Auto Engine Performance & Driveability

by

Chris Johanson
ASE Certified Master Technician

Publisher
THE GOODHEART-WILLCOX COMPANY, INC.
Tinley Park, Illinois

Library of Congress Catalog Card Number 96-47692
International Standard Book Number 1-56637-369-7
3 4 5 6 7 8 9 10 98 02 01 00

Important Safety Notice

Proper service and repair is important to the safe, reliable operation of motor vehicles. Procedures recommended and described in this book are effective methods of performing service operations. Some require the use of tools specially designed for this purpose and should be used as recommended. Note that this book also contains various safety procedures and cautions, which should be carefully followed to minimize the risk of personal injury or the possibility that improper service methods may damage the engine or render the vehicle unsafe. It is also important to understand that these notices and cautions are not exhaustive. Those performing a given service procedure or using a particular tool must first satisfy themselves that neither their safety nor engine or vehicle safety will be jeopardized by the service method selected.

This book contains the most complete and accurate information that could be obtained from various authoritative sources at the time of publication. Goodheart-Willcox cannot assume any responsibility for any changes, errors, or omissions.

Library of Congress Cataloging in Publication Data

Johanson, Chris.
Auto engine performance & driveability / by Chris Johanson.

p. cm.
Includes index.
ISBN 1-56637-369-7
1. Automobiles--Motors--Maintenance and repair.
2. Automobiles--Performance.
I. Title.
TL210.J65 1997 96-47692
629.25'028'8--dc21 CIP

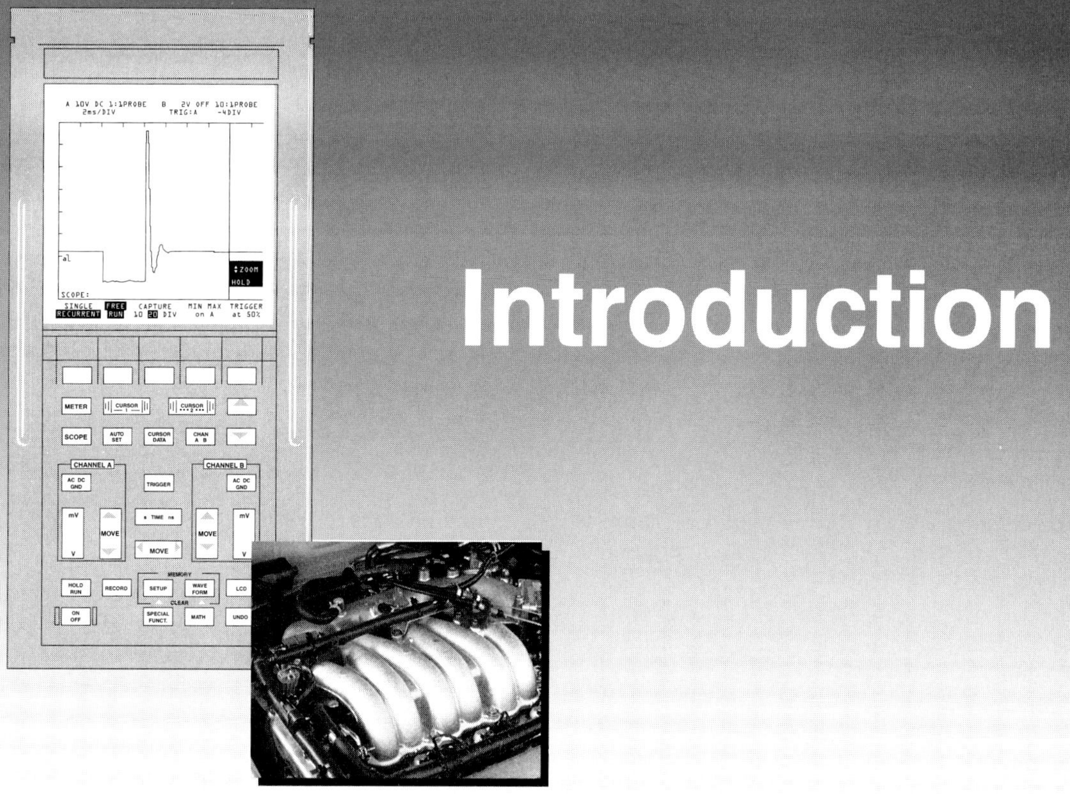

Introduction

Auto Engine Performance & Driveability is designed to help you diagnose and repair the complex engine and computer control systems on the modern automobile. As you may or may not know, major changes in service procedures for performance and driveability have occurred in the last 25 years. In the past, the cure for most automotive performance problems involved changing worn out ignition parts and readjusting the carburetor, simply referred to as a "tune-up." Today the tune-up is performed as maintenance, since most of the parts that were changed or adjusted are no longer used, improved so that they no longer need service or adjustment, or are computer-controlled. Most engine performance complaints are caused by the failure of a sensor, output device, or component, rather than worn spark plugs or other components that were replaced as part of a tune-up.

Troubleshooting also used to consist of simple procedures. Many problems could be found by looking at parts of the engine, and the number of possible problems was less than it is today. Even electrical tests could be made with simple equipment. Troubleshooting today's vehicles takes complex test procedures, much service literature, and elaborate test equipment. In addition, fuel economy and exhaust emissions from modern vehicles must be closely controlled to meet federal and state guidelines. How these changes came about, and what you can do to service driveability complaints, is explained in **Auto Engine Performance & Driveability.**

For you the technician, the primary effect these changes have had can be summed up as the need for more knowledge. Reading is a necessity for the modern technician to absorb the information in manuals, troubleshooting charts, and schematics. Some basic knowledge of mathematics and electrical theory is necessary to understand the operation of computer control systems, and how they affect the vehicle. Experimentation and random parts changing can no longer be performed. Both are simply too time-consuming and expensive on modern vehicles. The day of the backyard or shadetree mechanic is quickly drawing to a close.

You must be able to use service literature, test equipment, and most importantly, your reasoning and logic skills to find the cause of a problem. Simply being able to perform diagnostic tests is not enough; you must be able to evaluate what the test results mean. **Chapters 12** and **13** provide you with a seven-step diagnostic process that will help you locate the cause of performance problems in today's vehicles. This seven-step process can also be used in diagnosing *any system,* not just those that are automotive related. Once you learn this process, you will find it invaluable in other automotive classes.

Auto Engine Performance & Driveability has been written to help you develop those skills necessary to properly diagnose and fix driveability problems. Each chapter of the textbook begins with Objectives that provide focus for the chapter. Technical terms are printed in ***bold italic type*** and are defined when first used. Figure and Chapter references are printed in **bold type.** Warnings and Cautions are provided where a danger to life, limb, or property may exist in the system being discussed. Notes are provided to give hints to make routine tasks easier. Society of Automotive Engineers (SAE J1930) standardized sensor and electronics terminology is used throughout the text. Each chapter also includes a Summary, a "Know These Terms" section, and Review Questions.

One of the reasons you are holding this book is ASE certification, which is important for any technician. To help you in this area, ASE-type questions are included at the end of every chapter. One of the final chapters is devoted to a detailed explanation of the ASE test procedures, including how to register for and take the tests.

Student technicians can use **Auto Engine Performance & Driveability** to get a head start on understanding the latest automotive systems. Parts of this text review basic troubleshooting techniques, automotive electronics, engine computer systems, ignition systems, electronic fuel injection, emissions control systems, and nonengine related performance problems. The book also reviews material covered in courses on engines, electrical/electronics, and drive train systems. It will help prepare you to take the ASE Engine Performance test (A8).

Working technicians can refer to this text for information on the latest automotive designs, test equipment, and computer-specific troubleshooting techniques. Information on On-Board Diagnostics Generation Two (OBD II) computer systems is included for technicians needing to learn the operation of this new technology. Technicians who are familiar with only one make or type of vehicle can also use the information presented here to determine how all systems, components, and service procedures are similar, and where they are different.

Auto Engine Performance & Driveability is also a good textbook for review before entering a state sponsored auto emissions Inspection and Maintenance (I/M and IM 240) training class in areas where this testing is performed. Others who will find **Auto Engine Performance & Driveability** useful as a source of review material include experienced ASE certified Engine Performance technicians preparing to take the ASE Advanced Engine Performance Specialist test (L1). The last five ASE-type questions in **Chapters 14–21** are similar to the type of questions asked on ASE's most difficult certification test to date.

A **Workbook for Auto Engine Performance & Driveability** is also available. It contains chapters that correspond with the chapters in **Auto Engine Performance & Driveability** as well as hands-on jobs.

Chris Johanson

As vehicles have changed over the years, driveability problems have become more difficult to diagnose. (Chevrolet)

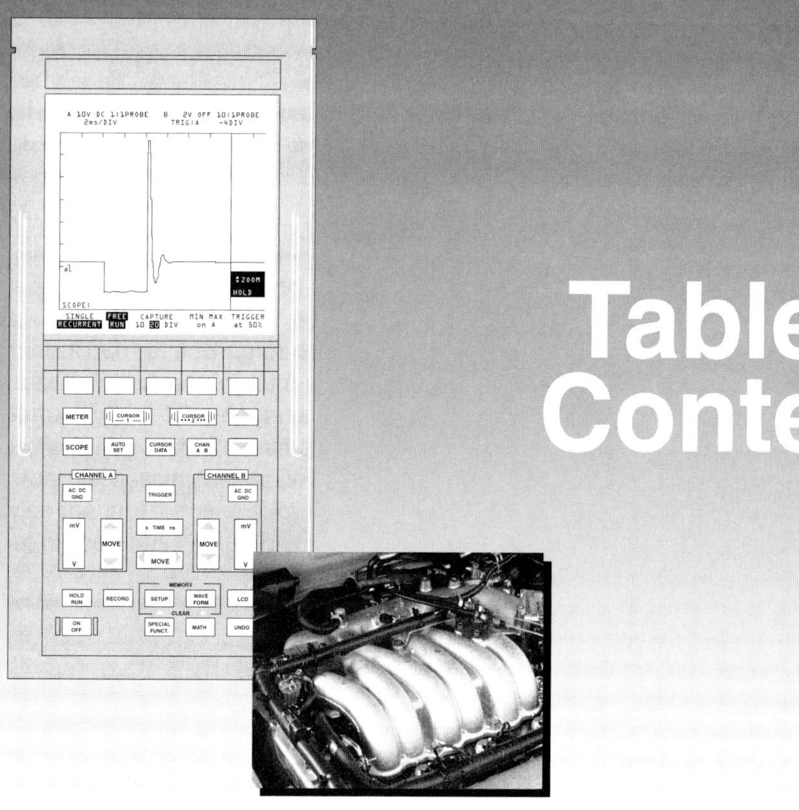

Table of Contents

Chapter 9: Fuel System Fundamentals 165

Chapter 10: Emissions Control and Exhaust System Fundamentals 189

Chapter 11: Drive Train and Vehicle System Fundamentals 213

Chapter 12: Basic Diagnostic Procedures 237

Chapter 13: Troubleshooting Techniques 253

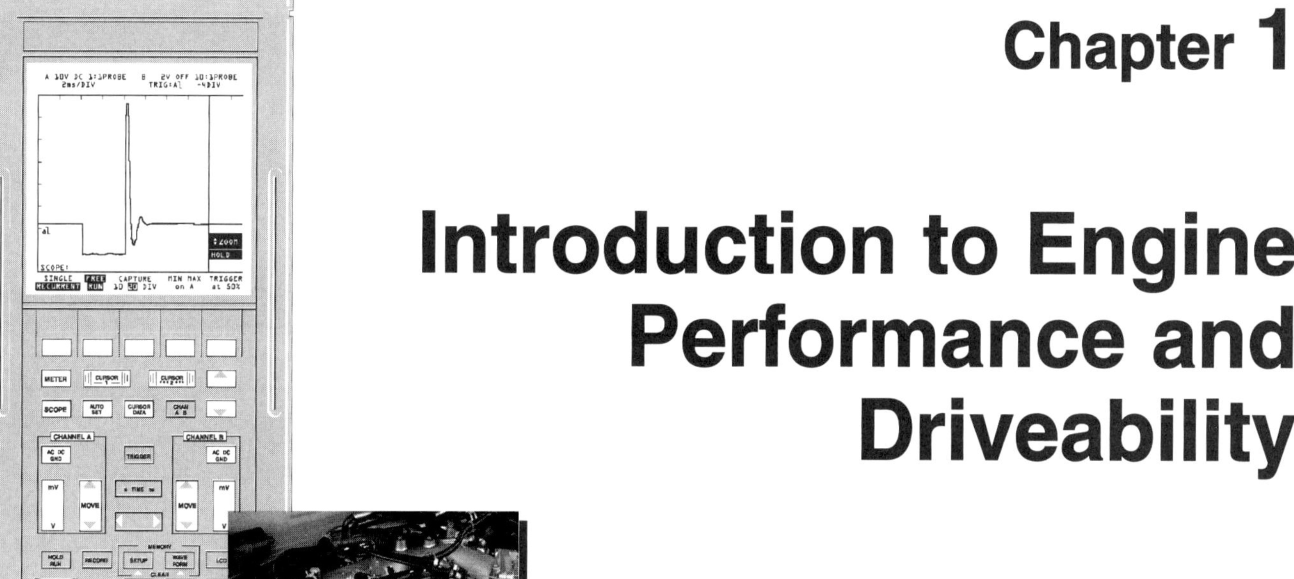

Chapter 1

Introduction to Engine Performance and Driveability

After studying this chapter, you will be able to:

- ❑ Identify the reasons for increasing the efficiency of automobiles and trucks.
- ❑ Identify the main methods of increasing the efficiency of modern vehicles.
- ❑ Identify the major uses of computer controls on modern vehicles.
- ❑ Identify what these changes mean for the modern automotive technician.

This chapter will identify and explain the reasons for the increasing complexity of modern vehicles and why more elaborate diagnostic procedures are necessary. You will also get an introduction to the concepts of engine performance and driveability. Studying this chapter will identify why these changes were needed and prepare you for the more detailed information in the next chapters.

Changes to the Automobile

The basic design of the automobile, the automotive culture, and driving habits have seen great fundamental changes in the last 25-35 years. In the 1960s, vehicles were simpler, larger, and very powerful. Solid-state electronic components under the hood did not exist and almost all cars and trucks were rear-wheel drive. There was virtually no fuel injection systems and carburetors were adjusted to compensate for any variations in driving conditions and state of engine tune-up. Fuel economy was unimportant to most drivers, since a dollar would buy about three to four gallons of fuel. Almost no one gave a thought to exhaust emissions.

This situation began to change around 1968, when emission controls were first introduced. These early control systems did not work very well and often became useless, either through lack of maintenance, or by deliberate tampering or removal. The combination of emission controls and large cars and engines reduced gas mileage to as low as six to eight miles per gallon on some vehicles.

The first oil embargo in 1973 caused a major increase in the price of fuel. This incident also made the driving public realize that crude oil was a finite resource that is constantly being depleted. During this time, consumer demand for smaller, fuel efficient vehicles increased. Some auto manufacturers were forced to begin making changes in their established methods of car design and production. This process was gradual, with many false starts and stop-gap measures, many of which proved to be unworkable.

Other factors that have helped to contribute to this process include increasingly strict regulations for exhaust emission levels and fuel mileage, a second oil embargo, and instability in several of the major oil producing countries. The gradual change in the attitudes, needs, and expectations of car buyers over the last two decades has also affected the redesign process. Modern vehicles are designed and used much differently than they were in years past. For example, many owners of light truck and sport/utility vehicles use them as cars.

Current Vehicle Designs

At the present time, all vehicle manufacturers have many methods for meeting the challenges of the modern automotive market. Although the internal combustion engine will be the dominant powerplant for the foreseeable future, it has and will continue to undergo modification to make it even more efficient. The drive train and other vehicle parts have also been redesigned, as has the shape and size of the entire vehicle. The next sections cover the major changes to the modern vehicle.

Redesigned Vehicle Bodies

To achieve better fuel economy, manufacturers began to reduce the weight of the average vehicle. In the 1960s, the average vehicle weighed about 4000 pounds (1814.6 kg). Modern vehicles have lighter bodies to consume less fuel. In many cases, the reduction in weight has been achieved by reducing vehicle size. Many vehicles with familiar names have been redesigned with smaller bodies, which average 3000 pounds (1360 kg), **Figure 1-1**. This overall reduction in size and weight has increased fuel mileage, while allowing good performance with smaller engines. Larger vehicles built today use aluminum, fiberglass, composite plastic, carbon fiber, and other lightweight materials in their construction to reduce weight.

Aerodynamic Designs

One of the major factors affecting fuel mileage is the amount of engine power used to overcome wind resistance at all speeds. All modern bodies have good ***aerodynamic*** shapes. This allows air to move over the body easily, with much less ***wind resistance***, or drag, than earlier models. Note that the older vehicle in **Figure 1-1A** has many corners and flat areas in its design. The vehicle in **Figure 1-1B** is curved and shaped to allow air to flow off easily. Sloped hoods and windshields, rounded surfaces, and fewer horizontal surfaces help to reduce drag.

The underside of the vehicle has been improved from an aerodynamic standpoint. Front-mounted air dams, or spoilers, reduce air flow under the vehicle, and increase the suction effect of air going through the radiator and engine compartment. See **Figure 1-2**. Underside surfaces are smoothed as much as possible to reduce wind resistance. The body shapes are also designed to use the air to help hold down the vehicle, increasing stability. All new body designs undergo extensive testing in wind tunnels and

Figure 1-1. *This figure shows the difference that years of vehicle redesign has brought about. A—This 1963 Grand Prix weighs about 4000 pounds and gets 12 mpg. B—This newer Grand Prix weighs 2500 pounds and gets about 30 mpg. (Pontiac)*

computer simulation to ensure that they produce as little wind resistance as possible, **Figure 1-3**.

Unibody Construction

Older cars used a separate body and frame in their construction. This type of design provided great strength, but also increased the vehicle's overall weight. Most cars today use unitized body or ***unibody*** construction, in which the frame and body assemblies are stamped from the same piece of sheet metal, welded together, or by a combination of these two techniques, **Figure 1-4**.

This design eliminates the need and weight of a separate frame and attaching hardware, while retaining the strength and rigidity of the heavier bodies. Unitized bodies are less likely to develop rattles. However, they are more prone to transmit noises from other parts of the vehicle. Light trucks, most sport-utility vehicles and some vans continue to use separate frames, but are becoming smaller and lighter than they were earlier.

Efficient Engines

Engines in the 1950s and 1960s were often oversized, having many more cubic inches or liters than necessary. This extra displacement resulted in loss of efficiency at low speeds, and, therefore, wasted fuel. At low speeds, the heat developed by the burning air-fuel mixture in a large displacement engine is wasted instead of producing pressure

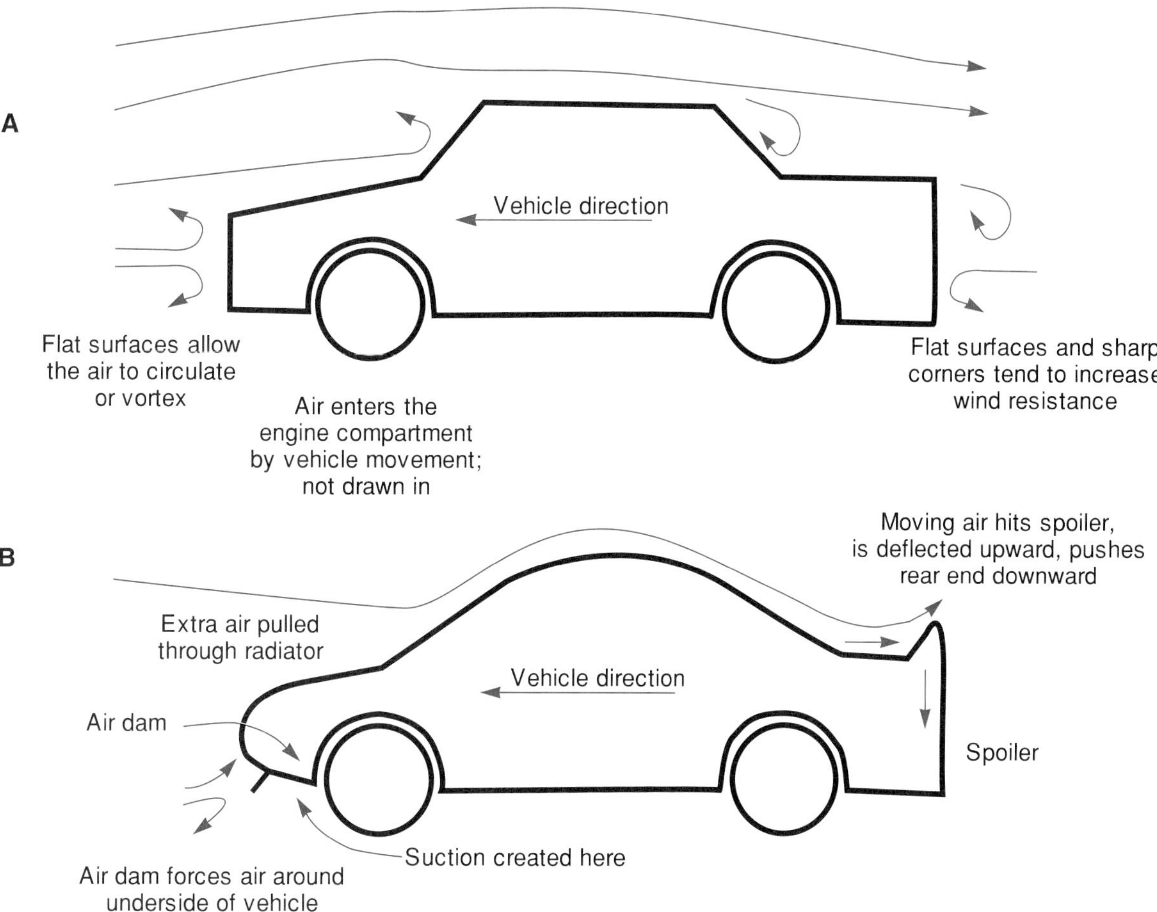

Figure 1-2. *Older vehicles had flat surfaces and sharp corners, which disrupts air flow. B—Aerodynamic body designs reduce wind resistance at high speeds, reducing the amount of power needed to move the vehicle, and increasing fuel economy.*

Figure 1-3. *Vehicle manufacturers perform numerous tests on new body designs before the first actual model is produced. A scale model of a sedan is tested for aerodynamic efficiency. The smoke indicates the path of air flow over the body. (Chrysler)*

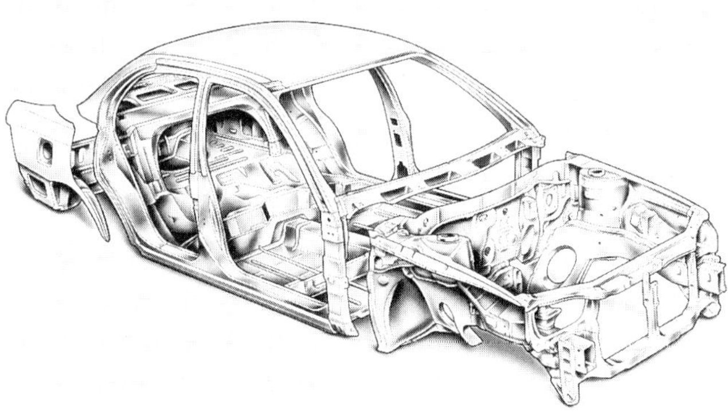

Figure 1-4. *Unibody construction, shown here, allows a lighter overall body weight than the separate frame and body design. Putting the frame and body into one unit also makes the body more rigid, increasing its strength.*

on the piston head. Large engines do not produce as much power as smaller engines for the amount of fuel consumed. Since the majority of modern driving is done at low speeds in urban areas, large amounts of fuel would be wasted by larger than necessary engines.

Modern engines have been reduced in size to improve efficiency. Smaller engines are used in most cars. Where the standard engine in an intermediate sized vehicle made in the late 1960s was a 350 cubic inch (5.7 liter) V-8, the standard engines in modern intermediate vehicles have four and six cylinders, ranging from about 130 cubic inches (2.2 liters) to 230 cubic inches (3.8 liters). Large luxury cars have seen engine sizes shrink from as much as 500 cubic inches (8.3 liters) to 276-300 cubic inches (4.6-5 liters). **Figure 1-5** illustrates the change in engine size.

Figure 1-5. *A—Older engines provided great power, but were not efficient. B—Today's smaller engines can produce as much power as the engines of the past, with better efficiency, thanks to improved manufacturing techniques and the use of electronics.*

More Horsepower from a Smaller Engine

The smaller engine size also decreased the amount of horsepower output. However, some of the lost high speed power was initially restored by installing carburetors with secondary systems. Fuel injection, camshaft with roller lifters, multiple valve per cylinder, and new engines with overhead camshafts began to restore some of the horsepower that was lost. In newer engines, multiport fuel injection, turbochargers, superchargers, more efficient transmission and differential gearing, and improvements in the electronic control of the fuel and ignition systems allows these smaller engines to produce as much or more horsepower than the older V-8 engines, while still getting good mileage at all other times. Some of the horsepower loss was due to the method manufacturers used to determine engine horsepower. In the 1960s, engine horsepower was measured at the engine flywheel. The horsepower of modern vehicles is measured at the driving wheels.

Improved Drive Train Design and Control

The transmission and drive train is used to multiply engine power and to transfer that power to the wheels. Unfortunately, it was also a major source of lost power and efficiency. The transmission and drive train has been improved to reduce weight, slippage, waste of engine power, and to reduce engine speed whenever possible.

Older transmissions had cast iron housings, which increased the weight of the unit. All transmissions and transaxles in use today have aluminum housings to further reduce overall vehicle weight. Older transmissions had a degree of slippage designed into them for smoother shifts. Modern transmissions and transaxles are manufactured to much closer specifications to reduce slippage.

Older automatic transmissions used two hydraulic inputs to determine when a gear shift was necessary. Many of today's automatic transmissions and transaxles are computer-controlled, using inputs from several engine and transmission mounted sensors to adjust output devices that control shift points, application of clutches and bands, and overdrive engagement. Other improvements include quicker, firmer shift patterns in automatic transmissions; shift lights to tell drivers when to shift a manual transmission for best efficiency; and reduced component weight.

Additional Transmission Gears

Since the job of the transmission is torque multiplication, manufacturers designed transmissions with more gears. The additional gears allow good acceleration, while still achieving maximum fuel economy. Where the manual transmission of the 1960s was usually a three-speed unit, today's manual transmissions offer as many as six speeds. The 1960s automatic transmissions had as few as two speeds, while the modern automatic transmission is often a four-speed unit. Five- and six-speed automatics are being planned for the near future.

The high gears of the four-speed automatic and the five- and six-speed manual is an overdrive ratio. See **Figure 1-6**. Modern transmissions, both automatic and manual, are designed with an overdrive gear for highway driving. This allows the engine to operate at a lower speed, while the vehicle continues to move at highway speed. The usual overdrive ratio is around 0.7 to 1.

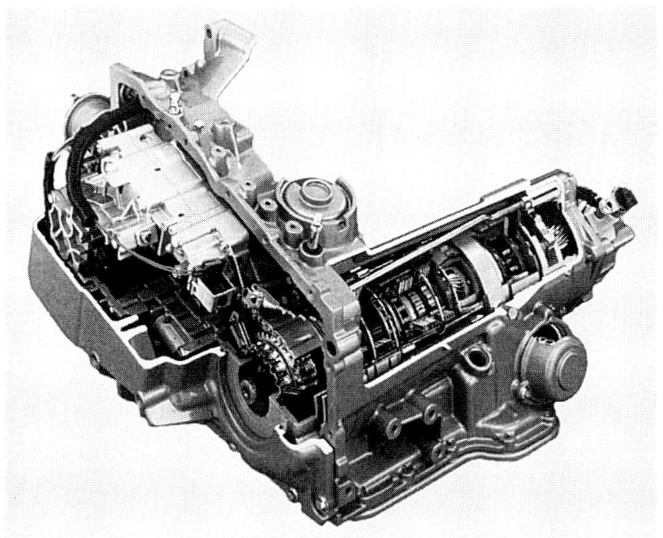

Figure 1-6. *The newest transmissions and transaxles employ aluminum parts, electronic controls, and construction to close tolerances to make them more efficient. (Chevrolet)*

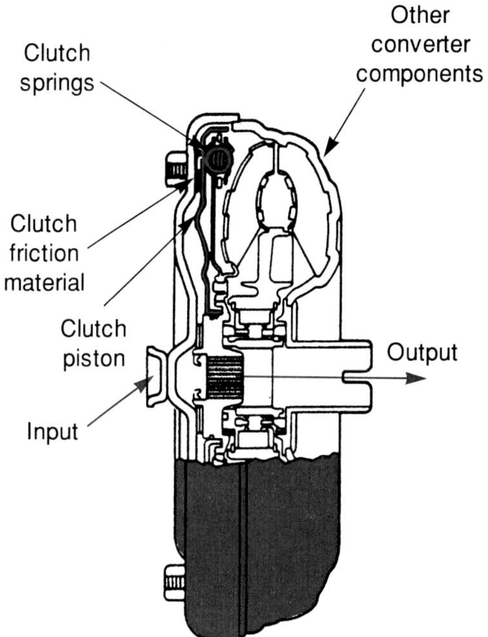

Figure 1-7. *The lockup converter uses a friction clutch to lock the input and output shafts together. The clutch is installed between the converter cover and transmission shaft and is applied by hydraulic pressure. The hydraulic pressure is controlled by solenoids energized by the vehicle computer. (Chrysler)*

Efficient Torque Converters

Many older automatic transmissions used inefficient fluid couplings. All vehicles with automatic transmissions produced since the late 1960s use torque converters. The torque converter was a major source of lost engine power at intermediate and high speeds. Also, the early torque converters were manufactured to allow much more slippage than modern units as a way to gain additional smoothness at low speeds.

More efficient torque converters are now being used on modern vehicles. Modern converters are designed to allow as little slippage as possible at any speed. In addition, most modern converters have an internal lockup clutch to completely eliminate slippage at higher speeds, **Figure 1-7**. The fluid drive design of the converter allows clutchless operation at low speeds, but is not needed once the vehicle is moving. The lockup clutch in the converter creates a direct mechanical connection, which bypasses the fluid drive. Most control systems apply the lockup clutch when the vehicle shifts into third or fourth gear.

Higher Final Drive Ratios

Higher ***final drive ratios*** are being used to increase gas mileage. The final drive ratio is the gear ratio produced by the ring and pinion gears in the rear axle assembly or in the front drive transaxle. As the ratio increases, the gear is said to be lower. As the number decreases, the gear is said to be higher.

Final drive ratios in older vehicles were often as low as 4.11 to 1, meaning that the driveshaft turned just over four times to turn the drive wheels once. These axle ratios could produce greater acceleration and low-speed power, at the expense of fuel mileage. Modern final drive ratios are closer to 2.5 to 1, reducing engine speed in relation to road speed. See **Figure 1-8**. The slower the engine turns at a given road speed, the less fuel it uses. When coupled with an overdrive transmission, the axle ratio can be selected to provide good fuel mileage, while delivering acceptable acceleration and power.

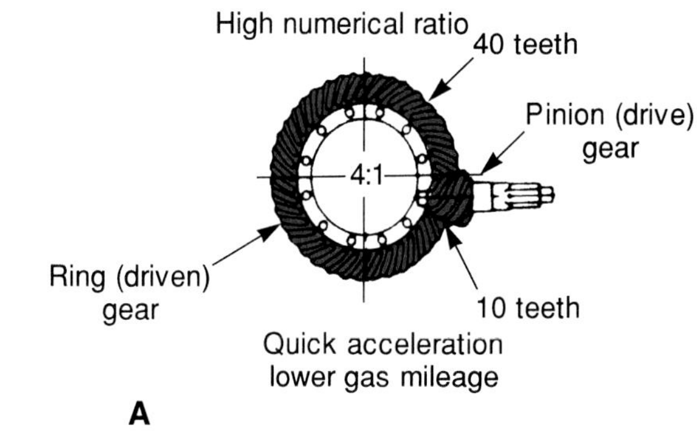

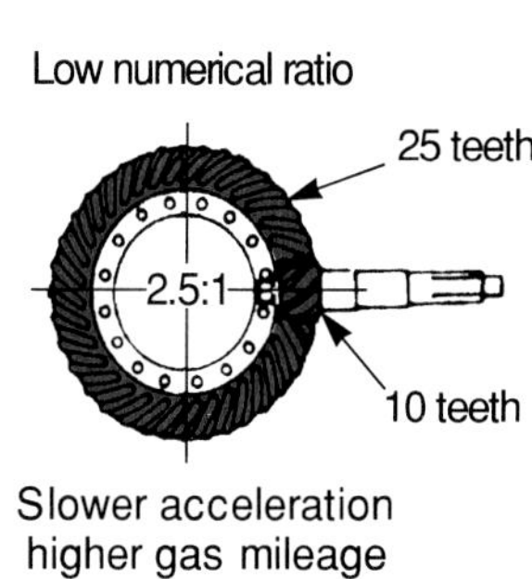

Figure 1-8. *A—Older final drive ratios were as high as 4:1 and higher. B—Modern final drive ratios are closer to 2.5:1 to reduce fuel consumption.*

Front-Wheel Drive

You learned earlier that older vehicles had rear-wheel drive. The majority of cars and minivans produced in the last 15-20 years use front-wheel drive. The front-wheel drive design saves weight, as well as providing better traction in all weather conditions. The heavy driveshaft and rear axle assembly are not used, creating more room inside of smaller car bodies, since the transmission and drive line "hump" is reduced or completely eliminated. A typical front-wheel drive vehicle is shown in **Figure 1-9**. Light trucks, larger vans, and a few cars continue to use rear-wheel drive.

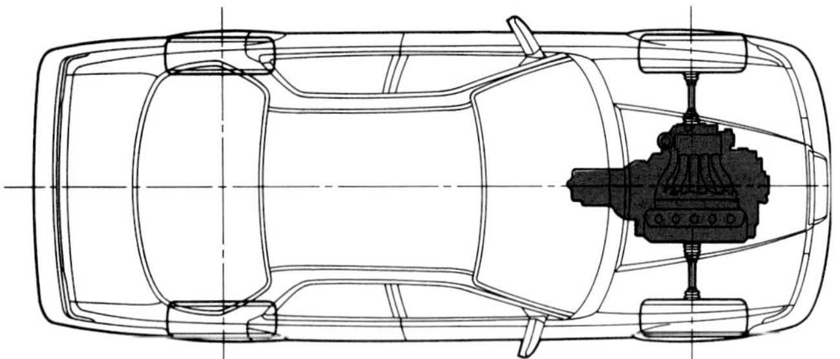

Figure 1-9. *Front-wheel drive assemblies, such as the one shown here, are used to reduce weight. Almost all automobiles produced today use front-wheel drive. (Acura)*

To reduce the amount of engine and drive train vibration transmitted through the body, many modern front-wheel drive vehicles use upper engine mounts and fluid-filled lower engine mounts. **Figure 1-10** shows an upper mount and lower mount.

Figure 1-10. *Older motor mounts, left, used on rear-wheel drive vehicles were usually installed on the bottom of the engine. Front-wheel drive and unitized body construction makes the use of upper motor mounts necessary. An example is the dog bone mount shown on the right.*

Computer Control

Computer control has been installed on all modern vehicles to monitor and control all engine functions, including spark creation and timing, fuel system operation, and emission control system operation. The computer shown disassembled in **Figure 1-11** contains many electronic components arranged into hundreds of circuits. The use of electronics to precisely control engine operation has been the single most important factor in the improvement of the modern vehicle in the areas of power, economy, and emissions. It has also contributed to the ease of maintenance by reducing the number of components that must be periodically replaced. It is also the major cause of the increased complexity in the modern automobile.

The on-board computer is an integral part of the modern powertrain and cannot be ignored during service. Many non-engine/transmission systems are now using on-board computers to control their operation. These systems include the air conditioner, air bags, anti-lock brakes, traction control, steering and suspension systems, anti-theft system, and cruise control. Computer usage in vehicles is growing at a tremendous rate. In 1990, the average number of vehicle functions that were monitored or controlled by the computer was less than 20%. Today, the number of computer-monitored or controlled functions is over 80% and is continuing to grow.

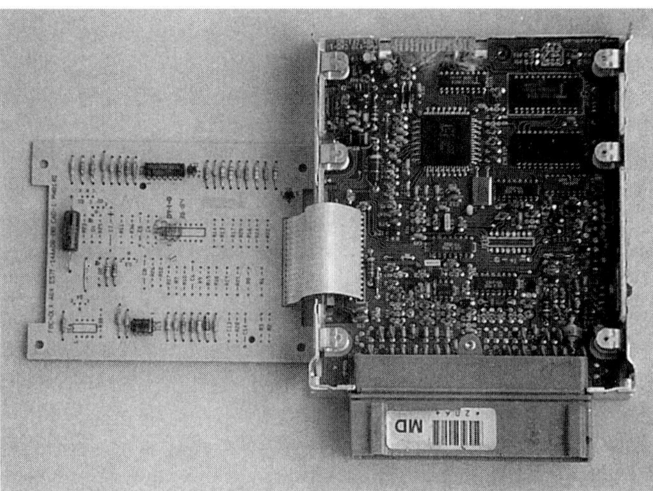

Figure 1-11. *The internal electronic components in this computer are the heart of the modern vehicle computer control system. They carefully and quickly control engine operation to provide the highest power and economy, with the lowest emissions.*

For instance, distributor vacuum and centrifugal units were used in older cars and trucks to advance the ignition timing in response to engine speed and load changes. These mechanisms were only approximately right most of the time. Today, the computer controls ignition timing based on sensor inputs that tell the computer exactly what is going on in the engine, drivetrain, and surrounding atmosphere. The computer then sets the ignition timing for maximum performance, and changes it constantly as conditions change.

The one change that is probably most noticeable when opening the hood of a modern vehicle is the reduction in the number of vacuum lines. A vehicle made in the 1970s and early 1980s had dozens of vacuum lines to the emission control devices. Computer control of the fuel and ignition systems has eliminated approximately 75% of these lines.

Automatic transmission shift valves were formerly moved by hydraulic pressures developed by the governor and throttle valves. This gave approximately correct shift points, subject to wear, disconnected components, and internal hydraulic pressure leaks. Shift valves are now moved by electric solenoids energized by the computer. This allows exact shift point changes under any conditions. Air conditioner compressor clutches, formerly engaged and disengaged by system pressures, are now operated by the computer.

Improvements in Fuel Delivery

Older vehicles used ***carburetors*** to deliver and regulate the engine's fuel and air intake. Carburetors had been in use since the earliest days of the gasoline internal combustion engine. The carburetor could be adjusted or modified to compensate for various driveability problems. However, these adjustments often contributed to poor gas mileage and higher exhaust emissions. Newer carburetors are non-adjustable and some have computer sensors and devices installed on or in them to monitor and control fuel flow.

With the improvements in electronics and computers, ***fuel injection***, which has always been used in diesel engines, began to see increased use in gasoline engines. Fuel injection has completely replaced the carburetor in all modern vehicles. The injection system pressurizes fuel and squirts it directly into the intake manifold of a gasoline engine, or into the cylinders of a diesel engine. Some gasoline fuel injectors are incorporated in a throttle body, which is mounted in the same location on the intake manifold as a carburetor.

The fuel is ***atomized*** (broken up into small particles) by the design of the injector nozzle, **Figure 1-12**. The operation of fuel injectors can be more precisely controlled by the on-board computer than the circuits of the carburetor. Therefore, the final air-fuel ratio can be controlled to meet the needs of the engine, while maintaining legal emission levels.

Turbochargers and Superchargers

Turbochargers and ***superchargers*** are used to increase the power of smaller engines. During normal driving, the engine operates in the same way as other engines, and delivers good fuel mileage. When more power is needed during periods of heavy loading or maximum acceleration, the turbocharger or supercharger increases horsepower by delivering more air and fuel into the engine under pressure. The turbocharger or supercharger can make the small engine produce as much power as a larger engine. Exhaust driven turbochargers, **Figure 1-13**, are the most common type, although belt-driven superchargers, **Figure 1-14**, are beginning to see increased use.

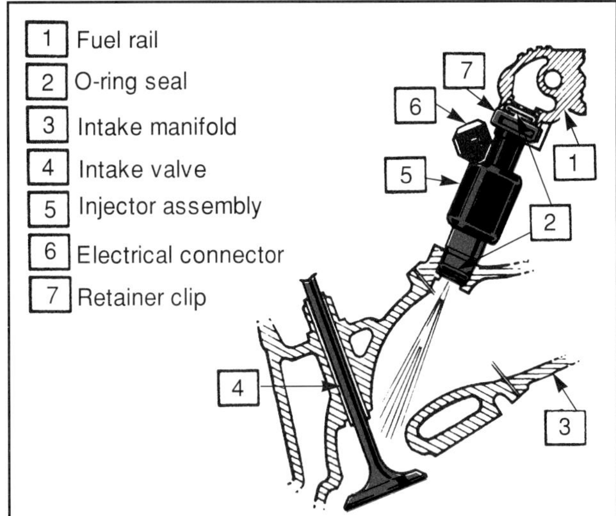

Figure 1-12. *The typical injector nozzle atomizes and delivers gasoline or diesel fuel to the engine. The setup shown here is typical of most gasoline fuel injectors. (General Motors)*

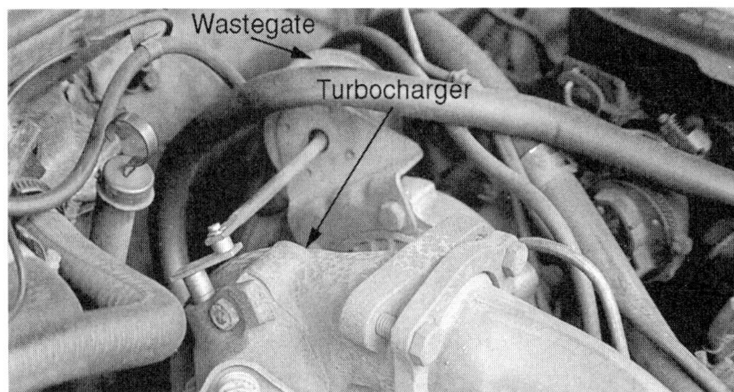

Figure 1-13. *Turbochargers are exhaust-driven air pumps. They pump extra air into the engine when extra power is needed. The wastegate prevents excess pressure in the cylinders, which could cause knocking.*

Emissions Control Systems

Unlike the earliest emission controls, which were bolted onto existing engines as afterthoughts, modern emission controls are an integral, engineered part of the modern vehicle. The first systems were mainly mechanical devices, which often failed, or were easily bypassed. Today, emission controls are included into the vehicle's design and often computer-controlled.

The use of the on-board computer as an emission control device has reduced the overall use of emission controls.

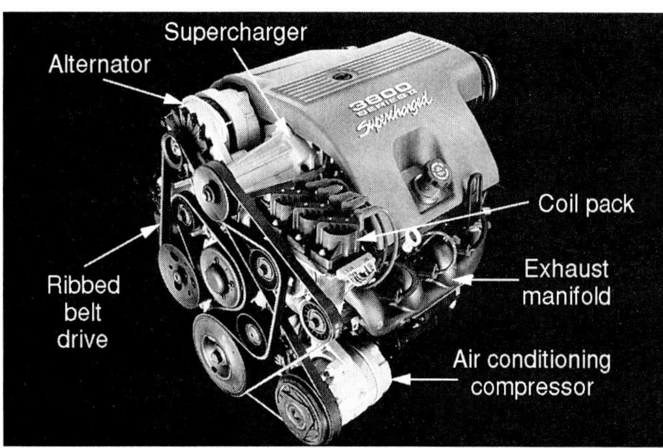

Figure 1-14. *Superchargers are belt- or gear-driven by the engine crankshaft, and deliver air at all times. They are most useful at higher engine speeds. Excess air is diverted to the air cleaner. (General Motors)*

A half-dozen types of vacuum advance restrictors have been eliminated by the use of computer-controlled ignition timing. Vacuum operated EGR valves have been replaced with solenoid-operated EGR valves controlled from the computer or eliminated altogether in some cases. Some engines no longer use an air injection pump system, which was quite common on many engines built in the 1970s and 1980s. Perhaps the biggest change is the use of the computer and oxygen sensor to replace the older methods of air-fuel ratio control. Instead of various carburetor modifications and the use of related devices to heat the air-fuel mixture, the input from the oxygen sensor is used by the computer to precisely control the air-fuel ratio.

Emission Control Troubleshooting Advances

Until the introduction of on-board computers, defective emission controls had to be diagnosed by the technician after the vehicle owner realized that there was a problem. In many cases, a defective emission control device remained non-functional for the life of the vehicle. The first spark and fuel control computer systems introduced in the late 1970s could sometimes fail without noticeably affecting vehicle operation.

After 1981, computer control systems were able to detect problems in their components and illuminate a dashboard light to warn the driver. The latest generation of computer controls will illuminate the dashboard light to warn not only of complete component failure, but will warn the driver when a component is beginning to fail, when an engine cylinder begins to miss, and when the air-fuel ratio begins to vary excessively.

Improved Manufacturing

When vehicles were manufactured years ago, design specifications and tolerances were not closely followed. Machine tools and other shop tooling was not checked regularly for damage or accuracy. This resulted in engine and vehicle parts that did not fit precisely, which lowered overall vehicle quality. These manufacturing practices, in an indirect way, contributed to reduced fuel mileage and increased emissions.

In order to install the electronic components, front-wheel drive trains, and other new technologies to improve the operation of the modern vehicle, it became necessary for auto manufacturers to also improve their production quality. Today, all parts are closely measured and compared to design specifications. Any part that is not within tolerance is rejected not as scrap, but is sent to be recycled. Building to closer tolerances in the case of engines and drive trains has resulted in better fuel mileage, less emissions, and greater horsepower output from a smaller package. By following design specifications and tolerances closely, a vehicle with a superior fit, finish, and performance is produced.

All tooling and machines are checked for accuracy and maintained on a regular basis. Robots and other automation are used on the assembly line to perform tasks that are tedious, boring, strenuous, or dangerous. Lasers and other equipment are used to make sure parts are within tolerances and in proper alignment before they are installed. Better manufacturing methods have resulted in fewer defects introduced at the factory, **Figure 1-15**.

Figure 1-15. *Modern manufacturing techniques result in a more precisely assembled vehicle. Each worker has access to a control panel that can be used to shut off the assembly line if a defect is found. (Saturn)*

Improvements have also been made in the quality and use of materials throughout the vehicle. Many chassis components are now made of aluminum instead of cast iron. The engine block, transmission case, master cylinder, air conditioner compressor, wheel rims, and many other engine and drive train parts are made from aluminum. Some engine and body components are made from plastic and carbon fiber for even more weight reduction. Many parts are now made from recycled plastics, such as antifreeze and soft drink bottles. Quality control is maintained throughout the vehicle's production, right up to when a new vehicle is delivered to the customer.

Driver Education

Drivers have been abandoning some of the wasteful practices of the past, either by a desire to conserve resources, or simple economic pressure. Most drivers have learned to reduce fuel consumption by reducing speed, removing unnecessary weight from their vehicles, and reducing trips by combining errands. Other fuel saving strategies for drivers include upshifting as soon as possible and selecting the highest gear possible for the particular driving conditions. The greatest change in driving habits has been the reduction of highway speeds over the last 20 years. Speed reduction has saved millions of gallons of fuel, and has resulted in a greater awareness of other things that could be done to increase fuel conservation.

What Is Driveability?

The term ***driveability*** has been adopted as a general term to cover smooth engine and vehicle operation without obvious defects. It does not include things that the driver cannot sense, such as high emissions levels or worn parts, although these are sometimes connected to a perceived driveability problem. The only time that high emissions levels will become a customer complaint is when the vehicle fails a state emissions test.

While driveability problems are often referred to as engine performance problems, they can be caused by a non-engine related system. Such a broad interpretation means that driveability can be affected by almost every vehicle system. A driveability problem, therefore, can be many things, including poor throttle response, surging, rough idle, hot and cold starting problems, pinging and detonation, dieseling, poor mileage, exhaust smoke, and a host of other symptoms. These problems can be caused by a defect in any of the basic engine systems, or a combination of defects. While the majority of driveability problems are caused by engine defects, many other vehicle defects can affect driveability. These include transmission and clutch problems, axle and drive line imbalance, brake pulsation, loose or bent suspension parts, tire imbalance, and air conditioner defects.

Since driveability problems are by definition those perceived by the driver as problems, there are no set driveability standards. A driveability complaint may be caused by a real defect or by unrealistic driver expectations. Therefore, the driveability technician is often called on to solve a problem using his own training, experience, and judgment, rather than by measuring a vehicle system against a set of specifications. Often, experienced technicians working in the field will refer to a performance related problem that has not been diagnosed as a driveability problem.

What a Tune-up Means Today

Twenty-five years ago, solving engine driveability problems usually consisted of simply replacing the ignition points, condenser, and spark plugs as part of a ***tune-up***. These parts, shown in **Figure 1-16**, wore out in about 10,000-12,000 miles (16 000-19 000 km). These components were estimated to be the source of about 85% of all driveability problems. In addition, the distributor cap, rotor, points, condenser, plug wires, advance mechanisms, and air cleaner element were checked and replaced if necessary. If necessary, the carburetor and manifold bolts were retightened, the valve train adjusted, and leaking vacuum hoses replaced. The initial timing and carburetor idle speed and mixture were checked and adjusted after the engine was started. Unless a more serious fuel, timing, or compression problem was found, these simple procedures were enough to get the vehicle back on the road and performing properly.

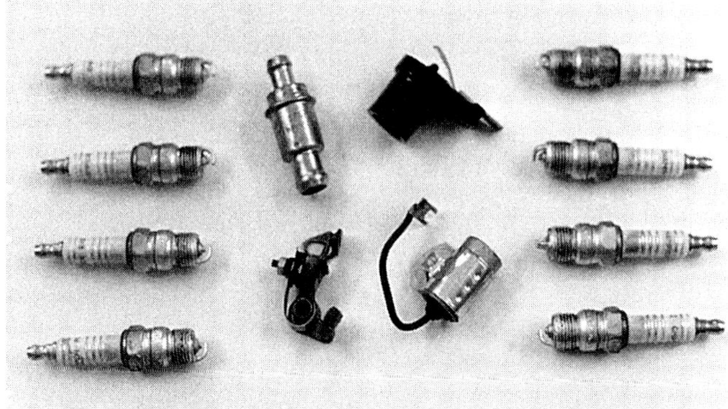

Figure 1-16. *All of the parts shown here were needed to tune up an engine 25 years ago. They usually required changing every 12,000 miles (19 200 km).*

Today these simple procedures are no longer effective or even possible. Most engines are precisely controlled for maximum performance. The computer and other electronic components used for control are not adjustable. The advantage of this is that the periodic tune-up is no longer necessary. There are no points and fewer or no moving ignition parts to wear out. Spark plugs last up to 100,000 miles (160 000 km) when fired by the higher voltages of the electronic ignition system. The fuel system is computer-controlled and is not adjustable. Only a few engines have adjustable valves, with the adjustment periods coming at much higher mileage.

The disadvantage is that extensive diagnostic routines and strategies are necessary to discover performance and driveability problems and to isolate any defective parts. Another problem is that engines with fewer cylinders tend to run more roughly than those with more cylinders, especially at low speeds in the higher gears. In addition, smaller cars with unitized bodies, firmer shifting transmissions, and front-wheel drives make any problem more obvious to the driver. Often, a tune-up performed on a modern vehicle that has a driveability problem will fail to correct the condition.

Troubleshooting

Troubleshooting older vehicles consisted of simple procedures. Many problems could be found by observation, and the number of possible problems was less than it is today. Electrical tests could be made using familiar equipment such as test lights and dwell meters. It was uncommon to see automotive technicians referring to service manuals. Repairs were more straightforward and new systems could often be figured out simply by examining them. There were many excellent technicians who were unable to read.

Troubleshooting today's vehicles takes complex diagnostic procedures, much service literature, and specialized test equipment. Today it is uncommon for a technician to get through even one day without referring to a service manual or other literature. Compare **Figures 1-17** and **1-18**. In the past, the service information for several vehicles could be contained in a single manual. Today, the amount of information needed to service a modern vehicle is so great that the service manuals for one vehicle often come in a volume set, with each volume dedicated to one or more systems.

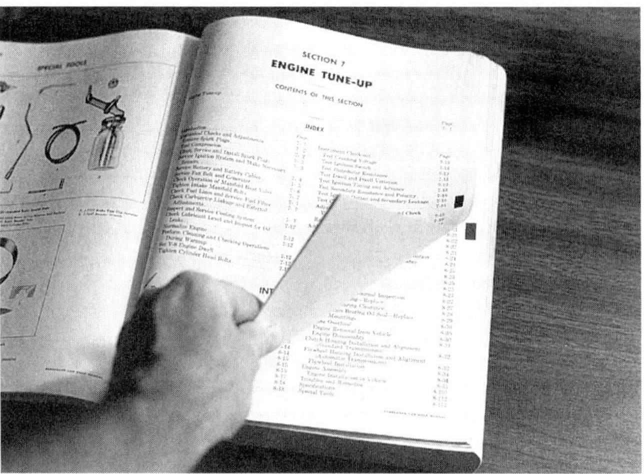

Figure 1-17. *Older service manuals had small sections on tune-ups, since the tune-up was a relatively simple process which almost always restored good performance.*

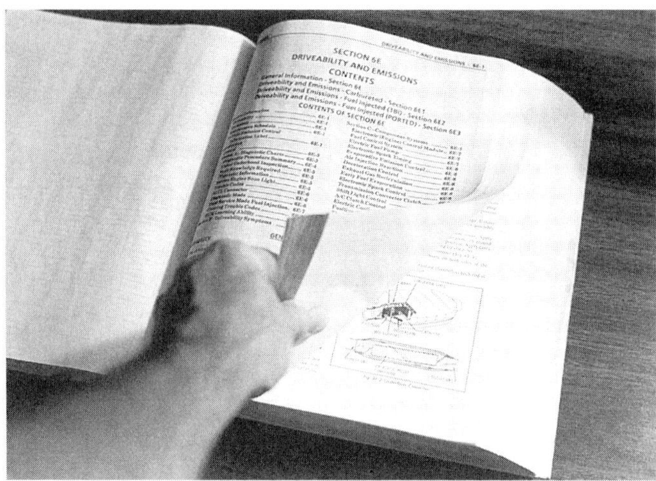

Figure 1-18. *Newer service manuals have a large section on driveability and performance diagnostics. This is necessary to cover all of the possible areas that could cause a driveability problem.*

It is no longer possible to properly diagnose vehicle problems by experimentation or random parts changing. Both are just too expensive on modern cars. The days of the backyard or shade tree mechanic are quickly drawing to a close. Specialized tools, test equipment, and other diagnostic equipment, as well as service manuals and other literature, are rapidly becoming a necessary expense.

Summary

The automobile has changed dramatically over the last 25–35 years, along with the rest of the automotive culture. Vehicles of the 1960s were larger and heavier, had large engines, and rear-wheel drive. As a result, they had poor mileage and high exhaust emissions. Various events, including oil shortages, technological advances, and mandatory emission controls have resulted in major changes to vehicles.

The easiest way to improve gas mileage is to reduce the size and weight of the vehicle. Bodies are aerodynamically designed to decrease the amount of engine power used to fight wind resistance. Oversized engines have also been eliminated.

The drive train has been improved to reduce slippage and waste of engine power, and to reduce engine speed whenever possible. More small cars and light trucks are manufactured with front-wheel drive, which eliminates the complexity and weight of the separate rear axle and driveshaft. Newer vehicles make extensive use of electronic control components; and are often equipped with fuel injection and turbochargers or superchargers.

The increase in the efficiency of modern vehicles has been accomplished through improvements in several areas, including the use of the latest electronic devices and metals technology. Electronic controls have replaced vacuum and mechanical devices. Emissions control systems have been improved and are computer-controlled. Some emissions control devices have been eliminated by computerized fuel and spark control. Another way to obtain efficiency is to educate drivers in how to conserve fuel.

All these changes mean that correct driveability diagnosis is much more important. One problem is that driveability means different things to different people. Twenty-five years ago, tuning an engine meant the tune-up; changing parts that were expected to fail when a certain number of miles were reached. On today's vehicles, parts do not need replacing at regular intervals and diagnosis is more complicated. Today's technician must have much more knowledge, access to service literature, and more complex test equipment to tune vehicles for driveability and performance.

Know These Terms

Aerodynamic
Wind resistance
Unibody
Final drive ratios
Carburetors
Fuel Injection
Atomized
Turbocharger
Supercharger
Driveability
Tune-up
Troubleshooting

Review Questions—Chapter 1

Please do not write in this text. Write your answers on a separate sheet of paper.

1. In what decade were the first emission controls introduced?
   (A) The 1950s.
   (B) The 1960s.
   (C) The 1970s.
   (D) The 1980s.
2. The truest statement about crude oil supplies is that they are:
   (A) constantly being depleted by use.
   (B) renewable resources.
   (C) mostly found in the United States.
   (D) never going to run out.
3. Match the letters that correspond with the era that the particular engine or vehicle characteristics were most common.

   ____ 1960s engine (A) 4000 pounds (1814 kg)
   ____ 1960s vehicle (B) 3000 pounds (1360 kg)
   ____ 1990s engine (C) 350 cu. in. (5.7 L)
   ____ 1990s vehicle (D) 2500 pounds (1125 kg)
   (E) 230 cu. in. (3.8 L)
4. Unibody construction means that the vehicle and frame are:
   (A) bolted together.
   (B) stamped out together.
   (C) welded together.
   (D) Both B & C.
5. In the 1960s, horsepower was measured at the ______.
6. The horsepower in modern vehicles is measured at the _____ .
7. Final drive ratios in modern vehicles is around:
   (A) 4.11:1.
   (B) 3.83:1.
   (C) 4.0:1.
   (D) 2.5:1.
8. The number of computer-monitored or controlled functions is over ______ .
   (A) 80%
   (B) 20%
   (C) 70%
   (D) 8%
9. Experienced technicians often refer to engine performance problems as ______ .
10. It is uncommon for today's technician to go through even one day without referring to a ______.

ASE Certification-Type Questions

1. All of the following have been used to increase engine and vehicle efficiency, EXCEPT:
   (A) smaller engines.
   (B) overdrive transmissions.
   (C) heavier bodies.
   (D) electronic engine controls.
2. Technician A says that unitized bodies develop fewer rattles than bodies that are bolted to a separate frame. Technician B says that unitized bodies transmit more rattles from other places in the vehicle than non-unitized bodies. Who is right?
   (A) A only.
   (B) B only.
   (C) Both A & B.
   (D) Neither A nor B.
3. Installing a higher final drive ratio in a vehicle will cause all of the following, EXCEPT:
   (A) decreased low speed pulling power.
   (B) increased low speed acceleration.
   (C) lowered engine RPM at cruising speeds.
   (D) lower fuel consumption at all speeds.

4. Technician A says that fuel injection is more efficient than carburetors in controlling air-fuel ratio. TechnicianB says that a fuel injected engine will produce more emissions than a carbureted engine. Who is right?
   (A) A only.
   (B) B only.
   (C) Both A & B.
   (D) Neither A nor B.
5. All of the following statements about driveability are true, EXCEPT:
   (A) there are no set driveability standards.
   (B) the driver will not notice most engine emissions problems.
   (C) driveability complaints are always caused by engine defects.
   (D) exhaust smoke can be a sign of a driveability problem.
6. Aerodynamic body shapes reduce ______.
   (A) wind resistance
   (B) top speed
   (C) fuel consumption
   (D) Both A & C.
7. Wind resistance is a source of power loss ______.
   (A) on windy days
   (B) at low speeds
   (C) at highway speeds
   (D) at all speeds
8. Older torque converters were a major source of lost engine power at ______ speeds.
   (A) low
   (B) intermediate
   (C) high
   (D) Both A & B.
9. A typical overdrive ratio is around ____ to 1.
   (A) 7
   (B) 3
   (C) 0.7
   (D) 0.07
10. Smaller vehicle bodies reduce vehicle ______ .
    (A) mileage
    (B) speed
    (C) smoothness
    (D) Both A & C.

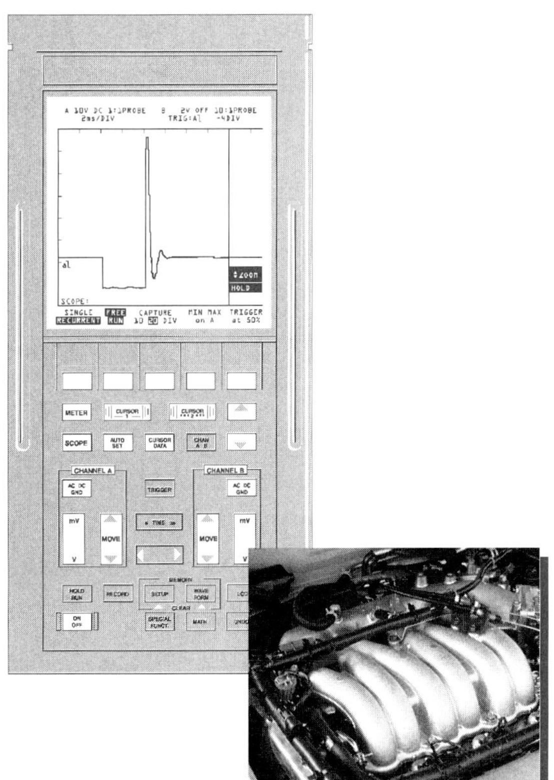

Chapter 2

Safety and Environmental Protection

After studying this chapter, you will be able to:

- Identify types of accidents that can occur when working on vehicles.
- Identify the two main classifications of accident prevention.
- Explain how proper shopkeeping can prevent accidents.
- Explain how proper work procedures can prevent accidents.
- Identify unsafe conditions in the shop.
- Identify unsafe work procedures.
- Identify environmental concerns that affect automotive shops.
- Identify ways in which automotive shops cause pollution.
- Identify proper waste disposal techniques.

This chapter will discuss the importance of safety in protecting the technician, the vehicle, and shop equipment. Proper safety procedures in all diagnosis and repair work will be discussed and stressed. An examination of unsafe conditions and work practices will be made and methods for correction given. You will also learn about the potential hazards to the environment by shop chemicals and the proper methods for storing and disposing of them.

The Importance of Safety

The most important part of any automotive diagnosis and repair is to work safely. An accident can result in injuries that keep you from doing your job or enjoying your time off, and may even kill. Even slight injuries are painful and annoying and may impair your ability to work and play. Even if an accident causes no personal injury, it can result in property damage. Damaged vehicles or shop equipment can be expensive to fix and could cost you your job. No mature and competent automotive technician wants to be injured or cause property damage.

However, even the most experienced technicians can become rushed and careless. For this reason, falls, injuries to hands and feet, fires, explosions, electric shocks, and even poisonings occur in auto repair shops. Carelessness in the shop can also lead to long term bodily harm from prolonged exposure to harmful liquids, vapors, and dust. Skin disorders, lung damage, and cancer can result from contact with these substances. For these reasons, safety must be kept in mind at all times, especially when it tends to be the last thing on your mind.

Causes of Accidents

Many ***accidents*** are caused when technicians try to take shortcuts instead of following proper repair procedures. Another common cause of accidents is failing to correct dangerous conditions in the work area, such as oil spills or tripping hazards.

Often, we hear of a chain of events, or a ***chain reaction.*** This is a series of actions or conditions which lead to other actions. Accidents are often caused by a chain of unsafe acts. An example of an unsafe act is removing the ground prong from the electrical plug of a portable drill, as shown in **Figure 2-1.** Some people do this because it is too much trouble to find a three-prong extension cord. Another unsafe act is using electric tools while standing on a wet floor, which many people do because it is too much trouble to clean up the spilled fluid, wear insulated shoes, or move to a dry area. Neither of these unsafe acts becomes an accident until someone tries to use the ungrounded drill while standing on a wet floor. If the drill motor develops a short to the drill body, current will flow through the user to ground, **Figure 2-2.** This causes severe electrical shock, with electrical burns, damage to internal organs, or even death.

In the example, no one deliberately set out to cause an accident. Instead, it was just too much trouble to do things correctly. The end result of these unsafe acts was an accident. To prevent accidents, the chain of unsafe acts must be broken.

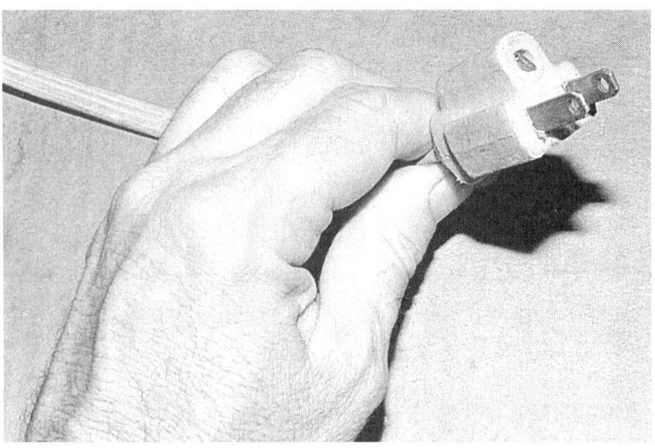

Figure 2-1. *Removing the ground prong from a three-way plug is very dangerous. Without this prong, electricity will take the next easiest path to ground, which may be through your body.*

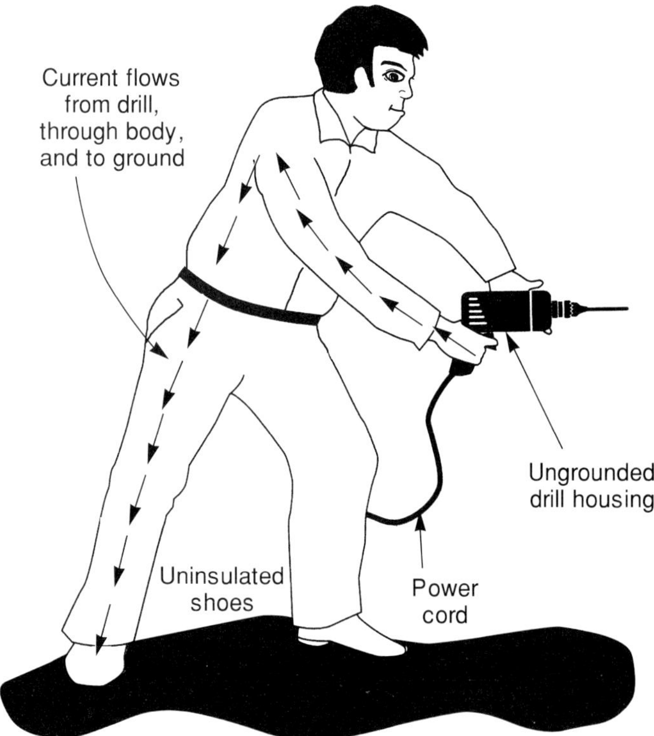

Figure 2-2. *Electricity always takes the path of least resistance to ground. If an ungrounded drill is used while standing on a wet floor, the person using the drill will receive a painful electric shock.*

Preventing Accidents

There are two major areas of unsafe acts: improper shopkeeping and improper procedures. ***Improper shopkeeping*** includes such things as failure to properly maintain tools and equipment; allowing old parts, containers, or other trash to pile up in the shop; and not cleaning up water, oil, or other liquid spills. ***Improper procedures*** include using the wrong tools or methods to perform repairs; not wearing safety glasses, safety shoes, or other protective equipment when necessary; using defective or otherwise inappropriate tools; and not paying close attention to the job at all times.

The best way to prevent accidents is to maintain a neat workplace and to use proper methods of repair. Some rules for preventing accidents are provided in the following sections. They are grouped into the areas of shopkeeping and procedures.

Proper Shopkeeping

The following rules apply to any shop. They do not mean just keeping the shop neat. Instead, they refer to the process of identifying and correcting unsafe conditions. Most of these rules will seem to be just a matter of common sense, but they are often disregarded.

Keep workbenches clean. This reduces the chance of tools or parts falling on your foot. In addition to preventing personal injury, this will reduce the chances of dropping parts on the floor, where they will be lost or damaged. A clean workbench reduces the possibility that critical parts will be lost in the clutter. It also reduces the chance of a fire from oily rags and paper debris.

Make sure that the shop has adequate lighting. Poor lighting makes it hard to see what you are doing, which can lead to accidental contact with moving parts or hot surfaces. Good lighting also makes the job easier to do. Overhead lights should be powerful and well-placed. Portable lights, or drop lights, should be operating and easy to use. Always use a rough service ***safety bulb*** in incandescent drop lights, **Figure 2-3.** These bulbs are more rugged than normal lightbulbs and will not shatter if they are dropped. Do not use a high wattage bulb in a drop light as they can create high temperatures, which will damage the light or cause burns if

Figure 2-3. *Rough service bulbs are clearly denoted on the top. They are also usually coated with a substance that helps to resist shattering.*

the housing is touched. Drop lights that use cool fluorescent bulbs are now available from some tool manufacturers.

Do not overload electrical outlets or extension cords by operating several electrical devices on one outlet, as shown in **Figure 2-4.** Only one electrical device, such as a drill or grinder, should be operated from each electrical outlet. Do not use more than one extension cord to power a high current device, such as an electric heater. Do not run electrical cords through water puddles, or use them outside when it is raining. Ensure that electrical cords and compressed air lines are in good condition. Avoid closing vehicle doors on electric cords or air lines. Do not drive a vehicle over an electrical cord or air line. Inspect air lines, couplings, electrical cords, and plugs regularly for leaks and damage.

Ensure that all shop equipment, such as grinders and drill presses, are equipped with ***safety guards.*** All of these devices are installed with guards by the manufacturer. These guards should only be removed for service operations, such as changing the grinding wheels. See **Figure 2-5.** When servicing any piece of equipment, be sure that it is unplugged. Read the equipment's service literature before beginning repairs. Return all tools and equipment to their proper storage places, **Figure 2-6.** This saves time in the long run, as well as reducing the chance of accidents, loss, or theft. Do not leave any pieces of equipment out where others could trip on them.

Keep tools and equipment in good repair. This includes such things as replacing damaged leads on shop equipment, checking and adding oil to hydraulic jacks, regrinding chisel tips, and replacing worn or damaged hand tools, **Figure 2-7.** Clean up spills immediately, before they get tracked all over

Figure 2-4. *Use only one plug per socket. This setup will result in an open circuit breaker or worse, an electrical fire.*

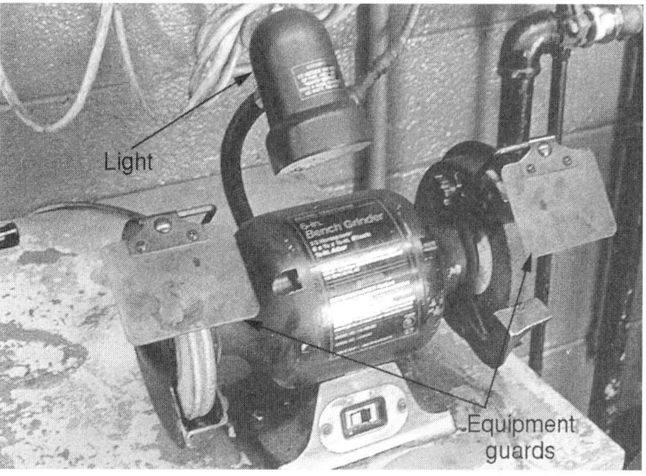

Figure 2-5. *This shop grinder has the necessary equipment guards in place. Be sure to wear eye protection when using any grinder.*

Figure 2-6. *A tool chest like this one will keep your tools stored where you can find them and safe from theft. The drawers of this box are open for display purposes. Never leave toolbox drawers open in this manner as the weight of the tools can cause the box to fall over.*

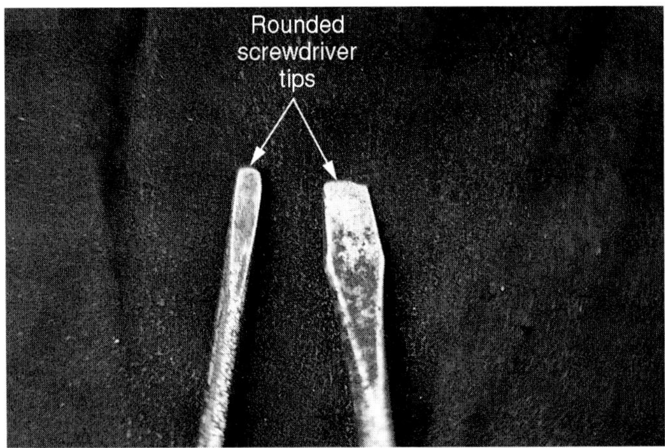

Figure 2-7. *Always replace any worn or broken tools immediately. In the past, screwdriver tips, such as these, were ground sharp and reused. Most tool manufacturers recommend replacing screwdrivers that are worn in this manner.*

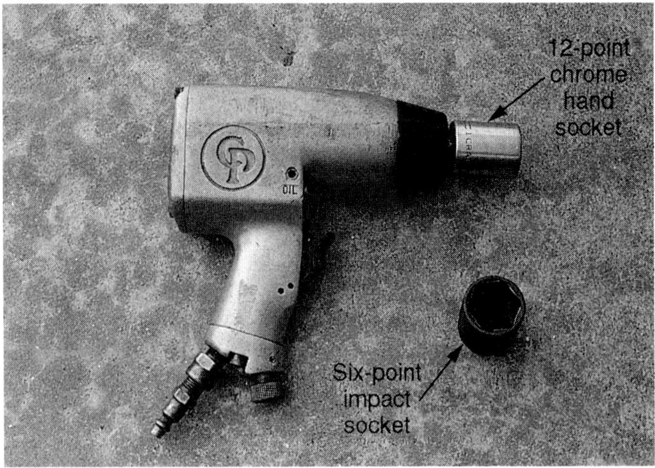

Figure 2-8. *Pneumatic tools are very powerful. Do not use hand sockets with air tools. Use only impact sockets, which are six-point and have a flat black color.*

the shop. Many people are injured when they slip on floors coated with oil, antifreeze, or water. Gasoline spills can be extremely dangerous, since a flame, cigarette, or the smallest spark can ignite the vapors, causing a major explosion and fire.

Proper Work Procedures

The following are some safe work procedures, which should be followed. Once again, these work procedures may seem to be simple. However, they are often disregarded, with tragic results. Study the proper procedure before beginning any job that is unfamiliar to you. Never assume that a procedure that you have used in the past will work on the vehicle you are facing. Always work carefully and avoid anyone who will not work carefully. Speed is not nearly as important as doing the job right and avoiding injury. Always lift with your legs, not your back. Make sure that you are strong enough to lift the object to be moved. If an object is too heavy to lift by yourself, get help.

Learn how to use new equipment properly before using it. This is especially true of air-powered tools, such as impact wrenches and chisels and large electrical devices, such as drill presses, boring bars, and brake lathes. These tools are very powerful and can cause damage or severe injury if they are used improperly. A good way to learn about new equipment is to read the equipment manufacturer's instructions. When working on electrical systems, avoid creating a short circuit with an unfused jumper wire or metal tool. Not only can this damage the vehicle components or wiring, it can develop enough heat to cause a severe burn or start a fire.

Use the right tool for the job. Using a screwdriver as a chisel or pry bar, or a wrench as a hammer, is asking for an accident or at least a broken tool. Never use a hand socket with an impact wrench, **Figure 2-8.** A hand socket can crack and shatter if used with an air tool.

Do not smoke in the shop. You may accidentally ignite an unnoticed gas leak. There are other less noticeable flammable substances around every vehicle. Batteries can produce explosive hydrogen gas as part of their normal chemical reaction. If a flame contacts escaping air conditioning refrigerant, it will produce poisonous ***phosgene gas***. A discarded cigar or cigarette can also ignite any oily rags or paper debris that may be lying around.

When using any type of vehicle lifting equipment, be sure to place the lift pads at the vehicle frame or other points that the manufacturer specifies as lifting points. Never raise a vehicle with an unsafe or undercapacity jack. Always support a raised vehicle with good quality jackstands. See **Figure 2-9.** Never use boards or cement blocks to support a vehicle. When using a post or hydraulic lift, be sure to place the lifting pads under the frame or on a spot that can support the vehicle's weight, **Figure 2-10.**

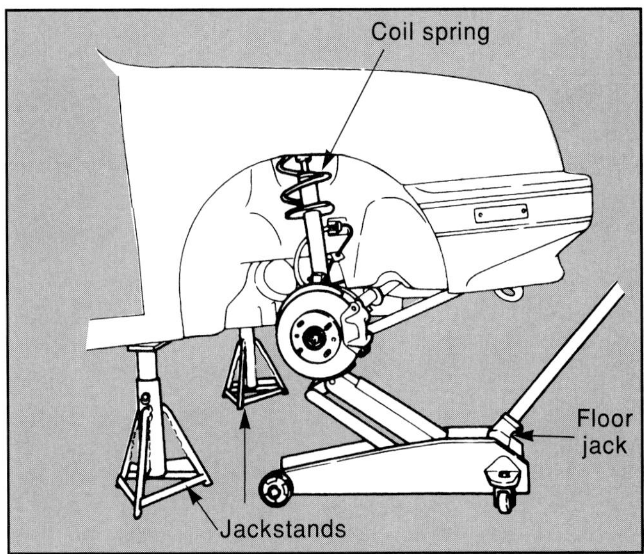

Figure 2-9. *Always use jackstands whenever lifting a vehicle with a floor jack. (Chevrolet)*

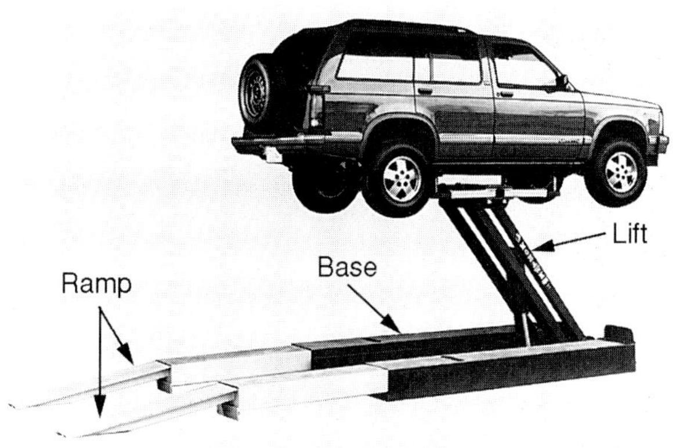

Figure 2-10. *While fixed post and hydraulic vehicle lifts are safer than using floor jacks, they can be dangerous if used improperly. (Mohawk)*

Do not run any engine in a closed area without good ventilation, even for a short time. ***Carbon monoxide*** can build up quickly, cannot be seen or smelled, and is deadly. When working on or near a running engine, keep away from all moving parts. Never reach between moving engine parts for any reason. Seemingly harmless parts such as the drive belts and fan can seriously injure you. Do not leave a running vehicle unattended. The vehicle may slip into gear or overheat while you are away. Whenever you are working on a running vehicle, set the parking brake.

When road testing a vehicle, be alert, and obey all traffic laws. Do not become so absorbed in the diagnosis process that you forget to watch the road. Be alert for the actions of other drivers. If you must listen or be involved in the diagnostic process during the road test, get someone to drive the vehicle for you.

Proper Clothing and Protective Equipment

Always dress appropriately and safely. Do not wear clothes with long, loose sleeves, open jackets, or scarves. They can get caught in moving parts and pull you into the machine or engine. Do not wear a tie unless your duty position requires it. If you must wear a tie, tuck it inside your shirt. If you wear your hair long, keep it away from moving parts by tying it up or securing it under a hat. Remove any rings or other jewelry. Not only can jewelry get caught in moving parts, it can cause a short circuit, which can result in severe burns or start a fire if caught between a positive terminal and ground.

Safety shoes, preferably with steel toe inserts, should be worn at all times. Most good quality safety shoes are constructed using materials that are oil and chemical resistant. Safety shoes have soles that are not only slip resistant and insulated, but also provide support and comfort. In any shop, there is a constant danger of falling parts or tools. Since such an incident can occur at any time, it is a good idea to wear safety shoes whenever you are in the shop.

Wear ***eye protection*** at all times while in the shop. In every shop, there are others working and you could become caught in a situation that may result in dirt, metal, or liquids being thrown into your face. This includes working around running engines; when using drills, saws, or grinders; and when working around batteries, parts cleaners, and hot cooling system parts. Two types of good eye protection are shown in **Figure 2-11.**

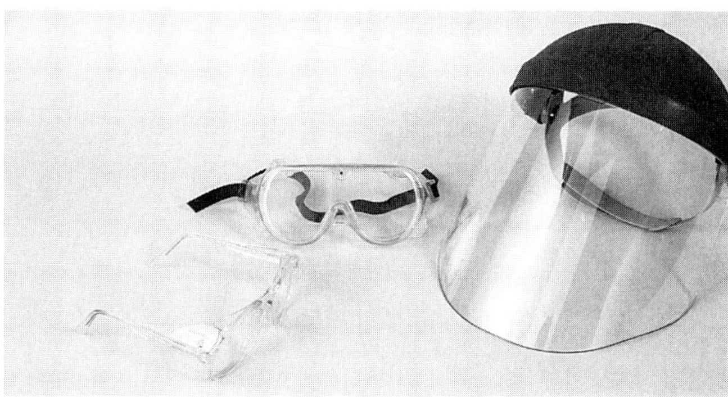

Figure 2-11. *Eye protection should be worn at all times. The eyeglass type is good for general work while the visor type is recommended for special operation, such as grinding and chiseling.*

Wear ***respiratory protection*** when working on brake systems or clutches. The dust from the friction lining material used in these devices contains ***asbestos,*** which can cause lung damage or cancer. Respiratory protection is also a good idea when working around any equipment that gives off fumes, such as a hot tank or steam cleaner.

Wear ***protective gloves*** when working with solvents, such as parts cleaning solvent. If you spill oil, gasoline, cleaning solvents, or any other substance on your skin, clean it off immediately. Prolonged exposure to these substances can cause severe skin rashes or chemical burns. All chemical manufacturers provide ***Material Data Safety Sheets,*** often called ***MSDS,*** for every chemical that they produce. These sheets list all of the known dangers of the chemical, as well as first aid procedures for skin or respiratory system contact. It is a good idea to obtain the proper MSDS and read it before working with any unfamiliar chemical.

However, it is still up to you to work safely and prevent accidents. Always use common sense when working on vehicles, and avoid people who do not.

The Technician and the Environment

In the last 10-20 years, the proper disposal of ***hazardous waste*** has become a major issue in the automotive shop. Pollutants, if allowed to escape without proper treatment, can cause major damage to the environment, **Figure 2-12.** Toxic materials lower the air and water quality, and can even affect food supplies. The effects of poisoned air, water, and soil may be noticed almost immediately, or may take decades to become apparent. The health and financial burdens of irresponsible waste disposal will grow

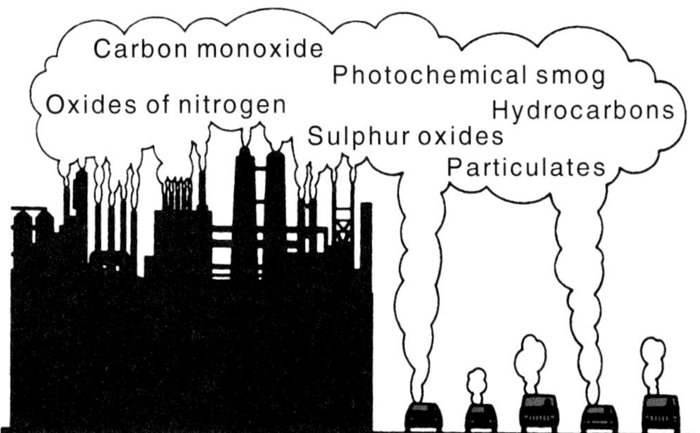

Figure 2-12. *Industrial and vehicle emissions contribute to environmental damage. It is the technician's responsibility to make sure each vehicle serviced produces as little pollution as possible. (Sun Electric Corp.)*

ever larger. Even if we escape the consequences, future generations will not.

The automotive service industry is a major emitter of solid and liquid wastes. Examples are used antifreeze and oil, cleaning solutions, tires, batteries, scrap parts, and various paper and plastic containers. Used motor oil and antifreeze can cause immediate damage to plants and wildlife, and can cause long-term damage to groundwater supplies. Solid wastes, such as used batteries, can contain acids and heavy metals that are extremely damaging to the environment. While some solid wastes such as scrap metal, paper, and tires are not immediately damaging, they are unsightly and increases the burden on local landfills. It is senseless to discard solid wastes that can be easily recycled.

In addition, improper servicing of engines and emission controls can increase air pollution through the production of excess carbon monoxide, hydrocarbons, and oxides of nitrogen. The thinking technician will always try to reduce environmental damage caused by careless waste disposal and service procedures.

Proper Waste Disposal

The lubricants, cleaners, and other materials used in the shop do not become waste until the shop is finished with them. However, when they become waste, they must be disposed of legally. Some basic rules for the environmentally responsible shop are given below. Note that some of these rules are legal requirements of federal, state, and local governments.

Do not pour used motor oil, transmission fluid, antifreeze, brake fluid, or gear oil on the ground or drains or leave open containers in the shop. For example, antifreeze contains ethylene glycol, which is poisonous and can cause damage to the environment and can poison any animal or person who accidentally drinks it.

Many ***recycling companies*** will pick up waste chemicals, such as used oil and antifreeze, to be processed for reuse. A 55-gallon drum or other secure storage container should be kept in the shop for this purpose, **Figure 2-13.** This is the most common way in which auto shops cause environmental damage. Improper disposal is illegal and you may be liable for cleaning up a contaminated area years after the actual disposal occurred. Some shop heaters utilize used motor oil as a fuel. Many states require that all chemicals, either new or used, be located in secure areas away from sources of sparks or flame and where leakage can be contained. In addition, new regulations require the use of above ground storage tanks in most areas.

Figure 2-13. *55-gallon drums, such as this one, are useful for storing used chemicals such as antifreeze. However, any drums or containers used to store hazardous waste should be clearly labeled and kept in an area where any leakage can be easily contained.*

Recycling Solid Waste

Recycle parts and scrap materials such as aluminum, iron, and steel whenever possible. Check with your local parts supplier to determine parts that can be returned for remanufacturing. Examples of parts that can be remanufactured are:

- ❑ Alternators.
- ❑ Starters.
- ❑ Carburetors.
- ❑ Fuel injection throttle bodies.
- ❑ Fuel pumps.
- ❑ On-board computers.
- ❑ Coolant pumps.
- ❑ Complete engines and transmissions.
- ❑ Power steering pumps.
- ❑ Brake shoes.
- ❑ Brake calipers.
- ❑ Brake boosters.

Catalytic converters should be turned in for recycling of the precious elements used in the catalyst. Used batteries should be collected and turned over for proper disposal. Tires, scrap metal, paper, glass, and plastic can often be recycled through your local sanitation service. If solid wastes cannot be recycled, do not dispose of them by illegal dumping or burning.

Proper Service Procedures

Servicing vehicles in such a way that they do not cause air polution is as important as proper disposal of toxic wastes. The following procedures should be followed to avoid damaging the atmosphere by improper service procedures.

Do not make improper adjustments or modifications to the emission controls of any vehicle, or modify any vehicle system without first determining the effect on emissions. Increases in power and economy are almost always at the expense of emissions, as well as engine durability and smooth operation. Keep in mind that any tampering that removes, disables, or defeats any emission control device is illegal.

Do not discharge any type of air conditioner refrigerant into the atmosphere. Refrigerants such as R-12 cause extensive damage to the ozone layer, leading to increased ultraviolet ray exposure as well as other environmental problems. Even so-called "environmentally safe" refrigerants, such as R-134a, still contribute to ozone layer depletion. Releasing refrigerants into the atmosphere is both illegal and needlessly expensive. Refrigerant recovery equipment must be available and at any shop that performs air conditioning service.

Additional information about waste disposal and vehicle emissions can be obtained from the Environmental Protection Agency. The EPA currently has ten regional offices and six field offices. For the address of the nearest EPA office, write:

Automotive/Emissions Division
United States Environmental Protection Agency
401 M Street SW
Washington, DC 29460

Summary

The most important part of driveability diagnosis and repair is to work safely. An accident may result in personal injury or damage to equipment or property. No mature and competent automotive technician wants to be injured or cause property damage.

However, even the most experienced technicians can become rushed and careless. For this reason, various kinds of accidents often happen in auto repair shops. Carelessness can lead to immediate injury, or to long term bodily harm. Safety must be kept in mind at all times. Many accidents are caused when technicians try to take shortcuts instead of following proper repair procedures. Another common cause of accidents is failing to correct dangerous conditions in the work area, such as oil spills or tripping hazards.

Accidents are often caused by a chain reaction beginning with one unsafe act. The act in itself may not result in injury at that time, but later, because of one or more added unsafe acts, may cause a severe accident. Not taking the time to do things correctly almost always results in an accident. To prevent accidents, the chain must be broken. There are two major areas of unsafe acts: improper shopkeeping, and improper procedures. The best way to prevent accidents is to maintain a neat workplace, and to use proper methods of repair.

It is up to the technician to work safely and prevent accidents. Always use common sense when working on vehicles, and avoid people who do not. The technician must prevent the escape of toxic materials into the environment. The most common ways in which the automotive shop causes pollution are improper waste disposal and improper vehicle service procedures. Toxic materials must be disposed of in ways that will not cause environmental damage. Scrap materials should be recycled whenever possible. Vehicles should be serviced in ways that will not cause the addition of pollutants into the atmosphere.

Know These Terms

Accidents
Chain reaction
Improper shopkeeping
Improper procedures
Safety bulb
Safety guards
Phosgene gas
Carbon monoxide
Safety shoes
Eye protection
Respiratory protection
Asbestos
Protective gloves
Material Safety Data Sheets (MSDS)
Hazardous waste
Recycling companies

Review Questions—Chapter 2

Please do not write in this text. Write your answers on a separate sheet of paper.

1. Some of the most common accidents in automotive shops are:
   (A) falls.
   (B) fires.
   (C) electric shocks.
   (D) All of the above.
2. Accidents are often caused by a ______ of unsafe acts.
3. The two major areas of unsafe acts are improper ______ and improper ______ .
4. The best ways to prevent accidents are to maintain a ______ workplace and to use proper methods of ______ .

5. Each electrical outlet should be used to operate how many tools?
   (A) 1
   (B) 2
   (C) 3
   (D) As many as the outlet will hold.
6. A high wattage bulb in a drop light can cause what type of personal injury?
   (A) Blindness.
   (B) Burns.
   (C) Electric shock.
   (D) All of the above.
7. Spilled ethylene glycol antifreeze can cause:
   (A) slipping.
   (B) poisoning.
   (C) environmental damage.
   (D) All of the above.
8. You should always wear respiratory protection when working on what two vehicle components?
9. Never use a ______ socket with an ______ wrench.
10. Creating a ______ with an unfused jumper wire or metal tool will cause damage to the vehicle's ______.

ASE Certification-Type Questions

1. Technician A says that accidents occur in auto repair shops when technicians try to take shortcuts. Technician B says that accidents occur in auto repair shops when technicians fail to correct dangerous conditions in the work area. Who is right?
   (A) A only.
   (B) B only.
   (C) Both A & B.
   (D) Neither A nor B.
2. If all tools and equipment are returned to their proper storage places, all of the following will occur, EXCEPT:
   (A) they will be hard to find the next time.
   (B) the chance of tool theft will be reduced.
   (C) the chance of tripping will be reduced.
   (D) the chance for tool damage will be reduced.
3. If the safety guards have been removed from a grinder, what should you do?
   (A) Be very careful when using the grinder.
   (B) Let someone else do the grinding.
   (C) Do not use the grinder until the guards are replaced.
   (D) Wear eye protection.
4. Always use a "safety" bulb in ______.
   (A) fluorescent drop lights
   (B) incandescent drop lights
   (C) every light in the shop
   (D) the ceiling lights
5. Wear eye protection when ______.
   (A) working around running engines
   (B) using drills, saws, or grinders
   (C) working around batteries
   (D) All of the above.
6. Technician A says that the automotive service industry is a minor source of pollution. Technician B says that solid wastes are a major source of automotive pollution. Who is right?
   (A) A only.
   (B) B only.
   (C) Both A & B.
   (D) Neither A nor B.
7. All of the following are examples of liquid wastes that must be disposed of, EXCEPT:
   (A) used motor oil.
   (B) contaminated gasoline.
   (C) refrigerant.
   (D) used antifreeze.
8. When does a substance become toxic waste?
   (A) As soon as it is manufactured.
   (B) When it is sold to the retailer.
   (C) When the shop begins using it.
   (D) When the shop is finished with it.
9. All of the following examples of solid waste can be recycled, EXCEPT:
   (A) used tires.
   (B) paper cartons.
   (C) used fuel filters.
   (D) defective aluminum parts.
10. Technician A says that the Environmental Protection Agency (EPA) is concerned only with air pollution caused by automobiles and light trucks. Technician B says that the EPA is concerned only with liquid and solid wastes that could contaminate groundwater. Who is right?
   (A) A only.
   (B) B only.
   (C) Both A & B.
   (D) Neither A nor B.

Chapter 3

Tools, Test Equipment, and Service Information

After studying this chapter, you will be able to:

- ❏ Identify and explain the uses of basic electrical test instruments.
- ❏ Identify and explain the uses of basic mechanical test equipment.
- ❏ Identify and explain the uses of input sensor testers and simulators.
- ❏ Identify and explain the uses of output device testers.
- ❏ Identify and explain the uses of starting and charging system testers.
- ❏ Identify and explain the uses of oscilloscopes.
- ❏ Identify and explain the uses of exhaust gas analyzers.
- ❏ Identify and explain the uses of service manuals.
- ❏ Explain how to use schematics.
- ❏ Explain how to use troubleshooting charts.
- ❏ Identify and explain the purposes of manufacturer hotlines.
- ❏ Explain the influence of computers in the automotive shop.

This chapter identifies and explains the purpose of tools and test equipment that will be used in driveability and performance diagnosis. This ranges from simple tools, such as gap gauges and jumper wires, to complex electronic test equipment used for checking the computer and other electronic components. Other equipment includes service literature and information hotlines. While some tests can be done with simple tools that are a part of every technician's basic tool collection, any shop that intends to perform driveability service must have test equipment capable of accessing information in the vehicle's computer as well as testing electronic components on all modern vehicles.

Many other tools and equipment are necessary in the course of diagnosing a driveability problem. These include engine oil, fuel, and automatic transmission pressure gauges, cooling system pressure testers, and other equipment. In addition to the tools and test equipment discussed here, the technician must also have a set of good quality hand tools. This chapter covers only the tools and test equipment that are used to check the engine, computer control system, and other components that are most directly related to vehicle driveability and performance.

Basic Diagnostic Tools

This section covers some basic tools that are used to perform diagnostic tests and make adjustments. These tools are a part of every technician's basic tool collection. Most of this equipment has not changed in many years, but is still used to diagnose and service parts on the latest vehicles.

Gap Gauges

All ***gap gauges*** are used to measure the clearance, or distance, between two surfaces. Many types of gap gauges are available for various uses. Gap gauges are used to measure the air gap of some electronic ignition pickup coils, spark plug gaps, and valve clearance as part of normal driveability service.

Feeler Gauges

Gap gauges known as ***feeler gauges*** are thin strips of steel or brass, available in various thicknesses. These thicknesses are measured in thousandths of an inch, or of a millimeter. Modern feeler gauges have both thousandths of an inch and millimeter markings. One thousandth of an inch, .001″, about .025 millimeters. Feeler gauges are often used to check valve clearance. ***Brass feeler gauges*** are used to set clearance on parts containing magnets. A steel gauge will stick to the parts, causing an incorrect reading. Brass feeler gauges resemble steel gauges, but can be identified by their distinct golden yellow color. Most brass gauge set sizes range from around .004″ -.015″ (.10-.38 mm). Feeler gauge strips are usually sold in sets to cover a range of uses. See **Figure 3-1.**

Wire gauges consist of wires calibrated to common spark plug gaps. They are sometimes used to gap spark plugs, especially used plugs. Used spark plugs develop a rounded surface as they operate and flat gauges will not accurately measure the gap. A wire gauge allows for electrode wear and the plug can be adjusted to the proper gap. Wire gauges are less common now, since used spark plugs are usually replaced rather than regapped.

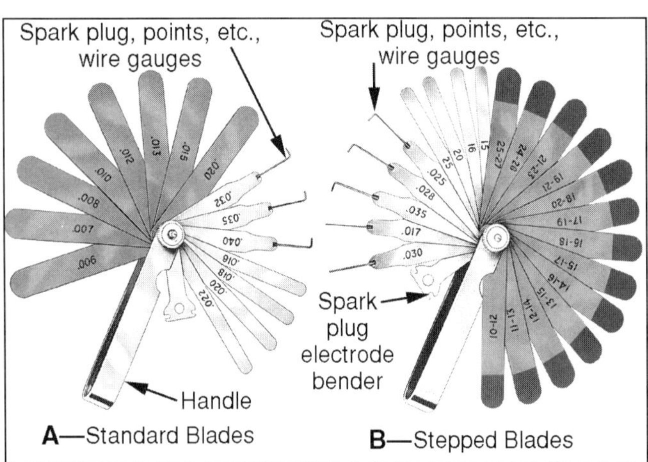

Figure 3-1. *Feeler gauges are used to measure gaps between surfaces. They are available in standard and metric sizes. (L.S. Starrett)*

Spark Plug Gauges

Spark plug gauges are a variation of the feeler gauge, designed especially for spark plugs. Spark plug gaps on modern vehicles vary from .025″ up to .095″ (.635 to 2.41 mm). A common type of plug gauge is the ***ramp gauge***, **Figure 3-2.** It provides a simple way of adjusting spark plugs. The gauge is inserted between the plug electrodes, and pulled along the ramp until it stops. It is then possible to read the gap size on the gauge. The ramp gauge eliminates the need for a large number of separate feeler gauges to set different plug gaps. Since the ramp gauge is made from a thick piece of steel, it is less likely to be damaged by the spark plug electrodes.

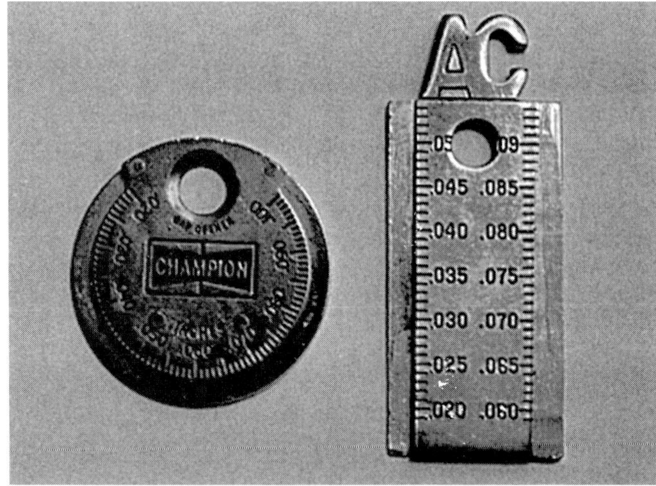

Figure 3-2. *Ramp gauges are used to measure spark plug gaps. They are very popular since they are easy to read and can check and adjust the gap of a spark plug in one operation.*

Vacuum Gauge

A ***vacuum gauge*** is a device that measures the difference in air pressure between the intake manifold and the outside atmosphere. This difference, called *manifold vacuum,* is a reliable indicator of engine load and condition. Under light loads, the engine will run relatively fast and try to pull in more air past the closed throttle plate. Vacuum will be *high.* Under heavy loads, the throttle plate will be wide open. Engine speed will be relatively low and the engine will be able to pull in all of the air it can use. In this case, vacuum will be *low.* A typical vacuum gauge is shown in **Figure 3-3.**

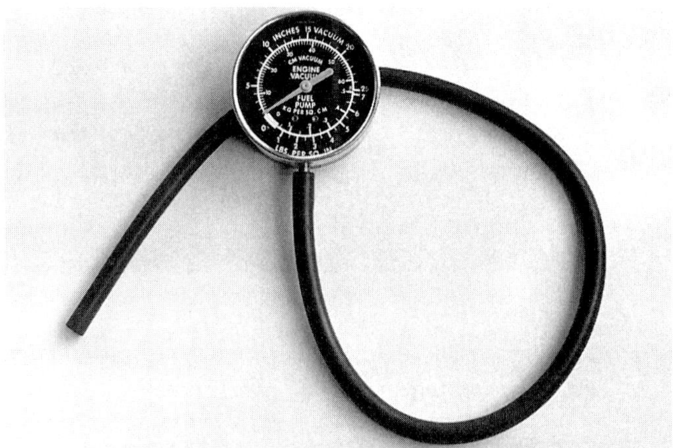

Figure 3-3. *Even on the most modern engine, a vacuum gauge is a valuable diagnostic tool. Manifold vacuum is a good indicator of how well an engine is running.*

A vacuum gauge can be used to help diagnose many internal engine problems. A burnt intake valve, for example, will show up as a sharp vacuum drop at a regular interval when the engine is idling. Retarded ignition or valve timing will show up as a low, steady reading. A reading that fluctuates from normal to about 6″ below normal may indicate an idle system problem or a vacuum leak. The vacuum gauge can also be used to test for proper vacuum to and from vacuum-operated accessories. Diagnosing engine problems using a vacuum gauge will be discussed in more detail in **Chapter 19.**

Vacuum Pump

The ***vacuum pump*** produces vacuum that can be used to test various vacuum-operated engine and vehicle accessories. It is frequently used to test for ruptured vacuum diaphragms. They can also be used to provide low pressure vacuum for other purposes, such as bleeding brakes. The vacuum pump is comprised of two sections: a simple hand pump for producing vacuum and a gauge for measuring vacuum. The pump creates the proper vacuum on the unit to be tested and the gauge measures the vacuum and indicates whether or not the unit can hold the vacuum.

The vacuum pump is often used with other equipment to test such things as vacuum diaphragms and MAP sensors. **Figure 3-4** shows a typical vacuum pump being used to check an EGR control valve diaphragm. Some vacuum pumps are part of the equipment in an engine analyzer. The pump is operated by an electric motor inside the analyzer.

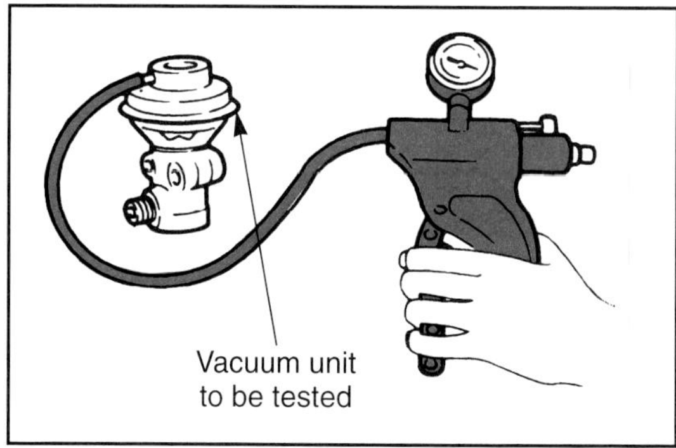

Figure 3-4. *Vacuum pumps are used to develop and measure vacuum. This is useful for checking vacuum-operated devices, such as ignition advance units and EGR valves. (Nissan)*

Compression Gauge

A ***compression gauge,*** **Figure 3-5,** is an air pressure gauge used to measure cylinder pressure. It is attached by a high pressure hose to the spark plug or injector hole of the cylinder to be tested. When the engine is cranked, the upward movement of the piston compresses the air in the cylinder. The compression gauge measures this compression in pounds per square inch (psi) or kilopascals (kPa). When using the compression gauge, all plugs or injectors should be removed to ensure a high enough cranking speed for accurate measurements. Use of the compression gauge is covered in more detail in **Chapter 19.**

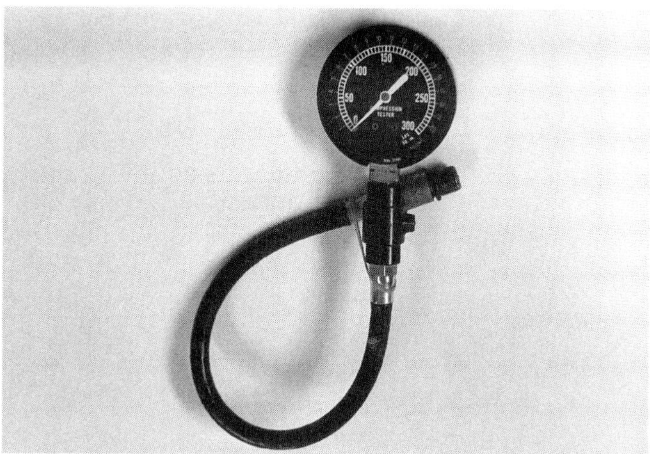

Figure 3-5. *A compression gauge is a pressure tester that is used to determine if the engine cylinders are developing the proper compression. Always remove all the spark plugs and disable the ignition system before making the compression test.*

Fuel Pressure Gauge

The ***fuel pressure gauge*** is a hydraulic (liquid) pressure gauge used to determine if the fuel pump is providing enough pressure and flow to keep the engine supplied with gasoline. A typical pressure gauge test kit is shown in **Figure 3-6.** Some gauges are supplied with a calibrated jar for measuring the flow rate of the pump. Pump flow rates should be obtained from the manufacturer service manual. Generally, a good mechanical pump should be able to deliver about one pint (.47 L) of fuel in 30 seconds at idle speed. An electric fuel pump in good condition should be able to pump one pint (.47 L) of fuel in 10-15 seconds.

Fuel pressure gauges are used on both carbureted and fuel injected engines. The normal fuel pressure on a carbureted engine, and some throttle body fuel injection systems ranges from 1.5 psi to 10 psi. (10.34 to 68.95 kPa). Multiport fuel injected engines can have pressures from 35 to 60 pounds (241.3 to 413.7 kPa). There are other types of gauges used to check engine and automatic transmission oil pressure and brake system fluid pressure.

PCV Tester

The ***PCV tester,*** **Figure 3-7,** is used to test the operation of the PCV system. Most PCV testers consist of a chamber with a ball inside. The PCV tester is placed over the oil filler opening as shown in **Figure 3-8.** If the ball is sucked toward the crankcase, this means that the PCV system is operating properly. If the ball is pushed away from the crankcase opening, the PCV system is not working properly or the piston rings are worn, causing excessive blowby.

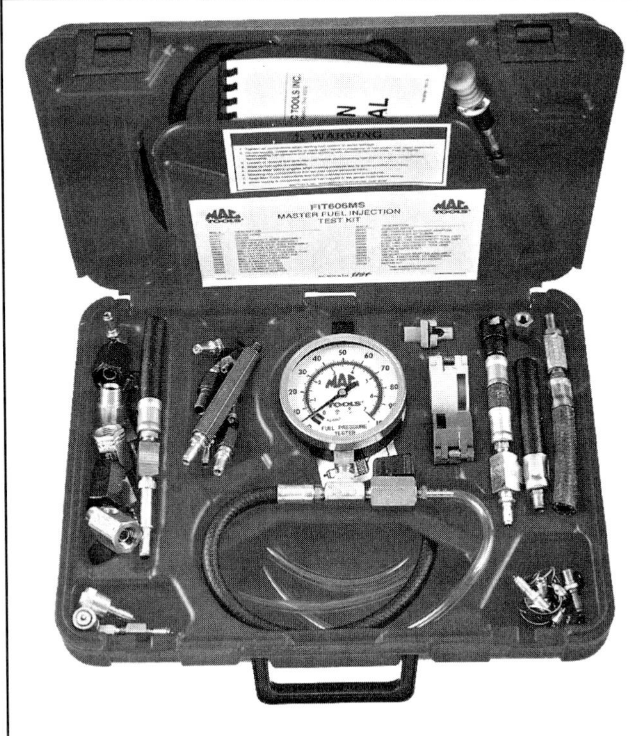

Figure 3-6. *The fuel pressure gauge is used to determine if the fuel system, particularly the fuel pump, is providing the necessary fuel pressure and quantity to operate the engine. (Mac Tools)*

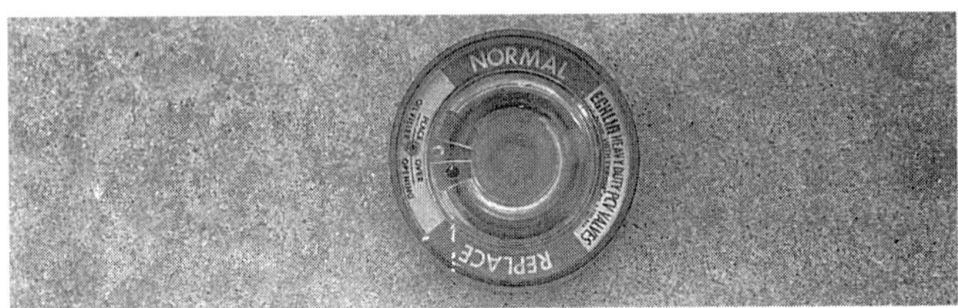

Figure 3-7. *PCV testers can be used on an idling engine to determine if the PCV system is operating correctly. They measure the vacuum developed in the crankcase as the PCV system draws blowby gases out of the engine.*

Figure 3-8. *To use the PCV tester, place it over the oil filler opening as the engine idles. If the PCV system is working correctly, the ball will move to the "normal" side of the chamber. If the engine has excessive blowby or the PCV system is not working properly, the ball will move to the "replace" side.*

Specialized Basic Tools

Since modern engines use electronic ignition, ***ignition spark testers*** are used to check for the presence of a spark, **Figure 3-9.** Many of these tools are simply spark plugs with a ground clip attached to the body. ***Fuel injection harness testers*** or so-called "noid lights" are available to test for pulse in fuel injection system wiring harnesses, **Figure 3-10.** There are hundreds of ***specialized tools*** that are needed by the driveability technician to properly diagnose and service

Figure 3-9. *Spark testers are used to determine if a spark is present in an electronic ignition system. This tool can be adjusted to increase or decrease the gap to help determine spark strength. (Mac Tools)*

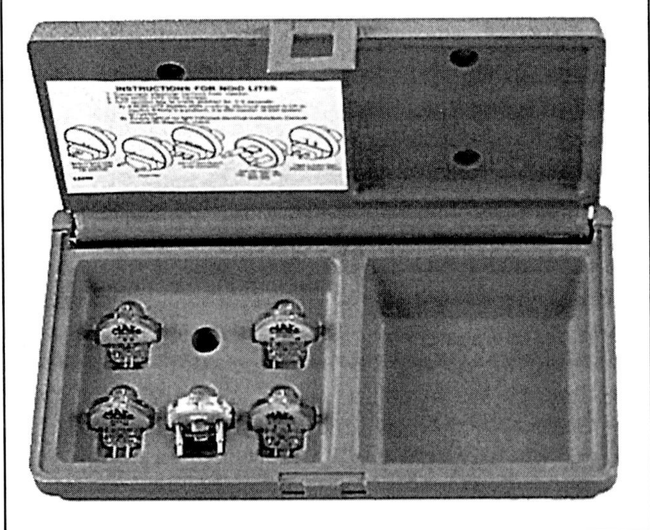

Figure 3-10. *Fuel injector harness "noid" lights, are used to quickly test for the presence of an injector pulse from the ECM. These lights reduce the chance of ECM damage from testing using a low impedance test light. (MAC Tools)*

the modern vehicle. As vehicles becomes more complex, even more specialized tools will be needed to properly diagnose and service the various systems.

Computer Memory Retaining Tool

Another specialized basic tool which is helpful to the driveability technician is the ***computer memory retaining tool,*** **Figure 3-11.** If the battery is disconnected from the vehicle, the ECM will lose all of its memory. This means that diagnostic codes, as well as learned programming is lost, and must be relearned as the engine operates. The vehicle may run poorly until the ECM relearning process is complete. To prevent loss of computer memory, the computer memory retaining tool is inserted into the vehicle cigarette lighter. A 9-volt battery maintains voltage to the ECM, keeping memory alive. Some memory tools use 110-volt shop current to provide the voltage signal. Memory tools should be plugged in before the battery is disconnected.

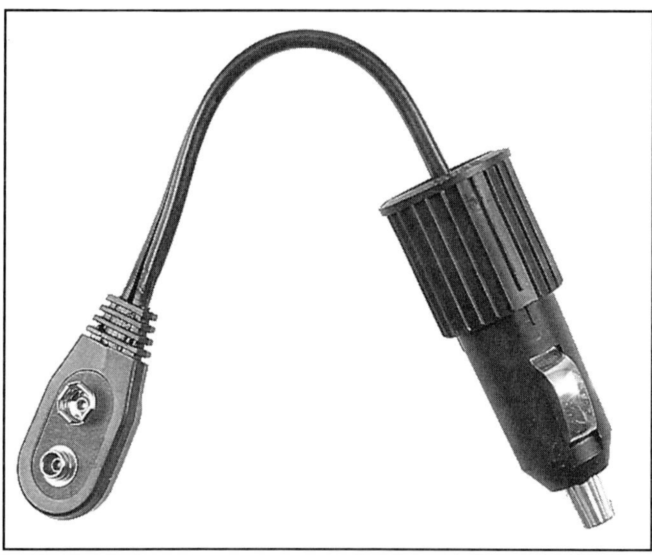

Figure 3-11. *A computer memory retaining tool can save the driveability technician time by preventing the need to reset idle speed and other ECM setting on some vehicles. (Mac Tools)*

Borescope and Stethoscope

The ***borescope*** is useful for locating defects or foreign objects in inaccessible places such as intake manifolds or plenums, cylinder head ports, exhaust pipes, and engine cylinders. The borescope is equipped with a light to illuminate the interior of the part being checked. Using a borescope will reduce the chance that a part is damaged by a foreign object, and also eliminates the possibility of disassembling an engine in good condition.

Many vehicle noises can be located with a ***stethoscope***. The stethoscope includes a long probe which is placed against the suspected part. Any noises are amplified and picked up by the earpieces. Some stethoscopes are electronic units capable of amplifying and filtering engine and drive train sounds.

Temperature Tester

To check the operation of vehicle temperature sensors, the technician must know the exact temperature of what the sensor is measuring. Temperature testers are designed to accurately measure the temperature of the engine coolant, incoming air, and exhaust manifolds. These can be compared to the readings provided by the sensor to determine the condition of the sensor. Many temperature testers use a probe which is placed directly on the unit to be checked, while others are able to take infrared temperature readings from a distance.

Devices are available to allow a standard multimeter to read temperature. They consist of a temperature probe and a device to convert the temperature reading to a value which can be read on the multimeter. See **Figure 3-12.**

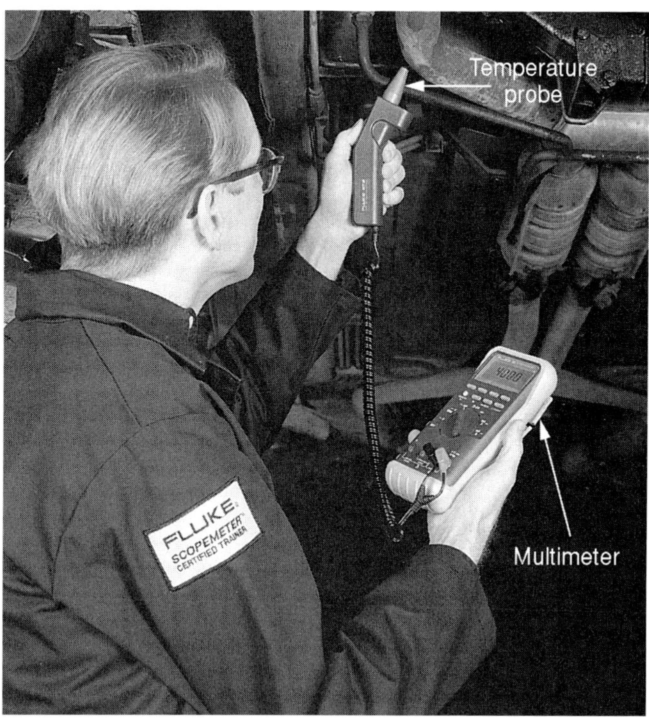

Figure 3-12. *A temperature probe can be connected to some modern multimeters. A temperature probe can allow the technician to safely check the exhaust system for restrictions by testing for variations in temperature. (Fluke)*

Hand Tools

Many common hand tools can make adjustments, diagnosis, and part replacement easier for drivability technicians. These tools include idle speed adjusters, specially shaped vacuum line plugs, fuel fitting removal tools, and special sockets for oxygen sensor removal. The drivability technician may have reason to use many other tools, such as carburetor and spark plug gap adjusters, steel and brass feeler gauges, micrometers, dial indicators, spark plug boot removers, and torque wrenches.

Basic Electrical Test Equipment

The following test equipment is electrical in nature and can be used to make many tests on modern engines and systems related to engine performance. Most of the equipment discussed here can be made or purchased as individual units, or as part of multiple use test equipment.

Jumper Wires

Jumper wires are short lengths of wire with alligator clips on each end, such as the one shown in **Figure 3-13.** They are used to make temporary electrical connections during vehicle testing procedures. Although they can be purchased, jumper wires can be made easily from a length of wire and alligator clips. Use copper wire that has a large enough gage to carry several amps of current, but small enough to be easily handled. 12 or 14 gage wire is a good size for most jumper wires. Cut about 2' to 3' (60.96 to 91.44 cm) of wire from the roll, then strip about .5" (12.7 mm) of insulation from each end of the wire. Crimp the alligator clips to the wire, making a secure joint at each end. For additional holding power and conductivity, the clips and wires can be lightly soldered together. It is also possible to use special connectors on the end of the wires for special uses. Some examples are shown in **Figure 3-13.**

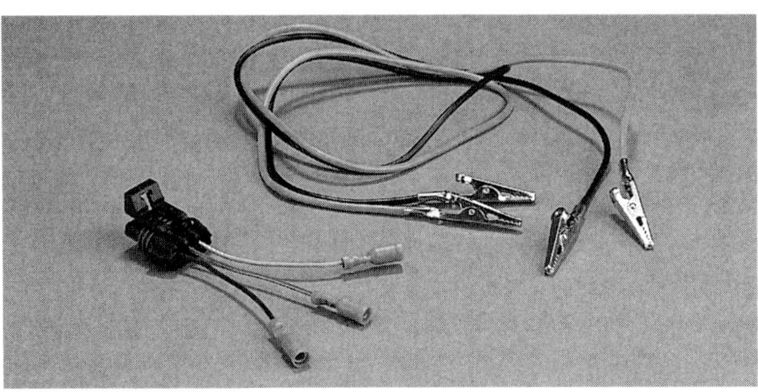

Figure 3-13. *Jumper wires with alligator clips are a handy tool for bypassing switches or relays and applying voltage or ground to various components. Some jumper wires can also be made from plug-in terminals cut from old wiring harnesses. This type of connector can be plugged into a device, such as a sensor, for testing.*

Note: It is a good idea to install an inline fuse or other circuit protection device in the jumper wire. This helps to protect the circuit wiring and any electronic components should a short circuit be created accidentally.

Another good source for jumper wires is the wiring harness in a wrecked vehicle. These can be found easily and inexpensively in most auto salvage yards. The advantage to having these is that they come with the original terminals, which ensures a good connection to the device to be tested. They are handy for testing sensors, fuel injectors, and other parts that are hard to reach with a meterís test leads. Remove the terminal from the sensor, injector, or part and then cut the terminal from the harness. Strip and crimp a terminal on the wire(s) as described earlier. Additional wire can be added if the connector is used in a hard-to-reach location. These are handy if you only work on a few models, such as at a dealer or specialty shop.

Jumper wires are often used to bypass electrical devices, such as relays and solenoids. Bypassing an electrical device will enable the technician to determine whether it is defective. Jumper wires can also be used to access the computer memory on many vehicles by grounding the diagnostic connector. They are useful for making an extended electrical connection, such as connecting an engine test lead to a test device inside the passenger compartment for a road test.

Caution: When using jumper wires, make sure that the current load does not exceed capacity. If the jumper wire insulation begins to discolor, blister, or smoke, remove the wire immediately using a shop towel or glove to prevent burns. Never grab a hot jumper wire with your bare hands. A severe burn could result. Never use a jumper wire to bypass a fuse or other circuit protection device.

Test Light

A ***test light*** is an electrical tester comprised of a 12-volt light bulb, two terminals, and connecting wiring. A test light is shown in **Figure 3-14.** Using a test light is a quick way to check for current in an electrical device or circuit. It is also used to test ignition wires for arcing to ground. However, test lights will not measure the amount of voltage or resistance.

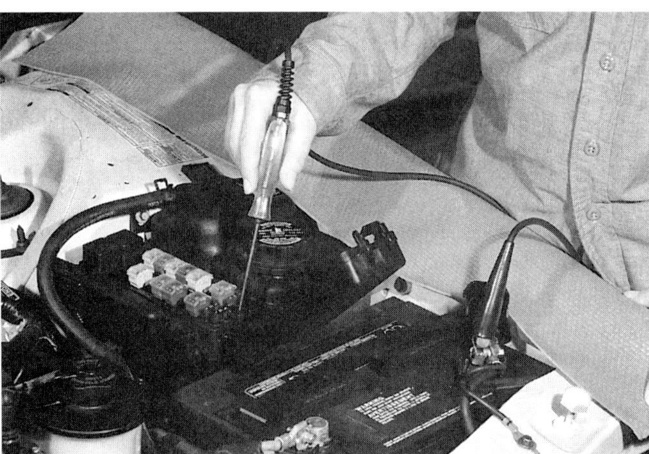

Figure 3-14. *This test light is being used to make a quick check for the presence of voltage. Other uses include testing ignition secondary wiring for leakage or shorts to ground. A variation of this is the powered test light, which is used to check for continuity.*

There are two types of test lights. The ***powered test light*** or self-powered, contains a battery and is used to check for a complete circuit (continuity) when no current is flowing in the circuit. The ***nonpowered test light*** is connected between a powered circuit and ground. It will light if voltage is present in the circuit. Proper use of the test light will be covered in later chapters.

Caution: Some computer and electronic components may be ruined by even the slightest amount of current flowing through a test light. A good quality test light has a minimum impedance (resistance to input current) of 10 megohms. This amount of impedance protects the tested circuit and any electronic components from excessive current flow. Also check the vehicle's service manual to make sure that you are using the proper test procedures.

Analog and Digital Meters

Meters are used to measure engine speed, dwell, resistance, voltage, current, and injector pulse. They are available in either analog or digital versions. ***Analog meters*** have a needle or pointer to indicate readings across a printed scale face. ***Digital meters*** display a number to indicate the reading. The numbers are formed by light emitting diodes (LEDs) or liquid crystal displays (LCDs). Meters are also available to measure other engine and vehicle functions.

Multimeters

Multimeter is a term for hand-held meters which can measure resistance, voltage, amperage, and other electrical properties. Some of the more elaborate multimeters can measure dwell, engine speed, sound levels, and temperatures. Some of the latest multimeters can measure the pulse timing and duration of fuel injectors, carburetor solenoids, distributor pickup coils, and Hall-effect switches and can display a computer's diagnostic codes as electronic pulses.

Multimeters can be either analog or digital. Most modern multimeters are the digital type like the one shown in **Figure 3-15.** Multimeters usually have red positive (+) and black negative (–) leads, and sometimes other leads for special functions. Polarity should be carefully noted when making some tests. Impedance is the resistance to current flow caused by the internal components of the meter. A high impedance meter can be used to take readings without the use of high current flowing through the meter.

Newer multimeters are designed to have high impedance. This is important when checking computer and other solid-state devices, since high current flow through the meter could damage the small circuits in the device being tested. With the exception of a few single-purpose meters, most meters sold today are high impedance multimeters.

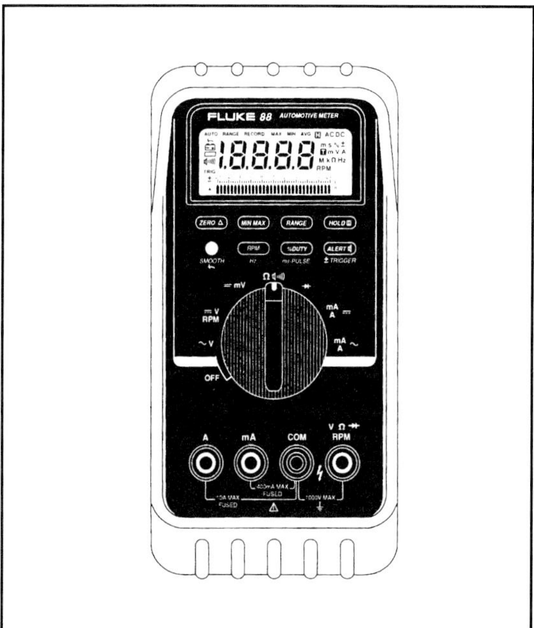

Figure 3-15. *Digital meters use a light emitting diode (LED) or a liquid crystal display (LCD) to display values as direct number readings. (Fluke)*

In this text, multimeters will be discussed as though they are the single-purpose meters described in the following sections.

Ohmmeter

Ohmmeters are used to measure resistance in an electrical device or circuit. Resistance is the opposition to current flow that exists in any electrical circuit or device. This resistance is measured in ohms. Ohmmeters can only check an electrical circuit when no current is flowing in the circuit. A typical ohmmeter connection is shown in **Figure 3-16.**

An ohmmeter has two leads, which are connected to each side of the unit or circuit to be tested. Polarity (direction of current flow) is not important when checking resistance, except in the case of diodes or some computer circuits.

Note: Always consult the manufacturer's service manual before using an ohmmeter to check any computer control circuit.

Most ohmmeters have selector knobs for checking various ranges of resistance values. Analog ohmmeters have a special knob to adjust the needle to zero before checking resistance. The needle should always be adjusted to zero whenever the range is changed.

Voltmeter

Voltmeters are used to check voltage potential between two points in an energized circuit. The circuit must have a source of electricity available before voltage can be checked. A typical hand-held voltmeter is shown in **Figure 3-17.** On some voltmeters, different scales can be selected, depending on the voltage type and level to be measured.

Always observe proper polarity when attaching voltmeter leads. When a voltmeter is used on a modern vehicle with a negative ground, the negative lead should always be connected to the frame or other ground, and the positive lead should be attached to the positive part of the circuit.

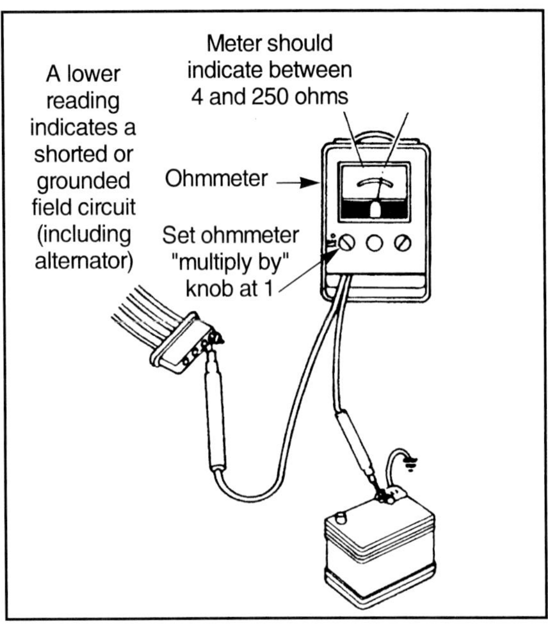

Figure 3-16. *Ohmmeters can precisely check the resistance of a circuit. They are always used on devices or circuits that have no electrical current present. (Ford)*

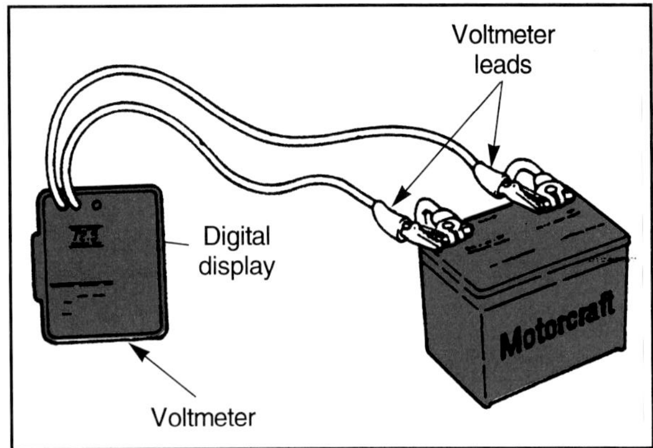

Figure 3-17. *Voltmeters measure the exact voltage potential in a circuit. To make a voltage check, electricity must be available in the circuit. (Ford)*

Ammeter

An ***ammeter*** is used to check the current flow in a circuit. Ammeters can also be used to check the current draw of motors, solenoids, and other electrical devices, when the technician suspects that they are drawing too much current. Most ammeters have two terminals for the positive lead. One is used to check low amperage, usually under 1 ampere. The other terminal is connected to the meter through a high amperage fuse and can be used to check circuits with current loads up to 10 amperes, **Figure 3-18.**

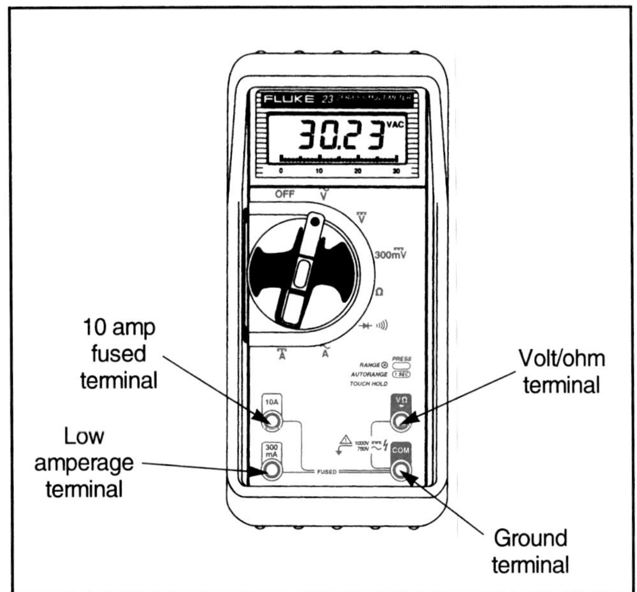

Figure 3-18. *The multimeter can be used to check amperage up to 10 amperes. Note the two amperage terminals on this digital multimeter. (Fluke)*

Heavy duty ammeters are used to check the amperage draw of starters or other motors, and when checking the charge level of a battery. These are usually part of a charging and starting system tester, which will be discussed later. Most ammeters used in automotive service are part of a multimeter.

Dwell Meter/Tachometer

The ***dwell meter/tachometer*** is connected to the primary side of the ignition system and measures engine speed. On older vehicles with point-type ignition systems, the dwell/tach can measure the time (expressed in degrees of distributor rotation) that the contact points are closed. The dwell reading is used to adjust the contact points. **Figure 3-19** shows a typical dwell meter.

Although point-type ignitions are no longer used, dwell developed by the electronic ignition module can be checked by attaching a dwell meter/tachometer to the coil connections. However, the dwell cannot be adjusted. On some late model vehicles with computer-controlled feedback carburetors, the dwell meter can be used to measure the pulse duration to the carburetor's mixture control solenoid.

Dwell/tachometers have at least two leads. One lead is connected to the distributor side of the ignition coil, and the other lead is connected to ground. Some dwell/tachometers

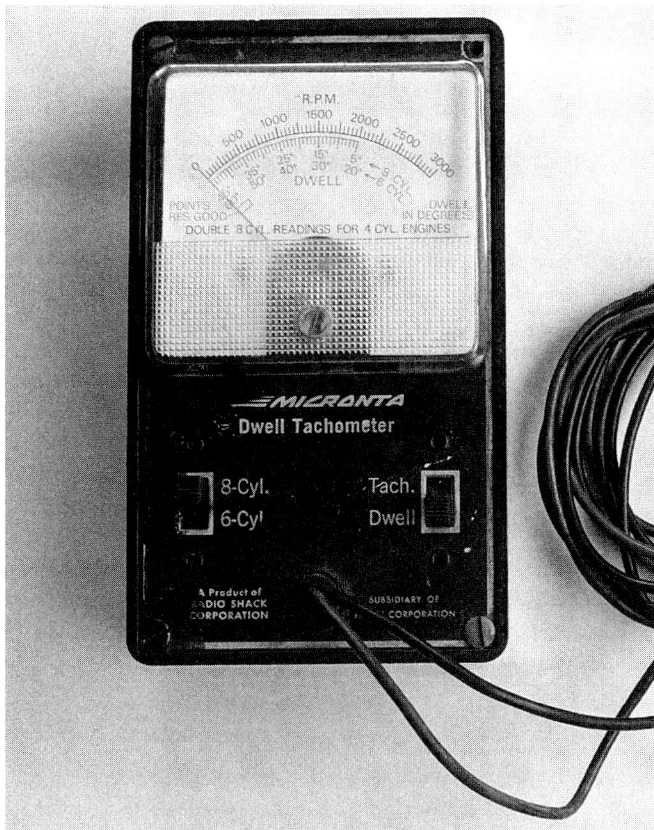

Figure 3-19. *Dwell meters are used to check the point setting on point type ignition systems. Other uses include checking the duty cycle on carburetor mixture control solenoids.*

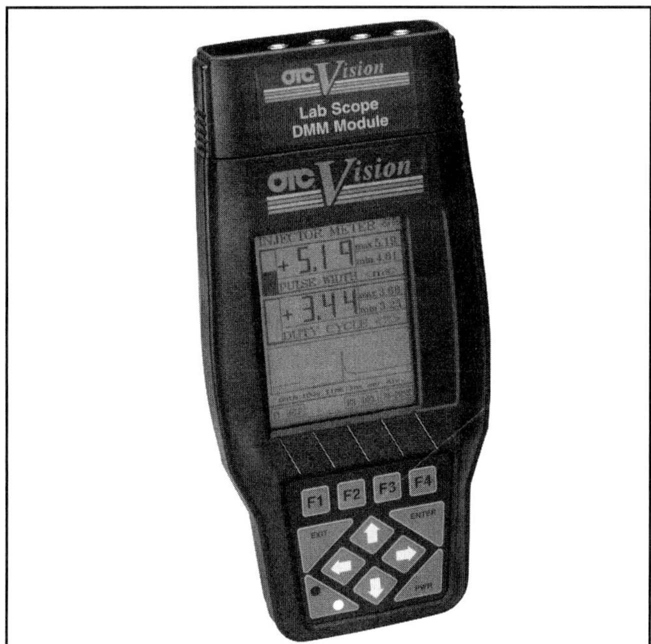

Figure 3-20. *Modern multimeters can have a scope feature that can be used to check for pulse and waveform in certain systems. (OTC Tools)*

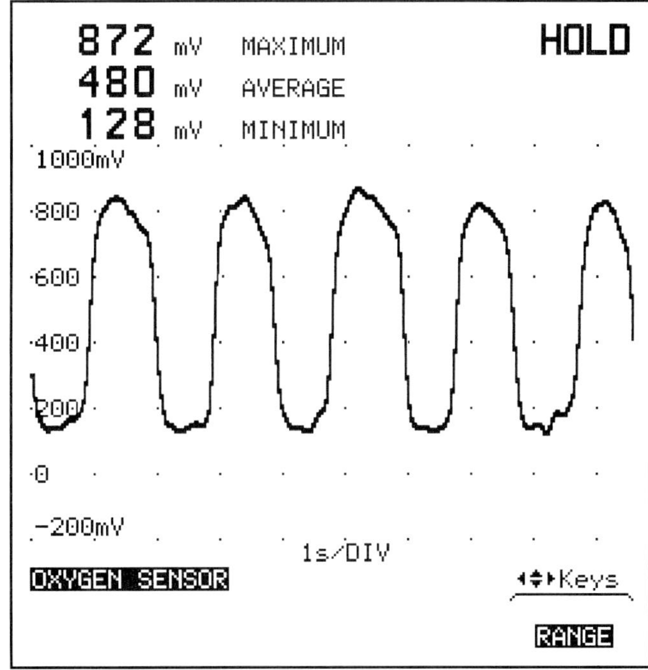

Figure 3-21. *Measuring waveforms can give a good indication of device operation. This is a typical waveform for an oxygen sensor. (Fluke)*

have an inductive pickup, which is placed over a plug wire for the tachometer reading. The tachometer section of the dwell-tach may have several ranges. The lower ranges are used for setting idle speeds, while the higher ranges are used to make various high speed tests.

Pulse and Waveform Meters

The ***pulse*** and ***waveform meter*** is used to measure the control voltages applied to such devices as carburetor solenoids and fuel injectors. In most cases, the pulse rate is controlled by the computer, and checking the pulse rate is a way of checking computer system operation. Pulse rate changes in response to changes in sensor inputs indicate that the input sensors, computer, and output devices are all working. Pulse meters are usually part of the more sophisticated digital multimeters, **Figure 3-20.**

The control voltage from the ECM has a particular shape or ***waveform.*** The waveform is always read across as the time and up as the amount of voltage. The shape of the waveform is a clue to the operation of the particular circuit. If the waveform does not match the standard pattern, a problem is most likely present. A typical waveform is shown in **Figure 3-21.**

Timing Light

A ***timing light*** is a type of rapidly flashing light called a strobe light. The bright, quick flash of the strobe light is used to synchronize, or match the operation of the engine compression and ignition systems. The timing light is connected to the number one spark plug wire. On older timing lights, an adapter was placed between the spark plug and wire. Most modern timing lights use an inductive pickup that is clamped over the wire to sense the magnetic field around the wire. This connection causes the strobe light to flash every time the number one spark plug fires. This matches the light to the ignition system.

A mark on the rotating vibration damper or crankshaft lines up with the zero mark on a stationary part of the engine every time the number one cylinder comes up to the top of its compression stroke. When the timing light is pointed at the timing mark, the flash of the strobe will appear to stop or "freeze" the mark. If the light freezes the mark next to the zero on the stationary plate, the number one plug is firing at the top of the cylinder's compression stroke. The distributor can be loosened and turned until the timing marks appear in the proper position. This adjusts the ignition timing. On some ignition systems controlled by the computer, the timing cannot be adjusted. However, timing can sometimes be checked as an aid to diagnosis. **Figure 3-22A** shows a typical timing light.

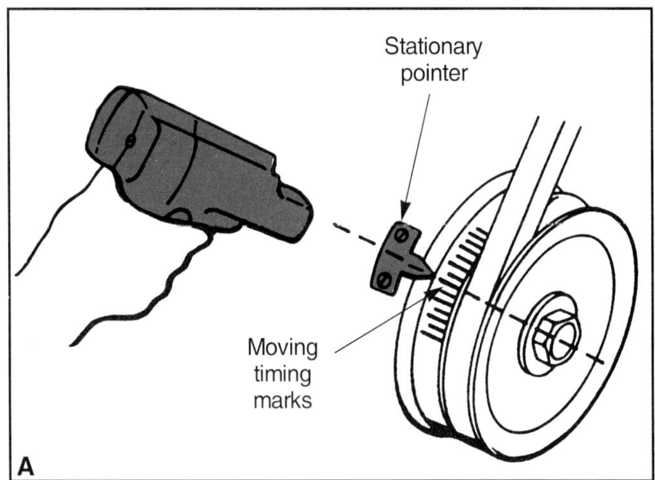

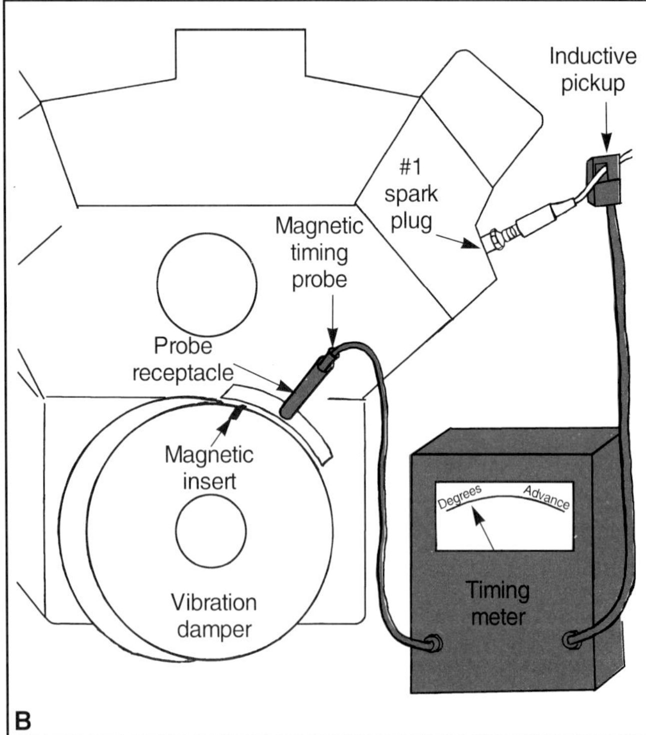

Figure 3-22. *A—Timing lights have been used for many years to set ignition timing. The best lights have a built-in advance meter. B—The timing meter uses a magnetic probe that detects exact crankshaft position, and therefore, the exact position of the number one piston.*

Timing lights can be adjustable or nonadjustable. Nonadjustable lights flash only when the plug fires. The timing must be checked by observing the engine timing marks. This does not affect the ability to set the initial timing. However, it is usually not possible to accurately check total advance, since most engines are not marked with the 30° to 40° of advance that most ignition systems use.

Adjustable timing lights contain internal circuitry that can advance or retard the timing of the strobe flashes. This simplifies initial timing setting and allows total advance timing to be checked. An adjustment knob on the timing light can be turned to bring the timing mark to zero. The advance can then be read on a dial in the timing light housing. Another job that can be performed by the timing light is checking the spray pattern of throttle body fuel injectors. Since most injectors fire with the spark plugs, the pattern can be noted by pointing the light at the suspect injector as the engine operates.

Other Types of Timing Equipment

While the timing light is still the standard device for checking ignition timing, another type of timing tester has been available since the 1970s. This tool consists of a magnetic probe, an inductive pickup, and associated electronic circuitry in a small case. The degrees of timing advance or retard are read on a meter in the case. This kind of tester is usually called a ***timing meter.*** A timing meter is more accurate than a timing light, since there is no possibility of misreading the timing marks.

The timing meter is connected by inserting the magnetic probe in a receptacle built into the engine at the vibration damper, **Figure 3-22B,** and clamping the inductive pickup around the number one spark plug wire. The engine is started and a magnetic strip on the rotating vibration damper creates a signal in the magnetic probe. The meter receives signals from the magnetic probe and the inductive pickup around the plug wire. The meter circuitry measures the amount of time between the signal from the magnetic probe and the inductive pickup, and compares them with engine speed. It uses these readings to determine the engine timing in degrees.

A related timing device is used to adjust the fuel injector timing on diesel engines. The diesel injector begins the combustion process when it squirts fuel into the combustion chamber, just as the spark plug begins the combustion process in a gasoline engine. Therefore, diesel injector timing has the same importance as ignition timing on a gasoline engine. The diesel timing meter, **Figure 3-23,** is used to determine injector timing by detecting the light flash in the cylinder as the mixture fires.

A probe, called a ***luminosity meter,*** is installed in the glow plug hole and detects the firing of the cylinder. This tells the timing device when the injector is opening. The piston position is determined by an electrical signal from magnetic timing marks. On some meters, the cylinder flash is converted to a strobe light flash and timing is set by observing the marks on the engine flywheel. Comparing the flash and timing marks allows the technician to accurately time the engine. Some diesel timing meters use a transducer that measures engine timing by sensing the number of pulses in the number one cylinder's fuel injector line.

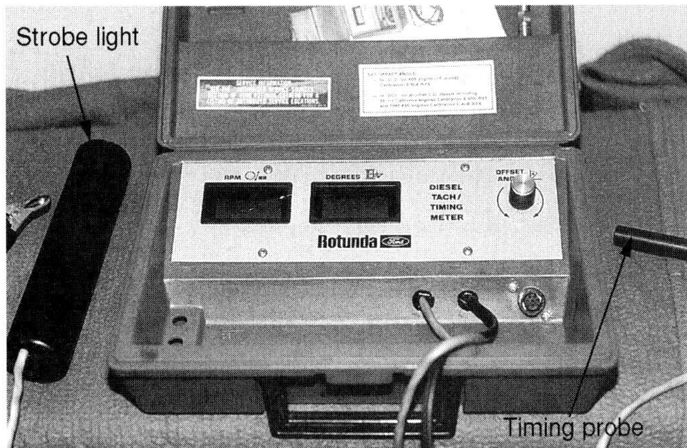

Figure 3-23. *To time a diesel engine, a luminosity probe is used. The probe detects the flash from the firing air-fuel mixture and converts this to a direct reading or to flashes of a strobe light that is directed to timing marks on the engine.*

Ultrasound Vacuum Leak Detectors

Ultrasound vacuum leak detectors, **Figure 3-24,** can locate vacuum leaks by sound. This procedure is much more accurate than other methods of leak detection. As air rushes through a vacuum leak, it creates a distinctive high pitched sound. The frequency, or wavelength of this sound is so high that it cannot be heard by the human ear. The circuits of the ultrasound leak detector are able to pick up, or "hear" this frequency. The ultrasound detector consists of a probe containing a microphone and internal circuitry capable of detecting the high frequency sound waves that indicate a vacuum leak.

Moving the probe of the ultrasound detector over the probable sources of a vacuum leak will allow the detector to locate the exact point of the leak. The detector will signal the leak with a high pitched sound, or by a display on its control panel. Some ultrasound detectors have a sensitivity adjustment, which allows for more accurate detection of varying types and sizes of vacuum leaks.

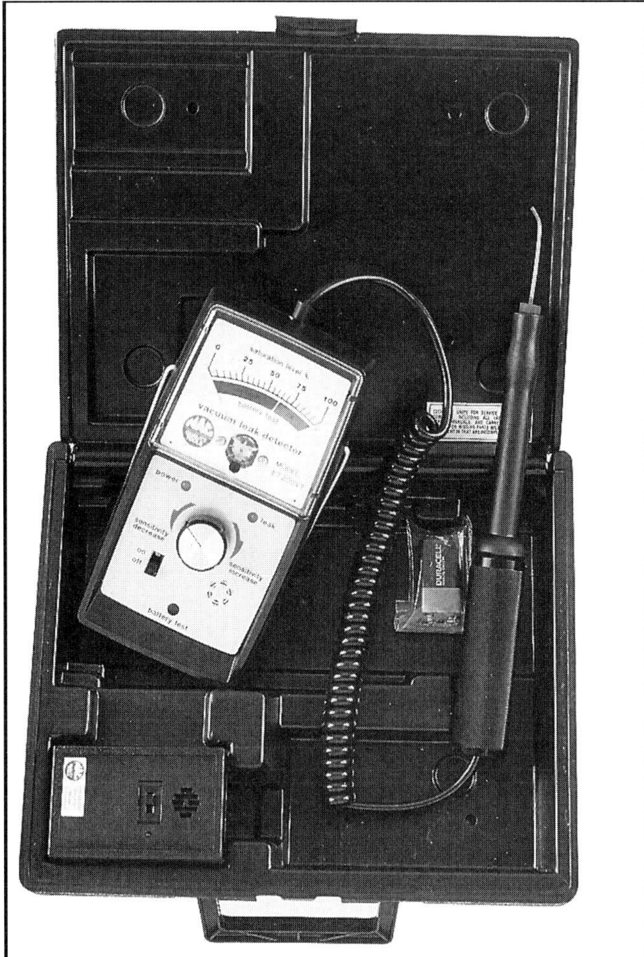

Figure 3-24. *Ultrasound leak detectors are useful for locating vacuum and other audible leaks. While most of these noises are too high pitched to be heard by the human ear, that are easily detected and can be pinpointed by this tool. (Mac Tools)*

Other Vacuum Leak Detectors

Another type of vacuum leak tester can locate vacuum leaks without the engine running. This tester operates by pumping a harmless vapor, or "smoke," into a closed area, such as the intake manifold. Within 20-30 seconds, the vapor will begin to exit through any leaks. Wherever vapor can get out, air can get in, indicating a vacuum leak. No vapor means that there are no leaks in the intake system.

Other Electronic Test Equipment

More than ever, technicians have a greater variety of electronic tools that they can use to assist them in diagnosis and repair. Many of these tools are highly specialized, and are able to perform direct tests on components that would have been testable by using awkward methods and in some cases, by replacement of the suspect part with a known good part.

Input Sensor Testers

Input sensor testers are specialized test devices dedicated to the computer control system input sensors. They can directly test the condition of input sensors by measuring their output voltages. To test the sensor, the simulator is attached to the sensor wiring and displays readings as the engine operates. Some sensors can be tested with the engine off. When necessary, the simulator can supply a reference voltage to the sensor. Typical single purpose sensor testers check the operation of throttle position sensors, MAP (manifold vacuum) sensors, oxygen sensors, and EGR valve position sensors. Some of these testers can also check the Hall-effect switches used in camshaft and crankshaft position sensors. A typical sensor tester is shown in **Figure 3-25.**

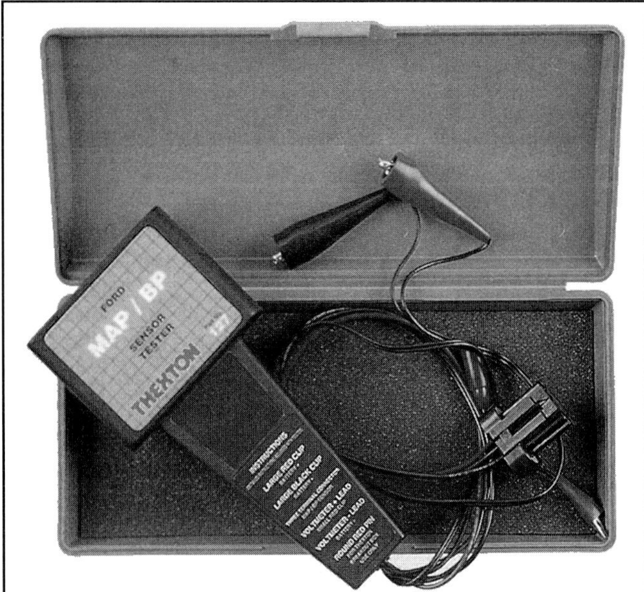

Figure 3-25. Dedicated testers are designed to check only one type of ECM component. This tool is used to check MAP and barometric pressure sensors. (Mac Tools)

Input Sensor Simulators

Sensor simulators are specialized test devices which can check the operation of the computer control system and test sensor operation indirectly by checking the operation of the ECM and related output devices. The simulator is used by disconnecting the suspect sensor and substituting the simulator. The simulator can be adjusted to change the input signal reaching the ECM. If the ECM sends the proper output signals based on the simulator inputs, the sensor can be assumed to be defective. If the ECM does not produce the proper outputs, it, rather than the sensor is defective. The simulator shown in **Figure 3-26** is a multi-function unit which can be used to simulate the oxygen, knock, mass airflow, MAP, temperature, and throttle position sensors.

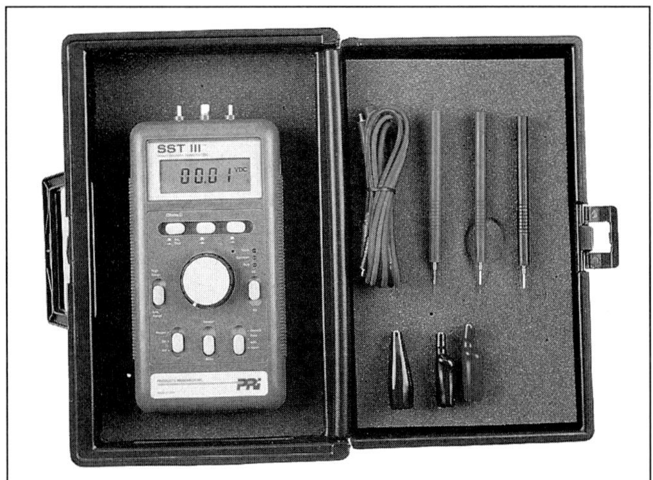

Figure 3-26. *Some tool manufacturers have designed sensor simulators that can duplicate the sensor readings of most engine and vehicle sensors. (Mac Tools)*

Output Device Testers

Output device testers are often used to test and adjust output devices such as idle air controls and the operation of idle speed control motors, idle bypass valves, ignition modules, distributorless ignitions, and diesel glow plug relays.

Idle Control Motor Testers

Some of the most common types of output device testers are ***idle control motor testers.*** A typical idle control motor tester is shown in **Figure 3-27.** This tester allows the idle speed and motor operation to be checked, and the motor adjusted if necessary. Some idle speed control testers can also be used to increase engine speed for checking timing advance and making other high RPM checks.

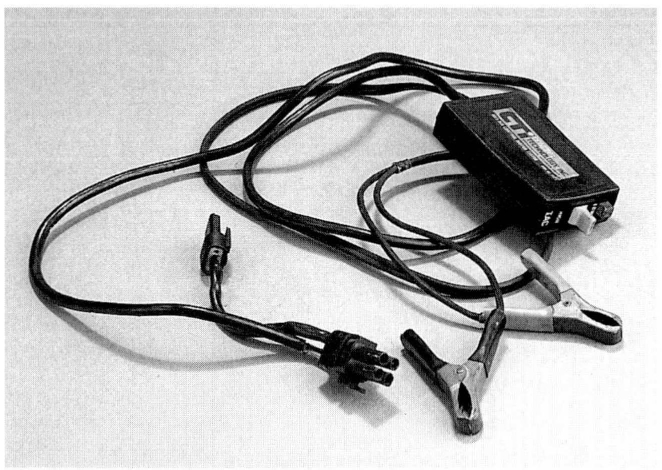

Figure 3-27. *This tester can be used to check the idle air control valve pintle for operation and binding.*

Ignition Module Tester

Ignition module testers are divided into two main classes, general testers which can test several kinds of modules, and the specialized, or dedicated, tester which can be used to check one type of module. Many of these testers can also test the pickup coil or Hall-effect switch. Other testers can test the components of distributorless ignition systems.

Fuel Injector Tester

Fuel injector testers can check the condition of the fuel injector windings as the injector operates. This is a more accurate method of testing than checking the injector with an ohmmeter, since the operation of the injector under normal current flow is being observed.

Glow Plug Circuit Tester

This tester checks the glow plugs by energizing each plug to simulate actual operation. By determining whether it heats up properly, and the accompanying voltage and current readings, the technician can get an accurate indication of glow plug condition.

Air-Fuel Monitor

The ***air-fuel monitor*** is a tool which can check overall system operation. The probe from the air fuel monitor is attached to the oxygen sensor and the monitor is placed under the hood. This allows the technician to closely monitor air-fuel ratio and quickly determine how adjustments and service operations affect engine operation. The air-fuel monitor can therefore check the entire system as the technician works on it.

Portable Diagnostic Testers

Portable diagnostic testers are small, hand-held testing devices used to check specific systems and components. They differ from multimeters in that they check the operation of individual systems, sensors and electrical parts, instead of universal qualities such as voltage or resistance.

Some portable testers are used to test an individual electronic component, with the component either installed or removed from the vehicle. These testers are attached directly to the leads of the device to be tested, usually with connectors identical to the one on the vehicle. Most of these testers display information by a series of lights, which illuminate to indicate whether the component is good or bad. These testers are usually dedicated to the component to be tested, and cannot be used for other tests. In some cases, the component is used on only one vehicle. Many vehicles use components that are similar in design, therefore, one tester can check the same component on several different vehicle makes.

Dedicated Scan Tools

Another type of diagnostic tester can retrieve computer trouble codes and display them in a readable form. It also allows the technician to clear codes and check for proper sensor operation. These are often called ***dedicated scan tools,*** since they are designed to work with, or be dedicated to the computer systems of only one manufacturer. All scan tools are designed to be used with the vehicle's ***data link connectors,*** or output terminals.

This type of portable tester usually contains a small LED or LCD screen for displaying the codes and other system information. The scan tool will also contain pushbuttons or switches for selecting different functions and possibly a numeric (number) keypad similar to keys on a pocket calculator. Many dedicated scan tools are able to test all the systems of one manufacturer, through the use of selector switches or special harness connectors. **Figure 3-28** shows a dedicated tester specifically designed to be used on one make of vehicle.

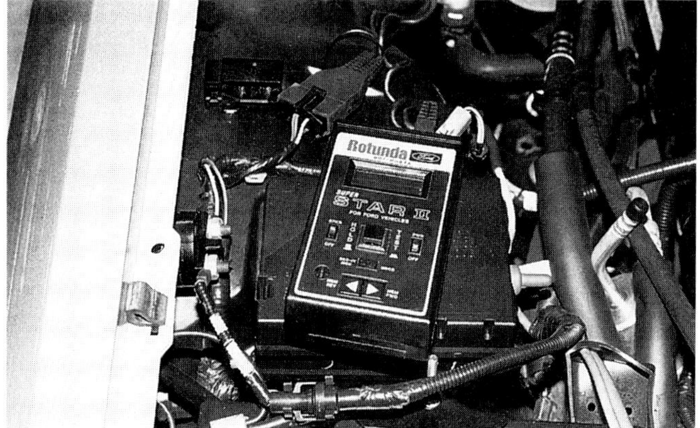

Figure 3-28. *Dedicated computer testers will check all the computer control systems of one manufacturer.*

Multi-System Scan Tools

When manufacturers began to computerize brake, transmission, body, steering and suspension, and airbag systems, it became necessary to expand the compatibility, as well as the diagnostic capabilities of the scan tool. Many of these systems require a scan tool not only to diagnose problems, but to perform certain repair procedures as well. These ***multi-system scan tools*** are able to retrieve trouble codes from more than one type of computer system on different lines of vehicles, and are sometimes referred to as multi-line scan tools.

A typical aftermarket multi-system tester is shown in **Figure 3-29.** It consists of a diagnostic unit and software packages, or modules. The modules are designed to interface with, or talk to, the various computers in a manufacturer's computer-controlled systems. Different modules are available for different systems and lines of vehicles. The module for a particular system or vehicle is plugged into the scan tool. The central unit contains basic circuitry for accessing the computer memory of several makes of vehicle, and a LED or LCD screen for displaying information. It is also equipped with wiring harnesses to connect with vehicle self-diagnostic connectors.

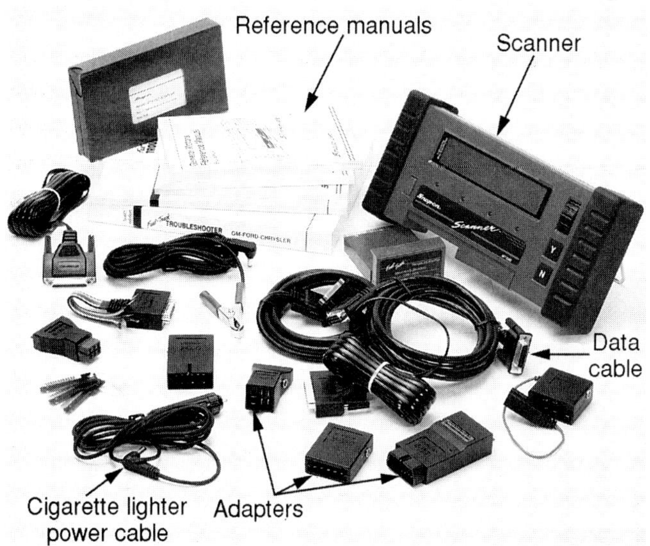

Figure 3-29. *Multisystem or generic scan tools can test the computer control systems of many different makes of vehicles. (Snap-on Tools)*

The module can read and interpret the information being sent from the computer control system. It translates the system information and displays it on the screen. To check a different make of vehicle, the memory section, **Figure 3-30,** is exchanged with the proper memory unit for the vehicle or system to be checked. Most tool manufacturers are now producing memory cartridges that can be used with more than one manufacturers' vehicle and their various computerized systems.

Figure 3-30. *The software cartridge can be changed in many multisystem testers to check different types of vehicles and systems. Information in the cartridges can be updated as new models are introduced and factory specifications are changed. (OTC Tools)*

Generic Scan Tools

The newest ***generic scan tools*** are able to access the computer systems of more than one manufacturer. These scan tools can perform system checks, tests, and can help the technician diagnose an intermittent performance problem by taking a "snapshot" of the sensor inputs and computer outputs while the problem is occurring. Some have multimeters built into them, **Figure 3-31.**

The newest scan tools have memory boards, rather than cartridges that can be updated by downloading new vehicle or system specifications through a computer from a CD-ROM disc. Scan tools have become such an essential part of engine performance and automotive repair, that some technicians purchase scan tools for their own use.

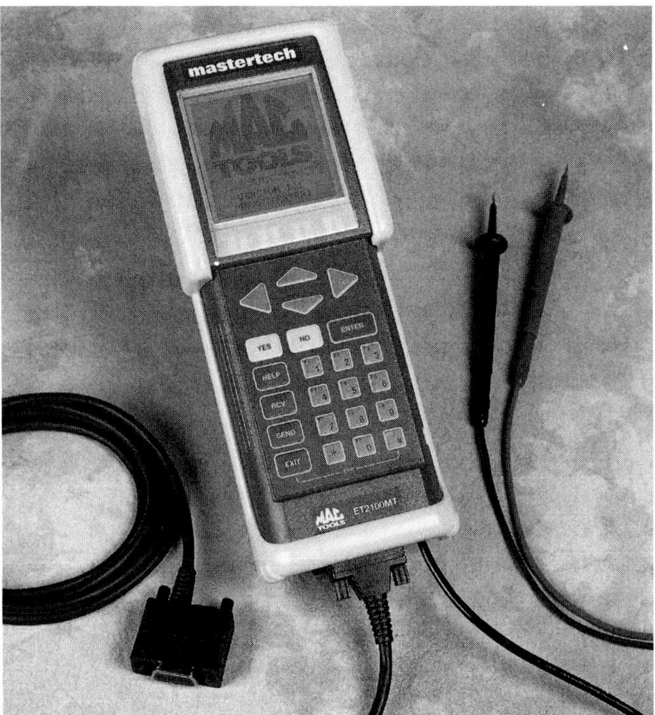

Figure 3-31. *Aftermarket generic scan tools provide multiple functions in a single tool. This scan tool also contains a fully functional multimeter. (Mac Tools)*

Breakout Boxes

With the increased amount of wiring in the modern vehicle, technicians needed a way to quickly test circuits from a central location. ***Breakout boxes*** are designed to allow the technician to use a multimeter to check voltage and resistance values in a system's various circuits, **Figure 3-32.** A breakout box is simply a box containing numbered test terminals and a harness designed to mate with a system's main wiring harness connector. The breakout box is connected to a system's harness, such as a computer wiring harness connector. The meter's terminals are then plugged into the various numbered terminals to check for proper voltage or resistance. A manual is used to outline the specific test procedures for each problem. This allows the technician to quickly isolate problems in complex electrical systems.

Large Testers

This section will cover the general types of large test equipment used to diagnose engine performance problems. The term large test equipment includes many kinds of testers. Some test equipment, like the oscilloscope and exhaust gas analyzer can be single pieces of equipment dedicated to one particular test, or part of a larger unit with other kinds of testers. Other large testers are combinations of

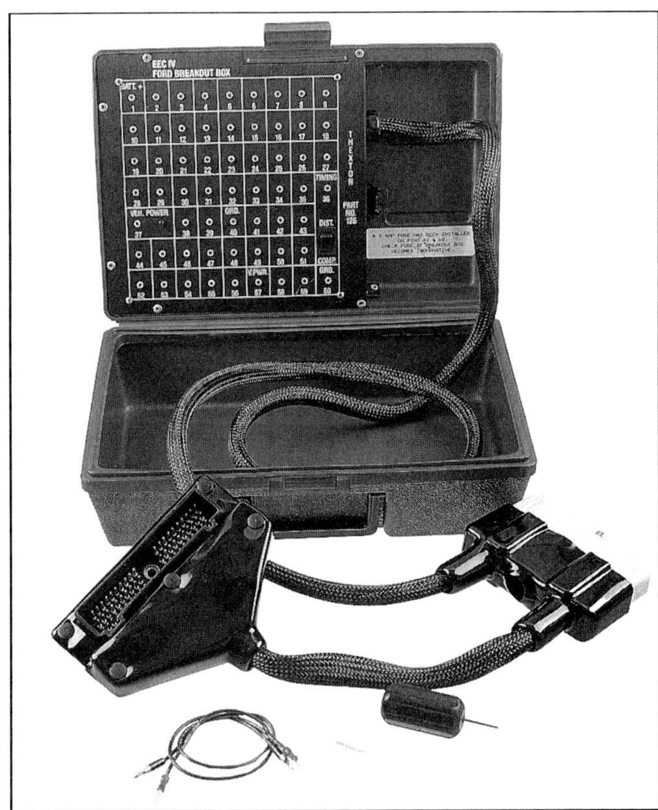

Figure 3-32. *Breakout boxes allow the technician to test the individual circuits in a particular system. This tool is very useful in diagnosing early computer control systems that do not have self-diagnostic systems. (Mac Tools)*

many testers in one large cabinet. This section will cover many variations of the large tester, but cannot be an exact guide to every type of tester. Always obtain and carefully study the tester manufacturer's instructions before attaching any test equipment to an engine. Also read and follow any tester maintenance instructions to ensure that the tester continues to operate satisfactorily.

Starting and Charging System Testers

Starting and charging system testers can check the battery's state of charge to determine whether the battery is defective. They can also test starter operation and current draw. The starting and charging system tester contains a voltmeter and ammeter The ammeter measures the current draw of the starter during cranking and the voltmeter measures the system voltage and any voltage drop. This will enable the technician to determine if the battery, starter, and related vehicle wiring are in good condition and cranking the engine fast enough for easy starting.

Most testers designed specifically to test starting and charging systems have a ***variable load*** or ***carbon pile*** for placing an electrical load on the battery and charging system. If the battery voltage drops too low for a given load, the battery is undercharged or defective. If placing a load on the battery of a running engine does not cause the alternator output to increase, the charging system is defective. The loading device is operated by the technician through a control knob on the instrument panel or automatically by the tester itself. **Figure 3-33** shows a typical starting and charging system tester.

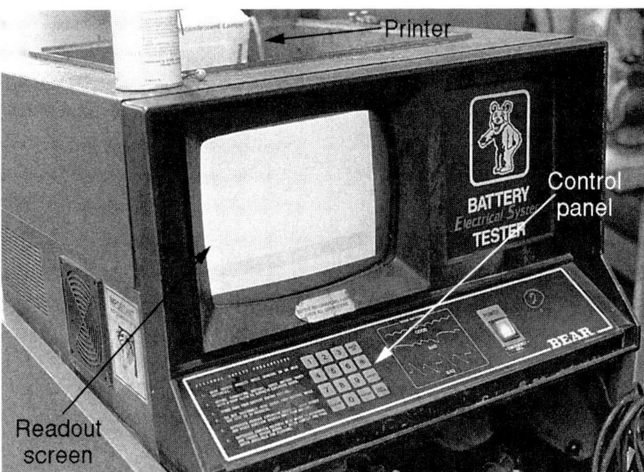

Figure 3-33. *Starting and charging system testers are often used in the course of driveability work. This tester can check all aspects of the starting and charging system and provide a printout of the results.*

Starting and charging system testers usually have three major leads. Heavy positive and negative cables are attached to the battery and allow heavy current to flow. They must be heavy because of the high-current draw of the electrical loading device. In most cases, these leads also measure the voltage present at the battery terminals. An inductive pickup is placed over the battery's negative cable. Its function is to measure the current draw of the starter during starting tests and the current produced by the alternator in charging tests.

Two smaller leads are used to make special voltage tests, such as alternator field current, or to measure voltage drop across electrical connections. These testers can also be used to check the electrical system wiring to isolate a high resistance connection. These tests help the technician to determine if engine driveability problems are caused by incorrect voltage levels or voltage spikes caused by the charging system.

Oscilloscope

The automotive ***oscilloscope*** is a device which converts the electrical activity in the ignition system into a visible pattern. Oscilloscopes are used to quickly identify defective ignition system components, such as worn out or fouled spark plugs, defective spark plug wires, carbon tracked caps or rotors, and shorted coils.

The basic oscilloscope is a cathode-ray tube (CRT), similar to the picture tube of a television set. Instead of the complex picture created by a TV tube, the oscilloscope CRT displays a single line. This line is varied by voltage fluctuations in the operating ignition system. The resulting waveform can be observed and interpreted as the condition of the ignition system. The complete waveform is more commonly called the scope pattern. The oscilloscope is equipped with various switches and controls for magnifying the pattern, switching between different types of patterns, or rearranging the pattern of all engine cylinders for easier comparison.

Electrical leads connect the CRT to the secondary and primary sides of the ignition system. The secondary pickup clamps over the coil secondary, or on top of the distributor cap. The primary pickup lead attaches to the negative primary connection on the coil. An inductive trigger pickup clamps over the number one plug wire, and a ground wire is attached to the engine block or other vehicle ground. Once these connections are made, the engine is started, and the scope pattern observed. Some oscilloscopes are also equipped with connectors for checking the starting and charging systems. A typical oscilloscope is shown in **Figure 3-34.**

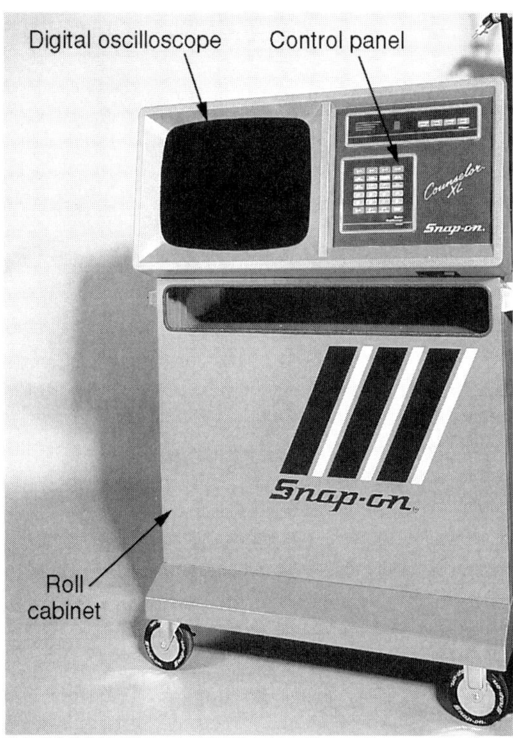

Figure 3-34. *The oscilloscope is part of most modern engine analyzers. It is useful in isolating many primary and secondary ignition problems as well as checking the operation of fuel injectors and the alternator. (Snap-on Tools)*

Exhaust Gas Analyzer

An ***exhaust gas analyzer*** measures the levels of certain gases in the vehicle exhaust. Keeping harmful exhaust gases at the lowest possible levels improves both gas mileage and emissions. The exhaust gas analyzer can be used to determine if an air-fuel ratio problem exists, to confirm that fuel or ignition system repairs have corrected a previous problem, and to make carburetor adjustments (when allowed). A typical exhaust gas analyzer is shown in **Figure 3-35.**

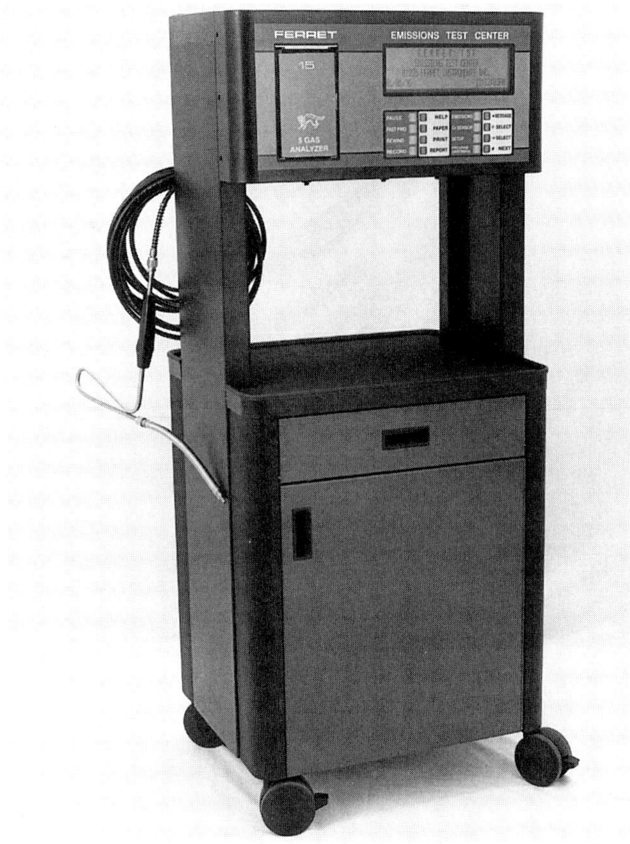

Figure 3-35. *Exhaust gas analyzers are helpful in determining many driveability problems, even those that are not noticeable to the driver. (Ferret)*

Exhaust gas analyzers are powered by shop electricity and the only connection to the vehicle is a sample probe inserted in the vehicle tailpipe. As the engine is operated at various speeds and loads, an internal vacuum pump draws some of the exhaust gases from the tailpipe, through a hose, and into the tester. Sensors in the tester measure the exhaust gas sample for the levels of four gases. Modern exhaust gas analyzers have five meters for measuring hydrocarbons (HC), carbon monoxide (CO), carbon dioxide (CO_2), oxides of nitrogen (NO_x), and free oxygen (O_2).

The ***HC meter*** measures unburned hydrocarbons, which is gasoline that has not been burned in the engine and is entering the atmosphere by way of the exhaust system. The ***CO meter*** measures a byproduct of combustion, carbon monoxide. The ***CO_2 meter*** measures carbon dioxide, a naturally occurring gas whose ratio to other gases on the exhaust system aids in diagnosis. The ***O_2 meter*** measures the amount of free oxygen in the exhaust. Measurement of CO_2 and O_2 is especially useful when checking vehicles with catalytic converters. The ***NO_x meter*** measures oxides of nitrogen, another combustion byproduct that is formed by extremely high combustion chamber temperatures.

Other Large Test Equipment

In addition to the oscilloscope and exhaust gas analyzer, many large testers contain additional types of test equipment, most of which was covered earlier in this chapter. Many large testers contain timing lights with total advance checking features, and sometimes timing meters. Other equipment commonly installed on large testers are tachometers, voltmeters, and ohmmeters. Some large testers also have ammeters, vacuum gauges, and compression gauges, and other attachments for testing engine compression in several ways. Older testers will be equipped with dwell meters, condenser checkers, and point resistance checking gauges. This type of tester is often called an ***engine analyzer.***

Some test gauges installed on large testers will be dedicated to a single purpose, such as a vacuum gauge or compression tester. Other gauges will be multiple-use types, such as combination volt-ohm-tachometers. To use multiple-use gauges, the needed test quality is selected by switches on the tester control panel.

Engine analyzers are usually enclosed in a single large cabinet. Due to the size of the cabinet, it is usually equipped with rollers for easy movement, or built into one service bay. **Figure 3-36** illustrates one popular tester. ***Dynamometers*** are used by vehicle manufacturers to test engine horsepower and torque, and by government and private repair facilities to operate vehicles during emissions inspection and maintenance testing programs, **Figure 3-37.**

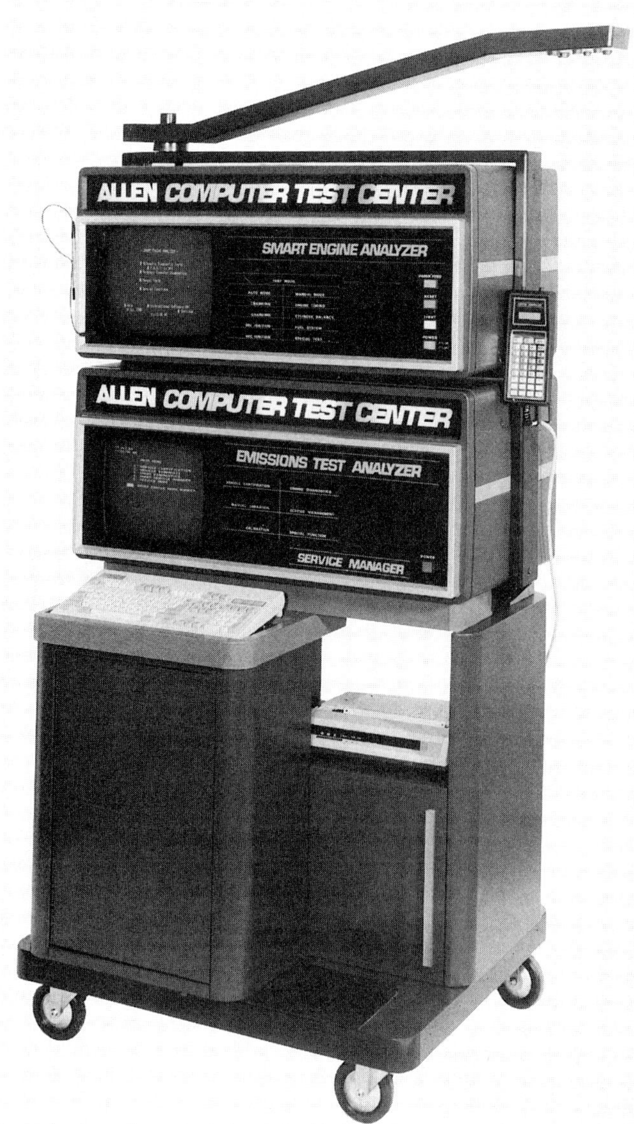

Figure 3-36. *Engine analyzers combine the features of many smaller testers in one package. The construction of the large tester allows the technician to make many ignition, fuel and compression system checks at the same time. (Allen)*

Engine Analyzer Functions

The testing capabilities and functions of engine analyzers are a combination of the features of many smaller testers. The difference is that these features are combined into one package. Engine analyzers can test the operation of gasoline engine ignition, fuel, and compression systems. Some engine analyzers can test engine starting and charging systems. Newer testers can test the operation and components of computer control systems. Some testers are equipped to perform tests on diesel engines, such as checking engine speed and injector timing.

Testing Ignition Systems

The oscilloscope portion of an engine analyzer is able to test the ignition primary and secondary systems, **Figure 3-38,** as well as some conditions in the charging and fuel injection systems. The oscilloscope is usually accompanied by one or more gauges for checking voltage levels, amperage draws, and electrical resistance. For most tests, the oscilloscope connections also provide the inputs to the gauges. Other leads are provided to make connections for tests that cannot be performed by using the oscilloscope leads. Older testers have a meter for checking contact point dwell and resistance, as well as condenser (capacitor) condition.

Figure 3-37. *Dynamometers are used in conjunction with exhaust gas analyzers and other test equipment. The chassis dynamometer allows tests to be performed with the vehicle operating under simulated driving conditions. (Chrysler)*

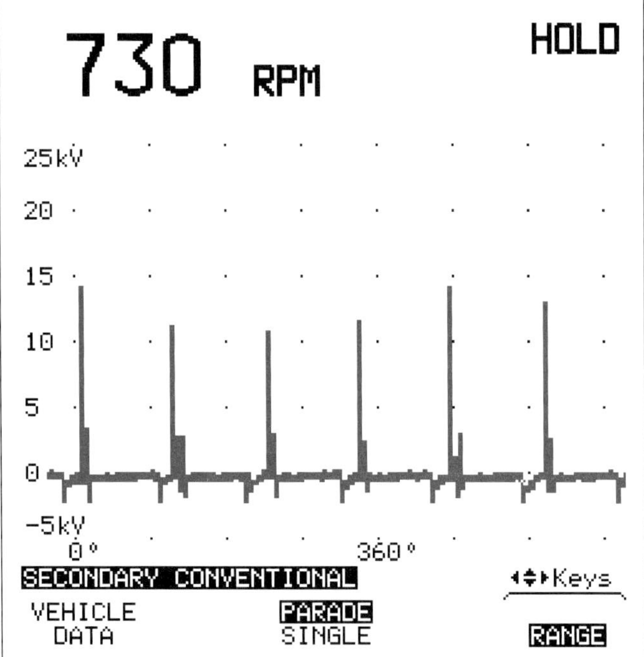

Figure 3-38. *The oscilloscope portion of the engine analyzer can be used to test the ignition primary and secondary system for proper operation. This waveform is the secondary ignition pattern for a six-cylinder engine. (Fluke)*

Testing Fuel Systems

The exhaust gas analyzer portion of an engine analyzer can check the air-fuel ratio of a running engine. The air-fuel ratio can be compared against specifications to check the operation of the carburetor or fuel injectors. The air-fuel ratios provided by exhaust gas analyzers can also be used to determine the operation of the oxygen sensor, computer, and fuel system output devices. The analyzer can also determine the operation of the emission control systems, and can even determine whether the initial timing is at the factory setting for producing lowest emissions. Some modern testers have devices for testing the duty cycle (percentage of time that the solenoid is electrically energized) of the solenoids used on carburetor mixture controls and fuel injectors.

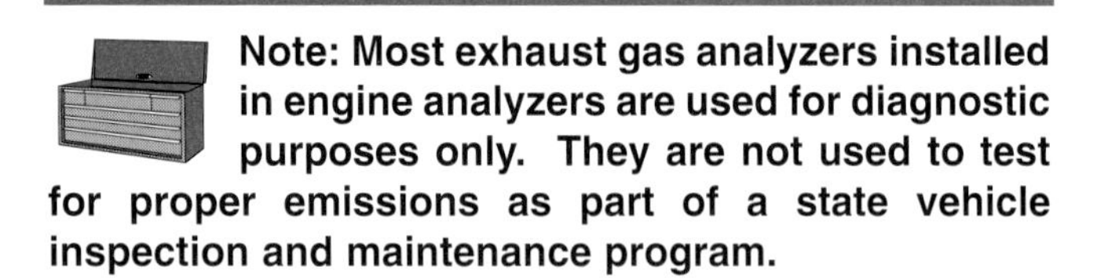

Note: Most exhaust gas analyzers installed in engine analyzers are used for diagnostic purposes only. They are not used to test for proper emissions as part of a state vehicle inspection and maintenance program.

Testing Compression Systems

Most engine analyzers are equipped with internal circuits that can momentarily prevent the spark plug from firing on individual cylinders. This feature is used to test the power output of individual cylinders. If a cylinder is producing power, preventing the plug from firing will cause engine rpm to drop. Weak cylinders will not cause the engine rpm to drop as much as other cylinders. Dead cylinders will not cause any rpm drop. By removing the spark on each cylinder in the engine, and comparing the rpm drops, weak or inoperative cylinders can be isolated quickly.

This test is often referred to as a *dynamic compression* or cylinder ***power balance test.*** It is useful for testing the mechanical condition of individual cylinders to determine if further testing of that cylinder is needed. This test can also check the balance between the barrels of multiple barrel carburetors and throttle body fuel injection systems. If an engine cylinder is not producing power, and the fuel and ignition components are operating properly, the cylinder can be checked for low compression with a compression gauge.

Many testers have air pressure operated devices which will check an individual cylinder for the exact cause of compression loss. A fitting is threaded into the spark plug opening with the cylinder on top of its compression stroke. Air pressure is then applied to the cylinder. A gauge, such as the one in **Figure 3-39**, will register the air leakage as normal or excessive. If air leakage is excessive, the air will be leaving the cylinder from the same place that compression is being lost. The technician can check for hissing at the intake manifold (leaking intake valves) or at the tailpipe (leaking exhaust valves). Excessive smoke from the crankcase breather indicates worn rings or a damaged piston, while bubbles in the vehicle's coolant are a sign of a blown head gasket or cracks in the head or block.

Testing Starting and Charging Systems

Some oscilloscopes and other engine analyzers are also equipped with gauges and connectors for checking the starting and charging systems. Engine analyzers that can test the starting and charging systems have a voltmeter and ammeter. The ammeter can measure the current draw of the starter during cranking and the voltmeter will measure the voltage drop of the battery. This will enable the technician to determine if the battery, starter, and associated wiring are in good condition and cranking the engine fast enough for easy starting. Most of these testers, however, do not have a loading device for checking alternator current output or battery condition under load.

The test results can be used to check the alternator output against manufacturer specifications, and to determine whether the voltage regulator is maintaining the proper charging voltage. The tester can also be used to check the electrical system wiring to see if engine driveability problems are caused by incorrect voltage levels or voltage spikes caused by the charging system.

Testing Computer Control Systems

Many of the newest engine analyzers are equipped with devices that can communicate with the computer on vehicles equipped with computer control systems. Almost all of the newest analyzers can retrieve trouble codes from the computer memory and monitor the operation of the computer system as the engine operates. Some testers can simulate the voltage inputs of certain sensors to determine

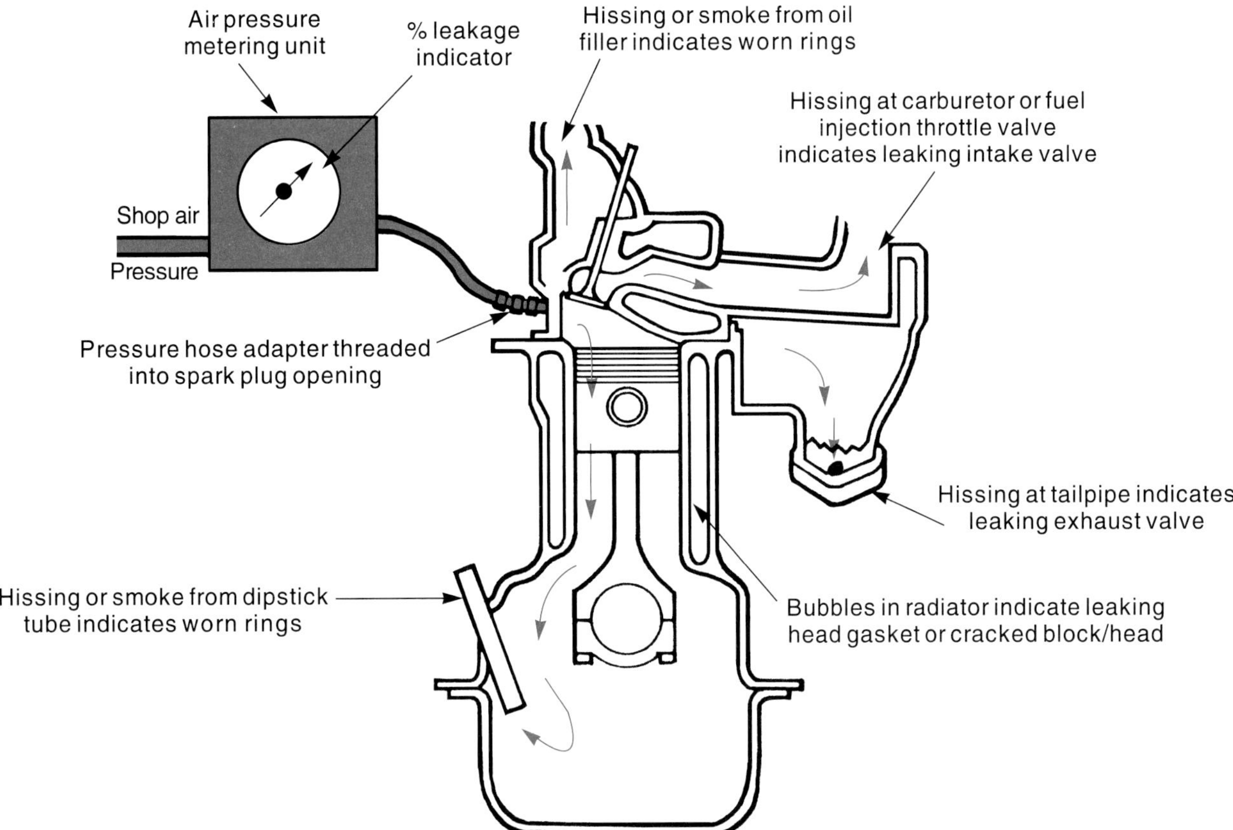

Figure 3-39. *Air pressure is used to determine where compression is being lost. This test is called a cylinder leakage test.*

whether the computer is responding properly to input sensor readings. Some testers can drive output devices to determine whether they are operating properly. Most of the analyzers that can communicate with the computer can also measure the voltage developed by some of the input sensors for comparison with manufacturer specifications.

Some engine analyzers can duplicate dedicated scan tools and perform various computer checking functions. Examples are converting temperature sensor voltage levels into temperature or pressure readings, driving output devices through specific operating cycles to test their condition, and determining the condition of the internal circuitry of the computer. Some testers are able to determine whether the computer and its programming (discussed in **Chapter 7**) are the correct ones for the engine and vehicle.

Service Information

Although it may seem strange to classify manuals and other ***service information*** sources under tools, they are often the technicians most valuable tools. Without these sources of information, your own knowledge and experience, or someone else's, all other tools are useless. Service manuals provide the technical information needed to successfully perform driveability diagnosis and parts replacement. They provide instant knowledge and experience for diagnosis and repair.

In the past, service information was generally ignored by both manufacturers and technicians. Vehicles were simpler in design and completely new systems were rare enough so that they could be learned about in the course of performing other repairs, or often ignored completely. Today, more information is needed for the technician to perform even the most basic service. The major classifications of service information are service manuals, schematics, troubleshooting charts, technical service bulletins, manufacturers hotlines, and computers. These types of service information are discussed below.

Service Manuals

The ***service manual*** is a book containing text and illustrations showing how to perform service operations. There are three major types of service manuals: the factory manual, the general manual, and the specialized manual. The factory or manufacturers manual is dedicated to one type of vehicle and contains information on every part of that particular vehicle. The factory manual is a useful source of detailed procedures for diagnosing driveability and other problems. It is also useful for providing detailed replacement and overhaul information for all vehicle components. Most modern factory service manuals now come in volume sets for one vehicle. The major drawback to the factory manual is its relatively high cost, compared to the limited range of vehicles it can be used with. **Figure 3-40** shows some of the factory manuals that are required for different makes of vehicles.

Figure 3-40. *Factory manuals are developed by the vehicle manufacturers and contain information about only one vehicle. Many factory manuals come in volume sets.*

Figure 3-41. *General manuals contain specific information about many kinds of vehicles. While they do not cover every part of a vehicle, they do cover the most common repair areas, making them a valuable resource in the repair shop.*

General Service Manuals

The ***general manual,*** **Figure 3-41,** contains the most commonly needed service information about many different makes of vehicles, such as engine, brake, and alignment specifications, fuse replacement data, and flasher locations. General manuals also contain procedures for preventive maintenance and minor repairs.

At one time, general service information for every vehicle could be covered in one manual. Today, due to the large number of different vehicles available, this is no longer possible. Modern general manuals are divided into automobile and light truck editions. Automobile manual publishers further divide their books into US, European, and Asian models. Some publishers are starting to offer their manuals in computerized format, which will be explained later.

The individual chapters of general manuals are grouped according to vehicle make, or several makes that are similar mechanically. Chapter subsections are devoted to particular areas of each make. General manuals also contain separate sections covering repair procedures which apply to all vehicles, such as engine overhaul, air conditioning service, and starter/alternator overhaul. The major disadvantage of these manuals is the necessity of eliminating most of the information on specialized vehicle equipment, sheet metal, and interior.

Specialized Manuals

Specialized manuals cover one common system of many types of vehicle. These manuals are often used to cover such topics as emission controls, automatic transmissions, computerized engine controls, electrical systems, brakes, or suspension systems. They combine some of the best features of the factory and general manuals. They provide detailed coverage for each make of vehicle, and cover all common makes. Some examples of these types of manuals are shown in **Figure 3-42.**

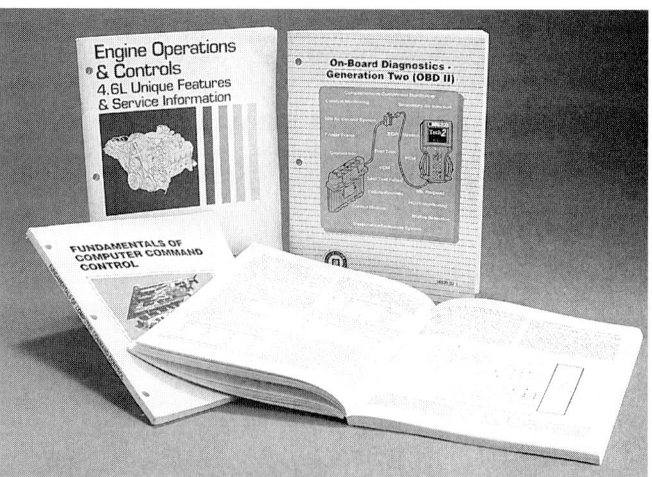

Figure 3-42. *Specialized manuals concentrate on one engine or vehicle system, such as the fuel, or computer control system. Specialized manuals are used in shops that specialize in one or a few areas of repair and are also a good source of information about new systems.*

Schematics

Schematics are pictorial diagrams which show the path of energy through a system. This energy can take the form of electricity, vacuum, air pressure, or hydraulic pressure. Most schematics used for driveability diagnosis show the flow of electricity, while a few show vacuum flow, or the inputs and outputs to a computer, **Figure 3-43.** Schematics do not show an exact replica of a system, but instead indicate the flow or process within the system. Some schematics show the exact flow of a form of energy while others show the general process of a particular system, **Figure 3-44.** Schematics are often included as part of a service manual, or may be supplied separately.

Tracing the flow through a schematic makes diagnosis easier by showing the exact path of electricity or other form of energy. Each line represents a single wire in the vehicle's wiring harness. The schematic lines are labeled with numbers to colors to correspond with a specific color, or color and color stripe combination on the actual wires. The path can be traced by carefully following the lines from component to component. Always carefully note the color designations of the wires and any stripes or bands to ensure that you are following the correct wire.

Troubleshooting Charts

Troubleshooting charts are summaries, or checklist versions, of the troubleshooting information about a particular vehicle or system. Although the information is found in a longer form elsewhere in the service literature, the troubleshooting chart allows the technician to quickly reference the problem, the possible cause, and the solution. **Figure 3-45** shows a typical troubleshooting flowchart. Some troubleshooting charts are arranged with the problem on the left-hand side of the page, the possible cause in the middle, and the corrective action on the right-hand side.

Technical Service Bulletins

Frequently, manufacturers issue ***technical service bulletins,*** or ***TSB,*** for newer vehicles to their dealership personnel. These bulletins contain repair information that is used to describe a new service procedure, correct an unusual or frequently occurring problem, or update information in a service manual. Many of the phone hotline and computerized assistance services receive these bulletins. They are a very good source of information to repair an unusual or frequently occurring problem. Subscriptions to these bulletins are also available through various services.

Manufacturer Hotlines

To further assist service personnel, various vehicle and aftermarket equipment manufacturers have established direct telephone links with service personnel. These telephone links are usually called ***hotlines*** or ***technical assistance centers.*** These hotlines connect with central information banks. Dialing one of these central information banks connects the technician to persons with access to troubleshooting and service information, or to computer memories called ***databases.***

The various types of hotlines reflect their purpose. Vehicle manufacturer hotlines are usually installed in dealerships handling that manufacturer's vehicles, and are intended for use only by the dealership service personnel. Some aftermarket part and equipment manufacturers provide toll free or "800" numbers for technicians in need of diagnosis or service information. Aftermarket hotlines are available to any technician and are provided as part of the marketing and advertising strategy of the aftermarket manufacturer. These numbers are available from the local outlets handling that brand of parts or equipment.

Computerized Assistance

A few years after computer control came to the automobile, manufacturers began to put computer driven analyzers into the service departments of their dealerships. These were simply computers that joined with a dedicated engine analyzer system designed to diagnose problems in that particular manufacturer's line of vehicles. Unfortunately, the first computer driven analyzers were not much better than traditional engine analyzers. Also, technicians were very apprehensive to use the new analyzers, since many of them had little or no exposure or training in computer usage.

Improvements in software and computer technology greatly improved the computerized analyzer. The newest computerized analyzers have user-friendly menus with touch screen capabilities or a standard computer mouse, which can be used to "point-and-click" on any particular menu selection. The newest analyzers can also be used on more than one manufacturers' line of vehicles. These computer driven analyzers, **Figure 3-46,** are able to communicate with the various computers on most late-model vehicles, as well as perform all standard diagnostic procedures. The analyzer's information can be updated through a computer-to-computer modem connection over the telephone to a central computer at the manufacturers' service headquarters.

One feature of these analyzers is that technicians seeking service information can access the manufacturer's main computer and type in the VIN number of the vehicle in question. The computer then provides a printout of all service information, including any technical service bulletins or recall campaigns applicable to that particular vehicle. Aftermarket tool manufacturers have begun to computerize their engine analyzers, as well as their alignment equipment and exhaust gas analyzers.

Unfortunately, most small shop owners cannot afford these large, expensive computerized diagnostic equipment

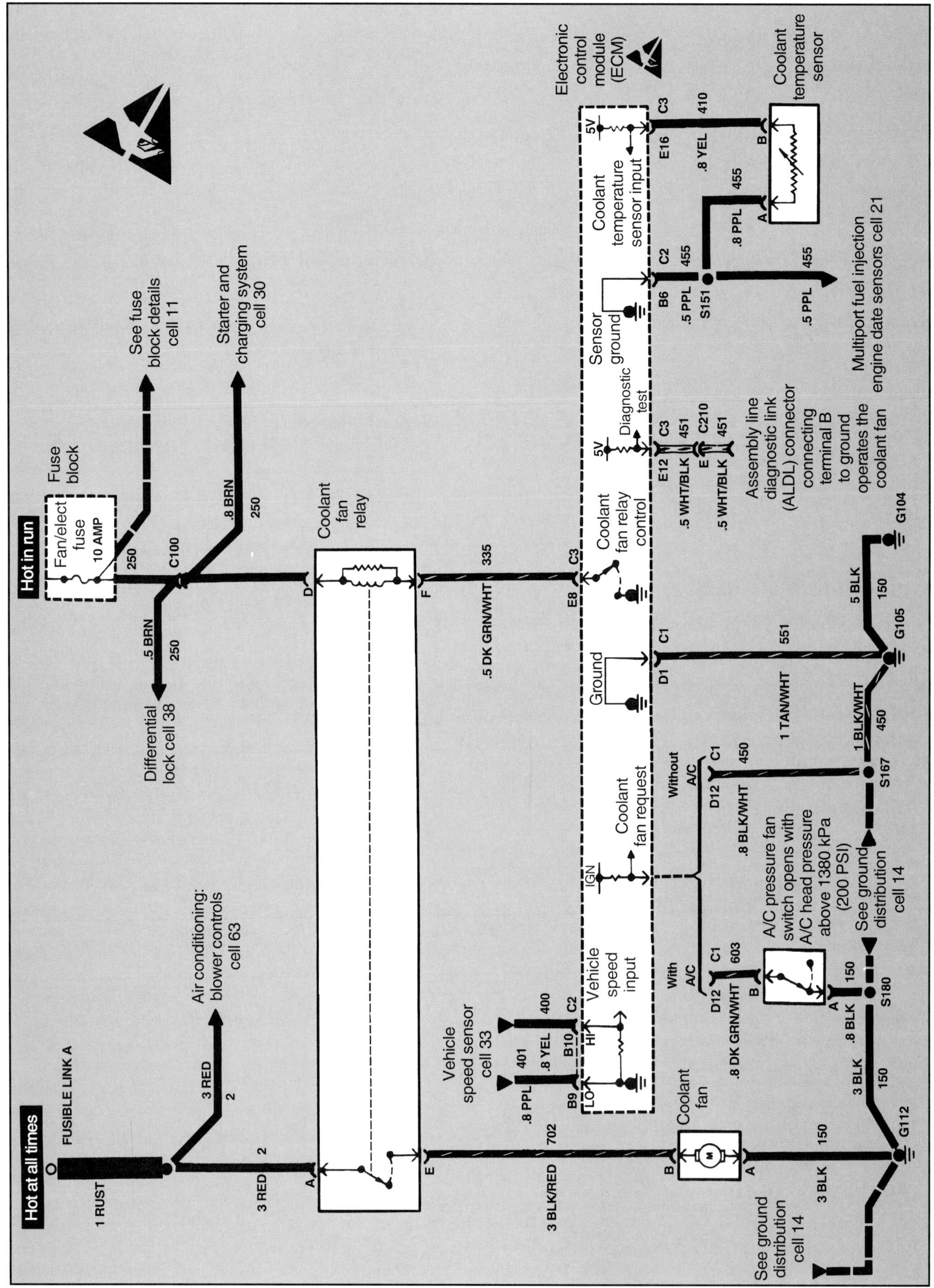

Figure 3-43. *Schematics show the exact flow of electricity or vacuum. These schematics are useful for determining the path that energy takes as well as the location of system components. (General Motors)*

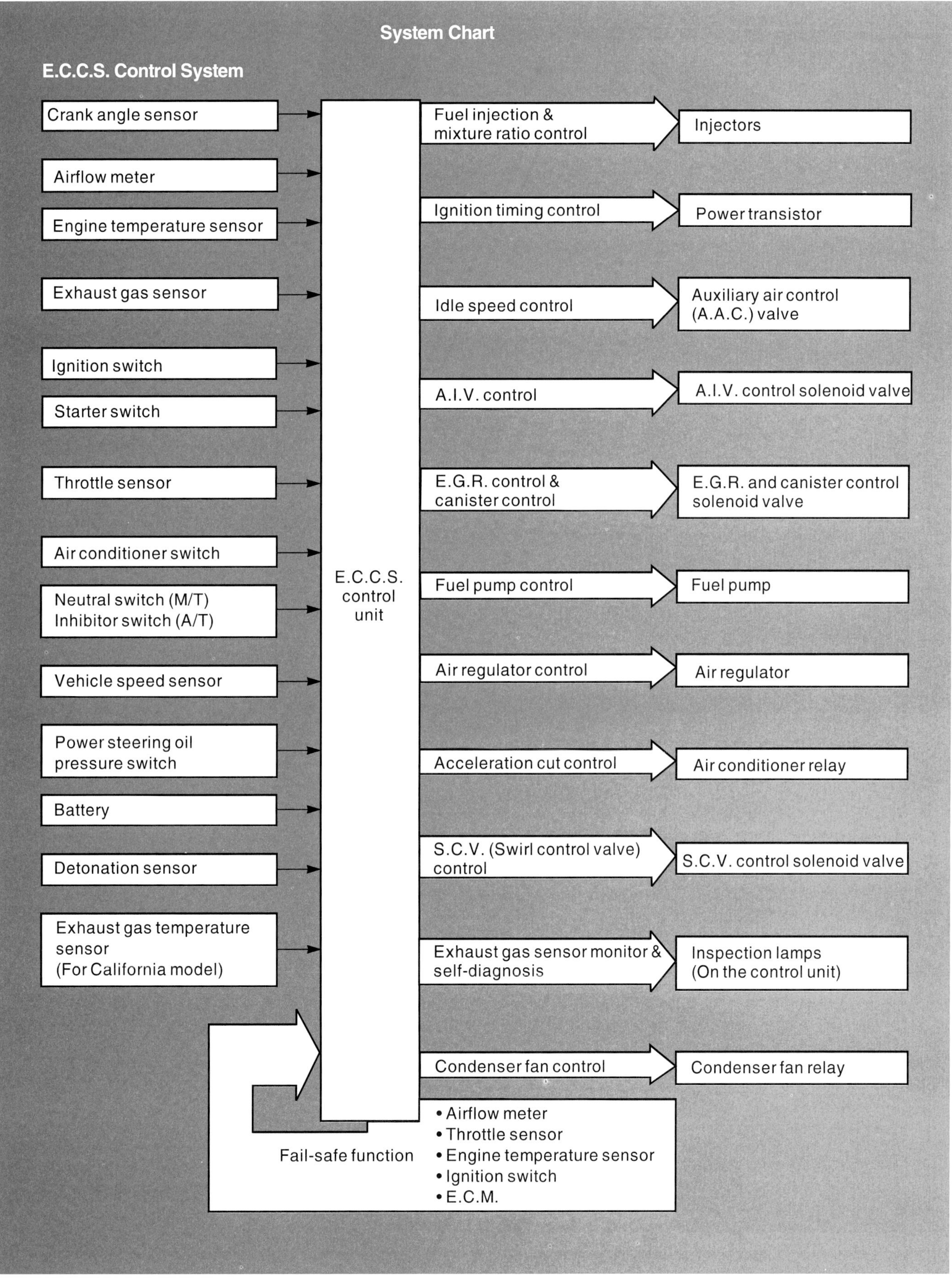

Figure 3-44. *Process flow schematics show how a process is completed, such as how inputs to the ECM result in changes to engine fuel and ignition system settings. (Fuji)*

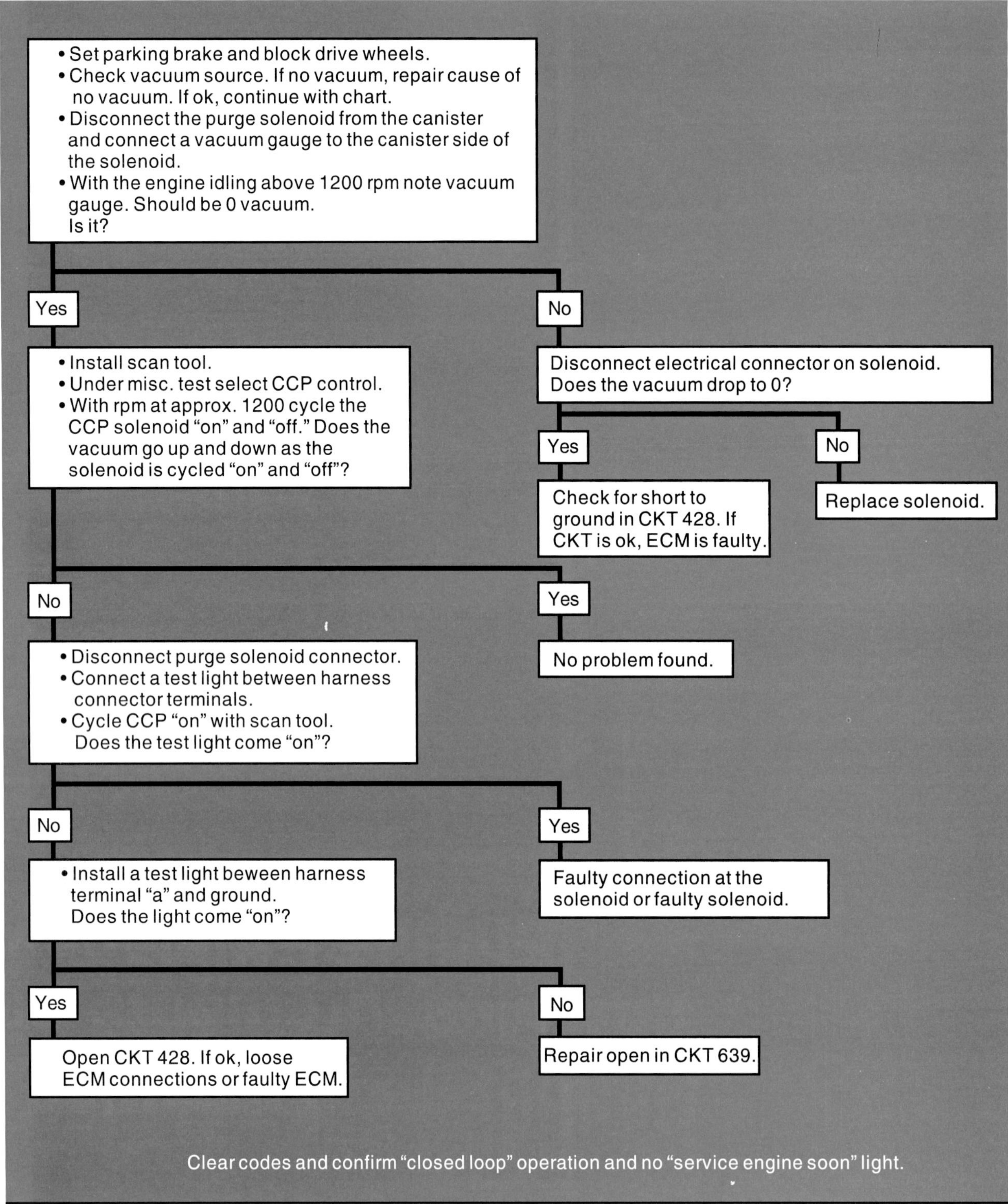

Figure 3-45. *Troubleshooting charts are used when trying to diagnose a problem. They contain information about the system in an easy-to-use form. (General Motors)*

nor are they allowed to interface with a manufacturer's central computers. However, some software and tool companies are beginning to offer add-on kits that contain diagnostic connectors, software, and related hardware that can be installed in a normal personal computer. These kits can turn a normal computer into a computer driven analyzer, with many of the same capabilities as the large, expensive computerized engine analyzers

Many shops now write their repair and purchase orders and keep their billing and accounting in order with computers. Service, parts, and labor time manuals are now available in CD-ROM format. One CD-ROM can provide the same amount of information found in a complete series of printed manuals. Some of the newest CD-ROM manuals show actual, step-by-step footage of certain repair operations.

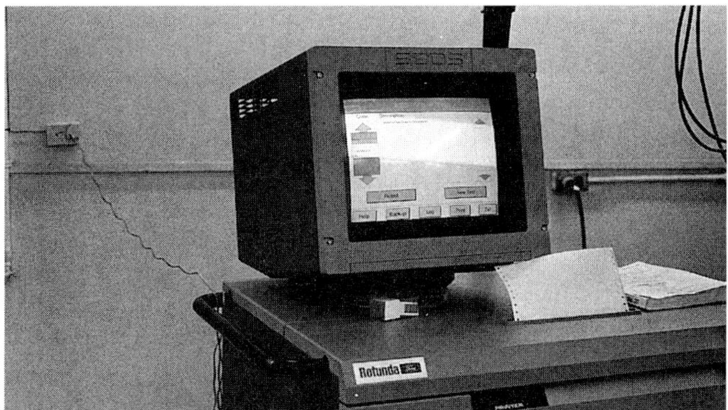

Figure 3-46. *Many analyzers are computer driven, and connected to a central computer located at the manufacturer's headquarters through a dedicated data telephone line.*

On-line Diagnostic Assistance

By using a computer on-line service, small shops can access any one of several automotive central information banks over the information superhighway. These banks can offer diagnostic tips, technical service bulletins, and other service information similar to the telephone hotlines described earlier. Many of these services have user-friendly interfaces that make them easy for anyone to use, **Figure 3-47.** Most of these on-line assistance centers are operated by aftermarket companies, private organizations, and individuals.

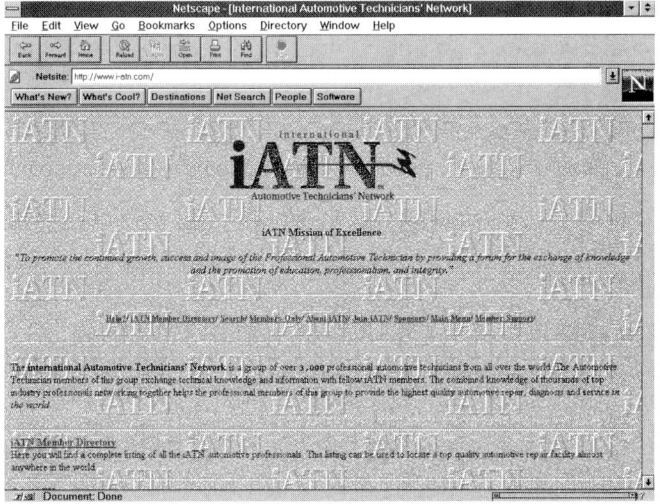

Figure 3-47. *Service information is available over the information superhighway. Some of the sites provide a forum for technicians to share information, ask questions regarding hard-to-diagnose problems, and listings of technical service bulletins in an easy-to-use format. This World Wide Web site is provided by the International Automotive Technicians' Network.*

Summary

The types of test equipment needed to diagnose driveability problems range from simple mechanical test equipment to complex electronic testers. Basic mechanical test equipment includes gap gauges such as feeler gauges, spark plug gauges, and wire gauges. Mechanical gauges that are helpful in driveability diagnosis include vacuum gauges and vacuum pumps, compression and fuel pump pressure testers, and PVC system testers. Computer memory retaining tools, borescopes, stethoscopes, temperature testers as well as various specialized hand tools are also handy.

Electrical testers that can be used for driveability diagnosis include such simple devices as jumper wires and test lights. Other electrical test equipment includes dwell meter-tachometers, ohmmeters, voltmeters, ammeters, and pulse meters and are usually combined into an all-purpose multimeter. Multimeters can be analog types with an indicator needle and scale or digital types which provide a direct readout of electrical values. Other electrical testers include timing lights, which may be adjustable or nonadjustable. Other timing tools are the timing meter, designed for use with late model engines having a timing receptacle and luminosity meters for testing diesel engine injector timing. An unusual but effective tester is the ultrasound vacuum leak detector, which can locate vacuum leaks by pinpointing the sound of air moving through the leak. Another type of vacuum leak detector operates by pumping smoke into the intake manifold.

Portable diagnostic testers are vital for checking the sensors, output devices, and ECM on computer-controlled vehicles. The system specific or dedicated scan tool is intended for a specific vehicle or line of vehicles. These testers are used to retrieve trouble codes from the computer and to make checks of related system voltage levels and sensor readouts. A multi-system scan tool can test more than one manufacturer's vehicles and computer systems. It contains removable memory chips which can be changed out to switch between different computer systems. Newer scan tools have memory boards that can be programmed using a personal computer or computerized engine analyzer. Other testers include input sensor and output device sensors and air-fuel monitors.

Types of large testers include starting and charging system testers, oscilloscopes, and exhaust gas analyzers. These testers are often combined with smaller test units and meters in a single large cabinet. The entire cabinet of test equipment is called an engine analyzer. Oscilloscopes present a graphic display of the operation of the primary and secondary ignition systems. Oscilloscope displays consist of a pattern that is common to all ignition systems. Some oscilloscopes can also be used to check the condition of the fuel injectors and alternator windings.

Exhaust gas analyzers measure the amount of certain gases in the exhaust stream. A sample of a vehicle's exhaust gas is drawn into the analyzer and checked for the presence of these gases. Gases measured by modern exhaust gas analyzers are carbon monoxide, carbon dioxide, hydrocarbons, and free oxygen. Other test equipment is usually installed on the engine analyzer to simplify checking the entire engine.

Functions of the engine analyzer include checking the engine fuel, ignition, and compression systems. Some analyzers can also check the starting and charging systems. Many modern engine analyzers are also equipped to retrieve trouble codes from the computer and perform other checks of the computer control system.

Service literature is as important as any other tool in diagnosing driveability problems. The most common type of service literature is the service manual. Factory service manuals provide detailed information on a single make and year of vehicle, while general manuals provide less detailed information covering many vehicle makes and years. Specialized manuals provide information on a single system, such as charging or fuel, but covering all makes of vehicles. Schematics are useful for tracing electrical, hydraulic, or vacuum system problems. They are provided as part of a service manual, or in separate manuals. Troubleshooting charts are distillations of information in other service literature, arranged in a form that makes possible problem causes easy to locate. Vehicle manufacturers and aftermarket part and equipment manufacturers often provide toll-free hotlines for information on driveability and other system problems. Other assistance is available on the information superhighway.

Know These Terms

Gap gauges
Feeler gauges
Brass feeler gauges
Wire gauges
Spark plug gauges
Ramp gauge
Vacuum gauge
Vacuum pump
Compression gauge
Fuel pressure gauge
PCV tester
Ignition spark testers
Fuel injection harness testers
Specialized tools
Computer memory retaining tool
Borescope
Stethoscope
Jumper wires
Test light
Powered test light
Nonpowered test light
Impedance
Meters
Analog meters
Digital meters
Multimeter
Ohmmeters
Voltmeters
Ammeter
Dwell meter/tachometer
Pulse meter
Waveform meter
Waveform
Timing light
Timing meter
Luminosity meter
Ultrasound vacuum leak detectors
Input sensor tester
Sensor simulators
Output device testers
Idle control motor tester
Ignition module tester
Fuel injector tester
Air-fuel monitor
Portable diagnostic testers
Dedicated scan tools
Data link connector
Multi-system scan tool
Generic scan tool
Breakout boxes
Starting and charging system tester
Variable load
Carbon pile
Oscilloscope
Exhaust gas analyzer
HC meter
CO meter
CO_2 meter
O_2 meter
NO_x meter
Engine analyzer
Dynamometer
Power balance test
Service information
Service manual
General manual
Specialized manuals
Schematics
Troubleshooting charts
Technical service bulletins (TSB)
Hotlines
Technical assistance centers
Databases

Review Questions—Chapter 3

Please do not write in this text. Write your answers on a separate sheet of paper.

1. Digital meters display a ______ to indicate the reading.
2. Is a timing meter more or less accurate than a timing light?
3. A luminosity meter is used to time ______ engines.
4. The vacuum gauge measures the difference in air pressure between the intake manifold and the:
   (A) outside atmosphere.
   (B) engine cylinder.
   (C) exhaust manifold.
   (D) All of the above.
5. High impedance means:
   (A) high price.
   (B) high resistance to current flow.
   (C) high screen visibility.
   (D) low screen glare.
6. Brass feeler gauges are used to set gaps on parts containing:
   (A) brass.
   (B) plastic.
   (C) iron.
   (D) magnets.

7. Jumper wires are used to make ______ electrical connections during vehicle testing procedures.

8. The nonpowered test light uses ______ power to light.
   (A) vehicle battery
   (B) internal battery
   (C) external power supply
   (D) None of the above.

9. Starting and charging system testers usually have:
   (A) a voltmeter.
   (B) an ammeter.
   (C) a variable load.
   (D) All of the above.

10. Oscilloscopes can quickly identify defective ______ system components.
    (A) fuel
    (B) ignition
    (C) compression
    (D) vacuum

11. Dynamic compression testers can check:
    (A) cylinder power output.
    (B) cylinder temperature.
    (C) carburetor balance.
    (D) Both A & C.

12. A ______ is a book containing text and illustrations showing how to service vehicles.

13. Specialized service manuals cover ______ common system(s) used on many types of vehicles.
    (A) one
    (B) two
    (C) three
    (D) all

14. ______ are pictorial diagrams which show the path of energy through a system.

15. A ______ can contain the same amount of information as a complete series of printed manuals.

ASE Certification -Type Questions

1. A dwell meter/tachometer can be used to ______.
   (A) measure engine speed
   (B) set dwell on electronic ignition systems
   (C) set dwell on point-type ignition systems
   (D) Both A & C.

2. Technician A says that component-specific testers usually have connectors that enable them to test almost every kind of electronic component. Technician B says that some portable testers are used to test an individual electronic component, with the component removed from the vehicle. Who is right?
   (A) A only.
   (B) B only.
   (C) Both A & B.
   (D) Neither A nor B.

3. Timing lights are used to synchronize the operation of the engine ______.
   (A) camshaft and crankshaft
   (B) fuel and compression systems
   (C) ignition and fuel systems
   (D) compression and ignition systems

4. The pulse meter can be used to check ______.
   (A) carburetor solenoids
   (B) fuel injectors
   (C) the ECM
   (D) Both A & B.

5. Most portable diagnostic testers are used with ______.
   (A) sensors
   (B) electrical parts
   (C) the ECM
   (D) Both A & B.

6. The exhaust gas analyzer is connected to the vehicle by a sample probe in the ______.
   (A) crankcase breather
   (B) intake manifold
   (C) tailpipe
   (D) exhaust manifold

7. All of the following can be used to make a jumper wire, EXCEPT:
   (A) 12-14 gage wire.
   (B) an inline fuse.
   (C) 18-22 gage wire.
   (D) alligator clips.

8. Computer memory retaining tools are being discussed. Technician A says that a 9-volt battery is sufficient to maintain computer memory. Technician B says that the computer memory retaining tool should be installed as soon as the battery is removed. Who is right?
   (A) A only.
   (B) B only.
   (C) Both A & B.
   (D) Neither A nor B.

9. Technician A says that a sensor simulator can check the operation of the ECM and related output devices. Technician B says that a sensor simulator can indirectly check sensor operation. Who is right?
   (A) A only.
   (B) B only.
   (C) Both A & B.
   (D) Neither A nor B.

10. Output device testers can be used to ______ output devices such as the idle speed air control.
    (A) diagnose
    (B) adjust
    (C) replace
    (D) Both A & B.

Vehicle manufacturers test fuels as much as they test the vehicles they are meant to power. Fuel is pumped into chill tanks to test how they will respond in extremely cold environments. (Chrysler)

Chapter 4

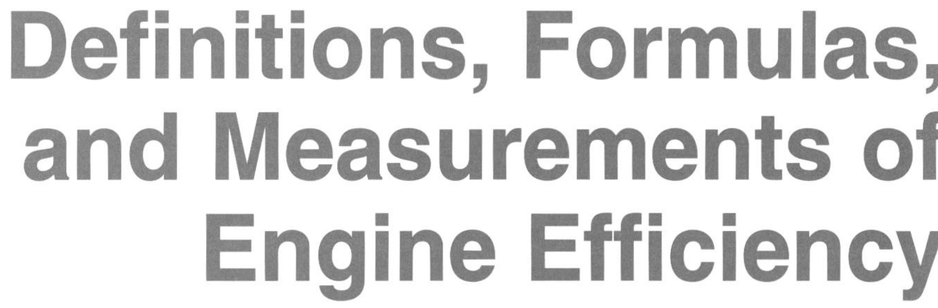

Definitions, Formulas, and Measurements of Engine Efficiency

After studying this chapter, you will be able to:

- Explain the difference between objective measurement and subjective judgment.
- Identify and explain rating systems for engine efficiency.
- Identify the English system of measurement.
- Identify the metric system of measurement.
- Identify and explain the use of common automotive formulas.
- Identify common linear measuring devices.
- Identify common pressure measuring devices.
- Identify common measuring devices for electrical properties.
- Identify common temperature measuring devices.
- Identify common engine measurement terms.
- Identify the various methods by which drivers measure efficiency.
- Identify the various methods technicians use to measure efficiency.
- Identify the various methods government agencies use to measure efficiency.

Engine efficiency is measured using many different factors. These can be set numerical standards used to determine the performance of an engine or vehicle. They can also be subjective, with each person or group comparing the performance to their own set of standards. The ability to use measurements, standards, common formulas, and make subjective judgments is critical for the driveability technician to successfully complete many diagnosis and repair operations. This chapter will discuss the most common measurement systems, including formulas to convert between the different systems of measurement and common measuring devices used by the driveability technician. This chapter will also discuss how drivers, technicians, and government agencies define performance and driveability.

Systems of Measurement

There are two major types of measurement, objective and subjective. We use both types of measurements in automotive troubleshooting and everyday life. ***Objective measurement*** relies on comparisons with universal standards, such as length, time, or temperature. ***Subjective judgment*** is based on the internal thought processes of the person performing the judgment. In this chapter, we will be dealing primarily with objective measurements and the systems used to make these measurements.

Objective Measurements

Objective measurements are based on universally agreed upon standards. Examples of such standards are ***length*** and ***width, volume, weight, temperature, time, speed, voltage, amperage,*** and ***pressure.*** An object to be measured can be compared against one or more of these standards for an exact determination of its condition. If a linkage rod is supposed to be 2.5" (63.5 mm) long, the technician can place it against a ruler, as shown in **Figure 4-1.** If it is the proper length, it can be reused. If it is not the proper length, it must be adjusted or discarded.

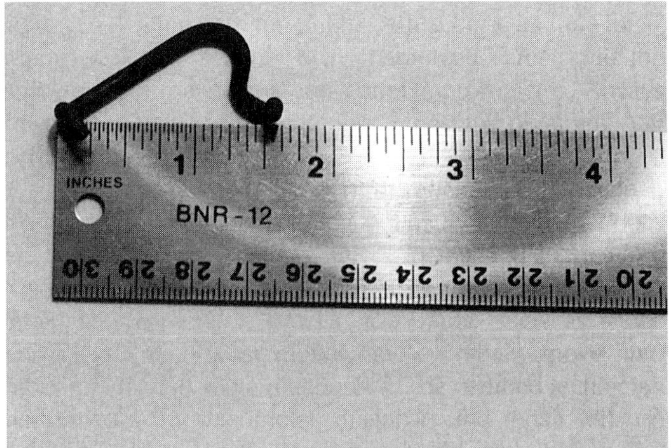

Figure 4-1. *Rulers are an often overlooked, but important tool in a technician's tool box. The carburetor linkage shown here is sometimes bent in an attempt to correct a problem.*

Measurements of many qualities, such as length, volume, speed, and temperature are all determined according to systems of measurement. The most common systems of measurement in the United States are the ***English*** and ***metric*** systems. The English or U.S. conventional system has been used in the United States for many years. The metric system is used exclusively in most other countries and somewhat in the United States. Since both systems will continue to be used interchangeably by auto manufacturers for many years, it is important to know how to convert from one to the other. Most of the conversions used in automotive work will concern speed, length and width, weight, volume, and temperature. Measurements of time and electrical properties are the same in both systems. Complete conversion charts are located in the back of this text.

Subjective Judgments

Subjective judgments are based on the individual driver's or technician's internal thought processes. They originate inside the mind, and cannot be proved or disproved according to any established standards. An example of subjective judgment is when a driver brings a vehicle in for repairs because "something feels wrong." Subjective judgment is also used when a technician looks inside a distributor cap and says that it is "okay." This measurement is based not on direct comparison with an established standard, but on the technician's concept of what the inside of an "okay" distributor cap looks like.

Subjective judgments have a vital place in automotive work, as they are closely connected to one's experiences. Subjective judgments are often necessary when a device cannot be tested against any objective standards. For example, the experienced technician uses his or her experience to determine that a spark plug, which had an acceptable firing line on the oscilloscope, has excess deposits which cause it to misfire under load. This is an example of subjective judgment.

However, subjective judgments are just what the term implies, judgments based on experiences with no established standard. Never substitute a subjective judgment for an objective measurement. For instance, a spark plug wire that looks okay (subjective) may turn out to have excessively high resistance when checked with an ohmmeter (objective). When in doubt, always substitute an objective measurement procedure for any subjective judgment whenever possible.

Process of Elimination

A combination of the two methods just described is utilized heavily in diagnosing automotive problems. Due to the nature and complexity of automotive systems, many components cannot be checked using normal testing methods. In these cases, they are isolated by the use of both objective measurements and subjective judgments. This is commonly referred to as ***process of elimination.***

For example, a dead cylinder can be caused by a misfiring spark plug, plug wire, faulty fuel injector, or defective primary wiring, all of which can be tested. However, a defective ignition module or on-board computer, which cannot be tested in most cases, can also be the cause of a dead cylinder. By using the process of elimination, a technician uses objective measurements to check all of the testable parts. Once all the testable parts have been checked, subjective judgment is used to analyze the results to isolate a non-testable part.

Automotive Formulas

The most common automotive formulas are those used to obtain engine power, electrical properties, and those used to convert from English to metric values. These types of formulas are discussed in the following sections. Other formulas for temperature-pressure relationships and electrical power consumption are sometimes used in automotive applications, but do not directly affect driveability and performance.

Horsepower

The rate at which work is performed is termed ***horsepower,*** or brake horsepower. To find horsepower, the total rate of work in foot-pounds accomplished is divided by 33,000. The formula is:

$$\frac{\text{Rate of work in ft.–lbs.}}{33{,}000} = \text{horsepower}$$

To find the rate of work, multiply the weight being lifted by the distance lifted per minute. The total formula would be:

$$\frac{\text{Distance moved} \times \text{weight}}{33{,}000} = \frac{D \times W}{33{,}000} = \text{horsepower}$$

Horsepower varies with engine speed. An early method of measuring crankshaft horsepower made use of a device called a prony brake, which is the origin of the term brake horsepower or BHP. The modern method of measuring horsepower is the ***dynamometer***. Dynamometers were discussed in **Chapter 3.**

Gross and Net Horsepower Ratings

The Society of Automotive Engineers (SAE) specifies ***gross horsepower*** as the maximum horsepower developed by an engine equipped with only basic accessories needed for its operation, such as the coolant pump, distributor, oil pump, and mechanical fuel pump. All other equipment is removed. The disadvantage of this method was that the engine horsepower rating was much higher than that available at the driving wheels, since power loss through the transmission and driveline was not taken into account. In addition, horsepower ratings could be artificially raised for advertising purposes by adjusting the timing and fuel mixture by hand as the engine operated.

Since 1972, the method of measuring engine horsepower is to place the entire vehicle on a chassis dynamometer and measure the output at the drive wheels. See **Figure 4-2.** The ***net horsepower*** is the maximum horsepower developed by the engine when equipped with all common accessories such as fan, alternator, catalytic converter, muffler, air pump, air cleaner, fuel pump, coolant pump, and all other devices that it would have when installed in the car. The net rating is more informative since it shows the horsepower that will actually be transmitted to the road. Test conditions are the same for both gross and net ratings. Some other nations use a din rating for net horsepower. The din rating is made under slightly different conditions and specifications, and will vary from SAE ratings.

Torque

Torque is a twisting or turning action developed at the crankshaft. It is measured in foot-pounds (ft.- lbs.) or Newton-meter (N•m). It is not the same thing as horsepower. Horsepower and torque both increase with engine speed. Horsepower increases as long as engine speed increases, until internal engine friction becomes excessive. As the pistons begin moving above a certain speed, excessive engine power is needed to overcome the friction between the rings and cylinder walls. Friction also increases in other moving engine and drive train parts as speed is increased.

Engine torque, however, is limited by volumetric efficiency or the amount of air-fuel mixture which can be drawn into the engine cylinders. Allowing the maximum amount of air and fuel to enter the cylinder produces the most push on the pistons and torque is greatest at this point. Any additional RPM increase will actually cause torque to diminish. **Figure 4-3** shows a chart relating the horsepower and torque of a typical V-8 engine at certain speeds. Note that both horsepower and torque do not remain steady, but rise and fall with engine speed. Also note that more engine rpm does not necessarily mean more power. After a certain rpm is reached, both horsepower and torque begin to fall off.

Figure 4-2. *Chassis dynamometers are used by manufacturers to measure horsepower, torque, and engine performance. Government agencies use dynamometers in emissions testing while some shops have dynamometers for diagnostic and testing purposes. (Chrysler)*

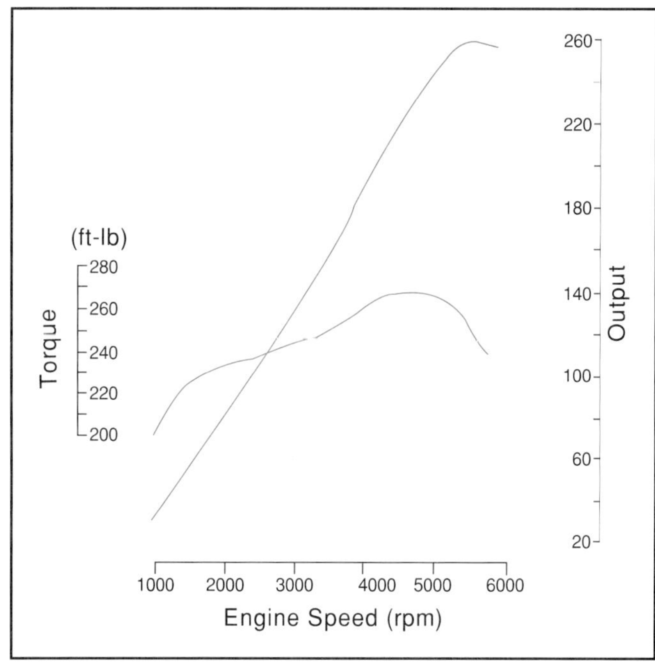

Figure 4-3. *An engine's horsepower and torque rise as engine speed increases. After an engine reaches a certain speed, torque and then horsepower, decreases. The beginning of the drop in horsepower and torque begins within 1000 rpm. (Lexus)*

Thermal and Mechanical Efficiency

If all of the heat in burning fuel could be converted into power, the engine would be 100% efficient. However, real engines are not this efficient. The average engine converts a very small percentage of heat into useful work. How much it converts depends on two kinds of efficiency: thermal efficiency and mechanical efficiency.

Thermal Efficiency

Engine ***thermal efficiency,*** sometimes called heat efficiency, is based on how much of the energy of the burning fuel is converted into horsepower. Heat generated by the burning fuel drives the piston down on the power stroke. However, not all of the heat is used to push on the piston. Some heat is lost to the cooling system, some to the lubrication system and a great deal to the exhaust system. The thermal efficiency of the average engine is around 25%, meaning that approximately 75% of the fuel's heat energy is lost.

Mechanical Efficiency

Mechanical efficiency is the ratio of power developed within the engine and actual brake horsepower delivered at the crankshaft. Mechanical efficiency is usually around 90%, with the other 10% lost to friction within the engine. Engine friction is not constant but increases with speed.

Practical Efficiency

Vehicle owners and technicians are primarily interested in horsepower delivered at the drive wheels. Every gallon of gasoline that enters the engine has the ability to do a certain amount of work, but many factors rob energy before it gets to the drive wheels. Losses include heat loss from the burning fuel, friction in the engine, and friction in the drive train. As little as 15% of the available energy in the fuel may be delivered to the drive wheels. This is the practical efficiency.

Volumetric Efficiency

Volumetric efficiency is the measure of an engine's ability to draw the air-fuel mixture into the cylinders. It is determined by the ratio between what is actually drawn in and what could be drawn in if all cylinders were completely filled. As engine speed increases beyond a certain point, piston speed becomes so fast and the intake stroke is of such short duration that less air and fuel can be drawn in. Since torque is greatest when the cylinders receive the maximum amount of air and fuel, torque will drop off as volumetric efficiency decreases.

Volumetric efficiency is influenced by engine speed, throttle valve opening, intake system design, valve size, amount of valve lift and duration, exhaust system design, and atmospheric pressure. It can be improved by supercharging or turbocharging; straighter, smoother, and longer intake manifold designs; larger intake valves; better exhaust flow, and other modifications to the induction and exhaust systems.

Engine Displacement

Engine displacement is only incidentally related to the outside physical dimensions of the engine. Displacement is the total piston displacement of all cylinders in the engine. Piston displacement is the total number of cubic inches (or liters) of space in the cylinder when the piston is at the bottom of the cylinder. This displacement depends on cylinder bore and piston stroke. The bore is the diameter of the cylinder. The stroke is the distance that the piston moves from the top of the cylinder, called top dead center (TDC), to the bottom of the cylinder, called bottom dead cylinder (BDC). Modern engines usually have a stroke that is shorter than bore diameter. This is referred to as an over square engine. By reducing the stroke, piston speed is decreased, prolonging the life of cylinder walls, pistons, and rings.

Calculating Displacement

If all other factors (such as compression ratios, combustion chamber design, and fuel system) are equal, the bigger the engine displacement, the more power the engine produces. To calculate piston displacement, begin by finding the area of the cylinder. This is done by squaring (multiplying by itself, called D^2) the cylinder bore. Then multiply the answer by 0.7854. This gives the area of the cylinder. Next, multiply the area by the total piston travel (stroke) from TDC to BDC. This answer is then multiplied by the number of cylinders and you have the total piston displacement in cubic inches. The formula is:

$$\text{Displacement} = 0.7854 \times D^2 \times \text{Travel} \times \text{Number of Cylinders}$$

For example, you may want to find the displacement of a V-8 truck engine having a cylinder diameter (bore) of 4.000 inches and a piston travel (stroke) of 3.480 inches.

$$0.7854 \times D^2 \text{ (16 square inches)} \times \text{Travel (3 inches)} \times \text{Number of Cylinders (8)} = 349.8486 \text{ cu. in.}$$

This is close enough to 350 cu. in., to tell you that you have a 350 cubic inch or 5.7 liter engine. This formula can also be used to find out how many cubic inches an engine gains when it is bored oversize. If a 145.78 cubic inch (2.389 liter) four-cylinder engine was bored .030 over, add .030 to the original bore (3.5 inches) to get a new bore of 3.53. The travel is 3.78 so you can use the formula:

$$0.7854 \times D^2 \text{ (12.46 square inches)} \times \text{Travel (3.78 inches)} \times \text{Number of Cylinders (4)} = 147.976 \text{ cu. in.}$$

This is about 148 cubic inches or about 2.4 liters.

Note: To change cubic inches to liters, divide the cubic inch figure by 61.024. This gives an approximate reading in liters.

Compression Ratio

Compression ratio is the relationship between the cylinder volume when the piston is at the top of its stroke (TDC), compared to the cylinder volume when the piston is at the bottom of its stroke (BDC). Modern compression ratios range from 8 to about 9.5 to 1 for gasoline engines, 17.5 to 22.5 to 1 for diesel engines. Although raising compression ratios increases power, compression ratios have dropped in the last 25 years. This was done to allow engines to run on lower octane unleaded fuel without knocking, and to reduce the formation of exhaust pollutants.

Obtaining Electrical Properties

The most commonly used electrical formula is ***Ohm's law.*** Ohm's law is often useful for determining electrical properties. For most vehicles and test equipment, all of the needed electrical properties are furnished by the manufacturer. In some cases, however, it may be necessary to use Ohm's law to determine a needed electrical property. The Ohm's law triangle, **Figure 4-4,** is a graphic representation of Ohm's law. To obtain the needed value, the following formulas can be used:

Volts = amperes times ohms or $E=I\times R$

Amperes = volts divided by ohms or $I=E\div R$

Ohms = volts divided by amperes or $R=E\div I$

Practical examples of how Ohm's law is used are discussed in **Chapter 6.**

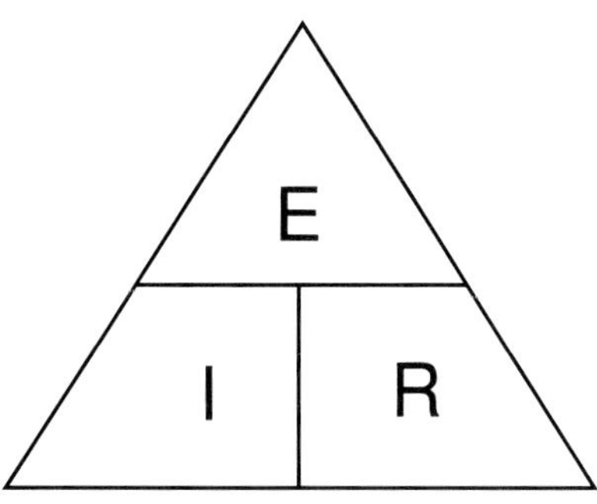

Figure 4-4. *Ohm's law can be used to find an unknown quantity when two other values are known. To use this triangle, cover the unknown value with your thumb and compute using the two remaining values.*

Temperature Conversion Formulas

The most commonly used English-metric formula in automotive applications is the temperature conversion formula. This formula allows the technician to convert from the English ***Fahrenheit*** (F) standard to the metric ***Celsius*** (C) standard. It consists of the following equation:

$$°C = °F - 32 \times 5/9$$

$$°F = °C \times 9/5 + 32$$

Use of this equation involves multiplying by a fraction, and subtracting or adding 32. Study the following temperature conversions:

The outside temperature is 80°F. To convert to Celsius, begin by subtracting 32 from 80, leaving 48. Then multiply 48 by 5/9. The easiest way to do this is to divide 48 by 9, getting an answer of 5.3. Then multiply 5.3 by 5, giving an answer of 26.5°C.

$$80°F - 32 \times 5/9 = 26.5°C$$

The temperature of an engine is 100°F. To convert to Celsius, subtract 32 from 100, leaving 68. Then divide 68 by 9, getting 7.5. Then multiply 7.5 by 5, getting 37.5°C.

$$100°F - 32 \times 5/9 = 37.5°C$$

The temperature of an engine is 100°C. To convert to Fahrenheit, begin by multiplying 100 by 9/5. The easiest way to do this is to multiply by 9, giving an answer of 900. Then divide 900 by 5, giving an answer of 180. Finally, add 32, giving an answer of 212°F.

$$100°C - 32 \times 9/5 = 212°F$$

Remember that to get °C from °F, always subtract 32 first before multiplying by 5/9. To get °F from °C, always multiply by 9/5 and add 32 last. Other formulas are often used to convert length, weight, or pressure from metric to English, or from English to metric. Often, however, it is easier to simply look up the corresponding value on a conversion chart.

Measuring Devices

The various types of measuring devices used in automotive work may already be familiar to you. You have also studied some of these in **Chapter 3.** This section will briefly review these devices and how they fit into the overall picture of driveability service.

Physical Measuring Devices

Physical measuring devices can be divided into several subgroups, including devices for measuring linear distances, pressure, electrical properties, and heat.

Linear measuring devices include ***rulers, scales, feeler gauges, calipers,*** and ***micrometers.*** The similarity between these devices is that they measure a linear distance. This distance can be the width and length of a piece of gasket material, the size of a spark plug gap, or the length of a linkage rod. Various linear measurement devices are shown in **Figures 4-5** through **4-7.**

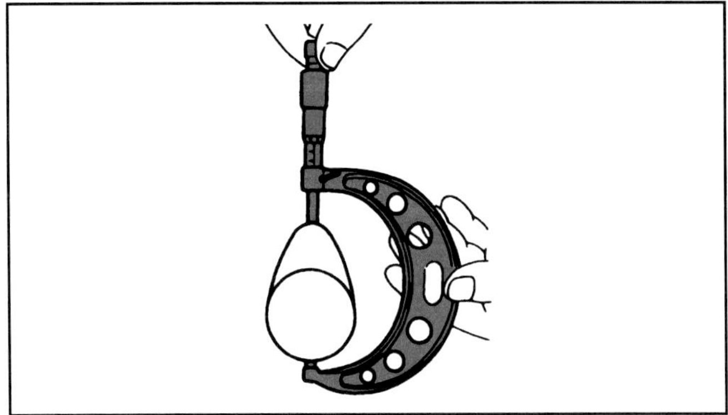

Figure 4-5. *Driveability work includes making many measurements. A micrometer can be used to precisely measure parts, including engine components. (Nissan)*

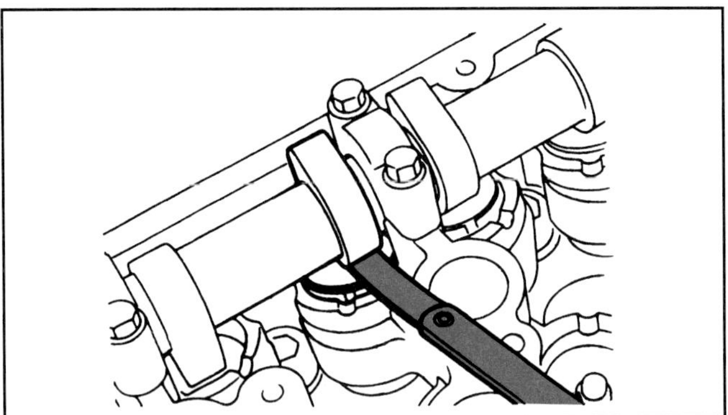

Figure 4-6. *Feeler gauges are used frequently in driveability work. Here, a feeler gauge is used to check the valve adjustment on an overhead camshaft engine. (Nissan)*

Figure 4-7. *Dial indicators are used to check for excess play and lift. This setup is used to check the end play on a crankshaft. (Chrysler)*

Pressure measuring devices include compression, vacuum, fuel, oil, and transmission pressure gauges. These testers are used to measure the force placed on a specific area, such as a square inch or millimeter, by pressurized air, vacuum, fuel, or oil. Electrical measuring devices include voltmeters, ammeters, ohmmeters, and multimeters. They measure the properties of an electrical device, using the universal measurement values of volts, ohms, and amperes. All of these devices were discussed in **Chapter 3.**

Heat is measured by various types of ***thermometers,*** and heat sensitive semiconductors. A conversion chart can be used to switch between Fahrenheit and Celsius scales when necessary. **Figure 4-8** is a conversion scale for Fahrenheit and Celsius temperatures. The measurement of heat is also a matter of checking the actual condition of something against a standard scale.

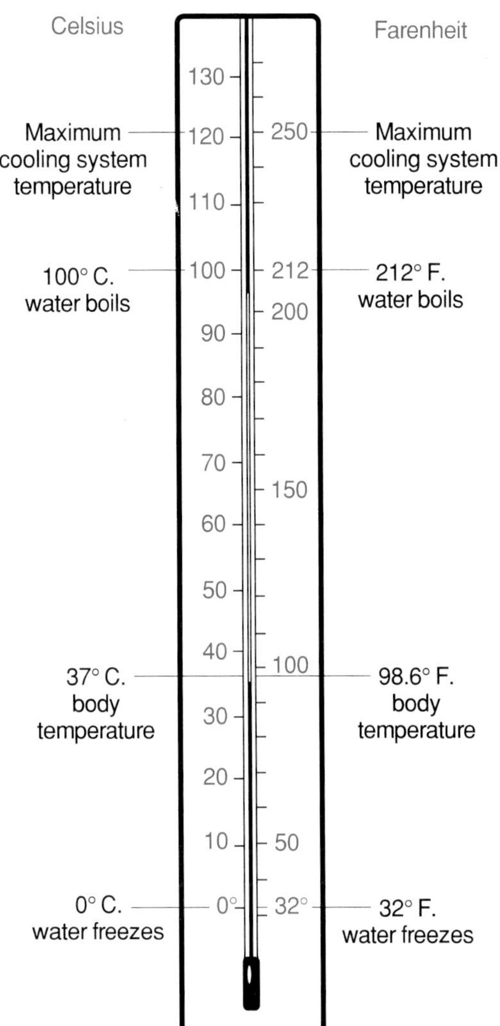

Figure 4-8. *A temperature conversion chart is useful to quickly find the Celsius or Fahrenheit equivalent. Other conversion charts and formulas are located in the back of this text.*

Vehicle Driveability Measurements

Two types of measurements are needed when investigating and correcting driveability problems. One type of measurement involves checking vehicle and system performance. The other type of measurement involves checking the status and condition of individual components. These types of measurements are explained in the following sections.

Vehicle and System Performance

Vehicle performance is simply defined as how the vehicle performs. This includes the engine, drive train, and all related subsystems such as the fuel or ignition systems and their individual parts. The methods of measuring performance are often vague, based mainly on subjective information. Many people see the best performing vehicle as the one that gets the best fuel mileage, while others think the best performing vehicle is the fastest, or the smoothest riding. Environmentalists will define a high performance vehicle as one that produces the least amount of pollution, while automotive technicians often view the best performing vehicle as the one that is the easiest to repair or has the fewest problems.

System performance is more objective and easier to measure. It is simply the measure of inputs causing outputs. For example, when you put your foot on the accelerator pedal of an operating vehicle, the engine will speed up. This process involves some complex operations on the part of various engine systems, but the end result is the same: increased pressure on the accelerator pedal equals increased engine speed.

System performance is easily measured when dealing with a simple part, but difficult when the system is complex. For example, a relay that is supposed to close when electric current is sent to one terminal can be easily checked by observing it as current is applied. However, a computer that is supposed to produce complex outputs based on many inputs cannot be checked by watching it. The performance measurement of a computer must be done by checking the operation of many components, using both objective and subjective processes.

Component Measurements

Individual components are measured according to clearly defined standards for which exact scales exist. The most common type of measurement, one that almost every person uses every day, is time. Time is sometimes measured on modern vehicles, for example, to check that the operation of various cold startup devices ends after so many seconds of engine operation. It is also possible to check the time interval, or "time-out" of the open loop mode of some engine control computers and the operating time of diesel glow plugs.

Physical dimensions of many components are measured as part of test procedures. These measurements of physical properties are often of linear size, that is, how long or wide the object being measured is. Some measurements, such as carburetor float level or spark plug gap, are made with a ruler or thickness gauge. These are examples of devices which measure length and width. Most of these measurements have been standardized over hundreds of years. An inch or a millimeter is the same size on any make of vehicle.

Electrical devices are measured for several established electrical properties, which are resistance, amperage, voltage, and wattage. Multimeters, oscilloscopes, and other electrical testers are all used to measure the same electrical properties using the same scales. These measurements are standardized worldwide. Therefore, a 12-volt Toyota or Mercedes Benz has the same voltage as a 12-volt Chevrolet or Ford.

Engine Performance Standards

One of the measurement areas of great interest is ***engine performance.*** These measurements fall into two general categories. The engine performance category includes horsepower and torque. Measurements of the entire vehicle include acceleration, top speed, mileage, emissions, noise, and driveability. Other performance measurements such as handling and ride quality are defined more in terms of the driver's expectations rather than with clearly defined tests and measurements.

Engine performance is defined by most drivers in terms of acceleration or top speed. Persons driving recreational vehicles or towing trailers may measure engine performance in terms of pulling power. Horsepower and torque ratings are the primary methods of comparing the relative power of various engines.

Measuring Horsepower and Torque

In the past, manufacturers obtained horsepower and torque ratings by attaching the engine to a dynamometer. The engine power output was measured with no engine-driven accessories such as air conditioning compressors or alternators in place. Modern vehicles are tested on a chassis dynamometer with the same timing and fuel settings that it would have while being driven.

The net horsepower and torque ratings also take into account the multiplying effect of the torque converter and transmission and the ratio created by the ring and pinion gears. This measurement is more accurate in determining the actual power potential of the vehicle. A 1960s V-8, for example, might produce 265 gross horsepower, but only 150 horsepower is actually measured at the drive wheels. A typical 1990s four-cylinder engine can produce 150 net horsepower, but would have been rated much higher in gross horsepower.

Acceleration and Speed

Acceleration and speed are measured by ***speedometers*** and other timing devices. The most common standard is the English system's miles per hour and the metric kilometers per hour. Modern ***analog speedometers*** (needle) have markings for both systems, **Figure 4-9.** ***Digital speedometers*** have a dashboard switch which allows the driver to change between English and metric measurement, **Figure 4-10.** It is also possible to mentally convert between systems when driving an older vehicle that does not have dual markings.

Figure 4-9. *Analog speedometers have been in use for many years and are still installed on most cars and trucks. Notice that the speedometer face has both mile and kilometer per hour scales. (Chrysler)*

Acceleration is the rate at which speed increases. It can be easily measured by comparing the speed of a vehicle at the beginning and end of a time period. A vehicle that can start from 0 miles or kilometers per hour and accelerate to 60 miles per hour (93 kph) in 10 seconds has better acceleration than a vehicle that takes 15 seconds to reach the same speed.

The time that a vehicle takes to cover a specific length of road is its ***average speed.*** A vehicle that covers one mile in 60 seconds has an average speed of 60 miles per hour. A vehicle that covers one mile in 30 seconds has an average speed of 120 miles per hour. A vehicle that covers 1 kilometer in 60 seconds has an average speed of 60 kilometers per hour.

Mileage

Mileage has assumed increasing importance as the price of fuel has increased. Auto manufacturers advertise their vehicle's fuel mileage as prominently as they once advertised their horsepower. Mileage is closely related to the vehicle's overall efficiency and emissions levels. It is also greatly affected by driving conditions and driving habits.

Vehicles built within the last 20 years have been certified as to a general range of mileage by the ***Environmental Protection Agency (EPA).*** This mileage is posted on new vehicles before they are sold, **Figure 4-11.** It is used mainly to establish the relative fuel efficiency between different vehicles and classes of vehicles. The fuel mileage numbers affect the technician when the actual mileage of a particular vehicle fails to approach the EPA mileage figure. Although this number can be affected by many outside variables, it is often helpful when diagnosing an excessive fuel consumption problem.

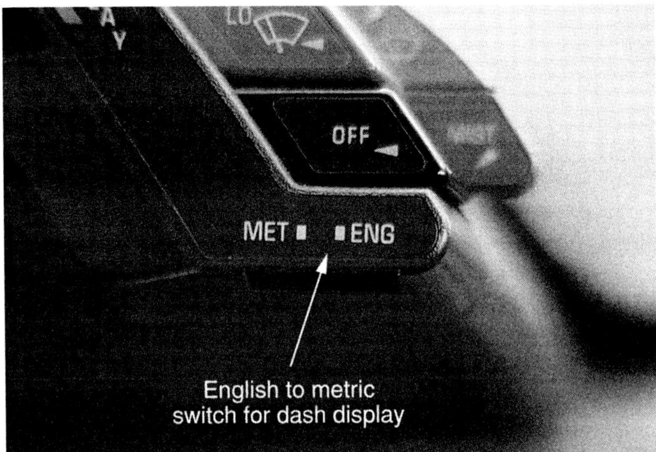

Figure 4-10. *Vehicles with digital speedometers have a switch to change the digital readout from miles per hour to kilometers per hour.*

Mileage is measured using the English miles per gallon or the metric kilometers per liter. This is the number of miles or kilometers that the vehicle will operate on one gallon or liter of fuel. A vehicle that travels 30 miles on one gallon has better mileage than a vehicle that travels 20 miles on one gallon of the same fuel. In the metric system, the mileage is determined by figuring the number of kilometers per liter. A vehicle that travels more kilometers on a liter of fuel has better mileage than one that travels fewer miles.

Emissions

At one time, ***exhaust emissions*** were neglected or ignored when engine and vehicle efficiency was being measured. Today, due to federal and state regulation of vehicle emissions, they have equal importance with other measurements of engine efficiency. The average person is now aware of the damage to the environment caused by engine emissions.

Modern vehicles must meet well-defined standards for emissions, **Figure 4-12,** as well as standards for mileage. These standards are expressed as parts per million, percentage, or grams per mile of ***hydrocarbons, carbon monoxide, oxides of nitrogen,*** and other substances. Agencies in the United States that set and monitor emissions standards include the Environmental Protection Agency (EPA), the California Air Resources Board (CARB), and other state and local agencies. In addition, a modern vehicle with legal emissions is also giving its best in power, smoothness, and mileage.

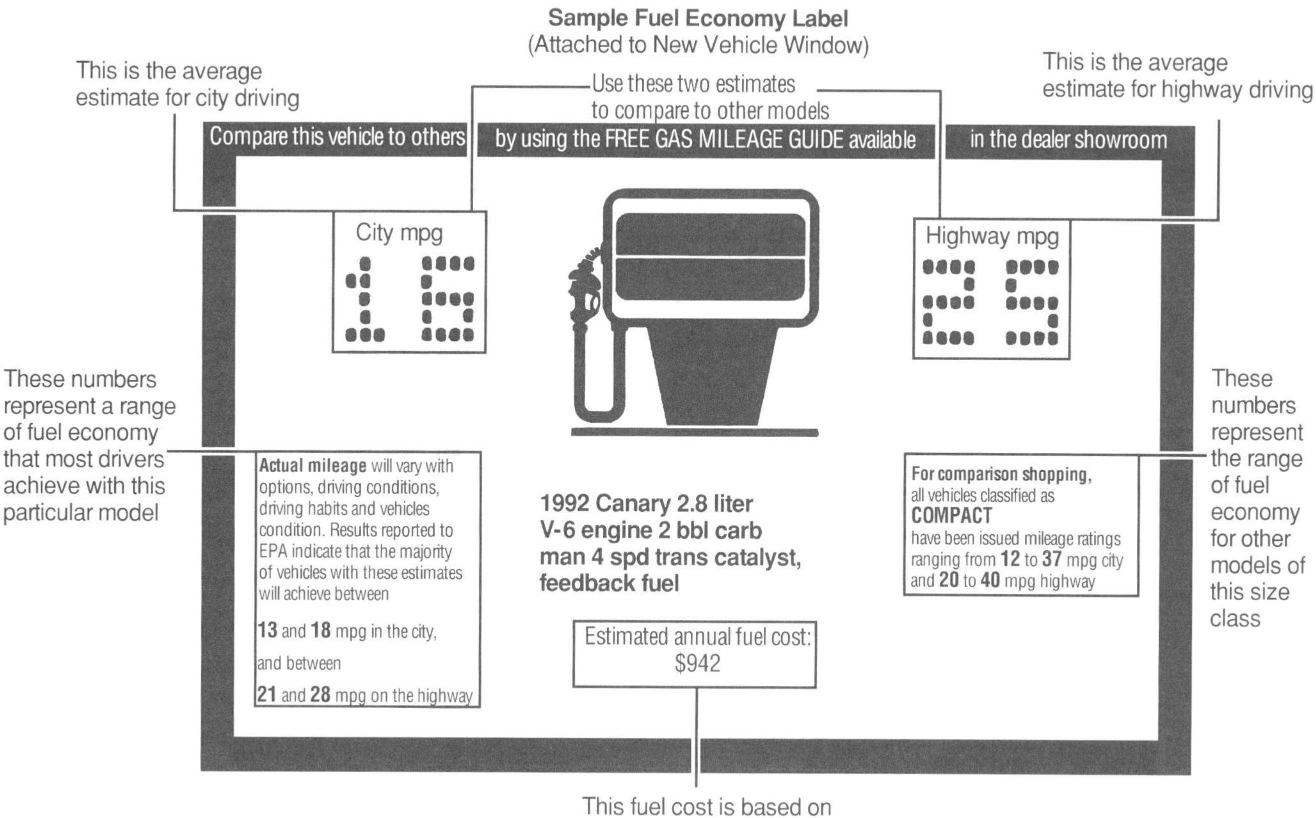

Figure 4-11. *All new vehicle manufacturers are required to affix an fuel economy label to every new vehicle sold in the United States. In addition to certifying the vehicle's fuel mileage, the sticker also shows the estimated annual cost for fuel. (U.S. Government Printing Office)*

Driveability

Driveability is the most common term used to describe such things as smoothness, lack of vibration, and the quick throttle response expected of a modern vehicle. Driveability also includes starting ease of a hot or cold engine and quick warm-up. Some common problems often grouped under driveability include pinging and detonation, run on (dieseling), and exhaust smoke. These problems are often caused by many of the same components that cause other engine problems.

There is no set scale for driveability standards, since they are usually set more by feel or expectations rather than any definable quality. Driveability problems are defined more in terms of the driver's or technician's impressions, rather than how the vehicle scores on any set tests.

Groups that Measure Engine and Vehicle Performance

There are three main groups that assess engine and vehicle performance. The vehicle owner's assessment of how the vehicle is performing, which is usually the most basic and inaccurate, but is sometimes the most important. The automotive technician, who has the tools and experience to make more accurate objective and subjective measurements of vehicle performance. And finally, the federal, state, and local government mandated assessments of vehicles in the areas of safety, mileage, and emissions. Each groups' individual criteria for engine and vehicle performance is discussed in the following sections.

Driver Assessments

The average driver's perception of good driveability and performance is somewhat subjective and often inaccurate, since most drivers are not equipped with the necessary test equipment or a thorough understanding of how the vehicle operates. If a vehicle will start and run without any obvious faults, many drivers are satisfied, **Figure 4-13.**

Only a few drivers equate efficiency with smoothness, acceleration, or mileage. A few owners closely monitor the operation of their vehicles and can detect slight variations in performance. Some drivers check the fuel mileage regularly. However, the majority of drivers are content to simply put in fuel and go.

However, even the most unconcerned driver cannot ignore basic driveability problems such as stalling, hard starting, overheating, hesitation, or detonation. The majority of drivers can often spot less obvious driveability problems,

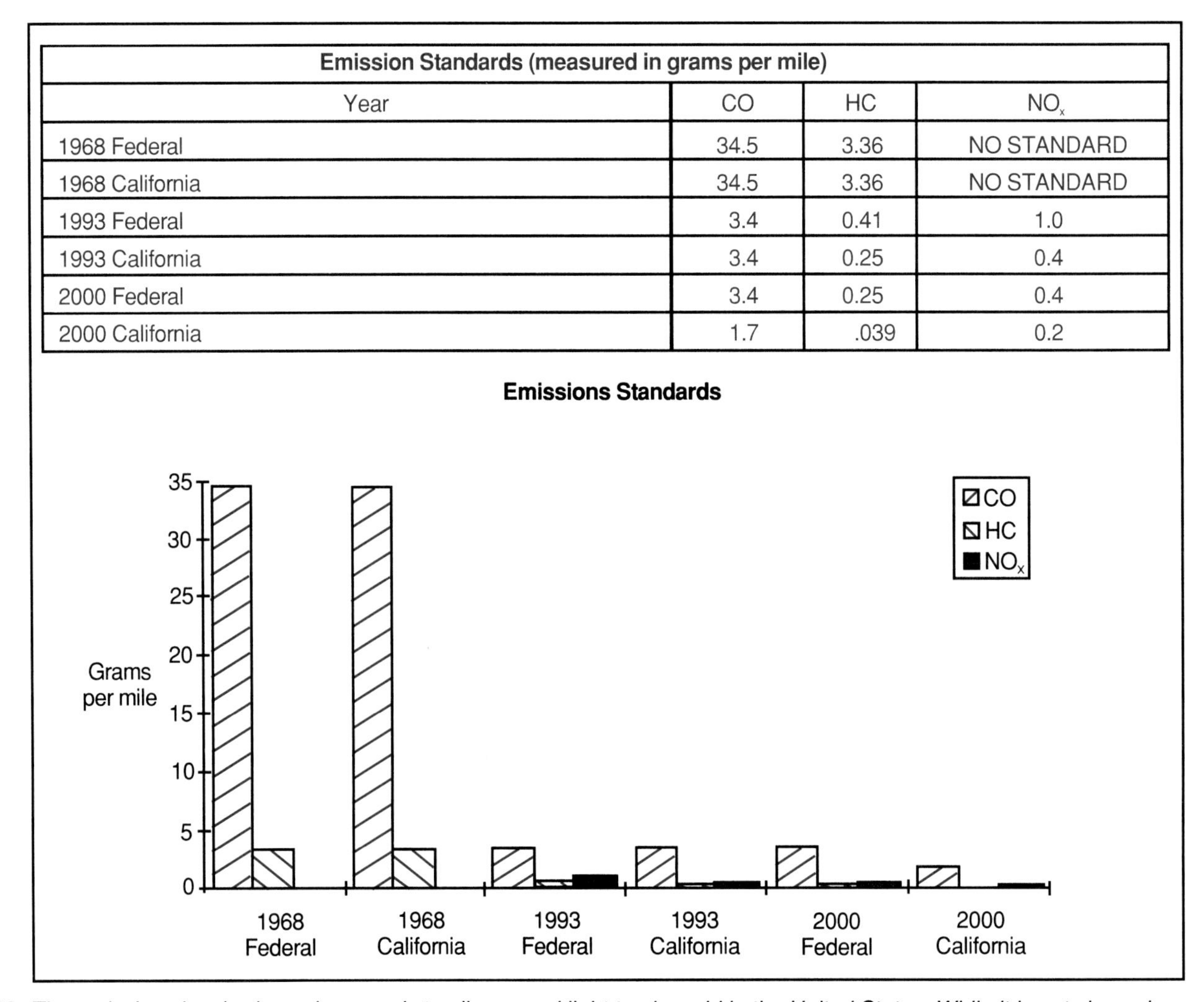

Emission Standards (measured in grams per mile)			
Year	CO	HC	NO_x
1968 Federal	34.5	3.36	NO STANDARD
1968 California	34.5	3.36	NO STANDARD
1993 Federal	3.4	0.41	1.0
1993 California	3.4	0.25	0.4
2000 Federal	3.4	0.25	0.4
2000 California	1.7	.039	0.2

Emissions Standards

Grams per mile
35
30
25
20
15
10
5
0
1968 Federal
1968 California
1993 Federal
1993 California
2000 Federal
2000 California
CO
HC
NO_x

Figure 4-12. *The emissions levels shown here apply to all cars and light trucks sold in the United States. While it is not shown here, emissions levels in the 1970s and 1980s decreased year by year. Some states, such as California, enact stricter emissions standards for vehicles sold in their respective states. (U.S. Environmental Protection Agency)*

Figure 4-13. *Of all the standards and measurements applied to vehicles, the most important one is often applied by the vehicle's owner. If a vehicle does not operate properly, is not repaired correctly the first time, or does not perform well, the owner will most likely be dissatisfied with the vehicle. (Chrysler)*

such as excess fuel consumption, surging, or high exhaust emissions. Almost all drivers have some idea, often mistaken, about their vehicle's fuel mileage. Most people underestimate the amount of driving that they do, and therefore, tend to underestimate their fuel mileage.

Since most emission problems have no obvious effect on driveability or vehicle costs, most drivers give no thought to their exhaust emissions unless the exhaust is smoking excessively or is producing offensive fumes. If their vehicle runs predictably and has reasonably good fuel mileage, most owners are satisfied, no matter how bad the emissions.

The vehicle owner has a very large financial and often an emotional interest in the vehicle. This makes it difficult for the owner to make good judgments about the condition of the vehicle. The cost of repairs is often a factor, as is the vehicle's status as an extension of the owner. Many owners are satisfied with a vehicle that has a poor running engine, because they do not realize how well the engine could run. And a great number of drivers are content to put up with a loss of mileage or power, because they are afraid of the expense of bringing the vehicle in for repairs. The driver's view of engine efficiency is most likely to be inaccurate, influenced by the lack of test equipment, an incomplete understanding of vehicle operation, complacency, or a distrust of auto technicians.

Technician Measurements

Since the technician has no financial or emotional investment in the vehicles brought in for service, he or she is able to make better subjective judgments about them. In addition to this, the technician has the proper test equipment to make sound objective measurements. The technician also knows how the various engine and vehicle systems operate together and what measurements are needed to determine overall efficiency. Therefore, the technician is in a better position to detect and correctly diagnose a problem, **Figure 4-14.**

Figure 4-14. *Since technicians are not financially or emotionally involved with most vehicles, they can make more subjective, and often much better judgments than vehicle owners.*

The technician can precisely measure such things as ignition timing and advance, air-fuel ratio, computer control system operation, and engine and vehicle mechanical condition. The technician can determine exactly what is happening in the engine and often in other parts of the vehicle. Therefore, the technician can precisely determine the efficiency level of the vehicle and take the proper steps to restore it to normal operating efficiency. The other test equipment that the technician brings to the vehicle is the experience gained from working on other vehicles with similar problems.

Mandated Measurements

Some objective measurements are made because government regulations demand that they be made. The government definition of efficiency is very different from the driver's or technician's definition. Mandated measurements that affect the procedures discussed in this text are those for emission controls and mileage. Many local, state, and federal laws affect the design and operation of engines and the vehicles in which they are installed.

The federal government enacts mandates which require manufacturers to construct their vehicles to meet certain average fuel economy and safety standards. You learned earlier that vehicles built within the last 20 years are certified as to a general range of mileage. However, each auto manufacturers' complete line of cars and light trucks sold in the United States also must meet an average fuel economy number, set by the federal government. This number is called the ***Corporate Average Fuel Economy,*** or ***CAFE,*** **Figure 4-15.** This average was first mandated in 1978 and increased in following years. The percentage increase is based off of the average economy numbers of all vehicles built in the 1974 model year. However, the average number and percentage has not changed since 1985.

Corporate Average Fuel Economy (CAFE) Standards		
Model Year	Miles per Gallon	% Improvement
1974	12	None
1978	18	50%
1979	19	58%
1980	20	67%
1981	22	83%
1982	24	100%
1983	26	116%
1984	27	125%
1985–1997	27.5	129%
1998–	27.5*	129%*

*CAFE number subject to change by Congress or DOT

Figure 4-15. *Each manufacturer's vehicles must meet the Corporate Average Fuel Economy (CAFE) number mandated by the Federal government. For example, a manufacturer might have to sell 2-3 economy cars for each sports car sold in order to meet this number. (U.S. Department of Transportation)*

The construction, operation, and effectiveness of the engine, drive train, related systems, and vehicle integrity is checked by conducting many tests, including crash tests. If an engine, drive train, vehicle chassis, or system fails to meet emissions requirements or has the potential to create a danger to passenger and/or vehicle safety, the government can compel the manufacturer to conduct a recall campaign.

The federal and some state governments have enacted strict emissions laws concerning vehicles made, serviced, and operated in the United States. Emission controls are often checked as part of state programs to reduce air pollution. These checks are made as part of a regular inspection program (the state inspection sticker or "tag"), or by random or "spot" checks of vehicles operated on public roads. In most cases, the vehicle is inspected to ensure that all original equipment emission controls are installed and in apparent working order and by sampling exhaust emissions at the tailpipe.

Some state inspection programs also test for the presence, operation, and effectiveness of safety equipment such as brakes, horns, vehicle window glass, and lights. Therefore, as well as satisfying the customer, the technician has to be prepared to satisfy all of the applicable emissions and safety laws on every vehicle leaving the shop.

Summary

Most driveability diagnosis involves taking various measurements. There are two basic kinds of measurement, objective and subjective. Objective measurement is the process of measuring something against accepted standards. Examples are measuring the size, temperature, or voltage of an automotive component. Subjective judgments are made on the basis of internal feelings. This type of measurement is less specific than other types, and often varies between different people.

The most common types of objective measurements are in either the English or metric systems. The metric system is slowly replacing the English system, meaning that it is often necessary to convert from one system to the other. Formulas are used to find an unknown quantity when other quantities are known. Horsepower is a measure of the rate at which work is performed and is measured by dynamometers. Gross horsepower figures are those made with only the basic accessories needed for engine operation. Net horsepower ratings are taken with all normal accessories on the engine. Torque is a measure of the engine's twisting or turning force at the crankshaft. Horsepower and torque are not the same. Horsepower increases with RPM until maximum is reached. Torque increases to the point where volumetric efficiency is greatest and any additional speed will cause it to diminish.

Engine efficiency is based on thermal and mechanical efficiency. Most engines have about 25% thermal efficiency and about 90% mechanical efficiency. Energy is lost to the cooling, lubrication, and exhaust systems as well as to engine, drive train, and wheel friction. In terms of practical efficiency, only about 15% of the total potential energy of the fuel is used to drive the wheels. Volumetric efficiency is a measure of how well the engine is able to draw in the air-fuel mixture. It varies with engine speed, and is influenced by a number of factors.

Engine size is usually given in cubic inches or liters. The number of cubic inches or liters of cylinder space depends on piston displacement. Cylinder bore and piston stroke determine engine displacement. Modern compression ratios are lower than previously for use with unleaded gas and to reduce NO_x emissions.

Ohm's law can be used to find one of the three basic electrical properties: voltage, amperage, and resistance, when the other two are known. Conversion formulas are used to convert between English and metric values. Measuring devices are used to make objective measurements. They always measure a physical quantity, such as size, pressure, voltage, or temperature. Measuring devices can be simple, such as a ruler, or complex electronic testers.

Vehicle driveability performance falls into two categories, vehicle and system performance checking, and individual component checking. Driveability testing involves how the vehicle or a subsystem operates and a combination of objective and subjective tests.

Engine performance standards is comprised of many different factors. One measurement that most people agree on is performance. Performance is defined by most drivers in terms of acceleration or top speed. The engine performance category includes horsepower and torque ratings, and is the most exact measurement of engine performance. Engine power is measured by placing the entire vehicle on a chassis dynamometer, and measuring the horsepower and torque at the drive wheels.

Acceleration and top speed are measured by speedometers and other timing devices either in English or metric measurements. Acceleration is the rate at which speed increases. It can be easily measured by comparing the speed of a vehicle at the beginning and end of a time period. The time that a vehicle takes to cover a specific length of road is its top speed.

Another factor in efficiency testing is mileage. Mileage is closely related to the overall efficiency and emissions levels of the vehicle. It is also affected by driving conditions and driving habits. Mileage is measured using the English system of miles per gallon, or the metric system of kilometers per liter. Vehicles built within the last 20 years have been certified as to a general range of mileage by the Environmental Protection Agency (EPA).

Emissions control has assumed equal importance with other measurements of engine efficiency. This is due to federal and state regulation of vehicle emissions and increased awareness of environmental problems. Modern vehicles must meet well-defined standards for emissions as well as standards for mileage. Various state and federal agencies monitor emissions.

Driveability is another factor in efficiency testing. Driveability includes smoothness, lack of vibration, and quick throttle response. Driveability also includes starting ease and smooth operation during engine warm-up. Other problems often grouped under driveability include pinging and detonation, run on (dieseling), and exhaust smoke. Driveability problems are defined more in terms of the driver's or technician's impressions rather than how the vehicle scores on tests.

There are three groups that make assessments of engine and vehicle performance: the owner's assessment of how the vehicle is performing; the automotive technician's assessment of vehicle condition, and mandated government measurement of safety, mileage, and emissions.

The average driver's perception of driveability and performance is inaccurate, since most drivers are not equipped with any test equipment, or a good understanding of how the vehicle operates. Different people have different ideas about what factors are the most important. Some see the best performing vehicle as the fastest, the one that gets the best mileage, the one that is the least harmful to the environment, the smoothest driving, or even the easiest to repair.

Since the technician is more objective, he or she is able to make more accurate judgments about a vehicle's driveability. Some measurements, such as safety checks, are made because government regulations demand that they be made. Many local, state, and federal laws affect the operation of engines and the vehicles in which they are installed.

Several states and the United States government have enacted strict emissions laws concerning vehicles made and serviced in the United States. Federal law also mandates fuel mileage for modern vehicles, as well as various safety laws. Emission controls are often checked as part of state programs to reduce air pollution.

Know These Terms

Objective measurement
Subjective judgment
Length
Width
Volume
Weight
Temperature
Time
Speed
Voltage
Amperage
Pressure
English system
Metric system
Process of elimination
Horsepower
Dynamometer
Gross horsepower
Net horsepower
Torque
Thermal efficiency
Mechanical efficiency
Volumetric efficiency
Engine displacement
Compression ratio
Ohm's law
Fahrenheit
Celsius
Rulers
Scales
Feeler gauges
Micrometers
Calipers
Thermometers
Vehicle performance
System performance
Engine performance
Speedometers
Analog speedometers
Digital speedometers
Acceleration
Average speed
Mileage
Environmental Protection Agency (EPA)
Exhaust emissions
Hydrocarbons
Carbon monoxide
Oxides of Nitrogen
Driveability
Corporate Average Fuel Economy (CAFE)

Review Questions—Chapter 4

Please do not write in this text. Write your answers on a separate sheet of paper.

1. A system of measurement that relies on comparisons with universal standards is an ______ measurement.
2. What two common measurements are the same in both the English and metric systems?
3. When a technician looks at a spark plug and determines that it has excessive deposits, he or she is using ______ judgment.
4. When in doubt, the good technician will always substitute an ______ measurement for a ______ judgment.
5. Match the letter on the right to the corresponding measurement tool on the left.

___ Ohmmeter	(A) Linear measuring device
___ Thermometer	(B) Pressure measuring device
___ Ruler	(C) Electrical measuring device
___ Micrometer	(D) Heat measuring device
___ Compresssion tester	

6. 200°F is how many °C?
7. 132°F is how many °C?
8. 109 C is how many°F?
9. If a 12-volt circuit is drawing 6 amperes, what is the resistance of the circuit?
10. If a coil has 3 ohms of resistance, and the circuit voltage is 12 volts, what is the current draw?
11. If a sensor circuit has a resistance of 12 ohms, and a current draw of .5 amps, what is the voltage?
12. How is performance defined?
    (A) Acceleration.
    (B) Pulling power.
    (C) Horsepower.
    (D) All of the above.
13. The modern way of measuring engine power is to place the ______ on a dynamometer.
    (A) engine
    (B) engine and transmission
    (C) entire vehicle
    (D) None of the above.

14. Fuel mileage for new vehicles is mandated by:
    (A) the federal government.
    (B) state governments.
    (C) vehicle owners.
    (D) Both A & B.

15. Emissions levels for new vehicles are mandated by:
    (A) the federal government.
    (B) state governments.
    (C) vehicle owners.
    (D) Both A & B.

ASE Certification-Type Questions

1. The automotive technician can precisely measure the following things, EXCEPT:
    (A) timing and advance.
    (B) exhaust emissions levels.
    (C) owner knowledge level.
    (D) computer control system operation.

2. Horsepower is a way to measure the ability to perform ______.
    (A) thermal efficiency
    (B) mechanical efficiency
    (C) work
    (D) friction

3. Technician A says that the average engine has a thermal efficiency of about 90%. Technician B says that the mechanical efficiency of an engine is reduced by internal friction. Who is right?
    (A) A only.
    (B) B only.
    (C) Both A & B.
    (D) Neither A nor B.

4. Net horsepower is determined using a ______.
    (A) thermal heat formula
    (B) engine dynamometer
    (C) chassis dynamometer
    (D) Either B or C.

5. The practical efficiency of the average vehicle is about ______.
    (A) 15%
    (B) 25%
    (C) 75%
    (D) 90%

6. Technician A says that torque is limited by practical efficiency. Technician B says that torque is limited by volumetric efficiency. Who is right?
    (A) A only.
    (B) B only.
    (C) Both A & B.
    (D) Neither A nor B.

7. The horsepower rating that is most useful to vehicle owners and technicians is the ______ horsepower.
    (A) gross
    (B) net
    (C) din
    (D) brake

8. All of the following can affect volumetric efficiency EXCEPT:
    (A) valve size.
    (B) engine speed.
    (C) compression ratio.
    (D) atmospheric pressure.

9. Total cylinder displacement is measured in ______.
    (A) cubic inches
    (B) liters
    (C) millimeters
    (D) Both A & B.

10. The most inaccurate measurement of a vehicle's efficiency is made by ______.
    (A) the federal government
    (B) state governments
    (C) technicians
    (D) vehicle owners

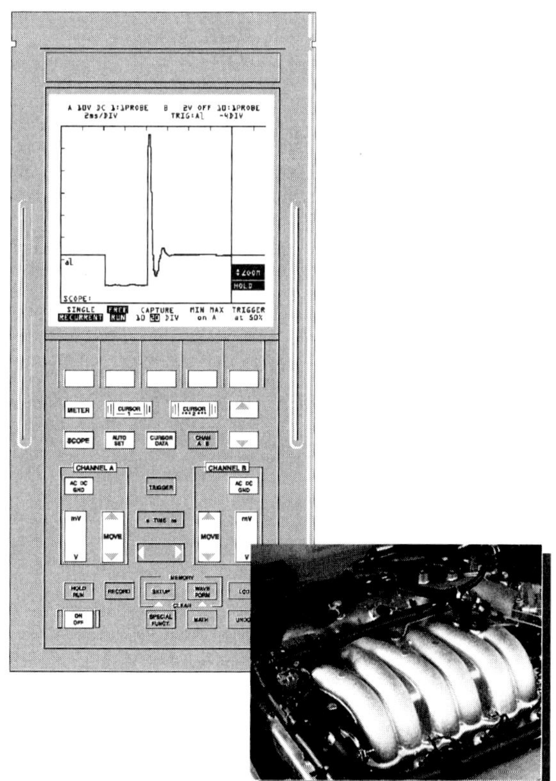

Chapter 5

Engine Operating Fundamentals

After studying this chapter, you will be able to:

- ❑ State the major action that takes place during each stroke of a four-stroke engine.
- ❑ Identify and explain the purposes of the major components of the engine compression system.
- ❑ Identify and explain the operation of a rotary engine.
- ❑ Identify and explain the operation of a Miller-cycle engine.
- ❑ Identify and explain the operation of a diesel engine.
- ❑ Identify the operating principles of liquid cooling systems.
- ❑ Identify the operating principles of air cooling systems.
- ❑ Identify and explain the purpose of the cooling system parts.
- ❑ Identify the purposes of the engine lubricating system.
- ❑ Identify and explain the purpose of the lubrication system parts.
- ❑ Identify common oil classification systems.

This chapter will identify and review the operation of the engine and its internal components. Engine problems can have a major effect on vehicle driveability. For example, without good compression or proper valve timing, the engine will not operate properly or at all. This chapter will cover the major engine components, review the operation, and how misadjustment or defects in each of these components can affect driveability.

Compression System

The ***compression system*** is comprised of the engine parts that develop engine compression and allow the burning air-fuel mixture to become mechanical energy. The parts of the compression system are sometimes called the engine internals. The next paragraphs are a brief review of the operation of the compression system. They are followed by a description of the individual components. The information presented here is true for any type of engine, whether it is a four-, six-, or eight-cylinder engine, and whether the cylinders are arranged inline, in a V-shape, or an opposed piston design.

Four-Stroke Engine Operation

The development of power in the typical automotive engine is a four-stroke process, with the piston moving up in the cylinder twice and down twice. Some diesel engines use only two-strokes rather than four. Refer to **Figure 5-1** as you read the following paragraphs.

On the ***intake stroke,*** the downward movement of the piston draws the air-fuel mixture into the cylinder through one or two open intake valves. When the piston reaches the bottom of the intake stroke, the intake valve(s) closes. The

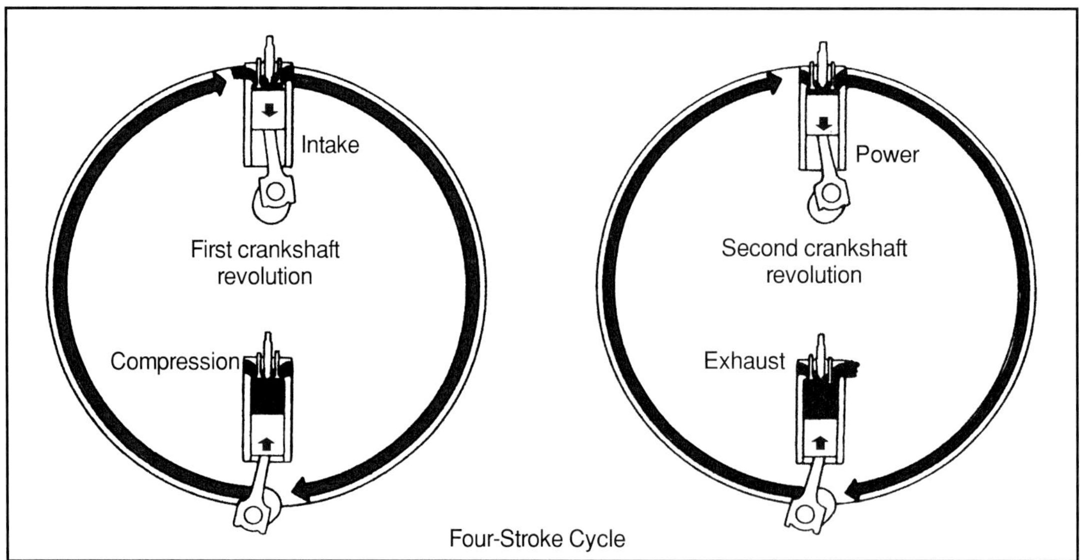

Figure 5-1. *The compression system is made up of the internal engine parts that develop and maintain pressure on the air-fuel mixture (or air in diesel engines). The cycle shown here is typical of all four-stroke cycle engines.*

piston then moves up in the cylinder, compressing the air-fuel mixture. This is called the ***compression stroke.*** When the air-fuel mixture in the cylinder is ignited, the burning fuel and oxygen produce heat. The heat causes the gases in the cylinder to expand, forcing the piston down under pressure, which is called the ***power stroke.***

The downward motion of the piston is transferred to the connecting rod, which in turn pushes the crankshaft. The connecting rod and crankshaft arrangement changes the downward piston movement into crankshaft rotation used to drive the vehicle. The piston reaches the bottom of the cylinder, and the exhaust valve opens. The piston then starts the ***exhaust stroke*** by moving up the cylinder, pushing the exhaust gases out the open exhaust valve. The cycle then repeats. The four strokes of the piston are the reason the automotive engine is called a four-stroke engine.

Engine Components

The following paragraphs describe the major components of the engine compression system. Although the exact design and number of engine components varies with the type of engine and the number of cylinders, the basic components and their functions are the same as described in the following sections.

Engine Block

The ***engine block*** is the basic support and attaching point for all other engine parts. Engine blocks are made by pouring molten cast iron or aluminum into molds made of a special type of sand. After the metal cools, the sand is washed out and the block is machined to allow other parts to be installed or attached, **Figure 5-2.** Major parts installed in or on the block are the pistons, crankshaft, camshaft, cylinder heads, and manifolds. Many modern engine blocks are machined to accept one- or two-piece brackets that hold

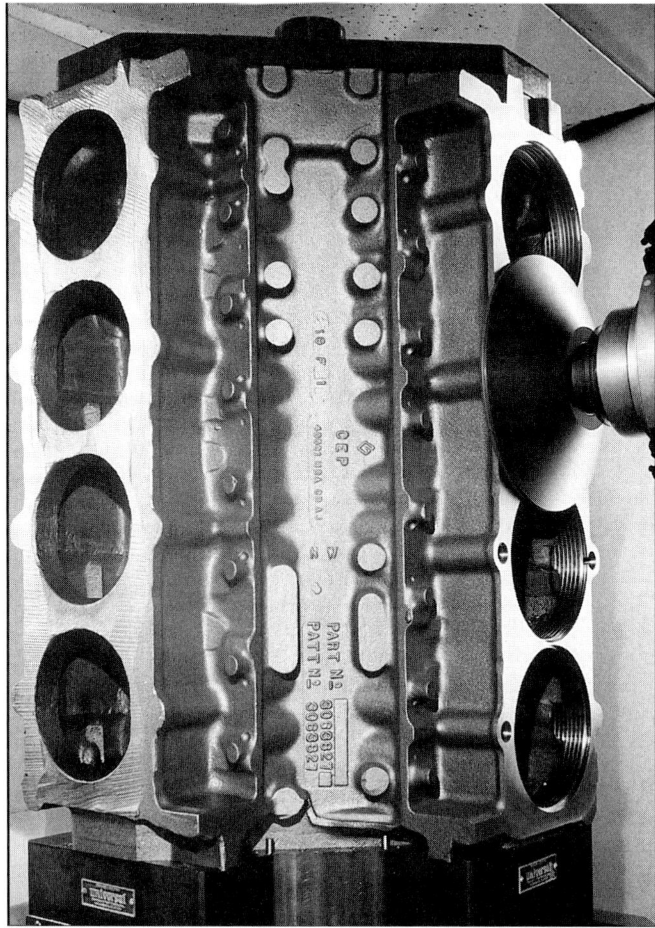

Figure 5-2. *The engine block is a large iron or aluminum casting machined to accept and support the other engine parts. (LeBlond Makino)*

the alternator, air conditioning compressor, power steering pump, and other engine-driven accessories. This eliminates the need for separate brackets, which are often sources of vibration and noise.

Pistons and Rings

Modern ***pistons*** are made of aluminum to reduce weight. Most engine pistons have two compression rings and one oil ring, as shown in **Figure 5-3.** ***Compression rings*** seal in the pressure created during the compression and power strokes. If this pressure was allowed to leak out, the engine would not start or would have severe power and driveability problems. The compression rings are installed at the top of the piston. A film of oil between the compression ring and cylinder wall seals in compression gases. This oil film is only about .001″ (.0025 mm) thick, but if it is removed, the engine will not develop enough compression to start.

The ***oil control ring*** is installed below the compression rings to prevent excessive oil consumption. During the piston's intake stroke, vacuum in the combustion chamber tries to draw oil from the crankcase. To reduce oil loss, the oil control ring scrapes most of the oil from the cylinder wall when the piston is moving down in the cylinder. A small amount of oil passes by the oil control ring to help the compression rings seal against the cylinder wall.

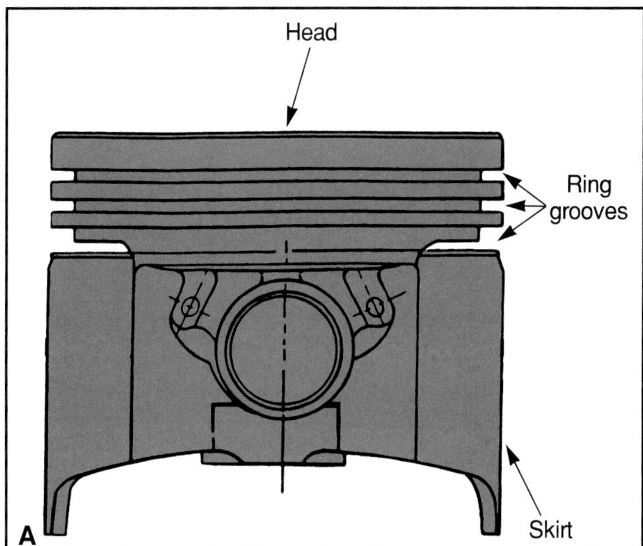

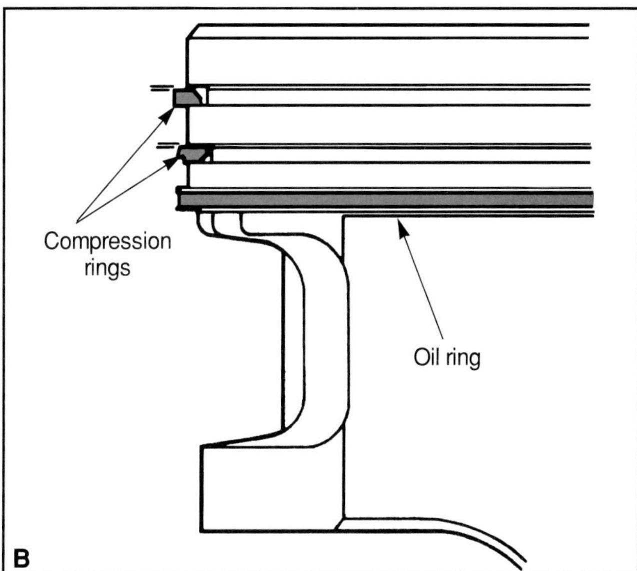

Figure 5-3. *The pistons and rings are the parts that most people associate with the compression system. A—This shows the various parts of the piston. B—Compression and oil control ring installation. Most spark ignition engines use this type of ring installation. (General Motors, Ford)*

Connecting Rods and Crankshaft

The ***connecting rods*** are forged steel rods that connect the piston to the crankshaft. They convert the straight-line motion of the piston into rotary motion at the crankshaft. Each connecting rod is connected to a piston by a ***piston pin.*** The rod is attached to the crankshaft by a bearing cap and bearing inserts that surround the crankshaft journal. The piston pin and crankshaft bearings allow the rod to move in relation to both the piston and crankshaft. Refer to **Figure 5-4.**

The ***crankshaft,*** **Figure 5-5**, is attached to the engine block by ***bearing caps*** and ***bearings,*** which surround the crankshaft journal. This design allows the crankshaft to rotate inside the bearings with a minimum of friction. The bearing caps are held to the engine block by 2-4 bolts carefully installed and torqued.

Cylinder Heads

The ***cylinder head*** contains the combustion chamber for each cylinder and forms the top of the cylinder. Cylinder heads contain the intake and exhaust valves and, in some cases, the camshaft and lifters. They also contain oil galleries, water jackets, and openings to allow the passage of intake and exhaust gases. Cylinder heads are made from either cast iron or aluminum. A sheet metal, cast aluminum, or plastic valve cover is installed over the upper valve train components. **Figure 5-6** shows a typical cylinder head and combustion chamber.

Cylinder Head Gasket and Retainers

The cylinder heads must seal in the coolant of the cooling system and must also contain the pressure of the exploding fuel. Thin steel, copper, and fiber ***head gaskets*** are used between the head and engine block. The cylinder head and head gasket are bolted to the block with ***head bolts.*** A few engines use studs and nuts, rather than head bolts.

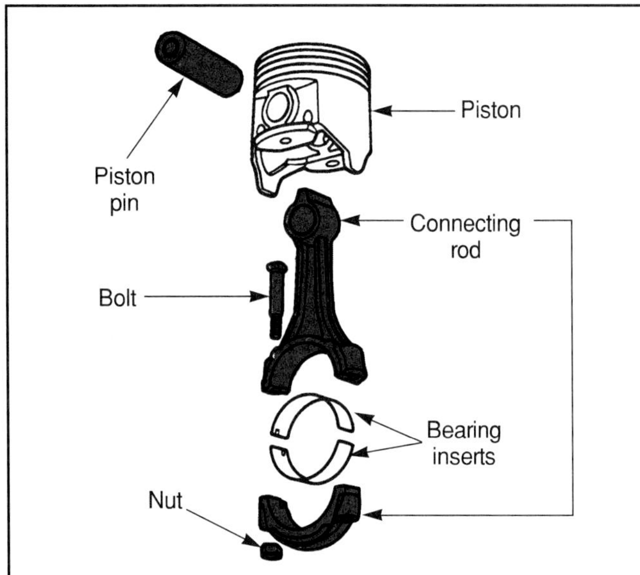

Figure 5-4. *The connecting rods attach the piston to the crankshaft. They convert the piston's up-and-down motion into rotation at the crankshaft. (Ford)*

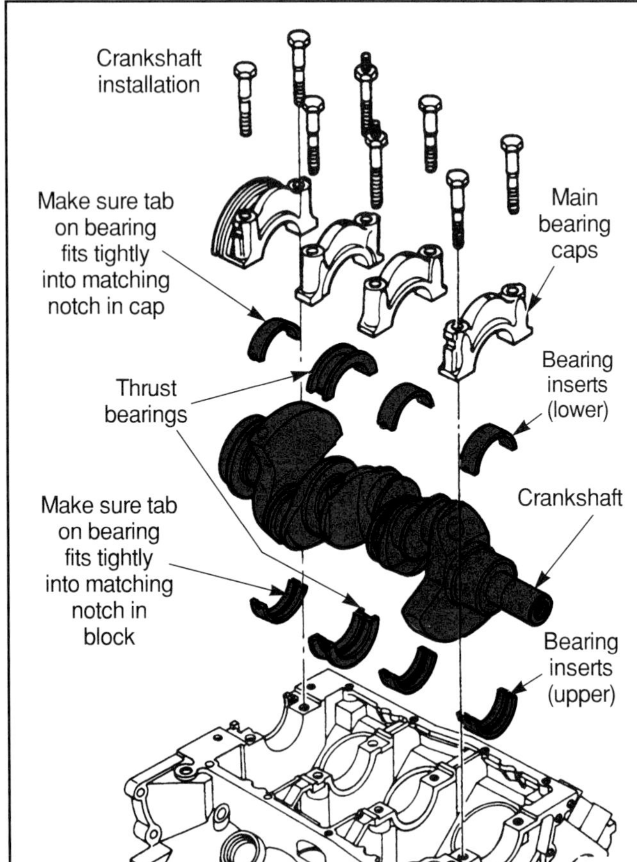

Figure 5-5. *The crankshaft is a rotating engine part that absorbs the power delivered by the pistons through the connecting rods, and delivers this power to the drive train. (Ford)*

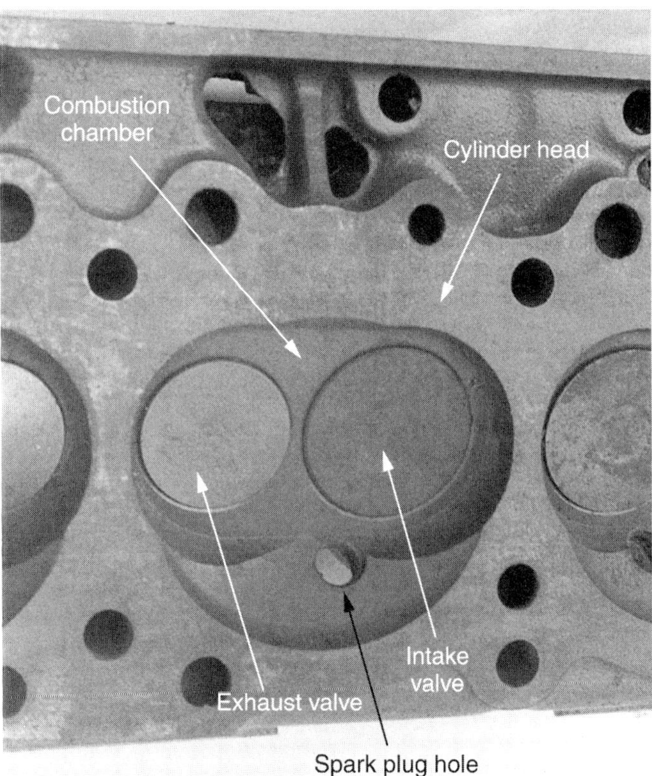

Figure 5-6. *The cylinder head contains the valves and acts as a cap on the top of the engine. Along with the cylinders, they form the combustion chamber.*

Valves and Related Components

One or more ***intake*** and ***exhaust valves*** are used to control the flow of air and fuel into and out of each cylinder and to seal it during the compression and power strokes. They are occasionally called *mushroom* valves due to their resemblance to a mushroom. Intake and exhaust valves are identical in shape, but intake valves are usually larger. Valves are operated by the valve train components.

The ***valve spring*** holds the valve against its seat, keeping it closed. Valve springs are always installed so that they are slightly compressed. This ensures that the valve closes tightly. The spring is held to the valve by ***valve spring retainers.*** The retainer is a cap which covers the spring. A locking device, usually called a ***split keeper,*** locks the cap to the valve stem. The valve and spring assembly is shown in **Figure 5-7.** The assembly is held together by valve spring pressure.

The valve stem rides against either an integral or a removable ***valve guide***. The valve guide keeps the valve steady and gives it a smooth surface to slide on. The ***oil seal*** at the top of each valve stem prevents engine oil from entering the combustion chamber. Without a seal, oil would be pulled between the valve stem and guide and into the combustion chamber. The valve seal can be either an umbrella type or an O-ring installed between the valve stem and valve retainer.

The valve must be in contact with its ***seat*** long enough to transfer heat to the head. This contact is most critical for exhaust valves, since they must transfer exhaust heat to avoid melting. Some valves are filled with metallic sodium to further aid in heat transfer. The intake valves are cooled by the incoming air, and are not as prone to damage. The contact is controlled by the ***valve clearance,*** which is the amount of looseness in the valve train between the camshaft and valve stem. Valve clearance can sometimes be adjusted. Clearance has more effect on valve life than engine performance.

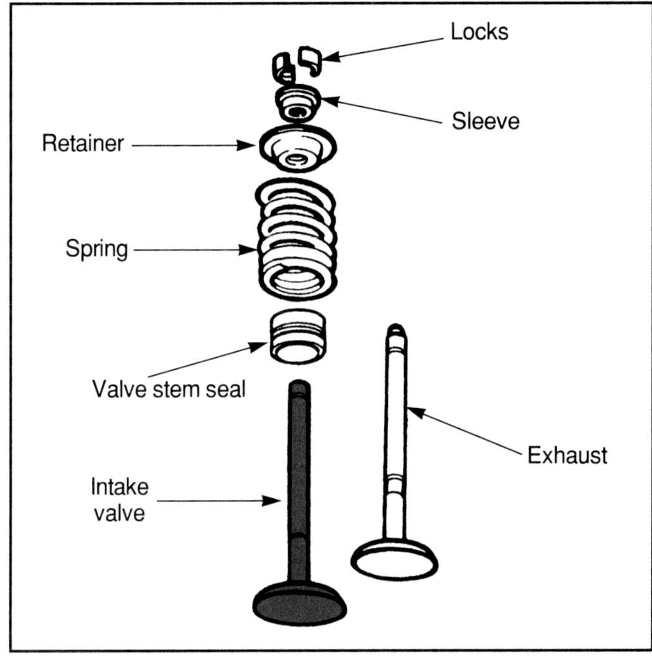

Figure 5-7. *The valve assembly consists of the valve itself, plus the valve spring or springs, retainers, retainer-to-valve stem locks, and oil seals. (Ford)*

Valve Train

The ***valve train*** is the group of components that opens the intake and exhaust valves. Valve train operation is similar in both overhead camshaft engines and engines having the ***camshaft*** in the block. The camshaft is turned by the engine crankshaft. The connection from the crankshaft to the camshaft can be two or more timing gears or sprockets and a timing chain or belt. As the camshaft rotates, the camshaft lobes push the ***valve lifters,*** which in turn overcome valve spring pressure and push the valves open. On overhead camshaft engines, the cam lobes usually push directly on the valve rocker arm, **Figure 5-8.** There is one camshaft lobe for each valve.

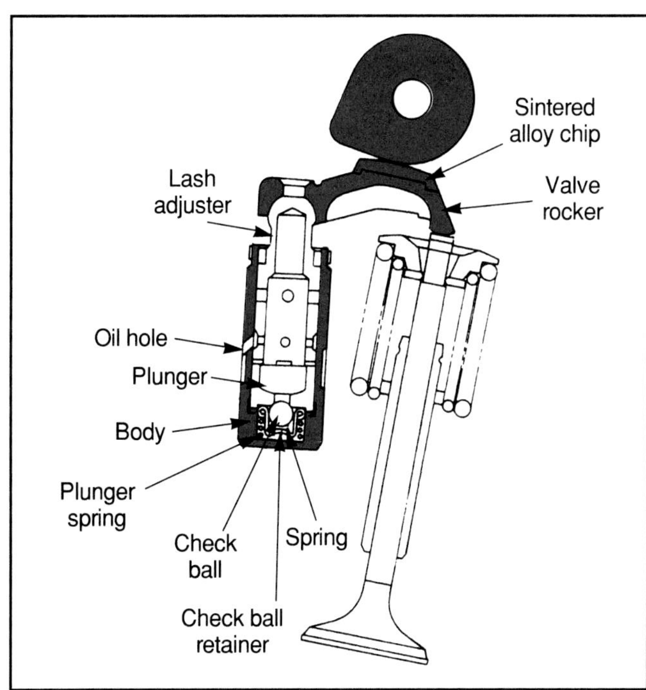

Figure 5-8. *The overhead camshaft valve train is quickly becoming the valve train of choice. Since it eliminates the pushrods, it allows quicker valve response with less play and friction between the components. Overhead camshafts are either belt- or chain-driven. (Subaru)*

Valve Timing

The valves must open and close in the proper relationship to the movement of the piston, or the engine will not run. This relationship is called ***valve timing*** and is determined by the relative positions of the crankshaft and camshaft. The valve must also open wide enough and long enough to allow the air-fuel mixture to get into the cylinder and for the exhaust gases to get out. ***Lift*** is how wide the valve opens, while ***duration*** is the amount of time that the valve stays open. Lift and duration are determined by the shape of the camshaft lobes. Valve timing, lift, and duration have a big effect on engine driveability. ***Overlap*** is the time that both intake and exhaust valves are open.

The crankshaft always turns two complete revolutions for every one revolution of the camshaft. This is because any cylinder in a four-stroke cycle engine, whether gasoline or diesel, requires two complete revolutions of the crankshaft to complete all four cycles. However, each valve in the engine opens only once during all four strokes. Every camshaft lobe, and therefore the camshaft, can only rotate once during the same four cycles when the crankshaft rotates twice. To accomplish this, the driving gear on the crankshaft always has half the number of teeth as the driven gear on the camshaft. This means that the crankshaft gear must turn two complete revolutions to turn the camshaft gear one revolution.

If the camshaft is installed in the block, the valve lifters are installed into machined bores in the block. The ***lifter bores*** provide enough clearance for the lifter to move up and down, while preventing side movement.

Valve Lifters

Valve lifters can be mechanical or hydraulic. Mechanical, or solid lifters transmit the motion of the camshaft lobes, and must be periodically adjusted. Hydraulic lifters are self-adjusting. A typical hydraulic lifter is shown in **Figure 5-9.** The outer lifter body contacts the camshaft lobe, and the inner piston contacts the pushrod or rocker arm. The space between the lifter body and under the inner piston is filled with engine oil. This oil is supplied by the lubrication system through a small passage.

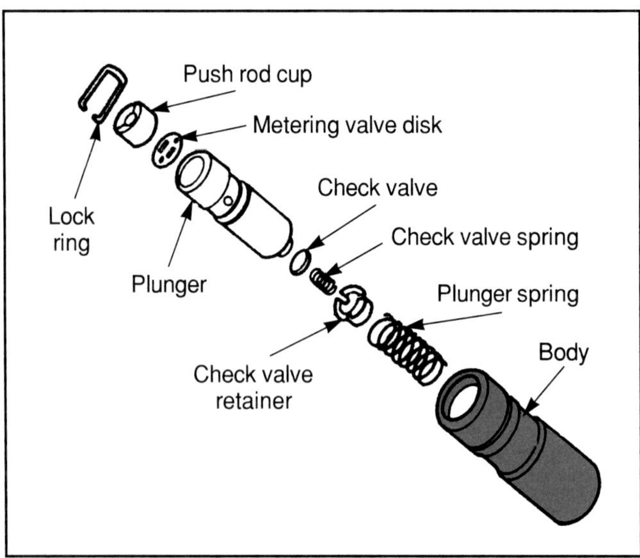

Figure 5-9. *Hydraulic lifters use the engine's oil pressure to automatically eliminate play from the valve train. Hydraulic lifters are used on almost all modern engines with an in-block (OHV) camshaft. (Ford)*

When the camshaft lobe pushes the lifter body upward, the oil passage into the lifter is sealed off. Since the oil cannot escape or compress, the hydraulic lifter becomes a solid unit and opens the valve. When the cam lobe allows the lifter to move down, the lifter oil passage is again open, and oil can flow into the lifter from the lubricating system passage. Engine oil pressure pushes the inner piston upward to remove any valve train clearance, but does not have enough power to open the valve.

The friction between the lobes and lifters in overcoming valve spring pressure is the highest friction in the engine and can cause the camshaft and lifters to wear out rapidly. To reduce friction, the camshaft lobes are tapered and the lifters offset. This causes the lifters to rotate as they are pushed up by the lobe. This rolling action between the lobe and lifter helps to reduce friction and wear. To further reduce friction, many late model engines use ***roller lifters.*** A roller is installed on the bottom of the lifter and turns with the lobe as the camshaft rotates.

Pushrods and Rocker Arms

Pushrods are used only on engines with the camshaft in the block. They transmit the lifter motion to the rocker arm. Many pushrods are hollow, and oil from the lifter flows through them to lubricate the rest of the valve train. ***Rocker arms*** are pivoting levers which convert the upward movement of the pushrod or lifter into downward movement of the valve. On some overhead cam engines, the rocker arm transfers the motion of the cam lobe directly to the valve.

On overhead cam engines, the camshaft is installed on top of the cylinder head, and opens the valves from above. In some overhead cam engines, the lifter is placed between the cam lobe and the rocker arm. In many cases, the overhead cam rocker arm is operated directly by the camshaft. A ***hydraulic lash adjuster***, similar in operation to a hydraulic lifter, maintains the proper valve clearance. See **Figure 5-10.**

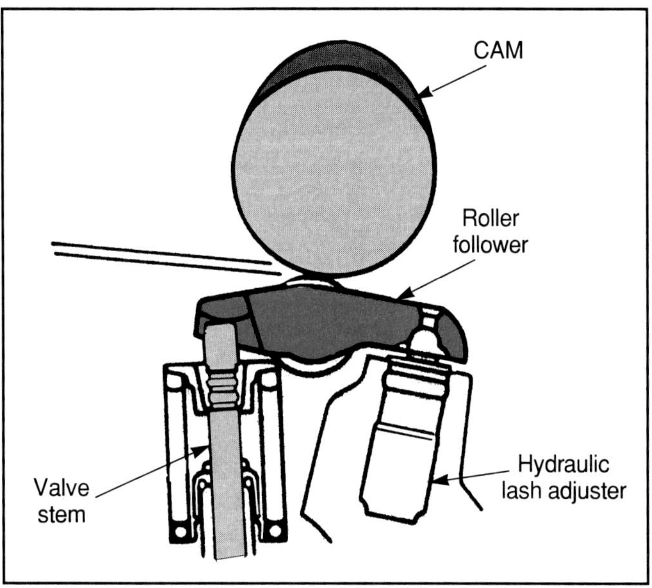

Figure 5-10. *Lash adjusters are used on overhead camshaft valve trains. They use the same principle as the hydraulic lifter, but they are installed on the rocker arm or on the opposite side of the valve stem. (Ford)*

Valve Timing Devices

To maintain the relationship between the valves and pistons, one of three types of ***camshaft drives*** is used. A few vehicles use the ***gear drive,*** **Figure 5-11.** In this drive two gears mesh to transmit power from the crankshaft to the camshaft. Note that the number of teeth on the crankshaft gear is exactly half the number of teeth on the camshaft gear. This causes the camshaft to turn at exactly half of crankshaft speed.

On the majority of overhead valve engines with the camshaft in the block, the ***chain drive*** is used. The chain drive, shown in **Figure 5-12,** transmits power by use of a chain and two gears. Note that the crankshaft gear is smaller and has half the teeth of the camshaft gear. A few overhead camshaft engines use a chain drive.

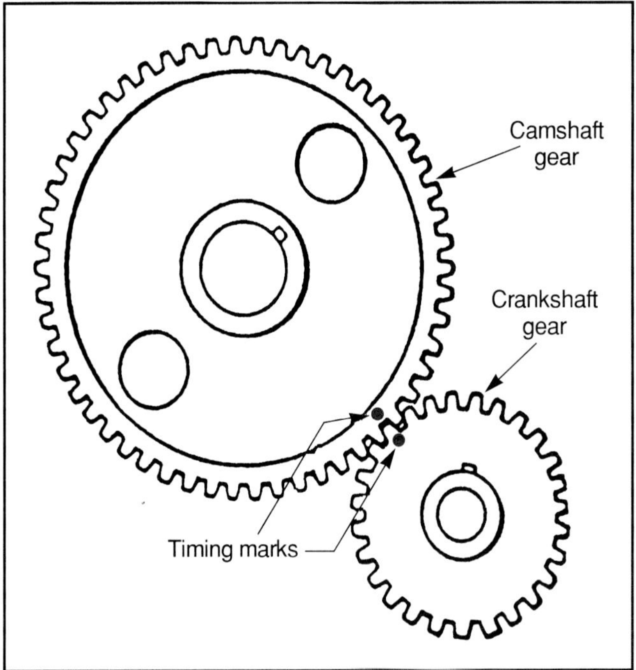

Figure 5-11. *A few spark ignition engines use gear drive. While gear drive provides excellent timing, it produces noise that is objectionable to most drivers. Gear drive is more common in diesel engines.*

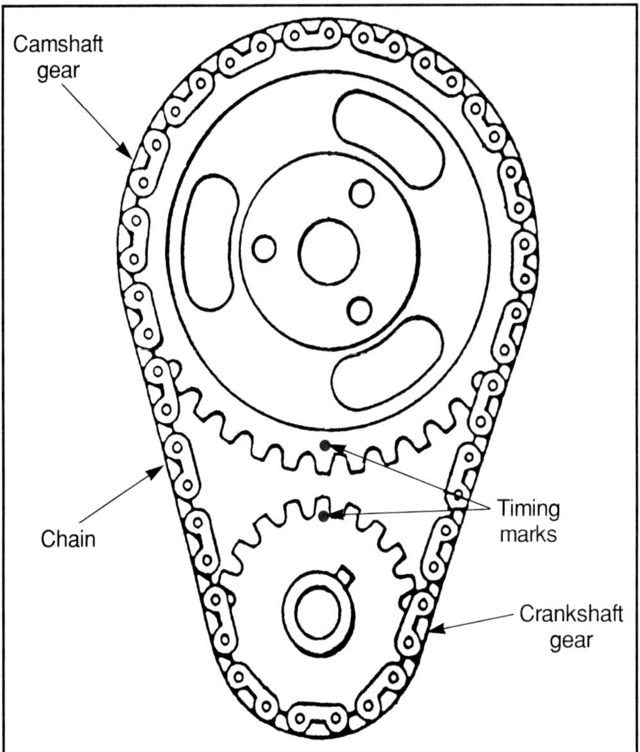

Figure 5-12. *Chain drive is the most commonly used timing methods. Note the timing marks on both the camshaft and crankshaft timing gears.*

Most overhead camshaft engines use a ***belt drive,*** **Figure 5-13.** In this design, the belt is driven by a crankshaft gear, or sprocket. The belt then drives the camshaft sprocket. The crankshaft sprocket has exactly one-half the teeth of the camshaft sprocket. On these designs, the belt often drives the distributor water pump, and/or oil pump. Some overhead camshaft engines use a combination chain and belt drive system.

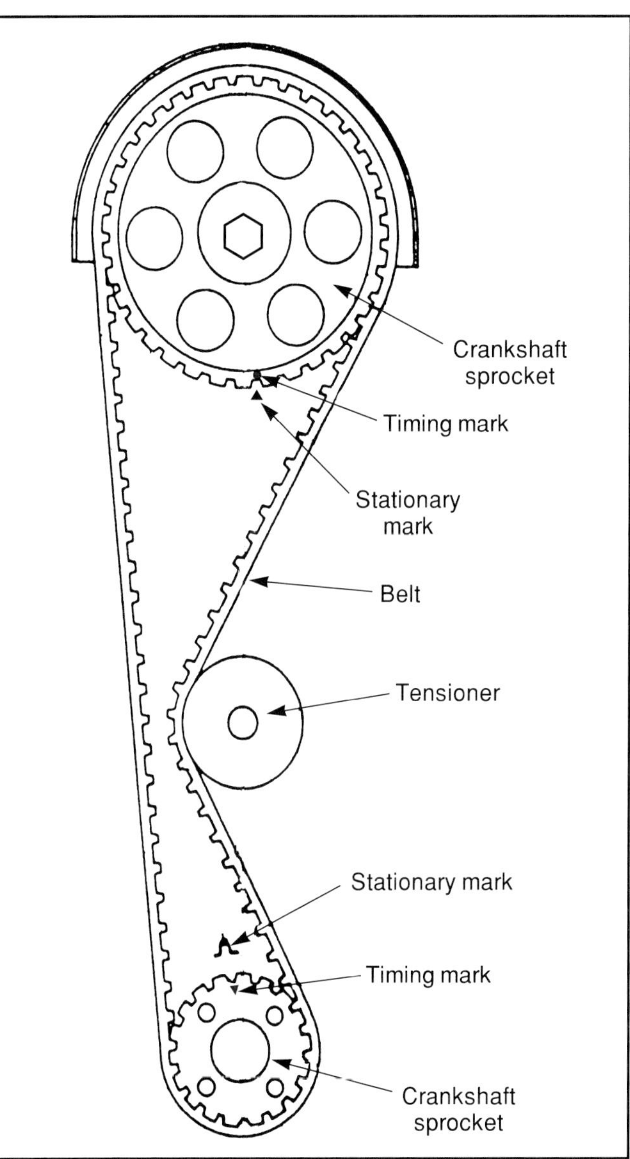

Figure 5-13. *Timing belts are primarily used on overhead camshaft engines. Some dual overhead camshaft engines use a combination belt and chain system to drive the camshafts.*

Harmonic Balancers

When the engine cylinders fire, power is transmitted through the crankshaft. When it receives this power, the front of the crankshaft tends to move before the rear, causing a twisting motion. When the torque is removed, the partially twisted shaft will unwind and snap back in the opposite direction. This unwinding action, although minute, causes what is known as torsional vibration. To stop the vibration, a *vibration damper*, sometimes called a ***harmonic balancer***, is attached to the front of the crankshaft. It is built in two pieces, connected by rubber plugs, spring-loaded friction discs, or a combination of the two.

When a cylinder fires and the shaft speeds up, it tries to spin the heavy section of the damper. When this occurs, the rubber connecting the two parts of the damper is twisted. The shaft does not speed up as much with the damper attached. The force necessary to twist the rubber and to speed up the heavy damper wheel smoothes out crankshaft operation. The unwinding force of the crankshaft cancels out the twist in the opposite direction. On some newer engines, the crankshaft pulley is an integral part of the balancer, **Figure 5-14.**

Vibration is also absorbed by the engine flywheel. The flywheel used with manual transmissions is heavy to absorb vibration and heat. Automatic transmission flywheels is a lightweight steel stamping, as the torque converter absorbs most of the vibration.

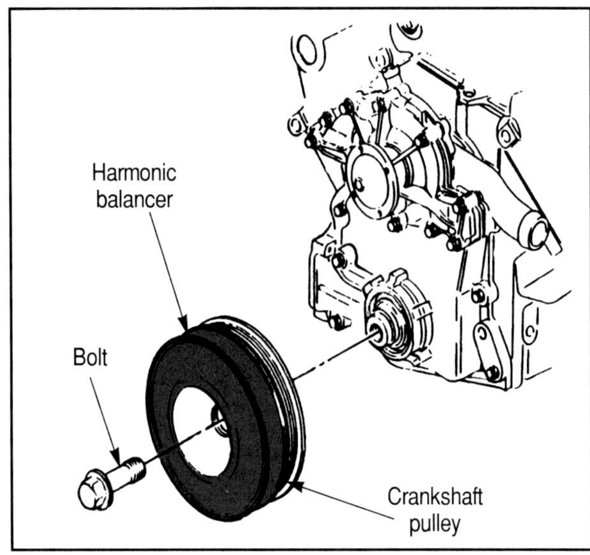

Figure 5-14. *Most modern crankshaft pulleys are one piece and simply bolt to the crankshaft. Others are two piece, with the harmonic balancer pressed onto the crankshaft. (General Motors)*

Balance Shafts

Inline and V-type engines are subject to vibrations at twice crankshaft rotational speed. For many years, this vibration went unnoticed, since most engines did not achieve the high engine speeds needed to cause this vibration to occur. As vehicles became smaller and engines operated at higher rotational speeds, vertical and torsional engine vibrations became more apparent. These vibrations are caused by secondary forces from piston motion.

In some late model engines, one or more ***balance shafts*** are added to counterbalance vertical and torsional vibrations. A balance shaft has offset bob weights that rotate in direction opposite to the crankshaft. These shafts are either turned by the camshaft through direct gearing or by the crankshaft through a belt or chain. Balance shafts help to provide a smoother idle and less vibration from the engine. **Figure 5-15** shows a balance shaft system used in a V-6 engine.

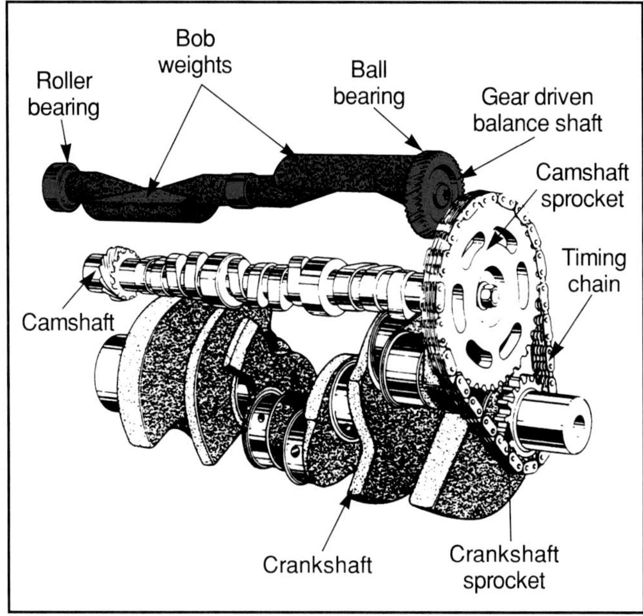

Figure 5-15. *Balance shafts cancel the vibrations produced by the crankshaft and the pulses of the engine firing. Note how the bob weights are opposite the crankshaft throws. (General Motors)*

Other Types of Engines

So far, we have discussed gasoline powered, overhead valve and camshaft engines. While these are the most common designs used in modern engines, other designs are used. These engines and their operating characteristics are explained in the following paragraphs.

Diesel Engines

Diesel engines operate on the same principles as gasoline engines, and contain many of the same parts. The difference between the diesel engine and the gasoline engine is that the diesel has much higher compression and does not use an electric spark to fire the fuel mixture. Instead, the tremendous heat created by compression causes the fuel to ignite. To ensure that the fuel ignites at the proper time, it is injected into the combustion chamber by a high pressure injector. The injector is both fuel and ignition system.

The major difference between diesel and gasoline engines is the heavier construction of the diesel engine components. Diesel pistons, connecting rods, crankshafts, and other parts must be larger and stronger to withstand diesel combustion chamber pressures. Some diesel engine cylinder heads have a ***precombustion chamber*** which often contains the fuel injector and glow plug, **Figure 5-16.**

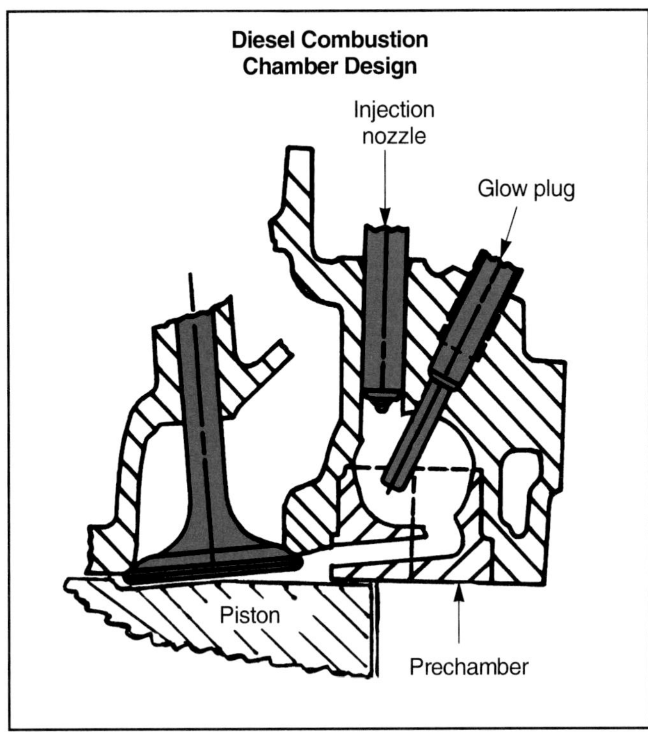

Figure 5-16. *Diesel engines inject the fuel directly into the combustion chamber. The glow plug warms the cylinder for better cold starting. The precombustion chamber is used on some engines while others use direct injection. (General Motors)*

Rotary Engines

In the past, the ***rotary engine*** has been used on cars and trucks, outboard motors, and large stationary engine-driven air compressors. Rotary engines are still found on a few imported vehicles and is sometimes referred to as a Wankel engine. The common automotive rotary engine consists of a chamber containing a three-sided rotor. The edges of the rotor are sealed to prevent leakage between the rotor sides. Movement of the rotor draws air into the engine and compresses it. The compressed mixture is fired, producing heat and pressure to move the rotor. A cutaway of a turbocharged rotary engine is shown in **Figure 5-17.** Most rotary engines contain several of these chambers for smooth operation. Rotary engines have fuel and ignition systems similar to those used on piston engines. Cooling, lubrication, emission, and exhaust systems also resemble piston engine equipment.

The advantages of rotary engines are a simpler design and they can reach higher engine speed than piston engines because of the lighter weight of the rotating parts. The major disadvantage of rotary engines is caused by the large surface area of the combustion chamber section. This reduces combustion chamber temperatures, causing lower gas mileage and higher emissions. Some rotary engines have a second spark plug which fires after the main spark plug. The extra plug ignites unburned fuel to reduce emissions and increase power output.

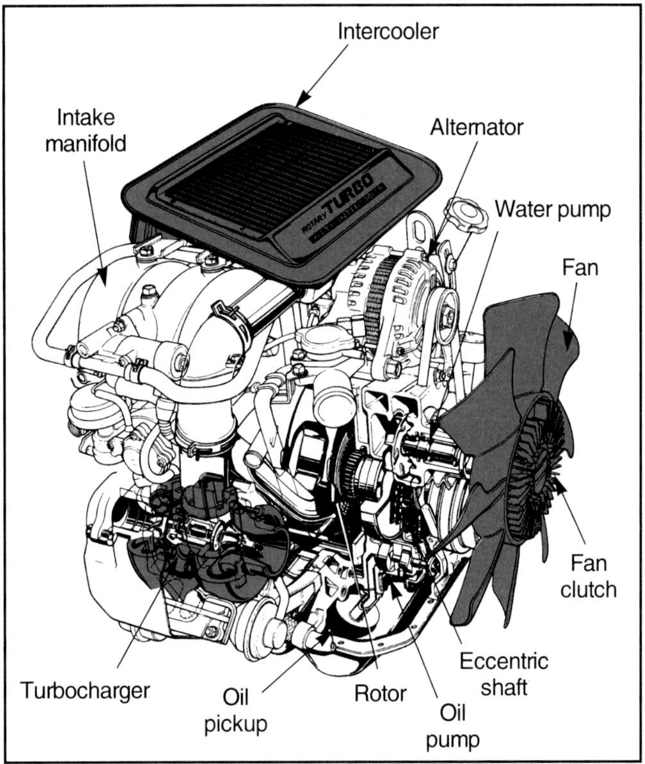

Figure 5-17. *Rotary engines use rotors instead of pistons to compress the air-fuel charge and produce the rotational force used to move the vehicle. Very few vehicles use this type of engine. (Mazda)*

Miller-Cycle Engine

In a conventional engine, the intake valve is open during the piston's intake stroke and closes almost as soon as the piston begins to move upward on the compression stroke. However, on a ***Miller-cycle engine,*** the intake valve is held open for a much longer time (degrees of camshaft rotation) as the piston comes up on the compression stroke, **Figure 5-18.**

To prevent the air-fuel mixture from exiting through the intake valve, a supercharger raises the intake manifold pressure. This forces more of the air-fuel mixture into the cylinder. The supercharged air passes through an intercooler, which reduces its temperature. This allows higher compression with less chance of preignition. The Miller-cycle design is said to produce more power and economy over a wider rpm range. The Miller-cycle and other supercharged engines are only used on a few cars, but may see more widespread use in the future.

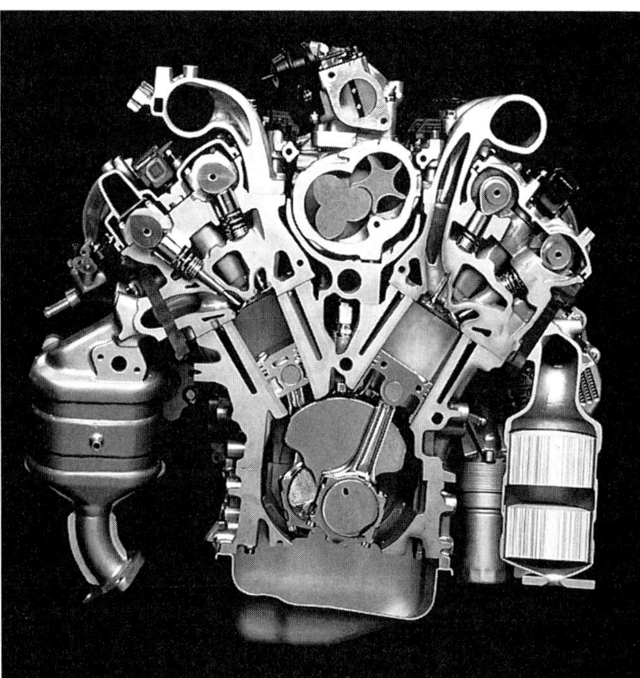

Figure 5-18. *Miller-cycle engines use a supercharger to force more air-fuel mixture in the cylinders. The intake valves in this engine are open longer to permit a greater amount of air-fuel mixture for better power. (Mazda)*

Other Engine Systems that Can Affect Driveability

The following sections will cover other engine systems which can affect driveability. These systems are all part of the overall basic engine, and will affect engine operation if they are not operating properly. These systems are the cooling and lubrication system.

Cooling System

The ***cooling system*** is a set of components that remove unwanted engine heat and regulates engine temperature. A cooling system is needed because not all of the heat of combustion is used to create the pressure that moves the pistons. The unused heat must be removed to prevent engine damage. Even a slightly overheating engine will wear due to the tighter than normal clearances of the moving parts. An excessively hot engine will tend to ping, diesel, be hard to start, and if it has a carburetor, suffer from vapor lock and percolation. An extremely hot engine will develop enough heat to melt its exhaust valves. This is usually called valve burning.

Some of the engine heat is removed along with the exhaust gases and some radiates out of the engine block and heads. The rest of the heat must be removed by the cooling system. The amount of heat removed must be controlled so that the engine does not run cooler than its normal operating temperature. An engine that runs too cold wastes fuel, drives poorly, pollutes the air, and wears out quickly.

All cooling systems remove excess heat from the engine and transfer it to the surrounding air. The two main kinds of air cooling are liquid cooling and direct air cooling. In liquid cooling, the heat is absorbed by a liquid, which then transfers it to the air. In direct air cooling, the heat is transferred directly to the air.

Liquid Cooling

The ***liquid cooling system*** is almost universally used in modern cars and trucks. In this system, the engine has many internal passages for coolant. The coolant is pumped through the cooling system passages by a belt-driven pump. As the coolant circulates through the passages, it picks up heat from the engine metal. The coolant then flows from the engine into the radiator. As the coolant travels through the radiator, heat is transferred from the coolant to the air passing through the radiator. A fan draws extra air through the radiator at low speeds. At higher speeds, the movement of the vehicle forces air through the radiator. The cooled liquid then returns to the engine to pick up more heat. **Figure 5-19** shows a modern cooling system.

Engine Coolant

The ***engine coolant*** is the medium of heat transfer. It must be able to absorb and release heat, and must not damage any other cooling system parts. Engine coolant is a mixture of antifreeze and water. Modern antifreeze contains a chemical called ***ethylene glycol.*** Also included are corrosion inhibitors to reduce rust and corrosion of the engine block and radiator. Small amounts of water soluble oils are added to lubricate seals and moving parts.

Pure ethylene glycol freezes at about 9°F (-12°C), and water freezes at 32°F (0°C). When ethylene glycol and water are mixed, however, the freezing point of the mixture is lower than either liquid alone. A 50:50 mixture will freeze at about -35°F (-37°C). Most engine manufacturers recommend a 50:50 mix of ethylene glycol antifreeze and water. A mixture of two-thirds ethylene glycol and one-third water will not freeze until the temperature reaches -67°F (-55°C).

A mixture of antifreeze and water has a higher boiling point than plain water. A 50:50 mixture will not boil until its temperature is about 11°F higher than the boiling point of water. An antifreeze/water solution also gives added boil over protection for summer driving. Up to a 70:30 mixture of antifreeze and water is sometimes used in severe climates and operating conditions.

Water Pump

The coolant or ***water pump*** is a cast iron or aluminum housing bolted to the engine block. The rotating portion of the pump is called the impeller. The impeller is a flat plate with curved blades, as shown in **Figure 5-20.** This type of

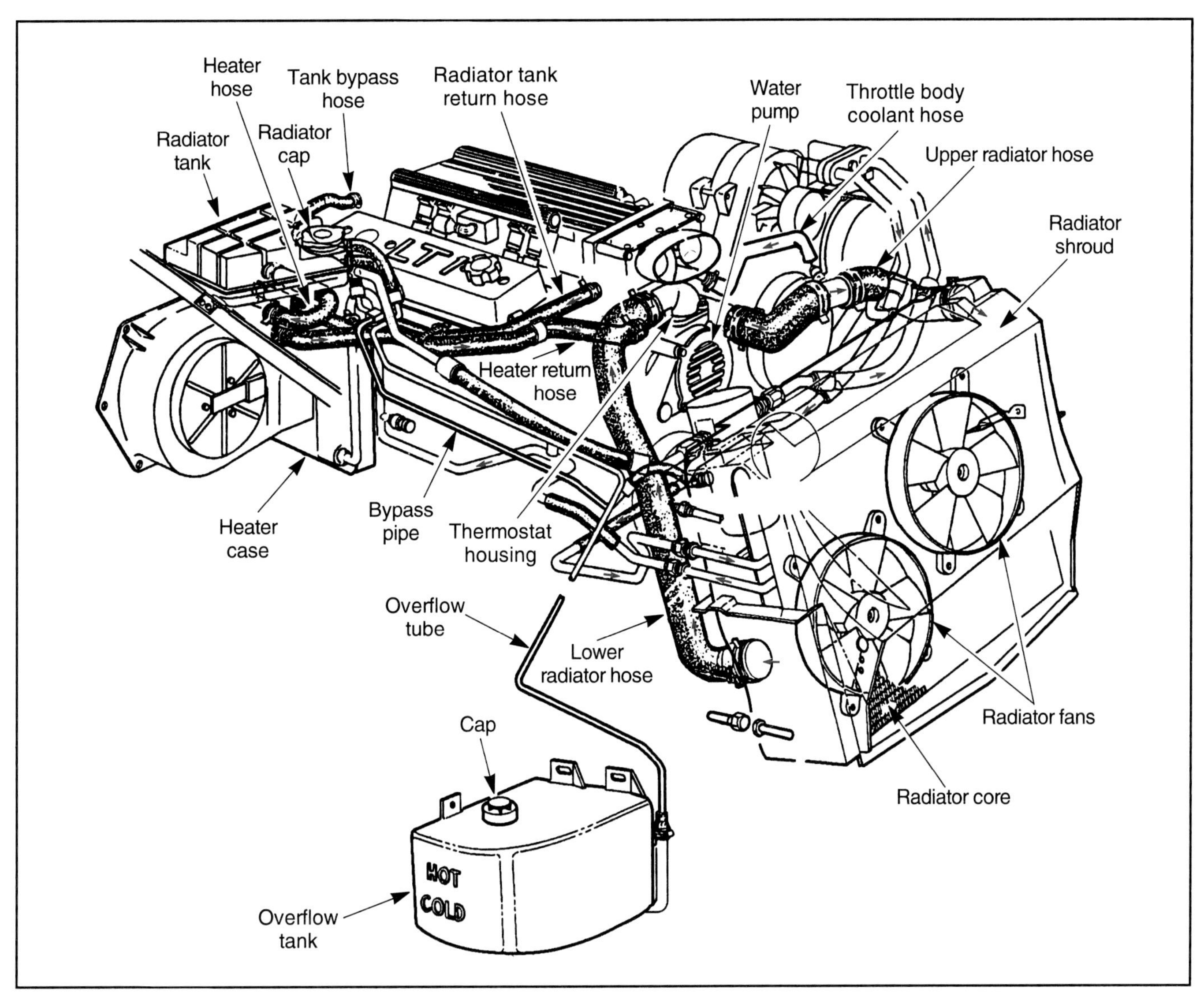

Figure 5-19. *Components of a typical cooling system. The arrows show the path of coolant to and from the radiator. (General Motors)*

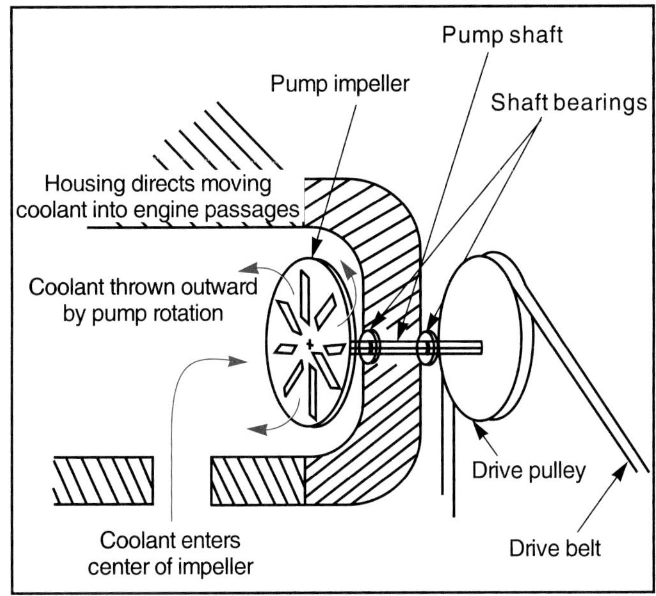

Figure 5-20. *The water pump uses an impeller to move the coolant through the engine. Water pumps are either belt- or gear-driven.*

pump is known as a centrifugal pump, capable of circulating several hundred gallons of coolant per hour. Pressure developed by this type of pump is about one or two pounds. The impeller is belt-driven from the crankshaft, and is supported by two sealed bearings. A seal where the impeller shaft exits the pump housing prevents coolant loss through the rotating shaft.

The pump intake is connected by a flexible hose to the bottom of the vehicle radiator. Coolant is drawn into the center of the rotating pump by suction. The coolant is thrown outward by centrifugal force. Passages in the pump and engine direct the coolant through the block and heads.

Coolant Passages

Internal ***coolant passages*** are cast into the block and heads when they are manufactured. These are sometimes referred to as *water jackets,* though they do not carry water. Coolant is pushed through internal passages in the engine and absorbs heat before exiting through the top of the engine. Passages near the hottest parts of the engine, such as the valves and cylinder walls, are designed so that more coolant flows through them. Some internal coolant passages are sealed with round stamped metal seals called core plugs or ***freeze plugs,*** **Figure 5-21.**

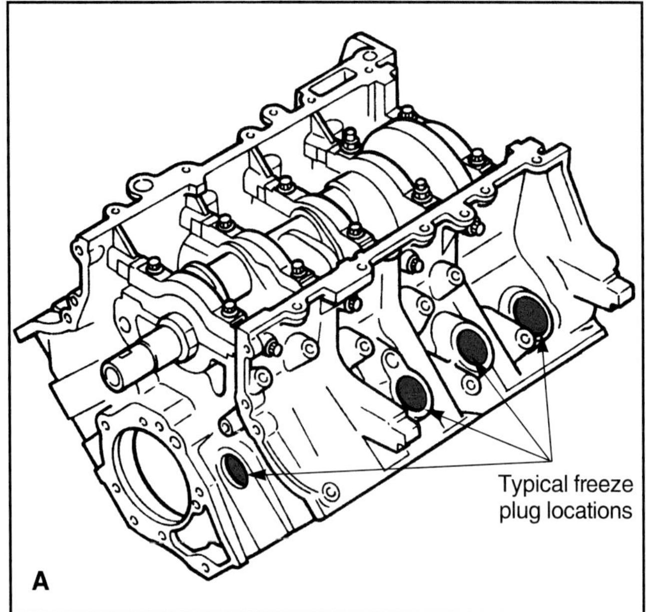

Figure 5-21. *A—Several freeze plugs are installed in various locations in the front, sides, and rear of the block. The holes in the block are left from where sand was drained from the block after casting. B—Freeze plugs are a source of coolant leaks. They are also referred to as core plugs (Ford)*

Radiator

The ***radiator*** is a heat exchange device made up of internal coolant tubes and external fins. As the coolant flows through the tubes, heat is transferred to the fins. The fins then transfer the heat to the air passing through the radiator. The radiator in most vehicles is capable of removing more heat than the engine can produce. Actual radiator efficiency is dependent on the flow rates of the coolant and the outside air temperature.

Older radiators were called ***top-flow radiators,*** which are designed so that the coolant flows from the top of the radiator to the bottom. Newer radiators are called ***cross-flow radiators,*** designed so that coolant flows from one side of the radiator to the other. Compare **Figures 5-22A** and **5-22B.**

Tanks on each end of the radiator direct coolant into the radiator tubes or to an outlet which leads back to the engine. Radiators used on cars with automatic transmissions have a heat exchanger mounted in the radiator to cool the transmission fluid. Transmission hydraulic pressure forces the fluid through the heat exchanger. The fluid gives up its heat to the engine coolant. The transmission cooler is always mounted in the radiator tank that feeds coolant back into the engine.

Radiator Fan

The ***radiator fan*** draws extra air through the radiator to aid in heat transfer at low speeds. Belt-driven fans are usually installed on the end of the water pump shaft, so that the same belt that drives the pump also drives the fan. Many belt-driven fans have a ***fluid clutch*** installed between the drive pulley and fan assembly, **Figure 5-23.** The fluid clutch allows the fan to free wheel at higher speeds when it is no longer needed. Some fluid clutch fans contain a thermostat which prevents the fan from operating until the engine warms up.

Electric radiator fans are used on front-wheel drive vehicles when the engine is mounted transversely in relation to the radiator. See **Figure 5-24.** Electric fans are sometimes used on vehicles with longitudinally placed engines to reduce engine drag. Fan operation is controlled by either thermostatic switches installed in a passage of the cooling system, or through the engine control computer. Electric fans remain off until the coolant temperature reaches a certain point. Electric fans will be on any time that the air conditioner compressor is operating, no matter what the coolant temperature. Manufacturers have begun producing a fan motor that operates on power steering hydraulic pressure.

Radiator Pressure Cap

The radiator ***pressure cap*** allows cooling system pressure to build up by holding a seal against the radiator filler neck. The seal is held in place with a spring that is calibrated to produce the proper system pressure. As the engine heats up, excess coolant pressure overcomes spring pressure and opens the valve. When the excess pressure is released, the spring closes the valve. This prevents coolant loss and maintains pressure in the cooling system. See **Figure 5-25.**

The cooling system is pressurized to prevent the coolant from boiling by raising the boiling point of the coolant. The boiling point of any liquid goes up as the pressure is increased. For every 1 pound (7 kPa) increase in pressure, the coolant boiling point increases by 3°F (1.8°C). Therefore, a 15 pound (105 kPa) pressure cap will raise the boiling point of the coolant by 45°F (27°C).

Coolant Recovery Systems

Coolant recovery systems are designed to keep the cooling system as full as possible at all times. The coolant recovery system consists of a plastic tank, connected by a hose to the radiator overflow neck, as shown in **Figure 5-26.**

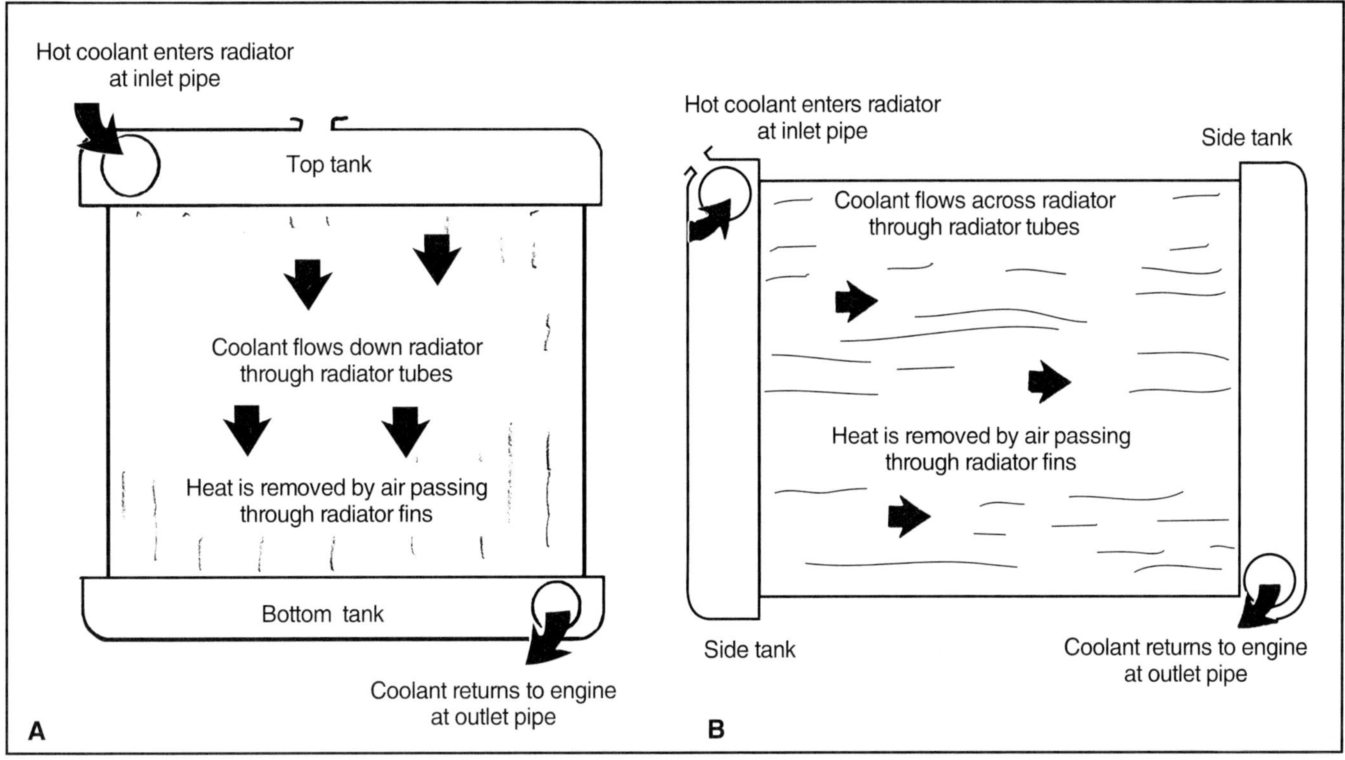

Figure 5-22. *A—Downflow radiators were used on older vehicles but can still be found on a few modern vehicles. B—Cross-flow radiators have become the radiator of choice by most manufacturers.*

Figure 5-23. *Most older rear-wheel drive vehicles use belt-driven fans. The fan clutch allows the fan to freewheel at highway speed, allowing more available power from the engine.*

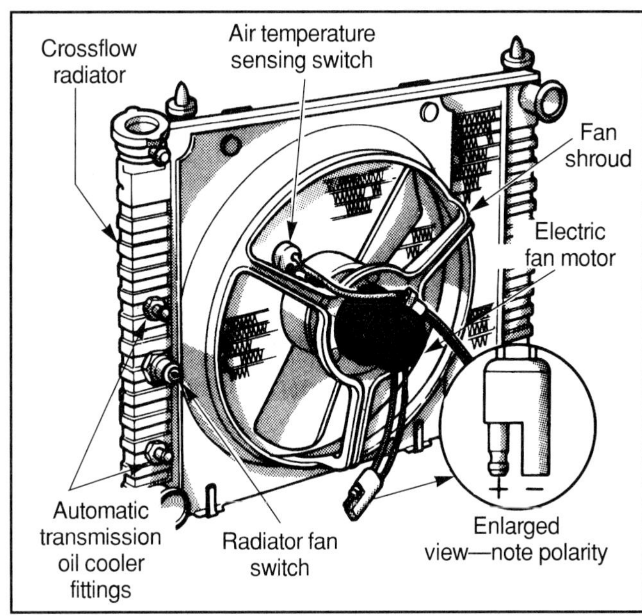

Figure 5-24. *One or more electric fans are used on all front-wheel and many newer rear-wheel drive vehicles. The electric fan allows for better control of fan operation as well as more available horsepower from the engine. (Chrysler)*

The plastic tank is called a ***recovery tank.*** Coolant that is pushed out of the radiator as the engine heats up enters the recovery tank. When the engine cools off, the cooling system loses heat and pressure and develops suction. This suction draws the coolant back into the cooling system. The system uses a special type of radiator cap. The cap is sealed so that only coolant can be drawn back into the radiator. If additional coolant is needed, it is added at the recovery tank.

Thermostat

The ***thermostat*** forces the engine to warm up quickly by keeping coolant from circulating through the radiator when the engine is cold. The thermostat consists of a

Figure 5-25. *Radiator caps are not a frequent source of driveability or engine problems; however, a faulty cap can cause diminished overheating protection. Never open a hot radiator cap.*

Figure 5-26. *The coolant reservoir bottle is usually mounted in a conspicuous location and can be used as a quick diagnostic tool. Bubbles in the reservoir bottle while the engine is running is a good indication of severe engine problems.*

heat-sensitive material, such as wax, sealed in a chamber with a piston at one end. The piston is attached to a valve, which opens or closes the thermostat to control coolant flow, **Figure 5-27.**

The thermostat is located at the engine outlet gooseneck, where coolant leaves the engine on its way to the radiator. When a cold engine is started, the wax is contracted and holds the piston in the closed position. As the engine warms up, hot coolant begins to circulate under the thermostat. This causes the wax to expand and push the piston, which opens the valve. When the thermostat opens, the coolant begins to circulate through the radiator. The thermostat has no effect on engine temperature once it passes the thermostat's wide open temperature.

Older thermostats were designed to open when coolant temperature reached 180 F (82 C). Thermostats used on newer engines will not open until the coolant temperature reaches higher temperatures, as much as 200 F (93 C). If the vehicle is operated in very cold weather, the thermostat may not open fully.

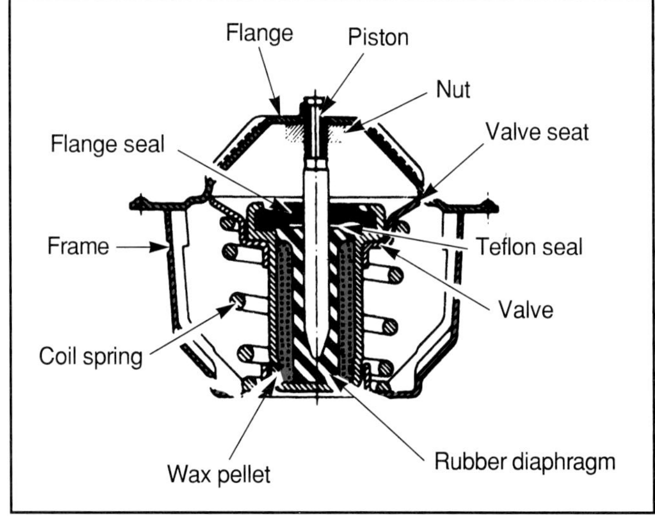

Figure 5-27. *Parts of a typical thermostat. Modern engines use a thermostat with an opening range between 190-195°F (87-91°C).*

Hoses and Tubing

The engine moves slightly in relation to the stationary radiator, and must be connected by flexible ***hoses.*** These hoses are usually made of rubber or neoprene, molded around a fiber mesh. The hoses are installed on the inlet and outlet with clamps. Some engines have a bypass hose which allows the coolant to circulate inside the engine until the thermostat opens. The bypass hose prevents damage to rapidly heating parts, such as exhaust valves or cylinder walls, before the thermostat opens. On other engines, the bypass is built into the engine casting or the heater core, and hoses are the bypass system.

Some cooling systems utilize fixed metal tubing. These are either connected directly or uses a flexible hose to the engine. This tubing is made of soft steel or aluminum and is used to route coolant along straight runs. Many cooling systems also have one or more small bleeder valves, which are used to bleed air from the cooling system during service.

Engine Belts

To drive the water pump, as well as the other engine accessories needed to operate the vehicle, belts are necessary. **Figure 5-28** shows ***V-belts*** being used to drive a coolant pump and other accessories. The belt is driven by a pulley on the front of the crankshaft. This same belt may be used to drive other engine accessories. A modification of the V-belt is the ***serpentine belt,*** **Figure 5-29.** It is a wide belt with several ribs that act as several small V-belts.

All belts must be in good condition and properly adjusted. Excessive tightness will place a heavy load on the coolant pump and possibly the crankshaft bearings, causing

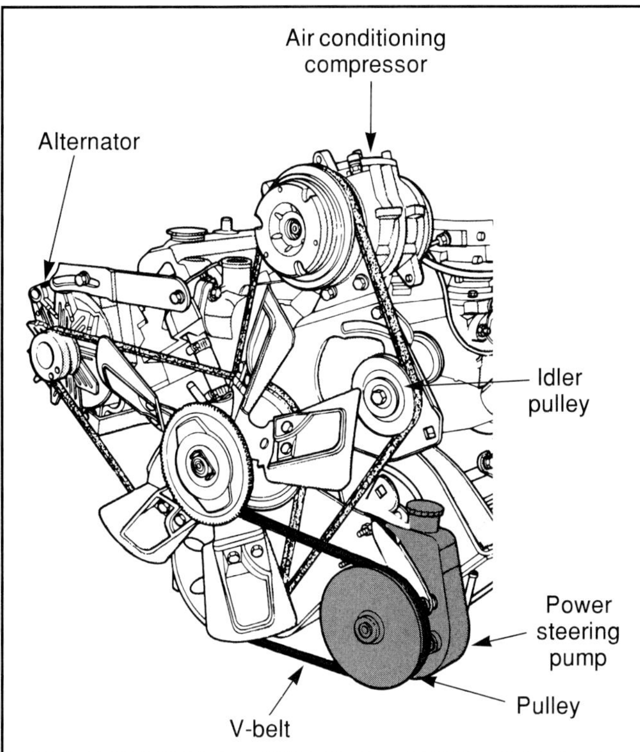

Figure 5-28. *Older engines use 2-5 V-belts to drive the engine accessories. Some of these belts are cogged or ribbed. (Chrysler)*

Figure 5-29. *Most modern engines use one or two serpentine belts to drive all of the accessories. These belts are ribbed in most cases. (Ford)*

premature wear. Belt looseness will permit slippage, which will reduce pump and fan speed, possibly causing overheating and battery discharge. A loose belt will have a tendency to whip or flap, which can cut hoses or cause intermittent loading of the bearings.

Direct Air Cooling

Direct air cooling is a method of transferring engine heat directly to the surrounding air. The air cooling system uses a fan to force air around and past the cylinders and cylinder heads, which are the hottest parts of the engine. The fan is driven from the engine crankshaft by a belt.

The cylinders and heads of an air cooled engine are made with fins to present a larger heat transfer surface. The engine is surrounded by a sheet metal shroud to direct more air over the hottest engine parts. See **Figure 5-30.** Most air-cooled engines have a separate oil cooler. Hot oil is pumped through the cooler and gives up its heat directly into the air stream. Although many engines of the past had air cooling systems, air cooling is generally confined to small engines, such as those used on lawn mowers, and is rarely used on late model vehicles.

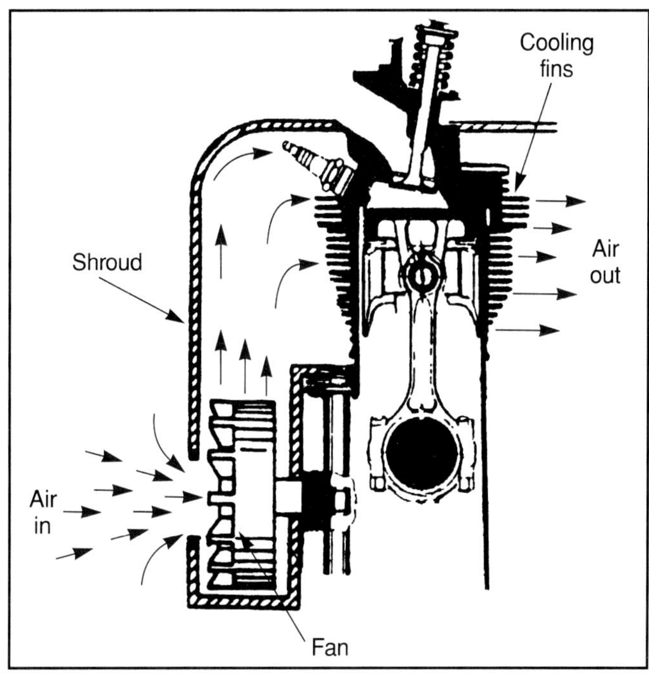

Figure 5-30. *Direct air cooling pushes external air over fins cast on the engine block and components to dissipate heat. This design is only used on a few automotive engine designs.*

Lubrication System

The ***lubrication system*** circulates engine oil to internal engine parts. Engine oil serves several purposes, and must be delivered to the right place at the right time, in the proper amounts. It might seem that the oiling system has no effect on driveability, but it is a vital part of the engine, and some oiling defects can result in driveability symptoms. Also, some driveability problems can affect the lubrication system.

Not all moving parts of the engine are subjected to equal stresses. The amount of oil needed varies between engine parts. Some parts of the engine are lubricated by *pressure,* while others are lubricated by *splash.* The parts that need constant, consistent lubrication are supplied with oil under pressure. The camshaft and crankshaft bearings are pressure lubricated. Hydraulic lifters must have a supply of pressurized oil to work properly. On some engines, pressurized oil drips through nozzles onto moving parts such as timing gears.

Some engine parts can be lubricated by oil that is splashed onto them. Oil is thrown upward by the rotation of the crankshaft in the oil pan, and strikes the engine cylinder walls, piston skirts, and piston pins. Some splash also reaches the camshaft and lubricates the cam lobes and valve lifters. Modern automotive engines are lubricated by a combination of pressure and splash. **Figure 5-31** is a combination ***pressure/splash lubrication system,*** typical of most modern engines.

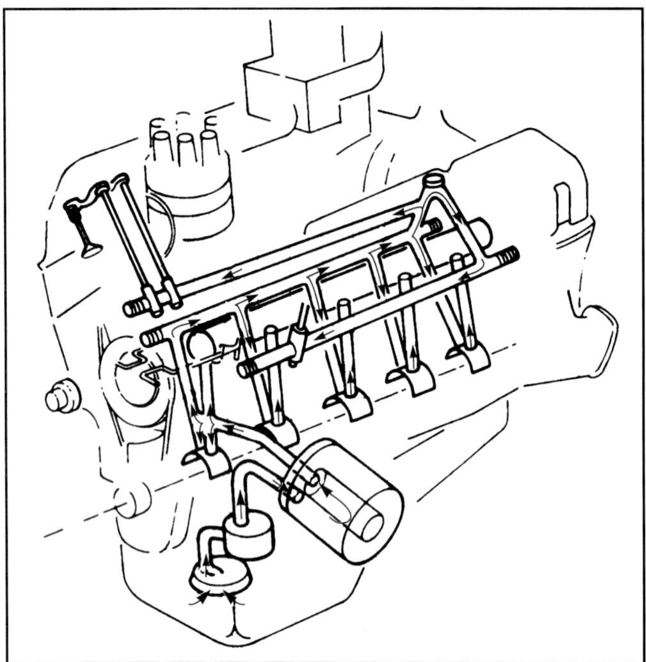

Figure 5-31. *The modern engine uses pressure and splash to lubricate the moving engine components. The lubrication system also serves to cool the internal engine components. (Ford)*

Lubrication System Components

The components explained in the following sections are common to all lubrication systems, and are used to produce and regulate oil pressure. They work together to circulate the oil continuously through the moving parts of the engine.

Engine Oil

The ***engine oil*** (often called motor oil) is what actually does the job of lubrication between moving parts. Modern oils are made to do a good job of lubrication, sealing, cooling, shock absorption, and cleaning. They also prevent sludge formation.

Lubrication is the most obvious job of the oil. Oil prevents friction by forming a layer between moving parts. All engine parts, no matter how highly machined, have microscopic high spots. On any moving parts, these spots will contact each other and begin to wear and overheat from friction. The oil separates the moving parts, preventing the high spots from touching. The oil acts like a set of microscopic ball bearings, allowing the parts to slide against each other, and friction is greatly reduced.

Cooling is one of the less obvious jobs of the oil. It is difficult to remove heat from the piston heads, since they cannot be cooled by the cooling system. At the same time, the pistons are among the hottest parts in the engine, and will overheat and melt if the heat is not removed. Even a slightly overheating piston can develop a localized hot spot that will cause detonation.

Oil splashed up under the piston head absorbs heat from the piston. Oil splash also removes heat from the cylinder walls, rods, crankshaft, and other engine parts that cannot be directly cooled by the engine cooling system. The oil itself is cooled by the cooling system and by direct heat transfer when it is in the oil pan. Sealing is another task of the oil. A thin film of oil between the piston rings and cylinder wall seals in engine compression. If this oil is not present, compression leakage will prevent the engine from running.

Shock absorption is a job performed by the oil for some moving parts, such as connecting rod journals and rod bearings. The oil cushions the shock when the rod changes direction at high speeds. This extends bearing life and prevents engine knocking. Other moving parts, such as the piston skirts and cam lobes, are also cushioned.

Cleaning is also an important job of the oil. Engines constantly collect impurities. Unburned gasoline, carbon from the combustion process, and water vapor get into the crankcase. These impurities can form sludge and varnish deposits in an engine. These deposits can cause the engine to overheat, burn oil, and wear out prematurely. Modern oil contains detergents that prevent engine sludge formation by picking up impurities and holding them in suspension. When the oil circulates through the oil filter, the impurities are trapped.

Grades of Oil

The American Petroleum Institute (API) classifies oil according to various factors that affect an oil's ability to prevent friction and deposits in an engine. The API service classification for most modern engines is SH for gasoline engines and CG4 for diesel engines. Oils marked SA, CA, SB, CB, SC, CC, SD, CD, CE or CF are for use in older engines only. Older engines can use the newer oils. The API service classification is printed on the oil container, **Figure 5-32.**

In addition to the API grades, the Automobile Manufacturers Association (AMA) classifies oils according to vehicle manufacturers test criteria. AMA classifications are called GF (for gasoline fueled) classifications. At the present time, GF-1 is the accepted standard, to be replaced by GF-2 in the near future. Another classification is the EC class, which defines an oil's ability to improve fuel

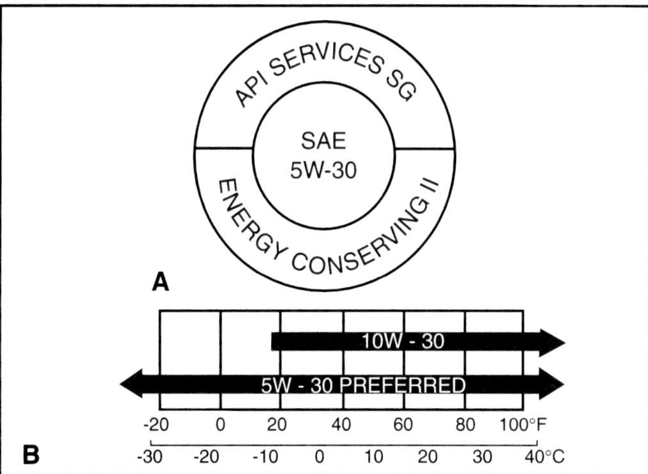

Figure 5-32. *A—Oil containers carry an American Petrolium Institute (API) marking indicating the oil viscosity and classification. B—Oil temperature range chart.*

economy. EC testing is performed by the American Society for Testing and Materials (ASTM). Present EC classes are EC-Iand EC-II. API accepts the ASTM classes, and they are often listed on motor oil containers.

Oil Pan

Oil that drips or is squirted out of any part of the lubrication system eventually drains into the ***oil pan.*** The oil pan is usually made of stamped sheet metal and functions as a reservoir for engine oil. The thin metal of the oil pan helps the oil to lose the heat that it picked up in the engine. The air passing underneath the vehicle removes this heat from the oil, **Figure 5-33.**

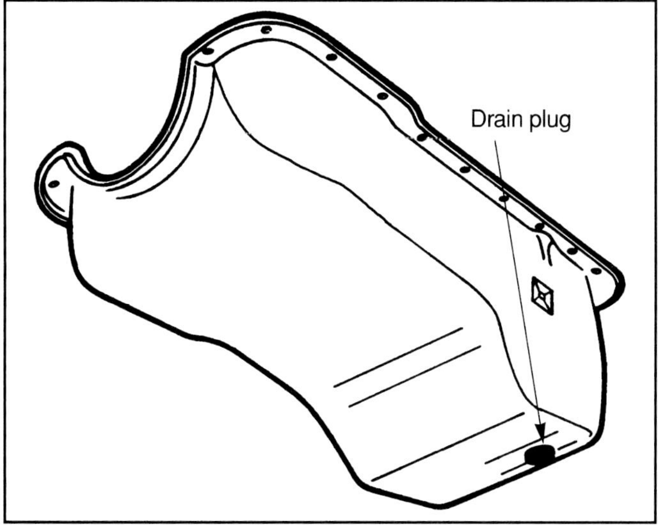

Figure 5-33. *The oil pump serves as the reservoir for the engine oil as well as providing a surface for heat dissipation. Most pans are single sump, however, a few engines use either a dual or dry sump pan. (Ford)*

Most oil pans have a drain plug at the lowest point of the pan. Some models have two drain plugs. Some newer vehicles are equipped with an oil pan made of cast aluminum with cooling fins cast into the bottom. The cast pan also adds rigidity to the engine block.

Some manufacturers are experimenting with an oiling system known as a dry sump. In a dry sump system, oil that reaches the bottom of the pan is immediately pumped to a separate oil reservoir, leaving very little in the pan. The advantage of this system is that it would allow the engine to be set lower, which would allow lower hood profiles. Dry sump oiling is a proven system that has been used in auto racing for many years.

Oil Pickup Screen

The oil pickup screen prevents any large particles, such as dirt, sand, or metal shavings from entering the engine oiling system. The oil screen is installed on the intake side of the oil pump, **Figure 5-34.** It is always located at the lowest point in the oil pan so that it is always covered by oil. This keeps the oil pump from drawing in air if the oil level drops because of oil consumption or sloshing during turns or hard braking.

Since it traps only large particles, the oil screen will usually not become plugged until the engine reaches very high mileage. Most oil screens cannot be removed and cleaned unless the oil pan is removed.

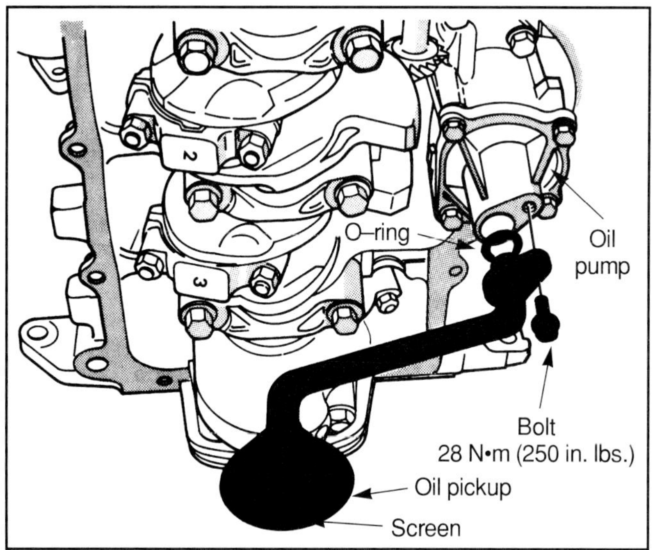

Figure 5-34. *The oil sump pick-up is either bolted or pressed onto the oil pump. The sump pick-up reaches into the bottom the oil pan sump to draw oil into the engine. (Chrysler)*

Oil Pump

The engine ***oil pump*** is driven by a gear on the camshaft or crankshaft. The oil pump drive is also the distributor drive on many engines. Most oil pumps are mounted near the bottom of the engine and connect to the camshaft drive gear through a shaft. Oil pumps are installed inside the oil pan. The oil pump speed varies with engine speed, since the pump is driven by the engine.

Engine oil pumps are always constant displacement types. A constant displacement pump delivers the same amount of oil with each revolution. **Figure 5-35** illustrates the gear and rotor versions of the constant displacement oil pump. The pump draws oil from the bottom of the oil pan when the pump gears move apart. The oil is carried around the housing in the spaces between the gears. When the gears come together, they squeeze the oil out the discharge port. The faster the pump revolves, the more oil it delivers. The gear type is the most common, however, more engines are using the rotor pump, **Figure 5-36.**

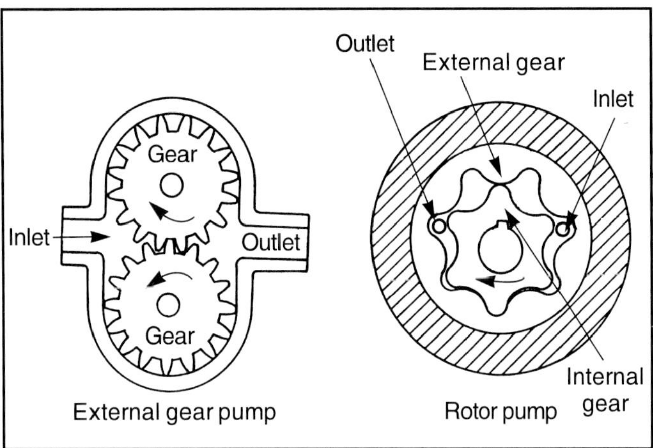

Figure 5-35. *The external gear pump is used in many distributor driven oil pumps and is mounted in the bottom of the engine. The rotor pump is usually mounted in the front of the engine and is crankshaft driven.*

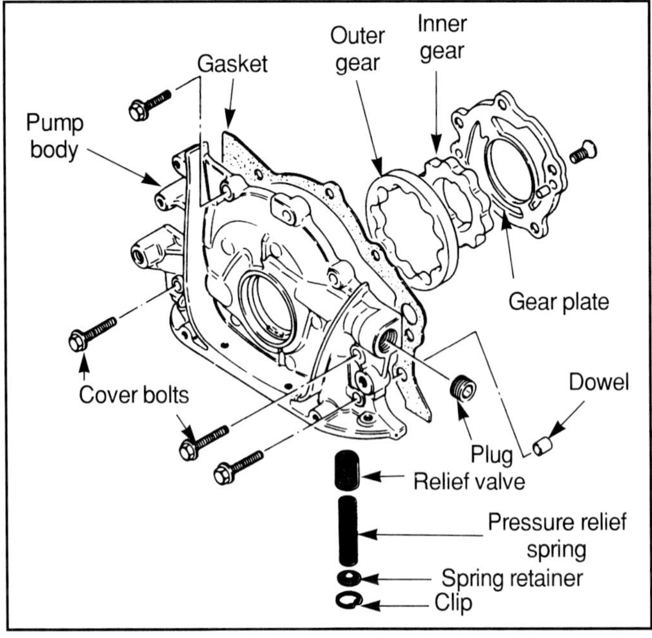

Figure 5-36. *Exploded view of a rotor driven oil pump. Note the pressure relief valve. (General Motors)*

Pressure Regulator

The oil pump has enough capacity to deliver sufficient oil pressure at idle speeds. At higher engine speeds, the pump will produce too much oil pressure, which could rupture seals or filter elements, affect hydraulic lifter operation, or cause oil burning.

Oil pump output is controlled by a ***pressure regulator.*** The pressure regulator consists of a valve which is held closed by a spring, **Figure 5-36.** Oil pressure from the pump pushes against the valve on the opposite side from the spring. When the pump pressure becomes too high, the spring is compressed, the valve opens, and excess oil pressure is dumped back into the sump. Engine oil pressure is usually regulated to about 35-45 pounds per square inch (241-310 kPa). Minimum pressure at idle should be 15-20 pounds per square inch (103-137 kPa). A typical pressure regulator is shown in **Figure 5-37.**

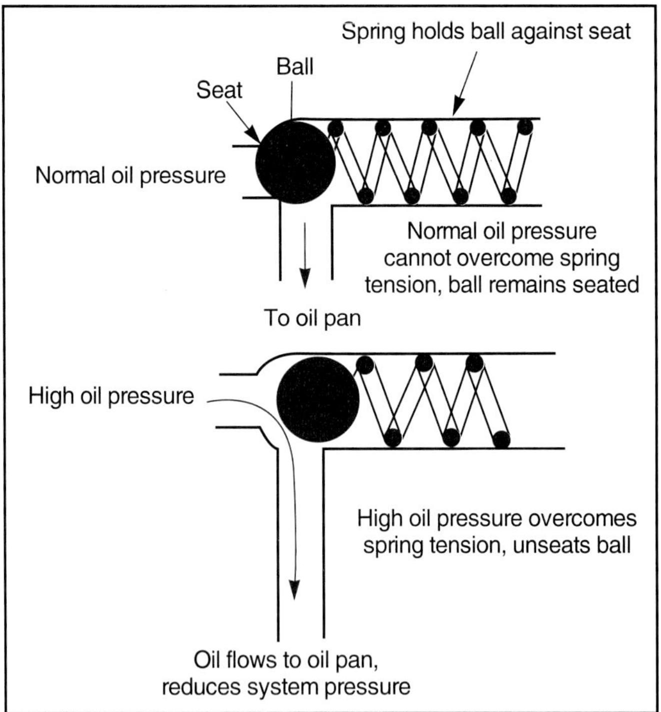

Figure 5-37. *The pressure relief valve allows engine lubrication to take place should the oil filter become clogged. Normal oil pressure will not unseat the valve, however, high oil pressure caused by a clogged oil filter will.*

Oil Filter

The modern ***oil filter*** consists of a stamped metal housing enclosing a pleated paper filter element. It is always installed after the oil pump. Oil under pressure flows into the filter, through the pleated paper, and out again. The housing is designed to withstand oil pump output pressure. The oil filter contains an internal bypass. The bypass allows oil to flow if the filter element becomes clogged.

The paper in the filter element will trap very small dirt and metal particles. An oil filter that does a good job quickly becomes clogged with trapped particles, and is replaced during an oil change as part of normal engine maintenance. A typical oil filter, such as the one shown in **Figure 5-38,** can be removed from the engine by unscrewing it from the mounting pad. A few are mounted off the engine on the firewall or elsewhere in the engine compartment. One type of oil filter is installed in the oil pan, and does not have a housing. The element resembles a small air cleaner.

Figure 5-38. *Oil filters come in many sizes and shapes. While it is a very simple and inexpensive part, it is very important to good engine performance.*

Oil Galleries

Oil galleries are internal engine passages which carry the pressurized oil. They are cast or drilled into the engine block and heads. Galleries extend from the pump and filter to the crankshaft and camshaft bearings, valve lifters, and rocker arm shafts. Galleries are drilled in the crankshaft to allow pressurized oil to reach the connecting rod bearings. Removable plugs are located at the rear of some engine blocks to allow the galleries to be cleaned during an overhaul.

Gaskets, Seals, and Sealant

Gaskets and seals are used in the intake system to seal air and fuel in, and to prevent vacuum leaks. Preventing fuel and air leaks is vital to driveability, performance, and emissions. Gaskets and seals must stand up to high engine temperatures, but be pliable enough to seal at various operating temperatures. Most ***gaskets*** are made of heavy paper or rubber, sometimes coated with other materials for maximum durability. See **Figure 5-39.**

Seals are usually made of neoprene to hold up against heat and chemical attack. The most common type of seal used in the fuel system is the O-ring, used to seal in much the same way that gaskets seal. The O-ring is pressed between two mating parts, and prevents leaks by conforming to the surfaces of the parts. O-rings are used in systems throughout the vehicle. Other seals are made to seal moving parts. They have a lip which prevents leakage while allowing the shaft to rotate. The design of the seal allows it to prevent leaks without being quickly worn out by the rotating shaft. See **Figure 5-40.**

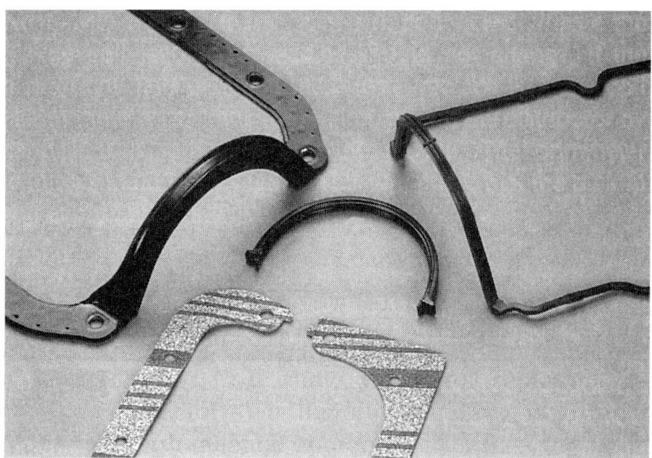

Figure 5-39. *The modern engine uses various gaskets made from cork, rubber, Viton, and other materials. Some gaskets are available with a bead of sealer added to the gasket surface for maximum sealing. (Fel-Pro)*

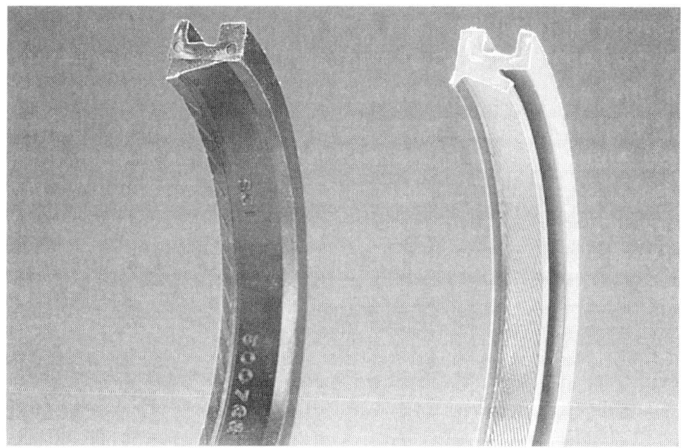

Figure 5-40. *Seals are used to seal moving parts, such as the crankshaft. They are not a frequent source of driveability problems. (Fel-Pro)*

Many modern engines use liquid ***sealant*** material that is applied to the normal gasket surfaces, and forms a seal when the parts are tightened together. These sealants are used with or in place of gaskets on many engines and transmissions. Two of the most popular are room temperature vulcanizing (RTV) and anaerobic sealants, **Figure 5-41.**

Figure 5-41. *Sealants are used alone or in conjunction with a gasket. Any sealer used on a computer-controlled engine should be approved for oxygen sensors as some sealers can contribute to oxygen sensor poisoning. (Fel-Pro)*

Summary

The three systems having the most effect on engine operation and driveability are the fuel, ignition, and compression systems. The compression system develops the compression pressure necessary to develop engine power. The four-stroke cycle operates through two revolutions of the crankshaft. The intake stroke draws air and fuel into the cylinder, or only air on a diesel engine. The compression stroke compresses the air-fuel mixture. When the mixture ignites, it pushes the piston down for the power stroke. On the exhaust stroke, the upward movement of the piston pushes the exhaust gases out of the cylinder. The cycle then repeats.

The major components of the piston engine are the engine block, pistons and rings, connecting rods and crankshaft, cylinder head, valves, and valve train. The valves are driven by the crankshaft through a set of gears, a gear and chain arrangement, or sprockets and a drive belt. Diesel engines have the same parts as gasoline engines, but operate at much higher compression pressures. Miller-cycle engines have a slightly different operating principle to the normal piston engine. Rotary engines have fewer parts and operate on a simpler principle. A rotor is turned by combustion action, and turns the crankshaft. There are no valves or pistons in a rotary engine.

The cooling system removes unwanted engine heat and regulates engine temperature. The two main kinds of air cooling are direct air cooling and liquid cooling. In direct air cooling, the heat is transferred directly to the air. In liquid cooling, the heat is absorbed by a liquid, which then transfers it to the air.

The liquid cooling system is almost universally used in modern cars and trucks. Coolant is pumped through engine passages by a belt-driven pump. The coolant picks up heat from the engine and then flows into the radiator where it transfers the heat to the air passing through the radiator. The cooled liquid then returns to the engine to pick up more heat.

Engine coolant is a mixture of antifreeze and water. Modern antifreeze is a chemical called ethylene glycol. Most engine manufacturers recommend a 50:50 mix of ethylene glycol antifreeze and water. A mixture of antifreeze and water also has a higher boiling point than plain water.

The water pump is a cast iron or aluminum housing bolted to the engine block. The radiator is made of internal coolant tubes and external fins. As the coolant flows through the tubes, heat is transferred to the fins, which then transfer the heat to the air passing through the radiator. Radiators used on cars with automatic transmissions have a heat exchanger mounted in the radiator to cool the transmission fluid.

The radiator pressure cap allows cooling system pressure to build up by holding a seal against the radiator filler neck. The thermostat forces the engine to warm up quickly by keeping coolant from circulating through the radiator when the engine is cold. Direct air cooling is a method of transferring engine heat directly to the surrounding air. The air cooling system uses a fan to force air around and past the cylinders and cylinder heads, which are the hottest parts of the engine. The fan is driven from the engine crankshaft by a belt. The cylinders and heads of an air cooled engine are made with fins to present a larger heat transfer surface.

The lubrication system circulates engine oil to internal engine parts. Engine oil serves several purposes and must be delivered in the proper amounts. The oiling system is a vital part of the engine, and poor oiling can eventually result in driveability symptoms. The amount of oil needed varies between engine parts. The parts that need constant, consistent lubrication are supplied with oil under pressure. On some engines, pressurized oil drips through nozzles onto moving parts. Some engine parts are lubricated by oil that is splashed onto them.

The engine oil (motor oil) is what actually does the job of lubrication between moving parts. Modern oils are made to do a good job of lubrication, friction reduction, cooling, sealing, shock absorption, and cleaning. They also prevent sludge formation. The American Petroleum Institute (API) classifies oil according to its ability to prevent friction and deposits in an engine. The API service classification is printed on the oil container. Vehicle manufacturers also have a grading system. Another classification system deals with the fuel efficiency of motor oils.

The oil pan is usually made of stamped sheet metal and functions as a reservoir for engine oil. The oil pickup screen prevents any large particles from entering the engine oiling system. The engine oil pump is driven by a gear on the camshaft, or by the engine timing belt. Oil pump output is controlled by a pressure regulator. When the pump pressure becomes too high, the spring is compressed, the valve opens, and excess oil pressure is dumped back into the sump. The oil filter consists of a stamped metal housing enclosing a pleated paper filtering element. It is always installed after the oil pump. Oil galleries are internal passages which carry the pressurized oil throughout the engine. The engine is sealed by gaskets, seals, and sealants.

Know These Terms

Compression system
Intake stroke
Compression stroke
Power stroke
Exhaust stroke
Engine block
Pistons
Compression rings
Oil control ring
Connecting rods
Piston pin
Crankshaft
Bearing caps
Bearings
Cylinder head
Head gaskets
Head Bolts
Intake
Exhaust valves
Valve spring
Valve spring retainers
Split keeper
Valve guide
Oil seal
Seat
Valve clearance
Valve train
Camshaft
Valve lifters
Valve timing
Lift
Duration
Overlap
Lifter bores
Roller lifters
Pushrods
Rocker arms
Hydraulic lash adjuster
Camshaft drives
Gear drive
Chain drive
Belt drive
Harmonic balancer
Balance shafts
Diesel engines
Precombustion chamber
Rotary engine
Miller-cycle engine
Cooling system
Liquid cooling system
Engine coolant
Ethylene glycol
Water pump
Coolant passages
Freeze plugs
Radiator
Top-flow radiators
Cross-flow radiators
Radiator fan
Fluid clutch
Pressure cap
Coolant recovery system
Recovery tank
Thermostat
Hoses
V-Belts
Serpentine belt
Direct air cooling
Lubrication system
Pressure/splash lubrication system
Engine oil
Oil pan
Oil pump
Pressure regulator
Oil filter
Oil galleries
Gaskets
Seals
Sealant

Review Questions—Chapter 5

Please do not write in this text. Write your answers on a separate sheet of paper.

1. Name the four strokes in a four-stroke engine.
2. Hydraulic lifters are self ______.
3. Many modern engine blocks and cylinder heads are made of ______ to reduce weight.
4. A film of ______ between the piston ring and cylinder wall seals in compression.
   (A) metal
   (B) oil
   (C) unburned fuel
   (D) None of the above.
5. The connecting rods convert the straight-line motion of the piston into rotary motion at the:
   (A) camshaft.
   (B) crankshaft.
   (C) push rod.
   (D) rocker arm.
6. List two devices used to reduce engine vibration.
7. The major difference between the diesel engine and the gasoline engine is that:
   (A) the diesel has much higher compression.
   (B) the diesel uses an electric spark to fire the fuel mixture.
   (C) the diesel has always been fuel injected.
   (D) Both A & C.
8. A unique feature of the Miller-cycle engine is that:
   (A) the exhaust valve stays open for an extended period.
   (B) it is equipped with a turbocharger.
   (C) the intake valve stays open for an extended period.
   (D) Both A & B.
9. Rotary engines have fuel and ignition systems similar to those used on:
   (A) gasoline engines.
   (B) diesel engines.
   (C) natural gas engines.
   (D) None of the above.

10. The engine block has ______ installed in it to prevent engine damage if the coolant should freeze.

11. Most modern vehicles are _____ cooled.
    (A) air
    (B) liquid
    (C) oil
    (D) thermostatically

12. Pressurizing the cooling system ______ the coolant boiling point.

13. Engine oil is used to ______ various parts of the engine.
    (A) cool
    (B) lubricate
    (C) seal
    (D) All of the above.

14. Valve lifters are usually lubricated by:
    (A) pressure.
    (B) splash.
    (C) combination pressure/splash.
    (D) None of the above.

15. Name two sealants used in modern engines.

ASE Certification-Type Questions

1. Technician A says that an engine must have sufficient compression before it will start and run. Technician B says that compression is developed on the power stroke of a four-cycle engine. Who is right?
    (A) A only.
    (B) B only.
    (C) Both A & B.
    (D) Neither A nor B.

2. All of the following statements about oil control rings are true, EXCEPT:
    (A) oil control rings are installed below the compression rings.
    (B) there should be no oil at the compression rings.
    (C) the oil ring scrapes oil from the cylinder walls when the piston is moving down.
    (D) defective oil control rings will cause high oil consumption

3. Technician A says that the intake valves are cooled by incoming air. Technician B says that the valve must be in contact with its seat long enough to transfer heat to the head. Who is right?
    (A) A only.
    (B) B only.
    (C) Both A & B.
    (D) Neither A nor B.

4. Most modern cylinder heads contain all of the following EXCEPT:
    (A) intake valves.
    (B) exhaust valves.
    (C) EGR valves.
    (D) camshafts.

5. The valve train is the group of components that opens the ______.
    (A) intake valves
    (B) exhaust valves
    (C) EGR valve
    (D) Both A & B.

6. A hydraulic lash adjuster, similar in operation to a hydraulic lifter, maintains the proper valve clearance on ______ engines.
    (A) diesel
    (B) rotary
    (C) overhead cam
    (D) All of the above.

7. Technician A says that overhead camshafts are usually driven by a belt. Technician B says that on overhead valve engines with the camshaft in the block the camshaft is always directly driven by two gears. Who is right?
    (A) A only.
    (B) B only.
    (C) Both A & B.
    (D) Neither A nor B.

8. Most vehicle manufacturers recommend a ______ mix of water and antifreeze.
    (A) 10:90
    (B) 50:50
    (C) 30:70
    (D) 80:20

9. Oil screens are placed in the oil pan at the inlet to the oil pump. What are they designed to screen out?
    (A) Large dirt or metal particles.
    (B) Small dirt or metal particles.
    (C) Additives.
    (D) Whatever the oil filter does not catch.

10. Technician A says that engine oil pumps are constant-speed pumps. Technician B says that engine oil pumps are constant displacement pumps. Who is right?
    (A) A only.
    (B) B only.
    (C) Both A & B.
    (D) Neither A nor B.

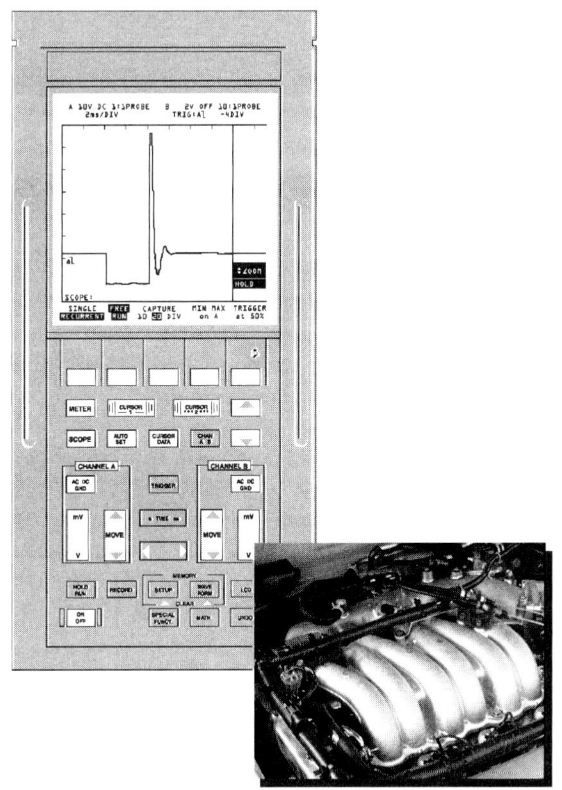

Chapter 6

Electricity and Electronics Fundamentals

After studying this chapter, you will be able to:

- Explain the electron movement theory or electricity.
- Identify basic electrical circuits.
- Identify basic electrical measurements.
- Identify and explain the purpose of vehicle wiring and connectors.
- Identify and explain the purpose of common vehicle electrical devices.
- Explain the construction of common semiconductor devices.
- Identify the uses of common semiconductor devices.
- Identify the major components of vehicle starting and charging systems.

It would be impossible to diagnose any problem in the modern vehicle without a thorough knowledge of electricity and electronics basics. This chapter is also intended as a review of the basic principles of electricity and semiconductor electronics, material you should have learned in earlier courses. Also in this chapter is an overview of vehicle starting and charging systems. Since defects in the starting and charging systems can affect driveability, the student should be familiar with their operating principles.

Electricity and Electronics Basics

As a driveability technician, you will be working with electricity and electronic components almost constantly. There are very few systems on the modern automobile that do not contain some sort of electronic sensor, switch, or output device. The following sections outline the basic principles that govern the operation of all electrical equipment. These principles apply to any make of vehicle.

Atoms and Electricity

Everything is made of atoms. Every atom has a center of ***protons*** and ***neutrons.*** The neutrons have no charge and the protons have a positive charge, making the center of the atom positively charged. Revolving around this center are negatively charged ***electrons.*** See **Figure 6-1.** In the ideal atom, the number of electrons is exactly equal to the number of protons in the center. In actual practice, the number of electrons varies, and there is movement of electrons between atoms. Electricity is the movement of large numbers of electrons from atom to atom.

Conductors and Insulators

Some atoms easily give up or receive electrons. These atoms make up elements that are good electrical

conductors. Examples are copper, gold, and aluminum. Materials whose atoms resist giving up or accepting atoms are called ***insulators.*** Glass and plastic are good insulators. Some materials can alternate between conducting and insulating. These materials are discussed in the electronics section later in this chapter.

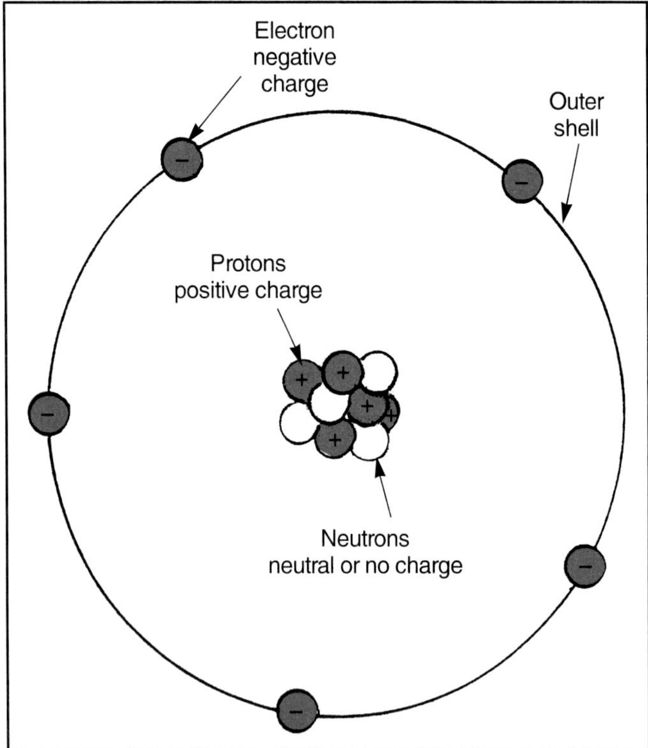

Figure 6-1. *An atom is made up of protons, neutrons, and electrons. The number of electrons in the atom's outer shell, or valence ring, determine its conductivity or resistance. The fewer electrons an atom has in its outer shell, the greater its conductivity.*

Electron Flow

As mentioned earlier, electricity is the movement, or ***flow*** of electrons. The path through which the electrons move is called a ***circuit.*** Electrons will not flow in a circuit unless two conditions are met:

- There are more electrons in one place than another. In a vehicle, this difference in the number of electrons is created by the battery and alternator.
- There is a connection between the two places. This connection is composed of the vehicle wiring and the electrical unit being operated. It must be made of materials that are good conductors.

A typical automotive circuit is the starting circuit. In the starting circuit, the electrical path is from the battery through the negative battery cable, starter, engine block, and positive cable, returning to the battery, as shown in **Figure 6-2.** This is called a ***complete circuit,*** since the electricity makes a loop through the battery, cables, starter, block, and back to the battery. The path for the electrons to return to the battery is as important as the path from the battery to the electrical device.

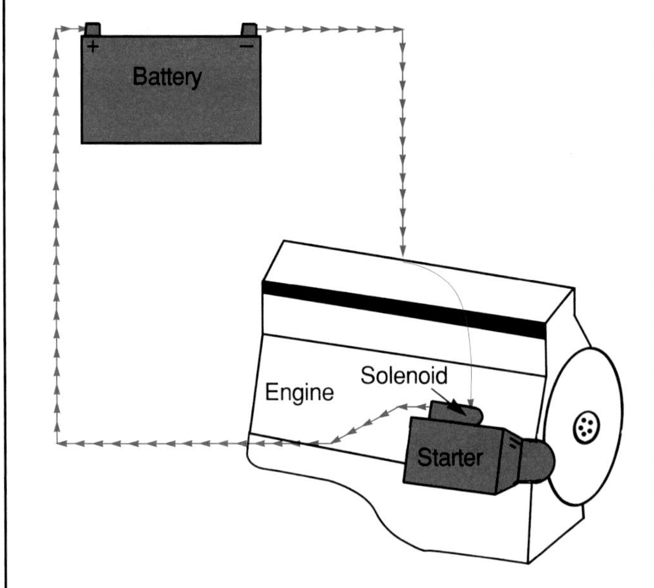

Figure 6-2. *Electricity makes a complete loop from the battery's negative terminal to its positive terminal. Remember the electron theory states that electrical energy flows from negative to positive.*

Electrical Measurements

The flow of electricity through a circuit depends on three electrical properties, all of which can be measured. The number of electrons flowing past any point in the circuit is called ***current.*** Current is measured in ***amperes***, usually shortened to *amps.* The higher the ampere rating, the more electrons are moving in the circuit.

The electrical pressure created by the difference in the number of electrons between two terminals in the circuit is referred to as ***voltage.*** It provides the push that makes electrons flow and is measured in *volts.* ***Resistance*** is the opposition to the flow of electrons in a conductor. All conductors, even copper, gold, and aluminum, have some resistance to giving up their electrons. Resistance is measured in ***ohms.***

Ohms Law

Sometimes, certain electrical properties of a circuit are not known. For example, you may want to know the total amperage draw of a set of aftermarket fog lights. If the lights have a resistance of 20 ohms and the vehicle has a 12 volt electrical system, you can use the voltage and resistance values to calculate the amperage.

$$\frac{12 \text{ volts}}{6 \text{ ohms}} = 2 \text{ amps}$$

If you know the amperage and voltage, and want to calculate the resistance in the above illustration, use the following equation to calculate the unknown resistance.

$$\frac{12 \text{ volts}}{2 \text{ amps}} = 6 \text{ ohms}$$

If you know the amperage and resistance, calculate the voltage by using the following formula.

$$2 \text{ amps} \times 6 \text{ ohms} = 12 \text{ volts}$$

These formulas are known as ***Ohms law.*** **Figure 6-3** is a graphic representation of Ohms law, sometimes called the Ohms law triangle. When applying Ohms law to find an unknown electrical property, remember to use the two known properties to find the unknown value.

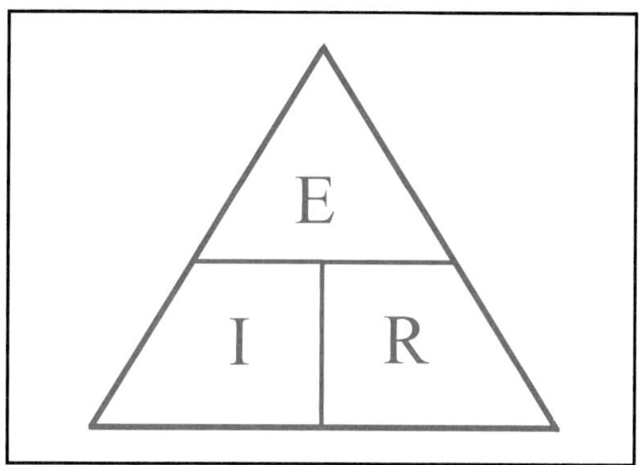

Figure 6-3. *The Ohm's law triangle is easy to use. To find the unknown value, simply cover the letter representing the unknown value and divide or multiply the remaining values.*

Direct and Alternating Current

In a car or truck, the battery always has two terminals: positive and negative. One of these terminals has more electrons than the other. There is always a shortage of electrons at the positive terminal, and an excess of electrons at the negative terminal. This current flows in only one direction, from negative to positive, **Figure 6-4.** This is called a ***direct current (dc)*** system. Since the battery generates direct current, almost every system on the vehicle operates on direct current.

In the electrical systems used in homes, schools, and offices, the flow of electrons changes direction many times every second. These are called ***alternating current (ac)*** systems, **Figure 6-5.** Some vehicle systems use devices that generate low voltage alternating current. These devices include some ignition system pickup coils, vehicles and wheel speed sensors. Since alternating current is a constant uninterrupted wave, it is better suited for use in these devices, which monitor movement or speed. However, since the vehicle battery cannot be charged by alternating current, most automotive electrical systems use direct current.

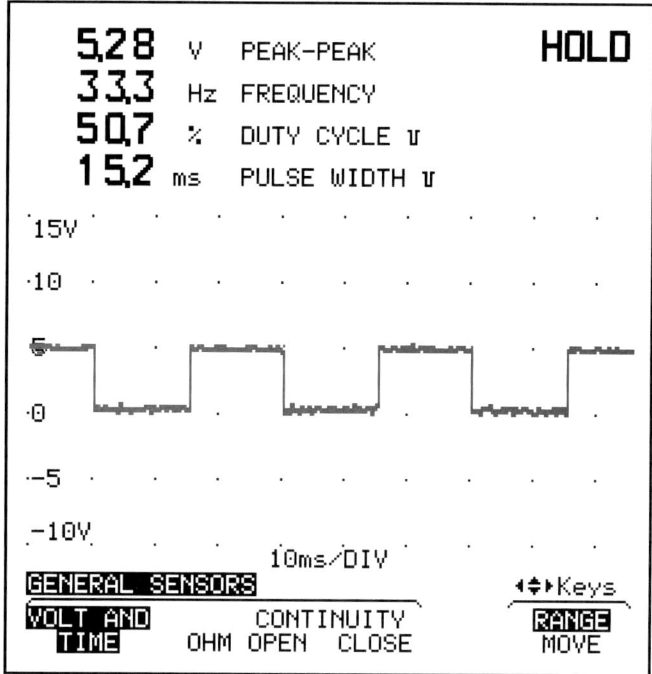

Figure 6-4. *Since the battery generates direct current, it is used in most automotive systems, including sensors and primary wiring. This waveform is typical of direct current. (Fluke)*

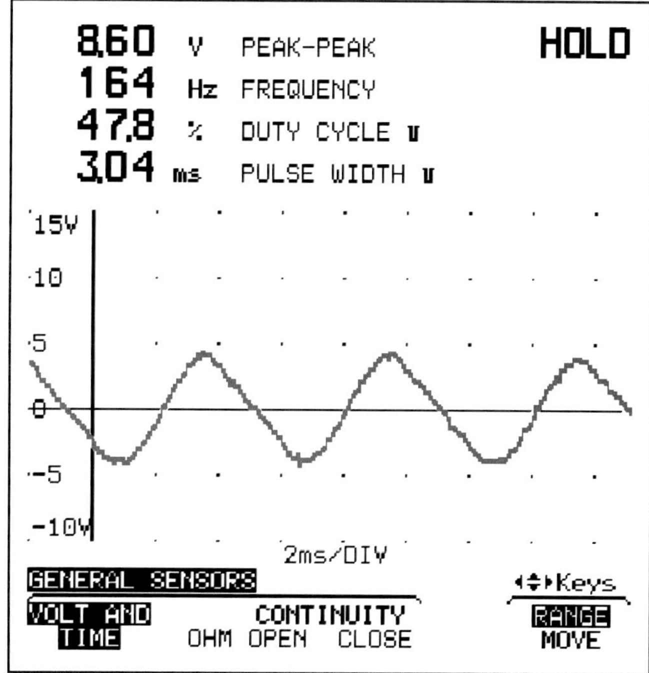

Figure 6-5. *While limited in its use in automotive systems, alternating current is used in applications where direct current is not practical. The waveform generated by alternating current is referred to as a sine wave. (Fluke)*

Warning: Always respect electricity. While most automotive electrical systems carry relatively low voltage, current can be high enough to cause serious injury or death.

Types of Electrical Circuits

There are three types of automotive circuits. Every wire in a car or truck is part of one of these types of circuits. The three types of circuits are series, parallel, and series-parallel. The circuit in **Figure 6-6** is the simplest type of automotive circuit, the ***series circuit.*** This series circuit consists of the vehicle battery, a switch, a lightbulb, and connecting wiring. Electrons flow from the battery through the wiring, switch, and bulb, and back to the battery. The wire in the bulb, called a *filament,* is made of a type of resistance wire that glows as the electrons pass through it. The same current (number of electrons) flows through every part of the series circuit.

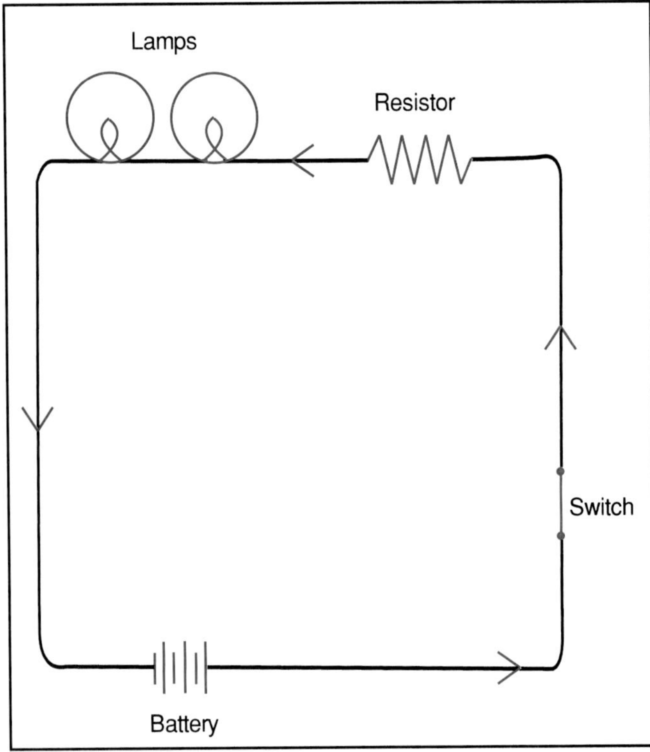

Figure 6-6. *The series circuit is the simplest circuit found in the modern vehicle. In some cases, the circuit will only have one load. A disadvantage of the series circuit is if an open occurs in one of the loads, all devices downstream of the load will not receive power.*

The circuit in **Figure 6-7** is a ***parallel circuit.*** In a parallel circuit, current flow is split so that each electrical component has its own current path. Different amounts of current will flow in different parts of the circuit, depending on the resistance of each part. The ***series-parallel circuit*** has some components that are wired in series and some that are wired in parallel, as shown in **Figure 6-8.** All current flows through some parts of the circuit, while the current path is split in other parts.

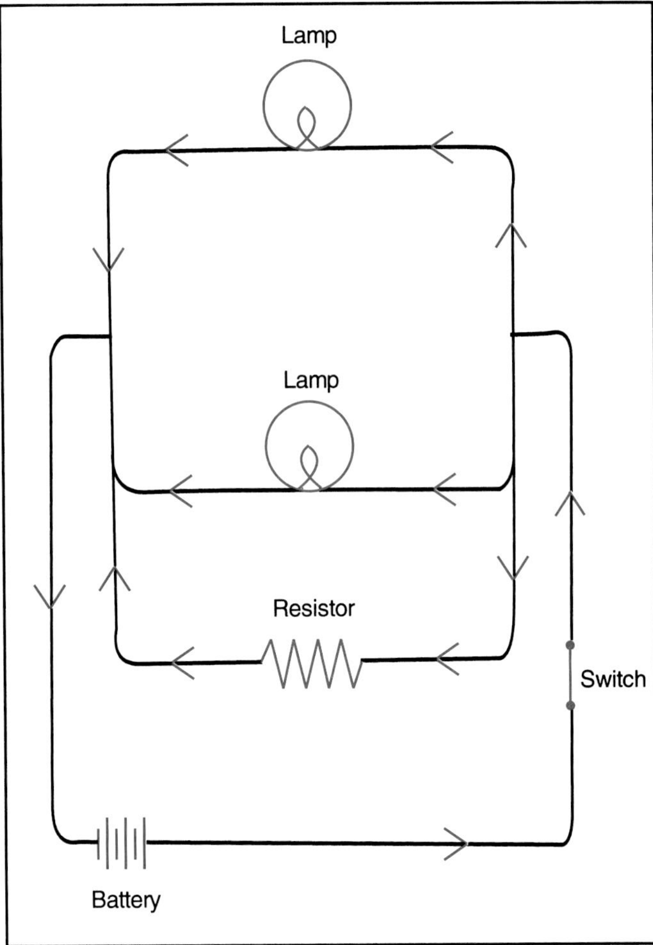

Figure 6-7. *Parallel circuits are used when it is necessary for the circuit to operate with one or more opens in the circuit.*

Circuit Defects

There are three main types of wiring defects: shorts, grounds, and opens. A ***short circuit*** is caused when the wire insulation fails, is damaged, or removed, allowing the wire to contact the frame, body, another circuit, or a part of the vehicle, **Figure 6-9.** If the circuit is not protected by a fuse, circuit breaker, or fusible link, a short could lead to damaged wiring or components, or a fire.

A ***grounded circuit*** is caused when a circuit's ground wire is shorted. In many ways, it is similar to a short circuit except that, a grounded circuit will have little or no noticeable effect on some circuits. However, if normal operation occurs when a circuit is grounded, a grounded wire may cause unwanted circuit or system operation.

An ***open circuit*** is a circuit which is not complete. Current cannot flow in an open circuit, as in **Figure 6-10.** Common causes of open circuits are loose or corroded connections, damaged or disconnected wires, open circuit protection devices, and defects in electrical components such as switches, bulbs, and fuses. A related problem is a high resistance electrical connection, usually caused by

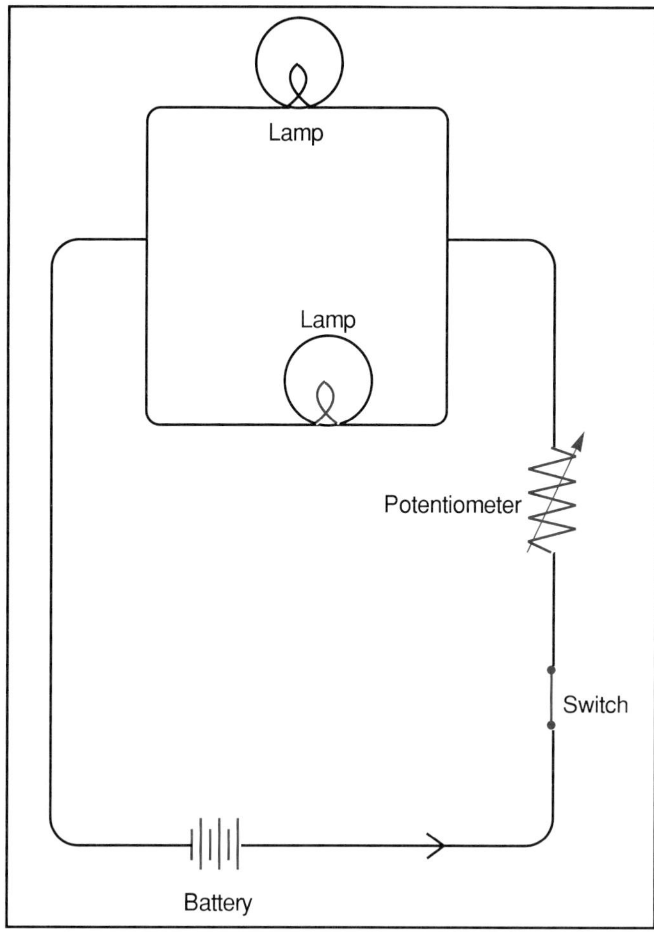

Figure 6-8. *Series-parallel circuits are found when power or ground is utilized by multiple loads. A common use is when two or more loads must be controlled by a single switch or potentiometer.*

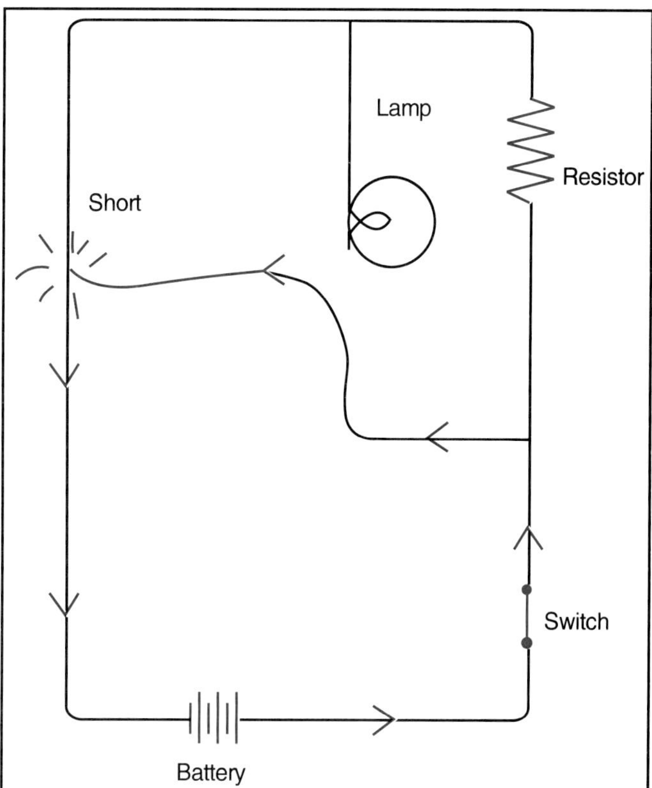

Figure 6-9. *A short circuit can be caused by many factors. If the short is located in the ground wire, the circuit is sometimes considered to be grounded.*

corrosion or overheating. The high resistance may cause the circuit to stop operating, or possibly start a fire at the connection.

A fourth type of failure is *intermittent circuit failures.* Intermittents can involve all three types of common circuit failures, however, a failure is classified as intermittent when it does not occur all the time. This type of failure is common and often, the most difficult to diagnose and repair.

Electromagnetism

The relationship between electricity and magnetism must be understood to service the modern vehicle. When a material is ***magnetic,*** the electrical charges of its electrons are aligned to create a force that extends outward from the material. Magnetic fields have definite North and South poles. This property of magnetic fields is called ***polarity.*** Like poles repel each other and unlike poles attract. It is usually not necessary for the technician to determine the North and South poles when servicing a magnetic unit.

A magnetic material attracts iron and metals that contain iron. Some materials are naturally magnetic while

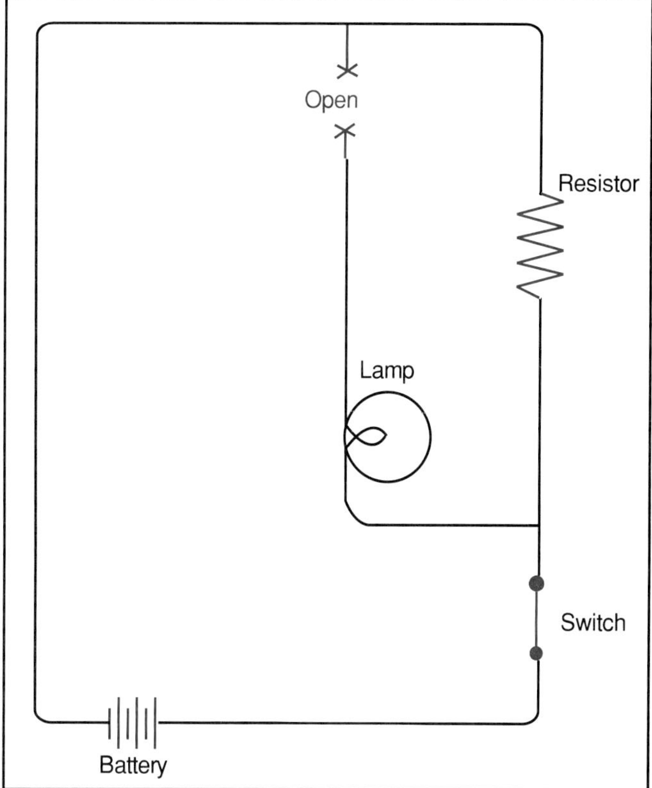

Figure 6-10. *An open in a circuit will cause the load to not function. However, it may or may not cause the circuit protection device to open.*

others can be magnetized by electricity. When current flows in a wire, the electrons start moving in the same direction. This alignment of electrons creates a magnetic field around the wire as long as current is flowing. When the wire is wound into a coil, the magnetic fields of each wire loop combine to create a very strong magnetic field. The combination of the coil winding and a metal core is called an ***electromagnet,*** **Figure 6-11.** The iron core helps to increase field strength and may be moveable. The magnetic field created may be used to move linkages or open electrical contacts. Electromagnets are the basic component of solenoids, relays, and starters.

When a wire moves through a magnetic field, or a magnetic field moves through a wire, the effect of the field on the electrons in the wire causes them to begin moving. This creates current to flow in the wire. This process is called ***induction,*** and is how current is produced in alternators. In modern alternators, the wires are stationary and the magnetic field moves.

Another common use of induction is the transformer. A transformer always has two wire coils. Electricity flows through one coil to produce a magnetic field. When the current flow in the coil is cut off, the magnetic field collapses through the second coil, inducing current flow. The difference in the number of coils, or turns, of wire between the two coils allows the voltage to be reduced or increased. This is the principle of the ignition coil, discussed in more detail in **Chapter 8.** The relationship between electricity and magnetism is called ***electromagnetism.*** Electromagnetism is used to operate many electrical devices on the vehicle. These devices will be discussed later in this chapter and throughout this text.

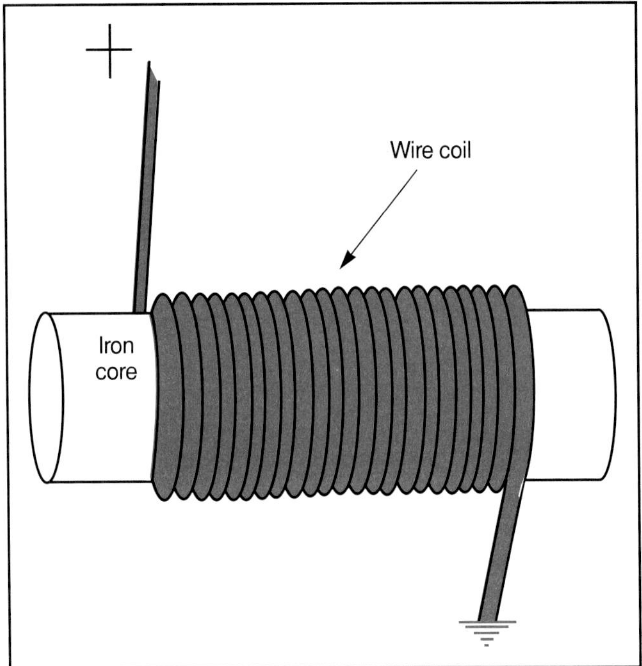

Figure 6-11. *A coil of wire wrapped around an iron core. When power is applied to the wire, the core becomes an electromagnet. This is the basis for motors, relays, solenoids, and ignition coils.*

Automotive Electronics

So far, we have reviewed the basic principles of electricity used in the modern automobile. These basic principles are also used in almost every electronic device in existence. The following sections will cover electrical principles and electronic devices that are common to modern vehicles.

Vehicle Wiring and Electrical Components

The modern automotive electrical system is a complex arrangement of wiring and electrical components. The electrical system must produce electricity and deliver it to the proper places, protect circuits from damage, reduce or increase voltage, change electricity into light, motion, or heat, and control the movement of liquids and gases. The construction, operation, and use of devices to accomplish this are discussed in the following sections. The operation and function of electronic devices will be discussed later in this chapter.

Wiring

Most automotive ***wiring*** is made of copper, aluminum, or aluminum coated with copper. Wires are plastic-coated, wrapped in plastic or fabric looms, and installed as assemblies called ***harnesses,*** **Figure 6-12.** For easier circuit tracing, automotive wiring is ***color-coded.*** This is done by giving the insulation of every wire a specific color. Modern vehicles have many wires and it is necessary to increase the number of available colors by adding a stripe or band of a contrasting color to the original insulator. Service literature contains schematics, discussed later in this chapter, that show all wires and their colors.

Many modern vehicles used printed circuits to connect the dashboard and other contained electrical units. These printed circuits consist of a plastic sheet on which electrical conduit has been etched or printed, **Figure 6-13.** Most printed circuits are replaced as an assembly when defective.

On most vehicles, there is no return wiring from various electrical units back to the battery. The vehicle frame forms the return wire, or ***ground wire,*** as shown in **Figure 6-14.** On all modern vehicles, the negative terminal is the ground terminal. This means that the battery, charging system, and all other electrical devices have their negative terminal connected to a common negative ground connection, usually the vehicle frame and body. On most vehicles, the battery negative cable is attached to the engine block. The body and chassis may be grounded directly to the battery through a smaller cable attached to the negative post, or grounded to the engine block by one or more ground straps.

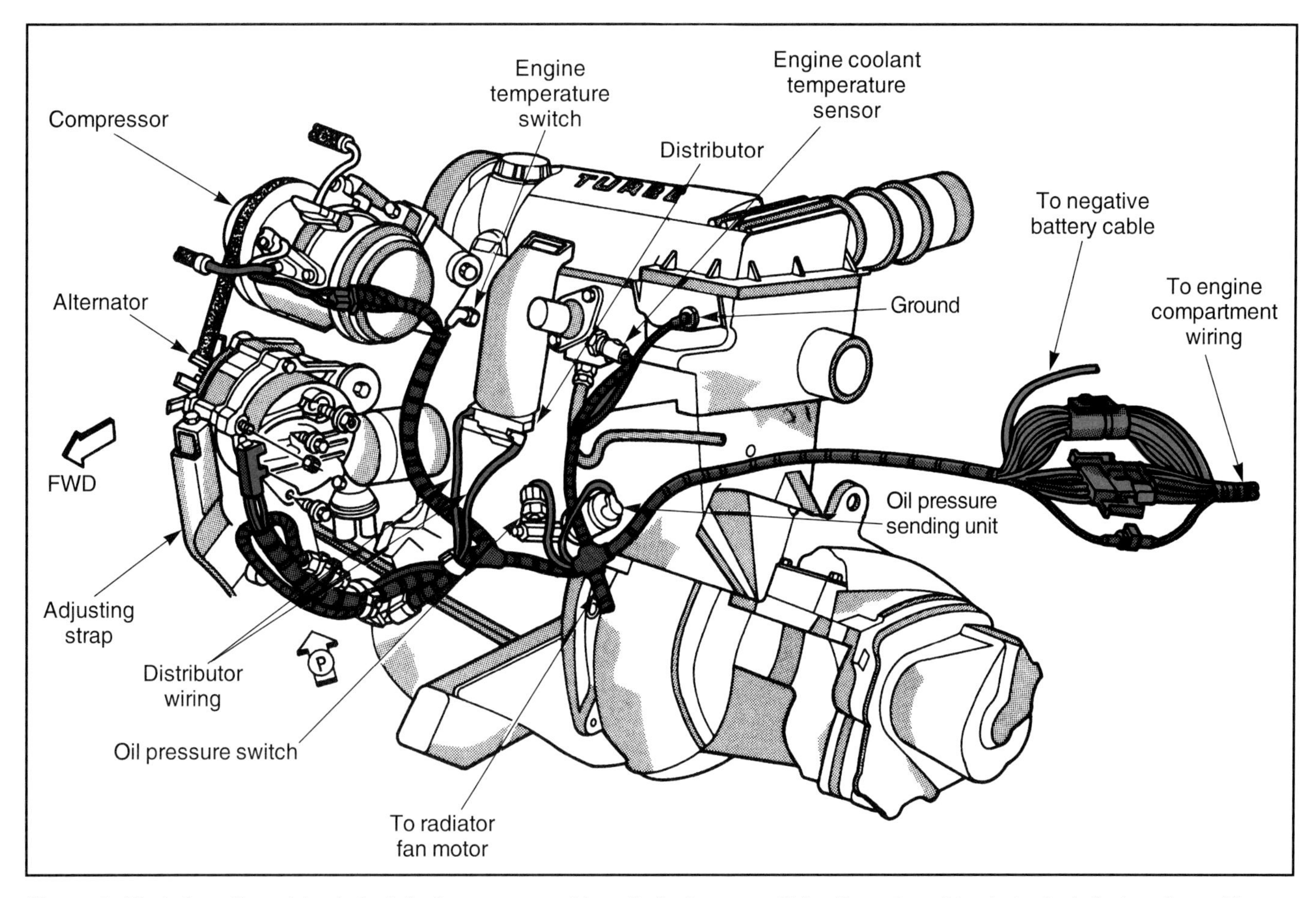

Figure 6-12. *Automotive wiring is installed as an assembly called a harness. This allows the wiring to be installed easily and keeps it in a neat package that cannot be easily damaged. (Chrysler)*

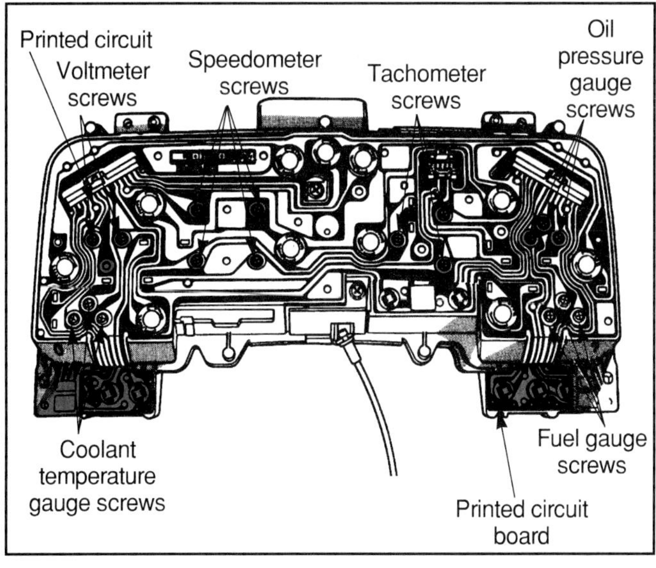

Figure 6-13. *Printed circuit boards are used in ECMs, many driver-operated system controls, and dashboard instrument clusters. Instrument clusters have used printed circuits for many years. (Chrysler)*

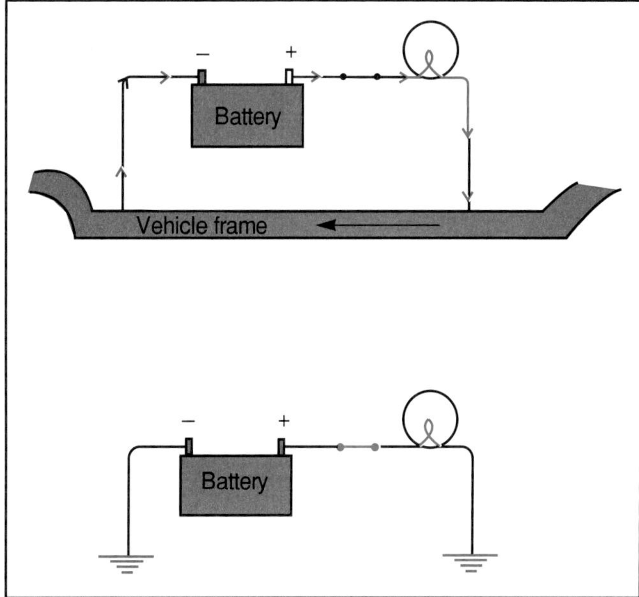

Figure 6-14. *The negative ground electrical system utilizes the frame, which is the unitized body of most modern vehicles, as the path from the negative battery terminal.*

Wire Size

All wires, no matter how well they conduct current, have some resistance and will lose some electrical power as heat. It is important that the wire in a circuit be large enough to carry the rated amperage without overheating. At the same time, the wire should not be unnecessarily large to reduce bulk, weight, and cost. ***Wire gage*** is the rating system used to measure wire diameter.

Wire gage can be measured in ***American Wire Gage (AWG)*** or metric sizes. Note that the gage refers only to the thickness of the wire itself, without insulation. The larger the gage number, the smaller the wire. A typical wire gage chart is shown in **Figure 6-15.** Note that the chart shows the highest amperage that can be carried by a particular size wire without overheating. The largest gage wires are used to connect the battery to the vehicle starter. To further reduce resistance losses, wires carrying high amperages are designed to be as short as possible.

Plug-In Connectors

The majority of connectors on modern vehicles are plug-in types. A ***plug-in connector*** has male and female ends, which are plugged into each other. Connectors having more than one wire are called multiple connectors. Many service manuals refer to a specific connector by the number of wires it contains, such as 12-wire connectors, 23-wire connectors, and so on, **Figure 6-16.**

Due to the shape of the connector, or by the use of special aligning lugs and slots on the connector, modern plug-in connectors cannot be assembled incorrectly. Harnesses with multiple connectors that plug into a single device are sometimes color-coded to prevent an accidental misconnection. Modern plug-in connectors are often thoroughly sealed to keep out moisture and corrosion. Since many vehicle electronic components operate on very low voltages, a small increase in resistance due to moisture and corrosion can affect circuit operation.

A few connectors may be screw terminals, or bolts which pass through a terminal eye to form a connector, **Figure 6-17.** These connectors are usually used to connect a ground strap to the vehicle body, or on circuits with very high current loads, such as the starter and charging system.

Wiring Schematics

A ***schematic,*** or ***wiring diagram,*** is a drawing showing electrical units and the wires connecting them. Schematics also show wire colors and terminal types. Use of the schematic allows the technician to trace out defective components in the wiring system. Many vehicle manufacturers break down the overall vehicle wiring into separate circuit diagrams, as shown in **Figure 6-18.**

Schematics use symbols to represent electrical devices. There is some variation in the use of these symbols. Some schematics have a combination of manufacturer-specific and standardized symbols. **Figure 6-19** illustrates some symbols widely used in automotive electrical diagrams.

Fiber Optics

One disadvantage of the modern electrical system is the great quantity of wiring needed to monitor and operate each system. The more systems a vehicle has, the more wiring needed. All this wiring, along with terminals, looms, and mounting components adds weight, which decreases fuel efficiency. This weight would be even greater if the chassis were not used as a ground wire. While the weight of

Circuit Amperes	Total Circuit Watts	Total Candle Power	Wire Gage (For Length in Feet)											
12V	12V	12V	3′	5′	7′	10′	15′	20′	25′	30′	40′	50′	75′	100′
1.0	12	6	18	18	18	18	18	18	18	18	18	18	18	18
1.5		10	18	18	18	18	18	18	18	18	18	18	18	18
2	24	16	18	18	18	18	18	18	18	18	18	18	16	16
3		24	18	18	18	18	18	18	18	18	18	18	14	14
4	48	30	18	18	18	18	18	18	18	18	16	16	12	12
5		40	18	18	18	18	18	18	18	18	16	14	12	12
6	72	50	18	18	18	18	18	18	16	16	16	14	12	10
7		60	18	18	18	18	18	18	16	16	14	14	10	10
8	96	70	18	18	18	18	18	16	16	16	14	12	10	10
10	120	80	18	18	18	18	16	16	16	14	12	12	10	10
11		90	18	18	18	18	16	16	14	14	12	12	10	8
12	144	100	18	18	18	18	16	16	14	14	12	12	10	8
15		120	18	18	18	18	14	14	12	12	12	10	8	8
18	216	140	18	18	16	16	14	14	12	12	10	10	8	8
20	240	160	18	18	16	16	14	12	10	10	10	10	8	6
22	264	180	18	18	16	16	12	12	10	10	10	8	6	6
24	288	200	18	18	16	16	12	12	10	10	10	8	6	6
30			18	16	16	14	10	10	10	10	10	6	4	4
40			18	16	14	12	10	10	8	8	6	6	4	2
50			16	14	12	12	10	10	8	8	6	6	2	2
100			12	12	10	10	6	6	4	4	4	2	1	1/0
150			10	10	8	8	4	4	2	2	2	1	2/0	2/0
200			10	8	8	6	4	4	2	2	1	1/0	4/0	4/0

Figure 6-15. *This wire gage selection chart can be used to measure and select wire for a new component installation or replacement for damaged wiring. (Belden)*

Locking tab only on one side

View A

Locking tabs are on both sides of the connector

View B

Insert a flat bladed screwdriver in the locking tab and twist, grasp the wire and pull to separate

Flat bladed screwdriver

Locking tab only

Insert a flat bladed screwdriver in the locking tab and twist. Grasp the wires and pull until the locking tab is on the ramp. Turn the connector over and repeat the procedure on the opposite side of the connector. Grasp the wires and pull apart.

Place a thumb under the locking tab and push up. Grasp the wires and pull to separate

View A

View B

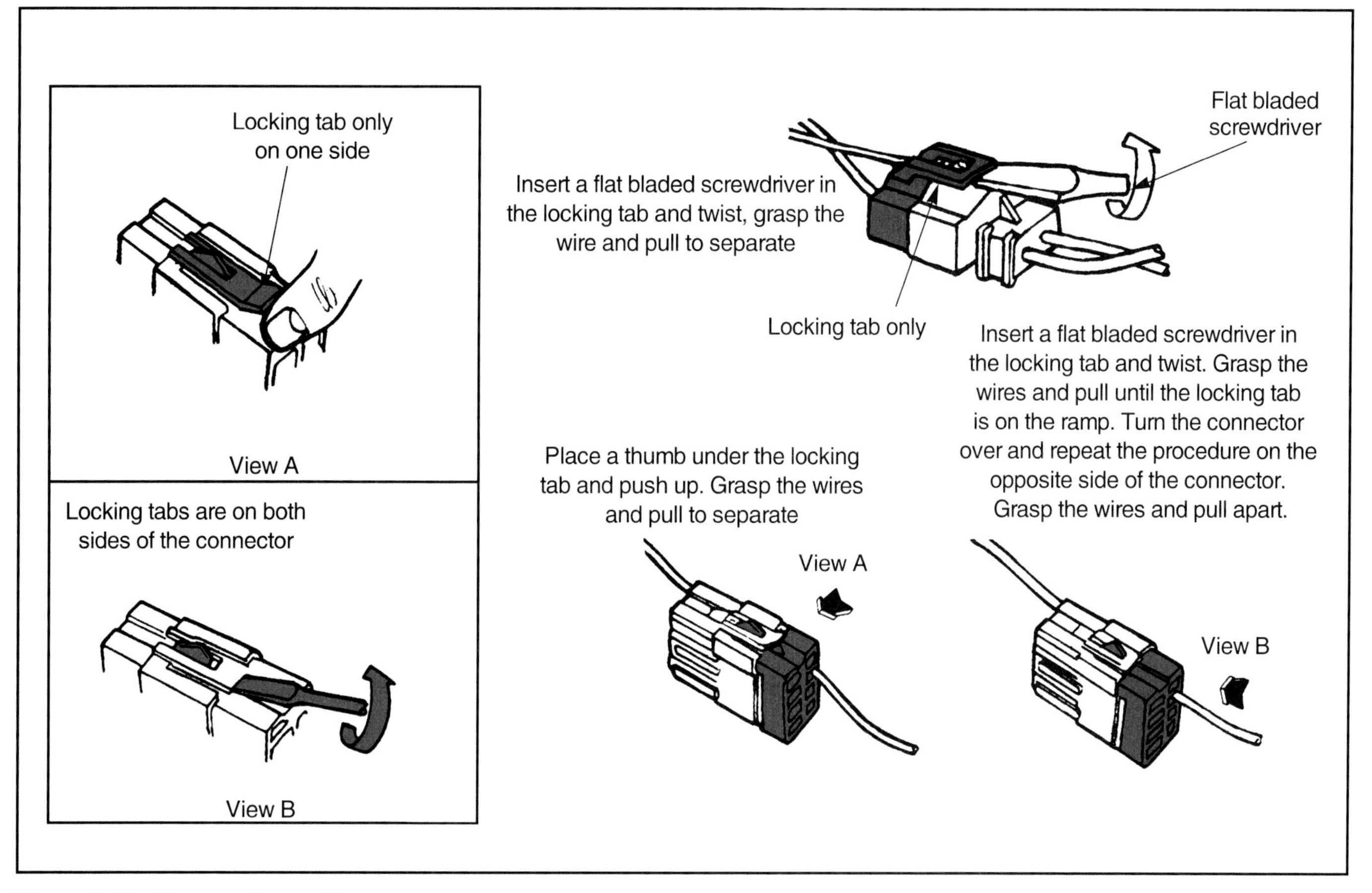

Figure 6-16. *Most modern vehicles use plug-in connectors. It is important that these connectors not be damaged during service. Any damage will allow dirt, water, and corrosion to enter the circuits. (Ford)*

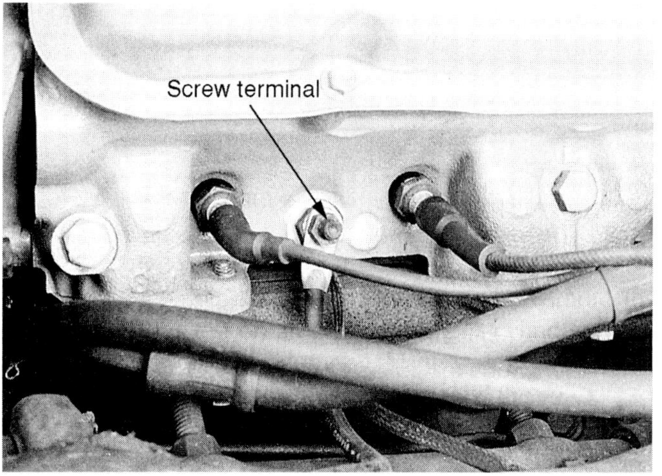

Figure 6-17. *Screw terminals are used to connect power or ground cables to components. This is a negative battery connection to the engine on one application.*

electrical wiring is not really a problem, engineers are always looking for ways to make the modern vehicle more efficient. As you learned earlier in this text, less weight means a more efficient vehicle.

A few of the latest vehicles are beginning to use ***fiber optics*** in place of separate wires to illuminate the instrument cluster. A fiber optic is a thin fiber strand made of glass, plastic, or other transparent material. While automotive use is currently limited to light transmission, fiber optics are best known for their ability to carry data transmissions. Fiber optics have been in use for years by the telecommunications industry and is beginning to be utilized by some television cable companies. In the near future, the maze of wires under the hood and dashboard of the current vehicle may be replaced by a few fiber optic cables.

Circuit Protection Devices

To protect vehicle circuits from damage due to excessive current flows, ***circuit protection devices*** are used. These include fuses, fusible links, and circuit breakers. Excessive current flow can be caused by short circuits, or defective components that draw too much current. All electrical circuits except the starter and alternator output will have a circuit protection device.

Fuses are made of a soft metal that melts when excess current flows through it before damage to system components or circuit wiring occurs. The majority of fuses are installed in a ***fuse block*** located under the dashboard, in the engine compartment, or in the glove box, **Figure 6-20.** Many modern vehicles have more than one fuse block. A melted, or blown fuse must be replaced. On many vehicles, heavy duty fuses called *Maxifuses* or *Pacific* fuses are used. These fuses are designed to carry large amounts of current.

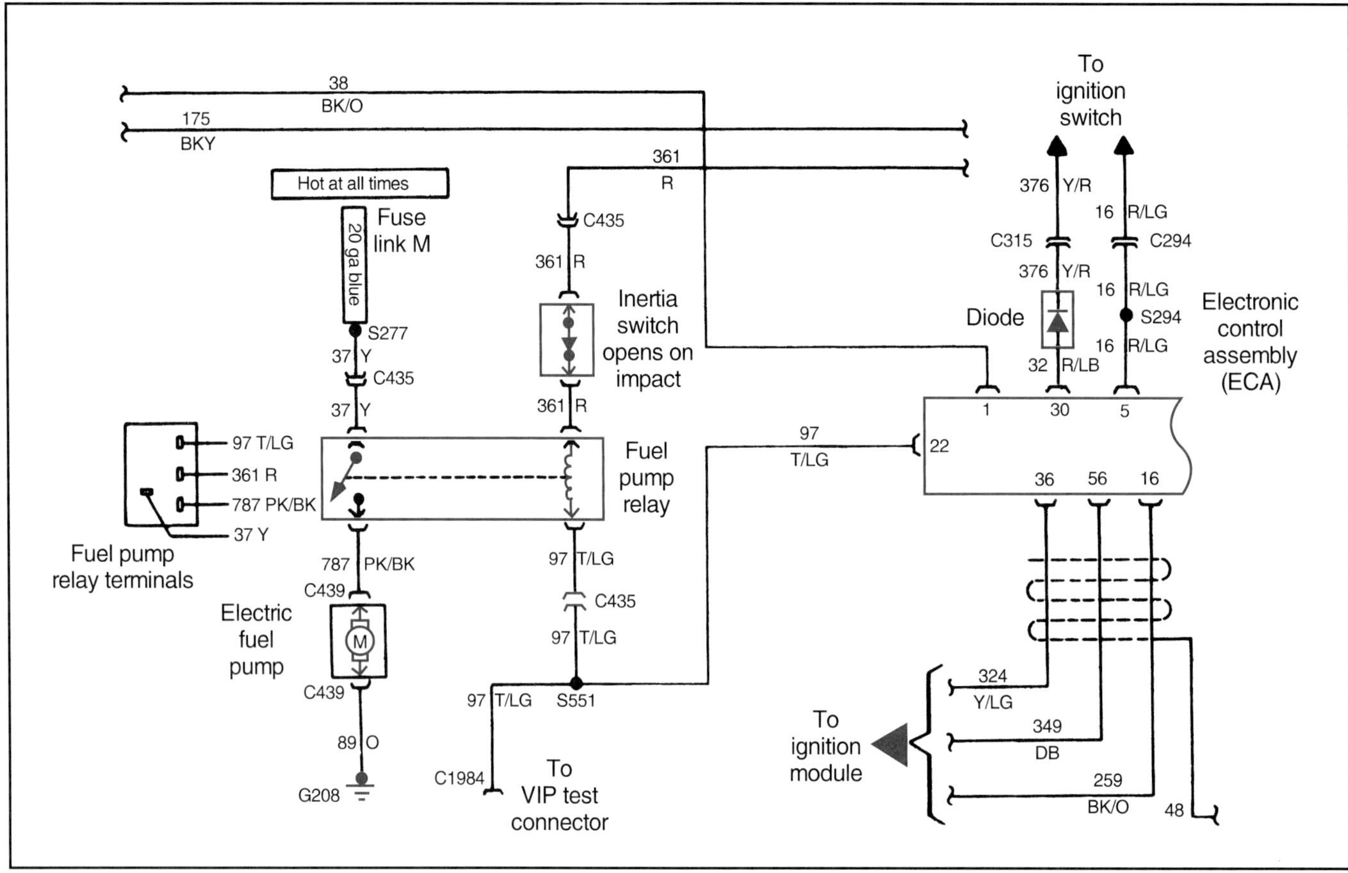

Figure 6-18. *A wiring schematic is a road map of the vehicle's electrical system. You should never attempt to diagnose an electrical system without one, even if you are familiar with the circuit in question. (Ford)*

Circuit breakers consist of a contact point set attached to a bimetallic strip. The advantage of the circuit breaker is that it can reset itself. The bimetallic strip will bend as it heats up. When it becomes hot enough, it bends enough to open the point set and break the circuit. When the strip cools off, it straightens out and allows the points to close. A typical circuit that uses a circuit breaker is the headlamp circuit. Many small electric accessory motors contain internal circuit breakers.

A ***fusible link*** is a length of wire made of soft metal. It operates in the same manner as a fuse, melting when excess current flows. Fusible links are usually installed in the main wiring leading from the battery, terminal block, or starter solenoid to the main electrical circuits. A circuit may be protected by both a fuse in the fuse block and a fusible link ahead of the block. If a fusible link burns out, it must be replaced, which involves cutting out the burned link and soldering or crimping a new link in the circuit. On many vehicles, fusible links are being replaced by Maxifuses.

Terminal and Junction Blocks

With the number of systems requiring battery power for operation or memory retention, connecting all of these systems directly to the battery terminal is impractical. It is far easier to connect a separate lead from the battery to a separate terminal. All systems that require battery power can then be connected to this terminal. This is called a ***terminal block.*** Fusible links not connected to the starter solenoid are often connected to a central terminal block. Either positive or negative connection can be achieved by using a terminal block.

As mentioned earlier, the modern vehicle usually has more than one fuse block. Sometimes, the fuse block contains more relays and other electrical components than fuses. They are often referred to as a ***junction block.*** Junction blocks are usually located in the engine compartment, **Figure 6-21.** Sometimes the block is located under the dashboard, away from the main fuse block. These blocks are sometimes called *relay centers.*

Switches

To control the flow of electricity through a circuit, some sort of ***switch*** is used. Some of these switches have simple on-off positions, while others, such as windshield wiper and blower motor switches, have several positions to place the circuit in various operating modes. Most switches are manually operated by the driver. Examples are vehicle headlight, ignition, windshield wiper, and air conditioner/heater switches. Other switches are operated as a by-product of other driver actions, such as the brake and backup light

Legends of Symbols Used on Wiring Diagrams			
+	Positive	→)—	Connector
—	Negative	——→	Male Connector
	Ground	)——	Female Connector
	Fuse		Denotes Wire Continues Elsewhere
	Gang Fuses with Bus Bar		Denotes Wires Going to One of Two Circuits
	Circuit Breaker		Splice
	Capacitor	J2 2	Splice Identification
	Ohms		Thermal Element
	Resistor	TIMER	Timer
	Variable Resistor		Multiple Connector
	Series Resistor		Optional Wiring with Wiring without
	Coil		"Y" Windings
	Step-up Coil	88:88	Digital Readout
	Open Contact		Single Filament Lamp
	Closed Contact		Dual Filament Lamp
	Closed Switch		L.E.D.ÑLight Emitting Diode
	Open Switch		Thermistor
	Closed Ganged Switch		Gauge
	Open Ganged Switch		Sensor
	Two Pole Single Throw Switch		Fuel Injector
	Pressure Switch	#36	Denotes Wire Goes Through Bulkhead Disconnect
	Solenoid Switch	#19 STRG COLUMN	Denotes Wire Goes Through Steering Column Connector
	Mercury Switch	INST PANEL #14	Denotes Wire Goes Through Instrument Panel Connector
	Diode or Rectifier	ENG #7	Denotes Wire Goes Through Grommet to Engine Compartment
	Bi-Directional Zener Diobde		Denotes Wire Goes Through Grommet
	Motor		Heated Grid Elements
	Armature and Brushes		

Figure 6-19. *In order to use wiring schematics, you must understand the symbols in the schematics. This chart shows the electrical symbols as used by one manufacturer. (Chrysler)*

switches. Some switches are activated by engine or transmission operating conditions. Examples are oil pressure and coolant temperature switches.

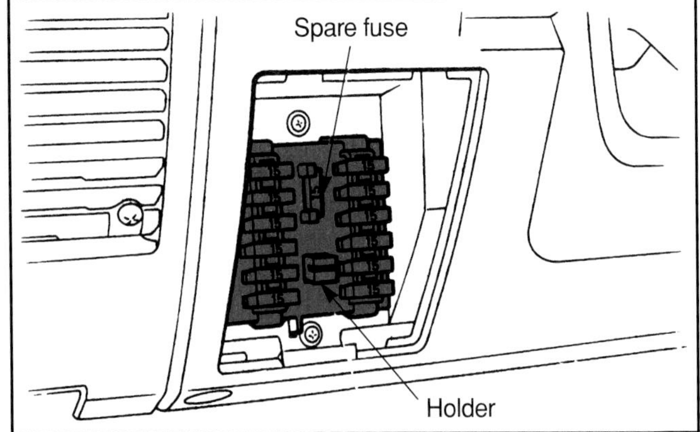

Figure 6-20. *The fuse box contains the circuit protection devices for many of the vehicle's electrical circuits. Most modern vehicles have more than one fuse block. (Ford)*

Figure 6-21. *Junction blocks contain high amperage fuses, circuit breakers, and relays used by systems close to the block's location. This junction block is located in the vehicle's engine compartment.*

Solenoids and Relays

Relays and solenoids are electromagnetic control devices. Electricity creates a magnetic field, which causes movement of a metal part. In a ***relay,*** the magnetic field closes one or more sets of electrical contacts, causing electrical flow in a circuit. This is useful when switching high-current devices, such as fan motors or resistance heaters, without excessive lengths of heavy wire. In a ***solenoid,*** the magnetic field performs a mechanical task, such as opening or closing a valve or moving drive linkage. Note that most solenoids not mounted on the starter are actually relays, since they control electrical flow. For convenience, they are referred to as solenoids, **Figure 6-22.**

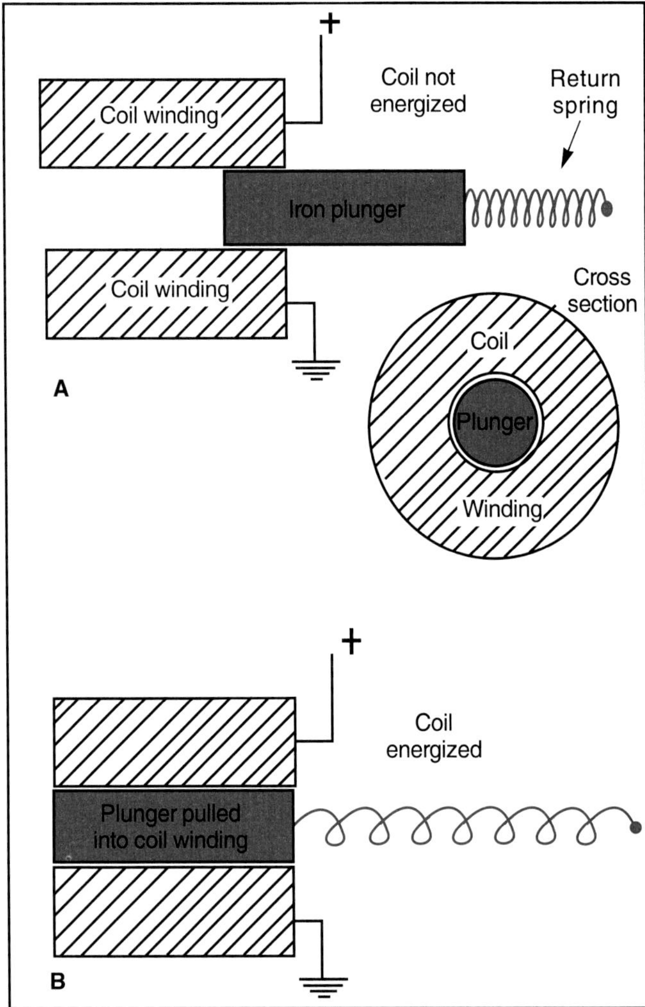

Figure 6-22. *A—A coil or wire and a iron core can be used to perform many functions. When the coil is not powered, a spring pulls the core back out of the coil. B—When the coil is powered, it pulls the core in. Attaching the core to a lever or plunger allows the coil and core to perform work. This is the operating principle behind the solenoid.*

Motors

To turn electricity into rotation, ***motors*** are needed. The most commonly used motors consist of a central ***armature*** made of many loops of wire. Surrounding the armature are ***field coils.*** The field coils produce a magnetic field while current flows through the armature windings. By controlling the direction of the current flowing through the armature windings, the armature can be made to turn inside of the field windings. Current direction is usually controlled by the use of a commutator and brushes. Refer to **Figure 6-23.**

Electric motors are found throughout the vehicle, including the starter motor, heater blower motor, windshield wiper motor, power window, seat, and antenna motors, and some small motors which control the ABS (anti-lock brake) system and some air conditioner duct doors. Many motors,

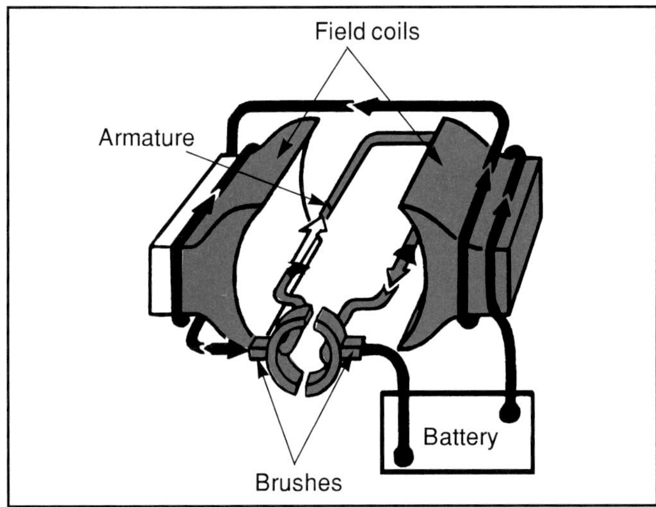

Figure 6-23. *In a motor, power is fed through the brushes to the armature. At the same time power is also fed through the field coils. When the field coils are energized, it causes the armature to turn.*

such as those used in cooling fans, rotate whenever they receive current. Other motors are designed to rotate an exact amount to precisely position a part, These are called stepper motors or servos.

Resistors and Resistance Heaters

Resistors are placed in a circuit to reduce current flow. They are made of carbon or various metals that create resistance to the flow of electrons. Resistors are commonly used to reduce voltage, motor speeds, and to protect other circuit components.

Resistors become hot, and this heat must be eliminated to keep it from destroying the resistor. Most resistors have some provision for absorbing heat. Sometimes the resistance heat is put to work in ***resistance heaters.*** Uses for resistance heaters include opening circuit breakers, opening electric chokes, and removing frost from windows and mirrors.

Capacitors

To damp out voltage fluctuations and control electronic frequencies, ***capacitors*** are often used. The capacitor serves as a trap for voltage surges (sometimes called spikes) by attracting excess electrons. Capacitors consist of a ***dielectric*** (non-conducting) material between a positive and a negative pole. The dielectric material absorbs electrons by becoming polarized, that is, the electrons in the dielectric material rearrange themselves into positive and negative areas similar to a magnet. This causes the capacitor to filter out voltage surges before they can effect radio and computer circuits. The condenser was used on older point-type ignitions to reduce point arcing and prolong point life. Almost all electrical devices that consume large amounts of current contain capacitors to reduce voltage surges.

Semiconductor Electronic Components

The technician who wishes to service modern vehicles must be familiar with electronic components and how they operate. The following section is a brief review of the basics of semiconductor electronic components.

Semiconductor Materials

All modern automotive electronics depend on the use of materials known as ***semiconductors.*** Semiconductors are made of silicon or germanium with a small amount of impurities to cause them to be either conductors or insulators, depending on how voltage is directed into them. Semiconductors change at the atomic level, depending on how their electrons are affected. Common semiconductors are diodes and transistors. These devices are small and can be combined into compact computers which perform the same number of calculations as a room full of older electronic devices.

Diodes

Diodes are semiconductors that allow current to flow in only one direction. Diodes are one-way check valves for electricity, becoming a conductor or an insulator, depending on which way the current tries to flow. Diodes are made up of two layers of semiconductor material, one positive and one negative, as shown in **Figure 6-24.** When current tries to flow in one direction, the area where the positive and negative materials meet becomes an insulator, and no current can flow. When current tries to flow in the other direction, the diode becomes a conductor and current begins to flow.

Diodes are used in alternators, where the alternating current produced by the alternator must be ***rectified***, or changed, to direct current to charge the battery. For this reason, diodes are sometimes called rectifiers. Note that diodes have two leads, one input and one output.

Zener Diode

Another type of diode is the ***Zener diode,*** which will not allow current to flow until a certain voltage is reached. When the triggering voltage is reached, the diode will become a conductor, allowing current to pass, **Figure 6-25.** Zener diodes are often used in electronic voltage regulators. The Zener diode is placed in the circuit so that current tries to flow through in the direction that is normally blocked. As long as the charging voltage is in the normal range, the diode keeps current from flowing. If charging voltage becomes too high, the Zener diode will open, allowing current to flow into other voltage regulator circuits which reduce alternator output.

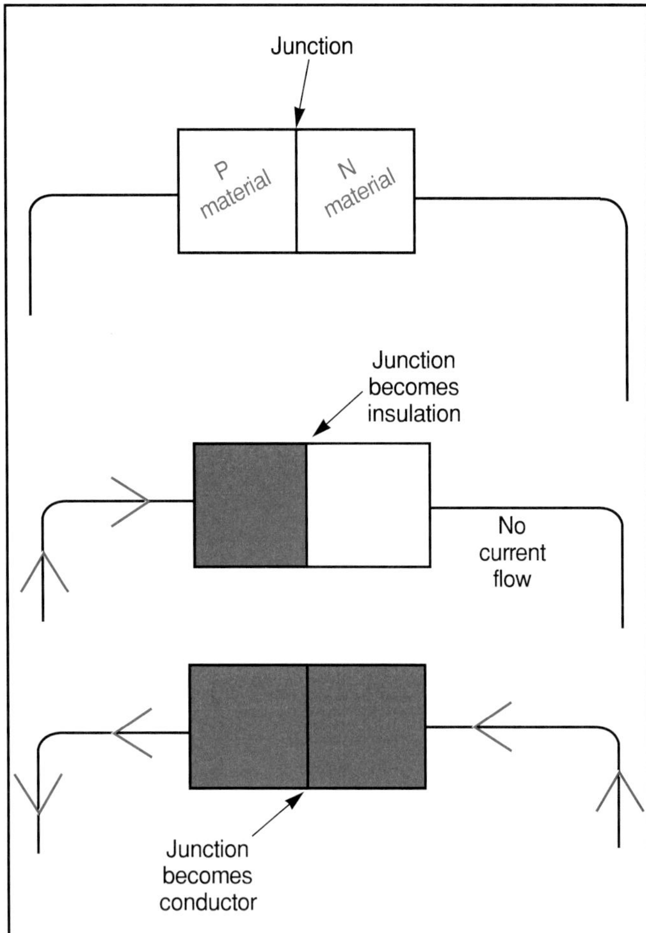

Figure 6-24. *The semiconductor material in a diode allows power flow in one direction but not the other. Typical automotive applications include alternators, air conditioning compressors, and anti-lock brake systems.*

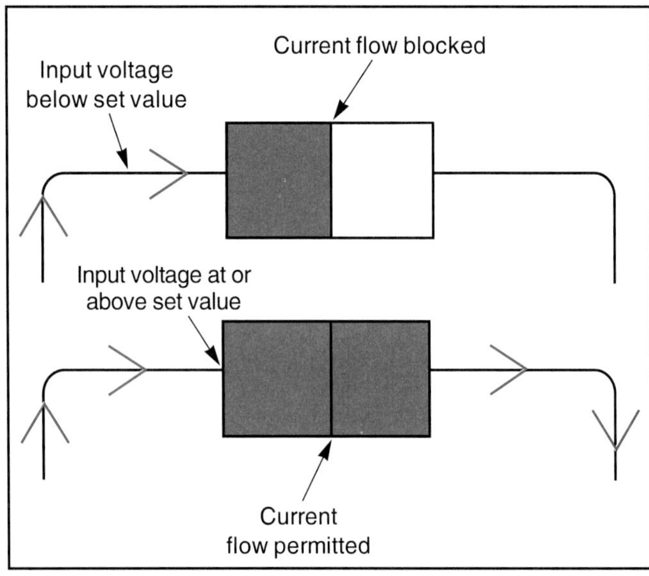

Figure 6-25. *The Zener diode is unique in that it uses voltage to vary current flow.*

Two related types of diodes are the ***light emitting diode***, or ***LED,*** and the ***photo diode.*** The LED produces light when current passes through it. The photo diode can absorb light and convert into an electrical signal. The uses of these two types of diodes will be discussed in later chapters.

Other kinds of diodes, called ***thyristors*** or ***silicon controlled rectifiers,*** are composed of two diodes connected together to form a unit with three junctions. Current flow through thyristors can be controlled by an external source. Thyristors are often used in situations where heavy, varying current and voltage must be controlled. Thyristors, unlike other diodes, have three leads.

Transistors

A ***transistor*** is a semiconductor device that is used as a switch or amplifier. A transistor can carry heavy current, but is operated by very low current. Transistors are comprised of either two outer layers of positive or negative semiconductor material with a center section of negative or positive material, depending on the application. The three materials are fused together, as shown in **Figure 6-26.** The transistor can be thought of as two diodes, with a common center section. Transistors always have three terminals.

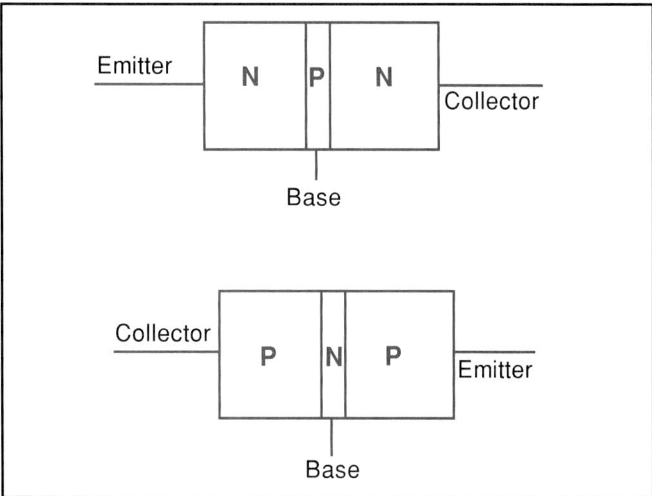

Figure 6-26. *Transistors are used in the same manner as a mechanical switch or relay. Depending on the application, its composition can differ.*

When the transistor is wired as shown in **Figure 6-27,** the battery current can flow into the transistor through terminals. The majority of the current enters at the collector terminal. A very small amount of current enters at the base terminal. When current flows through the base, the transistor becomes a conductor, and the much heavier collector current can pass through the collector circuit. All of the current leaves through the emitter circuit.

When the base circuit is broken, the transistor becomes an insulator and collector current instantly stops. A transistor with no moving parts is much more reliable than a set of mechanical points, or a vacuum tube, and transistors have largely replaced these older components, **Figure 6-28.**

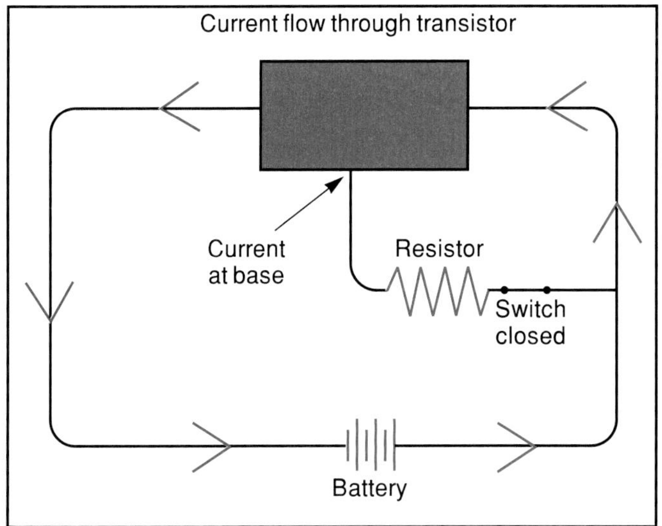

Figure 6-27. *When voltage is applied to a transistor's base, it closes the "switch" (transistor), allowing current to flow through the circuit.*

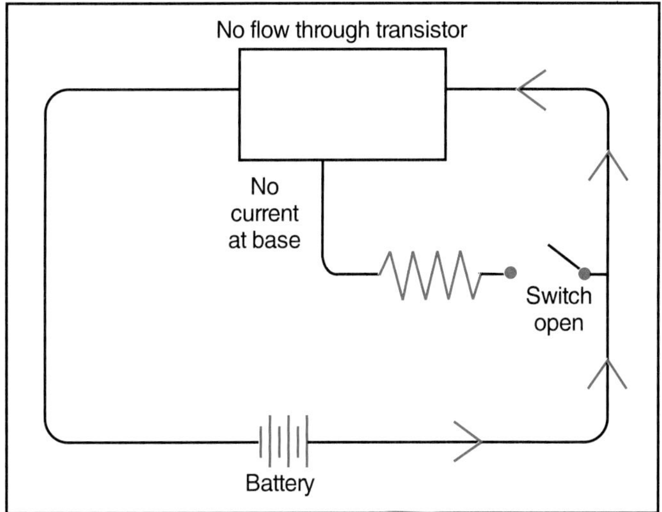

Figure 6-28. *If current is removed from the transistor's base, the "switch" (transistor) closes, or becomes an insulator, preventing current flow.*

Many modern transistors are not separate units, but are part of integrated circuits, discussed in the following sections. Transistors used in integrated circuits are too small to be seen by the naked eye. There are many variations of the basic transistor, such as the field-effect transistor. Large transistors designed to carry heavy current loads are called ***power transistors***. Power transistors, such as the one shown in **Figure 6-29,** produce large amounts of heat, and are usually placed away from other electronic components. Power transistors may have external fins to dissipate heat.

Figure 6-29. *Power transistors have uses in automotive applications, including lighting modules, amplifiers, and ignition modules, as shown here.*

Other Electronic Components

The following electronic components, although not semiconductors, are widely used in modern electronic devices. Their resistance to heat and vibration, as well as their small size, have made them a vital part of modern automotive electronics.

Chip Capacitors

Older capacitors were relatively large can-shaped devices, such as the condenser used on older point-type ignitions. The modern ***chip capacitor*** is made from an element called tantalum (Ta). Tantalum is combined with other materials to create a capacitor that does the same job as older capacitors, but is the size of a matchhead. Chip capacitors are soldered, or sometimes glued onto the circuit boards.

Quartz Crystals

Every modern calculator, clock, and computer relies on a ***quartz crystal*** as a timing device. Quartz is a naturally occurring mineral which can be made to vibrate at a certain frequency. These vibrations will not vary and can be modified by other electronic components to produce a timing signal which can be used to keep time or to monitor the elapsed times of other electronic processes. Other electronic devices can be arranged to create a timing circuit, but the quartz crystal is usually the simplest and most reliable.

Integrated Circuits

Many diodes and transistors can be combined onto a large complex electronic circuit by etching the circuitry on a small piece of semiconductor material. A circuit made in this manner is called an ***integrated circuit (IC).*** The IC also contains resistors, capacitors, and other electronic devices. Modern ICs, sometimes called ***chips,*** can control many vehicle functions formerly done by mechanical devices, such as alternator output.

The arrangement of ICs into a device which can make decisions based on input is often called a ***logic circuit.*** The modern computer contains many ICs, combined into a series of logic circuits that control engine functions, transmission shifting, anti-lock brake operation, air bag operation, suspension firmness, and traction control operation, **Figure 6-30.**

Static Electricity

While most electronic components used in vehicles are designed to withstand heat and some physical shock and abuse, other components, especially computer circuits, are sensitive to damage by ***static electricity.*** A technician can become charged with static electricity simply by sliding across a seat, which is a common occurrence when servicing most vehicles.

If you become charged with static electricity and touch a computer or other static sensitive component, the charge will arc to ground through the component, destroying it in the process. Fortunately, static electricity can be discharged simply by grounding yourself on the vehicle chassis or any metal part connected to the vehicle chassis before handling any static sensitive component. Most of these components and schematics that contain static sensitive components are marked, **Figure 6-31.**

Electromagnetic Interference

With the ever-increasing number of wires and circuits in the modern vehicle, the chances for problems increase. A problem that has existed in vehicles since the introduction of electronic ignition systems and has gone relatively unnoticed for the most part is starting to occur more frequently in newer vehicles. This problem is electrical malfunctions caused by ***electromagnetic interference (EMI).***

Electromagnetic interference is usually caused by voltage spikes and stray magnetic fields created by defects in such circuits as the charging system. Devices such as police and CB radios, cellular telephones, and other electronic accessories can also generate EMI during normal operation. In the past, electromagnetic interference problems were limited to background noise that was sometimes heard through the radio speakers. In the modern vehicle, EMI can create false signals to the ECM, resulting in sensor misreadings, unusual output device operation, and erroneous diagnostic codes, **Figure 6-32.**

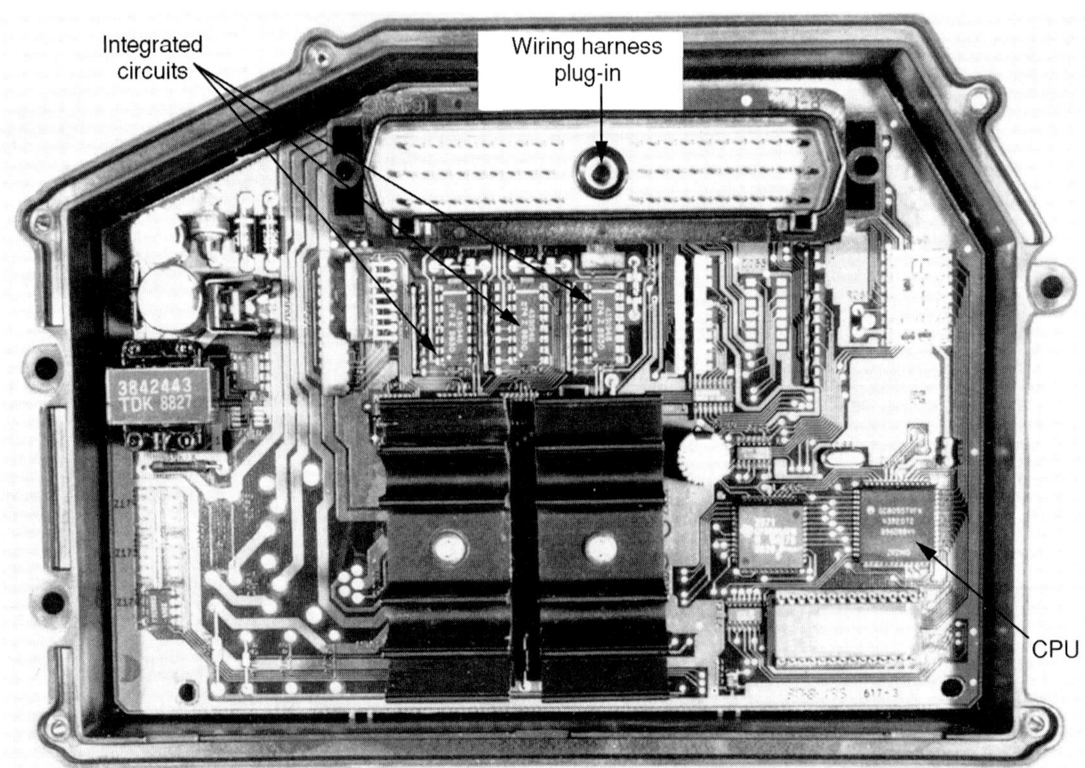

Figure 6-30. *The ECM is made up of many integrated circuits mounted on a printed circuit board. The CPU is also an integrated circuit, but is more complex. (Chrysler)*

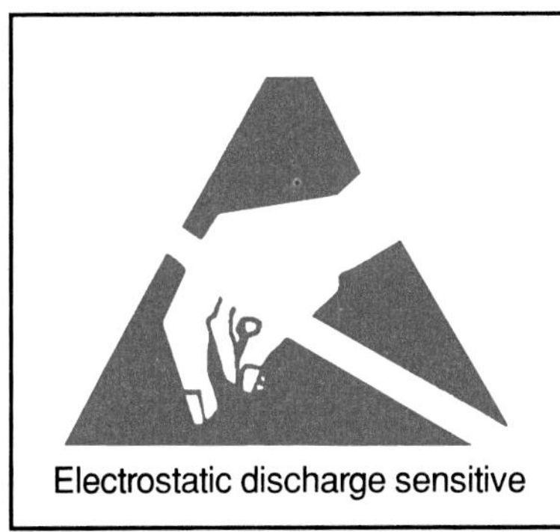

Figure 6-31. *The electrostatic discharge sensitive (EDS) symbol is placed on the packaging, in the service literature, and sometimes on the part itself. This symbol notifies the technician that static discharge procedures should be followed. (General Motors)*

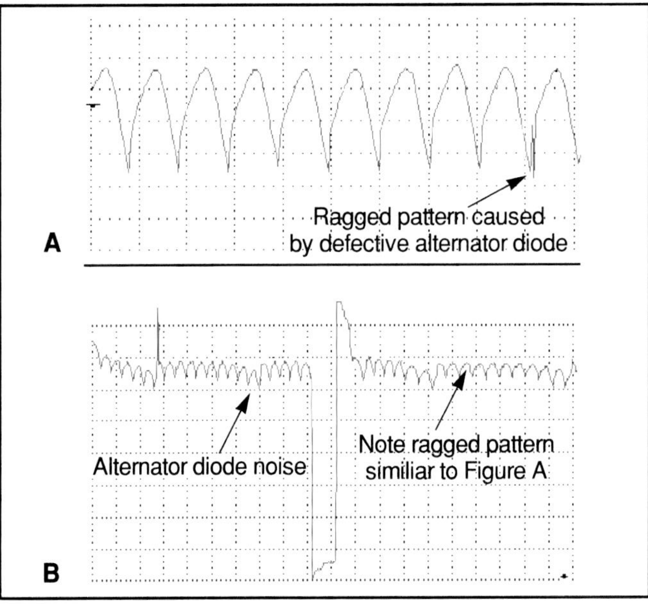

Figure 6-32. *Electromagnetic interference caused by defective or improperly installed components can cause minor and major problems in many vehicle systems. A—This wave is caused by a defective alternator diode. B—This waveform is the fuel injector pulse pattern. Note the unusual hump wave pattern that is similar to the alternator wave pattern; this is caused by electromagnetic interference. (IATN)*

Vehicle Starting and Charging Systems

The vehicle starting and charging systems are vital to the basic operation of the vehicle. They can also affect driveability. The starting and charging system can cause problems that affect engine operation, or even prevent the engine from starting. The battery, starter, and associated cables and controls provide the means to start the engine. These systems can cause starting problems, but are usually not related to driveability problems other than no-crank conditions. The following sections are a brief overview of battery, starter, and charging system operation.

Battery

The ***battery*** creates the energy required to operate the starter and ignition system during cranking, **Figure 6-33.** The battery is a set of positive and negative plates made of lead compounds, surrounded by a liquid solution of sulfuric acid and water called ***electrolyte.*** The battery creates electricity by chemical action. The battery provides current to the starter during cranking and powers lights and other electrical components when the engine is not running. The battery is usually mounted on the side of the engine compartment.

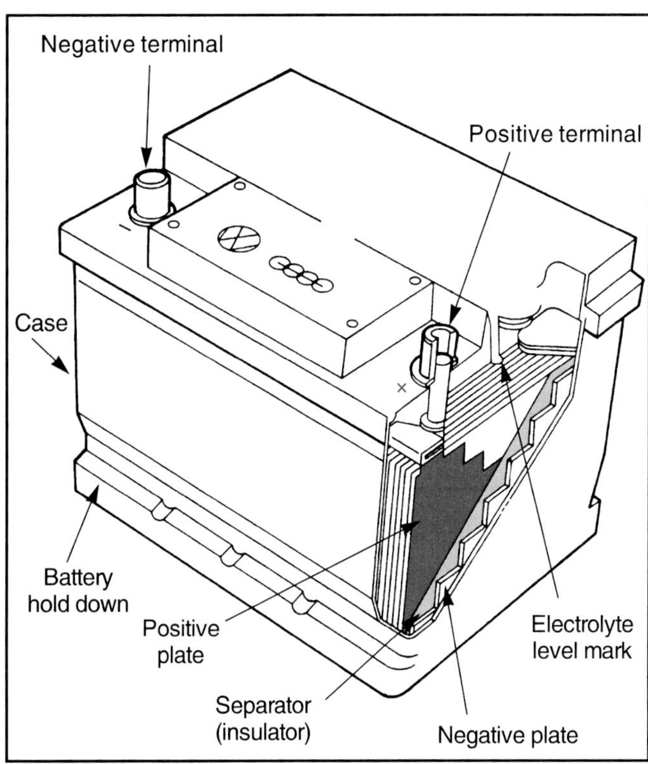

Figure 6-33. *Batteries come in many shapes and sizes, but their internal composition is quite similar. This cutaway shows the cell arrangement for a typical automotive battery. (Audi)*

Another function of the battery is to absorb extra current and damp out some of the voltage fluctuations, or spikes, that are created by other parts of the electrical system. This helps to protect the electronic components installed in the vehicle. Since it has separate positive and negative plates, the battery can operate only on direct current.

Battery Charging

Since the battery provides current to the starter during cranking, and powers the vehicle lights and other electrical components, it will lose all energy if it is not recharged when the engine is running. Due to its construction with separate positive and negative plates, the battery can be recharged only by direct current. The battery can also be recharged by an external battery charger.

Starter Circuit

The ***starting system*** turns the engine at a sufficient speed for start-up. The ***starter*** is a small, high powered electric motor designed for short bursts of operation. A typical starter can produce as much power as larger electric motors, but can only be operated for short periods without overheating. This is why most engines should not be cranked for more than 30 seconds at one time. If the starter is cranked for 30 seconds, it should be allowed to cool off for at least three minutes before cranking the engine again. The starter is installed on the engine in a position that allows it to turn the ring gear attached to the engine crankshaft, **Figure 6-34.**

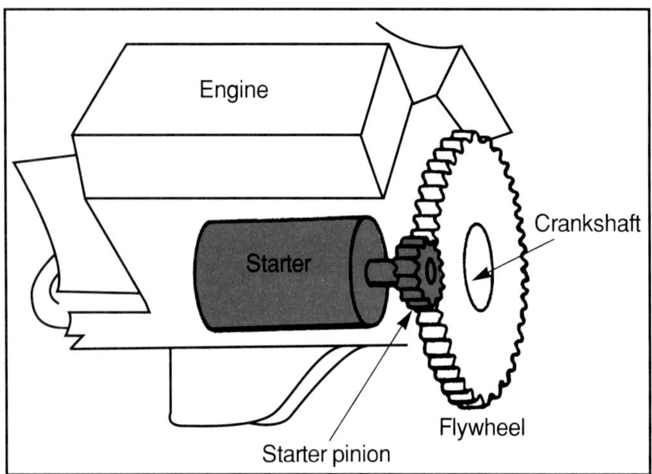

Figure 6-34. *In all cases, no matter what design is used, the starter's job is to engage a ring on the flywheel to start the engine.*

The ***starter solenoid*** may be mounted on the starter, or on the engine compartment sheet metal. The starter solenoid is energized by current flow through the ignition switch and a ***safety switch.*** The ignition switch and the other starting system components are mounted on the vehicle chassis.

Turning the ignition key completes the electrical circuit through the neutral safety switch. The automatic transmission/transaxle safety switch prevents starting in any gear except neutral or park, **Figure 6-35.** Safety switches on most manual transmission/transaxle vehicles prevent starting with the clutch engaged. When the starter solenoid is energized, the starting circuit is closed, and electrons begin to flow from the battery. This flow causes a chemical reaction inside the battery, that produces more electrons. The electrons flow through the battery cables to the starter solenoid and into the starter. The flow of electricity through the starter creates a magnetic field that causes the starter armature to turn.

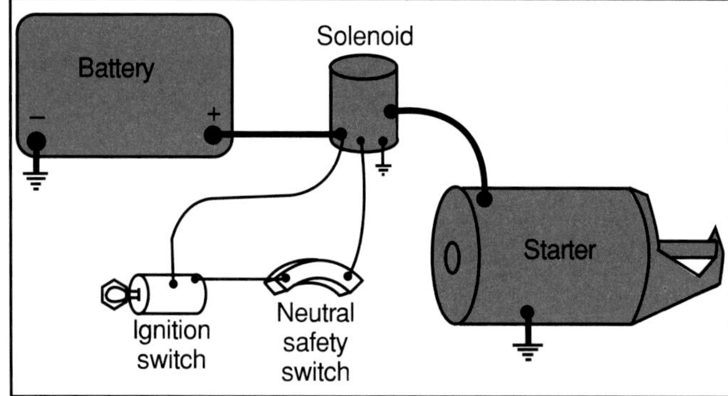

Figure 6-35. *The neutral safety switch prevents engine start-up while the transmission is in gear. Vehicles with manual transmissions use a similar switch connected to the clutch pedal.*

The ***drive gear*** on the end of the starter armature engages the ***ring gear*** on the engine crankshaft. The armature turns the drive gear and, therefore, the ring gear and crankshaft. Some starters have a reduction gear that increases starter torque by reducing armature speed before engaging the ring gear. During cranking, current also flows to the ignition system to provide high voltage to the spark plugs.

Starter Drives

The starter must be engaged with the engine flywheel for cranking, and disengaged once the engine starts. To engage the modern starter with the flywheel, two kinds of ***starter drives*** are used. The most common type of drive on modern vehicles makes use of the starter solenoid. The solenoid is mounted on the starter, and is connected to the starter drive gear by linkage, **Figure 6-36.** Starters with this type of drive are usually called ***linkage drive starters.*** The drive gear, or pinion, is able to slide on the starter armature shaft. Matching grooves on the shaft and pinion gear allow the pinion gear to slide while still transmitting power from the armature. The solenoid contains an internal return spring that keeps it in the released position when the starter is not being operated.

When the driver turns the ignition key, the solenoid is energized. As the solenoid pulls in to engage the electrical contacts, it also moves the linkage, which slides the pinion gear into engagement with the flywheel ring gear. When the engine starts, and the driver releases the ignition key, the solenoid is de-energized. This allows the solenoid return spring to pull the pinion gear out of engagement with the flywheel ring gear, as well as releasing the electrical contacts.

The other type of starter drive uses armature rotation to engage the pinion gear with the ring gear. This type of starter is often called an ***inertia starter*** or Bendix drive starter. Like the linkage drive starter, the inertia starter drive gear is able to move on grooves cut in the armature shaft and inside the gear. These grooves are in the form of a spiral, so that the drive pinion rotates slightly as it turns. A light spring keeps the pinion in the disengaged position when the starter is not being operated. **Figure 6-37** shows this type of starter.

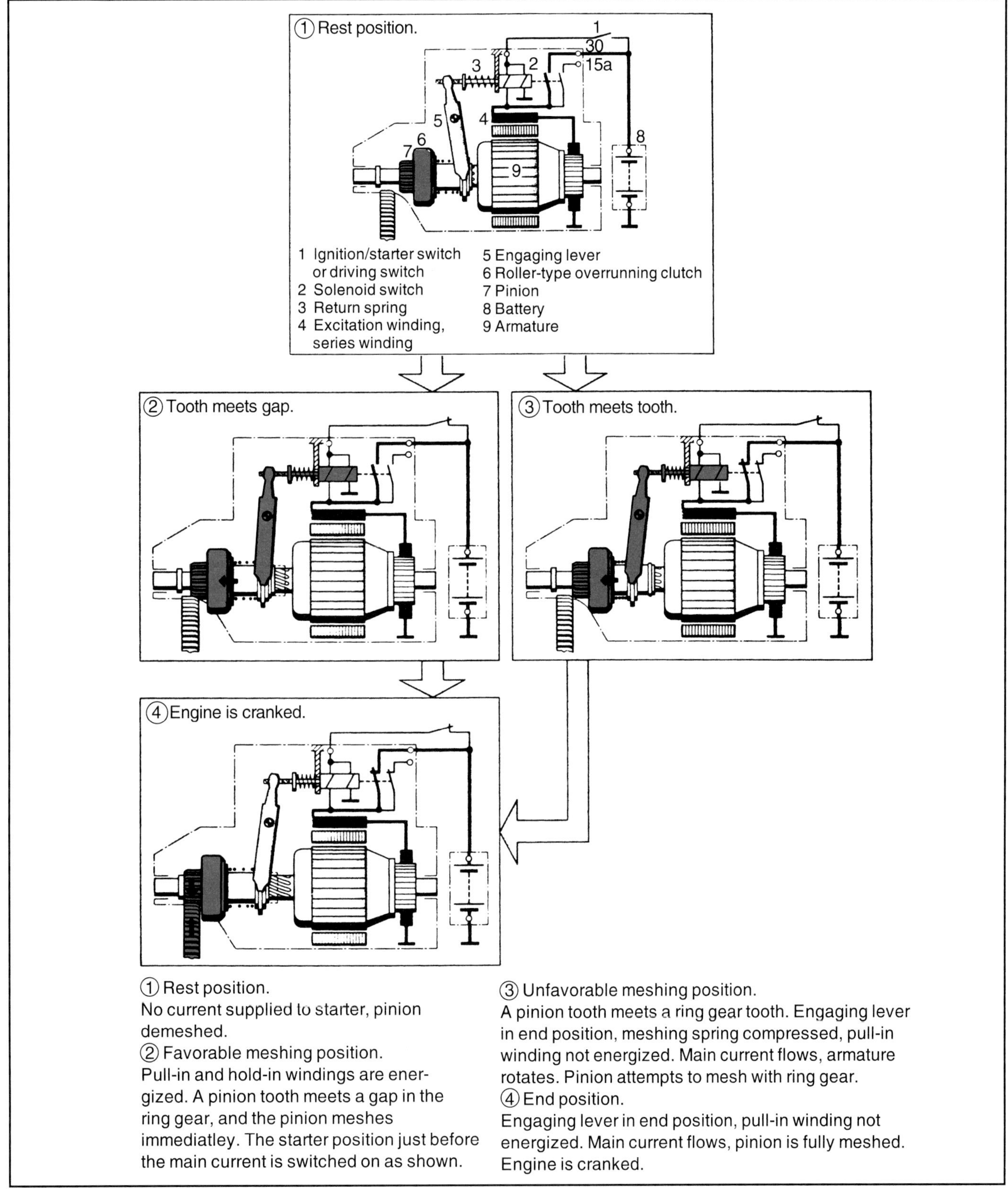

Figure 6-36. *This cutaway shows the operation of a linkage drive starter. The linkage is operated by the starter solenoid mounted on the motor. (Bosch)*

When the starter is energized, the armature begins spinning. The pinion gear, instead of spinning with the armature, tries to remain stationary. The spiral shape of the grooves causes the pinion gear to slide forward until it contacts the ring gear. Once the pinion and ring gear teeth are engaged, the pinion turns the ring gear and cranks the engine. After the engine starts, the ring gear turns faster than the pinion gear. The rotational speed pushes the pinion gear back along the spiral grooves in the armature shaft, and out of engagement.

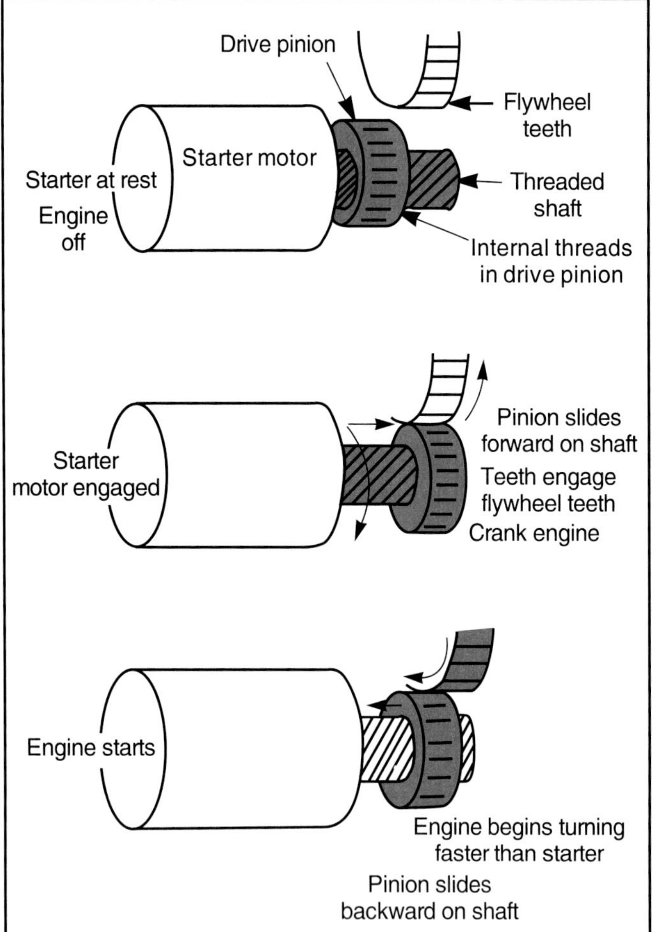

Figure 6-37. *The inertia drive starter uses the motor's rotational speed to force the drive pinion into engagement with the flywheel ring gear. In most cases, this type of starter uses a remote mounted starter solenoid.*

Charging System

The job of the ***charging system*** is to produce electricity to recharge the battery and to provide extra electricity to operate the ignition system and other electrical equipment when the engine is running. The charging system is comprised of the battery, the alternator, and the voltage regulator.

The charging system can affect the operation of the rest of the vehicle. Undercharging will allow the battery to become weak, making the vehicle hard to start. Overcharging can damage the battery and other vehicle components. Overcharging or undercharging can affect the operation of the ECM by creating voltages that are beyond the normal range of ECM operation.

Alternator

The ***alternator*** is an electrical device that changes rotational movement into electricity by using magnetism. The alternator consists of a set of rotating windings, called a ***rotor,*** and a set of stationary windings, called a ***stator.*** The alternator rotor is turned by the engine crankshaft through a drive belt and is magnetized by battery current delivered through slip rings and brushes, **Figure 6-38.**

The rotor's magnetic field has distinct North and South poles. As the rotor turns, the movement of the North and South poles produces a magnetic field in the stator windings. This magnetic field creates alternating (back and forth) current in the stator. Power to turn the alternator comes from the engine through the alternator drive belt. Some engine power is required to turn the alternator rotor, but the amount is small, usually less than one horsepower for a 60 amp alternator and about two horsepower for a 100 amp alternator.

Since the battery can only be recharged with direct current, diodes are installed in the alternator to change alternating current into direct current. The diodes are attached to the stator output, and arranged into a circuit that causes current to flow in only one direction, changing alternating current into direct current, **Figure 6-39.**

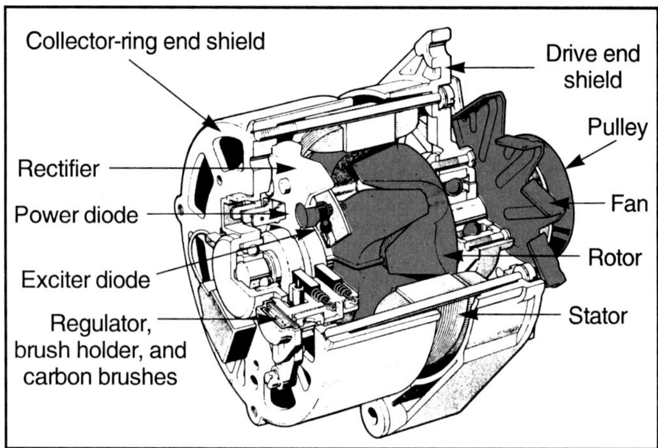

Figure 6-38. *This cutaway shows the internal components of a typical alternator. (Bosch)*

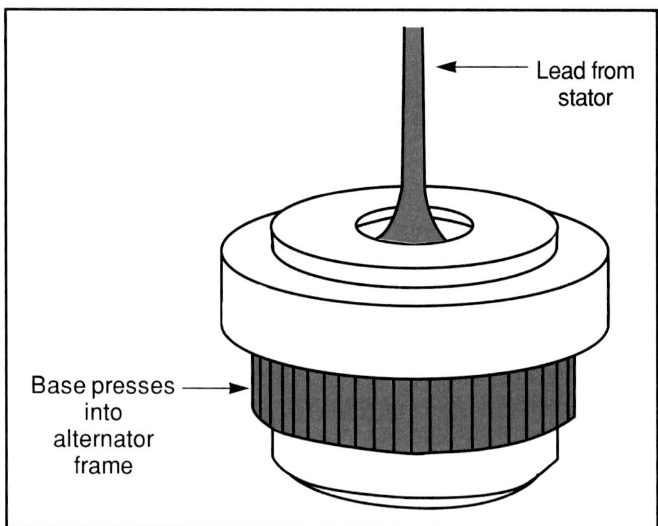

Figure 6-39. *Diodes are used inside the alternator to changer alternating current to direct current. Most alternators have 6-8 diodes.*

Voltage Regulator

The ***voltage regulator*** controls the alternator output by varying the strength of the rotor's magnetic field. The voltage regulator ensures that the battery is not over or undercharged, and that voltage does not become so high that it damages other vehicle components. It does this by controlling the size of the rotor magnetic field. Current flow into the rotor windings passes through the voltage regulator.

Older voltage regulators, such as the one illustrated in **Figure 6-40,** were electromechanical relays, making use of an electric coil and point set to control voltage. Modern regulators are electronic and voltage is sensed by a Zener diode, **Figure 6-41.** When voltage becomes too high, the Zener diode opens, allowing current to flow in other electronic components. These components reduce current flow to the rotor. Since the regulator components are small electronic devices, the modern regulator is often placed inside the alternator, or installed on the back.

Older alternators could often be repaired by replacing parts such as the bearings and brushes. Older regulators could be adjusted to keep voltage levels in the proper range. Many modern alternators, and almost all regulators, cannot be repaired and are replaced as a unit when they become defective.

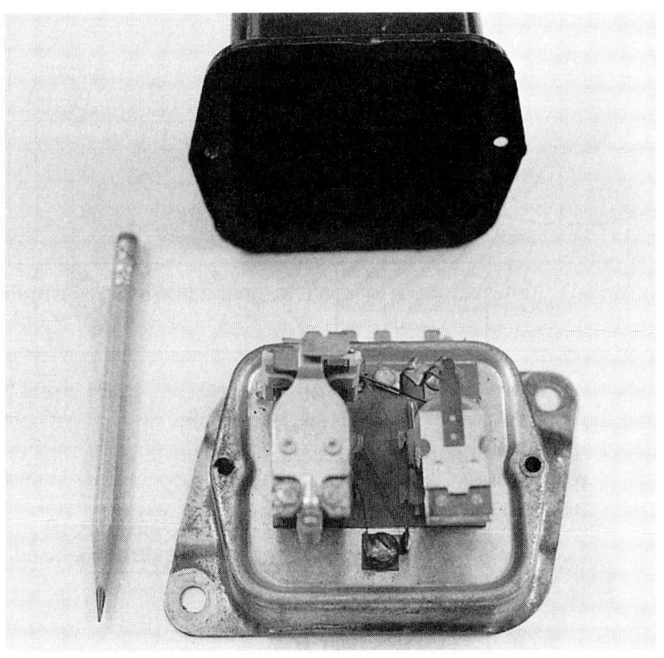

Figure 6-40. *For many years, electromechanical voltage regulators were used to control alternator output. The regulator contained coils that were affected by the alternator's voltage output.*

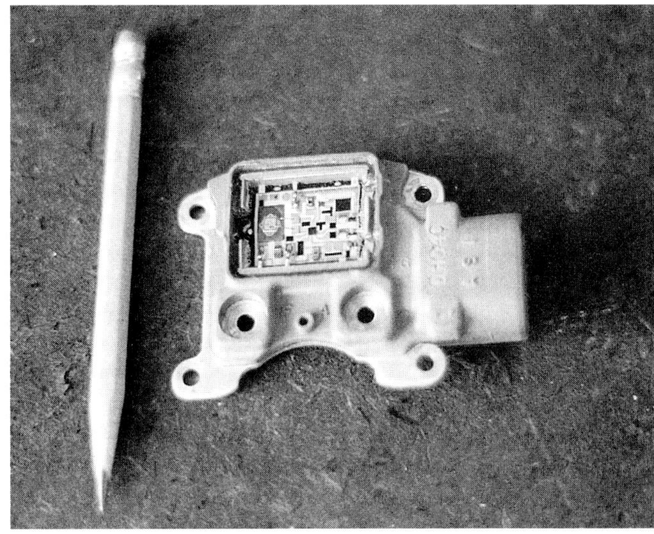

Figure 6-41. *Modern voltage regulators are electronic devices. This voltage regulator mounts on the rear of the alternator body. Other electronic voltage regulators are placed inside the alternator.*

Summary

Every atom has a positively charged center. Negatively charged electrons revolve around the center. Electricity is the movement of electrons from atom to atom. Materials whose atoms easily give up or receive electrons are called conductors. Materials whose atoms resist giving up or accepting atoms are called insulators. The path through which the electrons move is called a circuit. Electrons will not flow in a circuit unless there are more electrons in one place than another and a path between the two places exists. The three basic electrical properties are amperage, voltage, and resistance. Unknown electrical properties can be calculated using Ohms law.

The two types of current flow are alternating and direct. Most automotive electrical systems use direct current. The three types of automotive circuits are series, parallel, and series-parallel. Magnetism can be used to make electricity and electricity can be used to create magnetism. This relationship between electricity and magnetism is called electromagnetism.

Most automotive wiring can be copper, gold, or aluminum. The wire is plastic-coated, color-coded, and installed into harnesses. Wire gage is the rating system for wire diameter. Most electrical connectors are plug-in types. A connector with more than one wire is called a multiple connector. A schematic, or wiring diagram, is a drawing of electrical units and connecting wires. Schematics allow the technician to trace out defective components in the wiring system.

Circuit protection devices include fuses, fusible links, and circuit breakers. Switches control the flow of electricity through a circuit. Relays and solenoids are electromagnetic control devices. Relays control electrical flow, while solenoids cause physical movement. Motors turn electricity into rotation. Resistors are used to reduce current flow through circuits. Resistors produce heat which must be removed. Resistance heaters make use of this heat to open contacts or defrost glass and mirrors.

All modern automotive electronics depend on the use of materials known as semiconductors. Semiconductors can be either conductors or insulators, depending on how voltage is directed into them. Common semiconductors are diodes and transistors. Diodes allow current to flow in one direction only. A special type of diode is the Zener diode, which will allow current to flow in both directions when a certain voltage is reached. Transistors are used as switches or amplifiers. Transistors carry large amounts of current, but are operated by low currents. Transistors can perform switching operations with no moving parts. Chip capacitors perform the same jobs as older capacitors, but are much smaller. The vibrations of quartz crystals are used to create timing circuits. Transistors, diodes, capacitors, and resistors are etched onto small piece of semiconductor material to form an integrated circuit, or IC. ICs are also called chips.

The charging and starting system is composed of the battery, starter, and connecting cables. The battery changes chemical energy into electricity when the starter or other system begins operating. When current flows through the ignition switch and safety switch to the starter solenoid. The starter solenoid is energized and directs current to the starter motor. The starter begins turning and the drive gear engages the ring gear to turn the engine.

Once the engine begins running, it drives the alternator to recharge the battery. The alternator uses a spinning magnetic field to induce alternating current in the windings. Alternating current is converted to direct current by the diodes. Alternator output is controlled by controlling the alternator field strength. This is performed by the voltage regulator. Modern voltage regulators are electronic units using at least one Zener diode.

Know These Terms

Protons
Neutrons
Electrons
Conductors
Insulators
Flow
Circuit
Complete circuit
Current
Amperes
Voltage
Resistance
Ohms
Ohms law
Direct current (dc)
Alternating current (ac)
Series circuit
Parallel circuit
Series-parallel circuit
Short circuit
Grounded circuit
Open circuit
Magnetic
Polarity
Electromagnet
Induction
Electromagnetism
Wiring
Harnesses
Color-coded
Ground wire
Wire gage
American Wire Gage (AWG)
Plug-in connector
Schematic
Wiring diagram
Fiber optics
Circuit protection devices
Fuses
Fuse block
Circuit breakers
Fusible link
Terminal block
Junction block
Switch
Relay
Solenoid
Motors
Armature
Field coils
Resistors
Resistance heaters
Capacitors
Dielectric
Semiconductors
Diodes
Rectified
Zener diode
Light emitting diode (LED)
Photo diode
Thyristors
Silicon controlled rectifiers
Transistor
Power transistors
Chip capacitor
Quartz crystal
Integrated circuit (IC)
Chips
Logic circuit
Static electricity
Electromagnetic interference (EMI)
Battery
Electrolyte
Starting system
Starter
Starter solenoid
Safety switch
Drive gear
Ring gear
Starter drives
Linkage drive starter
Inertia starter
Charging system
Alternator
Rotor
Stator
Voltage regulator

Review Questions—Chapter 6

Please do not write in this text. Write your answers on a separate sheet of paper.

1. Glass and plastic are good ______, while copper and aluminum are good ______.
2. Briefly explain electromagnetic induction.
3. A material is a good conductor if it gives up ______ easily.
   (A) atoms
   (B) protons
   (C) electrons
   (D) neutrons

4. The number of electrons flowing past a point in a circuit determines the ______.
   (A) amperage
   (B) voltage
   (C) resistance
   (D) heating value
5. Capacitors are used to reduce voltage ______.
6. Wire gage is a way that wire ______ is rated.
   (A) length
   (B) material
   (C) insulation
   (D) diameter
7. As wire diameter decreases, its gage number ______ .
8. Power transistors are designed to control large amounts of ______.
   (A) voltage
   (B) current
   (C) resistance
   (D) heat
9. When a vehicle battery is discharging, the flow of electrons causes a chemical reaction which produces ______.
   (A) heat
   (B) resistance
   (C) oxygen
   (D) more electrons
10. The battery can absorb extra current, and therefore damp out some of the ______ that are created by the electrical system.
    (A) voltage fluctuations
    (B) voltage spikes
    (C) parasitic draws
    (D) Both A & B.
11. Since it has separate positive and negative plates, the battery can operate only on ______ current.
12. If a starter is cranked for 30 seconds, it should be allowed to cool off for at least ______ minutes before cranking the engine again.
    (A) 2
    (B) 3
    (C) 5
    (D) 10
13. The starter drives of many modern starters are operated by the starter ______.
14. The rotating part of an alternator is called the:
    (A) rotor.
    (B) stator.
    (C) armature.
    (D) diodes.
15. Alternating current from the alternator is changed into direct current by the:
    (A) rotor.
    (B) stator.
    (C) armature.
    (D) diodes.

ASE Certification-Type Questions

1. All of the following statements about atoms are true, EXCEPT:
   (A) the center of an atom is made up of protons and neutrons.
   (B) the neutrons have a positive charge.
   (C) electrons revolve around the nucleus.
   (D) the electrons have a negative charge.
2. Technician A says that current will only flow through a complete circuit. Technician B says that current will flow only if there is resistance in the circuit. Who is right?
   (A) A only.
   (B) B only.
   (C) Both A & B.
   (D) Neither A nor B.
3. When voltage and another electrical value are known, voltage is always ______ the other value to get the unknown value.
   (A) multiplied by
   (B) divided by
   (C) added to
   (D) subtracted from
4. Technician A says that direct current flows in only one direction. Technician B says that alternating current can be used to charge a battery. Who is right?
   (A) A only.
   (B) B only.
   (C) Both A & B.
   (D) Neither A nor B.
5. All of the following statements about magnetism are true, EXCEPT:
   (A) electricity flowing in a wire can produce a magnetic field.
   (B) when a magnetic field moves through a wire, electricity is produced.
   (C) a coil creates a weaker magnetic field than a single wire.
   (D) the iron core inside of a wire coil is used to increase field strength.

6. All of the current flows through every part of a(n) ______ circuit.
   (A) series
   (B) parallel
   (C) series-parallel
   (D) open
7. Circuit protection devices include all of the following, EXCEPT:
   (A) fuses.
   (B) circuit breakers.
   (C) resistors.
   (D) fusible links.
8. All of the following statements about resistors are true, EXCEPT:
   (A) resistors reduce current flow in circuits.
   (B) resistors produce excess heat.
   (C) resistors can be used to increase motor speeds.
   (D) resistors can be used to open circuit breakers.
9. Technician A says that a Zener diode allows current to flow in one direction at all times. Technician B says that a Zener diode allows current to flow in both directions when a certain voltage level is reached. Who is right?
   (A) A only.
   (B) B only.
   (C) Both A & B.
   (D) Neither A nor B.
10. Diodes are used to change alternator output into ______.
    (A) magnetism
    (B) alternating current
    (C) direct current
    (D) motion

Computer control is used to keep this V-12 engine and automatic transmission assembly operating as a unit. (BMW)

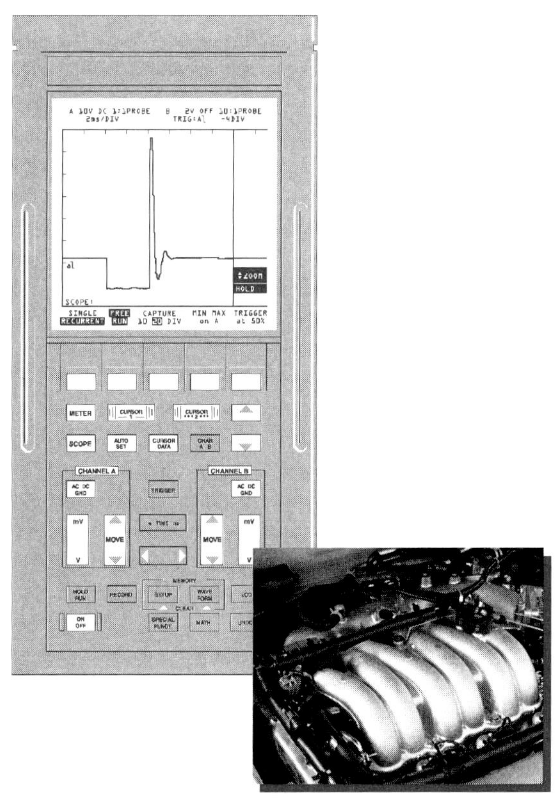

Chapter 7

Computer Control Fundamentals

After studying this chapter, you will be able to:

- ❑ Explain how efficiency is obtained by electronic engine controls.
- ❑ Explain the operation of the electronic control module.
- ❑ Identify the major internal components of the electronic control module.
- ❑ Identify and explain the operation of major input sensors.
- ❑ Identify and explain the operation of major output devices.
- ❑ Explain the basic control loop of the computer control system.
- ❑ Explain the major differences between OBD I and OBD II systems.

The typical automobile of 30 years ago averaged about 10-15 miles per gallon of gas, while modern vehicles average 25 mpg or more, with less pollution and better performance. Some techniques have been in existence for many years, and are now being reused on modern vehicles. However, many of these improvements involve the latest electronic devices. This chapter will discuss on-board computers and how they monitor and control the various engine, drive train, and vehicle systems. This chapter will not only provide you with a review of computer control systems, but also will prepare you for many of the subsequent chapters in this text.

Electronic Engine Control Systems

The most important advance in the areas of driveability, fuel economy, and performance is the development of the ***on-board computer*** to control the engine and drive train systems. The on-board computer system is usually called the ***electronic engine control system.*** These systems are comprised of electrical, electronic, and mechanical devices, that monitor and control engine operation. Electronic engine control systems always have a central computer that interacts with the other components. Some systems have a second computer which handles some of the electrical devices that draw heavy current. Other systems use the second computer to operate the fuel injection system or control automatic transmission operation.

Modern on-board computers control parts of the fuel, ignition, emissions, drive train, and accessory systems such as the air conditioner. A few of the latest computer control systems also operate parts of the valve train for maximum power. Although the engine control computer is often connected to other vehicle computers and receives information from them, it is primarily used to control engine and vehicle systems. Vehicle manufacturers often refer to computerized engine controls as emission control devices. While they do help to clean up the exhaust gases, they do so by improving overall engine efficiency. The computer and its related parts are constantly tuning all engine systems and components for optimum performance.

The basic components of the computer control system consists of three main parts: inputs, the computer, and outputs. The heart of this system is the computer. It receives inputs from components called ***input devices,*** or ***sensors,*** uses internal logic circuitry to decide on the actions to be taken, and sends commands to components called ***output devices.*** The outputs cause changes in engine and/or drive train operation, which are picked up as new readings by the input sensors. **Figure 7-1** illustrates this cycle.

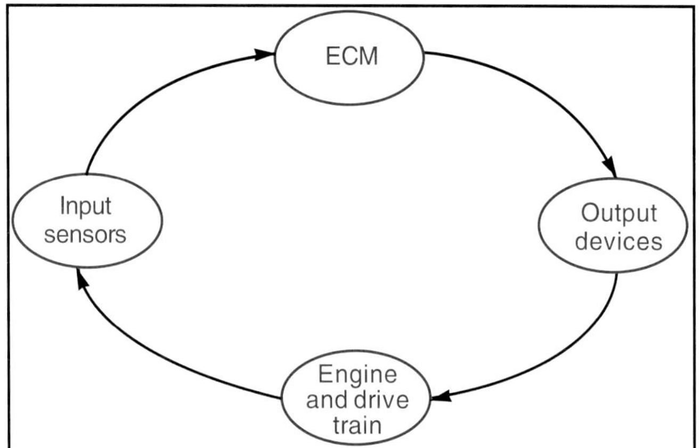

Figure 7-1. *A typical computer control system cycle or "loop." This loop cycle applies to all automotive computer control systems.*

The computer control system works in a continuous cycle or loop, with the inputs, computer, and outputs all affecting each other. The actual engine and drive train components being affected constitute a fourth part that completes the cycle.

Electronic Control Module

The ***electronic control module*** or ***ECM*** processes and interprets the electrical inputs from the input sensors through the use of tiny electronic components arranged so that they use logical processes. All information processing in a computer is a binary operation. This means that all information in a computer is a series of on-off signals, no matter how complex the information or signal. The manipulation of these signals is what operates the computer.

All ECM's are divided into several internal circuits that process the sensor inputs and issues commands to the appropriate output devices. A schematic of a simple ECM system is shown in **Figure 7-2.** Although each of the internal ECM circuits operates independently, they exchange information and share sensor inputs.

The engine control computer in various makes and models of vehicles is called by different names, according to individual manufacturer preferences. Some of these names include:

- On-board computer
- Microprocessor
- Controller
- Engine Control Module (ECM)
- Electronic Control Unit (ECU)
- Electronic Engine Control (EEC)
- Powertrain Control Module (PCM)
- Vehicle Control Module (VCM)

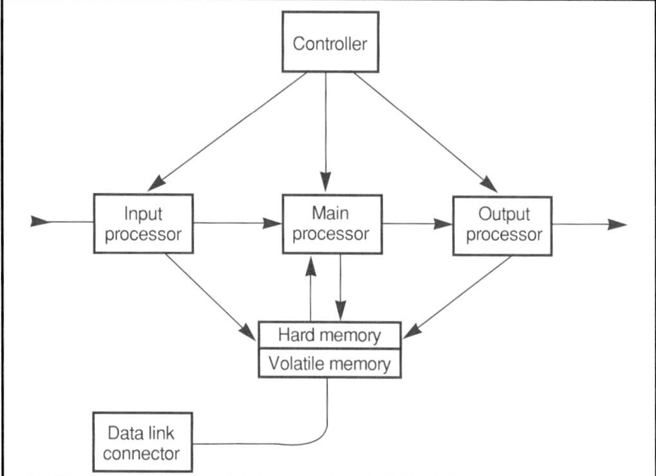

Figure 7-2. *The general relationship of computer control system components. All vehicle computer control systems operate in this manner. The types and number of sensors and output devices varies.*

The term ***Powertrain Control Module,*** or ***PCM,*** is beginning to see greater use by many manufacturers. In many newer vehicles, these computers are interconnected and exchange information. Since there are over two dozen motor vehicle manufacturers, the number of names for the computer can cause confusion. In this book, we will always call the engine control computer the *ECM* for *Electronic Control Module.* Some manufacturers use the term ECM to describe their engine control computers. Calling engine control computers ECMs makes it easier to separate them from other vehicle computers and the computers used in diagnostic equipment. In addition, electronic control module is an accurate functional description of all automotive computers.

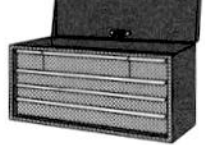

Note: The computers which control engine operation on most vehicles should not be confused with those that operate the anti-lock brake system (ABS), cruise control, air bags, anti-theft system, climate control system, or suspension.

Vehicle Control Module (VCM)

The latest version of engine control computer is the ***VCM.*** This acronym is an abbreviation for ***Vehicle Control Module.*** A vehicle control module differs from other on-board computers in that it also monitors and sometimes controls other vehicle systems besides the engine and transmission, such as the anti-lock brakes.

An advantage of the VCM is the elimination of other computers from the vehicle, which reduces weight in the form of wiring and other hardware. Other advantages includes centralized control of many vehicle systems, reducing the possibility of crossed signals and interference as well as increased capability, since a more powerful computer is needed to control all of these systems.

ECM Communications

Signals from the input sensors and to the output devices are analog signals or signals composed of variable voltage levels. The analog signals to and from the sensors vary from millivolts to battery voltage, depending on the system, and are usually referred to as ***reference voltages.*** Computers are digital devices, working off of a series of on and off signals that operate a binary computer. This information is referred to as ***serial data*** or the *datastream.*

The ECM operates output devices by turning them on and off. Output device operation is controlled by varying the length of time the device is operated, either by applying voltage or supplying ground. This length of time that the signal or ground is provided is called ***pulse width,*** **Figure 7-3.** The ECM also provides external communications to other on-board computers and diagnostic equipment, which is discussed in the next section.

Multiplexing

As explained earlier, the ECMs in some vehicles communicate with each other. They transfer information via a path called a ***data bus.*** This data bus is simply a roadway for data transfer between the ECMs and other control modules. **Figure 7-4** shows such a data bus network for one vehicle. By using a few wires to create the bus, weight is reduced by eliminating the need for separate wires to each ECM. This network is referred to as ***multiplexing.***

Multiplexing is the use of a single wire or pathway to transmit two or more messages at the same time. The same wire can be used to transmit any number of messages if the complete circuit for each message is separate. The simplest type of multiplexing is a battery ground cable which carries the electrical impulses of the ignition, headlight, and radio circuits while keeping each separate. Multiplexing was first used to make maximum use of undersea telephone cables. It has been used in sound systems and other applications for many years.

To prevent the information from overlapping, each signal has an identification code. The circuits inside each ECM will read the code before processing the signal. If the signal does not match the ECM's identification code, it is not processed. This identification code can also be used to prioritize signals. If two ECMs attempt to send a signal to a third ECM, the signal with the highest priority code is processed first.

In the future, fiber optics may be used to create the vehicle data bus. Since fiber optics have the ability to transmit multiple signals simultaneously, information can be transmitted, received, and processed at a much faster rate.

ECM External Communications

There are two types of external serial communication used in automotive computer systems. Most modern vehicles use a system called ***universal asynchronous receive and transmit (UART).*** This signal is used for communication

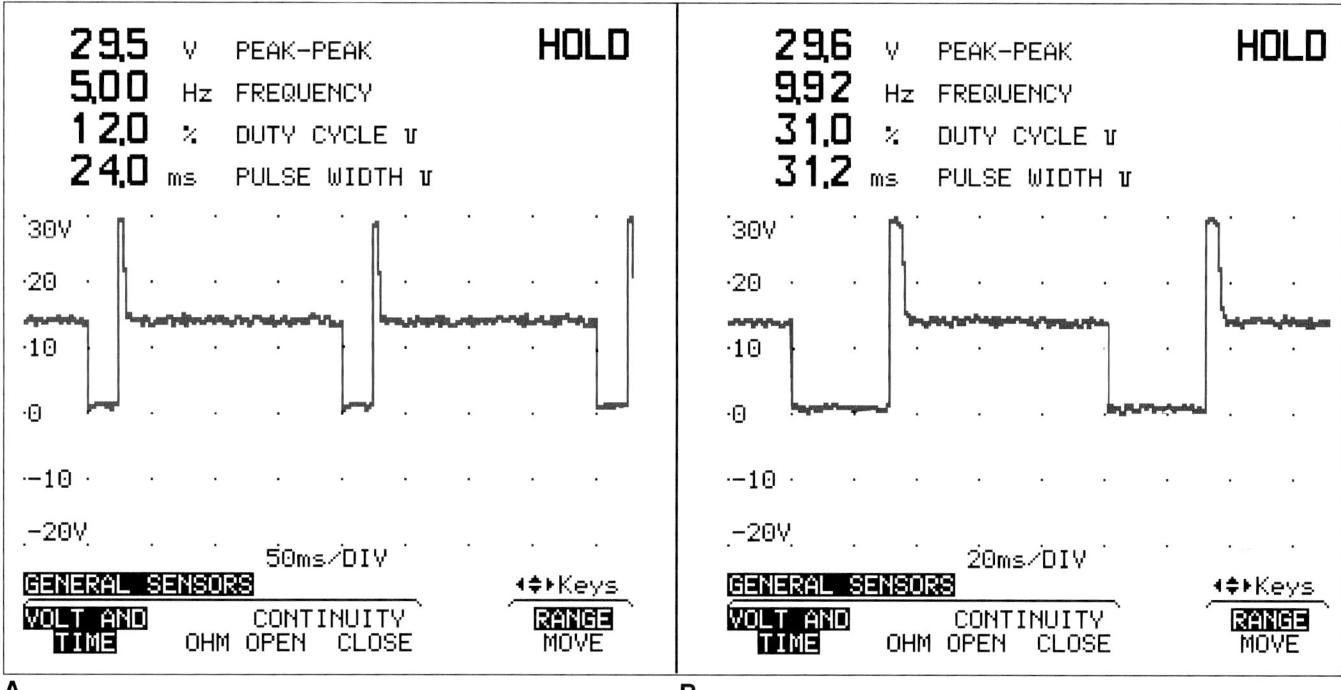

Figure 7-3. *Varying the length of the pulse width allows the ECM to control the operation of many devices. A—Narrow pulse width. B—Wide pulse width. Note the difference in the length between the pulses. (Fluke)*

between the ECM, off-board diagnostic equipment, and other control modules. UART is a data line that varies voltage between 0-5 volts at a fixed pulse width rate **Figure 7-5.** Some of the newest vehicles use UART, but also depend on the use of ***Class 2 serial communications.*** Class 2 data is transferred by toggling the line voltage from 0-7 volts, with 0 being the rest voltage, and by varying the pulse width. The variable pulse width and higher voltage allows Class 2 data communications to better utilize the data lines.

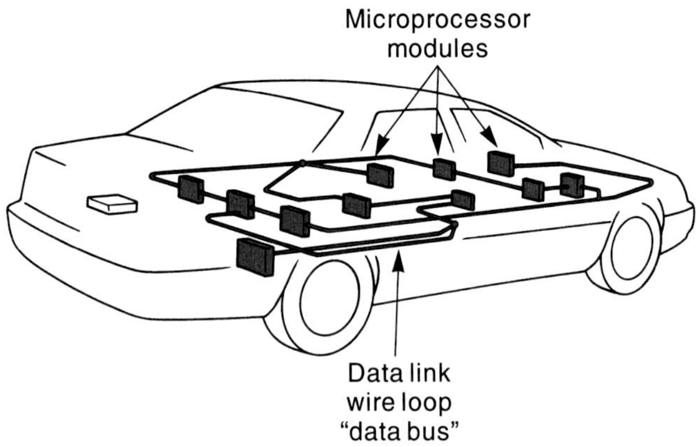

Figure 7-4. *The ECM's in modern vehicles communicate with each other along a network or data bus. (Cadillac)*

ECM Internal Components

There are several main circuits in the ECM. These circuits are called the input processor, the memory, the central processing unit or CPU, the output processor, and the controller. Although these circuits are not individually serviced, and are all replaced when the ECM is replaced, it is important to have some idea of how they work. The following sections explain these circuits in detail.

Input Processor

The ***input processor*** takes the analog voltage from the input sensors and converts them into digital pulses that the computer circuits can work with. Most sensors can send information at a faster rate than the ECM can process. The input processor reduces the amount of incoming information to a manageable level by averaging a series of analog voltage pulses into a single digital pulse.

The input processor also serves as a protective device. Early ECMs were easily damaged if a sensor or wire became shorted or grounded. Newer input processors limit the amount of current entering the ECM, preventing shorts and voltage surges from damaging the ECM.

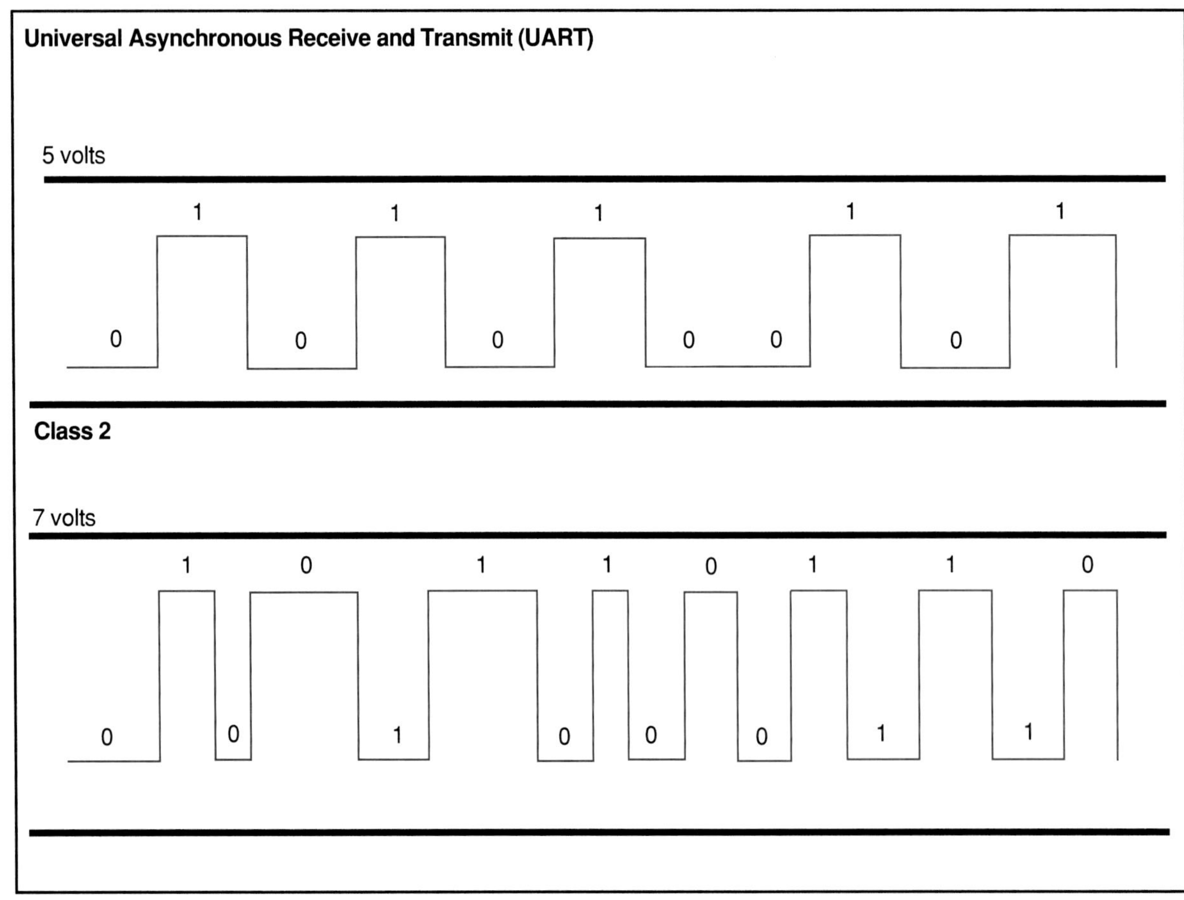

Figure 7-5. *Universal asynchronous receive and transmit signals are similar to a reference voltage signal and communicate by simply toggling the voltage on and off. Class 2 serial communications communicates by varying the pulse with and toggling voltage. This allows for communications at a faster rate, up to 41.6 kilobytes per second in some automotive computer control systems.*

Computer Memory

The memory contains the preset values for a properly running engine. There are two kinds of memory in the computer, fixed and volatile. The fixed memory circuits in most ECMs cannot be changed. However, the fixed memory in some ECMs is replaceable, while others can be modified with the ECM in the vehicle. The volatile memory is able to receive and store information from the sensors and output devices. Volatile memory is always affixed to the ECM and cannot be replaced.

Fixed Memory

Fixed memory is used to retain the operating instructions for the ECM and the vehicle's base engine or system information. There are four types of fixed memory. ***Read Only Memory*** or ***ROM,*** contains the basic instructions that tell the ECM how to operate. If power to the ECM is disconnected, the information is not lost. ROM is installed at the factory and, if defective, requires ECM replacement.

Programmable Read Only Memory or ***PROM,*** contains the operating parameters for the engine, drive train, or system. This information is used by the ECM as a reference for proper input sensor, output device, engine, drive train, and vehicle operation. If power to the ECM is removed, the information in the PROM is not lost. In some cases, the PROM can be removed if it is defective or an updated PROM is needed to correct a driveability problem. In some older and most newer ECMs, the PROM is permanently affixed to the circuit board. Most computer-controlled vehicles use PROMs in their engine ECMs.

Some ECMs use ***Eraseable Progammable Read Only Memory*** or ***EPROM.*** These are affixed to the ECM circuit board and can be reprogrammed by exposing the ECM to an ultraviolet light to clear the EPROM and programming in the new information. Unfortunately, this can only be done by the ECM manufacturer, which makes it necessary to replace any ECM that has a defective EPROM or if updated programming is needed to correct a problem.

Many of the newest ECMs use ***Electronically Eraseable Programmable Read Only Memory (EEPROM)*** or ***Flash Eraseable Programmable Read Only Memory (FEPROM).*** This type of memory, while affixed to the ECM, can be easily programmed or updated using computerized equipment. All or part of a vehicle's operating information is downloaded or "burned into" the EEPROM or FEPROM. Some EEPROM and FEPROM equipped ECMs can be programmed away from the vehicle.

Volatile Memory

The information in ***volatile memory*** changes with vehicle operation and indicates ongoing changes in the control system or other engine systems. Volatile memory is used to store diagnostic codes and temporary information on vehicle operating conditions from the input sensors. The incoming information can be used by other computer circuits as part of the ECM's "learning" ability. If power to the ECM is disconnected, all information stored in volatile memory is lost.

Note: Some newer ECMs can retain volatile memory for several days without battery power.

All ECMs have ***Random Access Memory*** or ***RAM.*** While the vehicle is operating, information from the sensors, such as minimum intake air flow rate and diagnostic trouble codes, is stored in RAM and constantly updated. Some ECMs can compensate for wear and changes in sensor output and output device response. For example, if an oxygen sensor begins to respond at a reduced rate of speed, the ECM stores this information in what is known as ***Keep Alive Memory*** or ***KAM.*** KAM is sometimes referred to as *Keep Alive* or *Nonvolatile RAM.* The ECM can access this information and use it to adjust the output devices to compensate for the sensor's reduced output. This is usually referred to as an ECM's ***adaptive strategy.***

Central Processing Unit (CPU)

The ***central processing unit*** or ***CPU*** actually controls ECM operation. It directs the operation of the other circuits, telling them what tasks to undertake next. It can be thought of as the ECM's manager, directing the flow of information through the various circuits. The CPU also determines the overall status of the engine and vehicle and compensates for short- and long-term changes in engine operation. This allows the ECM to keep input readings very close to the settings in its memory.

The CPU takes the information from the input processor and compares it with information in memory. If the input signals match the data in memory, the CPU takes no action. If the inputs do not match the preset data, the CPU instructs the output processor to change the operation of the output devices until the input signal matches the data in memory. Some inputs are simply for reference, such as intake air temperature and atmospheric pressure. The CPU compensates for these input readings by modifying the operation of other systems. Some outputs, such as air-fuel ratio, are directly and aggressively adjusted by the CPU.

Output Processor

The ***output processor*** or ***output driver*** takes the ECM digital commands and converts them back into analog electrical signals to the output devices. The output processor often acts like a relay, using the small voltage inputs from the main processor to control large current flows into solenoids or motors. The output processor also protects the ECM from voltage spikes or shorted output devices. On a few vehicles, the output processor is a separate unit, sometimes called the output or power computer. Since most output devices are operated by simply turning them on and off, the output signal, or pulse width, is varied in length.

In some cases, the ECM makes use of ***quad drivers.*** These are output chips that contain power transistors and other components that can absorb high current in the event that an output device or connection is shorted to ground. This prevents damage to the ECM internal components.

Other ECM Components

All ECM's have a chip that acts as an internal clock. This internal clock times engine and closed loop operation and is used by the ECM to coordinate its internal operations. Some ECM internal clocks monitor the actual time of day as long as they are connected to battery power. Most ECMs have a backup processor that has preset engine operating information. Should the ECM malfunction, this chip, often called a ***Mem-Cal,*** allows the vehicle to operate in what is commonly referred to as the limp-in mode. While the vehicle will not run efficiently, it can be driven to a shop for service.

Computer Control System Specific Functions

While monitoring input sensors and adjusting output devices, the ECM performs many functions important to vehicle operation and safety within its circuits. The following sections describe some of these specific internal ECM functions.

Short- and Long-Term Fuel Trim

The ECM interprets sensor readings and adjusts the fuel system to compensate. The fuel adjustment on older carburetors was dictated mostly by float level. Fuel adjustments on feedback carburetors could be seen by monitoring mixture control solenoid operation with a scan tool or a dwell meter. However, this is difficult to see in electronic fuel injection systems, since injector pulse width adjustments are an internal function of the ECM and occur very quickly.

The ECM makes adjustments to the injector pulse width by constantly checking the input sensor data in short term memory and making adjustment based on this information. This is called ***short-term fuel trim.*** Some manufacturers refer to this as an *integrator.* Short-term fuel trim adjustments are made during closed loop operation in response to temporary changes in engine operation, such as increased load when driving up a hill.

The ECM can also make longer, semi-permanent adjustments to the air-fuel ratio in response to engine operation. This adjustment is called ***long-term fuel trim,*** sometimes referred to as *block learn.* Long-term fuel trim adjustments are made when the ECM determines that the vehicle is operating under set conditions for an extended period of time, such as hot or cold weather, high altitudes, or an on-going driveability problem. Some scan tools have the capability to monitor short- and long-term fuel trim. The scan tool interprets both fuel trim readings as a count number or percentage. The ECM in some systems can monitor the fuel trim to detect problems.

Idle Speed Control

The ECM will keep the amount of air that flows past the throttle valve during idle as the proper rate necessary for good idle speed operation. It works off of a minimum number or percentage for flow rate. While at idle, the rate of air flow will not be less than this number or percentage. This is called ***minimum idle speed control*** or *minimum idle air rate.* On most engines, this is set automatically by the ECM when the vehicle reaches a certain speed and operating conditions. A few vehicles require the idle speed control to be set by the technician, usually with a scan tool or other means.

Engine and Vehicle Protection

Most ECMs are designed to protect the engine and vehicle from damage due to excessive engine and/or vehicle speed. Manufacturers add programming to the ECM that will limit the input of or shut off the fuel pump, fuel injectors, or ignition module if engine speed becomes excessive, usually between 6000-7800 RPMs. In some cases, the ECM will not allow the vehicle to exceed a predetermined maximum speed. The ECM will also shut down the engine during extreme deceleration conditions, such as a collision.

Clear Flood Mode

Some ECMs will shut down the injectors if the throttle position sensor indicates that the throttle valve is being held past a certain angle (meaning the accelerator pedal is depressed) by the driver during start-up. This is called ***clear flood mode.*** Clear flood is a feature programmed into a fuel injected engine's ECM that allows technicians to clear the cylinders should they become flooded with fuel. Some technicians will also use the clear flood mode while testing the ignition or starting system when engine operation is not desired. It also prevents engine damage if the driver has a tendency to depress the accelerator while cranking the engine. Clear flood will be discussed in more detail in **Chapter 17.**

Caution: Not all ECMs have clear flood capability. Check the vehicle's service manual before performing this procedure.

Voltage Monitoring

The ECM monitors battery voltage in order to compensate for electrical loads placed on the charging system. If the ECM detects a drop in battery voltage, engine idle speed is increased slightly to compensate. This increases alternator output in response to electrical loads. Some ECMs also have voltage regulator control and can increase alternator output or shut down the alternator for a time if excessive charging is detected. If the battery voltage drops below 10-10.5 volts or goes above 14.5-15 volts for an extended period, the ECM will store a trouble code.

Overheating Protection

The ECM monitors engine coolant temperature in order to adjust inputs to the output devices to compensate for engine temperature. Another feature programmed into most ECMs is overheating protection. On most engines, the ECM will shut off the air conditioning compressor, turn any electric cooling fans on, and increase the engine idle speed to force more coolant through the engine and, if equipped, a belt-driven fan's rotational speed.

A few ECMs provide overheating protection by selectively disabling cylinders in the engine to redistribute heat. If the ECM detects an overheating condition, it will shut off the fuel injectors and spark to selected cylinders in an alternating pattern. This decreases the build-up of engine heat and allows the vehicle to be driven for a period of time.

Input Sensors

The input sensors are the ECM's "nerves." Input sensors monitor a wide variety of engine and drive train functions. A brief description of input sensor operation is provided in this section. More detail about certain specialized sensors will be given in later chapters. Input sensors and switches monitor conditions in the engine and drive train, outside atmospheric conditions, and direct or indirect driver inputs. There are many types of input sensors. Sensors are grouped according to the vehicle systems that they affect.

Engine Sensors

Engine sensors are installed on the engine and measure the actual conditions in the engine. Typical engine sensors measure engine temperature and speed, vehicle speed, crankshaft position, camshaft position, manifold vacuum, intake air flow, intake air temperature, barometric pressure, exhaust gas oxygen, and whether the engine is knocking. **Figure 7-6** shows the location of some common engine sensors.

In many engine sensors, a reference voltage from the ECM is passed through a resistor unit, and modified as the resistance of the sensor varies with engine conditions. Some of these sensors contain a contact unit that slides along a resistor, **Figure 7-7.** Changes in the resistor length change the amount of current flow. The ECM reads the variation in current as an input signal. For example, the resistance in some airflow sensors varies according to the position of the resistor unit. The resistance of manifold vacuum and EGR position sensors varies according to movement within the resistor unit. These units are sometimes called *rheostats.*

In other resistance sensors, changes in temperature affect the resistance reading. Temperature affects the resistance of the sensor material, and therefore, the ability of current to flow through the sensor. This changes the voltage reading through the sensor. The ECM reads this change in voltage as a change in temperature. Intake air temperature and engine coolant temperature sensors are typical of heat-sensitive resistance sensors. A typical coolant temperature sensor is shown in **Figure 7-8.** Some vehicles are equipped with an ***exhaust temperature sensor.*** This sensor is used to tell the ECM when exhaust temperatures are high enough for the oxygen sensor readings to be reliable, **Figure 7-9.**

Some sensors are called ***transducers.*** They consist of an iron plunger inside a small wire coil. Current flows through the coil when the engine is operating, creating a magnetic field. When the plunger is moved by outside force, such as a vacuum diaphragm or driver-operated linkage, it changes the strength of the magnetic field in the coil. Changes in the magnetic field cause a change in the amount of current flowing through the coil. The ECM can read this change as a change in position of the monitored device. **Figure 7-10** shows a typical transducer.

Atmospheric Sensors

These sensors measure atmospheric conditions outside the vehicle. Atmospheric sensors include those that measure air temperature, barometric pressure, and sometimes humidity. Air temperature sensors work like coolant temperature sensors. In some cases, both the coolant and air temperature sensors on a particular vehicle are identical.

Intake Air Temperature Sensors

The ***intake air temperature sensor (IAT),*** monitors the temperature of the air entering the engine. An IAT's resistance changes in response to air temperature. When the air is warm, less fuel is needed to operate the engine. Intake air temperature sensors are also known as ***manifold air temperature (MAT) sensors,*** **Figure 7-11.** On some vehicles, the air temperature sensor is part of the mass airflow sensor.

Barometric Pressure Sensors

Some materials are pressure sensitive and are able to change their resistance as the pressure placed on them changes. These materials can vary the strength of the voltage signal passing through them as the pressures placed on them change. Sensors using this type of pressure sensitive materials include barometric pressure and manifold vacuum sensors.

Barometric pressure sensors measure atmospheric pressure. Since the air pressure is different at sea level than it is in the mountains, a vehicle that is normally operated at sea level would perform differently at higher elevations. Barometric pressure sensors send a voltage signal to the ECM, which compares the input to that from the manifold vacuum sensor and adjusts the air-fuel ratio, spark timing, and other outputs to maintain good performance. They operate in the same way as manifold vacuum sensors. In most cases, the manifold vacuum and barometric pressure sensors are installed in the same housing. These are called ***manifold absolute pressure*** or ***MAP sensors.*** The schematic for one of these sensors is shown in **Figure 7-12.**

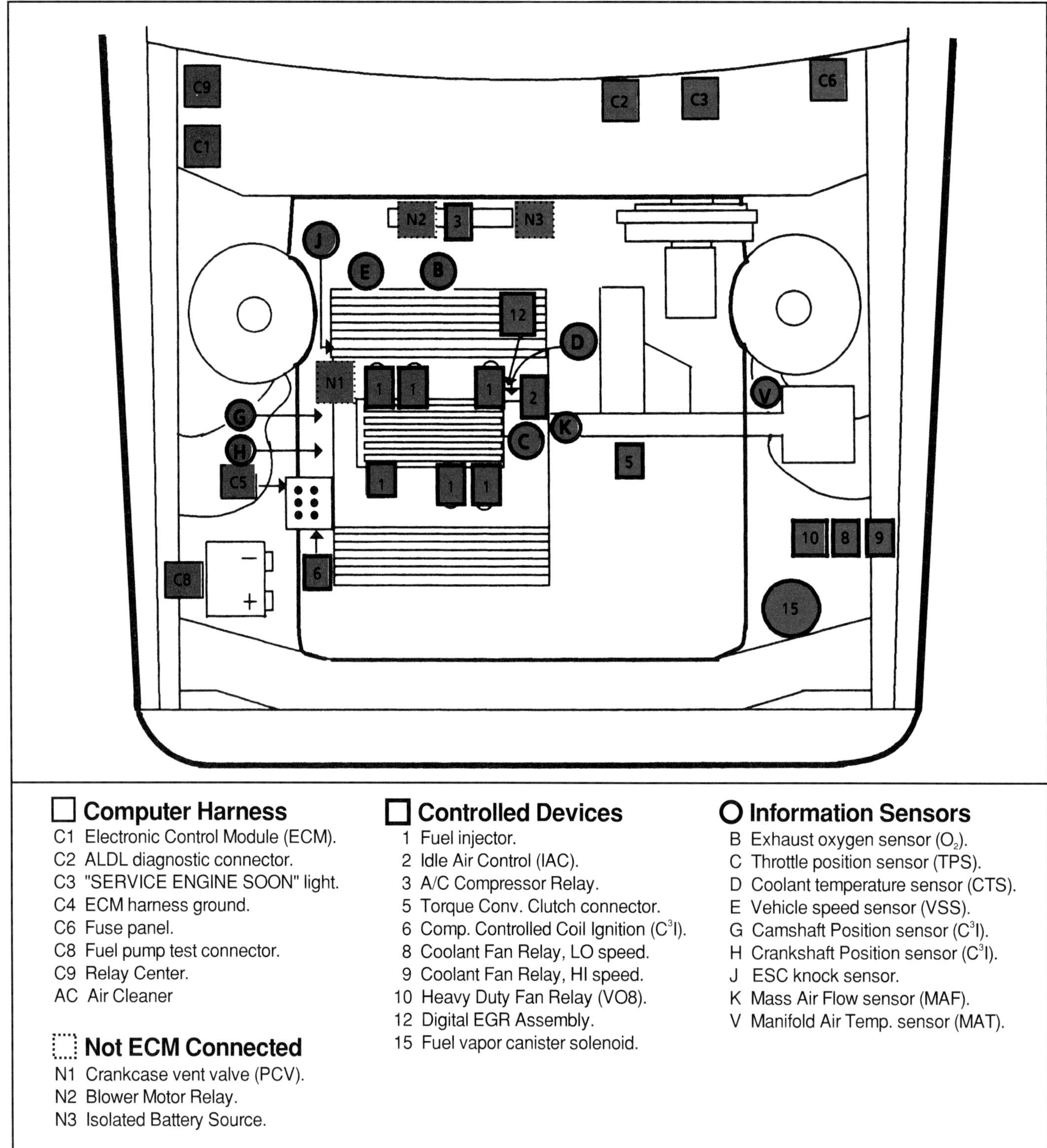

Figure 7-6. *Service manual contains diagrams showing the location of input sensors, output devices, the ECM, and other components of the computer control system. (General Motors)*

Mass Airflow Sensors

Many multiport fuel injection systems have a ***mass airflow sensor (MAF)*** that precisely monitors the amount of air entering the engine. The ECM uses the mass airflow sensor input to determine how much fuel to inject to match the amount of air entering the engine. Three types of airflow sensors are used with computer-controlled multiport fuel injection, the heated element, the air valve, and the Karman vortex. Another type of airflow sensor is used with continuous injection systems, but it directly controls fuel flow and does not send a signal to the ECM.

The *heated element sensor* consists of a wire or film made of temperature sensitive resistance material. This element extends into the intake air passage and contacts the stream of incoming air. A heated wire sensor is shown in **Figure 7-13.** The ECM directs a reference voltage through

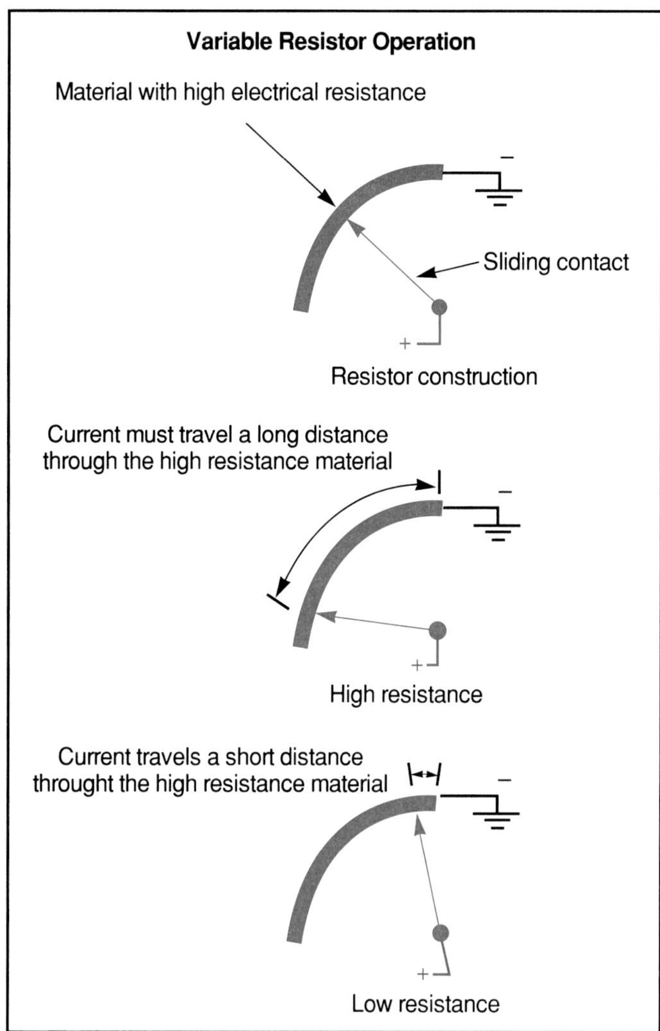

Figure 7-7. *A variable resistor is used to measure the position of a part, such as the throttle plate. It consists of a resistor and a sliding contact. The position of the contact on the resistor determines how much current can flow through the switch. The ECM reads the current flow change as a position signal.*

the element, causing current to flow. As current flows through the element, it heats up and its resistance increases. Since the reference voltage from the ECM is constant, increasing resistance decreases the amount of current flow, and therefore the voltage returning to the ECM. Air flow through the air passage removes heat from the element, reducing its resistance and increasing the voltage signal. Therefore, the voltage signal goes up when airflow is increased and goes down when airflow is reduced. The ECM reads the voltage variations and calculates the amount of air flowing into the engine.

The *air valve* consists of a moveable flap which extends into the air stream in the intake air passage. The flap is spring loaded so that air flow is opposed by spring pressure. As air flow into the engine increases and decreases, the flap moves according to the ratio of air flow versus the fixed spring tension. The flap is attached to a rheostat. The rheostat receives a reference voltage from the ECM. Variations in air flow

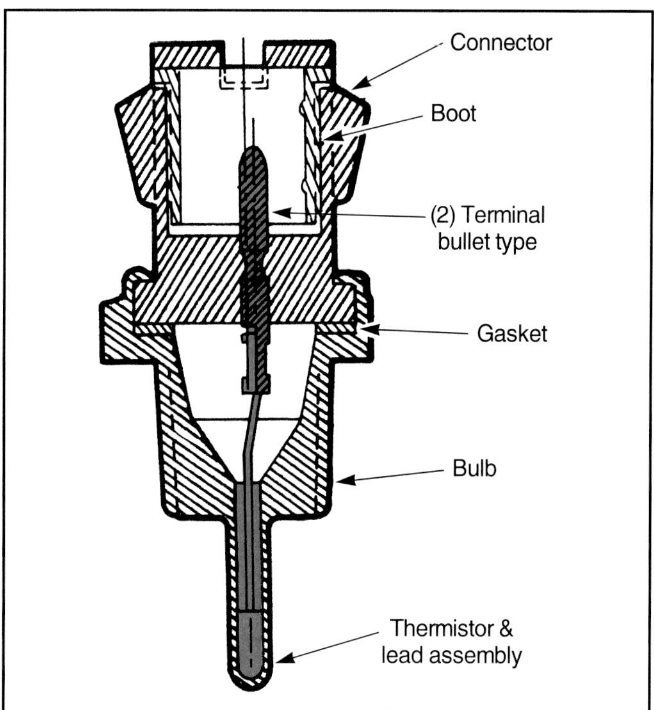

Figure 7-8. *Cutaway of a engine coolant temperature sensor. As temperature increases, the resistance of the material in the sensor decreases. (Ford)*

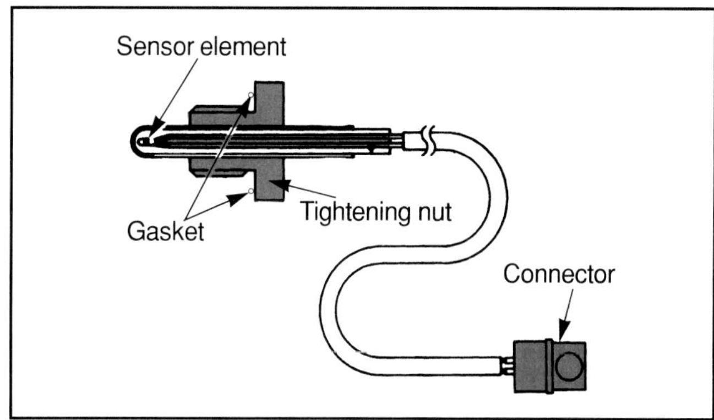

Figure 7-9. *Some vehicles use an exhaust temperature sensor to monitor when exhaust temperatures are right for accurate oxygen sensor readings. Most manufacturers do not use this sensor. (Subaru)*

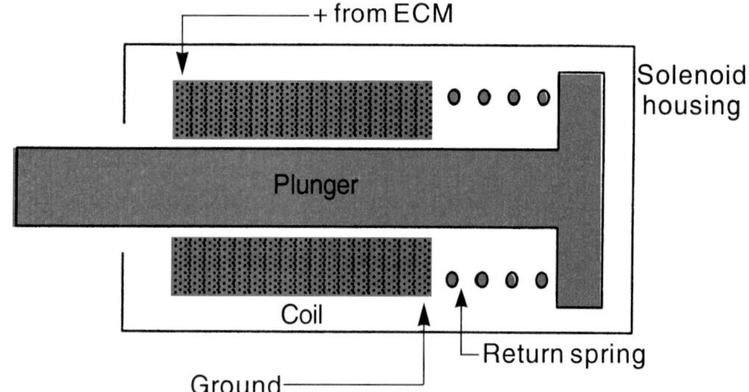

Figure 7-10. *A transducer is simply a wire coil with a metal plunger. Movement of the metal plunger causes fluctuations in the coil's magnetic field. This results in a voltage change that can be read by the ECM.*

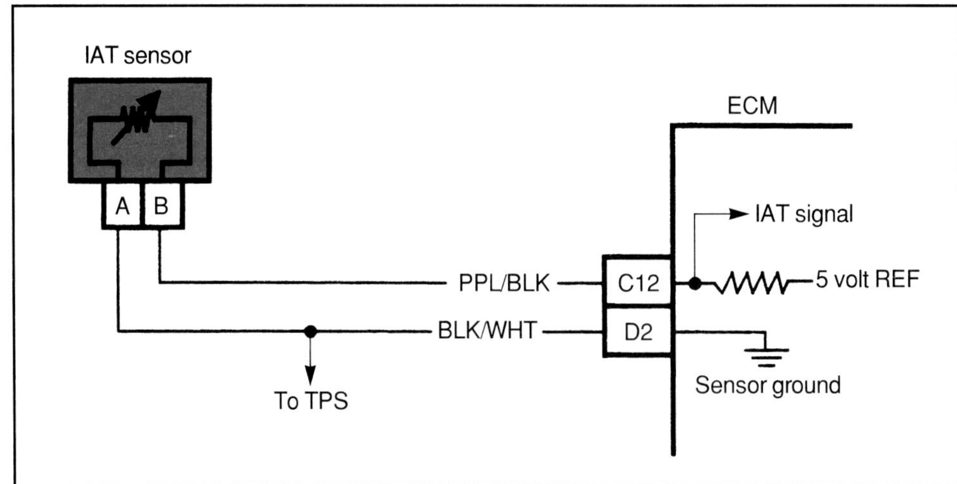

Figure 7-11. *The intake air temperature (IAT) sensor allows the ECM to compensate for air temperature when determining the correct air-fuel mixture. (General Motors)*

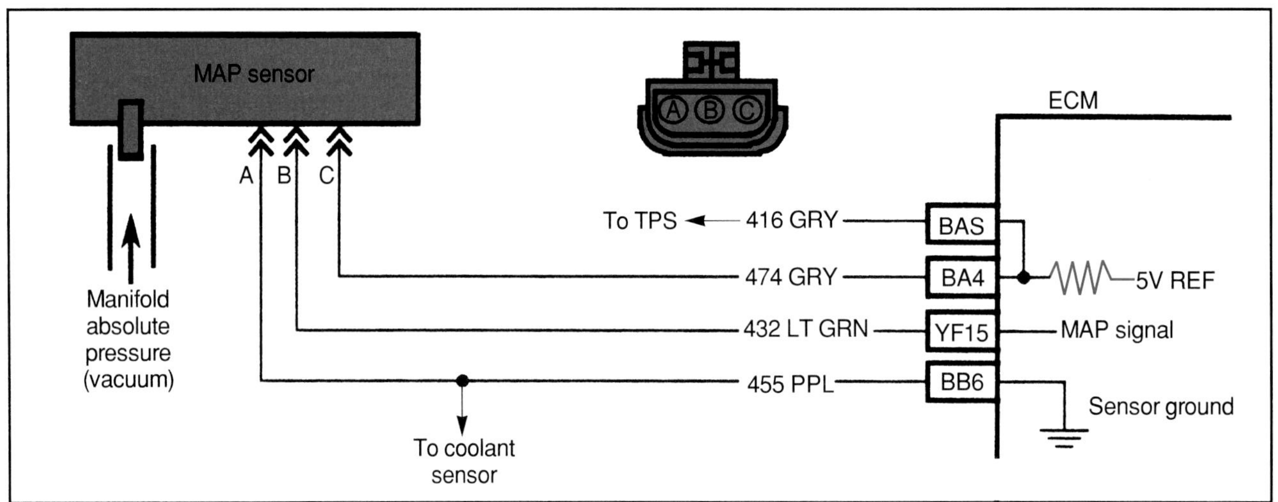

Figure 7-12. *Manifold absolute pressure (MAP) sensors are used to determine the difference between manifold and outside air pressure. (General Motors)*

cause the flap to move, which in turn moves the rheostat contact arm. This increases or decreases the voltage signal to the ECM, which reads it as changes in air flow.

The *Karman vortex airflow sensor* consists of a metal-foil mirror, a light emitting diode (or LED), and a light receptor called a photo transistor. The light produced by the LED is reflected from the metal-foil mirror into the photo transistor. The metal foil mirror is placed in the intake air passages, where the incoming air causes it to vibrate. Increases and decreases in air flow change the rate of vibration, and therefore, the amount of light which reaches the photo transistor. The photo transistor changes this light into a voltage signal that can be read by the ECM as airflow rate.

Many vehicles do not use mass airflow sensors since they are expensive and delicate. They are not used on some multiport engines and all throttle body injected vehicles. Therefore, some vehicle computer systems eliminate the

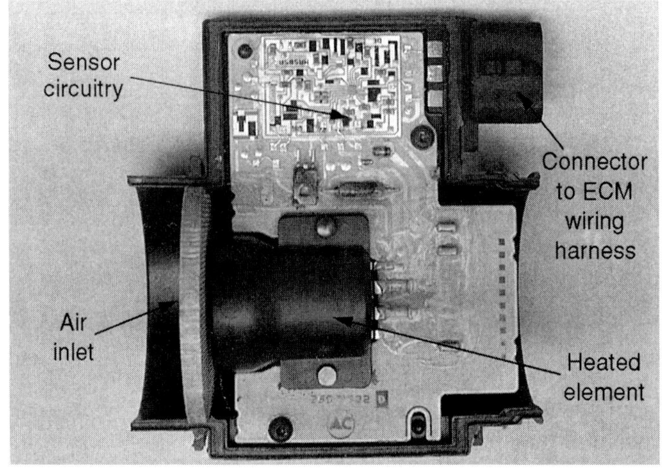

Figure 7-13. *Cutaway of a mass airflow sensor (MAF) shows the internal electronics of this important sensor.*

mass airflow sensor by monitoring other sensor inputs and calculating the amount of air entering the engine. These are known as speed density systems. In a typical speed density system the ECM computes air flow from inputs provided by the MAP sensor, engine speed sensor, throttle position sensor, vehicle speed sensor and intake air temperature sensor. This process will be discussed in greater detail in **Chapter 9.**

Engine Coolant Temperature Sensors

To properly adjust the fuel mixture and ignition timing, and to operate the emission controls, the ECM must know engine temperature. In addition, the ECM relies on engine temperature to decide whether to place the system in closed or open loop, and to turn the radiator fans and air conditioning compressor on or off. Since the temperature of the engine coolant is a good indicator of overall engine temperature, every computer control system has at least one ***engine coolant temperature (ECT) sensor.*** A few systems have a second sensor to more closely monitor engine warm up.

The engine coolant temperature sensor consists of a heat sensitive resistor material. Unlike most resistors, the material's resistance decreases with increases in temperature. The ECM sends a reference voltage to the sensor which passes through this material before returning to the ECM as a voltage signal. When the coolant and sensor are cold, the material has a very high resistance. Electricity flowing through the sensor meets this resistance and the voltage signal returning to the ECM is low. As the coolant warms the material, its resistance decreases. This increases the voltage signal returning to the ECM. The ECM reads these variations in voltage as engine temperature. In other systems, the sensor current flows directly to ground and the ECM reads increasing current flow as low voltage. The construction of a typical temperature sensor is shown in **Figure 7-14.**

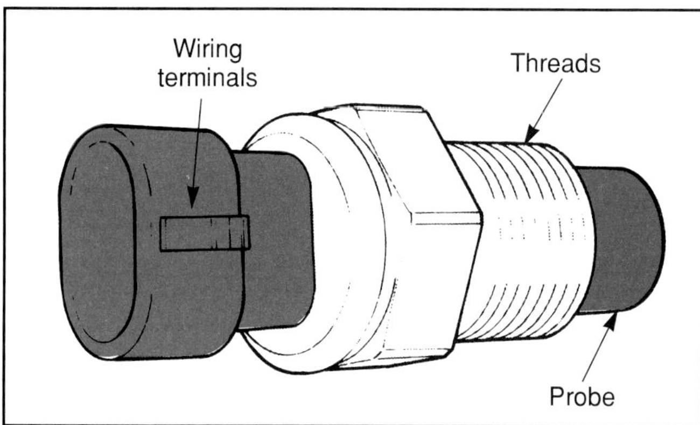

Figure 7-14. *Parts of an engine coolant temperature sensor. They are normally installed in the intake manifold or in the thermostat housing. (Chrysler)*

Engine Speed and Position Sensors

Some sensors develop a voltage signal by the creation or variation of a magnetic field. If a magnetic field is used to create the voltage signal, no outside voltage source is needed. In other sensors, a voltage signal created in the ECM is modified by the operation of the sensor.

Magnetic field sensors utilize a magnetic field generated by moving parts. In a typical system, a toothed wheel passes near a small pickup coil, creating a magnetic field. This magnetic field produces a small voltage signal to the ECM, **Figure 7-15.** Some sensors create a signal by interrupting an existing magnetic field. As shown in **Figure 7-16,** either a shutter or rotating magnets are used to affect the magnetic field. Interrupting the magnetic field creates a voltage fluctuation in the circuit, which is read by the ECM.

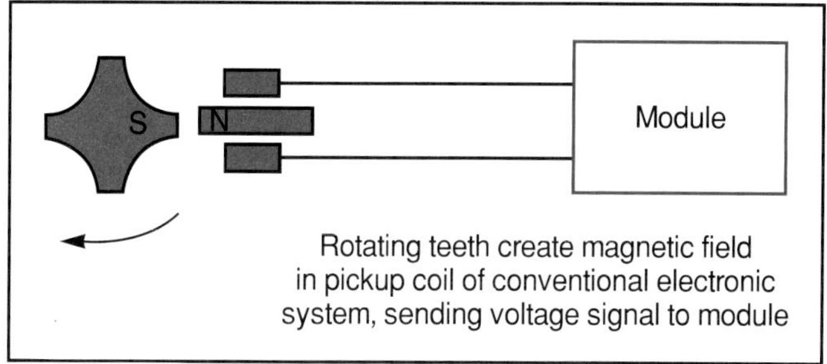

Figure 7-15. *The pickup coil is comprised of a small wire that responds to changes in nearby magnetic fields by producing a small voltage, which can be read by the ECM.*

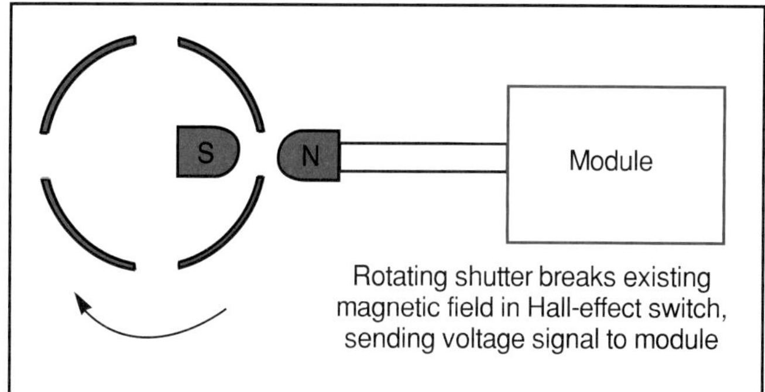

Figure 7-16. *The Hall-effect switch shown here is a wire coil with current flowing through it. Current produces a magnetic field around the coil. When anything affects the field, it changes the voltage of the current flowing through the coil, sending a signal to the ECM.*

In some systems, the input signal is generated as a byproduct of the operation of the ignition system. In these cases, the signal generated by the distributor mounted pickup coil or Hall-effect switch is used to trigger the coil discharge, and also as an input to the ECM. These sensors are discussed in more detail in **Chapter 8.**

Crankshaft position and ***camshaft position sensors*** used with distributorless ignition systems are installed near the crankshaft or camshaft, **Figure 7-17.** A disk with a series of magnetized teeth is installed on the crankshaft or camshaft. As the shaft(s) rotates, the disk passes near the sensor, creating a magnetic field. This magnetic field creates a voltage signal at the ECM. The ECM can process this signal to determine engine speed. The ECM can also determine the crankshaft and camshaft position and therefore, the position of each piston, by reading the pattern of voltage signals.

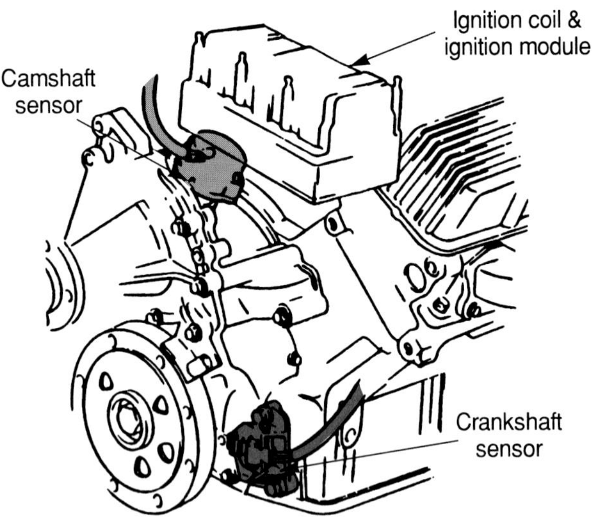

Figure 7-17. *One type of crankshaft position sensor. Also note the camshaft position sensor used with some engines. (General Motors)*

Light Sensors

Some sensors use light emitting and collecting devices to measure speed or locate the position of rotating parts. The majority of these sensors are used in ignition distributors. A few are used in vehicle speed and airflow sensors.

A typical light sensor contains three major components. Light is produced by a light producing device called an *emitter.* This light is directed at a light absorbing device called a *detector.* A rotating device called an *interrupter* periodically blocks the light beam between the emitter and detector.

Emitters can be light emitting diodes (LEDs) or infrared emitting diodes (IREDs). The detector is a light sensitive semiconductor, often called a photodiode or phototransistor. Current flows in the detector whenever the engine is running. Light striking the detector affects the amount of current flowing through the semiconductor material. The ECM monitors the current flow and reads variations as an input signal. The interrupter is attached to a rotating part of the vehicle, such as the speedometer cable or camshaft. Depending on the design, the interrupter blocks the light beam in one of two ways, as shown in **Figure 7-18.**

In **Figure 7-18A,** the interrupter is equipped with small mirrors spaced at intervals on the interrupter disc. When light from the emitter strikes a mirror, it is reflected into the detector. As the disc turns, the mirror moves out of position, and emitter light is no longer reflected into the detector. This causes a change in current flow in the semiconductor material. Further disc movement causes the next mirror to line up with the emitter and collector, again changing current flow.

In **Figure 7-18B,** the interrupter disc is constructed with shutter blades at intervals on its outer edge. When one of the shutter blades passes between the emitter and detector, the

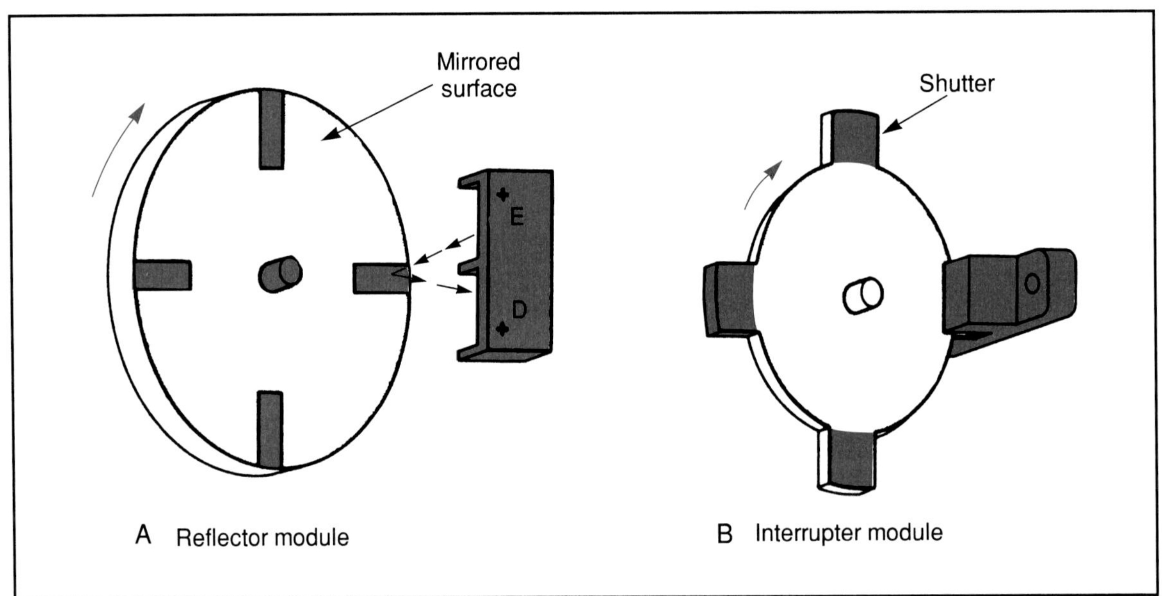

Figure 7-18. *Typical light sensor system. A—Light from the emitter hits the light sensor mirror and bounces back to the collector. B—The shutter assembly blocks the light path. The resulting signal can be used to send back speed and position to the ECM.*

light beam is broken. This causes a change in current flow in the semiconductor material. When the shutter blade passes, the light beam is restored, and current flow changes again.

Throttle Position Sensors

The ***throttle position sensor (TPS)*** measures the amount of throttle opening, using a variable resistor or transducer. These sensors are mounted on the outside of the throttle body on fuel injected engines and some carburetors. Some carburetors have the throttle position sensor located inside the float bowl.

The throttle position sensor is sent a voltage signal by the ECM. The voltage is modified by the variable resistor in the TPS. The lowered voltage signal from the TPS is interpreted as throttle position by the ECM. This process is shown in **Figure 7-19.**

Some throttle position sensors are simple on-off units, usually designed to tell the ECM that the throttle is in the idle position. Note that the idle switch is part of the throttle position sensor in **Figure 7-19.** Some throttle position sensors are adjustable while others are simply attached and are not adjustable.

Knock Sensors

Knock sensors contain a small crystal that reacts to pressure changes. Many knock sensors modify an incoming voltage, however, some actually produce a small electric current. The crystal is sensitive to certain types of vibrations, such as those produced by engine knock. When engine knocking is detected, the crystal modifies or creates electric current that is picked up by the ECM and tells it that the engine is knocking. See **Figure 7-20.**

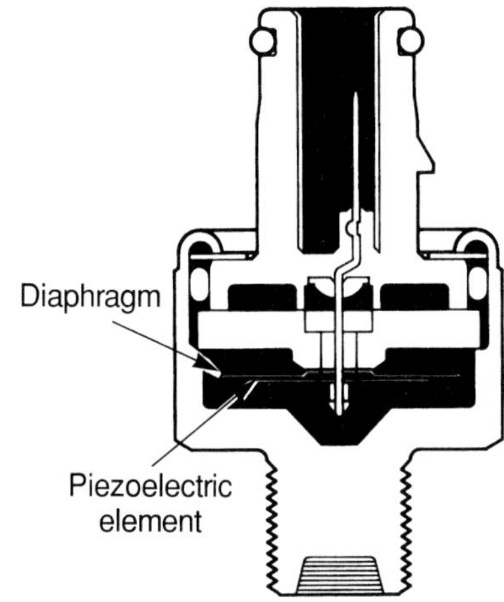

Figure 7-20. *Cutaway of a knock sensor shows the piezoelectric element, which vibrates and creates a small current or modifies reference voltage.*

Oxygen Sensor

Using a chemical reaction to produce an electric current is a common process. This is the source of electricity in all batteries. The reaction between various chemicals produces extra electrons, which will create an electrical potential, or voltage, at the battery terminals. This same principle is used in the ***oxygen sensor (O_2),*** **Figure 7-21.** Many automotive computer control systems rely heavily on the oxygen sensor for much of the engine's operating information.

The sensor is made of materials that react with oxygen to produce extra electrons, usually *zirconia.* The zirconia element is connected to internal electrodes, usually made of

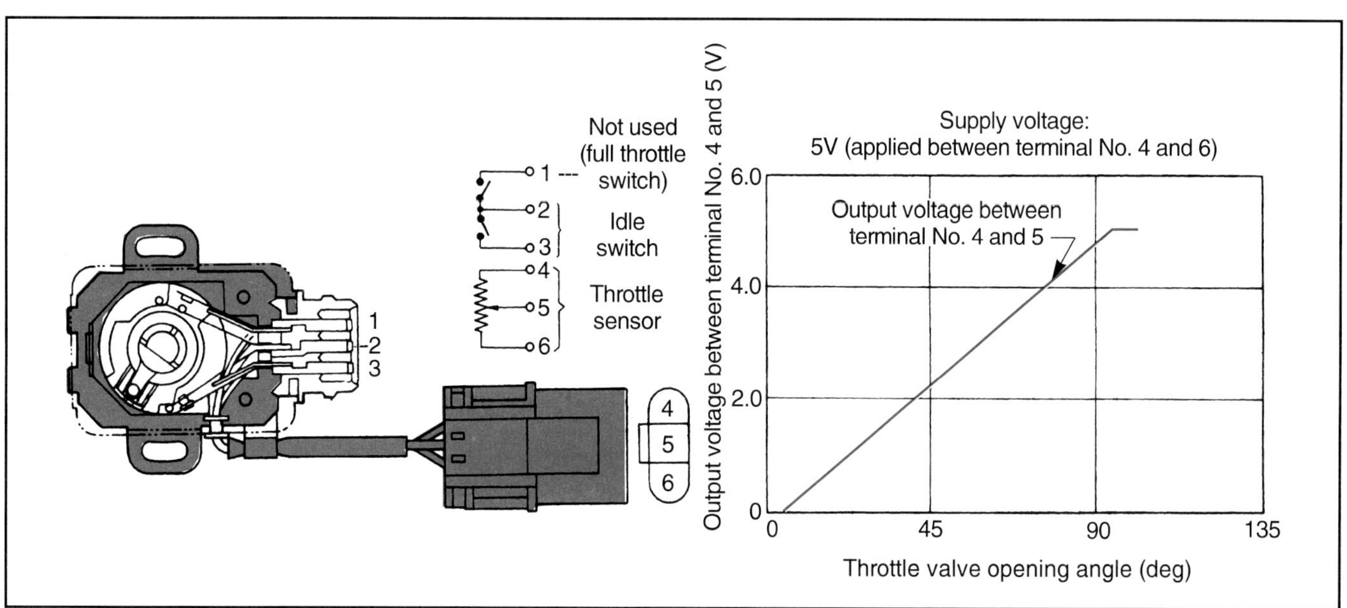

Figure 7-19. *Throttle position sensors are used to monitor the throttle position as it changes with driver input. (Subaru)*

platinum. The entire assembly is housed inside of a steel shell. The difference in the amount of oxygen in the exhaust gases and the outside air produces a weak electrical current, which is sent to the ECM. Variations in the oxygen content of the exhaust gases change the strength of the electrical signal. The ECM can read changes in the signal voltage as changes in the air-fuel ratio.

Variations of this sensor include oxygen sensors with ***titania*** cores. These sensors modify voltage, instead of producing it. This gives a titania O_2 sensor the ability to send relative information to the ECM immediately, rather than having to wait for engine or sensor warmup. Most newer and all OBD II equipped vehicles use ***heated oxygen sensors (HO_2S),*** which have an internal element that preheats the sensor during vehicle warm-up. This allows the computer control system to enter closed loop operation much more quickly.

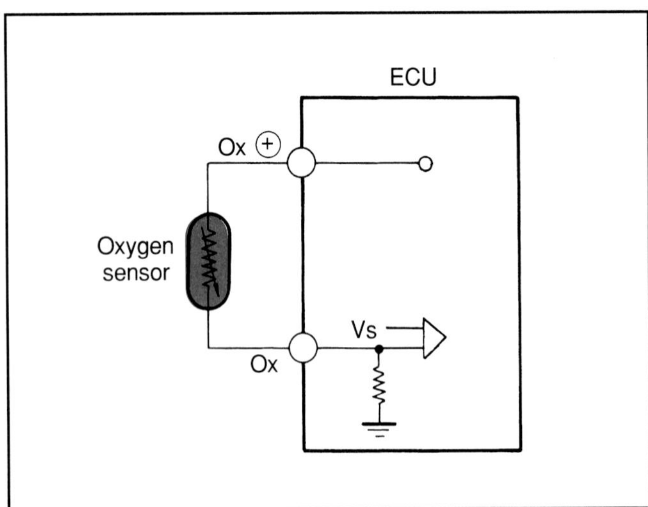

Figure 7-21. *The oxygen sensor is monitored closely by the ECM to determine engine operating condition. If this sensor is defective, it can affect the operation of the entire vehicle.*

Oxygen Sensor Location

Some vehicles have only one oxygen sensor, which is mounted either in the exhaust manifold or in the exhaust pipe near the manifold. Many vehicles with V-type engines have two oxygen sensors. All vehicles built after 1995 are equipped with at least two oxygen sensors. One is mounted before the catalytic converter and monitors O_2 levels in the exhaust gases as they enter the converter. The other sensor, which is referred to as a ***catalyst monitor,*** is mounted behind the catalytic converter. This sensor monitors O_2 levels of the exhaust gases as they leave the converter. The second O_2 sensor also acts as a backup should the first one fail.

Multiple oxygen sensors will be designated by engine bank (or side) and sensor position in relation to the engine. For example, one oxygen sensor might be designated as Bank 2, Sensor 1. This sensor would be located on the engine side opposite the number one cylinder and is the sensor on that side nearest the engine. The farthest number sensor (usually Bank 2, Sensor 2 or 3, depending on the number of oxygen sensors used) is always the catalyst monitor, **Figure 7-22.**

Drive Train Sensors

Drive train sensors monitor conditions in the transmission and other drive train components. Most drive train sensors are driver-operated. A driver-operated sensor is any sensor that detects the operation of a device, system, or component commanded by the driver. These sensors are usually installed on the throttle linkage, brake pedal, shift lever, and other drive train and chassis systems.

Vehicle Speed Sensors

Most ***vehicle speed sensors (VSS)*** are mounted in the transmission/transaxle case at a point where it can monitor output shaft speed, **Figure 7-23.** Most vehicle speed sensors are permanent magnet generators. Other speed sensors include photoelectric sensors mounted in the speedometer housing. Unlike most sensors, speed sensors produce alternating current.

Some vehicle speed sensors are mounted on the drive wheels to allow the ECM to control wheel traction or adjust engine settings for lowest emissions. The same wheel sensors are often used as inputs to the anti-lock brake computer. Most vehicle speed sensors use the same principles as engine speed sensors.

Reed Switches

Reed switches are often used to measure speed by measuring rotation. A typical reed switch is shown in **Figure 7-24.** The reed switch consists of two thin metal blades, or reeds, that contact each other inside a closed chamber. Reed metal resembles aluminum foil and can flex without breaking. A small amount of current flowing in the reeds creates a magnetic field, which attracts the reeds to each other. Any outside magnetic field, however, will cause the reeds to repel each other and move apart.

Reed switches are placed near a magnet attached to a moving part to be monitored. Movement across the magnet causes the two reeds to move apart, and then spring back together. This starts and stops the current flowing in the reed switch. This current variation can be read by the ECM. Reed switches are usually installed on the speedometer cable where it enters the speedometer head. These switches are used to control the cruise control as well as to provide inputs to the ECM.

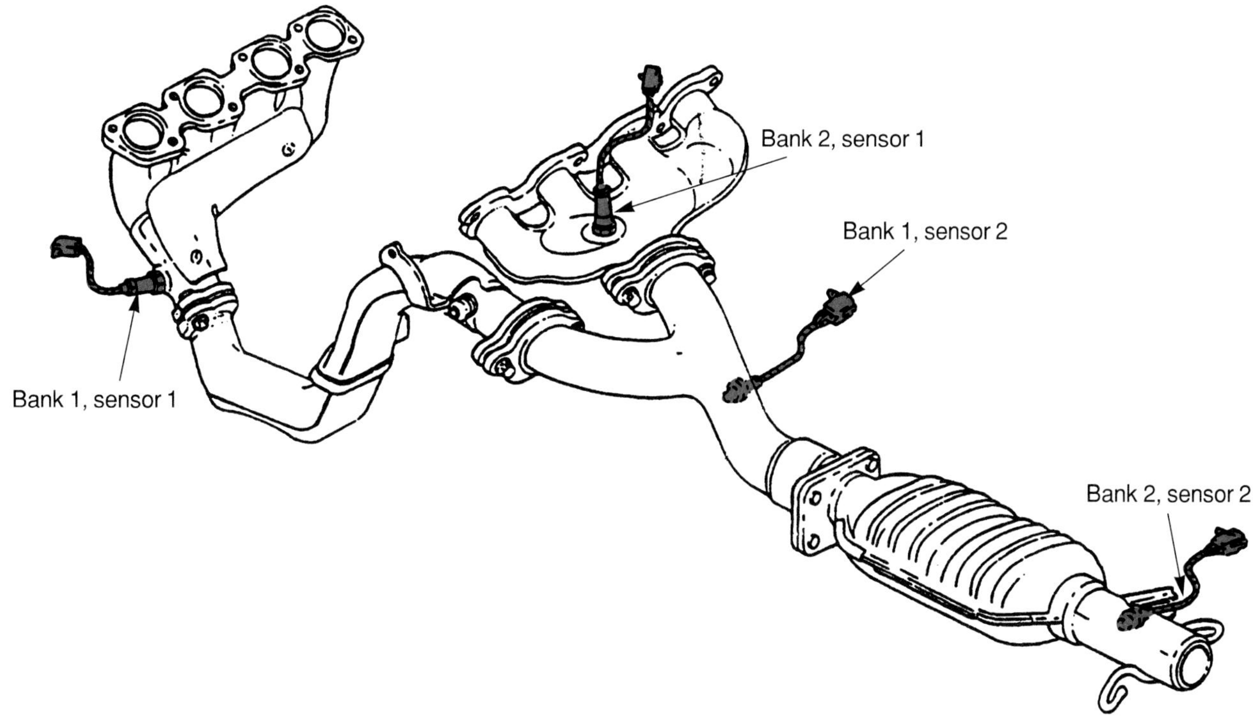

Figure 7-22. *Oxygen sensor location in a modern exhaust system. Bank 1 is the side of the engine that has the number one cylinder. Sensor 1 is the sensor closest to the number one cylinder. (General Motors)*

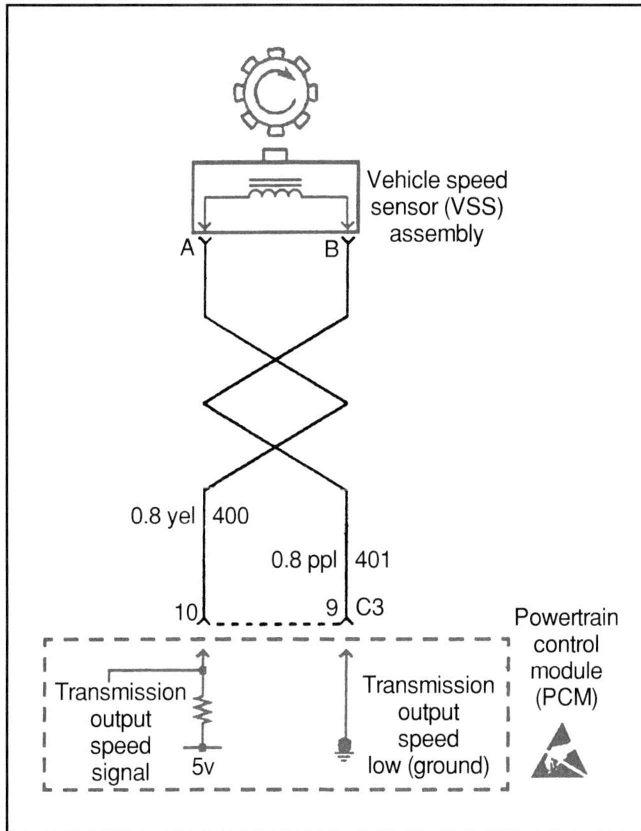

Figure 7-23. *The vehicle speed sensor (VSS) allows the ECM to monitor vehicle speed. Speed sensors are also located on each wheel or in the rear differential of many vehicles. (General Motors)*

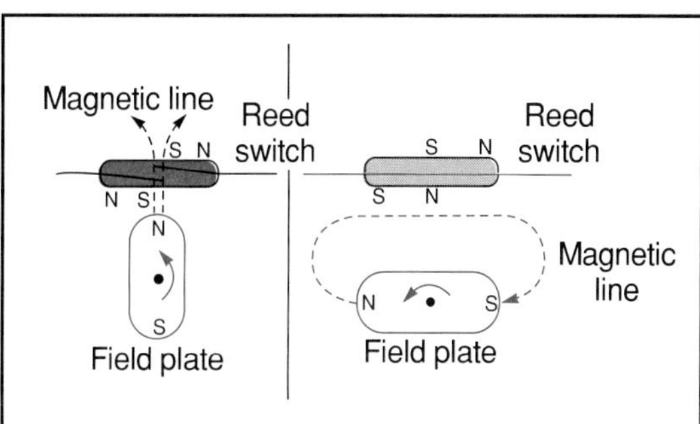

Figure 7-24. *Reed switches are magnetically operated. They are used to measure the speed of a part to determine RPM. (Nissan)*

Brake and Transmission Switches

Brake light and neutral safety switches are on-off switches, which are the simplest type of inputs to the ECM. The switch can be either on, allowing current to flow, or off, preventing the flow of current. They tell the ECM whether the brake is being applied or a particular transmission gear is selected.

Most automatic transmission sensors are pressure switches, such as those shown in **Figure 7-25.** Most manual transmission sensors are mechanical switches operated by the shift linkage, **Figure 7-26.** Additional switches are sometimes installed on the transfer case of four-wheel drive units.

Accessory Equipment Sensors

Accessory equipment sensors are used to monitor such conditions as power steering system pressure and whether the air conditioner or cruise control are on or off. This allows the engine to adjust idle speed, spark timing and other functions to compensate for the additional load placed on the engine.

Voltage Regulator

The alternator voltage regulator acts as a sensor, sending a very low amperage output voltage signal directly to the ECM. In some cases, the ECM reads the voltage directly from the alternator output terminal to the battery. The ECM can then adjust idle speed to compensate for low or high voltage conditions

Figure 7-25. *Pressure switches are installed on the case or valve body of automatic transmissions and transaxles to sense internal fluid pressure. When hydraulic pressure applies a band or clutch, the fluid pressure also activates the pressure switch, which signals the ECM that a shift has taken place.*

Power Steering Pressure Switch

Some computer control systems have a pressure switch installed on the power steering high pressure hose, **Figure 7-27.** This switch is used to determine whether the wheels are being turned to lock. This allows the ECM to compensate by raising engine idle and deactivating the air conditioning compressor. This switch operates in the same manner as the pressure switches installed in the automatic transmission.

Air Conditioning Compressor Relays and Cruise Control

The ECM receives an on-off signal from the air conditioner or cruise control through a wire harness connected directly to the air conditioner compressor clutch relay or cruise control module. Some driver-operated on-off switches deliver a signal to the ECM to indicate that a device or system is functioning. An example is the air conditioner compressor relay, which energizes the compressor clutch coil. The relay and the clutch in some cases is wired as an input connection to the ECM. When the relay energizes the compressor clutch, a low voltage signal informs the ECM that the air conditioning compressor is on.

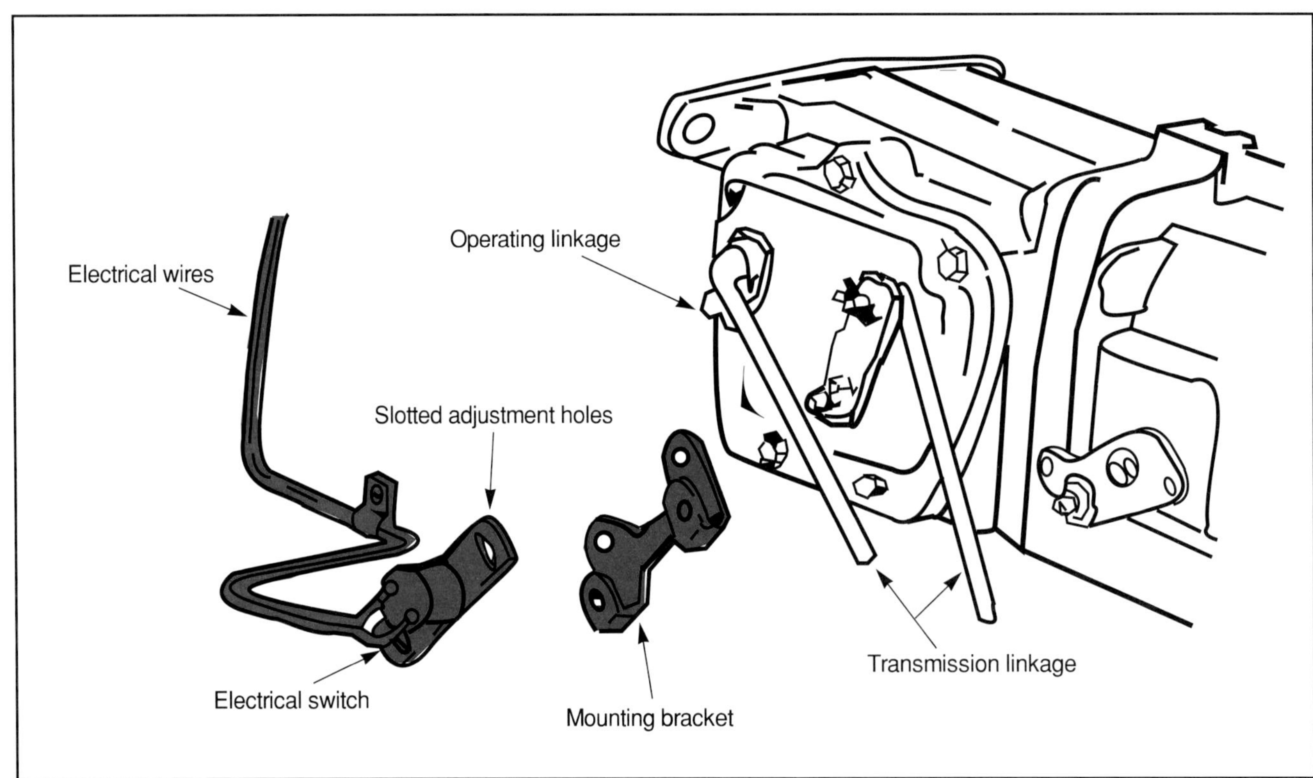

Figure 7-26. *The manual shift sensor is used to allow the ECM to monitor what gear the transmission is in. The ECM uses this information to adjust engine output.*

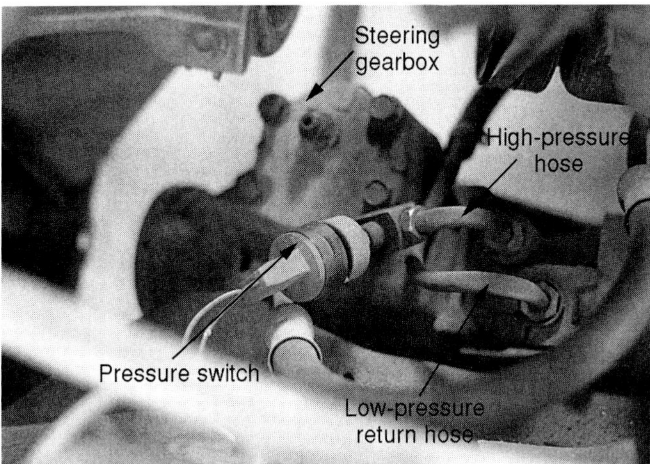

Figure 7-27. *The power steering pressure switch tells the ECM when the steering wheel is locked. The ECM can then increase engine speed to compensate for the extra load.*

Air Conditioning High and Low Pressure Sensors

Air conditioning systems have used high pressure switches for years, **Figure 7-28.** Their primary function is to monitor system pressure and to shut off the compressor if system pressure becomes too high. On some vehicles with computer control systems, the high pressure switch allows the ECM to monitor high pressure and compensate for high pressure. Many newer computer control systems monitor an air conditioning system low system pressure sensor. This sensor notifies the ECM that the air conditioning system is low on refrigerant. The ECM then deactivates and prevents the compressor clutch from engaging.

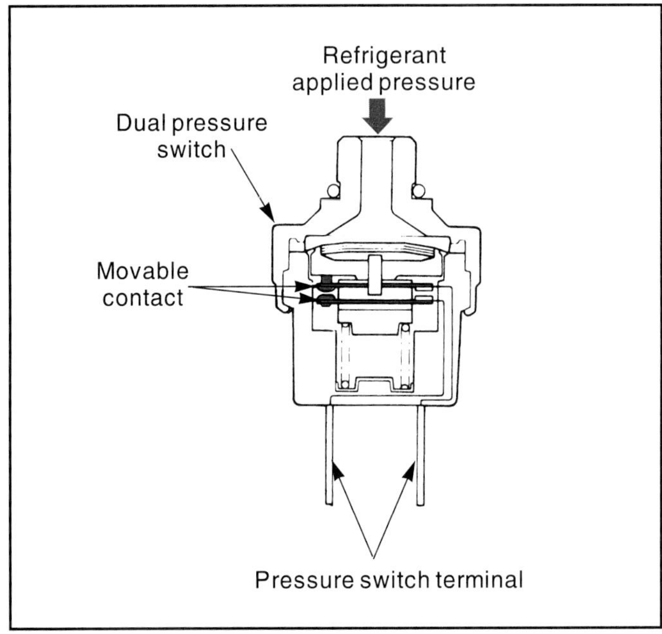

Figure 7-28. *Air conditioning pressure switches are ECM monitored for system operation, overpressure, and low refrigerant level.*

Output Devices

Output devices are electromechanical or electronic parts that carry out the commands of the ECM. Output devices can be solenoids, electric motors, or electronic devices. These are explained in the following sections. Output devices will be discussed in more detail in later chapters.

Solenoids

The most common output devices are electric ***solenoids.*** A solenoid consists of an iron plunger, operated by a magnetic coil. When the coil is energized, a magnetic field is created and moves the plunger. When the coil is de-energized, a spring returns the plunger to its original position. Often a solenoid is used to increase engine idle speed when the air conditioner is in operation. The plunger is positioned to contact a part of the throttle linkage when the coil is energized. The plunger opens the throttle to increase idle.

Many solenoid plungers are attached to flow control valves. The valves can control the flow of fuel, air, exhaust gases, or transmission fluid. Some solenoids are pulsed on and off many times per second. Examples are fuel injectors, and idle air control solenoids. Other solenoids operate less frequently, remaining on or off for seconds or minutes. Examples are the air pump switching solenoids, evaporative emissions canister solenoid, digital EGR solenoid, and torque converter control solenoid.

Idle Air Control (IAC) Solenoids

The ***idle air control (IAC) solenoid*** or valve allows air to bypass the throttle valve into the intake, **Figure 7-29.** The IAC valve is open as long as the throttle plate is closed. When the throttle plate opens, the ECM instructs the IAC valve to close the bypass port, since air through this port is not needed while driving. All of this happens very quickly. Due to the small size of the passages in the bypass, this valve is subject to sticking or clogging at times.

Relays

Electrical ***relays*** are also solenoids. Relays are used when an electrical component consumes more power than the ECM can safely deliver, **Figure 7-30.** In these cases, a small amount of current from the ECM energizes the relay solenoid, closing a set of contact points. A large amount of current can flow through the point set to the electrical component. Electrical components controlled by the ECM through a relay include the air conditioner clutch and radiator fan motor.

Electric Motors

Electric motors are sometimes used as output devices. These motors are usually direct current (dc) motors, similar to the starter motor and smaller motors that operate electric windows and power antennas. The electric cooling fan motor, **Figure 7-31,** is an example of an ECM-controlled motor.

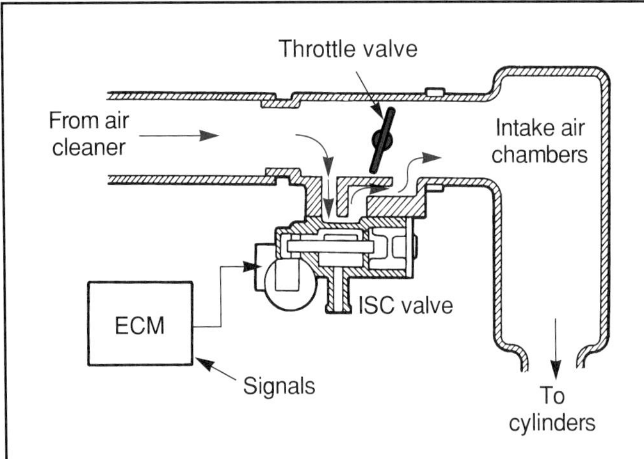

Figure 7-29. *The idle air control (IAC) solenoid allows air to bypass the throttle valve when the throttle valve is closed. When the throttle valve is opened, the valve shuts off.*

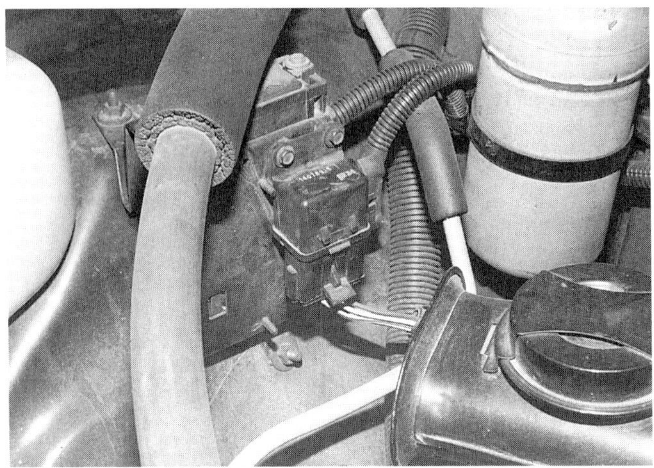

Figure 7-30. *The modern vehicle uses many relays to control high amperage power. This relay controls the air conditioning compressor clutch.*

Small electric motors are often attached to the throttle linkage to control idle speed. They open and close the throttle through a system of gears, that allow the motor to extend or retract a plunger. A motor-driven plunger can provide more precise idle speed control than a solenoid operated plunger. These motors are often called stepper motors or servo motors. They are designed to operate through an exact amount of travel, rather than simply rotating as is the case of fan motors. Some motors are reversible and can be driven forward and backward, according to the current delivered to them.

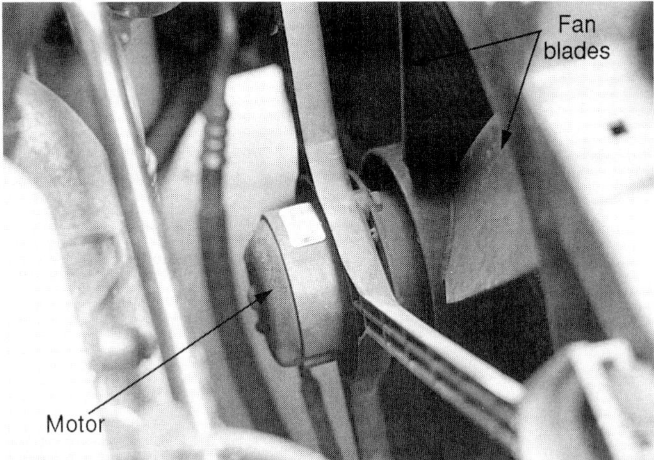

Figure 7-31. *Electric cooling fans can be controlled by the ECM to turn on and off and whether to run at high or low speed.*

Electronic Devices

Electronic output devices can be ignition modules or other solid-state components. These devices are similar to relays, since low current from the ECM controls high-current flowing in the electronic component. Many of these electronic components are input sensors as well as output devices. An example is the ignition module used on many vehicles. The module controls the timing and strength of the ignition coil output, based on commands from the ECM. It also sends engine RPM and crankshaft position signals to the ECM. See **Figure 7-32.**

Control Loop Operation

A ***control loop*** can be thought of as a constantly recurring cycle of causes and effects. The purpose of a control loop is to maintain a certain condition, even when other conditions are constantly changing. Many control loops are relatively simple. An example of a simple control loop is the engine oil pressure regulator valve. When oil pump output exceeds the pressure regulator spring setting, the spring is compressed and the valve opens. Oil escapes past the valve into the oil pan, lowering oil pressure. When pressure is lowered, the spring closes the valve. Refer to **Figure 7-33** for the oil pressure control loop. This action keeps oil pressure at the same level no matter how fast the oil pump is turning.

In the computer control system, the control loop is much more complex, but the basic principle is the same. The control loop tries to keep the air-fuel ratio as close to 14.7:1 as possible under all engine operating conditions. It does this by receiving inputs from the sensors, processing them, and issuing commands to the output devices. The control loop is from input sensors to ECM to output devices, to engine/system, and back to input sensors. A simple loop

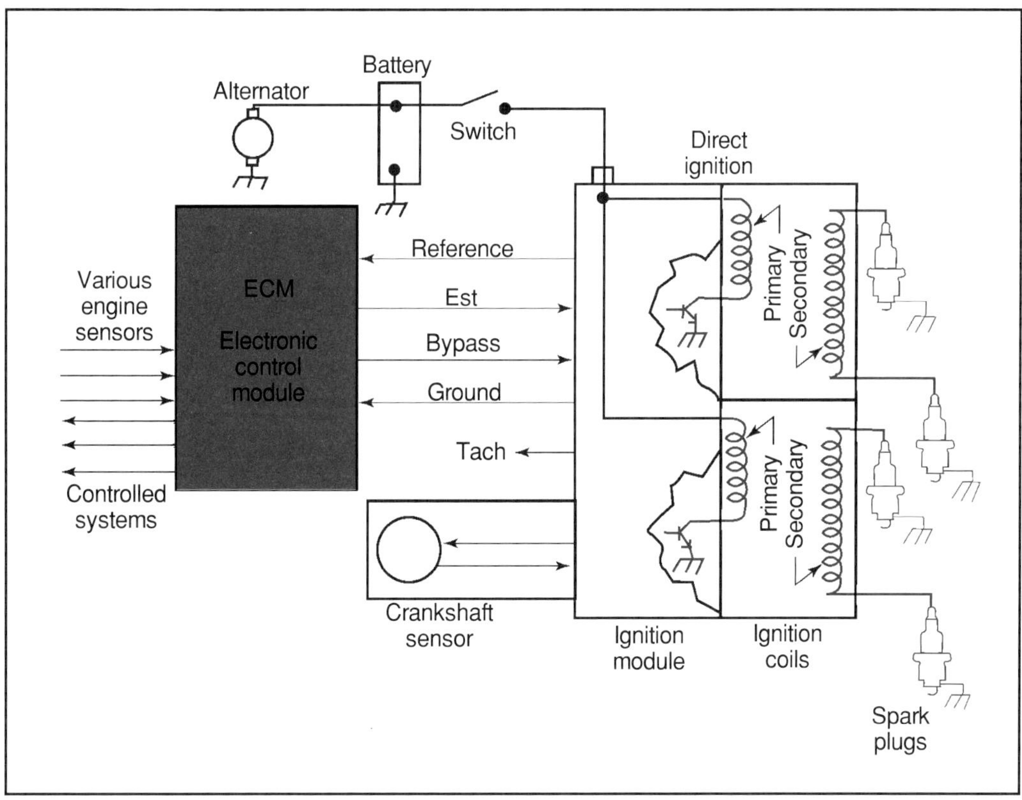

Figure 7-32. *The ignition module is one of the few automotive devices that functions both as an input sensor and an output device. More information on the ignition module is located in* ***Chapter 8.*** *(AC-Delco)*

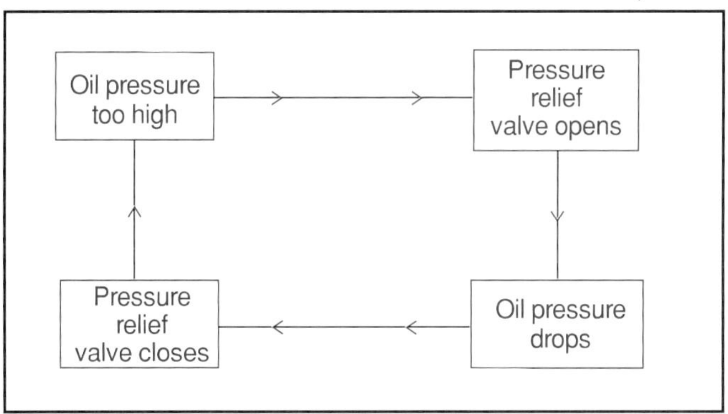

Figure 7-33. *The oil pressure regulation system operates in a loop, similar to the computer control system. Study how each function leads back to the beginning.*

would be from the oxygen sensor, to the ECM, to the fuel injectors, to the engine, and back to the oxygen sensor. Another simple loop would be from the knock sensor to the ECM to the ignition module to the spark plugs to the engine and back to the knock sensor. A third loop would be from the engine coolant temperature sensor to the ECM to the radiator fan to the engine and back to the temperature sensor.

The ECM uses its calculating ability to combine all of these simple loops into one complex loop that considers the input from all sensors, makes decisions, and sends commands to all output devices, **Figure 7-34.** When a vehicle is first started, it operates using preset parameters in the ECM while monitoring sensor input. This is called ***open loop*** operation. Once the vehicle reaches its normal operating temperature, and there are no problems with any sensor, output device, engine, or ECM, the system enters ***closed loop*** operation. During closed loop operation, the ECM monitors all sensors for changes, but pays particular attention to the oxygen sensor in most systems. The ECM adjusts air-fuel ratio based on oxygen sensor input.

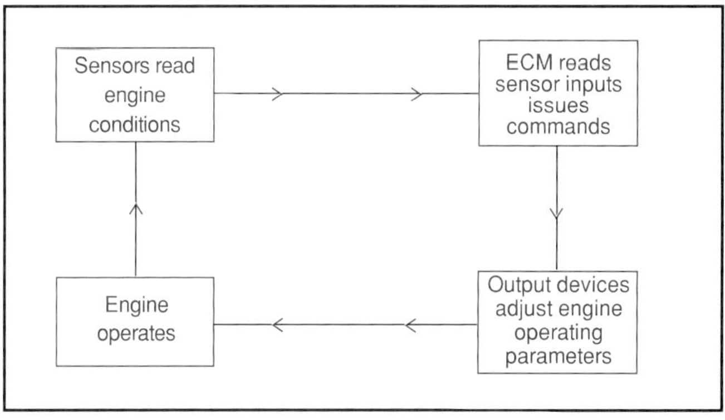

Figure 7-34. *Computer control loop. This is the order of operation during closed loop monitoring. Compare this to the loop in* **Figure 7-1.**

Data Link Connector

To access the ECM memory, a special wiring harness with a ***data link connector*** is connected to the ECM's memory circuits, **Figure 7-35.** Some ECMs have provisions for retrieving trouble codes without a special tester, either by flashing a light installed in the vehicle dashboard or on the ECM or by producing pulses which can be read on an analog or some digital multimeters. These methods can be used by grounding a contact on the diagnostic connector, putting the ECM in the self-diagnostic mode. Some ECMs can be placed in the diagnostic mode by pressing a certain sequence of buttons on the instrument cluster or by turning the ignition key on and off in rapid sequence.

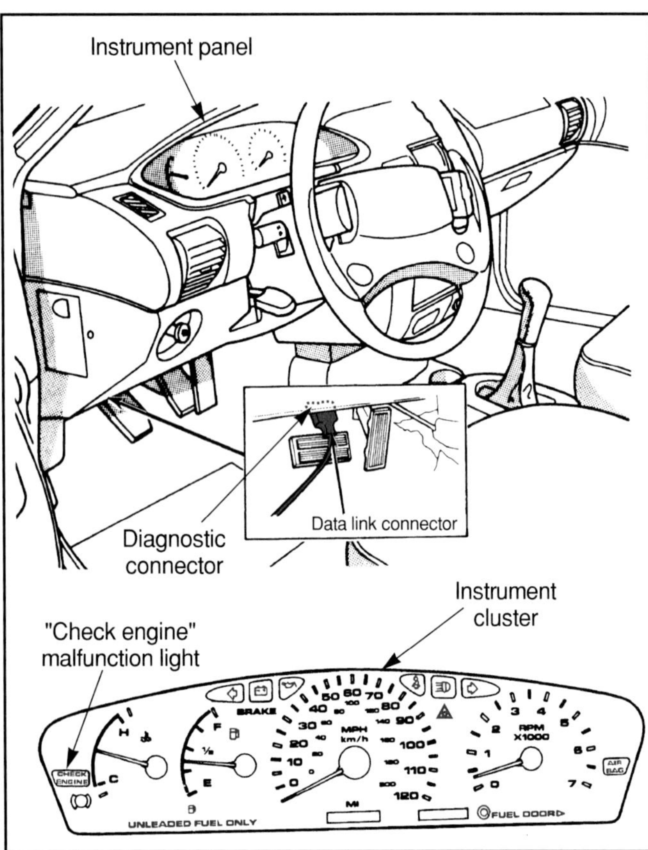

Figure 7-35. *The data link connector can be located almost anywhere on the vehicle in OBD I systems. However, OBD II data link connectors are located within easy access of the driver's seat. (Chrysler)*

However, it is much easier and more accurate to use a scan tool. Most scan tools will display the trouble codes as numbers and save then for easy recall. This makes it easier to perform testing based on one or more trouble codes. Many scan tools are able to perform other diagnostic tasks by sending a series of commands to the ECM and output devices and by monitoring the resulting system operation. Many scan tools can read voltage levels to and from input sensors to determine whether the sensor is operating properly. Some of the latest scan tools have a freeze frame mode, which can take a "snapshot" of sensor and output device readings when a malfunction occurs.

On-Board Diagnostics Generation One (OBD I)

Computer-controlled vehicles have had on-board diagnostics of some form since they were first installed in vehicles. As technology improved, the self-diagnostic systems also improved and many other sensors and output devices were added. These early systems are sometimes referred to as ***on-board diagnostics generation one*** or ***OBD I.***

Most ECMs have built-in ***self-diagnostic capabilities.*** The self-diagnostic system not only checks the ECM for proper operation, but monitors the sensors and output devices to ensure that there are no problems. If a sensor or device becomes defective, is disconnected, or provides readings that are out-of-range due to a non-computer related problem, the ECM stores a ***diagnostic trouble code.*** On most OBD I systems, this code is a two or three digit number that corresponds to a certain malfunction condition. The ECM simply illuminates the dash-mounted malfunction indicator lamp when a code is present.

On-Board Diagnostics Generation Two (OBD II)

OBD I systems could detect when a sensor malfunctioned or when a driveability condition caused a particular sensor's readings to go out of specifications. When the California Air Resources Board (CARB) mandated that all vehicles sold in that state after 1988 have on-board diagnostic systems, requirements for a new and more advanced on-board diagnostics system were also recommended. This new system would not just detect sensor malfunctions and driveability problems when they occurred, but would also detect potential problems before they affect emissions or even become noticeable to the driver. This system is called ***on-board diagnostics generation two,*** or ***OBD II.*** Auto manufacturers began phasing in OBD II systems in certain new models in 1994. All 1996 and newer vehicles sold in the United States are equipped with OBD II.

Note: Some manufacturers may use other names for their version of the OBD II system. OBD II is the term used by the Society of Automotive Engineers (SAE) in their standards for enhanced diagnostics systems, and will be used in this textbook to refer to all such systems. Basic operating principles will be the same for all systems, no matter what they are called.

OBD II Components

Vehicles equipped with OBD II are similar in many ways to other vehicles with older diagnostic systems. They use many of the same sensors, both have diagnostic connectors, and both are designed to aid the technician in diagnosing driveability problems. However, the OBD II equipped vehicle has several additional parameters that it must meet. These include:

- ❑ Redundant sensors for each monitored system. For example, many OBD II vehicles will use both a MAF sensor and a MAP sensor.
- ❑ Most OBD II vehicles will have multiple heated oxygen sensors before the converter.
- ❑ All OBD II vehicles will have a post-converter heated oxygen sensor (catalyst monitor).
- ❑ Most EGR valves will be electronically operated and equipped with a pintle position sensor.
- ❑ Less use of throttle body injection. Most engines will be using sequential multiport fuel injection.
- ❑ Evaporative emissions systems will be equipped with diagnostic switches for purge monitoring. More advanced systems will include test fittings, vent solenoids, and a fuel tank pressure sensor.
- ❑ All OBD II equipped vehicles must monitor the air conditioning system for low refrigerant levels, which would indicate a leak.

Service literature for some OBD II vehicles will have an emblem that indicates that the vehicle is so equipped. The emblem can be found on schematic pages and other charts, **Figure 7-36.**

Figure 7-36. *The service manuals for some OBD II vehicles will have a logo such as this on their schematics and other charts. This helps to identify the vehicle as OBD II compliant. (General Motors)*

OBD II ECM

The additional sensing and diagnostic requirements of OBD II necessitated a more powerful ECM with greater memory and processing capability. ECMs used in OBD I systems transmitted data at a rate of 8.2 kilobits per second (Kbps). OBD II ECMs use Class 2 data communications, which transmits data at a variable rate of 10.4-41.6 Kbps. The OBD II ECM also performs internal monitoring of the following:

- ❑ Cylinder Misfire
- ❑ EGR Flow Rate
- ❑ Fuel Trim.

OBD II PROMs

The OBD II protocol mandates that all PROMs be permanently affixed to the ECM. One of the reasons for this was the increasing popularity of so-called aftermarket "hot PROMs", which are designed to enhance a computer-controlled vehicle's performance. Often, this increased performance also produced increased exhaust emissions. Most newer ECMs use EEPROMs or FEPROMs that are write-protected and can only be reprogrammed using special equipment.

OBD II ECM Diagnostic Testing

The OBD II ECM will perform diagnostic tests to the various systems during operation. The ECM monitors all system components during normal operation. This is called ***passive testing*** and is similar to the way OBD I systems work. In addition, the OBD II ECM can perform ***active tests,*** in which the ECM orders a particular system or component to perform a specific function while monitoring performance. An active test is usually performed in the event a component or system fails a passive test, such as an out-of-range reading. The third test is called an ***intrusive test.*** This test is a type of active test that can affect vehicle performance or emissions. Many of these active and intrusive tests can be performed by the technician using a scan tool.

Malfunction Indicator Lamp

If a code is present, the vehicle's ***malfunction indicator lamp (MIL)*** will light to notify the driver that there is a problem. This lamp is usually amber colored and also called:

- ❑ ***Service Engine Soon lamp***
- ❑ ***Check Engine lamp***
- ❑ ***Power Loss lamp***
- ❑ ***Sensor lamp***
- ❑ ***PGM-FI lamp.***

This lamp may also have the initials "MIL," or the silhouette of an engine. On the older OBD I system, the light comes on if a defect or out-of-range condition is discovered in a sensor, an output device, or the ECU. On OBD II systems, the light will illuminate when a component becomes defective, but will also come on when the proper air fuel ratio is not being maintained, or when another output is not within specifications. The light can illuminate in one of two ways, on steady or flashing.

MIL On Steady

If the OBD II system detects a problem that does not have the potential to damage the catalytic converter, the MIL will burn steadily, indicating that the system should be checked as soon as possible. While the vehicle is driveable, prolonged operation with the MIL on could cause damage to the converter or other vehicle emission controls, as well as reduce mileage and driveability. Unlike the OBD I lamp, the MIL used with OBD II systems will not turn off immediately if the problem corrects itself. OBD II systems require at least three trips (start, warm-up, and stop) before the MIL will turn off.

MIL Flashing

A flashing MIL means that an ongoing engine misfire or other serious problem has occurred. Under severe misfire conditions, the computer may shut off the fuel injector and spark to a misfiring cylinder to protect the catalytic converter. The driver may notice or complain of a loss of power. In many cases, the light will stop flashing when the vehicle is restarted or the condition is no longer present. Even if the light does stop flashing, it will often remain illuminated. In any case, the vehicle should be brought to a service facility for diagnosis and repair as soon as possible. Some manufacturers recommend that a vehicle with a flashing MIL not be driven, but be towed to the nearest service facility.

Note: In some cases, the MIL will be illuminated because of an extremely low fuel level allowing air to enter the injection system, a missing fuel filler cap, or excess fuel entering the evaporative control system during refueling. In these cases, the problem is temporary and the OBD II system will turn off the MIL after a period of engine operation. However, if there is any doubt as to the cause of the flashing MIL, the vehlcle should be checked.

Data Link Connector Location and Design

In older systems, the data link connector could be located in the engine compartment or in many locations throughout the vehicle's interior. Also, each manufacturer used data link connectors that had different shapes. OBD II protocol requires manufacturers to place the data link connector in a location where it is out of visual sight, but is easily accessible from the driver's seat. The protocol also requires the use of a standardized 16-pin data link connector, **Figure 7-37.** The standardized data link connector and codes allows the use of a generic scan tool to read and clear the trouble codes on all vehicles.

Note: Some OBD I equipped vehicles use the 16-pin date link connector. These vehicles are not ODB II compliant.

Trouble Codes

OBD II systems utilize a five digit alpha-numeric trouble code system. There are two general categories of trouble codes. The first category consists of codes assigned by the Society of Automotive Engineers (SAE). A comprehensive list of these codes is located in **Appendix A** of this text. The second level is manufacturer specific codes, which are exclusive to a manufacturer's particular vehicle control system, **Figure 7-38.** Look in the appropriate service manual for these codes.

Within these general categories, there are four levels of diagnostic trouble codes (DTC). These are:

- *Type A codes* are emissions related and will illuminate the MIL. In the case of misfire or fuel trim DTCs, the ECM will flash the MIL whenever driving conditions cause this code to set in memory.
- *Type B codes* are also emissions related. However, the ECM will illuminate the MIL only when this type of code appears on two consecutive warm-ups.
- *Type C codes* are non-emissions related. They will not illuminate the MIL; however, they will store a DTC and illuminate a service lamp or the service message on vehicle equipped with a driver's information center.
- *Type D codes* are also non-emissions related, will store a DTC and will not illuminate any lamps.

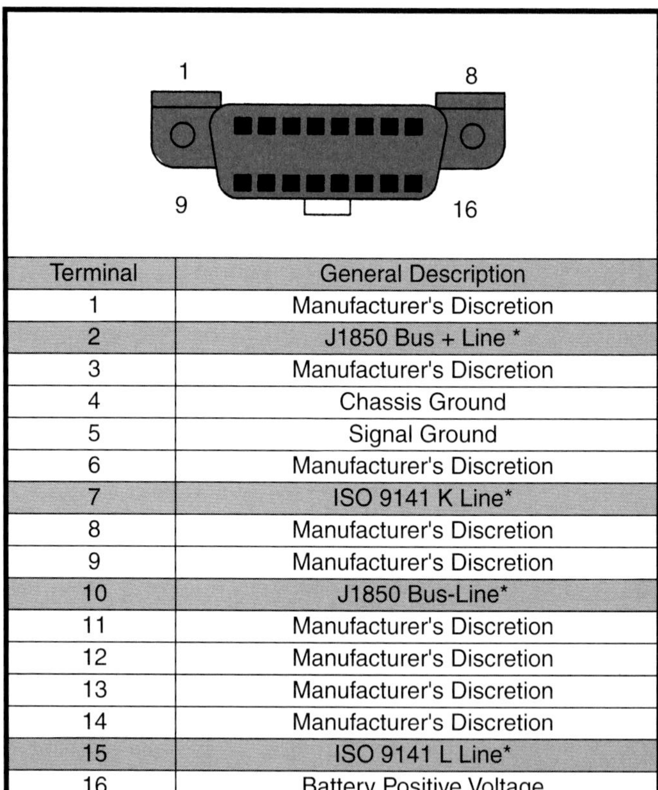

Terminal	General Description
1	Manufacturer's Discretion
2	J1850 Bus + Line *
3	Manufacturer's Discretion
4	Chassis Ground
5	Signal Ground
6	Manufacturer's Discretion
7	ISO 9141 K Line*
8	Manufacturer's Discretion
9	Manufacturer's Discretion
10	J1850 Bus-Line*
11	Manufacturer's Discretion
12	Manufacturer's Discretion
13	Manufacturer's Discretion
14	Manufacturer's Discretion
15	ISO 9141 L Line*
16	Battery Positive Voltage

*Pins 2, 7, 10, and 15 are used for external communications. On some vehicles, these pins may have alternate assignments.

Figure 7-37. *Sixteen-pin data link connector. This connector is used on some OBD I and all OBD II vehicles. Manufacturers usually install sensor leads for other ECMs to the terminals assigned "manufacturer's discretion." Terminals 2, 4, 5, 7, 10, 15, and 16 are assigned by SAE and cannot be changed. (Chrysler)*

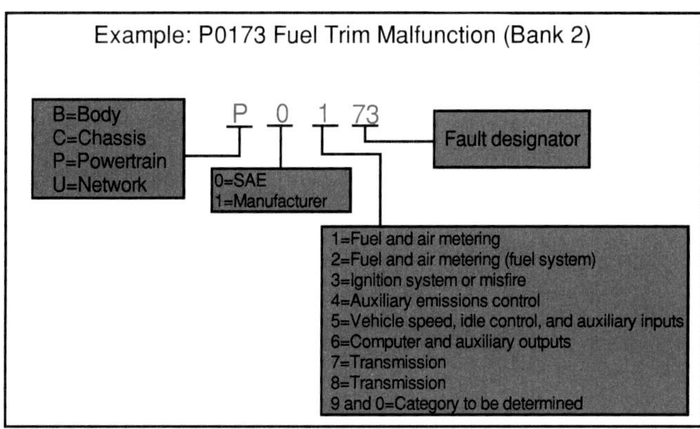

Figure 7-38. *OBD II codes are five digit numbers. The letter and first number is called the alpha-numeric discriminator. The second number indicates the nature of the code. The last two numbers is the fault designator and indicates which sensor or circuit is at fault and the type of problem.*

Failure Types

OBD II systems can also determine the type of problem a circuit has. OBD I systems could only report that a problem exists in a particular circuit, sensor, or device. These failure codes fall into four categories.

A ***general circuit failure*** is caused by disconnected or damaged wires and connectors, grounds and shorts, or a component that is constantly operating out of parameter. This type of circuit failure is often the easiest to locate, since it will set a code in almost all cases. These are similar to the types of hard failures detected by OBD I systems.

A ***low input failure*** is set when the ECM receives a weak or abnormally low voltage, current, or operation signal. This failure is caused by high resistance, poor electrical connection, or a contaminated or defective sensor. A ***high input failure*** is caused by sensor failure or a mechanical fault. This type of signal supplies the ECM with more voltage or current, or a false signal. These failures often trick the computer in an OBD I system into operating the outputs to lean or richen the air-fuel mixture to compensate for a false or artificial condition.

A ***range/performance failure*** occurs when a sensor is reading slightly lower or higher than normal. This can be due to a contaminated sensor, sensor wear, and poor electrical connections. This type of failure was often dismissed by technicians as an intermittent code or was never detected by self-diagnostics in OBD I systems.

Freeze Frame and Failure Record

When a Type A or B trouble code is set in the memory of an OBD II ECM, certain vehicle operations at the time of the code's occurrence will be stored in memory. This information is commonly referred to as the ***freeze frame*** record. Most ECMs will store the failure information for only one code that sets a DTC and illuminates the MIL lamp. If another Type B code occurs, this information will not be updated. If a Type A fuel trim or misfire DTC occurs, it will overwrite the freeze frame information unless a prior fuel trim or misfire code is already stored.

Some ECMs have a ***failure record*** that can be accessed in the case of multiple trouble codes. A fail record will be stored anytime a code is set. A failure record will be set when Type B, C, or D code is set. Type A codes are not stored in the failure record as the information for these codes is stored in the freeze frame. The failure record will be updated whenever a code is set. Most ECMs can store up to five failure records.

Summary

The increase in the efficiency of modern vehicles has been accomplished through improvements in several areas, including the use of the latest electronic devices. The most important advance in the areas of driveability, mileage, and performance is the development of the on-board computer to control the fuel and ignition systems. The on-board computer system is usually called the electronic engine control system. These systems are composed of electrical, electronic, and mechanical devices which control engine operation.

Modern on-board computers control parts of the emission control system, drive train, and accessory equipment such as the air conditioner. A few of the very latest computer control systems also operate parts of the valve train for maximum power. The computer and its related parts are constantly tuning all engine systems and components for best efficiency.

The basic components of the computer control system are the computer, the inputs, and the outputs. The computer receives electrical inputs, uses internal logic circuitry to decide on the actions to be taken, and takes these actions in the form of electrical outputs. The ECM processes the electrical inputs from the sensors. Through the use of tiny electronic components, arranged so that they use logical processes, the ECM interprets the incoming data. The ECM is divided into several internal circuits that process the input from a particular group of sensors, and issue commands to a particular group of output devices.

The main circuits in the ECM are the input processor, the memory, the main processor, the output processor, and the controller. These circuits are not individually serviced, and are all replaced when the ECM is replaced. The input processor takes the information from the input sensors and changes it into a form that the other circuits can use. The memory contains the preset figures for a properly running engine. There are two kinds of memory in the computer, fixed and volatile. The fixed memory circuits are installed at the factory, and cannot be changed. The volatile memory is able to receive and store information from the sensors and output devices. The volatile memory also stores information about malfunctions, in the form of trouble codes.

The main processor takes the information from the input processor and compares it with information in memory. If the input signals match the data in memory, the circuit takes no action. If the inputs do not match the preset data, the circuits affect the operation of the output devices until the input signal matches the memory data. The output processor takes the commands of the main processor and converts them into electrical signals to the output devices. The output processor uses the small voltage inputs from the main processor to control large current flows into solenoids or motors.

The controller directs the operation of the other circuits, telling them what tasks to undertake next. The controller also determines the overall status of the engine and vehicle, and can compensate for long-term changes in engine operation. This is called the "learning" ability of the ECM.

There are two generation of on-board diagnostics. Inputs to the ECM can be switches, resistance units, magnetic fields, chemical reaction sensors, pressure sensitive materials, light emitting and collecting devices, and transducers. Input sensors and switches monitor conditions in the engine and drive train, outside atmospheric conditions, and direct or indirect driver inputs. There are many kinds of input sensors, operated by various electrical, temperature, and chemical methods.

Engine sensors are installed on the engine and measure the actual conditions in the engine, such as temperature and RPM, crankshaft position, manifold vacuum, intake air flow, exhaust gas oxygen, and whether the engine is knocking. Other engine sensors may be used to measure the temperature of the incoming air, camshaft position, or exhaust gas temperature.

Drive train sensors monitor conditions in the transmission and other drive train components. A few vehicles use speed sensors mounted on the drive wheels to allow the ECM to control wheel traction. Accessory equipment sensors are used to monitor such conditions as alternator output voltage, whether the air conditioner or cruise control is on or off, and whether the wheels are being turned.

Atmospheric sensors measure conditions outside the vehicle. Atmospheric sensors include those that measure air temperature, barometric pressure, and sometimes humidity. Driver operated sensors are installed on the throttle linkage, brake pedal, and shift lever.

The outputs cause changes in engine operation, which are picked up as new readings by the input sensors. The actual engine and drive train components being affected constitute a fourth part that completes the cycle. The system works in a continuous cycle, with inputs, the ECM, and outputs all affecting each other. The ECM receives inputs from the input devices, processes these inputs, and sends commands to the output devices. Output devices are those electromechanical or electronic parts which carry out the commands of the ECM. Output devices can be solenoids, electric motors, or electronic devices.

The first computer control systems, or OBD I, was able to detect defective computer control system components and other serious problems. The OBD II system is a more sophisticated system which can more closely monitor engine conditions and tell when a part or system is deteriorating before it actually fails. The OBD II diagnostic system can provide much more diagnostic information than the OBD I system. In addition to the hardware of the OBD I system, the OBD II system has an extra oxygen sensor, located after the catalytic converter in the exhaust stream. The OBD II system is capable of detecting engine misfires before they are noticed by the driver. The shape and location of the OBD II diagnostic connector has been standardized.

Know These Terms

On-board computer
Electronic engine control system
Input devices
Sensors
Output devices
Electronic control module (ECM)
Powertrain Control Module (PCM)
Vehicle Control Module (VCM)
Reference voltage
Serial data
Pulse width
Data bus
Multiplexing
Universal asynchronous receive and transmit (UART)
Class 2 serial communications
Input processor
Fixed memory
Read Only Memory (ROM)
Programmable Read Only Memory (PROM)
Eraseable Programmable Read Only Memory (EPROM)
Electrically Eraseable Programmable Read Only Memory (EEPROM)
Flash Eraseable Programmable Read Only Memory (FEPROM)
Volatile memory
Random Access Memory (RAM)
Keep Alive Memory (KAM)
Adaptive strategy
Central processing unit (CPU)
Output processor
Output driver
Quad driver
Mem-Cal
Short-term fuel trim
Long-term fuel trim
Minimum idle speed control
Clear flood mode
Exhaust temperature sensor
Transducers
Intake air temperature sensor (IAT)
Manifold air temperature (MAT) sensors
Barometric pressure sensors
Manifold absolute pressure (MAP) sensors
Mass airflow sensor (MAF)
Engine coolant temperature sensor (ECT)
Crankshaft position sensors
Camshaft position sensors
Throttle position sensor (TPS)
Knock sensors
Oxygen sensor (O_2)
Titania
Heated oxygen sensors (HO_2S)
Catalyst monitor
Vehicle speed sensors (VSS)
Reed switches
Solenoids
Idle air control (IAC) solenoid
Relays
Electric motors
Control loop

Open loop
Closed loop
Data link connector
On-board diagnostics generation one (OBD I)
Self-diagnostic capabilities
Diagnostic trouble code
On-board diagnostics generation two (OBD II)
Passive testing
Active tests
Intrusive test
Service engine soon lamp
Malfunction indicator lamp (MIL)
Check engine lamp
Power loss lamp
Sensor lamp
PGM-FI lamp
General circuit failure
Low input failure
High input failure
Range/performance failure
Freeze frame
Failure record

Review Questions—Chapter 7

Please do not write in this text. Write your answers on a separate sheet of paper.

1. Electronic engine control systems are composed of ______, ______, and ________ which control engine operation.
2. ______ switches are often used to measure speed by measuring rotation.
3. In a coolant temperature sensor, changes in temperature affect the ______ reading.
4. An oxygen sensor is a ______ reaction sensor.
5. A knock sensor produces an electric current in response to a change in:
   (A) temperature.
   (B) speed.
   (C) pressure.
   (D) Both A & C.
6. The emitter, detector, and interrupter are part of a ______ sensor.
7. A sensor that consists of an iron plunger inside a wire coil is called a ______.
8. Most automatic transmission sensors are ______ switches.
9. Throttle position sensors are operated by:
   (A) engine temperature.
   (B) engine speed.
   (C) the driver.
   (D) vehicle speed.
10. The most common output devices are:
    (A) solenoids.
    (B) electric motors.
    (C) electronic modules.
    (D) None of the above.
11. Sometimes a solenoid is used to increase engine idle speed when the ______ is operating.
12. Electrical relays are energized by small amounts of electrical power from the:
    (A) ECM.
    (B) ignition module.
    (C) throttle position sensor.
    (D) MAP sensor.
13. Small motors are often attached to the ______ linkage to control idle speed.
14. The ability of the ECM to compensate for changes on sensor input is called _____ strategy.
15. The output processor is sometimes called an output ______.

ASE Certification-Type Questions

1. Technician A says that volatile ECM memory is called EEPROM. Technician B says that volatile ECM memory is called FEPROM. Who is right?
   (A) A only.
   (B) B only.
   (C) Both A & B.
   (D) Neither A nor B.
2. All of the following terms are associated with the KAM memory circuit, EXCEPT:
   (A) keep alive.
   (B) programmable read only.
   (C) nonvolatile RAM.
   (D) adaptive strategy.
3. Technician A says that using a scan tool is the easiest means of retrieving trouble codes. Technician B says that the OBD II diagnostic system can only be accessed by using a scan tool. Who is right?
   (A) A only.
   (B) B only.
   (C) Both A & B.
   (D) Neither A nor B.
4. In many sensors, a reference voltage is passed through a ______ to produce a final voltage reading.
   (A) diode
   (B) resistance unit
   (C) capacitance unit
   (D) PROM
5. All of the following are types of air flow sensors, EXCEPT:
   (A) capacitor discharge.
   (B) heated wire.
   (C) air valve.
   (D) Karman vortex.

6. The ECM reads engine ______ to decide whether to place the system in closed or open loop.
   (A) speed
   (B) oil pressure
   (C) temperature
   (D) air flow
7. All of the following are found on a OBD I (first generation) computer control system, EXCEPT:
   (A) engine misfire monitor.
   (B) exhaust gas oxygen sensor.
   (C) ECM.
   (D) data link connector.
8. Technician A says that a flashing MIL light on an OBD II system may be caused by a low fuel level. Technician B says that a flashing MIL light may be caused by a defect in the starting system. Who is right?
   (A) A only.
   (B) B only.
   (C) Both A & B.
   (D) Neither A nor B.
9. An example of an electronic output device is the ______.
   (A) ECM
   (B) ignition module
   (C) coolant temperature sensor
   (D) transmission pressure switch
10. Technician A says that many ECM-related electronic components are input sensors. Technician B says that many ECM-related electronic components are output devices. Who is right?
   (A) A only.
   (B) B only.
   (C) Both A & B.
   (D) Neither A nor B.

Most modern ignition systems have few moving parts or none at all. This distributorless ignition system (DIS) coil pack is one of two on a modern V-8 engine. (Jack Klasey)

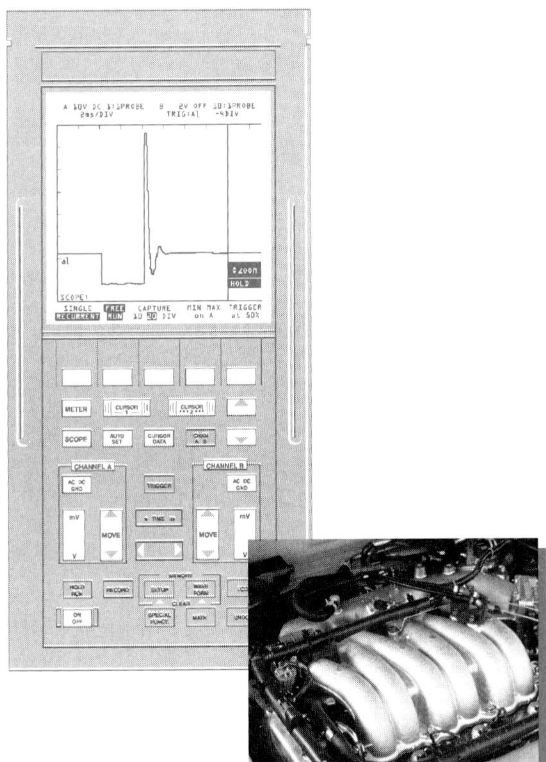

Chapter 8

Ignition System Fundamentals

After studying this chapter, you will be able to:

- Identify and explain the purposes of the ignition coil.
- Identify and explain the purposes of the ignition module.
- Identify and explain the purposes of the ignition triggering devices.
- Explain how the spark is developed in electronic and contact point distributor ignition systems.
- Identify and explain the purposes of distributorless and direct ignition systems.
- Identify and explain the purposes of vacuum and centrifugal spark advance systems.
- Identify and explain the purposes of computerized spark advance systems.
- Identify and explain the purposes of each of the major components of the ignition secondary system.

This chapter will identify the components of the ignition system. The condition of the ignition system is critical to good engine and vehicle driveability. If the ignition system has any type of defect, the engine will not run properly or at all. This chapter will cover the ignition components, reviewing the operation of each subsystem and how even minor misadjustments or defects in these subsystems can affect driveability. Before beginning this chapter, you may want to review the basic electricity and electronics material in **Chapter 6.**

Ignition System

The air and fuel cannot do any useful work unless they are made to burn, that is, ignited. On diesel engines, ignition is accomplished by high compression, which heats the air-fuel mixture in the engine cylinder to its flash point. On a gasoline engine, the ***ignition system*** produces a spark that ignites the compressed air-fuel mixture in the engine cylinder. The gasoline ignition system is comprised of a primary circuit that creates the high voltage needed to fire the spark plug, and a secondary circuit that distributes this high voltage to the right spark plug at the right time.

Warning: Modern ignition systems operate at very high voltages, up to 100,000 volts. A shock from these systems can injure or kill. Handle any ignition system with extreme care.

Electronic Ignition Systems

All ignition systems made since the mid 1970s use electronic components to create and time the spark. These systems are referred to as ***electronic ignition systems.*** These systems are capable of creating a spark as much as 100,000 volts. Older engines used a system known as the conventional or contact point system. Some modern engines use a variation of the electronic ignition system that eliminates the distributor and in some cases, the spark plug wires. All of these systems and their components will be discussed in this chapter.

Primary Ignition Circuit

The ***primary ignition circuit*** creates the high voltage spark from the battery's 12 volts. The primary circuits handle battery voltage or less, depending on the system. The modern primary system consists of a coil, an electronic control module, a triggering device, associated wiring, and a resistor (where used). The wiring in the primary ignition circuit is covered with a thin layer of insulation to prevent short circuits.

Ignition Coils

The ***ignition coil*** is a type of transformer; a device which converts electricity from low to high voltage. This voltage increase is the result of creating a magnetic field and then collapsing it. The ignition coil consists of two coils of wire, or windings, around a soft iron core. The ***primary winding*** is comprised of a few turns of heavy wire. The ***secondary winding*** is comprised of many turns of fine wire. The core is made of soft iron as this type of iron will not become permanently magnetized, which would upset the coil's operation.

To create high voltage, battery current passes through the primary winding when the ignition switch is turned on. The turns of heavy wire allow high current flow. This high current flow creates a magnetic field. When the current flow through the primary winding is stopped, the magnetic field collapses. As the field collapses through the primary winding, the voltage in the primary windings will be increased to about 200 volts. This is called ***self-induction,*** since the primary windings produce their own voltage increase.

As the magnetic field collapses through the secondary windings, it creates electricity that has very high voltage, from 60,000-100,000 volts, but with very little current. Since the secondary windings utilize very fine wire, it will not allow the energy from the collapsing magnetic field to develop high current. The difference in wire size and the number of wire turns between the primary and secondary windings causes the voltage created in the secondary winding to increase. This high voltage current is delivered through the secondary system to the spark plugs. The current discharges across the plug gap to ignite the air-fuel mixture.

Types of Coils

Coil shapes vary between vehicles and engines. The *round coil,* or oil-filled coil, was used for many years and can be found on a few vehicles built as recently as the early 1990s. The construction of a round coil is shown in **Figure 8-1.** Both the primary and secondary windings are wound around an iron core, then covered with an insulating paper or varnish. The secondary output is the insulated terminal at the top center of the coil. The coil is filled with oil or a material similar to paraffin that acts as insulation and as a heat absorber. The entire coil is housed in a type of plastic called Bakelite, or a metal casing made of steel or aluminum.

1 High-tension connection on the outside, 2 Winding layers with insulating paper, 3 Insulating cap, 4 High-tension connection on the inside via spring contact, 5 Case, 6 Mounting bracket, 7 Metal plate jacketing (magnetic), 8 Primary winding, 9 Secondary winding, 10 Sealing compound, 11 Insulator, 12 Iron core.

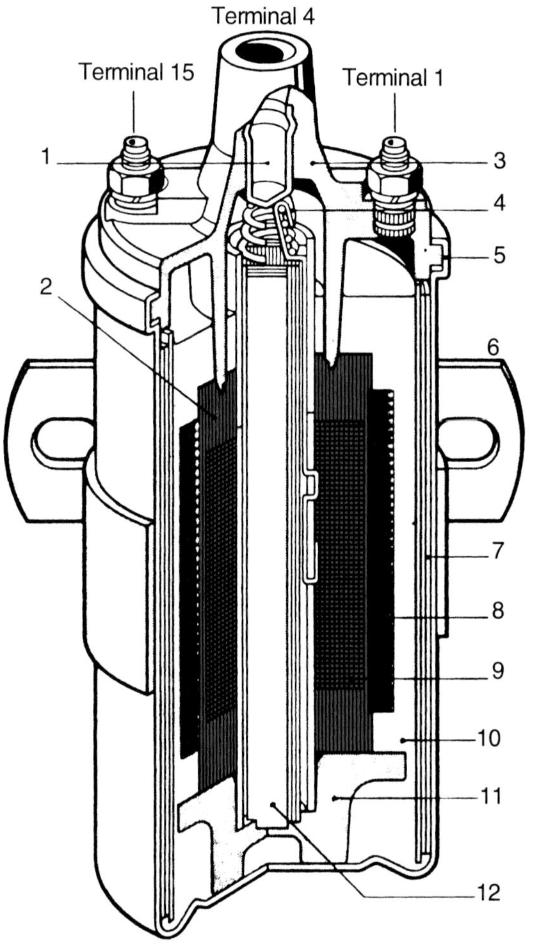

Figure 8-1. *The round coil was the only kind used on gasoline engines for over 50 years and is still found on a few vehicles. The small terminals are the primary circuit connections. The large terminal is the secondary circuit output terminal. (Bosch)*

Variations of the *flat coil* are used in most late-model engines, **Figure 8-2.** The flat coil consists of primary and secondary windings surrounded by a square iron frame, **Figure 8-3.** The windings are coated with epoxy to preserve the insulation between the primary and secondary windings. Many flat coils are also encased in heavy plastic to further insulate the windings. The positive and negative primary terminals are attached to the side of the coil. Some coils have a third wire that serves as a ground. The secondary output is at the center of the coil. Some coils are installed under a plastic cover on top of the distributor cap, **Figure 8-4.** The coil is not sealed and output goes directly to the rotor.

Coils used with distributorless ignition systems may be part of the control module assembly, **Figure 8-5.** There is

Figure 8-2. *The flat coil has replaced the round coil on most engines. It makes use of the same operating principles as the round coil.*

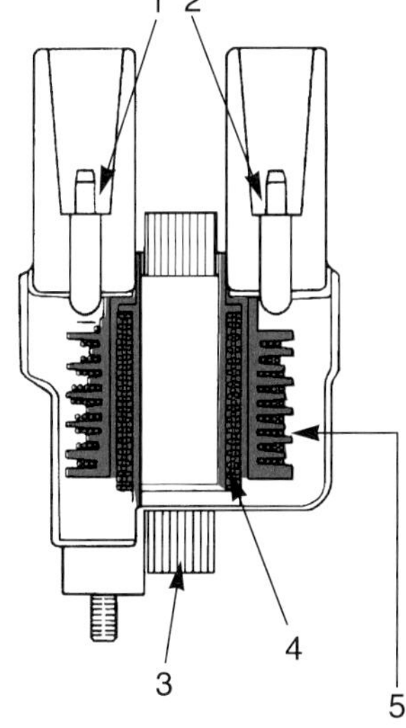

Figure 8-3. *The iron core extends around the coils of the flat coil to increase the magnetic field interaction. 1–Secondary output (+). 2–Secondary output (-). 3–Iron core. 4–Primary winding. 5–Secondary winding. (Mercedes-Benz)*

one coil for every two cylinders. In some cases, these coils can be replaced individually, while on other systems, the entire assembly must be replaced if one coil becomes defective.

Ignition Control Modules

The ***control module*** is the solid state device that processes the inputs from sensors and other ignition components and collapses the coil's magnetic field to produce a high voltage spark. Control modules are comprised of many solid state electronic components in a single sealed unit. Modules cannot be repaired and must be replaced if defective. Older control modules were large, as in **Figure 8-6.** Most newer modules are small, as in **Figure 8-7.**

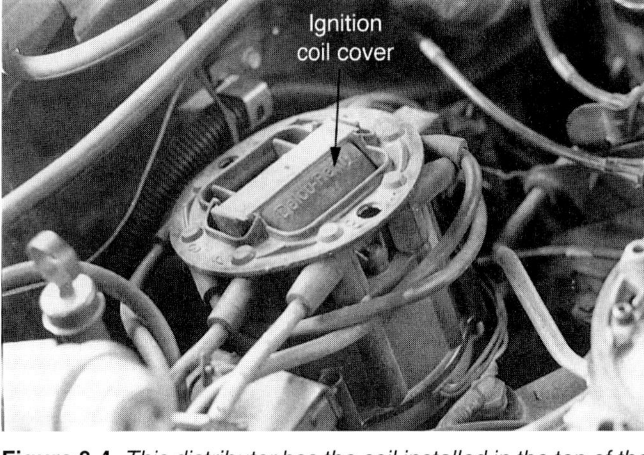

Figure 8-4. *This distributor has the coil installed in the top of the cap. Placing the coil here eliminates the need for a coil wire, reducing the chance of moisture shorting out the secondary connections.*

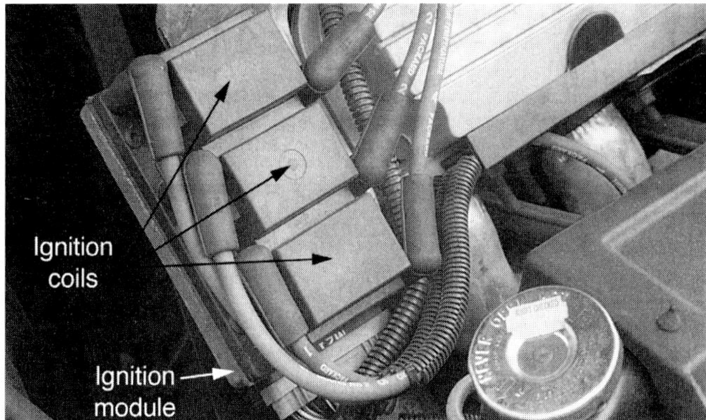

Figure 8-5. *There is one coil for every two cylinders in a distributorless ignition system. Each coil fires two spark plugs at the same time. One plug fires on the cylinder's exhaust stroke, but has no effect.*

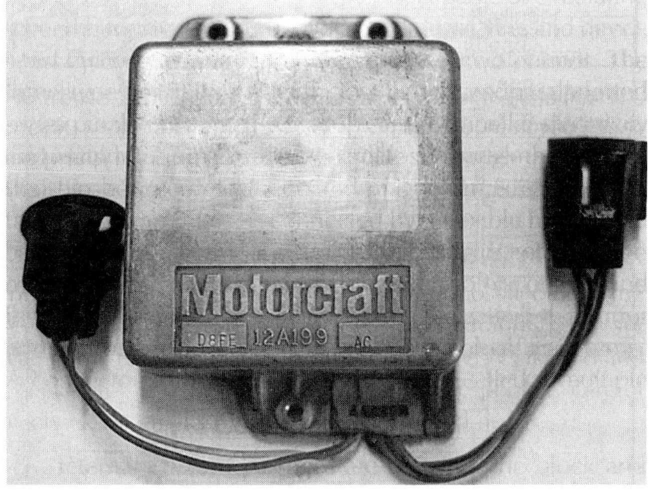

Figure 8-6. *This large ignition module is used on many older Ford vehicles. It is often mounted on the inner fender, away from as much engine heat as possible. In most cases, large modules like this were used on engines without computer-controlled spark timing.*

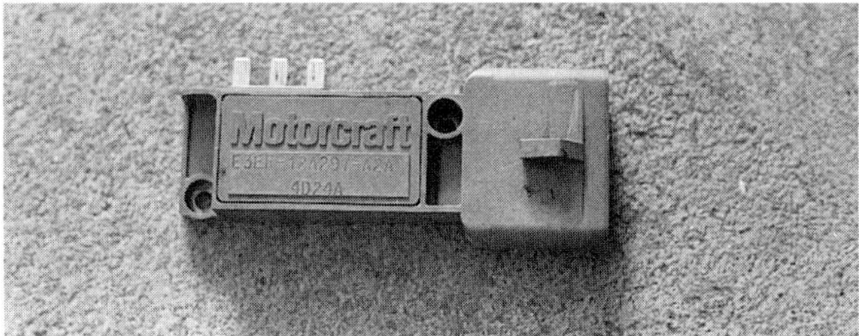

Figure 8-7. *This ignition module is mounted on the outside of the distributor. Some modules are mounted inside the distributor. This module is used on many engines with electronic engine controls.*

The module is often installed inside the distributor or on the outside of the distributor body. Some modules are located on the engine firewall or inner fender to remove them from engine heat. Other modules serve as the base for or are part of the ignition coil assembly.

The vehicle battery provides the module with low voltage electricity. Battery power is sent to the module only when the ignition switch is in the on or start position. The module passes this low voltage current onto the coil through a ***power transistor.*** The power transistor is a solid-state device that allows current to flow, based on electrical inputs from other module circuitry. Current flows from the power transistor and the coil primary winding to ground. This current flow creates a magnetic field around the coil windings, **Figure 8-8A.** At this time, the power transistor is a conductor, allowing full current to flow in the circuit.

A triggering device signals the ignition module circuitry, composed of diodes, transistors, capacitors, and resistors. When it is time to collapse the coil's magnetic field, the control module applies a voltage signal to the power transistor. The power transistor becomes an insulator. Since current cannot flow through an insulator, current flow through the coil primary circuit is stopped. When current flow stops, the magnetic field around the coil windings collapses, as shown in **Figure 8-8B.** Ignition modules are not serviceable. If a module is defective, it must be replaced.

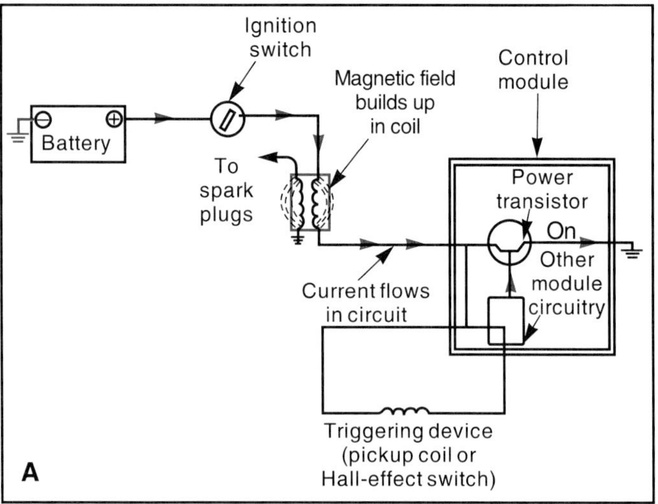

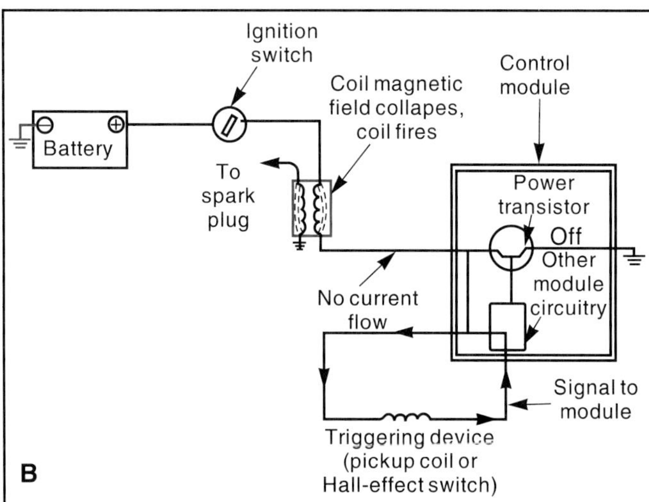

Figure 8-8. *A—Most modern ignition systems use a power transistor to control power flow through the coil. The power transistor is controlled by a small voltage signal from the pickup coil. In this figure, the power transistor is on and current flows through the coil, building up the magnetic field. B—In this figure, the power transistor is off, stopping current flow to the coil. This causes the coil's magnetic field to collapse and discharge through the secondary terminal.*

Triggering Devices

The creation of the secondary voltage is controlled by the breaking of current flow through the primary circuit. To cause the coil's magnetic field to collapse, the current flow through the primary windings must be interrupted instantly and cleanly with no jumps or arcs across space at the point of disconnection. Current flow is broken by the ignition control module, which uses transistors to operate the primary circuit, based on a signal from a ***triggering device.*** Three types of triggering devices are commonly used in modern electronic ignition systems. The pickup coil type will be discussed first, followed by the Hall-effect and the photoelectric switch.

Pickup Coil

Some electronic ignition distributors contain a small coil, called a ***pickup coil.*** The pickup coil was the first type of triggering device commonly used in electronic ignition systems and is still used in a few vehicles. The pickup coil is mounted on a stationary plate in the distributor. A pickup coil is shown in **Figure 8-9.** A toothed wheel is mounted on the distributor shaft. There is one tooth for each engine cylinder. This wheel may be called a *trigger wheel,* a *reluctor,* or an *armature,* depending on the manufacturer. The relationship of the pickup coil and wheel is shown in **Figure 8-10.**

Figure 8-9. *Magnetic pickup coils contain a small coil of wire and a magnetized iron core. These are sometimes referred to as permanent magnet generators.*

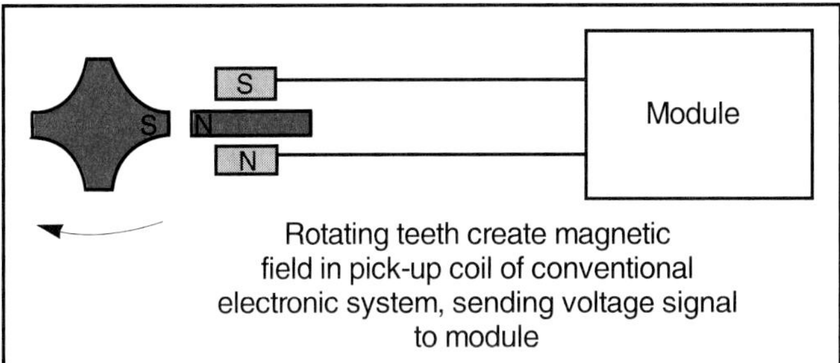

Figure 8-10. *The relationship between the pickup coil and trigger wheel is shown here. When the trigger wheel passes near the pickup coil, a magnetic field is created in the wire coil. This produces a small voltage that serves as a signal to the ignition module.*

The distributor shaft is driven by the engine, usually from the camshaft. As each tooth of the wheel passes the pickup coil, a magnetic field is created. The creation of this magnetic field produces a voltage signal that is sent to the control module. The module then interrupts current flow in the primary circuit, causing the output coil to fire. Some ignition systems may have as many as three coils, **Figure 8-11.** Multiple pickups indicate that the distributor is providing multiple engine speed signals to the ECM, ignition module, and in some cases, the cruise control.

Another type of pickup system uses a magnetic pickup coil with a notched wheel. When a notch in the wheel lines up with the pickup, the increase in the gap between the wheel and pickup affects the magnetic field, generating a voltage pulse to the ignition module or ECM. This system is usually used on a distributorless ignition system.

Hall-Effect Switch

In some electronic distributors, a magnetic field is created by current flow through an electromagnet with two separate poles. The electromagnet is called a ***Hall-effect switch.*** It is mounted on the distributor's stationary plate, **Figure 8-12.** Current to the Hall-effect switch flows through the control module. The distributor also contains a shutter assembly, sometimes called a chopper, which has one

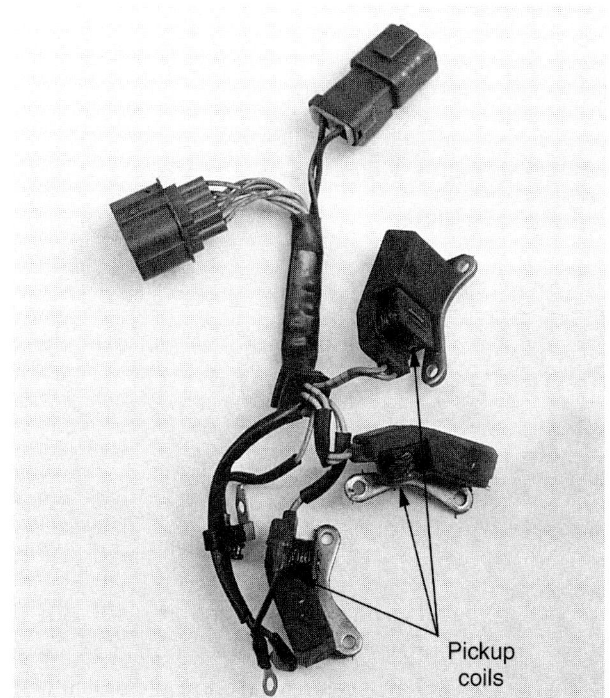

Figure 8-11. *All of these pickup coils are located in one distributor. Two of the pickups send separate electrical signals to the ECM. The ECM can then process each signal to determine engine speed and crankshaft position. The third sensor is used to trigger the ignition module.*

Figure 8-12. *The Hall-effect switch also has a coil of wire surrounding a metal core. However, current flows through the wire coil whenever the ignition is on. This creates a magnetic field between the two poles.*

shutter blade for each engine cylinder. The shutter assembly rotates with the distributor shaft. In many cases, the shutter assembly is part of the distributor rotor, as shown in **Figure 8-13.**

The magnetic field is interrupted every time a shutter blade passes between the two poles of the Hall-effect switch. This causes a voltage fluctuation, which is read by the control module. The module then shuts off coil primary current, causing the coil to fire. A typical Hall-effect assembly is shown in **Figure 8-14.**

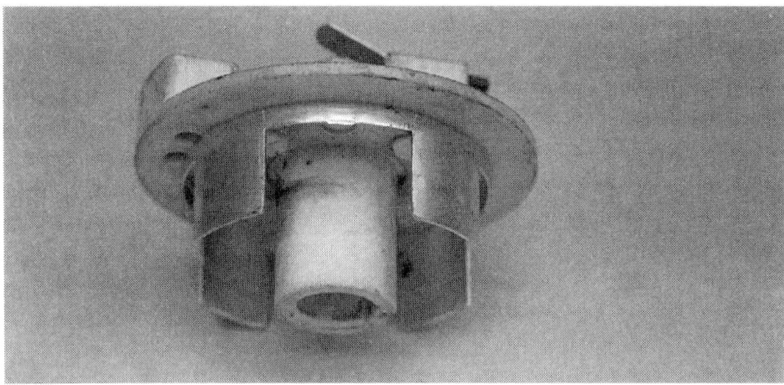

Figure 8-13. *The shutter assembly, such as the one shown here, is used to break the magnetic field. While most shutters are part of the distributor rotor, a few are installed directly on the rotor shaft.*

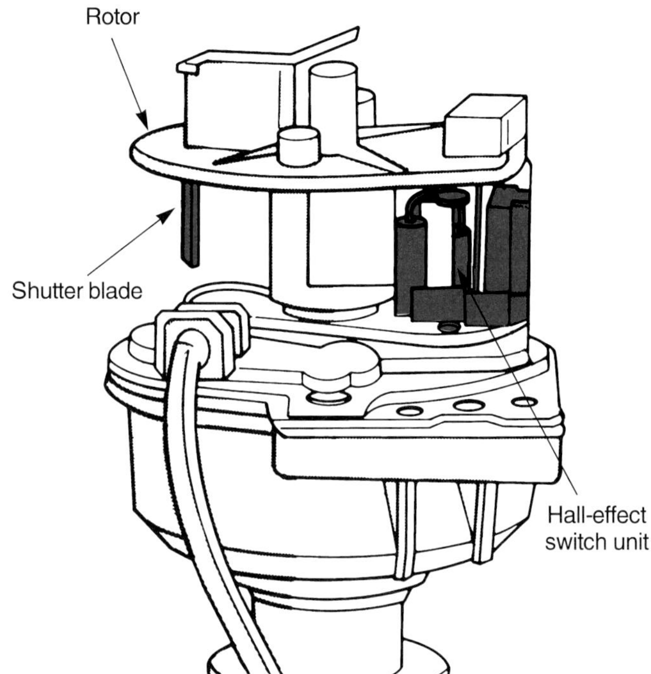

Figure 8-14. *The Hall-effect assembly is shown here, installed on a distributor. As the shutter turns with the rotor, it breaks the magnetic field in the Hall-effect switch. The resulting voltage fluctuation is sent to the ignition module. (Chrysler)*

Photoelectric Sensor

Some distributors generate a triggering pulse by the use of a ***photoelectric sensor.*** This sensor consists of three major parts, as shown in **Figure 8-15.**

- A light emitting diode (LED), sometimes called an emitter, which produces a light signal.
- A photodiode, sometimes called a detector, which receives the light signal from the LED.
- A rotating shutter assembly which breaks the light beam between the LED and photodiode.

The LED produces a light signal whenever the ignition is on. The shutter is attached to the distributor shaft and rotates with it. A series of openings, or slits, in the shutter make and break the light beam from the LED to the photodiode. The photodiode converts the making and breaking of the light beam into an electrical signal that is read by the ECM. The shutter assembly slits are arranged so that the ECM can determine crankshaft position, engine RPM, and also tell which cylinder is coming up on compression. On most systems, at least one slit is dedicated to each cylinder, with a master slit to determine when the firing sequence begins again. Many shutter assemblies are made with 360 slits, one for each degree of rotation, to ensure the fastest response to engine rpm changes. On vehicles equipped with OBD II, these speed inputs are monitored by the ECM for misfiring. This process is explained later in this chapter.

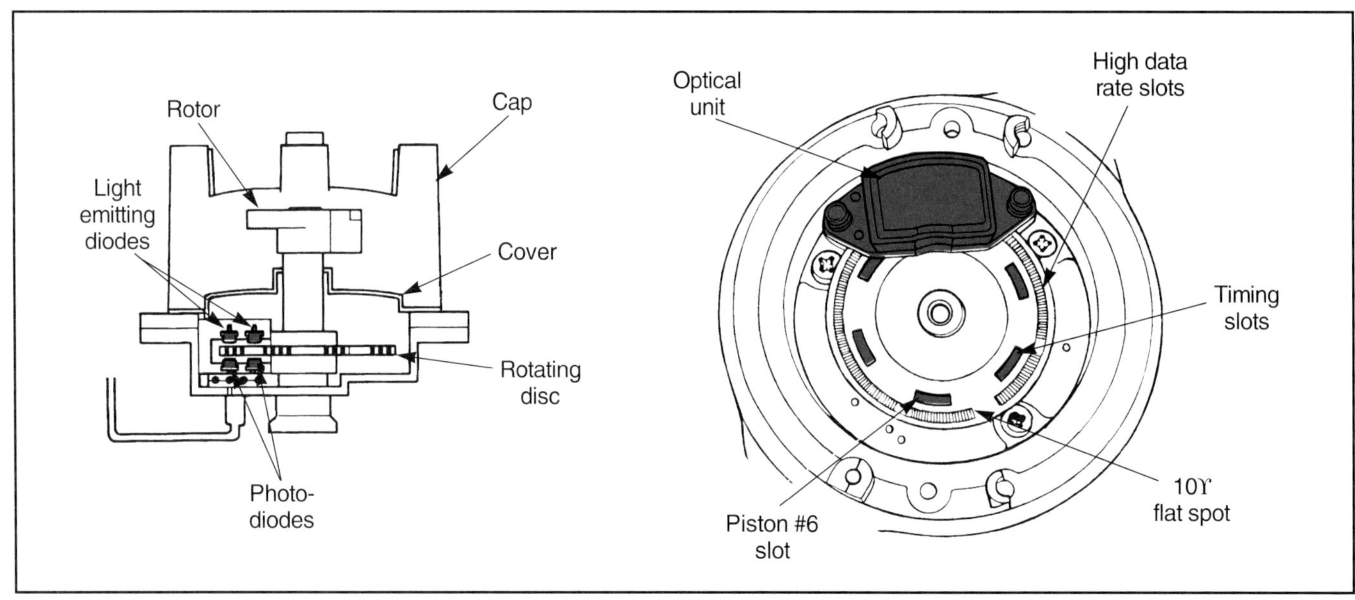

Figure 8-15. *A typical photoelectric distributor. Note the relative positions of the light emitting diodes, photodiodes, and shutter assembly. The position of the slots in the shutter control the reception of LED light by the photodiode. The photodiode converts the light to an electrical signal that is sent to the ECM. (Chrysler)*

Distributorless Ignition Systems

Distributorless ignition systems (DIS) are electronic ignition systems used on many modern engines. The distributorless ignition system has no distributor. **Figure 8-16** shows the schematic for the distributorless ignition system used on one engine. This system eliminates the distributor's moving parts that can wear out, even in electronic ignition systems.

The distributorless ignition uses a pickup coil or Hall-effect switch that performs the same job at the crankshaft as in the distributor—matching the firing of the spark plug to the piston compression stroke. Some distributorless systems have a second sensor monitoring the camshaft. The ignition sensor is placed on the engine near a rotating disc installed on the engine crankshaft, camshaft, or crankshaft pulley. The disc uses notches, embedded magnets, or a shutter assembly to produce a signal or to vary current flow in a sensor assembly. See **Figure 8-17.** In either system, the result is that an electrical signal is generated and sent to the ignition module whenever the crankshaft or camshaft is rotating. This signal allows the ignition module to determine the position of each piston in the engine. **Figure 8-18** shows the placement of typical distributorless ignition system sensors.

The design of the crankshaft sensor ring creates a pattern of sensor inputs that allows the ignition system to determine which piston is coming up on compression, **Figure 8-19.** Many engines utilize more than one crankshaft sensor ring. On these engines, as well as engines having crankshaft and camshaft sensors, the combination of sensor readings is used to determine piston position. The camshaft disc is cast as part of the camshaft, so the engine cannot be mistimed. The camshaft sensor input may also be used to determine engine speed. It is also used by the ECM to determine the amount of ignition timing advance.

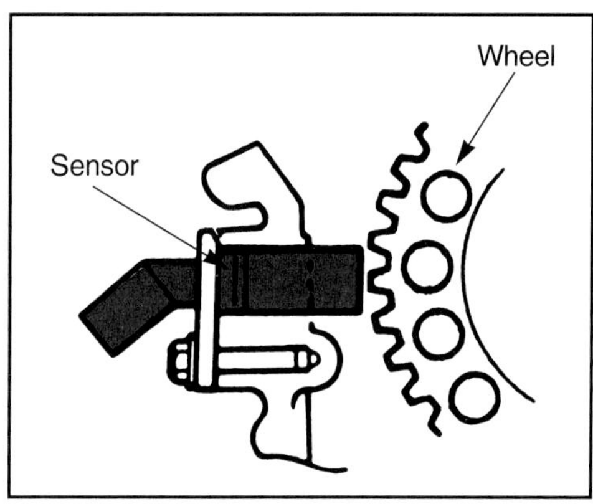

Figure 8-17. *DIS sensors can be pickup coils or Hall-effect switches. The pickup coil is placed near a magnetic strip or notch in a wheel attached to the crankshaft, crankshaft pulley, or camshaft. (Ford)*

The distributorless ignition creates a high voltage spark using ignition coils. There is one ignition coil for every two cylinders. A four-cylinder version has two coils, a six-cylinder has three coils, and a V-8 uses four coils. See **Figure 8-20.** Multiple coils are used since there is no distributor cap and rotor to distribute the spark from a single coil. In most systems the ignition coils are connected to two of the engine spark plugs through conventional plug wires.

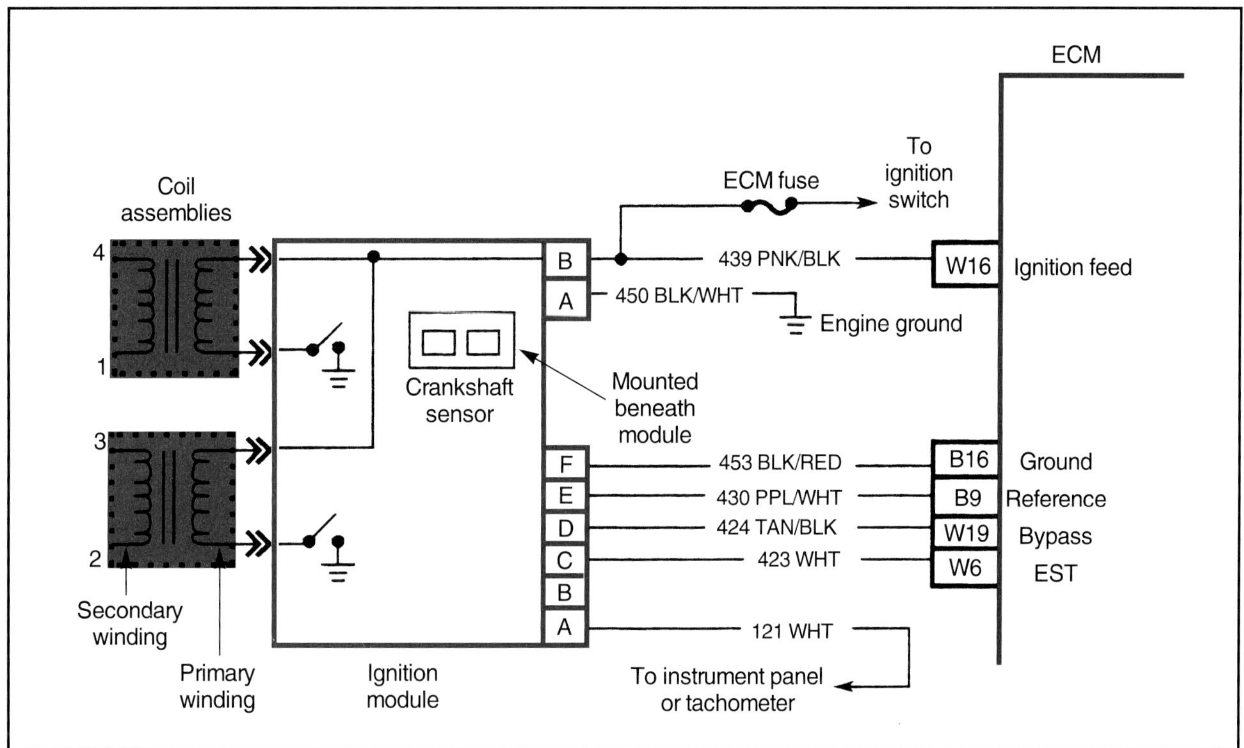

Figure 8-16. *In the distributorless ignition system, the triggering function is performed by sensors located on the engine crankshaft, and in many cases, the camshaft. One ignition module is used with one coil for every two cylinders. (General Motors)*

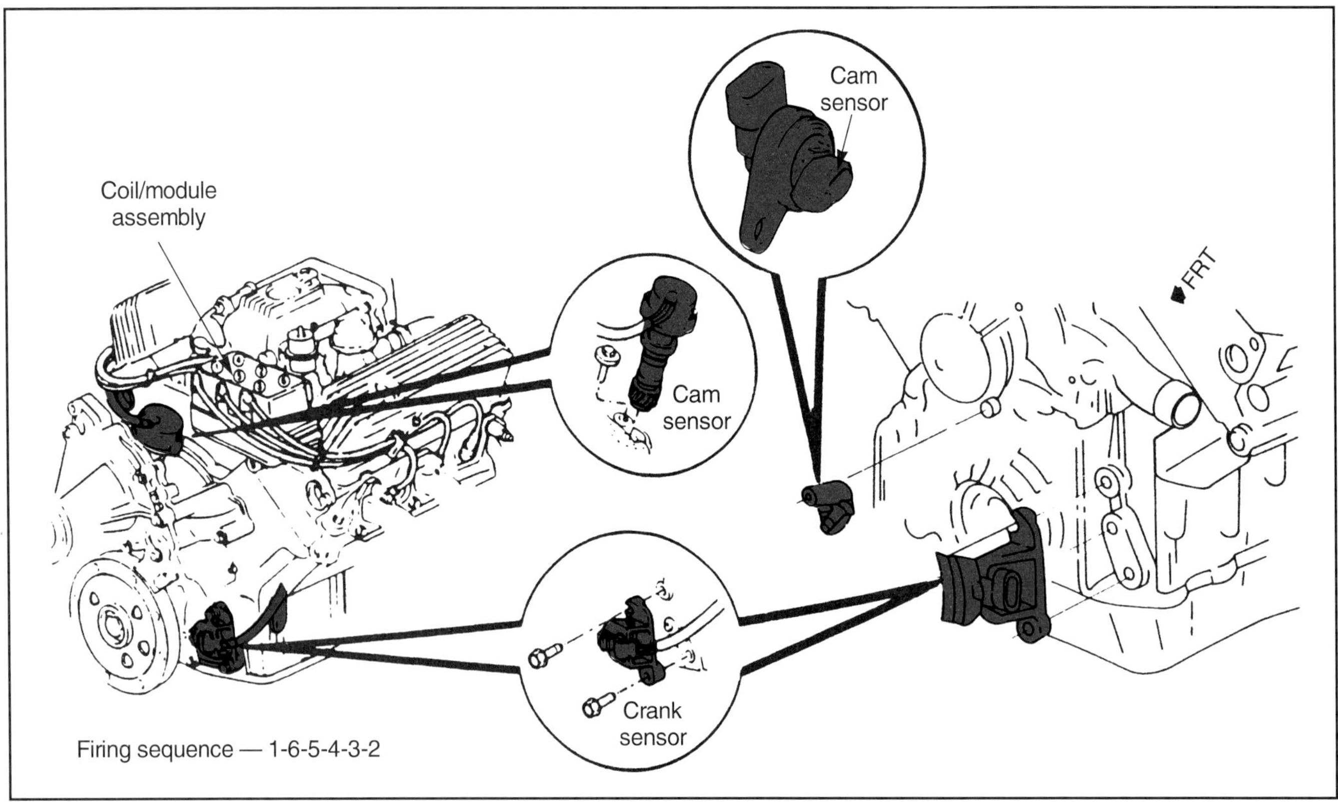

Figure 8-18. *Some of the possible locations of distributorless ignition system sensors. Some sensors can be simply removed and replaced. Others must be carefully adjusted after installation to prevent sensor damage. (General Motors)*

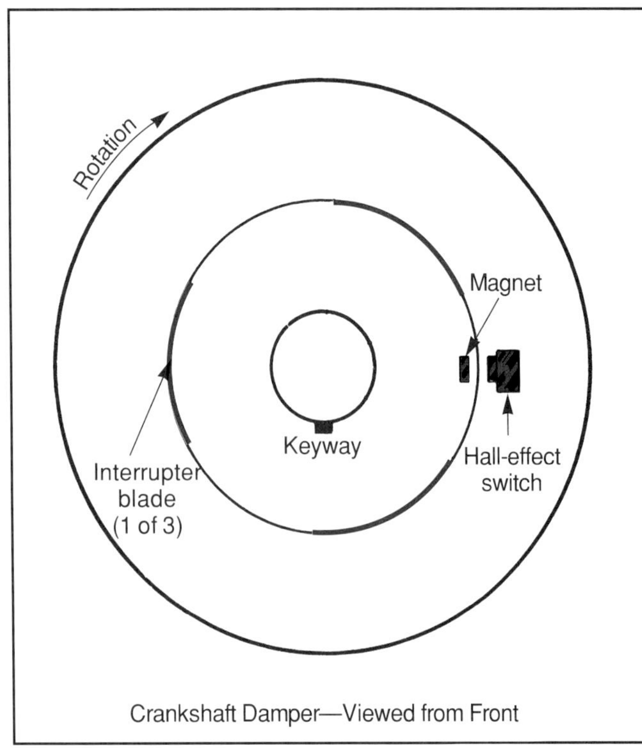

Figure 8-19. *The interrupter blades on this damper assembly are positioned to cause the Hall-effect switch to send a speed signal to the ECM. The basic operating principle is the same as a distributor equipped with a Hall-effect sensor. (General Motors)*

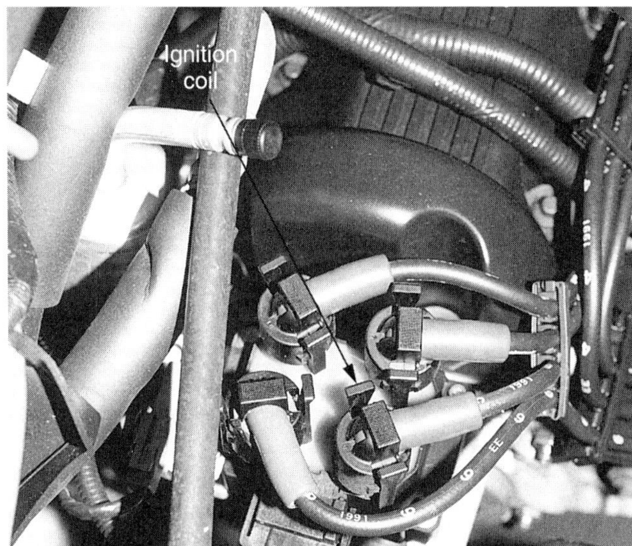

Figure 8-20. *This coil assembly is used with a DIS system. While it appears to be one coil, there are actually two coils in this assembly.*

When the coil fires, the spark exits one terminal, travels through the plug wire to fire the spark plug, and returns to the other coil terminal through the engine block, the other spark plug and plug wire, **Figure 8-21.** In effect, the coil fires both spark plugs at the same time. The firing order is arranged so that the coil fires one plug on the top of

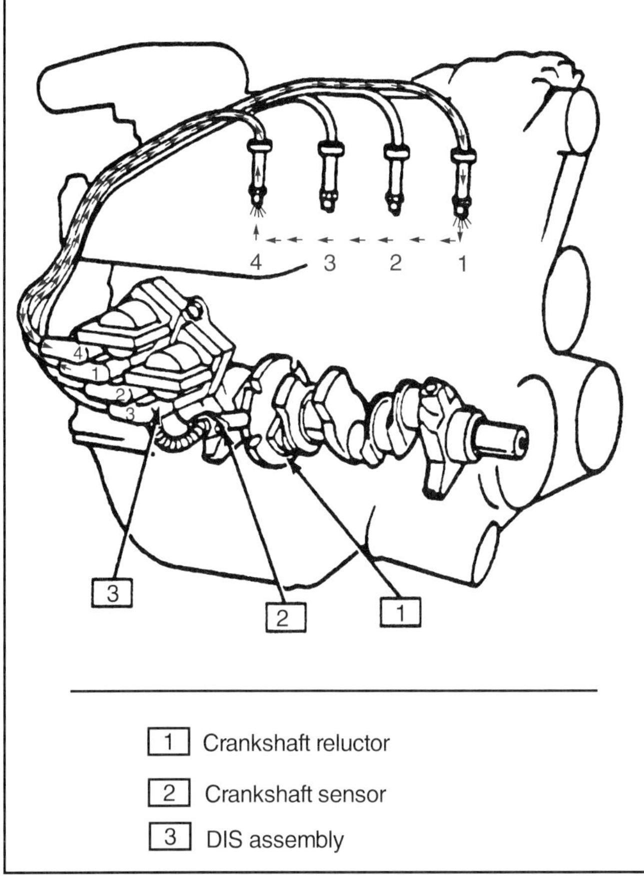

Figure 8-21. *When a distributorless ignition coil fires, the charge creates a complete circuit. Note how the power flows from cylinder one to cylinder four and back to the coil assembly. (General Motors)*

the compression stroke, and the other plug on the top of the exhaust stroke. The plug firing on the top of the exhaust stroke has no effect on the operation of the engine, and is often called the ***waste spark.*** It takes very little voltage to jump the spark plug gap on the exhaust stroke, as the coil is powerful enough to fire both plugs.

Direct Ignition Systems

Some distributorless ignition systems eliminate the spark plug wires, which can become a source of driveability problems over time. These systems fire the spark plugs through boots connected directly to the coils. **Figure 8-22** shows a typical ***direct ignition system.*** Some direct ignition systems that do not use spark plug wires are also referred to as *integrated direct ignition systems (IDI).*

The coils, spark plug boots, and module are installed in a single tower unit. The advantage of this system is that spark plug wires and boots are eliminated and all parts are located together in a sealed unit installed onto a cover plate. The only remotely mounted ignition component is the crankshaft position sensor. This particular system is used on four-cylinder engines with the spark plug mounted in the center of the combustion chamber.

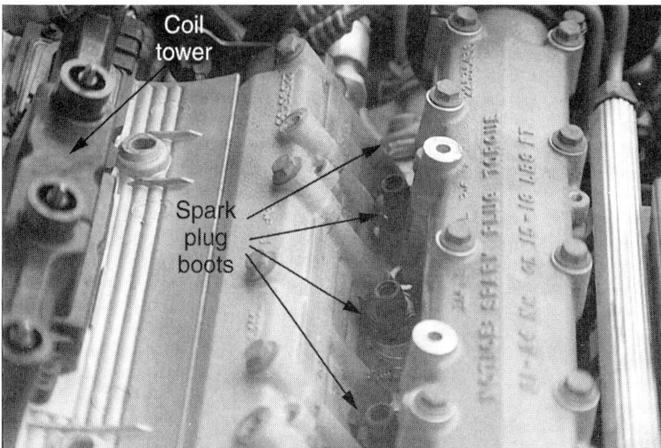

Figure 8-22. *Integrated direct ignition systems eliminate the spark plug wires, using boots to relay the secondary power to the spark plugs.*

Contact Point Ignition Systems

Vehicles manufactured before the mid 1970s were equipped with ***contact point ignition systems.*** The contact point ignition system was used in all gasoline engines until it was replaced by the electronic ignition system. Contact point ignition systems were unable to develop the high voltages needed to operate modern engines and required more frequent service. In the contact point system, the primary current flow was started and stopped by a set of contact points inside the distributor. The contact points opened and closed by rotation of a cam installed on the distributor shaft. A contact point distributor is shown in **Figure 8-23.**

The ***point set*** had a hard plastic projection, called a rubbing block, that contacted the distributor cam to open and close the points. High temperature lubricant was used on the block to reduce wear. The point contacts were held closed by spring tension built into the point assembly. When the rubbing block was on the low point of the cam lobe, the point contacts closed. As the cam rotated to the lobe's high

Figure 8-23. *The contact point distributor is used on older vehicles. In this system, primary current flow through the coil is controlled by a set of mechanically operated contact points. The points are opened and closed by a cam installed on the distributor shaft.*

point, the point contacts opened, breaking the primary circuit and producing voltage in the coil's secondary windings.

The ***condenser,*** also shown in **Figure 8-23,** was connected in parallel with the points. It absorbed voltage surges and prevented arcing when the points opened. The action of the condenser also caused the coil's magnetic field to collapse quickly, producing a stronger spark. Condenser capacity was closely matched to prevent pitting, which could quickly ruin the points. Since there are no points to pit, condensers are not used on electronic ignition systems. Most condensers are constructed of two sheets of very thin foil separated by two or three layers of insulation. The foil and insulation are wound together into a cylindrical shape and placed in a small metal case and sealed to prevent the entrance of moisture. The close placement of the foil strips creates capacitance, or the ability to attract electrons.

A variation of the contact point ignition system used the points to trigger a power transistor, which actually controlled the coil firing. Point-type ignition systems received battery current through a resistor. When the engine was being cranked, the resistor was bypassed to provide full battery current to the coil.

Point Gap and Dwell

The amount that the points open is called the ***point gap.*** The gap is set with the rubbing block on the high point of the cam. Proper adjustment allows the points to open and close, causing the coil to fire. The time that the points are closed is called the ***dwell,*** and is measured in degrees. See **Figure 8-24.** Dwell is sometimes referred to as cam angle. The amount of dwell is the time between each distributor shaft cam lobe that the points are closed. Dwell and point gap are related. Increasing one reduces the other. If the dwell is too small, the gap is too wide. If the dwell is too large, the gap is too small. The most accurate way to adjust contact points is by setting dwell using a ***dwell meter,*** discussed in **Chapter 3.** On electronic ignition systems, dwell can still be measured as an aid to diagnosis, however, it cannot be adjusted.

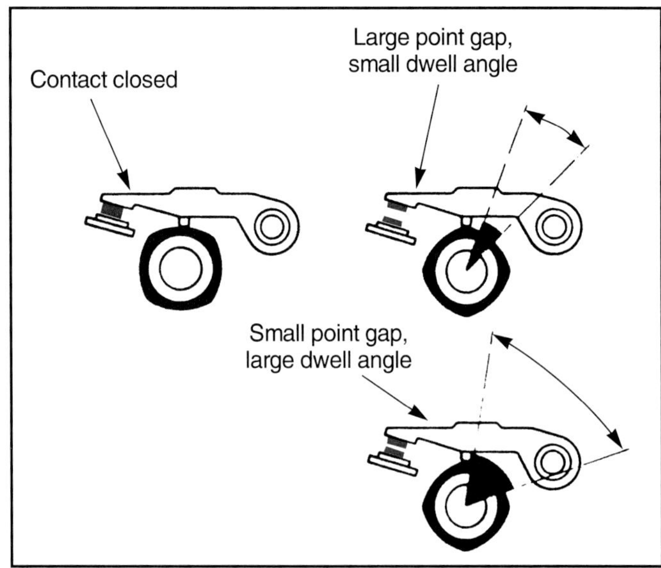

Figure 8-24. *The concept of dwell comes from the time when contact points were widely used. It is the amount of time that the points are closed and current is flowing in the coil primary winding. Dwell on contact point systems could be adjusted by moving the point set. While dwell cannot be adjusted on electronic systems, it is used as a diagnostic aid. (Bosch)*

Primary Wiring and Connectors

Primary wiring and ***connectors*** carry relatively high current loads at low voltages. They have large diameter conductors with light insulation. Primary ignition wiring resembles the other chassis wiring in the vehicle. Primary wires are selected by gage, a wire diameter classification system that uses a numerical code. The higher the number, the smaller the wire diameter. A size such as "0" indicates a large wire, such as a battery cable, while a number such as "18" indicates a small wire, such as a taillight wire. Most primary ignition wiring is from 8-14 gage, while modern battery cables are 4-6 gage.

Modern primary wiring connectors are often sealed plug-in units, such as that shown in **Figure 8-25.** These connectors reduce the possibility of bad connections due to dirt, water, or corrosion. Plug-in connectors are shaped to fit one way, which eliminates the possibility of reversing the coil connection, a common mistake made when installing older round coils. Wiring on older engines consisted of simple hook or eye connectors or plastic-encased male and female connectors, often called quick-disconnects.

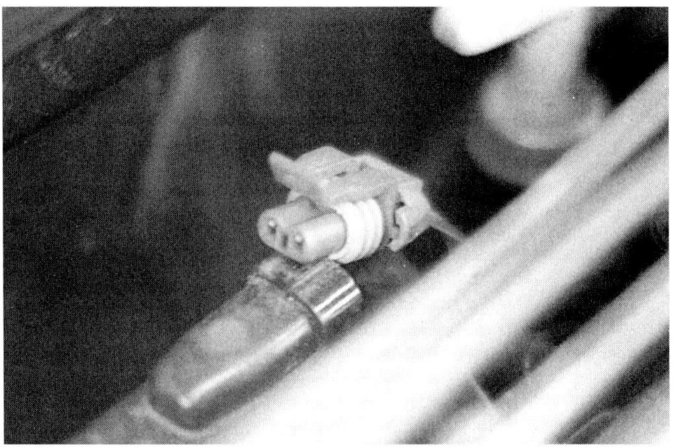

Figure 8-25. *Sealed plug-in wire connectors are used on modern engines due to the low voltages common in computer-controlled systems. The seal is important to keep out dirt and moisture, which can cause corrosion and increase circuit resistance.*

Resistors

Contact point and some electronic ignition systems include a ***resistor*** in their primary circuits. Electricity flows from the ignition switch to the resistor, which controls the amount of current reaching the coil. The resistor may be either a calibrated resistance wire built into the wiring harness or a temperature sensitive, variable ballast resistor.

The resistor lowers battery voltage to around 9.5 volts during normal engine operation. When the engine is started, the coil receives full battery voltage from a bypass wire. The bypass wire supplies the coil with full battery voltage from the ignition switch or starter solenoid while the engine is cranking. When the key is released, the circuit receives its power through the resistance unit. Some early transistor ignition systems use two ballast resistors to control coil voltage. From the resistor, the current travels to the coil. Most modern vehicles with electronic ignition do not use a resistor in the ignition circuit. The majority of modern electronic ignition systems use full battery voltage at all times.

Spark Advance Mechanisms

As engine speed increases, it is necessary to fire the mixture sooner. If this is not done, the piston would reach TDC and start down before the air-fuel mixture can be ignited. To properly fire the air-fuel charge, ***spark advance mechanisms*** advance or retard the ignition timing in relation to engine speed and load to prevent spark knock.

To start the engine, the spark must be near top dead center. After the engine starts, the spark must be advanced for best performance, mileage, and driveability. On older vehicles, vacuum and centrifugal advance mechanisms were used. On newer cars and trucks, the ECM adjusts the timing.

Centrifugal Advance

The ***centrifugal advance*** provides spark advance in relation to engine speed. At high speeds, the beginning of the combustion process must be advanced to match higher piston speeds. The centrifugal advance consists of two weights mounted on the distributor shaft, and held closed by springs. This is shown in **Figure 8-26.** The upper part of the distributor shaft is designed with play, so that it can be advanced by the weights. In effect, the distributor shaft is divided into upper and lower parts.

When the engine is idling, spring pressure keeps the two weights drawn together and the shaft remains at the position for low speed timing. As engine speed increases, the weights develop enough centrifugal force to overcome the springs and move outward, as shown in **Figure 8-27.** The weights are designed to act as levers, which advance the distributor shaft with rotation. Since the trigger wheel, chopper, or cam is attached to the distributor shaft, it advances also, causing the timing to advance. As engine speed decreases, the springs return the weights to the closed position.

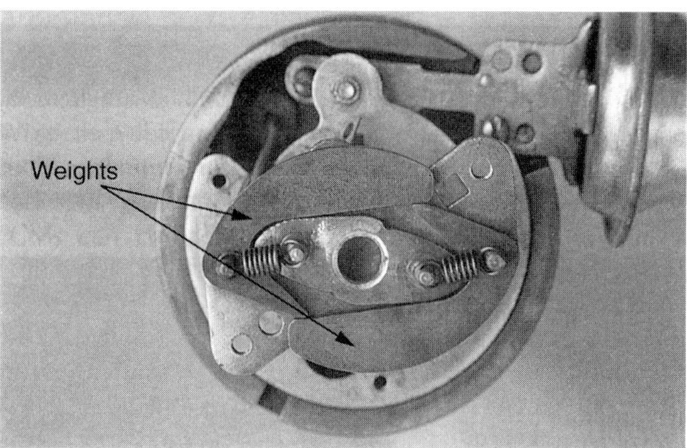

Figure 8-26. *Centrifugal advance mechanisms advance engine timing based on engine speed. The weights are shown in the closed position. This position is maintained by the return springs at low engine speeds.*

Figure 8-27. *At high engine speeds, centrifugal force throws the weights outward, overcoming return spring tension and advancing the ignition timing.*

Vacuum Advance Units

The ***vacuum advance unit*** advances the spark timing in relation to engine load. The vacuum advance unit is a spring-loaded sealed diaphragm, **Figure 8-28.** It is connected to the distributor plate holding the pickup coil or Hall-effect switch (or the point set on contact point ignitions). The plate is designed to move in relation to the rest of the distributor housing. It moves against distributor shaft rotation to advance timing and with rotation to retard timing. A hose connects the vacuum chamber to the engine intake manifold.

When the engine is running, manifold vacuum can move the diaphragm against spring pressure. This moves the linkage and distributor plate, varying the spark advance with changes in manifold vacuum. When manifold vacuum drops, the spring moves the plate toward the base timing setting. When vacuum is high, the plate moves toward the advanced position. This is shown in **Figure 8-28A.** Some vacuum advance units are dual diaphragm types, as in

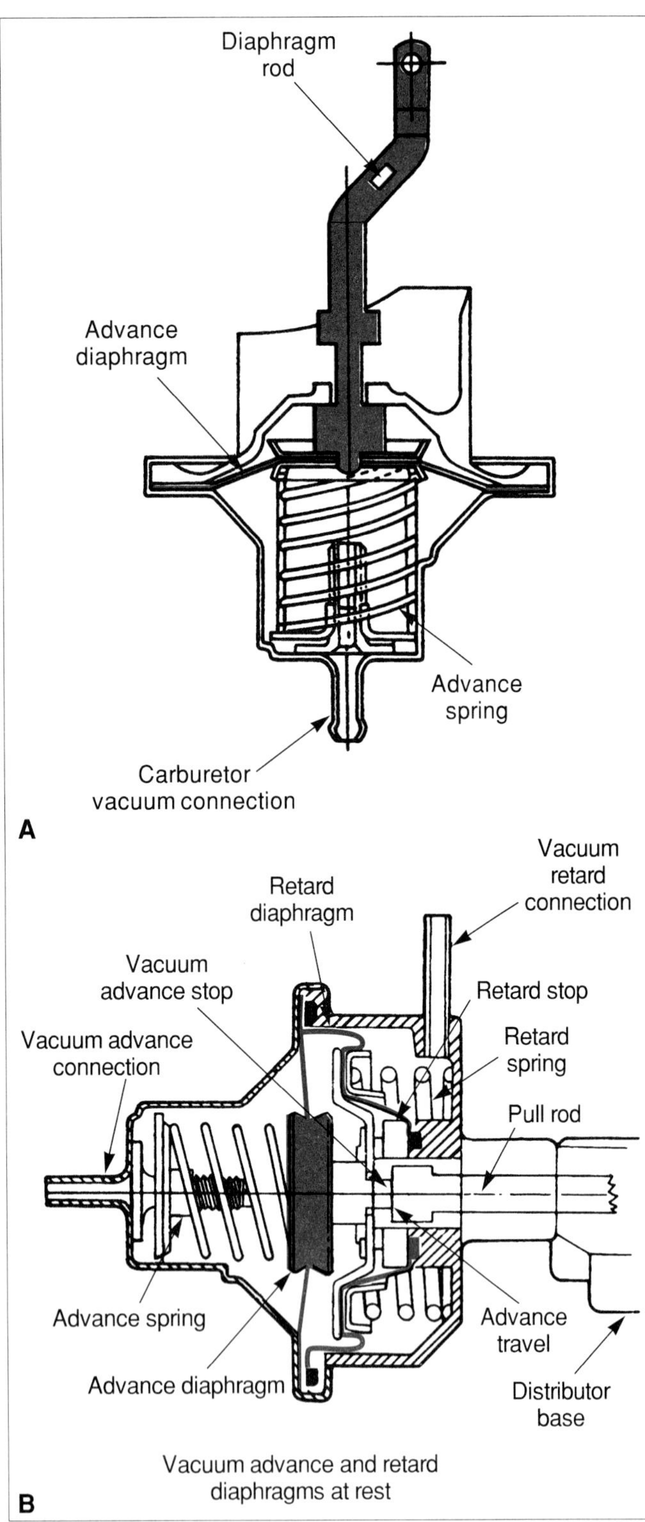

Figure 8-28. A—Cross-section of a vacuum advance. Vacuum enters at the vacuum port and lowers the pressure inside the diaphragm chamber. Air pressure moves the diaphragm and rod against spring pressure. B—Cross-section of a dual diaphragm vacuum advance. (Ford)

Figure 8-28B. Dual diaphragm vacuum advance units have manifold vacuum sent to the opposite side of the diaphragm to retard the spark at low speeds. At higher speeds, vacuum is allowed to enter the vacuum advance side of the unit. Because of the larger diaphragm area on the advance side, the diaphragm is pulled to the advance position when vacuum is equal on both sides. The placement of the vacuum advance unit on a distributor is shown in **Figure 8-29.**

On some early engines, the spark is *ported.* Ported spark was often used as an early emission control device. Instead of being connected to the intake manifold, the vacuum advance hose was attached to a port above the throttle plate. The vacuum advance did not receive vacuum until the throttle was opened above idle. Some ported vacuum advance units are routed through a ***thermal vacuum valve (TVV),*** **Figure 8-30.** This switch is connected to an engine coolant passage, usually in the intake manifold. When the engine is warming up, full vacuum is supplied to the advance unit, retarding engine timing and reducing exhaust emissions. When the engine heats up, the TVV sends vacuum to the EGR valve. Most late-model engines use full manifold vacuum to the vacuum advance diaphragm for best mileage.

Computer Advance

Computer advance is a precise way of controlling spark timing using the engine control computer. Computer controls are more efficient than vacuum and centrifugal advance mechanisms. They are able to change timing much faster than older systems and compensate for changes in engine operation or outside factors, such as air temperature. Computer control systems were discussed in **Chapter 7.**

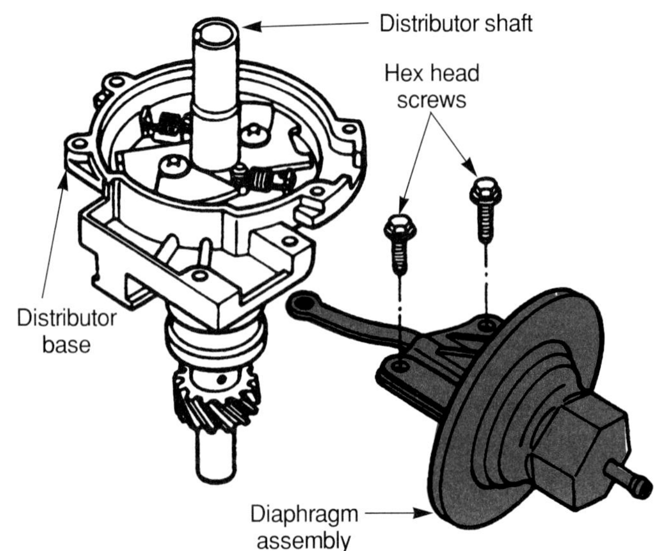

Figure 8-29. *Vacuum advance units are found on older engines as well as some newer ones. The vacuum advance is mounted so that the lever arm can move the internal distributor plate. (Ford)*

Figure 8-30. *On older vehicles, thermal vacuum valves were used to adjust ported spark. Thermal vacuum valves are used on newer vehicles with early fuel evaporation (EFE) systems.*

On a computer-controlled system, the ECM bases its timing settings on information provided by a distributor or engine-mounted triggering device, and other various engine and vehicle sensors. Typical inputs to the computer include engine temperature, engine and vehicle speed, manifold vacuum, throttle position, intake air flow, and intake air temperature. Modern ECMs also receive inputs from a knock sensor. The computer processes these inputs and controls the ignition module to vary the timing as required, **Figure 8-31.**

The module interrupts the primary circuit through the coil in the same manner as on conventional electronic ignition systems. However, the trigger input comes from the ECM circuitry instead of directly from the triggering device. By varying the timing of the triggering signal, the ECM advances or retards coil firing and therefore spark timing.

The ECM-determined advance and retard curves are similar to the curves of the older vacuum and centrifugal units. However, timing changes can be made at a much faster rate and are more precisely matched to engine speed, load, and temperature. Under normal operating conditions, the ECM advances the ignition timing as much as possible for maximum fuel economy without increasing emissions or causing detonation (pinging). Timing is constantly advanced or retarded as sensor inputs indicate changes in engine operating conditions.

During wide open throttle operation, the ECM may retard timing to a preset amount to reduce the chances of detonation and engine damage. In some cases, the ECM will disregard input from the knock sensor during wide open throttle operation for maximum power output. Unusual conditions, such as engine overheating or the failure of a computer system component, may cause the ECM to go into

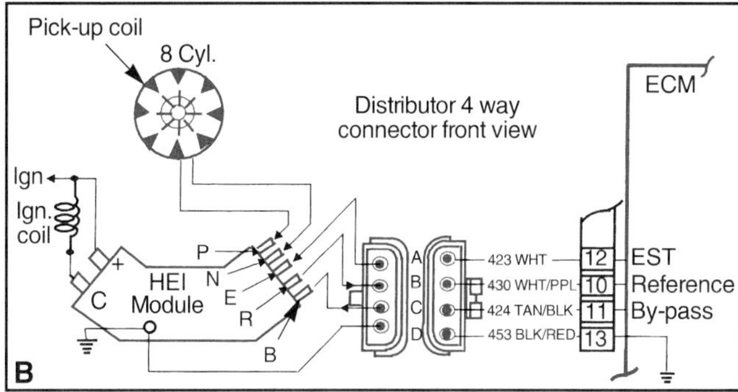

Figure 8-31. *Computer-controlled ignition systems do not use centrifugal weights or vacuum advance units, since the ECM can do a much better job. A—The ignition module as used in one distributor. B—Schematic of this computer-controlled ignition module. (General Motors)*

open loop mode and reduce timing to a preset amount. The advance is also stationary at a preset amount when the engine is cold and the ECM is in open loop mode.

The basic ignition system is similar to older electronic ignition systems. However, note that it has no vacuum or centrifugal advance, **Figure 8-31.** On some systems, there is no separate ignition module. Instead, the pickup coil or Hall-effect switch signals the computer directly. Most distributor ignition systems with computer-controlled advance have a provision for adjusting the base timing. This is usually accomplished by turning the distributor with terminals on the data link connector grounded or by disconnecting the distributor ECM input.

Some computer-controlled distributors consist of only a shaft, rotor, and cap. The spark is created and controlled by the computer, based on a signal from a pickup at the engine crankshaft. The distributor is used only to distribute the spark to the proper cylinder. On late model ignition systems with no distributor, the ECM together with a sensor mounted on the engine crankshaft (and sometimes a second sensor on the camshaft), produces and delivers a properly timed spark to the plug.

Many ECMs also control the amount of time that the primary current flows in the coil. This is done to reduce coil heating by producing only enough current flow to produce a spark hot enough to fire the spark plugs. When a hotter spark plug is needed, the computer increases primary current flow time. The time that current flows is called dwell, which was discussed earlier in this chapter.

Secondary Ignition System

Unlike the primary system, which creates and times the high voltage current, the ***secondary ignition system*** is used only to distribute the current from the coil to the spark plugs. The primary function of the secondary system is to ensure that the spark reaches the right plug at the right time, without shorting or arcing to ground. Defects in the secondary system usually cause engine missing, backfiring, or other symptoms that are similar in older and newer engines.

Many parts of the secondary ignition system contain small resistors or resistance increasing material. Increasing resistance reduces the size of the magnetic field around the secondary ignition components. Reducing the magnetic field reduces electromagnetic interference or static, in vehicle radios and tape/CD players, on-board computers, vehicle systems, and cellular telephones. Specific resistance reducing devices are discussed under each secondary system component.

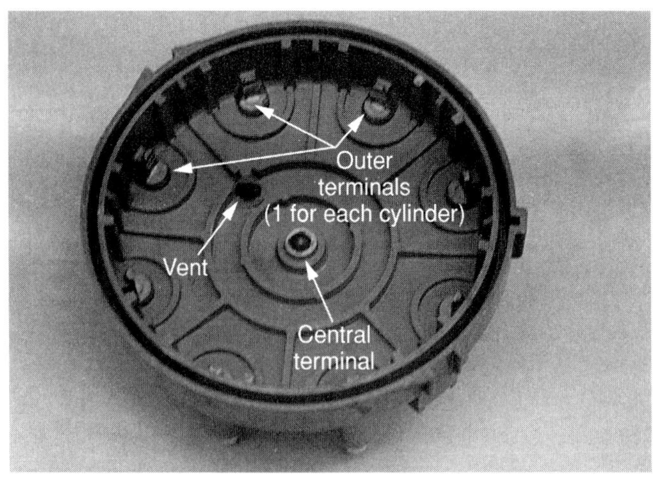

Figure 8-32. *The distributor cap works with the rotor to distribute the spark to the individual plugs. It is made of a plastic with high electrical resistance with copper or aluminum terminals. The small black vent allows moisture to escape.*

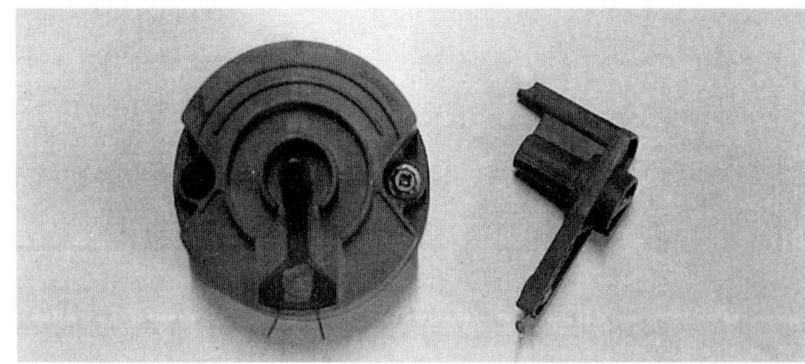

Figure 8-33. *Rotors come in many shapes and sizes. However, each rotor has the same job; working with the distributor cap to send the spark from the coil to the individual plugs. They can be pressed down on the distributor shaft or held with small screws.*

Distributor Cap

The ***distributor cap*** is constructed of electrically resistant plastic, sometimes mica-filled to reduce flashover. Some older caps are made of Bakelite, while most modern caps used with electronic ignition systems are made of more durable plastic. On some vehicles, the coil is installed in the distributor cap. The cap works in conjunction with the rotor to distribute high voltage with minimum loss. The cap contains one center terminal and one outer terminal for each engine cylinder. The terminals are made of copper, aluminum, or other electrically conductive material. Some modern distributor caps contain a vent to allow moisture to escape. See **Figure 8-32.**

Rotor

The ***rotor*** is a plastic device used to carry and distribute high voltage. See **Figure 8-33.** It is installed on top of the distributor shaft and turns with the shaft to distribute the spark to the proper distributor cap terminal. An electrically conductive center leaf contacts the center terminal of the distributor cap. The center leaf is attached to the outer portion of the rotor electrode. The outer portion of the electrode passes near the cap terminals, allowing the high voltage to jump from the rotor to the cap terminal. The rotor often contains a small resistor for static suppression. Some rotors used with Hall-effect switch distributors are equipped with shutters.

Secondary Wires

Secondary wires, commonly called ***spark plug wires,*** are used to conduct high voltage electricity from the coil to the spark plugs, as shown in **Figure 8-34.** The actual conductor is only a small part of the wire. The majority of the wire's diameter is made of heavy insulation such as rubber, neoprene, plastic, or other non-conducting materials. All modern secondary wiring is made of *fiberglass string,* impregnated with graphite or carbon. The graphite impregnated string has enough electrical resistance to reduce the magnetic field around the wires. Copper or aluminum core wire has not been installed on production vehicles for many years, although it is available as an aftermarket item. Most secondary wires are 6 to 8 millimeters thick.

Caution: Copper and aluminum core spark plug wires are designed for off-road use only. They should never be used to replace resistance wires as they create interference that can affect engine and vehicle operation.

Secondary wires have metal connectors on both ends. The internal end of the connector is attached to the conductor. The exposed end of the wire is designed to tightly connect to the distributor tower, coil tower, or spark plug, as applicable. Each wire end has a rubber or neoprene boot to keep out water and air. Some wire ends are specially made to fit certain engine designs, such as the one in **Figure 8-35.**

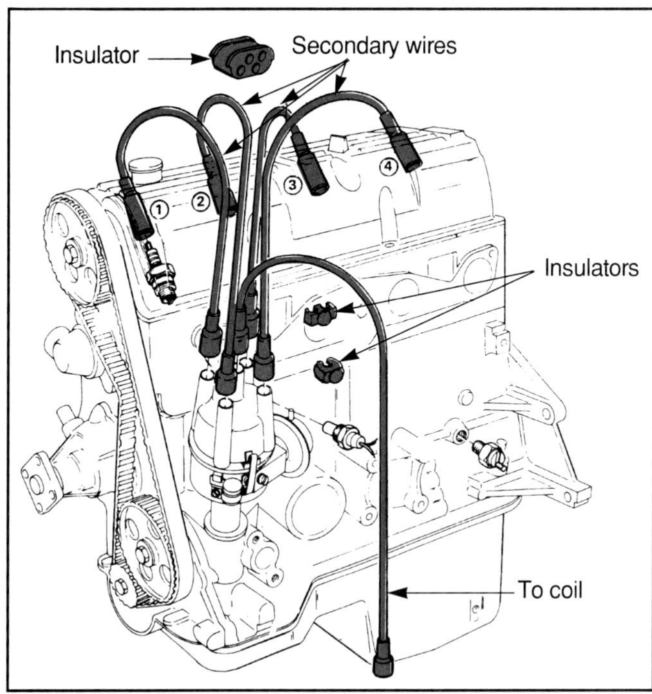

***Figure 8-34.** Spark plug wires carry the high secondary voltage from the distributor to the spark plugs. They come in several sizes, made out of carbon or silicon impregnated fiberglass string with a thick covering of insulation. (Ford)*

***Figure 8-35.** Some spark plug wires have special boots. These special boots are used when the plug is in a hard-to-reach spot. Since the spark plugs in many new engines are placed in the center of the combustion chamber, boots such as this one are becoming more commonplace.*

Spark Plugs

The ***spark plugs*** are ignition devices which provide a gap in the combustion chamber where the secondary current can jump. This jump causes a spark, which ignites the compressed air-fuel mixture. The spark plug also provides an insulated path for the secondary current into the combustion chamber. The spark plug must be able to withstand extremes of combustion chamber heat and chemical attack during this process. The typical modern spark plug may be expected to last from 15,000-30,000 miles (24 000-48 000 km) up to 100,000 miles (160 000 km). Modern spark plugs last longer due to the higher voltages produced by the electronic ignition system and the use of unleaded gas. If not changed periodically, however, they can cause performance problems.

A cutaway view of a typical spark plug is shown in **Figure 8-36.** It is comprised of the center electrode, ground electrode, and insulator. The center electrode conducts the spark into the combustion chamber. Typical materials used to make spark plug center electrodes include copper and nickel alloy. Platinum, although expensive, is sometimes used. Most center electrode assemblies contain a spring under tension to ensure physical contact between the electrode's metal parts at any temperature. Modern spark plugs have a small resistor built into the center electrode. The resistor reduces static interference caused by the magnetic field created by the spark jumping the gap.

The ground electrode is attached to the steel outer shell, which is threaded into the engine. The steel shell top is crimped over to bear against a seal. The crimping process grips the insulator tightly and also forms a pressure seal at both the top and bottom of the insulator, which prevents combustion leaks and arcing. The spark gap is formed between the center and side electrodes. Spark plug gaps can vary between .025″–.100″ (0.635-2.54 mm), depending on the engine and ignition system. Spark plug side electrodes come in a variety of designs and shapes.

The center electrode is insulated from the rest of the plug by a ceramic insulator. The insulator is made of a ceramic material that isolates the center electrode and steel shell. A common material used for making spark plug insulators is aluminum oxide. The aluminum oxide is fired at high temperature to produce a glassy smooth, dense, and very hard insulator.

Electrons will flow easily from a hot surface to a cooler one. Spark plugs are manufactured in many ***heat ranges.*** Heat range is determined by the diameter and length of the insulator as measured from the sealing ring down to the plug tip and to the cooling system, as shown in **Figure 8-37.** Spark plug heat range is vital to proper engine operation. A plug that operates without becoming hot will be fouled by combustion byproducts. A plug that is too hot will begin to glow. This will ignite the air-fuel mixture ahead of time, causing detonation and possible engine damage. Spark plugs are manufactured in many thread types and lengths. Among the most common thread sizes are 18 mm, 14 mm, and 10 mm.

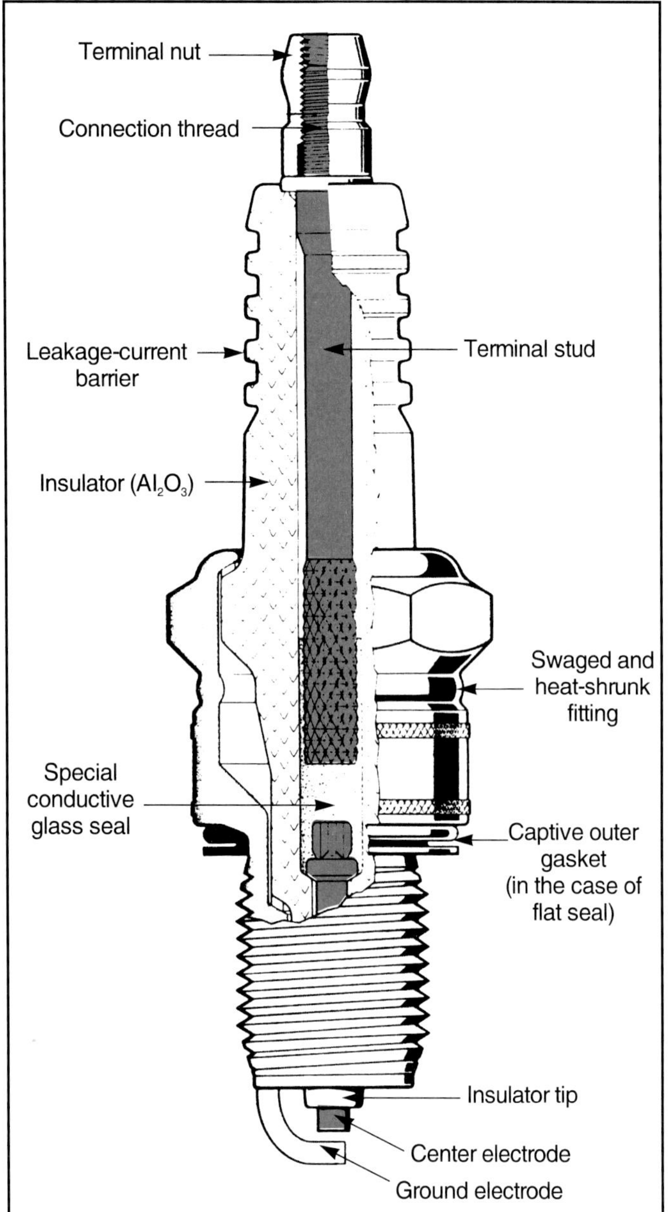

Figure 8-36. *This cutaway shows the internal components of a resistor spark plug. Most modern spark plugs are constructed in a similar manner. (Bosch)*

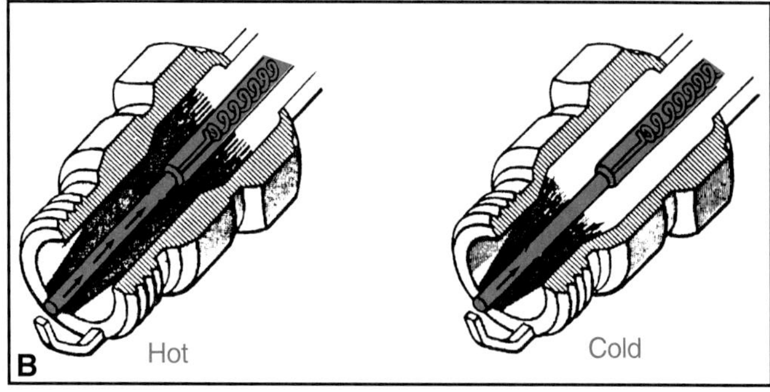

Figure 8-37. *A—Some spark plugs have an extended tip. This centralizes the flame and allows the movement of the gases in the combustion chamber to keep the electrode surface cleaner. B—Spark plug heat ranges are determined by how far the heat must travel from the plug tip to the plug body. The farther the heat must travel, the hotter the plug. (Ford)*

Ignition Timing

For correct engine timing, each cylinder should receive a spark at the plug electrodes as the piston nears the top of its compression stroke (a few degrees before TDC). This is made possible by driving the distributor shaft so that it turns at one-half crankshaft speed. The distributor shaft may be turned by one-to-one gearing with the camshaft or by using a timing belt driven by the distributor.

The distributor shaft gear is timed so that the spark is produced when the cylinder is ready to fire. The rotor will then point toward the cylinder's cap plug terminal. The wires are attached to the cap starting at the number one cylinder and follows the firing order in the direction of distributor shaft rotation. As the engine turns, the distributor shaft revolves. Each time the distributor shaft has turned enough to cause the rotor to point to a plug terminal, the ignition system produces a spark. If the plug fires later than the specified setting, the timing is said to be retarded. If the plug fires earlier than specified, the timing is referred to as advanced.

Most older engines, and many newer ones, have timing marks in the form of a line marked on the rim of the vibration damper, **Figure 8-38.** Some engines in front-wheel drive vehicles have timing marks on the flywheel. A pointer is attached to the timing cover. When the mark is exactly under the pointer, the engine is ready to fire number one cylinder. The spark will occur with the rotor pointing to the number one cap terminal. The timing is generally set by

Figure 8-38. *Many engines have timing marks cast on the engine crankshaft pulley. The mark is used to indicate engine timing on a scale located near the pulley.*

using a strobe lamp, which is a light that is operated by high voltage surges from the spark plug wire. The strobe lamp is usually referred to simply as a *timing light,* discussed in **Chapter 3.**

Many late model engines can be timed with a *magnetic timing meter.* This meter has a timing probe which is installed in a magnetic timing receptacle near the conventional timing marks or an inductive pickup that clamps over the number one spark plug. Once all connections are made, the engine is started, and timing can be read directly from the meter dial.

To adjust the timing, the distributor clamp is loosened and the distributor turned by hand. As it is turned, the timing mark moves. When turned in the proper direction, the mark will line up with the pointer. Once the two are aligned, the engine is properly timed and the distributor clamp can be tightened. This procedure will be discussed in more detail in **Chapter 16.**

Misfire Monitor

Engine misfiring due to ignition, fuel, or compression system problems can cause excessive emissions and performance problems. On OBD I equipped vehicles, any misfiring would go uncorrected until it became noticeable and the vehicle brought in for repairs. The OBD II protocol requires that the computer control system monitor the engine for misfiring. This system is called the ***misfire monitor*** and is an internal function of the ECM. The monitor is sensitive enough to detect a misfire within 200 crankshaft revolutions.

Misfire Monitor Operation

Each cylinder contributes to overall engine speed and power. When a misfire occurs, engine power drops, but more importantly, crankshaft rotational velocity decreases momentarily. To monitor the engine for misfiring, the ECM keeps a record of crankshaft revolutions. This is not a new concept, since ECMs have monitored crankshaft speed as part of normal engine sensor input for years. The ECM-monitored crankshaft readings are divided into separate, even samples, **Figure 8-39.** If a misfire occurs, one or more of the crankshaft speed samples will have an unexpected drop or fluctuation. The OBD II protocol requires misfire detection under the following conditions:

- Misfire sufficient to create exhaust emission levels exceeding 1.5 times federal standards. This is classified as a Type B misfire.
- Misfire sufficient to result in catalytic converter damage. This is sometimes classified as a Type A misfire.

When a misfire condition exists that will raise emissions levels, the ECM will set a trouble code and illuminate the MIL light. If the problem is intermittent, the light will go out, but will illuminate again if the misfire reoccurs under the same conditions and within a certain number of engine warm-ups. If a misfire condition that can cause catalyst damage exists, a trouble code is set and on most systems, the MIL light will flash. When the condition no longer exists, the MIL light will stay on steady. All OBD II systems have trouble codes for random or multiple cylinder misfires. **Chapter 7** discusses OBD II system operation in more detail.

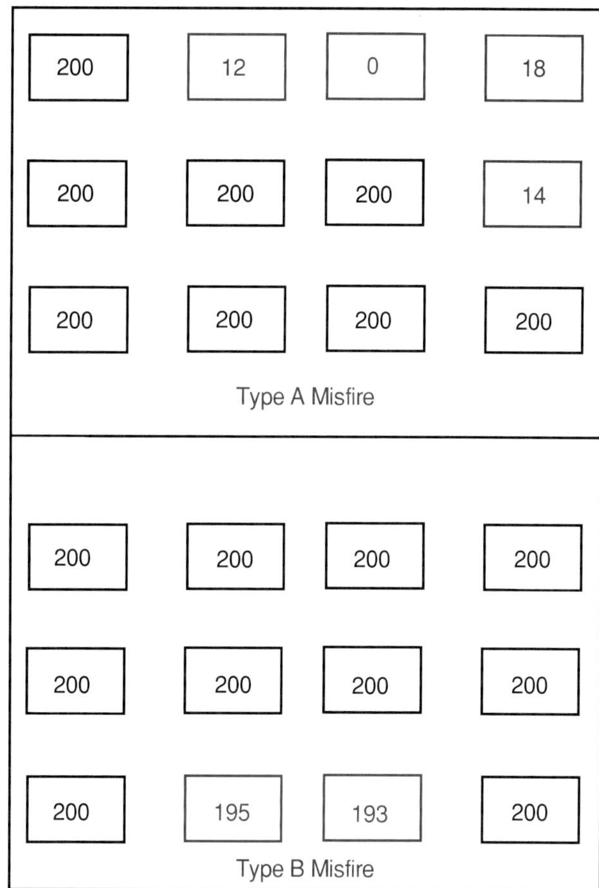

Figure 8-39. The OBD II misfire monitor separates crankshaft speed into even samples of 200 rpms. If a sample drops significantly, it will cause the ECM to set a Type A trouble code and flash the MIL light. If the misfire is relatively minor, the ECM will set a code and illuminate the MIL light. However, if the condition goes away, the MIL light will go out, but the code remains stored in the ECM.

Note: Some systems can set misfire trouble codes for individual cylinders.

However, a drawback is that rough roads can sometimes cause false misfire detection. Driving over a rough road can cause torque to be applied to the drive wheels, which can temporarily decrease the engine speed. To aid in reducing misdiagnosis, manufacturers add programming to the ECM to detect patterns consistent with rough roads. The second is for the ECM to monitor the anti-lock brake system's wheel speed sensors. The ECM on some vehicles equipped with overdrive automatic transmissions will disable the torque converter clutch when misfire is detected.

Summary

Fuel is ignited in the diesel engine by the heat created by high compression. The gasoline ignition system uses a high voltage spark to ignite the air-fuel mixture. It is comprised of primary and secondary systems. The primary system creates the high voltage spark using a coil, ignition module, and triggering device. Triggering devices can be distributor or engine mounted pickup coils, Hall-effect switches, or photoelectric cells. Coils are transformers that turn the battery voltage into high voltage to jump the spark plug gap. The ignition module converts the triggering device input into a signal that turns off the coil primary current.

Distributorless ignition systems have a triggering device installed on the camshaft or crankshaft instead of in the distributor. These systems have one coil for every two spark plugs. Older engines use contact point distributors and do not have an ignition module. A condenser is used to prevent point pitting. Older engines used spark advance mechanisms operated by centrifugal force and manifold vacuum. On newer engines, the spark is advanced by the ECM. The secondary system distributes the spark created by the primary system. It consists of a distributor cap, rotor, spark plug wires, and spark plugs.

Secondary components are made of materials which direct the high voltage coil output to the spark plugs without allowing any of it to leak away to ground. The distributor cap and rotor are made of high dielectric plastic. Plug wires are composed of carbon impregnated string surrounded by a thick insulator. Any part of the secondary system may contain resistors to reduce radio and computer interference.

Spark plugs are designed with a gap between the center and ground electrodes. The high voltage developed by the ignition system jumps this gap to ignite the air-fuel mixture. Spark plug gaps vary according to engine designs. Spark plugs are made in many sizes, threads, and heat ranges. Most plugs contain an internal resistor for radio suppression.

Know These Terms

Ignition system
Electronic ignition systems
Primary ignition circuit
Ignition coil
Primary winding
Secondary winding
Self-induction
Control module
Power transistor
Triggering device
Pickup coil
Hall-effect switch
Photoelectric sensor
Distributorless ignition system (DIS)
Waste spark
Direct ignition system
Contact point ignition systems
Point set
Condenser
Point gap
Dwell
Dwell meter
Primary wiring
Connectors
Resistor
Spark advance mechanisms
Centrifugal advance
Vacuum advance unit
Thermal vacuum valve (TVV)
Computer advance
Secondary ignition system
Distributor cap
Rotor
Secondary wires
Spark plug wires
Spark plugs
Heat ranges
Misfire monitor

Review Questions—Chapter 8

Please do not write in this text. Write your answers on a separate sheet of paper.

1. The coil converts ______ voltage electricity into ______ voltage electricity.
2. The ignition control module must be ______ if defective.
3. Some modern distributor caps contain a ______ to allow moisture to escape.
4. As each tooth of the trigger wheel passes the pickup coil, a ______ is created.
   (A) spark discharge
   (B) magnetic field
   (C) voltage signal
   (D) Both B & C.
5. On a Hall-effect distributor, the rotor is often used as a:
   (A) shutter.
   (B) module.
   (C) trigger wheel.
   (D) condenser.
6. The distributorless ignition system has a pickup coil or Hall-effect switch located near the:
   (A) camshaft.
   (B) crankshaft.
   (C) crankshaft and camshaft.
   (D) All of the above, depending on the engine manufacturer.
7. In the distributorless ignition system, there is one ignition coil for every _______ cylinders.
   (A) 2
   (B) 3
   (C) 4
   (D) All of the above, depending on the engine manufacturer.
8. In the contact point ignition system, the primary current flow is controlled by a set of contact points inside the:
   (A) engine.
   (B) module.
   (C) ECM.
   (D) distributor.

9. If the point dwell is too small, the gap is too ______. If the dwell is too large, the gap is too ______.
10. The vacuum advance is used to increase:
    (A) engine power.
    (B) top speed.
    (C) gas mileage.
    (D) All of the above.
11. The centrifugal advance operates at high engine:
    (A) vacuum.
    (B) speed.
    (C) temperature.
    (D) Both A & B.
12. The ______ was the first kind of triggering device used on electronic ignition systems.
13. The usual distributor mounted pickup coil has one tooth for each engine ______.
14. The Hall-effect shutter may be part of the distributor ______.
15. Crankshaft mounted triggering devices are ______ types.
    (A) magnetic pickup
    (B) Hall-effect
    (C) photodiodes
    (D) Both A & B.

ASE Certification-Type Questions

1. All of the following are part of a photoelectric sensor assembly, EXCEPT:
    (A) LED.
    (B) magnet.
    (C) photodiode.
    (D) shutter.
2. Technician A says that the pickup coil rotates with the distributor shaft. Technician B says that the Hall-effect switch is stationary on all types of systems. Who is right?
    (A) A only.
    (B) B only.
    (C) Both A & B.
    (D) Neither A nor B.
3. Technician A says that a distributorless ignition system may have more than one coil. Technician B says that a distributorless ignition system may have more than one pickup. Who is right?
    (A) A only.
    (B) B only.
    (C) Both A & B.
    (D) Neither A nor B.
4. Self-induction occurs in the ______ coil windings.
    (A) primary
    (B) secondary
    (C) Both A & B.
    (D) Neither A nor B.
5. The waste spark fires on the ______ stroke.
    (A) intake
    (B) compression
    (C) power
    (D) exhaust
6. Technician A says that the ignition module may be mounted inside the distributor. Technician B says that the ignition module may be mounted under the ignition coil. Who is right?
    (A) A only.
    (B) B only.
    (C) Both A & B.
    (D) Neither A nor B.
7. A wire that carries current to a single lightbulb is most likely to be a(n) ______ gage wire.
    (A) 0
    (B) 6
    (C) 18
    (D) 100
8. The purpose of the dual diaphragm distributor is to reduce ______.
    (A) emissions
    (B) mileage
    (C) top speed
    (D) detonation
9. The input from the knock sensor helps the ECM to compute ______.
    (A) waste spark
    (B) ignition timing
    (C) ignition dwell
    (D) coil output
10. A rotor is not used on a vehicle with a(n) ______ ignition system.
    (A) electronic
    (B) contact point
    (C) distributorless
    (D) Both A & C.

Partial cutaway of a V-10 engine. Note the position of the fuel injectors and fuel rail. (Ford)

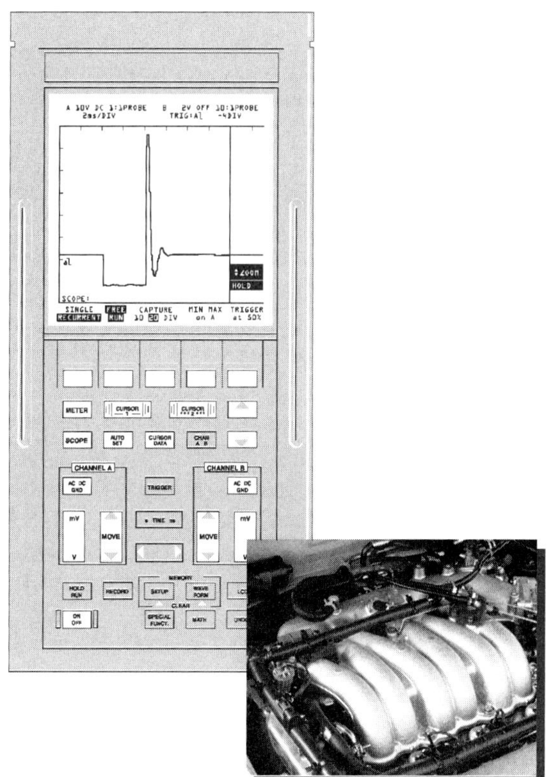

Chapter 9

Fuel System Fundamentals

After studying this chapter, you will be able to:

- Identify and explain the purposes of fuel tanks.
- Identify and explain the purposes of fuel pumps.
- Identify and explain the purposes of fuel filters.
- Identify and explain the purposes of major components of fuel injection systems.
- Identify and explain the purposes of fuel pressure regulators.
- Identify the differences and similarities between pulsed and fixed injectors.
- Identify the differences and similarities between throttle body and multiport fuel injection systems.
- State the major differences between gasoline and diesel engine fuel injection systems.
- State the purposes of intake manifolds and plenums.
- State the purposes of air filters
- State the purposes and designs of superchargers and turbochargers.

Fuel, ignition, and compression are necessary to the operation of any internal combustion engine. If any one of these three systems is not operating properly, the engine will run poorly or not at all. This chapter discusses the components and operation of the fuel system, and how misadjustment or defects can affect driveability. Fuel system diagnosis and repair will be discussed in **Chapter 17.** Carburetors have been used for almost 100 years and fuel injection has been widely adapted to and has replaced the carburetor in gasoline engines during the last 10 years. This chapter concentrates on fuel injection systems and the typical fuel delivery systems found on most vehicles. Specific information on carburetor operation and service is discussed in **Appendix B.**

Fuel System

The ***fuel system*** is a set of components that delivers fuel to the engine cylinders. The fuel system has several purposes. It must deliver the proper amount of pressurized fuel to the engine cylinder, closely control the air-fuel ratio, and atomize the fuel so that it will mix with the air and become a burnable mixture. Fuel system condition is very critical to good engine and vehicle driveability. Improper operation of any fuel system part can upset the operation of the engine, causing serious driveability problems. However, before discussing the fuel system, we should understand exactly what fuel is and what fuels are acceptable for use in automotive engines. Fuels are discussed in the next section.

Automotive Fuels

Fuel can be defined as any kind of chemical compound that will burn and produce energy. Most common fuels are compounds of hydrogen and carbon, usually called ***hydrocarbons.*** Hydrocarbon fuels can be solids, such as coal and oil shale; liquids such as kerosene, gasoline, and diesel fuel; or gases, such as propane and butane. All of these hydrocarbon fuels contain large amounts of ***energy*** that can be easily released as heat. This heat can then be used to accomplish a task.

The most commonly used fuels in internal combustion engines are gasoline and diesel fuel. These fuels are liquids at normal temperatures. They must be vaporized (turned into a gas) and mixed with air before they will burn and operate the engine. This is done in the fuel system by breaking the fuel into small particles. This process is called ***atomization.*** The atomized fuel is mixed with air before the cylinders in gasoline engines, or in the cylinders of diesel engines. Manufacturers are working on improving flex fuel vehicles that can operate on more than one type of fuel. Compressed and liquefied natural gas are beginning to see wider use, especially in fleet vehicles.

Gasoline

Gasoline is a compound of the elements carbon and hydrogen. The gasoline mixture must be made carefully to ensure that it remains a liquid in the fuel tank, while still able to vaporize before it reaches the engine cylinders. The amount and type of additives varies between grades of gasoline, as does the ***octane rating,*** a measure of the gasoline's ability to resist knocking. The following additives are blended into the distilled gasoline at the refinery:

- ❑ Alcohols, such as denatured ethanol, as an octane enhancer, product extender, and to reduce Carbon Monoxide emissions.
- ❑ Anti-icers to prevent fuel system freeze-up in cold weather.
- ❑ Anti-oxidants to minimize gum formation in stored gasoline.
- ❑ Corrosion inhibitors to reduce fuel system corrosion.
- ❑ Detergents and fluidizer oils to remove or eliminate intake valve and fuel system deposits.
- ❑ Lead replacement additives to minimize exhaust valve recession.
- ❑ Metal deactivators to minimize the effect of metal-based components that may be present in the refined gasoline on the fuel system.

To burn in the engine, gasoline must be ***vaporized,*** that is, changed from a liquid to a vapor. In cold weather, gasoline is harder to vaporize, especially when the engine is cold. Therefore, gasoline blends are changed to increase their vaporization rate in the fall and winter. Since gasoline vaporizes easily in hot weather, vapor lock could be a problem if the gasoline blend was not changed to reduce vaporization rate in the spring and summer. In addition, the final gasoline blend depends to some extent on where the original crude oil came from. Crude oil varies between oil fields, and differences in sulfur content, viscosity, and volatility cannot be totally eliminated in the refining process.

Since gasoline is such a complex, variable product, it can have an effect on driveability. A common problem is caused by a sudden rise in temperature in the spring when fuel tanks are still full of "winter" gasoline, or an early drop in temperature in the fall when "summer" gasoline is still in use. In some parts of the country where ozone levels are excessively high, reformulated gasoline—often called ***oxygenated gasoline***—is sold. Oxygenated gasoline reduces ozone formation but may cause a vehicle to operate differently for a time than with traditional gasoline until the ECM can adjust for the reformulated fuel.

The driveability technician must be sure to take the gasoline into account when diagnosing an engine performance problem. You should also remember at all times that gasoline can become contaminated with dirt and water at any point along the distribution route from refinery to fuel tank. Carburetors were able to handle small amounts of water and dirt, but fuel injection systems are much more likely to be affected by impurities.

Diesel Fuel

Diesel fuel is also a hydrocarbon fuel. Unlike gasoline, however, it is a more chemically stable fuel that does not vaporize as easily. For this reason, as well as the high compression necessary for engine operation, diesel fuel must be injected directly into the cylinder. The injection process violently agitates the fuel, vaporizing it and mixing it with air. Diesel fuel also contains additives to assist vaporization, reduce combustion chamber deposits and fuel system corrosion, and reduce the possibility of fuel line icing. The ***cetane number*** of diesel fuel refers to its burning ability, and has no relation to the *octane* ratings of gasoline. The cetane number is determined by comparing the burning characteristics of a sample of diesel fuel against a test fuel called cetane.

Diesel fuel is also classed according to its viscosity, from D-1 (lightest) to D-6 (heaviest). The two commonly available classes of diesel fuel are D-1 and D-2. Most automobile engines are designed to run on D-1, available at most gas stations. In some places, D-2 may be available. D-2 is acceptable for use in warm weather, but should not be used when temperatures will drop below 32°F (0°C). Do not try to operate an automotive diesel on grades D-3 through D-6, as these are too heavy (high viscosity) for automotive use. Automotive diesel fuel is the same as diesel used in marine engines and off-road equipment. However, diesel fuel made for non-highway use has a dye added to prevent its use in automotive diesel engines.

The most common impurity which causes diesel fuel injection problems is water. Most diesel engines have a separate water trap in the fuel lines. This trap should be checked before starting any other diagnostic operations on a diesel engine. If water stays in the fuel system long enough, bacteria and mold can begin to grow at the meeting point of the water and fuel. This bacteria can produce large amounts of slime that will quickly clog the fuel filters. It is very important that only clean, fresh, water free diesel fuel be used.

Fuel Delivery System

Fuel systems can be divided into two main groups: those using fuel injection and those with carburetors. However, both carbureted and fuel injected vehicles use many similar components in their fuel delivery systems. Fuel delivery and fuel injection system components are discussed in the following sections.

Fuel Tanks

The ***fuel tank*** is used to store fuel and, sometimes, fuel vapors until they can be burned in the engine. Tanks are made from sheet steel or plastic, with internal baffles to prevent fuel sloshing. All tanks contain a combination fuel gauge sending unit and pickup tube. Some sending units also contain an electric fuel pump. A few tanks have separate fuel pump and gauge assemblies. The tank is vented through the vapor control system so that air can replace the fuel as it is pumped to the engine. **Figure 9-1** illustrates the connections to and from the fuel tank. Some fuel tanks have valves and baffles that prevent fuel siphoning. Some newer fuel tanks have valves that monitor system pressure. Other tanks have fuel pressure sensors that may artificially pressurize or draw a vacuum into the tank to check for leaks.

The fuel tank can be mounted in many places on the vehicle. The most frequent location used is the rear of the vehicle, either in front or behind the rear axle. Other locations include above the rear axle, behind the passenger compartment, or along the vehicle's frame. Some vehicles, like vans and pickup trucks, can have more than one fuel tank. The overall design features of a vehicle often dictates the location of the tank. A filler neck, reaching to a convenient spot, is attached either directly to the tank or by a neoprene hose.

Fuel Lines

Fuel lines connect the fuel tank to the rest of the fuel system. Most fuel lines are designed to carry liquid fuel; however, other lines carry vapor back to the fuel tank. Some fuel lines are made of galvanized steel coated internally with a corrosion inhibitor that will not react with the fuel. Flexible rubber or neoprene is used in the lines that must move with the engine. Often, flexible lines are used to connect components due to their ease of installation and sealing. Most of the flexible fuel lines used on carbureted engines are low pressure neoprene lines with clamp type fittings. These lines are designed to carry fuel at no more than 7-10 psi (48.26-68.95 kPa).

Many vehicles with fuel injection use fuel lines similar to carbureted vehicles. However, they use connectors with ***tube fittings*** or ***push-on connectors,*** rather than clamp fittings, **Figure 9-2.** Some fuel injection systems operating under high pressure use lines made of braided metal with high pressure tubing fittings, sometimes referred to as a *banjo fitting.* Since fuel injectors are much more susceptible to clogging from rust in the fuel system, most newer fuel injected vehicles have nylon or plastic fuel lines, along with plastic fuel tanks, that will not corrode.

Other parts of the fuel delivery system are used to assist the carburetor or fuel injectors. They deliver the fuel to the engine from the tank and ensure that the incoming fuel is kept free of contaminants.

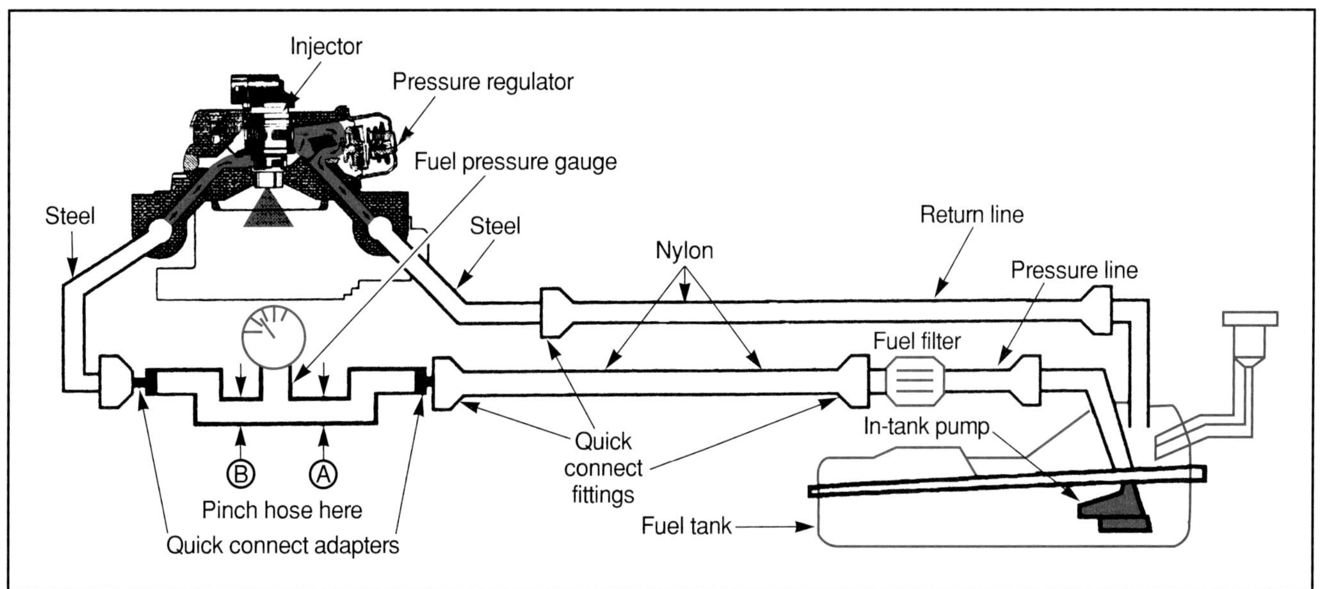

Figure 9-1. *The components of the modern fuel system. While this system uses some metal fuel lines and a metal fuel tank, most modern fuel systems are made of plastic from the fuel tank to the throttle body or injector rail. (General Motors)*

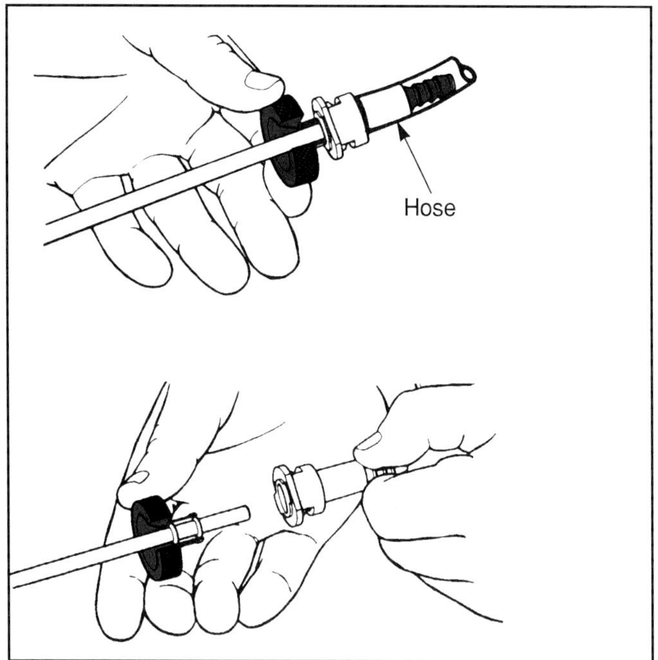

Figure 9-2. *Most modern fuel lines use push-on connectors. Some can be removed with your bare hands, while others require a special tool for removal. (Ford)*

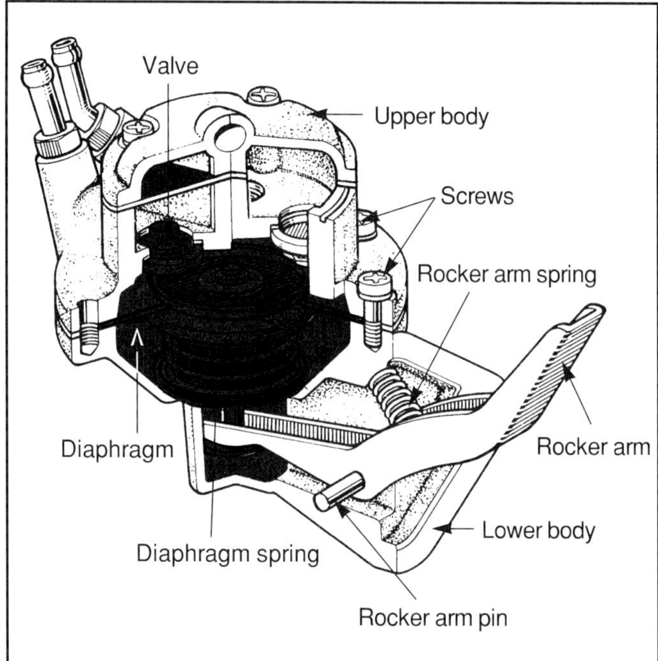

Figure 9-3. *Most mechanical fuel pumps are used on carbureted engines. A unique feature of this pump is that it is rebuildable, however, most fuel pumps are now simply replaced. (Toyota)*

Fuel Pumps

The ***fuel pump*** is a device for transferring fuel from the tank to the injectors at the proper flow rate and pressure. Both mechanical and electrical fuel pumps are used on modern engines. Most carbureted engines have a mechanical fuel pump, and all vehicles using fuel injection have at least one electric fuel pump. Some fuel injected engines have two fuel pumps, one of which may be mechanical. The first pump draws fuel from the tank, while the second pump provides the high pressure needed at the fuel injectors.

Mechanical Fuel Pumps

Mechanical fuel pumps, such as the one shown in **Figure 9-3**, are diaphragm types with two one-way check valves. When the diaphragm moves away from the check valves, fuel is drawn from the tank through the inlet check valve, while the outlet check valve remains closed. When the diaphragm moves in the opposite direction, it pushes the fuel out through the outlet check valve, while the inlet check valve closes. The check valves ensure that the fuel always flows through the pump in the same direction. This type of fuel pump is usually operated directly from a lobe on the camshaft, an eccentric lobe bolted to the front of the camshaft gear, or through a pump rod. Typical mechanical fuel pump placement is shown in **Figure 9-4.** The majority of modern mechanical fuel pumps are nonrepairable, and must be replaced as a unit. If the pump is defective, it is simply discarded.

Figure 9-4. *One possible location for a mechanical fuel pump. Locations can vary by engine and manufacturer.*

Electric Fuel Pumps

Electric fuel pumps are operated from the vehicle electrical system whenever the ignition switch is on. Some pumps are energized by the engine control ECM before the engine is started. ECM systems also energize the pump through the oil pressure switch as a backup. This allows the pump to operate when the engine is running fast enough to develop oil pressure. Some pumps are diaphragm models operated by an electric solenoid instead of a cam lobe. Electric fuel pumps can use positive displacement vanes or rotors to pump the fuel, or a non-positive impeller pumps, similar in operation to a water pump. Electric pumps can be located in the fuel tank or in the line from the tank to the engine. **Figure 9-5** shows an in-tank fuel pump. The motor is cooled by the surrounding gasoline.

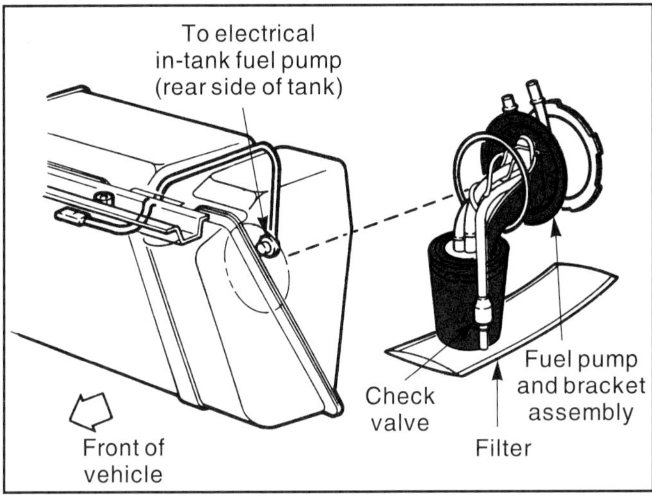

Figure 9-5. *Electric fuel pumps are usually located inside the fuel tank and may be a part of the fuel sender assembly. (Ford)*

Electric fuel pumps produce the same pressure no matter what the engine speed. The electric fuel pump is designed to produce maximum pressure whenever it is energized and usually produces more pressure than the mechanical pump. Pump pressures on fuel injection engines are usually somewhere between 30-60 psi (206.8-412 kPa). These higher pressures are necessary on most fuel injected engines, although some throttle body injection systems operate on pressures as low as 6-15 psi (36-103 kPa). Electric fuel pumps can fill and pressurize the fuel lines before the engine is started. This makes it easier to restart a vehicle when the fuel system has been serviced or has run out of fuel.

Fuel Filters

The ***fuel filter*** is a device that removes impurities from the fuel before it reaches critical parts. Gasoline and diesel fuel often contains dirt, water, or rust particles as a result of storage and shipping contamination, or condensation and rust from the vehicle's fuel tank. These contaminants can be drawn into the fuel system. Once there, they will clog or corrode the small passages of the carburetor or fuel injectors and wear the moving parts of the fuel pump. Any of these conditions can cause driveability problems. If the fuel filter does not do a good job of filtering, or becomes clogged, it can cause performance problems or complete fuel system failure.

Various types of fuel filters are used on modern vehicles. Filters can be sealed units which are replaced as a unit, **Figure 9-6**, or may be a reusable housing that has a replaceable element. All filters need periodic replacement. They can be located in several places in the fuel system, including the carburetor, fuel pump inlet, or in the fuel line, either in the engine compartment or under the vehicle. Some fuel systems, especially diesel fuel systems, have more than one filter.

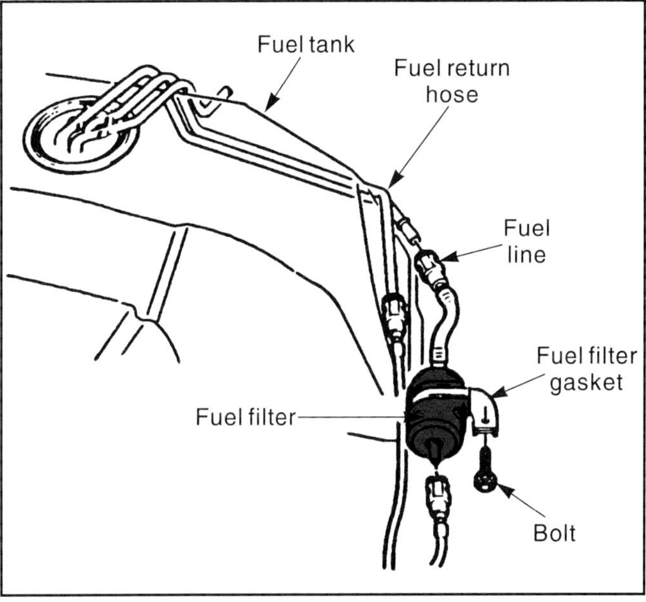

Figure 9-6. *Fuel filters can be located anywhere from the fuel tank, as shown here, forward to the engine compartment itself. This filter uses plastic fuel lines. (General Motors)*

Filters are constructed of special fuel-resistant medium, such as paper, ceramic, or sintered bronze that will allow gasoline or diesel fuel to pass through, but will trap water and particles of dirt and rust. Most fuel tank pickup tubes are also equipped with fiberglass or plastic screens to trap large pieces of debris and some water. This tube can enter from the top, side, or bottom of the tank. The tube end is generally located about 0.5″ (12.7 mm) from the tank bottom. This allows considerable water and sediment to form before being drawn into the pickup tube. Diesel fuel filters usually contain a water trap. This trap can often be drained without removing the filter from the vehicle.

Fuel Injection

Fuel injection has always been used on diesel engines and has completely replaced the carburetor on late model gasoline engines. Fuel injection precisely meters fuel into the engine, providing more accurate control of the air-fuel ratio than is possible with carburetors. Modern ***fuel injection systems*** are controlled by the engine computer, making them very accurate. The system tries to maintain as nearly a perfect air-fuel mixture as possible. A ***stoichiometric*** air-fuel mixture will burn very cleanly with a minimum of pollutants. The ideal mixture is 14.7 parts of air to 1 part of fuel by weight.

Fuel Injectors

Fuel injection systems consist of one or more fuel supply devices called ***fuel injectors.*** A typical injector is shown in **Figure 9-7.** Injectors are operated by fuel pressure and a control system. Some injectors are designed to deliver fuel to the intake manifold whenever the fuel pump is running. Other injectors are opened and shut by electrical current acting on a ***solenoid*** inside the injector. The solenoid moves an internal injector component called the ***pintle,*** which seals the injector opening or ***nozzle.*** The pintle is the control valve for the injector. The pintle is closed by ***spring pressure*** whenever the solenoid is de-energized.

The length of time that the injector remains open is very short, usually measured in thousandths of a second. The injector in **Figure 9-7** contains a solenoid, pintle, and spring. Fuel injection systems rely on high pressures to inject fuel into the combustion chamber, **Figure 9-8.** All injectors squirt fuel in a cone-shaped pattern for maximum atomization. The fuel injection process vaporizes the fuel so that it can readily mix with the incoming air.

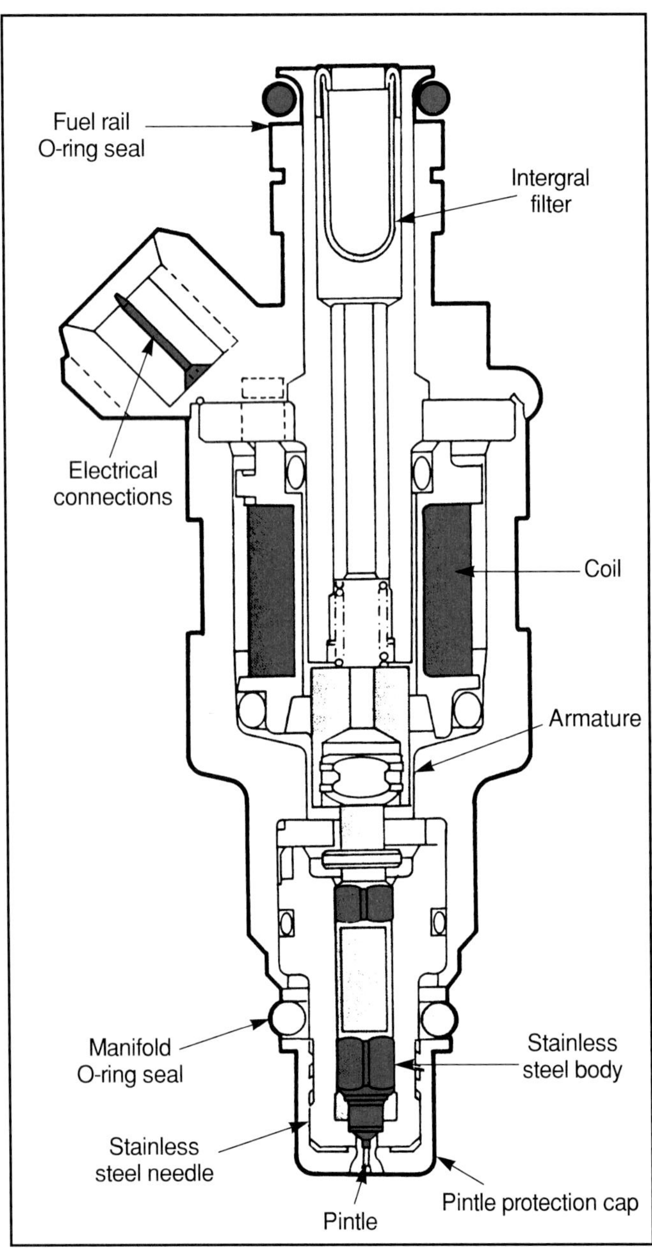

Figure 9-7. *Cutaway of a multiport fuel injector. Study all of the parts carefully. (Ford)*

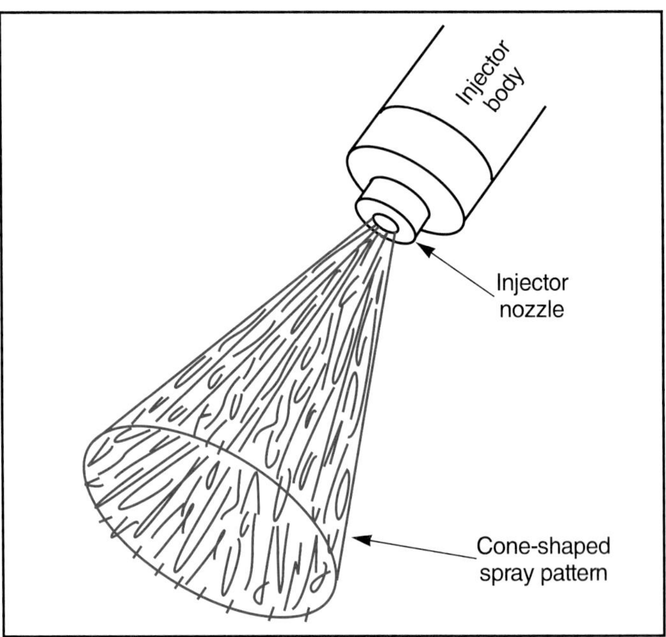

Figure 9-8. *Injectors spray fuel in a cone-shaped pattern. This pattern can be affected if the injector becomes dirty.*

A fuel injection system allows the air-fuel ratio to be precisely controlled, either by varying the length of time that the injector is open, or by varying the fuel pressure at the injector. On most modern systems, the length of time that the injector is open can be changed by varying the length of the electrical signal from the ECM. Injectors used on this type of cylinder always contain an internal electrical solenoid. The ECM bases its electrical signal to the injector(s) on inputs from various engine and drivetrain sensors, **Figure 9-9A.** This is discussed in more detail later in this chapter.

On some systems, the pressure of the fuel going to the injector is varied to match the throttle opening. The injectors are always open, and fuel pressure is varied to match the fuel flow with the air flow. These systems contain a mechanically operated pressure control valve. This valve increases pressure based on the amount of air flowing into the engine. Most late model versions of these systems have the fuel pressure modified by an electronic control system, which bases its pressure decisions on inputs from various sensors. A modern system is shown in **Figure 9-9B.**

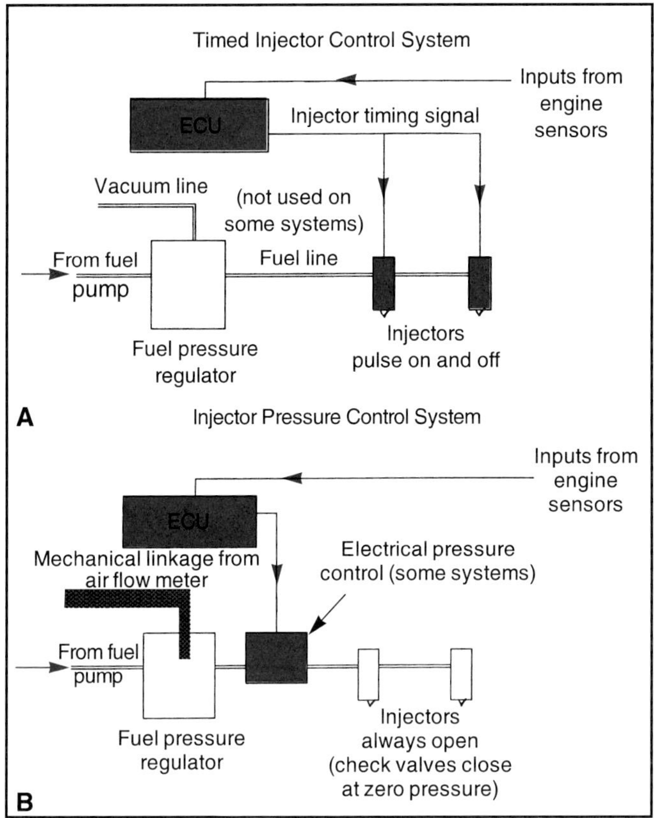

Figure 9-9. *On all modern systems, the amount of fuel that the injectors can deliver is controlled by the ECM, based on inputs from sensors. A—The injectors are pulsed on and off by the ECM to control mixture. B—The injectors are always open and injector system pressure is controlled by the ECM.*

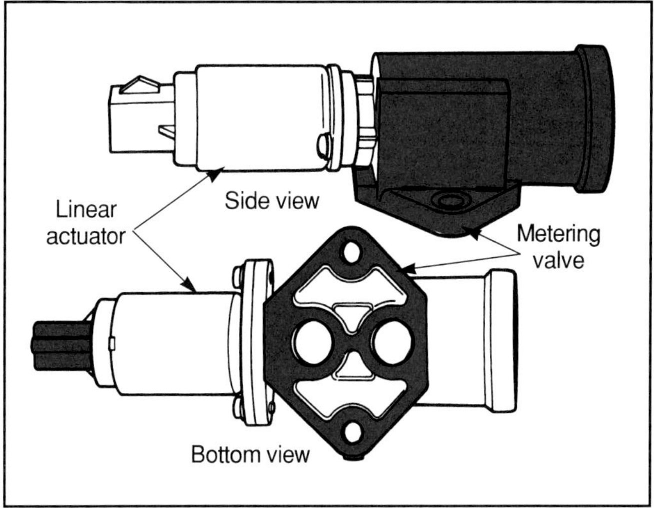

Figure 9-10. *An idle air control motor as used on a multiport fuel injected engine. Designs and sizes can vary. (Ford)*

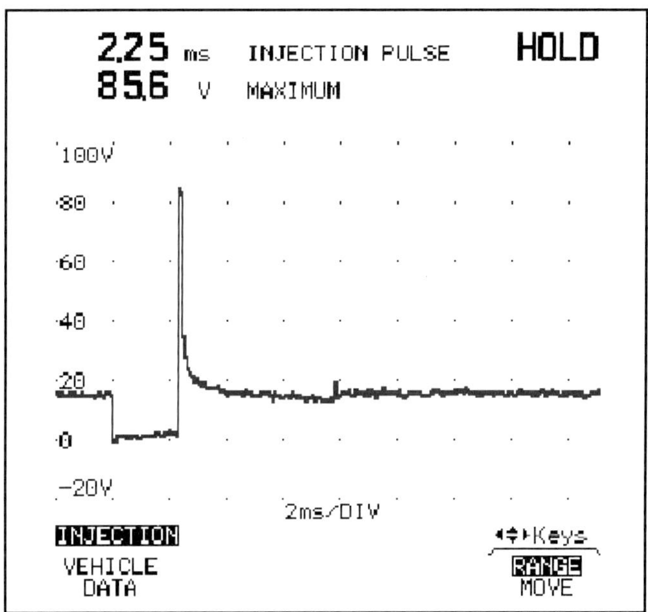

Figure 9-11. *Injector pulse width is varied to control fuel injector opening in some systems. Waveforms can be important clues to diagnosing a possible fuel injection problem. (Fluke)*

Idle Speed Control

Gasoline fuel injection systems inject fuel into the air intake system before the intake valve. Air flow into the fuel injected engine and idle speed is controlled by a plate-type throttle valve, identical to the throttle valve of a carburetor. Air flow past the closed throttle valve is often through a small passage controlled by an ***idle air control motor,*** or by an idle bypass valve or solenoid. See **Figure 9-10.** On some vehicles, a second set of throttle valves is used to increase air-fuel mixture control. At low speeds, only one set of throttle plates is open. This increases air speed through the intake runners, allowing better fuel atomization. When full power is needed, the second set of throttle plates opens, allowing maximum air to enter the engine.

Injector Pulse Width

Injector ***pulse width*** refers to the amount of time during which the injector electronic solenoid is energized, causing the injector to open. Pulse width is a measurement of how long the injector is kept open; the wider the pulse width, the longer the injector stays open. The amount of fuel delivered by a given injector depends upon the pulse width, **Figure 9-11.** Pulse width is controlled by the ECM.

Fuel Injection System Classification

Fuel injection systems can be divided into two main groups, throttle body injection systems and multiport injection. Vehicles for years used both types of fuel injection. Today, most vehicles are manufactured with various kinds of multiport injection systems. **Figure 9-12** is a comparison of the two types of fuel injection. These systems are referred to by various names, usually related to their operation. A throttle body or plate is used on all fuel injection systems.

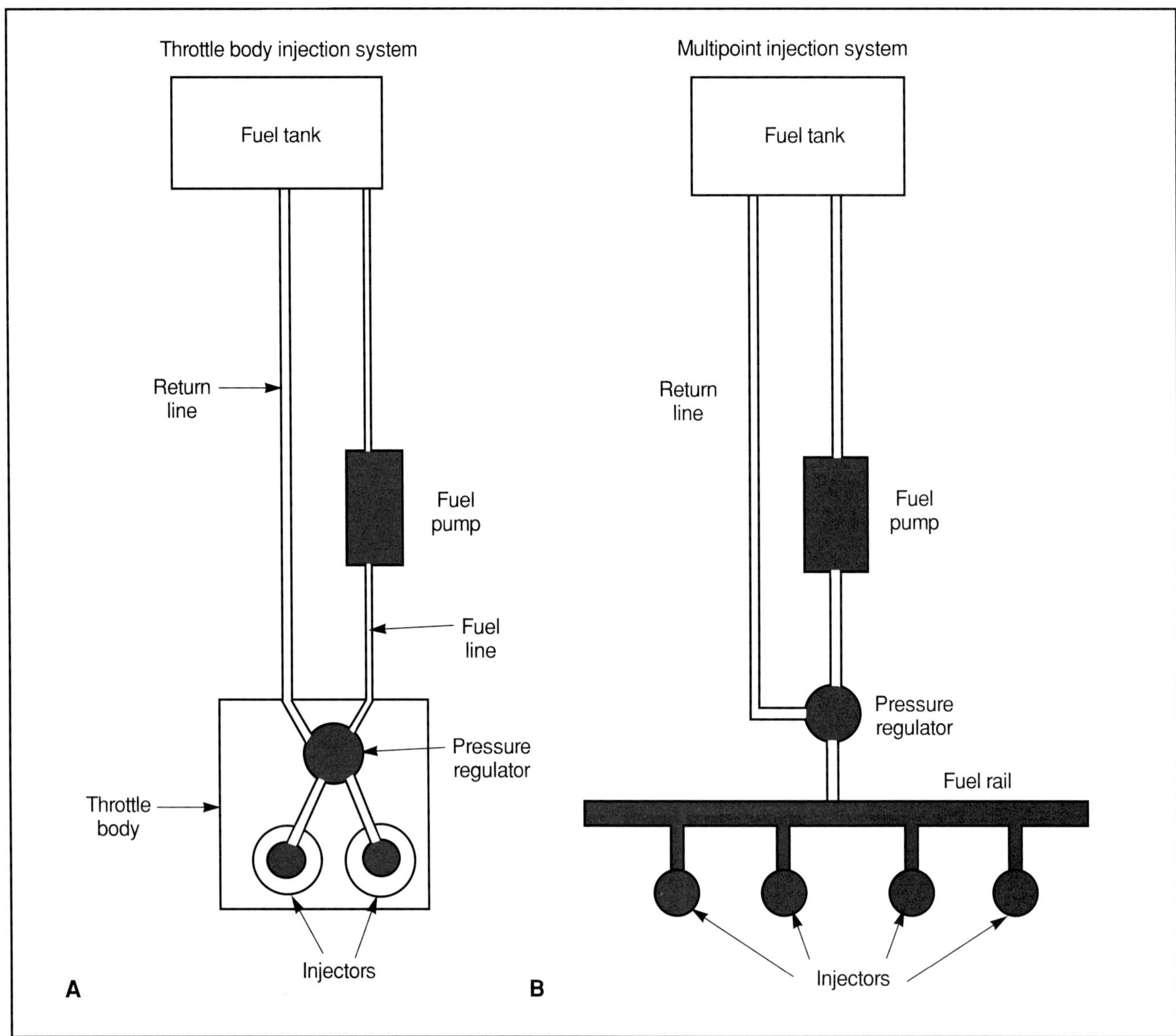

Figure 9-12. *Component schematics of the two fuel injection systems in use today. A—Throttle body fuel injection system. B—Multiport fuel injection system.*

Throttle Body Injection Systems

When the engine has a single injector assembly containing one or two fuel injectors, it is mounted at the entrance to the intake manifold, where the carburetor was formerly placed. These systems are usually referred to as ***throttle body injection (TBI).*** These systems are sometimes called by other various names, including:

- Single-point injection
- Central fuel injection (CFI)
- Electronic fuel injection (EFI)

Throttle body fuel injection assemblies consist of one or more aluminum or pewter castings. These castings are precision machined to reduce or eliminate air and fuel leaks. Refer to **Figure 9-13.** The injector assembly has two fuel line connections. One line supplies pressurized fuel to the assembly while the other line returns excess fuel to the fuel tank. Internal passages in the assembly direct pressurized fuel to the injectors. The injector solenoids are energized by electrical input from the ECM and closed by spring pressure. When the injectors are open, pressure developed by the fuel pump causes fuel to be sprayed into the air horn ahead of the throttle valve. The fuel is atomized and mixed with air entering through the throttle valve. Some throttle body injection assemblies have a Schrader type valve (similar to a tire valve) for pressure checking and relieving system pressures.

Most throttle body injectors are pulsed on and off at the same rate as the firing of the spark plugs. This is usually done by using the ignition triggering sensor impulse as a fuel injector triggering impulse. Matching the pulse rate of the fuel injectors to the firing impulses of the ignition system allows the pulse rate to vary with engine speed. The length

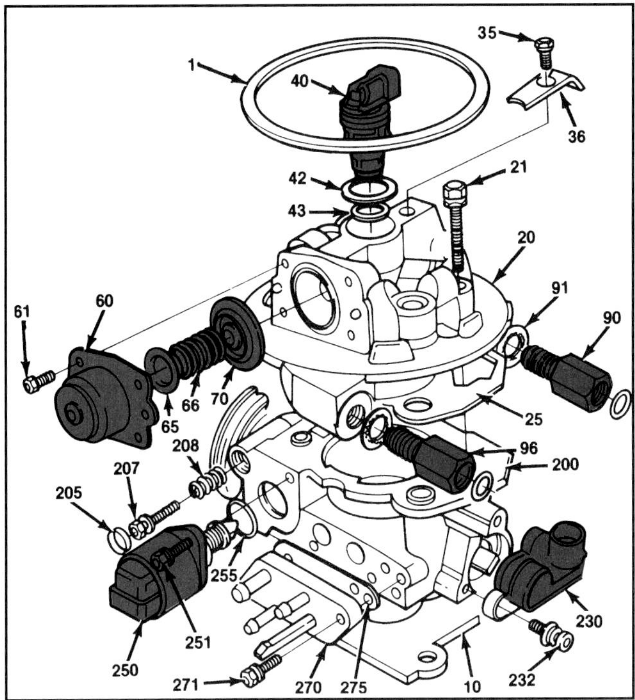

Figure 9-13. *An exploded view of a throttle body injection unit. 1—Air filter gasket. 10—Gasket flange. 20—Fuel meter assembly. 21—Screw and washer assembly. 25—Fuel meter body-to-throttle body gasket. 35—Injector retainer screw. 36—Injector retainer. 40—Fuel injector. 42—Fuel injector upper O-ring. 43—Fuel injector lower O-ring. 60—Pressure regulator cover assembly. 61—Pressure regulator attaching screw. 65—Spring seat. 66—Pressure regulator spring. 70—Pressure regulator diaphragm assembly. 90—Fuel inlet nut. 91—Fuel nut seal. 96—Fuel outlet nut. 200—Throttle body assembly. 205—Idle stop screw plug. 207—Idle stop screw and washer assembly. 208—Idle stop screw spring. 230—Throttle position sensor (TPS). 232—TPS screw and washer assembly. 250—Idle air control valve (IACV). 251—IAC attaching screw. 255—IAC O-ring. 270—Tube module assembly. 271—Manifold attaching screw. 275—Manifold tube gasket. (AC-Delco)*

of time that the injector is open, and therefore the air-fuel ratio, is controlled by the ECM based on inputs from various sensors. A pressure regulator is installed on the throttle body to control the pressure at the injectors. Excess fuel returns to the tank through the return line. The fuel pressure of a throttle body injection system ranges from 6-15 psi (41-103 kPa).

Multiport Fuel Injection

When the system uses one injector per cylinder, the individual injectors are located inside the intake manifold or in the cylinder head, close to the intake valves. These systems are called ***multiport fuel injection.*** Some typical names for multiport systems include:

- Sequential Fuel Injection (SFI)
- Tuned Port Injection (TPI)
- Multi-point Injection (MPI)
- Port Fuel Injection (PFI)
- Digital Fuel Injection (DFI)
- Electronic Fuel Injection (EFI)

Typical multiport injectors are shown in **Figure 9-14.** A metal ***fuel rail*** connects the injectors to the fuel delivery system.

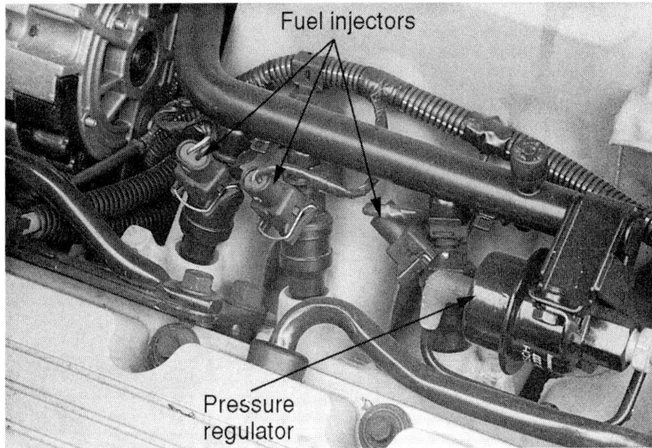

Figure 9-14. *Multiport systems use one injector per cylinder. The injectors are installed on the intake manifold and are actuated by the ECM.*

The most common type of multiport injection system uses solenoid operated injectors. The typical injector is normally held closed by spring pressure. An electrical signal from the ECM opens the solenoid and allows fuel to spray into the intake manifold runner just ahead of the intake valve, **Figure 9-15.** Some injectors are timed to the ignition system so that they open just before the intake valve opens. Others are fired in banks of two or three, with the pulse rate controlled by the engine speed. The time that the injectors are open is controlled by the ECM to maintain the correct air-fuel ratio. A fuel pressure regulator and fuel return line are installed on the fuel rail to maintain the correct pressure in the system. Most multiport systems contain a Schrader type valve for checking fuel system pressure and removing pressure before servicing the system. Pressures on multiport fuel injection systems range from 30-60 psi (206-412 kPa).

On fixed or continuous flow multiport fuel injection systems, the injectors are open whenever the engine is running. Check valves at the end of the injector nozzles are opened by fuel pressure when the engine starts. The check valves close when the engine is turned off to keep fuel from dripping into the manifold and flooding the engine. The check valves have no effect on injector operation when the injection system is pressurized, except to vibrate slightly when fuel passes through them. This vibration helps to atomize the fuel. On this type of system, air-fuel ratios are controlled by varying the fuel pressure. This is accomplished by a mechanical air flow sensor, discussed later in this chapter. Most modern fixed injection systems use additional ECM based controls to further control the air-fuel ratio.

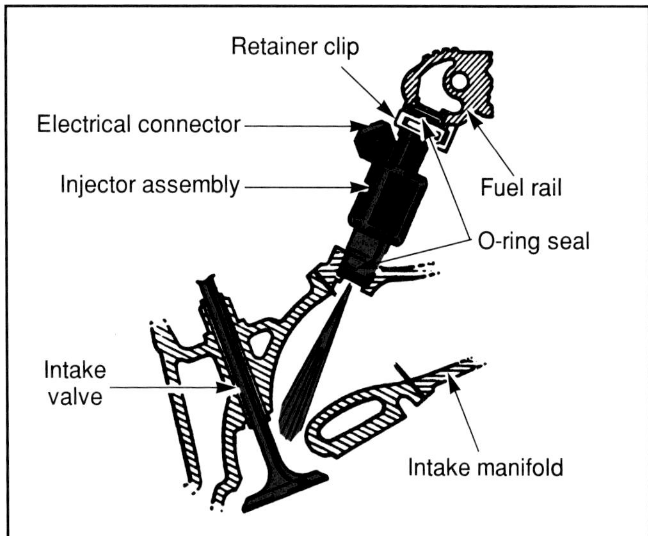

Figure 9-15. *Air and fuel is mixed in the intake manifold before it is drawn into the cylinder by the piston. This allows for maximum atomization. (General Motors)*

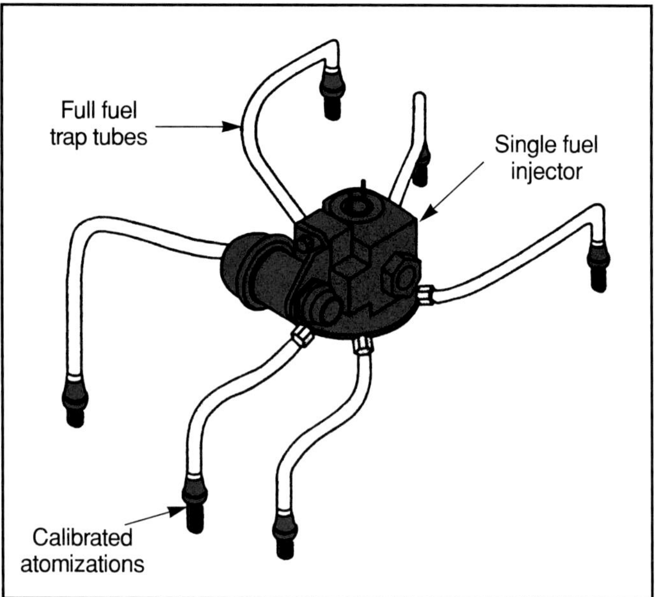

Figure 9-16. *Central port fuel injection systems use only one injector, but directly feeds fuel to each cylinder. While this system may look inefficient, it works quite well.*

A version of the multiport injector system uses one central fuel injector with lines connecting it to outlet nozzles at each intake passage. This system, shown in **Figure 9-16,** combines the simplicity of the single injector with the more precise fuel control of the multiport system. It is usually referred to as ***central point fuel injection (CPI).***

Diesel Fuel Injection Systems

Diesel engines always have fuel injection systems. Heat caused by the high compression ratios is what causes the diesel fuel to begin burning. Therefore, the fuel cannot be compressed along with the air inside the diesel engine cylinder or it would ignite before the piston reaches the top of its compression stroke. Therefore, the diesel injector is designed to squirt fuel directly into the cylinder at the top of the compression stroke. Diesel injectors operate under very high pressures (up to several thousand psi) and are opened and closed by mechanical linkage. See **Figure 9-17.** Diesel injectors are mechanically operated, therefore, they are not directly controlled by an ECM-based control system. However, some other systems on diesel engines are ECM controlled.

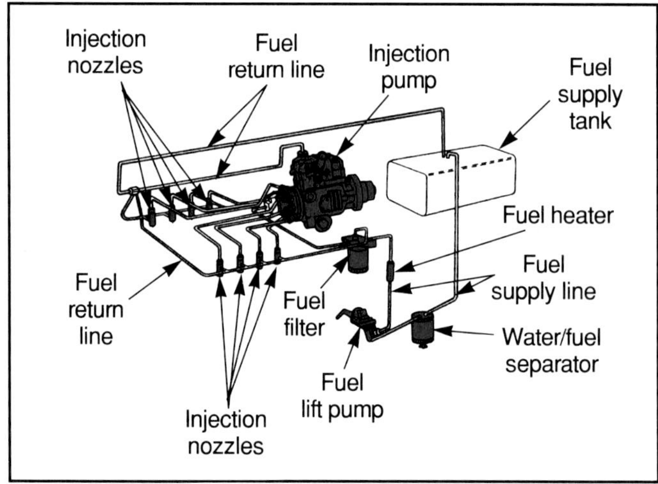

Figure 9-17. *Diesel injectors are either mechanically or electrically operated. Mechanical operation can be by camshafts similar to those used to open the valves, or driven by the engine crankshaft. (Ford)*

Due to the severe heat and pressures involved, diesel fuel injectors are made of heat treated steel, manufactured to extremely close tolerances. Even slight defects or dirt buildup can affect the operation of a diesel injector. Therefore, clean fuel is very important. Many diesel cylinder heads contain a precombustion chamber that creates turbulence to aid in fuel mixing and ignition.

Glow Plugs

Since the diesel fuel is ignited by compression heat, a cold diesel engine is extremely hard to start. As compression heat is developed, it is absorbed by the cold piston and cylinder walls. One solution to this problem would be to leave the diesel engine running at all times, as is often done with railroad locomotives and stationary engines. However, this is not practical with automobiles and light trucks, so ***glow plugs*** are used to heat cold diesel engines. The glow plug consists of a resistance unit that becomes very hot when current passes through it, **Figure 9-18.** The glow plug is installed in the cylinder head, sometimes in a special precombustion chamber. Every combustion chamber has one glow plug. The glow plug control system directs high battery current to the glow plugs only when the engine is cold. This additional heat is sufficient to ignite the diesel fuel and allow the engine to start.

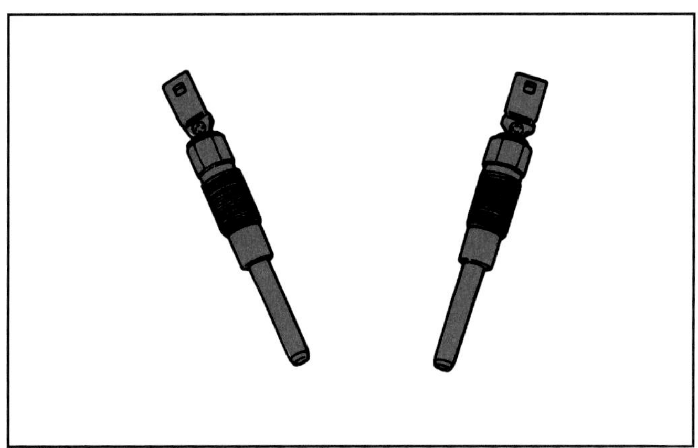

Figure 9-18. *Glow plugs are used as a cold start aid. They heat the combustion chamber in preparation for engine start-up. Some diesel engines use intake air heaters. (Ford)*

A few diesel vehicles mount glow plugs or other heating devices in the intake manifold. If the incoming air is heated before it enters the combustion chamber, there will be sufficient heat to ignite the fuel, even though the combustion chamber is cold. Most vehicles with diesel engines have two batteries to provide current for the starter and glow plug system.

The glow plug control system may be timed to turn off after a certain time has elapsed. On a few systems, the glow plugs are directly operated by an engine mounted temperature sensor. The glow plug system usually contains a dashboard light which illuminates to tell the driver that the glow plug system is operating. In most cases, the vehicle can be started when the glow plug light goes out. Some vehicles contain an interlock circuit which prevents engine cranking until the glow plugs have finished heating the combustion chambers.

Computer-Controlled Diesel Engines

On most modern diesel engine cars and light trucks, the glow plugs or air intake heaters are energized by relays controlled by the ECM. On most late model diesel engines, the EGR valve is controlled by the ECM. On some diesel engines, the fuel injector timing is also controlled by the ECM. The ECM usually operates one or more dashboard warning lights connected with diesel engine operation. Typical inputs to the ECM include throttle position, manifold vacuum, intake air temperature, transmission gear selected, engine RPM, and engine temperature.

Fuel Injection Pressure Regulators

Pressure in some throttle body fuel injection systems is as low as 7 psi (48.3 kPa), while some multiport systems can reach 60 psi (380 kPa) or more. Injection system pressure on most computer-controlled injection systems is regulated by a fuel injection ***pressure regulator.*** The regulator controls pressure by bleeding excess fuel back into the fuel inlet line or fuel tank. Pressure regulators used on multiport fuel injection systems are usually connected to intake manifold vacuum, so that changes in engine load can modify the fuel pressure setting. Most pressure regulators used on throttle body fuel injection systems are controlled by spring pressure alone.

On throttle body injection systems, the pressure regulator is mounted on the throttle body, **Figure 9-19.** Multiport system pressure regulators are attached to the fuel rail as in **Figure 9-20.** On almost all modern fuel injection systems, regulator pressure is preset during manufacturing, and there is no provision for pressure adjustment.

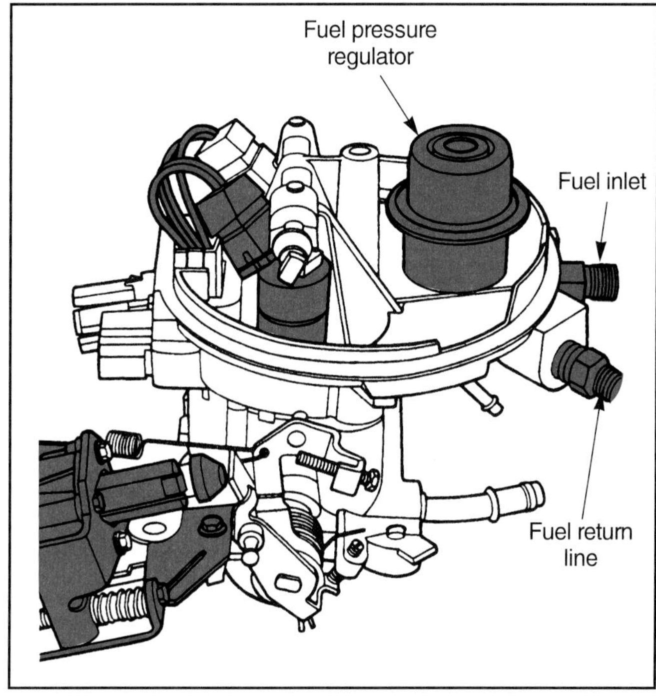

Figure 9-19. *Fuel pressure regulators limit the amount and pressure of the fuel entering the engine. This pressure regulator does not use vacuum assistance. (Ford)*

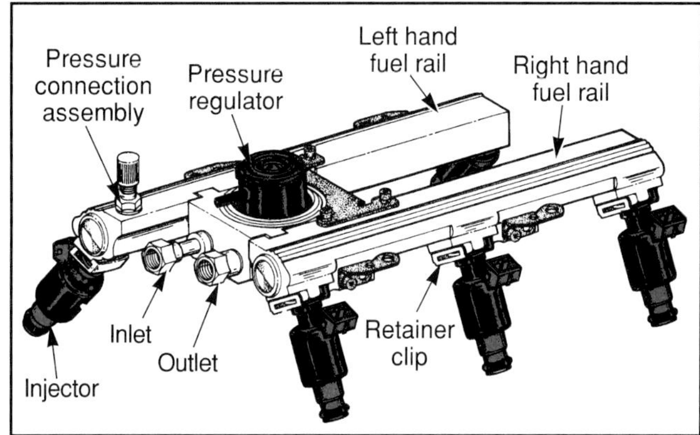

Figure 9-20. *A multiport fuel rail with pressure regulator as used by one manufacturer. This regulator uses manifold vacuum to assist the spring in fuel pressure adjustment. (AC-Delco)*

The fuel pressure regulator is one of the main components affecting the air-fuel ratio. It has an important part in controlling emissions, mileage, and driveability. Typical pressure regulators are shown in **Figure 9-21.** Note that the amount of fuel returned to the tank is controlled by a spring on the regulator in **Figure 9-21A.** In the regulator in **Figure 9-21B,** the spring is assisted by manifold vacuum. High vacuum caused by low engine load allows fuel pressure to unseat the spring and return to the tank easily, keeping fuel pressures low. Low vacuum caused by high engine loads allows the spring to apply full pressure to the fuel return valve, keeping fuel pressures high.

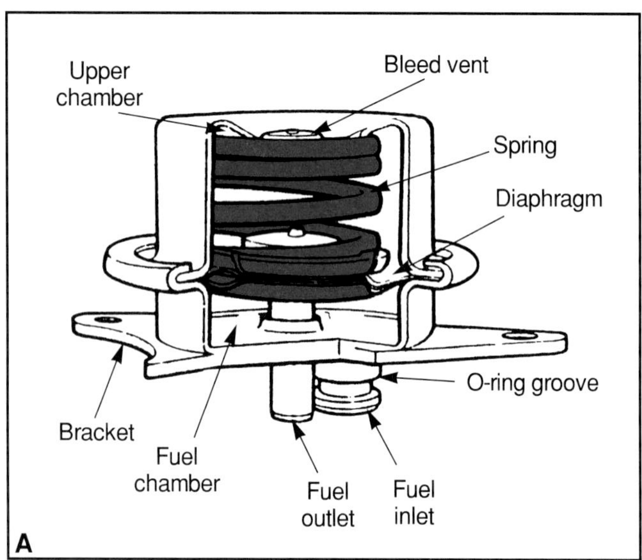

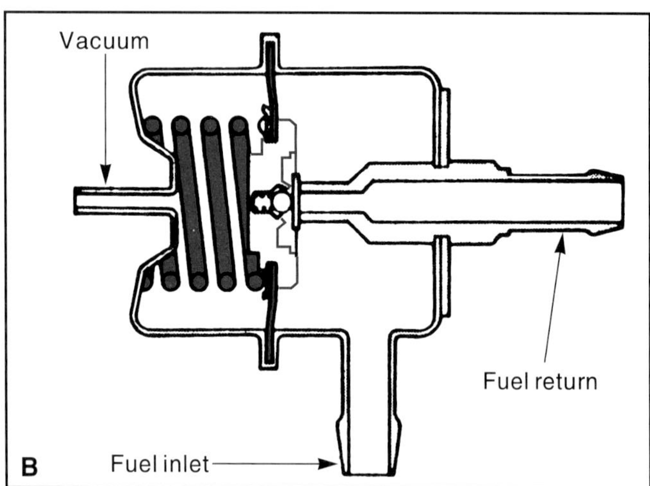

Figure 9-21. *Fuel pressure regulators are operated by spring pressure and sometimes, vacuum assist. A—Most throttle body injection systems use sealed pressure regulators. B—Most multiport systems use vacuum assisted pressure regulators. The regulator is connected to engine vacuum to vary power assist. (Chrysler)*

Fuel Injection Cold Start-up

In addition to the components described earlier, the fuel injection system can incorporate other units for special functions, such as cold starts. Some systems use a ***cold start injector*** to add extra fuel to the intake manifold when starting a cold engine. The cold start injector sprays extra fuel into the intake manifold for a temperature controlled period of time. The amount of fuel injected is controlled by a thermal switch which regulates the amount of time that the valve is energized, depending on the engine temperature. During cold start-up on some systems, an ***auxiliary air regulator*** admits additional air into the intake manifold to increase the idle speed. The auxiliary air regulator is controlled by a thermostatic switch located in the engine water jacket.

However, most systems rely on the ECM to prepare the engine for cold start-up. Most computer control systems will energize the fuel pump and pulse all of the fuel injectors for 2-5 seconds when the ignition key is turned to the run position and the engine is cold. This causes the fuel system to pressurize and injects fuel into the intake manifold, where it can help start the engine once the key is turned to the start position. Since most drivers immediately turn the key to start immediately, this additional pulse occurs while the engine is cranking, easing cold start-up.

Computer Control of the Fuel Injection System

Modern fuel injection systems are controlled by the ECM, based on inputs from various engine sensors. You have already learned sensor and ECM operation in this chapter and earlier in **Chapter 7.** All sensor readings are taken into account by the ECM in deciding how to adjust fuel system operation. However, it does look at some inputs more than others. In this section, you will learn how these sensors affect fuel injection system operation. You may want to review ECM and sensor operation in **Chapter 7** before continuing.

Airflow Sensors

One of the most important sensor input on some multiport injector systems is the intake air flow reading. Sensors that read the amount of air flowing into the engine are located ahead of the throttle valve, as shown in **Figure 9-22.** Airflow sensors are installed in the air intake system of some multiport fuel injection engines. To more precisely control the fuel injection system, the other sensors provide readings of engine temperature, outside air temperature, engine speed, throttle position, manifold vacuum, and exhaust gas.

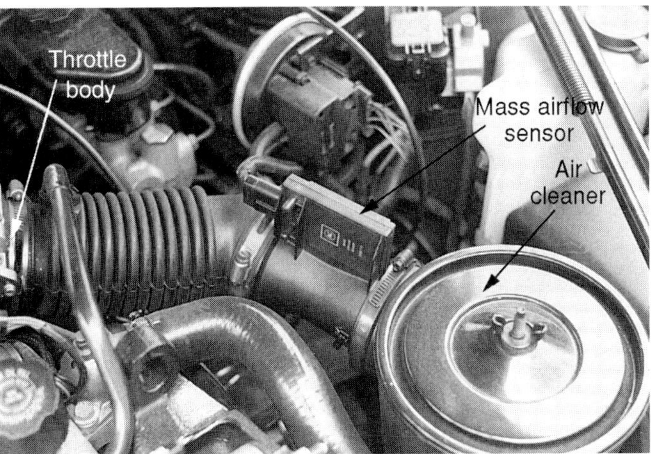

Figure 9-22. *Mass airflow sensors are used with multiport fuel injected engines. The sensor is installed between the air cleaner and throttle valve.*

The ECM processes these readings and matches injector fuel flow to keep the air-fuel ratio as close to 14.7:1 as possible, while maintaining good driveability and performance. There are three major kinds of air flow sensors, the mechanical, heated element and the Karman vortex.

Mechanical Airflow Sensors

Mechanical sensors can be flap meters or air vanes. The *flap meter* is a flat metal plate placed in the air intake system in the path of the incoming air. The plate is spring-loaded to remain in the zero-flow position when the engine is not running. When the engine starts, the plate is moved against spring pressure by the air flowing into the engine. The plate moves in proportion to the amount of air flowing against it. The plate is connected by linkage to a fuel pressure regulator, as shown in **Figure 9-23.** This system is always used with a fixed injector system.

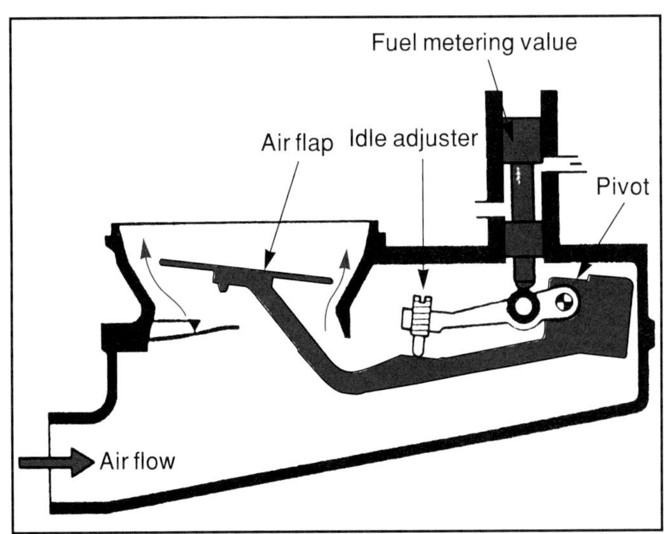

Figure 9-23. *Mechanical airflow meters are used on vehicles with pressure controlled fuel injectors. It directly controls how much fuel pressure goes to the injectors.*

The other type of mechanical airflow sensor, the *air vane,* is shown in **Figure 9-24.** This type consists of a vane that is placed in the air stream, with a lower projection into a sealed chamber. Air flow through the engine moves the vane. The bottom projection and chamber damp the vane movement, preventing rapid swings in air flow readings. This type of air flow meter is attached to a variable resistor, which sends a voltage signal to the ECM.

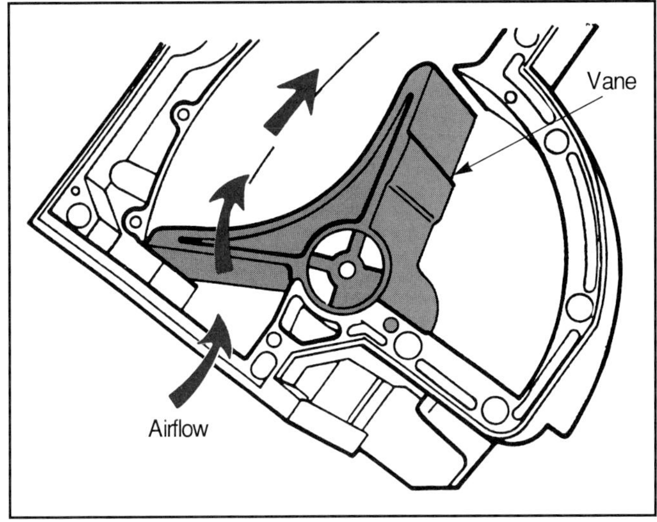

Figure 9-24. *The vane airflow sensor is similar to the flap type. However, the vane is attached to a variable resistor that sends a signal to the ECM.*

Mass Airflow Sensors

The second type of airflow sensor is the heated element, often called a ***mass airflow (MAF) sensor.*** The sensor consists of a wire or film with a closely calculated electrical resistance, placed in the intake system air stream. Electrical current flows through the element. Since current flow in the wire or film is always proportional to the temperature of the wire, the sensor circuitry can send an electrical signal to the ECM. The internal parts of a heated wire sensor are shown in **Figure 9-25.** Since this sensor measures air flow by how much heat it removes from the wire, it can partially compensate for changes in temperature and humidity of the incoming air.

When the engine is running, the air flow across the element removes heat, affecting its resistance. When resistance of the element changes, the current flow also changes. The sensor, therefore, sends an electrical signal to the computer that is directly proportional to the amount of air flow into the engine. If the air temperature or humidity changes, the amount of heat removed from the wire or film will also change. Cold, wet air will remove more heat, while warm, dry air will remove less heat. This allows the sensor to compensate for changes in both temperature and humidity. Most ECMs also process inputs from the air temperature sensor and MAP sensor to further refine the air flow reading.

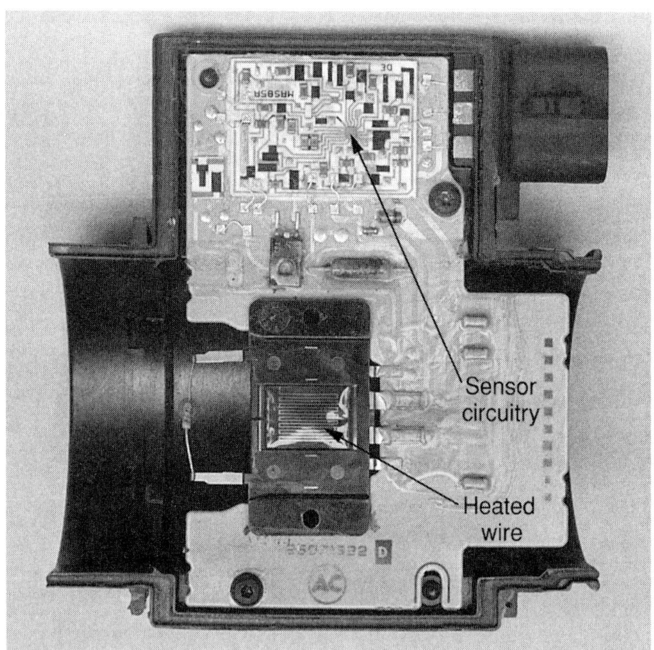

Figure 9-25. *The internal parts of a heated wire mass airflow sensor. Many sensors use a heated film versus a wire.*

Karman Vortex Sensors

The ***Karman vortex airflow sensor*** operates by sensing the vortex, or air vibrations, set up by air entering the engine. The Karman vortex sensor consists of an LED, a photodiode, and a lightweight foil mirror, **Figure 9-26.** In this design, the light emitted by the LED reflects from the mirror to enter the photodiode. The photodiode converts the light received into a voltage signal that is sent to the ECM.

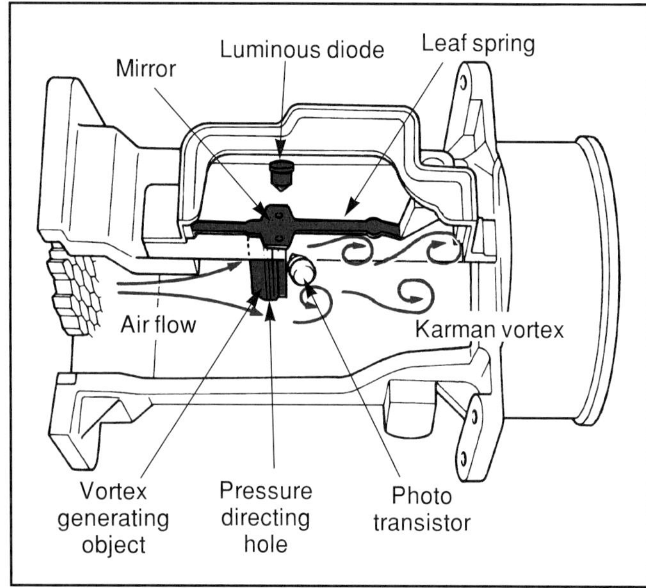

Figure 9-26. *Cutaway of a Karman vortex airflow sensor. Note how the vortex exerts pressure on the mirror, causing vibration. (Ford)*

As the engine draws in air, air flow through the sensor increases. This causes the foil mirror to vibrate. Vibration of the mirror decreases the amount of LED produced light that is reflected from the mirror to the photodiode. This in turn reduces the voltage of the signal sent to the ECM. As air flow increases, mirror vibration increases, and the voltage signal to the ECM varies in proportion. Other systems use the Karman vortex principle to produce a pressure differential that can be read by the ECM as an air flow signal, **Figure 9-27A.** In **Figure 9-27B,** the Karman vortex is used to produce sound waves which also can be interpreted as air flow.

Oxygen Sensors

Most fuel injection systems rely heavily on the oxygen sensor for much of the engine's operating information. The oxygen sensor monitors the amount of oxygen in the engine's exhaust gases. As the oxygen content in the exhaust gases changes, the voltage signal produced by the sensor also changes. The computer uses signals from the oxygen sensor to control the air-fuel mixture. The oxygen sensor is generally mounted in the exhaust manifold or a crossover pipe. OBD II systems use multiple oxygen sensors; at least one for each exhaust manifold, one before the catalytic converter, and one after the converter, **Figure 9-28.**

Speed Density

Airflow sensors are not used with throttle body fuel injection systems and some multiport systems. Instead, these systems rely on inputs from other sensors to determine the amount of air coming into the engine. This method is referred to as ***speed density.*** To determine the amount of air entering the engine, the ECM relies heavily on engine speed and MAP sensor inputs, however, the ECM uses inputs from all the sensors. Speed density has several advantages over airflow sensors. These include:

- Simplicity.
- Fewer driveability problems.
- Elimination of the complex and expensive airflow sensor.

The MAP sensor receives a vacuum signal from the engine whenever it is running. The MAP sensor sends a voltage signal to the ECM. Engine speed is monitored through the distributor, crankshaft and/or camshaft sensor. The ECM uses these signals to control the amount of fuel injected and the idle air control motor's pintle position. As the engine speed increases, the ECM receives a signal from the distributor or sensor, as well as a modified voltage signal from the MAP sensor. With these sensor signals, as well as the readings from the other sensors, the ECM "calculates" the approximate amount of air flowing into the engine and adjusts the fuel injection pulse width and ignition timing accordingly. While the airflow sensor may seem to be more accurate, speed density performs quite well.

Pressure sensor
Air
Karman vortex
To throttle body
Amplifier output
AFS output
Power supply (from control relay)
5V
Pressure sensor
Amplifier
Filter circuit
Square wave forming circuit
5V
Airflow sensor
Engine control unit
A

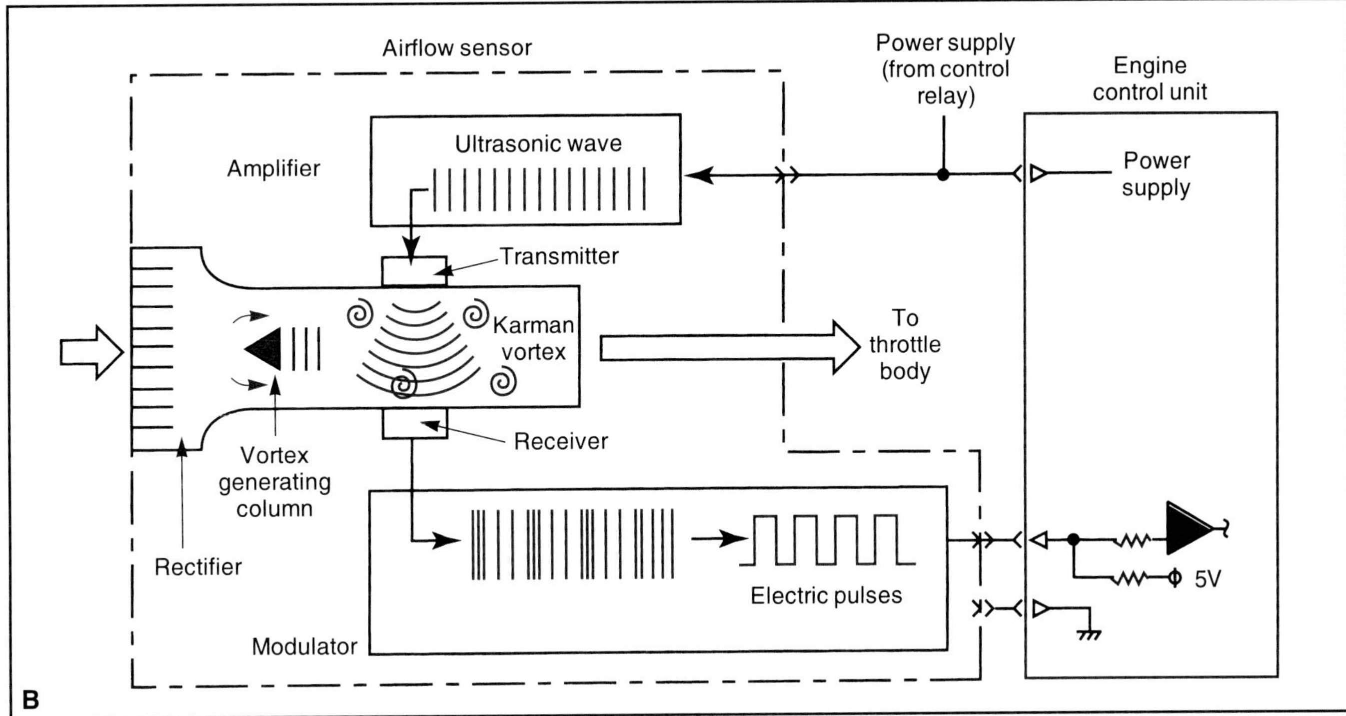

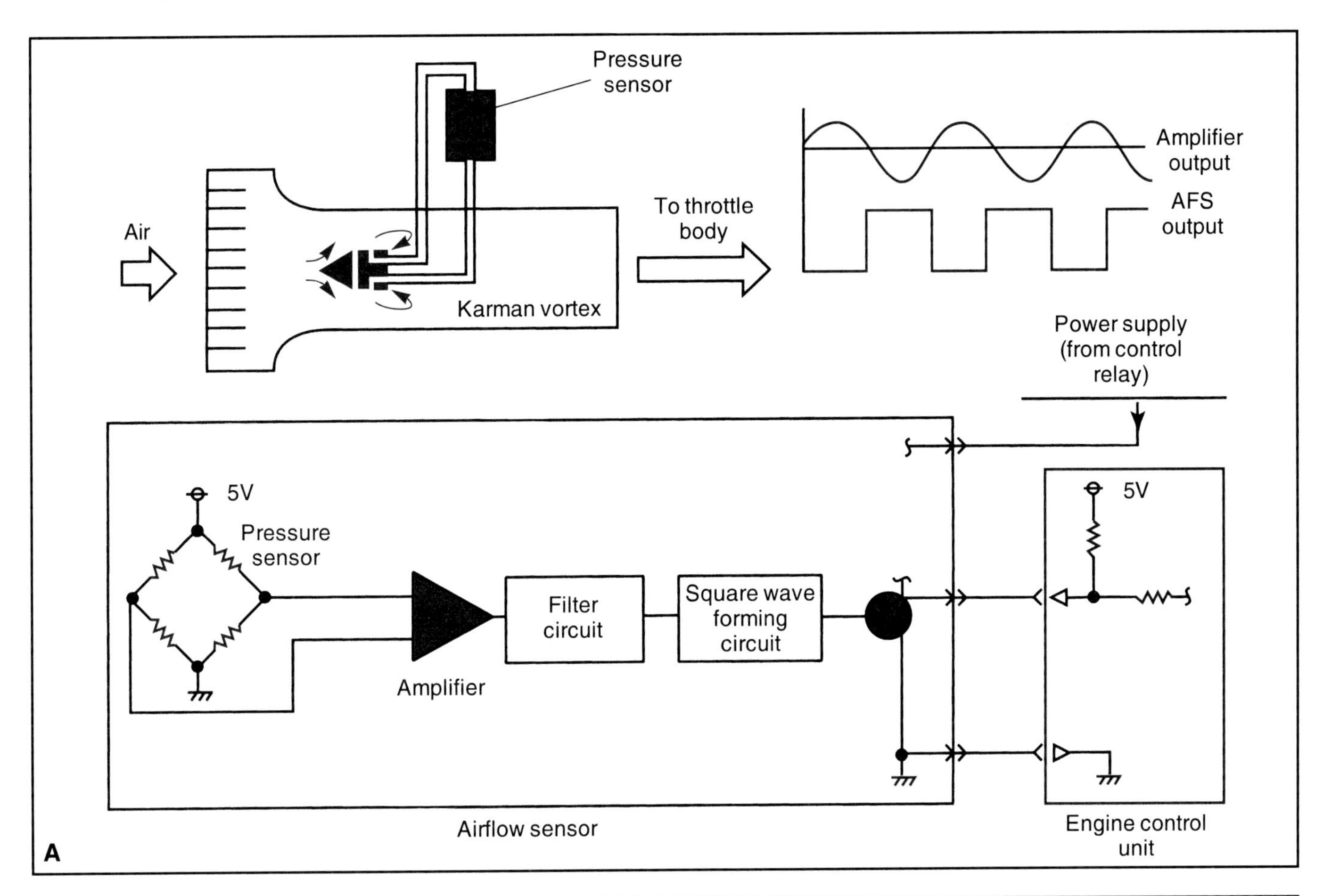

Figure 9-27. *The Karman vortex principle can be used to create pressure and sound waves. A—The number of pressure waves per a given amount of time increases as the air flow increases. B—The ultrasonic mass airflow sensor uses a transmitter that sends a sound to the receiver. As air passes through the sensor, it affects the sound waves. The receiver converts the sound variations into an electrical signal that is sent to the ECM. (Chrysler)*

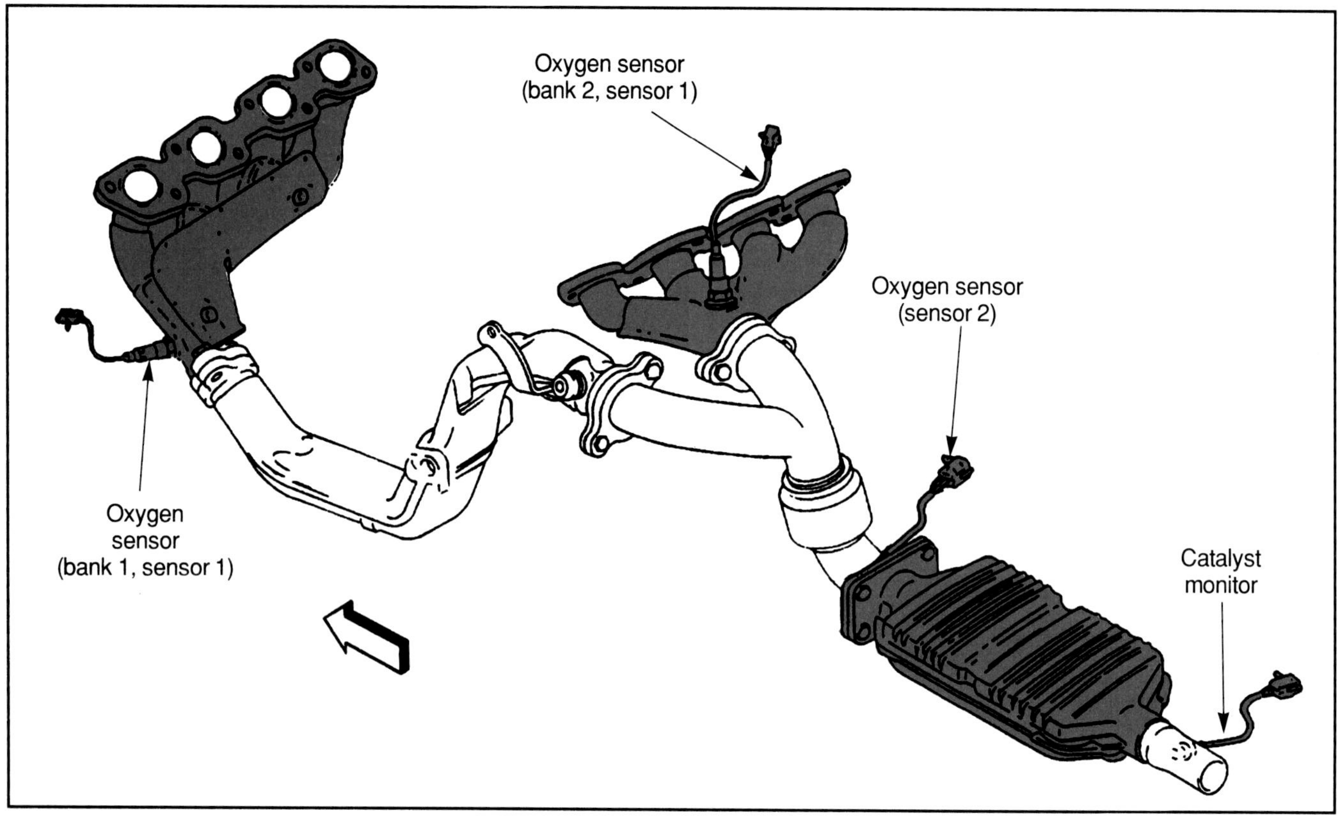

Figure 9-28. *Most newer vehicles use multiple heated oxygen sensors (HO_2S). The catalyst monitor is an oxygen sensor; however, it is used to monitor catalyst performance. (General Motors)*

Short- and Long-Term Fuel Trim

You learned earlier in this chapter that the ECM adjusts the fuel injection pulse to compensate for engine speed, load and other conditions. The term to describe this control is ***fuel trim.*** However, while this adjustment may appear to be a simple process, it actually involves a much more complicated dual process involving input sensors and the ECM, to control fuel injection on two levels.

The ECM normally adjusts for temporary changes in engine operation, such as climbing a hill, slow deceleration from high speeds, or the weight of additional passengers in the vehicle. This is known as ***short-term fuel trim.*** In the past, this was also referred to as the *fuel integrator* or simply as the *integrator.* The ECM also makes changes for more permanent conditions, such as altitude, external ambient temperatures, or an on-going driveability problem. This is called ***long-term fuel trim.*** This is also referred to by some manufacturers as *block memory, block matrix,* or *block learn.*

OBD II Fuel System Monitoring and Control

The OBD II system has provisions that monitor the fuel trim and fuel tank pressure. Much of the fuel system monitoring is done as part of internal ECM operation. The ECM must monitor the fuel system continuously for compliance with emissions standards.

Fuel Trim Monitoring

The ***fuel trim monitor*** system checks the ECM's changes in short- and long-term fuel trim during closed loop operation. If a problem occurs that creates emissions levels 1.5 times greater than federal standards on two consecutive drive cycles, a fuel trim problem is present and a diagnostic trouble code is set. This system monitors long-term fuel trim closely, as significant change in long-term fuel trim usually indicates a possible driveability problem.

To monitor fuel trim, the ECM monitors the averages of short- and long-term fuel trim values. The fuel trim monitor compares these averages against the rich and lean limits, which is set in the ECM's permanent memory. If either short- or long-term fuel trim number is within the limits, the system is considered to be functioning properly. A malfunction is indicated if *both* fuel trim values reach and stay outside their minimum or maximum limit values for a period of time. When a rich fuel trim diagnostic code is set in most systems, the ECM will test the evaporative emissions systems to determine if the canister is causing a rich condition.

Fuel Tank Pressure Sensor

On some vehicles, the evaporative emissions system uses a fuel tank pressure sensor mounted on the fuel tank, usually on top of the fuel pressure sending unit. The sensor is much like a MAP sensor, however, this sensor measures the difference between outside air pressure and the air pressure or vacuum inside the tank. On other systems, the fuel tank is pressurized or has a pump that applies vacuum to the tank. The amount of pump operation needed to produce pressure or vacuum is read by the ECM to determine if leakage is excessive.

Fuel Induction System

After the air-fuel mixture (or air only on a diesel engine) is metered by or through the throttle valves, it must be distributed to the cylinders. This is performed by the intake manifold. Some systems have turbochargers and superchargers to force additional air into the cylinders.

Intake Manifolds

Intake manifolds are passageways for the incoming air-fuel mixture (or air only on diesel engines). They are made from cast iron, aluminum, or plastic, **Figure 9-29.** The intake manifold is installed on the side of inline engines, and between the cylinder heads on V-type engines. Modern intake manifolds are designed to create, as much as possible, a straight path from the injectors or carburetor into the cylinder. Eliminating sharp corners and bends reduces unwanted flow restrictions in the manifold, and reduces the tendency of the gasoline to recondense as it makes changes in direction.

Some gasoline engine intake manifolds contain exhaust gas passages at the bottom of the manifold. These are called exhaust crossover passages. When the engine is cold, some of the exhaust gases are diverted to the crossover passages where they preheat the intake manifold to keep the vaporized fuel from recondensing. This allows the engine to run smoothly as it warms up. Other intake manifolds contain passages to allow exhaust gases to pass from the exhaust system into the EGR valve and intake manifold.

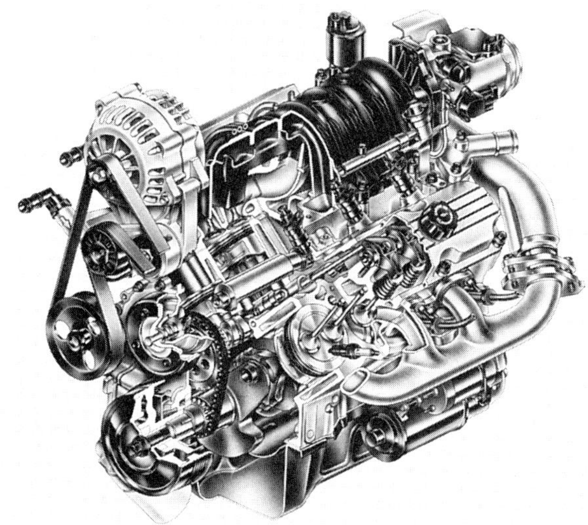

Figure 9-29. *Intake manifolds can be made of cast iron, aluminum, plastic, or a combination. (General Motors)*

Intake Plenums

The intake passage assemblies of many multiport engines are called ***plenums.*** The difference between a plenum and an intake manifold is that the manifold carries an air-fuel mixture, while the plenum carries only air. Most plenums are made of aluminum or plastic mounted to an aluminum manifold that bolts to the engine. Some metal plenums are polished smooth or have an internal Teflon® coating to reduce friction. Because of engine design factors and low hood clearances, many plenums are made in two or three sections, **Figure 9-30.** Many engine manufacturers still use the term manifold when referring to the intake passages of a multiport fuel injection system. The newest plenums contain computer-controlled valves that can be opened to allow additional air to enter the intake manifold, **Figure 9-31.**

Figure 9-30. *Intake plenums deliver air at high speed and turbulence to the intake valve area. This creates better fuel atomization.*

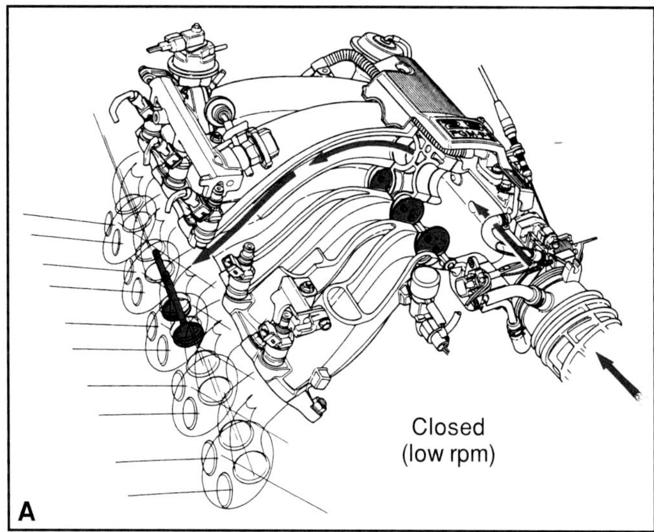

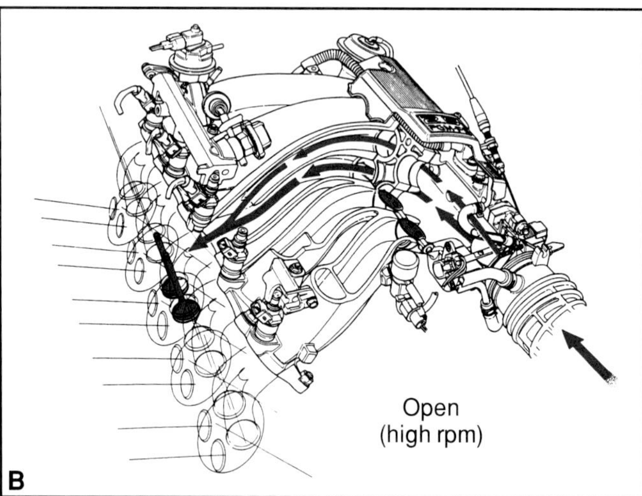

Figure 9-31. *Some intake plenums have additional features for better performance and power. This plenum and intake uses two air intake runners per cylinder. A—At low speeds, the extra valves remain closed. B—When extra power is needed, the ECM opens the valves and increases the fuel injector pulse width. This allows more air and fuel to reach the engine for maximum power. (Acura)*

Superchargers and Turbochargers

Superchargers and ***turbochargers*** increase engine performance by forcing more of the air-fuel mixture into the combustion chambers. A small engine with a supercharger or turbocharger can produce as much power at maximum throttle as a much larger engine. Superchargers and turbochargers are often used to increase the power of diesel engines. On diesel engines, they pump only air into the combustion chamber. Turbochargers have been used on production vehicles for years. Superchargers have been used in racing and on diesel engines for over 20 years and is now being used on newer engines.

Superchargers and turbochargers can increase manifold pressures by 7-16 psi (48.3-110.3 kPa) over atmospheric pressure. This pressure increase is called ***boost.*** The latest gasoline engine superchargers and turbochargers are always used with fuel injection. Many older systems are used with carburetors. The carburetor on older systems is usually mounted so that air flows through the carburetor before entering the compressor wheel. Superchargers and turbochargers can be thought of as pumps or compressors used to force more air into the engine. The major difference between the two is the drive method.

Superchargers

Modern gasoline engine superchargers are belt-driven by the engine crankshaft. Diesel engine superchargers are usually gear driven from the crankshaft. Superchargers take engine power, but begin operating at lower speeds because of their direct connection to the engine, **Figure 9-32.** It consists of a drive belt, drive and driven gears, and a centrifugal compressor wheel. The belt drives the gear and compressor wheel, which pumps air into the engine. The supercharger is usually installed on top of the intake manifold on V-type engines, and on the side of inline engines.

Turbochargers

Turbochargers use the pressure of the escaping exhaust gases as a power source. Turbochargers, therefore, use the exhaust gas pressure that is normally wasted, and do not reduce engine power. Turbocharger construction is shown in **Figure 9-33.** The turbine wheel is connected to the compressor wheel by a shaft. The shaft is supported by ball or tapered roller bearings to handle the high speeds of the shaft (as much as 150,000 rpm). The shaft bearings are pressure lubricated and cooled by the engine oiling system. The turbine wheel is installed behind the exhaust manifold. The compressor wheel is connected to the intake manifold passages after the air cleaner and before the intake valve. However, the engine must be running at high rpm before enough exhaust pressure is available to operate the turbocharger. Exhaust gases leave the engine at high speed, and spin the turbine wheel. The turbine wheel spins the compressor wheel, which forces more of the air-fuel mixture into the engine.

Wastegate

Excess boost on a supercharger or turbocharger will cause detonation, or engine knocking. Detonation occurs when the fuel mixture explodes in the combustion chamber, instead of burning smoothly. Detonation can destroy the engine if it is not controlled. The ratios of the drive and driven gears of the supercharger are designed to prevent overspeeding of the compressor wheel. Turbochargers are

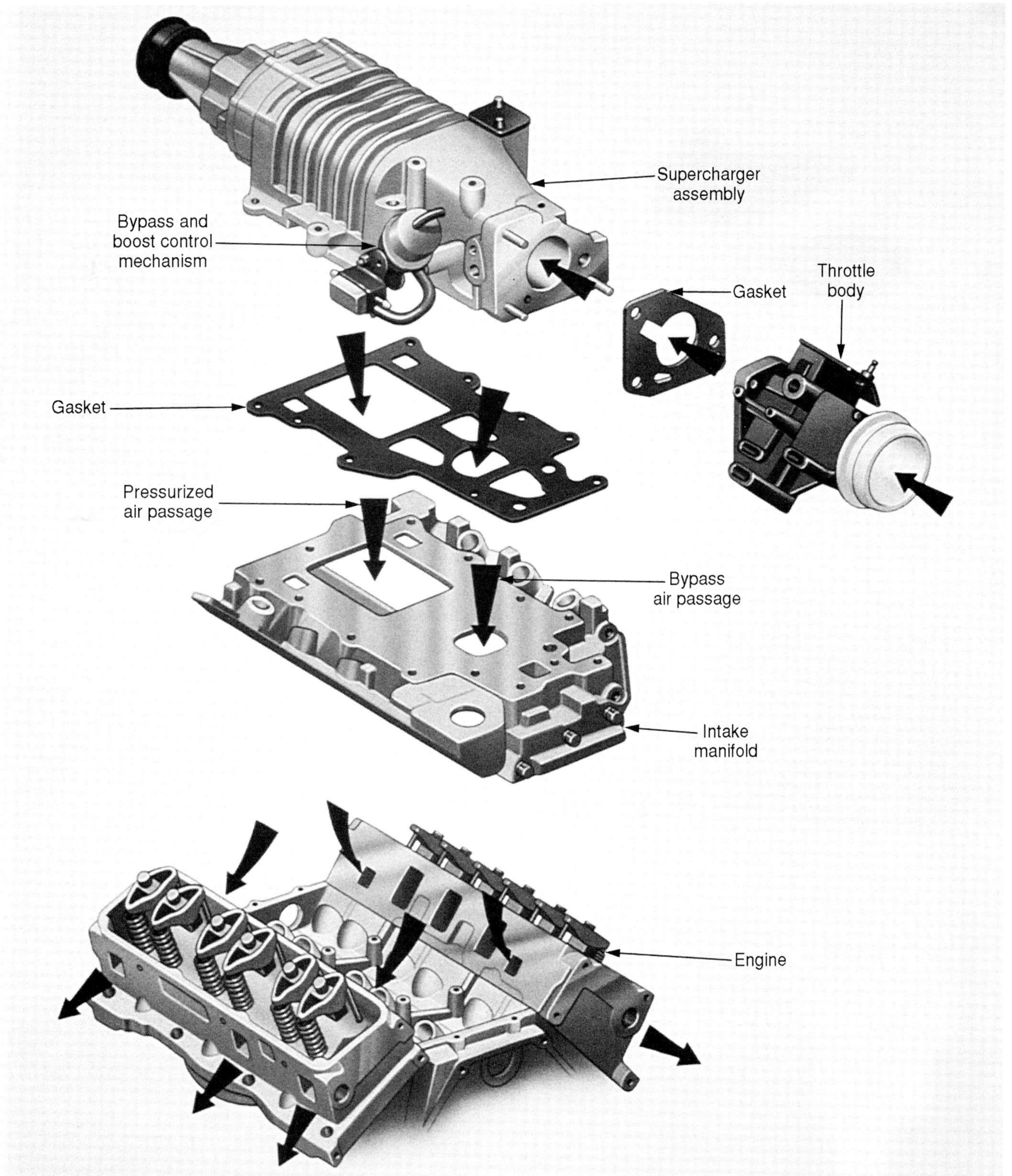

Figure 9-32. Superchargers are belt or gear driven by the engine crankshaft. Superchargers deliver power at all speeds, while turbochargers must develop speed before they become useful. (Pontiac)

equipped with a pressure reduction device called a ***wastegate.*** The wastegate, **Figure 9-34,** opens a passage that bypasses some of the exhaust gases around the turbine wheel. This lowers turbocharger speed if the boost goes too high. The wastegate is opened by a pressure operated diaphragm that responds to excessive intake manifold pressure. To reduce engine knocking, wastegates on many modern engines are controlled by the engine control computer. The engine control computer on many supercharged and turbocharged engines provides further control of detonation by retarding ignition timing when the engine begins detonating.

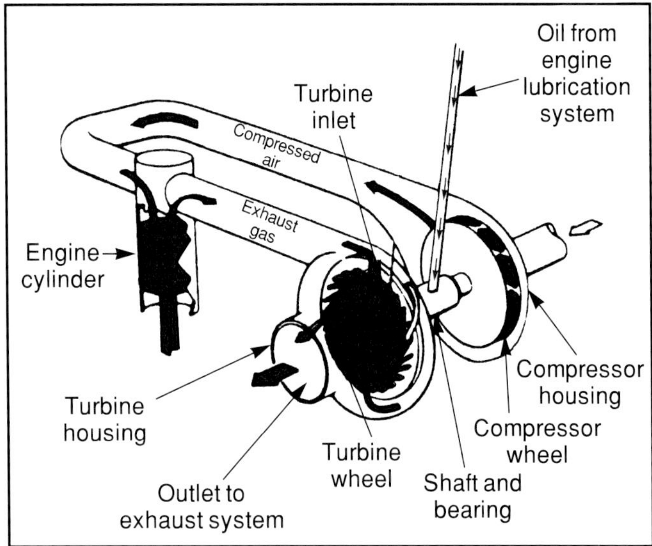

Figure 9-33. *Turbochargers are installed on the exhaust system, usually just after the exhaust manifold. The turbine wheels spin at very high speeds, pumping extra air into the engine.*

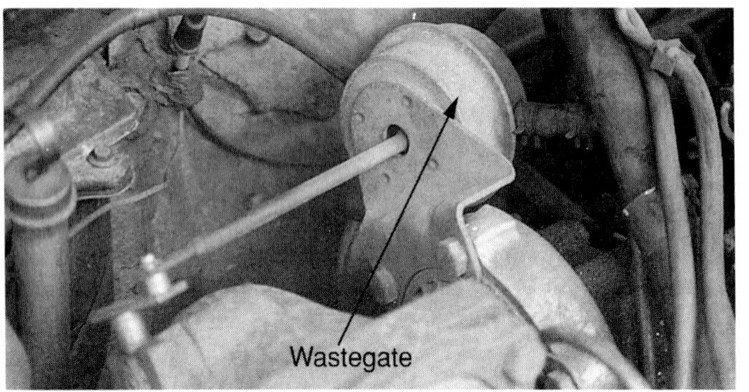

Figure 9-34. *A turbocharger wastegate installed on a V-6 engine. The diaphragm is attached to the intake manifold.*

Vacuum Hoses

The fuel and air induction systems are dependent on ***vacuum hoses.*** Small vacuum hoses carry the vacuum signal to the distributor vacuum advance diaphragm, the EGR diaphragm, MAP sensor, and the transmission modulator, when used. Larger vacuum hoses are the means of transferring crankcase gases through the PCV system into the engine. Hoses also allow the vapor control system to absorb and release gasoline vapors, and provide vacuum to the power brake booster.

Vacuum hoses are made of oil and fuel resistant neoprene. Larger hoses usually have a braided fabric core for added strength and resistance to collapse. Vacuum hose routings can be complex, and a single misrouted hose can cause major driveability and emissions problems. Most vehicles have a hose routing chart on or near the emissions sticker to aid in hose reinstallation. This chart should always be followed when reattaching hoses. In addition, vacuum hoses of similar size are usually marked with a color stripe to ensure that they are reinstalled properly. Some hoses are molded into a multiple connector to simplify installation, **Figure 9-35.**

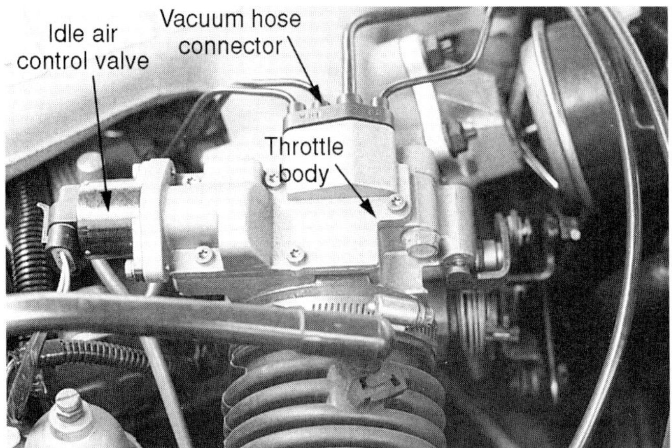

Figure 9-35. *While most vacuum lines are individual hoses, some newer engines use multiple hoses that mount in one central location.*

Air Cleaners and Filter Elements

The ***air cleaner*** is the part of the air induction system which cleans the incoming air. The air cleaner is composed of the air cleaner housing and the filter element. Filtering is necessary to remove the dust particles in the air. Although they are usually too small to be seen, these particles will rapidly wear the moving parts of the engine.

The ***filter element*** is made of a special paper containing microscopic openings. The openings allow air to enter, but catch the dust particles. Filter elements can be round or flat, **Figure 9-36.** Over time, an air filter can become clogged, resulting in reduced air flow. While the ECM can compensate for the reduced fuel flow, mileage and

Figure 9-36. *Air filters can be round, flat, oval or any number of shapes and sizes, depending on the application. They all perform the same function, keeping dirt out of the intake system.*

performance may suffer. A badly clogged air filter may cause the engine to run so rich that oil will be washed from the cylinder walls, causing rapid ring and cylinder wear.

The air filter is installed in a housing that directs the air flow through the filter element. Many modern air filter housings also contain a smaller filter which removes dirt from the air entering the engine crankcase. Modern automotive air cleaner housings can be located on top of the engine, directly over the intake manifold, or off to one side. The housings are connected to the engine by large hoses, called ducts. See **Figure 9-37.** The housing is constructed to silence the air rushing into the engine. Many air cleaner housings have ducts that draw cooler outside air into the engine from outside the engine compartment. Older air cleaner housings were made of sheet metal, while the latest housings are made of high strength plastic. Many air filter housings used within the last 25 years contain a thermostatic device which warms the incoming air.

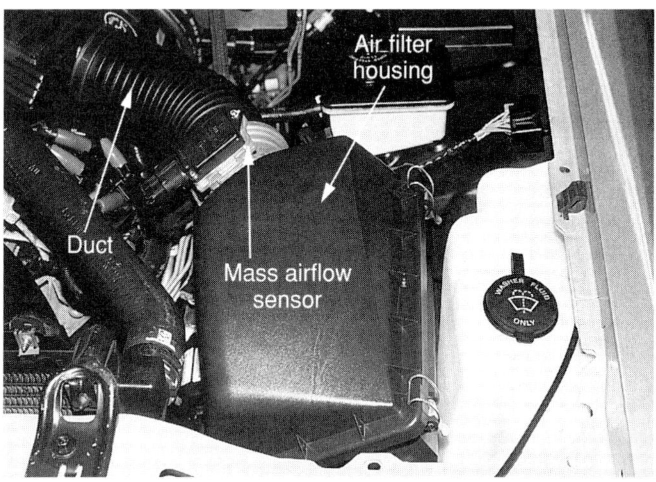

Figure 9-37. *Since most vehicles have very little clearance between the engine and hood, the air filter housing on most modern vehicles is installed on the fender or radiator support.*

Summary

Gasoline and diesel fuel are compounds of carbon and hydrogen, and is called hydrocarbons. Both types of fuels must vaporize before they can burn. The fuel tank stores fuel and vapor until they are used by the engine. Fuel lines connect the fuel tank to the rest of the fuel system. The fuel pump transfers fuel from the tank to the carburetor or fuel injection system. Fuel filters remove any dirt and water from the fuel system before it can cause problems.

The three systems having the most effect on engine operation and driveability are the fuel, ignition, and compression systems. The fuel system is responsible for turning liquid fuel into a burnable vapor, properly mixed with air in the right proportions. The two major types of fuel systems are carburetors and fuel injection systems.

Fuel injection is becoming more common on newer engines. Diesel engine fuel injectors squirt fuel directly into the combustion chamber, while gasoline injectors deliver fuel ahead of the intake valve. Types of fuel injectors include the pulsed type which is operated by an electric solenoid energized by the ECM, and the fixed type, which is open whenever the engine is running. Air-fuel ratios are controlled by the length of time that the injectors are held open on the pulsed system, and by fuel pressure on the fixed injector system.

The two major types of modern gasoline injection systems are the throttle body injector with one or two injectors in a central casting, and the multiport, with one injector for each cylinder. Injector system pressure is controlled by a pressure regulator. Some pressure regulators are modified by intake manifold vacuum.

The ECM receives inputs from airflow sensors, oxygen sensors, intake and coolant temperature, as well as other sensors to establish the fuel trim to maintain the proper air-fuel ratio. The three types of airflow sensors are the mechanical, mass airflow and the Karman vortex. Some systems eliminate the airflow sensor by calculating the airflow using the other input sensor readings. This is referred to as a speed density system.

Other critical fuel system components are air filter housings and elements, intake manifolds, and vacuum hoses. Superchargers and turbochargers are sometimes used to force more of the air-fuel mixture into the cylinders to increase power. Superchargers are belt or gear driven from the engine, while turbochargers are driven by exhaust gases. A wastegate prevents excessive pressure development by the turbocharger. This reduces the danger of detonation and engine damage. Many turbocharging and supercharging systems have a provision for retarding the spark by the ECM if detonation begins. The ECM input is made by one or more knock sensors installed on the engine. The modern engine contains vacuum hoses, usually color coded to simplify installation. Air filters are critical to trap dirt and debris before they can get into the engine.

Know These Terms

Fuel system
Fuel
Hydrocarbons
Energy
Atomization
Gasoline
Octane rating
Vaporized
Oxygenated gasoline
Diesel fuel
Cetane number
Fuel tank
Fuel lines
Tube fittings
Push-on connectors
Fuel pump
Mechanical fuel pumps
Electric fuel pumps
Fuel filter
Fuel injection systems
Stoichiometric
Fuel injectors

Solenoid
Pintle
Nozzle
Spring pressure
Idle air control motor
Pulse width
Throttle body injection (TBI)
Multiport fuel injection
Fuel rail
Central point fuel injection (CPI)
Glow plugs
Pressure regulator
Cold start injector
Auxiliary air regulator
Mass airflow (MAF) sensor
Karman vortex airflow sensor
Speed density
Fuel trim
Short-term fuel trim
Long-term fuel trim
Fuel trim monitor
Intake manifolds
Plenums
Superchargers
Turbochargers
Boost
Wastegate
Vacuum hoses
Air cleaner
Filter element

Review Questions—Chapter 9

Please do not write in this text. Write your answers on a separate sheet of paper.

1. Most common fuels are compounds of ______ and ______ .
2. Gasoline must be ______ and mixed with ______ before it will burn.
3. When the engine is ______, gasoline is harder to burn.
4. All fuel injectors squirt fuel in a ______ shaped pattern for maximum efficiency.
5. If an engine uses a single injector assembly, where will it be mounted on the engine?
   (A) Where the carburetor used to be.
   (B) At the entrance to the intake manifold.
   (C) Just before the intake valve.
   (D) Both A & B.
6. The multiport fuel injection system uses one injector per:
   (A) cylinder.
   (B) cylinder bank.
   (C) manifold runner.
   (D) engine.
7. Air flow into the fuel injected engine at idle is controlled by an:
   (A) throttle valve.
   (B) position motor.
   (C) idle air control solenoid.
   (D) All of the above.
8. Diesel injectors squirt fuel directly into the cylinder at the top of the ______ stroke.
   (A) intake
   (B) compression
   (C) power
   (D) exhaust
9. Fuel pressure regulators used on multiport fuel injection systems are usually connected to:
   (A) the engine oiling system.
   (B) atmospheric pressure.
   (C) EGR pressure.
   (D) intake manifold vacuum.
10. All cars using fuel injection have at least one ________ fuel pump.
11. Many modern air filter housings have a small extra filter which removes dirt being drawn into the:
    (A) intake manifold.
    (B) engine crankcase.
    (C) EGR system.
    (D) None of the above.
12. Why must the fuel tank be vented?
13. Gasoline engine superchargers are driven by a ______ from the engine crankshaft.
14. All turbochargers have a pressure reduction device called a ______.
15. The mass airflow sensor (MAF) is an electrical sensor that measures airflow by:
    (A) using air flow to generate electricity.
    (B) measuring the resistance of a heated wire.
    (C) measuring the temperature of a heated wire.
    (D) measuring the movement of a flap.

ASE Certification-Type Questions

1. The fuel system has the following purposes, EXCEPT:
   (A) delivering the proper amount of fuel to the engine cylinder.
   (B) controlling the air-fuel ratio.
   (C) maintaining proper cylinder temperature.
   (D) atomizing the fuel.
2. Technician A says that some oxygen sensors are installed in the exhaust manifold. Technician B says that some oxygen sensors are installed downstream of the catalytic converter. Who is right?
   (A) A only.
   (B) B only.
   (C) Both A & B.
   (D) Neither A nor B.

3. Technician A says that air-fuel ratio on a fuel injection system is controlled by varying the length of time that the injector is open. Technician B says that the air-fuel ratio on a fuel injection system is controlled by varying the fuel pressure at the injector. Who is right?
   (A) A only.
   (B) B only.
   (C) Both A & B.
   (D) Neither A nor B.
4. Technician A says that gasoline must be carefully blended to ensure that it vaporizes before it leaves the fuel tank. Technician B says that gasoline must be carefully blended to ensure that it vaporizes before it reaches the engine cylinders. Who is right?
   (A) A only.
   (B) B only.
   (C) Both A & B.
   (D) Neither A nor B.
5. Which of the following is least likely to be affected by small amounts of dirt or water?
   (A) Throttle body injection.
   (B) Multiport injection.
   (C) Central point injection.
   (D) Carburetors.
6. The length of time that a pulsed injector is open is determined by the ______.
   (A) ECM
   (B) ignition sensor
   (C) fuel pressure
   (D) air pressure
7. All of the following statements about fixed or continuous flow multiport fuel injection systems are true, EXCEPT:
   (A) the injectors are open whenever the engine is running.
   (B) check valves in the injector nozzles are opened by fuel system pressure when the engine is off.
   (C) the air-fuel ratios are controlled by varying the fuel pressure.
   (D) modern fixed injection systems use ECM controls to further control the air-fuel ratio.
8. Glow plugs are energized on ______ diesel engines.
   (A) cold
   (B) hot
   (C) continuously operated
   (D) Both A & B.
9. Technician A says that most fuel pressure regulators are preset by the manufacturer. Technician B says that fuel pressure regulators can be adjusted by removing an access plug. Who is right?
   (A) A only.
   (B) B only.
   (C) Both A & B.
   (D) Neither A nor B.
10. The Karman vortex sensor may have all of the following, EXCEPT:
   (A) an LED.
   (B) a magnet.
   (C) a photodiode.
   (D) a mirror.

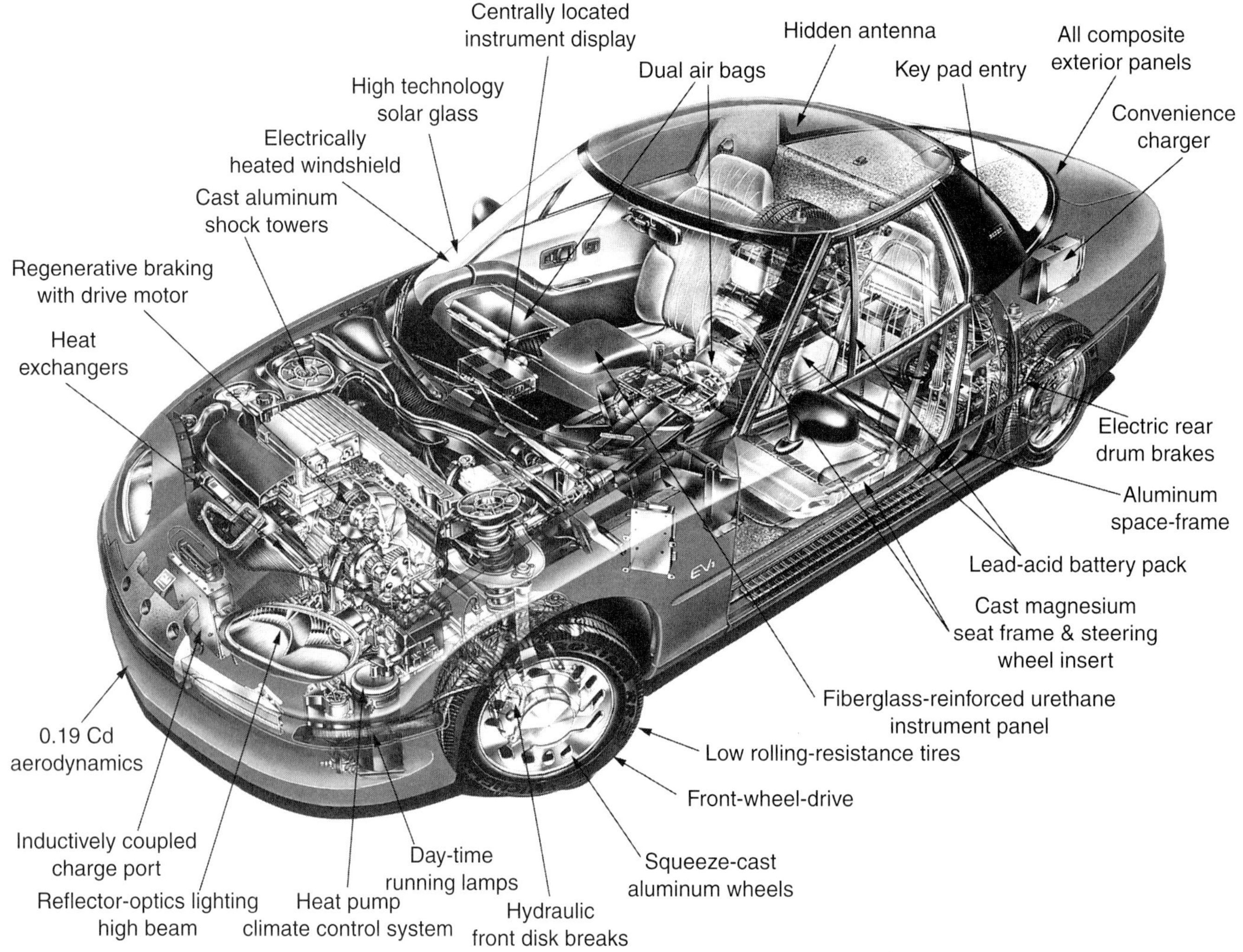

Cutaway of the EV1 electric car and its major components. (General Motors)

Chapter 10

Emissions Control and Exhaust System Fundamentals

After studying this chapter, you will be able to:

- ❑ Identify the three major pollutants produced by internal combustion engines.
- ❑ Identify the major types of internal engine modifications for emissions control and describe their purpose.
- ❑ Identify the major types of external emissions controls and describe their purpose and function.
- ❑ Explain how the use of emissions controls have been affected by the introduction of on-board computer systems.
- ❑ Identify the purposes of the exhaust system.
- ❑ Identify the major components of the exhaust system.

This chapter will concentrate on the emission controls and the exhaust systems used on the latest vehicles. Many of the emission controls are part of the overall engine control system, and are therefore, part of the computer inputs and outputs that you studied in **Chapter 7.** However, many are similar in operation to emission controls installed on older vehicles as far back as the late 1960s.

Many of the emission control systems discussed here are inspected as part of state emissions inspections. Other inspections are performed in some regions as part of federally mandated inspection and maintenance, or I/M programs. In addition, almost all emission controls can fail in a way that affects vehicle driveability. The driveability technician must know how these systems operate.

Common Pollutants

The three pollutants that the modern emission control system reduces are ***carbon monoxide,*** usually abbreviated as ***CO***; unburned ***hydrocarbons,*** abbreviated as ***HC;*** and ***oxides of nitrogen,*** abbreviated as ***NO_x.*** Carbon monoxide is a product of incomplete combustion, caused by rich air-fuel mixtures. Scarce oxygen causes the carbon atoms to combine with only one oxygen atom during the burning process, instead of the normal combustion cycle where one carbon atom combines with two oxygen atoms, forming carbon dioxide, (CO_2), a relatively safe compound. Carbon monoxide is odorless, colorless, and is a poisonous, potentially lethal gas.

Most HC is unburned fuel that passes through the engine and enters the atmosphere, as well as any lubricating oil that escapes from the crankcase as a vapor. NO_x is caused by excessively high combustion chamber temperatures. Excessively high combustion temperatures force the combination of oxygen with nitrogen, forming NO_x. In sunlight, oxides of nitrogen combine with unburned hydrocarbons to form the air pollution that we know as smog.

Modern Emission Controls

The first emission controls were grafted onto existing engines, without much knowledge of how they would affect other engine systems. Early emission control devices often lowered one type of emission while increasing another.

Air-fuel ratios were sometimes adjusted so lean that CO almost completely disappeared, but caused the engine to misfire, increasing HC levels. Increasing combustion chamber temperatures reduced the formation of HC, but created more NO_x and caused detonation. One emission control device often created a need for another device to overcome new driveability and/or emission problems.

Over the years, manufacturers have gradually solved these problems and designed a fully integrated emission system. Modern emission devices work together with each other and with the basic engine systems. Much of this has been accomplished by the use of engine control computers. On the modern engine, the ECM-controlled emission system works with the fuel system to maintain a ***stoichiometric*** air-fuel ratio as close to 14.7:1 as possible while starting, idling, accelerating, decelerating, and cruising in all weather conditions, **Figure 10-1.** The ECM keeps the ignition timing advanced as far as possible for mileage and power, but not far enough to cause excess emissions or spark knock. The action of the ECM, fuel system, cooling system, and emissions controls keeps the combustion chamber hot enough to lower HC levels, but not hot enough to cause excessive NO_x or pinging.

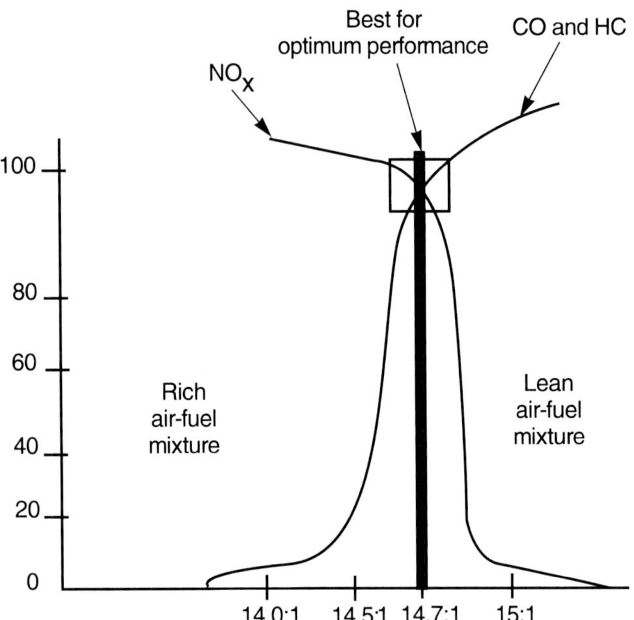

Figure 10-1. *A stoichiometric air-fuel ratio must be maintained for optimum fuel economy as well as exhaust emissions. A narrow window at 14.7:1 provides the best performance. Note how emissions increase as the air-fuel mixture goes lean or rich.*

Computer-Controlled Systems

The use of computer control systems has reduced the use of older emission controls. Compression ratios have increased and valve overlap has been reintroduced since the computer control system reduces pollution in other ways. Precise control of ignition timing by the ECM eliminates the vacuum advance unit and the various means of restricting vacuum to the unit. ECM control of automatic transmission shifting application and application of the air conditioner compressor clutch also lowers emissions.

The main advantage of the computer is that it has made electronic fuel injection possible. By eliminating the carburetor, many carburetor controls (mixture control solenoids, electric chokes, anti-dieseling solenoids) have been eliminated, along with the need for early fuel evaporative (EFE) systems and in most cases, thermostatic air cleaners. Eliminating these devices also eliminates the vacuum hoses that once covered the engine.

On older models, it was often possible to correct a driveability problem by adjusting the fuel system or ignition timing. It was also possible to correct driveability problems, however improperly, by disconnecting one or more of the emission control devices. Integrated design and computer control gives the technician a smaller range of options when dealing with emissions system problems. On newer vehicles, the emission controls must be working properly for maximum driveability as well as lowest emissions.

Internal Engine Modifications

In modern engines, internal parts are designed to balance the needs of performance, mileage, emissions, and driveability. This is in contrast to the design of earlier engines, which were intended to produce the greatest power possible. Fuel mileage was a secondary concern and emissions was an afterthought. Some of these modifications were developed many years ago and are still used on the latest vehicles.

It is important to understand that internal engine design plays an important part in driveability as they can drastically affect engine operation. While it is not practical, and sometimes illegal, to correct a driveability problem by substituting internal engine components, knowing about these modifications can make the job of diagnosing driveability problems easier. This is especially true if tampering is suspected. The most common of these internal modifications are explained in the following sections.

Combustion Chamber Shapes

The combustion chamber in modern cylinder heads is designed to reduce the amount of HC in the exhaust. Older combustion chamber designs, **Figure 10-2A,** used complex shapes. The goal was to create a high amount of turbulence in the combustion chamber, thoroughly mixing the air-fuel mixture. Most piston heads had areas that almost touched the top of combustion chamber when it was at the top of the cylinder. This added additional turbulence to the mixture as the piston came up on its compression stroke. These shapes produced more power with less detonation. However, they caused unburned fuel to condense and cling to the sides of

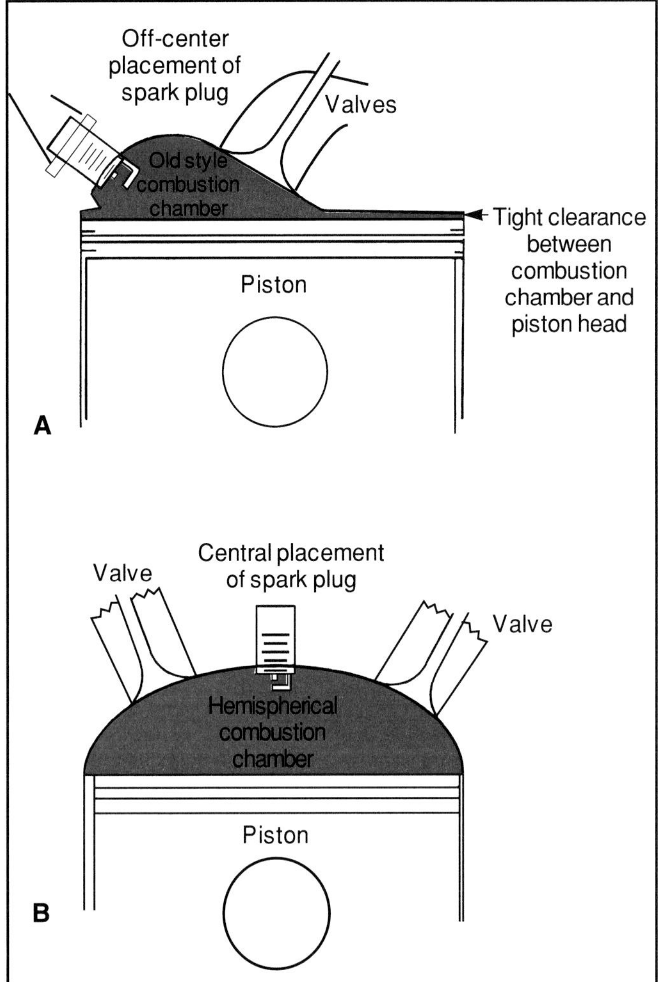

Figure 10-2. *A—Older head designs produces more power, but also raised hydrocarbon (HC) emissions levels. B—The hemispherical head, originally developed for racing, is the ideal design for lowest emissions and maximum power.*

the combustion chamber. This fuel would eventually find its way into the exhaust without burning.

The combustion chambers in newer heads have simple shapes and as small a surface area as possible to reduce the formation of HC. They also have a larger volume to lower compression ratios, discussed later in this section. Reducing the combustion chamber area reduces the opportunity for HC to become trapped and not burn. However, engines with the newer type of combustion chamber have less power and are more likely to detonate (ping). To reduce these problems, some cylinder heads are designed to place the spark plug in the center of the combustion chamber. This shortens the distance that the flame must travel, and ensures that most of the air-fuel mixture in the combustion chamber burns at the same rate.

Another method of reducing HC levels was the stratified charge combustion chamber. In this design, the chamber contained two levels of fuel, one very rich, and the other lean. The spark plug fired the rich mixture, which in turn ignited the lean mixture in the rest of the combustion chamber. This system ensured constant firing with very lean mixtures, and also reduced NO_x formation. This system was used in some older engines and has since been replaced by other designs.

The most efficient combustion chamber shape is the *hemispherical* design, **Figure 10-2B.** Hemispherical or "hemi" combustion chambers were originally developed in the 1950s for use in racing engines. The hemispherical design is seeing more wide spread use, especially in DOHC engines, due to its simplicity and ability to create a great amount of power. When the spark plug is installed in the center of the chamber, the firing process is nearly perfect.

Compression

Since the early 1970s, compression ratios have been lowered from as much as 11:1 to around 8:1. This allows modern engines to operate on lower octane unleaded gasoline without detonating or dieseling. Since lower compression also lowers the temperature of the combustion process, the oxygen and nitrogen are less likely to combine into NO_x. In addition, the lower compression allows the engine to operate on unleaded gas, which must be used if the vehicle has a catalytic converter. Compare **Figures 10-3A** and **10-3B**.

However, the reduction of compression has resulted in a loss of engine efficiency. This is because higher compression develops more heat, which increases the combustion pressure that can be exerted on the piston during the power stroke. The loss from lower compression must be overcome by other engine systems. Improved controls have allowed recent compression ratios to be increased to about 9.5:1 or 10:1 on some engines.

Number of Valves

All older engines, and some newer engines, had two valves per cylinder. To produce more power from smaller engines, most manufacturers are now designing cylinder heads with four valves per cylinder, **Figure 10-4**. These heads allow more of the air-fuel mixture to flow into the cylinder, and provide a low restriction escape for the exhaust gases. The use of four valves also allows the combustion chamber to be almost completely hemi spherical, improving combustion efficiency. Some manufacturers compromise by using two intake valves and one exhaust valve per cylinder. One engine has three cylinders with three valves per cylinder, making a nine valve engine, **Figure 10-5.**

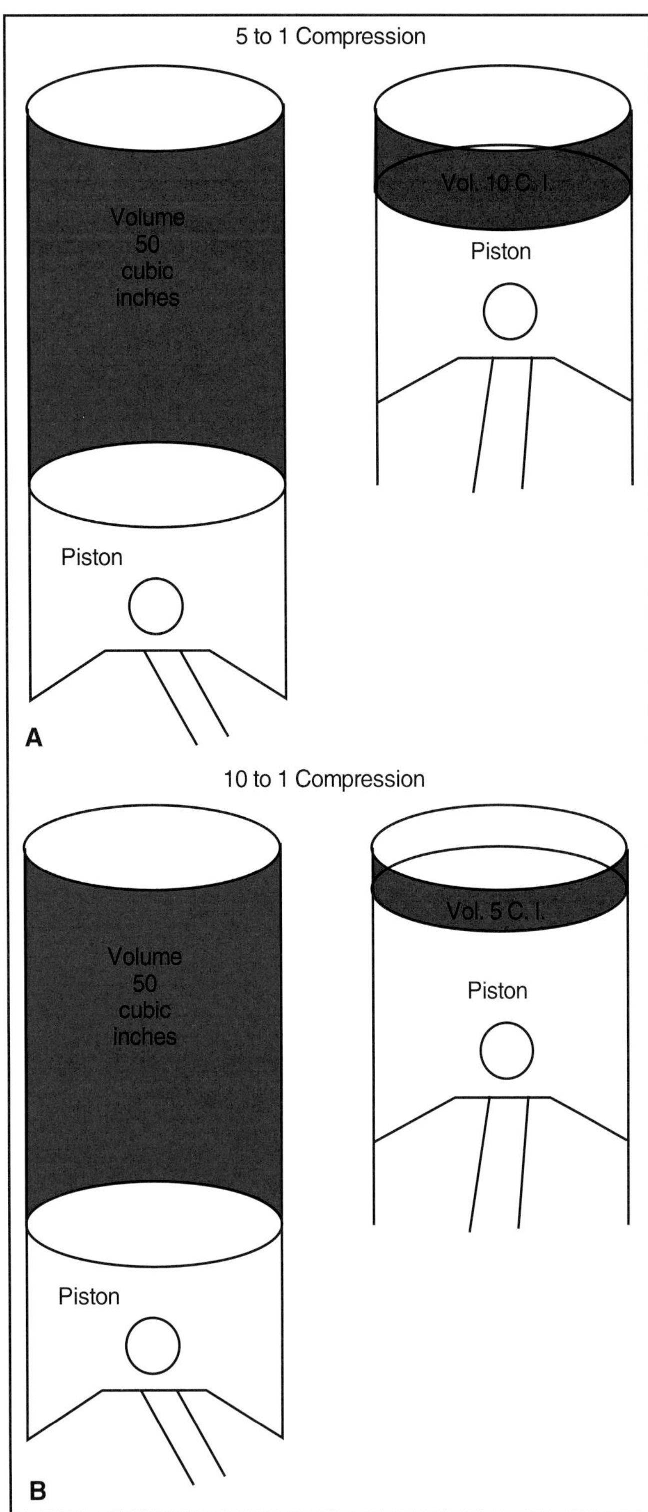

Figure 10-3. *Higher compression ratios produce more power. Modern ratios have been lowered to ensure that the engine will operate on unleaded gasoline and to reduce oxides of nitrogen.*

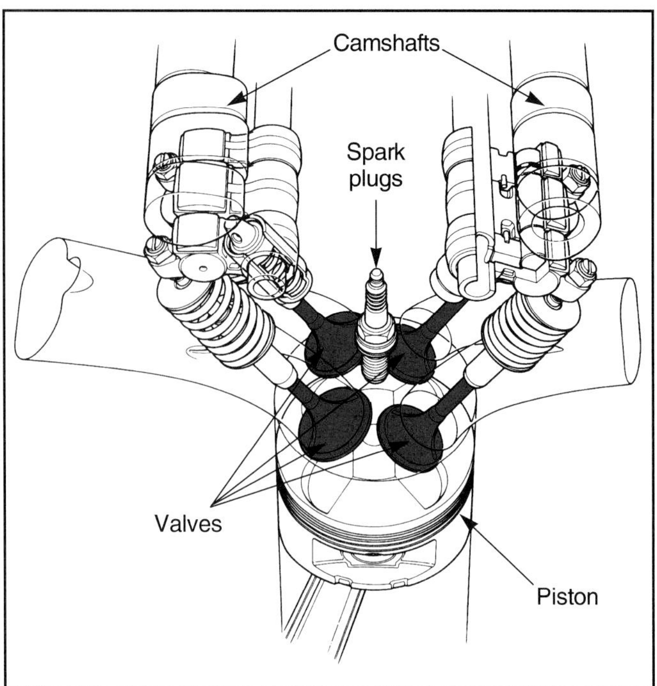

Figure 10-4. *Cylinder heads with four valves per cylinder are used frequently in modern engines as they permit better flow through the combustion chamber. (Acura)*

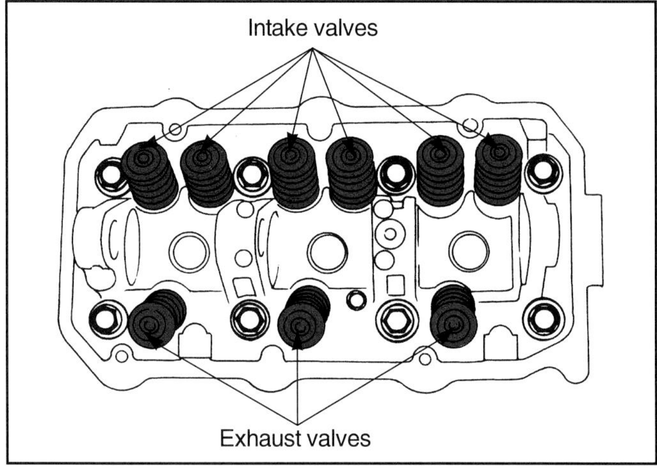

Figure 10-5. *Some small engines have three valves per cylinder. This improves flow at a lower cost than a four valve head. This head is also used in cases when the combustion chamber is too small to fit four valves. (Fuji)*

Camshafts

Modern camshafts have less ***overlap*** than older cam designs. Overlap is the time when both the intake and exhaust valves are open. This occurs during the end of the exhaust stroke, and the beginning of the intake stroke. Older cams were engineered with large amounts of overlap for maximum air flow on the intake stroke, and therefore, good high speed performance. However, excessive overlap caused the incoming air-fuel mixture to be diluted by exhaust gases at idle and low speeds. On older engines, this dilution was overcome by rich carburetor idle and off-idle mixtures. Since it is no longer desirable to have rich idle mixtures, valve overlap was reduced to prevent exhaust dilution and rough idle. Compare **Figures 10-6A** and **10-6B.** These "milder" camshafts reduce engine power to increase smoothness.

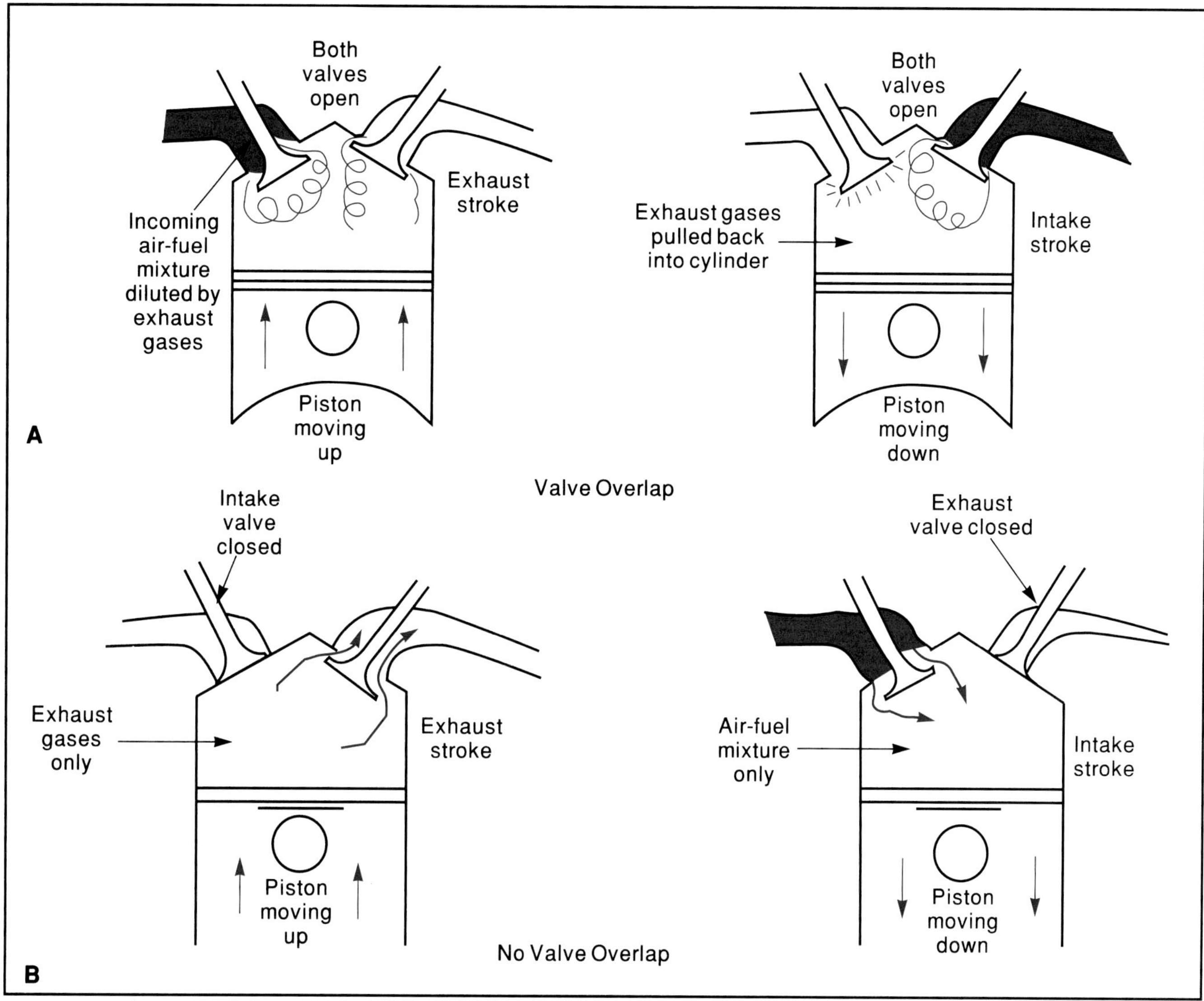

Figure 10-6. *A—Camshaft overlap provides more power at higher speeds, but can cause mild dilution of the incoming air-fuel mixture at idle. B—By reducing and eliminating overlap, high speed power is reduced, but the engine idles smoothly with less emissions.*

Thermostats

Cooling system thermostats used on modern engines do not begin to open until the coolant temperature has reached at least 190°F to 200°F (88°C to 93.5°C). Thermostats used on older vehicles opened at 160°F to 180°F (71°C to 82°C). The modern liquid cooling system warms up quickly, and operates at higher temperatures. This keeps the combustion chamber at higher temperatures for most of the time that the engine is operating. Higher temperatures reduce gasoline condensation on the combustion chamber surfaces and cylinder walls, and levels of HC will be lower, **Figure 10-7.**

External Systems and Components

Many external emission systems are installed on modern engines. Some of these systems go back many years. However, they are all part of an integrated emission control system on modern engines. These systems are discussed in the following sections.

Lean Feedback Carburetors

The carburetors on older vehicles operated on air-fuel ratios of about 12:1 for cruising, and about 8:1 for full

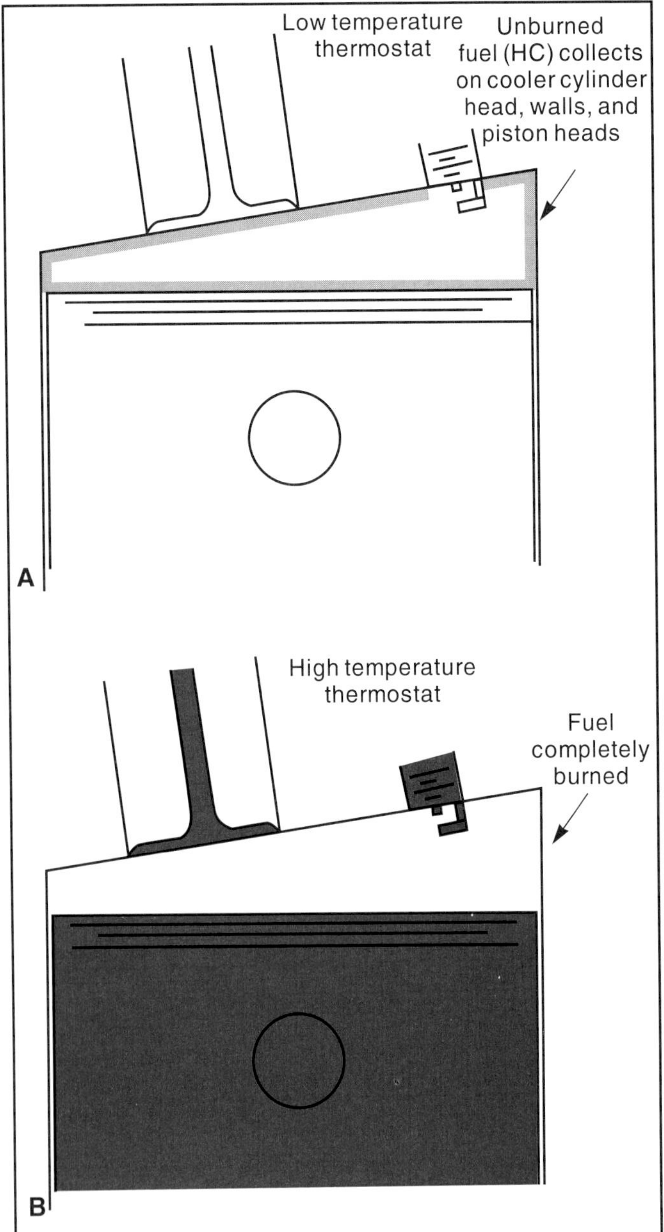

Figure 10-7. *A—Lower engine temperatures allow unburned fuel to collect on the combustion chamber walls. Some of this unburned fuel escapes the exhaust as hydrocarbon emissions. B— Using a high temperature thermostat raises the engine temperature, which prevents the fuel from condensing.*

power, **Figure 10-8.** Newer vehicles operate at a ratio as close to 14.7:1 as possible. This is the ideal ratio for mileage and emissions, but not necessarily for power and driveability. Newer carburetors have smaller main jets, reduced accelerator pump output, and power valve springs that do not allow the valve to open until the engine is under very heavy loads. The idle mixture screws are sealed to prevent misadjustment.

Figure 10-8. *For years, carburetors, such as this two-barrel, were the fuel delivery method of choice. They were simple and reliable, but inefficient.*

Most carburetors are monitored and controlled by the ECM, through a throttle position sensor and mixture control solenoid installed on or inside the carburetor, as well as other sensors. When the engine is warming up, or when wide open throttle operation is necessary, the air-fuel ratio may be richened by the choke, accelerator pump, or power valve. Rich ratios are used as little as possible. For an overall view of carburetor operation, refer to **Appendix B.**

Fuel Injection

Feedback carburetors can control the air-fuel ratio to a degree. However, the ultimate solution to air-fuel ratio control is electronic fuel injection. A fuel injector sprays a precise amount of fuel. The amount of fuel injected can be varied easily in response to external conditions, engine need, or driver demand. Using computers and sensor inputs, the air-fuel ratio can be precisely controlled to keep it as close to 14.7:1 as possible, **Figure 10-9.** Fuel injection has totally replaced the carburetor on all of the latest engines and was discussed in **Chapter 9**.

Ignition Timing Control

Ignition timing control has passed through three distinct phases in the last 30 years. The earliest timing control system was the combination vacuum and centrifugal advance for maximum power and economy. These systems were discussed in **Chapter 8.** The first emission-related timing control systems restricted vacuum advance to obtain the lowest HC emissions. On these systems, the advance unit was connected to manifold vacuum through a ported

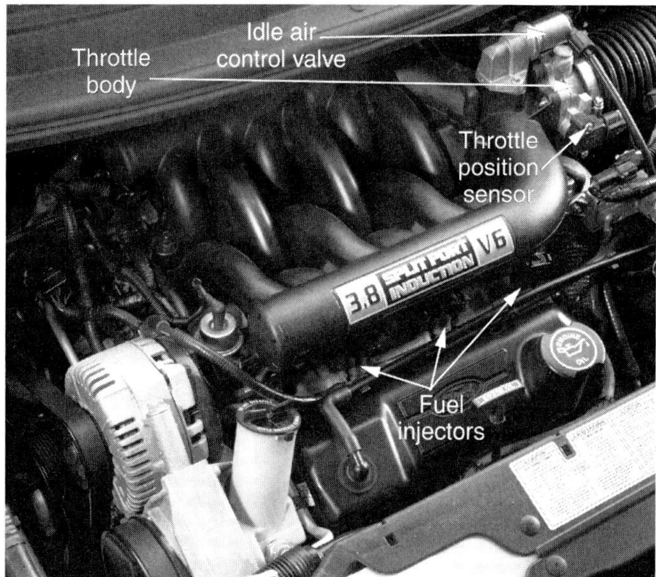

Figure 10-9. *Modern engines use fuel injection. This method allows the ECM to control how much fuel enters the engine. This provides maximum economy with low emissions. (Ford)*

opening on the carburetor. The ported vacuum connection ensured that no vacuum reached the advance unit at idle speeds. A restrictor was also inserted into the vacuum line. When the throttle was opened, ported vacuum was forced to pass through the restrictor. The restrictor prevented full vacuum from reaching the advance unit for as much as 30 seconds after the throttle was opened. This lowered HC emissions while giving good highway mileage. Typical vacuum restrictors are shown in **Figure 10-10.**

Figure 10-10. *Vacuum restrictors were used to limit the amount of vacuum that reached the distributor vacuum advance, as well as other vacuum operated devices.*

Some early emission control vacuum advance units were dual-diaphragm types. Manifold vacuum retarded the spark and venturi vacuum advanced the spark. The vacuum advance could not begin advancing the spark until venturi vacuum overcame manifold vacuum at high speeds. On some older vehicles, solenoid-operated valves were installed in the vacuum line from the manifold to the advance unit. These valves were energized by electrical switches on the transmission or speedometer. The vacuum advance could not receive full vacuum until the vehicle was in high gear or had reached a certain speed, **Figure 10-11.**

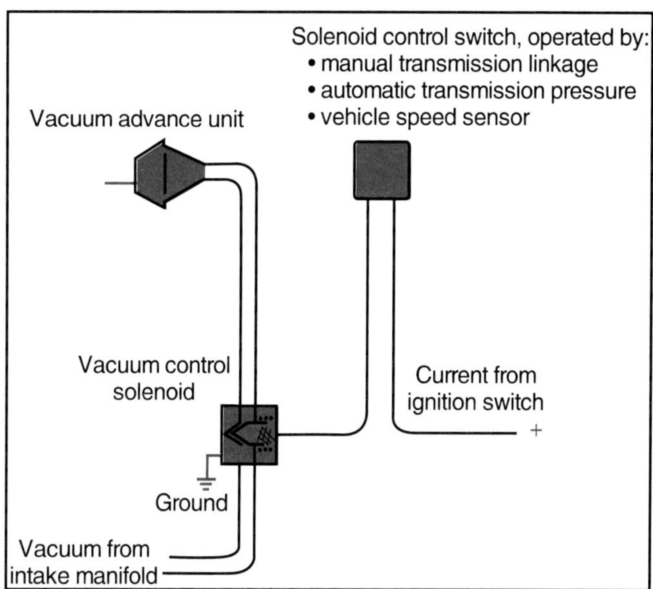

Figure 10-11. *The vacuum control valve is also used with the advance unit. The solenoid is energized when the vehicle reaches a certain speed or the transmission is shifted to high gear.*

The third generation of timing controls began when catalytic converters were introduced, about 1975. Full vacuum advance was used to improve gas mileage, and the catalytic converter was used to reduce the resulting emissions. Since the introduction of computer engine controls, ignition timing is precisely controlled by the ECM. The ECM advances the spark timing as much as possible for maximum performance and fuel mileage. Computer control has eliminated the need for centrifugal and vacuum advance.

Positive Crankcase Ventilation (PCV)

The ***positive crankcase ventilation system,*** or ***PCV,*** is a device that reduces air pollution and keeps the engine internals clean. The PCV system is necessary because piston rings, although providing a tight fit to the cylinder wall, do not seal perfectly. Combustion chamber pressures force some blowby gases into the crankcase. These blowby gases are composed of unburned gasoline, exhaust gases, and water vapor. These gases will cause engine damage if not removed. Positive crankcase ventilation is a controlled vacuum leak from the engine crankcase into the intake manifold. PCV systems are used on both gasoline and diesel engines. This is one of the first emissions control devices installed on modern engines. PCV valves have been used since the early 1960s.

In the typical PCV system, a large vacuum hose connects the crankcase to the intake manifold. A flow control valve called a ***PCV valve*** is installed in the hose. Intake manifold vacuum pulls crankcase gases into the engine, where they are burned. A typical PCV system is shown in **Figure 10-12.** Spring pressure tries to open the PCV valve, while manifold vacuum tries to keep it closed. The PCV valve is held closed when manifold vacuum is high, at idle and light loads. At this time, compression is low, and few blowby gases are produced. The slight amounts of blowby gases flow through a small hole in the valve.

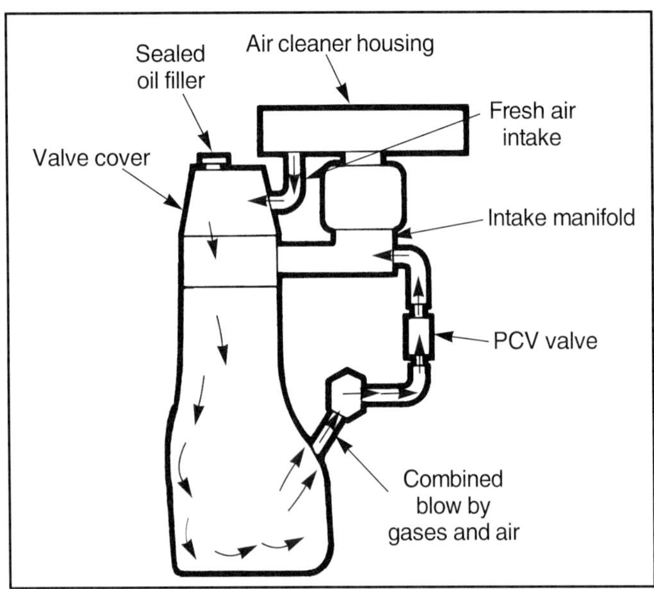

Figure 10-12. *The PCV system draws harmful gases out of the crankcase, while preventing them from entering the atmosphere.*

As engine load increases, compression and blowby gases begin to increase as manifold vacuum begins to decrease. As the vacuum drops, the valve is opened by spring pressure, and more blowby gases can flow to the intake manifold. If the engine backfires, the pressure surge in the intake manifold pushes the PCV valve shut. This prevents the flame from traveling back into the crankcase and causing an explosion.

Outside air is constantly being drawn into the crankcase because of the vacuum created by the PCV system. This incoming air must be filtered to prevent dirt from entering the crankcase. The crankcase inlet vent contains a small filter, usually installed in the air cleaner as in **Figure 10-13.** Some PCV vents are piped into the air intake system after the main air filter, and do not use a separate filter. Most of these are found on fuel injection systems with sealed intake ducts. Some PCV systems are cast as part of the intake plenum on a few multiport fuel injected engines. A few engines use an oil/air separator or other system that eliminates the PCV valve system.

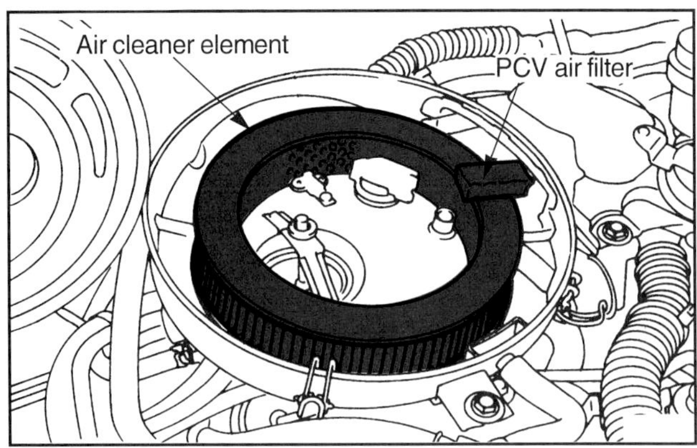

Figure 10-13. *The outside air must be filtered to prevent dirt and dust from entering the PCV system and causing engine wear. On newer systems, the PCV system uses air that has already been filtered by the air filter, and does not need a PCV valve filter. (Fuji)*

Heated Oxygen Sensors (HO_2S)

The first computer control systems use a single oxygen sensor that monitored the exhaust gas for oxygen content. However, the computer system had to be in closed loop mode before the ECM would recognize readings from this sensor. Also, the oxygen sensor had to reach an operating temperature of approximately 600°F (315°C) before it would operate properly. This usually does not occur until the engine is near operating temperature. On an extremely cold day, this temperature may not be reached for an extended period of time or at all.

In the early 1990s, manufacturers began to install oxygen sensors with built-in heating units that would heat the sensor to operating temperature in a short period of time. This allowed the computer system to enter closed loop operation sooner, optimizing vehicle operation. Many OBD I and all OBD II equipped vehicles have more than one oxygen sensor. One oxygen sensor is used as a catalyst monitor, discussed later in this chapter.

Thermostatic Air Cleaner

Thermostatic air cleaners use exhaust manifold heat to warm the incoming air when the engine is cold. The heated air enters the air intake system to help vaporize the gasoline and reduce recondensation. This is important on carbureted and throttle body fuel injection systems. Multiport injection systems do not use a thermostatic air cleaner, since the fuel is injected near the intake valves. The use of a thermostatic air cleaner allows the choke on carbureted engines to be opened sooner, and also reduces the possibility of carburetor icing. This lowers emissions and improves cold weather driveability.

Icing occurs when the lower air pressure in the carburetor venturi or throttle body causes water vapor to freeze in the throat. This ice formation can cause the engine to run roughly or stall in cold, humid weather. While it does not happen often, icing can also occur at the throttle plate of a fuel injected system under the right conditions. The problem is most likely to happen on a throttle body (TBI) fuel injection system, but can occur on a multiport system having a restricted area at the throttle valve. Therefore, thermostatic air cleaners are sometimes used on fuel injection systems.

A typical thermostatic air cleaner is shown in **Figure 10-14.** A metal pipe or heat-resistant flexible hose connects the intake snorkel to a metal baffle that fits closely over the exhaust manifold. Air is heated as it passes between the baffle and exhaust manifold. A sheet metal valve closes off the cold air intake and opens a passage for the manifold heated air. The valve is operated by a vacuum motor that is controlled by a thermostatic sensor. The sensor closes and sends intake manifold vacuum to the motor when the engine is cold and opens to bleed vacuum into the air cleaner housing as it warms up. The sensor is usually mounted in the air cleaner housing.

Some fuel injected engines warm the incoming air by directing coolant through a chamber at the bottom of the throttle body. Hot coolant on its way to the heater core warms the throat and valve, preventing icing and improving cold weather performance.

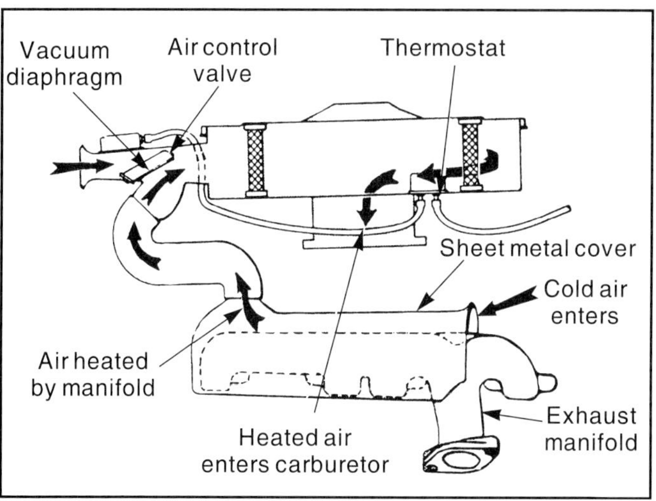

Figure 10-14. *The thermostatic air cleaner helps the engine warm up quickly and operate efficiently on a lean mixture. The heated air also reduces the chance of fuel condensation and carburetor icing. (Chrysler)*

Early Fuel Evaporation (EFE) Systems

Early fuel evaporation (EFE) systems are devices which quickly heat the intake manifold of a cold engine. If the intake manifold reaches normal operating temperature quickly, gasoline will not recondense on the manifold walls. The main advantage of the EFE system is that it allows the mixture to be leaned quickly while maintaining cold engine driveability. This reduces HC formation.

The first EFE systems were called *heat risers.* Heat risers are only found on vehicles over 20 years old. The heat riser was operated by a thermostatic spring on the exhaust manifold, similar to a choke thermostat, which closed a valve in the exhaust manifold. The closed valve forced the exhaust gases to pass directly under the intake manifold, heating it as soon as the engine started. Heat risers often became stuck and their use was discontinued when thermostatic air cleaners were introduced.

The vacuum-operated EFE is similar to the heat riser, with a valve mounted in the exhaust manifold. However, the thermostatic spring has been replaced with a vacuum motor, similar to the one used to operate the thermostatic air cleaner or EGR valve. The vacuum motor is controlled by a vacuum switch. All non- and some early computer-controlled engines use a switch called a ***thermal vacuum valve (TVV)*** or ***thermal vacuum switch (TVS).*** The TVV is installed in an engine coolant passage, usually in the intake manifold. It allows manifold vacuum to go to the vacuum motor when the engine is cold and shuts off vacuum as the engine warms up, **Figure 10-15.** An internal spring can then close the valve. EFE systems on most computer-controlled engines are operated by a vacuum solenoid that sends manifold vacuum to the motor when directed by the ECM.

A few vehicles are equipped with an electric EFE system. This EFE system consists of a grid made of resistive wire installed between the carburetor or fuel injector throttle body and the intake manifold. When the engine is cold, a thermostatic switch allows current to flow into the grid. The electrical resistance of the wire causes the grid to give off heat to the air-fuel mixture as it passes through the grid. The thermostatic switch shuts off current to the grid when the engine is warm.

Evaporative Emissions Control System

The ***evaporative emissions control system*** is a charcoal-filled vapor canister with attaching purge valves and hoses, **Figure 10-16.** It prevents gasoline vapors from escaping into the atmosphere from the fuel system. The pressure relief vents in the gas tank are connected by hoses to the canister. The canister contains charcoal that is able to absorb gasoline vapors. When the engine is turned off, any gasoline that evaporates from the tank or fuel system enters the canister. An additional line from the canister is connected to the fuel bowl on carbureted engines.

Whenever the engine is running, manifold vacuum draws air through the canister. Air flow through the canister draws gas vapors into the engine for burning. Flow is controlled by a vacuum valve that allows the engine to operate briefly before vapors can be drawn in. This reduces surging or stalling on a hot engine. Some carbureted engines have a solenoid on the fuel bowl vent. The solenoid is energized when the engine is running, and closes the vent valve. This is done to prevent air flow through the fuel bowl, which could result in excessive gasoline evaporation in the bowl. Some evaporative emissions systems have a diagnostic switch to use in determining purge solenoid operation.

Exhaust passage
Bi-metallic spring
A
Vacuum motor
Valve
B
A
B
Vacuum connections
Connection to coolant passage
C
Atmospheric vent
Rain cap
Blade type terminals or molded connector
Outlet port
Inlet port
Bleed
D

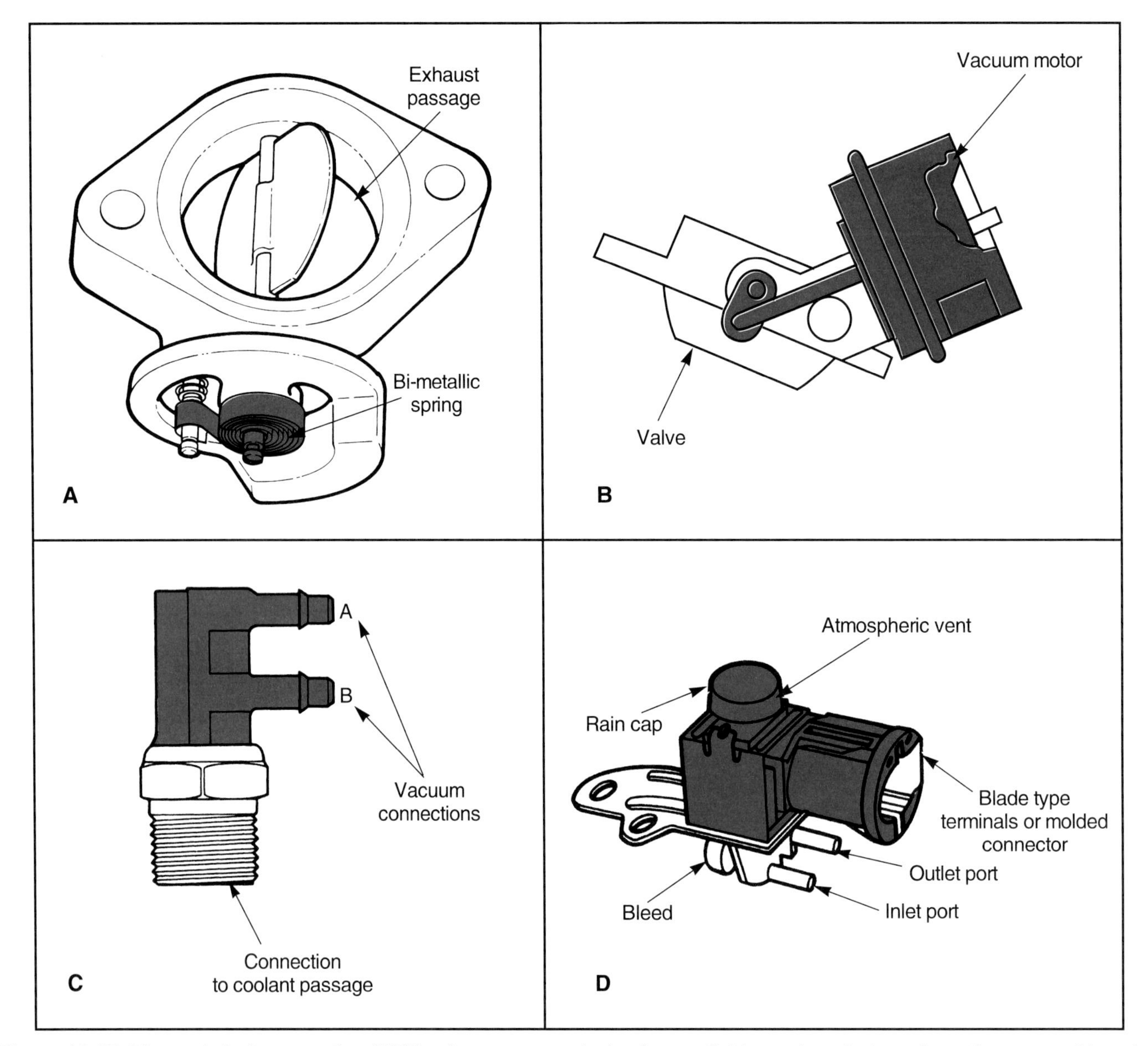

Figure 10-15. *The early fuel evaporation (EFE) valve warms up the intake manifold to reduce fuel condensation on a cold engine. A—The heat riser is operated by a spring. B—The vacuum motor opens the heat riser valve to allow exhaust gases to flow under the intake valve. C—Vacuum to the motor is controlled by a thermostatic vacuum valve. D—In some cases, an ECM operated vacuum solenoid is used. (Ford)*

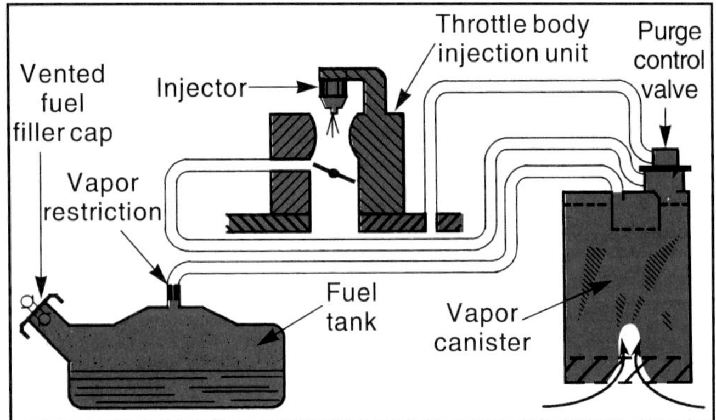

Figure 10-16. *The evaporative emissions system absorbs evaporated fuel and stores it in a charcoal-filled canister. The fuel is later drawn into the engine and burned. (General Motors)*

Enhanced Evaporative Emissions Control System

Some vehicles equipped with OBD II computer control systems use an enhanced evaporative emissions system, **Figure 10-17.** In addition to the components used on most evaporative emissions systems, an enhanced evaporative system also has a service port, fuel level sensor, fuel tank pressure sensor, and vent valve.

The ***service port*** is used to clean any carbon released into the system by the canister. Most service ports will have a Schraeder-type valve and a cap to keep the port free of dirt and grease. The ***fuel level sensor*** is monitored by the ECM to determine if the fuel level is sufficient for system diagnostics. The ***fuel tank pressure sensor*** mounts on the fuel tank or in one of the fuel lines. This sensor measures the

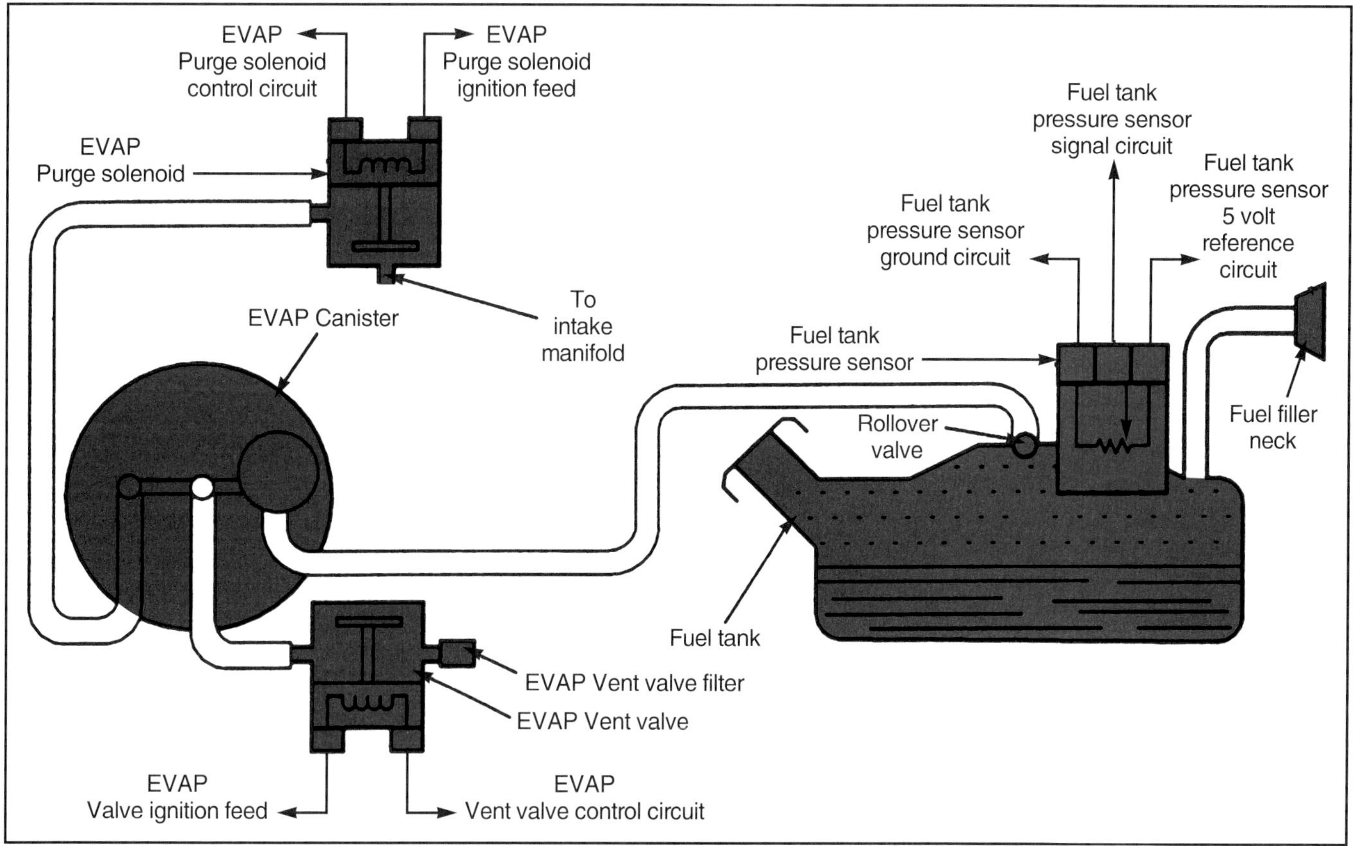

Figure 10-17. *OBD II equipped vehicles use an enhanced evaporative emissions system. The EVAP purge solenoid is ECM controlled to operate only under vehicle conditions when venting fuel vapor in the engine will not affect exhaust emissions. While this system may seem complicated, it uses many of the same components as the basic system shown in* **Figure 10-16.** *(General Motors)*

difference between the air pressure or vacuum in the fuel tank and outside air pressure. The ***vent valve*** allows fresh air into the system.

Enhanced Evaporative Emissions System Operation

The purge solenoid on the enhanced system is operated by the ECM. Purge valve operation is determined by the ECM based on readings from the MAF or MAP and intake air temperature sensors, along with the current fuel trim. However, purge valve operation will not begin until the following conditions are met.

- The engine must be in operation for about 2–3 minutes on a cold start or 30 seconds to one minute on a hot start.
- The engine must be at operating temperature.
- The computer system is operating in closed loop mode.

The ECM will perform leak detection diagnosis by closing the vent valve and applying a vacuum to the evaporative emissions system. Enhanced emissions system diagnosis and repair will be discussed in **Chapter 18.**

Air Injection Systems

Since the combustion process in the cylinder is not perfect, some unburned gas escapes into the exhaust system. If not reduced in some way, the excess gasoline would overwhelm the catalytic converter and escape into the atmosphere. To burn the gasoline before it reaches the converter and atmosphere, outside air is pumped into the exhaust manifold. The two types of air injection systems are the pulse air system and the air injection reactor.

Pulse Air Injection

Pulse air injection uses the pulses in the exhaust manifold to draw outside air into the manifold. As each exhaust valve closes, the sudden loss of air flow causes a negative pressure, or suction, in the exhaust manifold. The pulse air system consists of check valves placed in fittings at carefully calculated positions on the exhaust manifold. The check valves open when a vacuum pulse occurs in that section of the manifold, and close when the exhaust manifold is pressurized. Therefore, the opening and closing of the check valves draws air into the exhaust manifold, without allowing exhaust gases to escape.

Outside air is drawn into the pulse air system through a hose attached to the air cleaner housing. This ensures that the incoming air is cleaner than the outside air, and that the intake pulses of the system are silenced inside the air cleaner housing. The pulse air system does not use a belt-driven air pump or any control valves and is inexpensive to produce and takes up less space on the engine. Another advantage of the pulse air system is that it uses the natural pressure variations in the exhaust system, and does not consume any engine power. This system cannot be controlled as well as the belt-driven type, but is satisfactory for smaller engines where the air flow does not have to be closely controlled, and power losses must be kept to a minimum. These systems were gradually eliminated during the 1980s

Air Injection Pump Systems

The ***air injection reactor system*** uses a pump, sometimes referred to as a *smog pump*. It is a centrifugal pump that forces outside air into the engine exhaust manifold. Most air pumps are belt-driven from the engine. A few pumps are driven by electric motors controlled by the ECM. The pump, **Figure 10-18,** develops low pressure at no more than 10 to 15 pounds. This is sufficient pressure to force air into the exhaust manifold. High exhaust temperatures combined with the oxygen content of the outside air causes the unburned gasoline to vaporize and burn, completing the combustion process in the exhaust manifold. Flexible hoses and steel tubing connect the air pump to the exhaust manifolds and to the catalytic converter. The tubes at the exhaust manifold are made of heat-resistant steel, threaded into the exhaust manifold, **Figure 10-19.**

Air Injection System Valves

Several vacuum-operated control valves direct the output of the air pump system. ***One-way check valves,*** **Figure 10-20,** are used to prevent exhaust gases from backing up into the smog pump if exhaust gas pressures become higher than normal. High exhaust pressures could be caused by high engine loads, or exhaust restrictions, such as a plugged catalytic converter, muffler, or collapsed pipe. The check valves are usually installed where the flexible hoses attach to the exhaust manifold pipes.

Diverter valves are installed to prevent backfiring during deceleration by diverting the pump output to the atmosphere. The engine is more likely to backfire when decelerating, since higher intake manifold vacuum is drawing more fuel from the fuel system. This fuel could reach explosive proportions, and therefore, the oxygen supply is cut off by the diverter valve. Some air pump systems installed on fuel injected engines do not use a diverter valve, since the fuel supply is cut off during deceleration. The diverter valve is mounted on the pump or in the hoses between the pump and check valves. See **Figure 10-21.**

Most engines with a three-way catalytic converter and air pump are equipped with ***switching valves.*** This valve directs air to the exhaust manifold when the engine is cold. When the engine heats up, the switching valve redirects the air to the catalytic converter. This valve allows the incoming oxygen to combine most efficiently with the unburned gasoline. **Figure 10-21** shows a typical air pump switching valve. The air pump valves are controlled by vacuum diaphragms, usually operated by the ECM through solenoid valves. Air injection valves on OBD II vehicles are monitored for function and the presence of airflow in the exhaust stream.

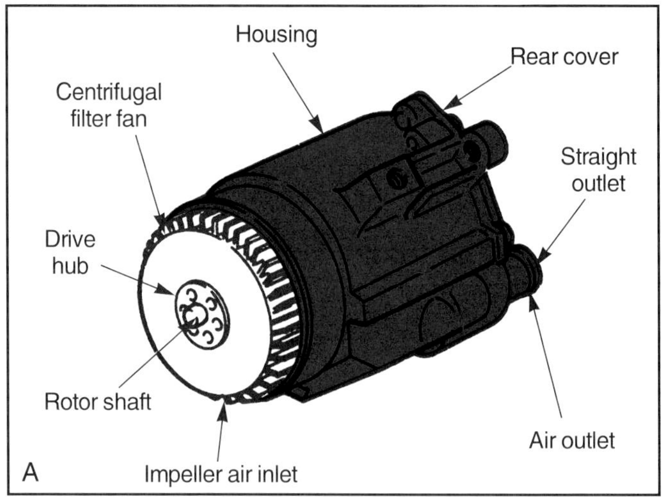

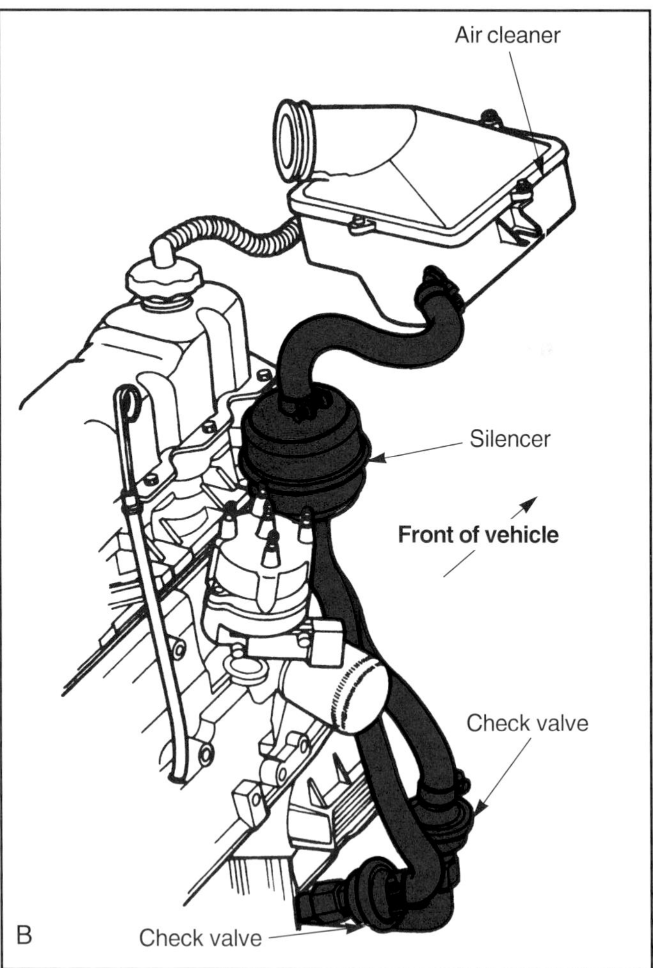

Figure 10-18. *A—A typical air injection reactor (AIR) pump is a low pressure high volume air pump which delivers air to the exhaust manifold. B—A pulse air injection system uses the scavenging effect of exhaust pulses to draw air through a check valve into the exhaust manifold. (Ford)*

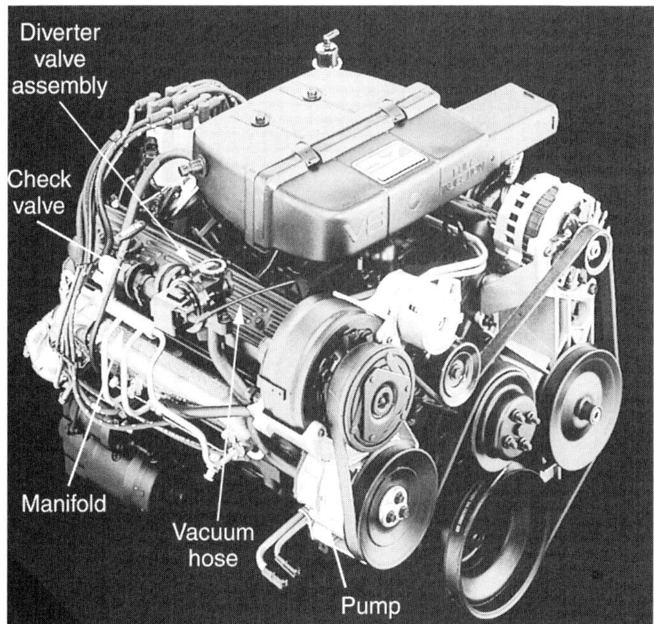

Figure 10-19. *Study the location of the various parts of the AIR system as they are mounted on this engine. (General Motors)*

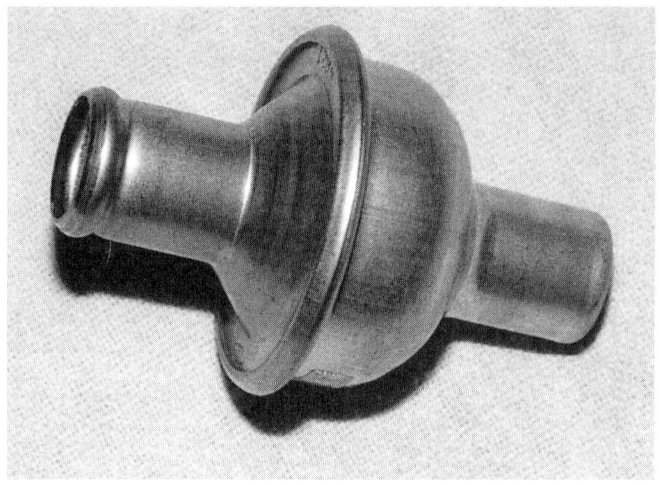

Figure 10-20. *The check valves open to allow air into the exhaust manifold, but do not allow exhaust gases to back up into the pump or valves.*

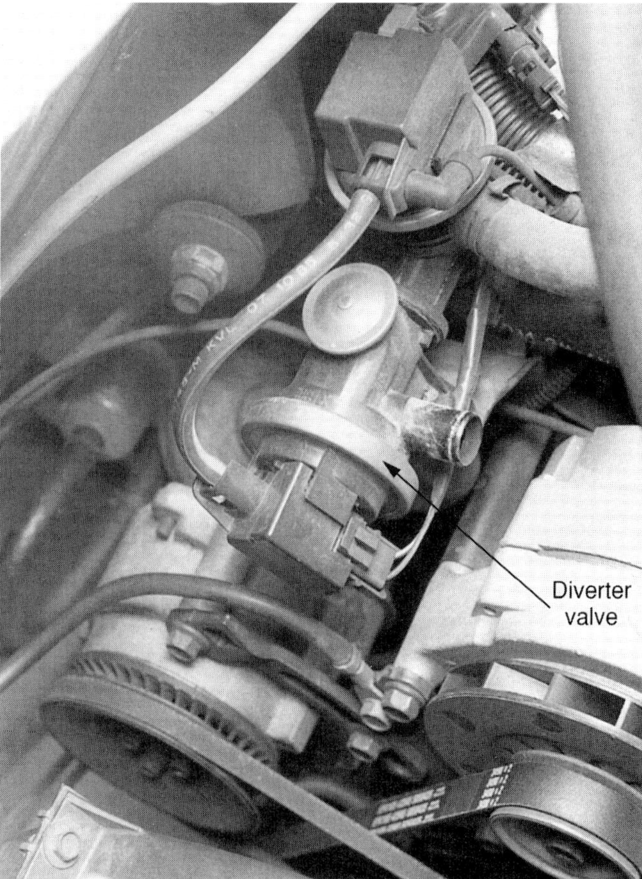

Figure 10-21. *The diverter valve is installed on the air pump outlet piping and is controlled by intake manifold vacuum. It prevents backfiring when the engine is decelerating. The switching valve switches air between the exhaust manifold and catalytic converter.*

Exhaust Gas Recirculation (EGR) System

As you learned earlier, high combustion chamber temperatures cause oxygen to combine with normally inert nitrogen to form oxides of nitrogen. If combustion chamber temperatures are reduced by reducing overall engine temperature, gasoline will condense on the cooler cylinder walls and head surfaces, raising HC levels. The ***exhaust gas recirculating (EGR)*** system is used to reduce NO_x formation without raising levels of hydrocarbon emissions.

Using an ***EGR valve*** directly reduces combustion temperatures by recycling exhaust gases into the intake manifold. Since exhaust gases do not contain enough oxygen to support combustion, the chemical reaction of the air-fuel mixture is reduced. This lowers the combustion temperature, reducing NO_x formation. At the same time, the heat in the exhaust gases keeps the combustion chamber temperature high enough to prevent HC formation. Most modern diesels installed in cars and light trucks have EGR valves, since they operate on much higher compression ratios and are more likely to produce NO_x.

The EGR valve is a vacuum, electrically, or computer-operated valve made of cast iron and steel to withstand the heat and corrosive effects of exhaust gases. **Figure 10-22** is a diagram of the EGR valve and passages. The intake and exhaust manifolds contain passages for the exhaust gases. The EGR valve is not needed at idle speeds and is closed to prevent rough idle.

The vacuum-operated EGR diaphragm is controlled by ported vacuum. The vacuum line often contains a vacuum restrictor or backpressure transducer to delay valve opening. **Figure 10-23** shows a typical EGR system. The vacuum signal to many newer EGR valves is computer-controlled, and EGR valve position is monitored by the ECM through a position sensor installed on the EGR valve. **Figure 10-24** is an EGR valve with a position sensor.

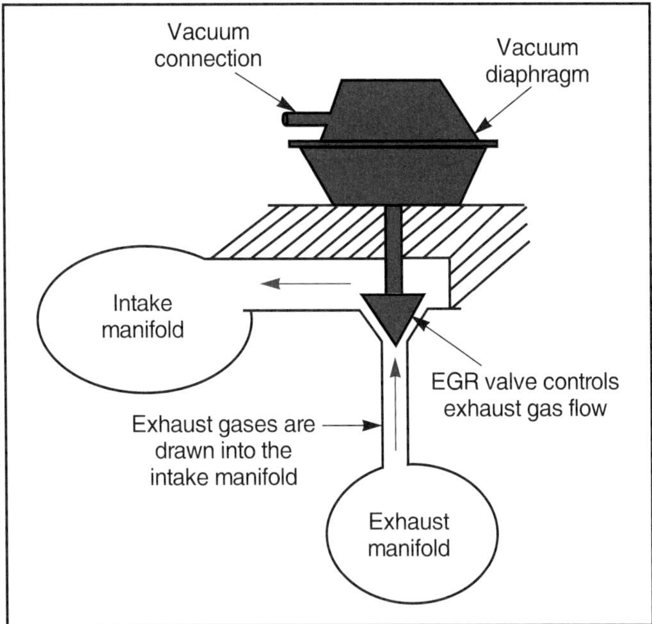

Figure 10-22. *Diagram of an exhaust gas recirculation (EGR) system. The EGR system recirculates some of the exhaust gas in the engine to reduce NO_X. A vacuum operated diaphragm controls the exhaust flow. Newer EGR valves are ECM controlled and monitored.*

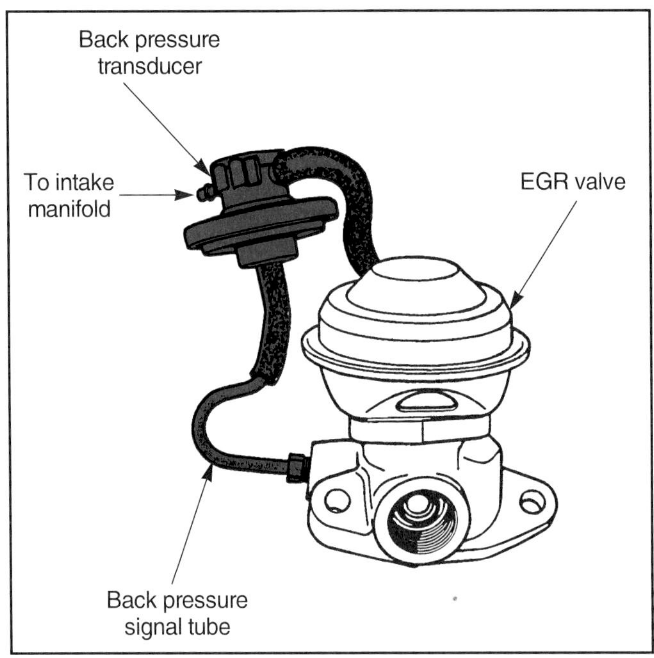

Figure 10-23. *This EGR valve uses a back pressure transducer to reduce valve opening at high speeds. (Chrysler)*

OBD II EGR Valve Flow Rate Monitor

OBD II equipped vehicles also monitor flow rate through the EGR valve. The EGR system in an OBD II vehicle has an additional sensor that monitors system pressure and flow. The monitor checks for abnormally low and high flow rates and changes in manifold pressure. Most

Figure 10-24. *This EGR valve is equipped with a pintle position sensor. The sensor signals the ECM, which operates a vacuum solenoid that controls EGR valve opening. This allows precise control of EGR valve movement in response to engine needs.*

OBD II EGR flow sensors measure the amount of exhaust gas flow by measuring the pressure differential (difference in pressure) on each side of the EGR valve. Other systems monitor changes in the MAP sensor, which monitors manifold vacuum, over several EGR valve actuations.

Emission System Tampering

Although the emissions and other engine control systems are critical for good vehicle operation, tampering remains a major cause of many driveability problems. There are two types of emission system tampering. *Accidental tampering* usually occurs unknowingly. For example, a technician accidentally misroutes a vacuum hose during the process of looking for the cause of a driveability problem. The other is *intentional tampering;* a vacuum hose is plugged to correct a driveability problem and in the process, creates an emissions or other driveability problem, leaving the original problem uncorrected. Emissions system inspection will be covered in more detail in **Chapter 18.**

Caution: *Never* intentionally remove or disable any emission control device under any circumstance. This is a violation of federal, and in most cases, state law. Do not modify any emission control system, component, or device unless the modification is manufacturer approved.

Exhaust System

The ***exhaust system*** is a set of pipes and sound deadening devices that remove exhaust gases and heat from the engine, **Figure 10-25.** The exhaust system operates in a way that will prevent damage to the vehicle or injury to the passengers from either heat or fumes. The exhaust system also reduces the noise of the escaping exhaust gases without causing excessive restrictions in their flow. Exhaust systems with catalytic converters are part of the emission control system and help to clean up the exhaust gases.

The exhaust system includes the exhaust manifold, exhaust pipe, catalytic converter, muffler, and tailpipe. Some models are equipped with more than one muffler. Many exhaust system components are connected together by flanged connections. Other exhaust system components have straight pipe ends. The pipes slip into each other and are tightened by muffler clamps.

Exhaust Manifolds

The ***exhaust manifold***, **Figure 10-26,** conducts the exhaust gases from the cylinder head to the exhaust pipes. It is usually made of cast iron for maximum durability and noise reduction. The exhaust manifold contains the early fuel evaporation valve, or on older engines, the heat riser. Only one exhaust manifold of a V-type engine contains an EFE valve. On engines equipped with an air pump, the exhaust manifolds are equipped with air pump inlet tubes that are threaded for maximum sealing. There is usually one inlet tube for each engine cylinder.

One or more oxygen sensors are installed in the exhaust manifold or in the exhaust pipes just after the

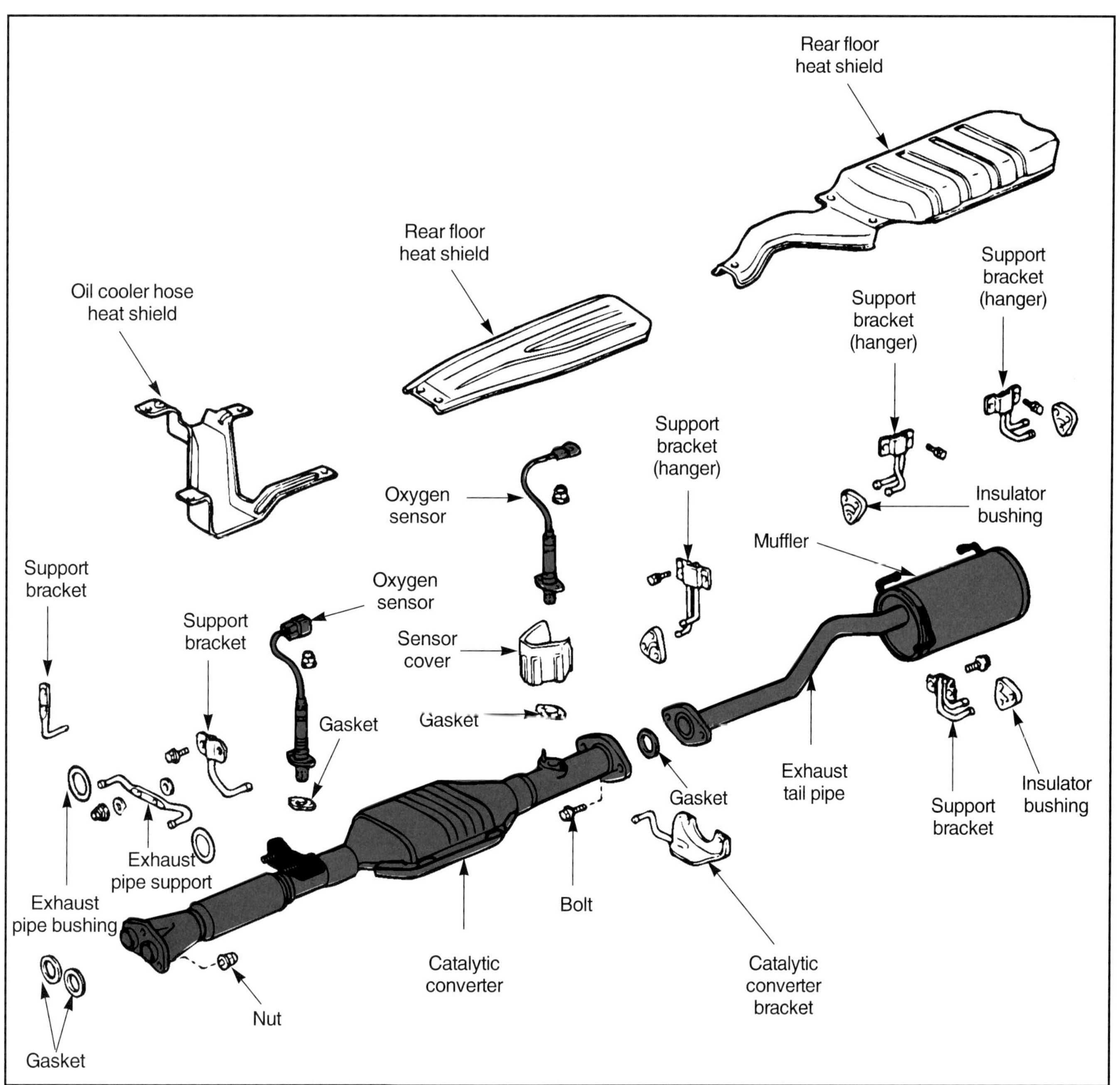

Figure 10-25. *Exploded view of a modern exhaust system. Note the two oxygen sensors mounted before and after the catalytic converter. (Toyota)*

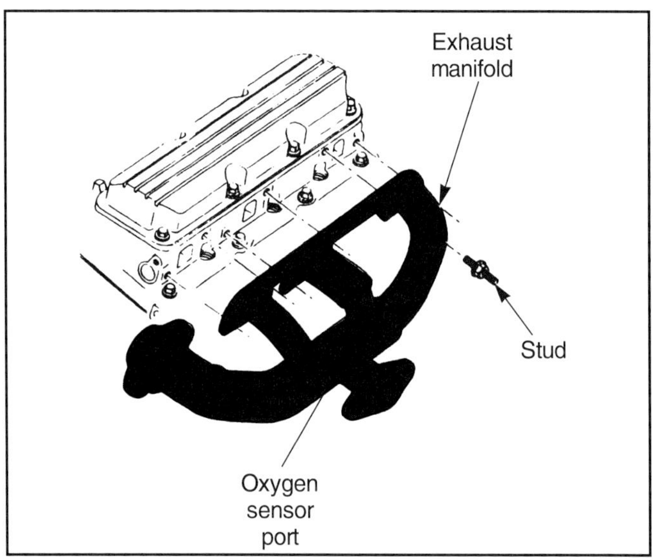

Figure 10-26. *The exhaust manifold often serves as the installation point for the oxygen sensor as well as routing the exhaust gases from the engine to the exhaust system. (General Motors)*

manifold. Turbochargers, when used, are installed in the exhaust manifold, or in a special exhaust pipe immediately behind the exhaust manifold. The turbine wheel is exposed to the exhaust gases, which spin it and the compressor wheel.

Catalytic Converter

The ***catalytic converter*** is a device that reduces the amount of harmful exhaust gases. Unless it becomes clogged, it has no effect on engine operation. The catalytic converter contains the elements platinum (Pt) and palladium (Pd). Platinum and palladium are ***catalysts,*** which is any material that causes a reaction in something else without being changed itself. These elements cause a chemical reaction which converts HC and CO into carbon dioxide (CO_2) and water (H_2O). While HC and CO are dangerous pollutants, CO_2 and H_2O are relatively harmless. Platinum and palladium are too expensive to be used in large amounts and are therefore lightly coated on many small ceramic pellets or a ceramic mesh screen. The pellet and ceramic mesh converters are shown in **Figure 10-27.**

Most current vehicles use a ***three-way catalytic converter.*** This type of converter reduces NO_x as well as HC and CO. It consists of two separate sections: a front NO_x reducing catalyst section, and a rear conventional converter section, with a center inlet for outside air, delivered by the air pump. This type of catalytic converter is always used with an air pump, covered earlier in this chapter. Air pump output is forced into the center inlet between the three-way catalyst section and the conventional converter section. See **Figure 10-28.**

The front section contains platinum, and another catalytic element called rhodium (Rh). These catalysts react with the exhaust NO_x, and partly with the HC and CO, breaking the NO_x into pure nitrogen and oxygen with some residual CO_2 and H_2O. This section of the converter is called the *reducing converter,* since it reduces the NO_x to its component elements. The exhaust gas then flows into the rear section of the converter, where it combines with the oxygen delivered by the air pump. This oxygen allows the two-way catalyst to finish converting the remaining CO and HC into CO_2 and H_2O. This part of the converter is sometimes called the oxidizing converter. The air pump system delivers air into the converter only when the engine is at operating temperature.

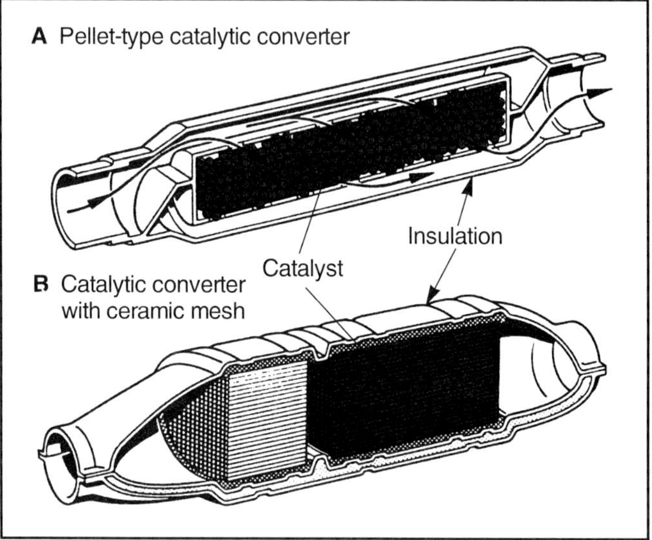

Figure 10-27. *A—Older catalytic converters use pellets which could be replaced in some cases. B—Newer converters use a ceramic catalyst-coated honeycombed mesh.*

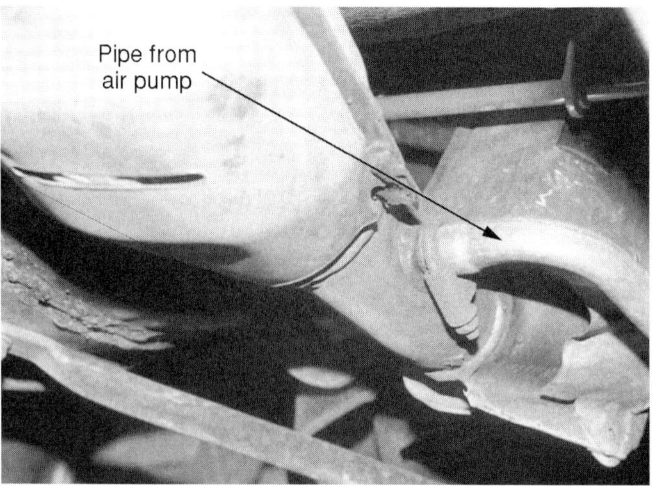

Figure 10-28. *Many converters have additional air pumped into the exhaust system from the AIR system. This increases the catalyzing action as well as catalyst life.*

Newer catalytic converters contain the element cerium (Ce). It is a base metal that has the ability to attract and release oxygen. The excess oxygen in the exhaust stream stabilizes the catalyst elements and enhances the converter's ability to convert exhaust gases into carbon dioxide and water. This extends the converter's effective life, since catalysts will deteriorate over time and use. The more oxygen the catalyzing elements receive, the better and longer they will be effective.

Since the catalytic converter operates by chemical reaction and is also receiving hot gases from the engine, it can easily reach temperatures up to 1600°F (872°C). Therefore, the internal and external parts of the converter are made of heat-resistant metals, such as stainless steel. Metal heat shields are also placed around the converter to protect the rest of the vehicle, and anything under the vehicle from the intense heat.

Vehicles with catalytic converters must be operated on unleaded gasoline, since the tetraethyl lead in leaded gas will coat the catalyst, rendering it inert and unusable. Once the catalyst stops working, cooler converter temperatures will cause the converter to accumulate deposits and plug up. Cooler exhaust system temperatures will also cause other exhaust system components to collect water and acid, quickly rusting them out. Since leaded gas has not been available in most areas for a long time, this is usually not a problem on most vehicles, unless the vehicle was operated on leaded gas sometime in the past.

OBD II Catalyst Monitor

All OBD II equipped vehicles utilize a ***catalyst monitor*** to ensure that the catalytic converter is operating properly and that excessively low or high oxygen levels (which could indicate a potential driveability condition) are not present. The catalyst monitor uses two or more heated oxygen sensors, **Figure 10-29.** One or more sensors are mounted in the exhaust manifold(s) or exhaust pipe upstream from the catalytic converter. A single heated O_2 sensor is mounted downstream from the converter, usually just to the rear of the converter's outlet.

A catalyst that is not working properly will not efficiently convert exhaust gases into harmless substances. The catalyst monitor measures the converter's ability to store and release oxygen. A good converter will have the ability to store a high amount of oxygen. When the catalyst begins to fail, its ability to store oxygen will decrease and will be picked up by the catalyst monitor. The waveform can be monitored to show how the monitor works, **Figure 10-30.**

Mufflers and Resonators

The ***muffler*** is always placed behind the catalytic converter. Conventional or "stock" mufflers consist of an airtight metal housing enclosing a series of internal pipes and resonance chambers. See **Figure 10-31.** The internal pipes and chambers reverse the exhaust flow several times, and reduce the speed of the exhaust gases. This reversal and slowing of the gases suppresses and cancels out sound waves and pressure surges from the engine. The disadvantages of this type of muffler are restriction of the gas flow, called

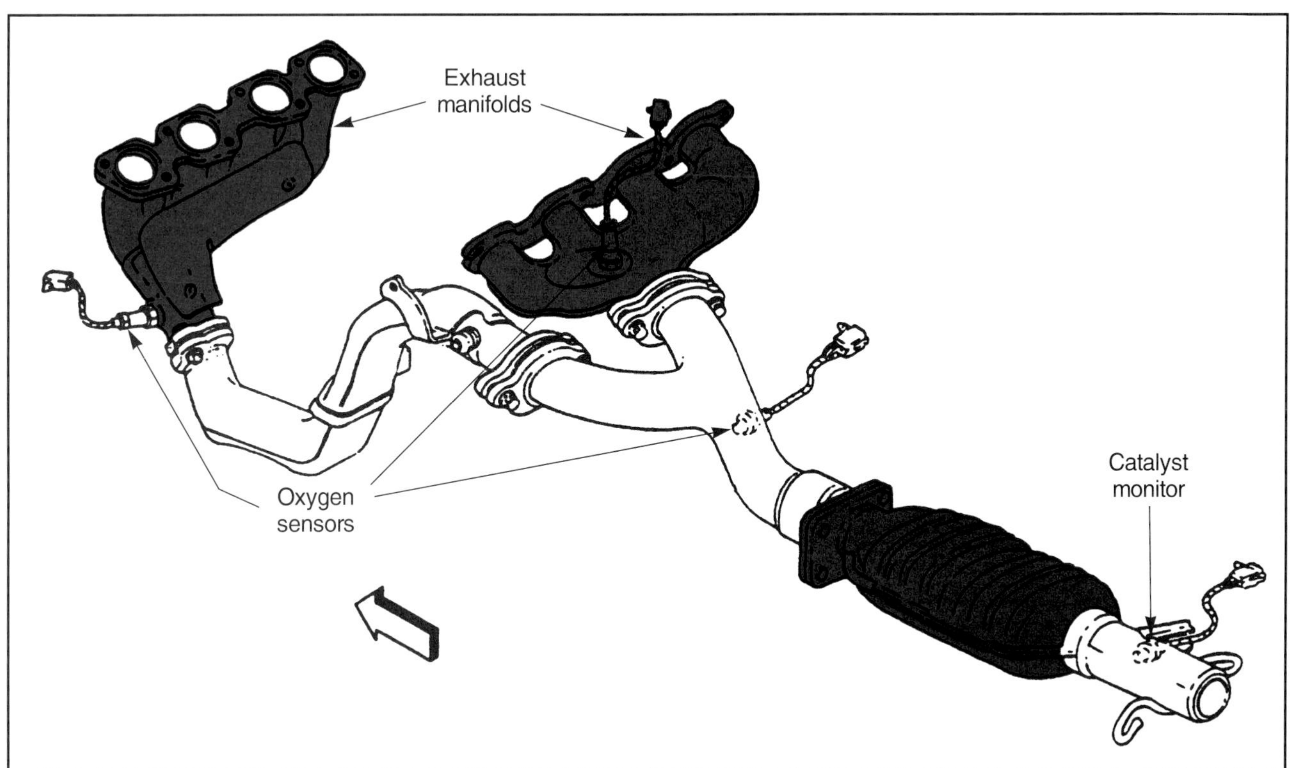

Figure 10-29. *Most newer vehicles use multiple heated oxygen sensors. The heated oxygen sensors allow the computer system to enter closed loop operation in a shorter period of time. A single oxygen sensor is mounted downstream of the converter to monitor catalyst operation. (General Motors)*

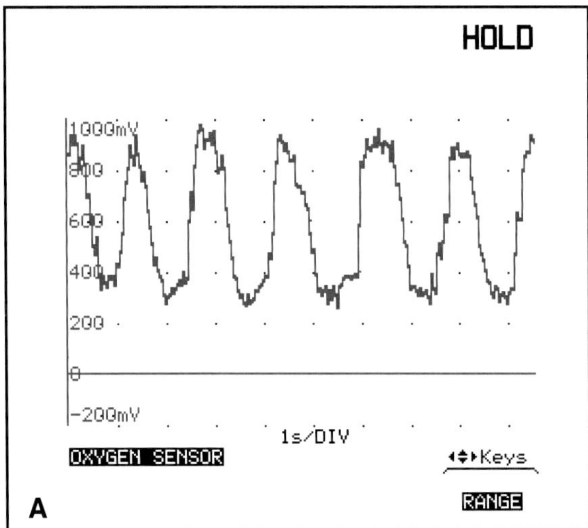

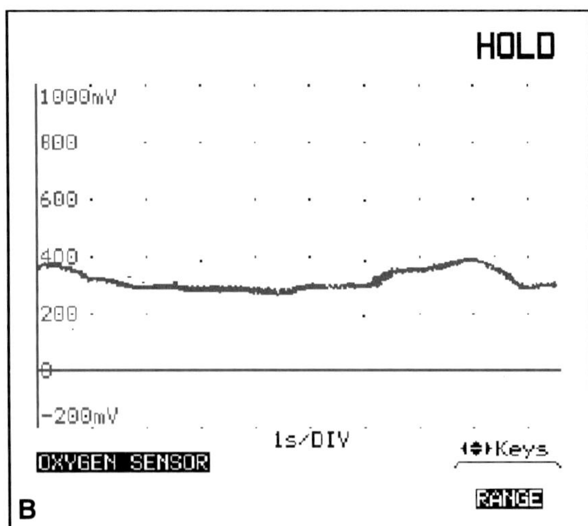

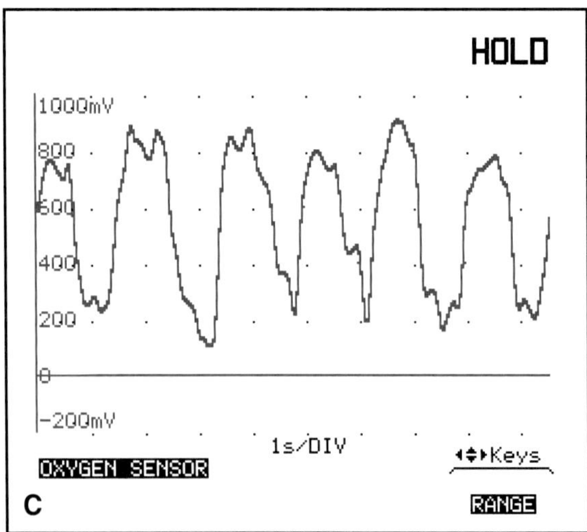

Figure 10-30. *The catalyst monitor checks the catalyzed exhaust for oxygen. This indicates the converter's ability to store and release oxygen. A—This waveform is typical of preconverter oxygen sensors. B—If the converter is operating properly and no other problems exist, the catalyst monitor waveform will be a somewhat flat line with very few waves and no real peaks. C—The catalyst monitor will indicate a defective converter. The wave will look much like a preconverter oxygen sensor. (Fluke)*

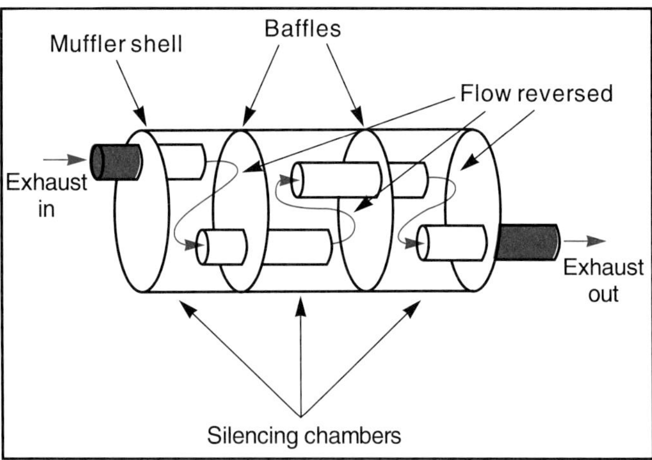

Figure 10-31. *A cutaway of a typical stock muffler. The baffles and chambers reverse the exhaust gas and reduce noise. Stock mufflers are very quiet and durable.*

back pressure, which is comprised of pressure waves in the exhaust stream caused by restrictions in the system, and the muffler's tendency to collect water and corrosive elements. Many engines, especially those used on luxury cars, have a ***resonator*** mounted behind the muffler. The resonator is a smaller muffler which absorbs any noise which escapes the first muffler. A small hole is often drilled at the lowest point on the housings of mufflers and resonators. This allows water that has collected in the housing to drain out before it can cause rust.

Some resonators, and many aftermarket mufflers, are *straight-through mufflers.* The straight through, or fiberglass muffler is often used to reduce back pressure. Straight through mufflers, **Figure 10-32,** have no internal baffles or chambers. They consist of fiberglass packed around a pipe containing many small holes or slots. Sound waves enter the fiberglass through the holes in the pipe, and are absorbed. The straight through muffler is cheaper and less restrictive. However, it does not do as good a job of muffling noise. The fiberglass packing also tends to burn out quickly, and does not last as long as the conventional muffler.

Exhaust Pipes and Tailpipes

Exhaust pipes are used to conduct the exhaust gases between the other exhaust system components. The pipes are steel tubes, bent to conform to the underside of the vehicle. Pipes that connect major components are called exhaust pipes. The final pipe in the system, which exits to the atmosphere, is called the ***tailpipe***.

Many exhaust and tailpipes are made of two tubes, one inside the other. This is known as double-wall construction. The double-wall construction deadens sound waves. Some exhaust pipes on luxury automobiles have a high-temperature plastic film sandwiched between the two metal walls for additional sound reduction.

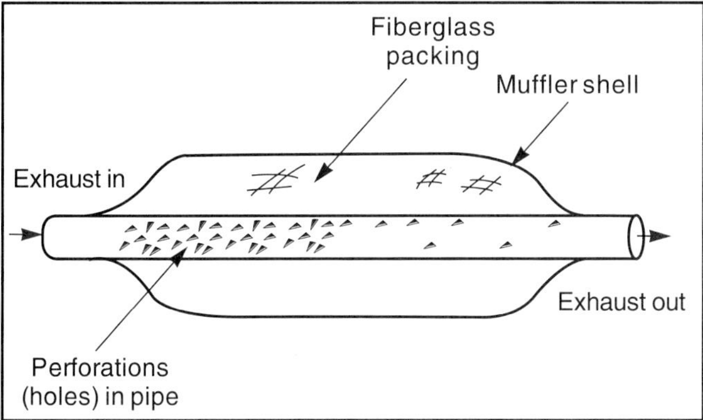

Figure 10-32. *A straight-through muffler reduces exhaust back pressure, but is also louder than stock mufflers. Also, the fiberglass packing in straight-through mufflers have a tendency to burn out quickly.*

Exhaust System Factors

Modern vehicles with fewer engine cylinders tend to be noisier than older V-8 engines. This is because the sound waves caused by the exhaust pulsations from each cylinder cancel each other out. The more cylinders, the more canceling effect. Since most vehicles today have four- or six-cylinder engines, modern exhaust systems tend to be noisier than older systems.

Some engines, especially high performance V-8 engines, have two separate exhaust systems, one for each cylinder bank. This is known as a ***dual exhaust system,*** and it increases the power of the engine by reducing the amount of back pressure in the system. Modern vehicles with dual exhausts will also have two catalytic converters. A dual exhaust system is noisier than a single exhaust on the same engine, because the canceling effect of multiple cylinders is reduced. To regain some of this canceling effect, equalize pressure between the two banks, or to provide an access point for an oxygen sensor in some cases, dual exhaust systems are sometimes equipped with an equalizer pipe.

The exhaust system has a harder job than its simple construction would indicate. If the exhaust gases cannot flow freely, the pistons must push against excess back pressure in the exhaust system to remove the exhaust gases from the cylinder. This excess back pressure can reduce engine power or cause overheating. In extreme cases, it can keep the engine from starting.

Exhaust gases contain corrosive combustion by-products and water. When these by-products enter the cooler exhaust system from the combustion chamber, they condense and settle in the low points of the exhaust system. These pools of corrosive water can quickly rust out the exhaust system. Pipes and mufflers are made of rolled steel and other quality metals that resist corrosion. Exhaust systems with a catalytic converter tend to last longer, since the extra heat produced by the converter causes the corrosive gases and water vapor to remain in the gaseous state, and exit the system. Dual exhaust systems are more prone to rust than single exhaust systems, since a dual system runs cooler than a single exhaust on the same size engine.

Non-Stock Exhaust System Alterations

Many people modify the exhaust system to gain extra horsepower. Vehicles with V-type engines and single pipe exhaust systems can sometimes be modified by replacing the single exhaust system with a dual exhaust system. However, this type of modification is expensive and is often limited by space under the vehicle. Another common modification is to replace the entire exhaust system with a system that has a larger pipe diameter. The most common modification is to replace the stock conventional muffler with a straight through or low restriction muffler, explained earlier. This type of muffler reduces the exhaust restriction, allowing the exhaust gases to flow more freely.

Another common modification to the exhaust system is to replace the stock cast iron exhaust manifolds with steel exhaust ***headers.*** Headers are designed to reduce exhaust back pressure, and therefore increase engine power. Headers for modern engines are manufactured with fittings, for installing the oxygen sensor(s) and the air pump nozzles when they are used.

The best brands of headers use the *scavenging* principle, in which exhaust pulses from each cylinder are routed to produce a suction effect that further reduces back pressure. While steel tube headers are usually installed after the vehicle leaves the factory, many vehicle manufacturers have redesigned their cast iron exhaust manifolds to obtain some of the free exhaust flow characteristics of the header. A few manufacturers have actually begun to equip their high performance engines with cast iron headers as original equipment.

Note: Many states and municipalities restrict certain non-stock exhaust parts such as headers, straight through mufflers, catalytic converters, and other system modifications to off-road use only due to emissions and noise. Non-manufacturer approved exhaust system modification or the installation of aftermarket parts in these states is considered a form of emission system tampering. A non-stock exhaust part approved for highway use will be able to fit all the applicable air tubes and sensors as well as the original factory part. Be sure to check all state and local regulations before replacing any exhaust system part. Also, the use of non-stock exhaust system parts or modifications could void any warranty that may apply to the vehicle.

Emissions Inspections

To help further reduce exhaust emissions in certain areas, inspection and maintenance or I/M programs require all light cars and trucks to be inspected and tested for exhaust emissions on a regular basis. The test procedure and test interval for each program varies from state to state. Most states simply check for the presence of the catalytic converter and fuel inlet restrictor and perform an exhaust gas analysis. In some areas, a more stringent program, called IM 240 is used, in states where high airborne pollution rates are a problem. More information on emissions testing programs will be given in **Chapter 18.**

Summary

The three pollutants that the modern emission control system must reduce are carbon monoxide (CO), unburned hydrocarbons (HC), and oxides of nitrogen (NO_X). CO is caused by rich air-fuel mixtures. HC is unburned fuel, as well as any lubricating oil that escapes from the crankcase as a vapor, that passes through the exhaust system and enters the atmosphere. NO_X is caused by high combustion chamber temperatures, forcing the combination of oxygen with nitrogen.

Modern emissions systems are fully integrated, making use of computer control. The ECM-controlled emission systems keep the air-fuel ratio as close to 14.7:1 as possible. The action of the ECM, fuel system, cooling system, and emissions controls keeps the combustion chamber hot enough to lower HC levels, but not hot enough to cause excessive NO_X or pinging.

On newer vehicles, the emission controls must be working for maximum driveability as well as lowest emissions. Internal engines have been modified to balance the needs of performance, emissions, and driveability. Some of these modifications were first developed many years ago, but are still used on the latest vehicles.

Many external emission systems are installed on modern engines. Newer vehicle carburetors have smaller main jets, reduced accelerator pump output, and power valve springs which do not allow the power valve to open until the engine is under very heavy loads. The idle mixture screws are sealed to prevent readjustment. Most carburetors are controlled by the ECM through a mixture control solenoid installed on the carburetor.

Early emission-related timing control systems were attached to the vacuum advance unit, and restricted vacuum advance to lower HC emissions. Since the introduction of computer engine controls, timing is precisely controlled by the ECM. The PCV system reduces air pollution and keeps the engine internals clean. Thermostatic air cleaners heat the incoming air when the engine is cold, using exhaust manifold heat.

EFE systems quickly heat the intake manifold of a cold engine, allowing the choke to be quickly opened to reduce HC levels. The vacuum operated EFE uses a vacuum motor combined with a thermostatic vacuum switch or ECM-operated solenoid. A few vehicles are equipped with a grid of resistive wire. It is installed between the carburetor or fuel injector throttle body and the intake manifold.

The evaporative emissions system prevents gasoline vapors from escaping into the atmosphere from the fuel system. Fuel that evaporates from the fuel tank or other fuel system components is stored in a charcoal canister. When the engine is running, these vapors are drawn into the engine. To burn raw gasoline before it reaches the atmosphere, outside air is often pumped into the exhaust manifold. The two types of air injection systems are the air pump and pulse air system. Using an EGR valve directly reduces combustion temperatures and NO_X formation by recycling exhaust gases into the intake manifold.

The exhaust system is a set of pipes and muffling devices which removes exhaust gases and heat from the engine. The exhaust system prevents damage to the vehicle or injury to its passengers from either heat or fumes. The exhaust system also muffles the noise of the escaping exhaust gases without causing excessive restrictions in their flow. Exhaust systems with catalytic converters are considered part of the emission control system, and help to clean up the exhaust gases. Turbochargers are installed in the exhaust manifold, or in a special exhaust pipe immediately behind the manifold. The turbine wheel is spun by the exhaust gases.

The exhaust system includes the exhaust manifold, exhaust pipe, catalytic converter, muffler, and tailpipe. Some models are equipped with a small second muffler called a resonator. Conventional (stock) mufflers consist of an airtight metal housing enclosing a series of internal pipes and resonance chambers that reverse the exhaust's flow and reduce its speed. This suppresses and cancels out sound waves and pressure surges from the engine. This type of muffler restricts the exhaust gas flow to some degree. Some resonators, and many aftermarket mufflers, are straight through mufflers that are often used to reduce back pressure. The straight through muffler is inexpensive and less restrictive, but does not do as good a job of muffling noise. This type of muffler does not last as long as the conventional muffler. Exhaust system pipes are used to conduct the exhaust gases between the other exhaust system components. The final pipe in the system is called the tailpipe. Many of these pipes are made of two tubes, one inside the other.

The catalytic converter is a chemical device that reduces the harmful content of the exhaust gases. Unless it becomes clogged, it has no effect on engine operation. The converter contains elements called catalysts, which cause a chemical reaction in something else without being changed themselves. Most current vehicles use a three-way catalytic converter. This type of converter reduces NO_X as well as HC and CO. It consists of two separate sections.

The converter is made of heat resistant materials, and is surrounded by baffles to protect the vehicle and surrounding area from the extreme heat. Vehicles with catalytic converters must be operated on unleaded gasoline, since lead will ruin the catalysts. If the catalyst stops working, the cool converter can accumulate deposits and plug up. Cooler exhaust system temperatures will also cause other exhaust system components to collect water and acid, quickly rusting them out.

Modern vehicles with fewer engine cylinders tend to be noisier than older V-8 engines, because the sound waves caused by the exhaust pulsations from each cylinder cancel each other out. The more cylinders, the more canceling effect. Since most vehicles today have four or six cylinder engines, modern exhaust systems tend to be noisier than older systems. Some V-type engines have two separate exhaust systems, one for each cylinder bank. This is known as a dual exhaust system, and increases the power of the engine by reducing the amount of back pressure in the system. Modern vehicles with dual exhausts will also have two catalytic converters. A dual exhaust system is noisier than a single exhaust on the same engine, because the canceling effect of multiple cylinders is reduced. Some vehicles are equipped with an equalizer pipe connecting the separate exhaust systems.

Many people modify the exhaust system to gain extra horsepower. The most common modification is to replace the stock conventional muffler with a straight through muffler. Another common modification to the exhaust system is to replace the stock cast iron exhaust manifolds with steel exhaust manifolds called headers. While steel tube headers are installed after the vehicle leaves the factory, many vehicle manufacturers have redesigned their cast iron manifolds to obtain some of the free exhaust flow characteristics of the header.

Know These Terms

Carbon monoxide (CO)
Hydrocarbons (HC)
Oxides of nitrogen (NO_x)
Stoichiometric
Overlap
Ignition timing control
Positive crankcase ventilation system (PCV)
PCV valve
Thermostatic air cleaners
Early fuel evaporation (EFE) systems
Thermal vacuum valve (TVV)
Thermal vacuum switch (TVS)
Evaporative emissions control system
Service port
Fuel level sensor
Fuel tank pressure sensor
Vent valve
Pulse air injection
Air injection reactor system
One-way check valves
Diverter valves
Switching valves
Exhaust gas recirculating (EGR)
EGR valve
Exhaust system
Exhaust manifold
Catalytic converter
Catalysts
Three-way catalytic converter
Catalyst monitor
Muffler
Back pressure
Resonator
Exhaust pipes
Tailpipe
Dual exhaust system
Headers

Review Questions—Chapter 10

Please do not write in this text. Write your answers on a separate sheet of paper.

1. Name the three pollutants that the modern emission control system reduces.
2. What air-fuel ratio does the emission control system try to maintain under all engine conditions?
   (A) 8:1.
   (B) 12:1.
   (C) 14.7:1.
   (D) 16.5:1.
3. For emission control purposes, combustion chamber shapes have been redesigned to:
   (A) increase volume.
   (B) reduce fuel condensation.
   (C) lower compression.
   (D) All of the above.
4. For smooth idle with lean mixtures, camshafts have been redesigned to reduce ______.
5. Modern cylinder heads have ____ valves per cylinder.
   (A) two
   (B) four
   (C) five
   (D) Both A & B.
6. Fuel vapors that are stored in the carbon canister are burned in the:
   (A) engine cylinders.
   (B) exhaust manifold.
   (C) catalytic converter.
   (D) Both A & C.

7. Match the following emission control device with the phrase that best describes it.

   ______PCV valve

   ______Thermostatic air cleaner

   ______ EGR valve

   ______Air pump

   ______Catalytic converter

   ______Thermostat

   (A) Prevents overheating
   (B) Assists fuel burning in the exhaust manifold
   (C) Cleans exhaust gases through chemical reactions
   (D) Helps engine to heat up quickly
   (E) Allows exhaust gases into the intake manifold
   (F) Reduces crankcase sludge.
   (G) Prevents carburetor icing

8. Engines with three-way catalytic converters and air pumps are equipped with ______ that direct air to the exhaust manifold when the engine is cold.

9. The EGR valve is used to reduce ______.
   (A) carbon monoxide
   (B) carbon dioxide
   (C) unburned hydrocarbons
   (D) oxides of nitrogen

10. Technician A says that some diesel engines have EGR valves. Technician B says that some diesel engines have catalytic converters. Who is right?
    (A) A only.
    (B) B only.
    (C) Both A & B.
    (D) Neither A nor B.

11. All of the following operate to control engine idle speed when the engine is running, EXCEPT:
    (A) idle speed up solenoids.
    (B) throttle position sensor.
    (C) air conditioning speed up solenoids.
    (D) idle air control valves.

12. Which of the following will result in a leaner air-fuel ratio at cruising speeds?
    (A) Lower fuel level in the bowl.
    (B) Smaller main jets.
    (C) More accelerator pump stroke.
    (D) Both A & B.

13. The evaporative emissions system is designed to trap ______ before they reach the atmosphere.

14. The evaporative emissions system on many OBD II vehicles have ______ .
    (A) a service port
    (B) a computer-controlled purge solenoid
    (C) a charcoal canister
    (D) All of the above.

15. OBD II vehicles use _______ to monitor the catalytic converter.
    (A) a pre-converter airflow sensor
    (B) a post-converter airflow sensor
    (C) a computer-controlled pressure sensor
    (D) a post-converter oxygen sensor

ASE Certification-Type Questions

1. Technician A says that modern engines have hotter thermostats for better control of emissions. Technician B says that thermostats on new engines open at higher temperatures than thermostats installed in engines many years ago. Who is right?
   (A) A only.
   (B) B only.
   (C) Both A & B.
   (D) Neither A nor B.

2. All of the following can cause ice formation at the throttle plate, EXCEPT:
   (A) high humidity.
   (B) high engine temperature.
   (C) low outside air temperature.
   (D) low air pressure.

3. The main advantage of electronic engine controls and fuel injection is that they have made it possible to eliminate the ______.
   (A) carburetor
   (B) catalytic converter
   (C) EGR valve
   (D) evaporative emissions system

4. Technician A says that lowering engine compression allows the engine to run on unleaded gas. Technician B says that lowering engine compression lowers engine efficiency. Who is right?
   (A) A only.
   (B) B only.
   (C) Both A & B.
   (D) Neither A nor B.

5. Technician A says that the spark timing on most modern engines is controlled by the ECM. Technician B says that most modern engines with vacuum advance mechanisms and catalytic converters restrict vacuum to the vacuum advance diaphragm. Who is right?
   (A) A only.
   (B) B only.
   (C) Both A & B.
   (D) Neither A nor B.

6. Technician A says that the PCV system was one of the earliest emission controls. Technician B says that the PCV system reduces pollution from the crankcase. Who is right?
   (A) A only.
   (B) B only.
   (C) Both A & B.
   (D) Neither A nor B.

7. Thermostatic air cleaners prevent the following problems, EXCEPT:
   (A) overheating.
   (B) icing.
   (C) hesitation when cold.
   (D) poor driveability during warmup.

8. All of the following applies to heat risers, EXCEPT:
   (A) they are an early form of early fuel evaporation (EFE) system.
   (B) they are operated by a thermostatic spring.
   (C) they are always used with a thermostatic air cleaner.
   (D) they reduce fuel condensation in the intake manifold.

9. Technician A says that the air pump diverter valves are used to prevent backfiring. Technician B says that the air pump switching valves are used to switch air flow between the exhaust manifold and catalytic converter. Who is right?
   (A) A only.
   (B) B only.
   (C) Both A & B.
   (D) Neither A nor B.

10. The air pump injection system and the pulse air injection system share the following components, EXCEPT:
   (A) check valves.
   (B) drive belt.
   (C) exhaust manifold connections.
   (D) tubing and hoses

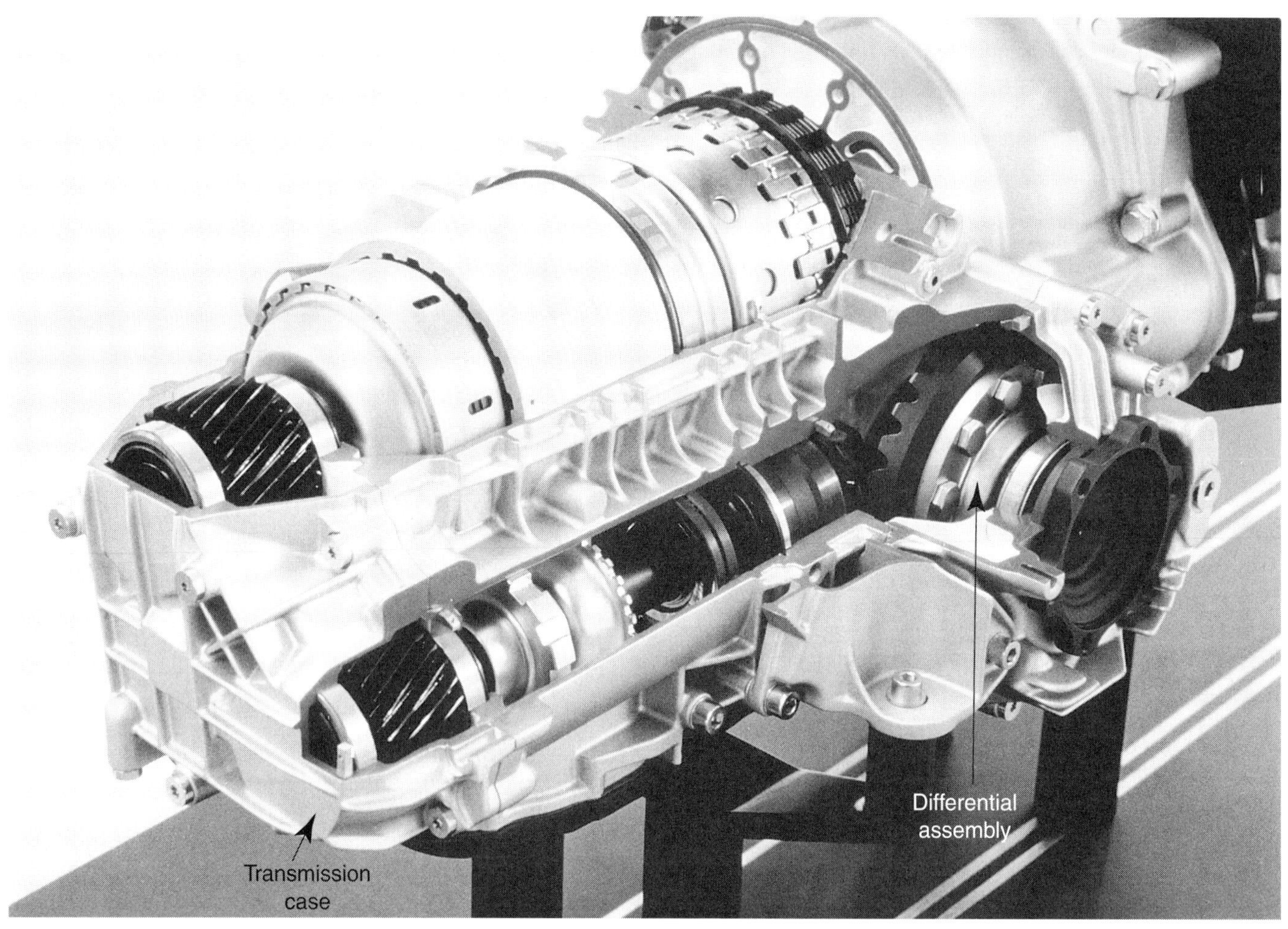

Cutaway of a five-speed automatic transmission. Although this transmission is set longitudinally, it drives the axles on a front-wheel-drive vehicle. (Audi)

Chapter 11

Drive Train and Vehicle System Fundamentals

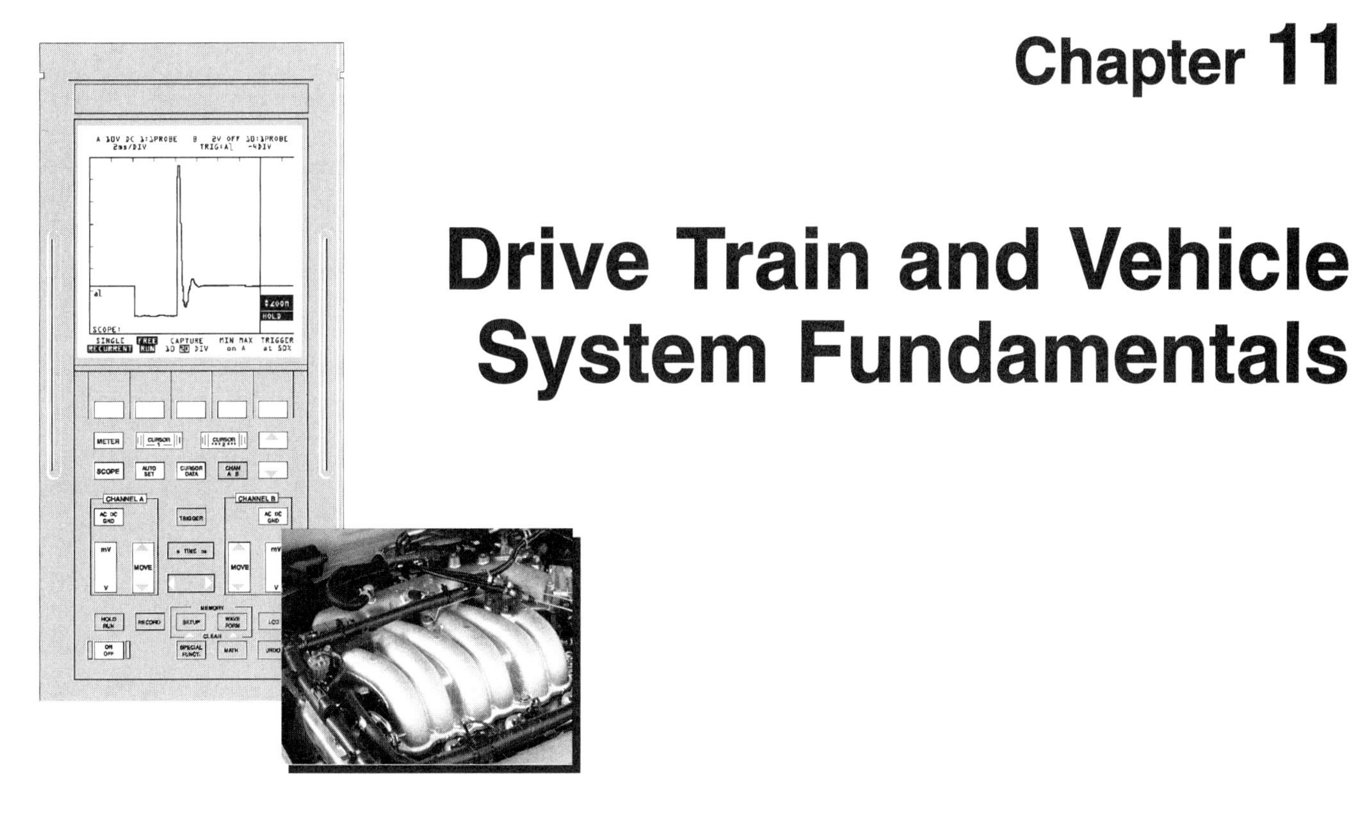

After studying this chapter, you will be able to:

- ❑ Identify drive train components having an effect on driveability and performance.
- ❑ Explain the basic purpose of both manual clutches and torque converters.
- ❑ Identify and explain the purpose of transmission/transaxle components.
- ❑ Identify the control system components of an electronically controlled automatic transmission.
- ❑ Explain how transmission problems can affect driveability.
- ❑ Explain how rear axle ratios can affect performance and gas mileage.
- ❑ Describe the air conditioning cycle.
- ❑ Explain how the air conditioner can affect vehicle operation.
- ❑ Identify the parts of the brake system.
- ❑ Identify the parts of the anti-lock brake system.
- ❑ Identify the brake problems that can be confused with engine driveability problems.
- ❑ Identify the parts of the suspension system.
- ❑ Identify the suspension problems that can be confused with engine driveability problems.
- ❑ Identify the parts of the steering system.
- ❑ Identify the steering system problems that can be confused with engine driveability problems.
- ❑ Identify tire problems that can be confused with engine driveability problems.

Driveability problems are not always caused by engine components. Various drive train and vehicle systems can affect driveability, performance, and mileage. While these parts do not directly affect the operation of the engine, they can affect the ability of the engine to deliver smooth power to the road, or can cause symptoms that may be mistaken for engine driveability complaints. Drive train problems can have the greatest impact on the vehicle's performance. The air conditioner compressor places a drag on the engine when it is operating. Some air conditioning defects as well as defects in the brakes, steering, suspension, and tires can also be mistaken for driveability problems. All of these systems and possible problems are discussed in this chapter. Understanding their relation to the engine and the vehicle is important to becoming a good driveability technician.

How Non–Engine Systems Can Affect Driveability

Most drivers do not believe a non-engine system, such as the air conditioner or brake system can affect driveability. Some technicians do not make this connection either. The modern vehicle is a series of systems, all working together. Even systems that are designed more for driver comfort than vehicle performance can affect driveability.

The following is an example of how a relatively minor system designed for driver comfort can affect vehicle performance. A sedan with an electric antenna is driven through an automatic car wash with the antenna up. Since most electric antennas are not very flexible, the retracting mechanism is usually damaged, preventing the antenna mast from retracting. Often, the motor will continue to run, draining the battery to the point that it is unable to start the vehicle. Although electric antennas are not discussed in this textbook, the example shows you how even minor, non-engine parts can drastically affect vehicle performance.

Drive Train Components

Major ***transmission, transaxle,*** and ***drive train*** components are discussed below, starting with the manual transmission clutch and automatic transmission torque converter. These components perform similar functions: forming a connection between the engine and the transmission and drive train. The major effect that the clutch or torque converter has on driveability is when they fail to transmit engine power, usually called slippage. The clutch and torque converter operate in different ways to accomplish the same thing. Another drive train component that can cause problems is the transmission itself.

Manual Clutch

Manual clutches are used to disconnect the engine from the road when the vehicle is stopped, or when shifting gears. All modern car and light truck clutches are known as ***single plate dry clutches***. The manual transmission clutch forms a connection between the engine and the road.

Warning: Clutch discs contain asbestos, a known carcinogen. Breathing dust from clutch materials can cause lung disease and cancer. Always wear a protective filter mask and avoid creating dust when working with clutches.

The single plate dry clutch consists of several major parts, as shown in **Figure 11-1.** The ***flywheel*** is attached to the engine crankshaft and acts as a mounting surface for the other clutch parts. The ***input shaft*** transfers engine power into the transmission. The ***friction disc*** is splined to the input shaft and is held between the flywheel and pressure plate to transmit power. The ***pressure plate assembly*** provides the spring pressure to tightly hold the friction disc when the clutch is applied. The ***throwout bearing*** overcomes spring pressure to release the friction disc. It is operated by the driver through linkage or a hydraulic cylinder and a ***clutch fork.***

The flywheel and pressure plate assembly are bolted together and revolve as a single unit when the engine is running. The transmission input shaft and the clutch disc are splined to each other, forming another unit which connects to the rear wheels through the transmission and differential. The input shaft and clutch disc turn whenever the vehicle is moving.

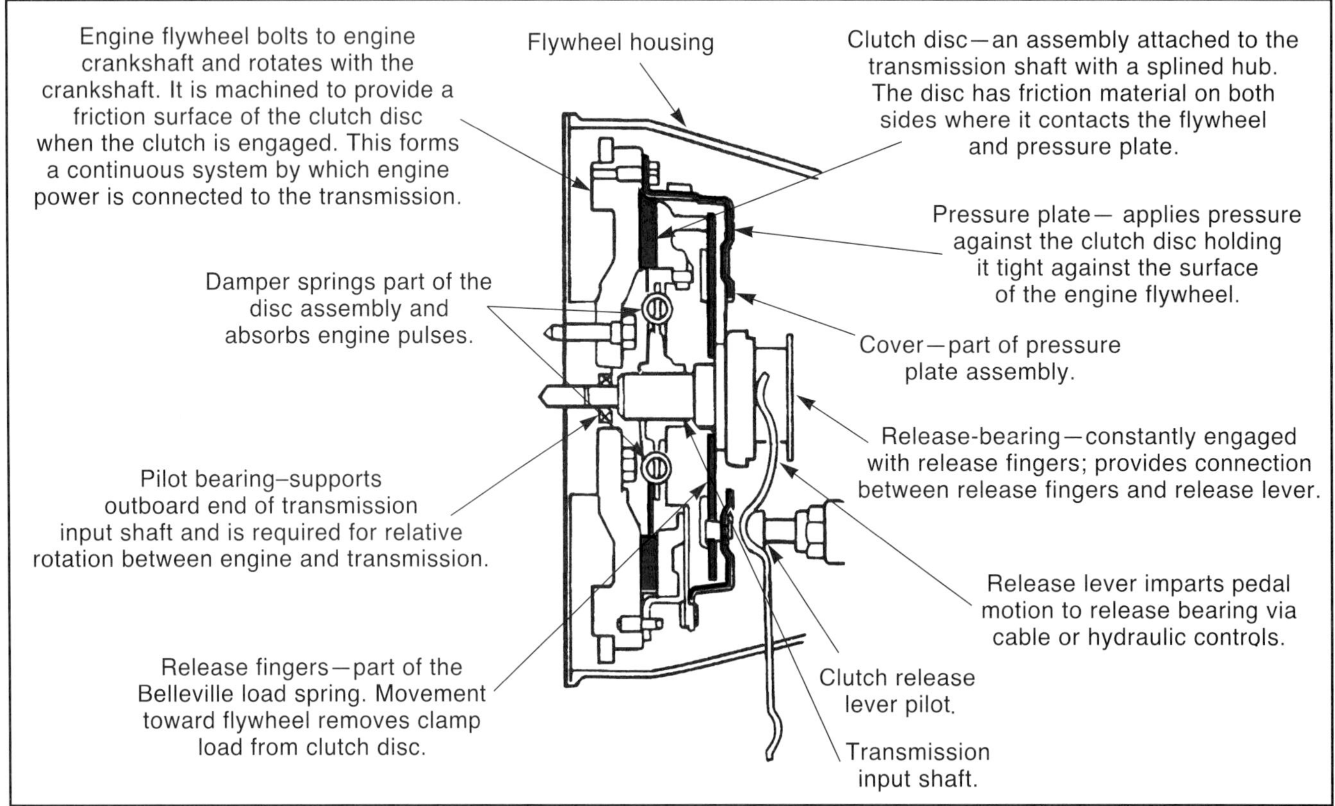

Figure 11-1. *The parts of a manual clutch work together to transfer power from the engine to the manual transmission input shaft. It is important for all driveability technicians to have a working knowledge of transmission, drive train, and vehicle systems. (Ford)*

When the pressure plate springs move the plate toward the flywheel, the disc is tightly pressed between them. The clutch disc is surfaced on both sides with a lining that allows it to grasp the metal surfaces of the pressure plate and flywheel without slipping. The disc then turns at the same speed as the flywheel and pressure plate. When the pressure plate moves away from the flywheel, the clutch disc is no longer pressed between the pressure plate and the flywheel. The clutch disc and transmission input shaft will stop turning, while the rest of the clutch assembly continues to revolve. Engaged and disengaged clutches are shown in **Figure 11-2.**

A defective clutch can slip, usually due to worn disc friction linings. Although this problem can sometimes be cured by adjustment, the clutch disc, and often other clutch parts, must usually be replaced to eliminate the slippage. If grease or oil gets on the disc friction linings, the clutch may grab when applied. This can often be confused with an engine miss. In many cases, the driver may not know how to operate the clutch smoothly, and may think that there is an engine problem. Letting the clutch out too quickly will cause bucking at low speeds, and letting the clutch out too slowly will cause slippage and eventual clutch damage.

Manual Transmissions and Transaxles

The manual transmission and transaxle are mechanical devices that provide different gear ratios for different vehicle conditions. See **Figure 11-3.** They are often referred to as *gearboxes.* Transmissions are used on rear-wheel drive vehicles, while transaxles are used on front-wheel drive vehicles. The transaxle incorporates the differential and final drive assembly in the same case as the transmission gears. All modern transmissions have at least one ***overdrive*** gear. Placing the transmission/transaxle in overdrive reduces engine speed and increases fuel mileage.

Manual transmission gears are selected by the vehicle operator. Manual transmissions do not cause driveability problems in most cases. Often, however, what appears to be a driveability problem is actually caused by a defective clutch or incorrect transmission operation. The driver may be using the wrong gear for a particular situation, changing gears too quickly, or engaging the clutch too quickly or slowly.

Automatic Transmissions and Transaxles

Automatic transmissions replace the manually operated clutch and gearshift with a system using planetary gears and a hydraulic control system. A typical automatic transmission is shown in **Figure 11-4.** An automatic transaxle is shown in **Figure 11-5.** ***Planetary gears*** are composed of a central sun gear, planet gears in a planet carrier, and an outer ring gear. Planetary gears are engaged at all times. Different gear ratios are provided by holding one part of the gearset stationary while driving the another part with engine power. The holding and driving actions are accomplished by holding members, which include bands, clutch packs, and

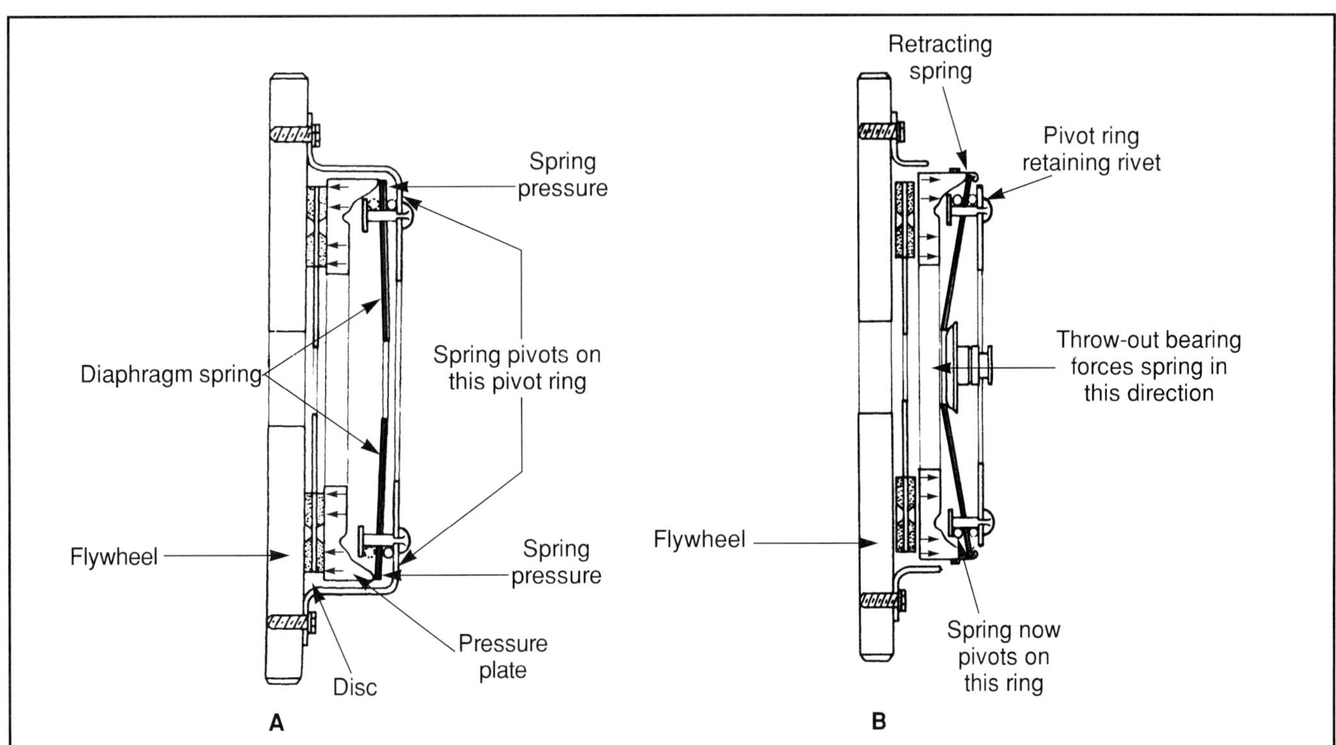

Figure 11-2. *Operation of a diaphragm spring clutch. A—The throw-out bearing has been released and the spring forces the clutch plate against the flywheel. B—When the throw-out bearing forces the diaphragm spring in, the clutch plate is released from the flywheel. (Chevrolet)*

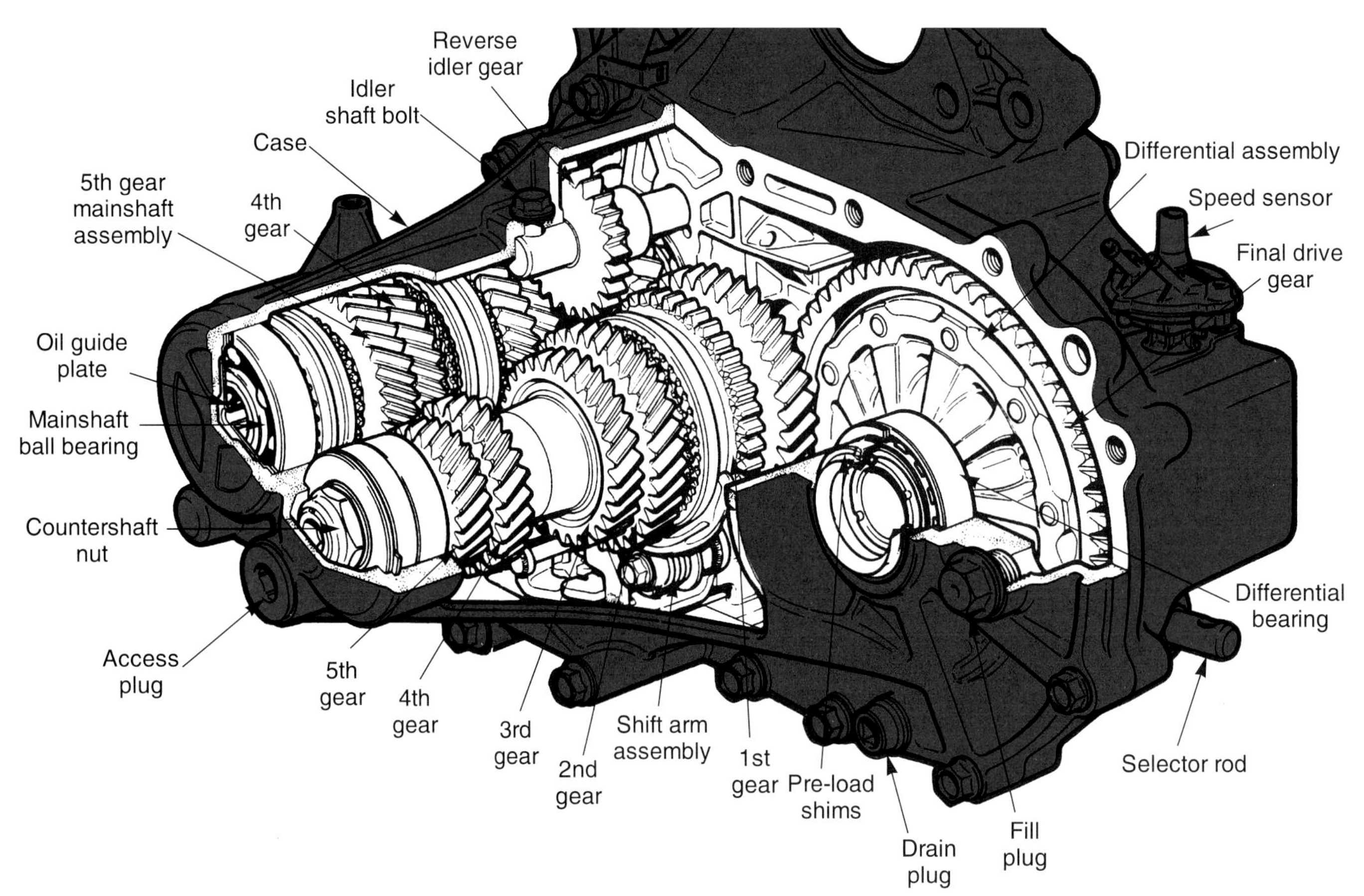

Figure 11-3. *Cutaway of a manual transaxle. Manual transaxle problems are rarely mistaken for engine performance problems. (Sterling)*

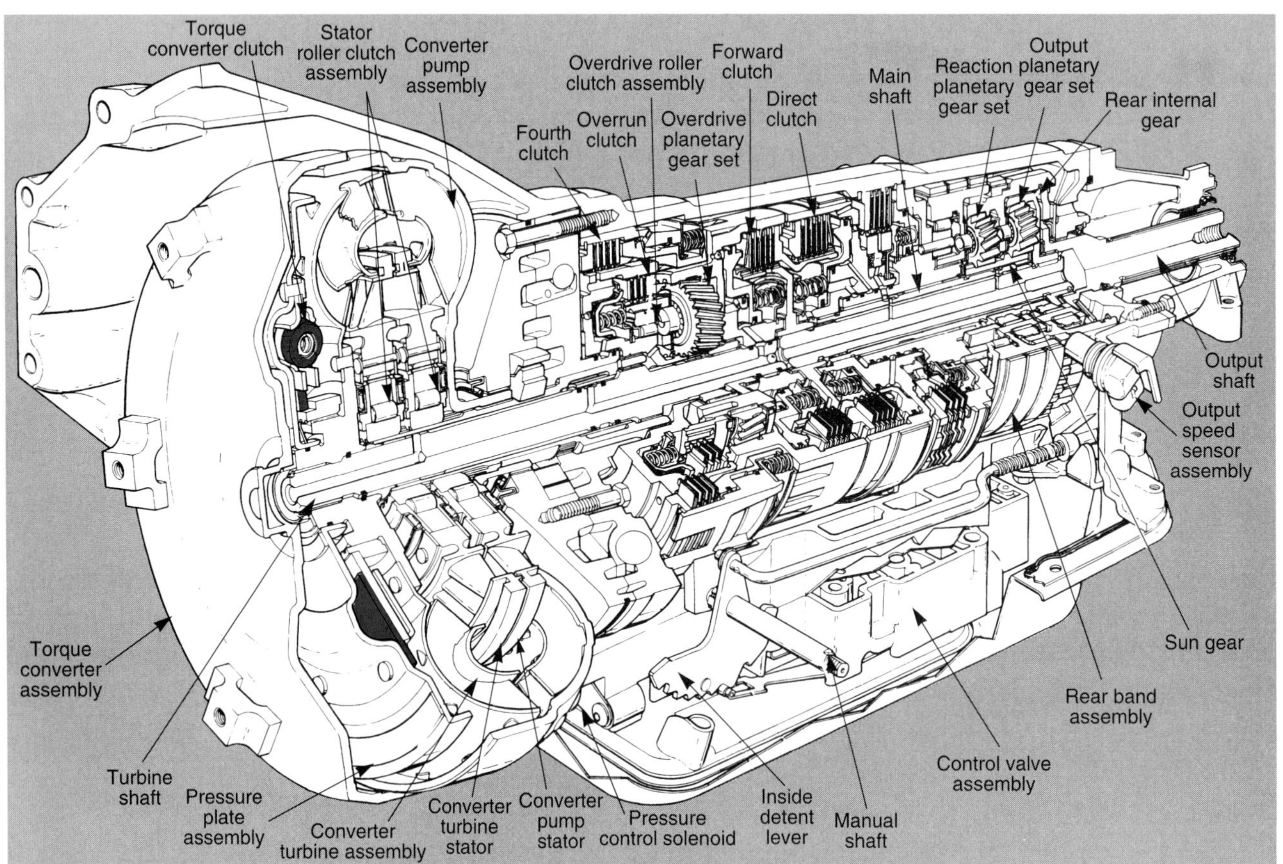

Figure 11-4. *Cutaway of a computer-controlled four-speed automatic transmission. Note the torque converter clutch. (General Motors)*

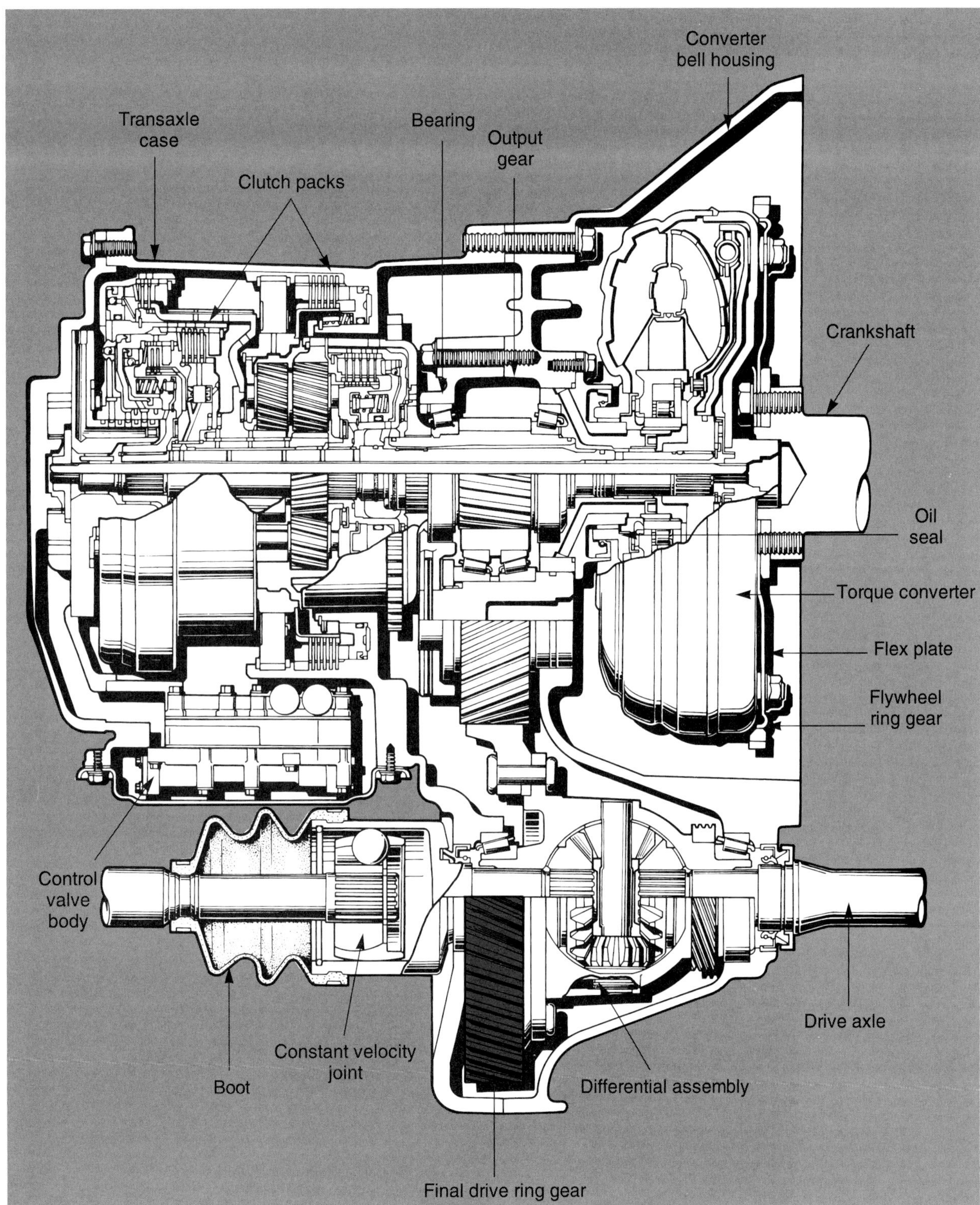

Figure 11-5. *Most automatic transaxles have an integral differential assembly. The differential in this transaxle is driven by a ring gear. (Ford)*

overrunning or one-way clutches. Bands are operated by hydraulic servos, and clutch packs are operated by clutch apply pistons.

The holding members are controlled by the action of the hydraulic system. A hydraulic system for a transmission with four forward gears is shown in **Figure 11-6.** Pressure is provided by an engine driven pump, and regulated by a spring loaded pressure regulator. The pump also fills the torque converter with fluid. The fluid is drawn from the transmission pan located at the bottom of the transmission or transaxle. A filter removes metal and sludge particles which may have entered the fluid. Fluid returning from the radiator cooler enters the pan to provide a supply of fluid for the hydraulic system.

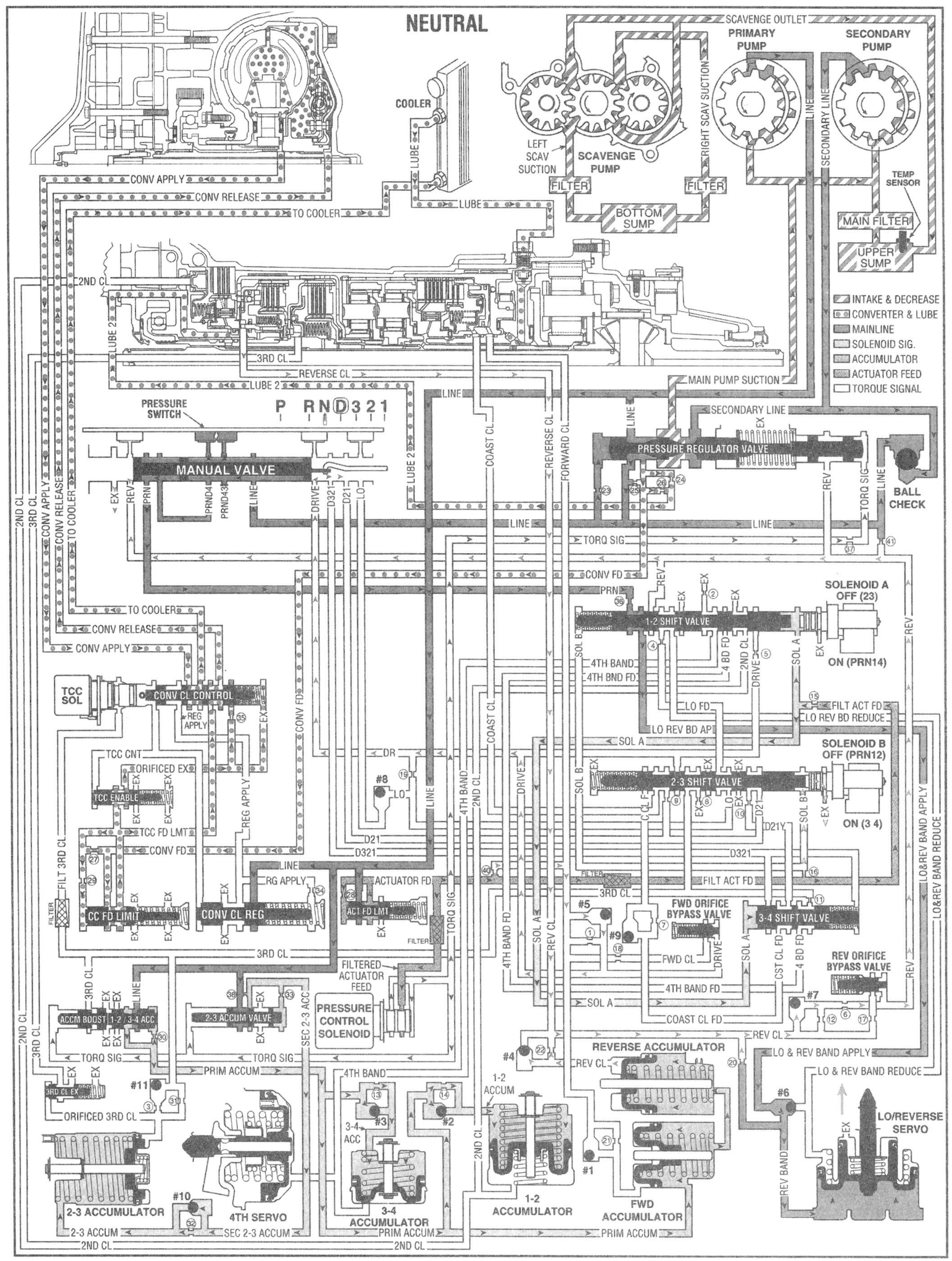

Figure 11-6. *While most modern automatic transmissions and transaxles are computer-controlled, they still rely on hydraulic shift valves to control the various clutches in the transmission unit. This is a hydraulic schematic for a four-speed automatic transaxle. (General Motors)*

Automatic Transmission Operation

The driver operates the manual valve that directs the fluid pressure to other parts of the hydraulic system. In the forward gears, shift valves direct fluid to the holding members based on pressure signals from the governor and throttle valves. The governor valve is turned by the output shaft, and increases pressure as the output shaft increases. The throttle valve is operated by the driver through linkage attached to the engine throttle plates.

As the vehicle speed increases, governor pressure overcomes throttle pressure and the shift valve moves into the upshifted position. Moving the shift valve directs fluid to the holding members needed to shift the planetary gears into the next higher gear. The speed at which the shift occurs is determined by the throttle pressure. Throttle pressure increases when the throttle is depressed, and decreases when the throttle is released. Therefore, the more the throttle is depressed, the later the upshift will occur.

When the vehicle slows down, the throttle pressure overcomes governor pressure and the shift valve returns to the downshifted position. Each shift valve is calibrated to move at different speeds and throttle openings. This is accomplished by changing the valve size, or by adding springs to the shift valve bore.

Other hydraulic system valves are used to modify shift timing, control shift feel, and provide a passing gear. Spring loaded accumulators are used to cushion transmission shifts by bleeding off some of the holding member apply pressure. Check balls are installed in many transmissions to further control shifting, prevent incorrect fluid flow, and cushion up and downshifts.

Automatic transmissions and transaxles may cause driveability problems. Defects can often occur in the hydraulic control system or in the clutches and bands that are applied to provide different gear ratios. Late or early shift points, overly soft or harsh shifts, slipping when hot, and slow engagement from neutral or park can often be confused with engine problems. Severe slippage is usually caused by burned clutches or bands, occasionally by a clogged filter, and is usually recognizable as a transmission problem. A sometimes unrecognized problem is a chatter caused by slight slippage of the holding members. This is often misdiagnosed as an engine miss.

In some cases, the transmission problem can be caused by a problem in the engine or the ECM. This is especially true of shift timing and quality, which is sometimes affected by engine vacuum or throttle linkage adjustment. Engine overheating can cause transmission overheating, since the transmission oil is cooled in the radiator.

Torque Converters

The ***torque converter*** transmits power through hydraulic fluid, usually called ***automatic transmission fluid (ATF).*** Torque converters are found only on vehicles with automatic transmissions and transaxles, installed between the engine flywheel and transmission. The torque converter consists of three major parts inside a sealed housing.

A set of vanes welded to the inside of the coupling housing turns when the engine turns. These vanes are called the ***impeller*** or pump. Another set of vanes is welded to a separate internal housing, which can turn in relation to the coupling housing. This is called the ***turbine.*** A set of curved blades called a ***stator,*** is mounted on a one-way clutch at the center of the torque converter, between the impeller and turbine blades. **Figure 11-7** shows the relationship of the torque converter parts.

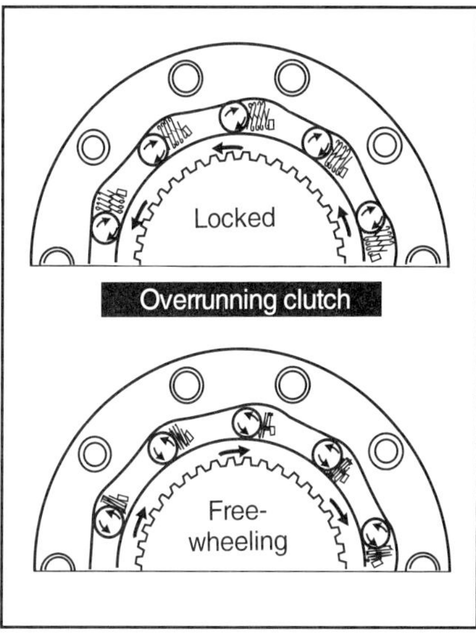

Figure 11-7. *Overrunning clutches are used to prevent rotation in one direction, depending on its position. Most overrunning clutches use spring-loaded balls or sprags. (Chrysler)*

The housing is filled with transmission fluid and the spinning impeller vanes begin to turn the fluid. Centrifugal force tries to throw the fluid outward, but it is redirected by the shape of the impeller housing, and is thrown toward the turbine. The impeller vanes also force the fluid to rotate at the same speed as the impeller. The fluid, therefore, leaves the impeller in a direction that is both outward and sideways. The power from the engine has been transferred from the impeller to the rotating fluid. The fluid then hits the turbine. The blades on the turbine are curved so that the fluid striking them is reversed. This transmits motion to the turbine, causing it to rotate. Engine power has now been transferred from the fluid to the turbine. When the fluid leaves the turbine, it hits the stator.

The stator is mounted on a one-way clutch, and can turn in one direction only. At low speeds, reversed fluid flow from the turbine causes the one-way clutch to lock so that the stator can function. At cruising speeds, the fluid flow from the turbine is no longer reversed, causing the stator to unlock and freewheel. One-way clutch action is shown in **Figure 11-5.** The stator blades turn the fluid around so that it hits the impeller in the direction of rotation, helping it to rotate. This has the effect of multiplying the engine power. **Figure 11-8** illustrates torque converter flow from impeller to turbine to stator and back to the impeller.

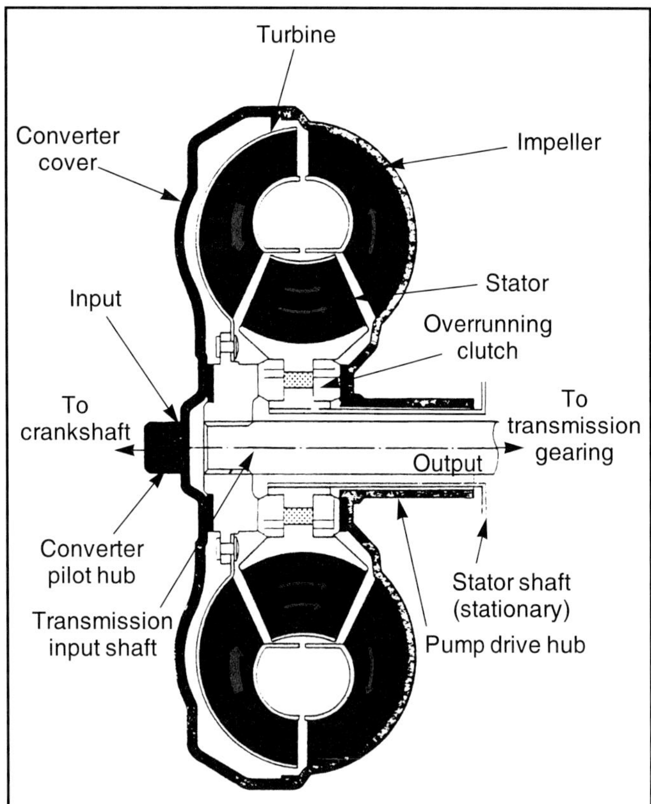

Figure 11-8. *Cutaway of a torque converter. Note the flow of fluid from the impeller, to the turbine, and through the stator. (Sachs)*

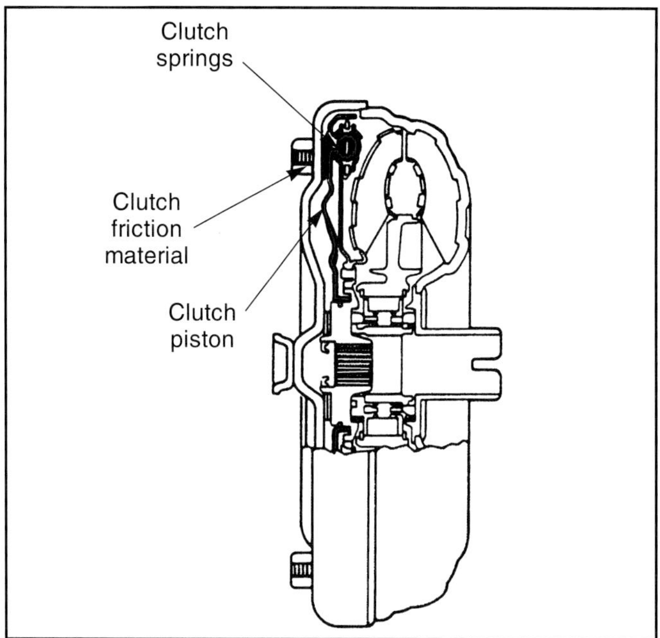

Figure 11-9. *Most lockup torque converters use an internal clutch plate to lock the converter in order to reduce slippage. (Chrysler)*

Torque converters multiply engine power due to the redirection of fluid by the stator at low vehicle speeds. The torque increase is about 2:1. This improves acceleration and mileage with no other modifications. As with any transmission, the torque increase is a trade-off with speed, so at maximum power output, the output shaft speed is about one-half of input. As vehicle speed increases, output shaft speed and torque multiplication changes in stages to about 1:1.

Converter slippage at idle and low speeds causes the transmission fluid to become very hot. For this reason, automatic transmissions are cooled by pumping fluid from the converter and transmission into a heat exchanger in the radiator. Auxiliary coolers are sometimes added to trucks and vehicles used in extreme conditions.

The stator one-way clutch rarely becomes defective, but occasionally will either fail to lock at high speeds, or fail to unlock at low speeds. If the clutch is unlocked at low speeds, acceleration will be sluggish. If the clutch stays locked at high speeds, top speed and fuel economy will be low, and the fluid may become overheated and cause transmission slippage.

Torque Converter Clutch

There is always some slippage and power loss in any torque converter. This slippage is reduced or eliminated by the use of an internal ***lockup clutch*** or ***torque converter clutch,*** to lock the turbine to the impeller. **Figure 11-9** shows a converter assembly with a lockup clutch. This is a hydraulically-applied clutch similar to a manual transmission clutch. The clutch is controlled by external hydraulic switches, or by the ECM on late-model vehicles.

One result of using the lockup clutch is that the driver feels more vibration from the driveshaft when the accelerator is pressed or released. This is because the cushioning effect of the torque converter fluid is lost.

Torque converter clutch problems usually appear in the form of driveline vibration when the lockup clutch is applied. If the torque converter clutch switch is defective, the lockup clutch may not engage or may slip in and out of engagement. If the clutch does not disengage when the vehicle is stopped, it will stall the engine. If the lockup clutch does not engage at highway speed, engine operation will appear to be normal, but mileage will be low. If the clutch slips in and out of engagement or is worn, vibration can result.

Computer-Controlled Transmissions and Transaxles

A computerized transmission/transaxle uses an ECM, which makes decisions concerning transmission shifts and pressures, torque converter lockup, and other tasks formerly done by hydraulic and mechanical controls. **Figure 11-10** is a schematic of an automatic transmission that is almost completely controlled by the ECM. The ECM makes all the decisions concerning shift patterns, operation of the converter clutch, and overdrive selection.

The transmission control portion of the ECM makes shift decisions based on inputs, in a manner similar to the fuel and ignition decisions made by the engine control

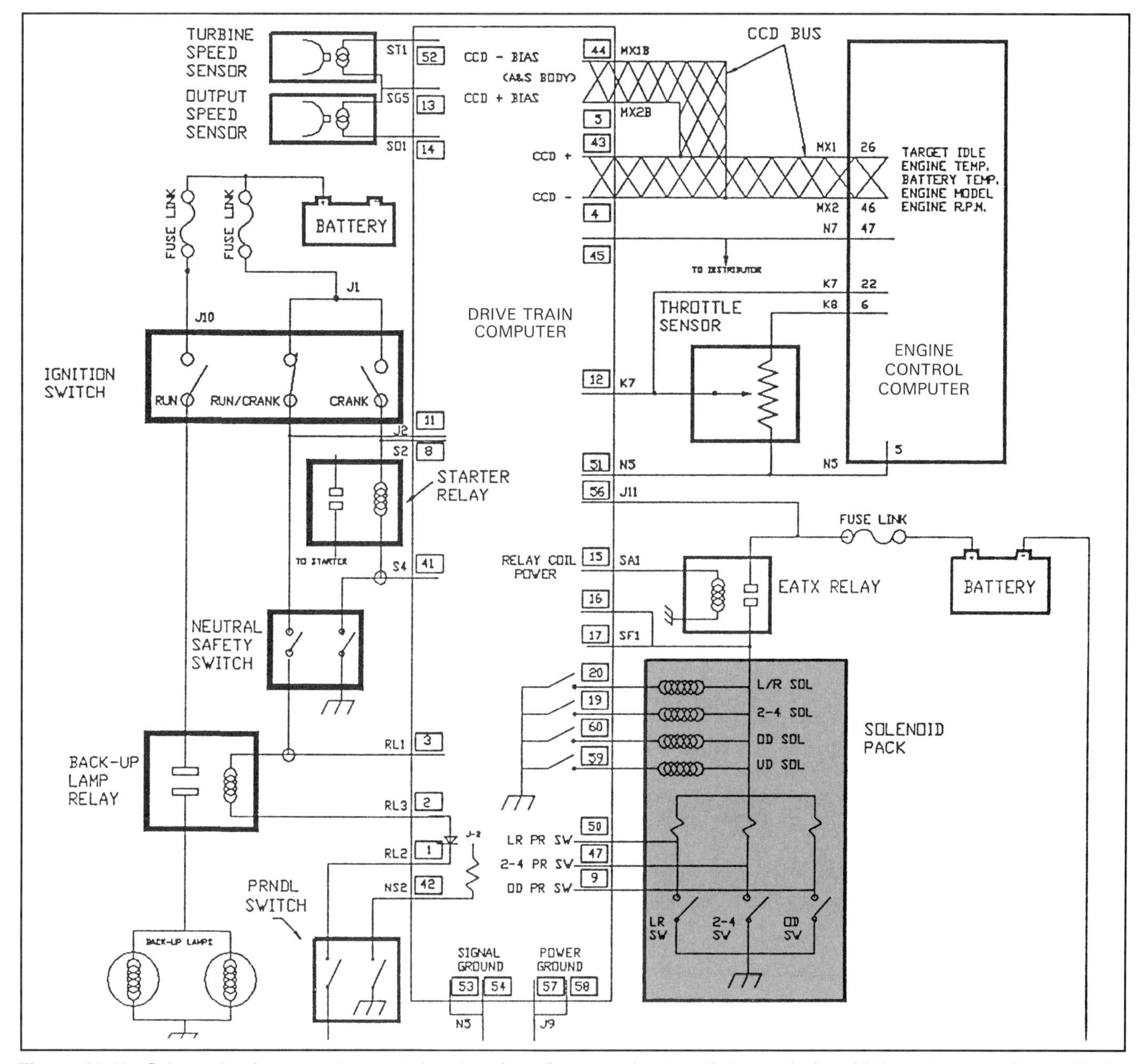

Figure 11-10. *Schematic of a computer control system for a four-speed automatic transmission. All the outputs are located in the solenoid pack installed in the transmission case. (Chrysler)*

portion. Typical inputs are engine speed, vehicle speed, and throttle position. These inputs are provided by sensors. The ECM may also receive input from electrical switches. Often, the same inputs are used to control both the engine and transmission. A valve body containing the shift solenoids for a late model four-speed automatic transmission is shown in **Figure 11-11.**

Figure 11-12 shows a drive train computer system. This system controls the operation of an automatic transmission. Other systems may simply control the converter lockup clutch, or provide a speed input to the ECM. There are many variations to the computerized drive train. In some cases, the ECM also controls the drive train, while in other cases, a separate drive train computer is used. ECM's used to control both engine and transmission operation are often referred to as a *powertrain control module (PCM)*. When a separate

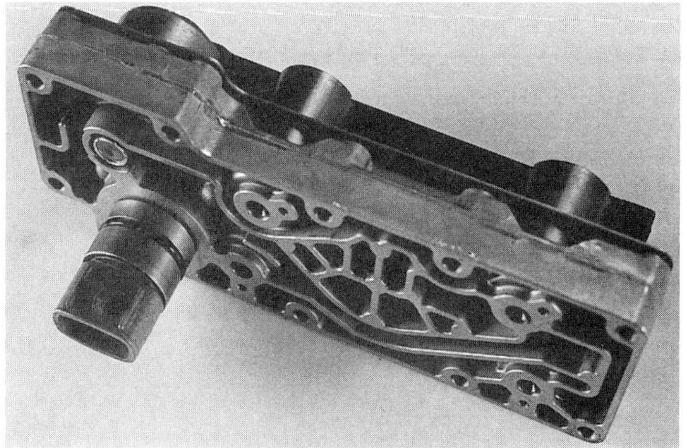

Figure 11-11. *The valve body contains the shift valves, check balls, and other components needed to control automatic transmission operation.*

computer is used, it may exchange information with the ECM, or it may be a separate unit.

Based on these inputs, the ECM makes decisions about what output commands to produce. The output commands are electrical signals to the output devices. In automatic transmissions, the output devices control parts of the transmission hydraulic system. The output devices shown in **Figure 11-12** are solenoids that control most of the transmission hydraulic system. The hydraulic system can be commanded to alter shift patterns, or to apply and release the converter clutch.

Figure 11-13 is a schematic of a computer control system used on many modern vehicles with four-speed transaxles. It consists of a control computer with inputs and outputs. The ECM and transaxle computer exchange information to control the operation of both the engine and transaxle. The ECM tells the transaxle computer how fast the engine is turning, engine temperature, and when the engine is idling. The transaxle controller also receives inputs from the throttle position sensor and two speed sensors. Internal transaxle pressure switches provide feedback from the hydraulic portion of the control system. In this system, any driveability problems that affect the transaxle can affect engine operation. An ECM-based transmission problem is likely if a computer-controlled transmission starts out in the wrong gear (usually second gear), or fails to change gears.

Computer-Controlled Transmissions in OBD II Equipped Vehicles

Most OBD II systems monitor and control the transmission and transaxle in the same manner as an OBD I system. However, OBD II will more closely monitor the transmission/transaxle operation as it relates to vehicle emissions. If a transmission problem that can affect engine performance and/or vehicle emissions exists, the ECM will store a type A diagnostic code. On some systems it will store a Freeze Frame and Failure Record.

Final Drive Ratios

The ***final drive ratio*** is the speed reduction ratio caused by the final drive gears in the drive train. Since the internal

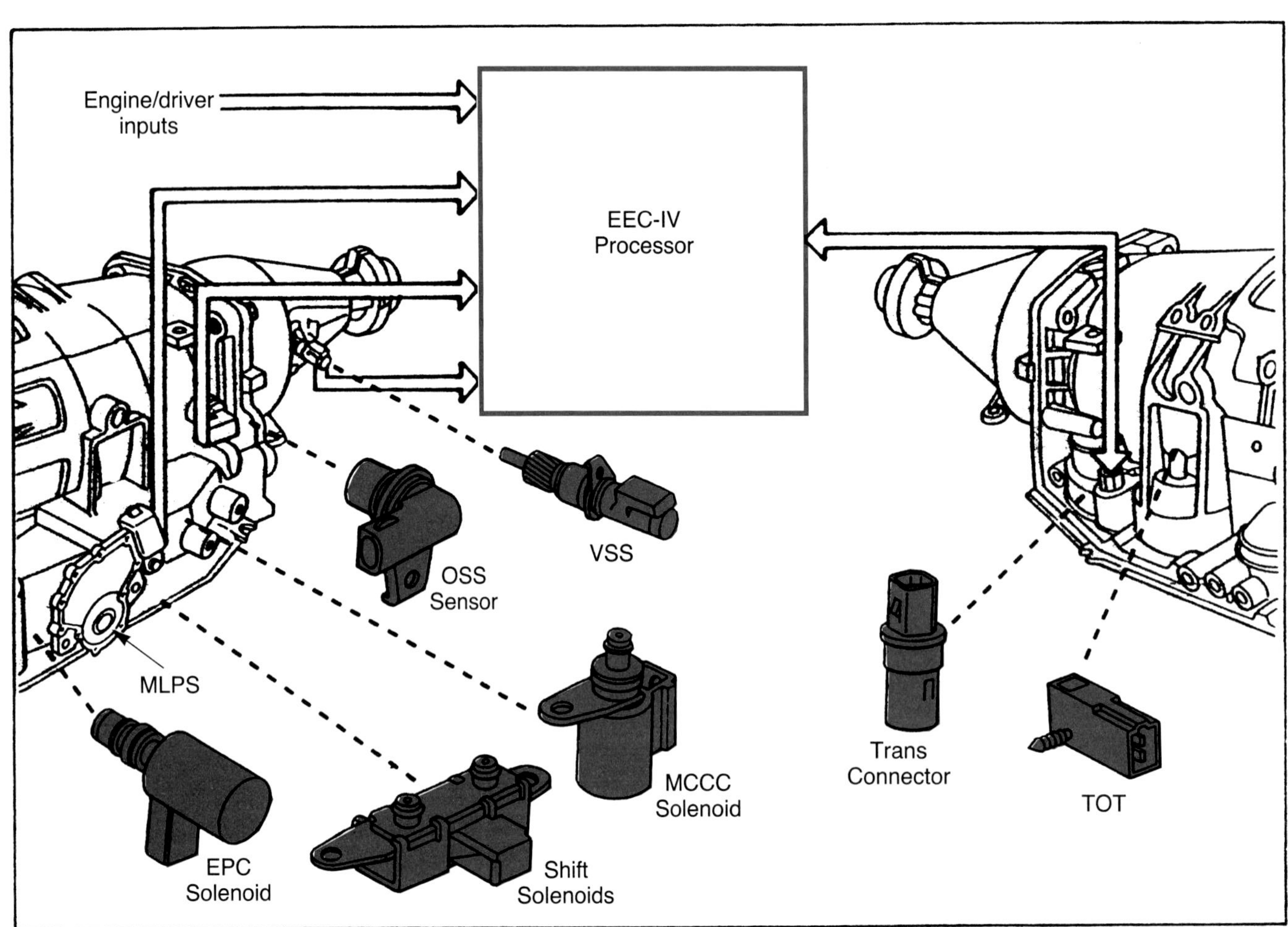

Figure 11-12. *Locations for the output devices for a computer-controlled transmission. Output device locations and type vary by manufacturer. (Ford)*

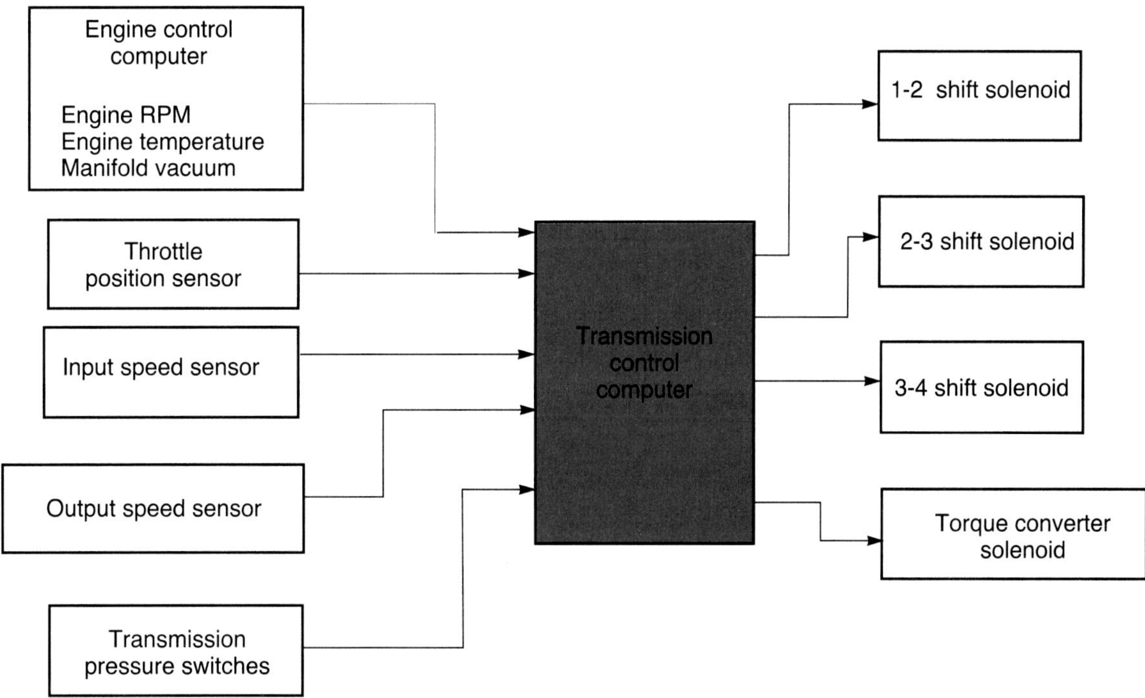

Figure 11-13. *This schematic shows the input sensors, output devices, and their electrical connections to the computer. On some vehicles, two ECM are used.*

combustion engine cannot drive the vehicle at a direct, or 1:1 ratio, speed is always reduced to increase torque and pulling power. The final drive unit is sometimes referred to as the ***differential,*** because the differential gears are installed in the same place in the drive train. The differential gears, however, are installed to allow the vehicle's drive wheels to rotate at different speeds for turning corners, and are not part of the final drive ratio gears.

The final drive ratio gears consist of a ***drive gear*** and a ***driven gear.*** The drive gear always has fewer teeth than the driven gear. The drive ratio is determined from the number of driven gear teeth divided by the number of drive gear teeth. If a drive gear with 9 teeth is used with a driven gear with 36 teeth, for instance, the drive ratio is 36 divided by 9, or 4:1. Common gear ratios today range from 1.8:1 to about 3.2:1 for best fuel economy. Ratios in older vehicles were usually lower (as much as 4.11:1) for greater pulling power and acceleration.

The final drive gear set on a rear-wheel drive vehicle is called the ***ring and pinion.*** See **Figure 11-14.** The ring and pinion is a matched gear set, installed in the rear axle of the vehicle. It is called a *hypoid gear set,* due to the design of the gears, which places the pinion gear below the centerline of the ring gear. The pinion is the drive gear, and the ring is the driven gear. The gearset runs in heavy oil for maximum life.

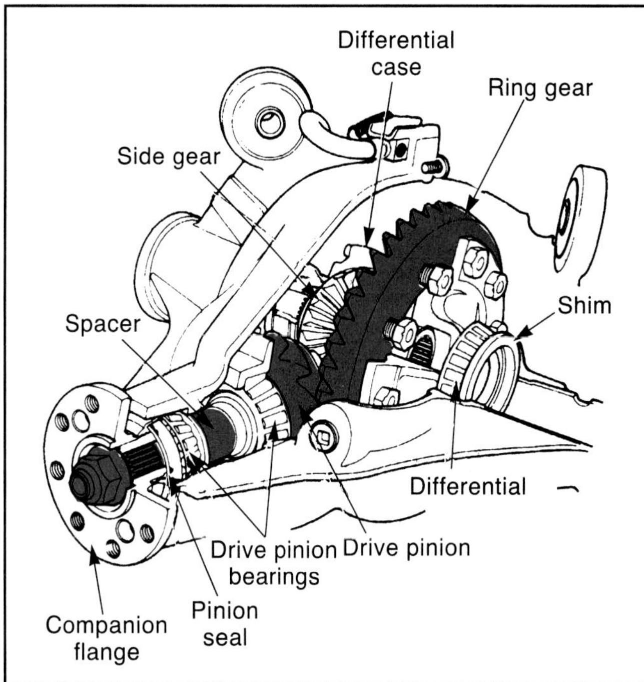

Figure 11-14. *The differential distributes the driving force from the driveshaft to the wheels. It also allows one wheel to turn faster than the other, which is necessary in turns. (Ford)*

Some transaxles also use a ring and pinion for gear reduction. When the engine is transverse (sideways) mounted, they are two flat helical gears, as shown in **Figure 11-15.** These gears are lubricated by the transmission fluid. If the engine is installed in the conventional (front facing) manner, the ring and pinion are identical to those used on rear drive systems, and are lubricated by a separate reservoir of heavy gear oil. Some front-wheel drive transaxles have a ***planetary gearset*** to produce the final drive ratio. This gearset is installed in the transaxle, just before the shafts for

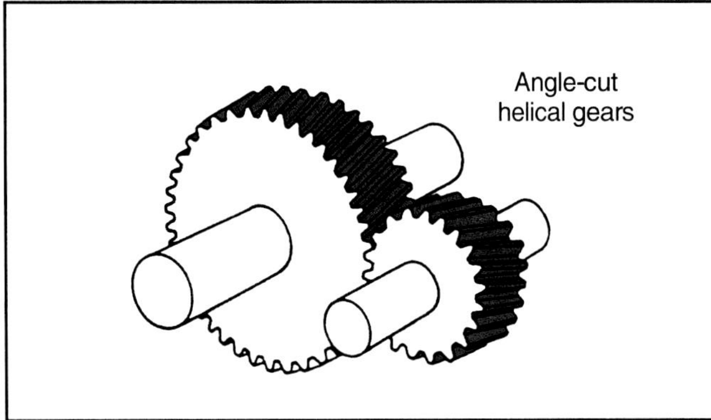

Figure 11-15. *Helical gears are used on all modern drivelines. They are cut on a slant to increase the gear contact area. This also makes them quiet and durable. (Deere & Co.)*

the drive axles. Planetary reduction provides the gear ratio. The gears are lubricated by the transmission fluid. See **Figure 11-16.**

Few driveability problems can be directly traced to the final drive ratio. If the complaint is low fuel economy or poor power, the technician should check that the right ratio is used on the vehicle. Some vehicles can be ordered with unusual axle ratios for trailer towing or highway driving, or a different ratio replacement rear axle or transaxle may have been installed. If the vehicle has a limited-slip or locking version of the differential assembly, wear, use of improper oil or lack of limited-slip additives can cause the differential assembly to make noise, buck, or shudder on turns. This problem will usually not be noticed during straight ahead driving.

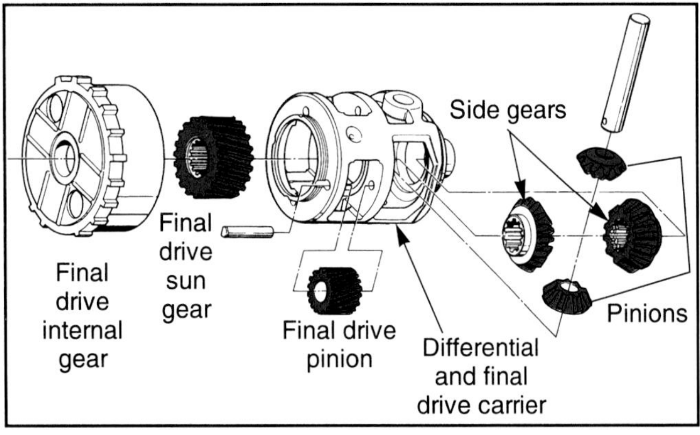

Figure 11-16. *Exploded view of a final drive assembly from an automatic transaxle. (General Motors)*

Driveline

The vehicle driveline consists of a single driveshaft on rear-wheel drive models, or two drive axles on a front-wheel drive vehicle. The rear-wheel driveshaft uses cross and roller type ***universal joints*** (usually called U-joints). The U-joints allow driveshaft angles to change as the rear axle moves up and down over bumps. A ***slip yoke,*** **Figure 11-17**, is used to

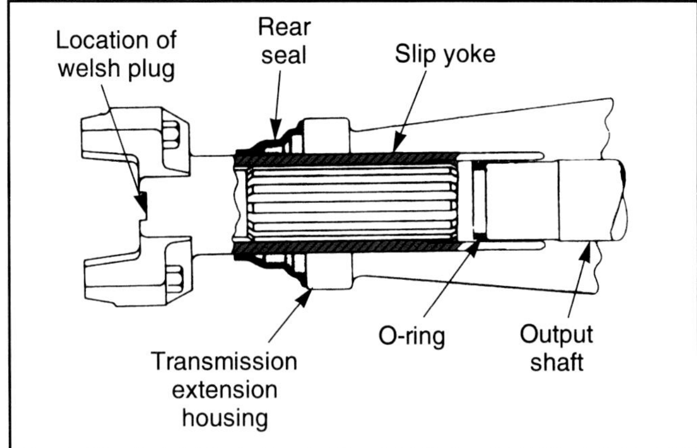

Figure 11-17. *The slip yoke allows the driveshaft to move in relation to the vehicle frame and differential. This prevents driveline vibration from unusual angles. Some driveshafts use two slip yokes. (Ford)*

allow the driveline length to change as the vehicle goes over bumps. One disadvantage of the U-joint is that it causes driveshaft speeds to fluctuate as the shaft is driven through an angle. Some driveshafts used on large vehicles have a *constant velocity universal* joint which consists of two conventional U-joints in a single housing. This reduces driveline fluctuations. Other driveshafts are two-piece units with three U-joints and a center support. Single and two-piece driveshafts are shown in **Figure 11-18.**

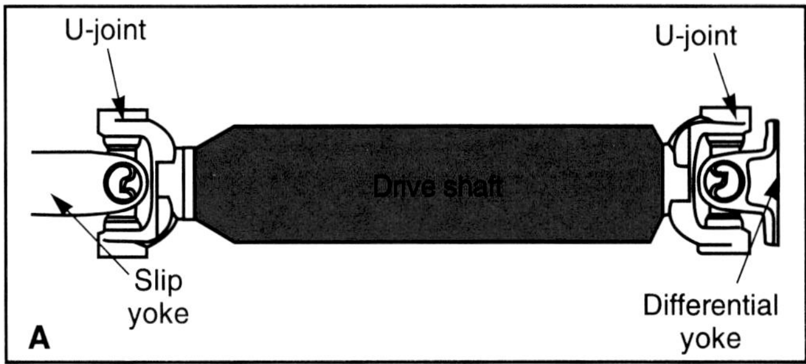

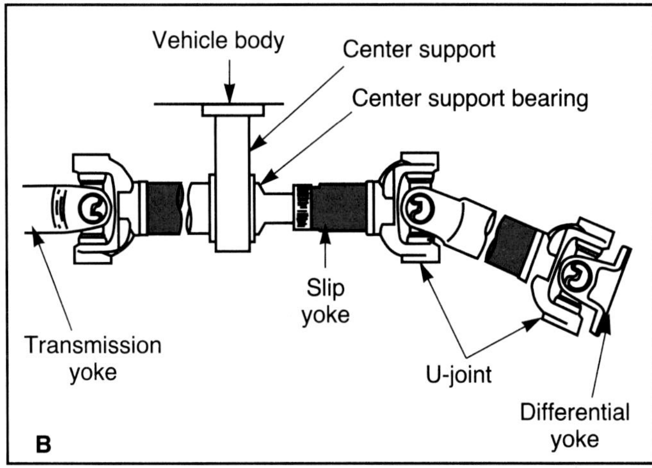

Figure 11-18. *A—A one-piece driveshaft is the most common. It can be made from iron, aluminum, or other materials. B—Two-piece driveshafts are commonly used on trucks and vans. The center support bearing is encased in rubber.*

Front-wheel drive axles use ***constant velocity joints,*** usually called ***CV joints.*** The use of CV joints allows drive axle angles to change with a minimum of vibration. This is because the CV joint design, unlike the U-joint, can transmit rotation with no speed fluctuation. The two major types of CV joints are the *Rzeppa joint* and the *tripod joint,* **Figure 11-19.** The CV joints are always sealed in a flexible boot. The boot keeps the joint lubricant from escaping, while protecting the joint from dirt and water. Some long CV axles are a two piece design with a support bearing.

Common driveability problems caused by the driveline are vibration and noises. These are almost always the result of worn U-joints or CV joints. The only fix for worn joints is to replace them. The driveshaft or drive axle may be bent if it has received an extremely heavy blow, such as running over a curb or other stationary object. A bent shaft will cause an obvious vibration that occurs at a much faster rate than a tire vibration. A bent drive shaft or axle should never be straightened. Always replace a bent shaft, or take it to a driveline service shop which can replace the bent section.

Worn front-wheel drive CV joints will usually make noises on turns. A worn outer or inner CV joint will usually make a clicking or knocking noise. A severely worn outer CV joint may vibrate or shudder, especially on turns. A worn inner CV joint will vibrate under engine load. If the U-joint angles of a rear-wheel driveshaft are excessive, the shaft will vibrate even when the U-joints are in good condition. Angles do not usually become excessive unless the transmission support mounts have collapsed, broken, or the mounting frame has been bent by a collision.

Air Conditioning and Heating System

The ***air conditioning system*** can affect driveability due to the load that the compressor places on the engine. Many late model compressors are designed to turn on and off (cycle), possibly causing complaints of rough operation or surging. This section will briefly explain the basic principles of the air conditioning system, and describe the common components of all air conditioning systems. Since all automotive air conditioners work according to the same physical and mechanical principles, this description will be applicable to air conditioners on all makes of vehicles. The air conditioner compressor is at least partially operated by the computer on many late model cars, so compressor problems can affect overall engine operation.

Basic Air Conditioning Cycle

All automotive air conditioners use a chemical compound called ***refrigerant.*** Older vehicles used ***R-12,*** which is also referred to as CFC-12, or dichlorodifluromethane

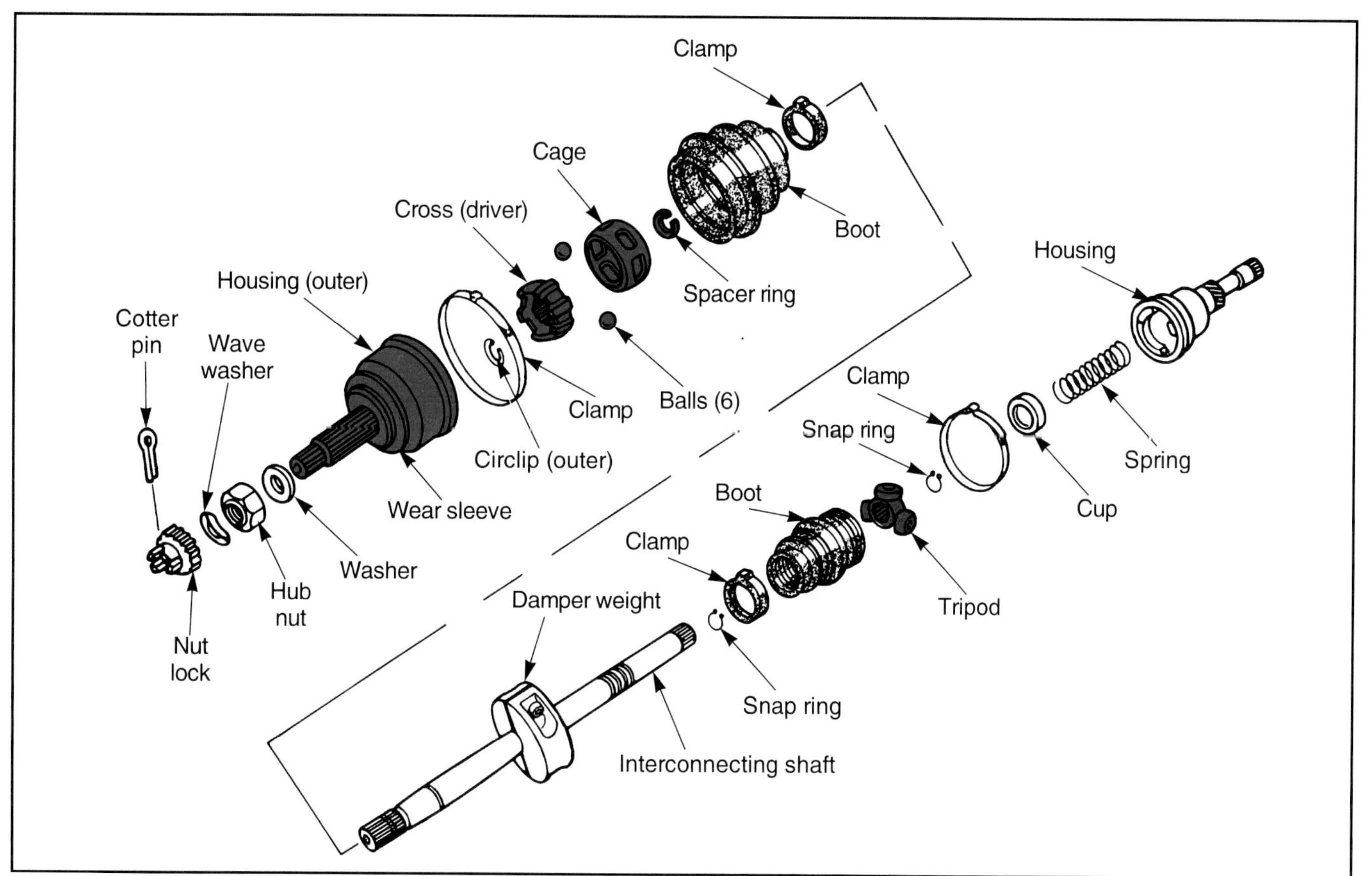

Figure 11-19. *Exploded view of a constant velocity (CV) axle assembly. Torn boots and worn joints from leaks are the most frequent causes of failure. (Chrysler)*

(CCl_2F_2). R-12 is no longer produced since it was determined that it contributed to the depletion of the earth's ozone layer. Most vehicles manufactured since 1992 use ***R-134a,*** which is also known as HFC-134a, or tetrafluroethane (CF_3CH_2F).

Changes in temperature and pressure cause refrigerant to change from a gas to a liquid and back again. Changing the pressure of the refrigerant allows it to pick up heat from the vehicle interior and release the heat to the outside air, even when inside and outside air is the same temperature.

At the beginning of the air conditioning cycle, liquid refrigerant is passed into a closed container called an ***evaporator.*** The evaporator pressure is low, and the refrigerant evaporates (changes to a gas). A fan forces air through the evaporator. As the refrigerant evaporates, it absorbs heat from the air. Moisture in the air condenses on the cold evaporator surfaces, and drips out under the vehicle. The fan forces the cooled and dried air into the passenger compartment.

The refrigerant is then drawn out of the evaporator by the ***compressor***. The compressor is a high pressure belt-driven pump. The compressor raises the refrigerant pressure and sends it to the ***condenser.*** Air is forced through the condenser by the engine fan, or by ram air when the vehicle is moving. This airflow removes some of the heat from the refrigerant and reduces its temperature. Since the refrigerant is under high pressure, the temperature at which it will condense (change from gas to liquid) is now higher than the outside air temperature. The refrigerant vapor condenses back into a liquid and gives up more heat.

The basic air conditioner cycle is shown in **Figure 11-20.** The system contains other parts which improve operation. The ***expansion valve*** is a device that allows the proper amount of refrigerant into the evaporator. Some systems use a ***fixed orifice tube*** to control the flow of refrigerant. The ***receiver-dehydrator*** or ***accumulator*** holds extra refrigerant and contains a desiccant (drying agent) which removes moisture from the system. Many older air conditioners use an additional valve that controls the refrigerant pressure in the evaporator. This valve allows the evaporator to provide maximum cooling without freezing the moisture that

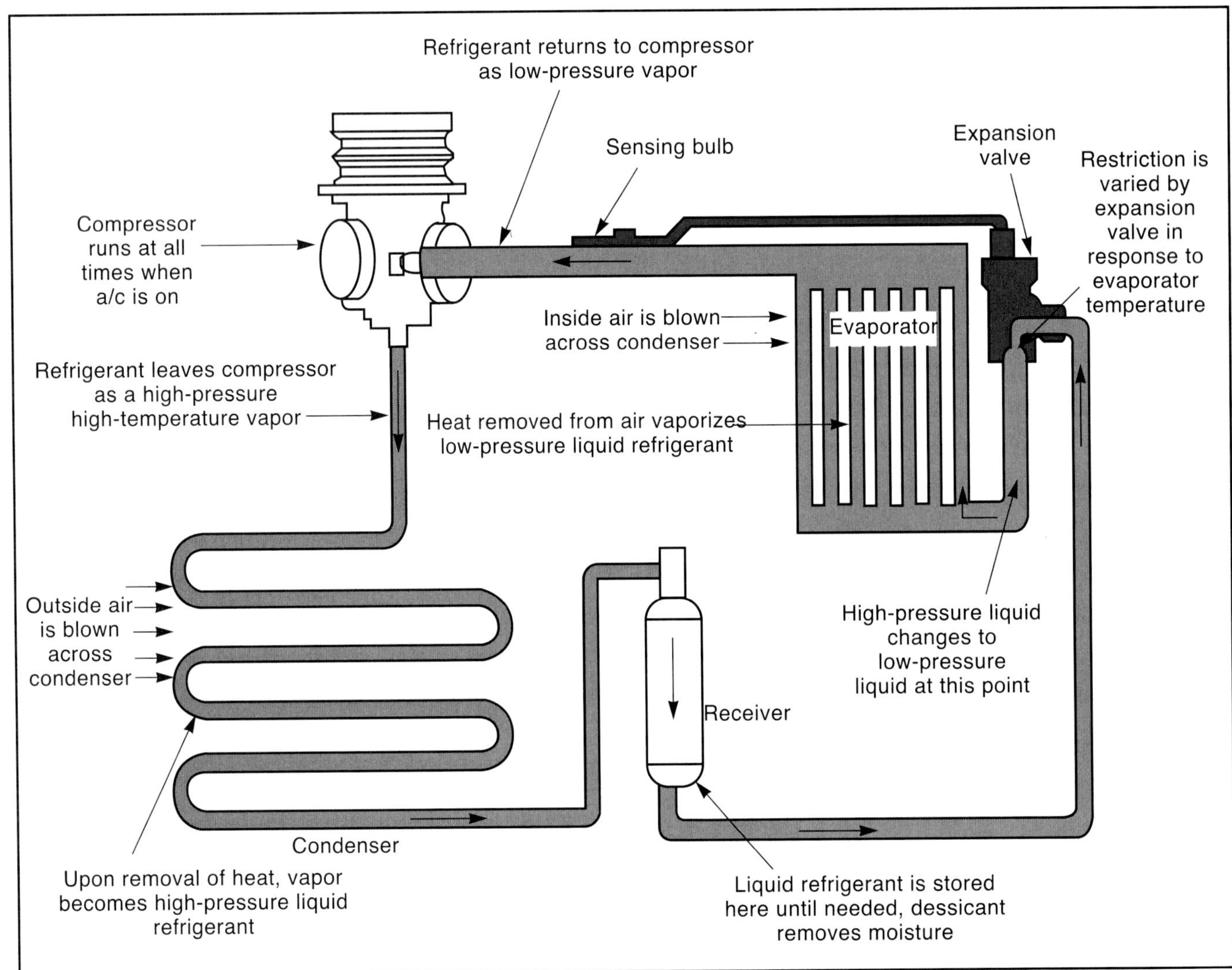

Figure 11-20. *Some systems use a compressor that is engaged whenever the system is operating. Follow the flow of refrigerant through the system.*

collects on it. The liquid refrigerant is then passed through the evaporator to begin the cycle again.

Newer systems, such as the one shown in **Figure 11-21,** control evaporator pressure by cycling the compressor, or by using a ***variable displacement compressor.*** Compressor cycling is accomplished by turning the compressor clutch coil on and off with a thermostatic or pressure switch on the evaporator outlet line. Variable displacement compressors are controlled by an internal valve that reduces compressor capacity when the evaporator pressure falls below a certain point. Variable displacement compressor clutches do not cycle on and off. Running the compressor at all times, however, places a greater load on the engine.

Computer-Controlled Air Conditioning

The air conditioning system is monitored by most computer systems for operation. When the air conditioning system is engaged on vacuum systems, a voltage signal is sent to the engine control ECM. The ECM, in turn, increases engine idle speed and makes other adjustments as needed. These systems contain the same basic components as a vacuum system, except that the control systems are operated by the ECM.

Newer computer control systems also control clutch operation. This allows the ECM to better manage compressor operation for maximum fuel economy. For example, it will delay compressor engagement for 1-3 seconds when the engine is first started to prevent stalling. The computer will also disengage the compressor clutch when the throttle is pushed to the wide open position. If the ECM detects a possible engine overheating condition, it will disengage the clutch to decrease the load on the engine.

Some newer systems and all OBD II equipped vehicles have a ***low refrigerant pressure sensor*** monitored by the ECM. The ECM will prevent compressor clutch engagement, and in some cases, set a diagnostic trouble code and illuminate the MIL lamp if low system pressure, which is an indication of a refrigerant leak, is detected.

Vehicle Heaters

The passenger compartment ***heater*** uses engine coolant as the heating medium. Coolant hoses are attached to the engine cooling system so that engine coolant pump operation forces coolant through the heater core. The heater core is a small radiator which gives up coolant heat as air is

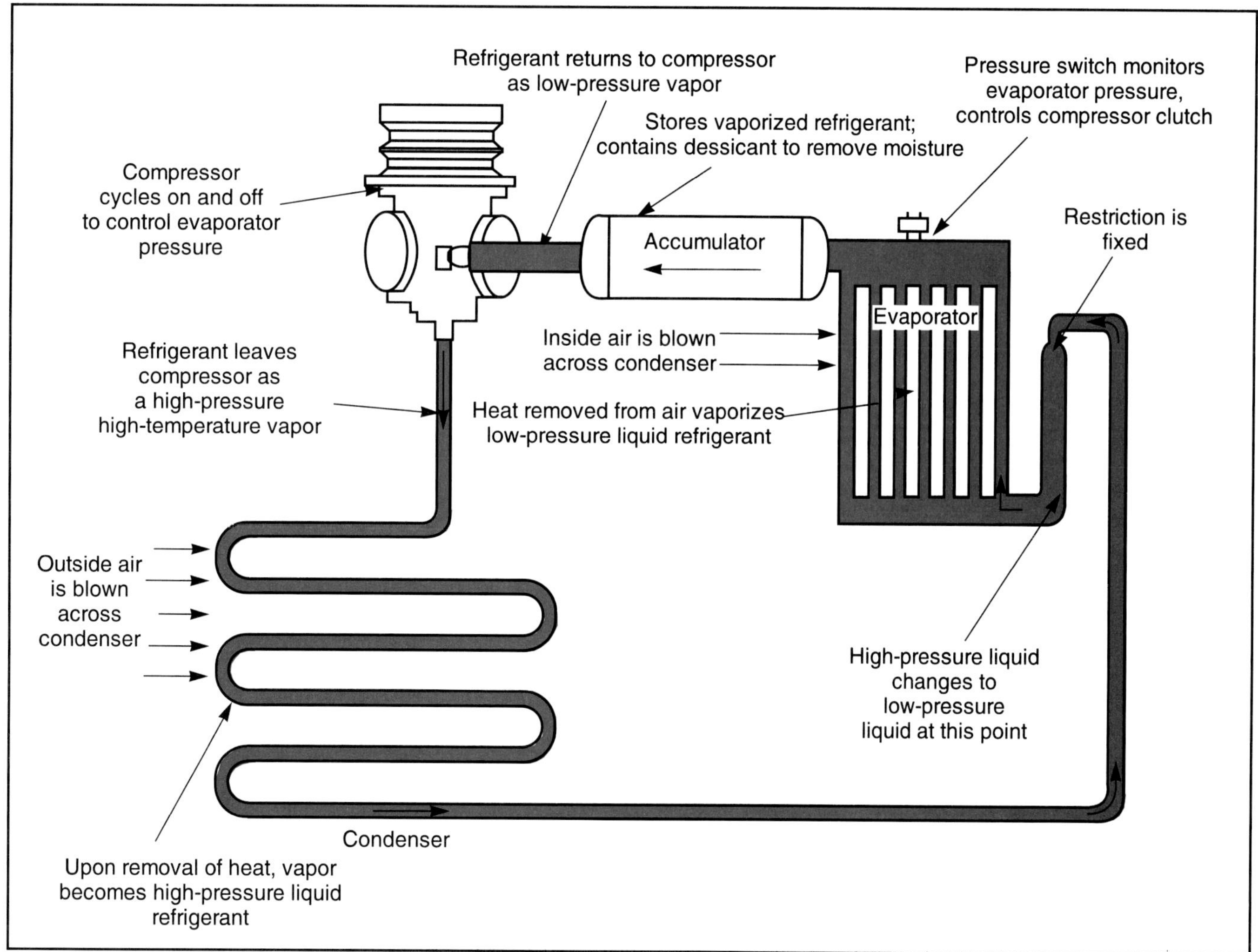

Figure 11-21. *Most systems are the cycling clutch-type. Air conditioning output is controlled by turning the compressor on and off.*

forced through it by the blower motor. Most late model heater systems have a heater shut-off valve in the inlet hose. This valve shuts off coolant flow when heat is not needed. The valve is usually operated by engine vacuum. A few shut-off valves are cable operated. **Figure 11-22** shows a vacuum-operated heater shut-off valve.

The most common problems caused by the heater system are coolant leaks and lack of heating. Leaks will usually be noticed as wet carpets, a film on the windshield, or a smell of coolant in the passenger compartment. Lack of cooling can be caused by a defective shut-off valve or by a clogged heater core. About the only other problem caused by the heater may be a complaint of a mysterious thumping noise as the heater shut-off valve opens or closes.

Figure 11-22. *The heater valve controls the flow of coolant into the heater core. It is also a source of coolant and vacuum leaks.*

HVAC Controls and Ductwork

The cooled or heated air must be distributed to the passenger compartment where it is most needed. The control system consists of various doors which direct the air to the floor, dashboard vents, or windshield as needed. The control system also varies blower speed and the temperature of the air leaving the system ducts. Most of these systems are manually controlled, but automatic controls are increasingly being used. A common result of a driveability problem is a lack of vacuum to the vacuum motors which operate the control doors.

Symptoms are inability to select airflow positions (air will usually come out of the heater vent only), or switching of airflow positions under low vacuum situations, such as hard acceleration. Vacuum problems are more often caused by a leaking or disconnected vacuum hose than an engine problem. Sometimes a rough idle or other driveability problem will be caused by a vacuum leak in the control system. In many cases, locating and repairing a vacuum leak in the heater-air conditioner control system will be more time-consuming than any other type of driveability service.

Brakes, Suspension, and Steering

Most driveability problems are caused by defects in the fuel and ignition system, engine internal parts, and drive train components. Other relatively common causes of driveability problems involve the cooling system, charging system, and electrical accessories such as the air conditioner. All of these systems directly affect the engine and drive train. They are responsible for the creation and delivery of engine power.

In addition to these primary systems, a small but significant portion of driveability problems are caused by vehicle systems that are essentially unrelated to the operation of the engine and drive train. Brakes and front end parts can cause vibration and noise problems that can affect driveability. Even the vehicle tires can cause driveability complaints. Certain defects in any one of these systems can reduce fuel mileage. Although the chance of these systems causing problems that may be misidentified as an engine or drive train problem is small, they are often worth investigating.

Brakes

The vehicle braking system is composed of two major subsystems, the ***hydraulic system*** and the mechanical or ***friction system.*** The hydraulic system is made up of the master cylinder, wheel cylinders, caliper pistons, hydraulic lines, and control valves. This system develops the hydraulic pressure needed to actuate the mechanical system.

The friction portion of the brake system is composed of the brake pads and shoes, drums and rotors, return springs, brake pedal, parking brake parts, and adjusting mechanisms. These parts are used to develop the friction that causes the vehicle to stop. The friction portion of modern brake systems can be divided into two major classes, disc brakes and drum brakes. ***Disc brakes*** operate when the disc pads contact the surface of the moving brake rotor, **Figure 11-23**. ***Drum brakes,*** **Figure 11-23,** operate as the shoe is pushed outward by hydraulic cylinders to contact the inner surface of the brake drum. The wheel bearings and spindles are generally associated with the brake system also. Applying the brake pedal creates pressure in the hydraulic system. This pressure is used to apply the brake pads and shoes to stop the vehicle.

Warning: Brake linings, like clutch discs, may contain asbestos, which can cause cancer if inhaled. Follow all precautions when working with brake linings.

Brake problems that affect vehicle driveability usually occur in one of three ways: noise, vibration, and resistance. Brake noises will most often occur only when the brakes are applied. Brake noises that go away when the brakes are applied are caused by a wear sensor contacting the rotor, which is the result of worn brake linings. Glazed linings can also cause brake noises, and sometimes vibration. Vibration

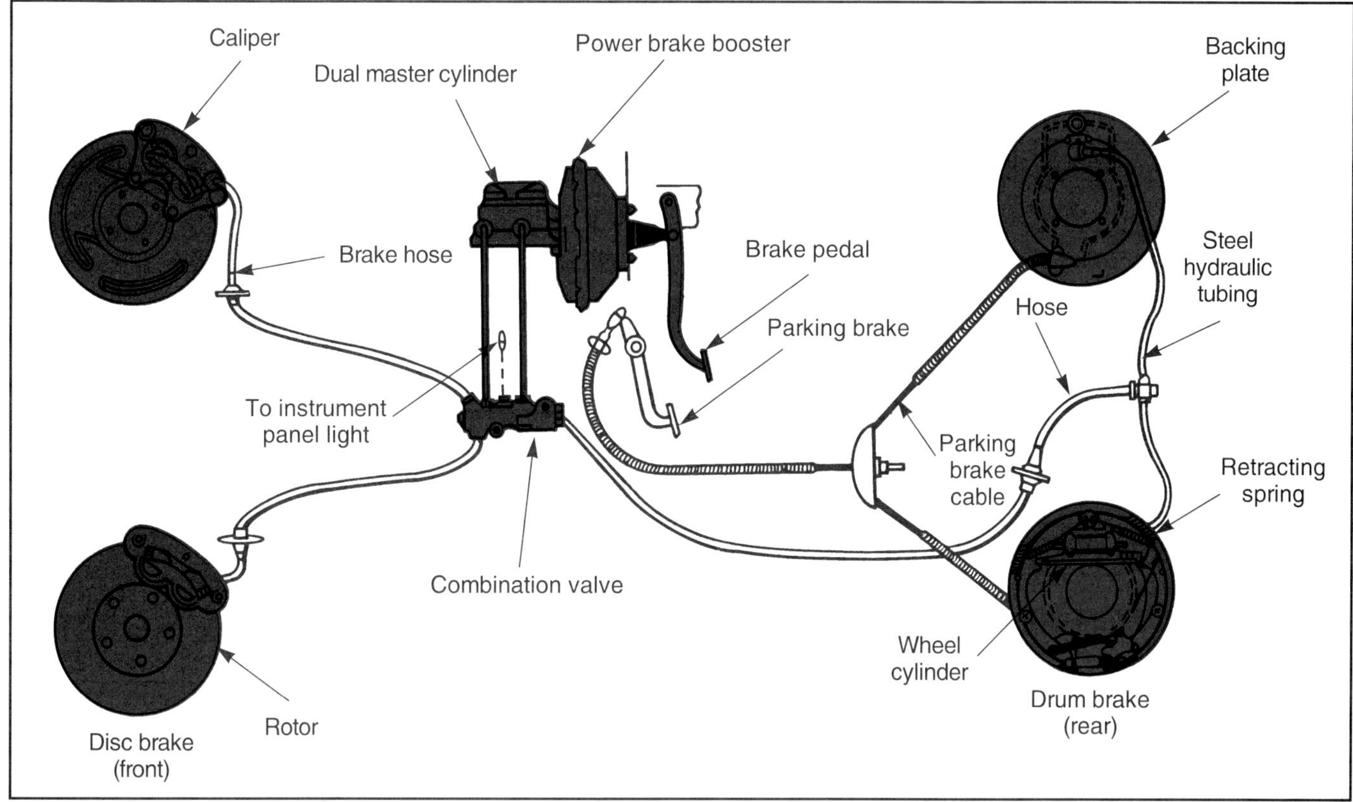

Figure 11-23. *The major hydraulic and friction components of the brake system are shown here. The disc brakes apply pads to a flat, turning rotor. The drum brakes contain wheel cylinders that push out on shoes that contact the inside diameter of a metal drum. (Bendix)*

when the brakes are applied is usually caused by an out-of-round rotor or drum, or by loose wheel bearings. This vibration will be noticed only when the brakes are applied.

Resistance caused by the brake system is usually called *dragging brakes.* Dragging brakes are partially applied at all times; sometimes caused by mechanical problems such as a sticking parking brake cable; misadjusted stoplamp switch, or by hydraulic system problems, such as swollen seals or defective control valves. Dragging brakes may not be noticed by the driver, but will reduce fuel mileage and performance, as well as prematurely wear the brakes.

Anti-Lock Brake and Traction Control Systems

In the past, drivers were advised to reduce the possibility of wheel skidding by "pumping" the brake pedal to keep the wheels from locking up. On newer cars and trucks, brake pumping is accomplished electronically by the ***anti-lock brake system, (ABS).*** The basic ABS system consists of a hydraulic control unit which controls the pressures to the brake calipers and/or wheel cylinders, an electronic control module which processes inputs and sends instructions to the hydraulic control unit, and wheel speed sensors which tell the electronic control module which wheel(s) are in danger of locking up. Some ABS systems also have G-force sensors. **Figure 11-24** shows the components of a typical ABS system.

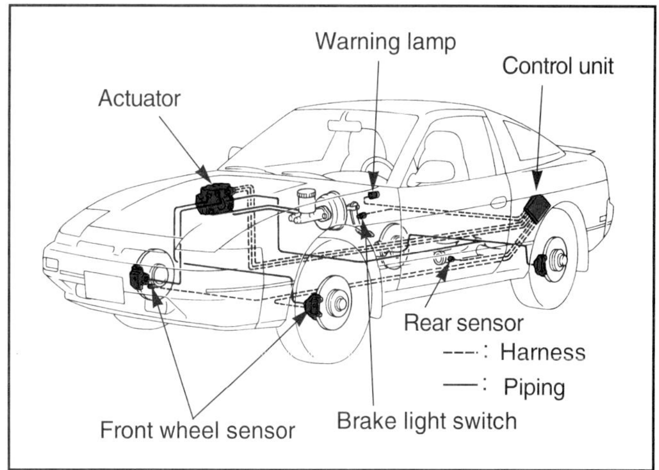

Figure 11-24. *The anti-lock brake system uses many of the same components as the base brake system. It will stop like a non-ABS equipped vehicle if a malfunction exists in the system.*

When the vehicle is braked normally, the ABS system does not operate. Under extreme braking conditions such as panic stops or stops on ice, the ABS electronic control module monitors the inputs from ***wheel speed sensors*** to determine whether any wheels are approaching the locked (no rotation) stage. When the ECM determines that a wheel is approaching lockup, it instructs the hydraulic modulator to seal off the brake hydraulic line to that wheel. If the ECM senses that further action is needed to prevent lockup, it instructs the hydraulic modulator to bleed off hydraulic pressure to that wheel. The hydraulic modulator performs these actions with a series of solenoid operated valves. Some modulator units also contain a small hydraulic pump to produce extra pressure if the bleeding process has reduced pressures beyond that provided by the driver's foot on the brake pedal.

Traction control systems (TCS) are similar in operation to ABS and often use the same components. The traction control system uses wheel speed sensors to determine when the wheels are losing traction during acceleration. It then uses the brake hydraulic system to apply the brakes on the wheel losing traction. Most traction controls also control the engine power output by reducing throttle opening or by retarding ignition timing. Some newer vehicles use the engine control ECM to monitor the wheel speed sensors. This type of ECM is sometimes referred to as a ***vehicle control module,*** (***VCM***).

Suspension Systems

The components of the ***suspension system*** are used to cushion road shocks and steer the vehicle. Springs, shock absorbers, struts, ball joints, and various bushings make up the suspension system. Modern suspensions can be the wishbone type, with upper and lower control arms, and a separate spring and shock absorber; or the MacPherson strut type, with the spring and shock absorber combined into one unit. Compare **Figures 11-25** and **11-26.**

Vehicle suspension systems contain many pivot points, such as ball joints, control arm bushings, and strut rod bushings. These parts can wear out and cause noise and handling problems. These have to be checked out carefully, since

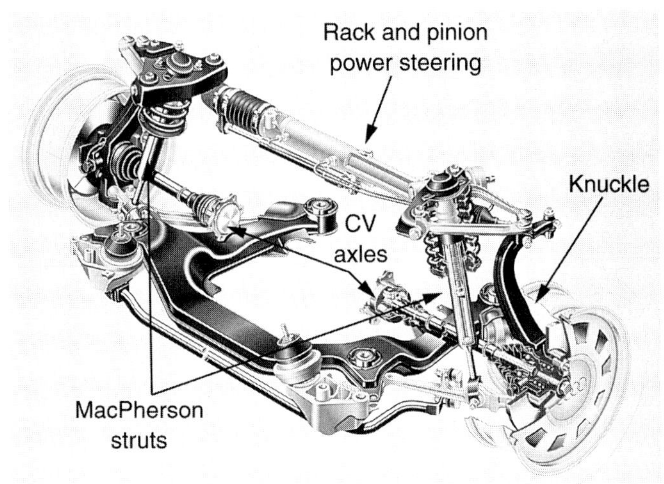

Figure 11-26. *MacPherson strut suspension systems have become the standard for most newer vehicles. The compact design reduces weight and allows for installation in a unibody frame. (Audi)*

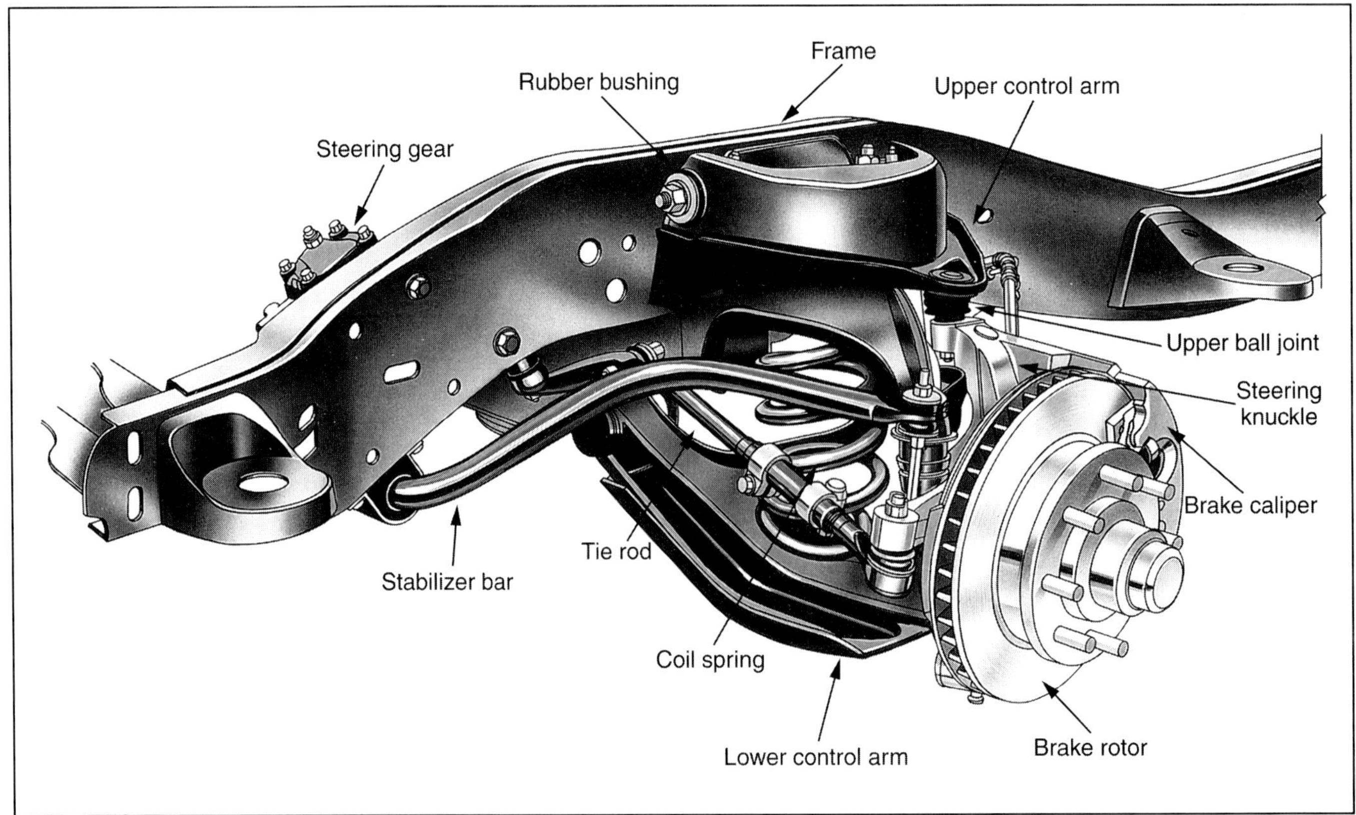

Figure 11-25. *The conventional suspension uses upper and lower control arms. The springs cushion the vehicle over rough surfaces. (Dodge)*

worn suspension parts can be a safety hazard. Note that some parts, such as ball joints, can be badly worn but not make any noise.

Shock absorbers and struts can wear out and lose their ability to cushion shocks and control spring oscillation. This is usually noticed as an excessively hard or bouncy, ride and leaning on turns. Springs can sag and affect ride height and the ability to absorb road shocks.

Some electronically controlled suspension systems consist of a vehicle mounted air compressor, rear-mounted air shock absorbers, and a computer control system. The control system adds or releases air from the shocks depending on ride height sensor input. Other systems affect ride by electronically varying the size of fluid orifices in hydraulic shock absorbers. Some systems have a provision for driver control of the ride firmness.

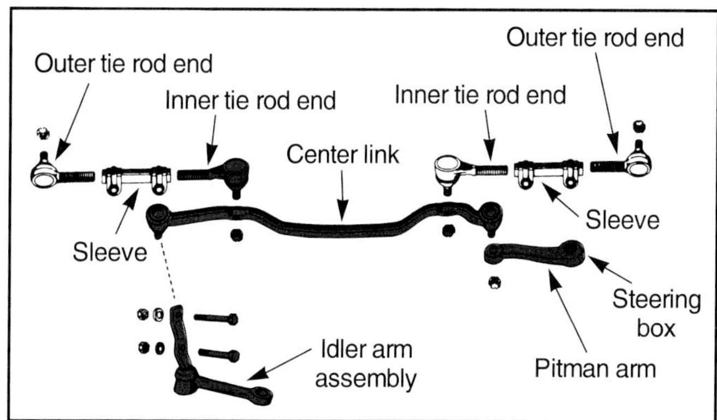

Figure 11-27. *Conventional suspension systems use a parallelogram steering system with a frame mounted manual or power steering gearbox. The type and number of components vary by vehicle. (Chrysler)*

Steering Systems

Many ***steering systems*** are the ***parallelogram linkage*** type, with a steering gearbox, drag link, tie rods, pitman arm, and idler arms, **Figure 11-27.** At one time, all steering systems were the parallelogram type. The steering gearbox is used much less than in the past, but can usually be found on light trucks, sport utility vehicles, and vans. The ***rack-and-pinion*** type, with a rack and pinion unit and tie rod ends is shown in **Figure 11-28.** This type of steering system is more common, even on larger vehicles. Both types of steering systems use spindles attached to the steering knuckles, as well as a steering wheel and steering shaft.

Power steering is often installed on late-model vehicles. It assists the driver, and is composed of a belt driven pump, a control and power unit, and hydraulic lines. The entire group of suspension and steering parts is often called the front end.

The most common driveability problems caused by defective front end parts are vibration and noise from loose parts. These defects will also cause handling problems and rapid tire wear. In most cases, front end misalignment will not cause vibration or noise, unless the misalignment is severe. The most common sign of misaligned wheels is steering pull to one side, or excessive or uneven tire wear. Another cause of pulling or uneven steering wheel feel is a defective power steering gearbox, or a loose and glazed power steering pump belt. A defective power unit in the gearbox or rack-and-pinion unit will also cause pulling.

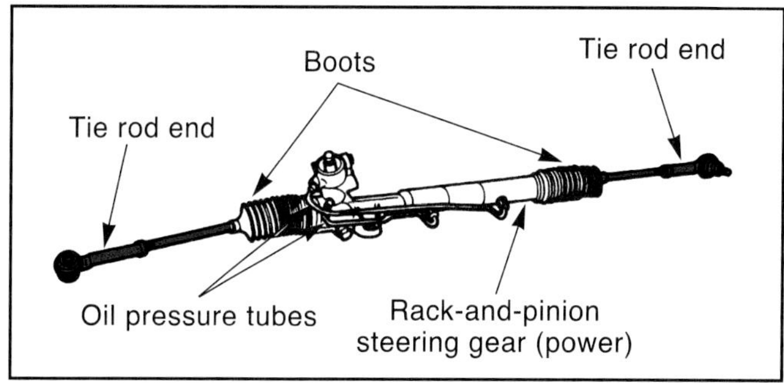

Figure 11-28. *Rack-and-pinion steering is used in almost all applications. The compact design reduces the number of parts that can possibly wear. The power steering assembly is enclosed inside the gear body. (Chrysler)*

Computer-Controlled Suspension and Steering Systems

The electronic steering system consists of a conventional steering system with an electronic flow control added. The computer receives a signal from a pressure switch installed on the power steering high pressure line. When the computer receives a signal from the pressure switch, it increases engine speed to compensate for the increased load of the power steering pump.

The electronic system is sometimes adjustable by the driver and operates by changing the size of a restriction in the hydraulic pressure system. Changing this restriction increases or decreases the flow of fluid between the power steering pump and the rack-and-pinion unit. Changes in flow translate into changes in steering effort. Some computer control systems use the pressure switch input to reduce engine load. When the wheels are turned at a severe angle, the ECM shuts off the air conditioning compressor and prevents the spark timing from advancing.

Steering and Suspension Alignment

The basic definition of ***alignment*** is the series of angles formed by the vehicle steering and suspension parts. These angles have a direct bearing on the steering and handling of the vehicle. Symptoms of misalignment include pulling to one side, road wander, poor recovery on turns, and tire wear. A common problem on front-wheel drive vehicles is a thumping noise in the rear of the vehicle caused by tire cupping. This is the result of incorrect alignment settings on the rear axle.

The vehicle may be brought in with an alignment problem mistaken for a driveability problem. Since the average customer has no idea how to tell an engine and drivetrain problem from an alignment problem, the technician must carefully determine the actual complaint. The driveability technician must learn how to locate such steering and alignment problems as wornout parts, undercarriage damage, sagging springs, loose wheel bearings, missing shims (when used), and misalignment.

Tires

Every automobile and truck is supported by four or more ***tires.*** The tires are the connection between the drive train, suspension and steering systems, and the road. Tires provide both traction and cushioning. Most modern tires are ***radial*** types for better traction and wear. A few ***bias-ply*** tires continue to be sold, mostly in the low-end grades.

In spite of, or perhaps because of their universal use, tires are the most neglected vehicle parts. Driveability problems caused by tires include vibrations and poor fuel economy. The most common cause of tire problems is underinflation and failure to periodically rotate and balance. Tire defects, such as belt slippage or cord separation, can cause vibration, pulling, and handling problems. See **Figure 11-29** for some common tire defects. Another common problem is the wrong tires being installed on the vehicle. Low profile tires (tires with a small aspect ratio or height) often reduce gas mileage and increase steering difficulty, while high profile tires with a small "footprint" (the area where the tire contacts the road) reduces both handling and traction.

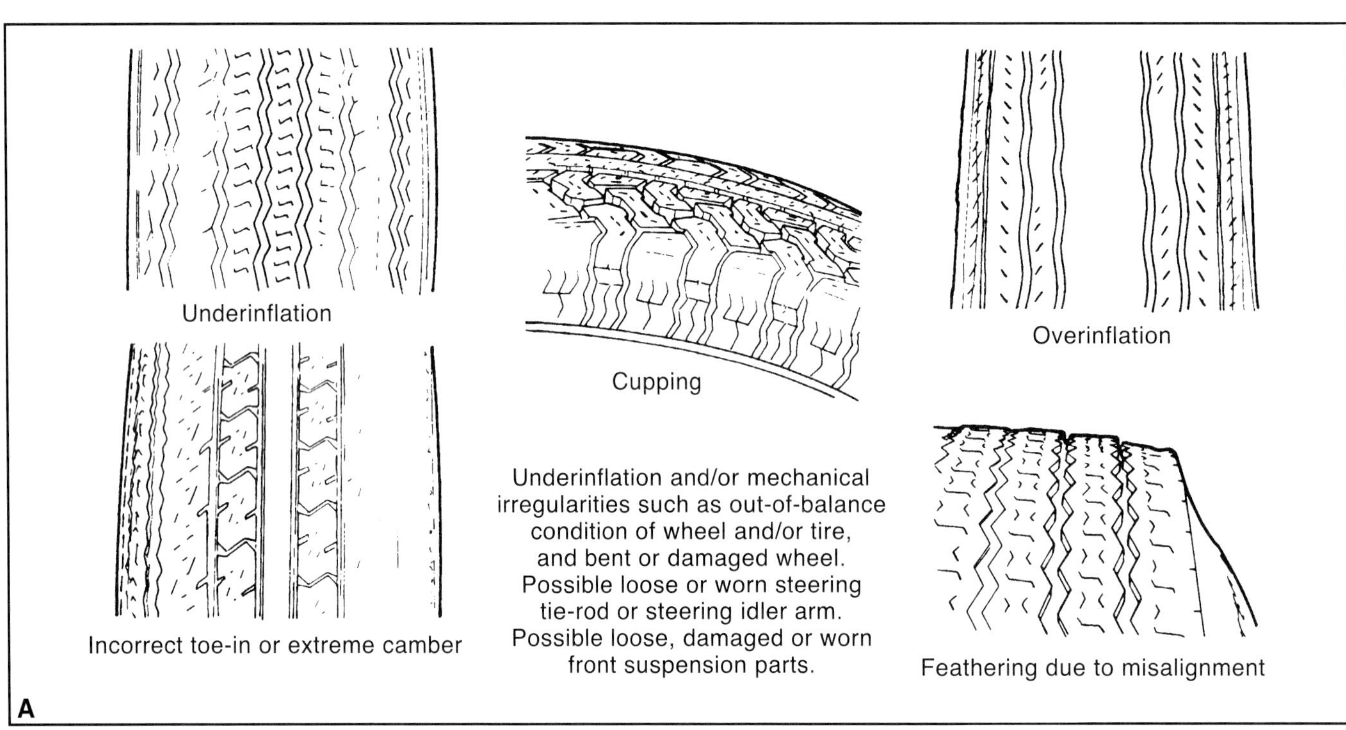

Figure 11-29. *A—Possible tire wear patterns and their causes. B—When tires are severely worn, they can cause a number of conditions, including tire separation and blowout. (Ford)*

Summary

Although most driveability problems are caused by defects in the fuel and ignition system, engine internal parts, and drive train components; some driveability problems are caused by vehicle systems that are essentially unrelated to the operation of the engine and drive train. Problems with the drive train components can have the most impact on the vehicle's driveability.

The manual transmission clutch and the automatic transmission torque converter perform similar functions, forming a connection between the engine, transmission, and drive train. The major effect the clutch or torque converter has on driveability is failure to transmit engine power, called slippage. Another component that can result in problems is the transmission, especially automatic transmissions that communicate with, or are controlled by, the ECM.

A defective clutch can slip. Usually, the clutch disc and other clutch parts must be replaced to eliminate the slippage. If grease or oil gets on the disc friction linings, the clutch may grab when applied. The torque converter can cause slippage, low top speed, and poor acceleration. Manual transmissions do not cause driveability problems in most cases. Usually the driver is using the wrong gear for a particular situation, changing gears too quickly, or engaging the clutch too quickly or slowly.

Automatic transmissions may cause driveability problems. Defects can occur in the hydraulic control system, or in the clutches and bands which are applied to provide different gear ratios. Late or early shifts, overly soft or harsh shifts, slipping when hot, and slow engagement from neutral or park can be confused with engine problems. Shift quality problems, such as rough downshifts or slipping upshifts, are often confused with engine driveability problems. In some cases, transmission problems such as slippage or improper shift timing can be caused by a problem in the engine or computer control system.

The final drive ratio is the speed reduction ratio caused by the final drive gears in the drive train. Few driveability problems can be directly traced to the final drive ratio. If the complaint is low fuel economy or poor power, the technician should check that the right ratio is used on the vehicle. If the vehicle has a limited-slip or locking version of the differential assembly, wear or the use of improper oil can cause the differential assembly to buck or shudder on turns. This problem will not be noticed during straight-ahead driving.

The air conditioner compressor places a drag on the engine when it is operating, and air conditioning defects can show up as driveability problems. Many late-model compressors are designed to cycle on and off, possibly causing complaints of rough operation or surging. Some new systems control evaporator pressure by using a variable displacement compressor, with a clutch that does not cycle on and off.

Brakes and front end parts can cause vibration and noise problems that can affect driveability. The vehicle tires can cause driveability complaints. Certain defects in any one of these systems can reduce fuel mileage. Although the chance of these systems causing problems that may be misidentified as engine or drive train problems is small, they are often worth investigating.

Know These Terms

Transmission
Transaxle
Drive train
Manual clutches
Single plate dry clutches
Flywheel
Input shaft
Friction disc
Pressure plate assembly
Throwout bearing
Clutch fork
Overdrive
Planetary gears
Torque converter
Automatic transmission fluid (ATF)
Impeller
Turbine
Stator
Lockup clutch
Torque converter clutch
Final drive ratio
Differential
Drive gear
Driven gear
Ring and pinion
Universal joints
Slip yoke
Constant velocity joints
CV joints
Air conditioning system
Refrigerant
R-12
R-134a
Evaporator
Compressor
Condenser
Expansion valve
Fixed orifice tube
Receiver-dehydrator
Accumulator
Variable displacement compressor
Low refrigerant pressure sensor
Heater
Hydraulic system
Friction system
Disc brakes
Drum brakes
Anti-lock brake system (ABS)
Wheel speed sensors
Traction control system (TCS)
Vehicle control module (VCM)
Suspension system
Steering system
Parallelogram linkage
Rack-and-pinion
Power steering
Alignment
Tires
Radial
Bias-ply

Review Questions—Chapter 10

Please do not write in this text. Write your answers on a separate sheet of paper.

1. The clutch friction disc is splined to the transmission ______ .
2. Bands are operated by hydraulic ______.

3. Holding one part of the gearset stationary while driving the other part with engine power is the power transfer method of the _____.
   (A) manual transmission
   (B) automatic transmission
   (C) ring and pinion type rear axle
   (D) differential assembly
4. When the pressure plate springs move the plate toward the flywheel, the clutch disc is:
   (A) applied.
   (B) released.
   (C) allowed to freewheel.
   (D) locked to prevent movement.
5. If grease or oil contaminates the clutch disc friction linings, the clutch may ______ when applied.
6. On late-model automatic transmissions, the application and release of the torque converter clutch is controlled by:
   (A) the ECM.
   (B) the driver.
   (C) engine vacuum.
   (D) Both A & B.
7. If a drive gear with 12 teeth is used with a driven gear having 48 teeth, the drive ratio is:
   (A) 2.5:1.
   (B) 3:1.
   (C) 4:1.
   (D) 1:3.
8. Many late-model air conditioning compressors are designed to turn on and off as they operate. What is this called?
   (A) Rotating.
   (B) Chopping.
   (C) Rectifying.
   (D) Cycling.
9. When refrigerant evaporates, it ______ heat.
10. Variable displacement air conditioner compressors are controlled by an internal valve which reduces compressor ______ when necessary.
   (A) speed
   (B) capacity
   (C) temperature
   (D) Both B & C.
11. The type of air conditioner which consumes the most engine power is the type where evaporator pressure is controlled by ______.
   (A) cycling the compressor clutch using a pressure switch
   (B) cycling the clutch using a temperature switch
   (C) using a variable displacement compressor
   (D) running the compressor at all times and using a valve between the evaporator and compressor
12. The vehicle braking system is composed of two major subsystems, the ______ system and the ______ system.
13. Applying the brake pedal creates pressure in the brake ______ system.
14. Dragging brakes are partially ______ at all times.
15. Tires are used to provide ______ and ______.

ASE Certification-Type Questions

1. On a hydraulically controlled automatic transaxle, shift valves direct fluid to the holding members based on pressure signals from the ________ valve.
   (A) governor
   (B) throttle
   (C) main pressure regulator
   (D) Both A & B.
2. All of the following statements about automatic transmissions and transaxles are true, EXCEPT:
   (A) planetary gears are in engagement at all times.
   (B) holding members are operated by hydraulic pressure.
   (C) bands and one-way clutches are holding members.
   (D) newer automatic transmissions have no hydraulic components.
3. Governor valve pressure increases as ______ increases.
   (A) input shaft speed
   (B) engine RPM
   (C) output shaft speed
   (D) Both A & B.

4. The final drive ratio is provided by ______ gears.
   (A) drive and driven
   (B) ring and pinion
   (C) planetary
   (D) All of the above, depending on the manufacturer.
5. Automatic transmission problems that can be mistaken for driveability problems include the following, EXCEPT:
   (A) late shifts.
   (B) overly soft shifts.
   (C) slipping when hot.
   (D) vibration and noises on turns.
6. Technician A says that the most common driveability problem caused by the front end is misalignment. Technician B says that the most common driveability problem caused by the front end is vibration and noise from loose parts. Who is right?
   (A) A only.
   (B) B only.
   (C) Both A & B.
   (D) Neither A nor B.
7. Technician A says that a common cause of tire problems is underinflation. Technician B says that a common cause of tire problems is failure to periodically rotate and balance tires. Who is right?
   (A) A only.
   (B) B only.
   (C) Both A & B.
   (D) Neither A nor B.
8. Variable displacement compressors are controlled by ______ .
   (A) the ECM
   (B) a pressure switch
   (C) an internal valve
   (D) calculated belt slippage
9. Technician A says that the vehicle alignment can cause a driveability problem. Technician B says that some driveability problems can be the result of tire defects. Who is right?
   (A) A only.
   (B) B only.
   (C) Both A & B.
   (D) Neither A nor B.
10. Technician A says that the ABS system operates when the vehicle is braked normally. Technician B says that most traction control systems use some of the vehicle ABS components. Who is right?
    (A) A only.
    (B) B only.
    (C) Both A & B.
    (D) Neither A nor B.

The engine compartment of a modern vehicle is a complex maze of wires, hoses, and parts. Locating problems in this maze is not difficult if you use patience and a step-by-step diagnostic process. (Ford)

Chapter 12

Basic Diagnostic Procedures

After studying this chapter, you will be able to:

- ❑ Define the concepts of strategy-based diagnostics.
- ❑ Identify and describe the seven steps of proper troubleshooting.
- ❑ Explain the importance of follow-up.
- ❑ Apply the seven step diagnostic procedure to specific driveability problems.

This chapter is intended as an overview of the techniques used by technicians to solve driveability and other problems. It will do this by outlining a strategic troubleshooting process of seven steps that will enable you to locate and correct almost any driveability problem. This chapter will also take you through several sample troubleshooting problems, showing how logical diagnostic procedures can be combined with an understanding of automotive systems to reach a solution. Studying this chapter will provide you with a valuable tool for finding solutions to driveability and other automotive problems.

Strategy-Based Diagnostics

In the past, it was fairly easy to find and locate a driveability problem, since most of the vehicle systems were simple and common to many, if not all vehicles. As vehicles became more and more complex, the methods used to diagnose them became obsolete and in some cases, inapplicable. Technicians who were used to using the older diagnostic routines, or no routine at all, began to simply replace parts, hoping to correct the problem, often with little or no success. Unfortunately, this process was very expensive, not only to the customer, but to the shop owner as well.

In response to this problem, a diagnostic routine involving the use of logical step-by-step processes to find the solution to a problem was devised for use by technicians. This routine is called ***strategy-based diagnostics.*** The strategy-based diagnostic routine involves the use of a logical step-by-step process, explained in the next sections. A flowchart of this process as recommended by one vehicle manufacturer is shown in **Figure 12-1.**

Seven Diagnostic Steps

The process of strategy-based diagnostics is a series of seven logical steps involving reasoning to reach a solution to a problem. This seven-step process, in the majority of cases, is the quickest way to isolate and correct a problem. The great advantage of this process is that it will work in the troubleshooting of *almost any vehicle system.* Variations of strategy-based diagnostics are used in many fields outside of automotive repair. **Figure 12-2** shows the seven-step process as it will be used in this text. Refer to **Figure 12-2** as you read the following sections.

Step 1—Determine the Exact Problem

The first step is to determine the exact problem. This means finding out the nature of the problem, what its

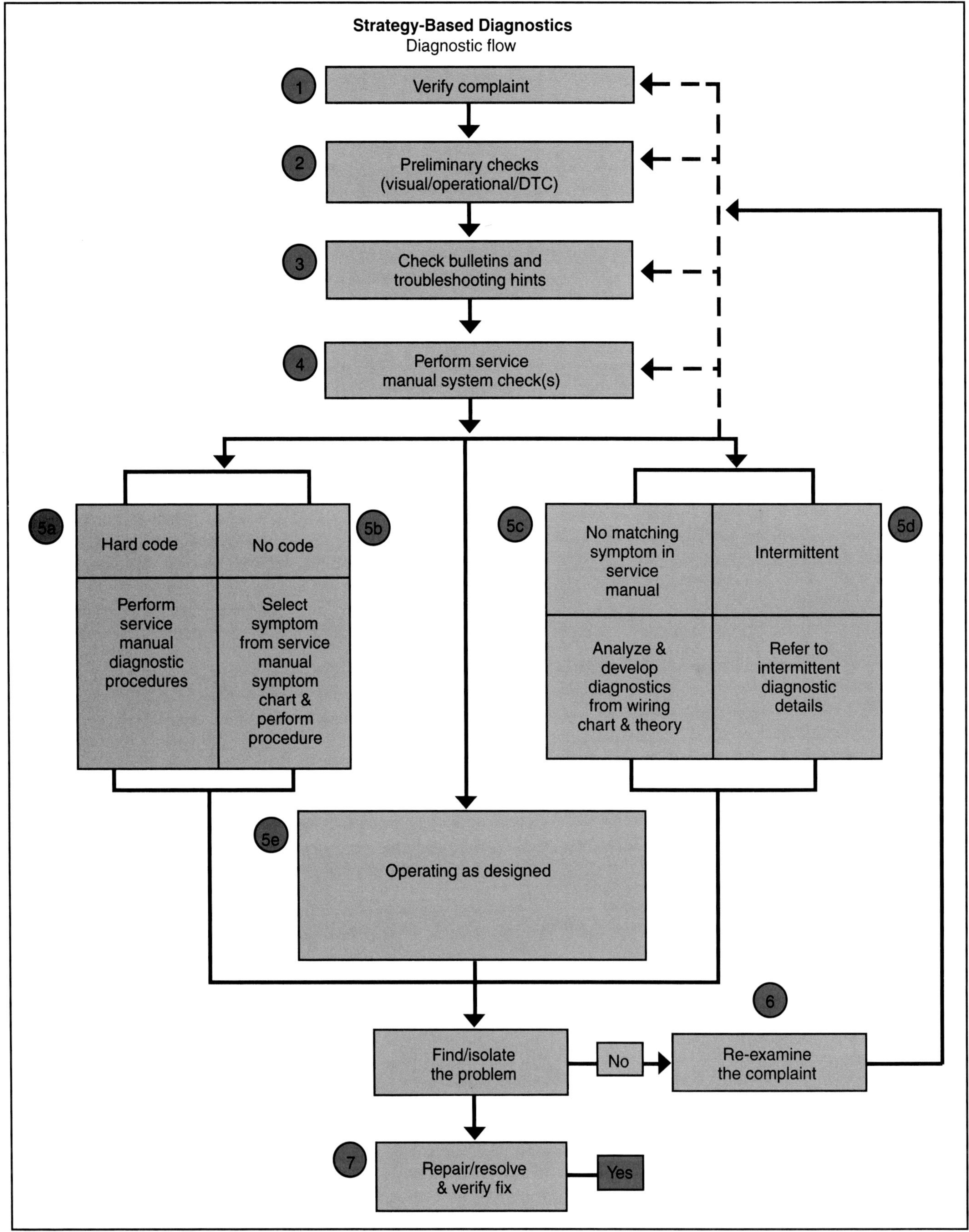

Figure 12-1. *Flowchart of the strategy-based diagnostic process as recommended by one manufacturer. Each manufacturer's recommended diagnostic process will vary slightly. Note the seven steps as they are shown in this chart. (General Motors)*

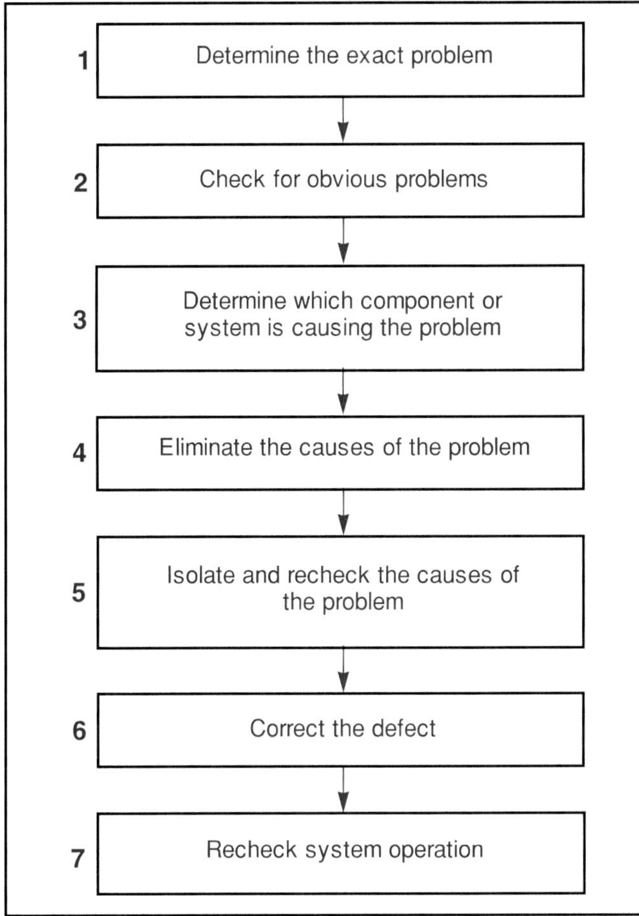

Figure 12-2. *The seven-step process shown here will be used in this chapter as well as the diagnostic and repair chapters to follow in this text. Following this process will save time in diagnosing a driveability or any other automotive problem.*

symptoms are, and how it affects the operation of the vehicle. This is mostly a process of listening to the driver's complaints and translating them into commonly accepted automotive terms.

Sometimes, the driver's complaint is normal vehicle operation mistaken for a problem. A good example is the normal operation of traction control systems. Some traction control systems use a throttle control stepper motor to decrease throttle angle, when necessary, to compensate for wet or icy road conditions. Some drivers may perceive this throttle control by the traction control system as an intermittent loss of power. If you do not listen or dismiss the driver's description of the problem and simply start looking for a problem that is simply normal vehicle operation, you will become very frustrated in a short period of time.

A very important part of listening to the complaint is understanding what the driver is trying to describe. The driver that can accurately describe a problem and use proper descriptive terms is very rare, so be patient and listen. Many drivers will give you a general description of what the problem is. Sometimes, the driver will try to duplicate the problem using hand gestures, body language, or imitating a noise that he or she hears.

Another part of this step is road testing the vehicle to confirm that the problem exists and that it is really what the driver says it is. Try to take the driver on the road test with you, especially for complaints involving vibration and noise. This way, you can be sure that you are addressing the driver's concern.

During this step, you should locate the service manual, along with any other service literature at your disposal. Working on a modern vehicle without a service manual is like driving on a long trip without a road map. Even if you have worked on a particular type of vehicle many times, there is always something new or something you may have forgotten that will be described in the vehicle's service manual. A service manual is as vital a tool as any in your toolbox. If you have access to technical service bulletins, you should also refer to these. Sometimes, a specific or unusual complaint is addressed by these bulletins. Using service literature is covered in more detail in **Chapter 13.**

Step 2—Check for Obvious Problems

Step 2 is visually checking for obvious problems that can be easily tested. This search should concentrate on areas that are most likely to cause the problem. For example, detonation or preignition is not going to be caused by the starter motor. A no-crank condition is not going to be caused by a defective spark plug wire.

The actual visual inspection begins with the road test and continues throughout the seven-step diagnostic process. Most visual checks take only a little time, and can often save a lot of time. Examples of this step are checking for a loose drive belt, disconnected vacuum hoses, or observing whether or not the dashboard malfunction indicator light is on. In many cases, these simple checks will reveal the problem, or the most likely place to start looking in Step 3.

Step 3—Determine Which Component or System is Causing the Problem

The third step is to set about determining which components or systems could cause the problem. This combines the information obtained in Step 1 with the knowledge of what automotive systems cause what symptoms. In a sense, this is a continuation of Step 2, but with simple and a few elaborate testing procedures. Simple tests include testing for fuel pressure, ignition spark, and injector pulse. Some of these tests are often performed by experienced technicians in Step 2. Elaborate procedures include such tests as exhaust gas analysis, oscilloscope testing, and scan tool testing.

This step is designed to eliminate systems and components that are operating normally and isolate those systems or components that are causing the problem. Instead of something obviously wrong, you are looking for something specific that could cause the problem. In performing this step, you will call upon techniques learned in other automotive courses along with those that you will learn in the remainder of this text. These steps will help you to eliminate components or systems that are not causing the problem, so that you can concentrate on finding the component or system that is causing the problem in Step 4.

Step 4—Eliminate the Causes of the Problem

In the fourth step, you begin eliminating possible causes of the problem, one by one. Always begin by checking the components or systems that are the most likely sources of the problem. This involves more specific system and component tests designed to narrow down the cause of the problem to a specific system or component.

This part of the diagnostic process starts at the most likely and easiest components to check. From there it continues on to the components that are less likely to cause the problem, and more difficult to check. For example, to check a no-cranking condition, you would first start at the battery and then eliminate the terminals, cables, starter solenoid, starter motor, alternator, neutral safety switch, and the engine itself as possible causes. This step is the most exhaustive, involving various, sometimes complex test procedures. You will eventually find the most likely suspect, to be further tested in Step 5.

Step 5—Isolate and Recheck the Cause of the Problem

In this step, the most likely cause of the problem is isolated and rechecked. This step requires reviewing the various test procedures that were performed in the last step, and determining whether the suspect component is likely to be the source of the problem. When you think you have the suspect part isolated, recheck the part using the specific test procedures outlined in the vehicle's service manual. The process of eliminating working parts in Step 4 should isolate most parts that cannot be tested, such as ECMs and most ignition modules.

Closely check and test the condition of the suspect part and review all the other possibilities to ensure that the wrong part is not identified. For example, if the cause of the no-cranking condition mentioned in Step 4 was isolated to the starter motor, you would perform additional tests to the starter motor itself to verify that it is the problem. In many ways this is the most important step, since it will give you confidence that your answer is correct, or lead you to check other possible causes.

Step 6—Correct the Defect

Step 6 is to correct the defect by making adjustments or replacing parts as necessary. If a part is adjustable, always try to adjust it before replacing it. If the part cannot be adjusted, or if it has a history of not maintaining an adjustment, it should be replaced.

It is important in this step to perform the repair properly, just as it was important to correctly diagnose the original problem. If you do not perform the repair in a proper manner, for example, replacing a defective fuel injector and failing to install an O-ring on the injector nozzle, you are taking the first step toward the vehicle coming back for a different and potentially more serious driveability problem.

Step 7—Recheck System Operation

The final step is to recheck system operation to determine whether the problem has been corrected. This is the most vital step, for it allows you to determine whether the previous steps accomplished the task of correcting the problem. Most times this involves road testing the vehicle again to verify that the complaint has been corrected. Sometimes, this step involves using some of the tests described earlier, such as performing an exhaust gas analysis when a vehicle has failed a state air quality test. If the problem has not been corrected, repeat Steps 1 through 6 until this step indicates that the problem has been fixed.

The Importance of Follow-Up

Once the seven-step diagnostic process has isolated and corrected the immediate problem, you will want to simply close the hood and get on to the next vehicle. However, it is worth your time to think for a minute and decide whether the defect that you found is really the ultimate cause of the problem.

For instance, if you were able to correct an engine miss by replacing a fouled spark plug, do not assume that the problem is corrected until you review the causes of spark plug fouling. If the plug itself is not obviously defective, then you must find out why it became fouled. One possibility is a leaking valve seal or worn rings that could allow excess oil leakage into the cylinder. If the oil leakage is not corrected, the new plug will soon foul out. Another possibility is that the plug may have been in the engine for so long that it can no longer fire properly. If the other plugs are not changed along with the fouled one, they will begin to fail also. In either case, the vehicle will be back soon, along with a dissatisfied driver.

Always try to find the ***root cause of failure,*** which is the actual cause of the problem, even when it appears to be simple. If a vacuum hose is loose, find out why. It may be too hardened or worn to stay on the fitting, or it may have been knocked loose by contact with a moving part or by a backfire through the intake manifold. Instead of just reattaching the hose, investigate the reason it became loose. Looking for the root cause of failure is important in order to avoid a comeback.

This process of making sure that the observed defect is the root cause of failure is commonly known as ***follow-up.*** Hidden causes of problems occur often and can cause a vehicle to return again and again with the same defective part(s). Some of these problems can be tricky, such as a dirty cooling system that causes coolant temperature sensors to fail again and again, or a rich mixture that causes several catalytic converters to overheat. This is where good observation skills and customer feedback can be helpful.

Documentation of Repairs

Part of the follow-up process includes writing on the repair order what the problem was, and what was done to correct the problem. This is called ***documentation*** and it is a vital part of the diagnostic process. Every repair order line should have three things. These three things are:

- ❑ What the driver's complaint was.
- ❑ The cause of the complaint.
- ❑ What was done to correct the complaint.

This type of documentation not only allows the driver to clearly see what was done to correct the vehicle's problem, it also supplies a good history of what has been done. If the vehicle should come back with a similar problem, it gives the technician working on the vehicle a place to start looking, without having to repeat some of the steps you took to find the problem. See **Figure 12-3.**

Using Strategy-Based Diagnostics to Diagnose Specific Problems

The following section will review examples of the strategy-based diagnostic process as applied to some specific problem areas. Some of these problems are similar to problems that you may have encountered in your past experiences, while others may be new. The symptoms mentioned are intended to aid in the diagnosis of the specific problem. Other symptoms are mentioned only if they will aid in finding the defect.

The problems presented here include hard starting, rough idle, stalling, hesitation, pinging, vibrations, intermittent problems, and poor fuel mileage. This section is not intended to be a complete list of every potential driveability problem. Instead it is supposed to indicate how the seven-step process of strategy-based diagnostics can be applied to troubleshoot specific driveability problems.

Note: Although driveability diagnosis might not seem to include finding causes of starting problems, the driveability technician will eventually be confronted with a vehicle having starting problems. Many of the same systems that cause performance problems can cause hard-start, no-crank, or no-start conditions. In addition, part of ASE Certification test A8, Engine Performance, contains starting and charging system problems.

Example 1—Cold Start Problem

Someone brings his car into the shop where you work. He says that he has trouble starting it in the morning. This vehicle is equipped with gasoline fuel injection, electronic ignition, and a computer control system.

Step 1

You ask the owner if it cranks slowly, or cranks okay but will not start. He says that it seems to crank fine, but does not start easily. Then you ask him if it cranks okay after it is warmed up. He says that it does. You ask if it has been doing anything else unusual lately, and he says that the engine also seems to hesitate until it warms up. Based on this information, you can assume that the battery and starter are okay, since they crank the engine well.

After a short test drive, you conclude that the engine runs well at normal operating temperature and that the malfunction indicator light stays off. You can, therefore, assume that the problem occurs only when the engine is cold, since it seems to operate well at all other times. Before starting a visual inspection, you locate the vehicle's service manual.

Step 2

Open the hood and check for obvious problems. In this case, you do not find any loose connections or hoses, signs of fuel leaks, or other problems.

Step 3

Think about what could go wrong with this engine. You can rule out most of the fuel, ignition, and compression system components, since they would cause the engine to run poorly whether it is hot or cold. Also, the majority of emission controls or other engine and vehicle systems will not affect engine starting.

Now you start thinking about the parts of the engine that can affect starting and cold operation. Spark plugs could be worn to the point that it prevents a good spark when the engine is cranking cold, while still delivering an acceptable spark after the engine starts and heats up. You notice in the service manual that this vehicle's fuel injection system has a cold start injector, which would only affect the operation of the engine when it is cold.

Step 4

In this step, you remove a few spark plugs and check their tips. In this case, the plugs look good, as though they have been recently replaced. A check with the driver reveals that the plugs were replaced about 5000 miles (8000 km) ago. This rules out the plugs, and you can now proceed to check the cold start injector.

To check the cold start injector, you must allow the engine to cool off for several hours, preferably overnight. Locate the cold start injector and remove the electrical connector, **Figure 12-4.** Using a high impedance test light, check to ensure that electric current is reaching the injector when the engine is cold. If the injector is receiving current, check that the injector coil has continuity by using an ohmmeter. This test reveals that the injector coil does not have continuity, therefore, it is defective.

Step 5

Before replacing the injector, you should reconsider what you have learned in Step 4. Is there any other problem that could cause this problem, or contribute to it? In this case, the answer is probably no. Since the cold start injector is defective, and a defective cold start injector could cause starting problems when the engine is cold, this is probably the cause.

Step 6

You obtain a new cold start injector and replace the old injector using the procedure outlined in the service manual. You relieve fuel system pressure before replacing the injector.

1501

Samples Auto Service
549 Cherry St.
Greenville, SC 24555

NAME	CUSTOMER'S ORDER NO.	DATE
John Ledbetter		9-22-96
ADDRESS: 9 Carrage Lane	ORDER WRITTEN BY	PROMISED: 9-22-96 A.M. / 5 P.M.
CITY, STATE, ZIP: Greenville, SC 24555		
HOME PHONE: 268-1234	BUS. PHONE: 555-1212 / EXT. 4325	ODOMETER: 43,126
YEAR, MAKE AND MODEL: Oldsmobile Cutlass Ciera '88		LICENSE NUMBER: JBX-999
SERIAL NUMBER: XY7322210B7	MOTOR NUMBER	TERMS: VISA

DESCRIPTION OF WORK	AMOUNT
☐ LUBE ☐ CHANGE OIL ☐ OIL FILTER ☐ TUNE-UP ☐ TRANS. ☐ DIFF.	
Check for poor acceleration	
No spark advance	
Replace knock sensor .5 hours	22 50
Replace knock sensor amplifier 1.0 hours	45 00

ESTIMATED COSTS

PARTS	LABOR	TOTAL
Knock sens	22.50	40.00

I hereby authorize the above repair work to be done along with the necessary materials. You and your employees may operate above vehicle for purposes of testing, inspection, or delivery at my risk. An express mechanics lien is acknowledged on above vehicle to secure the amount of repairs thereto. It is also understood that you will not be held responsible for loss or damage to cars or articles left in cars in case of fire, theft or any other cause beyond your control.

SIGNATURE ☒ RETURN PARTS ☐ DISCARD PARTS

REVISED ESTIMATE/ADDITIONAL WORK

PARTS	LABOR	TOTAL
K.S. amplifier	45.00	69.35

AUTHORIZED BY		
J. Ledbetter	☐ IN PERSON ☒ BY PHONE	
DATE: 9-22-96	TIME: 11:05	CALLED BY: Al	PHONE NUMBER

TOTAL LABOR	67 50
TOTAL PARTS	41 85
SHOP SUPPLIES	
GAS, OIL AND GREASE	0 0
SUBLET REPAIRS	0 0
EPA / WASTE DISPOSAL	0 0
	109 35
TAX	5 42
TOTAL	114 77

THANK YOU

QTY.	PART NO. AND DESCRIPTION	PRICE	
	Knock sensor	17 50	17 50
	K.S. amplifier	24 35	24 35
	TOTAL PARTS		41 85

(MAY BE CONTINUED ON OTHER SIDE)

REPAIR ORDER ______ DATE ______ JOB NUMBER ______ JOB NAME

Figure 12-3. *Good documentation of the repair is as important as diagnosing the problem and performing the repair properly. When documenting a repair, remember to include the driver's complaint, the cause of the complaint, and what was done to correct it.*

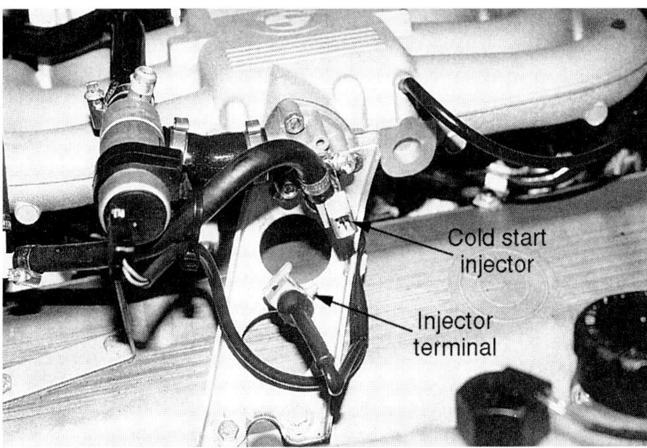

Figure 12-4. *A faulty cold start injector is frequently the cause of cold start and rich mixture problems on such equipped vehicles. Not all vehicles use cold start injectors.*

Step 7

Recheck your repairs. The best way to do this is to keep the car overnight, and see how it starts in the morning. If you still have a problem, repeat Steps 1 through 6 as necessary to locate the other problems.

Follow-up

In this case, there are no other systems that could cause the cold start injector to fail. In addition, there is evidence that the owner maintains the car properly. Therefore, follow-up should consist of reminding the owner to perform maintenance according to the manufacturer's schedule.

Example 2—Hesitation Problem

The owner of a pickup truck complains that it hesitates when accelerating. The truck's engine has a carburetor and an electronic ignition system. This truck has no on-board computer control system.

Step 1

You ask the owner if it has any other problems. She mentions that it is hard to start when cold, unless she pumps the gas pedal several times. Taking the truck on a road test, you verify the hesitation. After returning to the shop, you locate the appropriate service literature for this truck.

Step 2

Open the hood and check for obvious problems that might cause this condition. Keep in mind that it is very easy to confuse a miss with hesitation, so look for disconnected plug wires, or other problems that could cause a miss. In this case, you find no visible problems. Since the vehicle is not equipped with an on-board computer system, there are no codes to check.

Step 3

You think about what could cause the problem. The most common causes of hesitation on engines with carburetors are a bad accelerator pump, retarded timing, or worn out spark plugs. These conditions can also cause hard starting.

Step 4

You check the possibilities that you thought about in Step 3, starting with the easiest, in this case, the accelerator pump. You remove the air cleaner housing and operate the throttle linkage while looking in the air horn for a stream of gas from the accelerator pump, **Figure 12-5.** The stream seems to be weak and intermittent. This seems to be the most likely cause of the hesitation.

Figure 12-5. *Operating the carburetor while looking down the air horn can sometimes provide clues to the source of a fuel related driveability problem. Be sure the engine is not running when you do this, however.*

Step 5

Since the accelerator pump is working to some extent, you might want to check for other possible causes before deciding that it is the problem. You check the ignition timing and advance, and remove one or two spark plugs to check their condition. The checks show that the timing is correct and that the plugs are in good condition, but worn by high mileage. If you are still not convinced that the accelerator pump is bad, you could double-check for other ignition problems with an oscilloscope, and make a dynamic compression test to determine the mechanical condition of the engine. In this case, all other engine systems check out fine.

Step 6

After receiving approval from the owner, you overhaul the carburetor according to specifications in the service literature. You bench test the carburetor before reinstallation, making sure that the new accelerator pump and check balls produce a strong stream of gasoline.

Step 7

You recheck vehicle operation and it runs without hesitation.

Follow-up

In this case, the owner should be advised that the engine, while running fine for now, will require new spark plugs in the near future and possibly other maintenance. This example also illustrates the fact that there are still many vehicles on the road that are not equipped with on-board computer systems. While the seven step diagnostic process was intended to help in the diagnosis of today's complex vehicle systems, it is very useful in diagnosing older vehicles as well.

Example 3—Vibration Problem

A customer brings his car into your shop. He says that the engine seems to buck and shudder at highway speeds. This car is equipped with multiport fuel injection and a four-speed automatic overdrive computer-controlled transaxle.

Step 1

You ask the customer whether the problem occurs at any other speeds. He replies that it only occurs at highway speeds. Taking the vehicle for a road test, you verify that the vehicle does shudder badly at highway speeds. When you return to the shop, you note the speed at which the shudder occurred and locate the car's service manual.

Step 2

You open the hood and check for the sort of obvious problems that have been discussed earlier. In this case, you do not find any loose connections or hoses, signs of fuel leaks, or other problems.

Step 3

Think about what could go wrong. The problem would have to be in a system that only operates at highway speeds. An ignition, fuel, or compression problem would occur at other speeds, not just highway speeds.

You decide to take the vehicle for a second road test. During the road test, you notice that the shuddering stops when you gently apply the brake pedal or when you accelerate hard at highway speed. After pulling off the road, you shift the transaxle into third gear. When the vehicle is brought back up to highway speed, it does not buck or shudder. This would seem to indicate that the problem is drive train related.

Step 4

While driving back to the shop, you think about the operation of the overdrive gear, transaxle, and drive train. While an out-of-balance shaft would cause a shudder, it would do it all of the time, mostly under load. When the transaxle goes into overdrive, the torque converter clutch applies, locking the converter in direct drive, **Figure 12-6.** If the torque converter clutch was defective, it would not engage, stick in the released or engaged position, or slip in and out of lockup. A slipping converter clutch could be the cause of the shudder.

Following the manufacturer's service manual procedure, you first test the torque converter clutch solenoid, **Figure 12-7,** and find its resistance to be within acceptable range. By finding this information, you can say that the most likely location of the problem is the torque converter clutch itself, which is located inside the torque converter.

Step 5

Before declaring an expensive part such as the torque converter to be the problem, you should consider other possibilities. Since the torque converter solenoid is controlled by the ECM, which depends on input sensors, these systems should also be checked. However, in this case, the converter clutch solenoid was not found to be defective, and checks of other systems did not expose any problems. This would seem to indicate that the torque converter is the problem. Before calling the vehicle's owner, a transmission specialist should be consulted and additional tests performed on the torque converter and clutch solenoid to verify your diagnosis.

Step 6

Since the replacement of the torque converter requires transaxle removal, you turn the vehicle over to a transmission specialist after receiving customer approval.

Step 7

After the converter is replaced, you drive the vehicle and the transaxle operates normally with no shuddering.

Follow-up

This is an example of how a drive train problem can act like an engine driveability problem. While drive train problems are fairly easy to diagnose, they are often misdiagnosed as driveability problems. In this case, follow-up consists of a thorough road test to ensure that there are no other problems, which should be done anytime a drive train repair is performed.

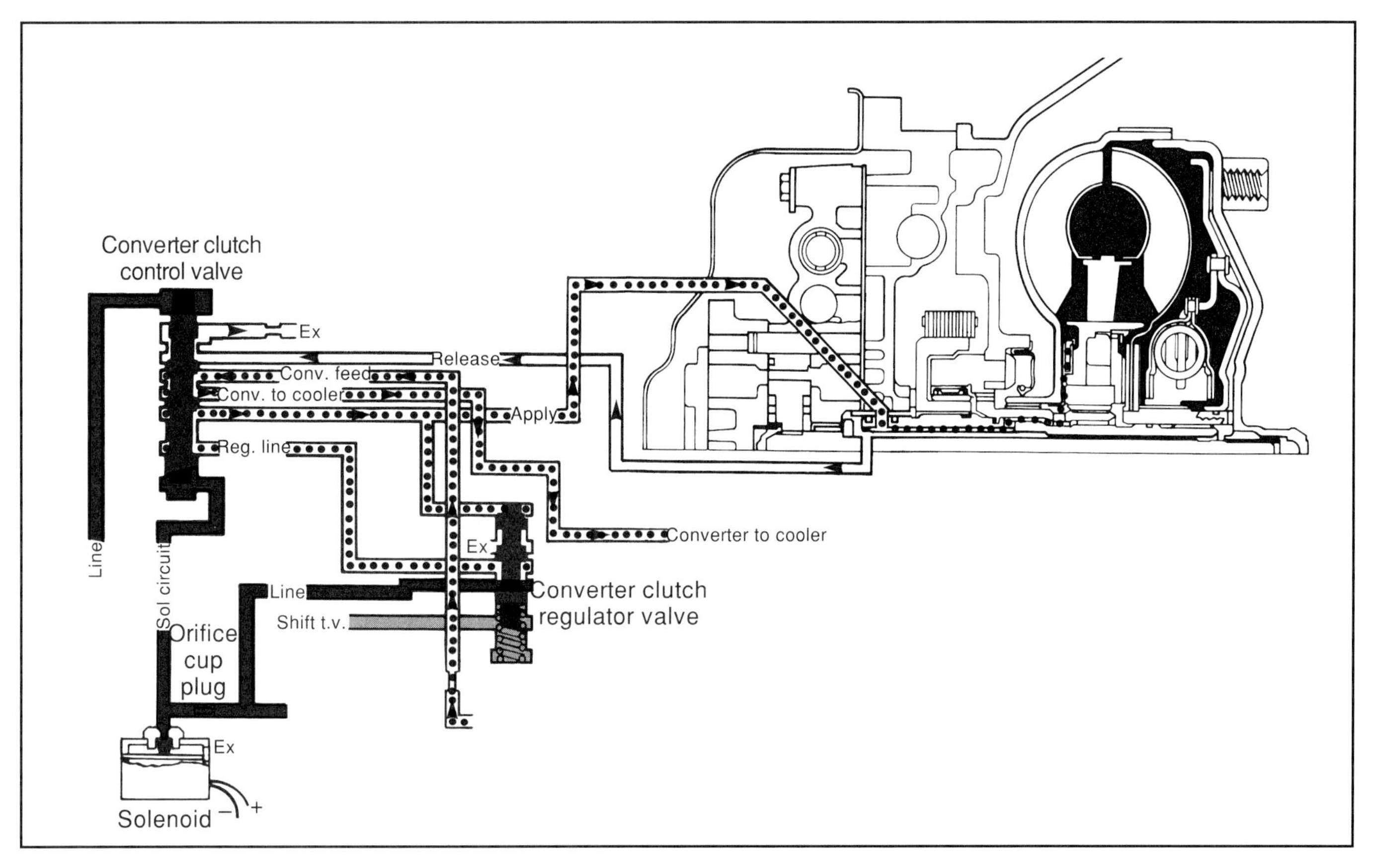

Figure 12-6. *The torque converter and converter clutch system is an unlikely suspect as the cause of a performance problem. However, they can cause problems, such as stalling, vibration, bucking, and poor fuel mileage. (General Motors)*

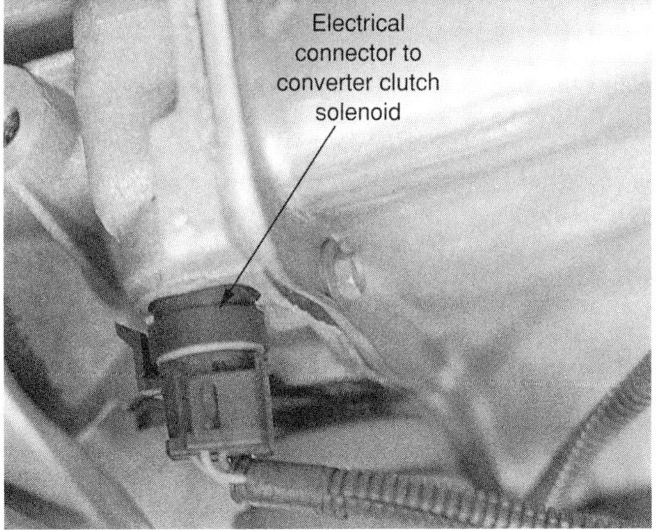

Figure 12-7. *Computer-controlled transmissions use internally mounted torque converter clutch solenoids. An external electrical connector, such as this one, is used to connect the internally mounted control solenoids to the ECM.*

Example 4—Stalling Problem

The owner of a front-wheel drive sedan complains that it stalls and fails to restart at times. The sedan has multiport fuel injection and distributorless ignition.

Step 1

You ask the owner if it has any other problems. She mentions that it usually stalls when coming to a stop and when turning corners. She says it often restarts immediately after it stalls, however, it sometimes will not crank after it stalls.

You road test the vehicle first. When you start the vehicle, the engine idle is high and unstable. The sedan stalls at the first stop you make, but restarts immediately. While making a sharp right turn, it stalls again. However this time, you notice that all of the electrical systems seemed to go dead when the engine stalled. Pulling off the road, you try to restart the engine. The engine will not crank on the first two attempts, but finally restarts on the third attempt. At this time, you notice that the sedan's digital clock has reset to 12:00 and all of the preset radio frequencies are gone. You decide to return to the shop before the engine stalls again. After you arrive at the shop, you locate the service manual for this vehicle.

Step 2

You have potentially three separate problems, or one problem causing all of the conditions. You think about what could cause the problem(s). The most common causes of a no-crank condition is a dead or undercharged battery, faulty starter, faulty starter solenoid, or transmission position switch. A stalling condition could be caused by a faulty ignition module, mass airflow sensor, or crankshaft sensor. The high and unstable idle could be caused by a misadjusted throttle position sensor, a sticking idle air control valve, or an internal ECM problem. While these components might cause one or more of these performance problems, they would not cause all of them. Open the hood and check for obvious problems that might cause all of these conditions. During the visual inspection, you find that the negative battery terminal is loose, **Figure 12-8.**

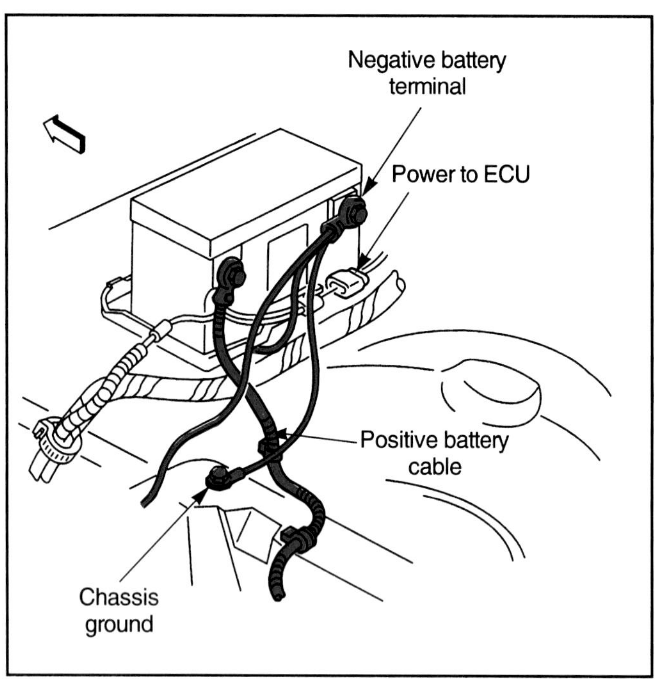

Figure 12-8. *The electrical system can be the cause of almost every possible driveability problem. If a vehicle has multiple driveability or system problems, a good place to start looking is the electrical system. (General Motors)*

Step 3

Think again about the problems as they occurred on the road test. Since no single malfunctioning component could cause all of the performance problems, this would indicate that the problem is in the electrical system. The loss of the digital clock time and radio presets is an important clue that indicates the existence of a loose connection in the electrical system, which could cause all of the problems.

Step 4

You check the negative battery terminal and cable again by twisting it. When the terminal and cable is twisted, electrolyte seeps from around the battery post and the underhood light flickers on and off.

Step 5

Since it is obvious that the vehicle is losing battery ground at times, this would account for the intermittent no-crank condition. It would also cause the other problems, including the unstable idle, since all of the ECM's learned memory is erased when ground is lost. This means that information stored in RAM or KAM, such as vehicle operating conditions, and most important, minimum idle speed control, is also lost.

On most engines, the ECM automatically resets minimum idle speed control the first time the vehicle is driven after power has been reconnected. However, some engines requires the idle speed control to be reset by the technician whenever battery voltage to the ECM is disconnected. If the idle speed control is not reset properly, it will cause the engine to idle high, erratically, and even stall when the throttle plate is closed suddenly, which happens when the driver quickly switches her foot from the accelerator to the brake pedal.

Step 6

Since the battery is leaking electrolyte, the negative battery cable is probably corroded internally. To ensure a good repair, you replace the battery and the negative battery cable after getting the okay from the owner. Since this engine requires the idle speed control to be reset when the battery is disconnected, you also do this at this time.

Step 7

You road test the vehicle. The idle is steady and the vehicle runs without stalling or losing battery power.

Follow-up

This is a case where a single, easy to overlook problem can cause more than one complaint. While a poor ground can cause relatively few problems in older vehicles, it can cause major problems in modern vehicles. When a leaking battery is replaced, you should also make certain that electrolyte has not damaged the battery tray or any chassis wiring.

Example 5—Rough Idle

A customer brings in an older minivan with a carburetor and electronic ignition. He says that it is idling rough and surging at cruising speed. He has decided that it has a bad fuel pump, and he wants you to replace it.

Step 1

You ask when the symptoms started. He says that it seemed to backfire through the carburetor one morning, and then it began idling rough. You ask if you can make a few checks before replacing the fuel pump, and he says okay.

Step 2

You look under the hood, and immediately spot a disconnected vacuum hose at the intake manifold,

Figure 12-9. This is an obvious problem, one that should be corrected before proceeding further. In this case, you can skip Steps 3 through 5 and go directly to Step 6.

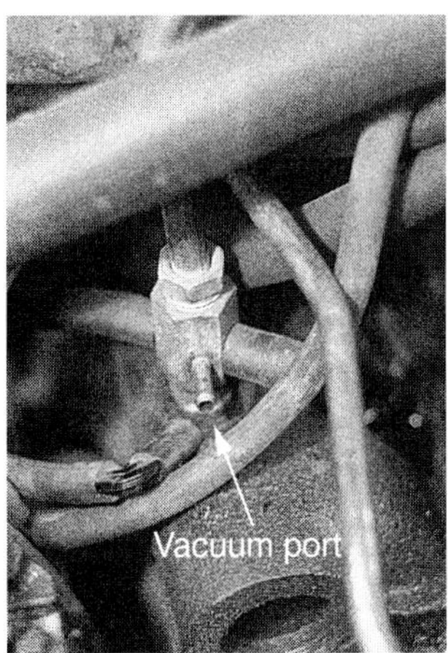

Figure 12-9. *Vacuum leaks, such as this disconnected hose, are simple problems that are frequently overlooked, even by experienced technicians. Always check the simple things first.*

Step 6

You reattach the hose, and ensure that it fits tightly on the manifold fitting.

Step 7

You recheck the engine. It now runs smoothly.

Follow-up

You have determined that the hose was the source of the rough idle and surging problem. However, something caused the hose to become loose. Since you already checked that the hose fit tightly on the manifold fitting, some outside force must have caused it to come loose. Remember that the owner stated that the engine seemed to backfire one morning. This is not a normal condition for any engine. Therefore, you should check the carburetor, choke, thermostatic air cleaner, and ignition system components for possible problems that could cause a backfire when the engine is cold.

Example 6—Computer Control Problem

A customer brings in a vehicle with a multiport fuel injected V-6 and an electronically controlled transmission. She says that the malfunction indicator light comes on sometimes, and the engine bucks and stalls.

Step 1

You ask if the engine is doing this all of the time. She says no, sometimes it seems to run fine. When you ask if she has noticed any other symptoms, she says that the transmission sometimes seems to shift too much, and that there is a slight hesitation on mornings when the temperature is relatively high. A short test drive verifies the customer's complaint. You locate the service literature before beginning the visual inspection.

Step 2

You look under the hood, and determine that nothing is obviously wrong with the engine. Next, you access the computer memory to retrieve the trouble code. The ECM has stored one trouble code, which indicates that the oxygen sensor is reading overly rich. Your first thought is that the sensor is defective. However, when you clear the computer memory and take a test drive, the code does not reset. This indicates one of two things: the oxygen sensor is intermittently defective, or something else is causing a rich mixture at the sensor.

Step 3

In this step, you need to consider all possible causes of the symptoms that you have determined so far: occasional rich oxygen sensor reading, along with occasional bucking, stalling, hesitation at some temperatures, and improper transmission shifting. Possible causes include defects in the fuel injectors, ECM, ignition system, or a compression system problem, such as a burned valve. The cause could also be a coolant sensor or airflow sensor that is out of range, but not badly enough to set a diagnostic trouble code.

Step 4

You proceed to check the fuel injection components, ignition system operation, and engine compression. Each system and all of their components checked good. Then you check the coolant temperature sensor and mass airflow sensor. The coolant sensor is in range for the present temperature of the engine, but the mass airflow sensor, **Figure 12-10,** does not check out good. You know from studying the service manual that a bad mass airflow sensor could cause the engine and transmission symptoms.

Step 5

Now that you have isolated the mass airflow sensor as the problem, rethink your original diagnosis. Could anything else cause the same symptoms, and what else could cause the fuel system to run rich? You know that the engine hesitation at some temperatures could be caused by an out-of-range coolant temperature sensor. After thinking about it for a while, you decide that the coolant sensor could be reading incorrectly at other temperatures. You then allow the engine to cool further, and recheck the coolant temperature sensor. In this case, it checks out as slightly off at a lower temperature.

Step 6

After getting the customer's approval, you replace the mass airflow sensor and coolant temperature sensor. You are

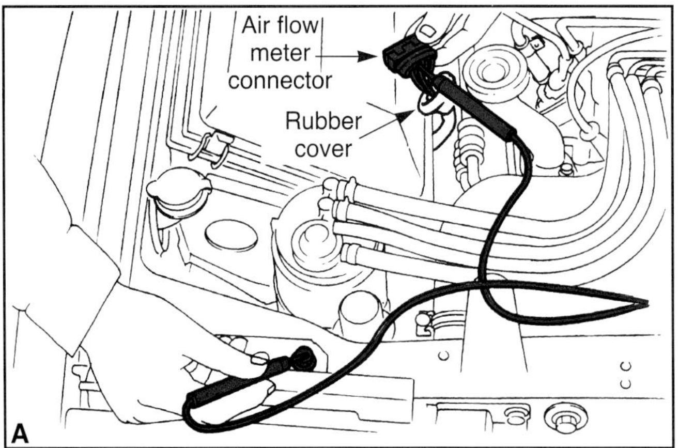

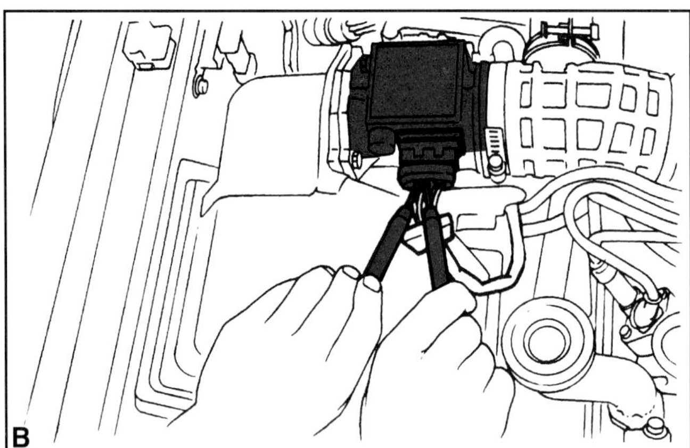

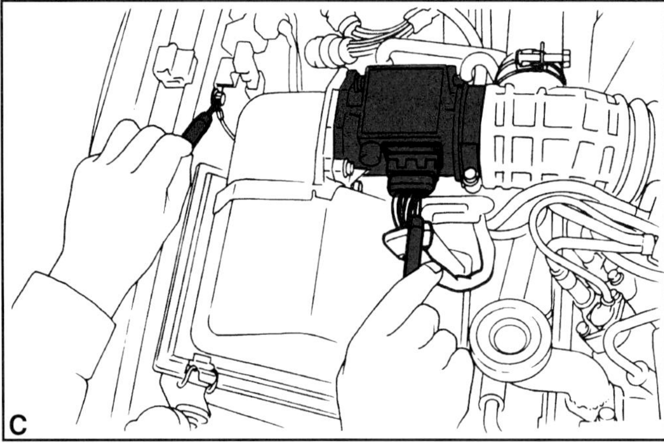

Figure 12-10. *The mass airflow sensor (MAF) is a relatively easy sensor to test and diagnose. A—Testing the connector for good power and ground. B—Testing the MAF sensor's output. C—Testing the sensor using a good body ground. Other tests will be discussed later in this text. (Nissan)*

careful to depressurize the cooling system before removing the temperature sensor. After removing the temperature sensor, you check it for deposits that could indicate a dirty cooling system. There are none in this case.

Step 7

You recheck the engine operation. The engine runs normally.

Follow-up

The follow-up procedure in this case does not involve making any more checks, since both sensors probably became defective without being affected by any outside factors. If you had noticed any cooling system deposits, it would have been a good idea to sell the owner on the advantages of flushing the cooling system and installing new antifreeze. Since this was an intermittent problem, the owner should be informed that the problem that she experienced could not be duplicated in the shop, and that she should bring the vehicle back if it exhibits any other problems.

Example 7—Spark Knock Problem

A vehicle with a computer-controlled engine is brought into your shop with a complaint of spark knock while accelerating. The owner says that it began "pinging" after another shop worked on it.

Step 1

You already have a solid lead in that the problem occurred after work was performed by another shop. You ask the owner why the first shop worked on it. He says that he brought it to that shop for a tune-up. When you ask what kind of fuel he uses, he says that he uses only premium, which has a higher octane than this engine requires. A short test drive verifies that the engine detonates badly in all transmission gears, and that the malfunction indicator light does not come on. After returning to the shop, you quickly find the vehicle's service manual.

Step 2

After locating the service manual for the vehicle, you open the hood and look for something obvious. In this case, nothing is out of place.

Step 3

You consider the conditions that can cause spark knock: low octane fuel, overadvanced timing, transmission upshifting too soon, defective knock sensor, or a disconnected or defective EGR valve. Since the engine is operated on premium gasoline, low octane should not be a problem. You also discount the transmission, since it seemed to shift at the right speeds when you drove the vehicle.

Step 4

You check the ignition timing and the operation of the knock sensor and both check out okay. You are able to verify that the timing is not overadvanced and that the knock sensor retards the spark when you tap lightly on the engine with a hammer. You check the EGR valve stem, **Figure 12-11,** and it does not move when the throttle is opened. Further examination reveals a steel check ball in the EGR vacuum line.

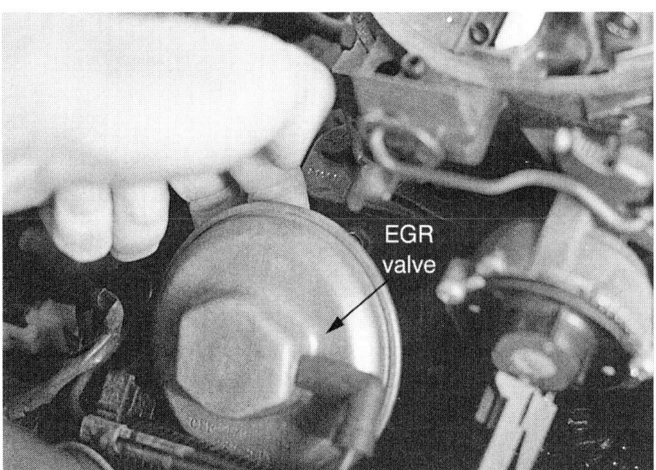

Figure 12-11. *Vacuum controlled EGR valves can be easily tested for movement by simply pulling up on the diaphragm. Electronic EGR valves cannot be tested in this manner.*

Step 5

You consider what you have learned so far. The disabled EGR valve is the most obvious problem. The other possibilities, such as timing and fuel, seem to be adjusted properly. In this case, it is obvious that the other shop disabled the EGR valve.

Step 6

You remove the check ball from the EGR valve line and reconnect it. You verify that the EGR valve stem moves when the engine is accelerated.

Step 7

You road test the vehicle, and discover that the pinging is now gone. However, the engine hesitates when first accelerated, indicating that the EGR valve was disabled to eliminate this problem.

Follow-up

To satisfy the owner and to find the root cause of failure, you should investigate the entire vehicle to determine and correct the actual problem. In this case, the EGR valve was disabled in an attempt to correct another driveability problem. This means performing the seven diagnosis steps again until the real problem is located. The owner should be cautioned against allowing the EGR valve or any other emission control device to be disabled or disconnected, as this is considered to be tampering, which is illegal.

This example gave you experience in what is being done by others to improve driveability. In this case, it was the wrong thing. In the future, you may want to check the operation of the EGR valve first, since this may save you some diagnosis time.

Example 8—Fuel Consumption Problem

A luxury vehicle with a V-8 engine and throttle body fuel injection is brought to the shop with a complaint of poor fuel mileage.

Step 1

You question the owner as to how he determined that the mileage is poor. He says that he documented and averaged the mileage through several full tanks of fuel, and that it appears to be about 5 mpg lower than it used to be. He also says that the malfunction indicator light has never come on, except when cranking. A road test reveals no obvious performance problems. This type of problem is a good example of when a service manual becomes a vital tool in the diagnostic process.

Step 2

You open the hood to look for obvious problems and find none. However, you note that the engine appears to be neglected, with no evidence of any recent maintenance. You also attempt to retrieve any trouble codes, although the malfunction indicator light has never come on during normal driving. As you suspect, there are no trouble codes stored.

Step 3

You consider the possible causes of the problem, and decide that the quickest method of finding out what is wrong is to check out the ignition system and the air-fuel ratio. If this does not reveal anything, you should check the drive train, brakes, alignment, and tires for problems.

Step 4

You attach the shop diagnostic equipment to check all of the engine systems. The timing and ignition components check out okay, and a power balance test indicates no weak cylinders. However, the long- and short-term fuel trim readings indicate that the air-fuel ratio is richer than normal, although not rich enough to cause the malfunction indicator light to come on. Removing the air cleaner improves the mixture, but only a little. You proceed to check all the possibilities for a rich mixture, including the oxygen sensor, coolant sensor, fuel pressure, and MAP sensor. All components check out okay, except the coolant temperature sensor, which indicates a much lower temperature than actual engine temperature, **Figure 12-12.**

Step 5

After the tests have been completed, you think about what you have discovered. It makes sense that the coolant sensor is telling the ECM that the engine is running much cooler than it actually has been, and the ECM has been richening the mixture to compensate. Since you cannot find another defective part, you decide to replace the sensor.

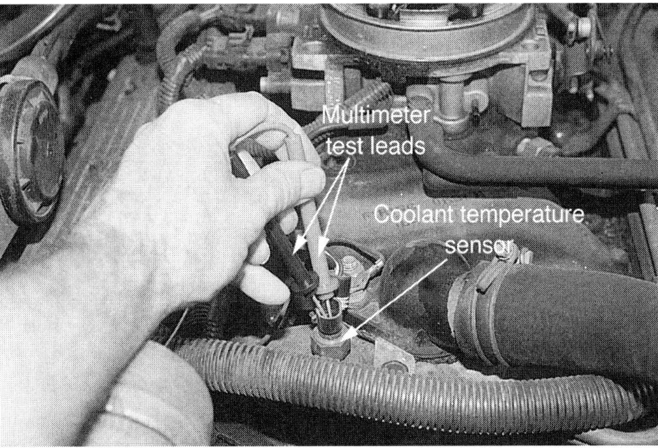

Figure 12-12. *Testing the coolant temperature sensor is a relatively simple procedure on most engines. This relatively simple sensor can cause a host of problems, including overheating and poor engine performance when hot or cold.*

Step 6

After removing cooling system pressure, you replace the coolant temperature sensor. When the sensor is removed, you spot heavy rust deposits on the sensor tip. This tells you that the sensor has been damaged by lack of cooling system maintenance.

Step 7

You attach the shop diagnostic equipment to examine long- and short-term fuel trim and determine that the air-fuel ratio is now normal.

Follow-up

After the air-fuel ratio has been returned to normal, an exhaust gas analysis should be performed to ensure that the rich mixture has not carbon-loaded the engine. In this case, the owner should be strongly advised to perform periodic maintenance with more regularity. In this particular case, you should also recommend a complete cooling system flush with new antifreeze, along with changing the engine oil, oil filter, air filter, PCV valve, and fuel filter. It may also be time to replace the spark plugs, plug wires, distributor cap and rotor. The owner should also be advised to continue to monitor fuel mileage to ensure that the fuel consumption problem has been corrected.

Example 9—Hard Starting Problem

A pickup truck with a diesel engine comes into your shop with a complaint of hard starting.

Step 1

You ask the owner to describe the problem, and she says that the engine cranks slowly at all times, and on several occasions it has needed a jump start from another vehicle. She states that the engine cranked much more easily when jump started, and that it runs well once started. When you ask if she has noticed any other problem, she says that the alternator light glows dimly at times when the engine is running. You start the vehicle and note that the engine barely cranked and that the alternator light does flicker at times.

Step 2

You remember that most vehicles with diesel engines are equipped with two batteries due to the high torque required to start the engine. If one of the batteries were sulfated, this would explain the hard starting, but does not explain the alternator light glowing dimly. You open the hood to check for obvious problems, and discover that the alternator drive belt is loose and glazed.

Step 3

From the information gathered in Step 2, as well as from the driver, you decide that the fuel injectors, compression system, and glow plugs seem to be in good working order. You decide to perform an electrical test before checking anything else.

Step 4

You perform an electrical system test, which indicates that the starter is in good condition, but the charge in both batteries is low. You also discover that the alternator is producing only about half of its rated current output.

Step 5

You analyze the situation and decide that the belt is the most obvious problem. Since the belt is glazed, you decide that it is worthwhile to replace it and recheck alternator operation.

Step 6

You install a new alternator belt and tighten it properly.

Step 7

You recheck charging system operation, and find that the alternator is producing within 5% of its rated current, and system voltage is within the proper range.

Follow-up

Although the alternator is now charging properly, the batteries are still weak. Considering the heavy electrical load placed on the batteries by the high compression diesel engine and the glow plug system, the batteries must be in top condition to properly start the vehicle. Therefore, both batteries should be recharged and retested before returning the vehicle to the owner.

Summary

Always use logical procedures when diagnosing a problem. Strategy-based diagnostics will quickly locate most problems. Step one is to determine the exact problem. Step two is to check for obvious causes of the problem. Step three is to determine which system or component could cause the problem. Step four is to make tests of the most likely systems. In Step five, the problem is isolated and rechecked to make sure that it really is the source of the trouble. In Step six, repairs are made. And in Step seven, the vehicle is rechecked to make sure that the problem has been fixed.

Another important part of diagnosis is follow-up. Always find out if the problem that you discovered is the real problem. Make sure that the part that failed was not damaged by some other vehicle system or operator action. Ensuring that the real cause of the problem has been found and corrected will eliminate much wasted time and customer dissatisfaction.

Always document all service performed to the vehicle. This includes maintenance as well as diagnosis and the repairs. All diagnostic information and repair steps should be clearly noted in writing. Parts required and labor times should also be included. Make sure that both the shop and the vehicle owner retain copies of the documentation. Accurate documentation allows the technician to quickly reference the vehicle, its original problem, and the repairs made. Proper documentation also helps to reduce the chance of misunderstandings between the shop and the vehicle owner.

Know These Terms

Strategy-based diagnostics
Root cause of failure
Follow-up
Documentation

Review Questions—Chapter 12

Please do not write in this text. Write your answers on a separate sheet of paper.

1. Diagnosis is a process of taking ______ steps to reach a solution to a ______.
2. Strategy-based diagnostics involves reasoning through a problem in a ______ of steps.
3. The ______-step process is usually the quickest way to isolate and correct a problem.
4. Talking to the vehicle driver is part of Step ___.
5. Checking for obvious problems is part of ______.
6. Test driving is part of step ___.
7. Checking for disconnected vacuum hoses is part of step ___.
8. When you reason as to what could be causing the problem, this is step ___.
9. Explain, in your own words, why finding starting problems is part of driveability diagnosis.
10. ______ is very important in the troubleshooting process.

ASE Certification-Type Questions

1. Technician A says that older diagnostic techniques may not work on newer vehicles. Technician B says that the easiest way to repair a newer vehicle is to replace parts. Who is right?
   (A) A only.
   (B) B only.
   (C) Both A & B.
   (D) Neither A nor B.
2. All of the following symptoms reported by the vehicle driver are actual driveability related problems, EXCEPT:
   (A) MIL light occasionally flashes when engine is running.
   (B) hesitation on acceleration when the engine is cold.
   (C) steering wheel vibration on acceleration.
   (D) fuel odors when the engine is not running.
3. An obvious problem is most often located ______.
   (A) visually
   (B) with a scan tool
   (C) by making a compression test
   (D) by making tests with a multimeter
4. What should the technician do first if a preliminary test indicates that the problem is in the ECM?
   (A) Replace the ECM and retest.
   (B) Replace or reprogram the ECM's PROM and retest.
   (C) Make further checks.
   (D) Nothing.

5. In Step 5, you ______ the part that is causing the problem.
   (A) replace
   (B) check
   (C) recheck
   (D) Both A & B.
6. Eliminating the causes of the problem one at a time is done in Step ______.
   (A) four
   (B) two
   (C) seven
   (D) one
7. Part of Step 6 includes ______.
   (A) making adjustments.
   (B) replacing parts.
   (C) diagnosing the problem.
   (D) Both A & B.
8. In Step 7, you recheck ______.
   (A) the defective part before replacing it
   (B) the new part before installation
   (C) the system and vehicle after repairs
   (D) Both A & B.
9. Good follow-up often reduces ______.
   (A) comebacks
   (B) dissatisfied customers
   (C) profits
   (D) Both A & B.
10. Proper documentation includes all of the following EXCEPT:
   (A) the description of the vehicle's problem.
   (B) a description of the diagnostic procedure.
   (C) what was done to correct the problem.
   (D) what was causing the problem.

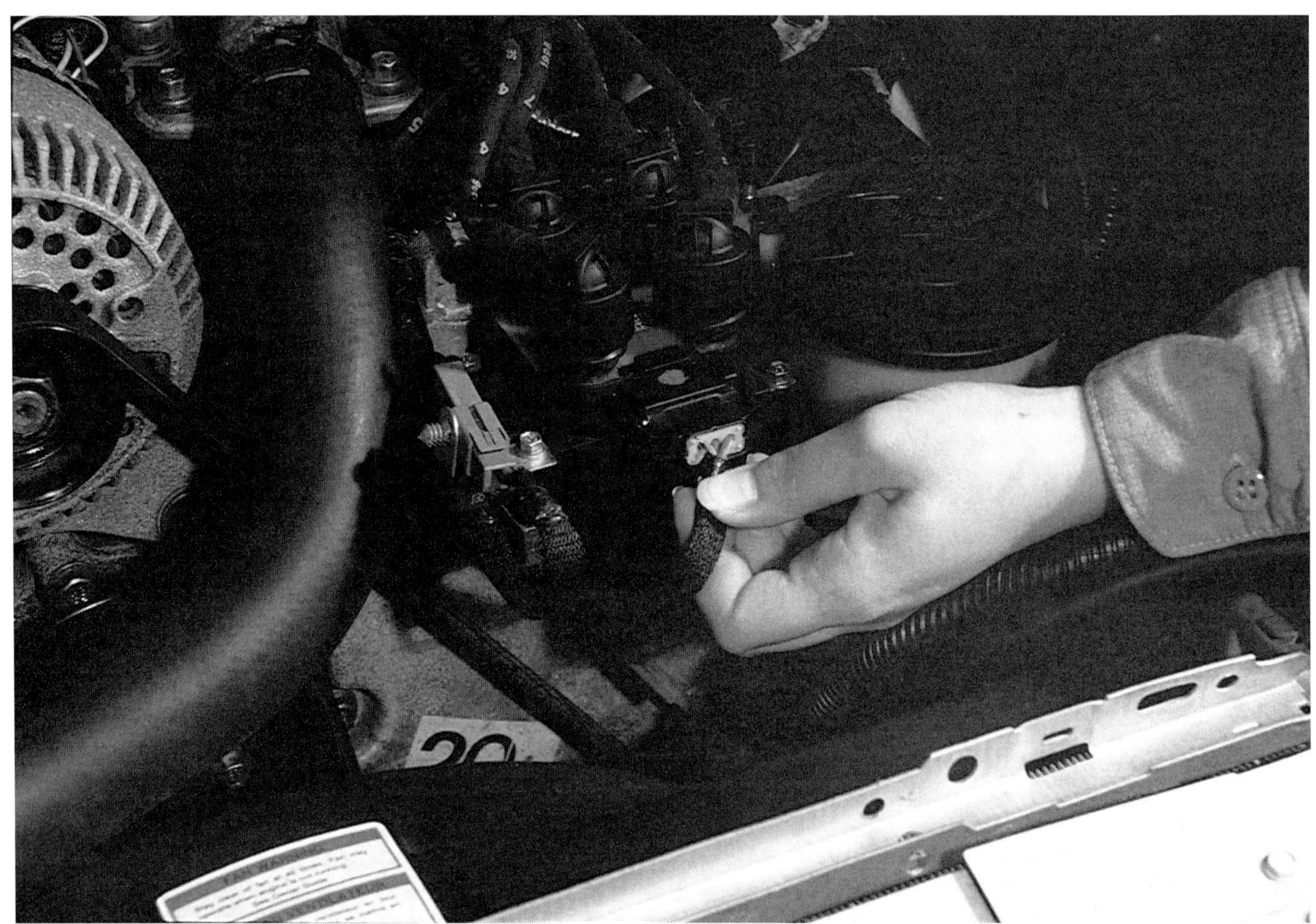

Loose wiring and dirty connections are a frequent cause of engine performance problems.

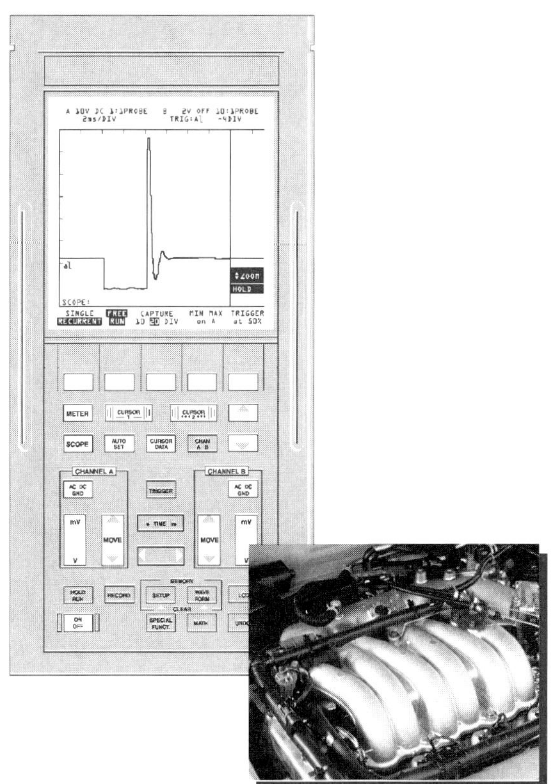

Chapter 13

Troubleshooting Techniques

After studying this chapter, you will be able to:

- ❑ Identify and explain the basic principles of driveability diagnosis.
- ❑ Evaluate driver input concerning vehicle problems.
- ❑ Make visual checks for underhood problems.
- ❑ Perform a road test.
- ❑ Identify and explain the basic methods of diagnosing driveability problems caused by engine systems.
- ❑ Identify and explain the basic methods of diagnosing driveability problems caused by the vehicle drive train.
- ❑ Identify and explain the basic methods of diagnosing driveability problems caused by computer control systems.
- ❑ Identify and explain the basic methods of diagnosing driveability problems caused by other vehicle systems.
- ❑ Identify methods of determining what further actions need to be taken.
- ❑ Identify and explain factors to be considered when deciding to adjust, rebuild, or replace parts.
- ❑ Identify steps in determining who to contact about needed repairs.

From studying the previous chapters, you know that driveability problems are not caused by a single area of the vehicle. You also know that all systems and components should be considered when trying to diagnose a driveability problem. In the earlier chapters in this book, you learned how the various engine systems, drive train systems, and other vehicle components can affect driveability. In **Chapter 12,** you learned the principles of strategy-based diagnostics. This chapter will begin the process of taking everything that you have learned previously, along with the information in the remainder of this text, and putting it together into a complete diagnostic and repair package.

Basic Diagnostic Procedures

Isolating and curing a driveability problem involves many procedures. Some of these procedures are diagnostic, while other procedures are service and repair operations. These procedures include checking the operation of vehicle systems to determine problem areas, which parts could be defective or out of adjustment, testing the condition and operation of the suspected parts, repairing or replacing these parts, and then rechecking the vehicle. This chain of procedures is part of a strategy-based diagnostic routine, **Figure 13-1,** which was discussed in **Chapter 12.**

It is relatively easy to change a part or make a simple adjustment to restore vehicle driveability. This is the sort of work that almost anyone could do with enough time and the proper tools. The hard part is discovering exactly what new parts or adjustments are needed. This discovery process is the heart of diagnosis, or ***troubleshooting.*** This is what separates the real automotive technician from the so-called technician who can only remove and replace parts and guess at whether it will solve the problem.

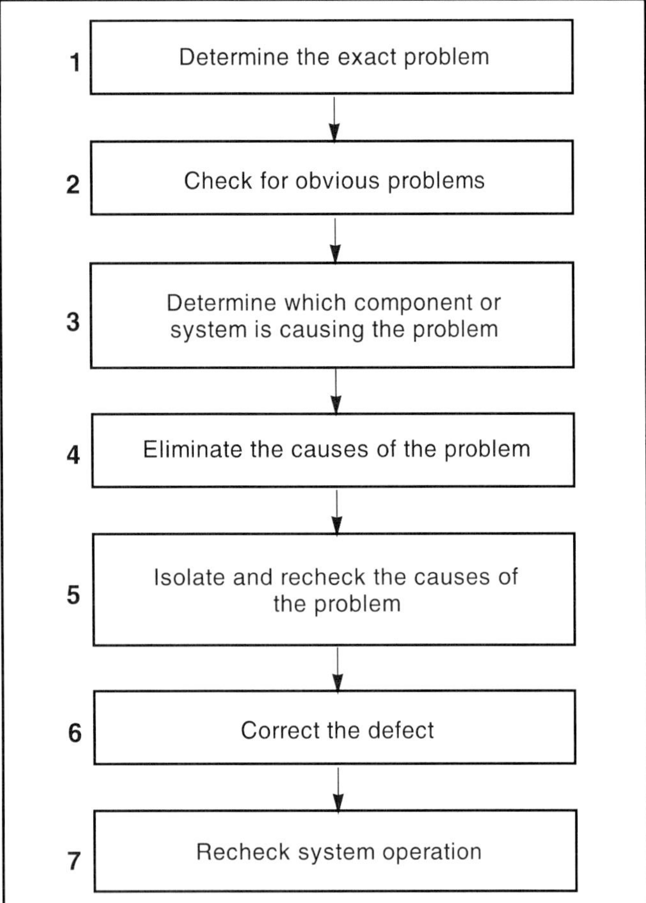

Figure 13-1. *By following a logical sequence, you should be able to diagnose most problems in a timely manner. If the purpose of this chart is not clear to you, review the material in Chapter 12.*

Learning to Diagnose

Some persons may seem to have a great natural ability to diagnose problems. What seems like a magical quality in some technicians is actually years of practice and experience in using their diagnostic ability. This ability is present in every person and can be improved with study and experience. The principles of strategy-based diagnostics will give you a framework in which to build upon.

You would not be reading this book if you did not have an interest in automobiles, and a desire to fix things. However, these two qualities alone will not make you a good driveability technician. To become an expert at troubleshooting, you will first need to increase your knowledge of how internal combustion engines and other systems operate and how their operation relates to vehicle performance. After you are well-versed in this knowledge, you will need to develop the ability to approach a problem, no matter how mysterious or frustrating, in a calm and logical way.

Evaluating Driver Input

Obtaining information from the driver is the first and the most important part of troubleshooting. Information from the driver will often allow the technician to bypass some preliminary testing and go straight to the most likely problem. In one sense, the driver begins the diagnostic process by realizing that the vehicle has a problem and deciding that it is serious enough to require service.

Always try to talk to the person who normally drives the vehicle. Try to get an accurate description of the problem before beginning to work on the vehicle. Since the driver is usually the vehicle owner, he or she can provide some idea of past service problems and any maintenance that has been performed or neglected. The driver's description of the problem often makes finding the solution easier. However, what the driver or owner says must be carefully evaluated. A tool used by some technicians is a ***driveability worksheet,*** which is given to the driver to fill out in his or her own words, **Figure 13-2.** This sheet also gives the technician a written description of the problem.

Assessing Driver Input

While taking into account what the driver says, try to estimate the driver's attitude and level of automotive knowledge. Because drivers are not usually familiar with the operation of automobiles, they often unintentionally mislead the technician when describing symptoms or may have reached their own conclusion about the problem. In describing vehicle problems, drivers have been known to use hand gestures, body language, and even simulate noises that they have heard. While this can be fun to watch, it is a part of the diagnostic process. Many times, important clues can be found simply by observing a driver's physical actions while describing a particular problem.

In many cases, the person bringing in the vehicle has already formed an opinion as to what is wrong. These opinions are a common occurrence, often based on a poor or incomplete understanding of vehicle operation, advice from uninformed friends, or the inability to fully comprehend the problem. The best course is to listen closely to the driver's description of the symptoms. Never accept a driver's diagnosis until you can verify it. Some drivers will be sensitive to even slight changes in driveability, and may be overreacting to a normal condition.

Often, the owner is concerned about the cost of repairs. Some will even downplay the symptoms, hoping for an inexpensive repair. Very few vehicle owners are unconcerned about the cost of vehicle repairs and maintenance. Do not give any type of uninformed estimate, even though you may have a good idea what the problem is. Giving an estimate without diagnosis is a mistake made by many technicians. This practice invites one of two things to occur. Either the repair that is recommended will not correct the

Driveability Worksheet

Name: ______________________________ Date: __________

VIN: ____________________ Year: ______ Make: ________ Model: __________

Style: ________ Color: ______ Engine: ________ Trans: ______ A/C: ______ P/S: ________

Describe the vehicle's problem as accurately as you can: ______________________________

____

____

____

Check the boxes that best describe when the vehicle's problem occurs.

☐ Hot ☐ Cold ☐ Constant ☐ Intermittent ☐ Recurring

Check the boxes that best describe the type of symptom(s) that occur.

☐ Stalling	☐ Lack of power	☐ Overheating
☐ Hard start	☐ Surge/Chuggle	☐ Noise
☐ Does not start	☐ Pinging	☐ Vibration/Harshness
☐ Rough idle	☐ Miss/Cuts out	☐ Fluid leak
☐ Incorrect idle	☐ Poor fuel economy	☐ Unusual odors
☐ Hesitation	☐ Indicator light(s) on	☐ Smoke

Other: ____

____

Background information (any service or repair work performed recently, other problems, etc.): __________

____

____

Figure 13-2. *Driveability worksheets are a good source of information in diagnosing most performance problems. The questions on the sheet encourage the driver to describe the problem in his or her own words.*

problem or it will frighten the driver, who may take his or her vehicle to another shop or decide not to have the repair done at all. Explain that the charge for diagnosing the problem is actually more cost effective than paying for a service, such as a tune-up, that in most cases, may not fix the problem.

Difficult People

In most fields of employment, including auto repair, the time will come that you will have to face an angry or difficult person. Unfortunately, this is an inevitable occurrence, no matter how hard you try to fix vehicles right the first time. Reasons for anger and hostility vary, and it should not be assumed that any person is naturally difficult. In many cases, the owner is distressed because a problem has not been solved during previous visits. Sometimes the owner is very hostile and may be difficult to communicate with.

Some drivers will become angry and hostile because they are afraid that repairs will be expensive or are simply annoyed at the inconvenience. A person's anger and hostility may also be due to factors outside of the vehicle's problem. In a few cases, the person may simply have a difficult personality. Always try to separate the vehicle's problem from the driver's problem.

Handling Angry and Difficult People

Dealing with difficult people involves understanding and tact on your part. When speaking to an angry person, talk in calm tones. While trying to calm the person down, avoid getting angry. Getting angry in response to a customer's anger will only make the situation more difficult to handle. While working on the vehicle, keep the owner informed of the vehicle's status as frequently as possible. This will give them the feeling that they are involved in the diagnostic and repair process. Once you find the problem, inform them of the cause and exactly what is needed to correct it. If practical, bring the owner to the vehicle to inspect the defective part.

Road Testing

The ***road test*** is the most common type of automotive test. It can reveal or confirm many types of driveability problems and will often indicate specific problem areas and what further tests need to be made. Sometimes, a quick road test will determine that a perceived driveability condition is normal, and that no further testing is needed. Some problems can only be found by taking a vehicle on a road test. Most vehicle problems will often reveal themselves in a road test taking no more than 15 minutes, saving hours of in-shop diagnostic time.

Try to duplicate the exact conditions under which the driver says the problem occurs. If the problem only occurs when the engine is cold, the vehicle should be left overnight and checked when it has completely cooled off. If the problem occurs when the engine has been sitting hot for a few minutes, it should be driven enough to warm up, and then rechecked after it has been sitting for the same amount of time. Unfortunately, duplicating some conditions is not always possible. For example, it is almost impossible to duplicate a spark knock condition that occurs only at high altitudes at a shop located near sea level with no hills or mountains in the vicinity.

Always try to road test the vehicle with the owner. This will ensure that you are both talking about the same problem, and will save valuable diagnostic time. It is very frustrating to find out that you have spent time diagnosing and repairing an obvious problem, when the customer wanted a less obvious problem fixed.

Safety Considerations

Before beginning a road test, make a few quick checks to ensure that the vehicle can be safely road tested. Walk around the vehicle's exterior and make a note of any damage that is present. Check each tire to ensure that they are inflated properly and in good condition. Also make sure that all safety-related equipment, such as the brake lights, turn signals, and horn are working properly, **Figure 13-3**.

Press the brake pedal, it should feel firm and high. Turn the steering wheel and ensure that the steering system does not have excessive play. Also make sure the vehicle has enough fuel to conduct a road test. If there is a problem with any of the vehicle's safety-related equipment, it should be addressed before any driveability complaint. Wear your seat belt at all times during the road test.

Drive slowly as you leave the service area to ensure that the brakes and steering are working properly, and that no obvious mechanical problems exist that could further damage the vehicle or cause personal injury. Drive the vehicle carefully and do not do anything that could be remotely constituted as abuse. Tire squealing takeoffs, speed shifts, fast cornering, and speeding can all be interpreted as misuse of the vehicle.

Note: Do not adjust anything in the passenger compartment, such as mirror, seat, and tilt steering wheel position, unless absolutely necessary. If the radio is on, turn it off so that you can listen for unusual noises.

While road testing, obey all traffic rules, and do not exceed the speed limit. It is especially important to keep in mind that you are under no obligation to break any laws to test a customer's vehicle. Also be alert while driving. It is easy to become so engrossed in diagnosing the problem, that you forget to pay attention to the road or the traffic around you. If it is necessary to monitor a scan tool's readout or look for a problem while the vehicle is driven, get someone (not the vehicle's owner) to drive the vehicle for you.

Road Testing Procedures

During the road test, try to duplicate all normal driving conditions, such as light and heavy acceleration, deceleration, braking, and different cruising speeds. Carefully note the response of the engine, as well as related systems.

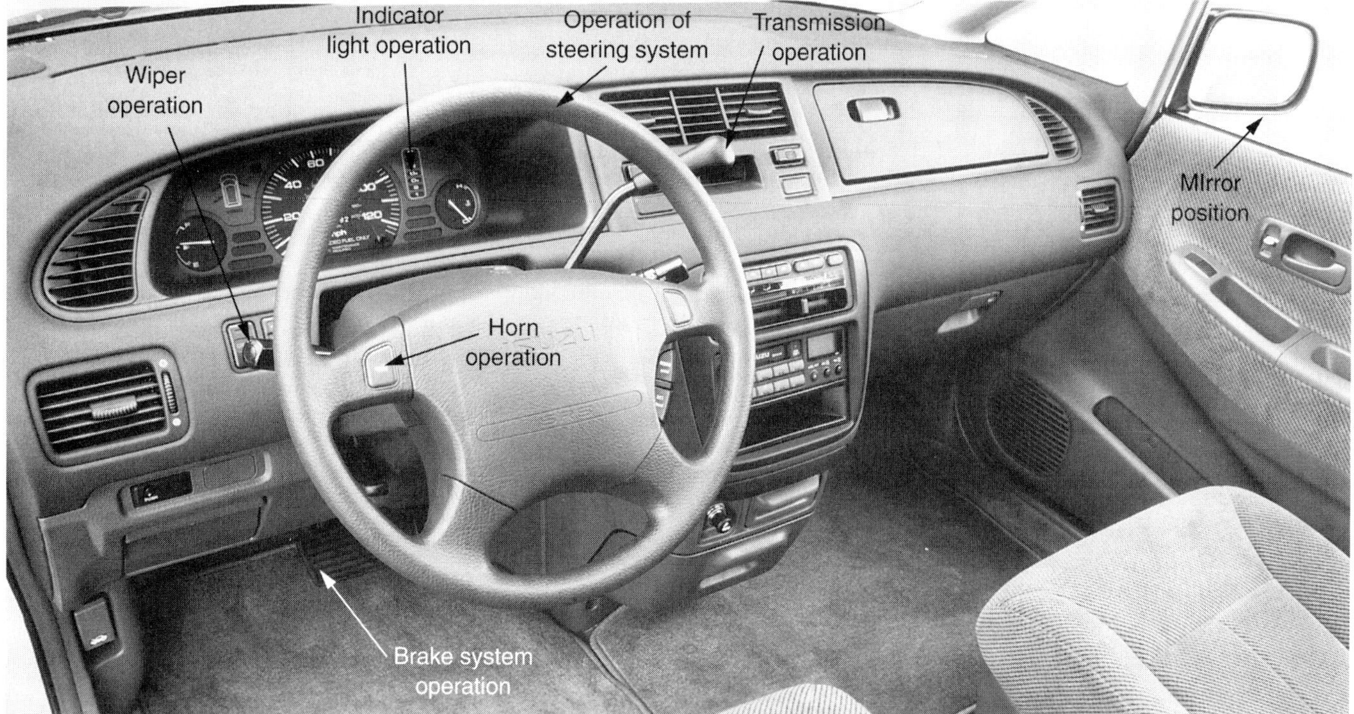

Figure 13-3. *Perform a preliminary check of vehicle controls before beginning any test drive. If a safety-related system, such as the brakes, is not working properly, the vehicle may be unsafe to drive. (Isuzu)*

Be alert for noises, vibrations, harshness, engine miss or hesitation, transmission shifts, and the operation of the brake and steering systems. You may want to bring a clipboard along to note any unusual conditions. If you are checking a problem that only affects the vehicle when cold, make all tests as quickly as possible, since modern engines reach normal operating temperatures quickly.

Once a specific problem is identified, you can pay more attention to conditions related to that problem. For instance, if you notice an engine miss, try to determine whether it occurs when accelerating lightly, under heavy acceleration, while cruising, when idling, or if it is constant. Also note if other vehicle components or driving conditions affect the problem. For example, does the problem occur only on a certain type of road surface, when the air conditioner is on, or when the power windows or seats are being operated, or when the vehicle is making a turn. Any of these situations could be related to the driveability problem. Note any and all of these factors before returning to the shop.

Diagnosing Intermittent Problems

If the problem does not occur, you must decide whether the vehicle really has a problem. It is tempting to dismiss the problem as the owner's imagination or as a normal vehicle operating condition, but the problem may well be real. ***Intermittent problems*** are the most difficult to diagnose, because they usually occur only when certain conditions are met. Intermittent malfunctions can be related to temperature, humidity, certain vehicle operations, or in response to certain tests performed by the ECM.

When dealing with an intermittent malfunction, always try to recreate the exact conditions in which the problem occurred. Unfortunately, most drivers do not relate intermittent problems to external conditions. Intermittent problems cannot always be duplicated, even in an extensive road test. If a road test of reasonable duration does not duplicate the problem, it is time to try other types of testing. It is *absolutely essential* that the principles of strategy-based diagnostics be followed closely when diagnosing intermittent malfunctions.

Performing Visual Inspections

Before performing any diagnostic tests, open the hood and check for visible problems, **Figure 13-4.** The actual visual inspection begins during the road test. In many cases, seemingly complicated problems are caused by a defect that can be easily seen, smelled, or heard. If the problem is not related to cold operation, leave the engine running at first, since underhood sounds and movement can provide vital clues. Be sure to stop the engine before investigating any component near hot or moving engine parts.

While under the hood, check the level and condition of the engine oil, coolant in the recovery tank, and automatic transmission fluid. Note any coolant, fuel, or oil leaks, as well as any disconnected hoses or air intake ducts. Check the air and PCV filters for clogging and make sure that the PCV valve and hose are not plugged. Check the condition of the drive belts, especially the alternator belt. If the engine is equipped with a serpentine belt, check the condition of the belt tensioner. Look for prior work done on the vehicle if no service history is available. Finally, take a close look for signs of abuse or tampering. Any engine components that have been removed or replaced by non-stock parts can affect engine operation.

Look for aftermarket add-on equipment that can affect vehicle operation. This includes alarm systems, CB and police radios, cellular telephones, fog lights, spot lights, fluorescent lights, car stereo amplifiers, compact disc players, trailer wiring harnesses and other external lighting and wiring. Any of these add-on components, if installed improperly, can affect engine and vehicle operation.

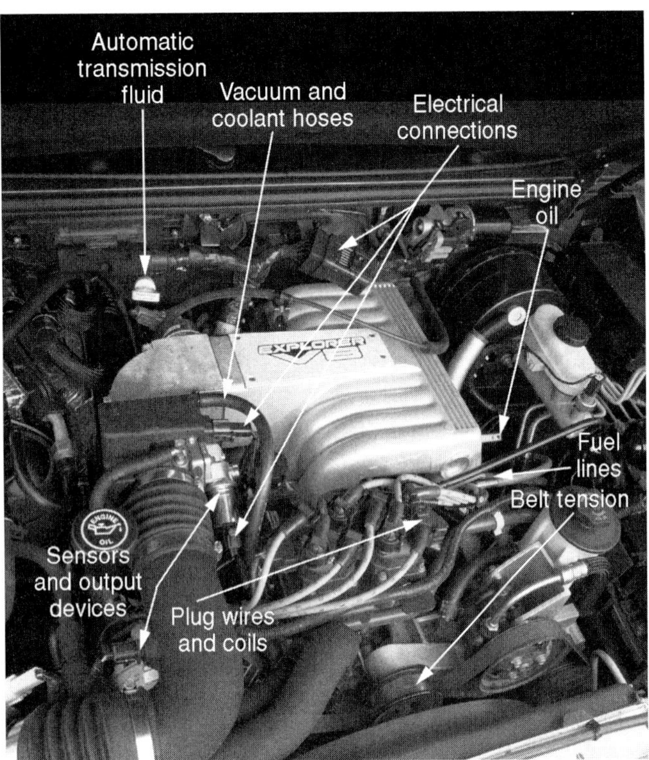

Figure 13-4. *Many driveability and performance problems are often found during the visual inspection. A quick visual inspection can often save you hours of performing needless tests. (Ford)*

Principles of Diagnosis

Although most of the basic principles of diagnosis could be called *common sense,* and hardly worth mentioning, they are often forgotten when a problem occurs. This is a good time to commit them to memory.

When presented with a driveability or other automotive problem, start by determining the exact nature of the problem. The information that you are given, either by the driver or your own testing, can be misleading at first. As discussed earlier, the driver's description may not accurately describe the problem. A problem described as a "miss" may be hesitation caused by a defective spark plug wire or surging caused by the air conditioner cycling on and off. Try to take the vehicle on a road test with the owner. If the driver requests a certain service be performed, ask the customer why he or she thinks that it needs the services requested. In some cases, the customer's description is correct, but you should always subject it to your own verification.

Make Sure Your Interpretation of the Information is Correct

Be sure that you correctly interpret all of the information that you get. In many cases, removing the symptom does not cure the problem. For example, if you find that the timing was set too far in advance on an engine vehicle that has begun idling rough, do not assume that the cure consists of resetting the timing. The misadjusted timing may be caused by a defective timing advance mechanism, a stretched timing chain or belt, or another problem. If you reset the timing, the engine might quit running rough, but it will leave your shop with the real problem uncorrected or in some cases, another problem.

Make sure that the presumed driveability problem is not caused by another vehicle system that can cause a similar problem. For example, a poor acceleration problem can be caused by a defective fuel pump or a slipping automatic transmission. If you concentrate on only one area because that is where the problem usually occurs, or "should" occur, you may miss an obvious defect in another system, and waste hours of diagnostic time. In all instances, you are looking for what is known as the ***root cause of failure.*** Understanding the function and possible defects in every part of the vehicle will allow you to consider all of the possibilities.

In some cases, experience, or information from other technicians or technical service publications may lead you to one particular area that is known to give trouble. For example, many computer systems in the past have developed problems due to certain defective sensors. Throttle position sensors, MAP sensors, and mass airflow sensors on certain vehicles have been prone to fail to the point where they have been the subject of manufacturer recall campaigns. In automotive and other areas of diagnostics, it is often quicker to start at a part that is known to cause the particular problem that you are facing. However, every suspect part still requires testing to verify that it is the root cause of failure.

Additional Diagnostic Tips

When diagnosing any problem, *always check the simple things first.* It will save a lot of time and annoyance to check all of the visible, obvious possibilities before going on to the difficult ones that require a lot of test equipment and time. You must check every part in a system, without assuming anything about the condition of any part. For example, do not try to solve a rough idle problem by working on the fuel or ignition systems until you are positive that all manifold vacuum hoses are in place and not split, **Figure 13-5.** Knowing that all the easy things are okay will allow you to zero in on the hard things. While diagnosing a problem, never assume that a particular part or system is functioning properly until it has been checked.

Figure 13-5. *As you learned earlier in this text, vacuum leaks are frequently responsible for many driveability problems. Check all vacuum hoses carefully.*

Sometimes you will hear of making an "educated guess." However, an educated guess is a reasonable decision, based on testing and the process of elimination. ***Uneducated guessing***, or jumping at the first possible cause that comes to mind, is a dangerous way of diagnosing problems. Unfortunately, it can quickly become a habit, done over and over no matter how many times it leads to disaster.

Remaining Calm

One of the hardest principles of diagnosis is to remain calm, no matter how much you would like to scream and throw things. Mastering your own emotions is often the hardest thing to do, especially if you meet with a series of dead ends while looking for a problem or are having to deal with an angry customer, but it is necessary. Nothing will be accomplished by losing your composure. If you lose your composure, you will waste valuable time and possibly upset the customer. If you have picked up a tendency to overreact to situations, you will have to unlearn this behavior and teach yourself to remain calm. Only a calm person can think logically.

Using Service Literature

Service literature is the source of all vehicle specifications, diagnostic procedures, expected test results, and repair procedures. As such, service literature is as much of a tool as a wrench or screwdriver. When working on the modern vehicle with its electronics and other complex equipment, service literature is often the most important tool in the shop. Service literature was discussed in detail in **Chapter 3**.

To refer to the appropriate service information, you must first decide which type of service literature that is needed. Service literature is available as manufacturer and general manuals; in the form of troubleshooting charts; and as electrical, vacuum, and information flow schematics. Always begin the diagnosis procedure by obtaining the proper service literature for the job that you are doing.

Service Manuals

In many cases, the service manual contains the diagnostic information that will enable you to pinpoint the problem. It will contain the service information needed to make repairs. Be sure to use the specific manual for the model and year of the vehicle that you are working on. Many changes are made to vehicles each year, and even a manufacturer's manual that is "close" may not be good enough. If the proper manufacturer's manual is not available, use a general repair manual which covers the model and year of the vehicle that you are working on.

Once you have obtained the correct manual, turn to the proper section for the specific problem the vehicle is experiencing. Since most manuals are divided into many driveability-related sections, it may be difficult to immediately locate the precise information that you are looking for. Some manuals have separate sections on driveability or tune-up. Most manuals are divided into fuel, ignition, emissions, and mechanical sections. Other manuals are divided into several volumes, with individual volumes for the body and chassis, and sometimes a separate section for wiring schematics. It may be necessary to try several sections before finding the proper information.

Once you have located the information that you need, use it to help in diagnosis. Some of the uses of manuals and other service literature include finding proper specifications; following a diagnostic routine; or studying the proper method of disassembling, reassembling, and adjusting components. Always read the information carefully to avoid skipping a step.

Since you will want to use the manual again in the future, take precautions to avoid damaging it. Do not place manuals on oily workbenches, or allow them to remain on the floor or any other area where they could be damaged. Close manuals after use and store them on a bookshelf or in a cabinet, **Figure 13-6.** This is not only to prevent damage to the manual, but is also a courtesy to other technicians. Do not use pencils or markers to trace over schematics in the manual, and do not tear pages out of the manual. If any procedure will damage the manual, make a copy of the needed pages, and use the copies.

Troubleshooting Charts

Troubleshooting charts cover a broad range of problems in an easy-to-use form. They are usually arranged in columns according to symptom. The condition that caused the symptom, or the area that should be investigated further is shown in the next column. Sometimes a third column is used to list the possible corrective actions that could be taken, **Figure 13-7.**

Using a troubleshooting chart is relatively easy. In some cases, you may have to rethink the symptoms, and decide which chart symptom they correspond to. For instance, what you or the driver may think of as a "flat spot" may be listed on the chart as "hesitation." Also, it is common to find

Figure 13-6. *Service manuals are as valuable as a wrench or any other tool. They should be cared for as well as you would care for any tool. Always return service manuals to the shelf or shop library once you are finished with them.*

listed on the chart as "hesitation." Also, it is common to find no reference to the problem that the vehicle is experiencing. Since no chart can cover every possibility, it may be helpful to look for what you know is a related problem, and investigate the causes listed there. For instance, if the chart has no reference to "surging," try the chart under "rough idle," since the two problems are often caused by the same defect. Do not inflict any unnecessary damage to the chart, since you or someone else will want to use it in the future.

Schematics

Schematics show the pathways for electricity or vacuum and how information flows to and from the ECM. Reading a schematic is a process of path tracing, from the beginning of the wire, hose, or information trail to its end.

Tracing the lines on an electrical schematic will enable you to locate problems caused by shorted or open wires, defective components, corroded connections, or other electrical problems. Tracing the hoses of a vacuum diagram will allow you to find loose or disconnected hoses, and defective vacuum devices. Tracing how information flows into and out of an ECM will allow you to determine which sensors affect the operation of which output devices. Care of schematics is essential. Do not use pencils or markers to trace over schematics. Instead, make copies of the needed pages, and trace the copies, as shown in **Figure 13-8.**

Restricted Exhaust System Check

Step	Action	Value(s)	Yes	No
1	Was the powertrain on-board diagnostic (OBD) system check performed?	—	Go to step 2	Go to powertrain OBD system check
2	1. Carefully remove the heated oxygen sensor (HO_2S). 2. Install the exhaust backpressure tester in place of the HO_2S. 3. Start the engine and idle. Is the reading greater than the specified value?	8.6 kPa (1.25 psi)	Go to step 4	Go to step 3
3	1. Increase the engine speed th the specified value. 2. Observe the backpressure gauge. Does the reading exceed the specified value?	2000 RPM 20.7 kPa (3 psi)	Go to step 4	Go to step 6
4	1. Check the exhaust system for a collapsed pipe, internal muffler failure, or any kind of heat distress. 2. If a problem is found, repair as necessary. Was a problem found?	—	Go to step 6	Go to step 5
5	Replace the catlytic converter. Refer to engine controls or exhaust of the appropriate service manual. Is the action complete?	—	Go to step 6	—
6	Operate the vehicle within the conditions under which the original symptom was noted. Does the system now operate properly?	—	System OK	Go to step 2

Figure 13-7. *Troubleshooting charts can help you narrow down the source of a vehicle's problem. Often, the charts will tell you the exact cause of the problem and the correction. (General Motors)*

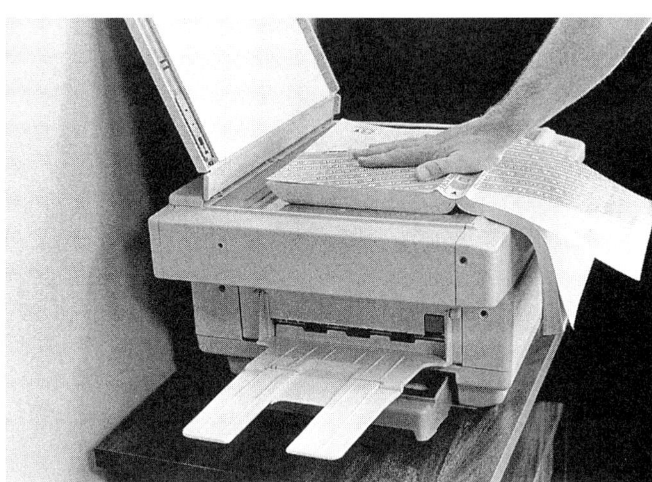

Figure 13-8. *If it becomes necessary to mark in a service manual, make a photocopy of the pages. Never write or mark pages in a service manual as it will quickly ruin the manual.*

Diagnosing Electrical Problems

Of all vehicle systems, this can be the most intimidating to diagnose as the modern vehicle uses wiring in almost every system. With over 80% of vehicle functions ECM monitored, controlled, or both, a minor electrical defect can become a major problem. However, diagnosing electrical problems can be accomplished by following a logical sequence of steps.

The most important aspect of electrical system troubleshooting is having complete wiring schematics for the vehicle and system. Make a photocopy of the schematic pages for the system to be diagnosed. Begin at the power source and physically trace the wiring back to the system ground connection. As each connector, length of wire, and component is tested and checked, mark off the component on the schematic photocopy. By using this system, you will eventually isolate the defective wire, connector, or component, **Figure 13-9.** Electrical system diagnosis is discussed in more detail in **Chapter 14.**

Diagnosing Charging and Starting Systems

The charging and starting systems can be checked by observing their operation as the engine runs. For example, the charging system can be checked for proper voltage and amperage output, and also for voltage spikes caused by an internal defect. Voltage spikes can be the source of many ECM problems. The starting system can be checked by observing cranking speed and performing a cranking amperage test.

If a starting problem is caused by a parasitic draw, start at the fuse block. Pull each fuse, one at a time, until the circuit causing the draw is isolated. Refer to the manual's schematics for the circuits powered by the particular fuse. Make a copy of the schematics for the affected circuits and trace each circuit until the problem is isolated. Charging and starting system diagnosis is also covered in **Chapter 14.**

Diagnosing Engine Problems

The fuel, ignition, and compression systems are the systems which allow the engine to run. They are the basis of engine operation and, therefore, can have the greatest impact on driveability. Other engine systems affect driveability by influencing the operation of these basic systems. The drive train or other vehicle systems affect driveability by degrading the output of the engine.

These systems should be checked first unless there is a very good reason to investigate another part of the vehicle. It is often difficult to determine whether the problem is in the fuel, ignition, or compression systems. For example, an engine miss could be caused by a fouled plug, defective fuel injector, or burned exhaust valve. Hesitation could be the result of a defective accelerator pump, retarded timing, or a jumped timing chain.

It is also possible for more than one part of a single system to cause the same symptom. For instance, an engine miss could be caused by a fouled spark plug, defective plug wire, or cracked distributor cap. These components are all part of the ignition system. Therefore, it is important to check all possible causes of a problem in the basic engine systems before making repairs. One common diagnostic method is to perform a ***dynamic compression test***, sometimes referred to as a power balance test, by shorting out each cylinder individually, **Figure 13-10.** This will quickly locate a cylinder that is not producing power. After the malfunctioning cylinder is found, the cause can be quickly located.

After the fuel, ignition, and compression systems have been tested, the emissions, cooling, lubrication, and exhaust systems can be checked. Testing these systems includes a simple visual inspection for disconnected hoses or wires. If a visual inspection does not reveal the problem, the operation of the entire system can be checked by various methods. It may be necessary to check the cooling system for internal leaks, or check the engine oil pressure by installing a pressure gauge. If the suspected system is controlled by the ECM, the diagnosis should include a thorough check for related trouble codes.

In some cases, the actual problem can only be determined by partial disassembly. Examples are removing a spark plug to check for deposits, removing an exhaust pipe to check for collapse, or removing the cylinder head to check for a burned valve. Obviously, some disassembly procedures are more involved than others. All other possibilities should be reconsidered and eliminated before proceeding with a major engine disassembly operation.

Diagnosing Computer Control Systems

Before beginning diagnosis on any vehicle equipped with an on-board computer, you must do three things: obtain the proper service manual and other information for the computer system, thoroughly check all non-computer systems, and check the ECM memory for trouble codes.

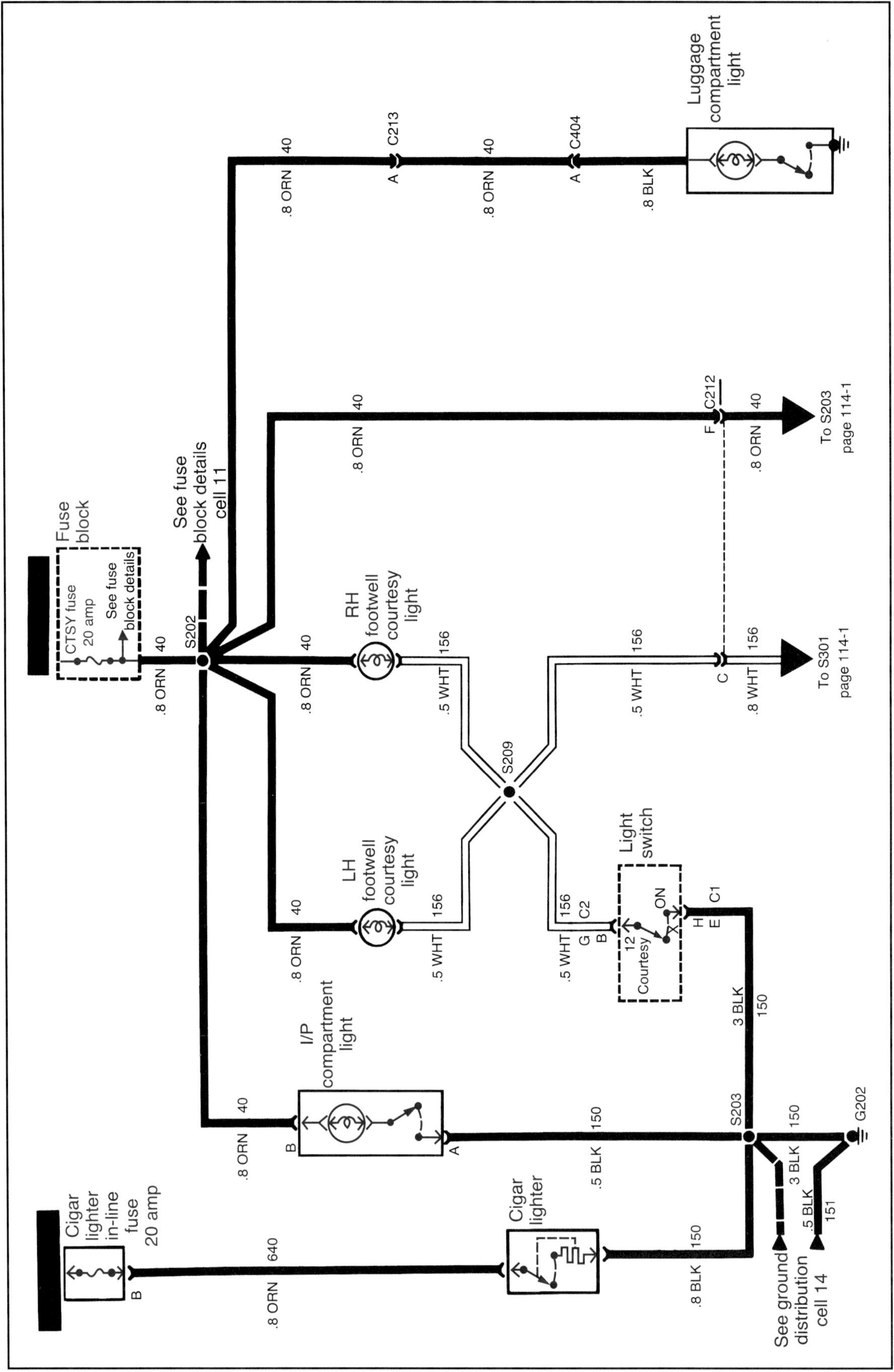

Figure 13-9. *By making a photocopy of a service manual page, such as a schematic, you can trace circuits to ground from the power source. Practice using a copy of this schematic. In this example, pretend the luggage compartment light is inoperative, however, the bulb is good and the rest of the lights in the circuit operate properly. Follow the circuit from the fuse to each lamp and power load. (General Motors)*

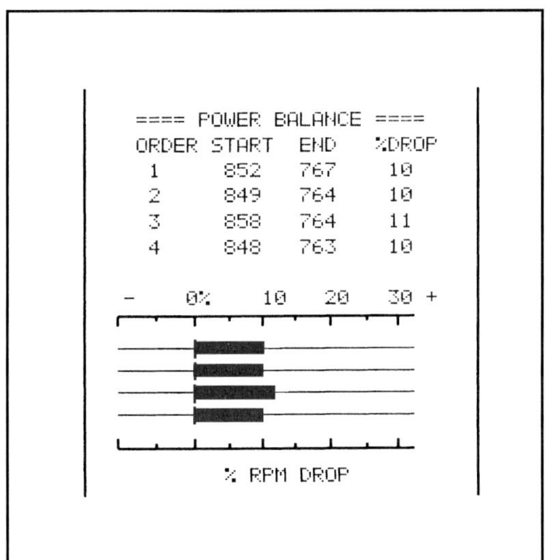

Figure 13-10. *Power balance tests can give a preliminary indication of the location of a misfiring or dead cylinder. However, this type of test should be followed by a compression and other diagnostic tests for a more accurate diagnosis. (Ferret)*

As you learned in **Chapter 7,** all computer control systems on newer vehicles have self-diagnostic capability. The ECM memory will monitor for any out of range readings from sensors, output devices, or the ECM itself. These readings are stored in the computer memory section as a trouble code. Some ECMs will also store a snapshot or freeze frame of the engine conditions when the trouble code was stored along with the number of keystarts or warm-ups that have occurred since the code was set. If the malfunction indicator lamp (MIL) is on or flashing, the ECM definitely contains trouble codes. **See Appendix C.**

If the MIL is not on, the ECM may still contain stored trouble codes. Therefore, always perform the code retrieval process before proceeding further. The stored codes may quickly pinpoint the trouble area. The trouble code retrieval process is discussed in **Chapter 15.** Once the trouble codes are identified, make further tests as identified in the manufacturer's service manual. In some cases, the defective component is identified by the trouble code. In other cases, a vehicle subsystem is isolated, and the entire subsystem must be diagnosed to find the defective part.

Computer systems with or without self-diagnostics often contain parts that cannot be tested, such as the ECM and ignition module. These components must be isolated by a process of elimination. Start by testing all parts that can be tested, such as ignition coils, primary and secondary ignition components, pressure switches, or fuel injectors. This may uncover the problem with no further checking. If all of the testable parts check good, a non-testable part is the most likely cause of the problem.

Procedures for Computer System Diagnostics

Many driveability problems on computer-controlled vehicles are caused by non-ECM defects. Always look for obvious problems first. If no obvious mechanical defects are found, retrieve any trouble codes from the ECM memory. However, do not assume that the presence of trouble codes is a sign of computer system problems. Trouble codes are often set in response to other engine and drive train defects. Find out what the trouble code indicates, and look for defects that could set that code. Many computer codes can be set by vacuum leaks, ignition secondary problems, chassis wiring problems, plugged fuel or air filters, EGR valve defects, or clogged fuel injectors.

If the procedures indicate that all of the basic engine and drive train systems are operating correctly, the computer control system may be at fault. Begin the testing procedure by checking all electrical connectors for tightness and cleanliness. Be especially careful when checking sensor connections, since most of them operate at very low voltages.

When checking for defective computer control components, check all of the input sensors and related wiring first, since defective sensors are the most common cause of problems. Check the output solenoids, motors, ignition modules, injectors, and other output devices next. Check the ECM last, since it cannot be tested. This sequence should be followed unless a part is visually observed to be defective, or when a check of the basic engine systems or system performance has uncovered a specific problem area. If a specific system or part has been pinpointed by earlier tests, time can usually be saved by checking it first.

Diagnosing Drive Train Problems

The engine creates and modifies the vehicle's power while the drive train delivers this power to the road. The clutch or torque converter, transmission or transaxle, drive axles, and final drive unit receive and transmit engine power, multiply it for additional performance, and reduce engine speed for economy, **Figure 13-11.** The drive train components, particularly the transmission, vary gear ratios to match engine power to the vehicle load and such conditions as road grade and traction conditions.

Figure 13-11. *The drive train and other vehicle systems should be scrutinized as carefully as the engine and its subsystems for possible problems that could contribute to the driveability problem. (Ford)*

The drive train components can affect driveability by incorrectly transmitting engine power. When the transmission or transaxle is fully or partially controlled by the ECM, it can cause control system problems. Defective pressure switches will send improper signals to the ECM, or defective solenoids will fail to perform the ECM output commands. Other drive train problems include noise, vibration, and harshness from worn parts.

It is usually difficult to diagnose drive train components, especially the transmission, by visual inspection. Defects such as worn out U-joints or CV joints, and oil on the clutch facings, can be easily spotted by raising the vehicle on a lift, while other problems will only show up when the vehicle is driven. A drive train problem can be difficult to distinguish from an engine problem, which is why all possibilities should be considered.

Manual Clutch and Transmission Diagnosis

Manual clutch problems can cause slipping, vibration, and chatter. If the vehicle has a manual clutch, it can be checked by slowly engaging it with the engine running, the vehicle stopped, and the transmission in first gear. See **Figure 13-12**.

Observe whether the clutch engages smoothly, with no jerks or chattering. Then place the vehicle in its highest gear with the vehicle stopped, and release the clutch. If the engine does not stall, the clutch may be slipping or misadjusted. However, manual transmissions and transaxles rarely contribute to a driveability problem. If a fault exists in a manual transmission or transaxle, it is almost always recognized as a drive train problem.

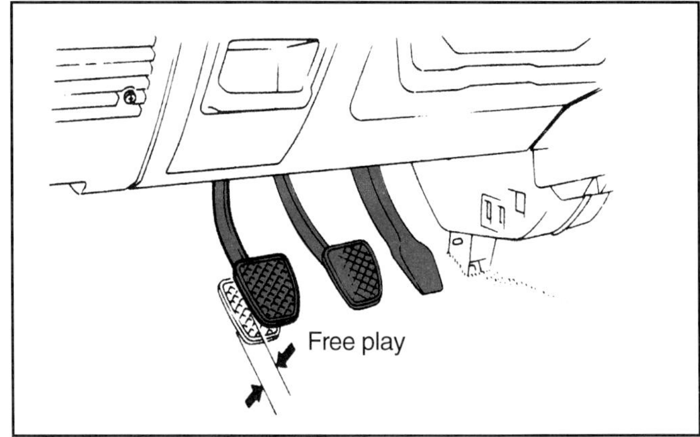

Figure 13-12. *While most manual transmission/transaxle problems are usually recognized as driveline problems, they are often mistaken for engine performance problems. To test for a slipping clutch, place the vehicle in its highest gear with the vehicle stopped and release the clutch. If the engine does not stall, the clutch is slipping or misadjusted. (Subaru)*

Automatic Transmissions and Transaxles

An automatic transmission can have a greater effect on driveability than a manual transmission. For example, mileage and performance can be affected when an automatic transmission fails to shift properly. After checking the level and condition of the transmission fluid, drive an automatic transmission vehicle enough to shift through all speed ranges. Carefully check that the shifts are smooth, without slippage, and that they occur at the proper times and speeds. See **Figure 13-13.** Try adjusting the throttle linkage, if

Hydra-Matic 4T60 Shift Speed Chart

	Ratios		1-2 min throttle	2-3 min throttle	3-4 min throttle	4-3 part throttle	3-2 part throttle	4-3 coast down	3-2 coast down	2-1 coast down
	Final drive	Sprockets								
AAH, ABH, AFH	3.33	37/33	13-16	22-25	39-50	53-55+	36-45	35-48	19-22	11-15
AJH	3.33	35/35	11-13	21-24	39-45	54-55+	41-48	36-43	18-21	9-11
ANH, ATH	3.33	37/33	13-16	24-27	45-55+	55+	45-54	40-53	21-24	11-15
BAH	3.06	37/33	12-14	18-20	43-52	55+	45-52	40-50	15-17	10-12
BDH	3.06	37/33	12-14	19-21	45-54	55+	46-54	42-52	15-18	10-13
BFH	3.06	35/35	12-15	20-22	45-53	55+	41-47	42-51	16-19	10-13
BHH	2.84	35/35	13-15	20-22	44-54	55+	47-54	41-52	16-19	11-13
BJH, BWH	3.33	37/33	12-15	20-24	46-54	55+	48-54	43-52	16-19	10-13
BMH	3.06	35/35	13-16	22-24	47-55+	55+	43-50	44-54	17-20	11-14
BPH	2.84	35/35	13-15	20-22	44-54	55+	47-54	41-52	16-19	11-13
CHH, YKH	3.33	35/35	12-15	21-24	43-50	55+	39-47	40-47	17-21	10-13
CJH, YLH	3.33	35/35	13-15	22-24	44-51	55+	40-48	41-49	18-21	10-13
LAH	3.33	35/35	12-15	21-24	43-50	55+	55+	40-47	17-21	10-13
LCH	3.33	35/35	12-15	21-24	43-50	55+	55+	40-47	17-21	10-13
LMH	3.33	35/35	12-14	20-22	41-47	55+	37-45	38-45	16-20	10-12
LNH	3.33	35/35	12-14	21-23	42-48	55+	38-46	39-46	17-20	10-13
PAH	3.33	35/35	12-15	20-22	46-54	55+	42-48	43-52	16-19	9-13

Notes:
1. All speeds indicated are in miles per hour. Conversion to km/h = MPH x 1.609.
2. Shift points will vary slightly due to engine load and vehicle options.
3. Speeds listed with + exceed 55 MPH.

Figure 13-13. *When an automatic transmission is suspected of slipping, note the speeds when each shift occurs. Check the recorded speeds against a shift speed chart, such as this one. Other charts can tell you which transmission clutches are applied, released, or held. (General Motors)*

equipped, and recheck the shift pattern. Also check that the torque converter clutch is applying when it is supposed to.

If the transmission problem cannot be corrected by simple adjustments, refer to the appropriate service manual for repair information. Basic transmission diagnosis is covered in **Chapter 20,** however, transmission and clutch repair and overhaul procedures will not be covered.

Diagnosing Other Vehicle Systems

Other vehicle systems that can affect driveability were discussed in **Chapter 11.** These systems include any vehicle part that is not included in the engine and drive train, but can have an affect on the performance, economy, or smoothness of the vehicle. Systems affecting driveability include the brakes, front and rear suspensions, steering system, air conditioner, and cruise control.

Many of these vehicle systems can be checked by inspecting the system's components. For example, the front end and steering linkage can be checked for worn parts and the brake rotors and drums can be measured for an out-of-round condition. See **Figure 13-14.** Look for obvious problems with any of the components. These include missing or loose drive belts on the alternator, air conditioner, or power steering pump. Look for loose electrical connections on the alternator or air conditioner. The air conditioning system can be checked to determine whether it has the proper charge of refrigerant, or for other problems that could cause the compressor to cycle excessively and affect engine operation.

Check for refrigerant leaks at the air conditioner fittings, **Figure 13-15.** These will appear as light oil smudges on the fittings or compressor seal, **Figure 13-16.** Leaks could lower

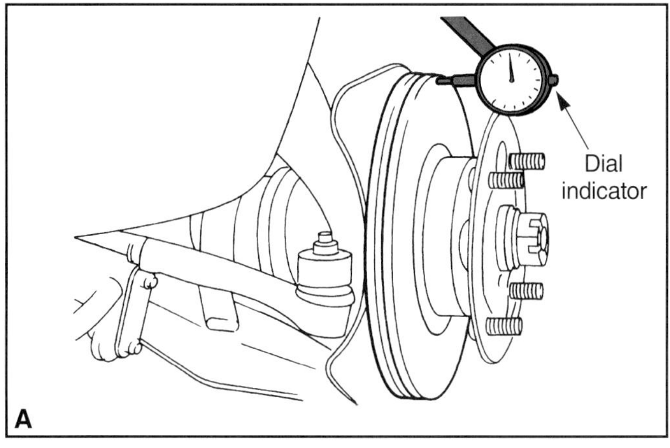

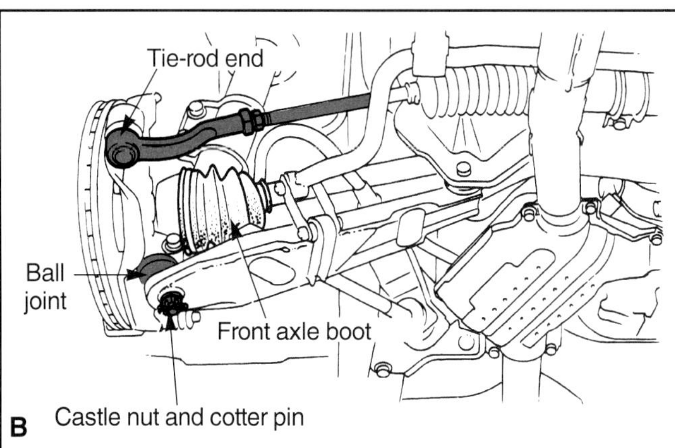

Figure 13-14. *The brake, suspension, and steering system can also cause problems that can look like an engine problem. A—A brake rotor that has excessive lateral movement or is warped will cause vibrations that can be mistaken for an engine or drive line problem. B—The drive axle, ball joints and other suspension components should be checked for play and wear. (Subaru)*

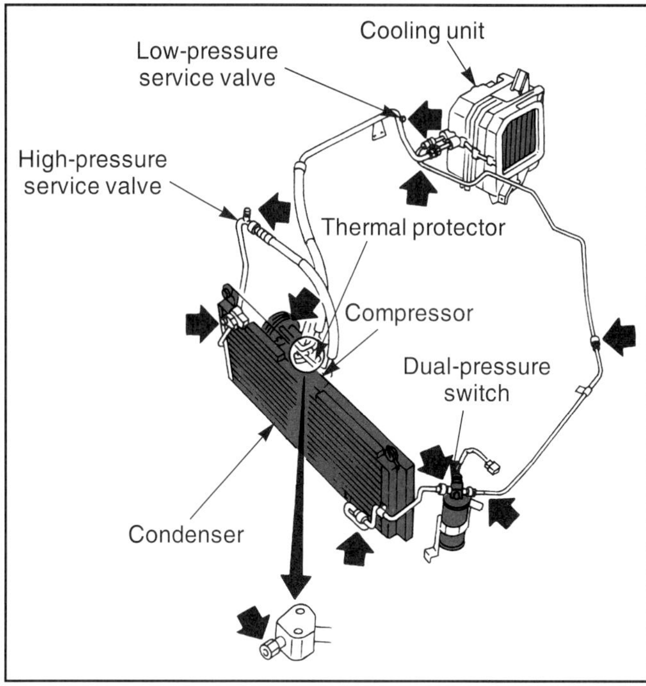

Figure 13-15. *The arrows on this air conditioning system component schematic show typical leakage points. Refrigerant oil on any of these points is a good indication of a potential refrigerant leak. (Subaru)*

Figure 13-16. *The compressor shaft seal is a common source of refrigerant leaks. The clutch itself can cause vibrations and noise that could be confused as an engine problem.*

the refrigerant level enough to cause rapid compressor cycling. A low refrigerant level will also set a diagnostic trouble code on some newer vehicles. These checks may allow you to quickly locate a vibration or other defect that acts like a driveability problem.

Whenever possible, the vehicle should be driven to detect problems. Problems in the suspension, brakes, or cruise control system are often quickly revealed during a road test. Review the road testing techniques and precautions listed earlier in this chapter. If the vehicle exhibits symptoms, or sets trouble codes that were not found when checking the engine systems in the shop, the problem could be in a system that operates only while the vehicle is moving.

Diagnosing Problems Caused by Aftermarket Equipment

Occasionally, add-on equipment will cause problems. While most add-on systems, such as alarm systems and fog lights, are carefully designed and well constructed to fit in with the factory electronic, computer control, and other on-board vehicle systems, problems can occur when it is improperly installed or a component in the system becomes defective.

If a suspected problem may be caused by a poorly installed add-on device, follow these guidelines for checking the system. First, examine the system for proper installation. This includes tight connections, components installed in locations that do not interfere with other vehicle systems, wiring cleanly routed, tied together, and preferably installed in plastic looms. Other checks are discussed in **Chapter 14.**

Deciding on Needed Work

Deciding on needed work is a process of interpreting the results of all diagnostic tests. It is simply a matter of taking all test readings and deciding what they mean. As discussed earlier, the test results can be simple observations of visible defects, detailed readings from elaborate test equipment, or any test procedure in between. Obviously, if you are investigating a vehicle with an engine miss, and find a disconnected spark plug wire, the source of the problem has been identified. However, most driveability problems are not that simple.

Before condemning any vehicle part based on test results, mentally review its interaction with the various engine and vehicle systems. Then decide whether the part in question can cause the particular test reading or symptom. For example, if charging voltage is low, the voltage regulator could be defective, but so could the alternator windings, diodes, or brushes. A low voltage reading could be caused by a discharged or defective battery, a slipping belt, or a poor connection at any one of several places in the vehicle wiring. Before replacing the alternator, you should always perform some more tests to ensure that it is the root cause of failure.

Troubleshooting charts and other diagnostic information can be a great asset to this process. If researched and prepared correctly, the troubleshooting chart will list all of the possible causes of the problem, allowing the technician to check everything in a logical sequence. Properly used, such information will speed up the checking and isolating process.

Always Perform Additional Tests

Additional testing is especially important when the suspected part is a solid-state, or otherwise untestable device. Most of these parts are too expensive to randomly replace without knowing for sure whether they are good or bad. Making further checks to confirm the problem is always a good idea, if only to increase the technician's confidence about finding the right part. Not many technicians are sorry that they made further checks, but a lot of them are sorry that they did not.

The following example illustrates the importance of additional testing. A single high firing line on an oscilloscope pattern is a common occurrence, usually indicating a defective or disconnected spark plug wire. However, a high firing line can also be caused by melted or widely gapped spark plug electrodes, or a missing or high resistance distributor cap contact. Do not be content to replace the wire and close the hood. Always check for other possible causes; in this case, by removing the distributor cap and affected spark plug to ensure that they are not the source of the problem. If the plug electrodes are melted or widely gapped, or if the contact inside the distributor cap is missing or has developed high resistance, the wire is not the problem. Another way is to recheck the firing line on the scope pattern after the new wire is in place. If the firing line is still high, either you have changed the wrong wire, or the cause is the plug or cap.

In another example, after entering the ECM self-diagnosis mode of a late-model vehicle, a technician finds an "oxygen sensor too rich" code has been set in the recent past. However, after the technician resets the ECM memory, the oxygen sensor code does not reset. Information from a scan tool indicates that the sensor is now working properly. It has not been proven that the oxygen sensor is defective. Before replacing the oxygen sensor, the technician should check for other engine conditions that could cause a momentary rich mixture that would set the sensor code. The technician should also investigate the fuel, ignition, and compression system components before deciding that the oxygen sensor is defective. If further tests indicate that all of the other components are operating properly, then the oxygen sensor can be replaced with confidence that it is the source of the problem. More often, additional tests will uncover another problem that was affecting the air-fuel ratio, and therefore, the oxygen sensor.

Deciding on the Proper Repair Steps to Take

The amount and type of needed corrective actions must also be determined. In some cases, the repair is as simple as reattaching a vacuum hose, removing dirt or corrosion from a sensor connector, tightening a belt, or resetting the ignition timing. In other cases, major unserviceable parts, such as the ECM, input sensors, output devices, fuel injectors, or the ignition coil must be replaced to correct the problem. To reduce the possibility of future problems, you should change parts that interact with the defective part. Examples are, replacing the rotor when the distributor cap is replaced, or replacing the fuel filter when installing new injectors. In all cases, the technician must thoroughly determine the extent of the repairs before proceeding.

Some parts, such as older carburetors, distributors, and alternators can be adjusted, have some components replaced, be rebuilt completely, or be replaced. Carburetors, for instance, contain many adjustable parts, while any defective parts in the alternator, and most distributors, must be replaced. Alternators are sometimes taken apart and repaired when the defective parts are relatively minor, such as brushes or bearings. Defective distributors and electronic voltage regulators, while theoretically serviceable, are almost always replaced today. This range of choices sometimes makes it difficult to decide what procedure should be performed.

Some engine parts, such as PCV valves, idle air control valves, or cooling system thermostats, may be cleaned up and returned to service. But when you consider the importance of the job that they do, and the high probability that they will fail again versus the cost of replacing them, it is dangerous to trust them. Factors that must be considered when deciding to adjust, rebuild, or replace a part are ease of adjustment, the need for special tools, cost of the replacement part, and the possibility that the old part will fail again.

Making Adjustments

If a part is easily adjustable, you can try the adjustment procedure before rebuilding or replacing. If adjusting the part does not restore its original performance, the part can still be rebuilt or replaced with little time lost. The important thing to remember is not to get too involved in trying to adjust a stubborn problem. For example, do not spend time trying to adjust the pickup coil air gap on a distributor with worn shaft bushings. The extra time spent will be lost if the component still requires overhaul or replacement. If there is any doubt about whether an adjustment has corrected the problem, replace the part.

Special Tools

Special tools are often needed to adjust or disassemble a complex assembly, and the cost of the tool may exceed the price of a complete replacement assembly. However, special tools can be used again for the same type of repairs in the future, and may be a good investment. The technician should also figure in the initial cost of the tool, versus the number of jobs that will be possible using that tool.

For example, a set of brass feeler gauges for adjusting distributor pickup coil air gaps is inexpensive, and will pay for itself on the first few jobs. A generic ECM scan tool is expensive, but may be a good investment if you expect to do a lot of diagnosis work on computer-controlled vehicles. If you expect to do a lot of the same type of repairs in the future, and the special tools are reasonably priced, they should be purchased.

Rebuild or Replace?

In cases of a defective component that must be rebuilt, the investment in materials and time must be weighed against the possibility that rebuilding the part will not fix the problem. At modern labor rates, it is often cheaper to install a new part than to spend any time on the old one. Many repair shops, and even some new vehicle manufacturers, are going increasingly to a policy of replacing complete assemblies. You must determine if rebuilding is cost effective.

In many cases, the customer will come out ahead with a new or remanufactured assembly instead of paying to rebuild an old part. The price of the new or remanufactured part is often less than the charge to rebuild the old part. These parts often come with a limited warranty from the manufacturer and the assurance that the part was assembled in a clean, controlled environment. The technician will often come out ahead, since the labor time saved rebuilding the old part can be devoted to other work.

Therefore, when deciding what to do to correct a problem, make sure that all parts that could contribute to the problem have been tested. In one form or another, every possible component and system should be tested. Then you can decide with assurance what components are defective.

Contacting the Owner About Needed Work

After determining what parts and labor are necessary to correct the problem and before proceeding to actually make repairs, contact the vehicle owner and get authorization to perform the repairs. *Never* assume that the owner will want the work done. The owner may not have sufficient money for the repairs, may prefer to invest the money in a newer vehicle, or prefer to have someone else perform the repair work. The defective part or problem may be covered by the vehicle manufacturer's warranty, the government-mandated emissions warranty, or a guarantee given by another repair shop or chain of service centers. In these cases, the vehicle must be returned to an approved service facility for repairs. If your shop is not one of these approved facilities, you cannot expect to be reimbursed for any more than diagnosing the problem.

If the vehicle is leased, the leaseholder is the actual owner. Depending on the terms of the lease, the leaseholder may be the only one who can approve any expenses in connection with the vehicle. Be especially careful if the vehicle is covered by an extended warranty or service contract. Extended warranties and service contracts are a form of insurance, and like all types of insurance, it is necessary to file a claim for any expenses. In some cases, the owner can file a claim after repairs are completed; while in other cases, approval must be granted from the insurer before the repair work can begin. Sometimes, the insurer will send an adjuster to inspect the vehicle before approval is given.

Before talking to the vehicle owner, leaseholder, or extended warranty company concerning authorization to perform needed repairs, you should make sure that you can answer three questions that will be asked. First, be prepared to tell exactly what work needs to be done, and why. Next, have available a careful breakdown of both part and labor costs. Third, be ready to give an approximate time when the vehicle will be ready. If you suspect a problem that requires further disassembly, such as a worn camshaft lobe, be sure that the customer understands that further diagnosis (and costs) may be needed before an exact price is reached.

Follow-up After Repairs

After making repairs, recheck the vehicle to make sure that the problem has been corrected. Check that the vehicle operates properly and that it meets emission standards when applicable. Keep in mind that on modern vehicles almost every system is connected to and affected by the rest of the vehicle.

Always recheck the vehicle, even when the defect is minor, such as an overheated electrical connector or a defective spark plug wire. Even a minor defect can have other consequences, or ruin other parts. For instance, a defective spark plug wire can cause a dead cylinder, leading to a rich mixture which overheats and destroys the catalytic converter. In other cases, the problem, although easily fixed, may have been caused by another vehicle defect. An overheated electrical connector may have been caused by excessive current flow due to a short elsewhere in the electrical system. If the cause of the excessive current flow is not corrected, the new electrical connector will be destroyed.

After work is completed, give the engine compartment a quick inspection for tools. Many technicians lose tools by accidentally leaving them in vehicles. Not only can this result in having to pay for replacing the tool, if the tool should fall into the cooling fan blades, it can be thrown outwards, possible causing damage to the fan, radiator, and other engine components. Finally, check the vehicle interior, steering wheel, and exterior body for oil, grease, and visible fingerprints. Your good repair will not be very good in the driver's eyes if they received their vehicle with grease on the seat and fingerprints on the steering wheel.

Summary

Isolating and correcting a driveability problem involves many procedures. Some of these procedures are diagnostic, or troubleshooting procedures. Other procedures are service and repair operations. These include checking the operation of vehicle systems to determine problem areas, determining which parts could be defective or out of adjustment, checking the condition and operation of the suspected parts, repairing or replacing these parts, and then rechecking the vehicle. Finding out exactly what new parts or adjustments are needed is the heart of diagnosis, or troubleshooting.

The ability to diagnose problems can be improved with study and experience. To become an expert at diagnosis, you first need to increase your knowledge of how internal combustion engines and their related systems operate, and you will need to develop the ability to approach a problem in a calm and logical way. When presented with a driveability complaint, start by determining the exact nature of the problem.

The first test involves driver input. What the driver says can help with diagnosis, but must be carefully evaluated. The driver may be unintentionally misleading you, or may be difficult to deal with. The technician should always make visual checks of the engine before proceeding further. Always open the hood and check for vacuum leaks, loose or overheated electrical connectors, or other obvious problems. Flex all electrical connectors while watching engine operation. If the engine operation changes, the connection is bad. Also look for other signs of problems, such as low fluid levels or lack of maintenance.

If a road test is necessary, the technician should always keep safety in mind. When making a road test, try to exactly duplicate the condition under which the problem seems to be occurring. Take notes if necessary. After the road test, compare your impressions with the information provided by the driver.

Make sure that the presumed driveability problem is not caused by another vehicle system that can cause a similar problem. Understanding the function and possible defects in every part of the vehicle will allow you to consider all of the possibilities. When diagnosing any problem, always check the simple things first. Check the obvious possibilities before going on to the difficult ones that require a lot of test equipment and time. You must check every part in a system, without assuming anything about the condition of any part. Knowing that all the easy things are okay will allow you to zero in on the hard things.

Experience, or information from other technicians or service publications, may lead you to one particular area that is known to give trouble. The basic systems should be checked first unless there is a very good reason to investigate another part of the vehicle. It is often difficult to determine whether the problem is in the fuel, ignition, or compression systems. Therefore, it is important to check all possible causes of a problem in the basic engine systems before making repairs.

After the fuel, ignition, and compression systems have been checked, the emissions, cooling, lubrication, and exhaust systems can be checked. Then check the drive train and related components. Before beginning diagnosis on any vehicle equipped with an on-board computer, you must do three things: thoroughly check all non-computer systems, check the ECM memory for trouble codes, and obtain the proper service manual and other information for the computer system. Other systems affecting driveability include the brakes, front and rear suspensions, steering system, charging system, air conditioner, and cruise control.

Deciding on needed work is a process of interpreting the results of all diagnostic tests and deciding what they mean. Before any vehicle part is condemned by test results, the interaction of the various engine and vehicle systems should be reviewed and a decision made on whether the part in question is the only possible part that can cause the test reading or symptom. Troubleshooting charts and other diagnostic information can speed up the checking and isolating process.

Additional testing is especially important when the suspected part is a solid-state device, or otherwise untestable. Therefore, when deciding what to do to correct a problem, make sure that all parts that could contribute to the problem have been tested. Then you can decide what components are defective. The amount and type of needed corrective actions must also be determined. In some cases, the repair is simple, while in other cases, major parts must be changed. Sometimes, the technician should change parts that interact with the defective part.

The technician must thoroughly determine the extent of the repairs before proceeding. Parts may be adjusted, rebuilt, or replaced. The technician must consider the ease of adjustment, the need for special tools, cost of the replacement part, and the possibility that the old part will fail again. If a part is easily adjustable, try to adjust it first, without getting too involved in the adjustment process. Special tools can be purchased to perform certain service operations, but their cost should be weighed against how useful they will be. When considering rebuilding a defective component, measure the investment in materials and time against the cost and the possibility that rebuilding the part will not fix the problem. At modern labor rates, it is often cheaper to install a new part than to spend any time on the old one.

After determining the parts and labor necessary to restore good driveability, and before making the repairs, the vehicle owner must be contacted to get authorization to perform the repairs. Never assume that the owner will want the work done. The problem may be covered by the manufacturer's warranty, the emissions warranty, or another kind of guarantee. This means that the vehicle must be returned to an approved service facility for repairs. If the vehicle is leased, the leaseholder may be the only one who can approve vehicle expenses. This may also apply in the case of extended warranties or service contracts.

Before talking to the vehicle owner about needed repairs, make sure that you can answer three questions: what work needs to be done, and why; the specific costs of parts and labor; when the vehicle will be ready. After all three of these things have been determined, the technician should call the customer and be ready to provide a detailed explanation of the needed repairs. If a problem that requires further disassembly is suspected, be sure that the customer understands that further diagnosis (and costs) will be needed before an exact price is reached.

Know These Terms

Troubleshooting	Root cause of failure
Driveability worksheet	Uneducated guessing
Road test	Dynamic compression test
Intermittent problems	Special tools

Review Questions—Chapter 13

Please do not write in this text. Write your answers on a separate sheet of paper.

1. To become an expert at diagnosis, the technician must learn to approach problems in a ______ and logical way.
2. When diagnosing any problem, always check the ______ things first.
3. If the suspected engine system is connected to the ECM, always check for ______.
4. When checking for defective computer control components, what order should they be checked in?

   ___ first (A) output devices

   ___ second (B) the ECM

   ___ third (C) input sensors
5. In some cases, ______ may enable you to bypass certain troubleshooting steps.
6. The ECM in systems without self-diagnostics must be checked by a ______.
7. Be especially careful when checking sensor connections, since they operate on very ______ voltages.
8. *True or False?* Whenever possible, make snap judgments about the cause of the problem.
9. A single defective part can cause ______ symptom(s).
   (A) only one
   (B) more than one
   (C) many
   (D) None of the above.

10. *True or False?* It is not necessary to recheck a problem after making repairs.

ASE Certification-Type Questions

1. Isolating and correcting a driveability problem involves ______.
   (A) diagnosis and troubleshooting
   (B) repair and service
   (C) diagnosis and repair
   (D) None of the above.
2. The easiest part of correcting a driveability problem is ______.
   (A) fixing what is wrong
   (B) finding out what is wrong
   (C) questioning the driver
   (D) selling the work to the driver
3. Technician A says that a difficult customer can be handled simply by ignoring them. Technician B says that getting angry in response to a customer's anger will make the situation better. Who is right?
   (A) A only.
   (B) B only.
   (C) Both A & B.
   (D) Neither A nor B.
4. Test results to be interpreted can include ______.
   (A) readings from test equipment
   (B) trouble codes
   (C) visible defects
   (D) All of the above.
5. All of the following are examples of simple checks EXCEPT:
   (A) checking vacuum hoses for cracks and leaks.
   (B) removing the spark plug to check compression.
   (C) inspecting the fuse block for open fuses.
   (D) checking the battery terminals for a tight, clean connection.
6. A good ______ will list all of the possible causes of a problem.
   (A) troubleshooting chart
   (B) schematic
   (C) technical service bulletin
   (D) None of the above.
7. Vehicle systems that can be checked by inspecting the system parts include the ______.
   (A) front end and steering linkage
   (B) U-joints
   (C) computer control system
   (D) Both A & B.
8. Technician A says that if a part can be cleaned up and reused, it should be. Technician B says that special tools are always cheaper than the cost of a replacement part. Who is right?
   (A) A only.
   (B) B only.
   (C) Both A & B.
   (D) Neither A nor B.
9. Technician A says that customers will sometimes come out ahead with a new part. Technician B says that it is cheaper to rebuild an old part in all cases, rather than paying for a new one. Who is right?
   (A) A only.
   (B) B only.
   (C) Both A & B.
   (D) Neither A nor B.
10. Technician A says that many technicians are sorry that they made further checks before replacing a part. Technician B says that most electronic parts are too expensive to randomly replace. Who is right?
    (A) A only.
    (B) B only.
    (C) Both A & B.
    (D) Neither A nor B.

Chapter 14

Electrical System Diagnosis and Repair

After studying this chapter, you will be able to:

- ❑ Evaluate if a potential problem is caused by an electrical fault.
- ❑ Perform a visual inspection of the electrical system.
- ❑ Use testlights and meters to check the electrical system and components.
- ❑ Perform electrical tests on various electrical devices.
- ❑ Perform repairs to the electrical system.
- ❑ Perform tests and service the charging system.
- ❑ Perform tests and service the starting system.

The electrical system is one of the greatest, and often the most overlooked contributor to driveability problems. This chapter explains how to perform basic electrical circuit testing and identifies ways of testing electrical components in systems that affect driveability. Studying this chapter will explain testing procedures that you will build on to learn how to diagnose problems in later chapters.

Identifying Driveability Problems Caused by the Electrical System

When approaching any driveability problem, the first area to check is the vehicle's electrical system. With computer control and electronics playing a major role in engine and drivetrain operation, a minor problem such as a loose ground connection can cause major driveability problems that act similar to problems caused by other systems. An electrical fault can cause problems in almost every vehicle system, therefore, it should be the first place to start looking for the cause of a driveability problem.

Unfortunately, there are only a few common driveability problems that are directly caused by the electrical system. These are limited to no-crank conditions. Typical electrical faults are shown in **Figure 14-1.** The electrical system faults shown in **Figure 14-1** will often create problems by causing another driveability related system, such as the fuel system, to malfunction. If a driveability problem seems unusual, there is no reference to it in the service manual, or the problem's occurrence coincides with another vehicle system's function or malfunction, an electrical fault is probably the source of the problem.

Preparing to Diagnose Electrical Problems

After verifying that a driveability problem exists that may be caused by the electrical system, the first step in diagnosis is to locate the service manual for the particular vehicle. The service manual contains the proper operating parameters for each of the vehicle's systems and components. Service manuals also contain the vehicle's wiring schematics. The schematics give you a road map in which

Figure 14-1. *A minor electrical defect can cause many problems on the modern computer-controlled vehicle. The systems pointed out here are the most common. Poor ground and power connections can occur throughout the entire vehicle. (Toyota)*

to look for the problem. If you do not use wiring schematics, you will be working in the dark, since the vast numbers of wires, connectors, and electrical components in even the simplest circuits can make proper diagnosis without a schematic almost impossible.

Using Schematics to Diagnose Electrical Problems

A ***schematic*** is a way of graphically representing the flow of electricity between electrical components. The path that the current takes through a single electrical unit or system is called a circuit. Following the current path is called ***circuit tracing.*** Using the correct schematic will allow the technician to visualize the flow of current through the wiring and components. It is not enough to simply have the proper schematics in hand, you must know how to properly use them. There are a few rules to using schematics and tracing wiring problems:

- ❑ Do not write on an original schematic. If you need to mark a schematic, make a photocopy of the page and mark the copy.
- ❑ If the suspected circuit shares a power source or ground with other unrelated systems, check to see if any of the other systems are also affected.
- ❑ Check the circuit's power sources, circuit protection devices, and ground connections first. These are the most common sources of electrical failures and are often overlooked, even by experienced technicians.
- ❑ If the power and ground connections are good, start at the power source and trace the wiring through the circuit protection devices, harness connectors, individual components, and finally to the circuit's ground connections.

However, the vast increase in electrical components and wiring in modern vehicles makes having one schematic for the entire vehicle impractical. Modern service literature contains many schematics, each covering one electrical circuit, or a few closely related circuits. For instance, the schematic in **Figure 14-2** illustrates only the current flow through the engine fuel injectors. Note that the electrical power passes through the ignition switch, through a fuse, several connectors, the injectors, and grounds through the fuel injector drivers in the ECM (called a Powertrain Control Module, or PCM, in this schematic). Also note that the schematic references other schematic pages when an electrical connection is made with another system.

The schematic in **Figure 14-3** is more complex. This schematic shows the operation of the basic ignition system. The Hall-effect crankshaft and camshaft sensors produce speed and position signals which feed into the PCM.

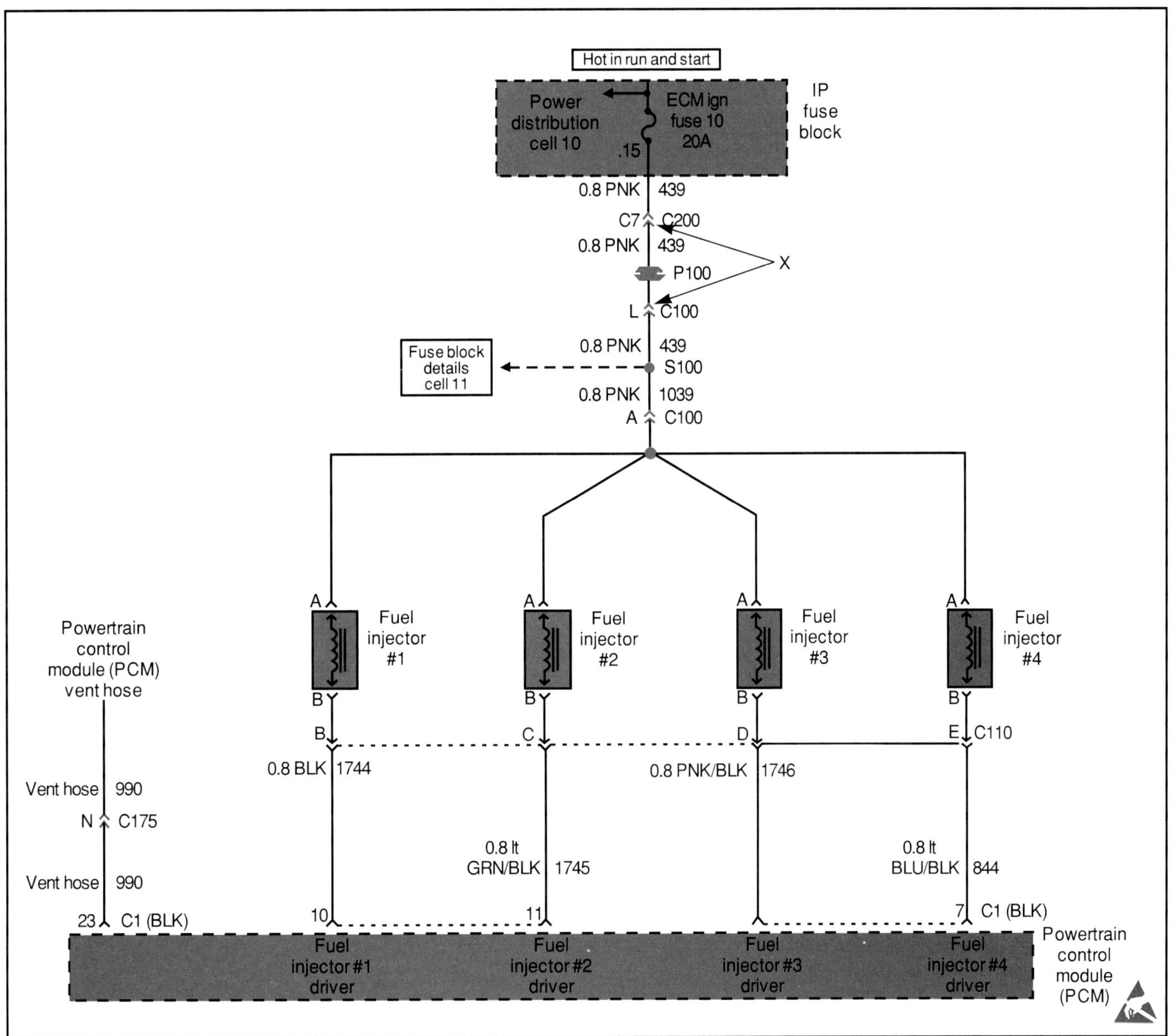

Figure 14-2. *Current is fed directly to the injectors from a fuse that is hot in start and run. The opening and closing of the injectors is controlled by providing a path to ground, which is done by the ECM. (General Motors)*

The knock sensor also provides input to the PCM. The PCM controls the firing of the plugs by controlling the flow of current into the ignition control module. The module receives current from the ignition switch through a 20 amp fuse. If the ignition system is not operating, this schematic will tell the technician which components to concentrate on first. For instance, finding out that the module is powered through a fuse should lead the technician to check the fuse condition before proceeding to anything more complex.

The schematic for a V-6 oxygen sensor circuit in **Figure 14-4** calls for further knowledge of the vehicle system. Note that the schematic shows four heated oxygen sensors (HO_2S). Two are installed in the left and right side exhaust manifolds, one in the exhaust pipes immediately ahead of the catalytic converter, and one behind the converter. The technician must consult other sections of the service manual to determine the exact placement and operation of these sensors.

Visual Inspection

As you learned in **Chapters 12** and **13,** a visual inspection should be performed before any diagnostic tests. A visibly damaged or defective wire, or device, is easy to spot. However, most defective electronic components will show no visible signs of damage. In some cases, the damaged component or wire is hidden in a loom, behind a panel or the dash, or under one or more components, which may or may not be related. Many electrical problems can be found by simply looking, listening, and smelling.

Look for visible evidence of electrical problems, such as loose, corroded, pinched, or burned wires, missing or

Camshaft position (CMP) sensor
Crankshaft position (CKP) sensor
Hall effect sensor
Hall effect sensor
Hot in run and start
Power distribution cell 10
ECM ign fuse 10 20 a
J5
IP fuse block
Tachometer input
Instrument cluster
8
0.5 WHT 121
B3 C200
P100
0.5 WHT 121
A C100
C B A
0.8 RED 631
0.8 PNK/BLK 632
S103
0.8 RED 631
C B A
0.8 PNK 439
C7 C200
P100
0.8 PNK 439
L C100
0.8 PNK 439
S100
Fuse block details cell 11
0.8 PNK 439
B C2
Ignition control module(ICM)
0.8 BRN/WHT 633
0.8 PNK/BLK 632
0.8 YEL 573
Ignition positive voltage
Ignition control A
Ignition control B
Ground
E D C1
A C2
08. WHT 121
0.8 PNK/BLK 632
0.8 PNK/BLK 632
16
0.8 RED/BLK 453
0.8 PPL/WHT 430
0.8 BLK/WHT 451
0.5 DK BLU 496
27 20 C3(BLU)
17 C2 (GRY)
0.8 RED 631
24
25 C1 (BLK)
32 C2 (GRY) 32
31 C3 (BLU)
KS signal
Tachometer output
CMP sensor signal
CMP/CKP sensor battery positive voltage (B+)
CMP/CKP sensor ground
CMP/CKP sensor ground
CKP 7x reference signal
Ignition control A
Ignition control B
Powertrain control module (PCM)
S106
1
Knock sensor (KS)
1 BLK/WHT 451
Ground distribution cell 14
0.8 TAN 413
G105

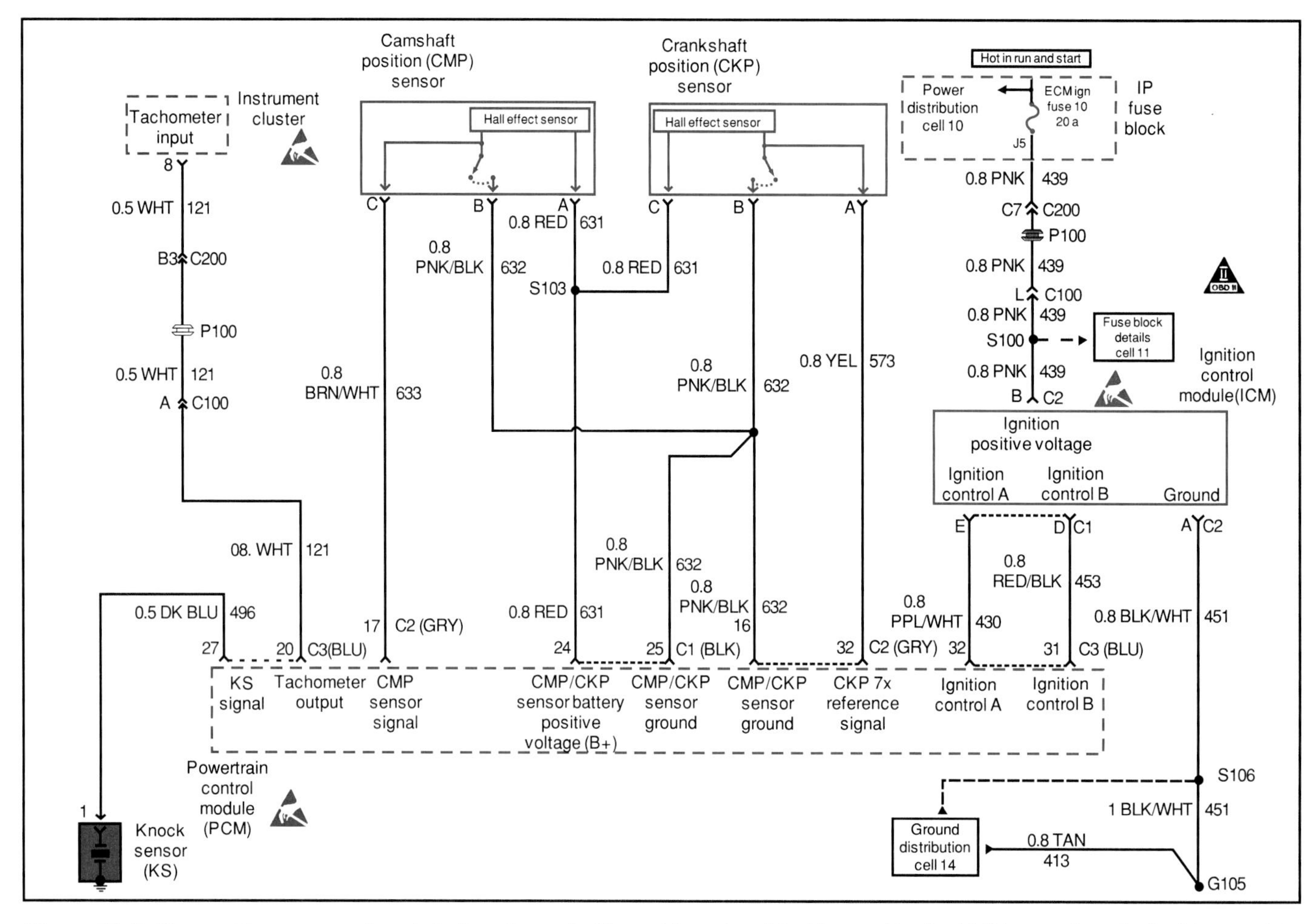

Figure 14-3. *The position sensors (crankshaft and camshaft) provide a signal that is used by the ECM as well as the ignition module. A defect in most of these circuits or components can cause the vehicle to stall or not run at all. (General Motors)*

damaged wire insulation, disconnected wires, overheated or loose electrical connectors, and shorted, disconnected, or otherwise inoperative electrical devices, **Figure 14-5.** Remember that many computer control sensors and output devices operate on very low voltages, and even slight problems in the wiring may cause inaccurate sensor readings. Arcing caused by loose electrical connections will often produce voltage spikes that can affect computer operation. Look for aftermarket lights, radios and CD players, alarm systems, cellular telephones, trailer brakes, and other nonfactory add-on equipment that may be affecting engine or drivetrain operation through improper wiring or electromagnetic interference.

Checking Circuit Protection Devices

If several circuits fail at the same time, the cause is usually a blown circuit protection device or a poor connection at a common power source, circuit splice, or ground. As you learned in **Chapter 6,** there is usually more than one fuse block in the modern vehicle and most of the fuses and circuit breakers are located in these blocks. Unfortunately, many technicians will often overlook circuit protection devices in their rush to solve a driveability problem.

Fuses and Circuit Breakers

Look for cracked or burned fuses. Sometimes a burned fuse will be clearly visible in the block, but in most cases, removal is necessary for proper inspection, **Figure 14-6.** If a fuse is damaged or burned out, it must be replaced. Circuit breakers will normally reset, but may be damaged by excessive current. Replacing a fuse or circuit breaker is easily accomplished; simply remove the damaged device from the block and replace it with a similarly rated fuse or circuit breaker, **Figure 14-7.**

Caution: Never use a higher rated circuit protection device or a wire to replace a fuse or circuit breaker. This can result in system damage and fire. If a new fuse burns out or circuit breaker opens after installation, find out why. This is an indication of a circuit drawing excessive current.

Fusible Links

Check the fusible links carefully. Most fusible links are located near the battery positive cable, attached to a body-mounted junction block, or at the starter solenoid. See **Figure 14-8.** All fusible links are connected directly to

Hot in run and start
Power distribution cell 10
Eng fuse 4 20 A
IP fuse block
L7
0.8 PNK 539
G C201
P100
0.8 PNK 539
N C100
0.8 PNK 539
Fuse block details cell 11
S123
0.8 PNK 539
G C180
0.8 PNK 539
S301
HO2S bank 1 sensor 1 signal
HO2S bank 1 sensor 1 low
HO2S bank 1 sensor 3 signal
HO2S bank 1 sensor 3 low
HO2S bank 2 sensor 1 low
HO2S bank 2 sensor 1 signal
HO2S bank 1 sensor 2 low
HO2S bank 1 sensor 2 signal
Vehicle control module (VCM)
19 25 22 26 7 21 24 20 C1 (BLU)
0.8 PPL/WHT 1665
0.8 PNK 539
0.8 PPL/WHT 1668
0.8 TAN/WHT 1669
0.8 TAN 1667
0.8 PPL 1666
0.8 TAN 1671
0.8 PPL 1670
0.8 PNK 539
0.8 TAN/WHT 1653
Heated oxygen sensor (HO2S) bank 1 sensor 1
Heated oxygen sensor (HO2S) bank 1 sensor 3
Heated oxygen sensor (HO2S) bank 2 sensor 1
Heated oxygen sensor (HO2S) bank 1 sensor 2
0.8 BLK 150
S300
0.8 BLK 150
H C180
0.8 BLK 150
S109
Ground distribution cell 14
0.8 BLK 150
S110
5 BKL 150
G102
Ground distribution cell 14

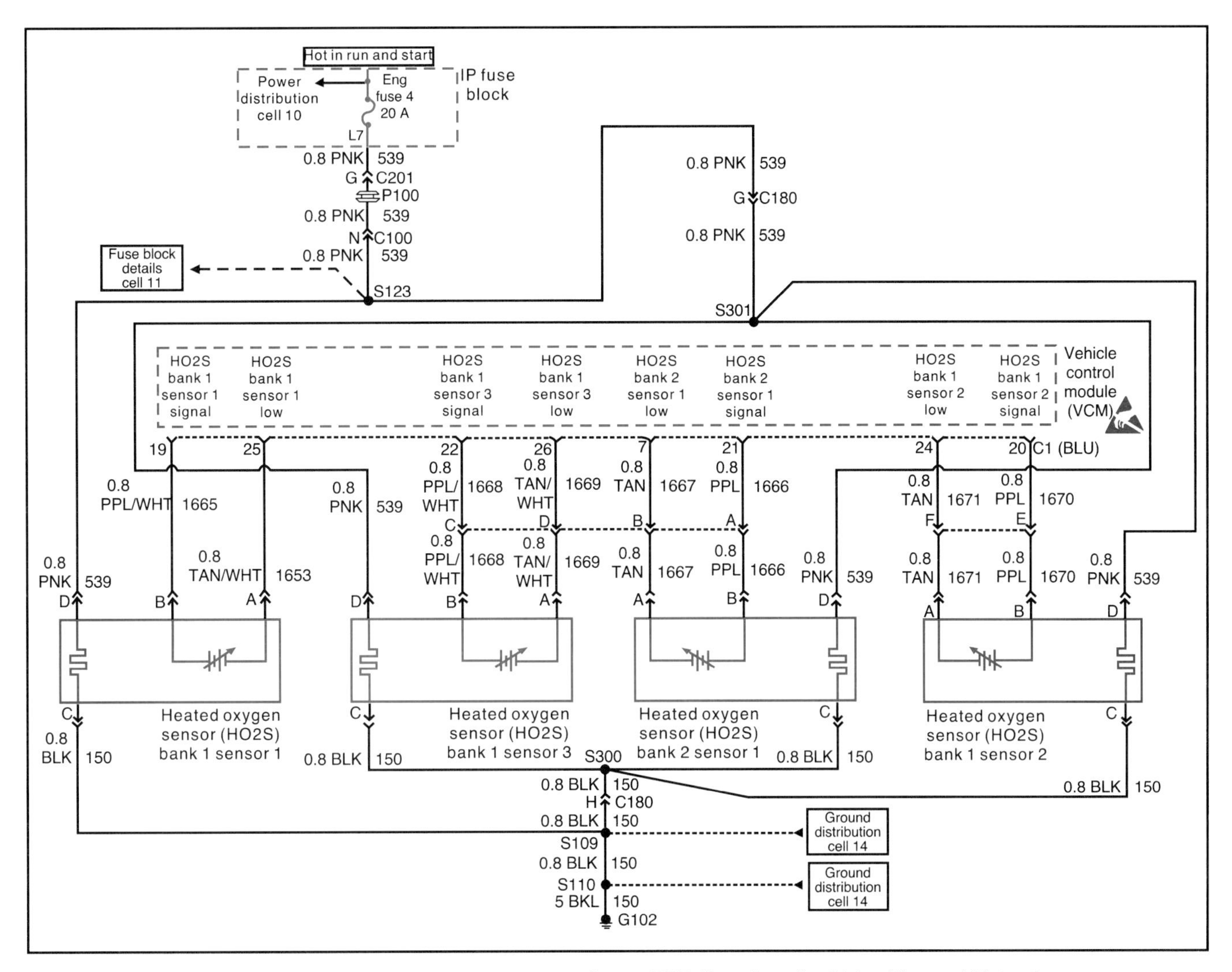

Figure 14-4. *This schematic shows the oxygen sensor setup for an OBD II equipped vehicle. (General Motors)*

Figure 14-5. *A visual inspection can often provide the answer to what initially seems like a complex problem. A—Loose and dirty terminals and connections, while more likely to occur in the engine compartment, can be found anywhere there is wiring. B—A disconnected part, such as this relay, is often the result of accidental or intentional tampering.*

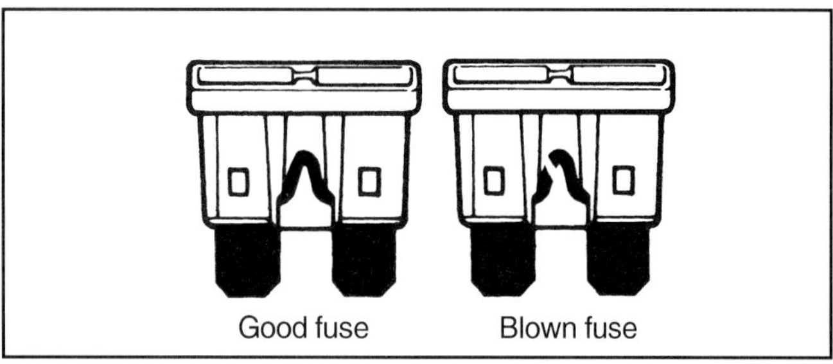

Figure 14-6. *The conductor for a good fuse will be intact and the insulator clear while a blown fuse will be dark and its conductor will be broken. (General Motors)*

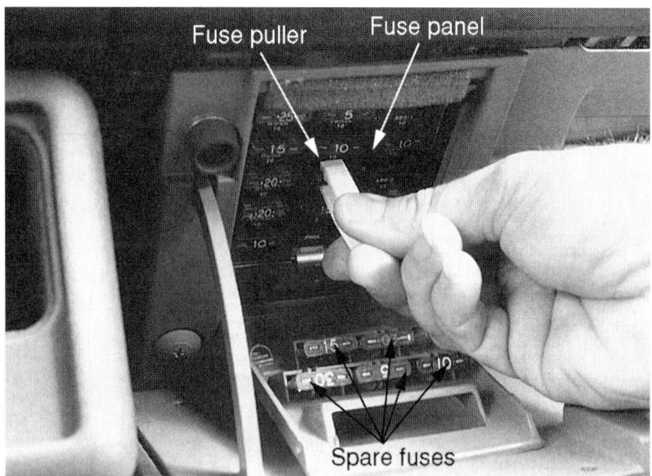

Figure 14-7. *Fuse blocks are usually located in an accessible place in or under the dash. Most manufacturers now supply additional fuses should one blow.*

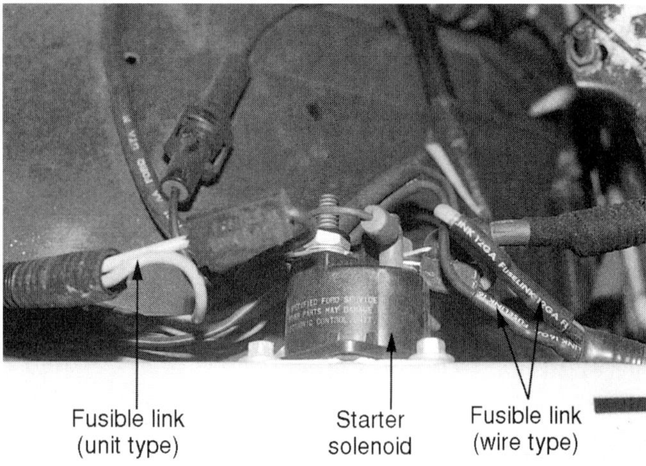

Figure 14-8. *Most fusible links are connected to the starter solenoid as shown here, or to a junction block located in the engine compartment.*

battery power in some way. A blown link is easily visible in most cases. If the insulation is discolored, blistered, or burned off completely, the circuit has overheated, and the link will melt through soon, if it has not already, **Figure 14-9.**

Sometimes a fusible link will burn out without melting the insulation, **Figure 14-10.** Twist and gently pull on the ends of any accessible links. If the link can be pulled apart or feels springy, the wire has probably come apart inside the insulation. Sometimes, the link melts at the end connection where the break is not visible and may appear to be in one piece.

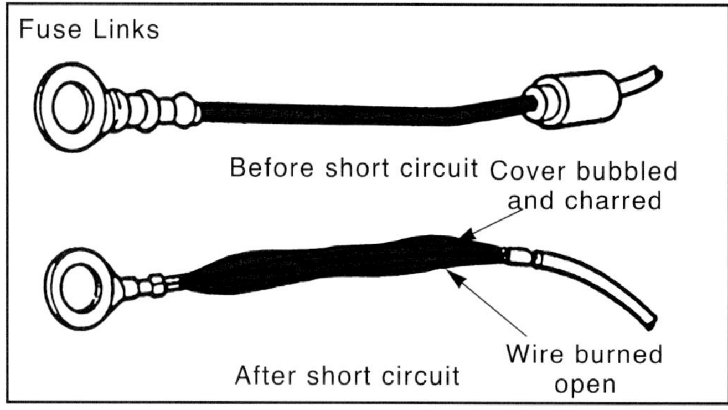

Figure 14-9. *Fusible links will usually show visible damage if they are subjected to excessive current. Note the differences between a good and blown fusible link.*

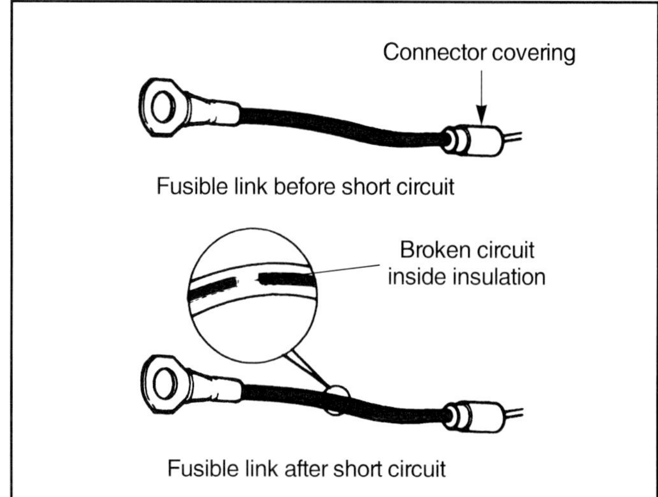

Figure 14-10. *Sometimes the fusible link will not show damage because the break is close to one of the ends. Be sure to feel the insulation for sponginess before declaring a fusible link good. (General Motors)*

Checking for Electrical Connection Problems

Defective electrical connections will often overheat, melting and discoloring the connector and the wire insulation. Any melted, distorted, or discolored electrical connectors and wires should always be investigated. In some cases, extra electrical connectors will be located in the engine compartment. Some of these connectors are there because wiring harnesses are standardized for a particular make of vehicle, with connectors for any available option, or optional equipment for the engine or a different engine. If the option was not installed on the vehicle, the connector will not be plugged into anything.

Some connectors are ***test leads*** that can be used to check for the presence of voltage in certain circuits or can be grounded to put a system into its test mode, **Figure 14-11.** Sometimes, however, loose connectors will be there because someone has disconnected an electrical device. When a loose electrical connector with no obvious mate is discovered, check the service manual to ensure that it is supposed to be disconnected on that particular vehicle or engine. On some vehicles, the test lead may be the computer system's data link connector, **Figure 14-12.**

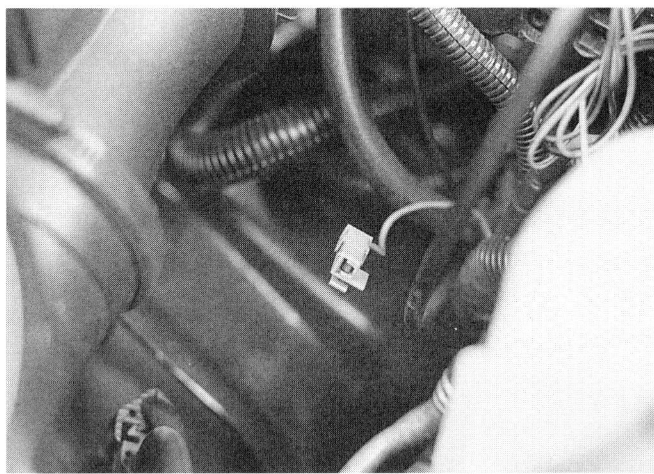

Figure 14-11. *A lead that is disconnected in the engine compartment may be a test lead. This particular terminal is used to check for voltage to the fuel pump. Grounding this lead would cause the fuel pump to run.*

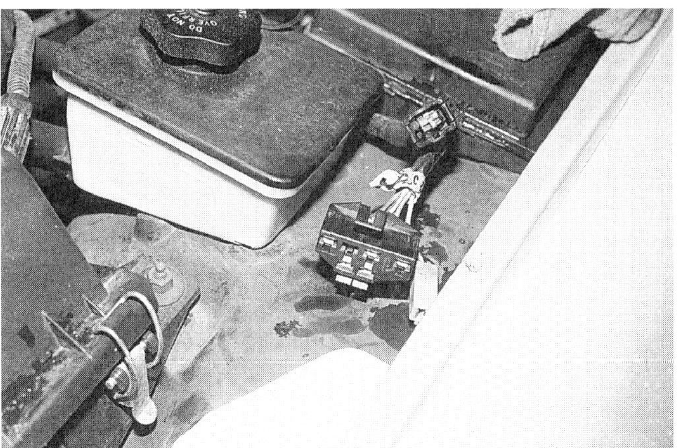

Figure 14-12. *Some OBD I vehicles have their data link connector located in the engine compartment. Applying voltage to or grounding the wrong lead on this connector can cause major computer damage. This is the reason that it is important to have a service manual on hand when diagnosing any vehicle.*

Flex Test

A common cause of many electrical problems is poor connections caused by dirt, grease, corrosion, overheating, or hidden short circuits. Even a small amount of dirt in a low voltage sensor circuit can change the wire's resistance and affect its input to the ECM. The ***flex test,*** **Figure 14-13,** is used to check the engine wires and electrical connectors for high resistance connections or internal shorts.

About half of all connection problems can be found by the flex test. With the engine or system running, wiggle, twist, tap, and generally flex all underhood wires and electrical connectors as you observe engine or component operation. If the engine, system, or component operation drastically changes, starts, or stops when a connector or wire is flexed, the connector should be taken apart or the wire stripped for further inspection.

Warning: Be careful when working near any running engine.

If the connector uses a rubber grommet, make sure that it is in good condition and in place on the connector. If the grommet is missing or in poor condition, replace it. This grommet prevents dirt, water, and other contaminants from entering the connector. See **Figure 14-14.**

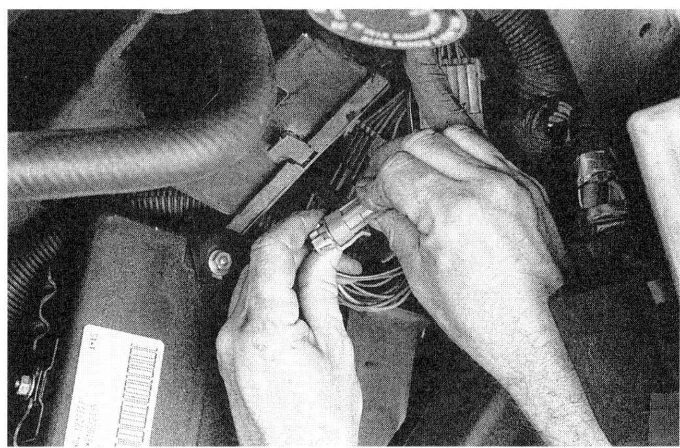

Figure 14-13. *The flex test will reveal problems caused by poor connections and dirt. If a device starts or stops operating when a connector or wire is flexed, you've found the problem.*

Checking for Electrical Device Problems

Watch and listen for proper device operation. Devices such as relays and solenoids will make audible noises if they are functioning. If the device has a lever or other moving component, observe the device's operation. Listen for unusual noises and visible signs of excessive heat and smell for burning. These are signs that a component is damaged internally. However, in most cases, an electrical component will show no visible signs of damage.

In many cases, the only way to determine whether a component is defective is to substitute a known good unit. When testing by substitution, however, the technician must keep several things in mind:

- ❑ Obtaining a known good part may be expensive and time consuming. In many cases, especially when substituting an electrical or electronic part, the part may not be returnable.

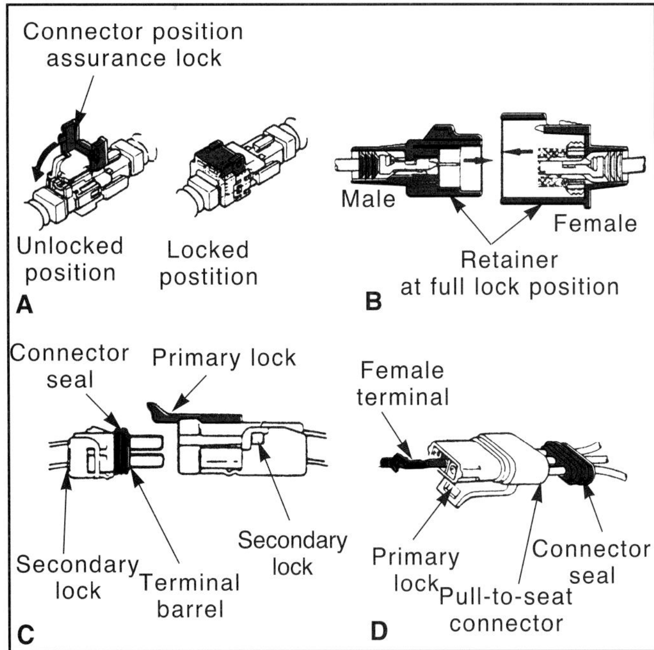

Figure 14-14. *The modern vehicle uses a variety of sealed, plug-in connectors. A—This connector uses a lock to ensure that it will not pull apart. This type of connector is used on air bag and other safety systems. B—Note the seals on the wires of both the male and female terminals. C—This is a weather pack connector. A missing seal is the most common cause of plug-in connector defects. D—Terminals can be pulled from the front and back of the connector, depending on the design. (Toyota and General Motors)*

- If the actual problem is a short or high amperage draw somewhere else in the system, the heavy current flow may ruin the substituted part.
- If the part requires a great deal of labor (time) to install, this labor will be wasted if the replaced part is not the problem.

The technician must be careful to eliminate all other sources of problems before substituting a part. Always consult the manufacturer's manual to determine whether a suspected part can be tested before replacing it. Technicians can partially offset the problem of obtaining substitute parts by acquiring a stock of commonly needed known good parts.

Test Equipment Used to Diagnose Electrical Problems

If a visual inspection has not revealed the problem, it will be necessary to begin tests using various tools and shop equipment. While test equipment can quickly reveal the source of the problem, they can also create problems if used improperly. The following sections will quickly review the most common types of tools used to diagnose electrical problems.

Note: Meters such as voltmeters, ammeters, and ohmeters, will be referred to by these names. As you learned earlier, these meters are often combined into a single, all-purpose multimeter.

Using Jumper Wires

Jumper wires are the simplest and easiest tools to use for electrical diagnosis. However, they can also cause a great amount of damage if used improperly. Unfused jumper wires and test probes are used to connect meters and other test equipment to various electronic components. They should be used to apply voltage to a circuit only when specifically recommended by the manufacturer, **Figure 14-15.** In all other cases, use a fused jumper wire as in **Figure 14-16.** This will prevent damage to the circuit and the jumper wire. Never use an unfused jumper wire to apply voltage to a circuit. Never leave a jumper wire in place for any period of time as it can start an electrical fire.

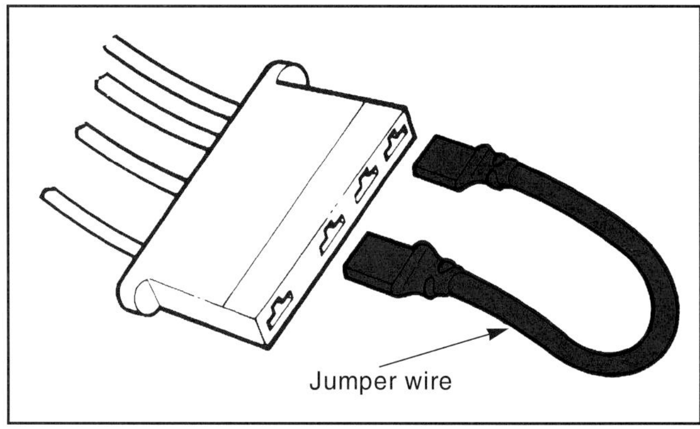

Figure 14-15. *An unfused jumper wire is used to apply voltage to a circuit. Be careful when applying voltage with any unfused wire. (Ford)*

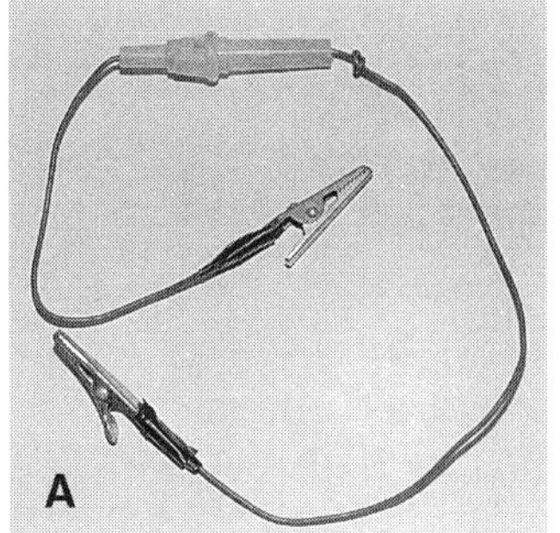

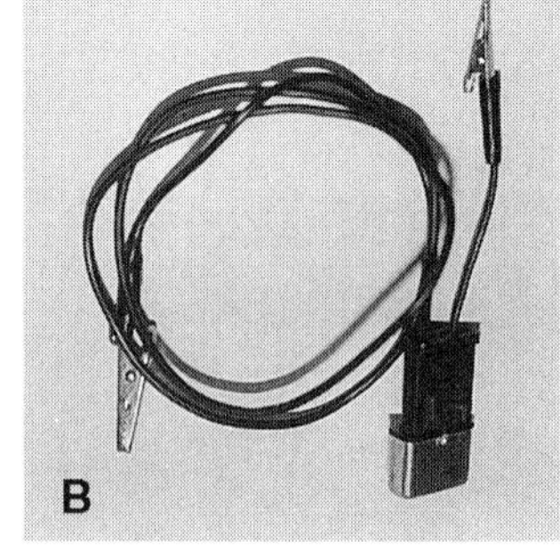

Figure 14-16. *Fused jumper wires should be used whenever power needs to be applied to any circuit. The fuse or circuit breaker will prevent vehicle damage. A—Jumper wire with inline fuse. B—This jumper wire can be used as a short finder.*

A jumper wire with a circuit breaker can also be used as a ***short finder.*** If the jumper wire is substituted for the blown fuse in a circuit, the short will cause the circuit breaker to open and close the circuit.

Caution: Be sure that the substituted circuit breaker is rated at the same or less amperage than the fuse for which it is being substituted.

Since current flow in a wire creates a magnetic field around the wire, a ***compass*** placed near the wire of the affected circuit will fluctuate (jump) every time that the circuit breaker opens and closes. Move the compass slowly over the wire harness, starting at the fuse, and heading for the component served by the fuse. When the compass stops fluctuating, current is not flowing in that section of the wire, and the compass has just passed the shorted section. This procedure can save much time as well as damage to wires and harnesses.

Using Test Lights

Both nonpowered and self-powered test lights are useful tools to test for the presence of voltage and circuit continuity. However, they are the single biggest contributor to electrical system and component damage due primarily to misuse by the technician. There are several rules to remember when using test lights including:

- ❑ Never pierce a wire with a test light or other probe. The probe will damage the wire and insulation. The resulting hole will also allow water, dirt, and other contaminants to enter, corroding the wire and changing its resistance. If a wire is pierced for any reason, place a small amount of silicone sealant on the pierced area. If the connector can be easily removed, a piece of shrink tubing should be placed and melted over the hole.
- ❑ Never use a test light to test a solid state component or ECM unless directed by the vehicle's service manual and only if the test light has a minimum impedance of 10 meg ohms.
- ❑ Never, under any circumstances, use a self-powered test light to test a solid state component.

Nonpowered Test Light

Nonpowered test lights can be used to test for the following:

- ❑ The presence of voltage in a circuit.
- ❑ A short to ground.
- ❑ The secondary ignition circuit for arcing to ground.

The secondary ignition procedure will be discussed in more detail in **Chapter 16.** As mentioned earlier, any test light used for driveability and other automotive electrical work must have a minimum 10 meg ohm impedance. The 10 meg ohm impedance protects the ECM and other sensitive electronic devices from damage as well as the technician from harm, **Figure 14-17.**

Using Test Lights

To use the test light to check for voltage, connect the alligator clip to a good ground. Touch the probe to the terminal or point to be tested. The test light should illuminate if voltage is present. If voltage is low, the light will glow dimly. The nonpowered test light can also be used to test for ground by connecting the alligator clip to battery voltage and touching the probe to the circuit's ground terminal. To test a power circuit for a short to ground, connect the test light to battery voltage. Probe the circuit's power terminal. If the test light illuminates, the circuit is grounded.

Caution: When testing a circuit for a short to ground, disconnect any electronic components.

Figure 14-17. *A test light is a powerful tool for finding many electrical problems. It can also cause damage if used improperly.*

Self-Powered Test Light

Self-powered test lights are useful for testing for continuity in a wire or circuit. This is a great time-saver, since a wire that is routed under the vehicle's dash, interior carpet or headliner, can be tested and eliminated if the test light illuminates. **Figure 14-18** shows a self-powered test light being used to test for continuity.

Using a Self-Powered Test Light

Disconnect all electronic components before testing any circuit with a self-powered test light. Connect the test light's alligator clip to one end of the circuit. Use the probe to touch the other terminal in the same circuit. If the circuit has continuity, the test light will brightly illuminate. If there is excessive resistance in the circuit, the light will glow dimly. If there is an open in the circuit, the light will not illuminate. The self-powered test light can also be used to check for shorts and grounds in the same manner as the nonpowered test light.

Using an Ammeter

Ammeters are used to test for the amount of current flowing through a circuit. This meter is used to determine if

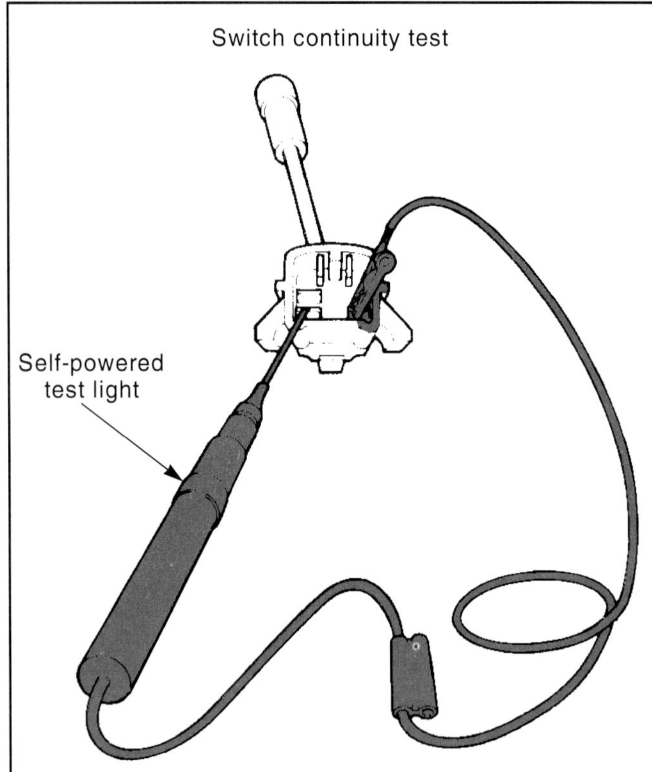

Figure 14-18. *The self-powered test light can be used to check switches and circuits for continuity. To test a switch, connect the test light as shown here and close the switch. If there is continuity, the test light will glow brightly. (Ford)*

a device or circuit is drawing excessive current. To use an ammeter, connect the ammeter in *series* with the circuit, **Figure 14-19.** Modern multimeters have an amp setting capable of handling amperage flows of up to 10 amps, **Figure 14-20.**

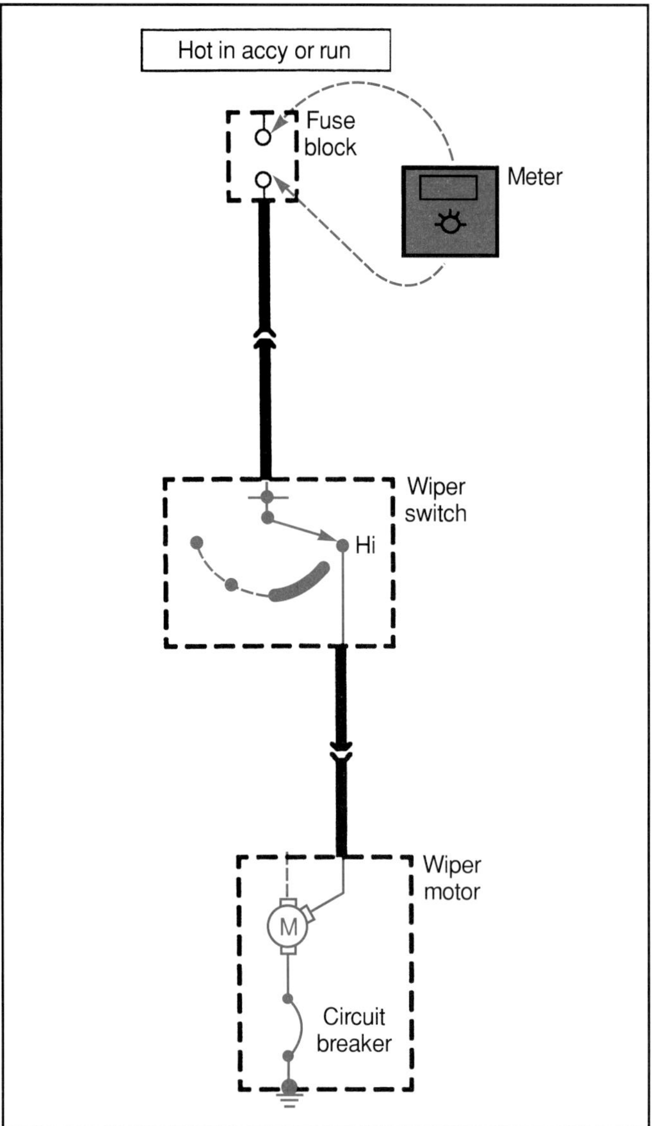

Figure 14-19. *When using an ammeter, always connect in series. (General Motors)*

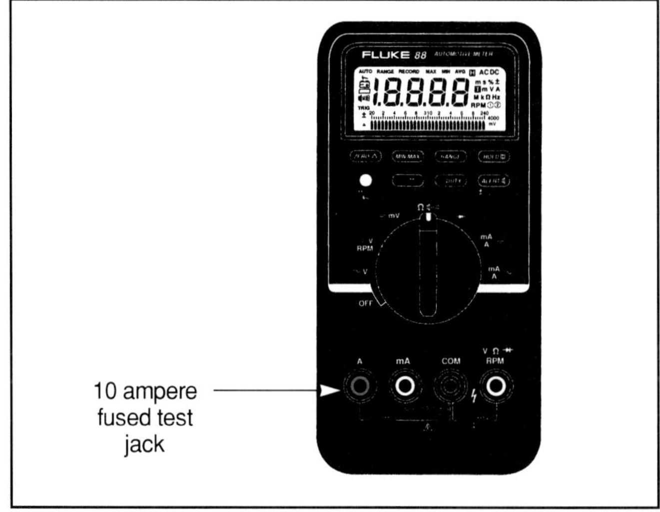

Figure 14-20. *Modern multimeters have a high amperage test jack that is fused for checking parasitic and other current draw problems. (Fluke)*

Using a Voltmeter

Voltmeters are used to check for the amount of voltage in a circuit, as well as the type. A good voltmeter for automotive use should be able to measure alternating as well as direct current. Voltmeters are connected in *parallel* with the circuit. The connections are made to the positive terminal connection and a ground. Many modern multimeters are equipped with an inductive pickup, **Figure 14-21.** This pickup is clamped over the negative battery cable. The pickup reads the magnetic field created by current flowing through the wire and converts it into a voltage or amperage reading.

Testing for Voltage Drop

A ***voltage drop test*** is used to test for excessive resistance in a circuit. It is very effective in checking components that cannot be tested by an ohmmeter. It also saves time, since no components have to be disconnected from the circuit. The voltmeter in **Figure 14-22** is connected to read the voltage across a battery cable connection as current flows

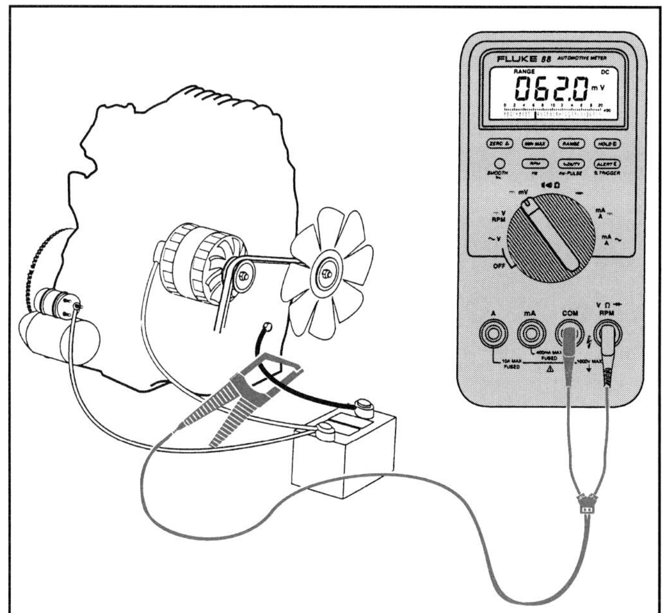

Figure 14-21. *Inductive pickups allow the multimeter to read a greater amount of voltage or amperage without subjecting the meter itself to high current or voltage. (Fluke)*

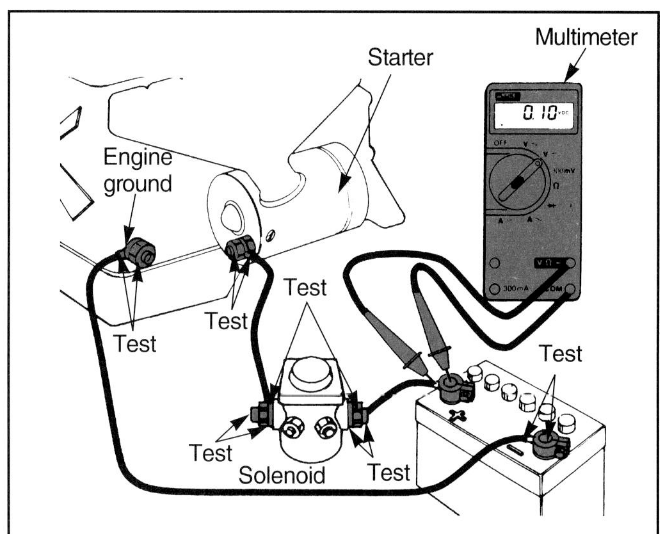

Figure 14-22. Voltage drops are determined by connecting a voltmeter between two points in a circuit. If voltage flows through the meter, it will be indicated in the display, meaning that there is a problem in the circuit between the points where the test leads were attached.

through it. If the connection has high resistance, current will try to flow through the meter, creating a voltage reading. Voltage higher than the specified figure means that the connection must be cleaned or replaced.

Using an Ohmmeter

The ohmmeter is the most commonly used meter for driveability testing. Since most input sensors are designed to change an ECM-supplied reference voltage, they are either variable resistors or thermistors. Ohmmeters are also used to check for resistance and continuity in wires, diodes, and other circuit components. To use an ohmmeter, connect it in *parallel* with the circuit to be tested. Ohmmeters can be used to check for continuity in the same manner as the self-powered test light. When checking a wire or part for continuity, the resistance should be at or near zero.

An analog ohmmeter can be used to check the operation of radio suppression and ignition capacitors, as shown in **Figure 14-23.** Set the ohmmeter scale to the highest position, and touch one probe to the capacitor body while touching the other probe to the capacitor lead terminal. The ohmmeter needle should jump toward the low resistance side, then return to infinity. If the needle does not jump, or goes to the low resistance side, the capacitor is defective. Before this test can be repeated, the capacitor lead must be touched to the body to discharge the capacitor.

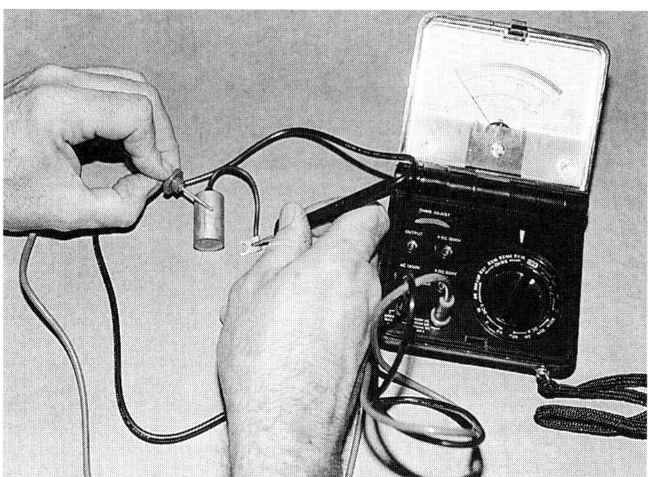

Figure 14-23. *Analog meters can be used for some tests, such as continuity tests in components such as capacitors. However, a high impedance, digital multimeter should be used whenever possible.*

Testing for Continuity

The great value of ohmmeters is that they can be used to check for circuit continuity. This is accomplished by first disconnecting power from the circuit. Then connect the leads of the ohmmeter to each end of the device or circuit to be tested, **Figure 14-24.** If the meter reads low or no resistance, the circuit or device has good continuity. Remember, you must use a meter that has a minimum impedance of 10 meg ohms when testing any automotive electrical system.

Pulse Meter

Another way of checking the operation of electrical devices is by using a ***pulse meter,*** included in many modern digital multimeters. The pulse meter displays a single line (called a waveform) which represents the electrical activity in a solenoid, motor, or electronic module, **Figure 14-25.** The waveform is similar to the familiar waveform of an ignition oscilloscope. Voltage is the vertical reading, while the horizontal reading is the passage of time. The device being tested must be electrically active, that is, being operated, before a waveform can be generated. There are many waveforms, corresponding to the electrical activity inside the device being tested. Most pulse meters can freeze the waveform for continued observation.

To diagnose a problem in a device by reading its waveform, the technician must know what the correct waveform looks like. The correct waveforms are often available in the vehicle service manual or the literature supplied by the pulse meter manufacturer. Once the correct pattern has been obtained, the technician can compare it with the waveform shown by the pulse meter, **Figure 14-25.** There are many waveforms showing many problems. In general, if the pattern displayed does not look like the standard pattern, something is wrong. The technician can then proceed with more detailed diagnosis. Many solid state components, such as ignition modules and ECMs, can be severely damaged by the careless use of test lights and multimeters. Specialized testers are often needed to check the operation of specific electronic devices or systems.

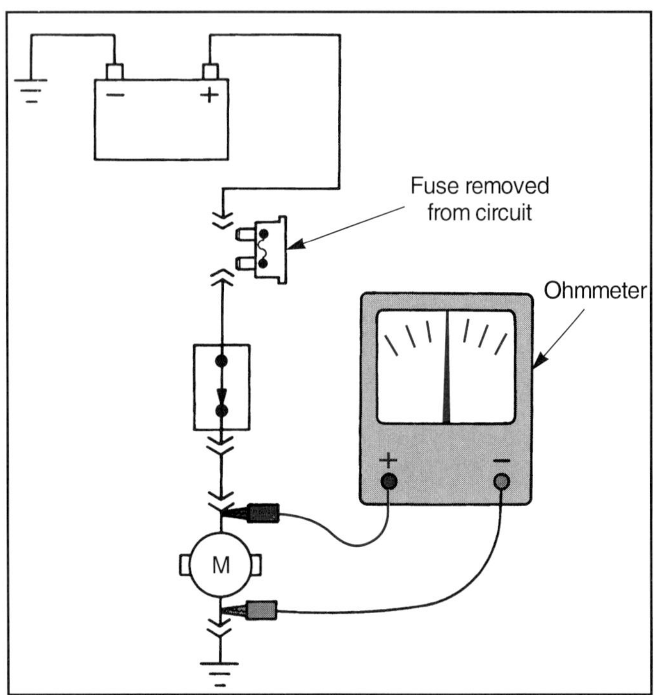

Figure 14-24. *Ohmmeters are the most frequently used meters in auto repair. They should be connected in parallel with the circuit. Be sure there is no power in the circuit when testing with an ohmmeter. (Chrysler)*

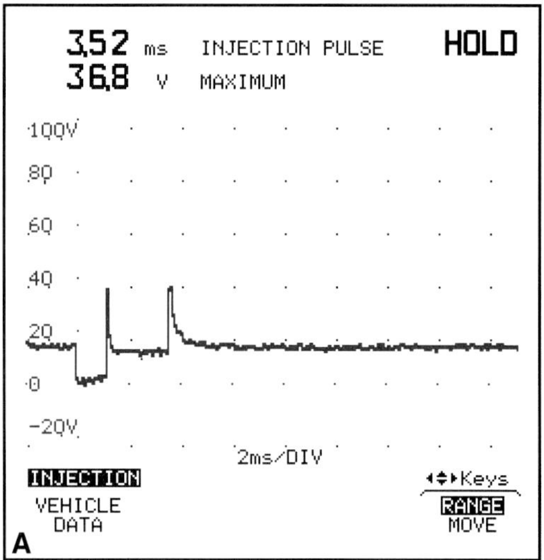

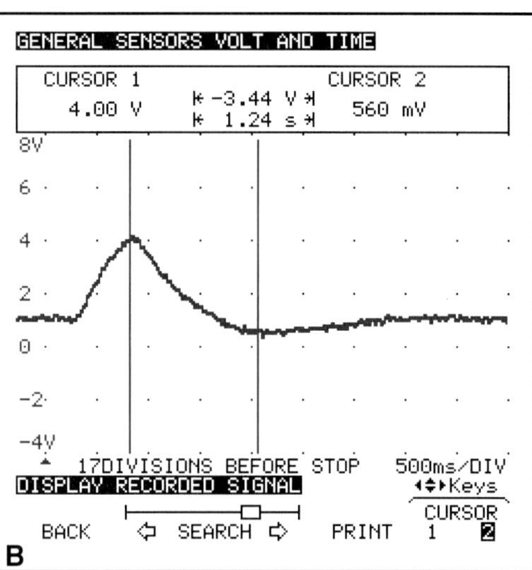

Figure 14-25. *Using waveforms to diagnose performance problems is an old, but proven method of diagnosis. Waveforms are used to diagnose ignition system problems. A—Fuel injector pulse. B—MAP sensor signal. (Fluke)*

Checking Specific Components

Caution: Many of the following procedures call for energizing a device with battery current. Always consult the service manual first before testing any electronic component to be sure that the device is designed to operate on full battery voltage. Some components are easily damaged if they are not tested properly.

Many vehicle components can be tested by using the tools you have just reviewed. Tests which can be made with more than one type of tester are identified in the following response of various electrical devices in response to changing engine conditions. Other tests can be made with the engine off. However, there are some vehicle components that require specialized electronic test equipment. Some electronic components cannot be tested with test lights or electronic meters.

Testing Relays

Relays are used to control power to a device that uses high current, such as an electric radiator fan. If you suspect a relay of causing a fault, the first step is to tap on the relay's body with a small screwdriver. If the system or component controlled by the relay starts or stops operating, the relay is sticking and should be replaced.

Further checks of relays can be made with jumper wires and a non-powered test light, as shown in **Figure 14-26.** The relay is energized by using the jumper wires while the test light is connected to the output terminal and ground. If the test light is illuminated when the relay is energized, the relay is good. Relay designs vary and a self-powered test light may be needed. Always consult the manufacturer's manual before testing any relay.

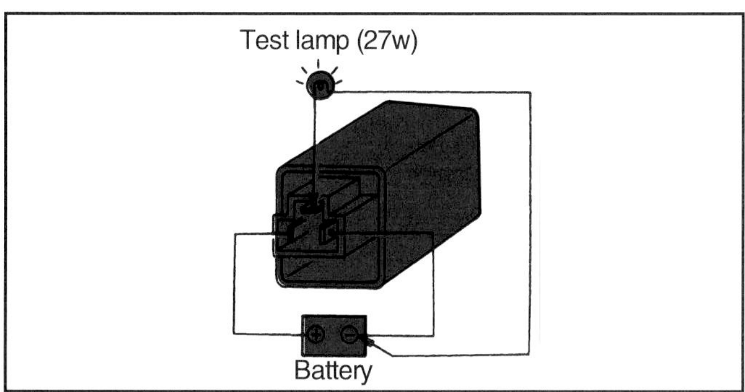

Figure 14-26. *Jumper wires and a non-powered test light can be used to check a relay for proper operation. If a relay works intermittently, replace it. (Nissan)*

Testing Ignition Coils

Ignition coils can be tested with an ohmmeter. In **Figure 14-27A,** the ohmmeter leads are connected to the primary terminals. This tests the coil primary winding for continuity and shorts. In **Figure 14-27B,** the ohmmeter leads are connected to check for continuity or shorts between the primary and secondary windings. In some cases, it may be necessary to lightly strike the coil with the handle of a screwdriver or other tool. This is done to check for an intermittent open or short.

Testing Solenoids

In most cases, solenoid operation can be observed, since most solenoids move an externally mounted lever or plunger. Using test equipment, solenoids can be checked by one of two methods:

- ❑ Making a continuity check with an ohmmeter or self-powered test light.
- ❑ Sending current through the solenoid and observing its operation.

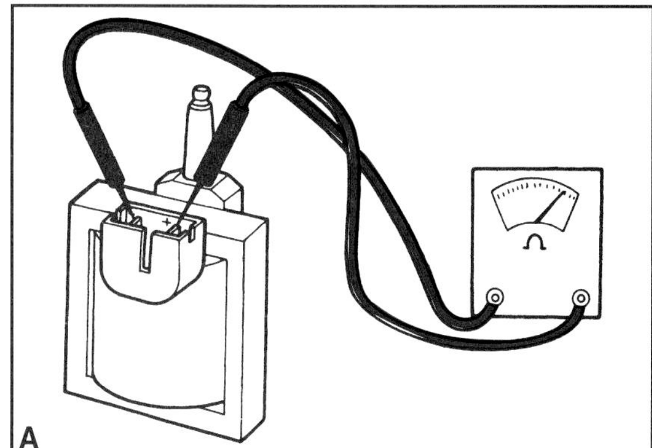

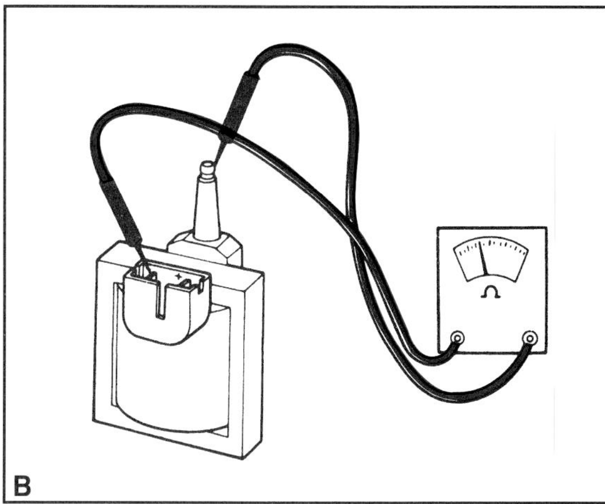

Figure 14-27. *An ohmmeter can be used to check an ignition coil for defects. A—Checking for shorts or opens in the primary windings. B—Checking for shorts between the primary and secondary windings. (Ford)*

To check continuity, connect an ohmmeter across the solenoid coil leads. The resistance reading should be between zero and infinity. Actual resistance will vary according to solenoid design. To check whether a solenoid is operating, use a fused jumper wire to apply power to energize the solenoid coil, **Figure 14-28.** When the coil is energized, you should hear a sharp click. If the solenoid has a visible plunger or valve, it should be seen to move.

Solenoids that control a ball check valve, such as the torque converter clutch solenoid in **Figure 14-29,** should be shaken to seat the ball before testing. If this is not done, the solenoid may not click when energized, leading to a false diagnosis. In some cases, a solenoid may quit working due to dirt or sludge build-up in the case of a transmission solenoid. Cleaning the solenoid may permit it to operate again. However, solenoids should be replaced if they are sticking.

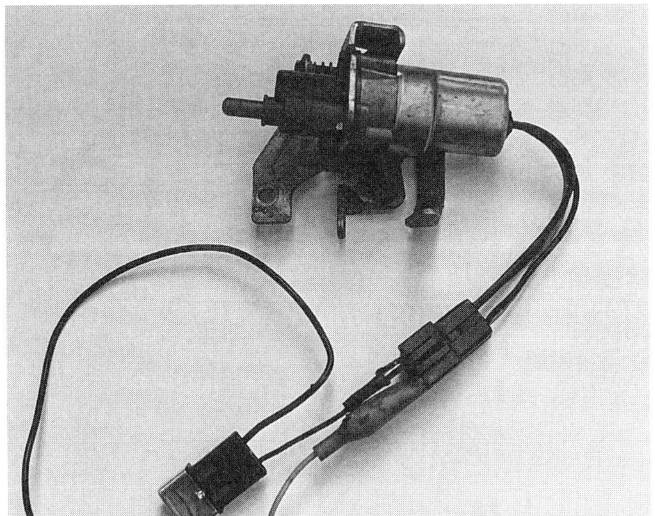

Figure 14-28. *To test solenoids and servo motors, apply power and inspect for proper operation.*

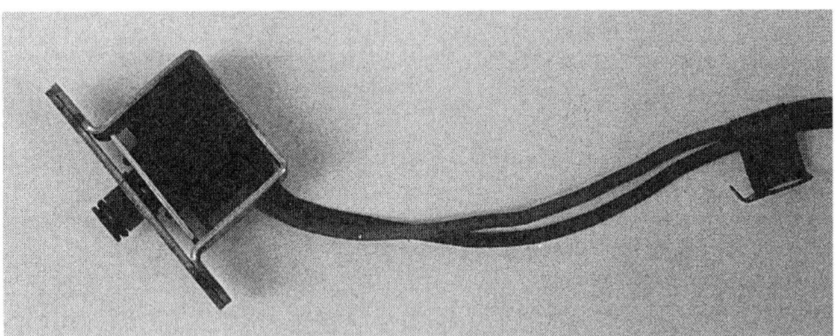

Figure 14-29. *Some solenoids do not have external moving parts and simply make an audible sound when operating.*

Testing Servo Motors

Like solenoids, servo motors can be tested by making a continuity check, or by applying current and observing motor operation. The type of motor will determine the exact testing procedure. Be sure to consult the manufacturer's manual before making any motor tests. Also make sure, when testing a reversible motor, to check for motor operation in both directions.

Testing Solid State Components

Although most solid state components require specialized test equipment, some basic checks can be made to some solid state components. Common checks include checking a terminal for proper grounding. This check must be made very carefully to avoid damage to the unit. In a few cases, the component can be checked for proper operation.

Testing Diodes

The simplest diode test is for continuity in one direction only. Attach the ohmmeter leads as shown in **Figure 14-30.** Test in one direction, then reverse the leads to test in the other. When testing a diode, polarity is important. If the diode shows low resistance in one direction only, it is good. If it shows continuity in both directions, it is defective. All of the diodes in an alternator should be checked individually, since one open diode cannot be isolated if the others are operating correctly.

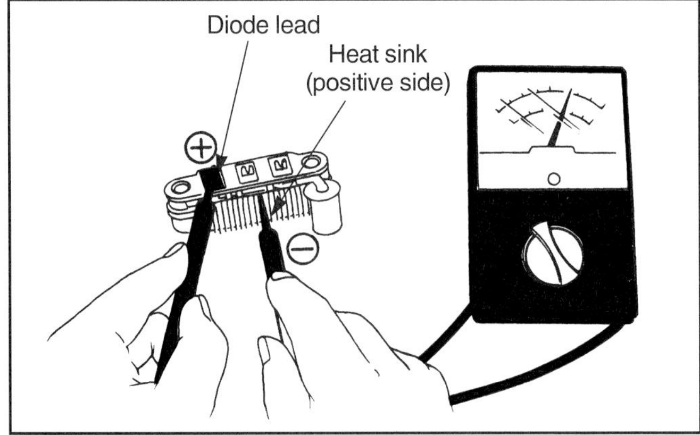

Figure 14-30. *Most diodes are tested by checking for continuity in one direction. This shows a test for continuity between two diodes on an alternator diode bridge. (OTC Tools)*

Testing Resistors

Resistance units are often used to produce heat or to send a temperature or position signal to the ECM. Checking most types of resistance units with a ohmmeter is relatively simple. **Figure 14-31** shows a diesel engine glow plug being checked with an ohmmeter. This checks not only the glow plug but the glow plug ground to the engine block. If the reading is not within specifications, the glow plug will not operate properly. Other resistors that can be checked with an ohmmeter are coolant and intake air temperature sensors. Unlike most resistors, however, temperature sensors show a decrease in resistance as temperature increases. Both the resistance and temperature of this type of sensor must be monitored to determine whether or not the sensor is defective.

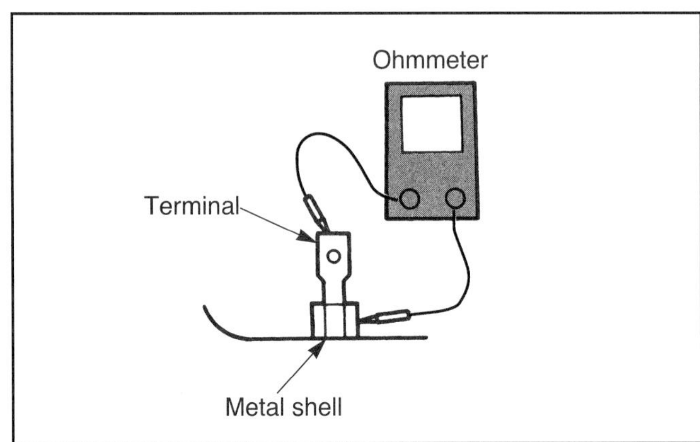

Figure 14-31. *An ohmmeter is used to check the resistance and ground condition of a diesel glow plug. (Ford)*

Some resistors depend on a sliding contact to vary resistance. Examples are throttle position sensors and vane-type mass airflow sensors. Sliding contact resistors should be checked with an analog ohmmeter, **Figure 14-32.** As the sensor is moved through its travel path, resistance should change smoothly. If the meter needle or reading jumps at any point, a bad contact is indicated and the sensor is defective.

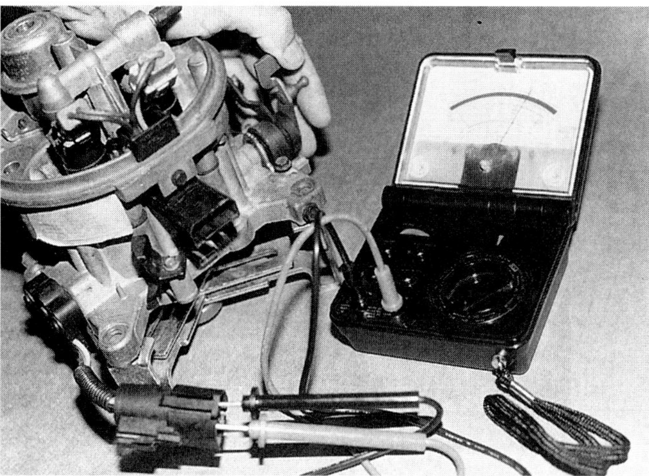

Figure 14-32. *An analog meter is useful to monitor the resistance change in throttle position sensors. Move the throttle while monitoring the meter.*

Problems Caused by Add-on Systems and Devices

Most add-on systems and devices are carefully engineered and constructed to fit and connect to a vehicle's electrical system without causing any problems. Unfortunately, these systems can cause a variety of problems, usually due to poor installation. A professional installation of an add-on system is barely noticeable and will have no effect on vehicle operation. Signs of a quality installation are wires carefully routed and often tucked into looms, components solidly mounted away from other vehicle components, and good connections, especially to battery power and chassis ground.

Poor connections to battery power and chassis ground are the most frequent mistakes made in an add-on system installation. For example, the power lead for an add-on device might be connected directly to the battery's positive cable terminal. While this fulfills the power requirements of the device, the battery terminal's connection is sometimes degraded as a result. This is especially true in the case of side terminal batteries. The additional connector allows corrosion to collect under the terminal, eventually resulting in a poor connection and a discharged battery. To correct this type of fault, replace the battery cable with one that has an additional lead or attach the device's power lead to a junction block connected directly to battery power. Some add-on devices may create electromagnetic interference, which can cause problems in a vehicle's computer-controlled systems.

Electrical Repair Operations

Once you have found and verified the source of the electrical problem, the next step is to repair the damage or replace the defective part. The next sections detail how to perform repairs on wiring, fusible links, and electrical connectors. Other electrical component repair operations were covered earlier in this chapter or will be covered in the appropriate chapters later in this text. Remember to follow the manufacturer's service manual recommendations when performing any electrical repair.

Repairing Damaged Wiring

Wiring can be repaired by replacing the damaged wire with a new wire of the same or larger gage. Wires are installed by soldering or the use of crimp connectors. ***Soldering*** is the recommended method of connecting wire, especially in the case of computer wiring and terminal connections. However, properly installed crimp connectors will satisfactorily connect most wires.

When preparing to replace a damaged section of wire, use wire that is of equal or larger gage. If you are installing a new section of wire, refer to a wire gage chart, such as in **Figure 14-33.** This chart will give you the correct gage size for the application.

Soldering

After cutting away the damages section, strip a section of insulation from each wire to be joined. The wires must be clean and free of damage for a good connection. Slip a piece of heat shrink tubing on one wire past the point where the wires are to be joined. Twist the wires together firmly. Plug in the soldering iron or gun and allow the tip to heat. When the soldering tool is at proper temperature, apply the tip to the wire. Touch the end of a strip of 60/40 rosin core solder to the wire, **Figure 14-34.** When the solder begins to melt readily into the wire, move the strip over the wire length.

If the solder is pasty looking, the wire is not hot enough and can result in a cold solder joint. Once the wire is soldered, allow it to cool for a time. Slide the shrink tubing over the wire and heat the tubing to melt it over. If shrink tubing is not available, use electrical tape.

Using Crimp Connectors

The best type of ***crimp connector*** uses heat shrink tubing. Select the proper type of connector for the job. Strip a section of wire as you did earlier. Place both ends of the stripped wire into the connector. Using a hand crimp tool, crimp the connector ends, making sure the wire does not slide out the end. If the wire comes out during crimping, the connector is too big.

After crimping, give the wires a slight pull to ensure that you have a tight connection. Heat the tubing to shrink it around the wire, which provides a waterproof seal. **Figure 14-35** shows the series of steps used to repair a wire with a heat shrink crimp connector.

Total Approx. Circuit Amperes	Total Circuit Watts	Total Candle Power	Wire Gage (For Length in Feet)											
12V	12V	12V	3′	5′	7′	10′	15′	20′	25′	30′	40′	50′	75′	100′
1.0	12	6	18	18	18	18	18	18	18	18	18	18	18	18
1.5		10	18	18	18	18	18	18	18	18	18	18	18	18
2	24	16	18	18	18	18	18	18	18	18	18	18	16	16
3		24	18	18	18	18	18	18	18	18	18	18	14	14
4	48	30	18	18	18	18	18	18	18	18	16	16	12	12
5		40	18	18	18	18	18	18	18	18	16	14	12	12
6	72	50	18	18	18	18	18	18	16	16	16	14	12	10
7		60	18	18	18	18	18	18	16	16	14	14	10	10
8	96	70	18	18	18	18	18	16	16	16	14	12	10	10
10	120	80	18	18	18	18	16	16	16	14	12	12	10	10
11		90	18	18	18	18	16	16	14	14	12	12	10	8
12	144	100	18	18	18	18	16	16	14	14	12	12	10	8
15		120	18	18	18	18	14	14	12	12	12	10	8	8
18	216	140	18	18	16	16	14	14	12	12	10	10	8	8
20	240	160	18	18	16	16	14	12	10	10	10	10	8	6
22	264	180	18	18	16	16	12	12	10	10	10	8	6	6
24	288	200	18	18	16	16	12	12	10	10	10	8	6	6
30			18	16	16	14	10	10	10	10	10	6	4	4
40			18	16	14	12	10	10	8	8	6	6	4	2
50			16	14	12	12	10	10	8	8	6	6	2	2
100			12	12	10	10	6	6	4	4	4	2	1	1/0
150			10	10	8	8	4	4	2	2	2	1	2/0	2/0
200			10	8	8	6	4	4	2	2	1	1/0	4/0	4/0

Figure 14-33. *This wire gage selection chart is used to select the proper gage wire by length. The lengths shown are for a single wire ground return. (Belden Mfg. Co.)*

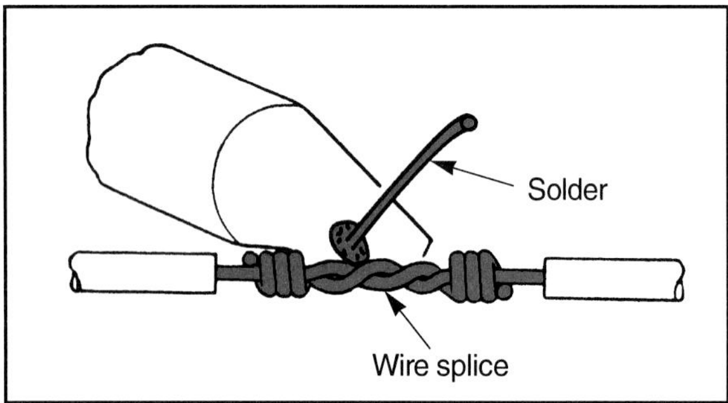

Figure 14-34. *When soldering, be sure to heat the wire and not the solder. Rosin core solder should be used for all electrical repairs.*

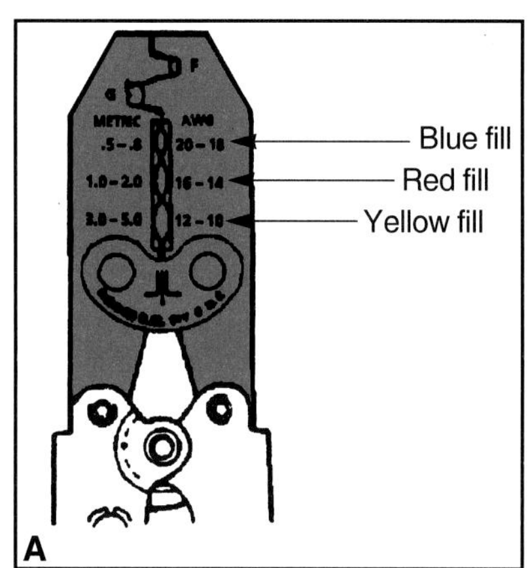

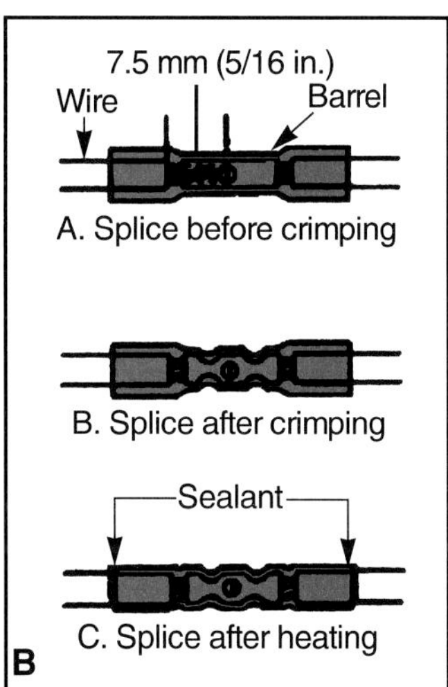

Figure 14-35. *A—Terminal crimping tool. Note the wire gage markings on the tool. B—Procedure for installing a crimp connector. (General Motors)*

Replacing Fusible Links

The process of replacing damaged fusible links is similar to the process used to replace damaged wires. However, a burned out fusible link must be replaced with another link having the same amperage rating. Never replace a fusible link with a length of standard copper, aluminum wire, or fusible link of a larger gage. See **Figures 14-36** and **14-37** for the installation steps used for replacing fusible links found in various parts of the vehicle wiring system. Note that the procedure varies with different fusible link locations.

Repairing Connectors

The amount of damage to the connector will determine what type of repair is to be performed. If the connector is not melted or physically damaged, it is possible to remove the defective terminal and either repair or replace it. If the connector is badly damaged or deformed, replace the connector. Replacement connectors are often available from the vehicle manufacturer. If a replacement connector is not available or the circuit passes through the firewall, it can be bypassed by routing a new length of wire around it, as shown in **Figure 14-38.**

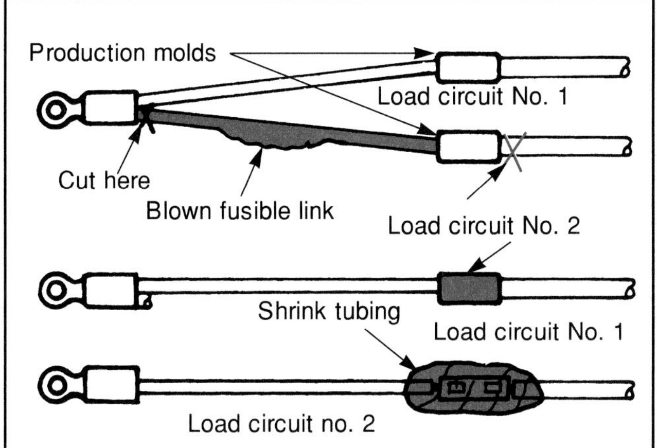

Figure 14-36. *Repair procedure for a blown fusible link. Be sure to replace the blown link with the same gage. (Ford)*

Repairing Plug-in Connectors and Terminals

If replacement connectors are available, they can be replaced by carefully removing the connector terminals. Removal and replacement of the connector terminals usually requires no more than a small screwdriver to unlock the tangs, as shown in **Figure 14-39.** Carefully unlock the terminal retainers. Retainers are usually a part of the connector body or a separate piece that plugs into the connector. Use a terminal removal tool or a small pick to depress the terminal locking tang while pulling on the wire.

Note: Depending on the design, the wire will slide out the front or rear of the connector. If the wire has a seal behind the terminal, carefully pry this seal out of the connector. This will ease wire and terminal removal. This type of terminal normally slides out the back of the connector.

Once the wire is free, reform the tang using a small pick and plug it into the corresponding spot in the new connector. Pull each wire one at a time so that they do not get confused and installed in the wrong location.

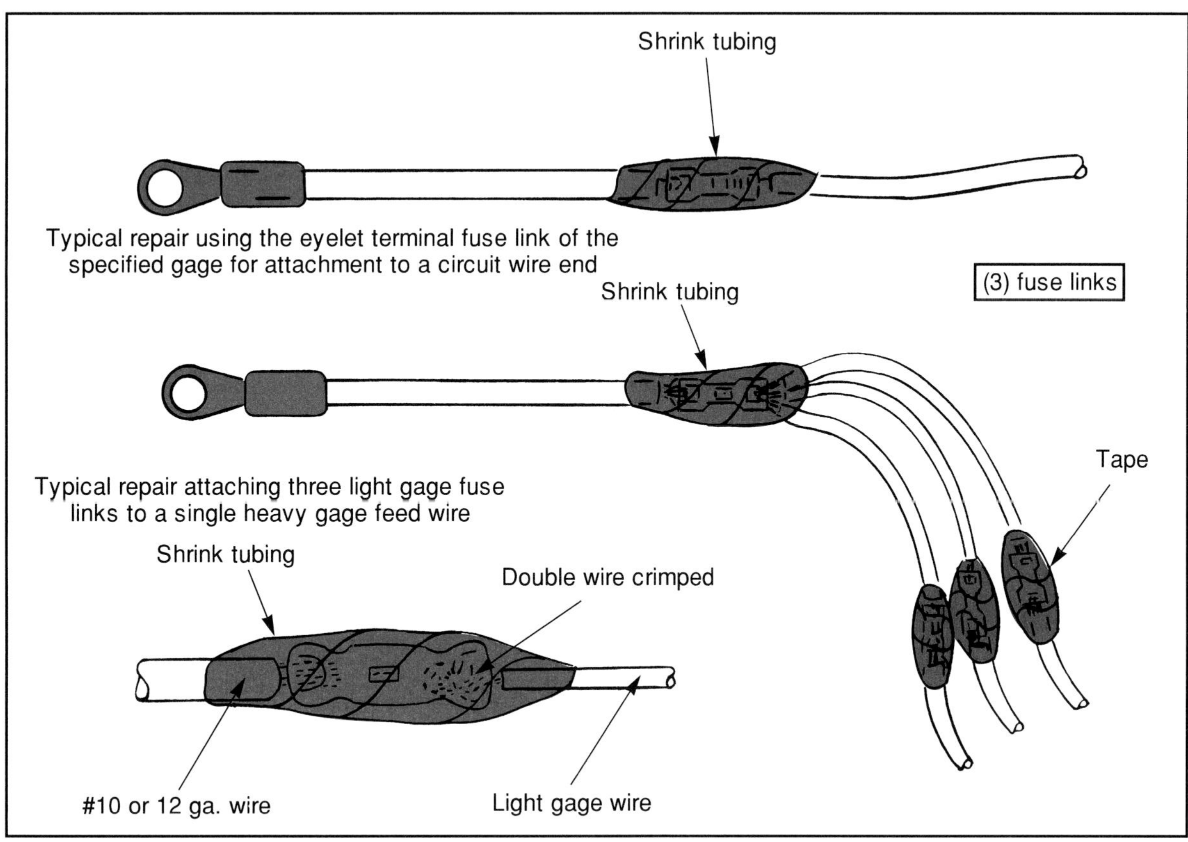

Figure 14-37. *Repair procedures for replacing single and multiple fusible links. Study the connection between the smaller gage fusible link and a larger wire. (Ford)*

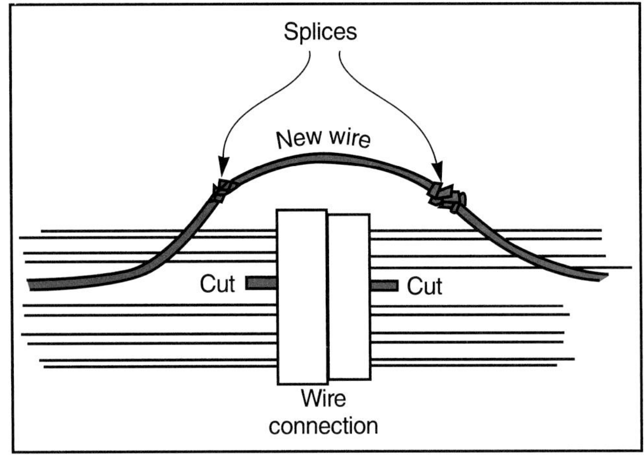

Figure 14-38. *Wires are sometimes bypassed and reconnected to repair a damaged connection.*

To replace a damaged plug-in connector terminal, cut the old terminal off, saving as much wire as possible. Save the old terminal to match up with the new one. If the wire has a seal, slide the old seal off and install a new one. Strip a section of wire. Locate the proper replacement terminal and crimp it to the wire. Apply a small bead of solder to the terminal and wire connection, **Figure 14-40.** Slide the repaired terminal back into the connector.

Checking Starting and Charging Systems

Common non-engine systems that can have an effect on driveability and performance include the starting and charging systems. Of all the parts in the starting and charging systems, the alternator is the greatest contributor to driveability problems. The battery, starter, and battery cable connections can also affect driveability and are discussed in this section.

Checking the starting and charging system will require a voltmeter and an ammeter. The best testing device is a combination unit, such as the one shown in **Figure 14-41**, which can check all parts of the starting and charging system. After the connections are made, start the engine while observing the meters. Excessive current draw or low cranking voltage indicates a problem with the battery, starter, or connections. Make further checks as necessary.

Battery

Starting problems are most often due to an undercharged battery. Causes of an undercharged battery can include:

- ❑ Parasitic draws.
- ❑ Charging system problems.

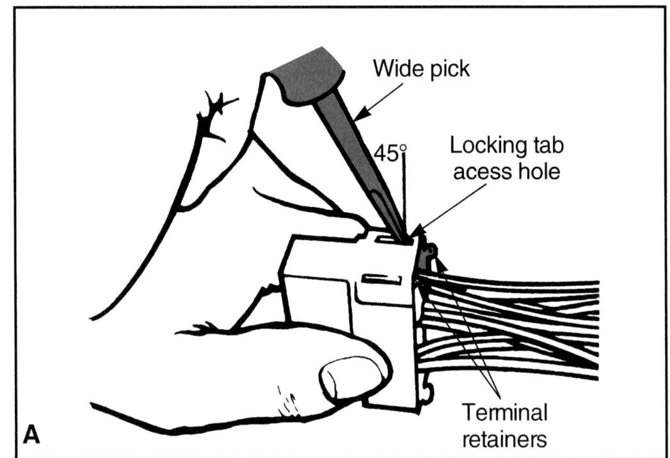

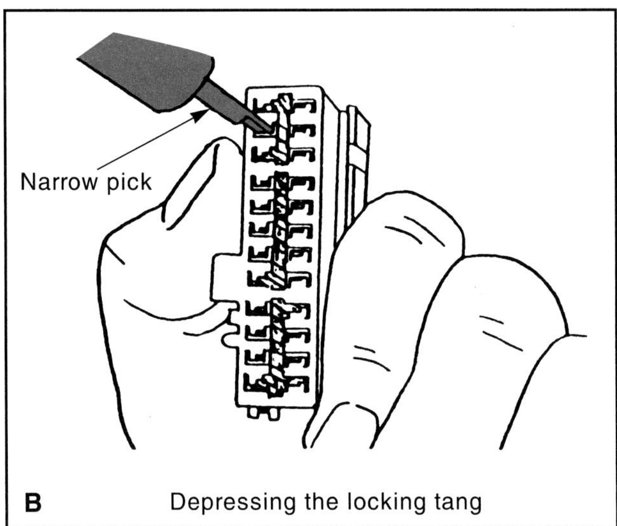

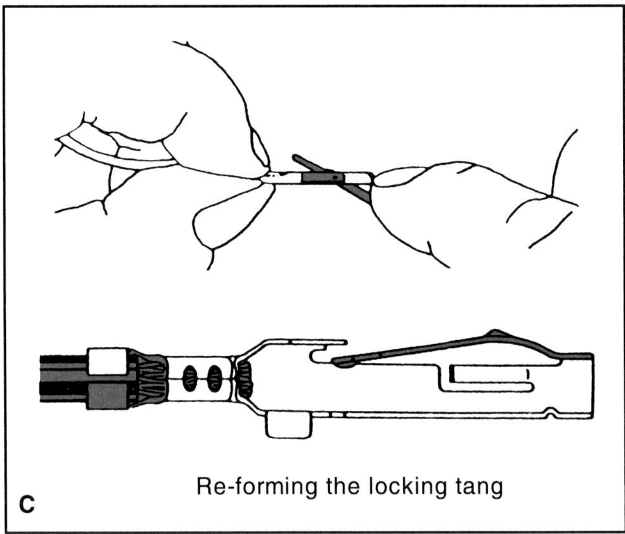

Figure 14-39. *Procedure for removing a plug-in terminal. A—Unlock the connector's terminal retainer. B—Use a narrow pick or terminal tool to unlock the tang. C—Before installing the terminal, reform the tang using the terminal tool or narrow pick. (General Motors)*

- ❑ Extended cranking conditions caused by a driveability problem.
- ❑ Excessive current draw by the starter motor.
- ❑ Poor battery cable connections due to corrosion.

The battery can become defective due to age, lack of maintenance on batteries with removable caps, excessive corrosion, overcharging, excessive cycling (discharging and recharging), and physical damage to the battery's terminals or case. Inspect the battery for electrolyte leaks and corrosion build-up. If the battery is leaking or corrosion is excessive, replace the battery.

Warning: Batteries produce explosive hydrogen gas as part of their normal chemical reaction. Do not smoke or create sparks or flame around a battery.

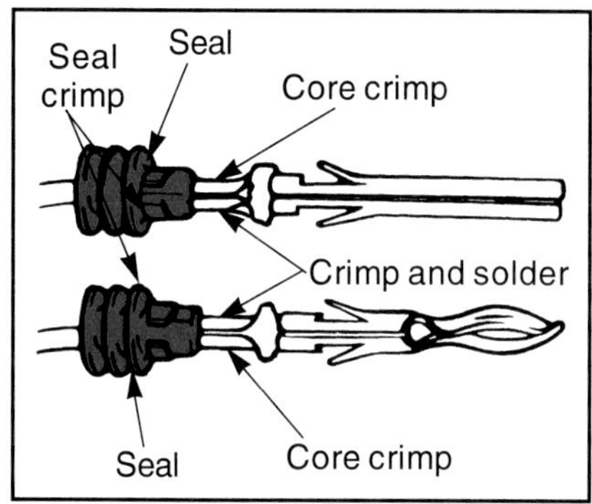

Figure 14-40. *Care should be used when repairing plug-in terminals. This type of terminal is crimped and then soldered for a good connection. (General Motors)*

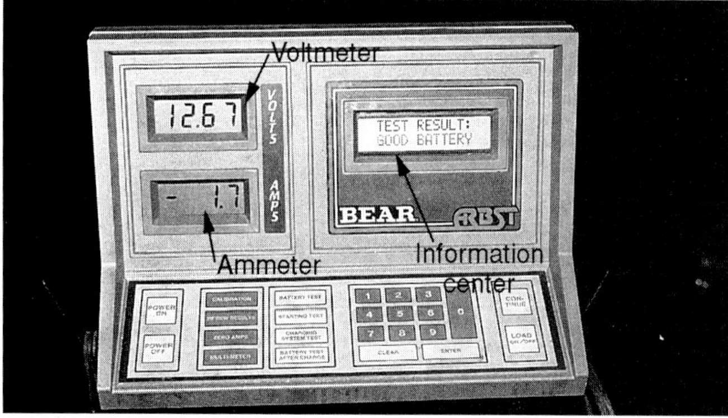

Figure 14-41. *A combination battery/charging/starting system tester is the best tool to use to check the charging and starting systems. This unit has a high amperage inductive clamp that can pick up amperage readings under 1 amp.*

Testing the Battery

Modern automotive batteries are usually sealed and can only be tested by performing an electrical check as shown in **Figure 14-42.** This type of check is often automated by the tester circuitry, and the technician has only to program in the cold cranking amps (CCA) of the battery to be tested. Older test equipment required a test sequence which involves placing an electrical load on the battery while checking the voltage. If placing a load on the battery equal to its CCA causes the voltage to drop below approximately 9.5 volts, the battery is either defective or is undercharged.

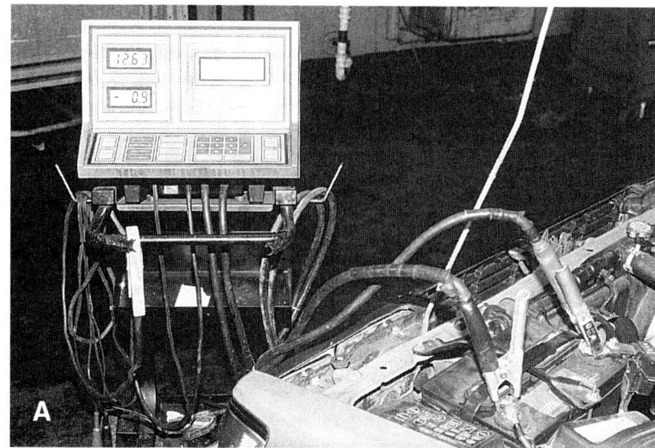

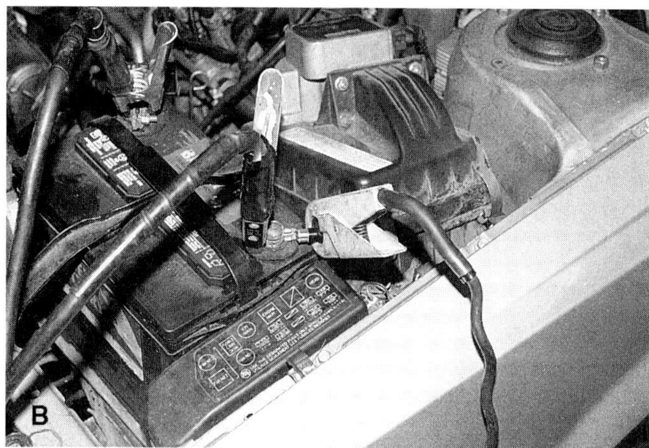

Figure 14-42. *A—Conducting a battery load test. The unit will do this test automatically. B—Connections to the battery for an amperage output test.*

Batteries with removable caps can be checked with a hydrometer, **Figure 14-43.** Squeezing and then releasing the hydrometer bulb draws some of the battery electrolyte into the tube. The small float inside the tube will float in the electrolyte. The specific gravity of the electrolyte is directly related to the battery's state of charge. The greater the battery charge, the higher the float will ride in the fluid. The float is calibrated and the state of charge can be read by matching the graduated lines with the fluid level.

Note: If a battery freezes in cold weather, it is almost always damaged internally. Some manufacturers recommend replacing the battery if it has been frozen. A frozen battery must be allowed to thaw completely before it can be tested.

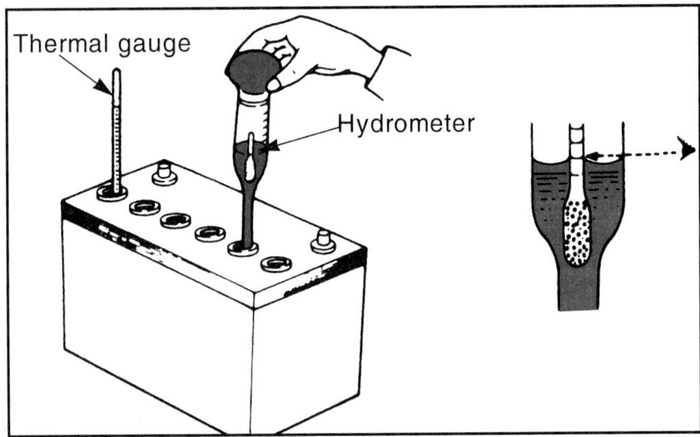

Figure 14-43. *Older batteries can be tested using a hydrometer. A reading of 1.260 indicates a full charge. (Subaru)*

Testing the Battery Cables

Some starting problems are due to high resistance caused by corrosion on and in the battery cable connections. The cables can be checked by testing for voltage drop. Refer to **Figure 14-44.** To check for voltage drop at a connection, attach the tester leads as shown, and then crank the engine. If the voltage reading is more than .5 volt, the connection has excessively high resistance.

If the battery terminal is damaged by corrosion, it can be replaced if the damage does not extend into the cable. If the cable has excessive resistance or visible damage from corrosion extends into the cable, replace the entire cable following the guidelines in the vehicle's service manual.

Testing for Excessive Parasitic Draws

If a battery has lost power to the point that it will not start the engine, and the charging and starting system are in proper working condition, the cause may be a ***parasitic draw*** from a system, component, or grounded wire. The ECM, radio, and any electrical device that contains a volatile memory draws power from the battery, however, the total of this draw is in milliamperes, and would take weeks to completely drain a fully charged battery. However, a draw of one-quarter ampere from the glove box light could drain the battery's power in a matter of days to the point that it would not start the engine.

Caution: Before conducting a parasitic draw test, be sure that any automatic headlight, electronic level control, anti-theft alarm, and other high amperage systems are shut off. If one or more of these systems engage when the meter is connected, damage to the meter will result. If the meter indicates a draw of over 10 amps when connected, disconnect the meter immediately, reconnect the battery, and perform a visual inspection of the entire vehicle.

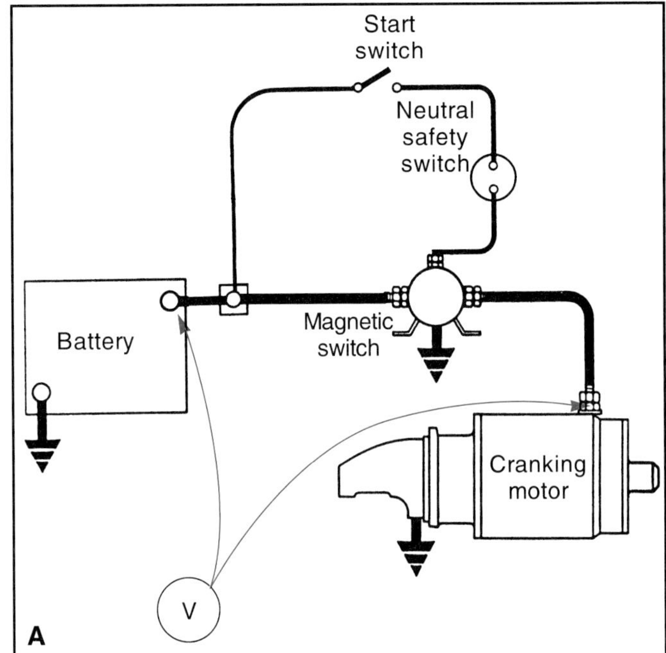

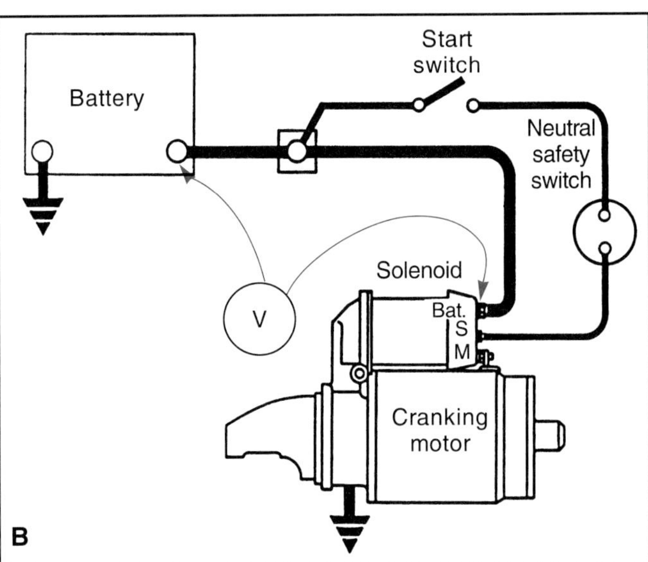

Figure 14-44. *A voltage drop test should be performed to check the battery cables. Voltage through the meter indicates that one of the cables between the two test points is defective.*

To test for a parasitic draw, disconnect the battery's negative cable and obtain an ammeter. Most ammeters have a jack that can be used to check for current above the milliampere range. It is connected through a slow-blow fuse, usually 10-15 amperes. Plug the meter's positive lead into the fused jack and the negative lead to the meter's common ground jack. Connect the ammeter in series by attaching the positive lead to the battery's negative cable and the meter's negative lead to the battery terminal, **Figure 14-45.**

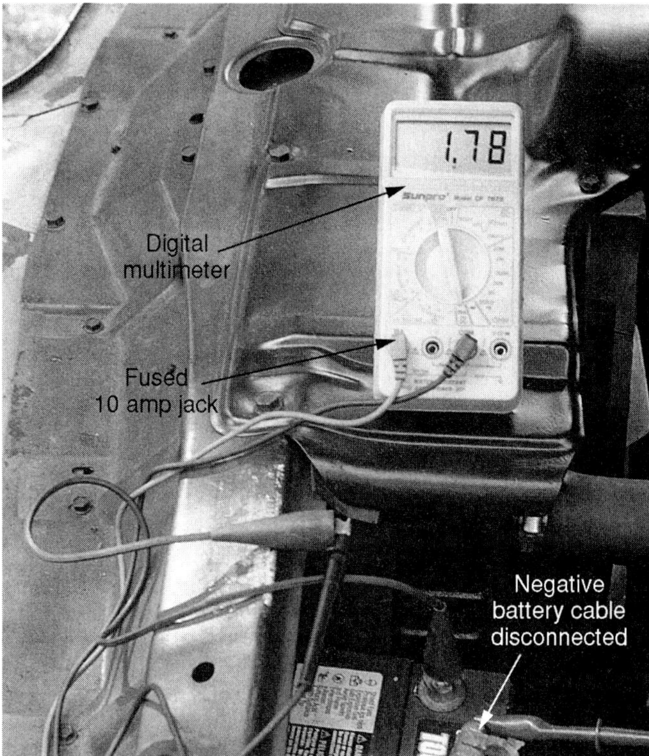

Figure 14-45. *Ammeter connection for a parasitic draw test. Note that the negative battery cable is disconnected. Be sure to shut off any high amperage circuits before connecting the meter.*

Caution: If the vehicle is equipped with an automatic headlight system, be sure that this system is shut off as the headlamps will illuminate as soon as the meter is connected. The power that this circuit will draw can damage the meter.

To ensure that the meter is connected properly, open one of the doors and check the meter. The interior lights should create a draw of between 1.5 and 4 amps, depending on the vehicle. After verifying that the meter is connected properly, close the door and recheck the meter. If the interior lights use an automatic timing relay, you may have to wait up to one minute for the meter to stabilize.

Normal parasitic draw is in the milliampere range. Typical parasitic draws for most on-board devices is shown in **Figure 14-46.** Check the amount of parasitic draw against manufacturers specifications. If parasitic draw is excessively high, position the meter so that it can be seen from the vehicle's interior. Start pulling fuses and circuit breakers, one by one, until the draw disappears. Replace each fuse or circuit breaker as you pull them. If the draw goes away when you pull a particular fuse, you have narrowed it down to the circuits powered through the just-removed fuse. However, if removing fuses and circuit breakers does not eliminate the draw, it is in a circuit or system connected through fusible link or directly to the battery.

Note: A 10 gage fused jumper wire with heavy-duty alligator clips can make this test easier.

	Current draw in milliamperes (mA)	
Component	Typical parasitic draw	Maximum parasitic draw
ABS-ECU	1.0	1.0
Alternator	1.5	1.5
Auto door locks	1.0	1.0
BCM	3.6	12.4
Chime	1.0	1.0
ECM	2.6	10.0
ELC	2.0	3.3
HVAC power module	1.0	1.0
Illuminated entry	1.0	1.0
Keyless entry	2.2	5.5
Radio	6.9	6.0
SRS	1.6	2.7
Theft deterrent system	0.4	1.0

Figure 14-46. *Typical parasitic loads in milliamperes of various on-board systems.*

Tracing Parasitic Draws

Once you have the parasitic draw narrowed down to a group of circuits, begin a visual inspection of wiring, connectors, and electrical components until you find the cause. Tracing parasitic draws is not as difficult as it might seem, if you have the correct schematic and patience. For example, pulling the courtesy (CTSY) fuse in **Figure 14-47** would eliminate a parasitic draw created by the luggage compartment light.

The first step is to reconnect the fuse and reposition the meter in a location where you can read it while looking for the source of the draw. Since this circuit is composed primarily of lights and door lock switches, the search begins with a visual inspection. In most cases, the source of a parasitic draw is found during the visual inspection.

Look at all of the lights, light switches, and related systems. Make sure all of the switches are open. If there are no lights illuminated, disconnect the audio alarm and passive restraint control modules. If none of these procedures isolates the draw, disconnect each component and test for shorts to ground. In some cases, you may have to physically search the wiring harness for the problem. This process is much more time consuming.

Battery Service

If the tests indicate that the battery is defective, it is simply replaced. Remove the negative battery cable first, then the positive. Remove the battery case or retaining device and lift the battery out. Install the new battery and retaining device. Install the positive cable first and the negative last. Anytime a battery is replaced, a test for excessive parasitic draw should be performed. If available, a memory saver, discussed in **Chapter 3,** should be installed in the cigarette lighter receptacle. If this is not done, be sure to reset the clocks, radio presets, and the idle speed on a fuel

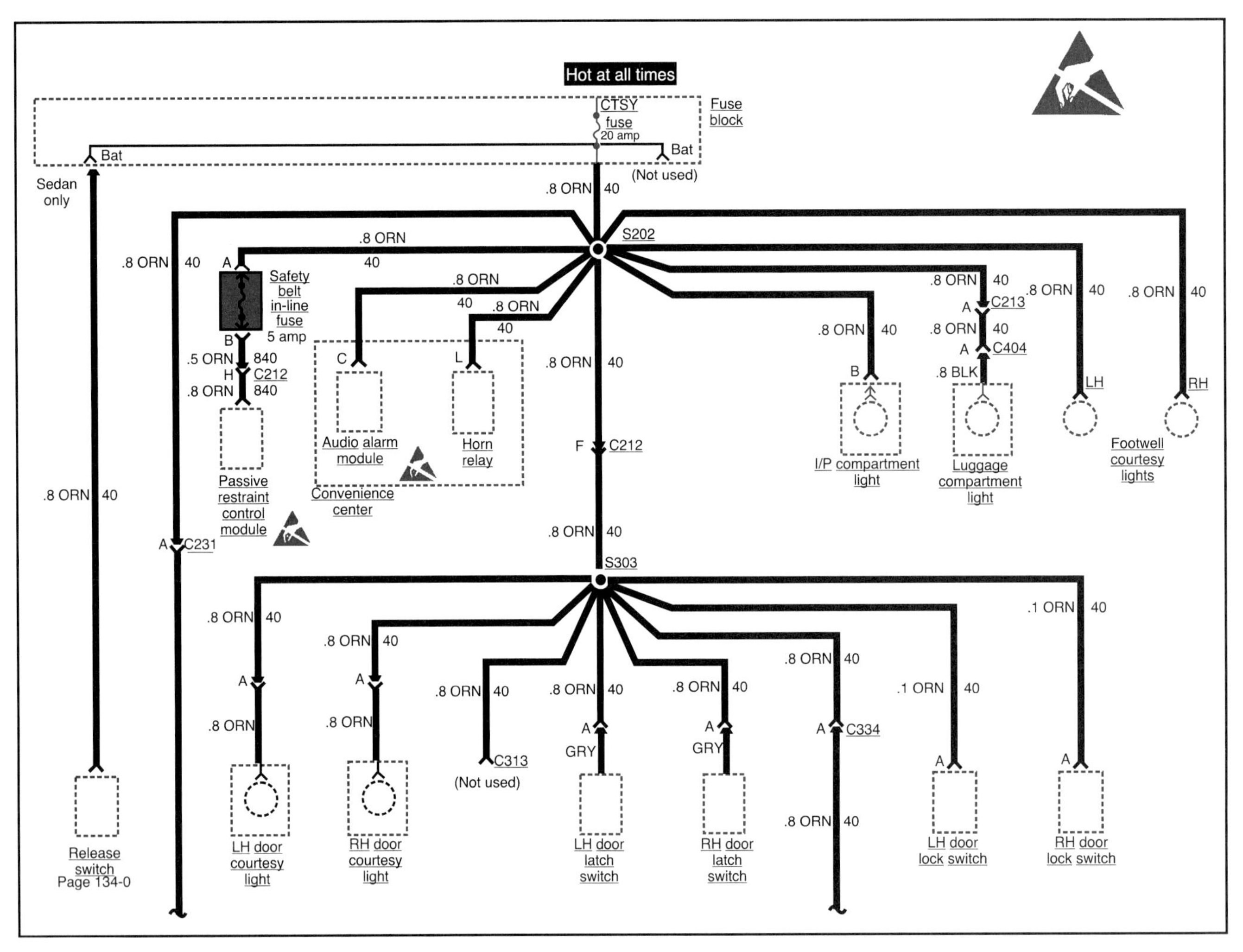

Figure 14-47. *When looking for a parasitic draw, start at the fuse block and pull fuses until you find the fuse that is supplying the circuit containing the draw. Once you find the circuit, begin checking circuit devices until you find the draw. (General Motors)*

injected engine before turning the vehicle back over to the customer. In most cases, simply driving the vehicle for approximately 10 minutes is sufficient for the ECM to reset engine operating parameters.

Warning: Batteries may leak electrolyte if tipped excessively or if the case is damaged.

Charging the Battery

A badly discharged battery should be recharged before it is installed in the vehicle. A badly undercharged battery will place a heavy load on the alternator, and will prevent the alternator from producing its rated voltage. To recharge a battery, connect the charger cables to the battery, as in **Figure 14-48.** Before charging a battery installed in a vehicle, remove the negative cable. Make the battery connections before plugging in and turning on the charger. Set the charging rate to as low a setting as possible. Fast (high current flow) charging can damage a battery, especially if it is badly discharged or cold.

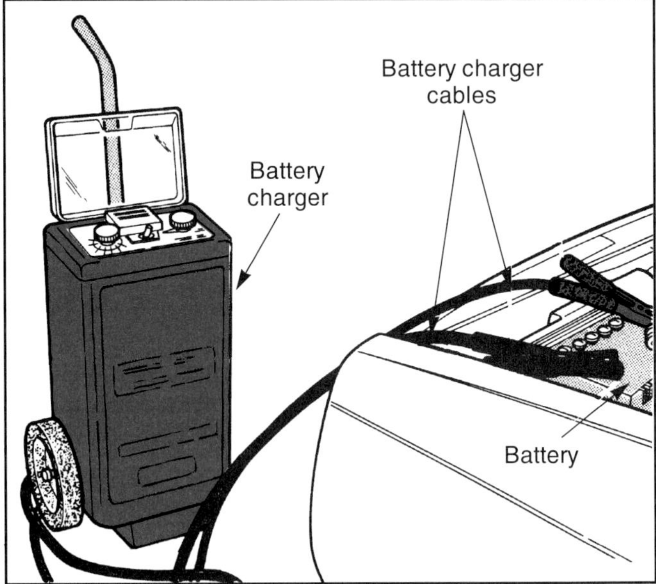

Figure 14-48. *When charging a battery in the vehicle, disconnect the negative battery cable to avoid voltage spikes that could damage the ECM and other electronic components. (Ford)*

Starting System

If the vehicle will not start, and the battery and cables check out good, the starter or solenoid may be defective. As a quick check of the starter and solenoid, turn on the headlights and attempt to start the vehicle. This test requires a fully charged battery. If the headlights dim severely when the ignition is turned to the start position, the starter is drawing excessive current and is defective.

Another possibility is that the engine has seized internally. If this is a possibility, try to turn the engine with a flywheel turner. If the engine will not move or moves with great difficulty, it has internal damage. If the engine can be turned with the flywheel turner, the starter is the problem. If the starter can be heard to spin, but will not crank the engine, the starter drive pinion may not be engaging with the flywheel, or the flywheel teeth may be stripped. In this case, the headlights will dim only slightly. Drive gear and flywheel problems can be checked visually by removing the flywheel dust cover.

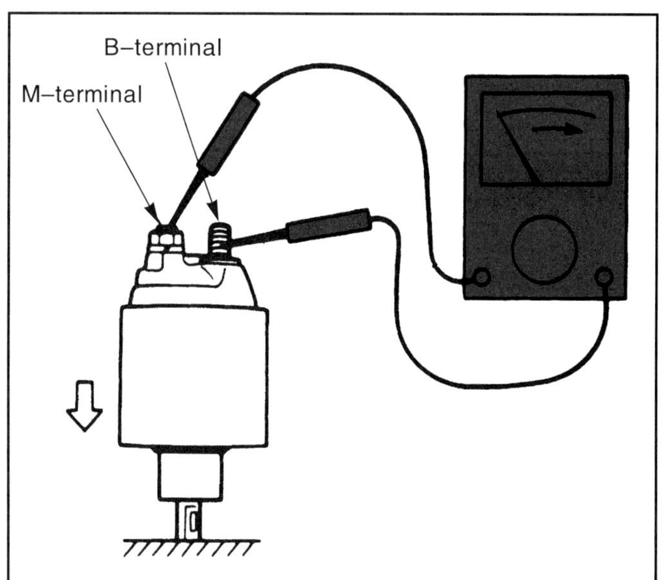

Figure 14-49. *Checking a starter solenoid for continuity. (Ford)*

Starter Load Tests

A starter ***load test*** can be made by connecting a high amperage ammeter to the positive battery cable. This can usually be done by most commercial battery/charging system testers. Place the amperage probe on the positive battery cable. Disable the ignition and fuel system and try to crank the engine for 15 seconds. Typical amperage draw for most starter motors is approximately one-half of the engine's cubic inch displacement. For example, if an engine displaces 302 cubic inches (5.0 L), the alternator should draw no more than 150-155 amps. Check the service manual for the proper specifications. If the draw is lower than specified, look for corroded battery connections and test for high resistance in the battery cables. If the amperage draw is high, the starter motor is defective.

Testing the Starter Solenoid

If the headlights remain bright when the ignition switch is turned to the start position, either the solenoid is defective or the ignition switch or neutral safety switch is not allowing current to reach the solenoid. In some cases, the ignition switch or neutral safety switch is simply misadjusted. To isolate the defective component, bypass the ignition and neutral safety switches by energizing the solenoid directly. A remote starter switch is best for this operation. Be sure that the transmission is in neutral or park before cranking. If the engine cranks, the solenoid is good, and the problem is in the ignition or neutral safety switch. If the engine does not crank, the solenoid is defective.

Most solenoids are part of the starter motor. However, some vehicles use a solenoid located near the battery. To replace this type of solenoid, simply disconnect battery power, disconnect the wires and cables from the solenoid and unbolt the unit from the vehicle. Installation is in reverse order. Starter mounted solenoids require starter motor removal. **Figure 14-49** illustrates one of the checks that can be made to the solenoid. If the solenoid winding is open or shorted to the case, it should be replaced.

Starter Motor Removal and Installation

To remove the starter motor, disconnect the negative battery cable. Raise the vehicle and remove any splash or heat shields and other parts that could make starter removal difficult. A few engines locate the starter under the intake manifold. Follow the service manual procedure carefully when replacing the starter motor on such an engine.

Note: In some cases, it may be necessary to loosen or disconnect motor mounts and raise the engine to remove the starter motor. Consult the service manual for the proper procedure.

Disconnect the positive battery cable, any auxiliary power wiring (these are often fusible links), and the starter solenoid wire. Unbolt the starter and remove it from the vehicle, being careful to remove any shims that may be located between the starter motor and the engine. Installation is in the reverse order of removal. Be sure the starter is properly located to engage the starter flywheel before lowering the vehicle. Reset any vehicle systems that have volatile memory, such as radios and the ECM.

Testing the Starter Motor

In most cases, technicians prefer to install a new starter or solenoid, rather than rebuild the old one. In some cases, it may be necessary to recondition the motor. Once removed from the vehicle, the starter and solenoid can be disassembled and checked with an ohmmeter. The starter (and sometimes the solenoid) can be disassembled to check the

the internal components. A detailed service manual should be used when attempting repairs to a starter.

The most common check to a starter once it is disassembled is a visual inspection of the brushes and commutator. Sometimes a slightly scored commutator can be lightly sanded to restore it to service. However, if the brushes are badly worn or the commutator shows signs of overheating or severe scoring, it is usually easier to replace the entire starter rather than replacing individual parts, **Figure 14-50.** Other tests can be made with an ohmmeter. Always follow the manufacturers instructions when making these checks.

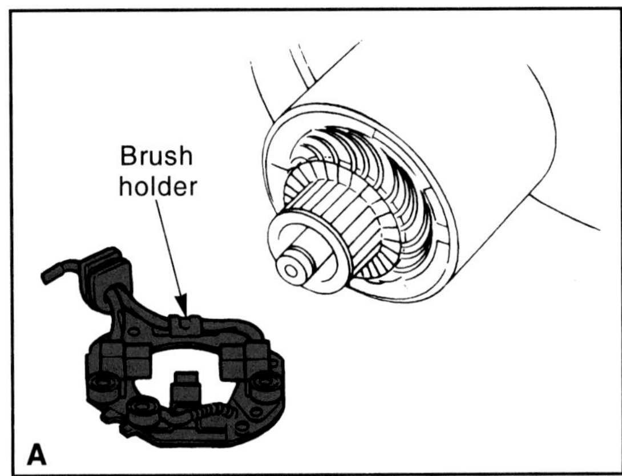

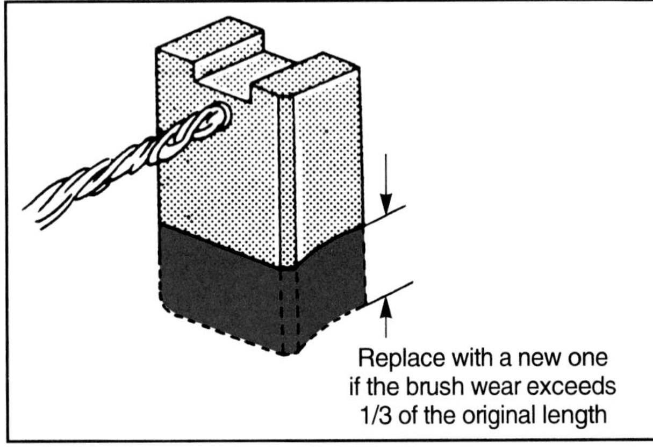

Figure 14-50. *The starter is a complex assembly that is often replaced rather than rebuilt at the shop. A—Starter brush holder. B—Typical brush.*

Alternator

Check the condition and tension of the drive belt before removing the alternator from the engine, since this is a common source of low charging rates. After the engine starts, check the alternator output and the charging voltage. In some cases, it may be necessary to place a load on the alternator to force it to produce its full output. After recording the voltage and amperage readings, compare them with the factory specifications, **Figure 14-51.** If alternator output is low, or voltage is incorrect, make further checks to isolate the problem.

One quick check is to bypass the voltage regulator, referred to as a ***full-field test.*** The alternator often includes the voltage regulator, either externally or built into the alternator's housing. To do this on a vehicle with an external regulator, disconnect the input wire to the alternator field terminal. Next, attach a jumper wire between the field terminal and the battery positive. Some alternators with internal voltage regulators can be full-fielded by inserting a small screwdriver in a small hole in the back of the alternator, **Figure 14-52.**

Figure 14-51. *Test results from an alternator output test. A—Results of voltage regulator control and output test. B—Results of peak amperage test. Some modern alternators can generate up to 140 amps. C—This unit monitors the diode pattern internally and can report any inconsistencies, which can cause charging problems.*

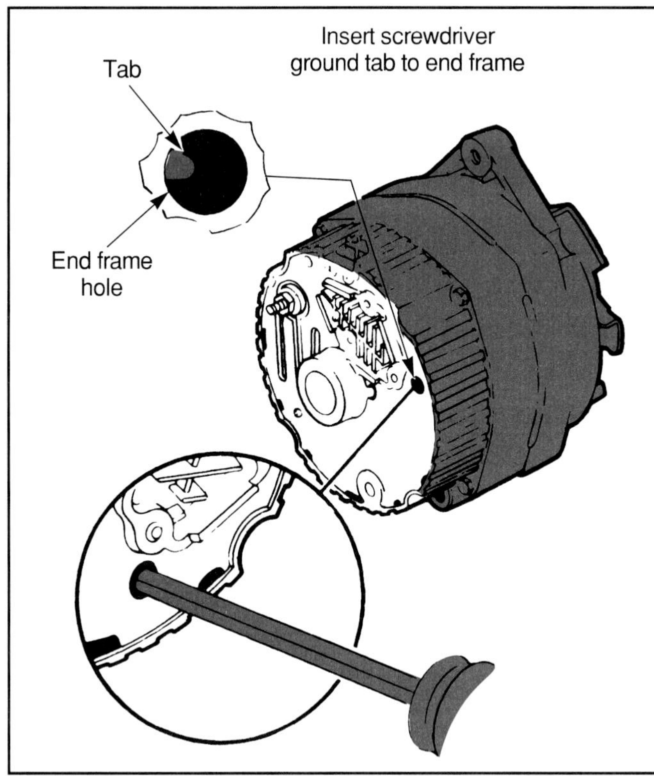

Figure 14-52. *To full-field an alternator, it is necessary to bypass the voltage regulator. On this alternator, a small hole is provided to insert a small screwdriver to bypass the regulator. (General Motors)*

Caution: Do not bypass the alternator's voltage regulator for an extended period of time. Bypassing the regulator causes the alternator to full-field (produce maximum current). Damage to the vehicle's electrical components can occur.

If the alternator begins charging, it is in working condition, and the problem is either in the voltage regulator or external wiring. If the alternator does not begin charging with the regulator bypassed, the alternator is defective. If the alternator has an internal regulator, consult the correct service manual for the exact regulator bypass procedures. Some oscilloscopes can also be used to check the alternator diodes and stator, **Figure 14-53.**

Alternator Removal and Installation

To remove the alternator, begin by disconnecting the negative battery cable. Depending on the engine, alternator removal can be very easy or somewhat difficult. Remove as many belts as necessary for alternator removal. Remove or relocate engine components, wiring, or brackets as necessary. Disconnect the wiring, unbolt, and remove the alternator from the engine. Installation is the reverse of removal.

Be sure that all belts are reinstalled and adjusted to the proper tension. Make sure that the lead to the battery is not grounded to a metal part and that any protective wiring

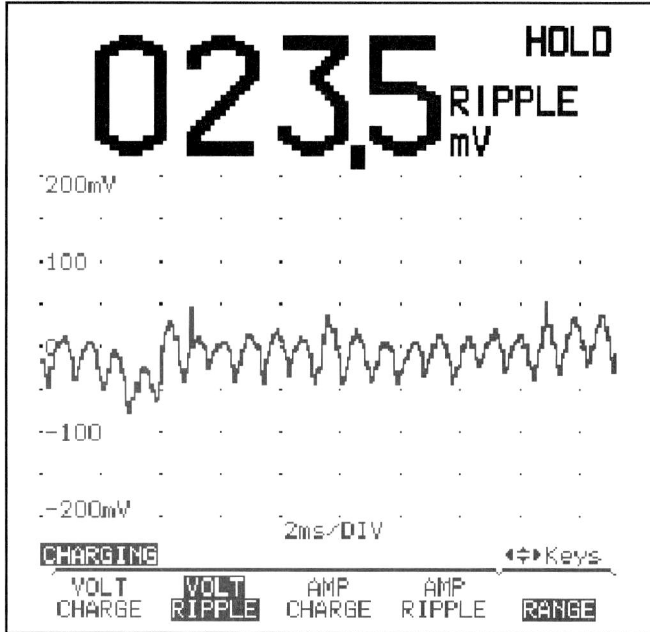

Figure 14-53. *Diode pattern as seen on an oscilloscope. This pattern indicates that the diodes are in good condition. (Fluke)*

covers are installed before reconnecting the battery. Reset the radio and road test the vehicle.

Testing Alternator Components

Most technicians simply choose to replace a defective alternator. However, some alternators can be disassembled to determine their condition and to replace defective parts. Place the alternator on a bench, mark a line on the alternator body with chalk, and remove the long bolts holding the two halves of the alternator together. If the front bearing is to be replaced, the rotor must be removed to release the rotor shaft from the front housing. After the alternator is disassembled, check the brushes for wear and use an ohmmeter to check the windings and diodes. These procedures are explained in the factory service manual. **Figure 14-54** shows some typical ohmmeter testing procedures.

Modern voltage regulators are sealed units and cannot be checked except by substituting a known good unit. This is relatively easy when the regulator is mounted outside the alternator, or on the engine compartment, but harder when the regulator is installed inside the alternator housing. If the regulator appears to be defective, replace it. Older electromechanical regulators can be adjusted by removing the cover and bending parts of the contact point linkage. However, this procedure is not always successful and the most effective repair is to simply replace the entire regulator.

Follow-up for Electrical Repairs

After any electrical repair, the affected system, device, or component should be operated for 10-15 minutes to ensure that the repair has taken care of the problem. If a

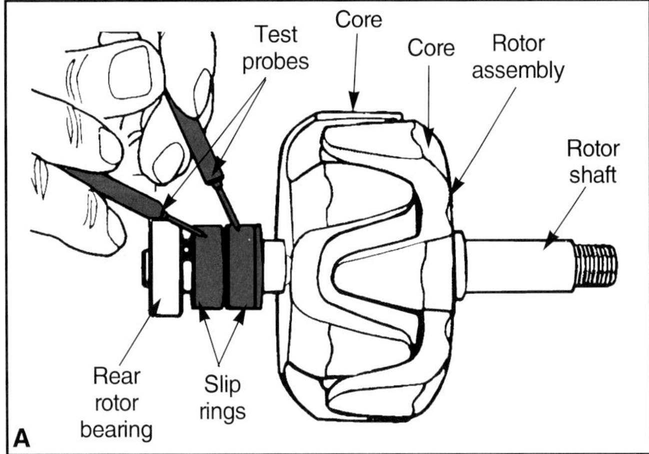

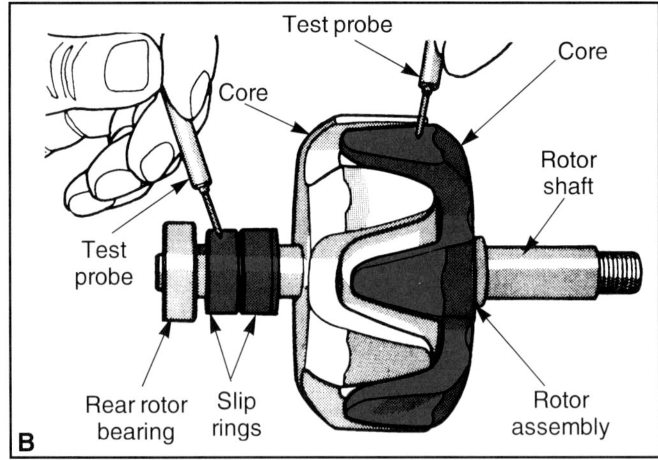

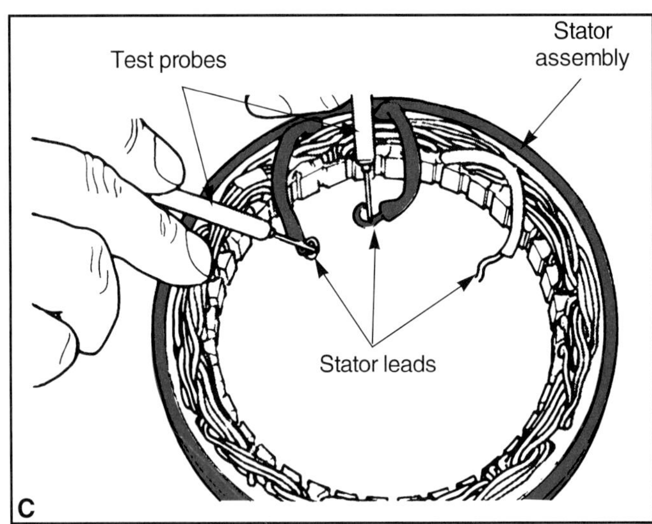

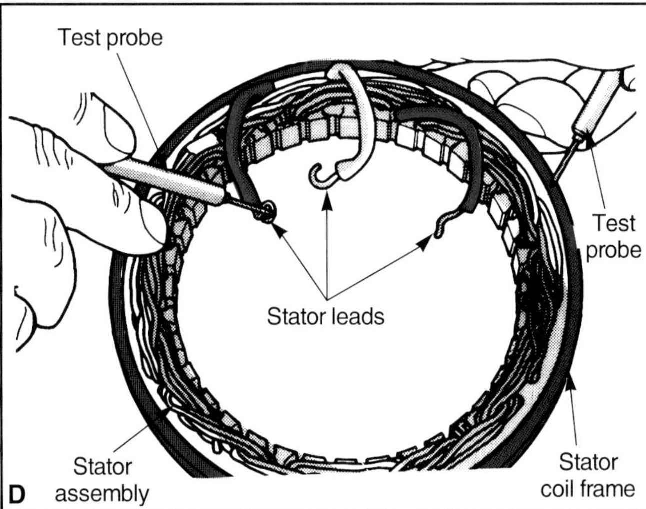

Figure 14-54. *Various meter tests on the stator and rotor assembly. All tests are done with an ohmmeter. A—Testing for continuity between the rotor windings and the slip rings. B—Testing for shorts and grounds between the slip rings and the rotor core. C—Testing for continuity between the stator windings. D—Testing for shorts and grounds between the stator windings and the stator body. (Chrysler)*

circuit protection device was replaced, operate all of the related devices. If a battery, starter, or alternator was replaced, check the entire electrical system to ensure that it is operating properly. Perform any other follow-up procedures recommended by the vehicle's manufacturer.

Summary

The vehicle electrical system and its components are often a cause of driveability problems, and is frequently overlooked. To properly diagnose any vehicle electrical problem, the technician must obtain the proper electrical schematics for the vehicle being worked on. A schematic is a way of graphically representing the flow of electricity between electrical components. The correct schematic will allow the technician to trace circuits easily.

Schematics allow you to trace circuits to arrive at a defective part or connection. Multimeters can check the operation of the oxygen sensor and ECM, and the operation of various other input and output devices. Fuses and circuit breakers are often overlooked when diagnosing a driveability problem. The technician should always locate the cause of fuse, circuit breaker, or fusible link problems instead of simply replacing the device. Many electrical problems are caused by defects in the wiring or connectors. A flex test may uncover wiring and connection problems. Many defective electrical components may be detected by observation or by substituting a known good part.

Equipment for making basic electrical tests includes jumper wires, test lights, multimeters, and specialized testers. Care should be taken when using any of these testers to avoid damage to electronic parts. Motors and solenoids can be tested with an ohmmeter, or by using battery current to operate the unit. Diodes and resistors can be tested with an ohmmeter.

Many add-on electrical devices can be sources of driveability problems. The most common cause of problems is improper installation. Wires, fusible links, and connectors can be repaired, often by crimping or soldering. Connectors can be replaced or bypassed.

The starting and charging system is often a source of driveability problems. Batteries can be checked by using an electrical tester, or with a hydrometer. Battery replacement or recharging should be done carefully to reduce the chance of damage to other vehicle components. Starter problems can be caused by bad cable connections, internal starter problems, or by defects in the solenoid, ignition switch, or neutral safety switch. Starters and alternators can be disassembled to check the condition of internal parts. Often it is cheaper to replace the entire unit rather than making repairs. Always recheck the operation of the vehicle electrical systems after making any electrical repairs.

Know These Terms

Schematic
Circuit tracing
Test leads
Flex test
Short finder
Compass
Voltage drop test
Pulse meter
Soldering
Crimp connector
Parasitic draw
Load test
Full-field test

Review Questions—Chapter 14

Please do not write in this text. Write your answers on a separate sheet of paper.

1. To verify that a driveability problem is in the electrical system, first locate the correct ______.
   (A) test light
   (B) service manual
   (C) multimeter
   (D) jumper wire
2. A drawing which graphically represents the flow of electricity between electrical components is called a ______.
3. In many cases, the only way to determine whether a component is defective is to substitute a ______.
   (A) fused jumper wire
   (B) jumper wire with a circuit breaker
   (C) known good unit
   (D) known defective unit
4. Sometimes a fusible link will blow out without damaging the ______. This sometimes makes it hard to locate the damaged section.
5. A short finder is a ______.
   (A) fused test light and a compass
   (B) high impedance multimeter and a jumper wire
   (C) jumper wire with a circuit breaker and a compass
   (D) Maxifuse
6. An inductive pickup ______.
   (A) is clamped over the wire
   (B) pierces the wire
   (C) is connected in series with the circuit being tested
   (D) is connected in parallel with the circuit being tested
7. Self-powered test lights can be used to test for ______.
   (A) voltage
   (B) resistance
   (C) amperage
   (D) continuity
8. The voltage drop test checks for excessive ______ in a circuit.
   (A) voltage
   (B) current flow
   (C) resistance
   (D) heat
9. Relays control the flow of electricity to devices that use a lot of ______.
   (A) heat
   (B) current
   (C) voltage
   (D) speed
10. Add-on electrical devices and systems will cause problems if they are poorly ______.

ASE Certification-Type Questions

1. Technician A says that the best modern schematics are able to show the entire vehicle electrical system. Technician B says that modern schematics often reference other schematics when they run out of space. Who is right?
   (A) A only.
   (B) B only.
   (C) Both A & B.
   (D) Neither A nor B.
2. The flex test can be used to check for all of the following, EXCEPT:
   (A) high resistance connections.
   (B) internal wire shorts.
   (C) high resistance in wire harnesses.
   (D) defective electrical components.
3. When should a wire be pierced with a probe or pin?
   (A) Whenever it is necessary.
   (B) When the wire connectors cannot be reached.
   (C) Never.
   (D) In dry climates only.

4. Parasitic draws occur when ______.
   (A) the engine is running
   (B) the key is in the *on* position
   (C) the key is in the *off* position
   (D) None of the above.
5. Defective fusible links can often be found by ______ .
   (A) visual inspection
   (B) the flex test
   (C) retrieving trouble codes
   (D) Both A & B.
6. Multimeters can measure ______ .
   (A) pulse width
   (B) vacuum
   (C) fuel pressure
   (D) All of the above.
7. Technician A says that solid state components cannot be checked electrically. Technician B says that some solid state terminals can be checked for a proper ground. Who is right?
   (A) A only.
   (B) B only.
   (C) Both A & B.
   (D) Neither A nor B.
8. Damaged electrical connectors and wires should be repaired in a way that ensures all of the following, EXCEPT:
   (A) reduced current flow.
   (B) prevention of dirt and moisture entry.
   (C) prevention of corrosion.
   (D) no increase in resistance.
9. Technician A says that motors can be tested by making a continuity check. Technician B says that motors can be tested by applying current and observing motor operation. Who is right?
   (A) A only.
   (B) B only.
   (C) Both A & B.
   (D) Neither A nor B.
10. To properly test a temperature sensor, what two things must be measured?
   (A) Resistance and temperature.
   (B) Current flow and temperature.
   (C) Resistance and voltage.
   (D) Current flow and voltage.
11. If a parasitic draw stops when a fuse is pulled, this means that ______.
   (A) another circuit is the source of the parasitic draw
   (B) a circuit protected by the fuse contains the parasitic draw
   (C) the fuse is the source of the parasitic draw
   (D) the parasitic draw is not fused
12. When disconnecting a battery's terminals, which of the following procedures should be used?
   (A) Disconnect positive first, reconnect negative last.
   (B) Disconnect negative first, reconnect positive last.
   (C) Disconnect positive first, reconnect positive first.
   (D) Disconnect negative first, reconnect negative last.
13. The vehicle headlights are turned on, and the ignition switch is turned to the start position. The headlights remain bright with the switch in the start position. All of the following could be the cause, EXCEPT:
   (A) excessive starter current draw.
   (B) defective starter solenoid.
   (C) defective ignition switch.
   (D) misadjusted neutral safety switch.
14. Technician A says that bypassing the voltage regulator checks the alternator. Technician B says that bypassing the voltage regulator checks the regulator. Who is right?
   (A) A only.
   (B) B only.
   (C) Both A & B.
   (D) Neither A nor B.
15. After the alternator is disassembled, the following can be checked with an ohmmeter, EXCEPT:
   (A) rotor windings.
   (B) stator windings.
   (C) brushes.
   (D) diodes.

Advanced Certification Questions

The following questions require the use of reference information to find the correct answer. Most of the data and illustrations you will need is in this chapter. Diagnostic trouble codes are listed in **Appendix A.** Failure to refer to the chapter, illustrations, or **Appendix A** for information may result in an incorrect answer.

16. A vehicle has experienced repeated battery discharging. Technician A says that an excessive parasitic draw could be the cause. Technician B says that excessive current from the alternator could be the cause. Who is right?
    (A) A only.
    (B) B only.
    (C) Both A & B.
    (D) Neither A nor B.
17. A voltage drop test is being performed on the circuit in **Figure 14-2.** A voltage drop is located between C200 and C100 as indicated by point X. Which of the following is the least likely cause of the voltage drop?
    (A) High resistance in the wire between C200 and P100.
    (B) High injector resistance.
    (C) An open in the wire at P100.
    (D) High resistance in the wire between P100 and C100.
18. A multiport fuel injected engine will not start. The ignition control module in the schematic shown in **Figure 14-3** was found to have no battery power. The problem could be caused by ______.
    (A) an open at connector G105
    (B) an open in RED/BLK circuit 453
    (C) an open at ECM IGN fuse 10
    (D) a defective ignition control module
19. An alternator rated at 140 amperes is producing 13 amps at 11.4 volts. The reading was taken at engine idle. There is no change when the alternator is full-fielded. Which of the following is the least likely cause?
    (A) A loose or badly glazed drive belt.
    (B) A defective voltage regulator.
    (C) An open diode.
    (D) A defective rotor or stator.

Engine Size: 350 cu. in. (5.7L)		
Cranking Test Results	Voltage	Amperes
Normal Draw	11.1 V	150-180 A
Actual Cranking Draw	9.7 V	442 A

20. A V-8 engine has an excessively slow cranking speed. The chart shown above are the results from a cranking test. All of the following could cause the test results shown above EXCEPT:
    (A) high resistance in the starter motor armature.
    (B) corroded battery terminals.
    (C) an open starter solenoid.
    (D) high resistance in the battery cables.

Computer control has allowed the use of older engines in newer vehicles. This V-8 was originally designed back in the early 1970s. (Chrysler)

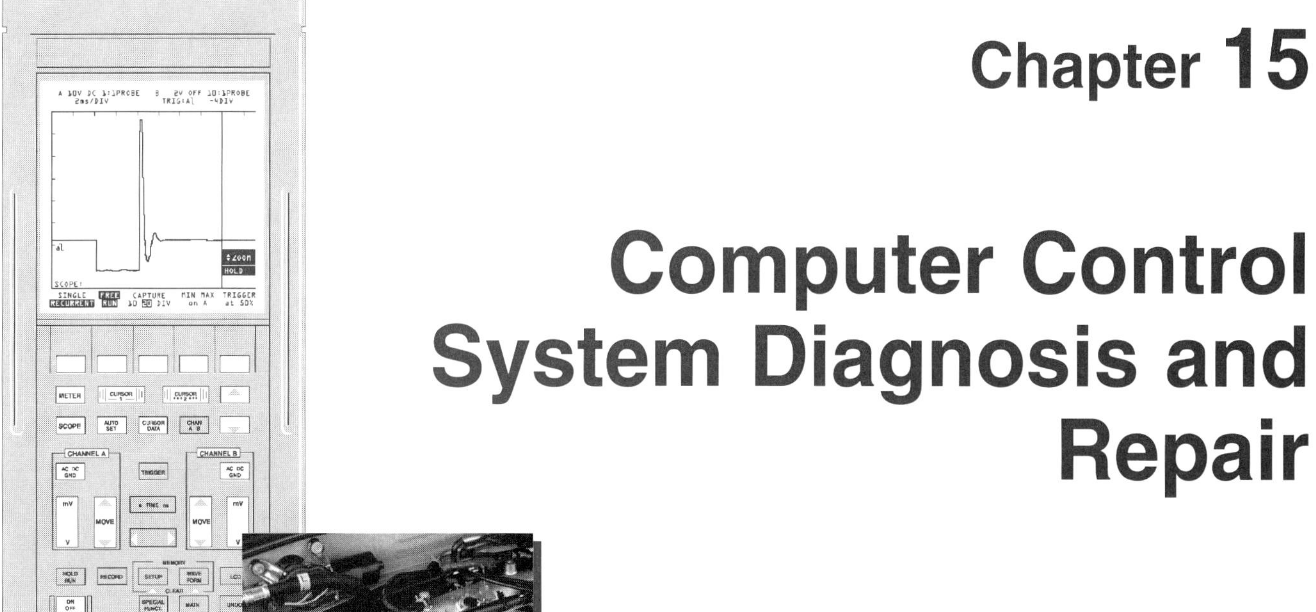

Chapter 15

Computer Control System Diagnosis and Repair

After studying this chapter, you will be able to:

- ❑ Identify methods of ECM trouble code retrieval.
- ❑ Retrieve ECM trouble codes.
- ❑ Identify trouble code formats.
- ❑ Interpret ECM trouble codes.
- ❑ Use a scan tool to check components and systems.
- ❑ Use a multimeter to check computer system components.
- ❑ Explain how to replace computer control system input sensors.
- ❑ Explain how to replace computer control system ECMs.
- ❑ Explain how to replace ECM memory chips.
- ❑ Explain how to program ECMs with eraseable PROMs

This chapter discusses diagnostic and repair procedures for the computer control system. This includes the ECM and the most commonly used input sensors. Service procedures for some input sensors and all output devices will be covered in later chapters. It is important that you have a good understanding of computer control diagnostic procedures. You will also use many of the test procedures that were covered in **Chapter 14.** Studying this chapter will explain the basic computer system testing procedures that you will build on as you learn to diagnose other problems in later chapters. You may want to review **Chapter 7** before beginning this chapter.

Beginning to Diagnose Computer Control Problems

The computer system always maintains some level of control over many vehicle systems. This level of control ranges from simple monitoring of operational state to aggressive adjustments to compensate for minor changes in vehicle operation. A minor or intermittent computer problem can have a dramatic effect on one or more of these systems. The types of driveability problems that the computer control system can cause vary from almost unnoticeable to completely disabling the vehicle.

Be sure to obtain the vehicle's service manual before you begin any diagnostic operation. Open the hood and check for visible problems before performing any other tests. Remember that computer input sensors operate on very low voltages and slight problems in the wiring may cause inaccurate readings. Any arcing or sparking caused by loose electrical connections will affect computer operation. Also check charging system voltage. Excessively high or low voltages will set codes and affect ECM operation. Use the electrical system diagnostic steps you learned in **Chapter 14.** The flex test is particularly effective in finding computer problems. Typical computer control problems and some of the systems they can affect are shown in **Figure 15-1.**

Check for signs of modification or tampering. This includes sensors and output devices that have been removed, unplugged, or disabled in some manner. If a visual inspection does not reveal any problems, the next

Computer control system
(defective ECM, input sensors or output devices; disconnected input sensor or output device; tampering)

Ignition system
(no or poor spark control, diesel glow plug control inoperative)

Fuel System
(wrong pulse width, wrong air-fuel ratio control, open injector, hard starting)

Catalytic converter
(catalyst monitor inoperative or contaminated)

Automatic transmission/Transaxle
(improper shifting, improper or no TCC lockup)

Cooling system
(radiator fan inoperative)

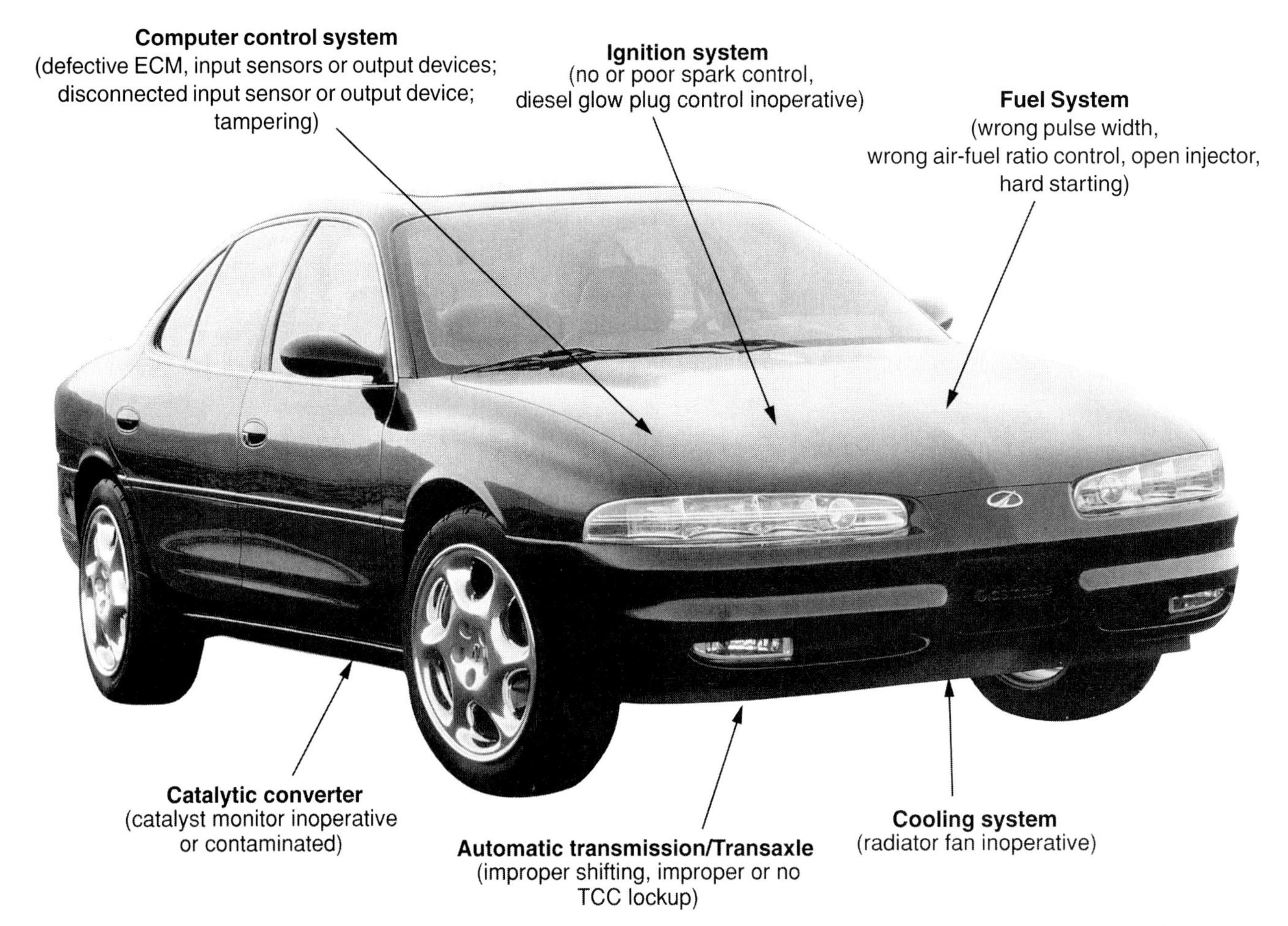

Figure 15-1. *The computer system monitors and controls systems that affect the entire vehicle. The problems shown here are only a few of the possible problems that can occur to systems controlled by the ECM. (Oldsmobile)*

step is retrieving any diagnostic trouble codes present in the ECM. Observe ECM controlled systems for proper operation. Remember that a problem may also be caused by normal ECM operation, such as an intrusive diagnostic test performed by an OBD II system's ECM. Other perceived problems include:

- ❑ The vehicle fails to exceed a certain speed or engine rpm. The ECM in most vehicles will not allow engine speeds over 6000-7800 rpm or the vehicle to exceed a certain speed.
- ❑ High idle caused by the operation of engine driven components, such as the air conditioning and power steering system.
- ❑ Hard starting in fuel injected vehicles caused by the driver depressing the pedal while cranking. This is caused by the ECM entering clear flood mode.
- ❑ These and other internal ECM functions were discussed in **Chapter 7.**

Retrieving ECM Trouble Codes

Before performing any involved diagnostic routines on computer-controlled vehicles, always check for and retrieve any trouble codes from the ECM's memory. If the dashboard-mounted malfunction indicator lamp is on when the engine is running, trouble codes are present in the ECM memory. The ECM's memory should always be checked anytime driveability work is performed, since trouble codes may be stored, even if the light is not on, **Figure 15-2.**

To retrieve trouble codes from the ECM, follow the procedures in the factory service manual. Basic procedures for code retrieval are discussed later in this chapter. On most older vehicles, the trouble codes can usually be retrieved

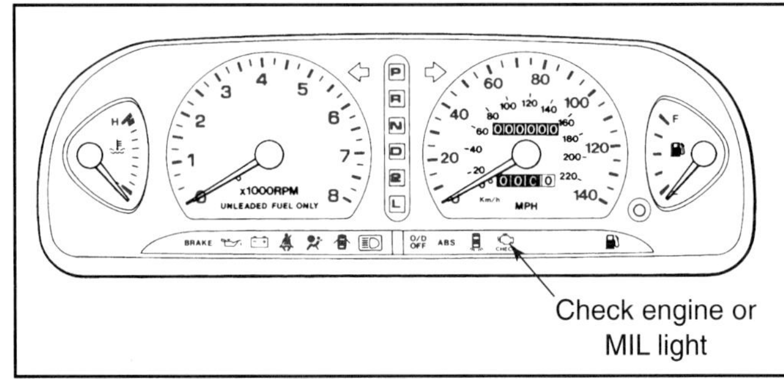

Figure 15-2. *The check engine or malfunction indicator light (MIL) is the first place to look when diagnosing any problem. Check the status of this light while test driving the vehicle. (Toyota)*

without any special equipment, although a code retrieval tool or scan tool will make the process easier. A *bi-directional scan tool* is the best for code retrieval, since it can provide information on system operation and has some test capabilities, **Figure 15-3.** The different types of code retrieval tools were discussed in **Chapter 3.**

In **Chapter 7,** you learned that for many years, each manufacturer used a different style of data link connector or code retrieval method. This made accessing diagnostic trouble codes a difficult process. These connectors could be located almost anywhere, depending on the manufacturer and vehicle model. However, the OBD II diagnostic protocol required manufacturers to use a standardized 16-pin data link connector located inside the vehicle's passenger compartment. Data link connectors used by various manufacturers are shown in **Figure 15-4.**

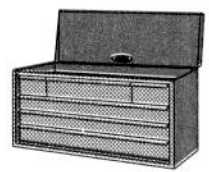

Note: The standardized 16-pin data link connector is used on some non-OBD II vehicles.

Some computer systems do not have data link connectors and use other methods to access the information in the ECM. In **Figure 15-5,** the information is obtained by turning a screw on the side of the ECM. This causes an LED to begin flashing to indicate trouble codes. On other systems, the information is displayed on the heater-air conditioning control head ac controls by pressing two control buttons at the same time, **Figure 15-6.**

The Code Retrieval Process

Trouble codes may be obtained from the ECM by using a ***two-part process.*** All methods of trouble code retrieval are variations of this two-part process, whether it is done manually by the technician or automatically by an electronic tool.

- ❑ Grounding a terminal to place the ECM in self-diagnosis mode.
- ❑ Reading the codes that the ECM displays.

Accessing Codes without an Electronic Retrieval Tool

Accessing trouble codes on OBD I systems without an electronic tool is very simple. The most common method is to ground one terminal of the data link connector and observe a series of flashes from a diagnostic light. The light may be the amber dashboard MIL light, or it may be a LED (light emitting diode) installed on the ECM case. The sequence of light flashes forms a numerical code. The factory service manual contains instructions for correctly reading the light flashes.

Another common method of reading trouble codes is to connect an analog voltmeter to the proper diagnostic connector leads. Another connector terminal is grounded and the trouble code output is read as a series of pulsations of the voltmeter needle, **Figure 15-7.** Always follow the manual procedure exactly, since the wrong sequence of steps can erase the trouble codes or possibly set false codes. These codes should also be written down for reference when consulting the service manual. Some digital multimeters can display codes as electronic pulses.

Data Scanned from Vehicle			
Coolant temperature sensor 204°F/95°C	Intake air temperature sensor –35°F/–37°C	Manifold absolute pressure sensor 1.2 volts	Throttle position sensor .45 volts
Engine speed sensor 920 rpm	Oxygen sensor .45 volts	Vehicle speed sensor 0 mph	Battery voltage 13.5 volts
Idle air control valve 33 percent	Evaporative emission canister solenoid Off	Torque converter clutch solenoid Off	EGR valve control solenoid 0 percent
Malfunction indicator lamp On	Diagnostic trouble codes P0110, P0111	Open/closed loop Open	Fuel pump Relay On
Brake light switch Off	Cruise control Off	AC compressor clutch Off	Knock signal No
Ignition timing (°BTDC)	Base timing: 8	Actual timing: 10	

Figure 15-3. *A scan tool can provide information as to the operational status of the engine and its related systems. Charts such as these will be used in this text to show the type of information presented by most scan tools. You must learn not only what type of information is available, but how to properly interpret the readings.*

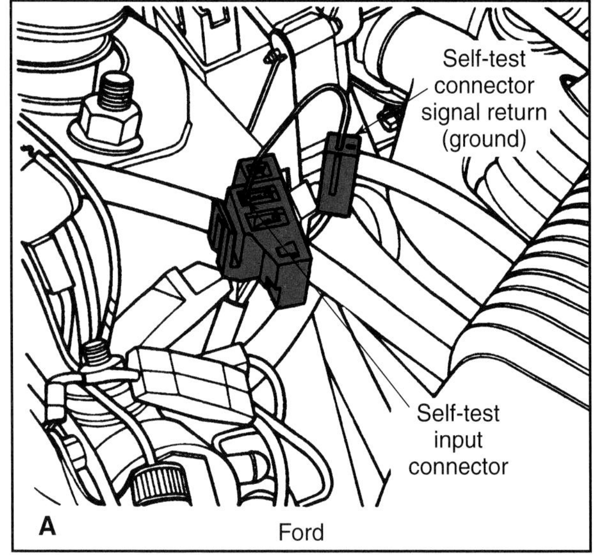

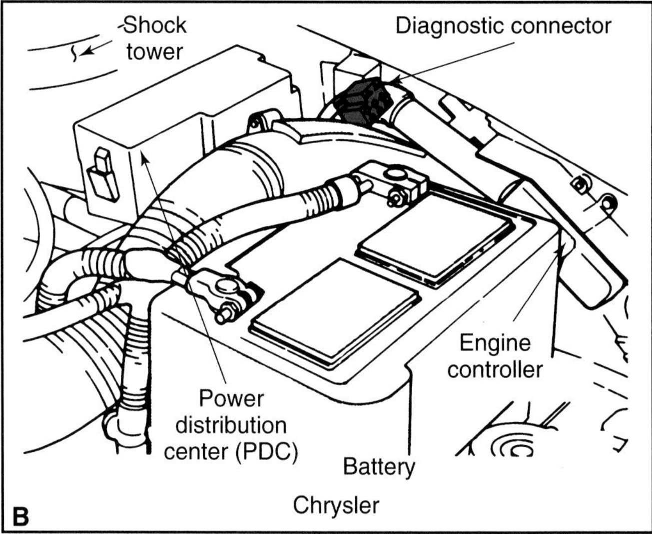

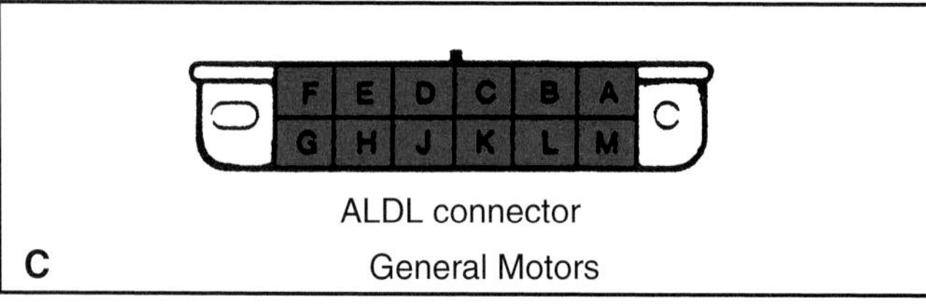

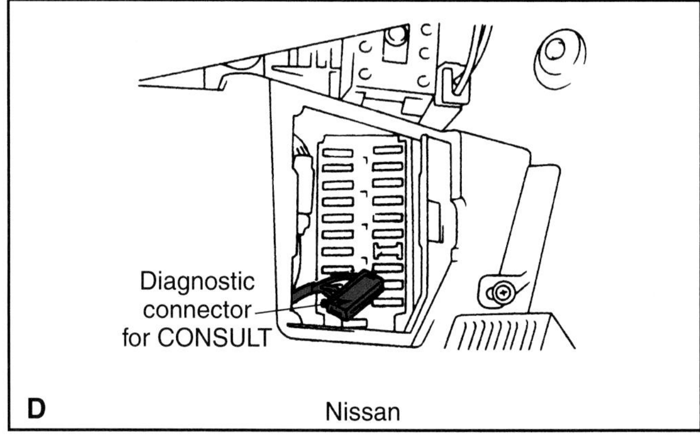

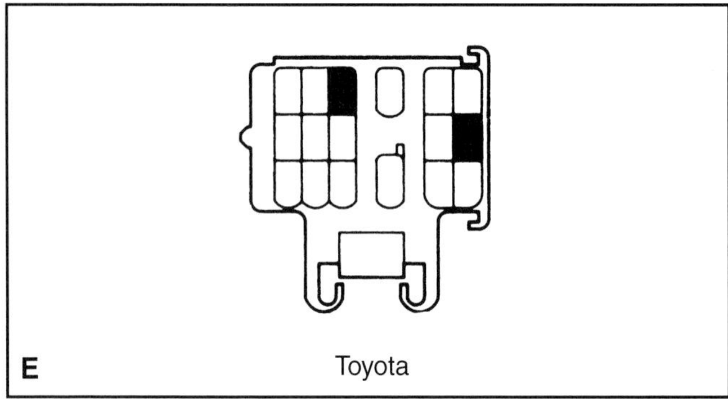

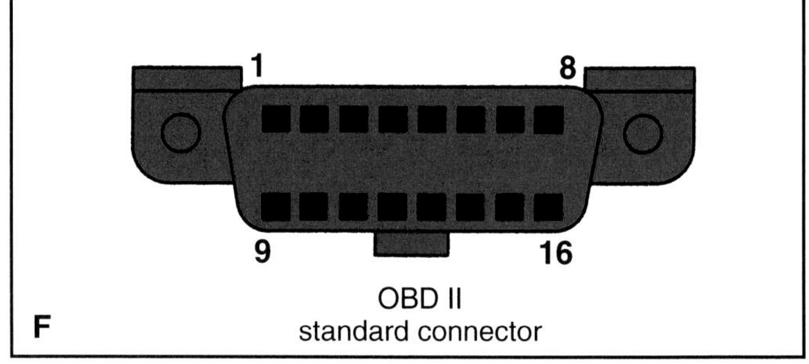

Figure 15-4. *Various data link connectors (DLCs) as used on modern vehicles. A—Ford EEC connector. B—Chrysler diagnostic connector. C—General Motors Assembly Line Diagnostic Link (ALDL). D—Nissan diagnostic connector. E—Toyota DLC. F—OBD II standardized data link connector. (Ford, Chrysler, Toyota, General Motors, Nissan)*

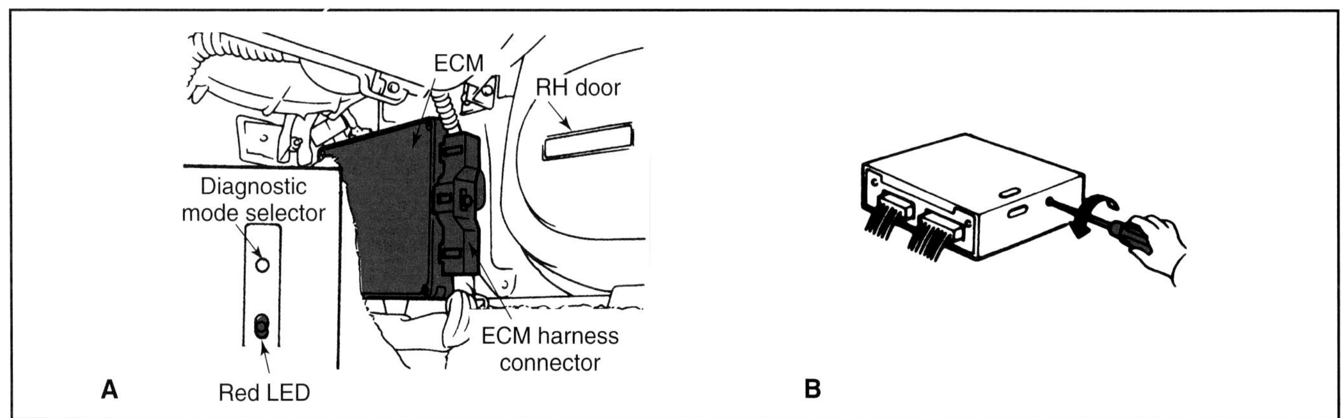

Figure 15-5. *Older diagnostic systems do not use a data link connector. Some use a switch or other method of activating the self diagnostics. This system uses a switch built into the ECM that is turned by a screwdriver.*

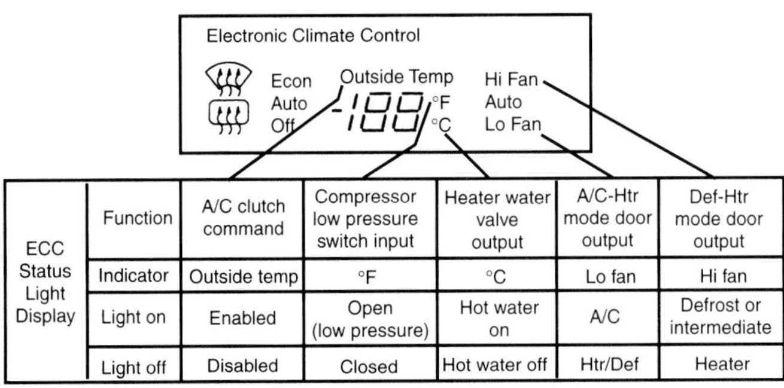

ECC Status Light Display	Function	A/C clutch command	Compressor low pressure switch input	Heater water valve output	A/C-Htr mode door output	Def-Htr mode door output
	Indicator	Outside temp	°F	°C	Lo fan	Hi fan
	Light on	Enabled	Open (low pressure)	Hot water on	A/C	Defrost or intermediate
	Light off	Disabled	Closed	Hot water off	Htr/Def	Heater

Parameters			
Number	Description	Range	Units
-00	ECC Fault Codes	00 - 55t	Code
-01	ECM Fault Codes*	00 - 55t	Code
-02	Program Number	0 - 255t	Counts
-10	Ignition Voltage	9.0 - 16.0	Volts
-11	Program Number * *	0 -100	%
-12	Vehicle Speed	0 -199	MPH
-19	Actual Blower Voltage	-3.3 - 18.0	Volts
-20	Commanded Blower Voltage	-3.3 - 18.0	Volts
-21	Engine Coolant Temperature	-40 - 151	°C
-22	Commanded Mix Valve Position	0 - 255	Counts
-23	Actual Mix Valve Position	0 - 255††	Counts
-24	Air Delivery Mode	0 - 6	Code •
-25	In-Car Temperature	-40 - 102	°C
-26	Outside Air Temperature	-40 - 93	°C
-28	Evaporator Inlet Temperature	-40 - 93	°C
-29	5.0 Liter vs. 5.7 Liter Engine (0000 - 5.0L, 01 - 5.7 L)	00 - 01	Code
-30	Ignition Cycle Counter	0 - 99	Counts
-31	PROM ID	0 - 9999	Code❏

❏ PROM ID code number identifies as individual calibration and is periodically updated
* 5.7 Liter engine only
* * "WARMER" increases program number;
"COOLER" decreases program number.
†† For parameter values greater than 199, the leading 2 is displayed as ¬ (i.e. 255 - ¬55)

Figure 15-6. *Some systems with electronic air conditioning or driver information centers have the ability to access the ECM by pressing two or more buttons on the control panel. (General Motors)*

Reading Trouble Codes

To read the flashes or meter pulses, connect the jumper wire, ground tool or meter and watch the light. For example if a code 26 was present, the light would flash two times, pause momentarily, and flash six times. The needle on an analog meter would move two times, pause, and then move six times. Some systems may repeat the same code up to three times before flashing the next code. When the first code is repeated, there are no more codes stored. Write down the numbers of all codes present.

A third method is to press buttons and/or turn the ignition switch on and off in a certain sequence. The ECM trouble codes on some vehicles can be accessed by turning the ignition switch on and off three times within five seconds. Some trouble codes can be accessed by pressing and holding down a certain sequence of buttons on the radio and/or air conditioner control head for a period of time. The code will appear as a number on the control head display. This feature is found only on a few vehicles with electronic entertainment and climate control systems.

Code Retrieval Using Electronic Tools

On OBD II equipped vehicles, the proper scan tool *must* be used to retrieve trouble codes. Do not attempt to retrieve trouble codes from an OBD II system by grounding a terminal, as this will damage the ECM. On OBD I systems, it is far easier and less troublesome to use a code retrieval tool or scan tool than to perform the procedures explained earlier. Code retrieval tools and scan tools display the code

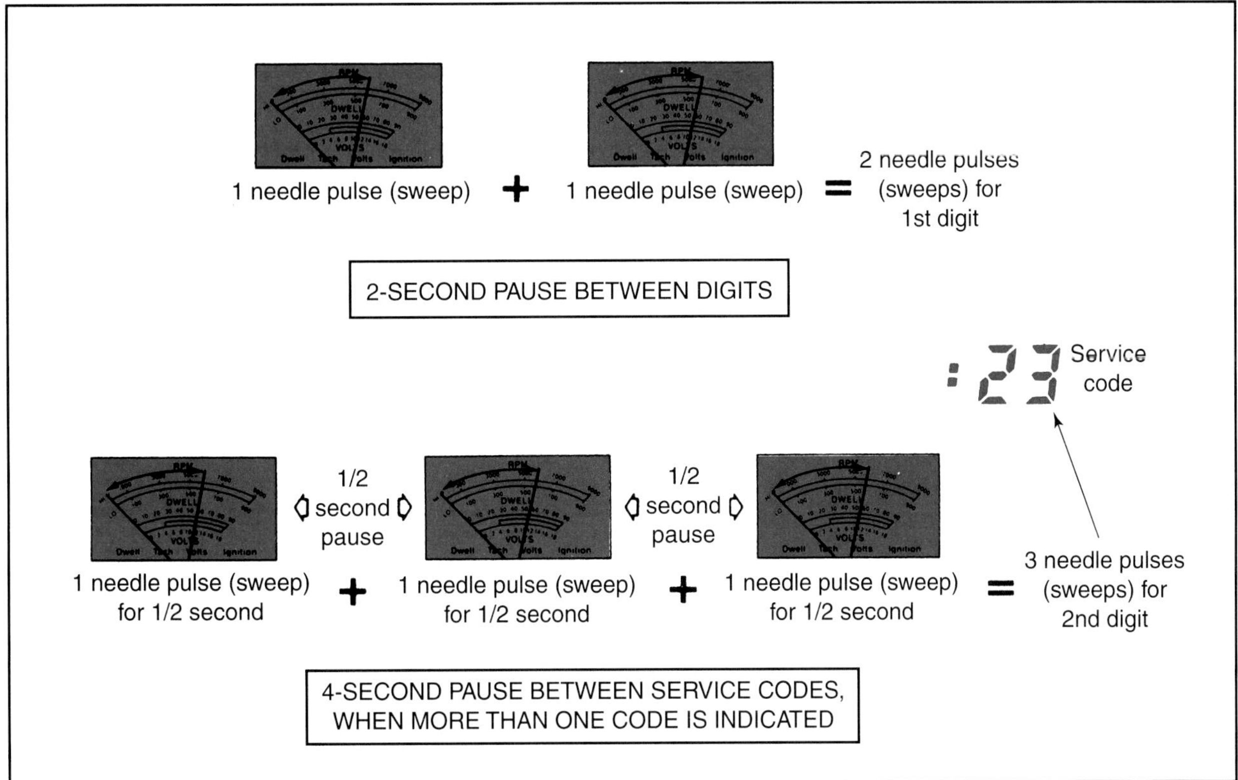

Figure 15-7. *An analog meter can be used to read the codes on most OBD I systems. Some digital meters can read codes as digital pulses. (Ford)*

numbers on a screen, and store them for future reference. They also perform the grounding step to place the ECM in the diagnostic mode. Typical tools are shown attached to vehicle data link connectors in **Figure 15-8.**

Scan tools can also show sensor and output data as well as perform certain output device tests. Always follow the manufacturer's directions exactly when installing the tool and retrieving the codes. Some scan tools have the capability to identify the type of ECM and/or PROM in use, **Figure 15-9.** See **Appendix C.**

Any trouble code number(s) present should be written down for later comparison with the corresponding numbers in the service manual to identify the faulty component or system. If you are unable to retrieve trouble codes or access sensor operating information, make sure that both scan tool and data link connector are connected to the ECM. If all of the connections are good, the ECM is most likely defective. This may also be an indication that someone may have tampered with the ECM or the system wiring.

Diagnostic Trouble Code Format

The trouble codes for most computer-controlled vehicles are displayed as or translated into two- or five-digit numbers, depending on the system, that correspond to a specific problem. Each trouble code corresponds to a spe-

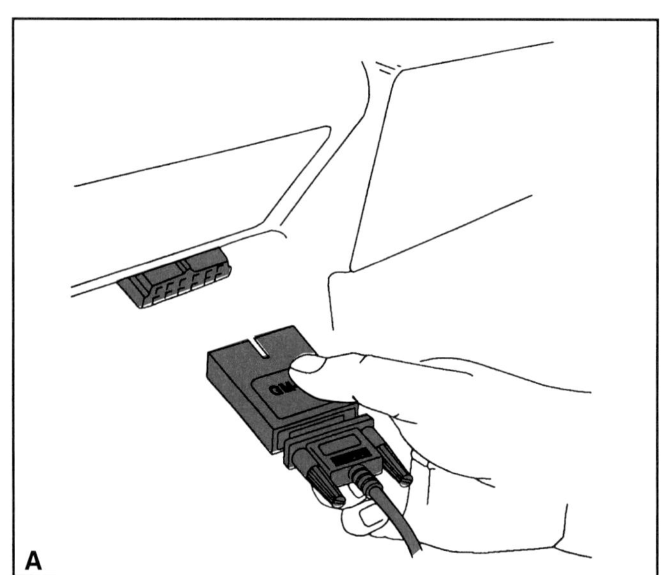

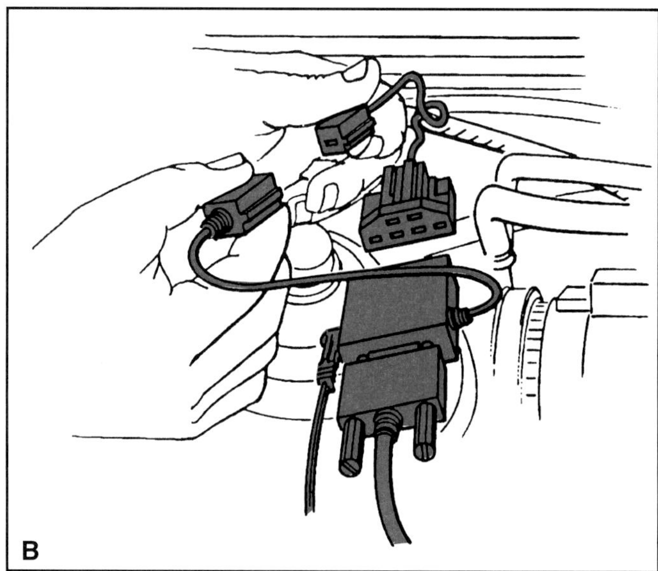

Figure 15-8. *The best way to read codes on any computer system is to use a scan tool. A—Connecting a scan tool to a GM ALDL connector. B—Some DLCs require a separate power lead, such as this Ford connector. (Snap-On Tools)*

Data Scanned from Vehicle			
Coolant temperature sensor 198°F/92°C	Intake air temperature sensor 77°F/25°C	Mass airflow sensor 2.7 volts	Throttle position sensor .78 volts
Engine speed sensor 930 rpm	Oxygen sensor .52 volts	Vehicle speed sensor 0 mph	Battery voltage 13.6 volts
Idle air control valve 38 percent	Evaporative emission canister solenoid Off	Short-term fuel trim 125	Long-term fuel trim 131
Malfunction indicator lamp Off	Diagnostic trouble codes	Open/closed loop Closed	Fuel pump relay On
PROM ID 5248C	Cruise control Off	AC compressor clutch On	Knock signal No
Ignition timing (°BTDC)	Base timing: 6		Actual timing: 19

Figure 15-9. *The ECM and PROM carry an identification code in its programming. This code can be accessed by a scan tool. The programming in OBD II systems can be identified by the date code in the program.*

cific problem area. Some trouble codes can often be eliminated if another code related to that system exists.

Note: A trouble code does not necessarily mean that the part or system is defective. For example, a constant rich reading from an oxygen sensor may mean the sensor is defective. It can also be caused by a defect that is causing the engine to run rich at all times

---

OBD I and OBD II Code Formats

OBD I systems normally display two-digit codes. The codes are correlated to a specific sensor or circuit. In a few cases, the code will specify a certain problem, such as low or high voltage. However, the code will simply indicate that there is a problem in the system.

OBD II systems display trouble codes as a three-digit number that follows an ***alpha numeric designator.*** The letter in the alpha numeric designator indicates the purpose of the vehicle system causing the problem; the number indicates whether the code is a generalized code assigned by SAE or a manufacturer specific code. In the three-digit number, the first number indicates the specific vehicle system, and the last two numbers are the ***fault designator,*** which indicates the sensor, output device, or ECM, and the type of fault. The two-digit and OBD II trouble code formats for a coolant temperature sensor problem are shown in **Figure 15-10.**

The three-digit trouble code format in OBD II systems does not conform to the OBD I two-digit format. *Do not* try to correlate the older fault codes with the OBD II fault designator. For example, there are only two or three codes used by most older computer control systems to indicate a possible oxygen sensor problem. OBD II systems have *thirty* or more possible generic codes for the oxygen sensor and may also have additional manufacturer specific codes, depending on the system. A complete list of the SAE generalized diagnostic trouble codes is located in **Appendix A** in the back of this text.

Interpreting Trouble Codes

Once the trouble codes are retrieved, they can be compared against the trouble code information in the service manual. Sometimes the same trouble code can be caused by one or more problems. Therefore, it is vital to correctly interpret the trouble codes. Trouble codes usually indicate one of three things:

- ❑ An engine problem is causing one of the sensors to transmit a voltage signal that is out-of-range (too high or too low).
- ❑ A sensor's ability to perform is starting to deteriorate.
- ❑ A part or circuit in the computer control system is defective.

Multiple codes should be addressed starting with the lowest code number and increasing. For example, if you

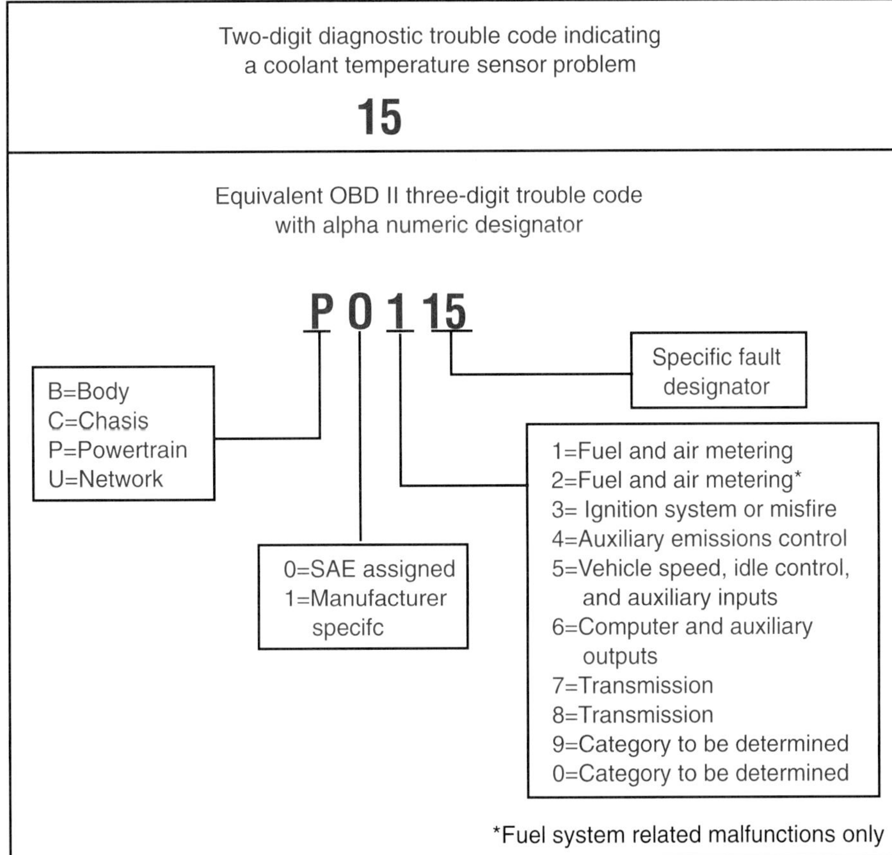

Figure 15-10. *OBD I and OBD II systems use two different diagnostic code systems. Both systems are distinct and any stored trouble codes in one system should not be correlated with the other system.*

were looking up codes P0175, P0304, and P0306, you would start with code P0175 first before looking up the others. In cases where multiple codes are stored, one or more codes are often eliminated using this process.

Once the problem area has been identified, you can proceed to concentrate on that area. If the trouble code indicates that the problem is a defective output device or ECM, you will be able to go to the specific part and determine whether the problem is a disconnected or defective part, a mechanical problem with the part, or a wiring problem. In the case of an out-of-range sensor reading, you can proceed to determine whether the problem is a defective sensor, or another engine condition that is causing the out-of-range reading.

Comparing Hard and Intermittent Codes

Two types of codes are stored in the ECM's volatile memory. They are called ***hard codes*** (permanent) and ***intermittent codes*** (temporary). Hard codes indicate an on-going problem, one that still exists on the vehicle. They are often the easiest to track down. Intermittent codes have been set by problems that occur occasionally, or only once, and are harder to isolate.

Note: Most ECMs will lose any stored codes if the battery is disconnected or goes completely dead. Therefore, trouble codes in a vehicle with a dead battery were likely set after the battery problem was corrected.

Separating Hard and Intermittent Codes

A hard code will illuminate or flash the indicator light whenever the engine is running. Intermittent codes will only turn the light on while the problem is occurring, after which the light will go out. Most systems require that you separate hard codes from intermittent codes. First, record all codes, then remove electrical power to the ECM for the period of time specified in the service manual to erase the stored codes.

This can usually be done by removing the negative battery cable or the vehicle fuse that supplies current to the ECM, **Figure 15-11.** Most scan tools have the capability to erase stored codes. Restart the engine and allow it to run for a period of time as specified by the manufacturer. After the engine has run long enough, stop it and re-enter the diagnostic mode.

Note: The ECMs in some OBD II equipped vehicles can retain trouble codes for several days without battery power. Removing battery power from the ECM on such equipped vehicles may not erase any stored trouble codes. Use a scan tool to clear stored trouble codes on OBD II systems.

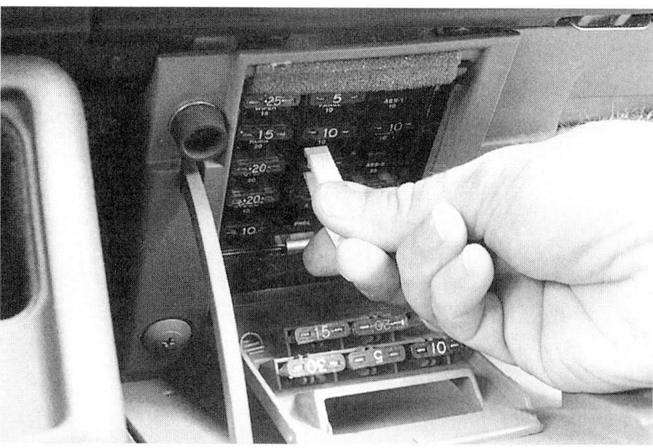

Figure 15-11. *The fuse that supplies power to the ECM is usually located in the main fuse block. Pull this fuse to clear any stored codes. Remember that this may not work with most OBD II systems.*

Hard Codes

Hard codes will usually reset almost immediately after the engine is started or when the ECM requests information from a sensor or tries to activate an output device. Hard codes are usually caused by a defective component, open or shorted circuit, or connector. Hard codes make it relatively easy to isolate the defect to a specific area, since the problem is always there.

Intermittent Codes

In many cases, an intermittent code may not mean that the sensor or output device is defective, only that it is responding to another problem. An example is an intermittent trouble code for a rich oxygen sensor reading. Although the oxygen sensor may be defective, it is more common for another component, such as a leaking fuel injector or defective airflow sensor, to cause the rich mixture problem. Problems that set intermittent codes take longer to diagnose.

Keep in mind that many sensors can go slightly out-of-range without setting a trouble code. The ECM in an older computer control system often will respond to the incorrect input in the same way that it responds to the correct input. OBD II systems are designed to monitor sensor performance and will usually set a diagnostic trouble code if a sensor goes slightly out-of-range, even if there is no apparent driveability problem. However, lack of a trouble code in any computer control system is not absolute proof that a sensor is good.

OBD II Trouble Codes

OBD II codes are very different from the two-digit codes used by older computer control systems. Each code correlates to a specific type of failure versus a general failure of a sensor, circuit, or output device.

In the older, or OBD I systems, the two-digit system limits the number of trouble codes to 100. In the three-digit

OBD II system there are 1000 potential codes. There are presently four letter codes used to begin the OBD II code sequence. Three letter codes cover the three major vehicle subdivisions. They are B (body), C (chassis), and P (powertrain). A fourth letter, U, designates internal computer communications network codes. The number immediately following the letter indicates whether the code is a general SAE code (0) or a manufacturer specific code (1).

With the four letters and two general/specific numbers, the total of possible OBD II codes is presently at 8000. In addition, OBD II scan tools display engine timing, temperature readings, and various sensor inputs. For instance, the OBD II scan tool can access the ECM and sensors to provide the technician with readings of engine idle RPM, as well as the desired (factory set) idle speed. The technician can instantly determine if idle speed is correct. Another feature will give the technician a series of numbers or percentages indicating whether the fuel trim (air-fuel ratio) is correct for both short- and long-term engine operation, **Figure 15-12.** Other inputs give exact engine timing in degrees, actual temperature reading from the engine temperature sensors, and voltage or duty cycle readings which indicate manifold vacuum, throttle position, EGR position, and many other operating conditions.

Short-term fuel trim	Long-term fuel trim
126	130

Figure 15-12. *The ECM uses the inputs to determine fuel trim for short- and long-term operation. The fuel trim is broken down into numbers or counts that can be read by a scan tool.*

Some scan tools are able to interface with the ECM to provide *snapshots* of engine operation. The snapshot is a method of recording engine operating conditions when a malfunction occurs. This allows the technician to determine exactly what is happening to cause the problem. To record the snapshot information, the technician only has to drive the vehicle with the scan tool attached until the malfunction occurs. The technician can then access the readings through the scan tool. Some vehicle ECMs save snapshot type information as soon as a malfunction occurs, and the technician can access it once the scan tool is attached. **Figure 15-13,** which shows information obtained while looking for the cause of a rough idle, is an example of the type of data available in a snapshot.

While the sophistication of the scan tool increases the amount of information available to the technician, it also makes it easier to misunderstand what the code means, or to look up the wrong code. Sometimes the sheer volume of information available from the scan tool can cause the technician to miss the actual cause of a problem. For this reason, the technician must carefully interpret all codes and other scan tool information before proceeding with diagnosis and repair.

Interpreting Scan Tool Data

One of the most important skills that any technician can develop is the ability to correctly interpret data presented by a scan tool. ***Scan tool data*** can be a tremendous help in diagnosing driveability problems. It can also mislead technicians if the data is interpreted incorrectly.

In order to correctly interpret scan tool data, you must know the normal operating parameters of the engine, as well

Snapshot Data Captured from Vehicle			
Coolant temperature sensor 212°F/100°C	Intake air temperature sensor 79°F/26°C	Mass airflow sensor 3.4 volts	Throttle position sensor 1.99 volts
Engine speed 2476 rpm	Oxygen sensor .31 volts	Vehicle speed sensor 40 mph	Battery voltage 13.2 volts
Idle air control valve 5 percent	Evaporative emission canister solenoid Off	Cooling fan On	EGR pintle position 11 percent
Malfunction indicator lamp On	Diagnostic trouble codes P0272, P0304	Open/closed loop Open	Fuel pump relay On
Transmission gear Drive	Cruise control Off	AC compressor clutch On	Knock signal Yes
Ignition timing (°BTDC)	Base timing: 7	Actual timing: 56	

Figure 15-13. *Data captured when a problem occurs can help you find the cause of a driveability problem. Some ECMs will capture this data automatically when a problem occurs. What type of problem does the data in this chart indicate?*

as all of the sensors and output devices. This text gives the typical operating parameters of most sensors and output devices. Check the vehicle service manual for the proper information for the vehicle you are working on. Service manual troubleshooting charts provide this information in an easy to use format. See **Appendix C** for more information on scan tool usage.

Computer Sensor and System Testing

A variety of test equipment can be used to check many computer control components. Some of these readings can be made using jumper wires, ohmmeters, or voltmeters. Tests which can be made with more than one type of tester are identified in the following individual test procedures. The meter will show the response of various computer control sensors and output devices to changing engine conditions. The following sections will also quickly review sensor and output device operation and discuss typical sensor readings.

Checking Oxygen Sensor Operation

There are two basic kinds of oxygen sensors, the electrically heated and the non-heated. Oxygen sensors with one or two lead wires are non-heated types. If an oxygen sensor has three or more wires, it is a heated type. Most vehicles made within the last several years are equipped with heated oxygen sensors.

If an oxygen sensor trouble code is set, and the "check engine" light is on, first check the harness for proper connection. Then make a quick check of the oxygen sensor by tapping on the exhaust manifold near the sensor with the engine running. If the light goes out, the sensor is defective.

Warning: Exercise caution when working around any exhaust manifold. Exhaust systems can reach extremely high temperatures very quickly.

If the light is still on, use a scan tool to access the input data from the oxygen sensor in question. Remember that some vehicles use more than one oxygen sensor. Connect the scan tool to the DLC and enter the vehicle's information. Make sure the engine is warm enough for the computer to operate in closed loop before beginning to monitor O_2 sensor output. If the engine is not in closed loop mode, run it at fast idle for a few minutes to raise engine and exhaust temperature. Allow the engine to stabilize. After the engine stabilizes, watch the readout, **Figure 15-14.**

The voltage will oscillate between 100 and 900 millivolts if the ECM is in the closed loop mode. The voltage readout will change very quickly with the amount of oxygen in the exhaust. The meter should show a minimum voltage of less than 300 mV, a maximum of more than 600 mV, and should average about 500 mV.

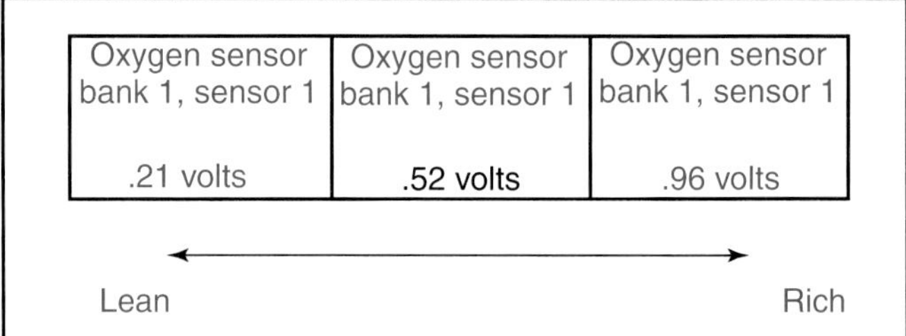

Figure 15-14. *Oxygen sensor voltage varies with the content of the oxygen in the exhaust gas. Be sure that you are monitoring the correct sensor on vehicles with multiple oxygen sensors.*

Oxygen Sensor Voltage Measurement

To measure the voltage output of the oxygen sensor, you must have a voltmeter or multimeter that can read very low voltages (in the 100 to 900 millivolt mV range). For this test to be accurate, the engine must be warm enough for the ECM to be in closed loop mode. The exhaust system temperature must be at least 600°F (350°C) before starting the test, unless the oxygen sensor is a heated type. When tested on the engine, the sensor must be heated either by its heating element (when used) or by exhaust heat. Heated oxygen sensors can be tested within seconds of start-up if the engine is warm. The service manual will tell you whether the sensor is the heated type. If removed from the engine, the sensor must be heated with a propane torch. In all cases, the manufacturer's instructions must be followed closely.

Caution: Oxygen sensors are sensitive to excess current flow. They can be destroyed by improper grounding, or the use of test lights and low impedence multimeters. Some manufacturers do not recommend meter tests of the oxygen sensor. Always consult the manfacturer's instructions before testing an oxygen sensor.

Begin by obtaining the correct specifications for the oxygen sensor that you are testing. Oxygen sensors use either Zirconia or Titania elements. A Zirconia type sensor will produce a voltage reading, while a Titania type sensor will produce a resistance reading. The majority of oxygen sensors in current use are Zirconia types. After determining the type of sensor, set the multimeter to the proper range, usually 2 volts for a Zirconia and 200KΩ for a Titania sensor.

After ensuring that the engine is warm enough, connect the meter to the oxygen sensor output wire as shown in **Figure 15-15.** Make sure that the meter is connected to the output wire and not to the sensor input or ground wires. One and three wire sensors are grounded through the sensor housing. On two and four wire sensors, one of the wires is a ground wire. Be very careful to make the proper connections, as slight voltage surges are sufficient to ruin an oxygen sensor. Set the meter to the proper range. Observe the meter for several minutes after the ECM enters the closed loop mode. Zirconia oxygen sensors should show a minimum voltage of less than 300 mV, a maximum of more than 600

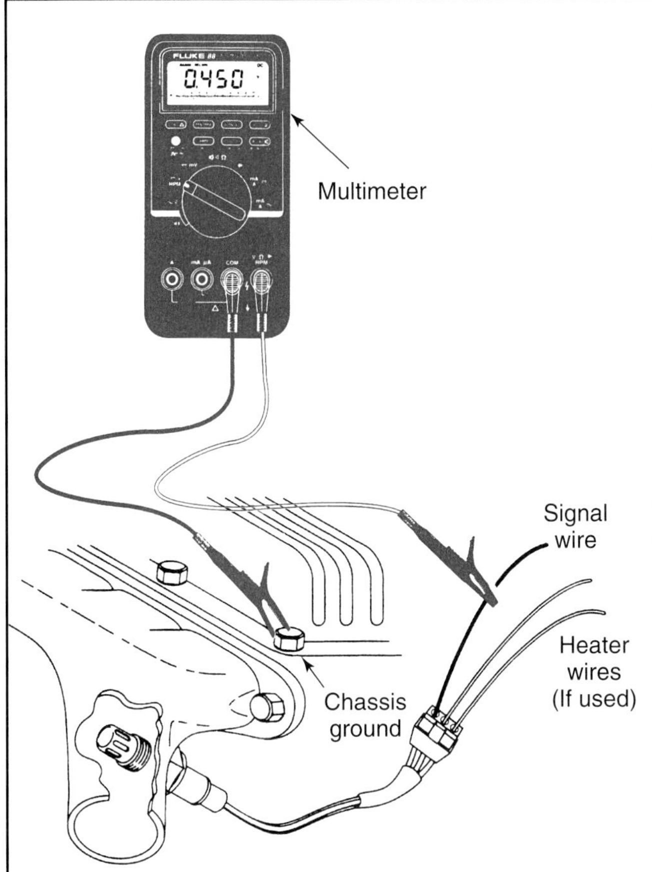

Figure 15-15. *A multimeter set to the voltage scale can check an oxygen sensor for proper operation. Be careful when working around hot exhaust parts. (Fluke)*

mV, and should average about 500 mV. Titania oxygen sensors should show low resistance when the air-fuel mixture is rich and high resistance when the mixture is lean.

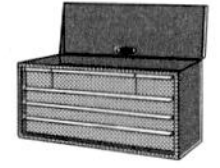

Note: Some manufacturers allow the oxygen sensor to be bypassed by grounding the sensor input to the ECM. The technician should only ground the O_2 sensor wire when specifically instructed by the service manual. Never ground any oxygen sensor wire without checking the service manual first.

Creating an Artificial Lean and Rich Mixture Condition

To test the oxygen sensor for response time, it is necessary to create an artificial lean and rich condition. To create an ***artificial lean mixture,*** disconnect the PCV hose or another large vacuum hose and cover it with your thumb. As you remove your thumb, a lean condition is created due to the vacuum leak. Partially cover the vacuum hose opening if the engine begins to stall. Then observe the meter for several minutes. The meter should show a lower than average voltage or high resistance as the oxygen sensor reacts to the lean mixture.

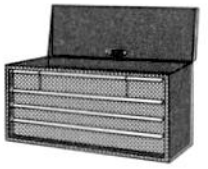

Note: The ECM on some fuel injected engines will react too fast for this test to be effective; the engine speed will just rise and fall.

After making this test, reconnect the vacuum hose and create an ***artificial rich mixture*** with a propane enrichment device, such as those used to adjust carburetors. Do not create a rich mixture with carburetor cleaner or gasoline, as this will simply overload the oxygen sensor. After the enrichment device is in place and operating, check the meter. Voltage should be higher than average, while resistance should be low, indicating that the rich condition is being read by the oxygen sensor. If the oxygen sensor does not react, reacts slowly, or indicates an out-of-range reading, the sensor is defective and should be replaced.

If you are using a propane torch to heat the sensor, a rich mixture can be simulated by completely surrounding the tip with the torch flame. The voltage of a Zirconia sensor should increase, while the resistance reading of a Titania sensor should decrease. When the flame is moved away from the sensor tip, the Zirconia sensor voltage should decrease, while the Titania sensor resistance should increase. If the sensor output responds quickly to changes in mixture ratio (within 1 to 3 seconds depending on the manufacturer), the sensor is good.

Testing Oxygen Sensor Heater Circuit

The heater circuit of a heated oxygen sensor must be checked for proper resistance to ensure that the heater resistor is not burned out or shorted. To test the heater, set the multimeter to the ohms range. Then connect the multimeter leads to the heater terminals of the oxygen sensor. Polarity is not important, but the leads should not contact the sensor signal terminal. If the resistance is within specifications, usually the heater circuit is good. Set the meter to the voltage scale and test the heater power terminals on the wiring harness. If 12 volts is present, the circuit is good.

Road Testing Oxygen Sensors

If the tests at idle are not conclusive, a road test while monitoring sensor operation with a scan tool or by using a voltmeter hooked up to the oxygen sensor can be done. In either case, the engine must be completely warmed up before the road test. If a scan tool is used, connect the tool to the DLC. Use an extension lead if the DLC is under the hood. If equipped, use the scan tool's snapshot feature to try to take a picture of sensor and vehicle operation when the malfunction occurs.

If a multimeter is used, start by connecting to the oxygen sensor through leads that will allow the meter to be placed inside the passenger compartment. If the meter that you are using has memory and averaging capabilities, follow the manufacturer's instructions to set the meter to record the average reading. Have someone drive the vehicle while you observe oxygen sensor operation. A high average voltage or low resistance reading means that the engine is running too rich. For example, the ECM is leaning out the mixture, but an injector is leaking. A low average reading or high resistance indicates too little fuel, such as when the ECM is richening the mixture, but cannot compensate for a vacuum leak. After the road test is complete, enter the meter's averaging function. This will allow you to get an average reading of the voltage levels.

Note: If the above tests are made with the engine running and the oxygen sensor disconnected, false trouble codes will probably have been set. After all tests are complete, be sure to clear the ECM memory.

Checking Mass Airflow (MAF) Sensors

Regardless of their design, all mass airflow, or MAF, sensors produce one of three types of outputs, depending on the manufacturer. These outputs are analog dc voltage, low frequency pulse, or high frequency pulse. Older MAF sensors are usually the dc voltage type, while many newer designs use frequency pulse.

This quick test will only work on MAF sensors that produce an analog dc output voltage. Check the proper service manual to determine if the sensor will respond to this test. Begin the test procedure by turning the ignition to on, but not starting the engine. Set the scan tool to monitor MAF sensor input, or the multimeter to measure dc volts and attach it to the MAF sensor as shown in **Figure 15-16A.** With the ignition key on and the engine off, voltage will be about 1 volt. Start the engine and observe the scan tool or voltmeter. Voltage should rise to about 2.5 volts, **Figure 15-16B.** Tap on the sensor with a small screwdriver handle. There should be no voltage fluctuation or engine misfires. If needed, repeat the test, tapping on the MAF sensor while heating it with a heat lamp. If the voltage fluctuates or engine operation changes dramatically (stalls or idles rough), the sensor is defective and should be replaced.

A variation of this procedure is shown in **Figure 15-17.** Instead of starting the engine, the technician blows through the MAF sensor with the ignition on. Voltage should rise, indicating that the sensor is responding to air movement. On some vane type MAF sensors, the sensor output voltage is determined by inserting an unsharpened pencil into the sensor. The sensor voltage output is measured with the ignition on and the engine not running. If the output voltage is not as specified with the pencil inserted as shown, the sensor is defective.

Testing Frequency MAF Sensors

The output of a frequency type MAF sensor can be measured by most scan tools, or by multimeters capable of reading RPM or duty cycles, **Figure 15-18A.** Consult the multimeter manual to determine the exact meter settings. The test procedure is similar to that of analog dc output MAF sensors. The frequency will be at a set value with the ignition on and the engine off. When the engine is started, increasing airflow through the MAF sensor will cause an increase in the frequency reading, **Figure 15-18B.** If the

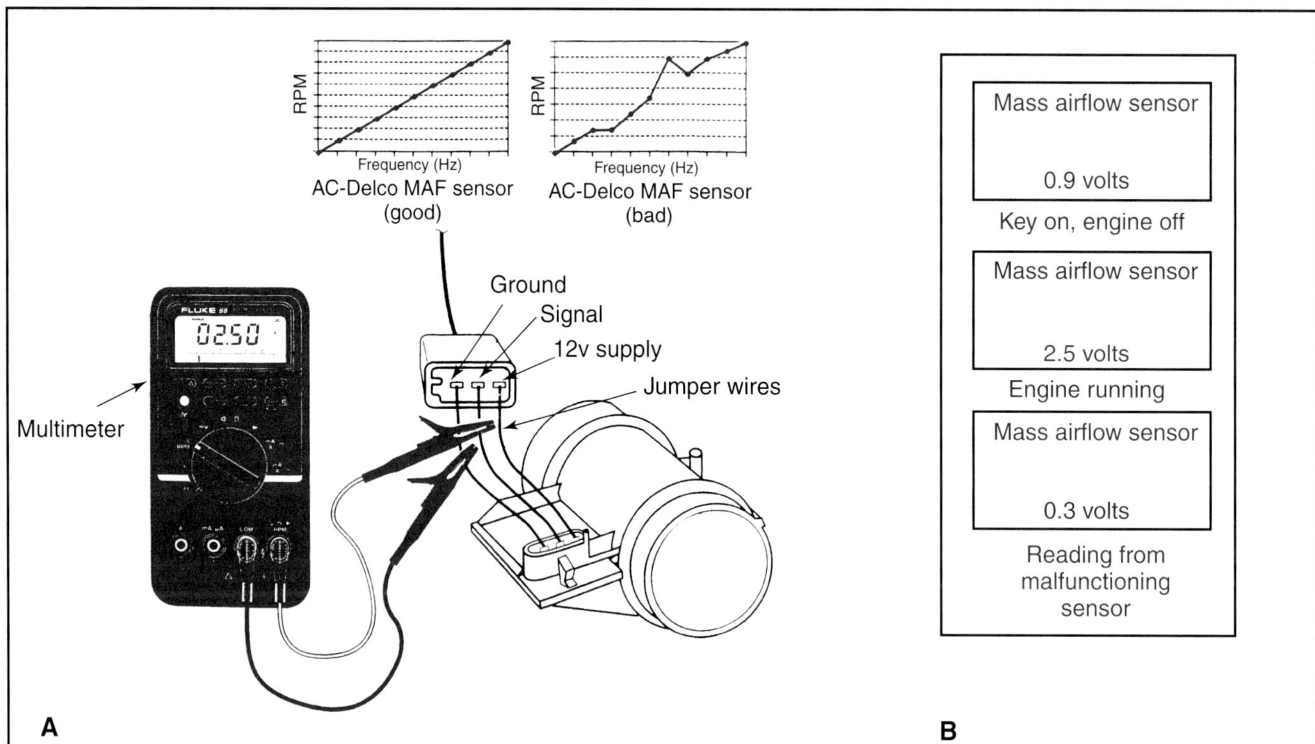

Figure 15-16. *A—A digital meter can measure mass airflow sensor operation. Do not use an analog meter to perform this measurement. B—A scan tool can also be used, however, any scan tool readings should be verified by using a meter. (Fluke)*

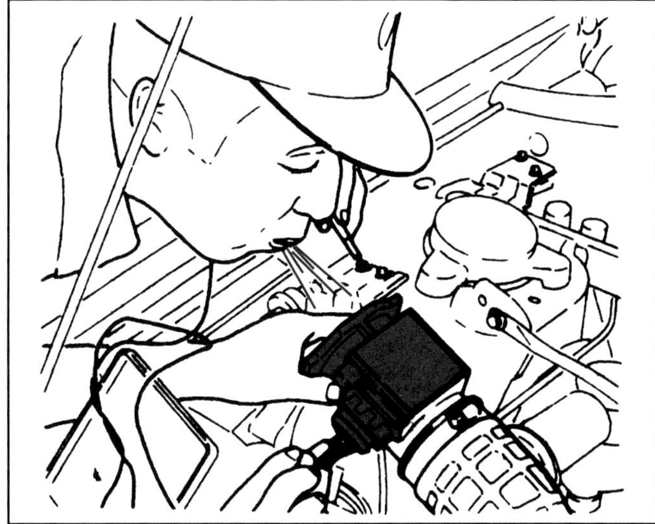

Figure 15-17. *Blowing through a MAF sensor will simulate the air that enters while the engine is running. (Nissan)*

frequency reading does not change or decreases, the sensor is probably defective.

Note: Always check carefully to determine which type of MAF sensor you are dealing with, since frequency type sensors closely resemble analog dc output types. If you are not sure, check the service manual.

Testing Manifold Absolute Pressure (MAP) Sensors

Manifold absolute pressure and other similar vacuum and barometric pressure sensors allow the ECM to compensate for manifold vacuum and high altitudes. This sensor is very important in vehicles that use speed density to calculate airflow. Before making this test, check the vacuum hoses for splits or clogging. Also check that the engine is providing sufficient manifold vacuum. Consult the service manual to determine whether this test can be made and for the exact procedures.

Begin testing by turning the ignition key on without starting the engine. Measure the dc voltage at the ECM and compare it to service manual specifications, **Figure 15-19.** Tap the sensor with a small screwdriver while watching for voltage jumps that could signal intermittent problems. Repeat the tapping procedure while heating the sensor with a heat lamp. Next, apply vacuum to the MAP sensor, **Figure 15-20,** while observing the voltage reading. If the voltage reading increases with increases in vacuum, the MAP sensor is probably ok. If the MAP sensor fails any of these tests, replace it.

A similar test can be made to the barometric pressure (BARO) indicator on some vehicles. Check the output voltage at the proper ECM terminals (ignition on, engine not running) and compare it to the specifications for the altitude in your area. A typical altitude compensation chart is shown in **Figure 15-21.** If the voltage is off, replace the sensor.

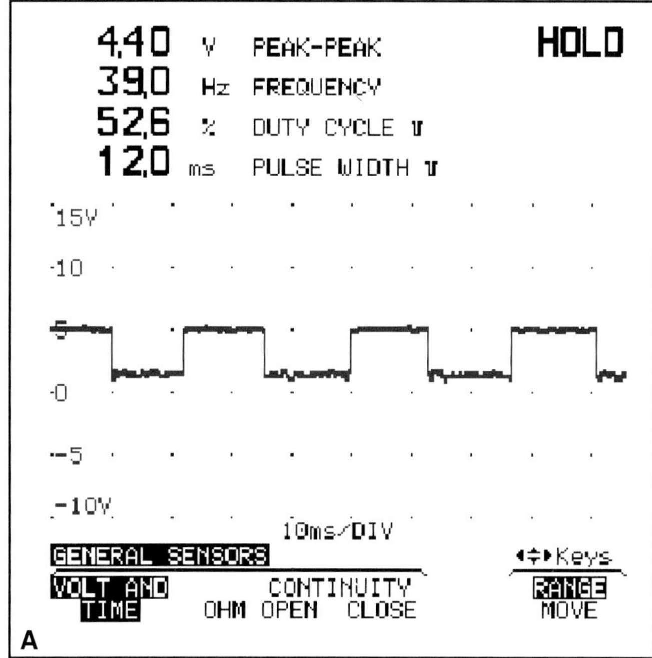

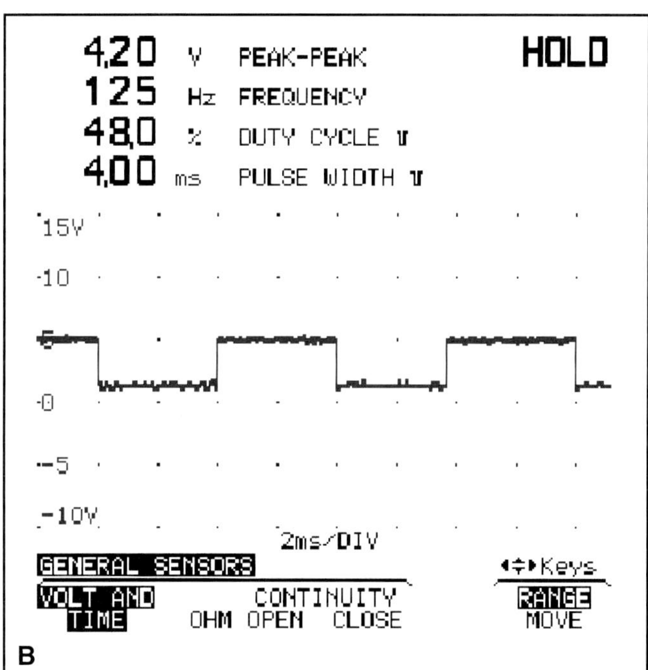

Figure 15-18. *A—The duty cycle of a frequency MAF sensor can be measured along with the waveform of the voltage output. B—Note the increased frequency and pulse width as the engine is started. (Fluke)*

Checking Temperature Sensors

Temperature sensors allow the ECM to compensate for changes in external air and internal engine temperatures. If a temperature sensor malfunctions, it will cause driveability problems that will usually occur when the engine is either

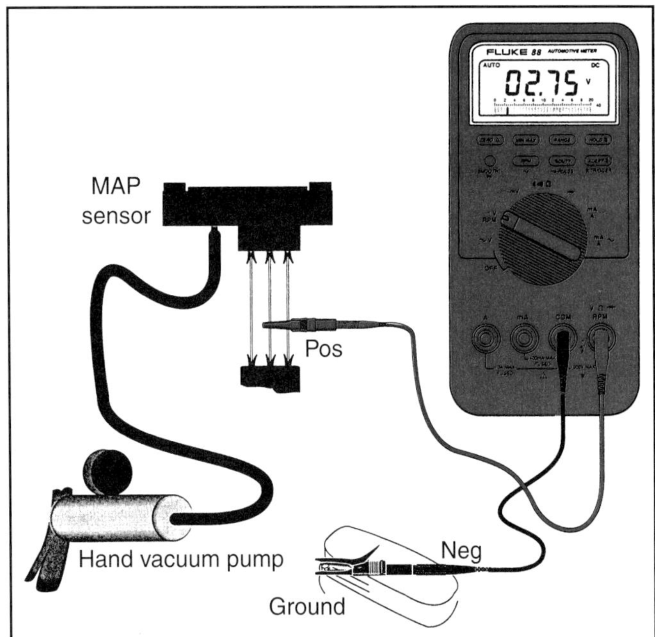

Figure 15-19. *Setup to measure dc voltage from the MAP sensor. (Fluke)*

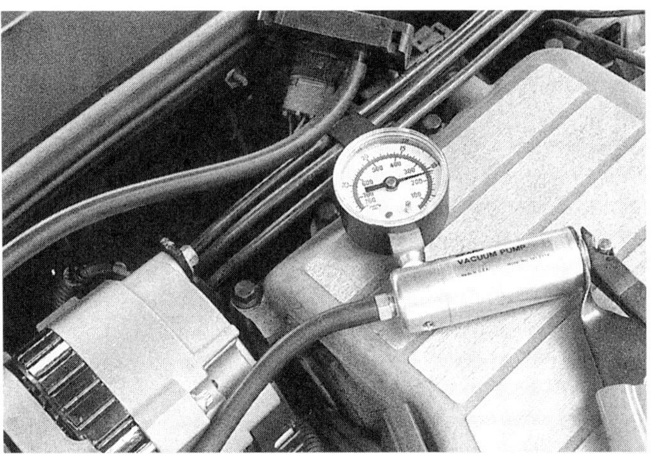

Figure 15-20. *Apply vacuum to the sensor to simulate engine vacuum. The readings should be similar to normal readings while the engine is at idle. Compare all readings to service manual specifications.*

Altitude		Voltage range
Meters	Feet	
Below 305	Below 1000	3.8–5.5v
305–610	1000–2000	3.6–5.3v
610–914	2000–3000	3.5–5.1v
914–1219	3000–4000	3.3–5.0v
1219–1524	4000–5000	3.2–4.8v
1524–1829	5000–6000	3.0–4.6v
1829–2133	6000–7000	2.9–4.5v
2133–2438	7000–8000	2.8–4.3v
2438–2743	8000–9000	2.6–4.2v
2743–3048	9000–10,000	2.5–4.0v

Low altitude = High pressure = High voltage

Figure 15-21. *BARO sensor output voltage varies with altitude. The lower the altitude, the higher the voltage.*

hot or cold. The following tests will work with engine coolant, intake air, and exhaust gas temperature sensors. Coolant temperature sensors are installed to contact the engine coolant, and will be located near the thermostat on most engines. Air intake temperature sensors may be installed in the intake manifold, or are often located in the air cleaner. Exhaust gas temperature sensors are located in the exhaust manifold.

This test measures the output resistance in relation to the temperature at the sensor tip. This will determine whether the sensor is correctly converting the temperature into a voltage signal during engine operation.

Note: Before performing this test on a coolant temperature sensor, check the level of coolant in the radiator. The sensor must be under the coolant level to operate properly.

Tests Using an Ohmmeter

The temperature sensor test requires a digital multimeter or ohmmeter. First, set the meter to measure resistance in ohms. Next, disconnect the sensor from the wiring harness and measure the resistance across the sensor leads and compare the readings with the proper manufacturer's specifications. These readings are temperature related, **Figure 15-22.** If the resistance measurements are incorrect, the sensor is defective.

If the resistance measurements are correct, the defect is either in the ECM or in the wiring. To check the wiring, reinstall the sensor connector, unplug the ECM, and measure the sensor's resistance at the ECM harness. If the resistance is now incorrect, check for a wiring problem. Remember that corroded or loose terminals will cause higher than normal resistance, while shorts to ground will cause lower than normal readings. If the resistance is correct, the problem is in the ECM. A heat gun can be used to raise the temperature to check for resistance variations with temperature.

Tests Using a Scan Tool

Another way to check the temperature sensor and wiring is by using a scan tool. See Appendix C. In order to measure the temperature sensor output using this procedure, you must have a scan tool that can read sensor inputs as temperature readings. This is possible on many scan tools by following the manufacturer's instructions.

Begin the test procedure by setting the scan tool to read the temperature as measured by the sensor. Remove the sensor's connector and note the temperature. The scan tool readings should indicate the sensor's maximum lowest reading, usually between -30 and -40 F (-34 and -40 C), **Figure 15-23.** Then, jumper the connector leads to simulate the sensor's highest possible reading, which can be up to 260 F (127 C) and higher. If the scan tool data shows the output to be correct, the sensor is the likely cause of the problem. If not, there is high resistance in a sensor connection, a shorted or grounded wire, or a problem in the ECM.

Temperature vs Resistance Valve (Approximate)

°C	°F	Ohms
100	212	177
90	194	241
80	176	332
70	158	467
60	140	667
50	122	973
45	113	1188
40	104	1459
35	95	1802
30	86	2238
25	77	2796
20	68	3520
15	59	4450
10	50	5670
5	41	7280
0	32	9420
–5	23	12,300
–10	14	16,180
–15	5	21,450
–20	–4	28,680
–30	–22	527,000
–40	–40	100,700

Figure 15-22. *Most temperature sensors are thermistors. The higher the temperature, the lower the resistance. (General Motors)*

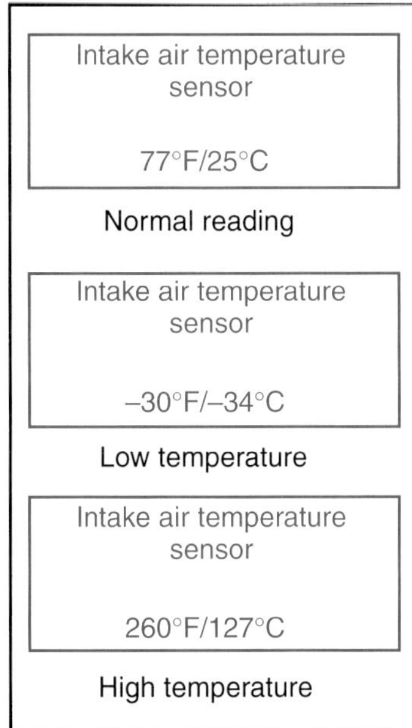

Figure 15-23. *Scan tool readings for the intake air temperature sensor. The high temperature reading may be lower than shown here, but will be much higher than normal ambient temperature.*

Checking Throttle Position Sensors

The throttle position sensor monitors throttle opening and closing so the ECM can adjust fuel and ignition spark timing to match driver demand. Typical TPS voltage can range from 0.2 volts at fully closed throttle to 4.5-5 volts at wide open throttle. However, this sensor is adjustable on most vehicles, so depending on the manufacturer and the engine, anything between 0.2 and 1.25 volts might be considered closed throttle. Throttle position sensors modify a reference voltage, therefore, an ohmmeter is used to test the meter. TPS resistance varies by manufacturer and application.

This test checks the internal resistance of the throttle position sensor at different throttle openings. Start the test procedure by disconnecting the throttle position sensor electrical connector. Connect an ohmmeter to the sensor as shown in **Figure 15-24.** Slowly open the throttle while watching the ohmmeter. The ohmmeter reading should increase smoothly without jumping or skipping. If it does jump or skip, the sensor is defective.

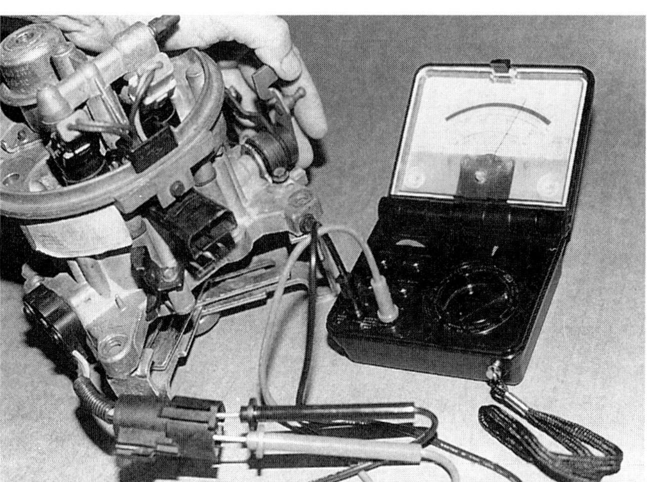

Figure 15-24. *An analog meter can be used to check the throttle position sensor for opens and poor internal connections. This test can also be done with the throttle body mounted on the engine.*

Some manufacturers recommend probing the TPS connector with a test light connected to either ground or battery voltage. This allows the technician to test the wiring for shorts or grounds and to eliminate the ECM as a possible cause of the problem. Follow the manufacturer's recommended procedure when using a test light to check a throttle position sensor. It is also possible to use a scan tool to check TPS operation. However, this should only be used as a quick check and any findings should be verified by the manufacturer's recommended procedure.

Testing Crankshaft and Camshaft Position Sensors

On most vehicles, the crankshaft and camshaft position sensors control both the fuel and ignition systems. A faulty crankshaft or camshaft sensor will produce either a no-start condition or intermittent stalling. The test procedures for crankshaft and camshaft sensors are similar to those for pickup coils, Hall-effect, and magnetic pickup switches mounted in the distributor. However, their location on or in the engine makes testing somewhat difficult. Visually inspect the sensor for a good connection to the harness and secure mounting to the engine. There are three quick tests that you can perform to check crankshaft and camshaft sensors.

Note: A problem that appears to be caused by a position sensor can also be caused by the ignition module, ignition coils, harness wiring, or the ECM. These generalized tests should only be used to narrow down a possible crankshaft or camshaft position sensor problem. If a problem in the position sensor circuits is suspected, it should be verified by using tests outlined in the vehicle's service manual.

The first test is to crank the engine while monitoring engine rpm using a scan tool. If the scan tool reads low or no rpms while cranking, this is an indication of a possible problem in the crankshaft or camshaft sensor circuit, **Figure 15-25.** The second test is to connect a spark tester to two or three adjacent spark plugs, one at a time, and crank the engine. If there is no spark, this is also a possible indication of a position sensor circuit problem. If some of the plugs wires produce a spark, the problem is most likely in the ignition module or coils.

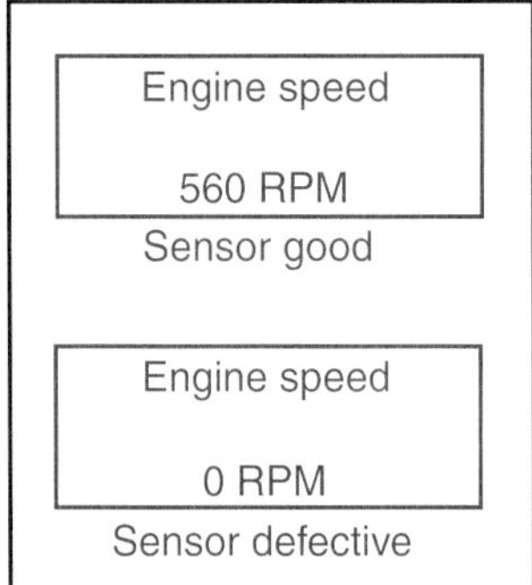

Figure 15-25. *A scan tool can be used to quickly check the crankshaft position sensor for operation. If the engine has a camshaft sensor, a scan tool reading of engine speed is not reliable as most ECMs take readings from both sensors to determine engine speed.*

The last test uses a injector harness "noid" light and a test light. First, use the noid light to test for injector pulse at two or three adjacent injectors. Pull the injector harness connectors, one at a time, install the light, and crank the engine. If injector pulse is present, the position sensors are most likely working and the problem is in another circuit. If there is no injector pulse, use the test light at each injector harness terminal for power and ground as instructed by the vehicle's service manual. Injector service will be covered in more detail in **Chapter 17.**

Testing Knock Sensor

The knock sensor allows the ECM to retard ignition timing when spark knock occurs. The knock sensor either produces a voltage that is sent to the ECM or modifies a reference voltage from the ECM. Check the service manual for the type of knock sensor you are testing.

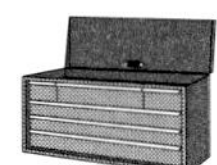

Note: Before performing the knock sensor test, make sure that an internal engine problem is not the cause of the knocking condition.

Testing a knock sensor is very easy. All that is needed is a scan tool or voltmeter and a small hammer. Connect the scan tool and set it to monitor knock sensor output. Turn the ignition switch to on, do not start the engine. Lightly tap on the engine block with the hammer.

Caution: Hammer on only cast iron parts or the block itself. Do not hammer on aluminum, plastic, or sheet metal parts.

If the sensor is working properly, the scan tool will show that the ECM is detecting the artificial engine knock, **Figure 15-26.** If the scan tool reading does not change, disconnect the knock sensor connector, connect the voltmeter, and repeat the hammer test. If the sensor produces a voltage signal, the problem is in the wiring or the ECM. A timing light can also be used to check the knock sensor on a running engine. Direct the light on the timing marks and tap on the engine. Tapping on the engine should make the timing marks move in the retard direction.

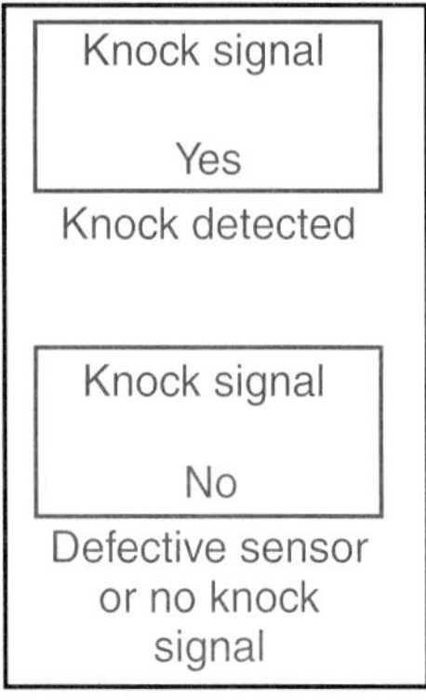

Figure 15-26. *The knock sensor signal will show up on most scan tool readouts as a yes (knock present) or no (knock not present).*

Testing Vehicle Speed Sensors

Most speed sensors produce an alternating current (ac) wave as they rotate. Therefore, the operation of many vehicle speed sensors can be measured by measuring the output in ac volts. The simplest way to check a speed sensor is to use the proper scan tool. Raise the vehicle and shift into drive. Modern scan tools can read the speed sensor output as miles per hour. If there is no reading, check the speedometer. If the speedometer is indicating speed, a defect exists in the wiring to the ECM or the ECM is defective. If the speedometer is not working, proceed to test the sensor directly with a multimeter.

To make this check, stop the engine and disconnect the speed sensor at the transmission or transaxle. Then set the multimeter to the ac volts range. Restart the engine and with the drive wheels off of the ground, shift into drive and accelerate. The multimeter should read increasing ac voltage as the engine speed is increased, usually over .5 volts. Some manufacturers call for checking the resistance of the speed sensor winding, but this is only done after other tests have indicated a sensor problem.

Replacing Computer Control System Components

This section gives general procedures for replacing many of the most widely used components of electronic engine control systems. Many electronic control system components can be replaced by unplugging the electrical connectors and removing the attaching screws. Some devices, such as the oxygen sensors, engine temperature sensors, and knock sensors, are threaded into the engine. They must be removed carefully to avoid stripping the threads or burns from hot engine parts or coolant. Before replacing any computer control system component, always turn the ignition switch to off and remove the battery negative cable. This will prevent any damage to the ECM, sensors, or other vehicle electronic devices from stray electrical charges.

Input Sensor Replacement

The following sections deal with replacing the input sensors. Note that not all types of sensors are covered in this section. Other input sensors not mentioned here will be discussed in the chapters with the systems they monitor. The many different types of computer control systems makes it necessary to consult the vehicle's service manual for the proper replacement procedures for a specific sensor.

Caution: As stated earlier, these procedures are general in nature and are meant to describe a typical replacement procedure. Always refer to the vehicle's service manual for the proper replacement procedure.

Oxygen Sensor

The oxygen sensor is probably the most delicate of all of the input sensors, and should always be handled carefully. Since it is installed in the exhaust manifold, it is subject to high temperatures and corrosion that can cause the threads to seize in the manifold. Many oxygen sensors require a special socket or tool for removal, such as the one shown in **Figure 15-27.** If the oxygen sensor will be reused, always use the special tool to remove it. If the sensor does not require a special tool, loosen it from the manifold using a socket or box end wrench that will contact all sides of the hex.

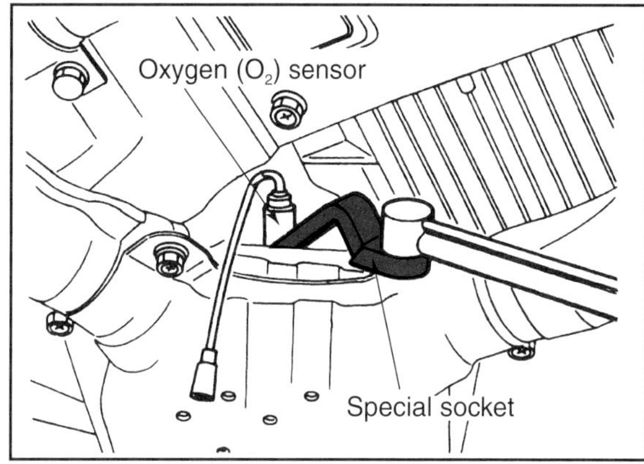

Figure 15-27. *Remove all oxygen sensors using the appropriate tool. Be sure you are removing the correct sensor on vehicles with multiple sensors.*

Caution: Make sure you are removing the correct sensor on vehicles with multiple oxygen sensors.

Once the sensor is out, check the manifold around the sensor fitting for cracks or pinholes. These will draw in outside air during engine operation, causing a false sensor reading. Replace the manifold or pipe if any damage is found. Inspect the oxygen sensor for indications of carbon loading, contamination, or poisoning. There are many substances that can cause an oxygen sensor to fail. These substances and sensor tip conditions that indicate their presence are shown in **Figure 15-28.**

Before installing the new oxygen sensor, check the service manual to determine whether the sensor threads should be coated with ***anti-seize compound.*** Some manufacturers recommend a special kind of high temperature sealant. Use only a light coating of sealant to avoid plugging the external air vents. Do not overtighten the oxygen sensor while installing it, since this may damage the brass shell. If anti-seize compound or sealant was used, wipe off any excess from the exhaust manifold and sensor after tightening.

Oxygen Sensor Tip Condition	Cause
Fluffy, black coating on the sensor tip.	Carbon loading due to rich air-fuel mixture.
Brown residue on the sensor tip.	Engine or transmission oil leaking into the exhaust.
White, glassy coating on the sensor tip	Silicon contamination from the use of silicon-based sealants or a coolant leak.
Shiny gray or black coating on sensor tip, not fluffy black. Looks like graphite coating.	Lead contamination from the use of leaded fuel. Will also cause damage to the catalytic converter and other systems.

Figure 15-28. *A normal oxygen sensor will have a light coating of carbon. This chart lists common causes of oxygen sensor contamination.*

MAF and MAP Sensor Replacement

The mass airflow and manifold absolute pressure sensors on most vehicles are retained by one or two bolts or clamps and are relatively easy to remove. Start by disconnecting the battery's negative cable. Remove any wiring and hoses from the sensor. Loosen the bolts or clamps, depending on the sensor and remove the sensor from the vehicle. When installing a MAF sensor, be sure you face the sensor's inlet in the correct direction. Tighten all retaining bolts or clamps. Reconnect all wiring and hoses next. Finally, reconnect the battery's negative cable.

Throttle Position Sensor Replacement

Misadjusted throttle position sensors affect ECM operation. They can go out of adjustment due to linkage wear, or changes in the sensor's electrical material. The sensor may also require adjustment when the throttle body is replaced. On most engines, the throttle position sensor is simply mounted with little or no adjustment. However, a specific adjustment procedure is required for some throttle position sensors. **Figure 15-29** shows a typical throttle position sensor mounted on the throttle shaft of a throttle body fuel injector assembly.

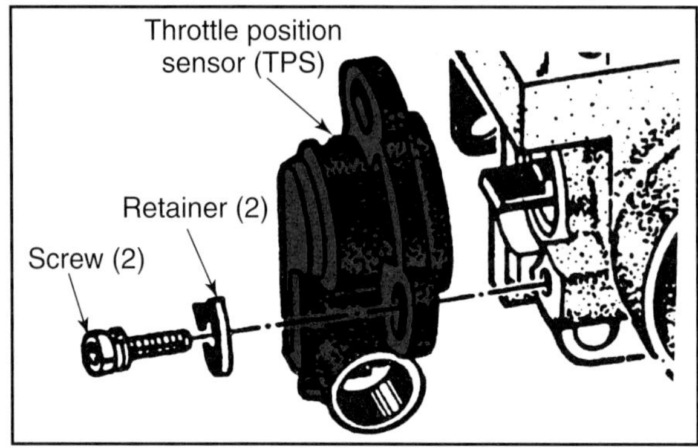

Figure 15-29. *The throttle position sensor mounts on the throttle shaft and is simply bolted on the throttle body in most cases. (General Motors)*

Note: Some throttle position sensors require special tools and/or adjustment procedures. Check the vehicle's service manual for the correct procedure.

Throttle Position Sensor Adjustment

To adjust the sensor, obtain the correct voltage specifications and set the meter to a range that will measure these voltages. Using jumper wires to the correct terminals, measure the input and output voltages. If the input voltage is incorrect, check the wiring and ECM and make repairs as necessary. If the output voltage is incorrect, first ensure that the throttle plate is in the fully closed position. Then loosen the throttle position sensor attaching screws and rotate the sensor until the proper value is shown on the voltmeter.

Temperature Sensor Replacement

Temperature sensors are used in many places on the engine and vehicle. With the exception of the coolant temperature sensor, they are relatively easy to replace and require no adjustment.

Coolant Temperature Sensor

Coolant temperature sensors are threaded into a coolant passage in the engine block or the radiator. They should not be removed until the engine cooling system is depressurized and drained below the level of the sensor, **Figure 15-30.** Always leave the radiator cap loose while changing any cooling system part to prevent pressure buildup, **Figure 15-31.**

Coolant temperature sensors use ***self-sealing pipe threads*** and are usually installed in the block. Most modern coolant temperature sensors can be removed by loosening and unthreading them from the engine using the proper size deep socket and ratchet handle. In some cases, the sensor can be removed with a box end wrench, **Figure 15-32.** A few sensors can only be removed with a special socket. Always disconnect the electrical connector from the wiring harness before installing the socket over the sensor.

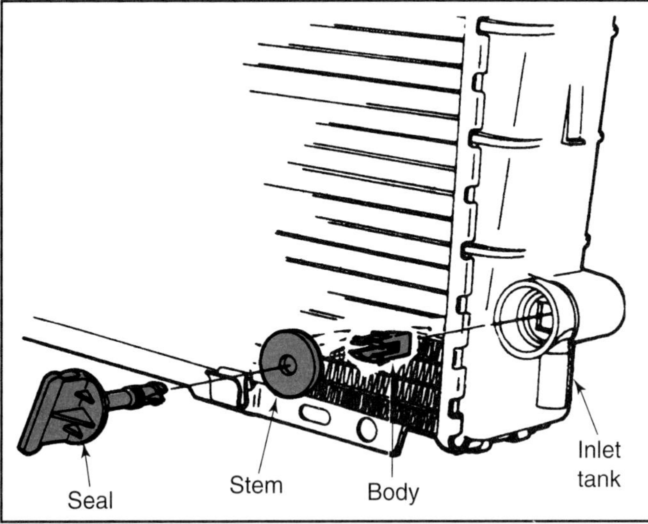

Figure 15-30. *Loosen the drain plug and drain the cooling system below the level of the coolant temperature sensor before removal. (Ford)*

Figure 15-31. *Loosen the radiator cap before servicing any part of the cooling system.*

Figure 15-32. *Some coolant temperature sensors can be located in an inaccessible location. A socket or wrench can be used to remove these sensors.*

Caution: Sensors installed in aluminum parts should not be removed until the engine has cooled for several hours. This cooling period is necessary to prevent damage to the threads in the aluminum part.

Before installing the new sensor, coat the threads with the proper sealant. When installing the sensor, do not overtighten, as this may damage the threads or distort the sensor shell. If the cooling system was drained, refill it before restarting the engine. Leave the radiator cap loose and recheck coolant level after about 10 minutes running time. A few front-wheel drive engines require a special fill procedure. The bleed valve shown in **Figure 15-33** can be loosened to remove air from the cooling system. Follow the manufacturer's recommended coolant fill procedure.

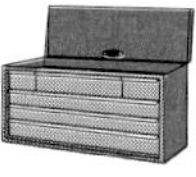

Note: When bleeding air from a cooling system, turn on the heater with the blower fan on high to assist in cooling system bleeding. This will also provide additional cooling and fluid circulation during the bleeding process.

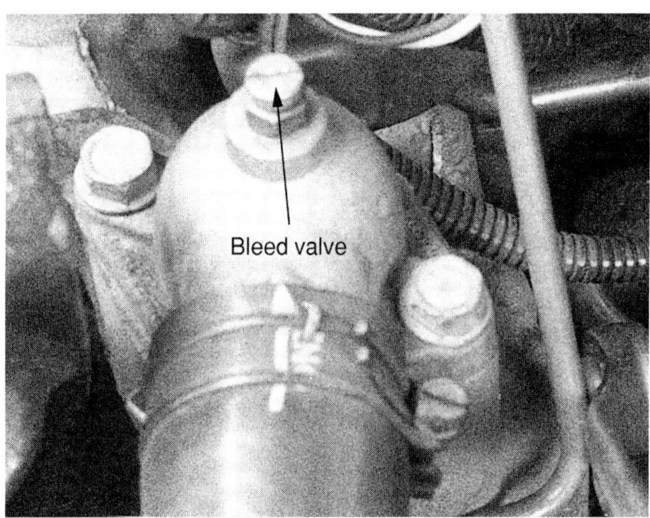

Figure 15-33. *After the sensor is replaced, it may be necessary to bleed air from the cooling system. Some engines are equipped with bleed valves. This one is located on the thermostat housing.*

Intake Air and Exhaust Gas Temperature Sensors

Intake air and exhaust gas temperature sensors are not under any pressure when the engine is not running. Intake air temperature sensors can be removed and replaced without special precautions. If the exhaust gas temperature sensor is installed in the exhaust manifold, use a special tool or a six-point socket to remove the sensor without damaging the shell.

Knock Sensor Replacement

Knock sensors are threaded directly into the engine block, cylinder head, or intake manifold, **Figure 15-34.** This allows them to easily record and transmit the sound of engine detonation. They can be removed and replaced without special precautions. Since the knock sensor shell is usually made of brass, do not try to remove it with an open-end wrench. Install and tighten the new sensor carefully. Use a socket or box end wrench to avoid distorting the shell. Then reinstall the electrical connector and check the operation of the knock sensor as explained earlier.

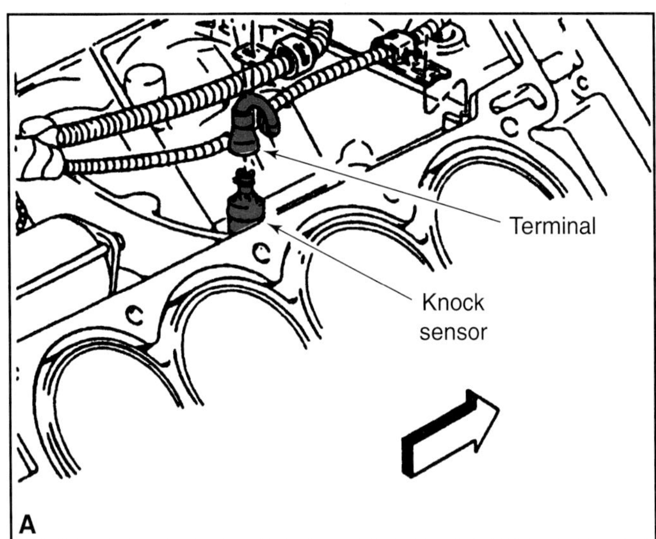

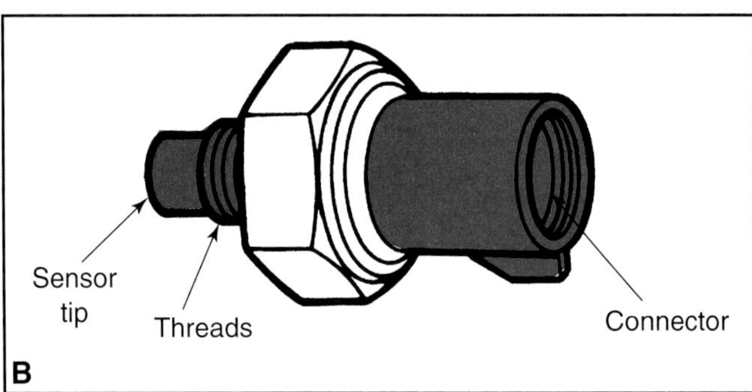

Figure 15-34. *A—Knock sensors are threaded into the engine. This one is located under the intake manifold. B—External parts of a knock sensor. (General Motors and Ford)*

Crankshaft/Camshaft Position Sensors

Most crankshaft and camshaft position sensors are installed on the engine block, timing cover, camshaft housing, or transmission housing near the flywheel. These sensors are held by a clamp and a single bolt or attached to a bracket held by one or more bolts. After the bolt(s) are removed, pull the position sensor from the engine. Then remove the electrical connector.

Check the end of the sensor to ensure that it has not contacted the crankshaft or camshaft sensor ring. If the end of the sensor has been damaged by contact with the sensor ring or other moving parts, determine the cause before installing a new sensor. To install the new sensor, make sure that any replacement O-rings or gaskets are in place. Then lightly lubricate the sensor tip with engine oil and push it into position. Reinstall the clamp and bolt, then install the electrical connector.

Caution: Some external crankshaft sensors require the use of a special alignment tool to properly position the sensor. If this tool is not used, damage to the sensor will occur. Consult the vehicle's service manual to ensure that the crankshaft sensor does not require the use of an alignment tool.

Adjusting Crankshaft and Camshaft Position Sensors

Many crankshaft position sensors are simply replaced and are automatically adjusted by the machining tolerances of the block and sensor. However, some crankshaft and camshaft position sensors must be adjusted to provide the proper signal to the ECM and to prevent damaging contact with rotating engine parts. The adjustment process usually involves the use of a special gauge, as shown in **Figure 15-35.** In this illustration a single pickup assembly is adjusted to the shutter assembly by using a spacer tool. This positions the shutter precisely between the two parts of the pickup assembly.

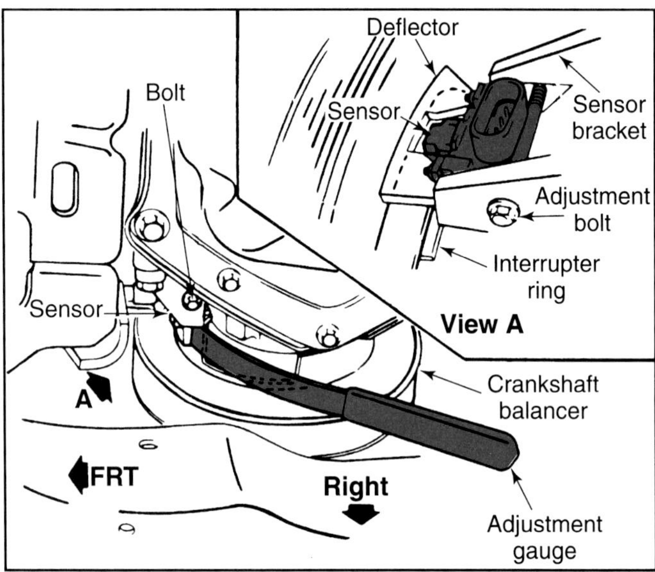

Figure 15-35. *Some crankshaft position sensors require adjustment after installation. (General Motors)*

In **Figure 15-36,** the Hall-effect switch has two pickup assemblies, one for crankshaft position and one for camshaft position. The crankshaft pulley has a double shutter assembly to operate each set of pickups. The special tool shown is used to match the rotating shutter assemblies to the stationary pickup coils.

A
Crankshaft position sensor tool
Sensor
Crankshaft
Pinch bolt
Pedestal
B
Crankshaft position sensor
Interrupter rings
Crankshaft position sensor tool
C
Harmonic balancer

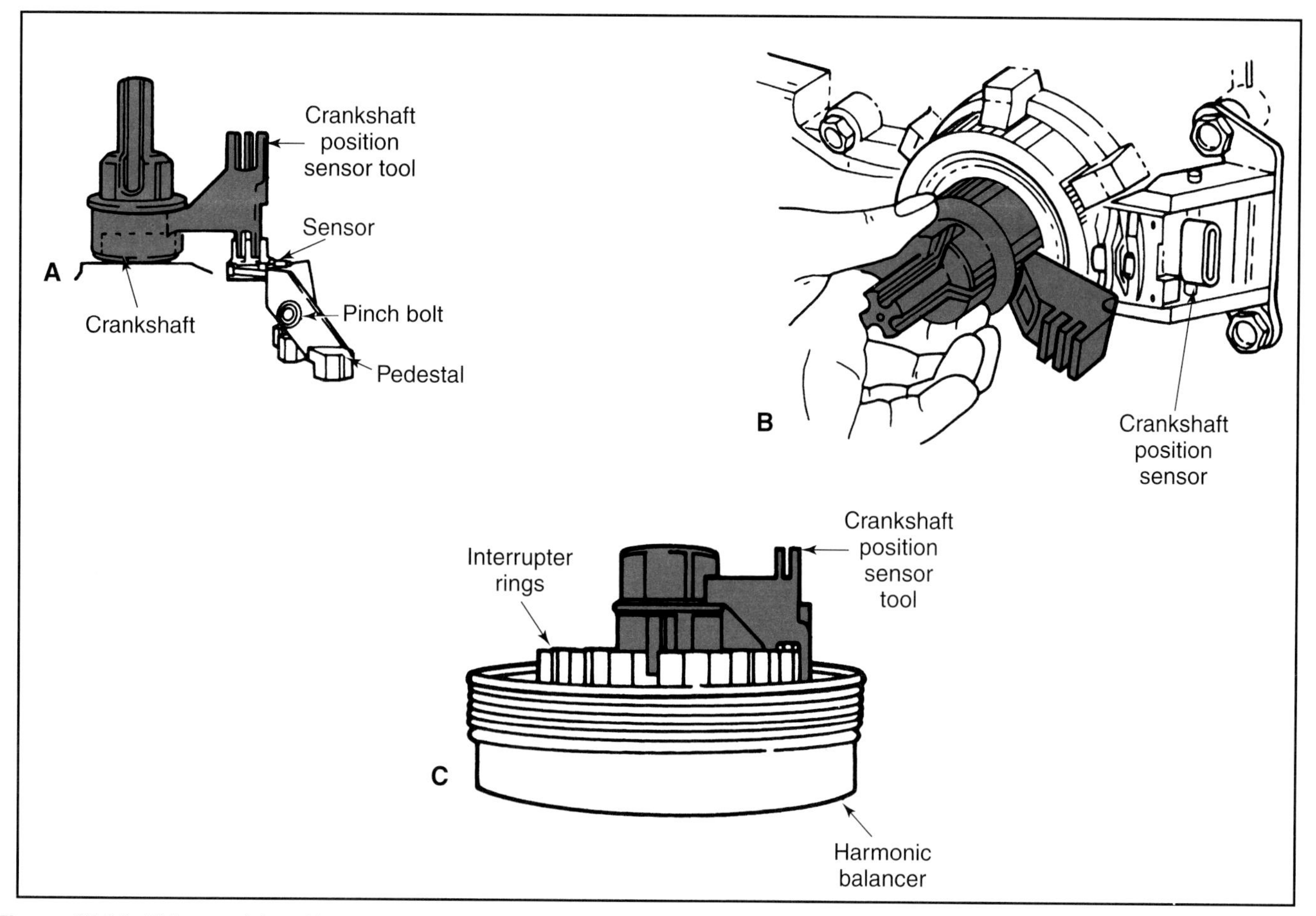

Figure 15-36. *This special tool is used to adjust the crankshaft sensor and can be used to check the interrupter rings on the crankshaft pulley. (General Motors)*

Caution: Always check the appropriate service manual for exact procedures before replacing or adjusting any position sensor or pickup. Severe damage can result if the pickup contacts any rotating engine part.

ECM Replacement

If the ECM fails, it must be replaced. Most ECMs are located inside the vehicle's passenger compartment so that they are protected from engine heat and the elements, but are still relatively accessible for testing and replacement. Most ECMs can be reached by removing an access panel, **Figure 15-37.** Dashboard-mounted ECMs can be reached by removing one or more of the dashboard panels, **Figure 15-38.** Other ECMs are located under a seat, and are protected by a dust cover panel, **Figure 15-39.**

Caution: Be sure to ground yourself before touching the ECM. Static electricity can easily destroy the internal ECM components. Even after grounding yourself, try not to touch any ECM terminals or internal parts. Static electricity and proper grounding procedures were discussed in Chapter 6.

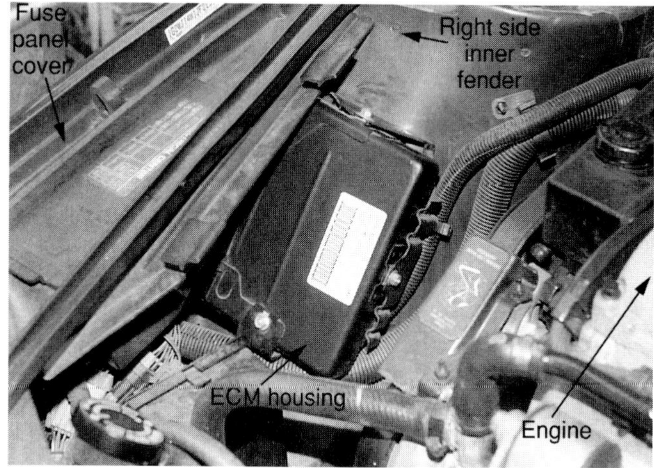

Figure 15-37. *ECMs can be located almost anywhere on the vehicle. One common location is in the engine compartment, under an access panel.*

ECM Removal

Once the ECM is located, it can be removed by disconnecting the wiring harness and removing the retaining bolts. Some ECMs are held in place by plastic clips. To remove the ECM, first disconnect the negative battery cable. Then remove any bolts or retaining clips, and pull the ECM out of its holder. Remove the electrical connectors from the

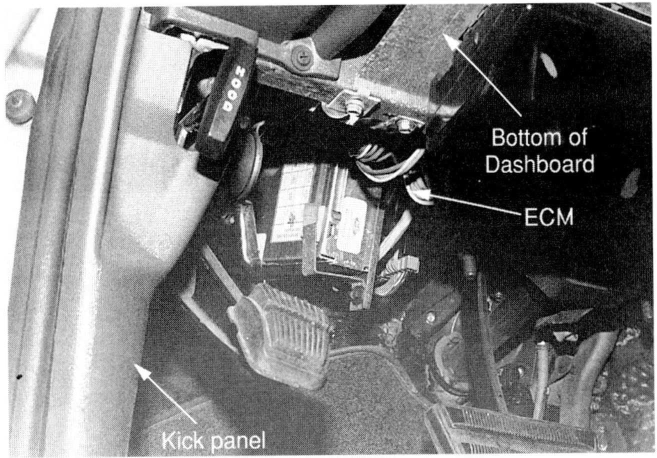

Figure 15-38. *Most ECMs are located under the dash.*

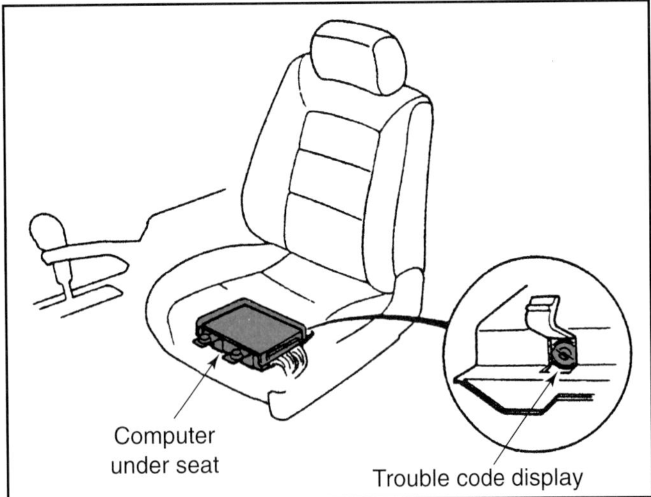

Figure 15-39. *Another common location for the ECM is under the seat. (Honda)*

ECM, and remove it from the vehicle. Always save the old ECM, since it usually has a substantial core deposit. Most ECMs are returned to the manufacturer to be rebuilt.

While the ECM is out, inspect the wiring harness connections for damage or signs of overheating. Other than removal of any PROM or Mem-Cal chips, disassembly of the ECM is not necessary. The other internal parts are not serviceable in the field and a visual inspection will not reveal anything in most cases.

ECM Memory Chip Replacement

Some ECMs are equipped with removable memory chips that contain the operating guidelines for the vehicle. These memory chips are often called a ROM (Read Only Memory) or a PROM (Programmable Read Only Memory). Some ECMs, especially those used with fuel injected engines, have two memory chips. The second chip is the Mem-Cal and provides a limp-in mode if the ECM fails. These chips can be removed from the ECM and replaced if they are faulty or incorrect. If the ECM is defective, the memory chip or chips must be removed and reinstalled in the replacement ECM.

To replace a ROM or PROM chip, make sure that the ECM is unplugged from all vehicle harnesses, then remove the access cover on the ECM case. See **Figure 15-40.** Note the position of any alignment marks or tabs on the chip and ECM case, then carefully pry the chip from the connector on the ECM board. Be careful not to bend the electrical contact pins or damage the connector receptacles. Special tools are available to safely remove and install the memory chip. Most ECMs and chips have an identification code and/or number that indicates when they were released. Check to ensure that the ECM or PROM is the exact one for the vehicle.

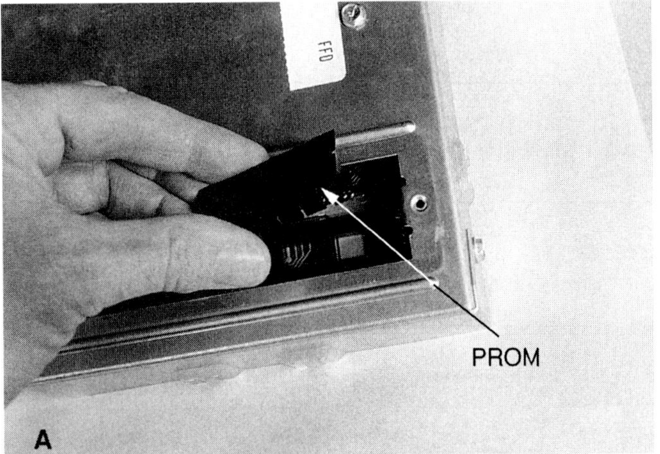

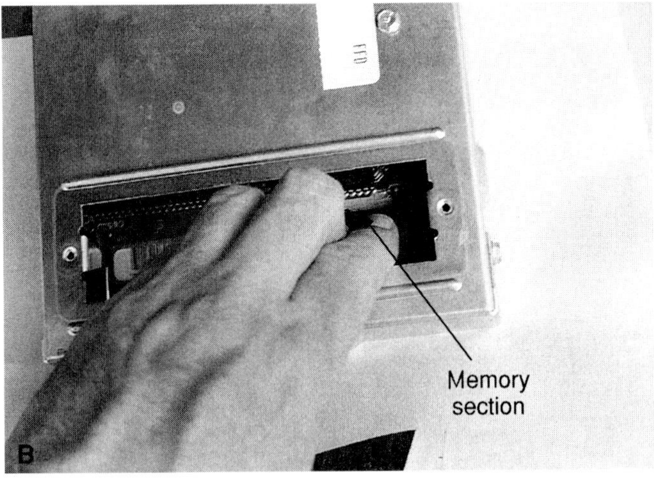

Figure 15-40. *A—An access cover can be removed to allow removal and replacement of PROMs on some ECMs. B—Be sure to follow all precautions regarding static electricity. Make sure the chip is installed securely.*

Caution: Some performance and emissions problems may be caused by the use of an incorrect or aftermarket PROM. If you discover the wrong PROM in an ECM, replace it with the correct chip and retest the vehicle. If the driveability complaint goes away, advise the driver. Use of the wrong PROM is considered a form of emissions system tampering in some cases.

Carefully place the new chip, or the chip from the old ECM into the recess, making sure that the chip pins align exactly with the receptacles. Also make sure that any chip and ECM alignment marks are in the proper position. Then gently push the chip down into the recess. Replace the access cover on the ECM and reinstall the ECM in the vehicle.

Knock Sensor Module Replacement

In some vehicles with EPROMs, the ECM contains a knock sensor (KS) module, **Figure 15-41.** This module is removed from the old ECM when replacement is necessary. The KS module is installed in the new ECM, ensuring that the proper timing retard is maintained for engine knock protection.

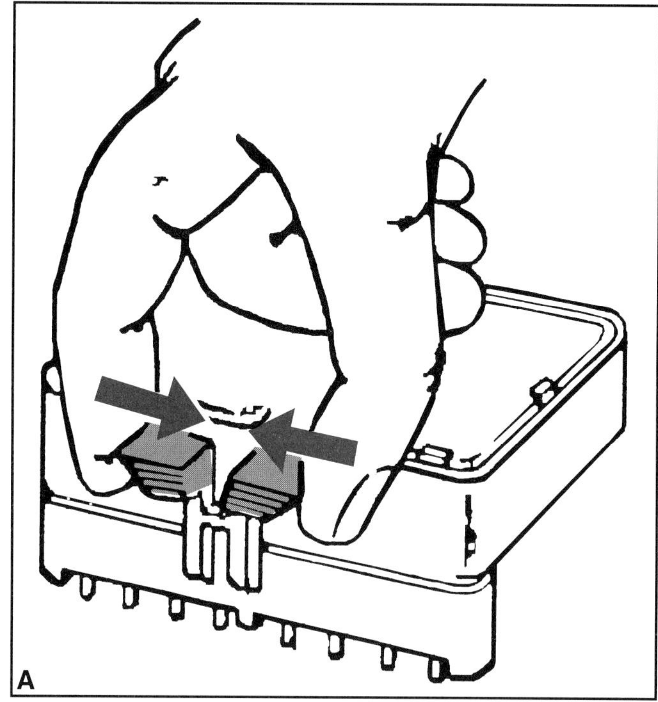

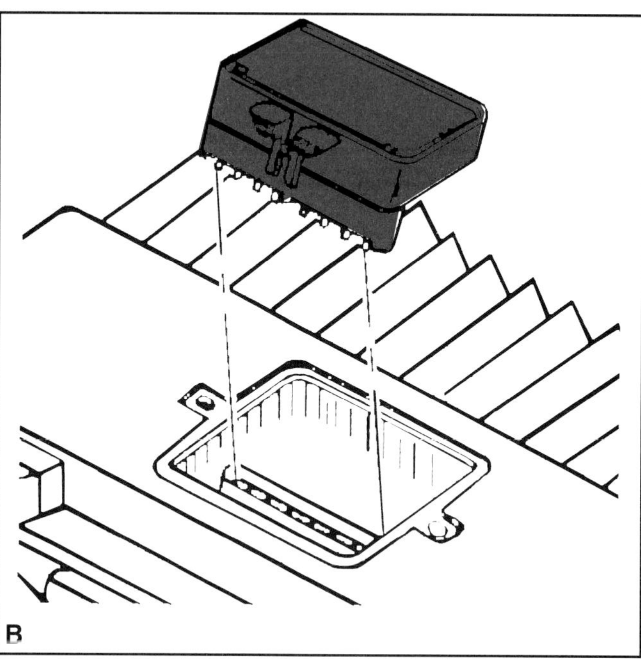

Figure 15-41. *A—Some newer ECMs use a knock sensor module. B—They can be removed and installed much like a PROM. (General Motors)*

ECM Installation

To install the ECM, attach the wiring harness and install the ECM in its holder. Install any covers or access panels that were removed. Connect the battery cable and start the engine. Note that many ECM memories have a learning mode, and must receive sensor inputs for a few minutes before they can begin to properly command the output devices. Therefore, it may be necessary to operate the vehicle for 10 to 15 minutes after installation. If this is necessary, it is often better to drive the vehicle to activate all sensors, rather than allow it to idle. If the ECM needs to be programmed, do this before starting the engine. Programming is covered in the next section.

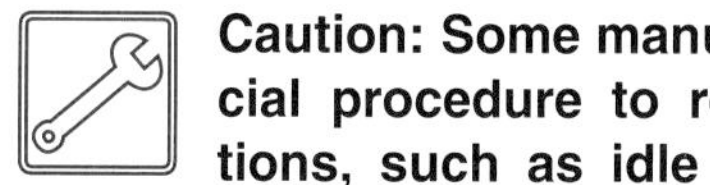

Caution: Some manufacturers require a special procedure to reset certain ECM functions, such as idle speed control. Follow these procedures exactly. Failure to do this can cause driveability problems possibly unrelated to the original complaint. If the engine has been operated without resetting the idle speed control, disconnect the ECM from battery power and perform the learning procedure. Some scan tools and computerized analyzers can perform the learning procedure automatically.

ECM Programming Using Computerized Equipment

Most newer ECMs use memory chips that are permanently soldered to the electronic board. These chips are called Electrically Eraseable Programmable Read Only Memory (EEPROM) or Flash Eraseable Programmable Read Only Memory (FEPROM). ECMs with EEPROMs or FEPROMs must be erased and programmed using electronic equipment. They can also be reprogrammed in order to correct certain driveability problems.

Flash programming can be done by downloading through a computerized diagnostic system or through a scan tool. Actual programming details vary between manufacturers, but the basic procedure begins with placing the ECM in the programming mode. One of three methods are used to program the ECM:

- Direct programming by a service computer or computerized analyzer.
- Indirect programming using a scan tool and a computer or computerized analyzer.
- Remote programming with the ECM off the vehicle.

Direct Programming

Direct programming is the fastest and simplest method. The new information is downloaded by attaching a shop recalibration device (usually a computerized analyzer) to the data link connector. The erasure and programming is done by accessing the proper menu and following the instructions as prompted by the computer. Then, the information (often contained in the computer analyzer's memory, on a CD-ROM disc, or through a connection to the manufacturer's database) as well as all settings is entered into the ECM through the connector. The shop recalibration device may be called a Service Bay Diagnostic System, Techline, Mopar Diagnostic System, or other names depending on the manufacturer.

Indirect Programming

To perform ***indirect programming,*** the proper scan tool must be available to connect to the programming computer and to the ECM as well as to reset some computer-controlled vehicle systems after programming. The programming computer resembles the personal computer or PC used in the home or it may be a computerized analyzer like the one used for direct programming.

In this type of programming, the information is downloaded from the computer into the scan tool and then downloaded from the scan tool into the ECM. The scan tool menu is accessed using the keypad. Most scan tools will use a high capacity cartridge to store the information. Some newer scan tools have sufficient fixed memory to hold the information and do not use a separate memory cartridge. In either case, follow the manufacturer's procedure as prompted.

Remote Programming

Remote programming is done with the ECM outside the vehicle. This procedure is used when changes needs to be made through a direct connection to a manufacturer's database. It can also be done in cases where normal direct and indirect programming is not practical or possible. Special connectors and tools are required for this type of programming. In most cases, this procedure is done only at vehicle dealerships.

Connections for ECM Programming

To begin programming the EEPROM-equipped ECM, make sure that the battery is fully charged. Recharge the battery if necessary, however, do not charge the battery during the programming procedure. Connect the service computer or scan tool to the ECM data link connector. Make any other vehicle connections as needed before proceeding with ECM programming.

Caution: Do not disconnect the scan tool or computer from the DLC during programming. Doing so will damage the ECM.

To start the programming sequence, it may be necessary to enter the engine type and vehicle identification number (VIN) or the type of vehicle and engine in a specific sequence. Once the vehicle information is entered, go to the programming software and follow the directions as prompted.

Depending on the manufacturer, it may be necessary to turn the ignition switch on or off during the connection and programming procedure. Double check any instructions on ignition switch position before making any connection or beginning programming or erasure. The next step is to determine the type of programming that is needed.

If a new ECM is being installed, only program that ECM. Do not attempt to program an ECM with information from the old ECM or an ECM from another vehicle. Any attempt to do this will cause a failed programming sequence code.

Note: In some cases, an erasure may need to be performed on a new ECM before initial programming can take place.

Reprogramming the ECM

If you are reprogramming an ECM that is to be reused, determine the date that the ECM's programming was downloaded or the current calibration number. If the information installed is the latest version, no further actions are required. If the latest information has not been installed, proceed with the reprogramming sequence. While most ECMs have internal circuitry that protects them from accidental reprogramming, be careful not to reprogram the wrong one.

Before reprogramming most ECMs, you must first erase the existing information. After this step is complete, select the updated calibration information from the reprogramming computer or scan tool menu. Then download the new information into the ECM. On some systems, the erasure step is not necessary as the new program will automatically overwrite the old information as it loads into the ECM.

Note: If the ECM cannot be erased or reprogrammed, check all connections first. Ensure that the correct ECM is being reprogrammed with the correct information and that all procedures are being followed. If the ECM still cannot be reprogrammed, it may need to be replaced.

Allow sufficient time for the programming to take place. Monitor the computer or scan tool to determine when the programming sequence is complete. Do not touch any connections until you are sure that the programming sequence is complete. After programming is finished, turn the ignition switch to the position called for and disconnect the computer or scan tool. After reprogramming is complete, use a scan tool to check the ECM and control system operation. While doing this, make sure that you have installed the proper program into the ECM.

ECM Relearn Procedures

Once the ECM is connected and programmed, when necessary, it must learn the vehicle's sensor inputs and to control output devices. This is usually done by driving the vehicle for a few minutes to allow the ECM to learn the sensors, how to adjust the output devices, and to adjust to any new EEPROM programming. On some vehicles, the idle speed control must be reset by using a computerized analyzer or scan tool. Some manufacturers recommend that the throttle body be cleaned before resetting the idle speed.

OBD II Drive Cycle

OBD II equipped vehicles must pass a drive cycle test after the ECM or the battery in an OBD II vehicle has been removed, codes erased, or as part of some states' I/M emissions testing procedures. The drive cycle test is performed before the actual testing begins. The test consists of attaching a scan tool and driving the vehicle for a set time. The drive consists of specific acceleration, cruising, and deceleration steps. The drive cycle is designed to tell the technician whether the OBD II system is operating and whether vehicle is operating efficiently enough to have a reasonable chance of passing an emissions test. **Figure 15-42** is a chart showing a typical drive cycle and the OBD II systems monitored.

Note that the entire drive test takes 12-15 minutes starting with a cold engine. If the vehicle fails the drive test, the technician can use the data amassed by the scan tool to quickly isolate the defective system or component. Once the system has been repaired, the drive test can be rerun to check OBD II system operation. IM 240 and other emissions testing procedures are covered in more detail in Chapter 18.

Follow-up for Computer Control System Repairs

After making repairs, you should erase any trouble codes from the ECM's memory. After erasing the trouble codes, road test the vehicle for at least 10 minutes to ensure that the ECM can compensate for any changes in engine operation, then re-enter the diagnostic mode to ensure that none of the trouble codes have been reset. If the repair was to correct a lean or rich condition, you should verify that the exhaust emissions output is within specifications.

Summary

To retrieve trouble codes, follow the exact procedure in the factory service manual. It is easier to use a scan tool than to read the dashboard light or a voltmeter. The next step is to separate the hard and intermittent codes. Interpret the codes to arrive at a defective system or part. Troubleshooting charts give a summary of possible problem causes, making the checking process quicker. There are many different types of computer control systems installed in various makes of vehicles. Therefore, it is necessary to consult the proper service manual for replacement procedures for specific sensors.

The oxygen sensor is probably the most delicate input sensor and should always be handled carefully. Since it is installed in the exhaust manifold, it is subject to high temperatures and corrosion. Many oxygen sensors require a special tool for removal. If the oxygen sensor will be reused, use the proper tool to remove it.

Typical OBD II Drive Cycle

Diagnostic Time Schedule for I/M Readiness	
Vehicle Drive Status	What Is Monitored?
Cold start, coolant temperature less than 50°C (122°F)	—
Idle 2.5 minutes in drive (auto) neutral (man), A/C and rear defogger ON	HO_2S heater, misfire, secondary air, fuel trim, EVAP purge
A/C off, accelerate to 90 km/h (55 mph), 1/2 throttle	Misfire, fuel trim, purge
3 minutes of steady state - cruise at 90 km/h (55 mph)	Misfire, EGR, secondary air, fuel trim, HO_2S, EVAP purge
Clutch engages (man), no braking, decelerate to 32 km/h (20 mph)	EGR, fuel trim, EVAP purge
Accelerate to 90-97 km/h (55-60 mph), 3/4 throttle	Misfire, fuel trim, EVAP purge
5 minutes of steady state cruise at 90-97 km/h (55-60 mph)	Catalyst monitor, misfire, EGR, fuel trim, HO_2S, EVAP purge
Decelerate, no braking. End of drive cycle	EGR, EVAP purge
Total time of OBD II drive cycle 12 minutes	—

Figure 15-42. *The OBD II drive cycle should be performed to prepare the vehicle for emissions inspection and to reset systems after the battery or ECM has been disconnected. (General Motors)*

Before replacing any computer control system component, always turn the ignition switch to off, and remove the battery negative cable. This will prevent any damage to the ECM, sensors, or other vehicle electronic devices. Many electronic control system components can be replaced by unplugging the electrical connectors and removing the attaching screws. Some devices, such as the oxygen sensors, engine temperature sensors, and knock sensors, are threaded into the engine. They must be removed carefully to avoid burns from hot engine parts or coolant.

Misadjusted throttle position sensors affect ECM operation. They can go out of adjustment due to linkage wear, or changes in the electrical material in the sensor. The sensor may also require readjustment when the carburetor or fuel injection throttle body is replaced.

Temperature sensors are used in many places on the engine. They are used to measure the temperature of the engine coolant, incoming air, or exhaust gases. Coolant temperature sensors are threaded into the radiator, or a coolant passage in the engine block, and should not be removed until the engine cooling system is depressurized and drained below the level of the sensor. The radiator cap should always be loose while changing any cooling system part to prevent pressure buildup.

Knock sensors are threaded directly into the engine block, cylinder head, or intake manifold for ease in recording or transmitting the sound of engine detonation. They can be removed without special precautions. Since the knock sensor shell is usually made from brass, do not try to remove it with an open-end wrench. Most crankshaft and camshaft position sensors are installed on the engine block or timing cover, and are held to the engine by a clamp with a single bolt.

If the ECM fails, it must be replaced. Most ECMs are located on the vehicle so that they are protected from engine heat and the elements, but are still relatively accessible for testing and replacement. Some ECMs are equipped with removable memory chips that contain the operating guidelines for the ECM and vehicle. These memory chips are often called a ROM (Read Only Memory) or a PROM (Programmable Read Only Memory). Some ECMs, especially those used with fuel injected engines, have two memory chips. The second chip provides a limp-in mode if the ECM fails.

Instead of changing the PROM on OBD II systems, the existing PROM is recalibrated using flash reprogramming. This is done by attaching the scan tool to the diagnostic connector and overwriting the old programming specifications with updated information.

Some state emissions testing procedures require that OBD II equipped vehicles pass a drive cycle test before proceeding with the actual emissions testing. This test is performed by attaching the appropriate scan tool and driving the vehicle through a series of acceleration, cruising, and deceleration steps.

Know These Terms

Two-part process
Alpha numeric designator
Fault designator
Hard codes
Intermittent codes
Scan tool data
Artificial lean mixture
Artificial rich mixture
Anti-seize compound
Self-sealing pipe threads
Flash programming
Direct programming
Indirect programming
Remote programming
Drive cycle test

Review Questions—Chapter 15

Please do not write in this text. Write your answers on a separate sheet of paper.

1. Which of the following should be done first?
   (A) Obtain the vehicle's service manual.
   (B) Scan for diagnostic codes.
   (C) Perform a flex test.
   (D) None of the above.
2. A temperature sensor can be checked with a(n):
   (A) voltmeter.
   (B) ohmmeter.
   (C) ammeter.
   (D) All of the above.
3. Trouble codes in most ECMs can be cleared by:
   (A) disconnecting the battery's negative cable.
   (B) using a scan tool.
   (C) removing a fuse to disconnect battery power from the ECM.
   (D) All of the above.
4. To install a coolant temperature switch, the cooling system must be ______.
5. Throttle position sensors can go out of adjustment because of:
   (A) linkage wear.
   (B) changes in the ECM.
   (C) changes in the sensor material.
   (D) Both A & C.
6. Throttle position sensors are adjusted by attaching what kind of meter to the sensor?
   (A) Voltmeter.
   (B) Ammeter.
   (C) Ohmmeter.
   (D) All of the above.

7. Which of the following can cause a false oxygen sensor reading?
   (A) A crack in the exhaust manifold.
   (B) Excessive anti-seize compound on the sensor threads.
   (C) Tight electrical connectors.
   (D) A discolored sensor shell.
8. The cooling system bleed valve is used to remove ______ from the cooling system.
   (A) coolant
   (B) air
   (C) sealant
   (D) rust
9. Knock sensors are threaded directly into the:
   (A) engine block.
   (B) cylinder head.
   (C) intake manifold.
   (D) All of the above.
10. When changing an ECM, the first thing to remove is the:
   (A) battery negative cable.
   (B) ECM retaining bolts.
   (C) ECM electrical harness.
   (D) ROM or PROM.

ASE Certification-Type Questions

1. Trouble codes can be read from ________.
   (A) the dashboard light
   (B) a voltmeter
   (C) a scan tool
   (D) All of the above.
2. Hard codes will cause the "check engine" light to be on __________.
   (A) intermittently
   (B) at all times
   (C) at no time
   (D) only when the key is on and the engine is off
3. An OBD II equipped vehicle has stored the diagnostic trouble code P1396. What is the best description of this code?
   (A) An SAE general code for a possible fuel and air metering problem.
   (B) A manufacturer specific code for a possible ignition system or misfire problem.
   (C) A manufacturer specific code for a possible computer system problem.
   (D) An SAE general code for a possible transmission problem.
4. The threads of most oxygen sensors are coated with ______.
   (A) anaerobic sealer
   (B) anti-seize compound
   (C) stainless steel
   (D) RTV sealer
5. Temperature sensors are often used to measure the temperature of all of the following, EXCEPT:
   (A) engine coolant.
   (B) incoming air.
   (C) exhaust gases.
   (D) engine oil.
6. An excessive rich reading from the oxygen sensor can mean all of the following, EXCEPT:
   (A) the oxygen sensor is defective.
   (B) the fuel system is running rich, causing the oxygen sensor to read improperly.
   (C) the gasoline is the wrong octane rating.
   (D) the air cleaner is plugged.
7. Technician A says that grounding one of the diagnostic terminals on an OBD II system will cause the engine to go into closed loop operation. Technician B says that a scan tool is needed to retrieve trouble codes from an OBD II system. Who is right?
   (A) A only.
   (B) B only.
   (C) Both A & B.
   (D) Neither A nor B.
8. All of the following statements about trouble codes are true, EXCEPT:
   (A) older (pre-OBD II) diagnostic systems used two-digit trouble codes.
   (B) OBD II data link connectors vary between manufacturers.
   (C) the OBD II system allows for 1000 potential trouble codes.
   (D) OBD II trouble codes do not correspond to older two-digit codes.
9. The output of a frequency-type MAF sensor can be measured by using a ______.
   (A) scan tool
   (B) multimeter which can read RPM
   (C) multimeter which can read duty cycles
   (D) All of the above.
10. All of the following statements about oxygen sensors are true, EXCEPT:
   (A) a Zirconia sensor produces a voltage reading.
   (B) a Titania sensor produces a resistance reading.
   (C) on one and three wire oxygen sensors, one of the wires is a ground wire.
   (D) an oxygen sensor with three or more wires is a heated sensor.

11. Technician A says that testing a BARO sensor requires a vacuum tester. Technician B says that testing a MAP sensor requires a vacuum tester. Who is right?
    (A) A only.
    (B) B only.
    (C) Both A & B.
    (D) Neither A nor B.
12. All of the following statements about speed sensors are true, EXCEPT:
    (A) the output of some vehicle speed sensors is measured in ac volts.
    (B) some vehicle speed sensors are checked by measuring resistance.
    (C) the output of some speed sensors is measured as a pulsed dc signal.
    (D) some speed sensors can be checked with a scan tool.
13. Throttle position sensor resistance varies with ______.
    (A) heat
    (B) RPM
    (C) throttle movement
    (D) All of the above.
14. Technician A says that tapping an engine block with a hammer or other tool will cause the ignition timing to retard. Technician B says that tapping the engine block with a hammer will cause the knock sensor circuit to operate. Who is right?
    (A) A only.
    (B) B only.
    (C) Both A & B.
    (D) Neither A nor B.
15. Technician A says that a drive cycle test is performed before the emissions are checked. Technician B says that a drive cycle test is performed after the ECM or battery has been disconnected. Who is right?
    (A) A only.
    (B) B only.
    (C) Both A & B.
    (D) Neither A nor B.

Advanced Certification Questions

The following questions require the use of reference information to find the correct answer. Most of the data and illustrations you will need is in this chapter. Diagnostic trouble codes are listed in **Appendix A**. Failure to refer to the chapter, illustrations, or **Appendix A** for information may result in an incorrect answer.

16. A throttle body fuel injected engine is not able to exceed 6500 rpm. Technician A says that the fuel pressure regulator could be the cause. Technician B says that the ignition pickup coil could be the cause. Who is right?
    (A) A only.
    (B) B only.
    (C) Both A & B.
    (D) Neither A nor B.
17. A multiport fuel injected vehicle has difficulty starting when warm. The scan tool data in **Figure 15-3** was obtained at idle. Which of the following is the *most likely* cause?
    (A) A disconnected intake air temperature sensor.
    (B) A defective oxygen sensor.
    (C) A misadjusted throttle position sensor.
    (D) A sticking idle air control valve.
18. A vehicle with distributorless ignition has an intermittent no-start condition. The ECM has stored diagnostic trouble codes P0335, P0336, and P0339. Which of the following is the *most likely* cause?
    (A) A faulty camshaft position sensor or circuit.
    (B) A faulty crankshaft position sensor or circuit.
    (C) A malfunctioning ignition coil.
    (D) A disconnected camshaft position sensor.

Data Scanned from Vehicle			
Coolant temperature sensor 206°F/96°C	Intake air temperature sensor 82°F/28°C	Manifold absolute pressure sensor .9 volts	Throttle position sensor .32 volts
Engine speed sensor 1150 rpm	Oxygen sensor .67 volts	Vehicle speed sensor 0 mph	Battery voltage 10.2 volts
Idle air control valve 59 percent	Evaporative emission canister solenoid Off	Torque converter clutch solenoid Off	EGR valve control solenoid 0 percent
Malfunction indicator lamp On	Diagnostic trouble codes P0562	Open/closed loop Open	Fuel pump relay On
Brake light switch Off	Cruise control Off	AC compressor clutch Off	Knock signal No
Ignition timing (°BTDC)	Base timing: 10		Actual timing: 16

19. A fuel injected sport-utility vehicle idles excessively high. The scan tool data shown above was obtained at idle. Technician A says the high idle could be caused by a defective alternator. Technician B says a dirty idle air control valve is causing the high engine idle. Who is right?
   (A) A only.
   (B) B only.
   (C) Both A & B.
   (D) Neither A nor B.

20. A vehicle has experienced repeated oxygen sensor failure. The oxygen sensor tip has a white glassy coating. Technician A says this is due to the use of a silicone-based engine sealant. Technician B says this is due to an on-going internal coolant leak. Who is right?
   (A) A only.
   (B) B only.
   (C) Both A & B.
   (D) Neither A nor B.

The modern vehicle has many complex systems. The diagnostic procedures discussed in this chapter can help make these systems easier to repair. (Jaguar)

Chapter 16

Ignition System Diagnosis and Repair

After studying this chapter, you will be able to:

- ❑ Use a scan tool to retrieve ignition system information.
- ❑ Use a timing light to check engine timing and advance.
- ❑ Adjust ignition timing.
- ❑ Use an oscilloscope to check the ignition system.
- ❑ Check ignition system components.
- ❑ Explain how to replace ignition coils.
- ❑ Explain how to replace ignition modules.
- ❑ Explain how to replace pickup coils.
- ❑ Adjust ignition system components.
- ❑ Explain how to remove, rebuild, and replace an electronic distributor.

This chapter identifies the various methods used to test the ignition system and explains how to replace various components. The ignition system must be working properly for the engine to perform at peak efficiency. Even a minor defect can cause serious performance and emissions problems. You will use many of the procedures and tests that you've studied in the earlier chapters. If necessary, review **Chapter 8** for basic ignition system operating principles.

Preparing to Diagnose the Ignition System

As in the previous chapters, start the diagnostic process by verifying that a driveability problem exists. Most driveability problems caused by the ignition system are usually noticeable, **Figure 16-1.** Either engine performance is reduced or the engine does not run at all. Once a problem has been verified, the next step is to locate the vehicle's service manual. However, unlike the earlier chapters, your search will concentrate on one specific system, rather than the entire powertrain.

Warning: Modern electronic ignition system voltages can exceed 100,000 volts. If proper procedures are not followed, equipment damage, personal injury, or death may occur.

Access to the proper service manual is important, since ignition systems vary by manufacturer and vehicle make, even by the type of engine in the vehicle. Flow charts and schematics, **Figure 16-2,** should be used throughout the diagnostic process.

Visual Inspection

After verifying that the vehicle's related primary wiring and computer control system is in good condition, you will perform two visual inspections of the ignition system, one with the engine off and one with the engine running. With the engine off, look at the ignition primary wiring, distributor (if equipped), ignition module (if visible) spark plug wires, spark plugs, coil(s), and any other ignition system components. If the engine has a distributorless ignition system, inspect the crankshaft and/or camshaft sensors.

Ignition System
Ignition triggering devices (inoperative, misadjusted, disconnected)
Electronic spark timing controls (defective, disconnected)
Ignition module (defective, disconnected)
Primary wiring (disconnected, improperly connected, high resistance)
Ignition coil (defective, disconnected, carbon tracks)
Distributor (carbon tracks, moisture, worn shaft, worn bushings, stripped drive gear or retaining pin)
Secondary wiring (broken, leaking insulation, disconnected, improperly connected)
Spark plugs (improperly gapped, fouled, cracked insulator, internally defective, incorrect, mismatched, worn)
Conventional spark timing controls (sticking, defective, linkage disconnected)
Ballast resistor (defective, disconnected)

Figure 16-1. *Problems in the ignition system can drastically affect engine performance. These are the most common ignition system defects. (General Motors)*

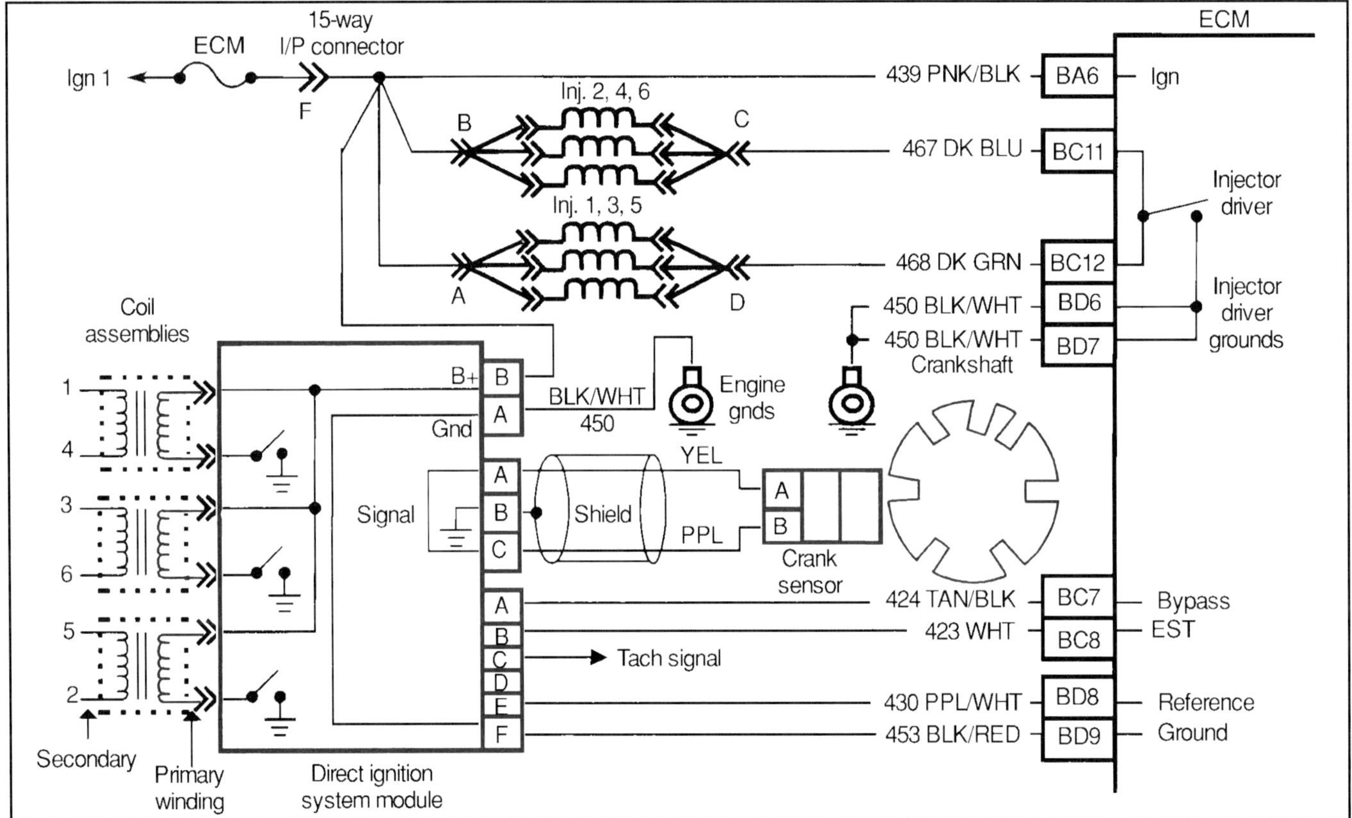

Figure 16-2. *Always refer to schematics and diagnostic flowcharts when looking for an ignition system problem. (General Motors)*

Check the distributor cap and ignition coil for ***carbon tracks***; this is a sign of arcing. If possible, check the harness connections at the ignition module. Inspect the terminal connections at the camshaft and crankshaft sensors. If the crankshaft sensor is mounted outside the engine, check for dirt, grease, or debris around the sensor. Check the spark plug wire connections at the distributor and/or ignition coil and at the spark plugs. Check spark plug wire routing, making sure that none of the wires cross, as this can cause misfiring and electromagnetic interference. If the problem is a no-start condition, proceed to spark testing after verifying that all ignition system components are present and connected.

If the first visual inspection does not reveal the problem, start the engine, if possible, and begin inspecting the system again. Look at the wires, spark plugs, and the distributor (if equipped) for arcing to ground and other electrical discharging. Perform the flex test on the primary wiring. If no problems are found, check the ECM for stored trouble codes, even if the malfunction indicator lamp is not on.

Warning: Do not perform the flex test on any secondary wiring while the engine is running.

Corona Discharge

During the visual inspection on some electronic ignition systems, you may notice a steady light around the spark plug insulators while the engine is running. This light will appear just above the point where the insulator enters the shell. This is called ***corona discharge.*** Corona discharge results from the formation of a high tension field around the spark plug insulators caused by the voltages of the modern electronic ignition system. You will probably notice nothing on most systems, as this discharge does not always occur and can only be detected in darkness.

Do not mistake corona discharge for a short to ground. Corona discharge will have no effect on engine performance. The discharge will sometimes leave a ring around the spark plug insulator. This ring is formed due to the repelling of dust particles by the high tension field. Many technicians mistake this ring as combustion gases blowing by the spark plug shell and insulator seal. A quick test involves the use of a test light, which is discussed later in this chapter.

Tests Performed during the Visual Inspection

The following tests are often performed by experienced driveability technicians during a visual inspection to quickly identify or eliminate the ignition system as the cause of a no-start or performance problem. Exercise caution when performing these tests as you will be dealing with open sparks and high voltages.

Spark Testing

Driveability technicians will often perform this test before a visual inspection when a vehicle has a no-start condition. The purpose of this test is to simply check for the presence of ignition spark. The advantage of this test is that you may be able to quickly identify or eliminate the ignition system as the cause of a no-start condition. Older contact point ignition systems could be checked simply by pulling a wire off the spark plug, holding the end close to a ground, and observing for the presence of a spark. However, the high voltages of modern electronic ignition systems make this method dangerous.

Warning: Do not spark test an electronic ignition system by pulling a spark plug wire. An electrical shock will result with possible injury or death.

Electronic ignition systems can be tested for spark by using a ***spark tester.*** This tool is shaped like a spark plug, however, it has a clip or lead that is attached to ground, **Figure 16-3.** To conduct this test, you will need a remote starter switch or someone to crank the engine. Unplug one spark plug wire and attach the wire to the spark tester. Clip the spark tester to ground. Crank the engine while observing the tester. Do not touch the spark tester during this time. If the wire produces a good strong spark, reconnect the wire and repeat the test with the other wires.

If all the wires tested produce a good spark and you find no other problems during the visual inspection, you can eliminate the ignition system as the cause of the problem. However, if one or more wires produce a weak spark or no spark, you have narrowed the problem down to the ignition system.

Figure 16-3. *The spark tester will indicate the presence of a spark. It will not, however, give an indication of the spark's intensity or voltage. (Jack Klasey)*

Checking the Secondary Ignition System Using a Test Light

Since the secondary wiring system carries the ignition system's high voltages, it is more likely to be the location of a problem. Unless you happen to select the misfiring cylinder(s) for spark testing, the method just described is very inefficient when the engine runs and the problem is possibly ignition related. A more effective way to find a defect in the secondary system is to use a test light. This test is often performed during the visual inspection as it can detect a wire that is becoming defective, but not enough to produce a visible arc, or a defective wire that will not arc at idle. This procedure uses a test light to check the wires for arcing. Due to the high voltages involved, a high impedance test light (minimum 10 meg ohms) must be used. Some tool manufacturers make special ignition probes for this procedure.

Caution: *Never* pierce a spark plug wire or boot with a test light probe. This will permanently damage the wire.

With the engine running, start at the distributor or ignition coil and slowly trace the test light along the spark plug wire within 1-2″ (25.4-51 mm) of the wire's surface, **Figure 16-4.** Continue along the wire's length, checking each wire in the system one at a time. Move the test light probe around each spark plug to make sure that any corona discharge is not an actual short. Finally, move the test light around the vicinity of the distributor and/or ignition coil. Some technicians will lightly mist the wires with water or a salt and water mixture to enhance any possible problems. However, this practice is not recommended. If the secondary system is good, there will be no arcing to the test light or to ground. If any wire or other part of the secondary system arcs to the test light, replace the defective part.

Figure 16-4. *A test light placed near a leaking spark plug wire will provide a ground path better than the wire, especially if the wire has high internal resistance. (Jack Klasey)*

Scan Tool Testing

Some scan tools can be used to test each cylinder for power. Normally this is done to isolate which cylinder is the cause of a misfire. This is called a ***power balance test*** and can vary by scan tool manufacturer. Some scan tools will deactivate a fuel injector or the spark plug for the particular cylinder to be tested. The power balance test is discussed in more detail in **Chapter 17.**

Note: Not all engines and vehicles can be power balance tested using a scan tool. Check the service manual before performing this test.

The scan tool can be used to check the knock sensor and electronic spark control for proper operation. Knock sensor testing was covered in **Chapter 15.** Vehicles with OBD II systems have additional tests that can be performed with the scan tool. These tests are discussed in the next section. More information on scan tools can be found in **Appendix C.**

OBD II Misfire Monitor

A misfire on any cylinder causes unburned fuel and excess oxygen to enter the exhaust system, eventually leading to catalytic converter overheating. The OBD II ***misfire monitor*** checks the engine for misfire at all times and illuminates or flashes the MIL if misfiring is excessive. The ECM counts crankshaft revolutions and monitors for any disruption in the sample. Some disruption due to rough roads and other external causes may register as a minor disruption (usually indicated by a number under 10). Counts of 15 or more is an indication of a minor or intermittent misfire. In cases of minor or intermittent misfire, a trouble code may or may not be set. Counts of 100 or more indicate a severe misfire. **Figure 16-5** is a visual depiction of the way that the ECM identifies and stores cylinder misfires.

Note: Since severe misfires can set up conditions that could damage the catalytic converter, some misfire codes are Type A codes and will cause the ECM to flash the MIL whenever the condition occurs.

On OBD II systems, a scan tool can be used to isolate the misfiring cylinder(s). When the scan tool misfire test screen is accessed, the tool will display a code that indicates which cylinder or cylinders is misfiring. The scan tool is also able to display the stored misfire messages for a set time in the past, **Figure 16-6.** After the misfire has been isolated to a particular cylinder or cylinders, other diagnostic steps can be taken to determine the cause of the misfire.

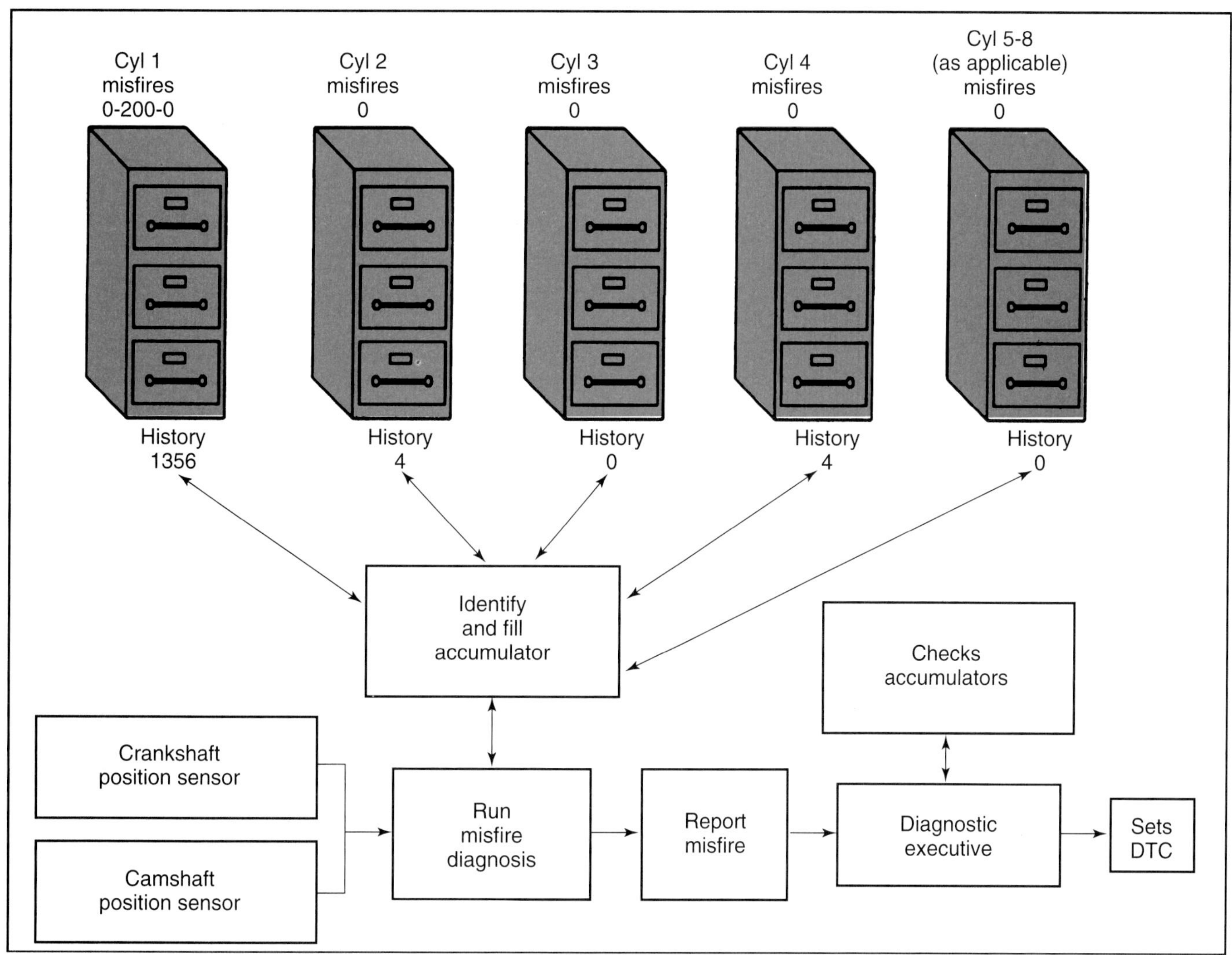

Figure 16-5. *The misfire monitor breaks the crankshaft speed into 200 rpm increments. A misfire will cause the crankshaft speed to decrease. When the ECM detects a decrease in crankshaft speed, it notes the cylinder that is firing when the misfire occurs. The number of misfires is counted and stored. Sufficient misfiring will cause the ECM to set a trouble code and illuminate, or in some cases, flash the MIL. (General Motors)*

Data Scanned From Vehicle			
Coolant temperature sensor 214°F/101°C	Intake air temperature sensor 92°F/33°C	Manifold absolute pressure sensor 2.1 volts	Throttle position sensor 1.55 volts
Engine speed sensor 2290 rpm	Oxygen sensor .76 volts	Vehicle speed sensor 28 mph	Battery voltage 13.9 volts
Idle air control valve 10 percent	Evaporative emission canister solenoid Off	Short-term fuel trim 140	Long-term fuel trim 117 percent
Malfunction indicator lamp Flash	Diagnostic trouble Codes P0301, P0351	Open/closed Loop Open	Knock signal Yes
Misfire cyl. #1 199	Misfire cyl. #2 0	Misfire cyl. #3 142	Misfire cyl. #4 2
Ignition timing (°BTDC)	Base timing: 8	Actual timing: 69	

Figure 16-6. *Misfire counts can be displayed on some scan tools.*

Checking the Ignition System Using Test Equipment

If a visual inspection and scan tool information does not reveal the problem, the next step is testing using electronic test equipment. The two main tools used for ignition system testing and service are the oscilloscope and timing light. The operation of these two devices is explained in the following sections.

Using the Oscilloscope

The ***oscilloscope,*** **Figure 16-7**, is used to spot defects in the ignition system. It is used to diagnose secondary system problems on modern electronic ignition systems, since the primary circuit is controlled by electronic circuitry and does not show up on the scope pattern. On older engines with contact point ignitions, the oscilloscope will reveal the majority of both primary and secondary system problems.

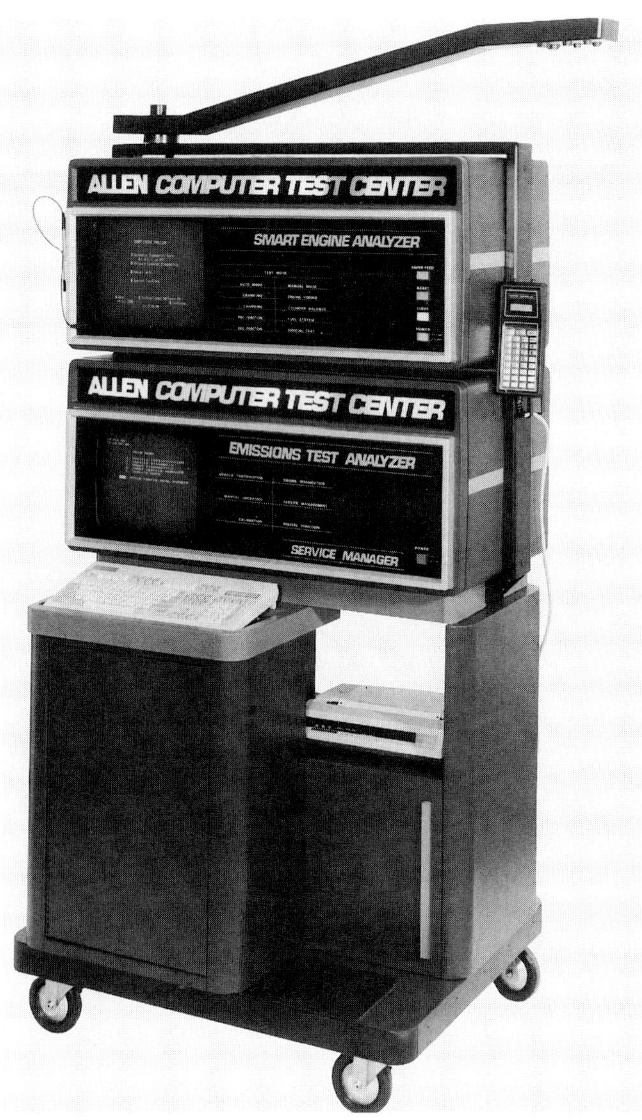

Figure 16-7. *Oscilloscopes can be powerful tools when you know how to use them properly. Oscilloscopes are usually combined with other testers and referred to as an engine analyzer. (Automotive Diagnostics)*

Note: To observe the oscilloscope pattern on engines with distributorless ignition, a special adapter may be needed.

Begin by connecting the oscilloscope to the engine according to the manufacturer's directions. Make sure that all cables are placed well away from any moving engine parts. Set the controls to display a secondary ignition pattern. Start the engine and observe the scope pattern. If any part of the scope pattern does not resemble the standard scope pattern, the ignition system may have a defect. Once secondary pattern observations are complete, switch to the primary pattern setting and observe the pattern. **Figure 16-8** shows the sections of primary and secondary ignition wave patterns. Look for possible problems in the primary system. Once a problem has been identified, make component checks as necessary. After making repairs, always recheck the scope pattern.

Secondary Pattern

The ***secondary pattern*** shows what is happening in the secondary, or high voltage, side of the ignition system. A typical secondary pattern is shown in **Figure 16-8A.** Since the pattern height is affected by ignition system voltage, the higher the voltage, the higher the line. The secondary pattern consists of three main sections discussed in the following paragraphs. Refer to **Figure 16-8A** while reading the following sections.

Firing Section

The ***firing section,*** or firing voltage, *(1)* is the portion of the pattern that shows the spark jumping the spark plug gap (and the rotor gap if applicable). Since the coil must produce a very high voltage to jump the plug gap, this section of the pattern is the highest, appearing as an almost straight line. This is called the firing line. The exact amount of voltage needed to fire the plug can be read on the scale, which is part of the CRT screen. Average firing voltage at idle is about 10,000 volts. The firing line may be too high, indicating a high resistance in the circuit, usually caused by an open spark plug wire or a worn plug with a wide gap. A short firing line is caused by a short circuit in the secondary delivery system, most likely a fouled plug, a grounded plug wire, or a carbon track in the distributor cap or on the ignition coil.

Once the spark jumps the plug gap, it requires less voltage to maintain the spark across the gap. Therefore, the next part of the pattern is lower than the firing line, but still higher than the rest of the pattern. This is the plug firing time *(2),* however, it is more commonly referred to as the spark maintenance line or ***spark line.*** The length of the maintenance line is matched to the height of the firing line. When the firing line is high, the maintenance line is short. When the firing line is low, the maintenance line will be long. This is because the available coil energy is a fixed amount, and whatever is not used to jump the gap is used to maintain the spark.

A

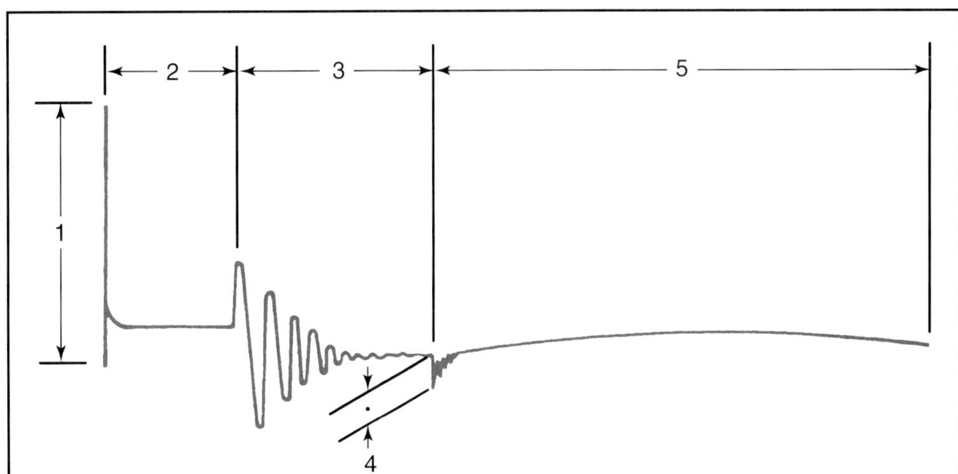

B

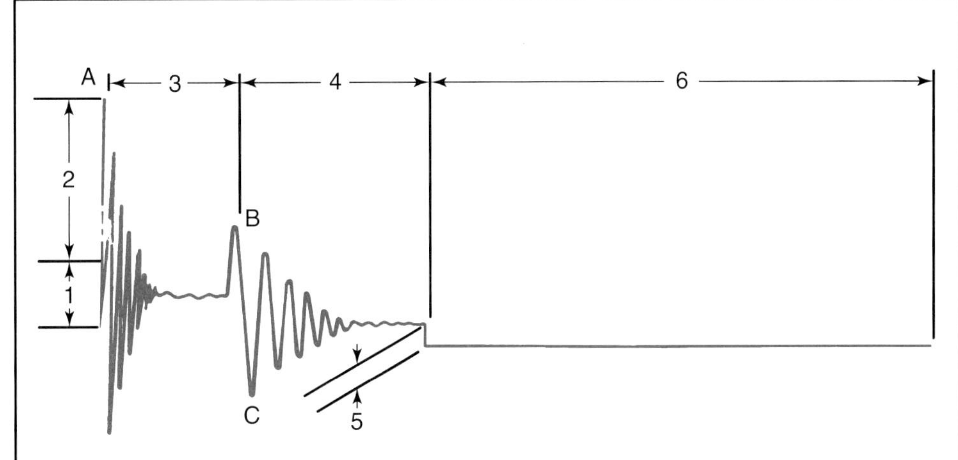

Figure 16-8. *A—The secondary pattern displays the high voltage waveform of the ignition system. This is the pattern you will be dealing with on most modern ignition systems. B—The primary pattern was important when diagnosing older contact point ignition systems. It is now used to double-check the secondary waveform. (Automotive Diagnostics)*

Intermediate Section

Eventually, the coil no longer has enough electrical energy to maintain the spark across the plug gap. However, some electrical energy is left over after the spark stops. The ***intermediate section*** is where this extra energy is used. The unused energy travels between the coil and various internal ignition components. This appears on the scope pattern in the form of oscillations, or up and down waves *(3)*. These waves will gradually reduce in size as the energy is used up. There should be at least four oscillations in the intermediate section. If there are less than four, the coil may be shorted.

Dwell Section

The ***dwell section*** shows the primary current beginning to flow in the coil's primary winding. On point-type ignition systems, the dwell section was used to diagnose points and condenser problems, such as arcing, bouncing, incorrect dwell setting, or high point resistance. On modern engines, the current flow is started by the switching on of a power transistor in the control module *(4 and 5)*. The coil magnetic field begins to build up in a series of oscillations as current flow continues. The oscillations become smaller as the magnetic field stabilizes. When the transistor shuts off, current flow stops. This triggers the collapse of the magnetic field, and the secondary spark is created. Dwell was discussed in **Chapter 8.** Electronic ignition systems have eliminated point-related problems, and the dwell section does not need to be as closely monitored.

Primary Pattern

The ***primary pattern*** shows what is going on in the primary, or low voltage side of the ignition system. A typical primary pattern is shown in **Figure 16-8B.** Note that the primary pattern is somewhat similar to the secondary pattern. The firing section shows the coil discharge as a series of oscillations as coil energy is used *(3)*. The intermediate section closely resembles that of the secondary section, since the unused coil energy is being dissipated through the primary ignition components *(4)*. The primary dwell section *(6)* has no oscillations since current is flowing through the primary side of the coil at a steady rate. On older point-type systems, the primary pattern was useful for finding point problems, but is now primarily a method of double checking problems found in the secondary. The primary pattern can be used to detect faulty operation of electronic fuel injectors when the proper connections are made.

Oscilloscope Patterns

If you are familiar with how to read the patterns, the oscilloscope will give you a powerful tool for finding ignition and other system problems. Some typical oscilloscope wave patterns and the problems that cause them is shown in **Figure 16-9.** Some of the newest multimeters have built-in oscilloscope functions.

Timing Lights and Timing Testers

Any type of ***timing light*** can be used to check and set the initial timing. However, to accurately measure the amount of timing advance, you will need an advance meter. ***Advance meters*** are built into many modern timing lights, and into ***timing testers*** using magnetic timing probes. To accurately check timing advance, an advance meter is vital. If your timing light does not have an advance meter, the total ignition advance can only be approximated. The purposes and designs of timing lights were discussed in **Chapter 3.**

Checking Ignition Timing

Always follow the vehicle and tool manufacturers' directions when using any kind of timing tester. There are differences in the timing procedure between vehicles. These include setting proper idle speed, whether to remove and plug any vacuum hoses, disconnecting any computer wiring to the distributor, and whether to unplug or ground the DLC or a special connector. See **Figure 16-10.** Small details like these will determine whether you check and set the timing accurately. In many cases, the underhood emission control sticker contains instructions for checking initial timing.

If you are using a timing light, begin by locating the timing marks on the engine. If the timing marks are dirty, clean them and use chalk or paint to highlight the rotating and stationary marks, **Figure 16-11.** If your timing light has a built-in advance meter mark the zero on the timing plate. If your timing light does not have an advance meter, mark the spot corresponding to the correct initial advance. If you are using a magnetic timing tester, it is not necessary to clean or highlight the timing marks.

If the engine is cold, warm it to operating temperature, either with a short drive, or by letting it idle until the upper radiator hose is hot to the touch. Once the engine is at operating temperature, shut it off and attach the timing light or magnetic probe of the timing tester. Disconnect any electrical connectors or vacuum lines as directed by the vehicle manufacturer's instructions.

Restart the engine and make sure that it is running at the correct idle speed for timing checking. Point the timing light at the timing marks, and turn the timing light control dial until the timing mark on the vibration damper appears at the TDC (zero degrees) mark, **Figure 16-12.** Then read the amount of advance or retard on the timing meter. If the timing light does not have an advance meter, note and compare the position of the moving and stationary timing marks. If you are using a timing meter, read the timing advance or retard on the meter. Compare the ***initial timing*** specifications on the underhood sticker to the actual timing. If the initial timing is correct, proceed to the next step. If the timing is off, it must be adjusted.

Checking Timing Advance

To check timing advance, shut off the engine but do not disconnect the timing light or timing meter. Reconnect any electrical or vacuum connectors that were removed as part of the initial timing check. Then restart the engine and note the timing. It should now be at a different setting than it was when making the initial timing check. If the timing has changed, this indicates that the advance mechanisms, whether vacuum or electronic, are working to adjust the timing.

Apply the parking brake firmly and raise engine speed to the rpm specified by the manufacturer for checking high speed advance. If using a timing light, you will observe the timing mark move in the advanced direction. Turn the advance meter control knob until the timing mark appears to return to the initial timing position. Then read the timing advance on the timing light dial. If using a timing meter, read the advance shown on the meter face. This is the ***total advance,*** comprised of the initial timing plus the automatic advance in response to the increase in engine speed. Compare the actual advance reading against the advance specifications for the engine. If you are using a timing light without an advance meter, or the advance specifications are not available, ensure that the system is advancing the timing roughly 35-45° at 2000 rpm.

Allow the engine to return to idle. If the engine has a computer control system with a knock sensor, watch the timing marks or timing meter at idle. Use a wrench or other light object to tap on the engine block or manifold near the location of the knock sensor as you observe the timing reading. Use a scan tool to monitor knock sensor operation during this test. Timing will retard between 10-20° if the knock sensor and computer spark control system are working, **Figure 16-13.**

If the timing is advancing and retarding according to specifications, the problem is not in the timing controls. Stop the engine and disconnect the timing light. If the initial timing or advance was not correct, record the actual readings and refer to the vehicle manufacturer's troubleshooting information.

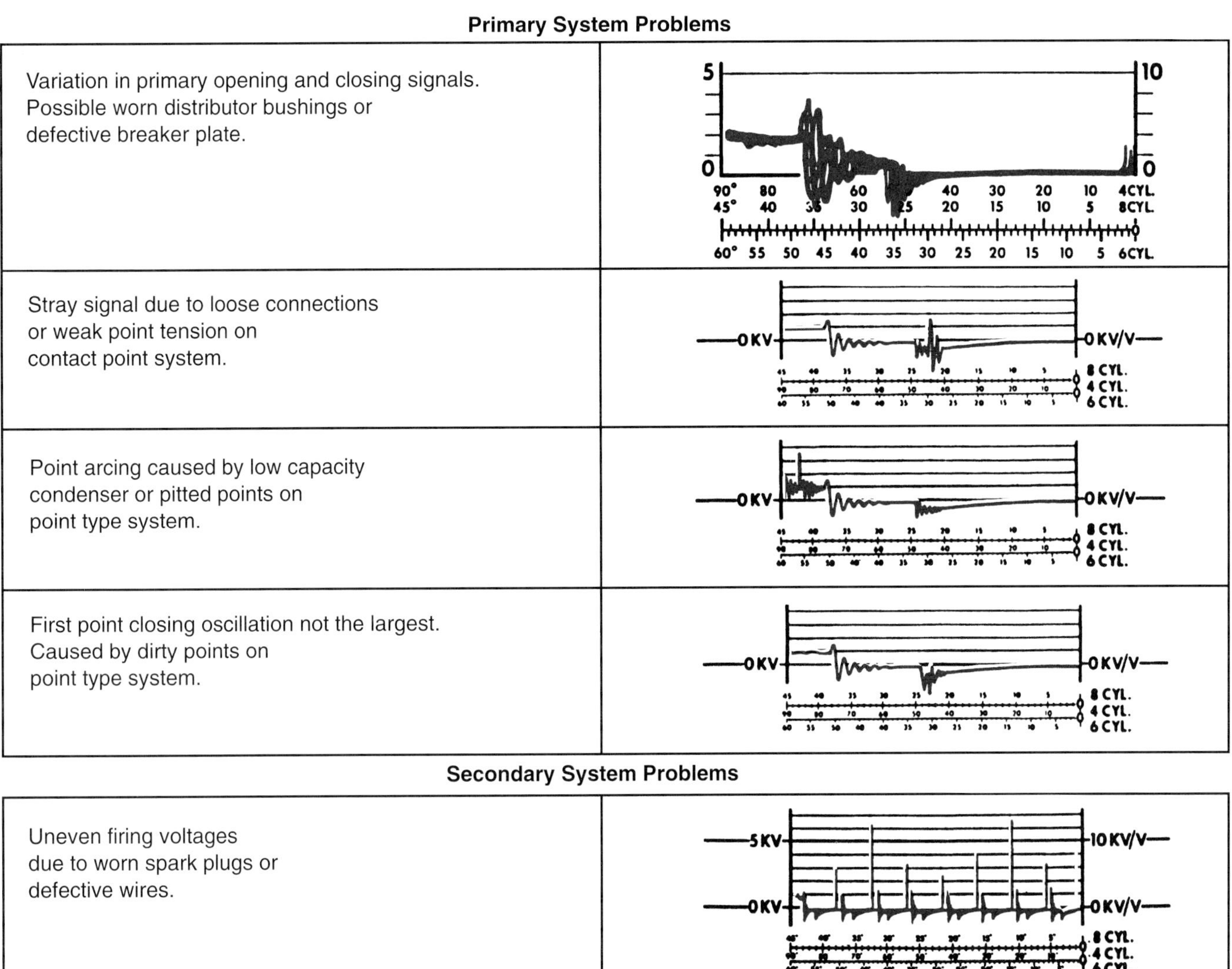

Primary System Problems

Variation in primary opening and closing signals. Possible worn distributor bushings or defective breaker plate.	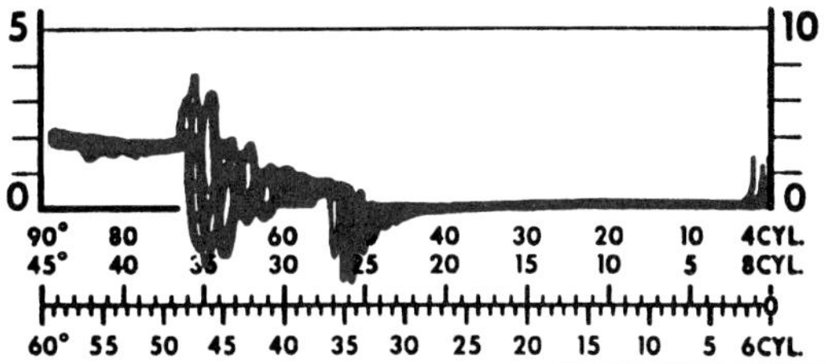
Stray signal due to loose connections or weak point tension on contact point system.	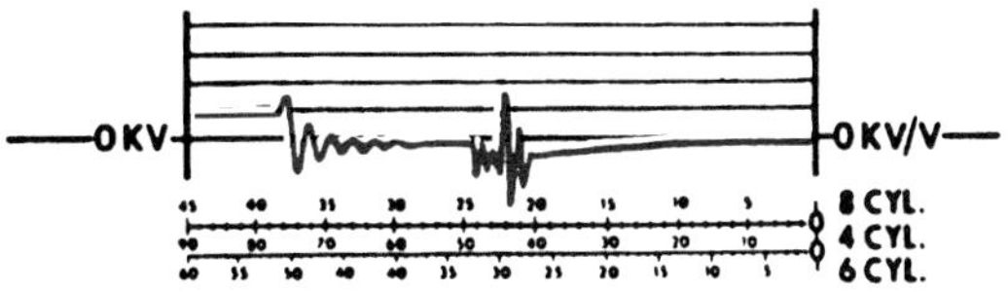
Point arcing caused by low capacity condenser or pitted points on point type system.	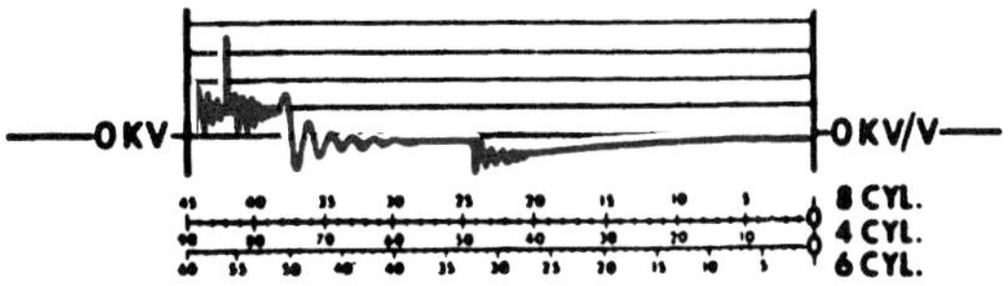
First point closing oscillation not the largest. Caused by dirty points on point type system.	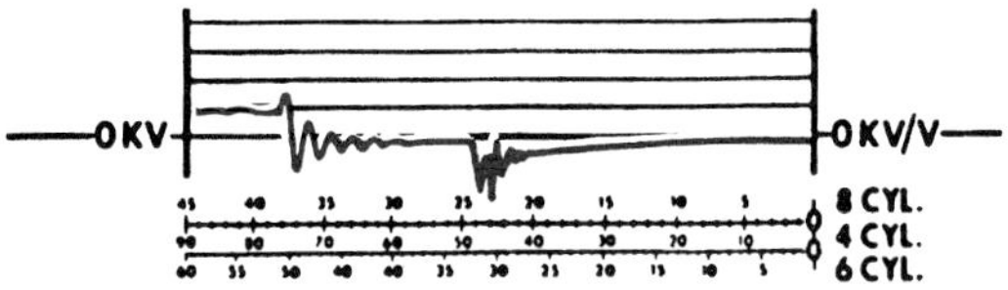

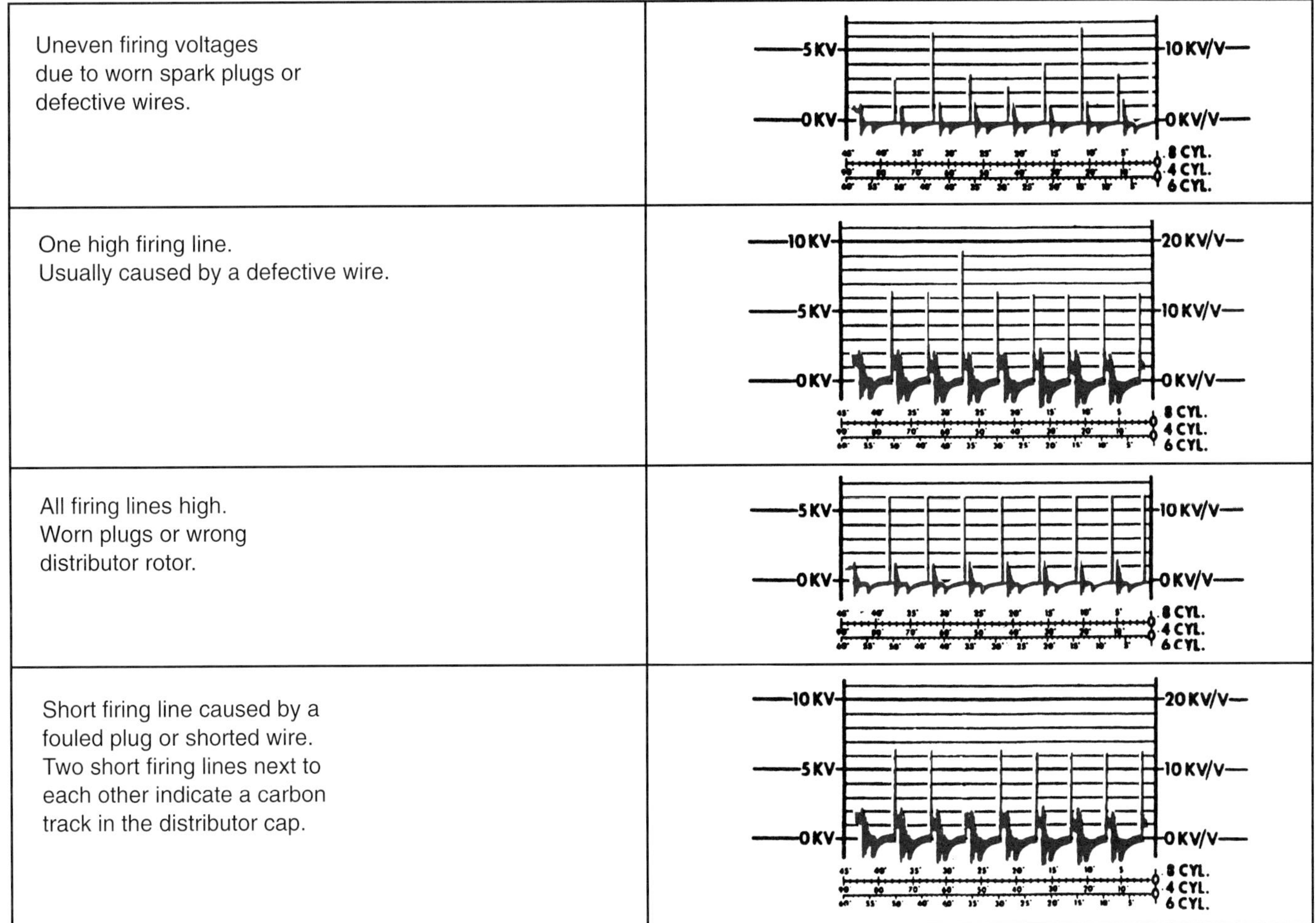

Secondary System Problems

Uneven firing voltages due to worn spark plugs or defective wires.	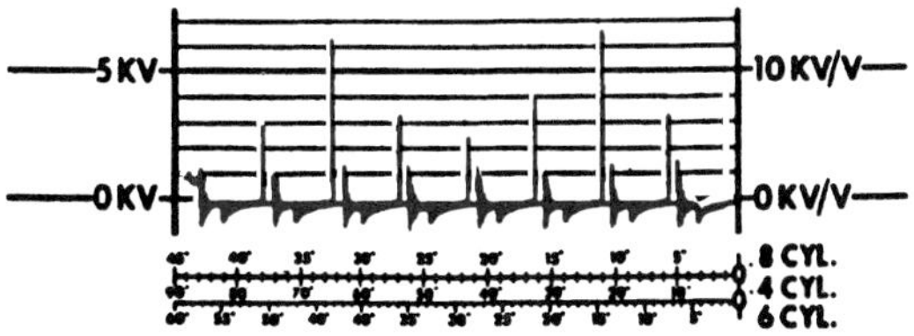
One high firing line. Usually caused by a defective wire.	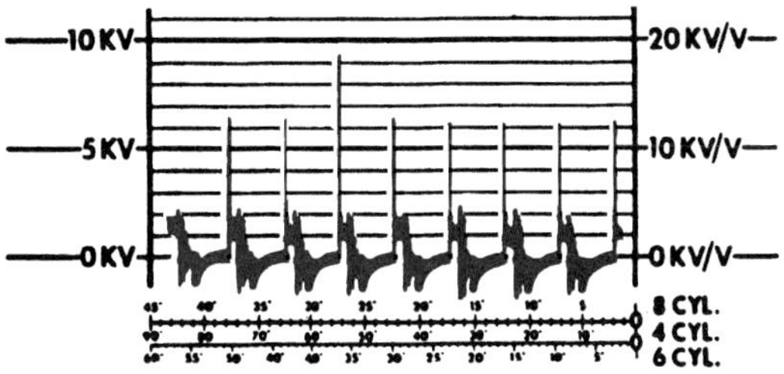
All firing lines high. Worn plugs or wrong distributor rotor.	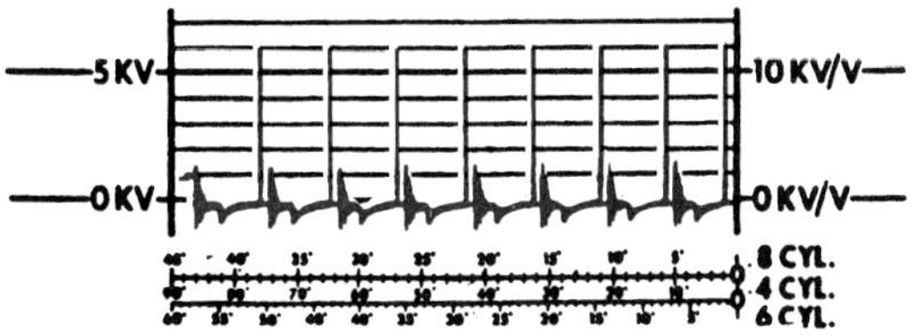
Short firing line caused by a fouled plug or shorted wire. Two short firing lines next to each other indicate a carbon track in the distributor cap.	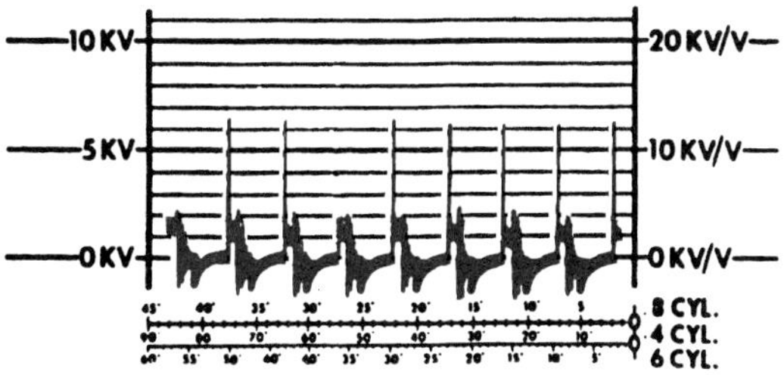

Figure 16-9. *These are some typical primary and secondary ignition system waveform patterns and the problems that cause them.*

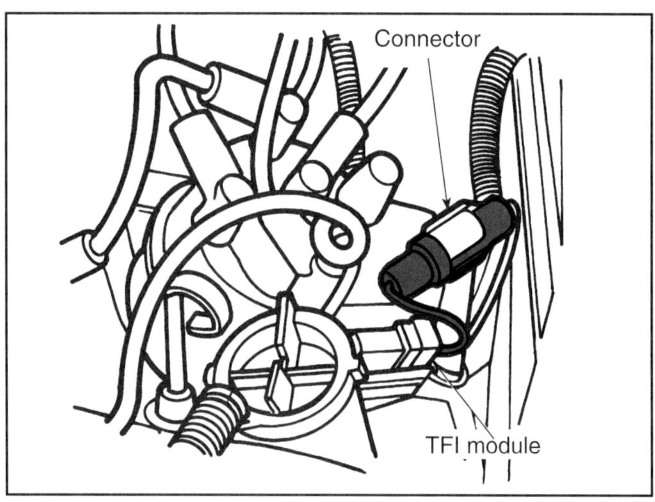

Figure 16-10. *Follow all procedures when setting engine timing on an electronic ignition system. Connectors may need to be disconnected or grounded. (Ford)*

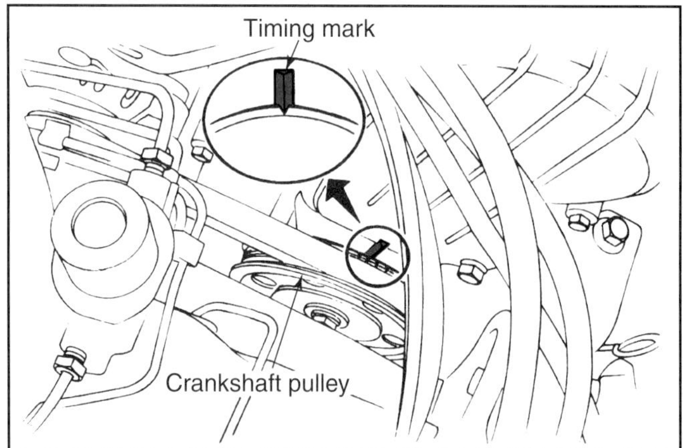

Figure 16-11. *The timing marks are usually on the crankshaft pulley and a fixed pointer on the engine. You will probably have to clean and mark the timing points on most engines. (Subaru)*

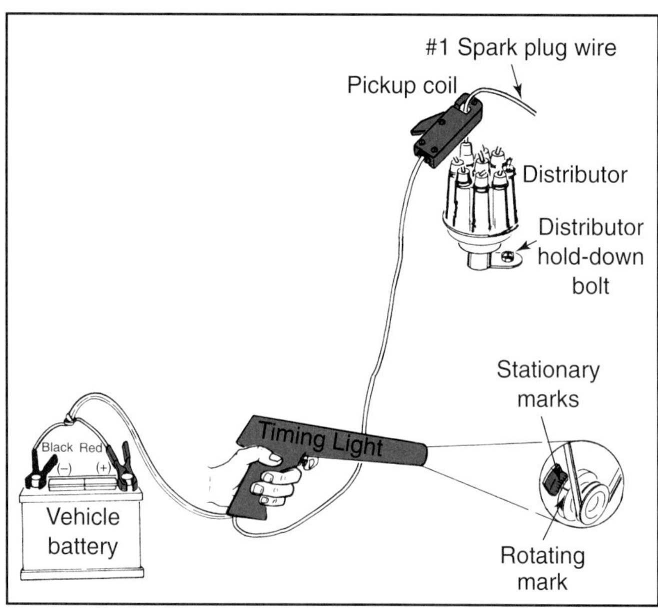

Figure 16-12. *Component setup for most inductive timing lights.*

Testing Ignition Coils

Coils can be tested by bypassing the rest of the ignition system. Start by disconnecting the coil primary leads and connecting jumper wires from the coil's primary connections to the positive and negative terminals of the battery. Remove the coil's secondary wire from the distributor and install a spark tester. When testing a coil installed in the distributor cap, run a jumper wire to the center connection inside the distributor cap, with the other end connected to a spark tester. Then make and break the coil circuit by removing the jumper wire from the battery's negative terminal. There should be a spark from the center wire.

Data Scanned from Vehicle			
Coolant temperature sensor 209°F/99°C	Intake air temperature sensor 89°F/31°C	Manifold absolute pressure sensor 2.8 volts	Throttle position sensor .73 volts
Engine speed sensor 880 rpm	Oxygen sensor .49 volts	Vehicle speed sensor 0 mph	Battery voltage 13.3 volts
Idle air control valve 41 percent	Evaporative emission canister solenoid Off	Short-term fuel trim 128	Long-term fuel trim 128
Malfunction indicator lamp Off	Diagnostic trouble codes	Open/closed loop Closed	Fuel pump relay On
Prom ID 5248C	Cruise control Off	AC compressor clutch On	Knock signal Yes
Ignition timing (°BTDC)	Base timing: 4		Actual timing: 23

Figure 16-13. *Scan tools should be used with oscilloscopes to diagnose computer-controlled ignition systems. Tapping on the engine with a hammer should cause the knock sensor to register a knock signal.*

On some vehicles, this test can be made by turning the ignition switch on and off. There should be a single spark from the coil when the ignition is switched off. If the coil will not produce a spark, check it for continuity and resistance with an ohmmeter. Place one lead on each coil primary terminal. The primary resistance should be near the service manual specifications, usually about 1 to 2 ohms, **Figure 16-14.**

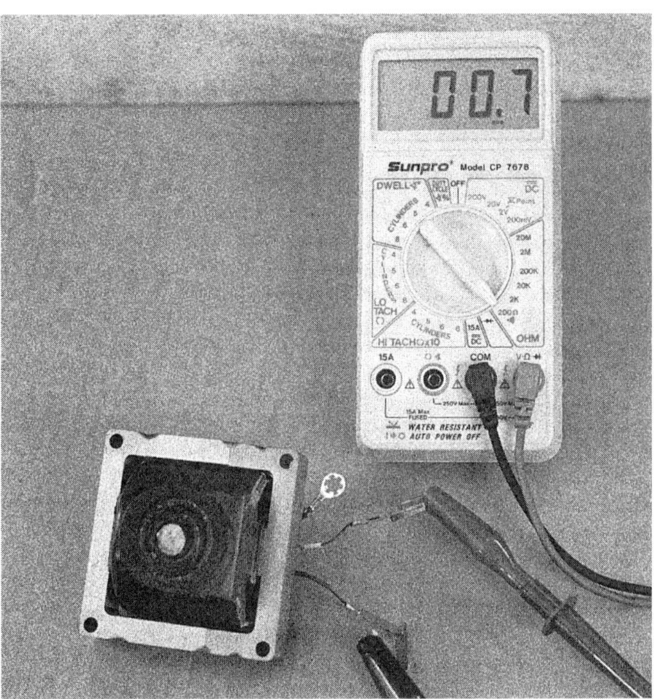

Figure 16-14. *Ignition coils can be checked with an ohmmeter. Note the connections to the primary coil leads.*

If the coil produces a weak or intermittent spark, place one of the meter leads on either primary terminal, and the other lead in or on the coil secondary tower, **Figure 16-15.** Secondary resistance should be near the factory specifications, usually between 8000 and 12,000 ohms.

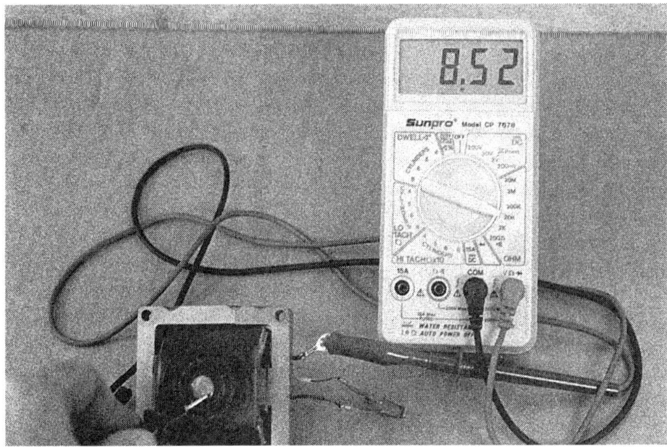

Figure 16-15. *Due to the number of fine wires in the circuit, secondary resistance is higher. Coil ground is the best terminal to use when checking the coil's secondary windings.*

On distributorless ignition systems, remove the spark plug wires from the suspect coil and connect the ohmmeter to the secondary towers. Have someone lightly tap on the coil with a screwdriver. If the resistance varies widely, is infinite, or below specifications, the coil is defective. If the coil fails any of these resistance tests, replace it.

Note: Ideally, ignition coil tests should be performed after the engine has been operated for several minutes.

Checking Spark Plug Wires with an Ohmmeter

Spark plug wire failure will usually show up as a constant or recurring engine miss. Often a plug wire will look good, but may be broken or corroded internally. If the test light does not show any obvious spark plug wire problems, the wires should be checked with an ohmmeter. Start by setting up the ohmmeter and removing the spark plug wire terminal ends from the spark plug and distributor cap. Place one ohmmeter lead on one end of the spark plug wire, and the other lead on the other end of the wire, as shown in **Figure 16-16.** If the cap has been removed from the distributor, place one lead to the proper contact inside the distributor.

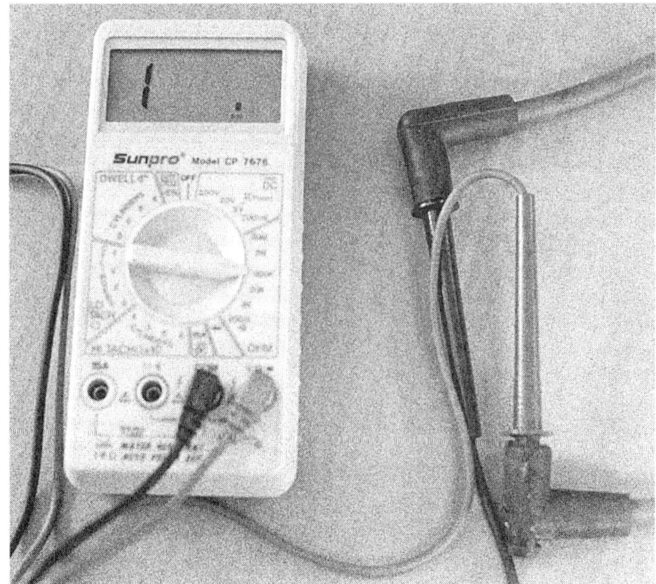

Figure 16-16. *Spark plug wires, no matter what type of system, should have no more than 4000 ohms resistance per foot of wire.*

The ohmmeter should read around 4000 ohms per foot of wire. A two-foot wire should read approximately 8000 ohms, and a three-foot wire should read about 12,000 ohms. The maximum that any wire should read is 20,000 ohms. If the reading is too high, replace the wire. You should also visually inspect the spark plug wires. Replace any brittle or damaged wires, or any wires with broken or corroded spark plug or distributor cap connectors. Damaged wires will allow the spark to ground on the engine block instead of jumping the spark plug gap.

Checking Integrated Direct Ignition System Coil Tower

Integrated direct ignition systems have no rotor or cap to develop carbon tracks. However, the high voltages used with these systems can cause ***flashover*** at the coil output towers, which can create carbon tracks on the coils and the tower. Always check the coils and towers carefully for signs of flashover or carbon tracking, visually or with an ohmmeter. Evidence of carbon tracking and flashover may be signs that a coil is leaking or the tower is defective. The integrated direct ignition system has almost no secondary system. The coil output travels directly from the coil towers to the plugs.

The coil tower can be checked for continuity between the coils and secondary conductors with an ohmmeter. Begin by removing the coil tower, then remove both coils and all the spark plug connectors from the tower, **Figure 16-17.** Connect the ohmmeter's positive test lead to one of the tower's coil terminals. Connect the negative lead to each spark plug terminal. Three terminals should give no reading, one should read near or full continuity. If there is no reading on any terminal for one or more coil terminals, replace the tower.

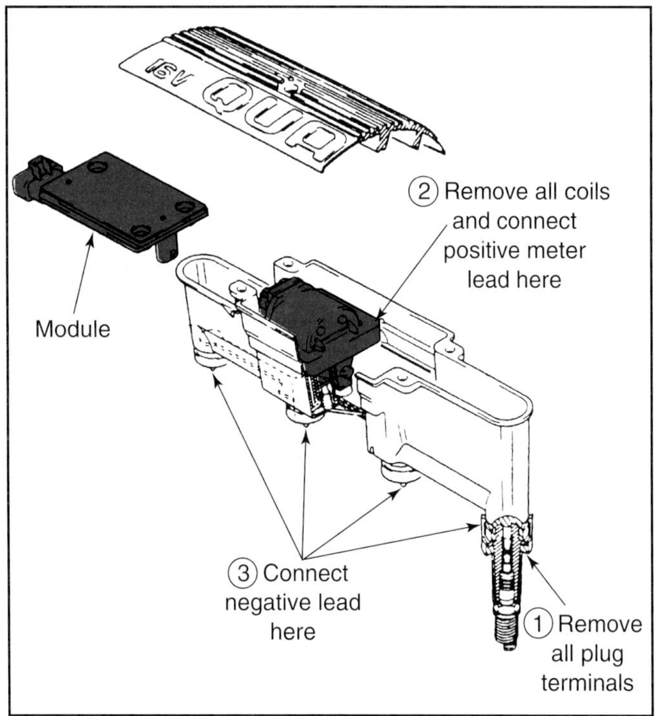

Figure 16-17. *Defects in the integrated direct ignition (IDI) coil tower will not always be visible to the eye. Remove all spark plug leads, all coils, and check each coil tower lead one at a time. (General Motors)*

Checking Spark Plugs

One of the most common checks is to directly observe the spark plug tips. Since the tips are directly involved in the combustion process, they are reliable indicators of what is going on in the cylinder. Begin by removing the plug wire from the plug to be checked. If the plug wire boot is stuck to the plug, lightly twist the boot to break it loose, or use a special boot removal tool, **Figure 16-18.** Note any obvious damage to the plug wire, such as burned insulation or damaged boots. Then install the proper size spark plug wrench over the plug, and loosen it by turning counterclockwise. If the plug is tight, it may be necessary to give the wrench handle a sharp jerk or tug to loosen it from the head.

Caution: If the cylinder head is made of aluminum, allow the engine to cool completely before removing the spark plugs. Attempting to remove the plugs from a hot aluminum head may strip the threads.

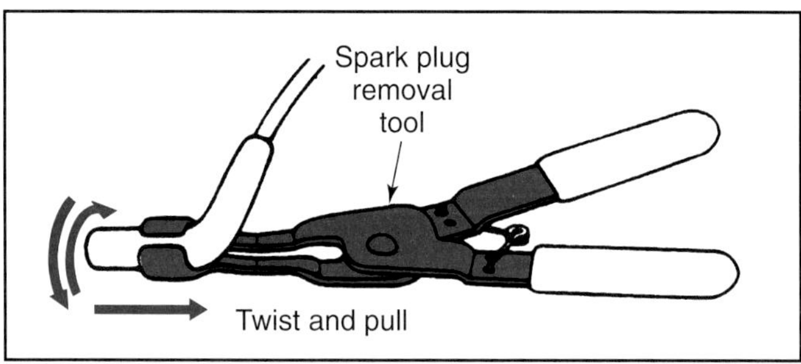

Figure 16-18. *A spark plug boot removal tool can make plug wire removal easier, especially with the limited space in most engine compartments. (Ford)*

After removing the plug, inspect the tip for unusual wear or fouling that indicates long plug usage, oil burning, or other problems. **Figure 16-19** illustrates some typical spark plug conditions, and the problems that they indicate. If the spark plugs have been used in a conventional ignition system for more than 12,000 miles (19 200 km), they are usually replaced. However, plugs with such low mileage can sometimes be cleaned, adjusted, and reinstalled in the engine. Spark plugs used in electronic ignition systems will usually run for longer periods before they need replacement. Often the plugs in an electronic ignition engine will last for 40,000-50,000 miles (64 000-80 000 km). Plugs with this much mileage are always replaced when they are removed from the engine. Some plugs are designed to last up to 100,000 miles (160 000 km).

It is sometimes helpful to check all the plugs with each other to determine certain engine problems. If you must remove all the plugs at the same time, label the wires with the corresponding cylinder number to ensure proper reinstallation. As you remove the plugs, place them on a flat

Normal Plug Appearance

A spark plug operating in a sound engine and at the correct temperature will have some deposits. The color of these deposits should range from tan to gray. The electrode gap will show growth of about .001 in. (0.025 mm) per 1000 miles (1600 km), but there should be no evidence of burning.

Fuel Fouling

Fuel fouling (dry, fluffy, black carbon deposits) can be caused by plugs that are too cold for the engine, a high fuel level in the carburetor, a stuck heat riser, a clogged air cleaner, or excessive choking. If only one or two plugs show evidence of fuel fouling, inspect the plug wires for those cylinders. Sticking valves can also cause fuel fouling.

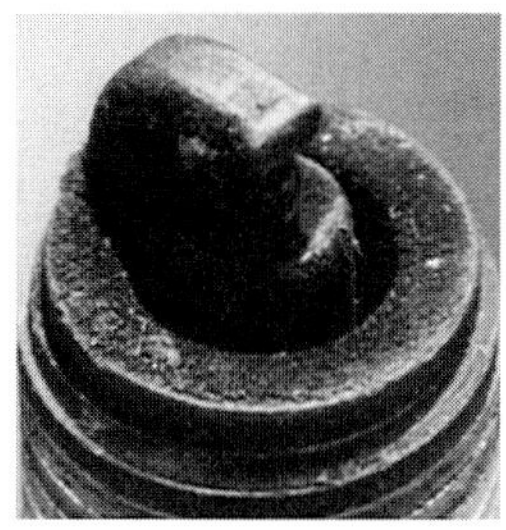

Oil Fouling

Oil fouling (wet, black deposits) is caused by an excessive amount of oil reaching the cylinders. Check for worn rings, valve guides, or valve seals. A ruptured vacuum pump diaphragm can also cause oil fouling. Switching to a hotter spark plug may temporarily relieve the symptoms, but will not correct the problem.

Splashed Fouling

Splashed fouling (plugs coated with splashes of deposits) can occur when new plugs are installed in an engine with heavy piston and combustion chamber deposits. The new plugs restore regular firing impulses and raise the operating temperature. As this occurs, accumulated engine deposits flake off and stick to the hot plug insulator.

Gap Bridging

Gap bridging (carbon-lead deposit connecting the center and ground electrodes) is not often encountered in automotive engines. Prolonged low speed operation followed by a sudden burst of high speed operation can form gap bridging. It can also be caused by excessive fuel additives.

Mechanical Damage

Mechanical damage can be caused by a foreign object in the combustion chamber. When a plug shows evidence of mechanical damage, all cylinders should be inspected. Valve overlap may allow small objects to travel from one cylinder to another.

Overheating

Overheating (dull, white insulator and eroded electrodes) can occur when the spark plugs are too hot for the engine. Cooling system problems, advanced ignition timing, detonation, sticking valves, and excessive high speed driving can also cause spark plug overheating.

Preignition

Preignition (fuel charge ignited by an overheated plug, piece of glowing carbon or hot valve edge before the spark plug fires) will cause extensive plug damage. When plugs show evidence of preignition, check the heat range of the plugs, the condition of the plug wires, and the condition of the cooling system. The engine should be checked for physical damage because it has been subjected to excessive combustion chamber pressure.

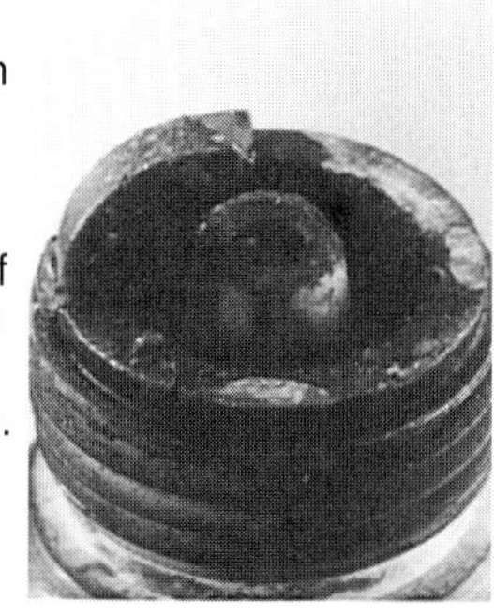

Detonation

Detonation can cause the insulator nose on a spark plug to crack and chip away. The explosion that occurs during heavy detonation creates extreme pressure in the cylinder. Detonation can be caused by an excessively lean fuel mixture, low octane fuel, advanced ignition timing, or extremely high engine temperatures.

High Speed Glazing

High speed glazing (hard, shiny, yellowish-tan, electrically conductive deposits) can be caused by a sudden increase in plug temperature during hard acceleration or loading. This condition often caused misfiring at speeds above 50 mph (81 km/h). If high speed glazing recurs, cooler plugs should be used.

Ash Fouling

Ash fouling (heavy white and yellowish deposits) is caused by the buildup of combustion deposits. The deposits may be caused by burning oil or fuel additives. Although ash fouling is not conductive, excessive deposits can cause spark plugs to misfire.

Worn Out

Extended use will cause the spark plug's center electrode to erode. When the electrode is too worn to be filed flat, the plug must be replaced. Typical symptoms of worn spark plugs include a drop in fuel economy and poor engine performance.

Figure 16-19. *The condition of the spark plugs can tell much about the operating condition of the engine. (Champion Spark Plug)*

surface in the order they came out of the engine. **Figure 16-20** shows a typical method using a board to hold the plugs. Note the condition of the plug tips. The tips should all be equally worn, and a light gray or brown color, with no signs of unusual conditions. If the plugs are in this condition, the compression, fuel, and ignition systems can be assumed to be operating properly, although some minor problems may be occurring. **Figure 16-21** illustrates some of the problems that can be determined by observing all of the plugs. Also check that the plugs are all the same make, number, and heat range.

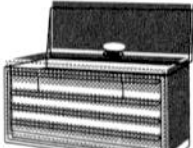

Note: Once all the plugs are out, you can proceed to check the engine compression. This procedure is outlined in Chapter 19.

Figure 16-20. *As spark plugs are removed, they should be arranged in a pattern corresponding to their location in the engine for later inspection.*

Fouled Plugs

V-8 V-6 4 Cyl.

Unbalanced mixture (multiple barrel carburetors or throttle body fuel injectors)

V-8 V-6 4 Cyl.

Plugged oil drain holes in head

V-6 6 Cyl.

Blown head gasket or cracked distributor cap

V-6 6 Cyl.

Engine not reaching normal temperature, or wrong heat range plugs

Overheated plugs

V-8 V-6 4 Cyl.

Engine cooling system problem or borderline lean mixture

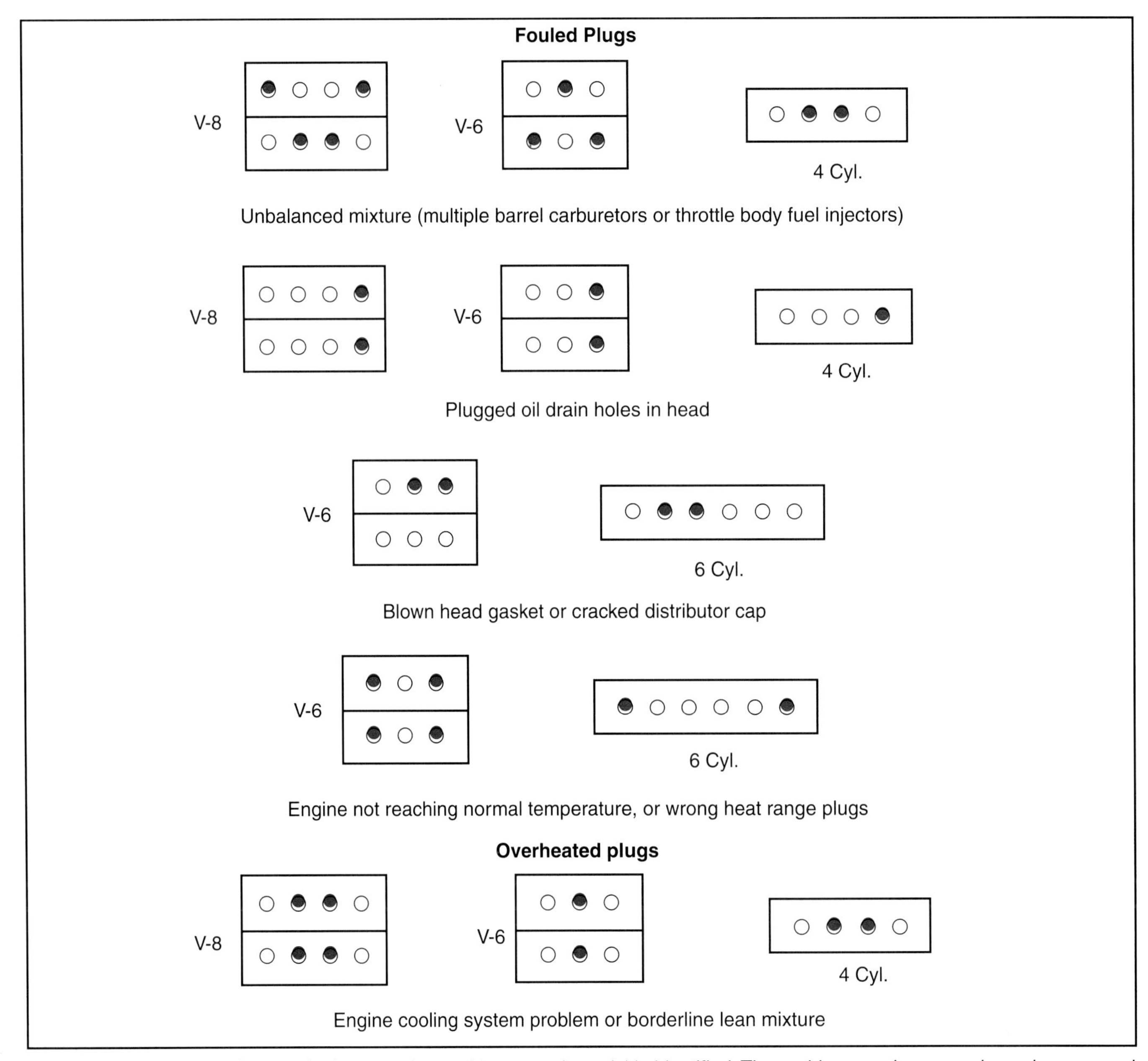

Figure 16-21. *By inspecting spark plugs, engine problems can be quickly identified. The problems and causes shown here are only a few.*

If the plugs are ok, you can either clean them and reinstall or replace them with new plugs. Make sure to check all spark plug gaps before installation. This procedure is also covered in **Chapter 21.**

Distributor, Cap, and Rotor Checks

The distributor cap, rotor, and other internal parts are common sources of trouble. Distributor checks are visual. Remove the distributor cap and check inside the cap for cracks, rust, or carbon tracks. Remove the rotor and check both sides for carbon tracks (paths of burned and carbonized material between two terminals) or rust, **Figure 16-22.** The rotor may develop a carbon track under the input (center) terminal due to high resistance plugs or wires, or long use. Some systems develop enough voltage to eventually burn a hole through the rotor, **Figure 16-23.** Engine oil inside the distributor is an indication that the distributor shaft is severely worn, although this rarely occurs.

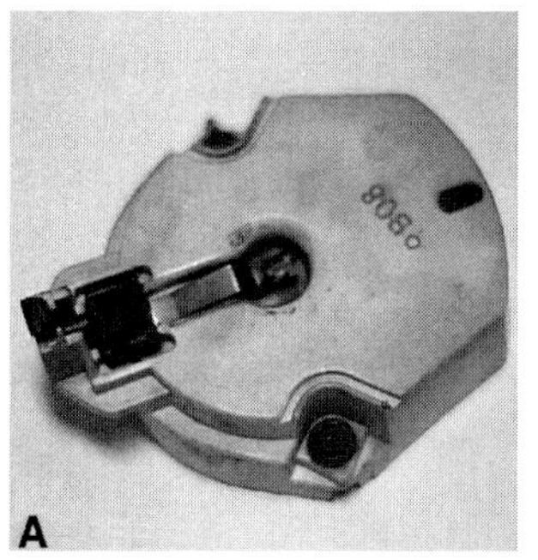

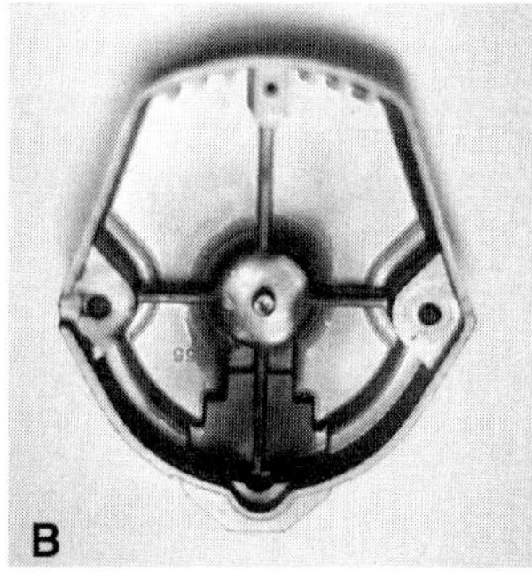

Figure 16-22. *A—The surface of the rotor should be checked for carbon tracks, wear, corrosion, and rust. B—Check under the rotor for evidence of burn through (hole burned through rotor plastic).*

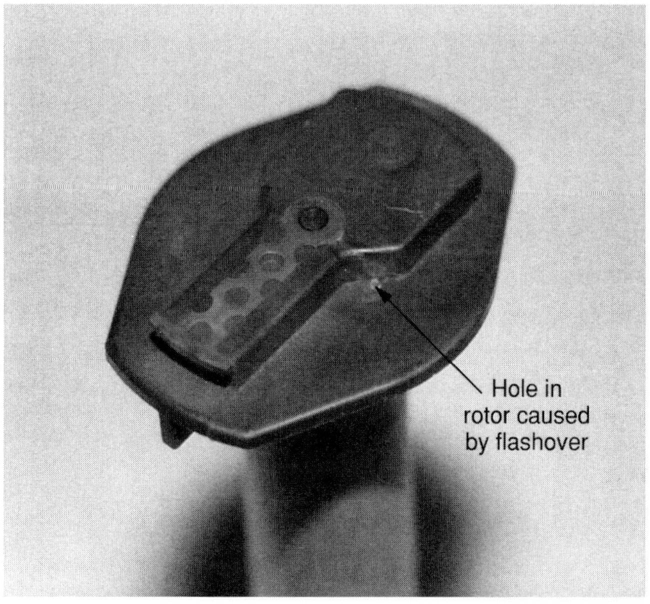

Figure 16-23. *Modern electronic ignition systems develop sufficient voltage to burn holes through the components. Lack of maintenance is the most frequent cause of this condition.*

Since the electronic ignition system produces much more voltage than the conventional system, the cap and rotor are more prone to carbon tracking. See **Figure 16-24** for typical locations for carbon tracks and other things to look for. Rust is a sign that moisture has entered the distributor, and the metal parts of the distributor should be checked for rust. Replace any distributor parts that appear to be worn or damaged.

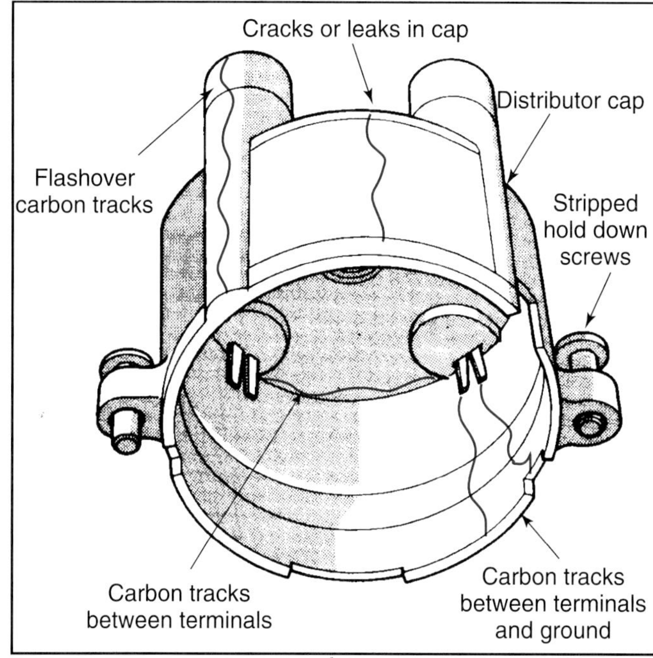

Figure 16-24. *Check the distributor cap for carbon tracks, flashover, and contamination. If any of these are present, replace the cap. (Chrysler)*

Checking Advance Mechanisms

If the distributor has vacuum and centrifugal advance mechanisms, you will need to check them for proper operation. Using a hand-held pump, apply vacuum to the vacuum advance unit, **Figure 16-25A.** Observe that the distributor plate moves freely in the opposite direction of shaft movement. If it does not move, the vacuum advance diaphragm is leaking, or the linkage is stuck.

To check the centrifugal advance, reinstall the rotor and turn it in the direction of distributor rotation, **Figure 16-25B.** It should move about one-half inch under light hand pressure, and return to its original position when released. If the rotor does not move, the centrifugal advance is stuck, and must be disassembled and cleaned.

Testing Ignition Pickup Devices

The two major types of sensor pickups are the magnetic pickup coil and the Hall-effect pickup. Refer to **Chapters 6-8,** and **15** for additional information on these devices. A digital multimeter should be used to test these devices as an analog meter cannot respond fast enough for

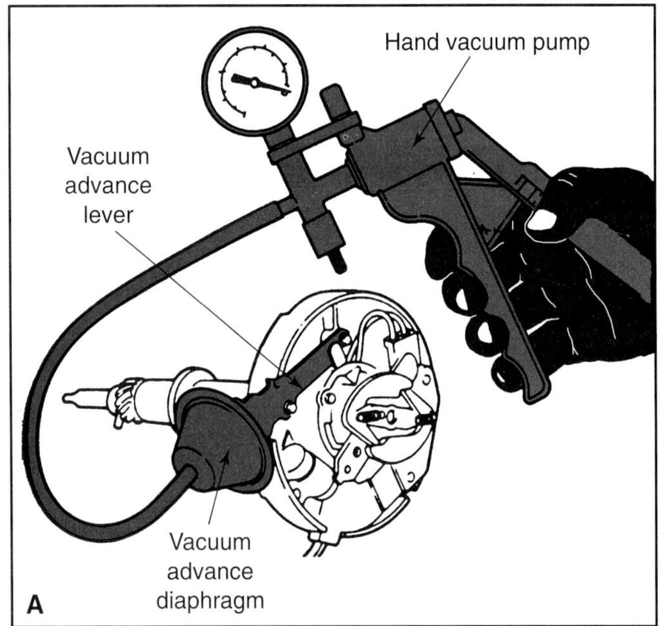

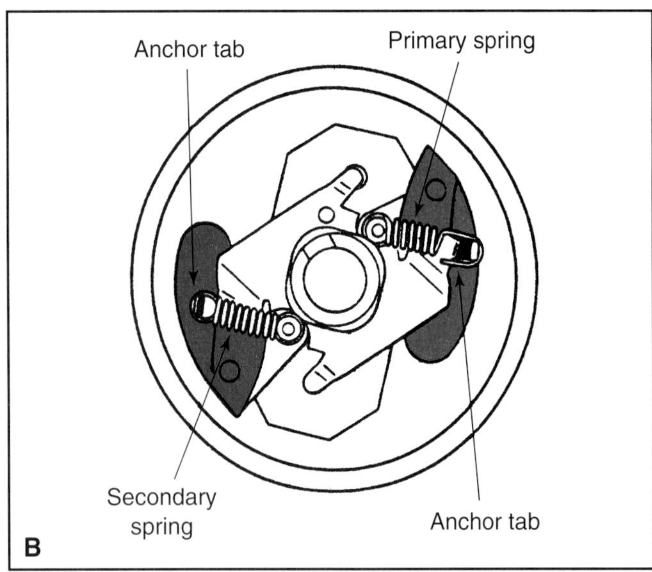

Figure 16-25. *A—Vacuum advance mechanisms can be checked by using a hand held vacuum pump. B—Centrifugal advance weights should move freely when turned. (General Motors and Toyota)*

some of the tests. Testing procedures for the pickups are similar, whether they are a distributor pickup or a position sensor mounted near the camshaft or crankshaft.

Testing a Magnetic Pickup Coil for a Sine Wave Signal

Start this test by turning the ignition key to on, but do not start the engine. Obtain a multimeter and set it to measure ac voltage. Connect the multimeter as shown in **Figure 16-26,** then disconnect the ignition system and crank the engine. Watch the multimeter while cranking. It should show about 0.1 volts ac or higher when testing distributor pickups and crankshaft position sensors that have many teeth on the trigger wheel, **Figure 16-27.**

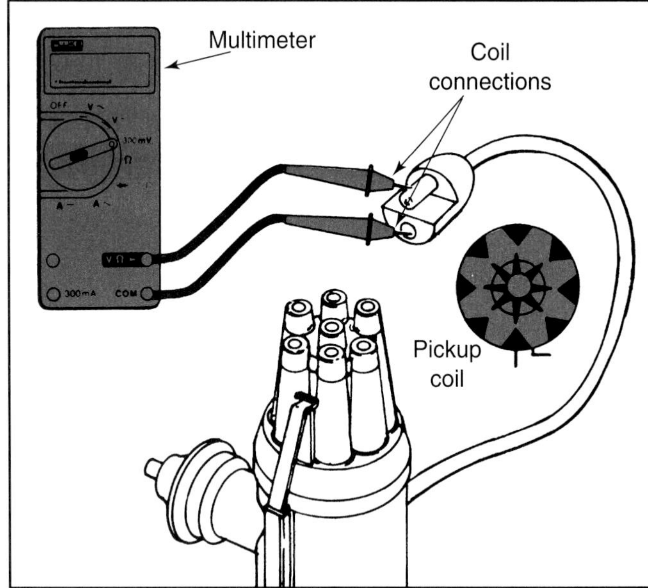

Figure 16-26. *Magnetic pickup coils produce ac voltage. A multimeter set to read alternating current should be connected directly to the pickup coil leads. (Fluke)*

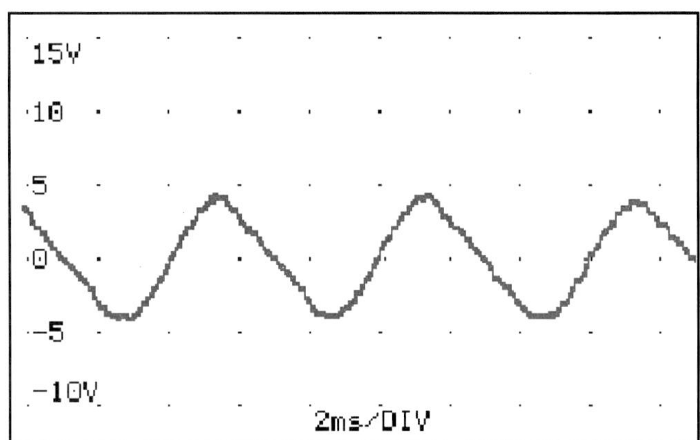

Figure 16-27. *Alternating current waveform generated by a magnetic pickup coil. (Fluke)*

On crankshaft position sensors with a small number of teeth on the trigger wheel, the reading will change slightly during cranking. It may be easier to switch the meter to the ohms scale and watch the ohms reading as the engine is cranked.

Testing a Hall-Effect Switch for a Square Wave Signal

This test should also be started by turning the ignition key to on with the engine off. Set the multimeter to measure dc volts. Disconnect the engine ignition and crank the engine. While cranking, watch for pulses each time the vanes pass the sensor. If you are using a multimeter with a duty cycle or pulse width feature, access it and note the duty cycle percentage. A good Hall-effect sensor will have a duty cycle or pulse width of about 50%, **Figure 16-28.**

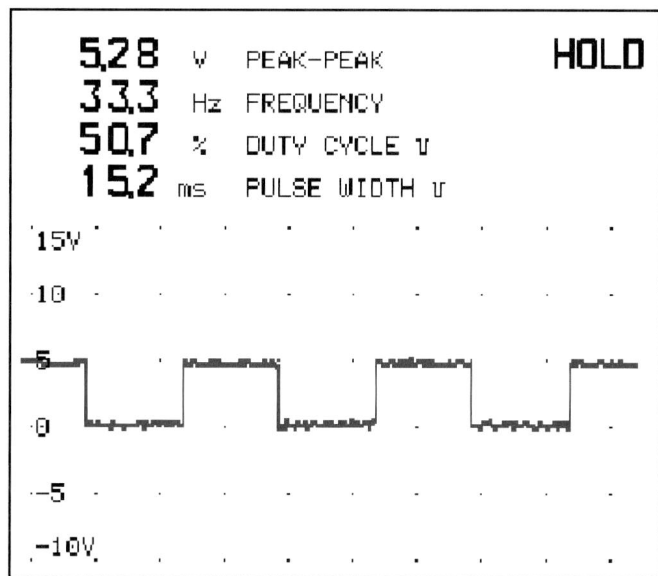

Figure 16-28. *Hall-effect switches generate dc voltage. Note the square waveform. However, the most important reading here is the duty cycle. The duty cycle of a Hall-effect switch should be about 50%. (Fluke)*

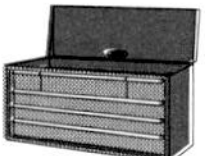

Note: Do not attempt this test with the engine running. At idle speeds or faster, the pulses occur too fast to be read.

Testing Pickup Coils and Hall-Effect Switches with an Ohmmeter

Pickup coils and some Hall-effect switches can be checked by using an ohmmeter. The pickup coil or Hall-effect switch does not have to be removed from the distributor to make this test. On some pickup coil systems, there is a procedure to adjust the air gap between the pickup coil and reluctor. Ordinarily, the air gap does not go out of adjustment, but should be checked if there is a problem with the pickup coil, or if the pickup coil is removed from the distributor, **Figure 16-29.**

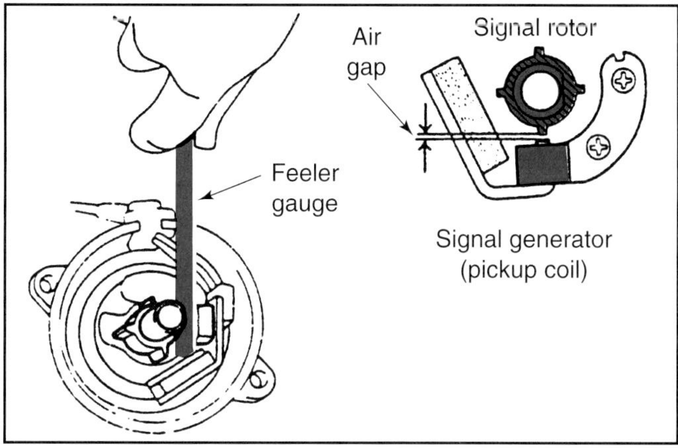

Figure 16-29. *Pickup coils have a small air gap between the rotor and generator. Check this gap with a nonferrous (brass) feeler gauge.*

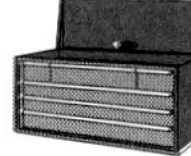

Note: Some Hall-effect switches and pickup coils can only be checked by substituting a known good unit. Consult the appropriate service manual.

Start by obtaining an ohmmeter. Then connect the ohmmeter leads on the pickup coil leads, and read the resistance. Compare the reading to the manufacturer's specifications. Resistance is usually between 300 to 1500 ohms. Next, with the ohmmeter leads still connected, wiggle the pickup coil leads. The reading should not change. If the pickup coil fails either of these tests, replace it. If the coil passes these tests, crank the engine, or turn the distributor by hand if it is removed from the engine. The ohmmeter reading should jump slightly. This will prove that the pickup coil is producing a voltage signal.

Checking Ignition Modules

Often, the process of eliminating other ignition parts will lead you to check this part of the electronic ignition system. While it is part of the ignition system, the ignition module is also an input sensor and output device of the ECM. While voltmeters, ohmmeters, and specialized tools can be used to test an ignition module, test procedures vary so greatly by manufacturer that describing a generalized procedure here is impossible. However, a few generalized tests can be described that can help isolate the ignition module.

Operate the engine or heat the module to its normal operating temperature. If the module can be reached while the engine is operating, tap lightly on the module with a screwdriver handle. If engine operation changes abruptly or stops, there is an intermittent connection in the module. If a coil in a distributorless ignition system is not firing, check the ignition module by removing the coil or connection and connecting a test light between the exposed module terminals. If the test light pulses when the engine is cranked, the module is good.

If the ignition module is separate from the coil, disconnect the coil's negative (ground) lead and place a test light in series with the negative wire and the battery's positive terminal. If the light pulses when the engine is cranked, the module is good. Sometimes the only way to verify an ignition module problem is to substitute a known good module.

Checking the Radio Capacitor

The radio capacitor on an electronic ignition system rarely becomes defective. However, it can prevent the engine from starting if it goes bad. Capacitors can be checked with an ohmmeter, as shown in **Figure 16-30.** Remove the lead from the ignition module and place one ohmmeter lead on the capacitor lead. Then place the other lead on the capacitor body. The ohmmeter reading should briefly jump, then return to infinity. If the ohmmeter reading

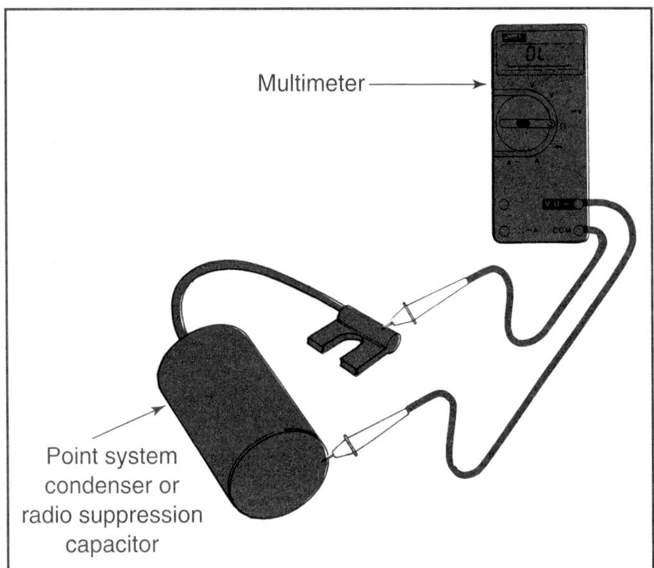

Figure 16-30. *Radio capacitors are rarely defective, but should be checked with an ohmmeter whenever the distributor or pickup coil is removed. (Fluke)*

does not jump, replace the capacitor. This test was originally used to check contact point system condensers.

Ignition System Service

Once the problem with the ignition system has been isolated, the process of repair begins. Most ignition system repairs are fairly simple. However, some repairs, such as distributor pickup coils, require complete distributor removal from the vehicle. Depending on the distributor's location, removal can be difficult. Crankshaft and camshaft sensors may require the removal of one or more engine components just to gain access. Replacement procedures for each part are explained in the following sections.

Setting Engine Timing

This procedure should be performed during the timing check if the engine timing is off. Be sure to follow all manufacturer's instructions before setting timing. Once all preliminary steps have been made, shut off the engine and loosen the distributor hold-down bolt. Restart the engine and turn the distributor while monitoring timing until the correct timing is reached, **Figure 16-31.**

Note: The timing on many computer-controlled ignition systems cannot be adjusted, including many systems with distributors.

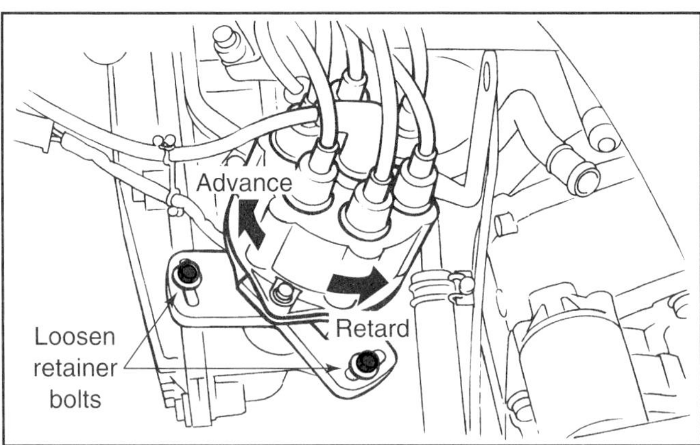

Figure 16-31. *Loosen the distributor hold-down bolts and turn the distributor in the direction needed to correct timing. Be sure not to loosen the bolts too much as they may come out and become lost in the engine compartment. (Subaru)*

Recheck the timing after adjustment. The timing, both base and advance, should be within specifications. If the timing is off, recheck all connection and grounds, and reset timing again. If the timing is still off after a second adjustment, check for problems in the advance devices. An engine that will not time properly may also have a stretched timing chain or loose timing belt.

Replacing Ignition Coils

Ignition coils are among the easiest devices to replace. The external round coil is usually held by a round bracket tightened by a single screw, **Figure 16-32.** Removing this screw and the primary and secondary electrical connectors allows the coil to be removed easily. External flat coils are also held by a bracket that can be loosened to remove the coil.

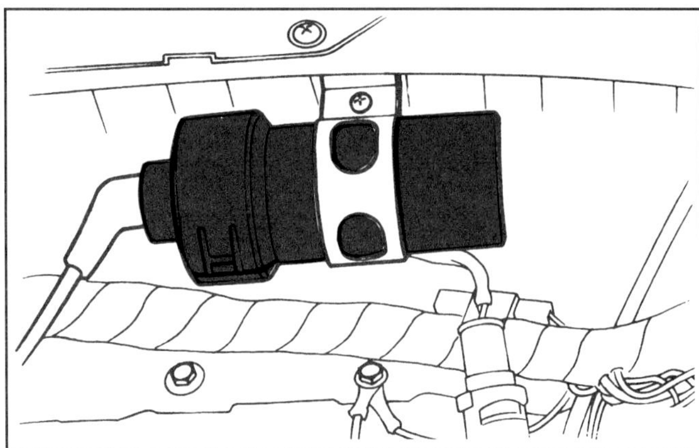

Figure 16-32. *External round coils and some remote square coils are held to the engine or vehicle body by one or more screws. (Subaru)*

Replace the external ignition coil by reinstalling it in the bracket and tightening the screw(s). Then reconnect the primary and secondary electrical leads, and installation is finished. If the coil does not have a keyed harness connector, make sure that the coil polarity is correct. The wire from the ignition switch should be attached to the coil plus terminal, and the wiring to the ignition module should be connected to the coil negative terminal. Modern block style coils are serviced in a similar fashion to the round coil.

Coils used with distributorless or direct ignition systems are held to the ignition module by one or more screws. Some coil assemblies are located in an inaccessible place which may require the removal of the ignition module and other components for coil service. Remove the spark plug wires and unbolt the coil. Using a rocking motion, carefully remove the coil from the module. Some coils have wire leads from the module. Carefully note the orientation of each wire and remove the coil. Coil installation is the reverse of this procedure, **Figure 16-33.**

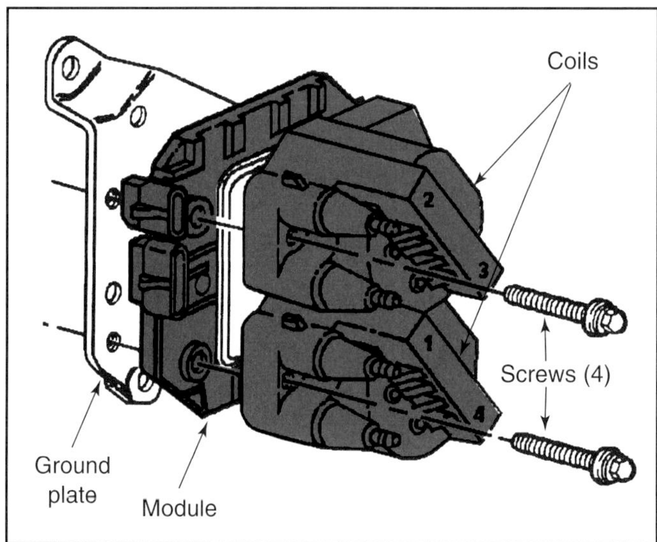

Figure 16-33. *Distributorless ignition coils are held by two or more screws. Note the ignition module installed under the coils. (General Motors)*

Coils Installed Inside the Distributor Cap

Coils that are installed in the distributor cap can be removed by unlatching the coil retainer on the top of the cap and removing the electrical connector, **Figure 16-34.** Unscrew the two or four screws holding the coil to the cap body, and pull the coil straight up. Ensure that the cap button and seal in the cap remain in place. Some replacement coil packages contain these parts.

Install a distributor cap mounted ignition coil by first aligning the coil and the primary electrical leads in the cap recess. Then reconnect the electrical leads and coil attaching screws. On coil-in-cap models, the ground wire should be reattached to one of the coil attaching screws. Then place the coil retainer over the coil and push gently until the retainer latches snap into place. Reconnect the electrical connector to the distributor cap.

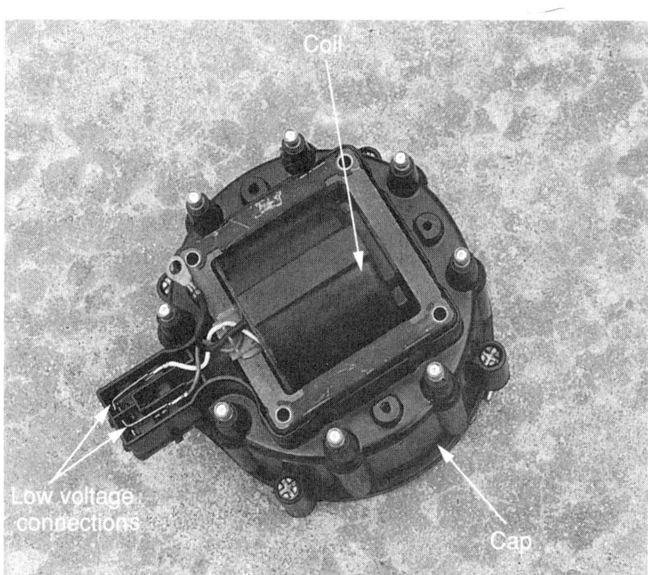

Figure 16-34. *Many electronic distributors have the coil installed inside the cap. Be careful handling the low voltage connections when replacing the coil or distributor cap.*

Ignition Modules

Ignition modules are installed in the distributor, on the distributor body, on or under the coil, or in the engine compartment, **Figure 16-35.** For all types, the removal process is basically the same. Before removing the module, ensure that the ignition switch is turned to the off position to reduce the chance of electrical surges. Remove the distributor cap on models having the module in the distributor. Then unplug the electrical connectors to the module, and remove the attaching screws.

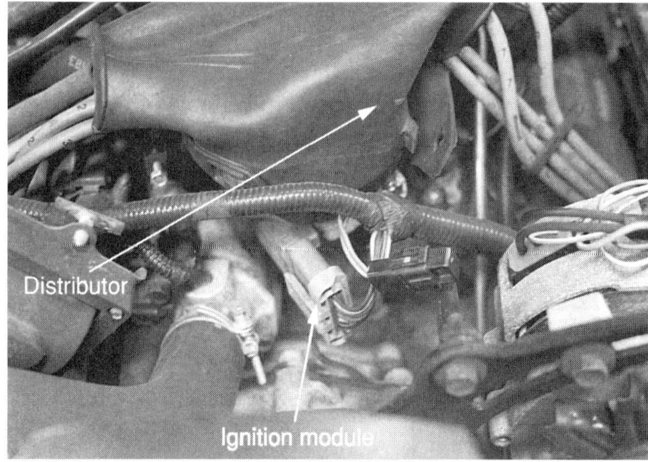

Figure 16-35. *Ignition modules can be installed in many places. This one is bolted to the side of the distributor.*

Before installing the new module, especially those placed on or in the distributor, find out whether dielectric grease should be spread between the module and the mounting surface to dissipate heat. To install the module, place it in position, and install and lightly tighten the attaching screws. Then replace the electrical connectors and distributor cap if necessary.

On distributorless systems, the module is often located under the coils. Loosen, relocate or remove any wiring, hoses, or engine components that prevent module removal. Remove the coils and disconnect the wiring harness from the module. In some cases, it may be faster to remove the coils and module as a unit. Remove the module attaching screws and remove the module from the vehicle. Distributorless ignition modules do not need dielectric grease in most cases. Be sure to reinstall the coils and ignition wires in the right location.

Distributor Advance Mechanisms

Although the ignition timing of most modern vehicles is controlled by the ECM, a few newer, and many older engines have mechanical and vacuum advance devices. Replacement of the mechanical advance may require that the distributor be removed from the engine and disassembled. Removing other mechanical advance units, and all vacuum advance units, requires partial disassembly of the distributor.

Mechanical Advance

To replace the mechanical advance on many distributors, the distributor breaker plate must be removed from the distributor. This is the plate that holds the pickup coil. A few distributors have the mechanical weights on top of the distributor shaft, where they can be reached after removing the rotor.

Once the weights are visible, remove the return springs and lift the weights from the pivot pins, **Figure 16-36.** Carefully check all parts for wear or corrosion. If the pivot pins or any of the parts which attach to the distributor shaft are worn, the distributor should be replaced.

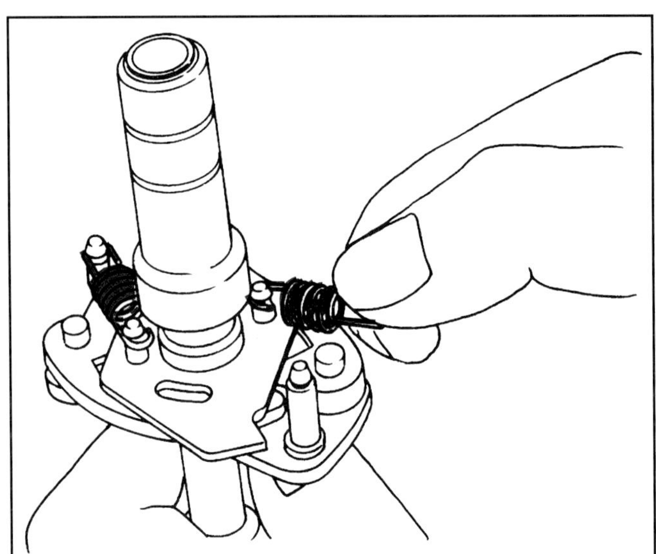

Figure 16-36. *Mechanical advance mechanisms are easy to replace. Remove the springs, then the weights. Do not drop the springs or the weights inside the distributor. (Subaru)*

To reinstall the weights, lightly lubricate them with a light film of distributor cam grease. Place the weights over the pivot pins and reinstall the springs. Check the mechanism for free movement, and make sure that the springs return the weights to the resting position. Then reassemble the rest of the distributor as necessary. If the breaker plate was removed, check the pickup coil air gap if adjustable. Start the engine and check advance operation.

Vacuum Advance

To replace the vacuum advance unit, remove the distributor vacuum line and distributor cap, and locate the vacuum advance unit attaching screws. Remove the screws and slightly twist the vacuum unit to disengage the link rod from the distributor breaker plate. Some rods are attached to the plate with a small clip, which must be removed before removing the rod, **Figure 16-37.**

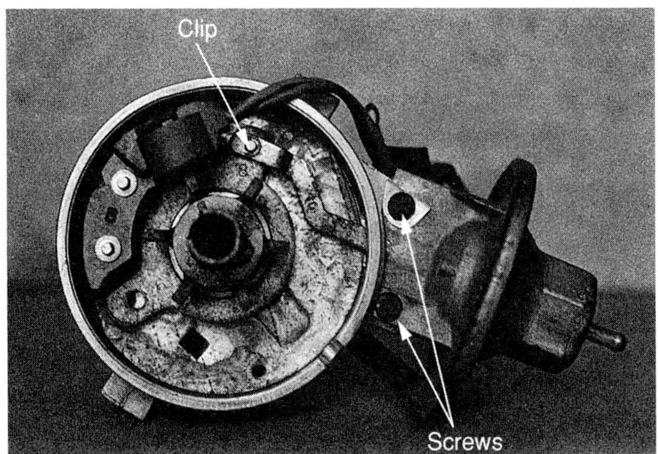

Figure 16-37. *Vacuum advance units are held to the distributor by two screws and a clip. Removal is easy in most cases and can be done with the distributor mounted on the engine.*

Place the new vacuum advance unit in the distributor and guide the link rod into engagement with the plate. Install the rod clip, if necessary. Then reinstall the vacuum advance attaching screws, and check the vacuum advance for free movement with a portable vacuum source. Reinstall the distributor cap and vacuum hose and check engine operation.

Distributor Pickups

The distributor pickup is usually attached to the distributor plate with one or two screws. Some distributors, such as the one shown in **Figure 16-38,** have a circular pickup coil that completely surrounds the distributor shaft. To replace most pickups, the distributor must be removed from the engine and disassembled. Refer to the distributor removal and disassembly section later in this chapter. On some pickups, start by removing the distributor cap and rotor, and removing the attaching screws holding the pickup in place. Then disconnect the electrical connector and lift the pickup

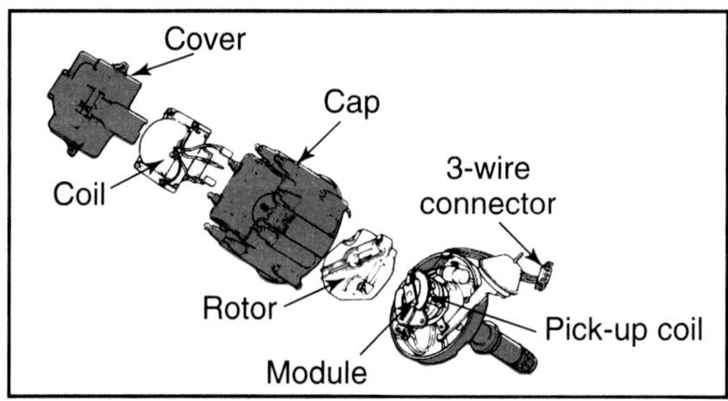

Figure 16-38. *The distributor pickup surrounds the distributor shaft in most cases. This design makes it necessary to remove the distributor to replace most pickup coils. (General Motors)*

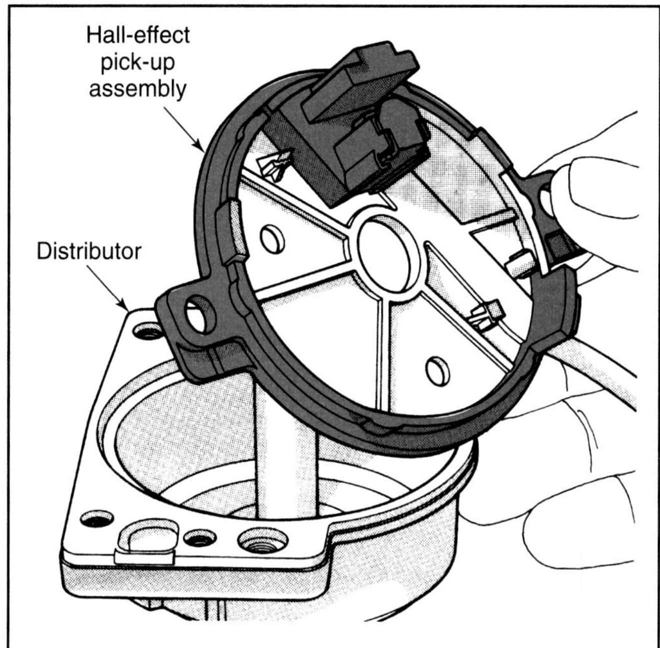

Figure 16-39. *A few pickup assemblies can be lifted out of the distributor. However, almost all pickups are held by one or more screws. (Chrysler)*

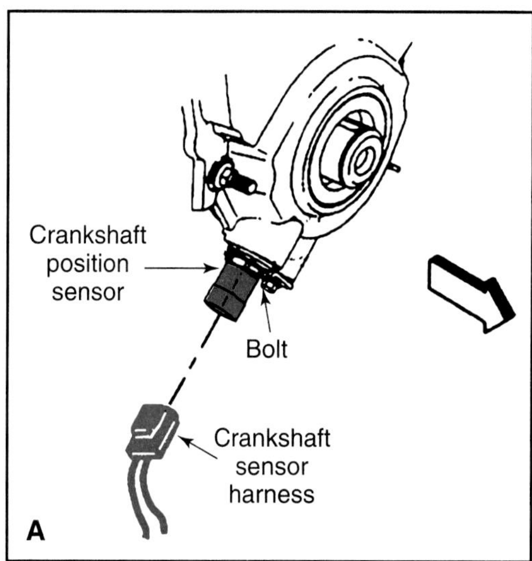

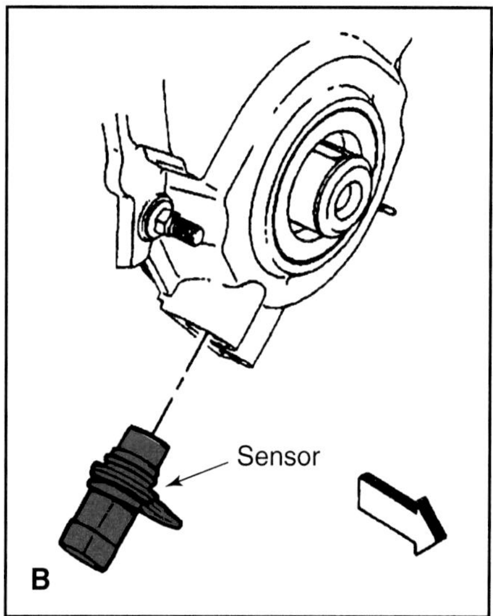

Figure 16-40. *A—Remove the harness connector before loosening the crankshaft sensor bolts. B—Crankshaft sensors are held by one or more bolts. (General Motors)*

from the distributor, **Figure 16-39.** Compare the old pickup with the replacement to ensure that it is correct.

Place the new pickup in position, then install and tighten the attaching screws. If the pickup air gap must be adjusted, follow the procedures in the service manual. Then reattach the electrical connector and reinstall the distributor rotor and cap. Crankshaft and camshaft position sensor service on some vehicles is similar, **Figure 16-40.** Service procedures for these sensors was discussed in **Chapter 15.**

Replacing Distributors

Distributors used in newer vehicles are usually not rebuilt. It is much easier to replace the old distributor with a new or rebuilt unit. However, distributors in good condition are often partially disassembled to replace internal parts, mostly in the primary ignition system. Examples of ignition parts installed in or on the distributor include reluctors, pickup coils or Hall-effect switches, and ignition modules. Some of these parts can be replaced without removing the distributor from the engine. In other cases, the distributor must be removed from the engine and disassembled for replacement. Sometimes the distributor is installed on the engine in an inaccessible location, or must be removed to gain access to other parts.

Distributor Removal

Before removing the distributor, disconnect the negative battery cable to prevent accidentally cranking the engine with the distributor removed. Then remove the distributor cap and carefully note the position of the rotor, **Figure 16-41.** Find a spot on the engine that the rotor points to, and mark this spot. Also note the position of the distributor housing on the engine. This will simplify proper ignition timing during reinstallation.

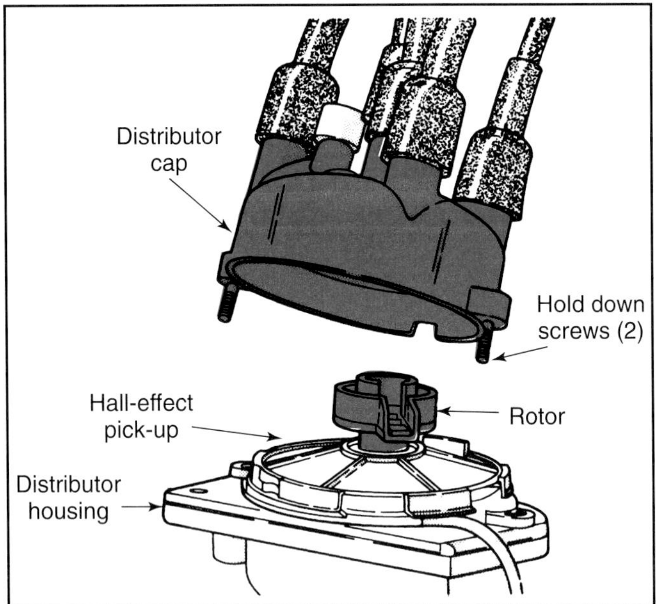

Figure 16-41. *The first step in distributor removal is to remove the cap. It may be necessary to remove one or more of the plug wires to remove the cap. Be sure to mark where the wires were connected. (Chrysler)*

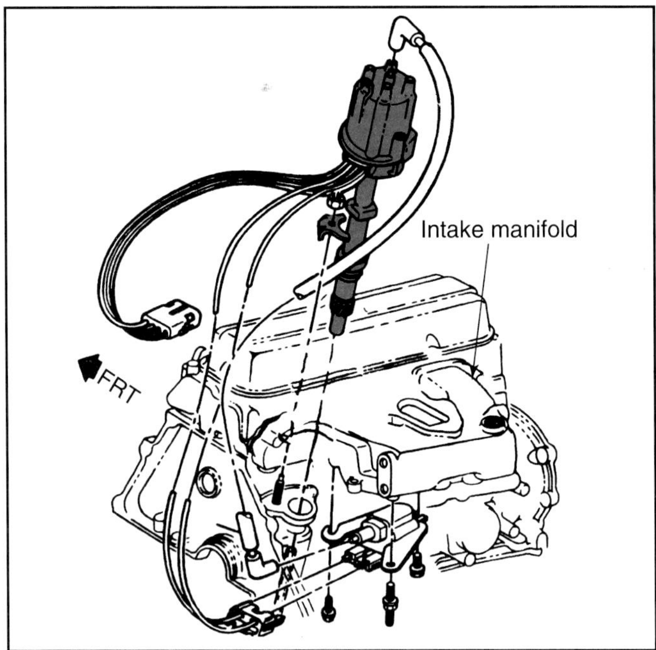

Figure 16-42. *One or two bolts normally hold the distributor to the engine. Try not to turn the distributor shaft during removal. (General Motors)*

Note: Some technicians prefer to crank the engine until it is on its number one compression stroke before removing the distributor.

Disconnect any electrical wiring to the distributor primary components, and the vacuum line if the distributor has a vacuum advance unit. Then loosen the bolt attaching the distributor hold-down clamp, and remove the bolt and clamp together. Pull the distributor away from the engine, **Figure 16-42.** You will note that on distributors with slanted (helical) drive gears, the rotor will move slightly as the distributor is pulled from the engine.

Try to pull the distributor out slowly, since many distributors also contain the oil pump drive rod. If the distributor is quickly pulled from the engine, the rod may be pulled out of engagement with the pump. If the distributor sticks in the engine, twist the distributor body lightly to loosen it. A light tap with a soft face hammer will also help loosen the distributor.

Distributor Disassembly

Remove the rotor, **Figure 16-43,** and place the distributor on a clean bench. Before proceeding further, check that the shaft and body are not damaged, and that the shaft bushings are not worn. If the distributor shaft must be removed, the drive gear must be removed first. Almost all distributors have a small roll pin holding the drive gear to the shaft. Note that some drive gears can be installed incorrectly. Temporarily reinstall the rotor and note the position of the gear against the rotor before removing the roll pin.

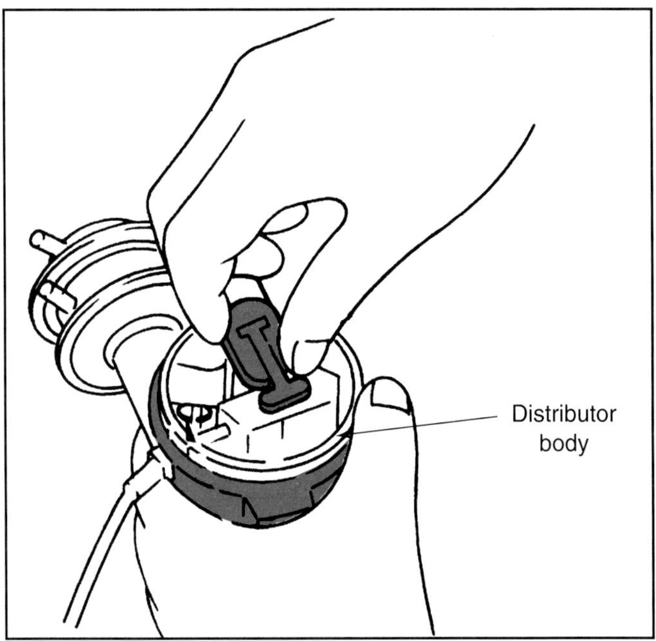

Figure 16-43. *The next step in distributor overhaul is to remove the rotor. Check the position of the rotor against the drive gear before removal. (Subaru)*

The roll pin can usually be removed by driving it through the shaft and gear, **Figure l6-44.** Lightly clamp the distributor in a vise with the roll pin facing up. Use a small pin punch and a brass hammer to avoid damage to the distributor parts. If the roll pin has a head on one end, drive the pin out from the other direction. Once the pin is out, pull the drive gear from the shaft, and then pull the shaft from the

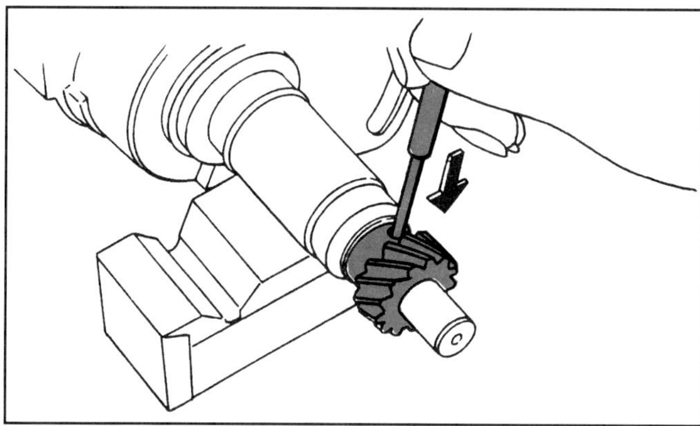

Figure 16-44. *Removing the roll pin will allow the shaft to be pulled from the housing. Exercise care in removing the pin; do not damage the drive gear. (Subaru)*

distributor body, **Figure 16-45.** Engine sludge often builds up on the drive gear end of the distributor, and light tapping may be necessary to remove the gear and shaft. Remove the ignition module, pickup coil, and all other distributor parts that will be replaced.

Note: It is a good ideas to replace the pickup coil if the distributor shaft is removed, even if the pickup coil is not defective.

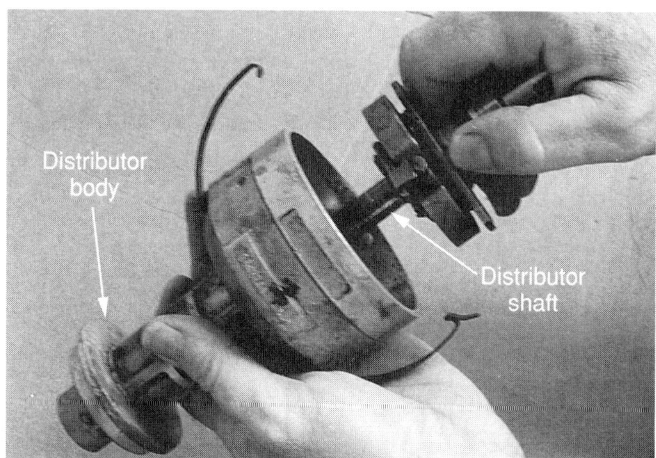

Figure 16-45. *After gear removal, pull the distributor shaft from the housing. This will permit replacement of the pickup coil, seals, and other distributor internal components.*

Check all the distributor internal parts, such as those shown in **Figure 16-46,** for wear, damage, or evidence of corrosion. On older distributors with vacuum or centrifugal advance mechanisms, check the mechanical advance weights for wear or sticking, and check the vacuum advance for a leaking diaphragm. If any parts are worn or damaged, they should be replaced. If there is any evidence of corrosion, water is entering the distributor and the distributor cap should be replaced. If oil is present in the distributor, the shaft body and shaft bushings may be worn and the distributor should be replaced.

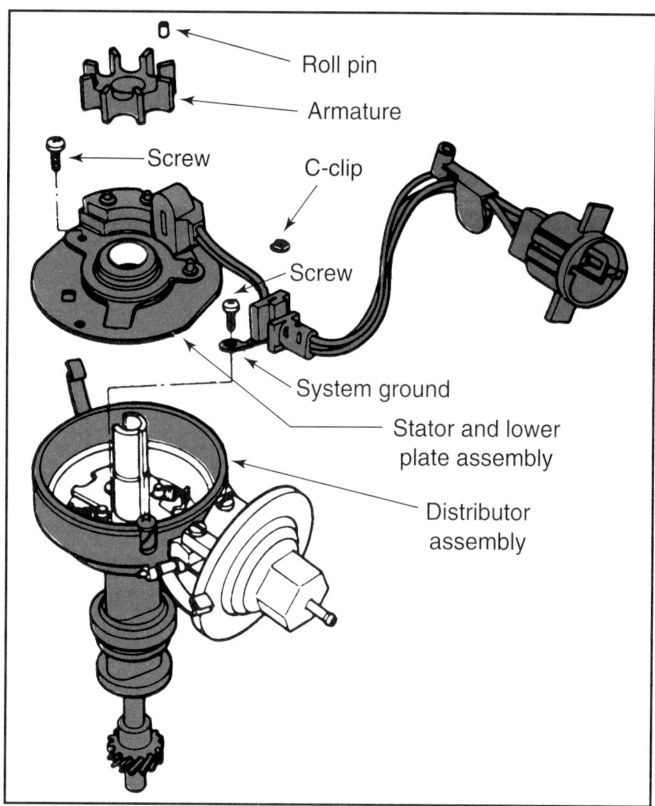

Figure 16-46. *Exploded view of an electronic ignition system distributor. The components shown here are typical of most computer-controlled distributors. (Ford)*

Clean all parts thoroughly in a mild solvent, and dry them with compressed air. Once the distributor housing is dry, lubricate the shaft bearings (and mechanical advance if used) with small amounts of the proper kind of lubricant. Check the manufacturer's manual to determine what kind of lubricant is recommended.

Distributor Assembly

Install new parts as necessary. Use a new roll pin to reinstall the drive gear on the shaft, **Figure 16-47.** If the

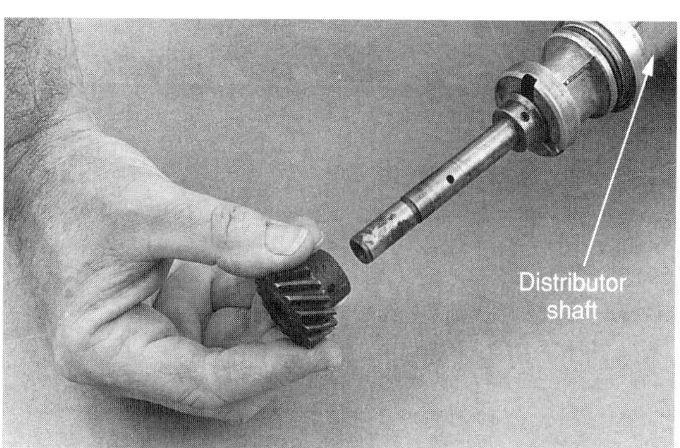

Figure 16-47. *Be sure to align the distributor rotor with the gear before installing the roll pin.*

distributor has a pickup coil with an adjustable air gap, adjust it before reinstalling the distributor in the engine. If the distributor uses a Hall-effect switch, turn the shaft a few times to ensure that there is enough clearance between the shutter and switch. If there is any contact, adjust as necessary, then install the rotor.

Before reinstalling the distributor, make sure that any distributor-to-engine gaskets or seals are in place. Align the rotor with the mark that was made on the engine. On distributors with slanted drive gears, the rotor will move slightly as the distributor gear meshes with the drive gear. Compensate for this before installation by turning the rotor slightly in the opposite direction.

Align the distributor body in its original position, then lightly push the distributor into the engine. Remember that on many engines, you are also engaging the oil pump drive rod as you engage the drive gears. Ensure that the rotor is in the proper position when the body is fully seated. If the body will not fully seat or the rotor is not in the proper position, remove the distributor, check for interference or misalignment, and try again.

Once the distributor is fully seated, reinstall the hold-down clamp and bolt, but do not tighten the bolt. Reinstall the primary wiring and vacuum advance line if necessary. Reinstall the distributor cap and the battery cable, then start the engine and allow it to warm up. Check and set timing according to the procedures in the manufacturer's service manual. These procedures were also outlined earlier in this chapter.

Replacing Diesel Engine Glow Plugs

Diesel engines will miss when cold when a glow plug fails to produce heat. The defective glow plug can be easily replaced. A press-in glow plug can be replaced by removing the electrical connector and unbolting the retainer clip. The glow plug can then be pulled out of the head. If the glow plug is threaded into the head, remove the electrical connector and unscrew the plug with the proper size socket and wrench.

The tip of the glow plug may break off and fall into the combustion chamber as the plug is removed. The tip must be removed to prevent engine damage. Disable the fuel injection pump and remove the fuel injector on the affected cylinder and crank the engine. The broken tip will be blown out of the injector or glow plug hole.

Caution: Wear eye protection and stand away from the engine compartment during this operation, until the engine is turned off.

Check the replacement glow plug to ensure that it is the correct part. To install the replacement plug, thread or press it into the head. Then either tighten the injector to the proper torque, or install the attaching clip. Install the electrical connector, start the engine, and check glow plug operation. Bleed the fuel system if necessary.

Follow-Up for Ignition System Repairs

After repairs are complete, clear any ECM diagnostic trouble codes that may have been set during the diagnosis and repair steps. Make sure all spark plug wires are correctly routed to prevent crossfiring or electromagnetic interference. Road test the vehicle a sufficient distance to ensure that there are no ignition or other driveability problems. It is sometimes a good idea to recheck ignition timing after the vehicle has been road tested. If the ignition problem was ongoing, it is a good idea to replace the spark plugs as part of a maintenance tune-up. This is covered in **Chapter 21.**

Summary

Before investigating the ignition system, make sure that the problem is not a normal condition, and that it is caused by the ignition system. You must have access to the proper service manual to check the ignition system. Begin by making a visual inspection for any problems. Check for a spark and for any discharge from the coil(s), cap and rotor, and wires. To isolate a misfire, a scan tool should be attached to the vehicle. After the scan tool has identified the misfiring cylinder, steps can be taken to isolate the cause. On older vehicles, the scan tool may not be able to detect which cylinder is misfiring, but will provide other information which will aid in diagnosis.

Two main tools for ignition testing are the oscilloscope and the timing light. Oscilloscope patterns can be read and used to determine the condition of the primary and secondary systems. The timing light is used to set ignition timing and determine whether the advance mechanisms are working. Many direct ignition systems have no provision for checking timing.

Ignition coils can be tested by bypassing the rest of the ignition system, or with the use of an ohmmeter. Spark plug wires can also be checked with an ohmmeter. Spark plug problems can be checked by observation of their firing tips. The distributor should be checked for carbon tracks and signs of moisture and oil. Mechanical and vacuum advance mechanisms, when used, should be checked for proper operation. Ignition pickups can be tested by observing whether they can produce a signal when the engine is cranked. Some pickups can be tested with an ohmmeter, while others can only be tested by substituting a known good unit. The ignition module should be tested according to manufacturer's procedures.

Ignition timing is usually set by using a timing light. The distributor hold-down bolt is loosened and the distributor turned until the timing is correct. Ignition coils are easily replaced by removing the electrical connections and brackets. The removal process for all ignition modules is similar. Replacing the mechanical or vacuum advance devices often requires removal and partial disassembly of the distributor. Pickup coils are installed in the distributor or on the engine block. Removal is usually simple, but the pickup may require adjustment.

On distributor equipped vehicles, the distributor can be removed after the position of the rotor is noted and the hold down clamp removed. Many distributors can be successfully overhauled. A few distributors are replaced as a unit. The distributor should be installed with the rotor in exactly the same position as it was originally.

Always follow up ignition system repairs by clearing trouble codes and road testing the vehicle. A timing light can be used to check the ignition timing, and the timing advance systems. The oscilloscope will determine the condition of the ignition system components, and can also be used to check the operation of the alternator.

Know These Terms

Carbon tracks	Intermediate section
Corona discharge	Dwell section
Spark tester	Primary pattern
Power balance test	Timing light
Misfire monitor	Advance meter
Oscilloscope	Timing testers
Secondary pattern	Initial timing
Firing section	Total advance
Spark line	Flashover

Review Questions—Chapter 16

Please do not write in this text. Write your answers on a separate sheet of paper.

1. What effect does corona discharge have on engine operation?
2. Name two tools that can be used to check the secondary wiring for the presence of a spark.
3. Why should the technician not test for a spark on an electronic ignition system by pulling a plug wire off of a spark plug?
4. The oscilloscope secondary pattern shows what is happening on the ______ voltage side of the ignition system.
5. Ideally, ignition coils should be tested ______.
6. On many vehicles, if the ignition switch is turned on and then off, the ignition coil will ________.
7. The best type of timing light has a(n) ______.
   (A) strobe lamp
   (B) ECM data link connector
   (C) non-inductive pickup
   (D) built-in advance meter
8. Dielectric grease is often spread between an ignition ______ and its mounting surface.
9. Name three places on the vehicle where the ignition module can be installed.
10. The air gap must be adjusted after replacing the distributor ______ on some electronic ignition systems.
    (A) pickup
    (B) cap
    (C) shaft
    (D) advance

ASE Certification-Type Questions

1. Technician A says that a misfire monitor is part of OBD II computer system. Technician B says that misfire information can only be retrieved by using a scan tool. Who is right?
   (A) A only.
   (B) B only.
   (C) Both A & B.
   (D) Neither A nor B.
2. All of the following statements about timing lights are true, EXCEPT:
   (A) a timing light can be used to set initial timing.
   (B) if a timing light does not have an advance meter, timing advance can only be approximated.
   (C) a timing light can be used to check initial timing.
   (D) all timing settings are the same on newer vehicles.
3. A power balance test can be made by some ______.
   (A) multimeters
   (B) scan tools
   (C) test lights
   (D) None of the above.
4. Carbon tracks or flashover can be present on all of the following parts, EXCEPT:
   (A) distributor caps.
   (B) spark plug tips.
   (C) rotors.
   (D) coils.
5. The condition of the spark plugs can indicate the condition of ______.
   (A) the fuel system.
   (B) the ignition system.
   (C) the compression system.
   (D) All of the above.

6. A distributor has experienced repeated pickup coil failure. When the distributor cap was removed, the inside of the cap was coated with engine oil. Which of the following is the least likely cause of the pickup coil failure?
   (A) Excessive distributor shaft play.
   (B) A leaking distributor shaft seal.
   (C) Excessive wear in the distributor body.
   (D) A leaking ignition module.
7. Technician A says that an oscilloscope pattern that does not resemble the standard pattern may indicate a problem. Technician B says that the primary pattern can often detect module and pickup problems that are not shown on the secondary pattern. Who is right?
   (A) A only.
   (B) B only.
   (C) Both A & B.
   (D) Neither A nor B.
8. All of the following spark plug conditions can cause performance problems, EXCEPT:
   (A) melted spark plug electrode.
   (B) a dark ring around the spark plug insulator.
   (C) wet oil on the spark plug tip.
   (D) a crack on the spark plug insulator.
9. To test a magnetic pickup coil that produces a sine wave signal, what range should the multimeter be set to?
   (A) Ohms.
   (B) DC volts.
   (C) AC volts.
   (D) Amps.
10. The ignition timing on an engine is advancing 12° at 2000 rpm. Technician A says that the advance mechanism is overadvancing the timing. Technician B says that the specifications should be consulted to determine the proper advance curve for this engine. Who is right?
    (A) A only.
    (B) B only.
    (C) Both A & B.
    (D) Neither A nor B.
11. The air gap must be set with a feeler gauge on some ______.
    (A) distributor-mounted magnetic pickup coils
    (B) distributor-mounted Hall-effect switches
    (C) crankshaft-mounted magnetic pickup coils
    (D) All of the above.
12. Distributor removal is often required for replacement of all of the following parts, EXCEPT:
    (A) distributor cap.
    (B) pickup coil.
    (C) distributor gear.
    (D) ignition module.
13. Before removing any distributor, carefully note the position of the ______.
    (A) timing marks
    (B) pickup coil
    (C) rotor
    (D) crankshaft pulley
14. Technician A says that the engine should not be cranked when the distributor is removed from the engine. Technician B says that many distributor components can be replaced without removing the distributor from the engine. Who is right?
    (A) A only.
    (B) B only.
    (C) Both A & B.
    (D) Neither A nor B.
15. Technician A says that the vehicle ECM should be cleared of trouble codes before returning the vehicle to the owner. Technician B says that the spark plugs should be changed whenever a vehicle is brought in with an ignition problem. Who is right?
    (A) A only.
    (B) B only.
    (C) Both A & B.
    (D) Neither A nor B.

Advanced Certification Questions

The following questions require the use of reference information to find the correct answer. Most of the data and illustrations you will need is in this chapter. Diagnostic trouble codes are listed in **Appendix A.** Failure to refer to the chapter, illustrations, or **Appendix A** for information may result in an incorrect answer.

Data Scanned from Vehicle			
Coolant temperature sensor 208°F/98°C	Intake air temperature sensor 78°F/26°C	Manifold absolute pressure sensor 2.2 volts	Throttle position sensor 1.76 volts
Engine speed sensor 2640 rpm	Oxygen sensor .62 volts	Vehicle speed sensor 55 mph	Battery voltage 13.7 volts
Idle air control valve 10 percent	Evaporative emission canister solenoid On	Torque converter clutch solenoid Off	EGR valve control solenoid 18 percent
Malfunction indicator lamp Off	Diagnostic trouble codes P0329	Open/closed loop Open	Fuel pump relay On
Brake light switch Off	Cruise control Off	AC compressor clutch On	Knock signal No
Ignition timing (°BTDC)	Base timing: 6		Actual timing: 21

16. A multiport fuel injected truck has an intermittent spark knock at speeds above 30 mph (48 kmh) under light to medium acceleration. The scan tool data shown above was captured at highway speed under light acceleration as the spark knock occurred. Which of the following is the *least likely* cause?
    (A) A poor connection at the knock sensor.
    (B) A spark plug wire shorting to ground.
    (C) High resistance in the knock sensor circuit.
    (D) A defective EGR valve.

17. A throttle body injected car equipped with OBD II and distributorless ignition idles rough, hesitates, and backfires under load. The ECM has stored trouble code P0302 and P0305. All of the following can cause these problems, EXCEPT:
    (A) a defective ignition coil.
    (B) a defective ignition module.
    (C) number 2 and 5 plug wires crossfiring.
    (D) number 3 and 6 plug wires crossfiring.

18. An integrated direct ignition system has experienced repeated ignition coil tower failure. Technician A says that a leaking coil could be the cause. Technician B says that a poor ignition system primary ground could be the cause. Who is right?
    (A) A only.
    (B) B only.
    (C) Both A & B.
    (D) Neither A nor B.

19. The direct ignition system in the schematic shown in **Figure 16-2** has no electronic spark timing (EST) control. Which of the following is the *most likely* cause of the problem?
    (A) An open ECM fuse.
    (B) An open in circuit BLK/WHT 450.
    (C) An open in circuit PPL/WHT 430.
    (D) An open in circuit 423 WHT.

20. A vehicle with an integrated direct ignition system runs rough intermittently and has a flashing MIL light. The scan tool capture in **Figure 16-6** was made under light acceleration. Technician A says that the cause could be a defective ignition module. Technician B says that the coil supplying cylinders 1 and 3 is defective. Who is right?
    (A) A only.
    (B) B only.
    (C) Both A & B.
    (D) Neither A nor B.

Testing a coolant temperature sensor with a digital multimeter. (Fluke)

Chapter 17

Fuel System Diagnosis and Repair

After studying this chapter, you will be able to:

- Check the operation of fuel injection systems.
- Check for correct air-fuel ratio.
- Use scan tools and multimeters to check fuel injection components.
- Use fuel pressure testers to check electric fuel pumps.
- Check operation of air control devices.
- Adjust idle speed of fuel injection systems.
- Clean throttle bodies and fuel injectors.
- Remove, service, and replace fuel injection components.

This chapter discusses diagnostic and repair procedures for the fuel system. Like ignition system diagnosis, you will be concentrating on one system. However, since parts are located over the length of the vehicle, diagnosing and servicing the fuel system will require you to inspect the entire vehicle. Once again, you will use many of the diagnostic and repair techniques that you learned earlier in this text. Remember that the modern fuel system is very dependent on the ECM and its related sensors and output devices for proper operation. Keep this in mind while you are studying this chapter. If necessary, review **Chapter 9** on fuel system fundamentals.

Preparing to Diagnose the Fuel System

Often, a fuel system problem will cause the engine to perform poorly or prevent it from running. However, if the engine runs, make a quick visual inspection for a fuel leak. If there is no evidence of a fuel leak, perform a short road test. Do not drive far from the shop, since a fuel problem that occurs when the vehicle is warm may cause the engine to stall and not restart. Ideally, this type of testing should be done on a chassis dynamometer.

Many vehicles with fuel system problems will exhibit symptoms that are sometimes mistaken for ignition system problems, **Figure 17-1.** If you suspect that the vehicle has a fuel system related problem, start by locating the appropriate service manual. You should also locate a fuel pressure gauge, since this tool will be used early in diagnosis to identify or eliminate the fuel system as the cause of a no-start condition.

Warning: Use appropriate safety precautions when diagnosing the fuel system. Remember that some fuel systems operate under high pressure and that gasoline is extremely flammable.

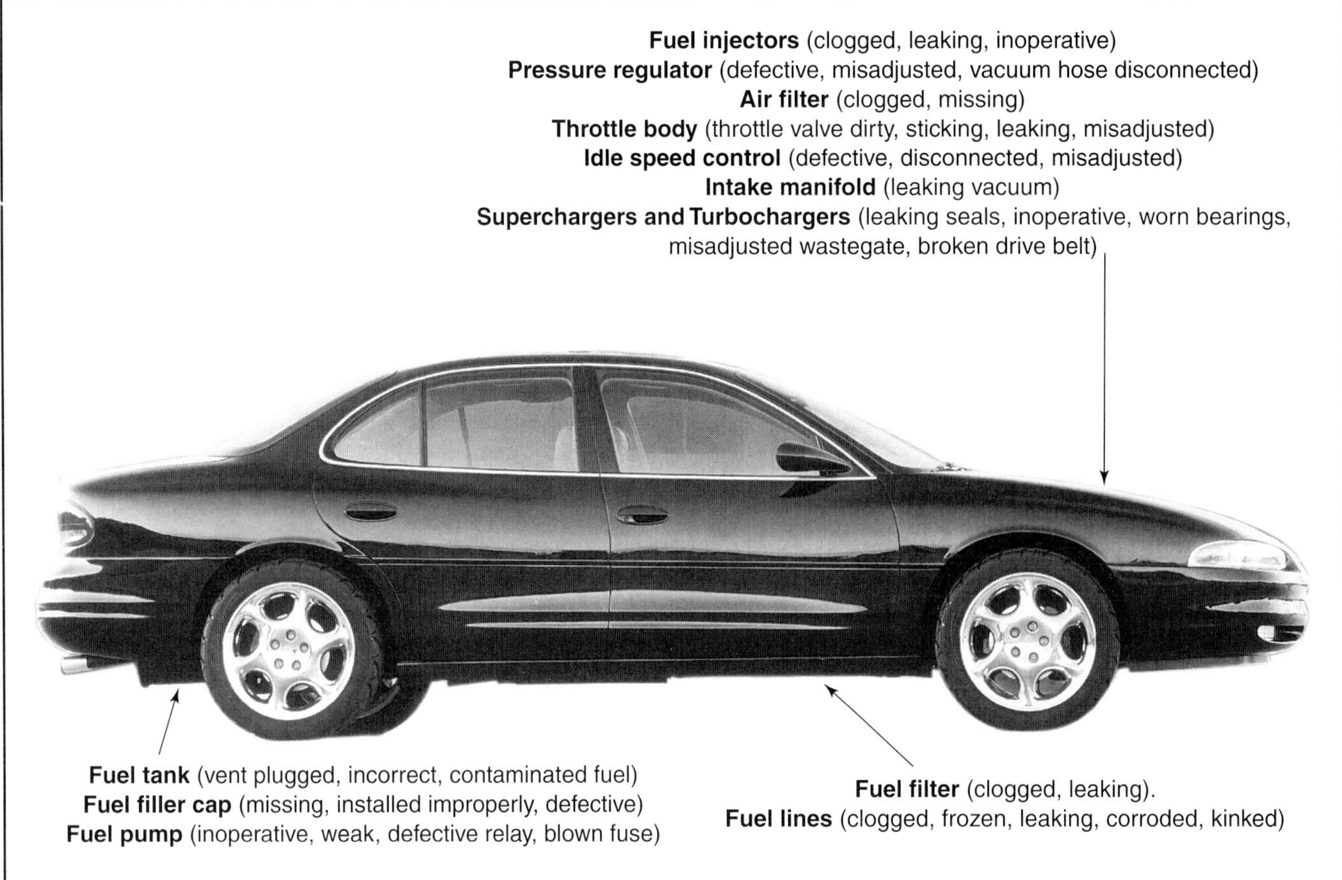

Figure 17-1. *Fuel system problems often prevent the engine from running, but can also cause economy and emissions problems. These are the most common fuel system problems. (Oldsmobile)*

Visual Inspection

Before inspecting and testing the fuel system, you should have already eliminated the electrical, computer control, and ignition systems as possible causes of the problem. If you have not, check these systems first before performing any fuel system inspection or test. Begin diagnosis by performing a visual inspection of the fuel system. **Figure 17-2** shows a typical fuel system with pressure, vapor, and vacuum lines.

Start at the engine where the fuel enters. This includes the fuel injectors, fuel rail, pressure regulator, all related sensors, output devices, and wiring. Look for leaks, loose fittings and connections, improper installation, modifications, damage, or tampering. Make sure all heat shields are in place and secure.

Next, make a visual check for vacuum leaks at the throttle body or intake manifold caused by loose bolts, damaged gaskets, or disconnected vacuum hoses. Also check any flexible ducts between the throttle body and the engine for holes or loose connections. Before checking other parts of the injection system, check the pressure regulator (if its operation was not checked as part of earlier testing). Ensure that the vacuum hose to the intake manifold, when used, is not disconnected, kinked, or split. Also make sure that the regulator is not leaking. Pull the oil dipstick and smell the oil. If you smell fuel, an injector may be stuck open, diluting the oil with fuel.

Check the idle control devices for obvious problems. The idle air control solenoid should be checked for sticking due to a buildup of deposits. If you suspect an injector problem, remove an injector, if possible, place it in a clear jar, and crank the engine to check. The injector should spray fuel in the correct pattern, and should pulse on and off as the engine cranks. If the engine has throttle body injection, remove the air cleaner and inspect the injector(s) while someone cranks the engine.

Check metal fuel lines for kinks, unusual bends (a bend where one is not required or a bend of more than 60-90°), leaks, loose fittings, or corrosion damage. Neoprene fuel lines, along with any clamps, should be checked for cracks, deterioration, loose clamps, and leaks. Most modern vehicles use plastic fuel lines and tanks, which eliminates the chance of fuel system damage by corrosion. However, plastic fuel lines are very susceptible to kinks, usually caused by careless handling. Loose connections are sometimes a problem. Check the fuel filter for damage or leaks and if it has been replaced recently. If the vehicle has a dual tank system, check the switching valve for proper operation. Follow the fuel lines back to the fuel tank. If the fuel pump is mounted outside the fuel tank, inspect it for leaks. If the weather is extremely cold (below 0°F or -18°C) the fuel lines

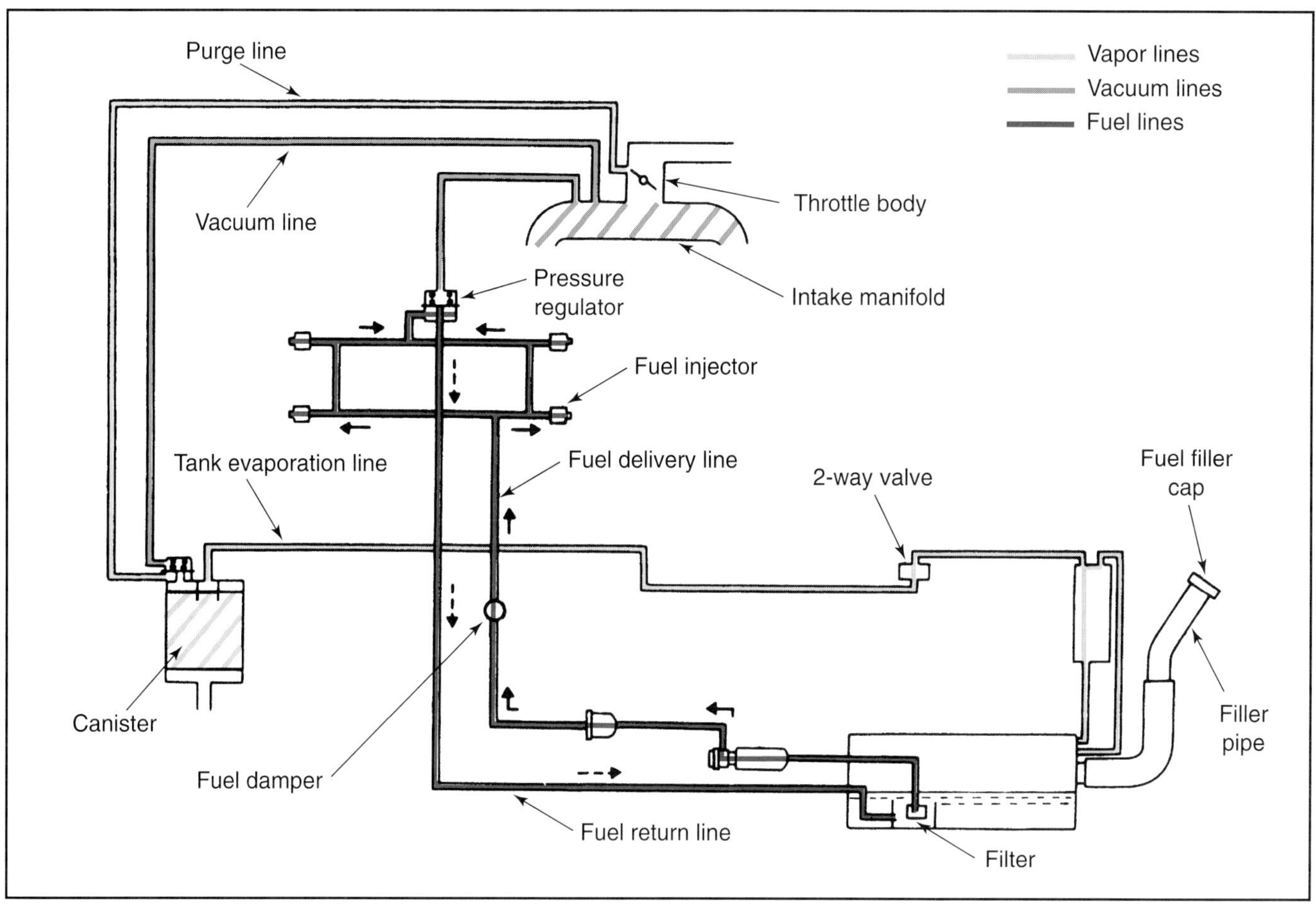

Figure 17-2. *The fuel system has lines, hoses, and components located throughout the vehicle. Parts of the system carry liquid fuel while others hold vapor or vacuum. (Subaru)*

may be frozen or have other problems. Bring the vehicle into the shop and allow it to warm up.

Check the fuel tank for physical damage and corrosion if the tank is metal. Check carefully for leaks, especially around the seams of metal tanks. Make sure that all lines, including fuel feed, fuel return, vapor, and fuel fill are installed and free of leaks or damage. All wiring for the fuel gauge sending unit and in-tank fuel pump should be connected. Make sure all fuel tank heat and rock shields, if used, are in place.

Fuel Pressure Test

Of all of the fuel system's many parts, the fuel pump is the most frequent cause of a fuel system problem. If the driveability problem seems to be related to an insufficient supply of fuel, such as losing power, hesitation or stalling on hard acceleration, stalling while cruising, or unexplained engine stall at various speeds, the fuel pump may be defective. The following tests will determine if the pump can deliver enough fuel at the proper pressure.

Checking Fuel System Pressure

Low pressure in the fuel injection system can cause various driveability problems, including poor cold performance, rough idle or stalling, hesitation, and surging at cruising speeds. Fuel pressure should be checked whenever a fuel injected engine has a driveability problem. Electric fuel pumps are used on fuel injection vehicles.

Begin by locating the test fitting on the fuel injection system and removing the protective cap. Relocate or remove any components that prevent attachment of the pressure gauge. Attach the pressure gauge to the test fitting, as shown in **Figure 17-3.** The gauge can be attached to most fittings by threading it onto the fitting. A few require the use of a quick disconnect coupler, similar to those used with air hoses. A few systems do not have a fitting and pressure can only be checked by removing a fuel line to attach the gauge. After installing the gauge, note whether the fuel pressure regulator is equipped with a vacuum hose to the intake manifold.

Turn the ignition key to on, but do not start the engine. The fuel pump should begin operating. Shut off the key and read the pressure on the gauge and compare it against the manufacturer's specifications. Pressure should come up quickly and reach the proper values. These specifications will range from 7-10 psi (48.26-68.94 kPa) on some throttle body fuel injection systems to over 60 psi (413.7 kPa) on some multiport injection systems. Monitor the gauge for 5-10 minutes. If pressure holds for 5-10 minutes with no decrease, continue with other testing. If the pressure drops, you will need to test for pressure drop, which is detailed in the next section.

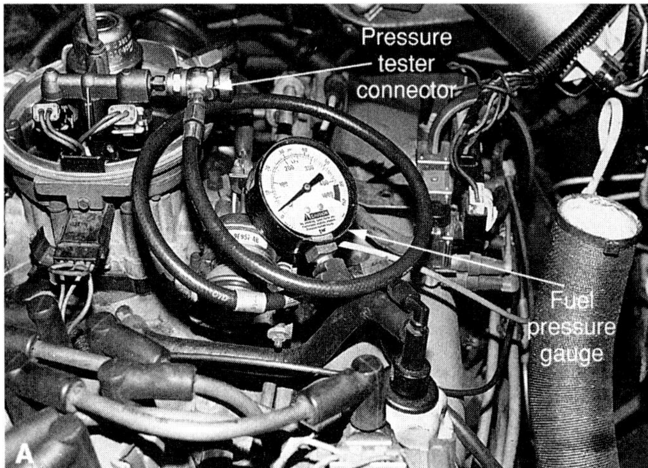

Figure 17-3. *A—Fuel pressure test setup on a throttle body injected engine. Often, a fuel line must be removed to attach the gauge on a TBI unit. B—Gauge setup on a multiport injected engine. These normally have a Schraeder valve located on the fuel rail.*

Note: Before proceeding with other tests, check for a plugged fuel filter if the engine does not seem to be getting enough fuel. Sometimes, this can save much diagnosis time. Fuel filter checking is covered later in this chapter.

Testing for Pressure Drop

If the pressure drops, turn the key off, relieve fuel pressure and turn the key on again. After the system is pressurized, clamp the fuel delivery line and shut off the key. If the fuel pressure remains steady the pump is defective, however, if pressure drops, remove the clamp and repeat the procedure. This time, clamp the fuel return line. If fuel pressure holds, the pressure regulator is defective. However, if pressure continues to drop, an injector is leaking. Remove the spark plugs, one at a time, to locate the leaking injector.

Caution: Never clamp a plastic fuel line. Install shut-off valves when performing a fuel pressure drop test.

After recording the reading with the engine off, start the engine and observe the pressure gauge. On systems equipped with a vacuum operated fuel pressure regulator, the pressure should drop after the engine starts. If the pressure does not drop, the regulator may be defective, or the vacuum line from the regulator to the intake manifold may be disconnected or restricted. Remove the vacuum line from the regulator, **Figure 17-4,** and note whether the pressure increases. If it does not, check the vacuum lines and regulator.

Figure 17-4. *During the fuel pressure test, remove the vacuum hose from the pressure regulator. This is also done when flushing injectors. Be sure to plug the vacuum hose while it is disconnected.*

On systems without a vacuum pressure regulator, the pressure should remain about the same. On all systems, the pressure may fluctuate slightly as the fuel injectors open and close. If the pressure is correct, stop the engine, remove the gauge from the fitting, and reinstall the protective cap. Reinstall any other components that were removed. If the pressure is not correct, refer to the vehicle manufacturer's troubleshooting charts.

If the gauge shows low or no pressure, turn the ignition key to off and turn off any noise producing accessories, such as the radio or air conditioner blower motor. Get someone to turn the ignition key to on while you listen for a whining noise from the pump. A pump whine indicates that the pump is working. However, if the pump noise appears to be excessive, the pump may be worn and should be replaced. If the pump does not make any noise, the pump may not be receiving power or may not be working at all.

Note: The whining noise will only occur for two to five seconds.

If the pump operates, loosen a fuel line at the pump or fuel filter, attach a line to an empty container and observe whether fuel is pumped out. If the pump passes these tests, the problem is a stuck pressure regulator, a plugged filter, or a restriction in the fuel line between the pump and the pressure test fitting.

Vehicles equipped with ***fuel pump priming terminals*** can be tested using another method. Simply apply power or ground, depending on the system, to the priming terminal and listen for fuel pump operation. Do not leave the priming terminal connected for more than 10 seconds. Check this terminal for a short to battery power or ground if the fuel pump is operating constantly, or the engine will not shut off with the key.

Note: If the engine is running, the electric pump is producing enough pressure for operation. Recheck the pressure gauge connections if no pressure is indicated on an operating vehicle.

Checking the Fuel Pump Power Circuit

A common source of fuel system problems is the fuel pump circuit and its relay. Most fuel pump circuits have a power fuse, **Figure 17-5.** This fuse often supplies power to the fuel pump. Always check this fuse before replacing the relay. If the pump does not appear to be working at all, bypass the pump relay momentarily with a jumper wire. Use the fuel pump priming terminal if the system has one. If the pump begins working and the relay is receiving battery power, the relay is defective.

Figure 17-5. *The fuel pump relay and power fuse are sometimes grouped together in a single housing. More often, however, they are located separately.*

Many fuel systems with electric pumps have a bypass to start the engine once oil pressure rises above 5 psi (34.5 kPa), **Figure 17-6.** A quick way to check the relay when the complaint is an extended cranking time when cold, is to watch the oil pressure gauge or light while cranking the engine. If the engine starts when the gauge begins to rise or the light goes out, the relay is probably defective or there is an open in the lead between the ECM and the relay.

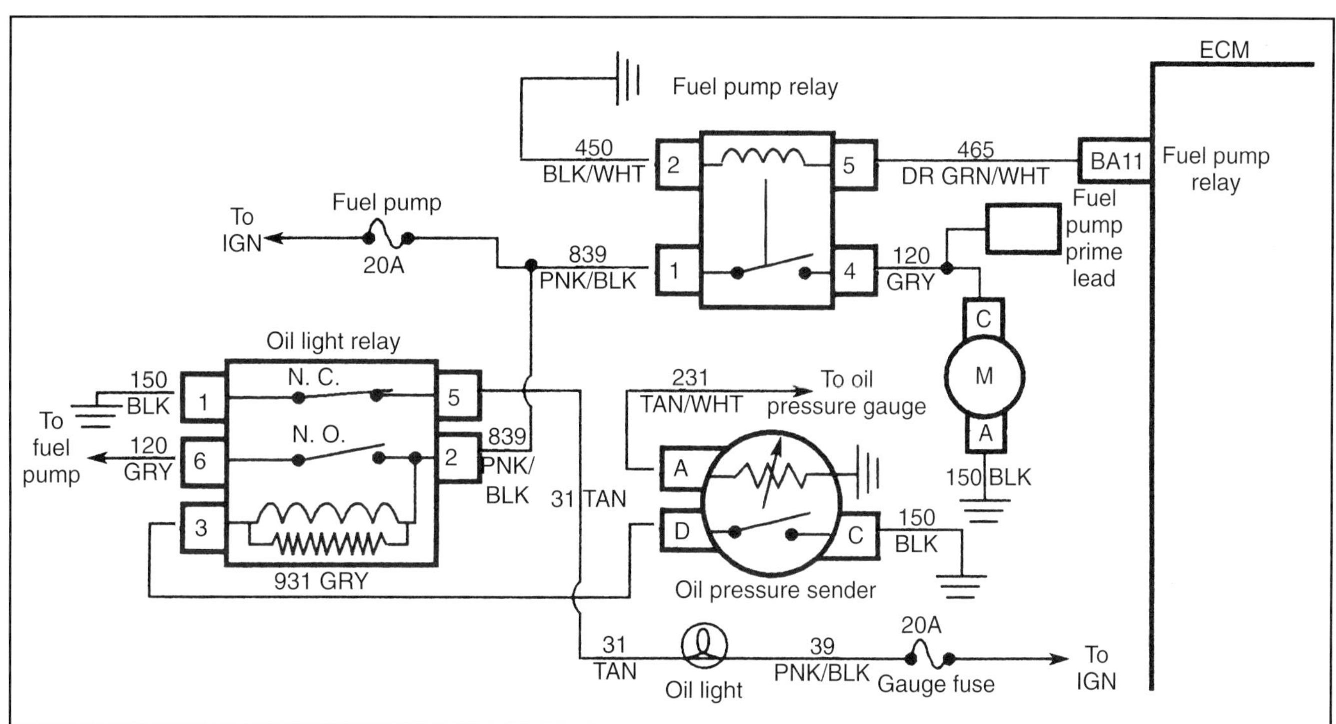

Figure 17-6. *Most fuel injection systems operate the fuel pump for 2-5 seconds to pressurize the fuel system when the ignition key is turned to run. A bypass through the oil pressure switch will operate the fuel pump once oil pressure is present. (General Motors)*

Note: Not all systems have an oil pressure bypass circuit. Check the service manual before conducting this test.

If the pump still does not start, tap it lightly with a soft-faced hammer or other tool (if the pump is accessible). If the pump starts, it was stuck, and should be replaced.

Adding Gasoline to Check a No-Start Condition

Some driveability technicians will add one gallon (3.8 L) of gasoline to the fuel tank when a vehicle has a no-start condition and is not showing fuel pressure, even if the gauge is showing the tank to have fuel. Occasionally, a no-start condition is caused by the driver operating the vehicle until it runs out of fuel. This test also checks the fuel gauge for proper operation.

Note: This test will only work on gasoline powered engines. If a diesel engine has been operated until it ran out of fuel, the fuel lines will contain air and must be bled.

After adding the gasoline, crank the engine several times in order to give the fuel an opportunity to fill the system. If the vehicle starts and the gauge was showing no fuel, you have found the cause of the no-start condition. If the gauge showed fuel in the tank and the engine starts, either the gauge or gauge sending unit is defective or the fuel tank is damaged internally. If the engine does not start, you have eliminated no fuel as a possible cause.

Basic Injector Checks

To quickly check to see that the injectors are operating, the following two methods can be used. More detailed methods of testing injectors and ECM pulses to the injectors are covered later in this chapter.

Listening for Injector Operation

This test is very effective on a running engine. Carefully touch a screwdriver tip on the injector as the engine idles, **Figure 17-7.** Place your ear on the screwdriver handle. If the injector is operating properly, you will hear a clicking noise as the injector pintle is moved by the solenoid. This test ensures that the injector is operable and that it is receiving a signal from the ECM. Repeat this procedure on all injectors. Any injector which does not make a clicking noise is not operating. If an injector is not operating, proceed to the next test to isolate the problem to the injector or the ECM. This test can be made on an engine as it is being cranked, but it may be hard to isolate the injector clicking from the noise made by the starter. Using a stethoscope may also help to isolate the sound.

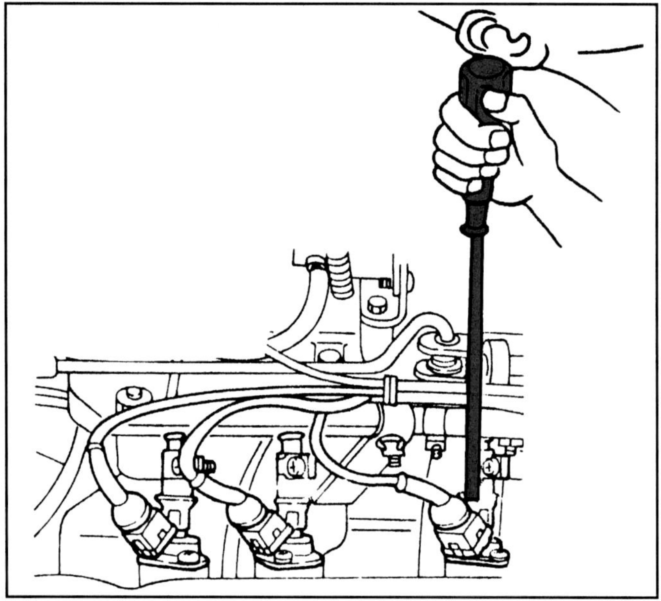

Figure 17-7. *A quick injector operation test is to listen for the telltale clicking of the injector pintle opening and closing.*

Testing for Injector Pulse

This test will allow you to quickly verify the presence of an injector pulse from the ECM. It is often performed by experienced driveability technicians during the visual inspection. A more detailed diagnostic procedure for testing injector pulse width will be discussed later in this chapter. You will need an injector harness "noid" light for this procedure. **Figure 17-8** shows the noid light attached to various fuel injector harness connectors. If earlier tests have narrowed the problem down to one or more cylinders, check the injector supplying the suspect cylinder first and then check the surrounding cylinders.

Remove one injector wiring connector from the injector. Install the noid light. Have someone crank the engine. If an injector pulse is present, the light will flash on and off, **Figure 17-9.** If the noid light flashes, reinstall the connector and test additional injector connectors. If the light flashes on each, the injectors are receiving a pulse and the problem is most likely in another part of the fuel system. If the light does not flash on one or more of the connectors tested, the problem could be located in the injector or ECM wiring, the ECM itself, or its related input sensors. This is why it is so important to check the electrical, computer, and ignition systems first. Be sure to check the service manual for the exact testing and service procedure.

Using Test Equipment to Diagnose the Fuel System

So far, you have used the fuel pressure gauge and injector noid light to test the fuel system. While these tools are very effective in identifying problems that could cause a no-start condition, they are not very effective in finding a

Figure 17-8. *An injector noid light can quickly identify the presence of an injector pulse. A—Multiport injector setup. Some multiport injectors may be difficult to reach. B—TBI setup.*

problem that occurs intermittently, or other problems that still allow the vehicle to run. Many tools are used to diagnose the fuel system. You have already used some of these tools, like the scan tool and the multimeter. The next sections will teach you the importance of the scan tool and multimeter in fuel system diagnosis. Other specialized tools will also be discussed later in this chapter.

Figure 17-9. *Injector noid lights are used to simply check for the presence of an injector pulse. Other equipment is needed to further diagnose an injector pulse for width, duty cycle, or any abnormal signals.*

Scan Tool Testing

Many fuel system components, as well as the overall fuel trim, can be tested with scan tools. A scan tool can provide the technician with many readings without the necessity of making any other test connections. A sample of the available scan tool information is shown in **Figure 17-10.**

See **Appendix C** for more scan tool information.

Data Captured from Vehicle			
Coolant temperature sensor 210°F/99°C	Intake air temperature sensor 52°F/10°C	Manifold absolute pressure sensor 2.1 volts	Throttle position sensor 0.39 volts
Engine speed sensor 860 rpm	Oxygen sensor .84 volts	Vehicle speed sensor 0 mph	Battery voltage 13.5 volts
Idle air control valve 31 percent	Evaporative emission canister solenoid Off	Short-term fuel trim 111	Long-term fuel trim 123
Malfunction indicator lamp On	Diagnostic trouble codes P0172, P0203	Open/closed loop Open	Knock signal No
Misfire cyl. #1 0	Misfire cyl. #2 0	Misfire cyl. #3 3	Misfire cyl. #4 1
Ignition timing (°BTDC)	Base timing: 8	Actual timing: 22	

Figure 17-10. *As you learned earlier in this text, scan tools are essential to proper diagnosis of modern vehicles. This chart shows typical fuel system readings, including short- and long-term fuel trim values.*

Long- and Short-Term Fuel Trim

One of the features available on OBD II systems is the ***short-term*** and ***long-term fuel trim monitor.*** Fuel trim is an ongoing record of changes in the air-fuel ratio, and the responses of the ECM and output devices. Observing the short- and long-term fuel trim allows the technician to tell whether the system is properly controlling fuel mixtures. However, long- and short-term fuel trim can be used to determine air-fuel ratio and whether or not the engine is running lean or rich. This is also called an integrator and block learn on some systems.

Most scan tools monitor short- and long-term fuel trim as a percentage or by counts or steps, based on a set number of 256, with 128 as the base line (14.7:1 air-fuel ratio). If the short-term fuel trim drops below 128, the ECM is compensating for a rich condition; above 128, the ECM is compensating for a lean condition. If percentages are used, 0% is the baseline and varies between -10%-10% when operating correctly. If the fuel trim monitor is below 0% the ECM is compensating for a rich condition. If it is above 0%, a lean condition is indicated. If the fuel trim monitor is below -10% or above 10%, a problem affecting fuel trim is present, **Figure 17-11.**

The ECM monitors short-term fuel trim and adjusts it aggressively. If the ECM notices frequent air-fuel changes or trends (going toward lean or rich operating conditions) in short-term fuel trim, the long-term fuel trim begins to become involved. The ECM will adjust long-term fuel trim either rich or lean, depending on the condition, and reset the short-term fuel trim to 128 or 0%, **Figure 17-12.**

Fuel Trim Monitor

The ECM in OBD II vehicles has a ***fuel trim monitor*** that continuously checks the short- and long-term fuel trim numbers or percentages. This fuel trim monitor is part of the ECM's programming and is an internal function of the ECM.

The fuel trim monitor averages the short- and long-term fuel trim values and compares them to set limits (usually 118-138 or -10%-10%). Both fuel trim values must be outside these numbers for a trouble code to be set. If only one number is out of range, a code is not set, since the ECM can still control fuel trim. The fuel trim is also monitored for problems that occur during certain periods of operation. If the fuel trim values reach and stay at their maximum allotted values for a certain period of time, a malfunction is present and the ECM flashes the MIL. The ECM, in some cases, will perform an intrusive self-test to determine if the evaporative emissions system is causing a rich condition.

Power Balance Testing

The scan tool can perform active and intrusive tests to the fuel system. An intrusive test is any system test that drastically affects engine or vehicle operation. Among the types of intrusive tests that can be performed to the fuel system include cylinder ***power balance tests.*** Power balance tests can be performed to isolate a misfiring cylinder and to determine if a misfire is caused by the fuel or ignition system.

A power balance test is performed by shutting off one injector at a time. Most scan tools have the capability of

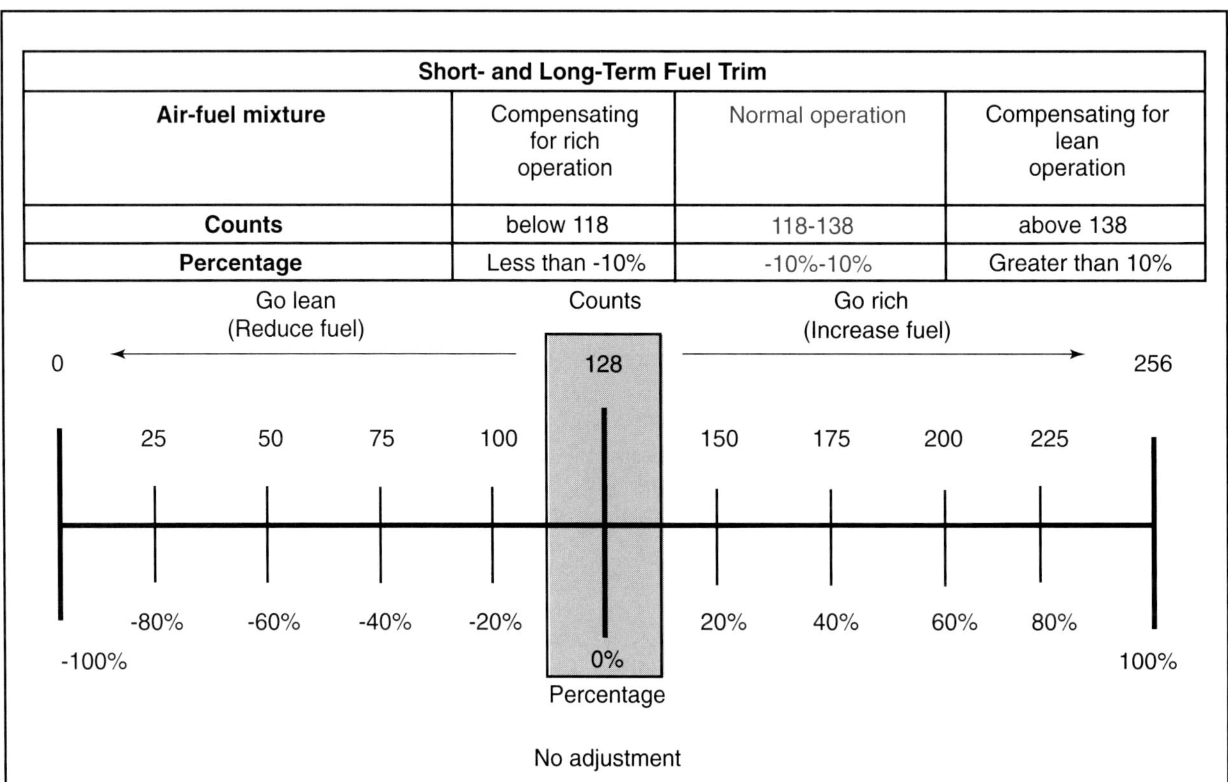

Short- and Long-Term Fuel Trim			
Air-fuel mixture	Compensating for rich operation	Normal operation	Compensating for lean operation
Counts	below 118	118-138	above 138
Percentage	Less than -10%	-10%-10%	Greater than 10%

Figure 17-11. *Fuel trim is the most frequent adjustment made by the ECM. The ECM tries to keep the fuel trim within a narrow area for maximum efficiency.*

Compensating for an on-going lean condition	
Before long-term fuel trim adjustment	After long-term fuel trim adjustment
Short-term fuel trim 109	Short-term fuel trim 128
Long-term fuel trim 128	Long-term fuel trim 122

Compensating for an on-going rich condition	
Before long-term fuel trim adjustment	After long-term fuel trim adjustment
Short-term fuel trim 142	Short-term fuel trim 128
Long-term fuel trim 128	Long-term fuel trim 133

Figure 17-12. *Short-term fuel trim tells you what type of air-fuel mixture is currently being delivered to the engine. Long-term fuel trim indicates particular trends that have been occurring. If the short-term fuel trim becomes excessively high or low for a time, the long-term fuel trim will adjust its value and reset short-term fuel trim to its default value (128 or 0%).*

performing power balance tests automatically. Select the menu for cylinder or injector testing and start the test. The scan tool will shut off the ignition spark to one cylinder for a few seconds. If the idle speed does not change when the spark is disabled, the cause of the misfire is located in that cylinder. Some scan tools will allow you to shut off the individual injectors. This feature can be used to verify the results from the power balance test.

Clear Flood

Some vehicles are equipped with an operational mode called ***clear flood.*** Clear flood will shut off the injectors if the ECM indicates that the throttle position sensor is being held past a certain angle, usually 60-80% or 3.0-4.5 volts. This prevents the injectors from flooding the engine with fuel.

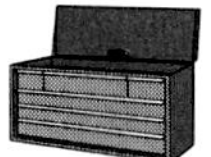

Note: Check the service manual before performing clear flood. Not all vehicles have this feature.

To clear flood an engine, with the engine shut off, depress the accelerator to the floor. Crank the engine for 15 seconds. Wait 30 seconds and then repeat. Although clear flood will clear an engine with minor flooding, if an engine has been flooded severely or for a period of time, it may still be necessary to remove and clean or replace the spark plugs. You can also use this procedure as a quick way to disable the injectors when testing for ignition spark.

Multimeter Testing

Fuel system testing often requires the use of multimeters. Multimeters can measure the resistance of various fuel system devices, as well as input and output voltages. The pulse and duty cycle features of the multimeter is needed to test injector pulse rate. Many modern multimeters have oscilloscope features that allow the technician to see the electrical pattern of what is happening in the injector windings. If necessary, review the features of multimeters in **Chapter 3.**

Using the Exhaust Gas Analyzer

A fuel system problem can sometimes go unnoticed until a vehicle fails an emissions test. If the air-fuel ratio is not properly controlled, the engine can develop driveability problems, such as poor performance, high exhaust emissions, and poor gas mileage. Using an ***exhaust gas analyzer*** is the only sure method of determining the air-fuel ratio. It is a good idea to perform an exhaust gas analysis after any fuel system repair work is completed. **Chapter 18** contains a description of exhaust gas analyzers.

If the vehicle is equipped with a computer control system, the following checks can be made. Start by creating a rich mixture. This can be done by briefly throttling (restricting) the air flow into the engine. Do not restrict the air flow enough to stall the engine, just enough to create a rich mixture. Do not spray gasoline or carburetor cleaner into the intake to richen the mixture, as this will overload the oxygen sensor. When the air flow is restricted, the exhaust gas analyzer should show a momentary rich mixture, and then begin to lean out as the ECM begins to compensate for the rich mixture signal from the oxygen sensor.

When this check is complete, remove the air flow restriction and operate the engine at high idle for one minute to clear any extra gasoline out of the intake passages. Allow the exhaust gas analyzer readings to stabilize. Create a lean mixture by removing the PCV hose or another vacuum hose. The exhaust gas analyzer should show a momentary lean mixture, then begin to move to the rich position as the ECM begins to compensate for the lean signal from the oxygen sensor. When the test is complete, reconnect the hose.

If the exhaust gas analyzer readings indicate that the computer control system is reading the changes in air-fuel ratio, and is changing fuel system settings to compensate, the system is in good condition. If the system does not appear to be reacting to air-fuel ratio changes, refer to the vehicle manufacturer's troubleshooting procedures to determine the cause. After making all needed tests with the exhaust gas analyzer, shut off the engine, and remove the analyzer probe from the tailpipe.

Fuel Delivery System Component Test

If the engine runs and your initial inspection and testing reveals that the fuel system is losing or has low fuel pressure, the most likely location for the root cause of the problem is the fuel delivery system. If the engine does not run and there is no fuel pressure, the problem is definitely located in the fuel system. While the fuel pump is the primary cause of most fuel delivery problems, the entire system must be considered. Failure to thoroughly check the entire system could result in a misdiagnosis.

Fuel Filters

A clogged fuel filter can cause stalling, hesitation, and poor acceleration as well as holding dirt that can possibly clog an injector. Before a fuel filter can be checked, it must be removed from the fuel system. The first step is to locate the filter. Many modern filters are located on the underside of the vehicle body, near the fuel tank. In most cases, the filter can be located by tracing the fuel line back from the engine, or if the vehicle is on a lift, forward from the fuel tank.

Once the fuel filter is located, it may be removed from the fuel lines by one of several methods. In many cases, the filter is held by a clamp, which may be held by a single bolt. Loosening the bolt will open the clamp enough to allow the filter to slide out, **Figure 17-13.** Some modern filters, such as the one in **Figure 17-14,** are equipped with built-in brackets, which are held to the vehicle with self-tapping screws. The bracket is usually replaced with the filter.

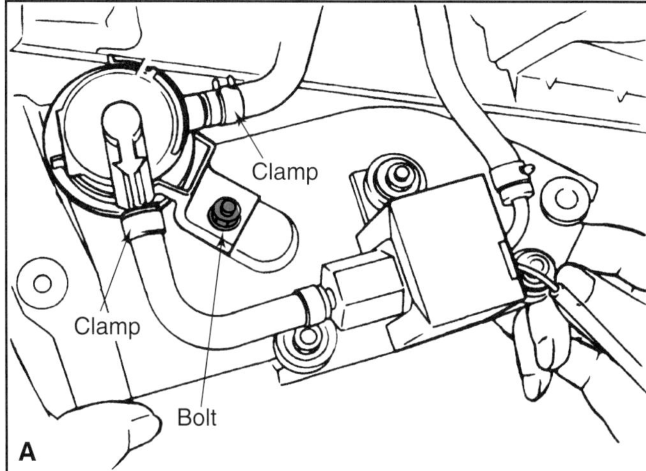

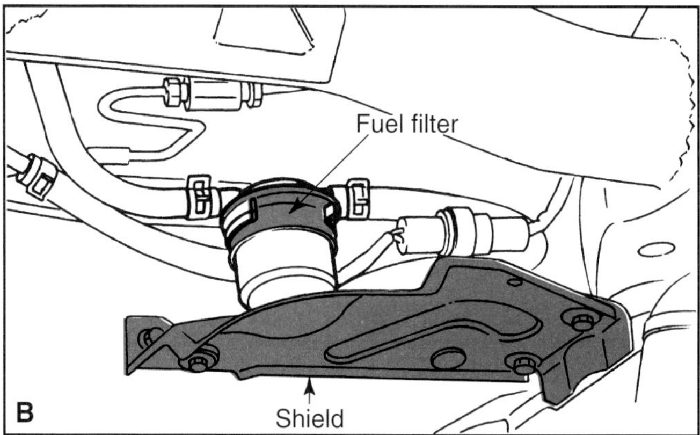

Figure 17-13. A—Fuel filters are often held by a clamp and one or more bolts. The bracket will expand when the bolt is loosened, allowing the filter to be removed. B—Some filters are protected by heat or rock shields. (Subaru)

Warning: Some filters are in the pressurized part of the fuel system. Pressurized fuel can spray out if not relieved before filter removal. Be sure to relieve fuel system pressure before loosening any fuel line fittings.

To remove filters that are attached by neoprene hoses, loosen the clamps and slide the filter and old hoses from the fuel lines. Twist the hoses slightly to loosen them from the lines before pulling on them. When installing the new filter, always use new hoses. When removing filters attached by tubing fittings, use a flare-nut wrench on each tube fitting to prevent twisting or rounding off the fuel lines, **Figure 17-15.** This is especially important when changing diesel filters, since the lines are internally coated and the coating could flake off if the lines are bent. Some filters are attached to the fuel lines with plastic clips or special spring-loaded fittings, **Figure 17-16.** Sometimes, a special tool must be used to remove the lines from the filter. Once the new filter is installed, the lines can be reconnected by pushing them onto the filter fittings.

After the filter has been removed from the vehicle, pour the gasoline out the inlet end. If large quantities of debris fall out along with the fuel, the filter may be plugged. Another way to check the filter is to blow through it. First, drain out the liquid gasoline and dry the filter thoroughly.

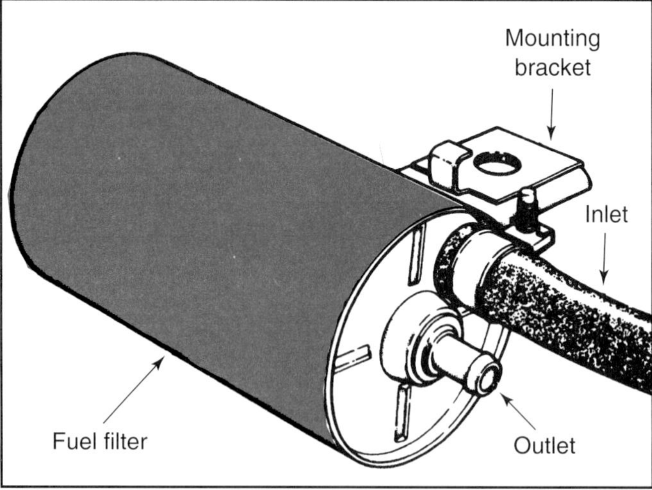

Figure 17-14. Some fuel filters have built-in brackets that are replaced along with the fuel filter. (Chrysler)

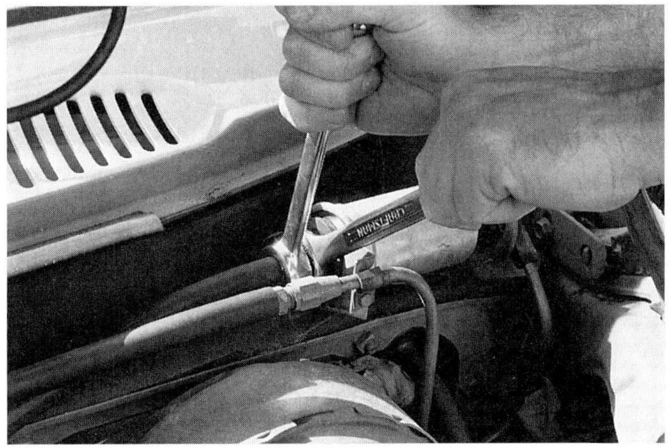

Figure 17-15. *Use a flare-nut wrench when removing any tube fuel line. The standard open-end wrench can be used to hold the filter.*

Warning: Gasoline fumes in large quantities can cause respiratory system damage. Be careful not to inhale gasoline vapors or ingest liquid gasoline.

After the liquid gasoline has been completely removed, install a piece of clean neoprene hose to the filter and attempt to blow through the filter. If any resistance is felt, the filter is at least partially plugged. It will be almost impossible to blow through a badly plugged filter. If the filter was badly plugged, be sure to perform a fuel quality test to ensure that dirty fuel is not the cause. If you find large quantities of metal particles in the filter, this indicates that the fuel pump is defective. Filter replacement is in the reverse order of installation.

Fuel tube or fuel system component
Pull tab
Quick-connect fitting
Fuel tube stop
Metal collar quick-connect fitting
Plastic collar quick-connect fitting
Step 1
Removal twist
Step 2
Blow
Step 3
Removal tool
or
Step 4
OIL
Step 1
Step 2
Step 3
Fuel tube
Removal
Quick-connect fitting
Push
Push
Push
Fuel tube
Installation
Push
Push
Plastic retainer

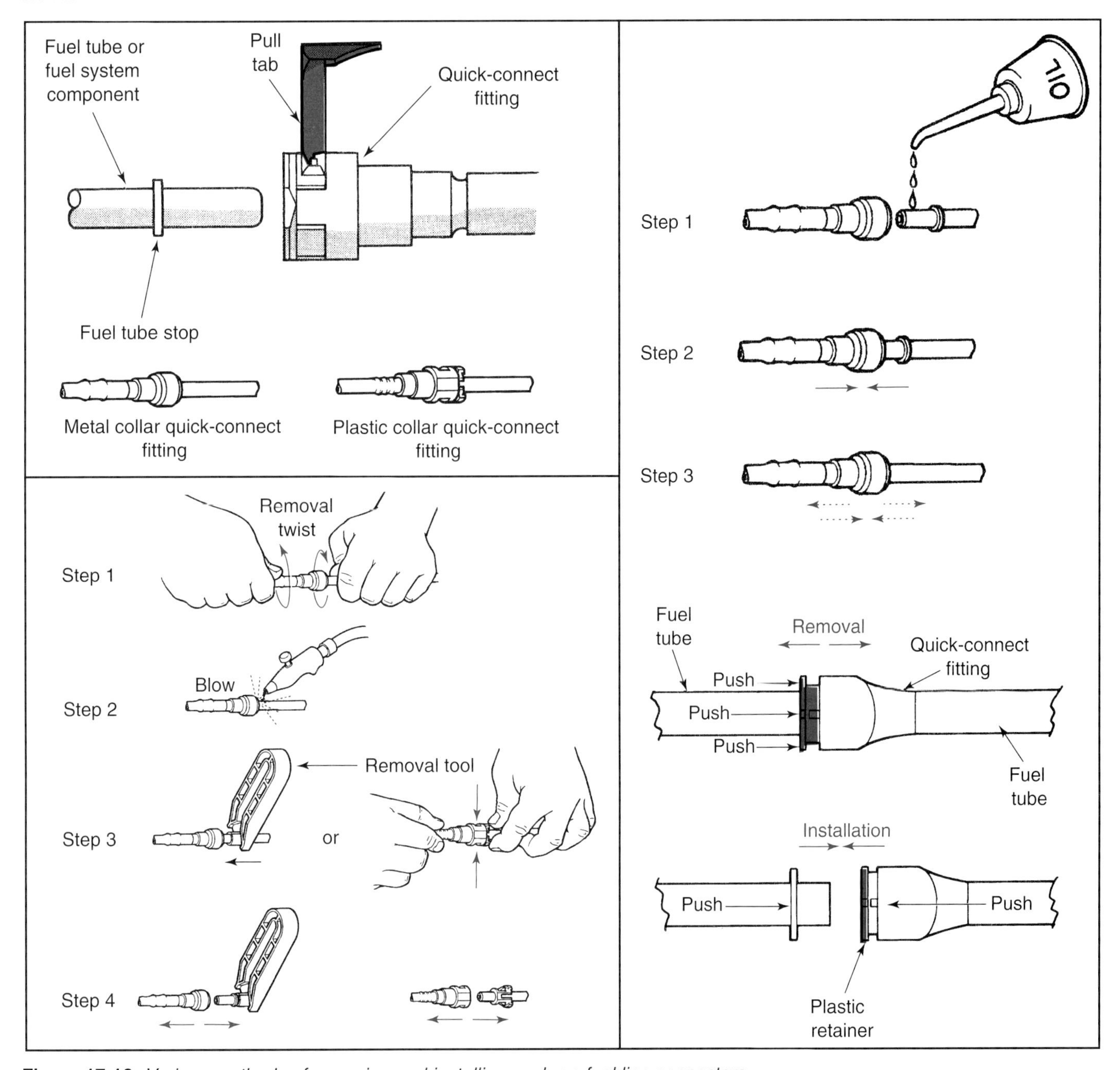

Figure 17-16. *Various methods of removing and installing push-on fuel line connectors.*

Diesel Fuel Filters

The fuel filters used on diesel engines operate and are checked in the same manner as gasoline filters. Many diesel filters have a ***water separator,*** which should be checked regularly for excess water accumulation. Some water separators can be drained through a petcock at the bottom of the filter assembly. Others must be replaced when full.

Checking the Fuel Lines

Most fuel line problems involve leaks, kinks, or restrictions due to external wear or damage caused by foreign debris and exposure to dirt and salt. Any leaks or exterior damage should have been found during the visual inspection. Other problems include misrouting near hot engine or exhaust parts, causing vapor lock. If water is present in the fuel lines, it may freeze in the lines during extremely cold weather. Check the lines carefully in spots where they are clamped to the frame. Vibrations can cause the lines to wear and leak at the clamps.

Checking Plastic Fuel Lines

While plastic fuel lines are not susceptible to corrosion damage by dirt or salt, they are more likely to be damaged by debris, or in most cases, mishandling. The most common damage is caused by a bend or kink placed in the line by careless handling. Damage to a plastic fuel line is usually obvious and can be spotted during a visual inspection. Check all fittings for proper installation. Any bends or kinks in the line cannot be straightened and become a permanent restriction. Most plastic lines are one piece and cannot be spliced if damaged, requiring complete replacement from the fuel tank forward to the engine. If a plastic fuel line fitting is not installed properly, a leak will result.

Checking the Fuel Tank

The fuel tank is probably the least likely cause of a performance problem. However, it is the most difficult part of the fuel system to inspect and test, since there is not an easy way to see into the tank except to remove it from the vehicle. If you have reached this point, you should have eliminated the rest of the fuel system, including the fuel pump, as the cause of the problem.

Often, a leak in a fuel tank will occur at the top of the tank, where condensed water collects and begins rusting. In other cases, the insulating material between the top of the tank and the body will retain water and corrode the tank. These leaks may not drip gasoline, but may prevent proper fuel tanks and system pressurization, affecting evaporative emission control operation.

The fuel tank pressure sensor used in some OBD II vehicles will detect loss of vacuum in the fuel tank. A code indicating low evaporative emissions system pressure may indicate a possible leak in the fuel tank. Also check for the presence of a fuel filler cap. On some OBD II systems, a missing or poorly installed fuel filler cap can cause the ECM to set a low system pressure code.

Checking Fuel Quality

When diagnosing the fuel system, the fuel is an often overlooked part. Although fuel quality is not often taken into account, even by experienced driveability technicians, it is sometimes the cause of driveability problems. In the past, fuel that had some water or excessive particles of dirt or rust did not affect engine performance drastically. However, fuel injectors are very susceptible to clogging by minute particles that would have gone unnoticed in a carbureted engine. Due to the fine air-fuel ratio control by the ECM, any water or an excessive amount of alcohol in the fuel will affect driveability.

Note: You should eliminate the other fuel system parts before checking fuel quality.

If a driveability problem caused by the fuel system is suspected, ask the driver when and where they normally purchase fuel, the fuel grade (octane for gasoline or cetane for diesel) and whether or not the problem started after the last fuel purchase. Sometimes a driveability problem can be corrected by a change in fuel grade or even where the fuel is purchased. If you removed the fuel filter earlier, you should have noted if the fuel that poured out was excessively dirty. *Do not* use this fuel as a judge of fuel quality. The sample must come from the fuel line before the filter.

Inspecting Gasoline Quality

Start by drawing a fuel sample for inspection and testing. Loosen the fuel outlet line from the fuel tank and draw approximately 200 ml (6 oz) of fuel into a clear plastic or glass jar. The fuel should be drawn from the bottom of the tank so that any dirt, particles, or water is drawn with the fuel into the sample jar. Examine the fuel sample, it should be clear with no floating dirt or particles. If there are particles of dirt, rust, the fuel is cloudy or the sample separates (water settles on its own level below the fuel), the tank should be drained and cleaned. If the tank is rusted internally, it should be replaced.

Alcohol in Gasoline Test

Most gasoline manufacturers add ***alcohols,*** such as methanol and ethyl alcohol to their gasoline to lower the price and to stretch their supplies when stockpiles are low. Excessive alcohol in gasoline can cause many driveability problems, including stalling, lack of power, hesitation, or no-start conditions. Most vehicle manufacturers allow a maximum fuel alcohol content of 10%.

Start by locating a 100 ml (3 oz) cylinder with 1 ml (.03 oz) graduations. Add 90 ml (2.7 oz) of gasoline from the sample to the cylinder, then add 10 ml (.3 oz) of water, place sample to the cylinder, then add 10 ml (.3 oz) of water, place a stopper on the cylinder and shake vigorously. Allow the cylinder to sit for 5-10 minutes after shaking. The water will separate from the gasoline to the bottom of the cylinder, along with any alcohol that is present. If the quantity of separated water and alcohol (the original 10 ml of water plus any alcohol) is above 20 ml (.6 oz), the alcohol content in the fuel is more than 10%, **Figure 17-17.**

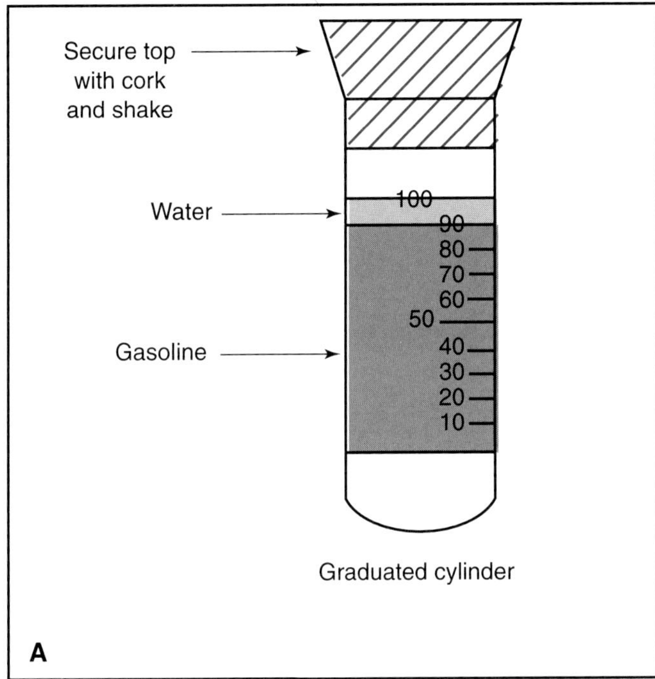

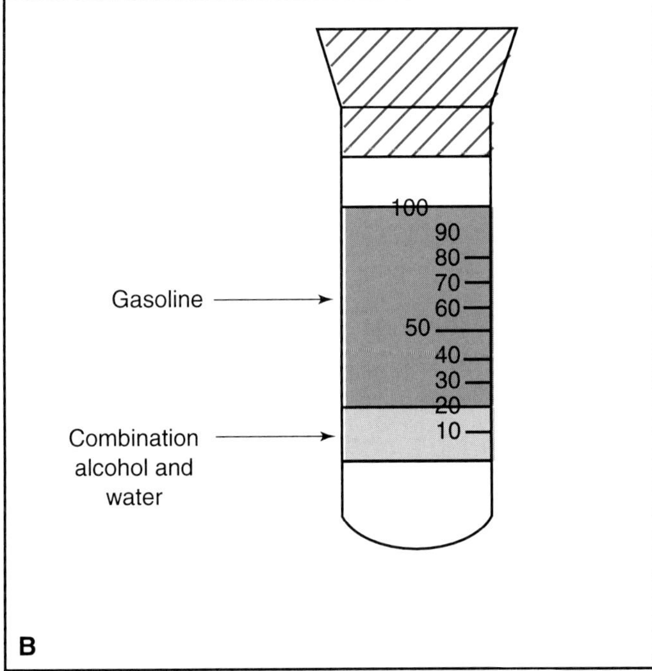

Figure 17-17. *A—To determine the quantity of alcohol in a sample of gas, it is necessary to conduct an alcohol in fuel test. Add 90 ml of gas to a 100 ml graduated cylinder, than add 10 ml of water, insert a stopper and shake. B—Any alcohol in the fuel will become suspended in the water and will descend to the bottom of the cylinder. There should be no more than 20% alcohol and water in the cylinder.*

Keep in mind that this test will only give you an approximation of how much alcohol is in the fuel. If you suspect that excessive alcohol is the cause of the driveability problem, advise the customer to try purchasing fuel from another supplier, especially if the driver purchases fuel from one particular supplier.

Diesel Fuel Quality Test

Fuel quality is sometimes a factor when a diesel engine has a driveability problem. Contaminated or low quality diesel fuel can cause hard starting, smoking, stalling, filter plugging, or internal engine damage.

To test diesel fuel quality, begin by drawing a sample of the diesel fuel into a glass or other clear cylinder and visually checking it for particles. There should be no foreign matter in the fuel. Next, allow the fuel sample to sit for about 20 minutes, and note whether any liquids have settled out, usually on the bottom of the test cylinder. The most common liquid contaminant is water. Water in diesel fuel is usually accompanied by particles. These particles are rust and hydrocarbon utilizing microorganisms that have begun to grow at the meeting point of the water and fuel. If the fuel has any contamination, the entire fuel system must be drained and flushed. If microorganisms have begun to grow, the entire fuel system must be treated with a ***biocide.***

After the fuel is visually tested for contamination, its ***cetane rating*** must be checked. Cetane rating is the burning ability of the diesel fuel, and should not be confused with the octane rating for gasoline. The easiest method of testing cetane is by using a ***cetane hydrometer,*** **Figure 17-18.** This hydrometer works in the same manner as the familiar battery hydrometer. To test the cetane level, draw a sample of clean diesel fuel into the hydrometer. Then read the number on the float where it meets the level of the fuel in the hydrometer barrel. Diesel fuel should have a minimum cetane rating of 40. If the cetane rate is lower, advise the driver to change diesel fuel suppliers.

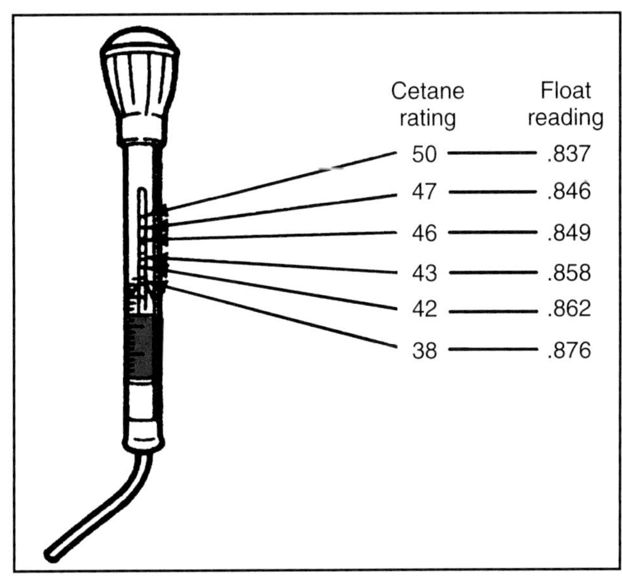

Figure 17-18. *Diesel fuel quality can be determined by finding its cetane number. This hydrometer will quickly give the cetane number of a sample of diesel fuel. (Ford)*

A problem that can affect diesel fuel in cold weather is ***waxing.*** In very low temperatures, poor quality diesel fuel can cloud and then solidify into a waxlike substance. Wax can form at any temperature, but is more likely the colder the fuel gets. This wax can clog filters and injectors, causing the engine to stall or not start at all. Often the vehicle is brought into a warm shop for diagnosis, the wax dissolves, leaving no clue to the real problem. To check for waxing, place a container of the suspect diesel fuel in a cold place, outside in cold weather, or in a freezer compartment. Wax will form obvious solid lumps in the fuel.

Fuel Injector and ECM Control System Diagnosis

An injector which is stuck open will cause a rich mixture, or flood the engine. This can be caused by dirt or an injector defect. The problem may be in the ECM rather than the injector itself. An injector which does not squirt any fuel will cause a miss on that cylinder. Also check the cold start injector, if the system has one. A cold start injector that sticks open will cause the engine to run rich when it is warmed up. If the cold start injector sticks closed, it will cause hard starting on a cold engine.

Basic check to determine whether the fuel injectors are operating were covered earlier in this chapter. This section will cover how to test the injectors to see how well they are working.

Note: Clean up any spilled gasoline.

Checking Fuel Injection Pulse Width

Measuring fuel injector pulse width or duty cycle (sometimes called fuel injector on-time) ensures that the ECM is correctly pulsing the injectors, based on sensor inputs. Earlier in this chapter, you learned how to check for the presence of injector pulse. For this test, you must have a pulse meter or a multimeter with a duty cycle feature. Set the multimeter to the *pulse width* function. Then start the engine and read the pulse width on the display screen.

Create an artificial lean fuel mixture by disconnecting the oxygen sensor and grasping the connector on the wiring harness side with one hand. If the vehicle has more than one oxygen sensor, determine which is the output wire for the oxygen sensor closest to the number one cylinder. If the oxygen sensor is the heated type, this test will still work. You must make electrical contact with the connector. Touch a good ground with your other hand. This procedure is safe, due to the low voltage and low current used in the oxygen sensor circuit. Read the injector pulse width from the display screen. If the ECM is operating properly, it will interpret the tiny current flow through your body as a lean condition due to high oxygen content in the exhaust and compensate with a rich command to the injectors. Pulse width should increase.

Note: On some vehicles, the oxygen sensor input lead to the ECM can be grounded to perform this test. However, the terminal should only be grounded if specifically called for by the service manual.

Next, create a rich mixture by removing your hand from the ground and touching a battery positive voltage point. This can be the positive battery post or any positive terminal. Read the pulse width on the display screen. The ECM should interpret this as a rich condition due to low oxygen in the exhaust and compensate with a lean command. Pulse width will decrease. If your scan tool has the capability of monitoring long- and short-term fuel trim, you can use this test along with the scan tool to check if the ECM is adjusting fuel trim and, therefore, injector pulse.

Observing Fuel Injector Scope Patterns

Much can be learned about fuel injectors by observing their electrical pattern on an oscilloscope. Many injector problems leave a distinct electrical signal which can be seen in their scope pattern. For instance, a sticking injector pintle will cause a low turnoff spike in the pattern when the current is turned off. This shows that the injector has a problem without having to remove it from the vehicle. These tests can be made with a multimeter having a scope pattern provision.

Monitoring Fuel Injector Pulse Width Using a Scan Tool

Modern scan tools can monitor the fuel injector pulse width and deliver the information to the technician. This information is provided in the form of a number, measured in milliseconds, that can be compared with normal readings to determine the condition of the injectors and its electrical system. Typical injector pulse width ranges from 1-4 milliseconds and will vary greatly, since the ECM's adjusts pulse width to maintain the proper air-fuel ratio. A low injector pulse width would indicate that the injector is not opening while a high pulse width number would indicate that the injector is not closing. **Figure 17-19** shows a typical scan tool reading of an injector pulse. The injector pulse width number normally displayed is the average pulse width time for all of the injectors. Many scan tools have the capability of monitoring the individual pulse widths from each injector. It is also an indicator of the type of air-fuel ratio control occurring in the engine.

Using a Timing Light to Check TBI Injector Operation

A timing light can be used to quickly observe the operation and spray pattern of TBI injectors. The test is made by attaching a timing light to the engine as if you were going to check engine timing and removing the air cleaner. Start the engine and point the timing light at the injector nozzle.

Data Scanned from Vehicle			
Coolant temperature sensor 212°F/100°C	Intake air temperature sensor 79°F/26°C	Mass airflow sensor 2.6 volts	Throttle position sensor .39 volts
Engine speed sensor 846 rpm	Oxygen sensor .55 volts	Vehicle speed sensor 0 mph	Battery voltage 13.4 volts
Idle air control valve 31 percent	Evaporative emission canister solenoid Off	Short-term fuel trim 127	Long-term fuel trim 128
Malfunction indicator lamp Off	Diagnostic trouble codes	Open/closed loop Closed	Fuel pump relay On
Injector pulse width 3 M/sec	Cruise control Off	AC compressor clutch On	Knock signal No
Ignition timing (°BTDC)	Base timing: 4	Actual timing: 17	

Figure 17-19. *A scan tool can be used to monitor the average injector pulse number. Often, the amount of time the injectors remain open reflects what is occurring with short- and long-term fuel trim.*

Since injectors are timed to match ignition system firing pulses, the timing light will flash when the injector sprays fuel. The flash from the timing light will illuminate the spray. In **Figure 17-20A,** the spray pattern of a single injector unit is illuminated by the flash from the timing light. In **Figure 17-20B,** one injector of a dual TBI system is being checked. On some low pressure systems, it is helpful to raise the idle to provide a more consistent flash. The spray should be strong and cone shaped. Weak, irregular, or ragged spray patterns indicate a problem in the injector or the control system.

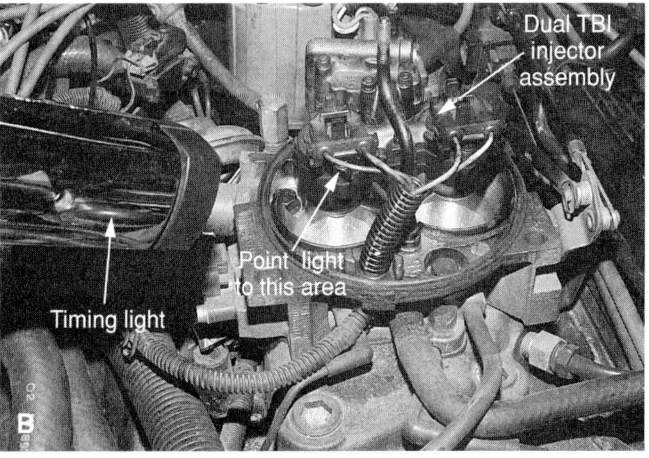

Figure 17-20. *A timing light can be used to watch TBI injector spray patterns. A—Single injector unit. B—Dual injector TBI assembly.*

Checking Idle Air Control (IAC) Motors

Checking idle air control (IAC) motors requires you to read the percentage (duty cycle) or ***counts.*** Percentages are generated by current control motors and can range between 0% (closed, throttle plate open) to 60-70% (maximum idle speed compensation). Counts are generated by IAC valves that use stepper motors and range from 5-50. These can only be monitored by a scan tool. A current control IAC can be checked by measuring the average dc voltage or the current draw. Before proceeding, check the proper service manual to determine which test a particular manufacturer specifies.

To begin the percentage duty cycle test, connect the multimeter to read voltage between the IAC motor's return wire and ground. Set the multimeter to the duty cycle position. Start the engine and observe the duty cycle percentage. A lower than normal duty cycle can indicate a sticking or otherwise defective IAC motor. Low duty cycle can also be caused by a high accessory load (such as the air conditioner compressor) or engine mechanical problems. Therefore, do not replace the IAC motor before checking other possible causes.

Using a Scan Tool to Measure IAC Counts and Percentages

As mentioned earlier, a scan tool is needed to check stepper IAC motors. It can also be used to check current control motors by checking the percentage. Monitor the scan tool's IAC reading, **Figure 17-21.** If the percentage or counts are excessively low, this could cause a extended crank or stall condition. Turn on the air conditioning compressor, the motor percentage or count should increase slightly, if not, check to see if the ECM is receiving a compressor request signal. If the scan tool shows that the compressor is on, the bypass chamber is dirty or the IAC motor is the problem. Compare any scan tool readings to the service manual recommended values.

Idle Air Control (IAC) Motor			
Throttle plate position	**Throttle plate open**	**Throttle plate closed**	**IAC open 100%***
Percentage	Idle air control valve 0 percent	Idle air control valve 31 percent	Idle air control valve 90 percent
Counts	Idle air control valve 0	Idle air control valve 24	Idle air control valve 50

*Throttle plate closed

Figure 17-21. *Idle air control values fluctuate greatly in response to engine load and demand. The numbers here are simply to give you an idea of what type of IAC values are typically seen during engine operation.*

Checking Turbocharger Boost Pressure

A general check of turbocharger operation can be made with the dashboard boost gauge. Start the engine and observe the gauge as you gradually accelerate the engine. The gauge should show the turbocharger ***boost pressure*** gradually rising to the maximum range and then remaining steady as the engine is accelerated further. If the boost pressure is within the proper range, and no other problems, such as pinging or low power are encountered, the turbocharger is probably okay.

However, most dash gauges are not accurate enough to make a precise check or to adjust boost pressure on those units where it is possible. A pressure gauge with a range of 0 to 20 pounds is necessary to accurately check boost pressure. The pressure gauge must be installed on a fitting on the intake manifold.

Caution: Do not attach the gauge to the turbocharger wastegate fitting.

Install a tachometer on the engine, and ensure that the engine is fully warmed up. After the pressure gauge and tachometer are installed, start the engine and increase engine speed until the turbocharger begins to operate.

Note: Watch the tachometer carefully to ensure that you do not exceed the maximum rpm limit for the engine. Most engines in good condition can speed up to about 4000 rpm without harm. The safe limit for the engine that you are working on should be determined before going above this rpm.

Increase the engine speed until the limit of boost pressure is reached. The wastegate should open somewhere within the range for maximum boost pressure. If the turbocharger pressures are below normal, either the turbocharger is defective, or the wastegate is opening too soon or stuck partially open. If boost pressures are higher than normal, the wastegate is stuck closed, the linkage is disconnected or jammed, or the actuator diaphragm or manifold hose is leaking. After correcting any problems and rechecking the boost pressure, remove the tachometer and pressure fitting. Some wastegates are adjustable. However, most are not and service is limited to replacing defective parts. Most newer wastegates are ECM-controlled.

Adjusting Fuel System Components

Once you have isolated and verified the problem, you will have to decide whether or not to adjust, if possible, or replace the component. Many parts of the fuel system can be adjusted to improve driveability. They include throttle position sensors, fuel injector pressure regulators, and fuel injection throttle valves. Note, however, that most components of the modern fuel injection system are controlled by the ECM and are not adjustable.

Throttle Position Sensors

Misadjusted throttle position sensors affect ECM operation. They can go out of adjustment due to linkage wear, or changes in the electrical material in the sensor. **Figure 17-22** shows a typical throttle position sensor for a throttle body fuel injection system. TPS adjustment and replacement was covered in detail in **Chapter 15.**

Fuel Injector Pressure Regulator

A few fuel injection pressure regulators have provisions for adjustment. To adjust the pressure regulator, connect a fuel pressure gauge to the pressure connector on the fuel system. Then turn the ignition key to on, but do not start the engine. Read the fuel pressure; if it is incorrect, locate the regulator adjusting screw. Loosen the locknut and turn the

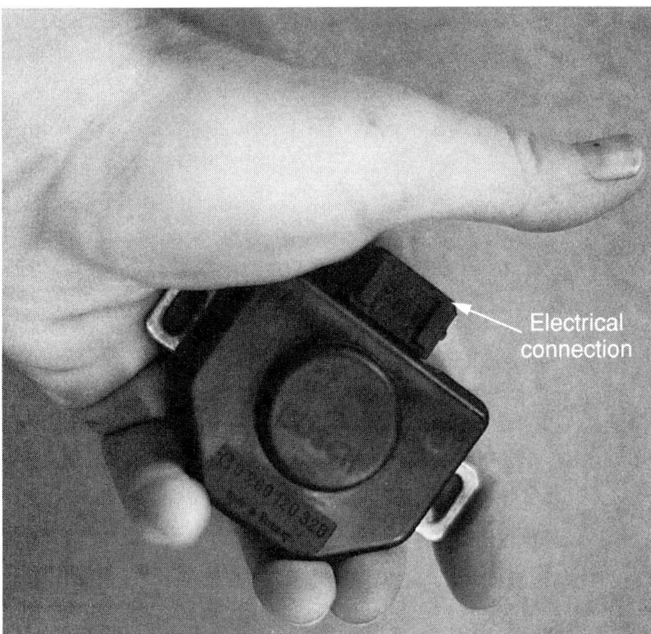

Figure 17-22. *Throttle position sensors come in many shapes and sizes. This TPS is for a throttle body injection system.*

adjuster until the proper pressure is reached. If the regulator has a vacuum line, start the engine and observe that the pressure remains in the proper range for a running engine. Then tighten the locknut. If the regulator cannot be adjusted to meet specifications, it must be replaced.

Fuel Injection Throttle Valve

The throttle valves used with many fuel injection systems are controlled by an idle motor or calibrated air bleed, **Figure 17-23.** In many cases, the throttle position cannot be adjusted, since the ECM will compensate for any changes. Sometimes, the idle speed stop screw can be adjusted for cold engine operation. Always follow the procedure given in the service manual to make this adjustment.

Figure 17-23. *Throttle valves are larger, but in many ways share similarities with the throttle valves used on carburetors. The idle speed stop screw can be adjusted, but is not recommended.*

Caution: Do not adjust the idle speed stop screw to compensate for low idle speed. The ECM will lower the IAC pulse width, creating other driveability problems.

Some throttle valves will build up deposits of gum or varnish, which cause the valve to stick, or the idle bleed system to clog, causing stalling or other problems. The valve can be cleaned with commercial spray carburetor cleaner, and readjusted if necessary.

Replacing Fuel System Components

Once a defective component has been isolated, replacement is usually a straightforward process of removing the old part and installing the new part. The vehicle should always be rechecked after replacing any part.

Fuel Lines and Fuel Tanks

Fuel lines and tanks are easily replaced, however, precautions must be taken to remove all pressure from the fuel lines before removing any filters. Remove the fuse that supplies the fuel pump with power or disconnect the fuel pump wiring harness. Crank the engine for 15 seconds. Leave the fuel pump disconnected to ensure the fuel system does not become pressurized.

Replacing Fuel Lines

Most fuel lines to and from the fuel rail or throttle body are formed lines and are simply replaced as an assembly. Disconnect the battery and loosen the fuel lines. Remove the fitting from the fuel tank first, since there may be still some pressurized fuel remaining. Check for O-rings as you remove the lines and save them for comparison with the new O-rings. Remove the fitting from the fuel rail next, then remove any clamps or retainers. Remove or relocate any necessary wiring or parts and carefully remove the line. Install the O-rings on the new fuel line and connect the fittings. Install any clamps or retainers last.

The metal fuel lines from the tank rarely need to be replaced. If they are damaged, they should be repaired using a flare fitting. Cut out the damaged section and flare in a new section. Use double-wall steel tubing only and make either a double-lap flare or an ISO flare. Plastic fuel lines cannot be repaired if damaged and should be replaced as an assembly.

Caution: Do not use copper tubing to replace fuel lines.

Removing Fuel Tanks

Fuel tanks are removed if the tank is leaking or damaged. However, the most frequent reason for tank removal is when the fuel pump must be replaced. If the fuel tank contains considerable fuel, it must be drained before removal. Raise the vehicle, disconnect the fuel lines, and cover with masking tape. Remove any wiring attached to the tank. Remove any components, such as heat shields, exhaust, and suspension parts that may block tank removal. With the help of an assistant, remove the tank support straps and lower the tank, **Figure 17-24.**

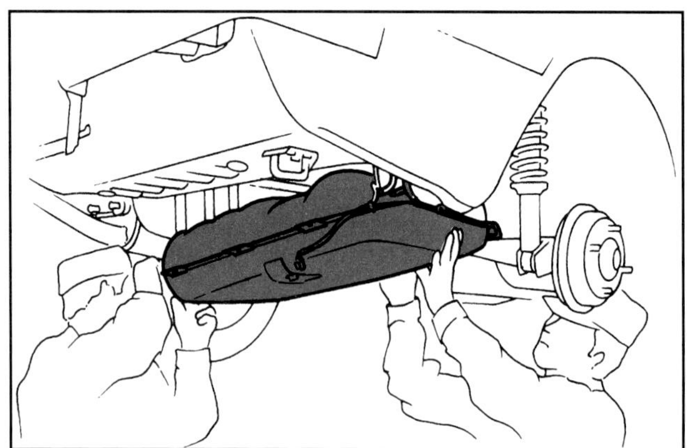

Figure 17-24. *Removing a fuel tank, even when empty, is a two person operation. The shape and bulk of the tank makes it almost impossible to be handled by one person. (Subaru)*

Replacing Fuel Pumps

Modern fuel pumps, whether mechanical or electric, are replaced rather than repaired. Electric fuel pumps can be mounted anywhere on the vehicle. Some pumps are installed in the fuel tank and must be replaced according to manufacturer's instructions. In most cases, the battery must be disconnected, and the fuel tank drained and removed from the vehicle. Then, the lock ring can be removed from the pump and sender assembly, and the pump withdrawn from the tank, **Figure 17-25.** When replacing this type of pump, always use new seals.

Caution: Always release the fuel system pressure and disconnect the electric fuel pump electrical connector before disconnecting any fuel system components.

Other electric pumps can be replaced easily after the battery negative cable is removed. First remove the pump electrical connections, **Figure 17-26.** Then remove the fuel lines and pump mounting hardware. Install the new pump in the same position as the old pump, and attach the mounting hardware.

Replace the electrical connectors and reattach the battery cable. Then turn the ignition switch to on and allow the

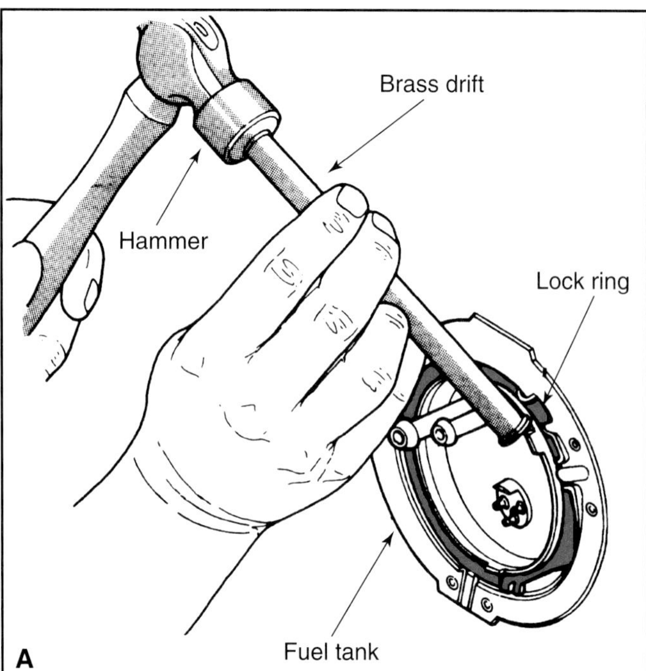

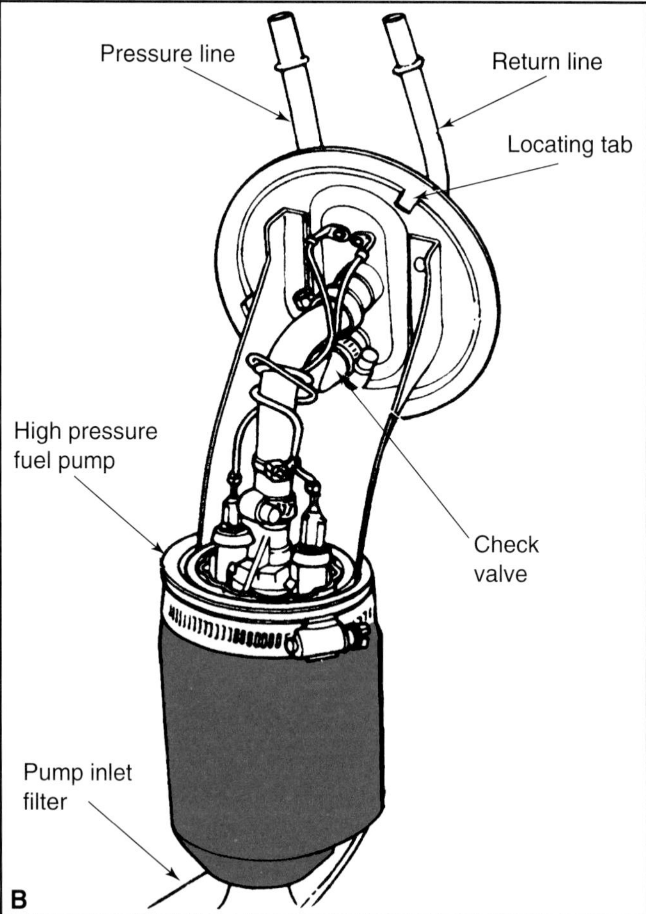

Figure 17-25. *A—The lock ring can be removed by using a hammer and brass drift. Special pliers are also available to remove this ring. B—Once the lock ring is removed, the fuel sender unit can be easily lifted out of the tank. (Chrysler and Ford)*

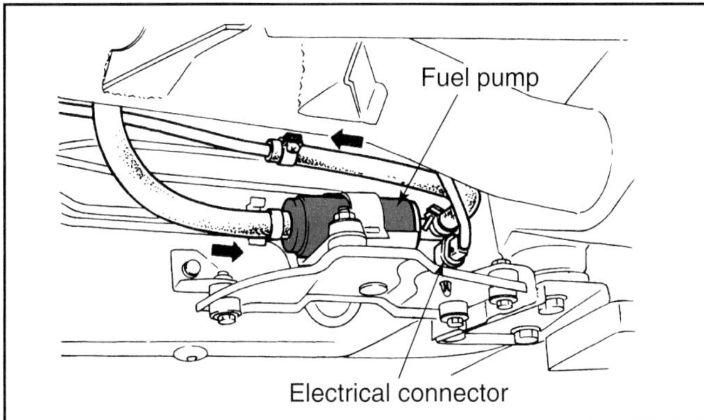

Figure 17-26. *Fuel pumps are sometimes located on the frame, near the fuel tank. (Subaru)*

pump to fill the lines. This may take as much as 30 seconds. Then start the engine and check pump operation.

Cleaning the Fuel Tank

If the tank contains excessive water, it should be cleaned by removing any remaining gasoline and water and adding a quart of nonflammable solvent. Hold a clean shop rag over the filler neck and sender hole. Tilt the tank several times to move the solvent, drain, and repeat. Blow the tank dry with compressed air. If the tank is corroded or damaged internally, the tank must be replaced.

Installing the Fuel Tank

With the help of an assistant, carefully raise the tank into position and attach the tank straps. Make sure all tank insulation strips are in place. Reinstall all fuel lines and wiring connections. Reinstall any other components that were removed. If the vehicle is equipped with a nonvented filler cap, make sure that it is in place and that the vent tube is open. Do not refill the tank until all components are in place and installed.

Cleaning and Servicing Fuel Injection System Components

Many times, fuel injection system components can be returned to service by removing deposits of solidified gasoline components (usually called gum and varnish), or by replacing defective components. These procedures are discussed in the following sections.

Cleaning Fuel Injectors on the Vehicle

Clogged fuel injectors will produce an irregular pattern that will not allow the fuel to be completely atomized. Badly clogged injectors will not allow enough fuel into the engine. In either case, misfiring will cause a complaint of poor driveability.

The internal parts of modern fuel injectors contain many small openings. For this reason, cleaning fuel injectors by soaking them in carburetor cleaner may not remove all deposits. In addition, soaking may damage internal seals and O-rings which are not designed to withstand contact with solvents. Fuel injector cleaners can be poured into the fuel tank to help prevent the formation of deposits, but are not strong enough to remove heavy deposits once they are formed. The most positive way to clean the injection system is by using a special injector cleaner device. The basic cleaning device consists of a pressurized can of fuel system cleaner which is attached to the pressure test fitting of the injection system, **Figure 17-27.** The fuel pump is disabled, and the can pressure pushes the cleaner solvent through the injector system. The engine can operate on the solvent without damage to itself or to the catalytic converter. Some cleaning systems use shop air to provide pressure.

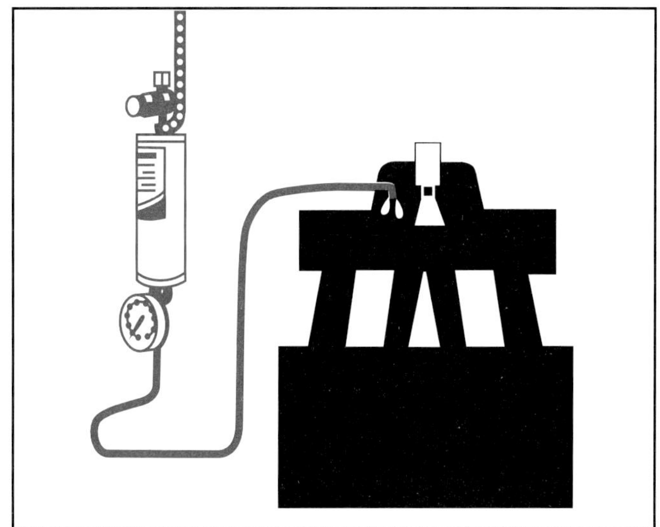

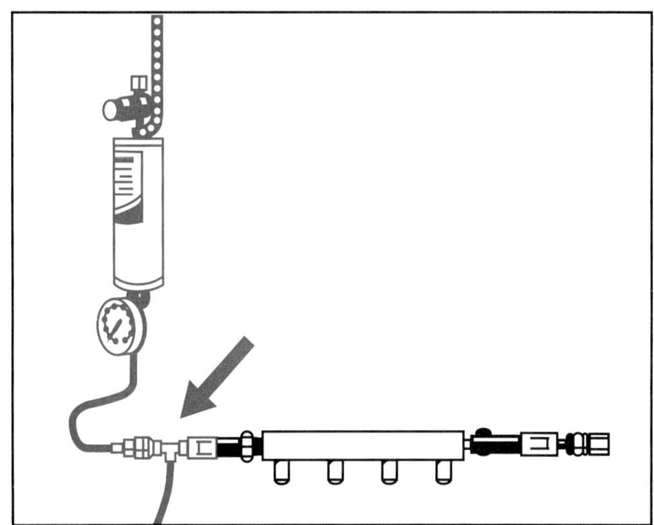

Figure 17-27. *A—Fuel injection flush setup for a TBI injection system. You may have to remove a fuel line to attach this type of setup. B—Flush setup for a multiport fuel injection system. The Schraeder valve makes flushing this type of system easy. (OTC)*

Caution: Some manufacturers no longer recommend injector cleaning. Check the service manual before performing any fuel injector cleaning.

To use the injector cleaner, first locate the fuel pump relay and disconnect it. Some relays have a fuse, which can be removed to disable the pump. On other vehicles, the electrical connector on the relay must be removed. Then remove the protective cap from the injection system fitting and attach the outlet hose from the injector cleaner. Attach a pressurized can of the cleaning solvent to the injector cleaner. **Figure 17-27A** shows the cleaner attachment on a TBI system, while **Figure 17-27B** shows the cleaner attached to a typical multiport injection system.

Slightly crack the flow control valve on the cleaner and start the engine. The engine will operate on the cleaner solution until no pressure is left in the solvent can. Allow the engine to idle as you control the flow of cleaner into the fuel injection system. The flow must be fast enough to keep the engine running, but not so heavy that large amounts of cleaner are bled off through the pressure regulator.

When the solvent can loses enough pressure, the engine will starve for fuel and die. Turn the ignition switch off and close the control valve. Disconnect the hose from the injector system fitting and reinstall the protective cap. Then reconnect the fuel pump relay and turn the ignition switch on. Allow the fuel pump to build up pressure, then crank the engine. After the engine starts, allow it to idle for a few minutes to purge any air and solvent from the injector system.

Cleaning Diesel Injectors

Diesel injectors usually do not become as clogged as gasoline injectors, since the higher pressures and resistance of diesel fuel to gumming reduce the chances of clogging. In many cases, diesel injectors wear out before they become clogged. If necessary, diesel injectors can be cleaned by removing them and soaking them in solvents. After soaking, the diesel injector should be completely disassembled and cleaned with compressed air. Many manufacturers recommend that clogged diesel engine injectors be replaced.

Throttle Body Cleaning

The two major types of fuel injection throttle bodies are the TBI, which contains the injectors, and the multiport, which controls air flow only. Both can be cleaned on the vehicle using a spray cleaner. Spray cleaning the throttle body is considered part of routine maintenance and is covered in **Chapter 21.** However, if the throttle body is coated with gum and varnish, it may be necessary to remove, disassemble, and clean the throttle body.

TBI Throttle Body Removal and Disassembly

To remove a TBI throttle body from the engine, begin by removing the air cleaner or intake ducts, disconnect all electrical connections at the throttle body, and remove all vacuum lines. Then remove the throttle linkage or cable, automatic transmission linkage, and throttle return springs. Remove the fuel lines and any inlet and outlet fittings. Then remove the throttle body attaching bolts, and remove the throttle body from the intake manifold.

Place the throttle body on a clean workbench. Start the disassembly procedure by removing any electrical devices that should not be immersed in cleaning solvent if you are going to repair the old throttle body. Examples are the idle speed motor, idle air bypass solenoid, and the throttle position sensor, **Figure 17-28.** Remove the air cleaner gasket, the pressure regulator assembly, and the injector(s), **Figure 17-29.** The throttle body can now be cleaned and rebuilt, or replaced with a new part.

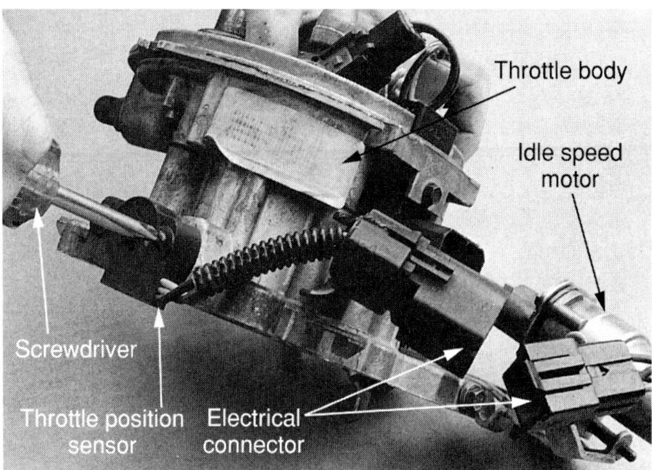

Figure 17-28. *Remove all external sensors and motors before disassembling the throttle body.*

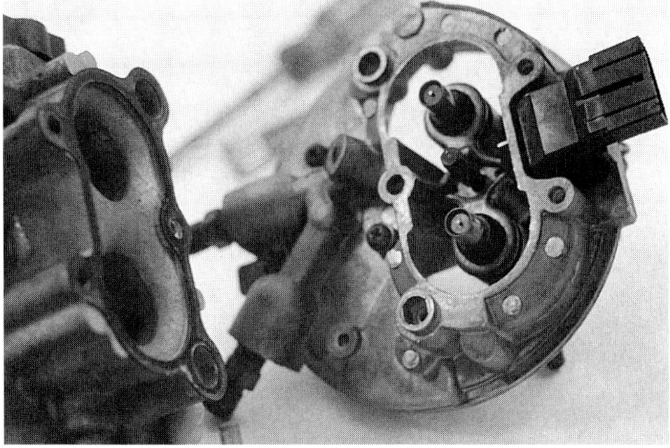

Figure 17-29. *Most valve bodies split easily after the fasteners are removed. Throttle bodies are much easier to disassemble than carburetors since there are fewer parts.*

The throttle body is usually made in two parts. The lower part contains the throttle valves while the upper part, called the air horn, contains the injectors and pressure regulator. To disassemble the throttle body, remove the attaching screws and split the throttle body. See **Figures 17-30** and **17-31.** Then remove the injectors, pressure regulator, and other components.

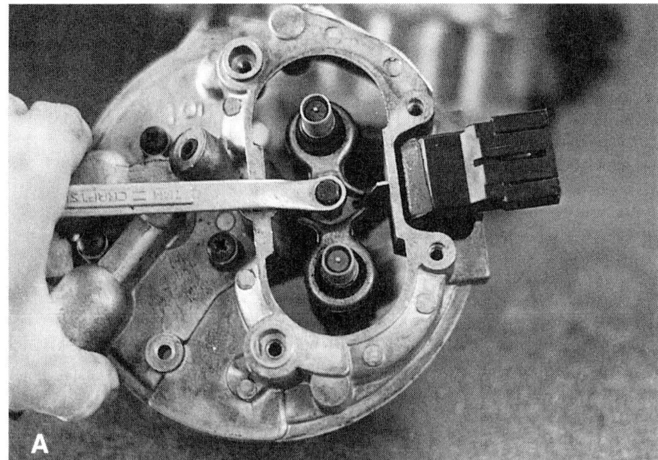

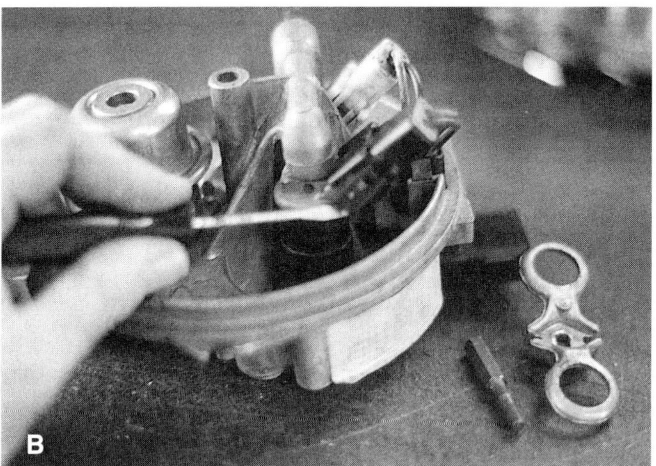

Figure 17-30. *A—A screw or bolt is used to hold most fuel injectors. B—Remove the wiring harness before removing the injector.*

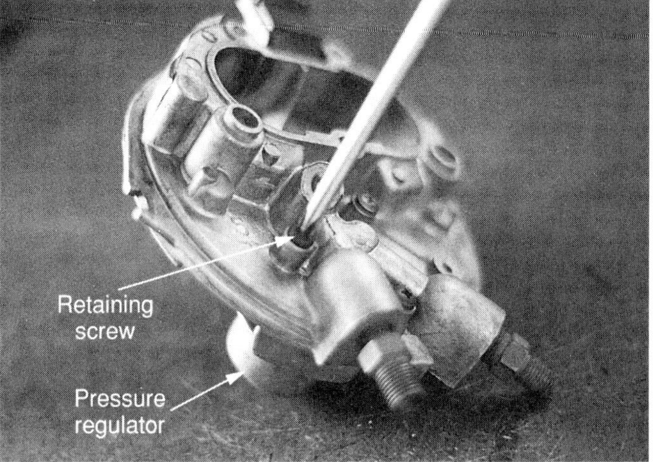

Figure 17-31. *The pressure regulator is held by one or more screws, often located outside the throttle body. This one is located inside the TBI air horn.*

To disassemble the air horn portion of the throttle body, remove the fuel injector cover if necessary. Then pry the injector from the throttle body. Do this carefully to prevent damage to the injector or throttle body casting. On some TBI systems, the pressure regulator can also be removed from the air horn. Soak the throttle body in cleaning solution for no more than 20 minutes. Soaking the parts longer than 20 minutes may damage the coatings on the body casting. Remove the castings and rinse with water. Blow out all passages with compressed air.

TBI Throttle Body Assembly and Installation

Obtain new gaskets and O-rings for the injectors and fuel lines. Install the injector(s) in the throttle body cavity. Line up any alignment marks on the injector and mating parts. Push down on the injector until it is fully seated. Be careful not to cut the O-ring during installation. Install the injector cover if used, and reassemble the two halves of the throttle body. Reinstall the idle control solenoid, throttle position sensor and all other components removed earlier.

Note: You may want to replace the injector(s) if the throttle body is from a high mileage engine.

To reinstall the throttle body, place it on the intake manifold, using a new gasket, and tighten the fasteners. Then install the fuel inlet and outlet fittings using new gaskets. Replace the throttle linkage, transmission linkage, and return springs. Reinstall the vacuum lines and electrical connections. Then start the engine and check for fuel leaks. If there are no leaks, observe general engine operation, and make any tests as necessary.

Multiport Fuel Injection Throttle Body

Servicing a multiport throttle body is somewhat less complicated, as you do not have injectors to remove. Start by removing the air intake hose, all electrical connectors, and linkage. Most multiport throttle bodies have coolant flowing through them to prevent icing, so drain the coolant down to a level below the throttle body. Some throttle bodies have the mass air flow sensor installed on the throttle body as a separate part. Often it is easier to remove the sensor with the throttle body. Unbolt the throttle body from the intake plenum, carefully pry the throttle body, and remove from the vehicle. Remove the throttle body-to-plenum gasket and clean the plenum mating surface.

The multiport throttle body is a much simpler assembly than the TBI unit, **Figure 17-32.** Start by removing all parts that will not be cleaned. This includes the throttle position sensor, idle air control motor, and mass air flow sensor, if attached, **Figure 17-33.** Remove all other parts such as the idle stop screw, idle air bypass chamber, vacuum and coolant ports, and any remaining gaskets. Soak the throttle body in cleaner for 20 minutes and clean with water. Then, thoroughly dry the parts with compressed air.

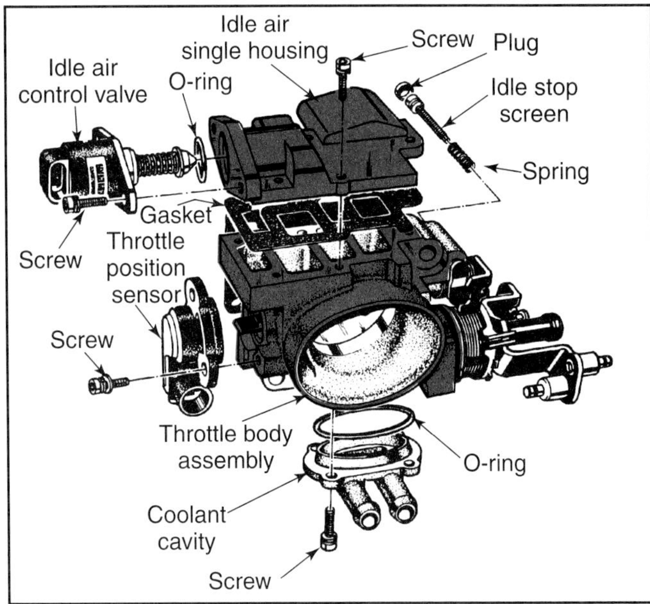

Figure 17-32. Parts of a multiport throttle body. These are even easier to service than the TBI units, since there are no injectors to remove. (General Motors)

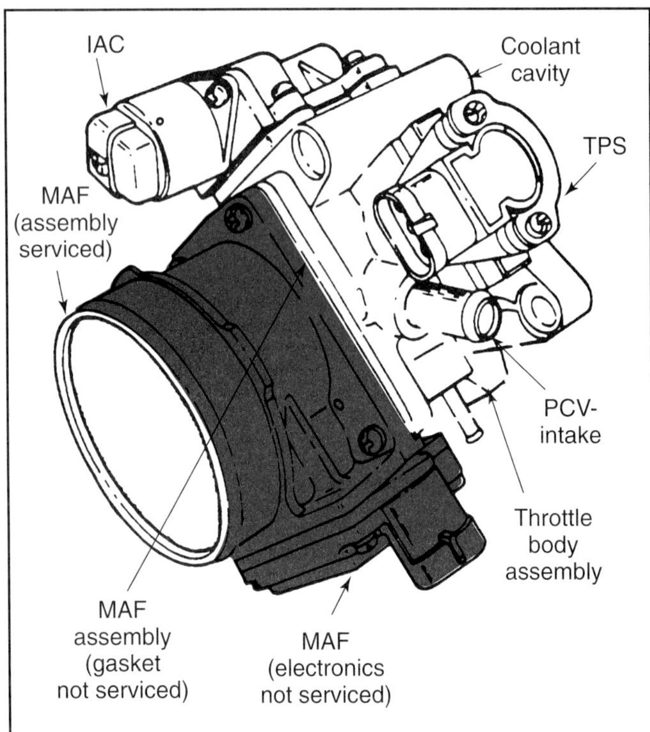

Figure 17-33. *Multiport throttle bodies use many of the same parts as TBI units. However, mass airflow (MAF) sensors are sometimes attached to them. (General Motors)*

Reassemble the throttle body using new gaskets. If neoprene hoses are used to supply the throttle body with coolant, now is a good time to replace them. Install the throttle body-to-plenum gasket and the throttle body. Reconnect all linkage, ECM connectors, and the air hose. Start the engine and listen for air and vacuum leaks. Reset minimum idle speed and road test thoroughly.

Multiport Fuel Injection Pressure Regulator

The pressure regulator is the major control of injection system pressure. Pressure regulators are installed on the fuel rail of multiport systems. See **Figure 17-34.** Most pressure regulators are preset and are not adjustable. These regulators are replaced when they are unable to properly control the fuel system pressure. If an adjustable pressure regulator cannot be successfully adjusted, it should also be replaced. To replace the pressure regulator, remove the air cleaner, hoses, or other components to gain access to the regulator. Bleed off fuel system pressure, then remove the vacuum line to the regulator, if it has one. Remove the fittings and fasteners that hold the pressure regulator to the fuel rail, and remove the pressure regulator from the rail. See **Figure 17-35.**

Figure 17-34. *Multiport pressure regulators are located on the fuel rail. Most are press fit and retained by one or more bolts.*

Install the new regulator, making sure that all seals and gaskets are in place, and install the fuel line fittings and attaching screws. Replace the pressure regulator vacuum line, and all components that were removed to gain access to the regulator. Then start the engine and check for leaks and proper fuel system pressure.

Multiport Fuel Injector Replacement

Since gasoline fuel injectors are manufactured to extremely precise tolerances, they cannot be overhauled. Some diesel injectors, however, can be repaired rather than replaced. When any injector service involves removing the fuel system tubing, avoid bending or flexing the tubing.

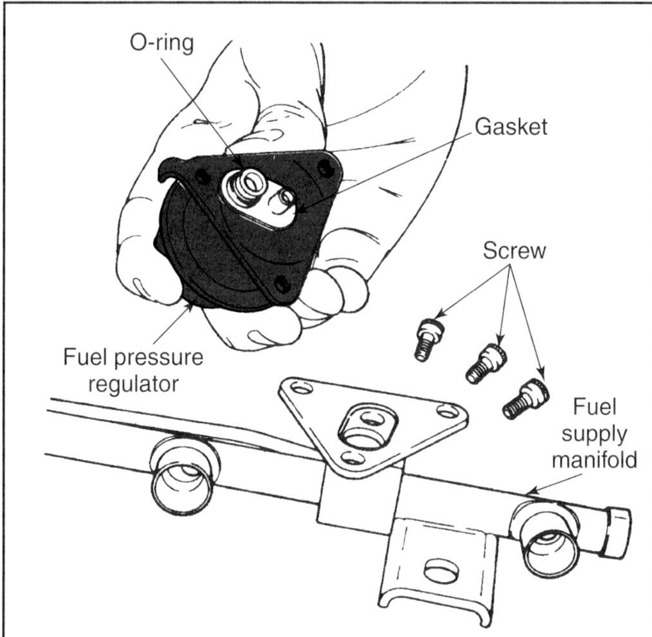

Figure 17-35. *When removing the old pressure regulator, be sure to check for the presence of an O-ring. Use a new O-ring if the regulator is to be replaced. (Ford)*

Most fuel injector tubing has an internal coating, which can flake off if the tubing is bent or flexed. If gasoline engine injectors cannot be restored to proper operation by the cleaning procedure described earlier in this chapter, they must be replaced.

Injector and Fuel Rail Removal

A typical injector installation is shown in **Figure 17-36.** Begin by removing the battery negative cable first and then removing engine parts as needed to gain access to the injector(s). On some multiport systems, it may be necessary to remove the intake manifold or plenum to gain access to the injectors, **Figure 17-37.** Also remove the injector electrical connectors.

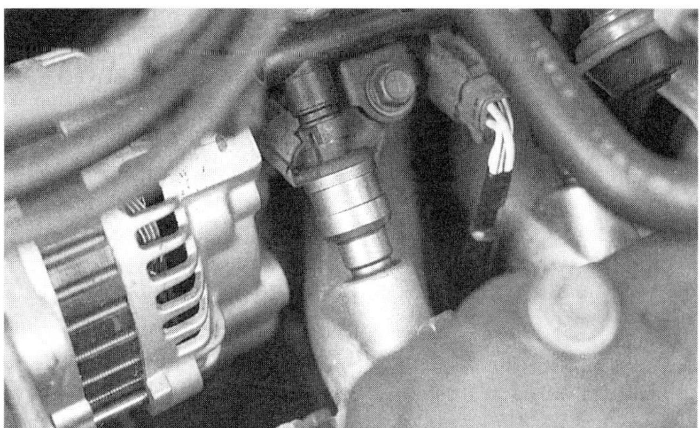

Figure 17-36. *Multiport injectors are press-fit into the manifold and the fuel rail retained by several bolts. Remove all wiring harnesses and any other parts before removing the injectors. Remove all multiport injectors, even if you are only concerned with one; this prevents fuel rail distortion.*

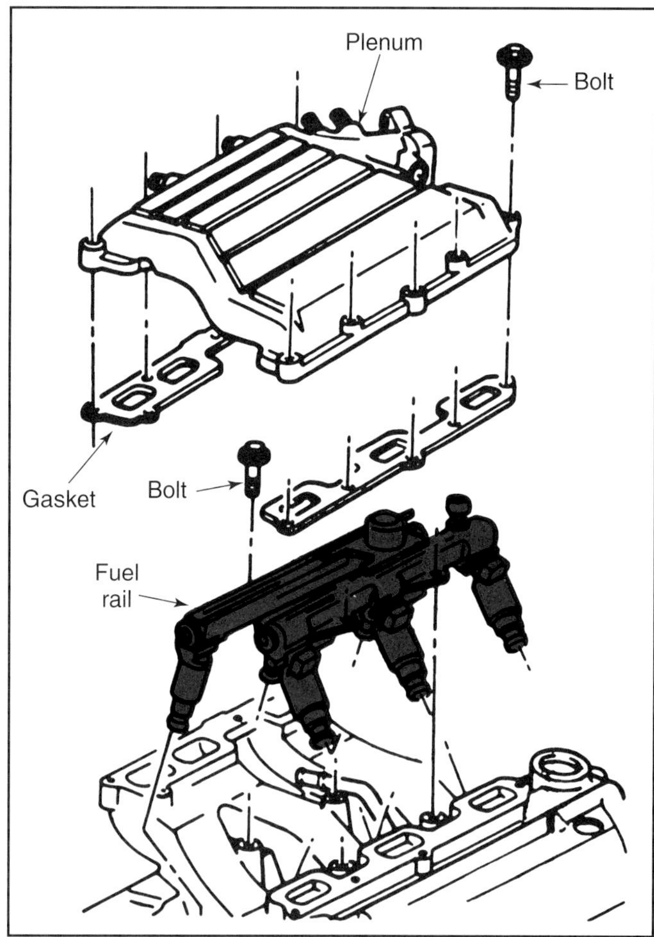

Figure 17-37. *In some cases, it may be necessary to remove the intake plenum to access the injectors. (General Motors)*

Note: Do not allow gasoline spills to remain on the engine or shop floor. Clean up gasoline spills immediately.

Remove the fuel rail by removing the supply and fuel return hoses, and the fuel pressure regulator vacuum hose. Remove any brackets holding the fuel rail to the valve cover or other engine components. Then remove the fuel injector to intake manifold attaching bolts and clips. Remove the fuel rail and injector assembly by pulling the rail straight up so that the injectors come straight out of the intake ports. Then remove the electrical connector if they were not accessible before removal, **Figure 17-38. Figure 17-39** shows the injectors and the fuel rail as an assembly. On some engines, the fuel rail is removed separately from the injectors and the injector is then pulled from the intake manifold, **Figure 17-40.**

Place the injectors and fuel rail assembly on a clean workbench. If the injector has a retaining clip, pull it from the fuel rail and injector. Pull the injector straight out of the fuel rail receiver cup, **Figure 17-41.** Check the injector O-rings for damage. If the O-ring is defective, it must be replaced. If you plan on reusing any injector, place the injector in a secure place so that the tip cannot be damaged.

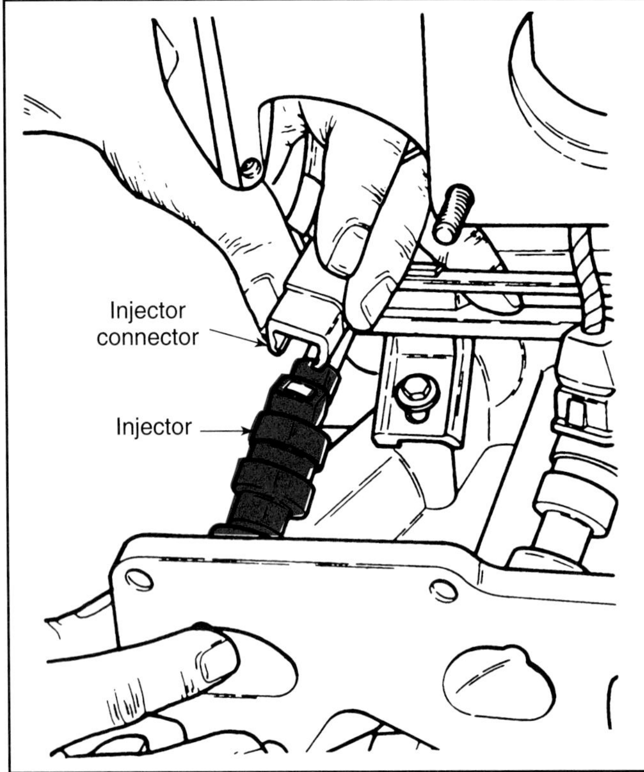

Figure 17-38. *Unplug the injector harness carefully. It is a good idea to mark each harness to the cylinder it is assigned to. (Ford)*

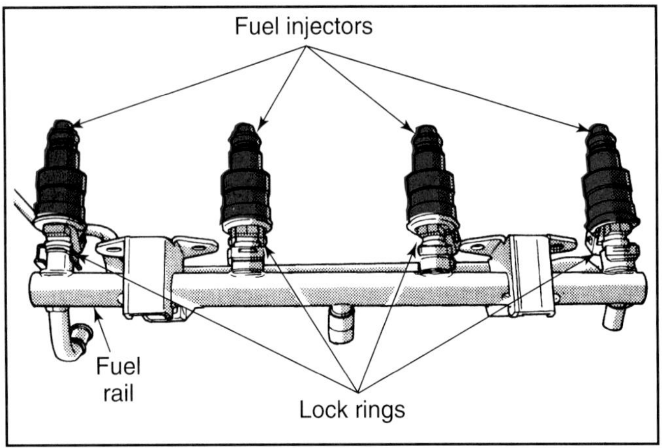

Figure 17-39. *Fuel injectors and fuel rail assembly. (Chrysler)*

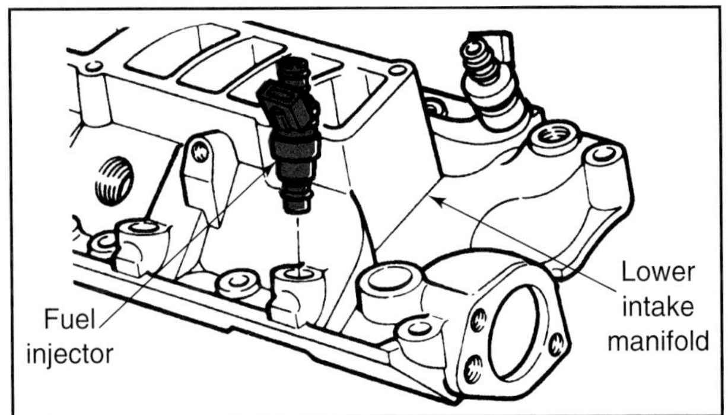

Figure 17-40. *Gasoline fuel injectors should pull from the engine with little difficulty. Do not use pliers or other tools to remove injectors unless instructed to by the service manual. (Ford)*

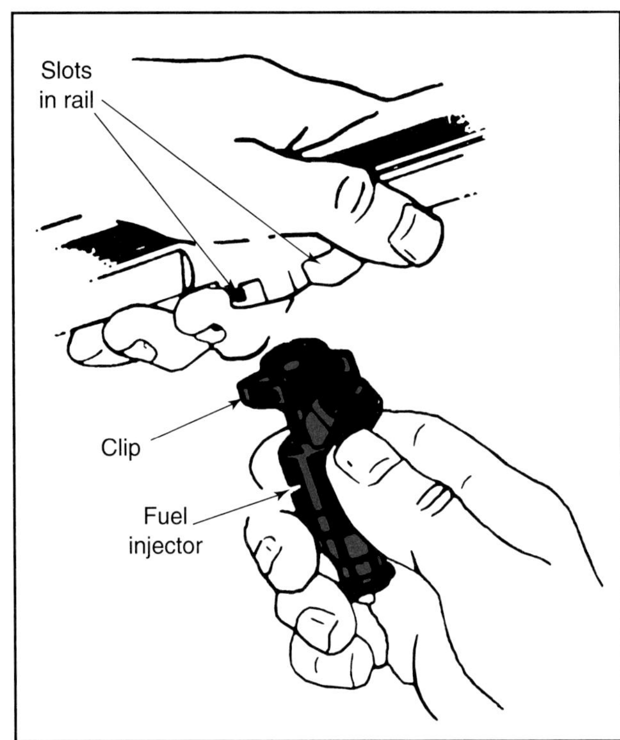

Figure 17-41. *Most injectors simply clip to the fuel rail. (General Motors)*

Injector Installation

Before installing new injectors, lubricate any O-rings with transmission fluid or clean motor oil so that the injector will be easier to install. This will also reduce the chance of O-ring damage during installation. Then install the injector into the fuel rail receiver cup.

Install the injector and fuel rail assembly into the intake ports. The fuel rail assembly must be drawn into the intake ports evenly, making sure each injector enters its own port. Make sure the injector terminal is facing in a direction that will allow a connection to the injector harness. When all the injectors are properly seated, install the injector hold-down clips and tighten the attaching bolts to the proper torque values.

Reconnect the injector electrical connectors, and install the pressure regulator vacuum line, if necessary. If the plenum was removed, reinstall it using new gaskets. Then attach the fuel supply hose to the fuel rail inlet. Reconnect the negative battery cable. Make a final check to ensure all fittings, fuel lines, ground straps, and wiring harnesses are located as they were originally.

Turn the ignition on and off twice to allow the fuel pump to fill the fuel lines. Start the vehicle and check the fuel injector system for leaks. If no leaks are found, check fuel system pressure and general injector system operation. If the system checks out, reinstall all remaining parts and road test.

Diesel Fuel Injectors

To replace diesel fuel injectors, release the fuel pressure by cracking a fuel line. Then loosen and remove the fuel line fitting from the injector(s) to be replaced. If the injector has a return line, remove it also. Pull the lines away from the injector, being careful not to bend or kink the tubing. Move the lines only far enough to provide clearance for removing the injector. Loosen and unscrew a threaded injector. Pull a pressed in injector from the cylinder head, **Figure 17-42**. If the injector is stuck, it can usually be removed by grasping it with pliers, or by using a small prybar to lightly pry up and loosen it. Repeat the removal process for all injectors that must be changed.

Thread or press the new injector in the head. Be careful not to damage the sealing washer during installation. Line up any alignment marks on a pressed injector before pushing the injector to its fully installed position. Tighten the threaded injector to the specified torque. Install the pressed-in injector retaining clip and tighten the attaching bolt to the proper torque.

Reinstall the tubing fittings, but do not tighten them. If only one or two injectors were removed, the engine can be started and the replacement injectors will be purged of air as the engine runs. If several or all injectors were removed, the injection system will contain too much air. The pump will not be able to develop sufficient pressure to start the engine and the system must be bled. With the fittings loose, crank the engine for approximately one minute, or until fuel begins squirting from the fittings. Then start the engine, following the usual diesel starting procedure, and allow it to run briefly. Tighten the tubing fittings and restart the engine. Then check the injector system for proper operation.

Warning: Wear eye protection and exercise care when bleeding diesel fuel injectors. The fuel injection system in a diesel engine operates at several thousand PSI and pressurized fuel can easily penetrate your skin.

Diesel Injection Pumps

Diesel injection pumps used in cars and light trucks are usually not field serviceable. The usual procedure is to remove the pump and replace it with a new or rebuilt pump. When removing the pump, note the orientation of the pump drive and driven gears. The replacement pump must be reinstalled in the same position as the old pump to maintain injector timing.

The diesel injection pump may contain the idle and timing adjustment devices. Actual procedures vary widely between manufacturers. Many modern injection pumps are at least partially controlled by the ECM. Always check the correct service manual before servicing a diesel injection pump.

Adjustment with an Exhaust Gas Analyzer

To adjust the idle with an exhaust gas analyzer, place the exhaust probe of the analyzer in the vehicle tailpipe.

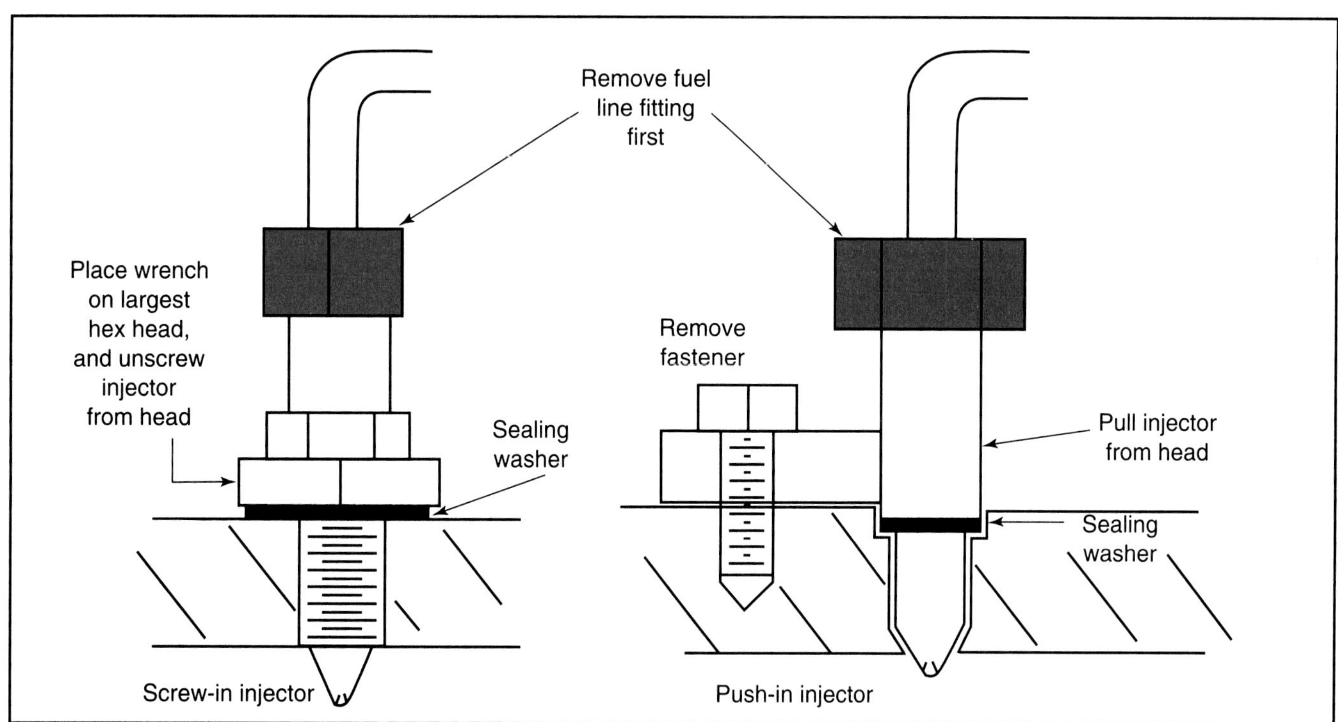

Figure 17-42. *Diesel fuel injectors are more difficult to remove. Since they are exposed to the high compression from the cylinder, it is often necessary to use special pullers to remove the injectors.*

Then start and calibrate the analyzer. The analyzer should be in its manual mode when used for calibration. Adjust the mixture to obtain the optimum level of emissions, which is much lower than that allowed by most local laws. The basic procedure for exhaust gas analyzer use is covered in more detail in **Chapter 18**. The emissions level varies with different model years, and the proper specification book should be consulted before adjusting the mixture.

Note: Although most states not require emissions testing on vehicles made before 1975, it is a good idea to use an exhaust gas analyzer when adjusting the air-fuel mixture on these vehicles.

Follow-Up for Fuel System Repairs

Road test the vehicle thoroughly. There should be no noticeable odors or fumes. Recheck the ECM for trouble codes that may have been set during the repair process. If possible, keep the vehicle overnight. This will allow any gaskets that were replaced to swell and seal more effectively. The idle speed settings can then be readjusted, if necessary. All sensor readings should be checked and verified for proper operating parameters, **Figure 17-43.** By keeping the vehicle overnight, any cold start circuits or devices can be rechecked with the engine cold. Perform an exhaust gas analysis to ensure the vehicle complies with local emissions standards.

Summary

A quick visual inspection and short road test are important first steps in fuel system diagnosis. An important beginning step in fuel system diagnosis is pressure checking. If the system does not produce sufficient pressures, the fuel pump may be defective, or the filter may be plugged. Excessive pressures indicate a defective pressure regulator. To check for lack of fuel add a small amount of gasoline to the tank. If the vehicle starts, low fuel was the problem. Also check the fuel pump relay if the pump is not operating.

Basic checks of the injectors include listening for injector clicking and measuring injector pulse. Scan tools can also be used to obtain information about individual components and overall system operation. Multimeters and exhaust gas analyzers are also useful when diagnosing fuel system problems.

To check the fuel filter it must be removed from the fuel lines. Filters are installed in the fuel lines in many ways, and in many locations. Once removed, the filter can be checked for clogging and the presence of dirt and metal particles. Also check the fuel tank and lines for damage and leaks.

Fuel quality should also be checked when no other problem is found. Remove a fuel sample and check for dirt and other contaminants. Check gasoline for excessive alcohol content, and check diesel fuel for the proper cetane rating.

Fuel injector and ECM diagnosis requires the use of a multimeter or scan tool. The injector and control system is checked by simulating false rich and lean mixtures. Changes in the pulse width can be read on the multimeter or scan tool. A timing light can be used to check TBI injectors for proper spray pattern. Idle air control motors can also be checked with a multimeter or scan tool. Turbocharger output is checked using a tachometer and boost pressure gauge. Throttle position sensors can also be checked with a scan tool, or the output voltage can be measured with a voltmeter. Voltage should vary smoothly as the throttle is opened.

Fuel pressure regulators are replaced if they are defective. Most throttle bodies have some provision for basic throttle opening adjustment, but the idle is actually controlled by the ECM. Some turbocharger wastegates are adjustable, but the majority are factory set with no adjustment device. Fuel system components are generally simple to replace. However, the technician should remember that gasoline is extremely dangerous. Always relieve pressure before removing any fuel system component.

Many fuel injection system components can be cleaned. Examples are injectors and throttle bodies. Injectors can be cleaned on or off the engine. TBI and multiport injection throttle bodies can be removed, cleaned, and overhauled. Regulators and injectors can be replaced as units when they are defective. Both procedures may involve some engine disassembly, such as removing the intake plenum and fuel rail. Diesel fuel injectors are replaced in a manner similar to gasoline injectors. Diesel fuel pumps are usually replaced as a unit instead of being serviced. Injection pumps must be carefully timed to the engine.

Know These Terms

Fuel pump priming terminals
Short-term monitor
Long-term fuel trim monitor
Fuel trim monitor
Power balance tests
Clear flood
Exhaust gas analyzer
Water separator
Alcohols
Biocide
Cetane rating
Cetane hydrometer
Waxing
Counts
Boost pressure

Scan Tool Values		
Scan Value	**Units Displayed**	**Typical Data Values**
Engine Speed	rpm	±50–100 rpm (± 50 in drive)
Engine Coolant Temperature (ECT)	F°/C°	185–221°F (85–105°C)
Intake Air Temperature (IAT)	F°/C°	52–176°F (10–80°C) (Dependent on underhood temp)
MAP sensor	kPa/Volts	20–110/.6–5 volts (Depends on vacuum and air pressure)
BARO	kPa/Volts	20–100/2.5–5 volts (Depends on engine vacuum and air pressure)
Mass Air Flow (MAF) sensor	Gm/Sec or Volts	3–9 or 2.5–4.0
Air Flow (Speed density)	Gm/Sec	3–9
Injector Pulse Width	M/Sec	1–5
Oxygen Sensor (O_2)	Volts	.1–1.0 and varying
Oxygen Sensor Heater	On/Off	On (Off once engine reaches operating temperature)
Throttle Position Sensor (TPS)	Volts	.3–1.5 (More with throtlle open)
Throttle Angle	Percentage (0–100%)	0
Idle Air Control (IAC)	Percentage or Counts	0–80 or 5–50
Transmission Position	P/N and RDL	Park/Neutral (P/N)
Short-Term Fuel Trim	Percentage or Counts	Varies
Long-Term Fuel Trim	Percentage or Counts	–10%–10% or 118–138
Open/Closed Loop	Open/Closed	Closed (May go open if code set or with extended idle)
Vehicle Speed Sensor (VSS)	mph	0 (Varies with vehicle speed)
Torque Converter Clutch (TCC)	On/Off	Off (On with TCC commanded)
EGR Solenoid	Percentage (0–100)	0
EGR Position	Volts	.3–2.5 volts
Knock Sensor	Yes/No	No
Spark Advance	Number of Degrees	Varies
Spark or Knock Retard	Degrees	0
Battery	Volts	13.5–14.5
Cooling Fan	On/Off	Off (On 225–230°F/106–110°C or Yes A/C Request)
Power Steering Switch	Normal/High Pressure	Normal
Transmission 2nd gear	Yes/No	No (Yes in 2nd, 3rd or 4th gear)
Transmission 3rd gear	Yes/No	No (Yes in 3rd or 4th gear)
Transmission 4th gear	Yes/No	No (Yes in 4th gear)
A/C Request	Yes/No or On/Off	No/Off (Yes/On when A/C is requested)
A/C Clutch	Yes/No	No (Yes when A/C is requested)
A/C Low Pressure	Yes/No	No
Brake Switch	Off/On	Off (On with brake pedal depressed)
Malfunction indicator lamp (MIL)	Yes/No	No
Fuel Pump	Yes/No or Volts	Yes or 13–14.5 volts
EVAP Canister Purge Valve	On/Off	Off
Misfire Monitor	0–200	0–5
Cruise Control	Off/On	Off (On when cruise control requested)
PROM/Program ID	Number/Date	Varies (Compare reading w/service literature)

Figure 17-43. *This chart shows scan tool values, the units displayed, and typical data values. The typical data values are for an engine running at idle, with a closed throttle, in park or neutral, in closed loop, with all accessories off. Most vehicles may show other scan values and ranges so check the service manual.*

Review Questions—Chapter 17

Please do not write in this text. Write your answers on a separate sheet of paper.

1. Many of the fuel system components are under high ______.
2. The fuel pressure may fluctuate as the ______ open and close.
3. If the engine is running, the ______ must be producing pressure.
4. A noid light will check the electrical condition of the ______.
   (A) injector
   (B) ECM injector driver
   (C) throttle position sensor
   (D) idle control valve
5. Fuel trim is a measure of the ______.
   (A) air-fuel ratio
   (B) manifold vacuum
   (C) injector pressure
   (D) catalytic converter operation
6. If you cannot blow through a fuel filter after all of the fuel has been removed, it is ______.
7. A driveability problem may occur if the gasoline has too much ______.
8. An IAC motor can be checked with a multimeter by readings its ______.
9. After a fuel system component is replaced, the electric fuel pump may take ______ seconds to repressurize the system.
10. Diesel injection pumps must be ______ to the engine for it to operate.

ASE Certification-Type Questions

1. A plugged fuel filter can cause all of the following, EXCEPT:
   (A) poor acceleration.
   (B) spark knock.
   (C) stalling.
   (D) hesitation.
2. A fuel pressure gauge is attached to the fuel rail of a fuel injected engine. When the ignition switch is turned on, no fuel pressure is indicated. All of the following could be the cause, EXCEPT:
   (A) clogged fuel filter.
   (B) clogged injector.
   (C) defective pump relay.
   (D) improper gauge connection.
3. On an engine with fuel injection and an electric fuel pump, fuel pressure will ______ when the engine starts.
   (A) go up
   (B) go down
   (C) not change
   (D) fluctuate at all times
4. If an electric fuel pump is not operating, bypassing the relay will determine whether the ______ is defective.
   (A) pump
   (B) relay
   (C) filter
   (D) Either A or B.
5. Fuel pressure rises on a running engine when the pressure regulator vacuum hose is removed. Technician A says that this indicates a defective pressure regulator. Technician B says that this indicates a plugged fuel return line. Who is right?
   (A) A only.
   (B) B only.
   (C) Both A & B.
   (D) Neither A nor B.
6. Technician A says that a defective injector may be stuck open by dirt. Technician B says that a defective injector may have failed internally. Who is right?
   (A) A only.
   (B) B only.
   (C) Both A & B.
   (D) Neither A nor B.
7. Pulse width is a measure of whether the ECM is operating the injectors based on inputs from the ______ sensor.
   (A) oxygen
   (B) throttle position
   (C) RPM
   (D) MAP

8. A modern scan tool can check all of the following, EXCEPT:
   (A) ignition timing.
   (B) injector pulse.
   (C) fuel pressure.
   (D) IAC operation.
9. Technician A says that multiport fuel injectors can be cleaned without removing them from the engine. Technician B says that a multiport injection system throttle body can be cleaned without removing it from the engine. Who is right?
   (A) A only.
   (B) B only.
   (C) Both A & B.
   (D) Neither A nor B.
10. All of the following are symptoms of injector problems, EXCEPT:
   (A) high current draw.
   (B) low turnoff voltage.
   (C) clicking.
   (D) ragged spray pattern.
11. The TBI throttle body contains all of the following parts, EXCEPT:
   (A) fuel filter.
   (B) injectors.
   (C) throttle valves.
   (D) pressure regulator.
12. Technician A says that injector O-rings should be lubricated with transmission fluid or clean motor oil so that the injector will be easier to install. Technician B says that injector O-rings should be lubricated with transmission fluid or clean motor oil to reduce the chance of O-ring damage during installation. Who is right?
   (A) A only.
   (B) B only.
   (C) Both A & B.
   (D) Neither A nor B.
13. Technician A says that if only one diesel engine injector was removed, the replacement injectors will be purged of air as the engine runs. Technician B says that if all of the injectors in a diesel engine were removed, the injection system must be bled. Who is right?
   (A) A only.
   (B) B only.
   (C) Both A & B.
   (D) Neither A nor B.
14. Technician A says that all fuel injection pressure regulators have a provision for adjustment. Technician B says that the throttle valves used with many fuel injection systems are controlled by an idle motor or calibrated air bleed. Who is right?
   (A) A only.
   (B) B only.
   (C) Both A & B.
   (D) Neither A nor B.
15. Before it can be cleaned by immersion in cleaning solvent, all of the following must be done to a TBI throttle body, EXCEPT:
   (A) it must be removed from the engine.
   (B) it must be disassembled.
   (C) it must be washed with water.
   (D) all electrical components must be removed.

Advanced Certification Questions

The following questions require the use of reference information to find the correct answer. Most of the data and illustrations you will need is in this chapter. Diagnostic trouble codes are listed in **Appendix A.** Failure to refer to the chapter, illustrations, or **Appendix A** for information may result in an incorrect answer.

16. The multiport fuel injection system in **Figure 17-6** is experiencing an extended crank condition when cold. Technician A says that a defective fuel pump relay could be the cause. Technician B says that an open in 465 DK GRN/WHT could be the cause. Who is right?
   (A) A only.
   (B) B only.
   (C) Both A & B.
   (D) Neither A nor B.

Short-term fuel trim	Long-term fuel trim
149	141

17. A vehicle with a multiport fuel injected engine has failed a state emissions test. A captured scan tool reading of short-term and long-term fuel trim is shown above. Which of the following is the *least likely* cause?
   (A) A defective oxygen sensor compensating for a lean condition.
   (B) A vacuum leak at the intake manifold.
   (C) Low fuel pressure.
   (D) A leaking fuel injector.

Data Captured from Vehicle			
Coolant temperature sensor 78°F/24°C	Intake air temperature sensor 72°F/21°C	Manifold absolute pressure sensor 1.1 volts	Throttle position sensor 4.5 volts
Engine speed sensor 0 rpm	Oxygen sensor 0.45 volts	Vehicle speed sensor 0 mph	Battery voltage 12.3 volts
Idle air control valve 34 percent	Evaporative emission canister solenoid Off	Short-term fuel trim 128	Long-term fuel trim 128
Malfunction indicator Lamp Off	Diagnostic trouble Codes	Open/closed loop Open	Fuel pump relay Yes
Brake light switch Off	AC request Yes	AC compressor clutch Off	Knock signal No
Ignition timing (°BTDC)	Base timing: 8		Actual timing: 8

18. A vehicle with the fuel system shown in the schematic in **Figure 17-6** will not shut off. Which of the following is the *most likely* cause?
    (A) The fuel pump prime lead shorted to battery power.
    (B) A short to ground in 150 BLK.
    (C) An open in 839 PNK/BLK.
    (D) The fuel pump prime lead shorted to ground.

19. The oil in a fuel injected engine is diluted by fuel and the exhaust has a rotten egg smell. The scan tool chart in **Figure 17-10** was captured at idle. Which of the following is the *most likely* cause of the oil dilution?
    (A) A stuck open injector in cylinder 2.
    (B) The circuit for the injector in cylinder 4 is shorted to ground.
    (C) The ECM is defective and is providing constant ground to the injector in cylinder 1.
    (D) A stuck open injector in cylinder 3.

20. A throttle body injected engine cranks, but will not start. The scan tool readings shown above was taken with the key on, engine off. A noid light test reveals no pulse to any of the injectors. Technician A says that a sticking AC compressor clutch is the cause. Technician B says that a defective or misadjusted throttle position sensor is the cause. Who is right?
    (A) A only.
    (B) B only.
    (C) Both A & B.
    (D) Neither A nor B.

Chapter 18

Emissions Control and Exhaust System Diagnosis and Repair

After studying this chapter, you will be able to:

- ❑ Use an exhaust gas analyzer to check air-fuel ratios.
- ❑ Diagnose problems in emissions system components.
- ❑ Diagnose EGR problems.
- ❑ Diagnose evaporative emissions control system problems.
- ❑ Perform an exhaust gas analysis.
- ❑ Replace emissions system components.
- ❑ Diagnose exhaust system problems.
- ❑ Replace exhaust system components.
- ❑ Perform an OBD II drive cycle.

This chapter discusses diagnostic and repair strategies for emission and exhaust system problems. While most exhaust emission problems are caused by other engine systems, it is important to know how to properly diagnose the emissions and exhaust systems. Since most exhaust system complaints are related to noise, the diagnostic procedures discussed in this chapter will concentrate primarily on emissions system components. Refer to **Chapter 10** if you need to review emissions system operation.

Preparing to Diagnose the Emission and Exhaust Systems

Most emissions system problems will have little or no noticeable effect on driveability. Emissions problems sometimes become known to the driver when the vehicle fails a state mandated emissions test. Some emissions complaints are related to the condition of the exhaust gas as it leaves the tailpipe. The driver may complain about the exhaust gas having an unusual color or the presence of an offensive odor. Some drivers may even complain about water exiting the tailpipe, which is a normal part of the catalyzing operation. If the vehicle has failed a state air quality test, try to obtain the test readout from the driver.

It is important to remember that most exhaust emissions problems are not caused by a defect in the emissions control system. More often, a defect in the engine, ignition, fuel, electrical, or computer control systems causes a problem that the emissions control system cannot completely compensate for, resulting in high exhaust emissions, **Figure 18-1.**

Most exhaust system complaints are not related to vehicle performance, but to audible noise. Audible noise is usually caused by leaks or defective or misaligned parts in the exhaust system. However, the exhaust system can

Figure 18-1. *These are the most common problems that occur in the emissions and exhaust systems. However, most emissions problems are caused by a malfunction in an engine-related system. (Oldsmobile)*

become restricted, which will affect vehicle performance. Also, an exhaust leak before the oxygen sensor and/or catalyst monitor will affect ECM input, possibly creating an artificial lean condition. Excessive exhaust noise will usually show up on engine start-up, making an extensive road test unnecessary.

Caution: Do not road test a vehicle with an exhaust restriction. This problem will show up as lack of power. If an exhaust restriction is suspected, perform a backpressure test before any road test. Road testing a vehicle with a restricted exhaust component could result in severe engine damage, fire, or explosion.

If the exhaust system sounds ok, bring the vehicle into the shop for inspection. Try not to road test the vehicle at this time. If a problem with the emissions system can be found by visual inspection, you do not want the engine heated to operating temperature. Some parts of the emissions control system, such as air injection pipes, can get extremely hot, requiring a cooldown period before they can be removed. As you learned in the earlier chapters, you should locate the appropriate service manual before starting the visual inspection.

Note: Most emissions problems are caused by a defect in the fuel, ignition, computer control, or electrical systems. Before inspecting the emissions or exhaust system, you should eliminate these other systems first.

Visual Inspection

Your inspection and testing of the fuel, ignition, electrical, and computer control systems may have revealed signs that the problem is in either the emissions or exhaust system. You may also be checking the emission and exhaust systems as part of follow-up for a repair performed on one of the other engine systems.

When visually inspecting the emissions system, check for evidence of tampering. This includes making sure that all emissions control devices are installed in their proper places and in good condition. Make sure that all electronic connectors and hoses are connected. Inspect the exhaust manifolds for cracks and leaks around the manifold and its gaskets (usually indicated by a line of discoloration on the manifold metal).

Raise the vehicle on a lift to inspect the exhaust system. Make sure that all heat shields and hangers are in place and tight. There should be plenty of clearance between the exhaust system and the vehicle components and heat shields. Inspect the system itself for damage from debris and corrosion. There should be no openings in the pipes, mufflers, resonators, or catalytic converters. Tap on the mufflers, resonators, and converters to check for internal damage. If the exhaust system is equipped with a catalyst monitor, make sure it is present and the connector is tight and properly routed.

Check for evidence of tampering with the exhaust system. This includes making sure that all exhaust components are in place. Add-on components, such as chrome exhaust tips, are considered acceptable in most areas so long as the original emissions and exhaust components remain in place and are not bypassed, or defeated, by the add-on component(s). Some states do permit some aftermarket replacement exhaust system components that enhance performance. However, these components must allow all original equipment emissions control components to be connected and must meet or exceed the original equipment emissions performance standards.

Many air quality inspection programs include a visual inspection for the presence of a catalytic converter. A common form of exhaust tampering involves removing the catalytic converter from the exhaust system and replacing it with a standard exhaust pipe. Sometimes, the converter is removed, hollowed out, and the shell reinstalled. These illegal procedures are done to gain better performance by reducing exhaust system backpressure. If the converter rattles or sounds hollow when you tap on it or an exhaust gas analysis shows an unexplained excessive rich mixture, lower than normal backpressure, or an excessive amount of pollutants, the catalyst element is either damaged or may have been removed.

Checking for Excessive Backpressure

While it is normal for all engines to have some backpressure, excessive backpressure can cause many driveability problems including loss of power, excessive fuel consumption, overheating, and severe engine damage if left uncorrected. Exhaust backpressure can be checked in one of two ways. One method uses a vacuum gauge and the other uses an exhaust backpressure gauge.

Using a Vacuum Gauge

The ***vacuum gauge*** is an established device for determining the internal condition of the engine. The vacuum gauge is attached to the intake manifold. With the engine running, the vacuum gauge reading will determine the overall condition of the engine, and may be useful in determining the condition of individual cylinders. The vacuum gauge can also be used to check for a restricted exhaust system. Attach the vacuum gauge and a tachometer and start the engine. Record the vacuum at idle. Set the parking brake and raise engine rpm to 2500 rpm. Allow engine speed to stabilize and read the vacuum gauge. The gauge reading should be steady and slightly higher than it was at idle. If the gauge reading is the same or lower, the exhaust system is probably restricted, **Figure 18-2.**

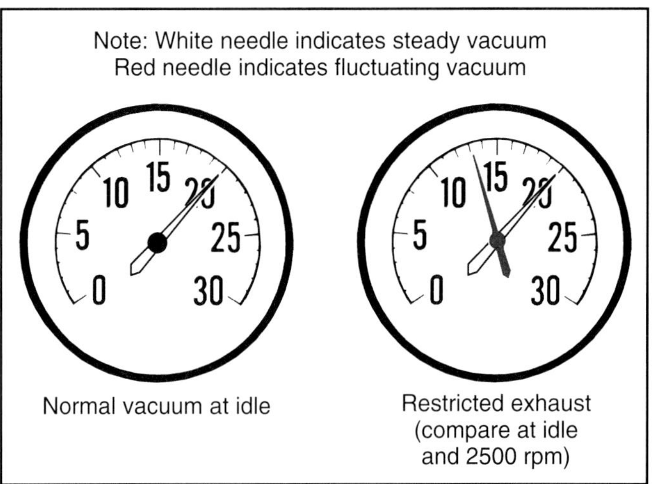

Figure 18-2. *The vacuum gauge can indicate the condition of the engine internally. A restricted exhaust is indicated if the gauge jumps to a normal reading and then slowly decreases, sometimes as low as 0 Hg.*

Exhaust Backpressure Gauge

A more accurate way of checking exhaust backpressure is to use a ***backpressure gauge.*** Some backpressure gauges are stand-alone tools, while others are simply adapters that can be used with a standard pressure gauge. Some backpressure gauges are designed to thread into the oxygen sensor's port. Remove the oxygen sensor and install the adapter. On vehicles with multiple oxygen sensors, use the sensor nearest the exhaust manifold. See **Figure 18-3A.** Other vehicles have air injection tubing and the gauge can be installed after removing the check valve, **Figure 18-3B.**

Note: Use anti-seize compound on the gauge adapter fitting.

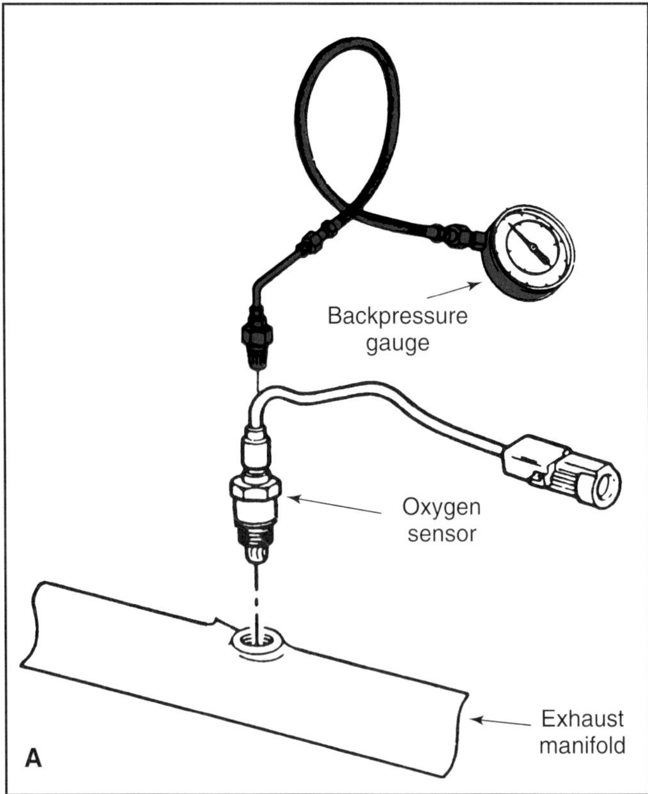

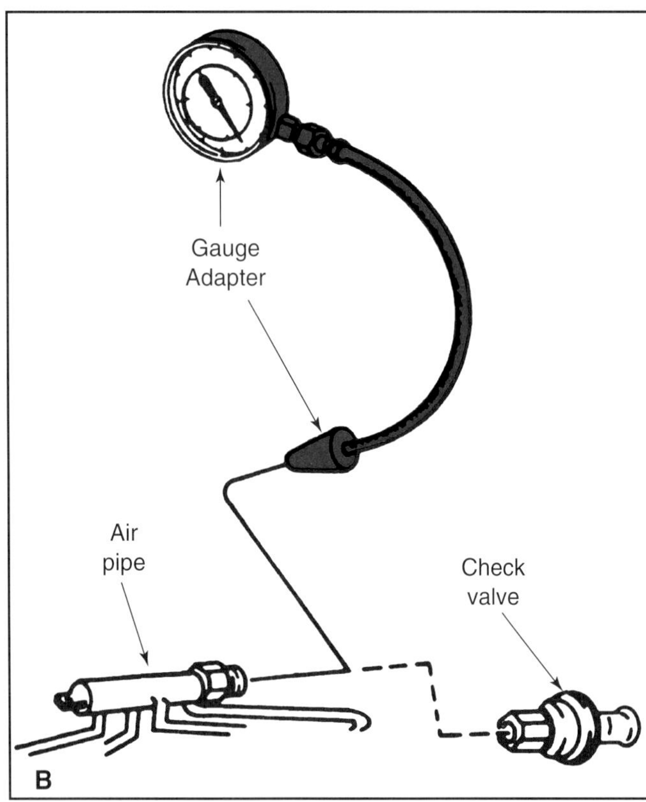

Figure 18-3. *A—A backpressure gauge is installed in the oxygen sensor port. Install it as close to the engine as possible at first and then move it to isolate a restricted exhaust component. B—If the vehicle is equipped with an air pump system, the backpressure gauge can be installed in the air pipe manifold. (General Motors)*

Start the engine and note the backpressure gauge. The gauge reading should not exceed 0.5-1.25 psi (3.4-8.6 kPa). Increase engine speed to 2500 rpm. The gauge reading should not exceed 2-3 psi (13.8-20.7 kPa). If exhaust backpressure is excessive, a restricted exhaust system is indicated. Inspect the exhaust pipes, resonators, mufflers, and catalytic converters for a restriction. If backpressure is normal, remove the gauge adapter, coat the oxygen sensor threads with sealant or anti-seize compound and reinstall.

Exhaust Gas Analysis

As stated earlier, an emissions problem is sometimes brought to the driver's attention when the vehicle fails a state mandated air quality (emissions) test. Since most parts in the emissions control system are not replaced as part of routine maintenance, most drivers do not notice the function of this system until there is a problem. If the exhaust system has normal backpressure and your visual inspection revealed no problems with any of the emissions or exhaust system parts, road test the vehicle thoroughly. This is to bring the engine and the exhaust system up to the proper operating temperature for an exhaust gas analysis test.

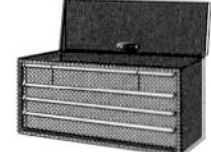

Note: Some vehicles that fail state air quality tests were not driven for a sufficient period of time before testing or the driver shut the engine off while waiting for testing.

Using the Exhaust Gas Analyzer

Using the ***exhaust gas analyzer*** is the only sure method of determining how efficiently the emissions control system, as well as the rest of the engine, is operating. If the air-fuel ratio is not properly controlled, the engine can develop driveability problems, as well as poor performance, high exhaust emissions, and poor fuel mileage. Often, an exhaust gas analysis will give you a good indication of where to look for a problem. There are several brands of exhaust gas analyzers and, therefore, no exact procedure for their use can be given. **Chapter 10** contains a description of exhaust gas analyzers and the types of gases they measure. The following information is a general guide only. Always follow the equipment and vehicle manufacturers' instructions exactly.

Note: Do not shut off the engine at any time during an exhaust gas analysis. An inaccurate reading will result. Make sure that the air conditioning system is off during testing.

After road testing, bring the vehicle directly into the shop; do not shut off the engine. Begin by starting and calibrating the exhaust gas analyzer in manual mode. Newer exhaust analyzers are very sensitive, so keep the probe away from the vehicle's tailpipe during calibration. Place the analyzer probe in the vehicle's tailpipe, and allow the engine to idle as the analyzer takes an exhaust gas sample. See **Figure 18-4.** Note the exhaust gas reading and compare it with manufacturer, federal, and/or state specifications.

EMISSION DATA		
Engine Speed		748 RPM
Engine Temp		192 °F
Hydrocarbons	HC	357 PPM
Carbon Monoxide	CO	0.93 %
Oxygen	O2	0.2 %
Carbon Dioxide	CO_2	14.4 %
Oxides of Nitrogen	NO_x	19 PPM
Air-Fuel Ratio	AFR	15.17
Lambda	λ	1.04

Figure 18-4. *Emissions testing is important to clean air and maintaining good performing vehicles. The readings indicate a lean air-fuel ratio. Depending on the vehicle's year of make, these readings could pass or fail. (Ferret)*

Inspection and Maintenance Programs

Some individual states have emission control test procedures, usually referred to as ***inspection and maintenance (I/M) programs.*** Existing I/M test procedures in most states usually consists of taking readings as the vehicle engine operates at idle, a set rpm, and/or vehicle speed. The program also requires a visual inspection for the presence of a fuel inlet restrictor and catalytic converter. This is often referred to as a *tailpipe emissions test.*

However, more detailed testing has been mandated by the Clean Air Act of 1990. New Environmental Protection Agency (EPA) regulations will cause a major change in the way that emissions tests are conducted in many states. Some states without emissions testing programs will begin testing soon, based on the new guidelines. The EPA has established three types of tests for checking the operation of emissions controls and the overall pollutant levels produced by the vehicle. The three tests are:

- ❑ The traditional tailpipe emissions test, modified to test the vehicle as it is accelerated and decelerated to mimic city driving conditions. This test is called the IM240 test.
- ❑ Purge flow test of the evaporative canister, designed to ensure that all unburned fuel collected by the evaporative canister can be drawn into the engine and burned.
- ❑ A pressure test of the evaporative system. This test ensures that the fuel tank and other fuel system components are not leaking.

Each test is relatively short, and the entire series takes no more than 10 minutes. These tests are discussed in more detail in the following paragraphs.

IM240 Emissions Test

The ***IM240 emissions test*** is similar to the traditional I/M test in that a probe is inserted in the tailpipe to record the emissions produced by the operating engine. Emissions monitored are the same as in earlier test procedures: hydrocarbons (HC); carbon monoxide (CO); and oxides of nitrogen (NO_X). However, unlike earlier stationary tests, the IM240 test is given as the vehicle is driven on a dynamometer and monitored by a computer driven emissions tester. The dynamometer places a load on the engine to simulate actual driving conditions. Emissions are read by the tester on a second-by-second basis. The emissions sensing equipment used with IM240 test setups is more sensitive than that used for older exhaust gas analyzers.

To perform an IM240 test, the vehicle is driven onto the dynamometer and an exhaust gas probe is inserted in the tailpipe. To test evaporative emissions purging, a flow meter is installed in the line between the evaporative emissions canister and the engine. Vehicles equipped with the OBD II diagnostic system may need to be driven through the drive cycle before the test is performed. The drive cycle is explained later in this chapter.

After all connections are made, the vehicle is operated through a specific cycle of acceleration, cruising, and deceleration which approximates normal city driving conditions. The entire cycle will take no more than 4 minutes, or 240 seconds. If the exhaust from a vehicle is exceptionally clean or dirty, the test will be concluded earlier.

Evaporative System Purge Test

Any gasoline that evaporates from the fuel system is drawn into the carbon canister, now called the evaporative canister. If this vapor was allowed to build up in the evaporative canister, it would eventually enter the atmosphere. Unburned fuel vapor is a major cause of pollution in many areas. The ***evaporative system purge test*** is conducted to ensure that evaporated gasoline from the fuel system is drawn from the evaporative canister into the engine.

This test is conducted as the IM240 test is being performed. A flow meter is installed in the line between the evaporative emissions canister and the engine is monitored by the analyzer. An acceptable flow rate tells the analyzer that that engine vacuum is drawing unburned fuel from the canister into the engine. Actual flow rates are usually much higher than the mandated flow rate. Current standards call for approximately .25 liters per minute, or 1 liter during the entire test. A properly operating purge system will flow as much as 10 to 25 times this amount.

Evaporative System Pressure Test

The ***evaporative system pressure test*** checks for any fuel system leaks that would allow unburned gasoline to escape into the atmosphere. To conduct the pressure test, the vapor line from the fuel tank must be disconnected at the evaporative canister. Then the fuel tank is pressurized to about .5 PSI using nitrogen. This pressurizes the entire fuel system. The analyzer monitors the pressure in the tank.

Note: Nitrogen is used as a pressurizing agent to reduce the possibility of fires. Compressed air should not be used for this test.

After the fuel tank is pressurized, the analyzer monitors the leak rate from the system. If the leakdown rate is not excessive the system passes. Current standards call for no more than a .25 PSI loss after two minutes.

Analyzing Exhaust Gas Readings

Remember that before the use of electronic engine controls, the carburetor and older fuel injection systems controlled air-fuel ratio based on the size of fuel passages, air pressure versus intake manifold pressure, air flow through the engine, and various mechanical settings. Therefore, exhaust emissions on these vehicles will be more likely to be incorrect and usually higher in pollutants such as hydrocarbons (HC) and carbon monoxide (CO).

On most late model vehicles, the ECM accurately adjusts the air-fuel ratio through the electronic fuel injectors. In addition, the catalytic converter on newer vehicles will attempt to clean up any rich mixtures. Therefore, problems causing incorrect air-fuel ratios will sometimes be masked and will not show up as clearly as on engines without converters. A defective, hollow, or missing converter will make a properly running engine seem much richer than it really is. A list of problems that can cause excessive exhaust emissions is in **Figure 18-5.**

Make any possible adjustments to the other systems to bring the exhaust emissions to specifications. This was covered in earlier chapters in this text. However, on most late model engines, no adjustments are possible. If the emissions levels cannot be corrected by adjustment, refer to the vehicle manufacturer's troubleshooting information. After making all needed tests with the exhaust gas analyzer, shut off the engine, and remove the analyzer probe from the tailpipe.

Understanding State Emissions Standards

Before you can compare exhaust gas readings, you must understand how to interpret the analyzer readings versus what is generated in a typical state air quality test. Depending on the type of testing system used, emissions can be measured in ***grams per mile (gpm)*** or in ***parts per million (ppm)*** and ***percentage (%),*** **Figure 18-6.** Most current state testing programs use the parts per million and percentage standard. A few states use grams per mile.

Hydrocarbons, oxides of nitrogen, and particulates are measured in parts per million. The lower the parts per million, the better the reading. Carbon monoxide, oxygen, and carbon dioxide are measured in percent of total gas. Carbon monoxide should be as low as possible, as this is an indicator of air-fuel ratio and how well the catalytic converter is working. Few states test for oxygen content, but this reading should fall within a certain range. Some states set a minimum requirement for carbon dioxide emissions, however, this percentage should be higher than the minimum requirement. In the other system, all quantities are measured in grams per mile. Vehicles with diesel engines are also tested for particulate emissions.

Depending on the state, only certain areas may be required to test vehicles for exhaust emissions, usually near larger cities and industrialized areas. A few states mandate emissions testing in all areas. The standard is also devised on a scale. Newer computer-controlled vehicles must produce less pollution than an older vehicle equipped only with a catalytic converter, EGR valve, and evaporative emissions canister.

Comparing Analysis Readings with State Inspection Readings

If the vehicle failed a state air quality test, compare the readings from the state test to your exhaust analysis, **Figure 18-7.** The two readings will be similar if the problem that occurred during the state test was present during your test. However, if the readings are dramatically different (the state reading shows only CO to be high and your reading shows HC to be high, for example), recalibrate the analyzer and retest the vehicle.

If the readings from the exhaust gas analysis you performed is correct and you can find no other problems, road test the vehicle again and retest in manual mode. If the exhaust emissions remain correct, the engine was probably not at operating temperature when the state test was done. If your shop is licensed to reinspect failed vehicles for exhaust emissions, you may do so at this time. Check local regulations before conducting a reinspection.

Other components used to check emission control devices include vacuum pumps, multimeters, and trouble lights. Since many of emissions controls are operated by the ECM, precautions should be taken to avoid damage. Do not use any test equipment on the computer-controlled components unless the manufacturer specifically recommends it.

Other Test Equipment

While the exhaust gas analyzer is a powerful tool for diagnosing emissions problems, other tools are necessary to locate which part is actually causing the problem. Earlier in this chapter, you learned how to use vacuum and backpressure gauges to check the exhaust system for restrictions. Scan tools can be used to test electronic EGR valves and to check the catalyst monitor on vehicles so equipped.

Excessive Hydrocarbon (HC) Reading: Excessive HC is usually caused by a problem that results in an incomplete burning of fuel. Sometimes accompanied by a "rotten egg" smell.

Poor cylinder compression	Open EGR valve
Leaking head gasket	Sticking or leaking injector
Ignition misfire	Improper fuel pressure
Poor ignition timing	Leaking fuel pressure regulator
Defective input sensor	Oxygen sensor contaminated or responding to artificial lean or rich condition
Defective output device	Fuel filler cap improperly installed
Defective ECM	

Excessive Carbon Monoxide Reading: Excessive CO is caused by a problem that results in a rich air-fuel mixture. However, excessive CO is often created by an insufficient amount of air or too much fuel reaching the cylinder. Will sometimes coincide with a high HC and/or low O_2 reading.

Plugged air filter	Higher than normal fuel pressure
Engine carbon loaded	Leaking fuel pressure regulator
Defective input sensor	Oxygen sensor contaminated or responding to artificial lean condition
Defective ECM	
Sticking or leaking injector	

Excessive Hydrocarbon (HC) and Carbon Monoxide (CO) Readings: When both HC and CO are excessive, this often indicates a problem with the emissions control system or an on-going problem, usually indicated by a rich air-fuel mixture, that has damaged an emissions control component. You should check all of the systems previously mentioned along with these listed below.

Plugged PVC valve or hose	Evaporative emissions canister saturated
Fuel contaminated oil	Evaporative emissions purge valve stuck open
Heat riser stuck open	Defective throttle position sensor
AIR pump disconnected or defective	

Excessive Oxides of Nitrogen (NO_X): Excessive oxides of nitrogen (NO_X) are created when combustion chamber temperatures become too hot or by an excessively lean air-fuel mixture.

Vacuum leak	Low coolant level
Leaking head gasket	Defective cooling fan or fan circuit
Engine carbon loading	Oxygen sensor grounded or responding to an artificial rich condition
EGR valve not opening	Fuel contaminated with excess water
Injector not opening	
Low fuel pressure	

Excessively Low Carbon Dioxide (CO_2) Reading: A low CO_2 reading is usually caused by a rich air-fuel mixture or a dilution of the exhaust gas sample. You should check all of the possibilities mentioned earlier, starting with the ones listed below.

Exhaust system leak	Sticking or leaking injector
Defective input sensor	Higher than normal fuel pressure
Defective ECM	Leaking fuel pressure regulator

Low Oxygen (O_2) Reading: Low oxygen readings are usually caused by a lack of air or rich air-fuel mixture, the same factors that can create an excessive CO reading. Note: Not all analyzers check the exhaust gas for O_2 content.

Plugged air filter	Higher than normal fuel pressure
Engine carbon loaded	Leaking fuel pressure regulator
Defective input sensor	Oxygen sensor contaminated or responding to artificial lean conditioning
Defective ECM	Evaporative emissions system valve defective
Sticking or leaking injector	

High Oxygen (O_2) Reading: A high oxygen level is an indication of a lean air-fuel mixture, dilution of the air-fuel mixture, or dilution of the exhaust gas sample by outside air. When a high oxygen reading is present, the CO reading is usually very low or does not register.

Vacuum leak	Defective input sensor
Low fuel pressure	Exhaust system leak near the tailpipe

Figure 18-5. *These are the most common causes of high emissions readings. Each section gives an explanation of how each excessive reading is created.*

Tested Gas	Standards (ppm &%)	Standards (gpm)
HC	220 ppm	.41
CO	1.2%	3.4
NO_x	Measured in gpm	0.4
Particulates	Measured in gpm	.08

Figure 18-6. *Emissions standards in parts per million, percentage, and grams per mile. These emissions standards vary, so check the standards that apply for your area.*

Gas	Standards (gpm)	Test Readings
HC	.41	1.13
CO	3.4	2.8
NO_x	.08	0.2

Figure 18-7. *Comparing test results to a set standard will give you a start on where to look for the problem.*

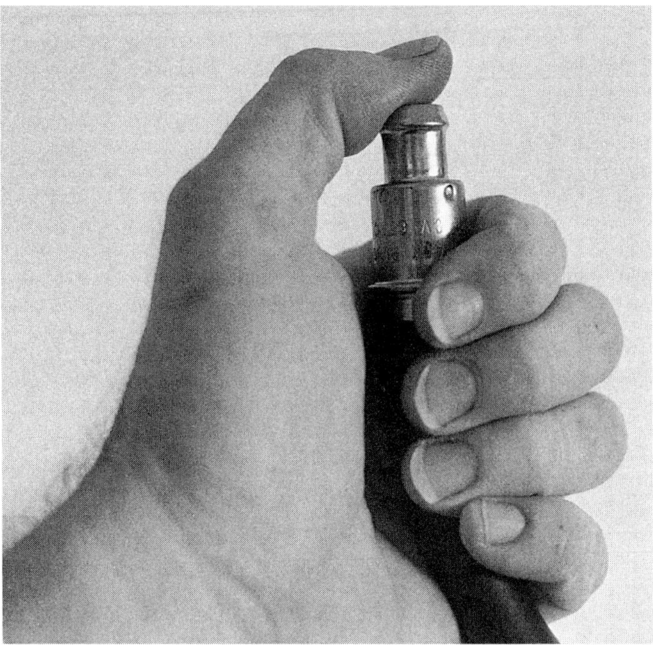

Figure 18-8. *A simple check for PCV valves is to remove the valve and place your finger or thumb over the opening. A strong vacuum should be felt if the valve is working.*

Checking Emission Control Components

Emission control components that can be checked include the PCV system, thermostatic air cleaner, evaporative emission control, EGR valve, and air pump. Many other emission control components are part of the fuel, ignition, or computer control system and should have already been checked as part of those systems.

Checking the PCV System

The PCV system is probably the easiest emission component to check. One simple test is to locate the PCV valve, remove it from the engine, and shake it. Listen for a rattling noise that indicates that the valve is not stuck. If no rattle is heard, the PCV valve is stuck and should be replaced.

A more accurate test can be performed on some PCV valves with the engine running. Start the engine and remove the PCV valve from the grommet that holds it to the valve cover or crankcase inlet hose (if it has not been removed already). Do not disconnect it from the hose leading to the intake manifold. Place your thumb over the valve, as shown in **Figure 18-8.** You should feel a strong vacuum, and the engine should slow down slightly when the vacuum is cut off. If this does not occur, the PCV valve is defective, or the hose is plugged. Another method of checking the PCV system is to use the PCV system tester described in the next section.

Note: Some PCV valves are not clearly visible when looking at the engine. Refer to the service manual for PCV valve location. Some engines do not have a PCV valve.

Fixed orifice PCV systems have been used in some vehicles in the past. In this type of system, manifold vacuum pulls blowby gases into the engine. However, the PCV valve is replaced with a calibrated opening, called an orifice. These systems have usually been abandoned because of problems with orifice plugging and oil pullover (oil being pulled into the intake manifold through the orifice). In this system, blowby gases pass through an oil separator before entering the intake manifold. In the oil separator, blowby gases flow direction is reversed, and the speed of the gases is reduced. Oil falls out of the air stream, and drains back into the crankcase through a tube at the bottom of the separator. To reduce clogging in cold weather, an ECM controlled heater is used to aid vaporization of the blowby gases before they pass through the orifice.

PCV System Testers

The PCV systems on some vehicles cannot be checked without the use of special testers. One kind of PCV tester is shown in **Figure 18-9.** These testers can be useful when the PCV valve is not easily accessible for checking by the other methods. To use the special tester, first remove the oil filler cap and plug the air cleaner's breather tube. Then start the engine and allow it to idle. Place the tester over the oil filler opening as the engine idles. If the PCV system is operating properly and the compression rings are not worn excessively, the float or ball in the tester will be pulled by vacuum to the "pass" side of the tester scale, **Figure 18-9A.** If the PCV system is plugged, or if compression rings are worn, allowing excessive blowby, the float or ball will be pushed by crankcase pressure to the "replace" side of the tester scale, **Figure 18-9B.**

Figure 18-9. *A—Place the PCV tester over the oil filler opening as the engine idles. If the PCV system is working correctly, the ball will move to the "normal" side of the chamber. B—If the PCV system is not working properly or blowby is present, the ball will move to the "replace" side.*

If the PCV valve is defective or the hose to the valve is clogged or collapsed, replace it. The relatively low cost of these parts makes cleaning unnecessary while also reducing the possibility of a comeback problem. PCV valve and hose replacement is the reverse of removal.

Checking Evaporative Emissions Systems

Evaporative emissions systems are relatively trouble free, but can develop four kinds of problems:

- Flow restrictions can be caused by clogged filters, inoperative flow control valves, or defective hoses.
- Vacuum buildup can be caused when the tank vent system fails.
- Canister can become saturated with fuel or charcoal may leak out.
- Leaks can occur anywhere in the system.

Evaporative emission system problems were more common on older vehicles with carburetors. Not only did they have a greater amount of vapors to absorb from the float bowl, flow failures could cause serious performance problems, even fuel starvation. Fuel injection engines produce fewer vapors, and are less likely to develop problems, **Figure 18-10.** However, any type of vehicle can develop evaporative emission system problems.

Flow Restrictions

The most common flow restrictions are caused by a canister vent filter that has clogged completely. Air must be drawn through the canister to allow the fuel vapors to be drawn out of the charcoal and into the engine. Filters that have clogged up will prevent operation of the purge system. Other problems are caused by collapsed or kinked hoses. Since these filters are often neglected, always assume that the filter has not been serviced and check it for clogging. Filter replacement is covered later in this chapter. The most common cause of hose problems are careless service procedures used by previous technicians. Often a hose is routed improperly and is kinked or pinched between other components. Sometimes a hose on an older vehicle will swell shut internally.

Checking Evaporative Flow Control Valves

Flow control valves are often defective. There are many variations of flow control valves, and the technician should refer to the proper service manual to diagnose a defective valve. Some systems also contain check valves to prevent the entry of liquid gasoline into the evaporative canister. In many cases, the valve is vacuum operated and controlled through a thermal vacuum valve. Using the proper service manual, check valve and temperature switch operation when the engine is cold and after it has warmed up. **Figure 18-11** shows a typical vacuum valve being checked with a vacuum pump. In **Figure 18-11A**, no vacuum is applied to the control diaphragm at the top. With no vacuum, no air should be able to flow between the second fitting and the two bottom fittings. This can be tested by trying to blow through the second fitting. When vacuum is applied to the control diaphragm, **Figure 18-11B,** air should be able to flow between the second fitting and the bottom fittings. If the valve allows air flow with no vacuum applied, or does not allow air flow when vacuum is applied, it should be replaced. The vacuum pump can be used to make other checks to this valve.

Testing Thermal Vacuum Valves

To test the thermal vacuum valve, apply vacuum to either port on the valve, as in **Figure 18-12.** If vacuum cannot be built up when the switch is cold, it has failed open. If vacuum can be developed when the engine is at normal operating temperatures, the switch has failed closed. In either case, the switch must be replaced. The service manual should be consulted to determine exact testing procedures.

Testing Purge Valves

Many purge valves on modern vehicles are operated by the ECM. On these systems, all of the vacuum operated

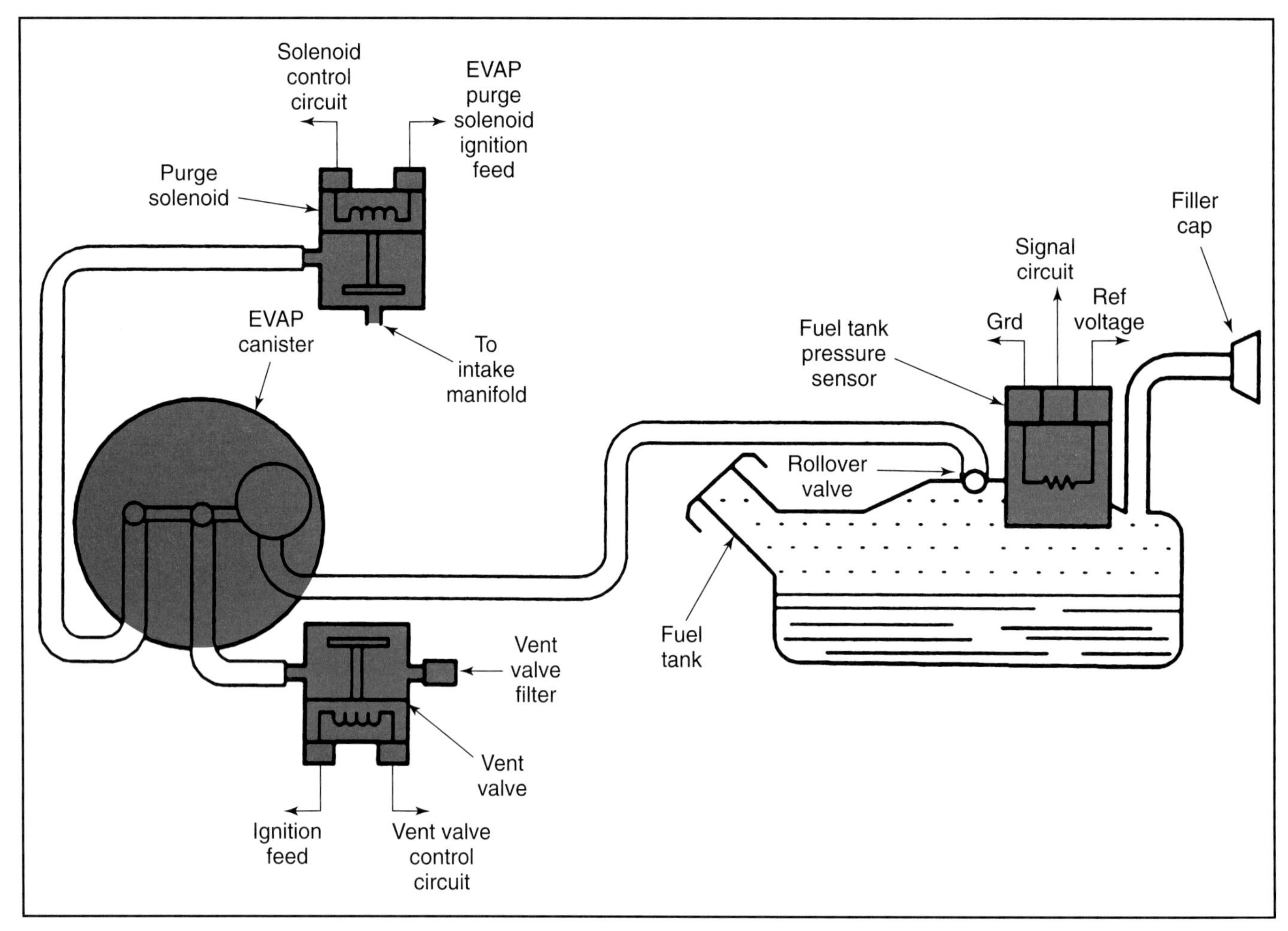

Figure 18-10. *Computer control and improved components have reduced the evaporative emissions produced by modern vehicles. This is the schematic for an enhanced evaporative emissions system. (General Motors)*

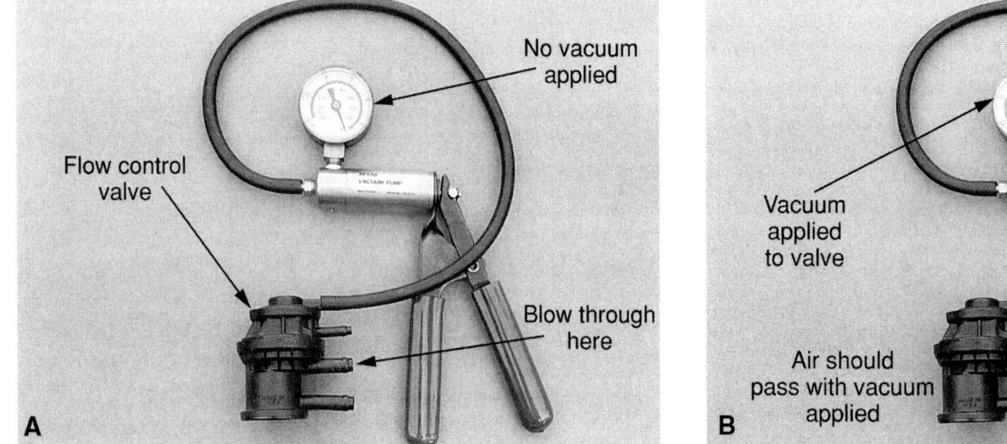

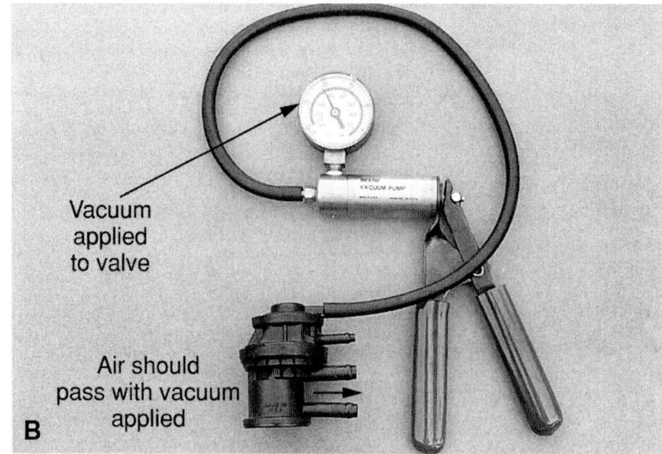

Figure 18-11. *A—Evaporative emissions flow valves can be checked with a vacuum pump. Some can be checked with a meter or scan tool. B—When vacuum is applied, air should be able to pass through the valve.*

controls are replaced by a single solenoid installed either in the purge line between the canister and intake manifold or on the canister itself. A resistance check of the purge valve solenoid can be made, as in **Figure 18-13,** or the actual operation of the valve can be monitored by use of a scan tool, **Figure 18-14.** Scan tools will give the solenoid's operational state or the percentage of vapor purge. Some solenoids can be checked by applying vacuum with a vacuum pump to determine whether the valve is being opened and closed by the ECM.

If carbon (actually charcoal) particles are found in the hoses, tank, valves, or elsewhere in the evaporative system,

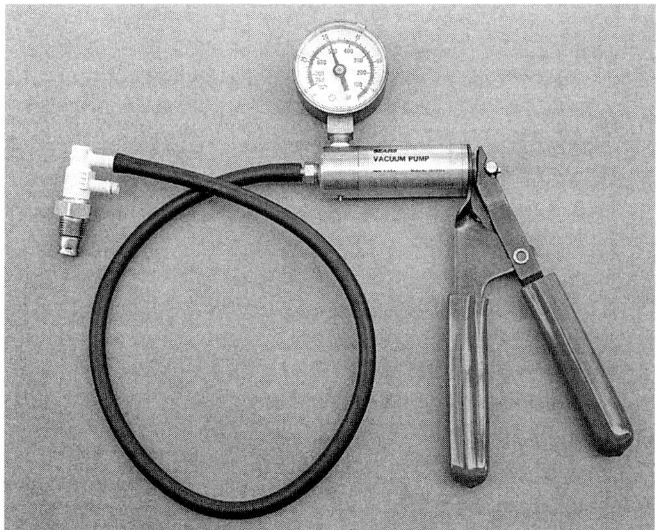

Figure 18-12. *Similar to the flow valves, thermal vacuum valves can also be tested with a vacuum pump. Also, apply heat to the probe tip to ensure the thermal valve is working properly.*

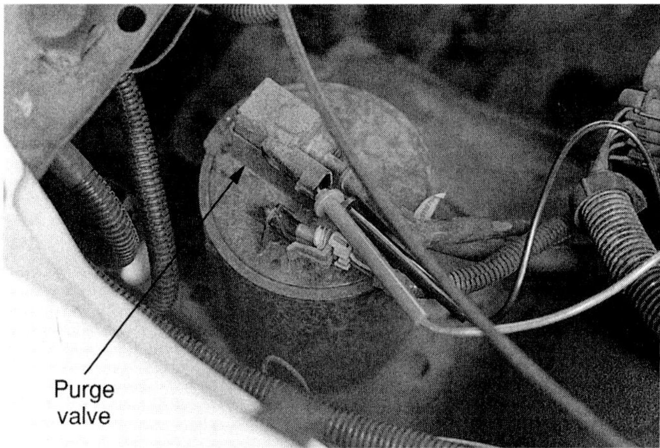

Figure 18-13. *Evaporative purge valve check using a multimeter. A test connector cut from the harness of a wrecked vehicle would make this test easier.*

Evaporative Emissions Purge Solenoid Operation		
State	Evaporative emission canister solenoid Off	Evaporative emission canister solenoid On
Percentage	Evaporative emission canister solenoid 0%	Evaporative emission canister solenoid 1–100%

Figure 18-14. *Evaporative emissions purge state can be monitored by a scan tool. Remember that purge normally occurs when the engine is running and the transmission is in drive.*

the evaporative canister has failed and carbon particles are escaping from the unit. The canister must be replaced. In cases of severe contamination, the entire system must be flushed to remove the carbon particles.

Vacuum or Pressure Buildup

On many vehicles, air can enter the fuel system through a check valve located in the fuel filler cap or through the emissions canister filter. These systems have a check valve between the canister and tank, and rely on the filler cap valve to admit air to the tank. The cap is designed to allow air to enter the fuel tank as the fuel is consumed. If the valve sticks, the fuel pump may not be able to draw fuel into the engine. If the canister filter is plugged, the purge system may draw liquid fuel into the canister. On some vehicles, a separate vent is used, but this is uncommon. If removing the fuel cap causes a rush of air into the tank, check the vent in the filler cap to ensure that it is not stuck closed. Proceed to check for a stuck vent and/or clogged canister filter.

Note: Air rushing out of the tank is a normal condition. The design of the vented cap will allow some pressure to build up in the tank.

If the vent is stuck closed, the fuel pump may be powerful enough to partially collapse the fuel tank. Therefore, if a venting problem is suspected, check the fuel tank for signs of collapse. On systems with OBD II vapor monitoring, the fuel tank is sometimes placed under a partial vacuum for test purposes. If the tank cannot hold a slight vacuum, the OBD II system will set a trouble code for a leak. Fuel tanks that are ECM tested under slight vacuum may have a non-vented fuel filler cap or a filler cap check valve which opens at a higher vacuum than those used on other systems. If the cap is opened while the vacuum test is going on, it will be normal to observe air rushing into the tank. Check the vehicle's service information carefully before deciding that there is a defect in the system.

Leaks

Any leaks of liquid gasoline from the fuel system will show up as liquid gasoline, or stains on or near the leaking component. Vapor leaks are harder to find. Often the most effective way of finding leaks is making a visual check for cracked or disconnected hoses, loose clamps, or damaged components. Locating leaks may require pressurizing the tank. Use nitrogen as the pressurizing agent, and pressurize the tank to no more than .5 PSI. The tailpipe probe from an exhaust gas analyzer can then be passed over the fuel system components. Any leak will cause an excess hydrocarbon reading on the analyzer.

Warning: Never pressurize the fuel tank with any gas except nitrogen. Use of another gas may cause an explosive mixture.

Checking EGR Valves

Vacuum operated EGR valves can be checked by observing the valve stem or feeling for diaphragm movement as the engine is accelerated, **Figure 18-15.** Set the parking brake and have someone depress the brake pedal before accelerating the engine. As the engine speed increases, the valve stem should be observed to move. If it does not, the valve is stuck, the hose is leaking or is plugged, or the vacuum diaphragm is leaking.

Note: If the EGR valve is controlled by a solenoid, this test may not give accurate results. Consult the service manual for exact testing procedures.

This test can also be performed with a portable vacuum pump. If the valve does not move when vacuum is applied, the valve is defective. To check the EGR pintle itself, the valve must be removed from the engine. The valve and passages can then be observed for sticking or excess deposit buildup.

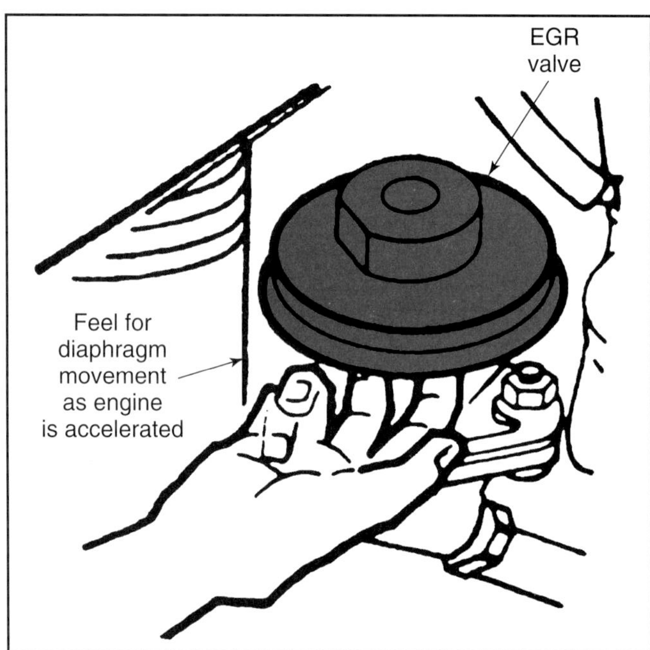

Figure 18-15. *Feel the EGR diaphragm for movement as the engine is accelerated. Be careful as the EGR valve gets very hot.*

Diagnosing EGR Control Solenoids

A common cause of problems on vacuum operated EGR systems is a defective control solenoid. The EGR solenoid controls vacuum flow to the diaphragm, based on an electrical signal from the ECM. Most EGR solenoids are normally open types, and will allow vacuum flow to the diaphragm when the ECM does not signal them to close. Therefore, most solenoid problems will cause the EGR valve to open when it is not desirable, leading to hesitation, surging, and stalling.

Some EGR solenoids can be checked by energizing them using battery current, **Figure 18-16A.** Other EGR solenoids can be checked by using a vacuum gauge and test light. In **Figure 18-16B,** the EGR diaphragm is activated using a vacuum pump. On this vehicle, if the EGR diaphragm moves when the ignition is on, and the shift lever is in park, either the EGR solenoid or ECM is defective. To check further, remove the input wire from the EGR solenoid and attach a test light to the solenoid terminals, **Figure 18-17.** If the light comes on with the transmission in park, the solenoid is receiving a close signal from the ECM. If the solenoid is receiving the close signal and still allows vacuum to reach the EGR diaphragm, it is defective. If the light does not come on, the problem is in the ECM.

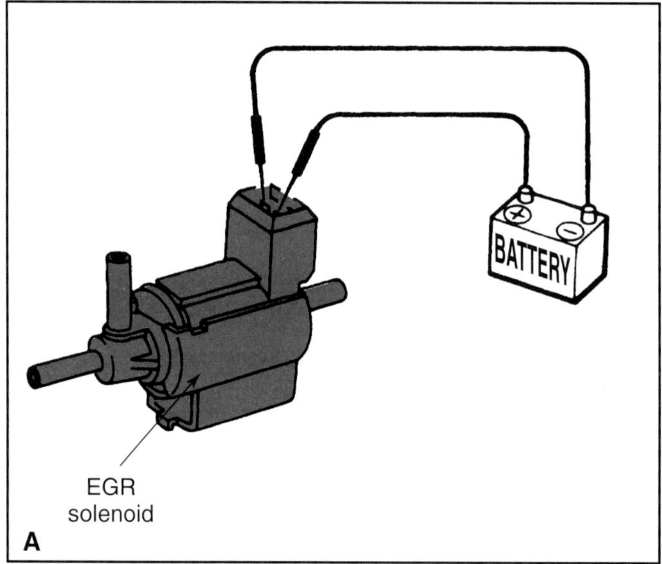

Figure 18-16. *A—Some EGR solenoids can be checked by jumping battery power to the terminals. Check the service manuals first before performing this test. B—A vacuum pump should be used to simulate engine vacuum to the EGR valve. (Subaru)*

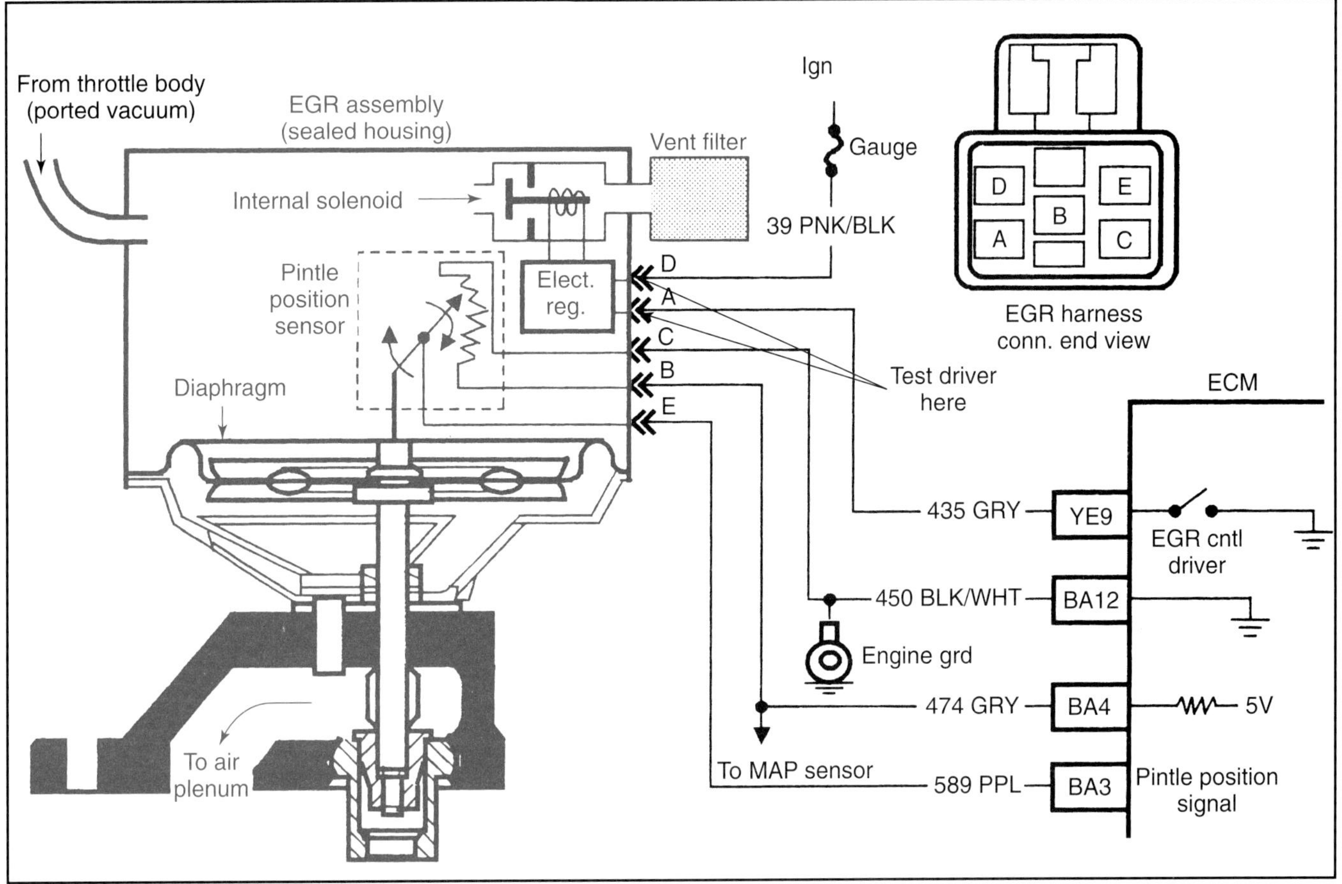

Figure 18-17. *Electronic EGR valves can be checked with a test light or by monitoring operation with a scan tool. Some scan tools can perform tests on the EGR solenoids. (General Motors)*

Checking EGR Position Sensors

If an EGR valve is equipped with a position sensor, the sensor operation can often be checked through the scan tool as the engine operates by monitoring the voltage or percentage, **Figure 18-18.** Another common test of EGR position sensors is to apply vacuum with a vacuum pump while observing voltage or resistance with a voltmeter. Check the shop manual for exact testing procedures and specifications. Generally, the electrical reading should change as vacuum is varied. If the sensor fails to produce the proper electrical reading at a specific vacuum, it should be replaced.

EGR Position Sensor Operation		
State	**Park, neutral and drive at idle**	**Driving except at idle**
EGR pintle (voltage)	EGR valve control solenoid 0.5 volts	EGR valve control solenoid 2.4–5 volts
EGR pintle (percentage)	EGR valve pintle 0 percent	EGR valve pintle 1–100 percent

Figure 18-18. *Scan tools can monitor EGR operation by monitoring solenoid voltage or pintle position. Note that there is low voltage and no pintle movement at idle.*

Electronic EGR Diagnosis

The pintle of an electronic EGR valve is operated by one or more electric solenoids or motors. The used of electrical devices means that the movement and position of the pintle can be directly read on a scan tool. The ECM calculates the needed movement from the signals received from the EGR valve itself. The condition of the EGR valve can be read and tested by a scan tool, **Figure 18-19.** Switch the scan tool to the intrusive test mode and test each solenoid. As each solenoid is energized, the engine will decrease in speed, idle rough, possibly set a code for excessive knock, and even stall. If engine operation does not change when an EGR solenoid is energized, either the valve, ECM, or wiring is defective.

EGR Flow Rate Monitoring

OBD II systems have a procedure for monitoring the flow rate through the EGR. On modern vehicles, less EGR flow is needed because close control of other engine operating parameters reduces NO_X formation. Therefore, the EGR is opened as little as possible to improve driveability. On some vehicles, flow rate is calculated by measuring the pressure drop across the EGR valve. This is done with a sensor similar in design to a MAP sensor. On other vehicles, the computer determines EGR flow rate by reading pintle position, manifold vacuum, and air flow, and calculates

Data Scanned from Vehicle During Intrusive Test			
Coolant temperature sensor 211°F/99°C	Intake air temperature sensor 82°F/28°C	Manifold absolute pressure sensor .6 volts	Throttle position sensor .44 volts
Engine speed sensor 713 rpm	Oxygen sensor .77 volts	Vehicle speed sensor 0 mph	Battery voltage 13.1 volts
Idle air control valve 32 percent	Evaporative emission canister solenoid Off	Torque converter clutch solenoid Off	EGR purge valve 0 percent
Malfunction indicator lamp Off	Diagnostic trouble codes	Open/closed loop Closed	Fuel pump relay On
EGR solenoid 1 On	EGR solenoid 2 Off	EGR solenoid 3 Off	Knock signal Yes
Ignition timing (°BTDC)	Base timing: 7		Actual timing: 21

Figure 18-19. *Intrusive tests can determine if the solenoids are defective. Here, EGR solenoid 1 is being tested manually by the scan tool. Idle speed will drop and the engine will run rough as each solenoid is tested.*

actual flow from one or more of these readings. On all systems with an EGR air flow measuring feature, air flow can be read by the scan tool.

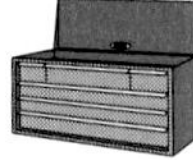

Note: When servicing any electrically controlled EGR system, make all electrical tests according to factory procedures. Failure to do so may destroy one or more of the ECM output drivers.

Checking the Air Pump and Valves

To test the air pump, the engine must be running. The pump should be quiet, with little, if any noticeable noise at idle. If the pump is noisy, remove the drive belt and retest. Race the engine while observing the diverter valve muffler. A low pressure gauge can be used to check the pump output. Pressure should be about 1 psi (6.9 kPa). If not, check the air filter for clogging. If the engine is raced and then the throttle is suddenly released, air should be heard and felt exiting through the diverter valve muffler, **Figure 18-20.** If air does not exit, the diverter valve may be defective.

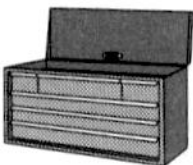

Note: The diverter valve may flutter when the engine is idled. This condition is normal.

If no air flow was felt from the muffler, remove a hose to the air manifold pipes (before the switching valve, if the system has one) and observe whether air is being pumped from the hose outlet. If air is coming out the hose, the pump is working. If no air is coming out, the pump is not delivering

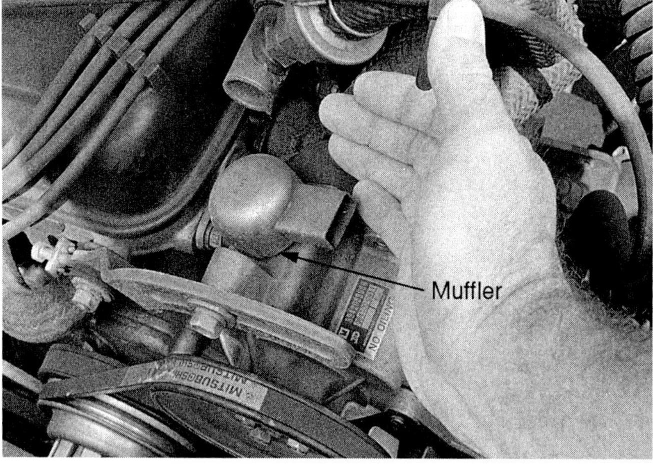

Figure 18-20. *Normal muffler operation will allow a small amount of air to exit the muffler. If air does not exit, the diverter valve or the pump is defective.*

air. Also remove and check the vacuum signal line to the diverter valve. There should be vacuum whenever the engine is running.

Note that on air pump systems with switching valves, the output air is delivered to the exhaust manifold when the engine is cold, and to the converter when the engine is warm. This can be confirmed by removing a hose past the switching valve. If the air pump system is controlled by the ECM, refer to the service manual for further testing procedures. Check valves can fail and allow exhaust gases to enter the air pump and control valves. If any evidence of exhaust gas build-up is found (exhaust deposits or corrosion) check the operation of the check valves. Check all hoses, lines, and manifolds for kinks, pinching, and leaks.

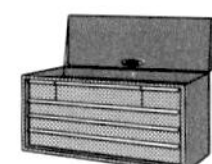

Note: Be sure to check the drive belt on the air pump for tension and condition.

Checking the Thermostatic Air Cleaner

To check the operation of the thermostatic air cleaner, observe the flap in the air cleaner snorkel when the engine is off. It should be open. Attach a manual vacuum pump to the air cleaner control vacuum hose. Apply vacuum to the hose and observe the flap, **Figure 18-21.** The flap should close. If it does not, the diaphragm is leaking or the control valve is not working. To ensure that the flap is opening, observe it after the engine is fully warmed up and running or heat the sensor with a heat gun. With the vacuum hose connected to the manifold, the flap should be completely open.

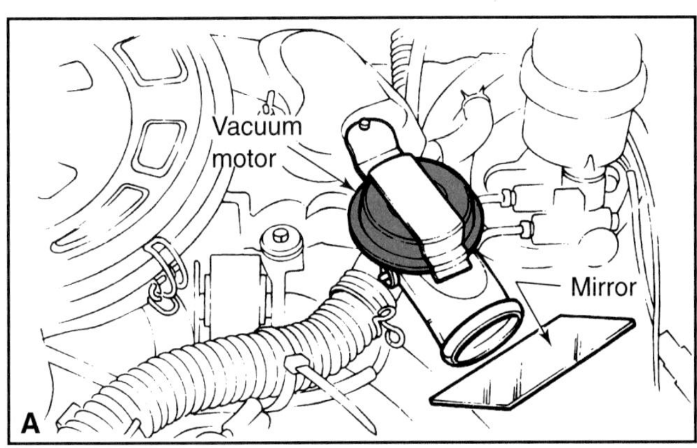

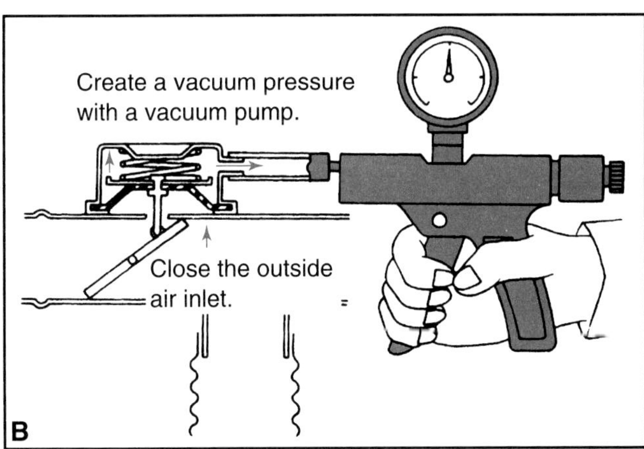

Figure 18-21. *A—Use a mirror to check thermostatic valve operation. The valve should close as the engine warms up. B—If the valve does not open or close, use a vacuum pump to test the motor. (Subaru)*

Checking EFE Systems

Two basic defects can occur on an EFE system: a sticking valve or a problem with the vacuum control system. If the exhaust gases cause the valve to stick, it can usually be freed up by soaking the shaft with penetrating oil and lightly tapping on the shaft. If a valve cannot be freed, or it keeps sticking, it should be replaced.

EFE vacuum control system problems can be caused by a leaking vacuum diaphragm or by failure of the thermal vacuum valve and related hoses. First, make a visual check of the vacuum lines to ensure that they are not split, collapsed, or disconnected. Then remove the vacuum line at the thermal vacuum valve and apply vacuum to the diaphragm with a vacuum pump. If the diaphragm does not move, it is defective (assuming that the valve is not stuck). If the diaphragm moves when vacuum is applied with the pump, but does not move when the engine is started cold, the diaphragm is good and either the thermal vacuum valve is defective or the hoses are misrouted. Before replacing the thermal vacuum valve, make sure that the coolant level is not low. If coolant does not contact the switch, it will not operate properly.

Emissions Filters

Emissions filters will usually not cause driveability problems unless they become completely plugged. If the PCV filter becomes completely plugged, it will cause the same symptoms as a plugged PCV valve, such as rough idle and poor gas mileage. If the evaporative emissions filter becomes plugged, the engine will not be able to draw air through the evaporative emissions system and may exhibit signs of a rich mixture. If an emissions filter problem is suspected, remove it and make a visual check for obvious accumulations of dirt or oil, **Figure 18-22.** Replacement is usually done as part of routine maintenance, which is discussed in **Chapter 21.**

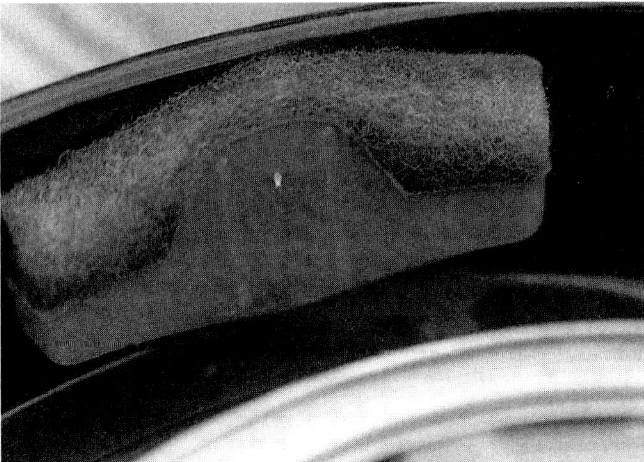

Figure 18-22. *A dirty emissions filter will cause driveability problems similar to a plugged PCV valve. Since this filter is not used in all vehicles and is not frequently mentioned, it is often missed during routine maintenance.*

Checking Exhaust System Parts

The most common driveability problem caused by the exhaust system is lack of power or low top speed caused by restricted exhaust components. If the exhaust system is restricted, the engine will use its power to clear out the exhaust gases instead of moving the vehicle. If you suspect a restricted exhaust, start by checking the manifold vacuum or exhaust backpressure at low and high engine speeds. This was covered earlier in this chapter. In most cases, exhaust system problems are related to exhaust gas leaks caused by system corrosion. **Figure 18-23** shows a corroded muffler. Corrosion usually starts at the exhaust component furthest from the engine, where low heat prevents the vaporization of water and acids.

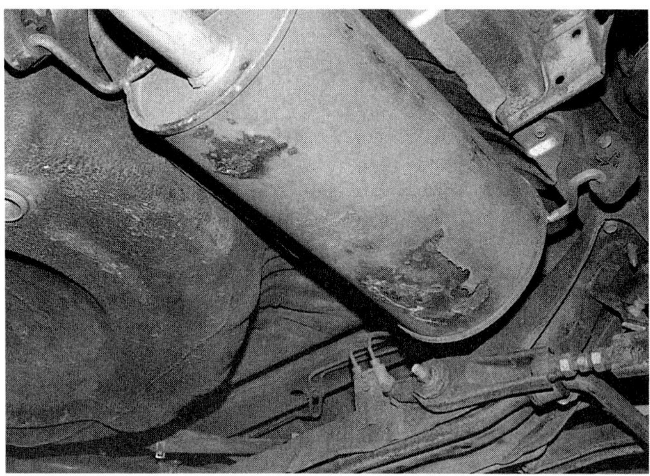

Figure 18-23. *Corrosion is the exhaust system's worst enemy. Corroded exhaust parts will change the exhaust gas content at the tailpipe, which will alter any readings by an exhaust gas analyzer. If the leak is before the catalytic converter, it can change the air-fuel ratio on some vehicles.*

If the vacuum or backpressure test establishes that the exhaust system is restricted, the problem can be isolated by removing exhaust system components and searching for the source of the restriction. As a quick check, an exhaust system joint can be loosened ahead of the suspected pipe, and the vehicle test driven a short distance. If the vehicle performs normally, the exhaust system is restricted somewhere past the loosened joint. Continue to loosen joints until you isolate the part.

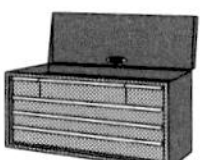

Note: Loosen the exhaust pipes connected to the catalytic converter first.

Possible Locations of Exhaust Restrictions

Several parts of the exhaust system can cause restrictions to exhaust gas flow, **Figure 18-24.** The diverter plate that is part of the heat riser or early fuel evaporation (EFE) system can stick in the closed position, causing a restriction inside the exhaust manifold. However, the system passages are designed to allow partial exhaust flow even when closed, and a closed plate is more likely to cause engine overheating, pinging, or fuel system flooding. On some V-type engines, a stuck EFE valve can cause the exhaust valves to burn on that cylinder bank.

After long use, the active elements in the catalytic converter can become coated with exhaust deposits. The

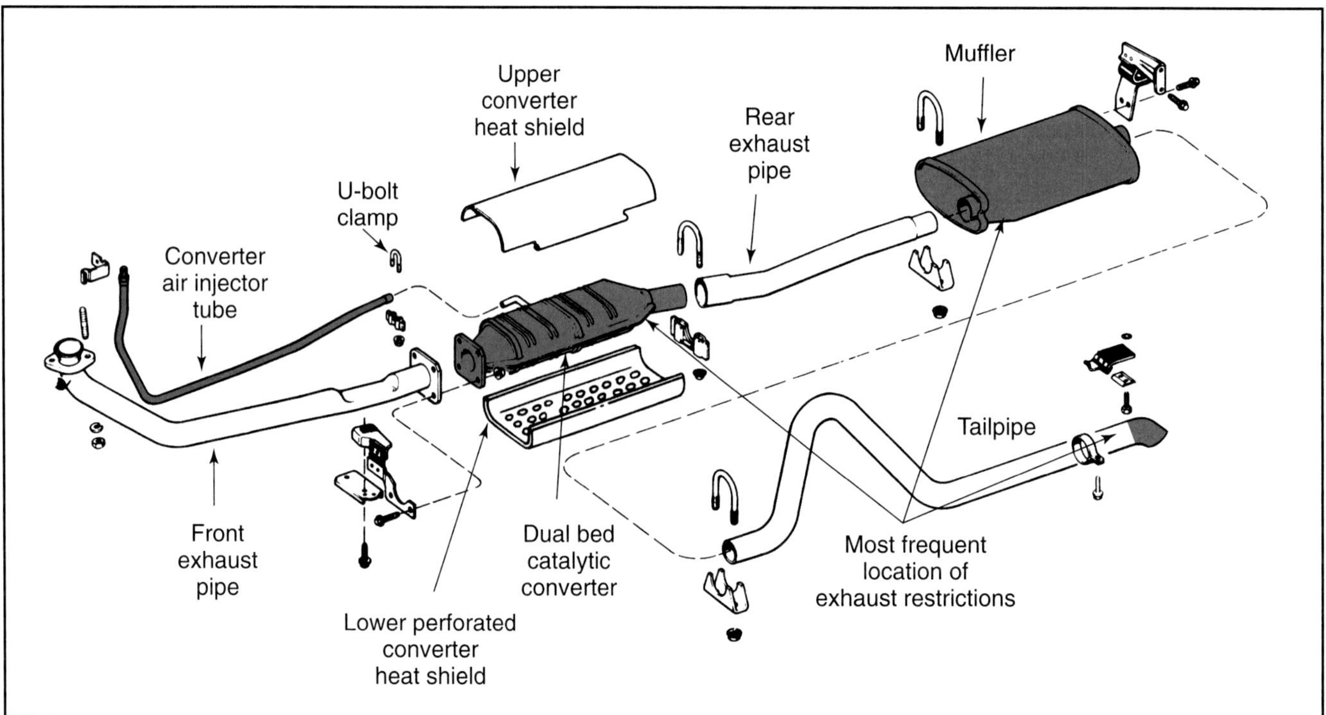

Figure 18-24. *Parts of the exhaust system are more susceptible to restrictions than others. The colored areas indicate the most frequent locations for exhaust restrictions. (Dodge)*

converter stops working, cools off, and becomes plugged with exhaust deposits. Usually, a converter that quits working can be identified by a rise in the exhaust emissions, especially hydrocarbons (HC). Also, a plugged or inoperative converter will be much cooler than a normally operating converter. While all parts of the exhaust system can cause a restriction, the catalytic converter is the most frequent cause of exhaust flow problems.

In some cases, the muffler can become plugged with soot or other exhaust debris, but this problem is rare, especially on newer vehicles with catalytic converters. Usually, a muffler that runs cool enough to build up this level of deposits will corrode and leak before restriction becomes a problem. Another problem is a clogged tailpipe from a restriction placed or wedged in the pipe or by a collapsed section. A collapsed pipe is obvious in some cases.

On many vehicles, double-wall pipes are used to reduce exhaust noise. These pipes are composed of two separate pipes, one inside the other, sometimes with a layer of high temperature plastic between the pipes for additional noise reduction. Sometimes, the inner wall of a double-wall pipe will collapse inward, restricting or almost totally blocking exhaust flow. If all other causes have been eliminated, the pipe can be probed with a wire after it is removed from the vehicle. In some cases, the only way that a collapsed pipe can be definitely located is by removing it and cutting it open.

Replacing Emissions and Exhaust Components

Once you have located the source of the problem, you can begin the process of parts replacement. Most emissions systems repairs are fairly simple, with very few, if any non-related parts to remove or relocate. While the exhaust system is simple, care must still be taken when dealing with this system, or other problems may be created.

Evaporative Emissions Canister

Some evaporative emissions canisters have an internal filter that cannot be replaced separately. If the filter cannot be replaced, or if the canister is otherwise defective, it must be replaced. Many canisters are located in the engine compartment, as shown in **Figure 18-25.** Others are located near the fuel tank or at the rear quarter panel. Consult the service manual if the location of the canister is not obvious. If the canister carbon has escaped into the rest of the system, it must be removed by blowing it out with compressed nitrogen.

Note: If carbon has been released, the evaporative emissions system must be purged before replacing any component.

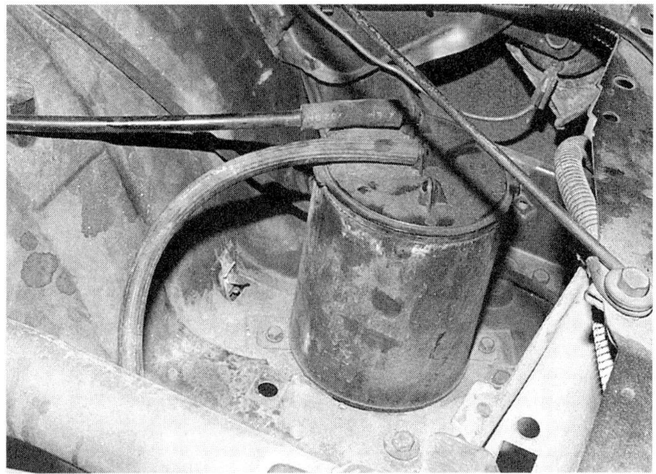

Figure 18-25. *Evaporative emissions canisters are normally located in an easy-to-reach location. This one is held by two bolts and has two hoses with no solenoid connections.*

After locating the canister, remove the vacuum and electrical connectors, being careful to mark any vacuum hoses if they are the same size. Then remove the mounting brackets and remove the canister. Compare the old and new canisters, and ensure that the new canister is equipped with a filter, **Figure 18-26.** Install by placing the new canister in its bracket and reinstalling the vacuum hoses and electrical connectors.

Figure 18-26. *The canister filter is easily replaced on most applications. The filter should be replaced if a new canister does not come with one.*

Thermostatic Air Cleaner

The thermostatic air cleaner contains two basic parts, the vacuum switch and the vacuum motor. The vacuum switch is usually installed on the air cleaner housing with one or more sheet metal clips that fit over the switch vacuum fittings. After removing the vacuum hoses, pry the clip from the vacuum fittings to release the switch. Then pull the switch from the housing, **Figure 18-27.** Note if a gasket is used between the switch and housing. Place the new

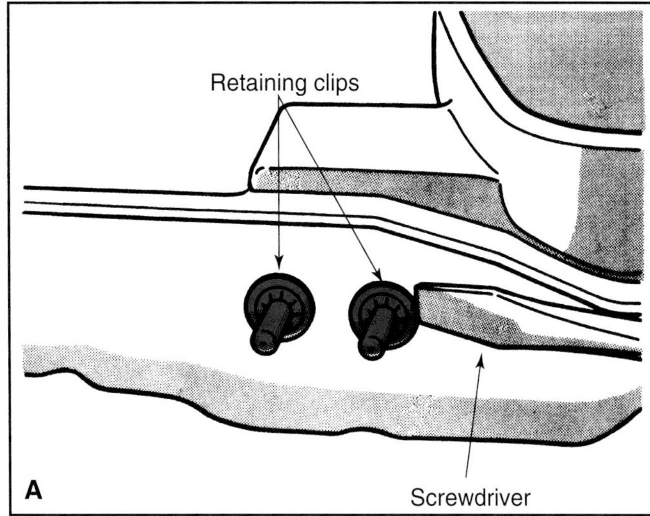

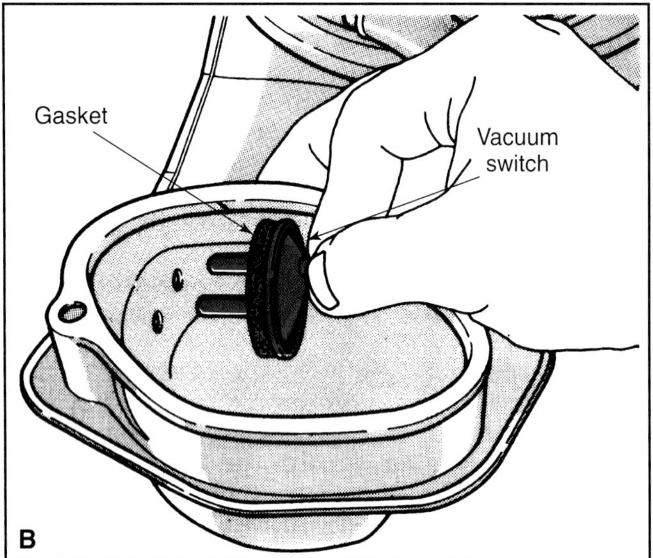

Figure 18-27. *A—Removing a thermostatic air cleaner vacuum switch is a simple operation. Most switches are held by one or two clips or bolts. B—Carefully remove the switch and replace it with a new one. (Chrysler)*

switch (and gasket if used) in the housing, and push the clip over the vacuum fittings. Then reinstall the vacuum hoses.

The vacuum motor in most thermostatic air cleaners can be changed without replacing the entire air cleaner assembly. Place the air cleaner in a bench vise or other holding fixture and remove the vacuum hose. Ensure that the air cleaner is placed so that the vacuum motor is upright. Remove the screws holding the vacuum motor to the snorkel, **Figure 18-28A.** If the vacuum motor is spot-welded or riveted to the snorkel, drill out the spot welds that hold the vacuum motor to the air cleaner snorkel. Use the smallest drill bit that will remove the weld. Lift the motor from the air cleaner snorkel and disengage the link from the vacuum motor to the damper flap by sliding it to one side of the flap, **Figure 18-28B.** Install the replacement vacuum motor in reverse order of removal. Screws are provided to attach the replacement motor to the snorkel.

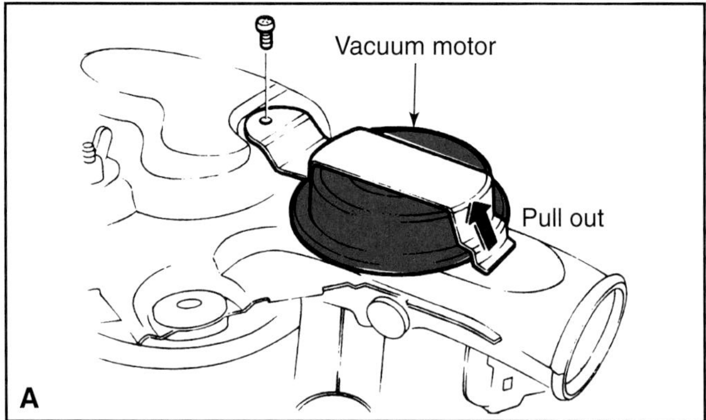

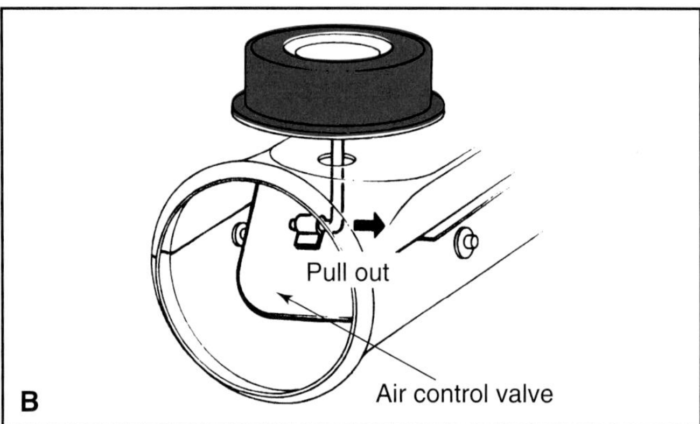

Figure 18-28. *A—The vacuum motor is held by one or two bolts or rivets. B—Make sure the link is removed from the flap before pulling the motor. (Subaru)*

After installing a new air cleaner thermostatic vacuum switch or motor, check thermostatic air cleaner operation using a separate vacuum source (hand pump) or intake manifold vacuum. Then check air cleaner operation with the motor connected to the system. The motor should close the flap when the switch is cold, and allow the flap to open when the switch is hot.

Early Fuel Evaporation Valve

To service the EFE system, the engine and exhaust system must be sufficiently cool. If the defect is in the valve or diaphragm, the vehicle must be raised on a lift. Remove the exhaust pipe at the exhaust manifold and slide the valve and diaphragm assembly of the manifold studs. If the diaphragm can be serviced, separate it from the valve assembly. If only the diaphragm is being replaced, make sure that the valve moves freely. Loosen it with penetrating oil and light hammer tapping if needed. Reassemble the valve and diaphragm as necessary, and reinstall the assembly on the engine. Be sure to check operation after the new parts are installed.

Thermal Vacuum Valve

To replace the thermal vacuum valve, depressurize and drain the cooling system to a level below the switch. Use a box-end or socket wrench to remove the switch. Specialized sockets are available to remove these switches. Compare the old and new switches. Coat the threads of the new switch with the correct type of sealant and install the switch in the coolant passage. Recheck operation after installing the switch.

Exhaust Gas Recirculation (EGR) Valve

The EGR valve is usually mounted on the top or side of the engine, where it can be easily removed. Refer to **Figure 18-29.** Start by removing the electrical and vacuum connections and then remove the bolts holding the valve to the intake manifold. Then pull the valve from the engine. Check the assembly for carbon deposits which may be holding the valve open. Some EGR valves can be cleaned and reused by striking the pintle with a soft hammer to knock loose any accumulated deposits, **Figure 18-30.** Be careful not to damage the valve during this operation. Also check the ports in the intake and exhaust manifolds for excessive carbon buildup and clean if necessary.

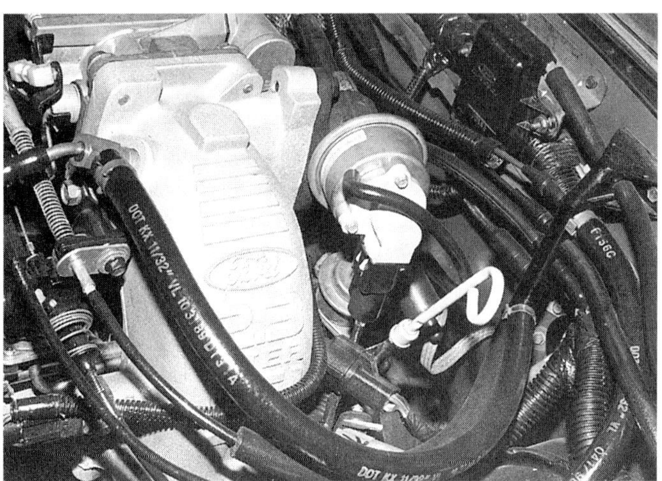

Figure 18-29. *EGR valves are normally bolted directly to the intake manifold or plenum on fuel injected engines. Some are remote mounted close to the engine and use a bolt-on pipe or passage.*

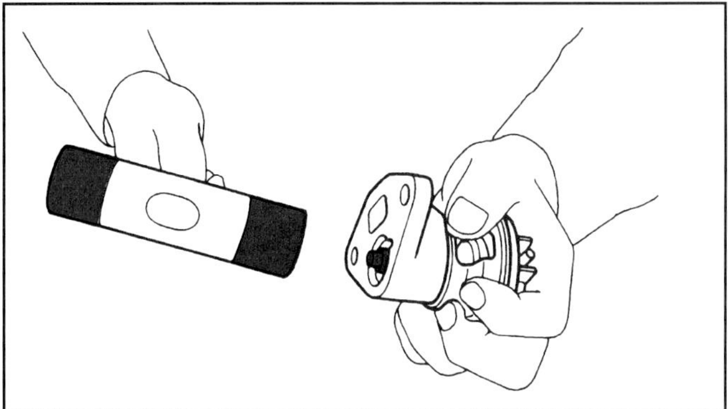

Figure 18-30. *If an EGR valve must be reused, it can be cleaned. If the pintle is sticking, clean the valve and tap lightly on the base with a hammer. (Subaru)*

Caution: If the EGR bolts are difficult to turn, spray the bolts with penetrating oil before continuing removal.

If the valve has a separate position sensor, remove the attaching screws and separate the sensor and valve body. Do not reassemble the sensor and valve body without using a new gasket. Before reinstalling the valve, check the EGR passages for carbon buildup, and clean them as necessary. Place a new gasket over the EGR opening, then reinstall the EGR valve and attaching screws. Reattach the vacuum hose and electrical connector. After reinstalling the EGR valve, recheck the operation of the system.

Servicing EGR Solenoids

EGR solenoids cannot be repaired. They are replaced as a unit. To replace the solenoid, disconnect the electrical connectors and vacuum lines. Then remove the fasteners holding the solenoid to the bracket. Compare the new and old solenoid assemblies to ensure that they are correct, and then install the new solenoid. Recheck EGR operation after the solenoid is installed.

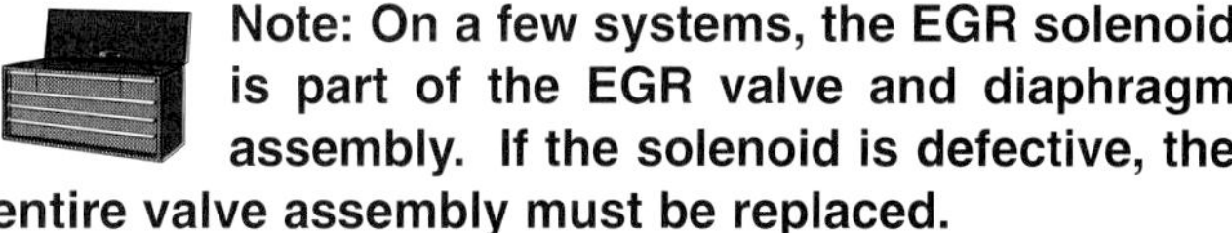

Note: On a few systems, the EGR solenoid is part of the EGR valve and diaphragm assembly. If the solenoid is defective, the entire valve assembly must be replaced.

Air Injection System

The air injection system is comprised of the air injection pump and drive belt, diverter valve or valves, check valves, and associated hoses and piping. Most of these parts are easily removed. The valves and hoses can be removed by loosening the hoses and any attaching hardware. The exhaust system check valve fittings and manifold tubing often corrode at the exhaust manifold. If they are hard to remove, soak them with penetrating oil and use a flare-nut or box-end wrench to loosen the threads. Always use new gaskets where applicable for reinstallation. An exploded view of the air injection manifold tubing and check valves is shown in **Figure 18-31.**

AIR Pump Replacement

The air injection pump is usually not repaired and is simply replaced when it is defective. Begin pump removal by loosening the attaching bolts enough to slide the pump and loosen the drive belt. Then remove the drive belt and hoses. Finish removing the attaching bolts and lift the pump from the engine. Before installing the new pump, check the hoses for cracks. Install the new pump by reversing the removal process. Tighten the belt to specifications and recheck pump operation.

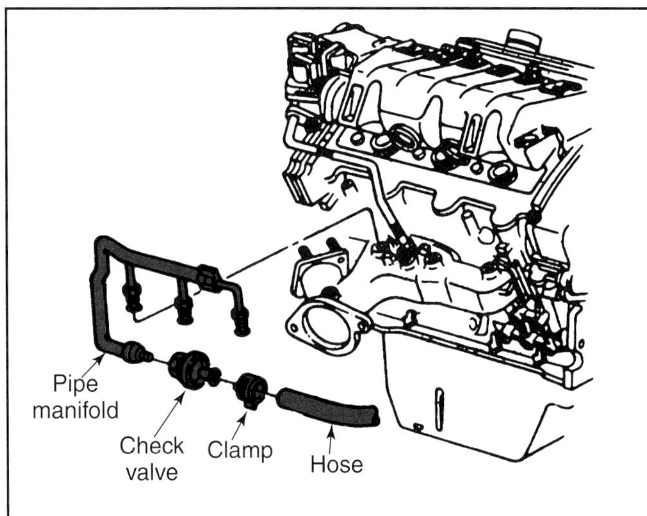

Figure 18-31. *Exploded view of the air injection tubing and valves. The check valve and manifold tubing are the parts that are the most frequently replaced in this system. (General Motors)*

Note: Some air pump pulleys are either pressed or bolted to the pump and are not replaced when the pump is defective. If the pulley is to be used with the new pump, remove the pulley before removing the old pump from the engine.

Figure 18-32 shows a pulley removed from the air pump. The centrifugal intake air filter cannot be removed without destroying it.

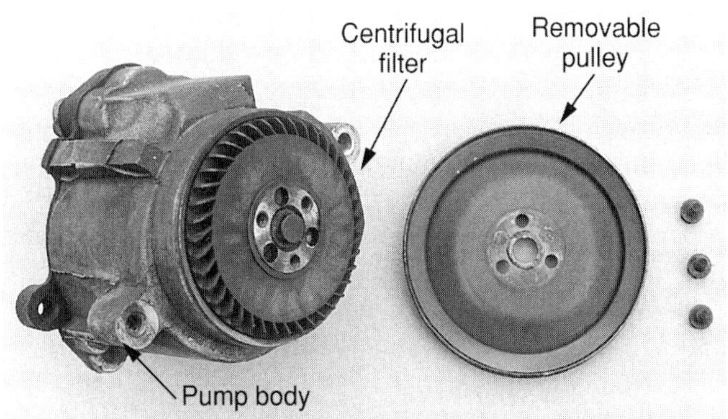

Figure 18-32. *Most air injection pumps have removable pulleys. The air filter on the front of the pump is serviceable, however, it is destroyed when it is removed from the pump.*

Diverter and Switching Valve Replacement

Diverter and switching valves may be mounted on the top or side of the air pump, **Figure 18-33.** Remove the electrical connections and hoses, then remove the screws holding the valve to the air pump. Then remove the valve from the pump. Use new gaskets when reinstalling the valve. Reinstall the diverter valve and tighten the attaching screws. Reattach the hoses and electrical connectors. Other diverter valves are separate from the air pump and are replaced by disconnecting them from the attaching hoses, **Figure 18-34.** After reinstalling any diverter or switching valve assembly, recheck the operation of the air pump system.

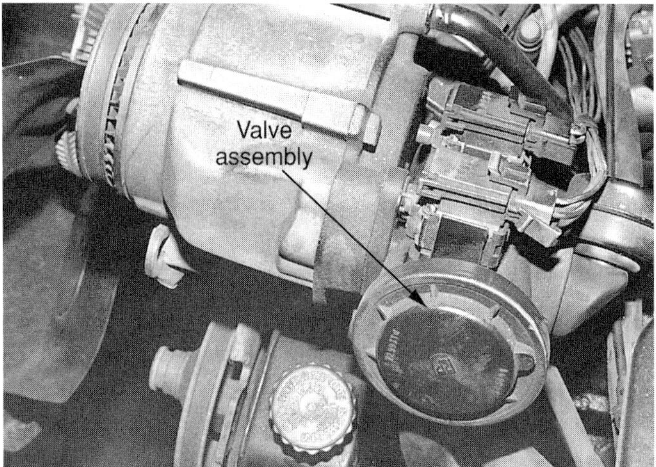

Figure 18-33. *Diverter valves are located on or near the pump. The connectors on the top of the valve are inputs to the ECM.*

Figure 18-34. *Soft neoprene hoses are used to connect remote mounted diverter valves. Replace the hoses and the clamps when replacing the valve.*

Replacing Exhaust System Components

Exhaust system components are relatively easy to replace. The vehicle should be on a lift to gain enough clearance to maneuver the pipes. Exhaust manifold and gasket replacement is discussed in **Chapter 19.**

Warning: Always wear eye protection, since exhaust systems contain large amounts of loose rust that will dislodge when the system is disturbed.

Most exhaust pipes slip together and are held and sealed by special clamps called ***saddle clamps,*** or simply muffler clamps. Some pipes have flanges on one end, which match a flange on the mating pipe. Flanges are held together tightly by flange collars and bolts. Some flanges use a gasket between the mating surfaces, **Figure 18-35.** Other flanges are bell-mouthed and do not use a gasket. Bell-mouthed flanges usually use separate collars on each pipe to allow the flanges to mate closely with each other at any angle. Flanges are often used to connect the exhaust pipes to the exhaust manifold. The exhaust system is attached to the underbody of the vehicle by flexible hangers, which allow the exhaust piping to move with the engine.

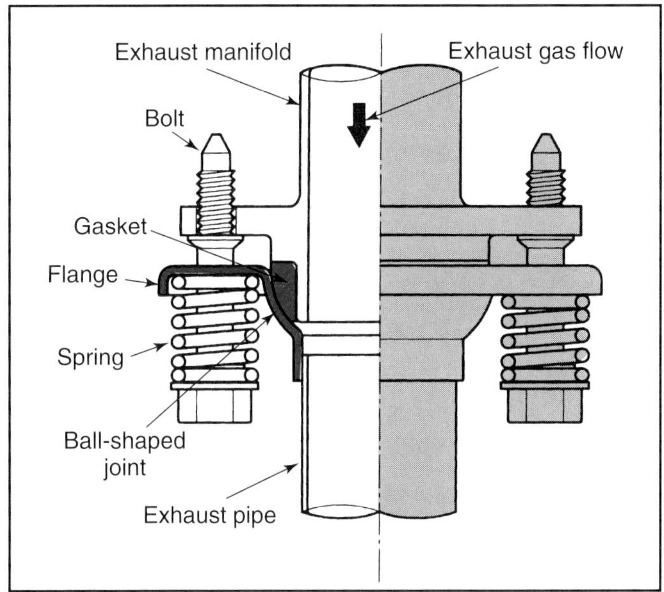

Figure 18-35. *Flange connections allow for movement between fixed and moving parts. The connection between the exhaust manifold and exhaust pipe is almost always a flange connection. (Toyota)*

Note: Use generous amounts of penetrating oil on all exhaust system fasteners.

To begin replacing exhaust system components, remove the exhaust hangers that hold the sections to be replaced. Then loosen and remove the pipe fasteners. If the fasteners are badly corroded, cut them off using a power cutting wheel, chisel, or torch. Since exhaust clamps are subjected to extreme heat and corrosive elements, it is recommended that the old clamps not be reused.

Removing Fitted Exhaust Components

When removing pipes that slip together, carefully cut or chisel the defective pipe or component from the pipe that is to be reused. Heat from a torch can also be used, if applied carefully. If the pipe to be replaced is the outside pipe, split the pipe down the entire length of the joint using a chisel. Then twist the outer pipe and pull it from the inner pipe. If the pipe to be replaced is the inside pipe or the joint is welded, use an exhaust pipe cutter to cut the pipe off just before the joint with the outside pipe. Then split and collapse the inner pipe inward, using a curved chisel. When the inner pipe is collapsed enough, twist and pull it from the outer pipe with a pair of adjustable pliers.

Caution: If a torch is used to heat or cut off an exhaust system part, be careful not to apply heat to the underside of the vehicle. The heat from the torch could start a fire inside the vehicle. Do not use a torch near the fuel tank, fuel filter, or fuel lines. Once a part has been heated, use heavy gloves or pliers to touch or grab the part.

When installing slip fit pipes and components, it may be necessary to expand the old pipe. This can be done by using an exhaust pipe expansion tool. Slip the new part onto the exhaust system and install new clamps. Tighten the clamp at about the midpoint on the sleeve where the two parts meet. Once installed, reinstall the exhaust system on the vehicle, along with any heat shields and other parts that were removed.

Note: While it is not necessary in most cases, some technicians will place one or more spot welds on the edge of an exhaust slip joint. This ensures that the two parts will not slip and come apart with exhaust system movements.

Removing Flanged Exhaust Connectors

If the mating pipes are flanged, remove the bolts from the collars. Then remove the collars (if separate), and lower the pipe from the vehicle. Once the old exhaust system component(s) have been removed, carefully clean the mating surfaces for the new pipes. If a flanged pipe uses a gasket, make sure that the old gasket material is removed from all pipes that will be reused. The new parts can be installed by first slipping them into position. Lightly attach the pipe clamps or flange bolts.

Before tightening any of the exhaust components, check the entire exhaust system for correct placement and alignment. Ensure that no part of the exhaust system contacts any heat shields or the underside of the vehicle. Once alignment is correct, tighten all of the clamps and hangers, then reinstall the pipe hangers that have been removed. Then start and road test the vehicle, making sure that no part of the exhaust system rattles against the body or heat shield.

Replacing Catalytic Converters

Replacing catalytic converters is similar to replacing mufflers and resonators. However, some extra safety precautions are necessary when removing converters. A properly operating converter becomes very hot during operation. The temperature of the converter can reach as much as

1600°F (871°C). Due to the mass of the catalyst elements, the converter cools much more slowly than other exhaust system parts. Therefore, extra care should be taken to avoid burns when working on or around the converter. Some converters are filled with coated pellets, which can be changed without removing the converter. The pellet changing procedure requires a special tool. However, most converters, even pellet types, are now simply replaced.

Since the mass of the converter makes it much heavier than other exhaust system parts, make sure that the converter is adequately supported before removing the hanger fasteners. Most converters have a heat shield that protects the area underneath the converter from excessive heat. Most heat shields are welded to the converter and are replaced when the converter is replaced. However, a few heat shields are bolted around the converter. These shields should be removed before removing the converter.

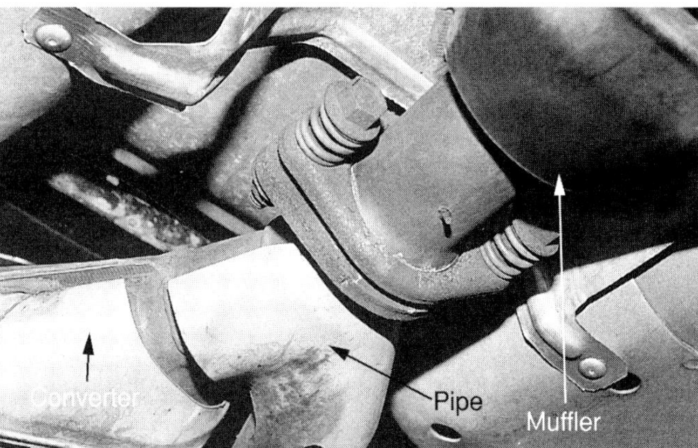

Figure 18-37. *Y-pipe assemblies used with V-type engines often have two or more converters. The Y-pipe is often replaced with the converters.*

Converter Removal

The converter can be removed by removing the pipe and hanger attaching clamps and then lowering it from the vehicle. The converter sometimes has flanged connectors and does not have to be chiseled or heated off of the exhaust system, **Figure 18-36.** Most exhaust system hangers are placed on pipes supporting the converter to isolate them from the converter's heat. Many modern converters are dual-bed types with a separate inlet connection for the air pump pipe. This pipe should be removed by loosening the clamp, heating the slip joint with a torch, and sliding the pipe from the converter. Do not damage these pipes, or they will require replacement.

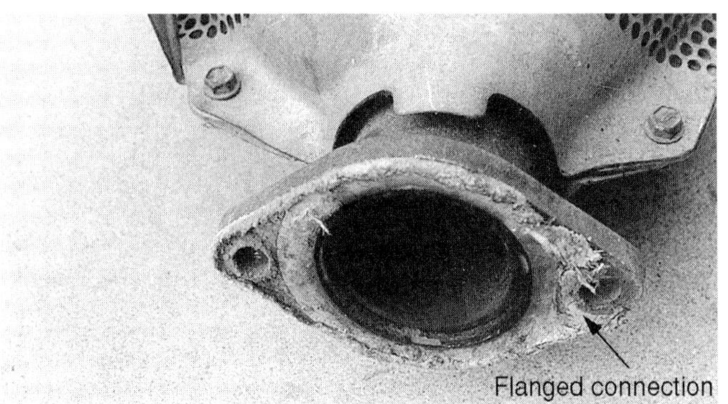

Figure 18-36. *Gaskets are normally used on flange connections to the catalytic converter. Some flange connections use a doughnut-type gasket.*

Some V-type engines have two converters installed in each side of the Y-pipe from the exhaust manifolds. These converters are usually welded in place and are changed by replacing the entire Y-pipe assembly. See **Figure 18-37.** The Y-pipe sometimes contains the oxygen sensors. The sensor electrical connectors should be detached before the pipe is removed.

Converter Installation

Before installing the new converter, install any accessory equipment, such as the oxygen sensors. To install the replacement converter, place it in position and attach all mating pipes or flanges. Then install the pipe and hanger fasteners. Ensure that the converter is properly aligned with the rest of the exhaust system and the vehicle body and tighten the fasteners, **Figure 18-38.**

The elements inside the converter are precious metals, such as platinum. These elements are extremely expensive and becoming increasingly hard to find. Used converters are now recycled to recover the platinum and other precious metals contained in the catalyst. Do not simply dispose of the old converter as scrap; find out where it should be turned in for a core credit.

Performing the OBD II Drive Cycle

The ***OBD II drive cycle*** is performed whenever the battery has been disconnected or before performing an emissions test. Some states require that the drive cycle be performed before an emissions test can begin. A scan tool is required for the drive cycle test. However, no such test is required on OBD I equipped vehicles. Each drive cycle is different for each vehicle so check the service manual.

Note: Carefully study the drive cycle procedure and scan tool operation before starting. You should be reasonably sure that you can complete the drive cycle from begining to end. If the drive cycle has to be aborted for any reason, the engine must be allowed to cool, which can cause a considerable delay.

To begin the drive cycle test, check that the coolant temperature is low enough to allow the ECM to start in the

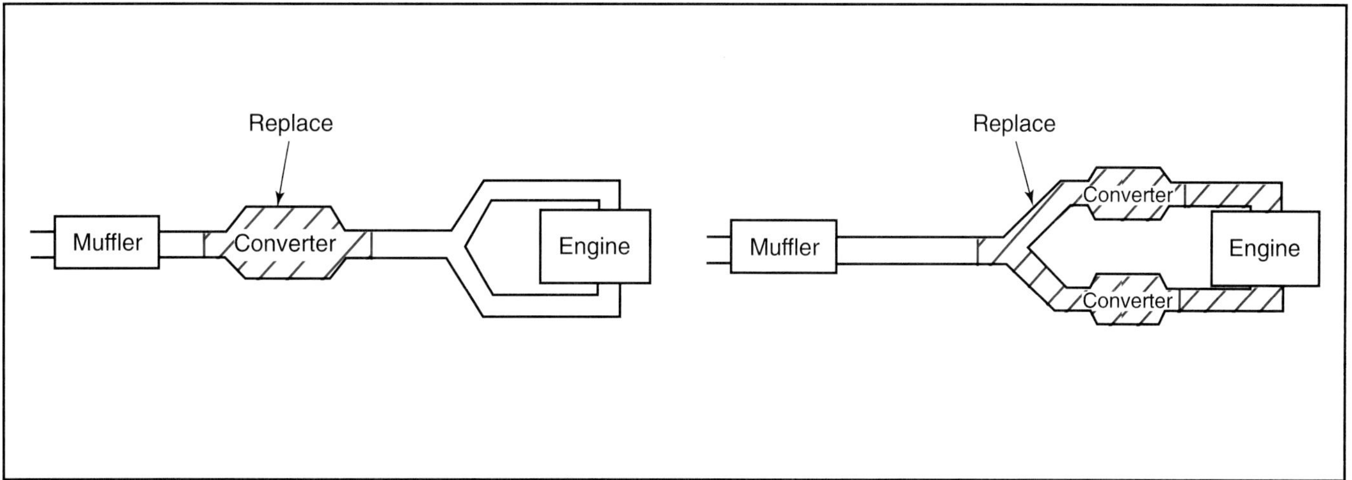

Figure 18-38. *The catalytic converter is replaced as a separate component of the exhaust system. In some cases, a section of pipe, such as a Y-pipe will be replaced with the converter.*

open loop mode. On most engines, coolant temperature should be below 120°F (49°C). Always check the service manual for the exact temperature at which the ECM will enter open loop. Be sure to allow for variations in the coolant temperature sensor calibration. In warm climates, the cool down period can take as long as six hours. Ideally, the vehicle should sit overnight before beginning the drive cycle.

After ensuring that the engine temperature is low enough to start the drive cycle, attach the scan tool. Set the scan tool to record the ECM status as the engine operates. Some scan tools have a dedicated drive cycle option. A typical drive cycle will cover engine warm-up, idling, accelerating, decelerating, and cruising, in a specific order, **Figure 18-39.** This order must be followed *exactly* as outlined in the service manual. Some scan tools will prompt the technician throughout the drive cycle. Start the vehicle and complete the drive cycle sequence as outlined in the service manual.

A typical drive cycle will take from 8-15 minutes to complete, depending on the manufacturer. Some drive cycles require the technician to turn the air conditioning on and off at certain times, and decelerate with the manual transmission clutch engaged or released, depending on the

OBD II Drive Cycle	
Drive Status	**System Monitored**
Cold Start, coolant temperature <122°F (50°C)	No system monitored, warm-up only
Idle 90 seconds in drive or neutral, A/C on	HO_2S heater, EVAP purge, secondary air, fuel trim, misfire
Accelerate to 45 mph (72 kph), 1/2 throttle, A/C off	Fuel trim, misfire, HO_2S monitor
2 minutes of steady throttle at 45 mph (72 kph)	Misfire, EGR, secondary air, fuel trim, EVAP purge
Decelerate to 20 mph (32kph), no braking	Fuel trim, EGR, misfire, EVAP purge
Accelerate to 50-60 mph (88-97 kph), 3/4 throttle	Fuel trim, EVAP purge, HO_2S, misfire
4 minutes of steady throttle at 55-60 mph (88-97kph)	Catalyst monitor, misfire, EGR, fuel trim, HO_2S
Decelerate to stop, no braking	EGR, EVAP purge
End of OBD II Drive Cycle	

Figure 18-39. *The OBD II drive cycle must be performed before an emissions inspection and after the battery has been disconnected. This particular drive cycle is not used by any manufacturer, however, it does represent the steps that normally occur in this procedure.*

portion of the cycle being performed. Some state emissions programs eliminate the warm-up portion of the drive cycle as it is impractical to allow the vehicle to cool off before testing.

Performing the drive cycle with the vehicle on a chassis dynamometer will allow the scan tool to gather readings in the shortest possible time. If the vehicle is driven on the road, it may be impossible to complete the drive cycle exactly as designed. For example, the drive cycle may require several acceleration and deceleration sequences, followed by driving on a road where 55 mph (88 km) can be maintained for several minutes. Fortunately, most scan tools take these factors into account and allow the cycle to continue once the proper conditions have been reestablished. Some scan tools can be paused when the drive cycle must be delayed. However, if the engine is turned off for any reason, the drive cycle must be restarted from the beginning.

The scan tool will indicate when the drive cycle is complete. Most scan tools will only indicate when the cycle is complete, *not* whether the vehicle passed or failed. If any malfunctions were detected during the drive cycle, they will be stored as trouble codes in the scan tool. Check for any stored trouble codes and make further diagnostic checks and repairs as needed. After repairs are complete, repeat the drive cycle to ensure that the vehicle is repaired. Do not attempt to perform an IM240 test until the drive cycle is successfully completed.

Follow-up for Emissions and Exhaust Repairs

After repairs, clear the ECM of any codes that may have been set during the diagnostic procedure. Road test the vehicle to ensure that it performs normally with no unusual noise or odor. Perform another exhaust gas analysis to verify good emissions system operation. If your shop has the proper permits, perform an air quality reinspection if the vehicle was in for failing a state air quality test. If necessary, conduct an OBD II drive cycle on those vehicles equipped with this diagnostic system. If the problem is not corrected, begin the diagnostic procedure again.

Summary

Most emission control system problems will only show up when the vehicle fails an emissions test. In many cases, the emission control devices are not the source of the problem, but are overwhelmed by other engine problems.

Before diagnosing emission problems, determine that the exhaust system has no obvious problems. Then visually inspect for defective components and signs of tampering. Raise the vehicle and check exhaust system condition. Check for exhaust restrictions if necessary. The two methods of exhaust restriction testing make use of a vacuum gauge or a pressure gauge.

The emission test procedure varies between states. New EPA regulations based on the Clean Air Act of 1990 mandate an emissions test that approximates 4 minutes of city driving. The test procedure is known as the IM240 test. At the same time as the IM240 test is being conducted, purge air flow from the evaporative canister is being tested. The fuel tank and other components are also tested for leaks.

If the vehicle fails an emissions test, it must be adjusted or repaired to bring it up to the proper specifications. If a vehicle is brought in for repairs after failing a state inspection, it should be retested before proceeding. Make sure that the readings obtained by the shop exhaust gas analyzer approximately match the readings of the state emissions test. To complete testing and emission control repair of the vehicle, other test equipment may be needed.

The PCV system can be checked by visual observation or by a special PCV tester which checks for air flow into the engine. The evaporative system can be checked for flow restrictions, vacuum buildup, and leaks. Many evaporative system problems are caused by plugged filters in the evaporative canister or restricted hoses.

EGR valves can clog up with exhaust deposits, usually on high mileage vehicles. Vacuum supply defects are the most common cause of EGR problems. The diaphragm may be defective, or the control solenoid may be defective. EGR position sensors, when used, can be checked with a vacuum pump and multimeter. Electronic EGR valves can fail electrically, or may have a defect in the wiring. On any type of ECM controlled exhaust gas recirculation system, be sure not to damage the ECM drivers while making electrical checks.

The air pump should be checked for proper output. If there is no output, check pump drive belt condition and tension before replacing the pump. Most other air pump system problems are caused by valve defects. Thermostatic air cleaners can be checked for proper closing when cold, and full opening when hot. Also check the EFE system to ensure that the valve closes when cold, and opens when hot. Checks to both of these systems can be made with a vacuum pump.

Common emissions filters are the PCV and evaporative emissions canister filters. These filters should be checked often and replaced when they show any signs of plugging. If an exhaust system restriction is suspected, the exhaust system may need to be disassembled to check for a restriction. Common sources of restrictions are stuck EFE valves, clogged converters or mufflers, kinked pipes, or double wall pipes that have collapsed internally.

Most emission control components are relatively easy to replace. The usual procedure is to remove electrical connectors and vacuum lines, remove any brackets, and remove the part. Always compare the old and new parts before installing the new part. Some emission control components, such as EGR valves and air injection system components, may require new gaskets. Installation of a new part is the reverse of removal. The OBD II drive cycle can be performed as necessary. Always follow the drive cycle procedure exactly.

Always recheck system operation after any new parts are installed. Clear all trouble codes from the ECM and road test the vehicle to recheck operation. If the original problem was a failed state emission inspection, recheck the vehicle using an exhaust gas analyzer.

Know These Terms

Vacuum gauge
Backpressure gauge
Exhaust gas analyzer
Inspection and maintenance (I/M) program
IM240 emissions test
Evaporative system purge test
Evaporative system pressure test
Grams per mile (gpm)
Parts per million (ppm)
Percentage (%)
Saddle clamps
OBD II drive cycle

Review Questions—Chapter 18

Please do not write in this text. Write your answers on a separate sheet of paper.

1. The entire series of IM240 tests should take no more than about ______ minutes.
2. The purge flow test ensures that all gasoline vapors are drawn into the ______.
3. To reduce the chance of fires, fuel tanks should be pressurized with ______ instead of compressed air.
4. Most EGR solenoids are normally ______ types, and are ______ by an electrical signal from the ECM.
5. On a properly operating EGR position sensor, the electrical value will vary as ______ varies.
6. The EGR valve diaphragm can be checked by:
   (A) applying vacuum.
   (B) applying battery positive current.
   (C) removing the EGR valve and visually inspecting it.
   (D) None of the above.
7. The air pump and its related control valves can be easily checked when the engine is:
   (A) running in park.
   (B) not running.
   (C) at 2000 rpm.
   (D) at wide open throttle.
8. The thermostatic air cleaner consists of a vacuum ______ and vacuum ______.
9. The air injection pump is usually ______.
   (A) repairable
   (B) replaced as a unit
   (C) rebuilt
   (D) Both B & C.
10. The exhaust and muffler pipes are held together and sealed by:
   (A) saddle clamps.
   (B) muffler clamps.
   (C) flanges and collars.
   (D) All of the above.

ASE Certification-Type Questions

1. Technician A says that a vacuum gauge can be used to check for an exhaust system restriction. Technician B says that a pressure gauge can be used to check for an exhaust system restriction. Who is right?
   (A) A only.
   (B) B only.
   (C) Both A & B.
   (D) Neither A nor B.
2. Technician A says that most existing state I/M programs test the vehicle as it operates at different speeds on a dynamometer. Technician B says that new emissions testing procedures call for performing the OBD II drive cycle first, even if the vehicle is not OBD II equipped. Who is right?
   (A) A only.
   (B) B only.
   (C) Both A & B.
   (D) Neither A nor B.
3. The IM240 test will take less than 4 minutes (240 seconds) under what conditions?
   (A) The vehicle emissions are extremely clean.
   (B) The vehicle emissions are excessive.
   (C) The OBD II drive cycle was performed first.
   (D) Both A & B.

4. Technician A says that fixed orifice PCV systems are being used widely on modern engines. Technician B says that PCV valves can be checked by shaking them. Who is right?
   (A) A only.
   (B) B only.
   (C) Both A & B.
   (D) Neither A nor B.
5. The evaporative emissions canister is saturated with fuel. Technician A says that a defective canister purge valve could cause the problem. Technician B says that the saturated canister could cause HC and CO levels to be excessively high. Who is right?
   (A) A only.
   (B) B only.
   (C) Both A & B.
   (D) Neither A nor B.
6. Evaporative emissions system flow restrictions can be caused by any of the following, EXCEPT:
   (A) clogged canister filter.
   (B) inoperative flow control valve(s).
   (C) disconnected hoses.
   (D) defective hoses.
7. EGR control solenoids can be checked using all of the following testers, EXCEPT:
   (A) flow meter.
   (B) vacuum pump.
   (C) test light.
   (D) battery and jumper wires.
8. All of the following statements about EGR flow rate monitoring are true, EXCEPT:
   (A) on modern OBD II vehicles, more EGR flow is needed.
   (B) the EGR is opened as little as possible to improve driveability.
   (C) on some vehicles, EGR flow rate is measured with a sensor similar to a MAP sensor.
   (D) EGR air flow can be read by a scan tool.
9. Technician A says that some PCV filters are not replaceable. Technician B says that some evaporative canister filters are not replaceable. Who is right?
   (A) A only.
   (B) B only.
   (C) Both A & B.
   (D) Neither A nor B.
10. The EGR valve and position sensor must be replaced ______.
    (A) as a unit
    (B) along with the intake manifold
    (C) individually
    (D) None of the above.
11. A defective EGR solenoid is usually ______.
    (A) repaired
    (B) replaced
    (C) cleaned
    (D) Any of the above, depending on the manufacturer.
12. Which of the following switches could be affected by a low coolant level?
    (A) EGR position switch.
    (B) Thermostatic air cleaner temperature switch.
    (C) EFE thermal vacuum valve.
    (D) Evaporative canister purge valve.
13. A restricted exhaust can cause ______.
    (A) low top speed
    (B) lean mixtures
    (C) improved gas mileage
    (D) All of the above.
14. Catalytic converters are ______ than other exhaust system components.
    (A) hotter
    (B) heavier
    (C) cleaner
    (D) Both A & B.
15. Engines with two converters are usually ______ engines.
    (A) inline
    (B) V-type
    (C) diesel
    (D) four-cylinder

Advanced Certification Questions

The following questions require the use of reference information to find the correct answer. Most of the data and illustrations you will need is in this chapter. Diagnostic trouble codes are listed in **Appendix A.** Failure to refer to the chapter, illustrations, or **Appendix A** for information may result in an incorrect answer.

16. The driver of a vehicle with throttle body injection complains of loss of power. The exhaust backpressure measured at the oxygen sensor port at idle is 7 psi (48.26 kPa) and 18 psi (124.1 kPa) at 2500 rpm. Which of the following is the *least likely* cause?
    (A) A restricted catalytic converter.
    (B) A restricted muffler.
    (C) A collapsed exhaust pipe
    (D) A cracked exhaust manifold.

Gases	Standards	State test readings	Shop test readings
HC (ppm)	below 220	458	9
CO (%)	below 1.20	2.37	0.01
CO_2 (%)	above 9	11.2	15.3

17. A vehicle has failed a state emissions test. You perform an exhaust gas analysis and compare the readings with the state test readings in the chart shown above. You should perform all of the following steps next, EXCEPT:
    (A) recalibrate the analyzer.
    (B) retest the vehicle immediately after driving.
    (C) take the vehicle on another road test.
    (D) turn the air conditioning compressor on and retest.

18. An exhaust gas analysis shows a fuel injected truck to have a high CO reading and a low O_2 reading. Technician A says that a defective evaporative emissions canister valve is the cause. Technician B says that low fuel pressure is the cause. Who is right?
    (A) A only.
    (B) B only.
    (C) Both A & B.
    (D) Neither A nor B.

Data Scanned from Vehicle

Coolant temperature sensor	Intake air temperature sensor	Manifold absolute pressure sensor	Throttle position sensor
212°F/100°C	82°F/28°C	1.8 volts	0.68 volts
Engine speed sensor	Oxygen sensor	Vehicle speed sensor	Battery voltage
680 rpm	0.53 volts	0 mph	13.4 volts
Idle air control valve	Evaporative emission canister solenoid	Torque converter clutch solenoid	EGR valve control solenoid
23 percent	Off	Off	23 percent
Malfunction indicator lamp	Diagnostic trouble codes	Open/closed loop	Fuel pump relay
On	P0300, P0403	Open	On
Brake light switch	Cruise control	AC compressor clutch	Knock signal
Off	Off	Off	No
Ignition timing (°BTDC)	Base timing: 8	Actual timing: 25	

19. A multiport fuel injected vehicle has excessive emissions and a rough idle. The scan tool readings shown above were captured at idle. When the engine speed is increased, emissions decrease to normal and the idle smooths out. Which of the following is the *most likely* cause?
    (A) A defective ignition coil.
    (B) An open injector harness wire.
    (C) A sticking PCV valve.
    (D) An EGR valve pintle not fully seated.

20. A multiport fuel injected vehicle has failed a state air quality test. The test results are shown in **Figure 18-7.** Technician A says that low fuel pressure can cause these readings. Technician B says that a leaking fuel pressure regulator can cause these readings. Who is right?
    (A) A only.
    (B) B only.
    (C) Both A & B.
    (D) Neither A nor B.

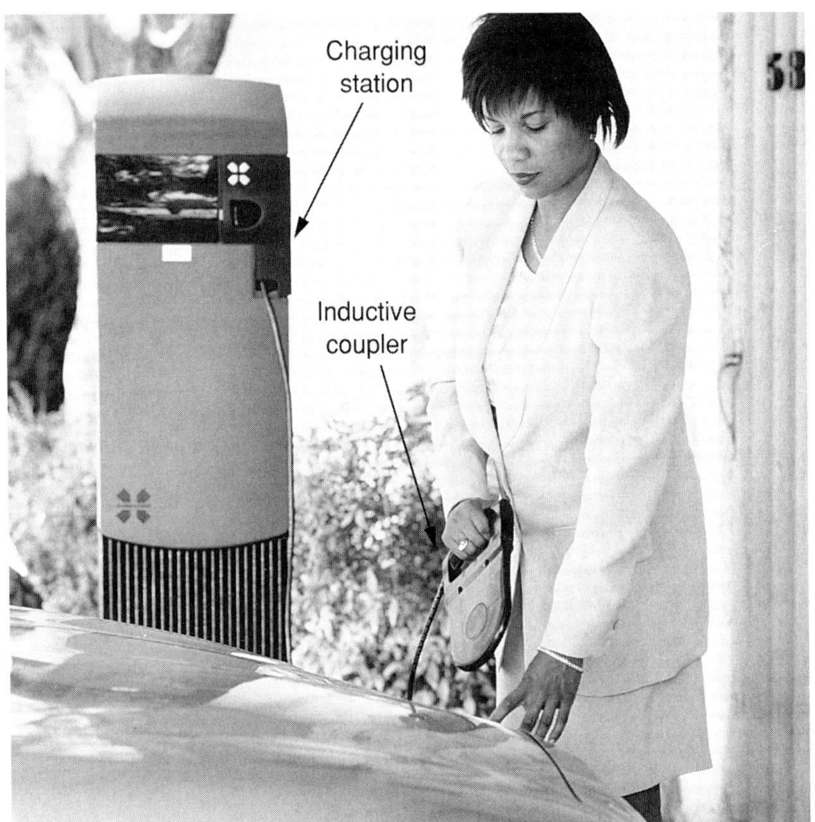

Above—Three-phase, alternating current induction motors are used to power electric vehicles, which are referred to as "zero emissions vehicles." Below—An inductive coupler and charging station is used to recharge the vehicle's batteries. (General Motors)

Chapter 19

Engine Mechanical Diagnosis

After studying this chapter, you will be able to:

- ❑ Diagnose and isolate engine noises.
- ❑ Use a vacuum gauge to detect engine problems.
- ❑ Use a compression gauge to detect engine problems.
- ❑ Check cooling system condition.
- ❑ Check turbocharger and wastegate operation.
- ❑ Check supercharger condition.
- ❑ Check engine oil pressure.
- ❑ Check valve train parts.
- ❑ Check valve timing.
- ❑ Adjust valve clearance.
- ❑ Replace cooling system parts.
- ❑ Replace turbochargers and superchargers.
- ❑ Replace timing gears, chains, and belts.
- ❑ Replace intake manifolds, plenums, and exhaust manifolds.

This chapter covers tests on engine components that greatly affect driveability. The contents of this chapter discuss diagnostic procedures with very few corresponding repair procedures. It is not possible to thoroughly cover all of the necessary procedures for internal engine repair in this text. However, some service procedures normally performed by driveability technicians is covered. Also, some of the previous chapters called for you to perform tests to check the condition of the internal engine components. These tests are outlined in detail in this chapter.

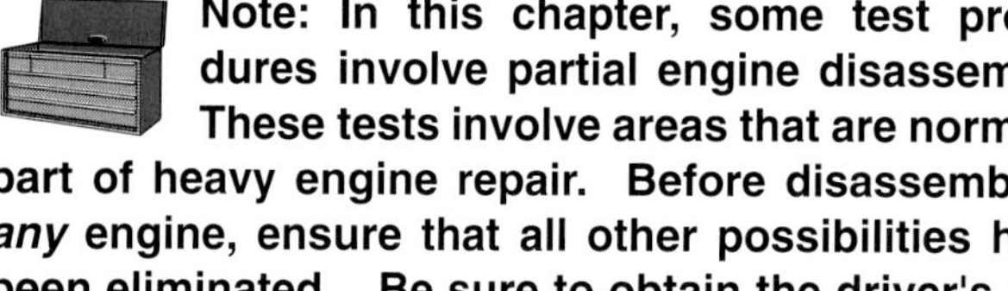

Note: In this chapter, some test procedures involve partial engine disassembly. These tests involve areas that are normally part of heavy engine repair. Before disassembling *any* engine, ensure that all other possibilities have been eliminated. Be sure to obtain the driver's permission before beginning engine disassembly.

Diagnosing Engine Problems

After all external systems (fuel, ignition, emissions, computer, electrical) have been tested and checked, the next system to be tested is the engine itself. In the process of testing the other systems, you probably performed tests, such as compression tests, to check the condition of internal engine components, **Figure 19-1.**

One of the greatest difficulties in diagnosing an internal engine problem is that most defective parts cannot be seen without disassembling some part of the engine. Always begin by examining the external parts of the engine, such as the intake and exhaust manifolds, intake plenum on multiport fuel injected engines, cylinder heads, engine block, and engine mounts. The tests in the following sections will usually indicate if an internal defect exists. Be sure to obtain the vehicle's service manual and any available service history before beginning any inspection and testing.

Figure 19-1. *The engine's internal parts can be a source of problems that can affect the entire vehicle. However, all external systems, such as the fuel, ignition, and computer controls, should be checked first. (Honda)*

Visual Inspection

If possible, begin by inspecting the engine while running. Listen for unusual noises, such as vacuum leaks, knocking, or tapping. If a knocking or tapping noise seems to diminish as the engine warms to operating temperature, it may be an indication that a lifter is bleeding down or there is excessive piston, piston pin, or connecting rod clearance. Check the exhaust for signs of smoking or excess water. Black smoke indicates a rich mixture. Blue smoke is a sign of oil consumption. Thick white smoke indicates water vapor, possibly from coolant entering the exhaust system.

Note: Be sure not to confuse the thin vapor-like white smoke that often occurs on cold humid days with the thick white smoke from coolant leaking into the exhaust.

Check the oil for signs of fuel or coolant contamination. Smell the oil for fuel, which can dilute engine oil and is also a sign of an on-going fuel system problem or rich air-fuel mixture condition. Coolant contamination will turn the oil white, with a similar appearance to a milk shake. Check for external oil and coolant leaks and clean the engine to identify the origin of any leaks, if necessary. Lift the vehicle to check the block and crankshaft seals. Be sure to check the seam between the cylinder head, head gasket, and engine block for leaks and seepage. Check the inside of the tail pipe. A pipe on an exhaust system with a catalytic converter will have a black coating on the inside. The tailpipe on a system without a converter will have a light gray coating. Varying colors or heavy build-up indicates that a problem exists.

If the engine does not crank or run, examine the engine for signs of external damage. Check the block for cracks and holes from damaged internal parts. Occasionally, a no-crank condition is caused by a seized engine. If the engine will not crank and the starter motor tests ok, attach a socket to a breaker bar and attempt to turn the engine clockwise. If the engine will not turn easily or does not turn at all, remove all the spark plugs and try to turn the engine again. If it still does not turn or turns with great difficulty, it is probably seized.

Inspect the cooling system by first checking the freeze plugs while inspecting the engine itself. All the freeze plugs should be present and in good condition with no coolant seepage or corrosion. Check the water pump for leaks and excessive shaft play. If the engine runs, listen to the water pump for excessive noise. Allow the engine to cool and check the engine coolant for oil contamination in the radiator by removing the radiator cap. Other tests will be discussed later in this chapter. If the vehicle is equipped with an electric cooling fan, inspect it for proper operation. Check the inside of the oil filler cap and the top of the dipstick for signs of excess sludge or varnish inside the engine.

Diagnosing Engine Noises

Any moving part on the vehicle can make noise. Before disassembling the engine, be sure that the noise is actually in the engine. Often what appears to be an engine noise is caused by another vehicle part or system. To isolate noises, it may be necessary to disconnect various components and operate the engine. For example, to determine whether a rattling noise is in the engine internals or the belt driven accessories (coolant pump, alternator, air pump, power steering pump, air conditioner compressor) remove the drive belts and briefly operate the engine. If the noise stops, it was coming from one of the belt-driven accessories. If the noise continues, it is being caused by an internal engine problem.

In another case, a noise that seems to be coming from the exhaust may be caused by a vibrating air injection check valve. This can be isolated by removing the hose between the pump and check valve. If the noise stops, it was caused by the check valve. If the noise continues, it is being caused by the exhaust system. In all cases, before replacing any part, make sure that it really is the source of the noise.

Cylinder Head and Valve Train Noises

Valve train problems usually create clattering or ticking noises. They are lighter sounding than the deep knocks made by worn rod or main bearings. Valve train noises are almost always caused by excessive play in the valve train components. Wear causes the clearance between the valve tip and rocker arm (or camshaft on some OHC engines) to increase to the point that the valve opening action becomes noisy. This is caused by the valve train parts building up excessive speed before making contact with the valve tip. Other causes of noisy valve trains are bent pushrods, collapsed lifters, or improper valve adjustment. Repairs to correct a noisy valve train can often be made without major engine disassembly. On some engines, parts may have to be removed to gain access to the lifters.

If the engine uses a timing chain, it may become worn enough to make a whirring or rattling noise as the engine operates. In some cases, the chain will become so loose that it strikes the timing cover or another stationary engine part, causing a severe knock. In most cases, however, when a chain becomes excessively loose, it will jump and stop the engine before it makes a significant amount of noise.

Engine Block, Crankshaft, and Piston Noises

Noises encountered in the engine block include knocking from the rod and main bearings and the wrist pins. Other problems occurring in this area include piston slap, excessive crankshaft play, and defects in the vibration damper or flywheel. Correcting noises in the block is always a major repair operation requiring engine disassembly.

Rod knock is a loud and deep knock that matches the speed of the engine. They can be isolated by disconnecting the spark plug wire of the suspected cylinder as the engine operates. If the rod bearing is loose, the knock will become quieter, or may even disappear entirely. Main bearing knocks are deeper in tone, and may only occur when the engine is placed under a load, such as hard acceleration. When a main bearing knock is suspected, always check the oil pressure. Loose main bearings usually leak oil and cause low oil pressure. Worn wrist pins make a knocking noise similar to a loose rod bearing. To isolate a loose wrist pin, disconnect the spark plug wire on the suspected cylinder. A wrist pin knock will become a double knock when the plug wire is removed from the plug.

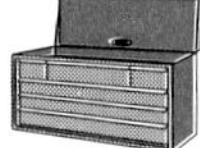

Note: The spark plug wire on the suspect cylinder should be removed with the engine off or the suspect cylinder should be disabled using a scan tool. Do not remove a spark plug wire with the engine running.

Piston Slap

A deep, muffled sound is often the result of a worn piston rocking in the cylinder. This is usually called ***piston slap.*** Piston slap can be isolated to a particular cylinder by disconnecting the spark plug wire of the suspect cylinder and driving the vehicle. If the sound stops as the engine is placed under a load, the defective cylinder has been located. To repair a cylinder with piston slap the engine cylinder must be bored out and an oversized piston fitted.

cylinder must be bored out and an oversized piston fitted. For maximum engine balance and smoothness, the other cylinders should be bored out the same amount.

Caution: Do not operate the engine for long periods with a plug wire disconnected. Unburned gasoline will overheat and destroy the catalytic converter. If possible, disconnect the harness to the cylinder's fuel injector.

Flywheel/Vibration Damper Knocks

Loose or defective flywheels can cause knocking noises. These noises are usually more random than those caused by loose rod bearings or piston slap. The most common cause of flywheel knocks is loose bolts, either at the crankshaft or the torque converter. They are usually more noticeable when the engine is sharply accelerated or decelerated. In some cases, a loose flywheel will only knock when the engine is turned off. If teeth on the flywheel are worn or broken, noise will be heard only when the engine is started.

Vibration dampers can become loose on the crankshaft, or the rubber ring holding the hub to the weight can break, allowing the weight to slap around inside the pulleys. If the pulleys are bolted to the damper, the bolts can become loose enough to cause knocks or rattles. Like flywheel noises, damper noises may be most noticeable when the engine is sharply accelerated or decelerated.

Excessive Crankshaft End Play

Excessive ***crankshaft end play*** allows the shaft to move back and forth. This can cause a knocking sound which varies with engine load. This noise can often be located on a manual transmission vehicle by pushing in the clutch. Pressure from the throwout bearing pushes the flywheel and crankshaft forward, taking up the end play. This noise is seldom encountered in automatic transmission vehicles, since the pressure in the torque converter hub pushes the converter and crankshaft forward whenever the engine is running.

Checking for Vacuum Leaks

You learned in **Chapters 14-18** that many driveability problems are caused by vacuum leaks. Check for obvious sources of leaks, such as disconnected or split vacuum hoses, loose manifold bolts, or damaged vacuum operated devices. See **Figure 19-2.**

Testing the Intake Manifold for Vacuum Leaks

Since a vacuum leak can lean the air-fuel mixture, its presence can be confirmed by richening the mixture through the leak. One technique to find a vacuum leak is to pour a light oil, such as penetrating oil, on the edges of the intake plenum and manifold where it meets the cylinder

Figure 19-2. *Vacuum leaks are frequently the cause of performance and emissions problems and are also overlooked by many technicians.*

head and increasing idle speed. If the engine runs more smoothly for a brief period and/or the exhaust smokes, the oil is entering the manifold through the leak. If you have an ultrasound vacuum leak detector, it can be used to quickly pinpoint the leak. Follow the equipment manufacturer's instructions exactly.

Using a Vacuum Gauge

You learned earlier that the vacuum gauge is used to determine if an engine problem exists. Begin by attaching the vacuum gauge to the intake manifold. With the engine running, read the vacuum gauge to determine the overall condition of the engine as well as the condition of individual cylinders.

Note: A common mistake of many technicians is connecting a vacuum gauge to ported vacuum. Make sure that you are connecting the gauge directly to manifold vacuum.

Interpreting Vacuum Gauge Readings

In general, low vacuum gauge readings point to a problem in the engine ignition or valve timing. Sharp drops in the vacuum reading indicate a problem in one or more cylinders or a restricted exhaust. A fluctuating reading indicates an air-fuel ratio or valve train problem. Refer to **Figure 19-3** for vacuum gauge readings and the problems that they indicate.

Note: Some vacuum gauge readings can be caused by external problems, such as a restricted exhaust, misadjusted fuel delivery, or improper ignition timing. Be sure to check these systems before suspecting an internal engine problem.

Vacuum Gauge Readings

Note: White needle indicates steady vacuum. Red needle indicates fluctuating vacuum.

Needle steady and within specifications.

Cause: Normal vacuum at idle.

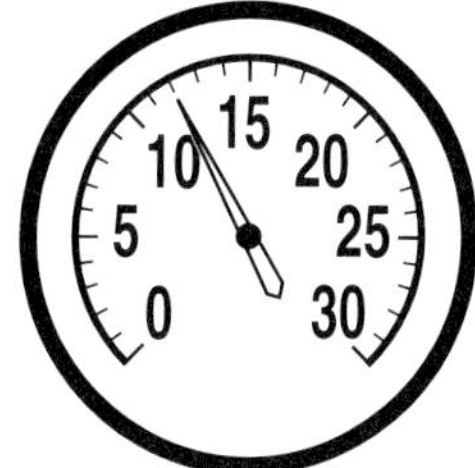

Needle very low and steady.

Cause: Vacuum or intake manifold leak.

Needle normal at idle, but fluctuates as engine speed is increased.

Cause: Weak valve spring.

Needle jumps to almost zero when throttle is opened and comes back to just over normal vacuum when closed.

Cause: Normal acceleration and deceleration reading.

Needle slowly drops from normal as engine speed is increased.

Cause: Restricted exhaust (compare at idle and 2500 rpm).

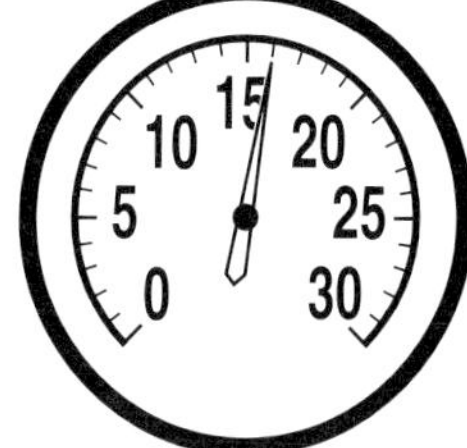

Needle steady, but low at idle.

Cause: Improper valve or ignition timing.

Needle has small pulsation at idle.

Cause: Insufficient spark plug gap.

Needle occasionally makes a sharp fast drop.

Cause: Sticking valve.

Needle regularly drops 4 to 8 inches.

Cause: Blown head gasket or excessive block-to-head clearance.

Needle slowly drifts back and forth.

Cause: Improper air-fuel mixture.

Needle drops regularly; may become steady as engine speed is increased.

Cause: Burned valve, worn valve guide, insufficient tappet clearance.

Needle drops to zero when engine is accelerated and snaps back to higher than normal on deceleration.

Cause: Worn piston rings or diluted oil.

Figure 19-3. *Vacuum gauge readings and the problems that cause them. Red needle indicates fluctuating vacuum.*

Checking Engine Compression

Compression tests are a sure way of determining internal engine condition. They are usually performed if other testing indicates one or more missing, or "dead" cylinders not caused by a fuel or ignition system problem. A compression test is easy to perform, once all the spark plugs are removed from the engine. However, it is a little harder to translate the readings into a diagnosis of what is wrong with the engine.

The tools needed for compression testing include a compression gauge with the proper adapter, and a pencil and paper. Begin by disabling the ignition system; a good way to do this is to disconnect primary power or ground from the ignition coil. Block the throttle valve in the wide open position. Thread the test adapter hose into the first spark plug hole, **Figure 19-4.** After the gauge is securely threaded into the plug hole, attach a remote starter switch to the starter circuit. Using the remote starter switch, crank the engine through at least four cycles of the cylinder being checked, or eight engine revolutions. Read the compression pressure on the gauge. Most compression gauges have a check valve that allows the gauge to hold the highest reading. Record this reading, then release the pressure and repeat the process with the next cylinder. Be sure to record the readings on all cylinders. Once all readings are taken, compare the results against each other, **Figure 19-5.**

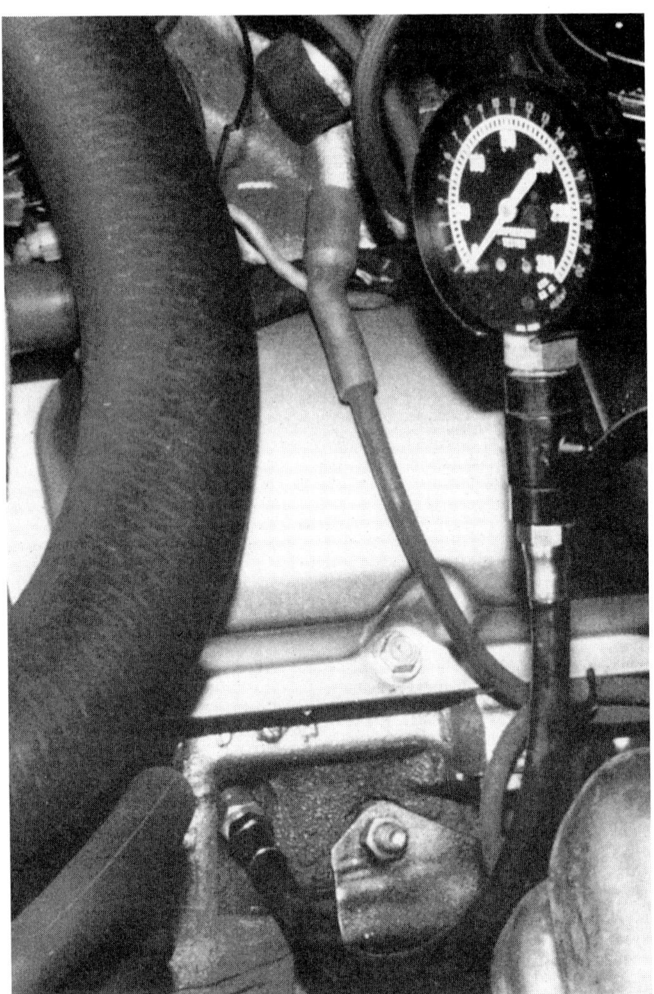

Figure 19-4. *Good engine compression is vital to efficient operation. The compression gauge threads into the spark plug hole and the engine cranked. This test gives a good indication of overall engine condition.*

Wet Compression Test

If the initial test shows engine compression to be low, perform a ***wet compression test***. Pour about one-half ounce of oil in the cylinder, crank the engine a few times, and recheck compression. If compression rises somewhat after the oil is added, the cylinders or rings are worn. If the compression stays low, a valve is burned or the head gasket is leaking. When this happens, the spark plugs will probably be fouled with soot from poor combustion.

Caution: Do not add an excessive amount of oil to the cylinder. Since fluids do not compress, an excessive amount of oil could damage the engine or compression gauge. Check the service manual for the exact amount of oil to be added for a wet compression test.

Analyzing Compression Readings

If all cylinders are low, which usually means at least 20% lower than specifications, the rings and cylinder walls may be excessively worn and the engine probably requires a complete overhaul. Incorrect valve train timing can also cause a low compression reading on all cylinders. If only one cylinder is low, having at least 20% less pressure than the average of the other cylinders, it may have a burned valve, damaged valve train part, or piston. If two cylinders next to each other are low, the head gasket may be leaking between the cylinders or the head or block is cracked, **Figure 19-5.** If the compression is near zero on one or more cylinders, this is an indication of severe engine damage, such as a hole in the piston, head, or block, or severe valve damage.

Sometimes, other test procedures indicate that one cylinder is not producing full power, and may have low compression. It is possible to perform a compression test on only one cylinder, but the readings will not be as accurate as those taken during a normal compression test, when all the plugs are out and the starter can turn the engine at a higher speed. If a one-cylinder compression test indicates very low or no compression, the rest of the engine should be checked to confirm that it is the only cylinder with low compression. If the compression check indicates pressure only slightly below normal, this is probably due to the slower cranking speed, and the problem is not in the cylinder.

Compression Gauge Readings		
Compression Test Results	**Cause**	**Wet Test Results**
All cylinders at normal pressure (no more than 10–15% difference between cylinders)	Engine is in good shape, no problems.	No wet test needed.
All cylinders low (more than 20% difference)	Burned valve or valve seat, blown head gasket, worn rings or cylinder, valves misadjusted, jumped timing chain or belt, physical damage to engine.	If compression increases, cylinder or rings are worn. No increase, problem caused by valve train. Near 0 compression caused by engine damage.
One or more cylinders low (more than 20% difference)	Burned valve, valve seat, damage or wear on affected cylinder(s).	If compression increases, cylinder or rings are worn. No increase, problem caused by valve train. Near 0 compression caused by engine damage.
Two adjacent cylinders low (more than 20% difference)	Blown head gasket, cracked block or head.	Little or no increase in pressure.
Compression high (more than 20% difference)	Carbon build-up in cylinder.	Do not wet test for high compression.

Figure 19-5. *Compression gauge chart showing the typical results of compression tests, problems that can cause each type of compression reading, and results from a wet compression test.*

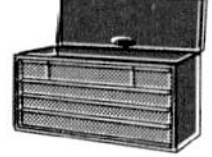

Note: Most causes of low compression require extensive engine repair. These procedures are not addressed in this text. Consult a text on engine overhaul.

Cylinder Leakage Test

While the compression test is good at identifying which cylinders are weak, the cylinder leakage test allows you to more accurately narrow down what the problem is, whether a valve, head gasket, or rings. Normally, cylinder leakage tests are only performed on those cylinders found to be weak in the compression test. However, some technicians will perform this test on all the cylinders. A cylinder differential pressure tool is used during this test.

The first step is to rotate the engine until the cylinder to be tested is at TDC on its compression stroke. Install the tool by attaching the adapter hose to the spark plug hole. Remove the oil dipstick, oil fill cap, radiator cap, and air cleaner housing. Connect shop air pressure to the tool and apply pressure to the cylinder, **Figure 19-6.** There should be no leaks out of any cylinders. The chart in **Figure 19-7** shows common causes of typical problems found in a cylinder leakage test.

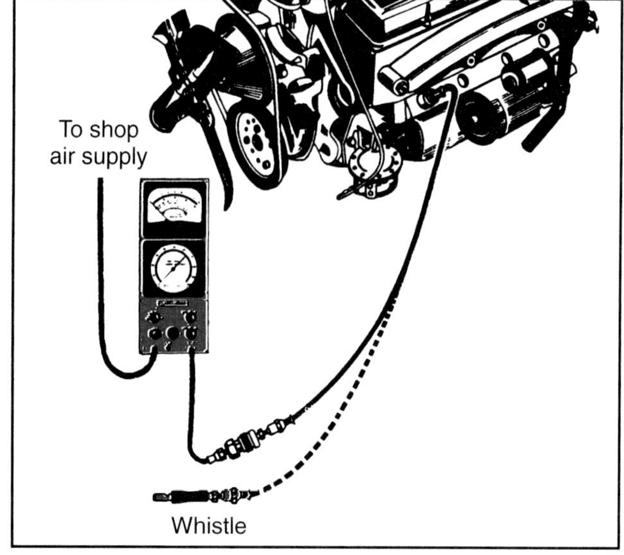

Figure 19-6. *A cylinder leakage tool tests the ability of the cylinder, valves, rings, and head gasket to seal. If air leaks out of a part of the engine, the type of problem can be quickly identified. (Sun)*

Checking the Cooling System

The ECM bases many of its output commands on input from the coolant temperature sensor. In addition, many emission controls and warm-up devices depend on inputs from thermal vacuum valves. Therefore, problems in the cooling system can cause ECM and emission control malfunctions, as well as the traditional engine problems caused by overheating. High NO_X emissions can result from the high temperatures. If the engine is pinging excessively, is hard to start when warmed up, or loses coolant, it may have an overheating problem. Undercooling or slow warm-up may not cause obvious problems, but can be the cause of

Cylinder Leakage Test Results	
Condition	**Possible Causes**
No air escapes from any of the cylinders.	Normal condition, no leakage.
Air escapes from carburetor or throttle body.	Intake valve not seated or damaged. Valve train mistimed, possible jumped timing chain or belt. Broken or damaged valve train part.
Air escapes from tailpipe.	Exhaust valve not seated or damaged. Valve train mistimed. Broken or damaged valve train part.
Air escapes from dipstick tube or oil fill opening.	Worn piston rings. Worn cylinder walls. Damaged piston. Blown head gasket.
Air escapes from adjacent cylinder.	Blown head gasket. Cracked head or block.
Air bubbles in radiator coolant.	Blown head gasket. Cracked head or block.
Air heard around outside of cylinder.	Cracked or warped head or block. Blown head gasket.

Figure 19-7. *Cylinder leakage test results. Most of these problems involve heavy engine repair to correct.*

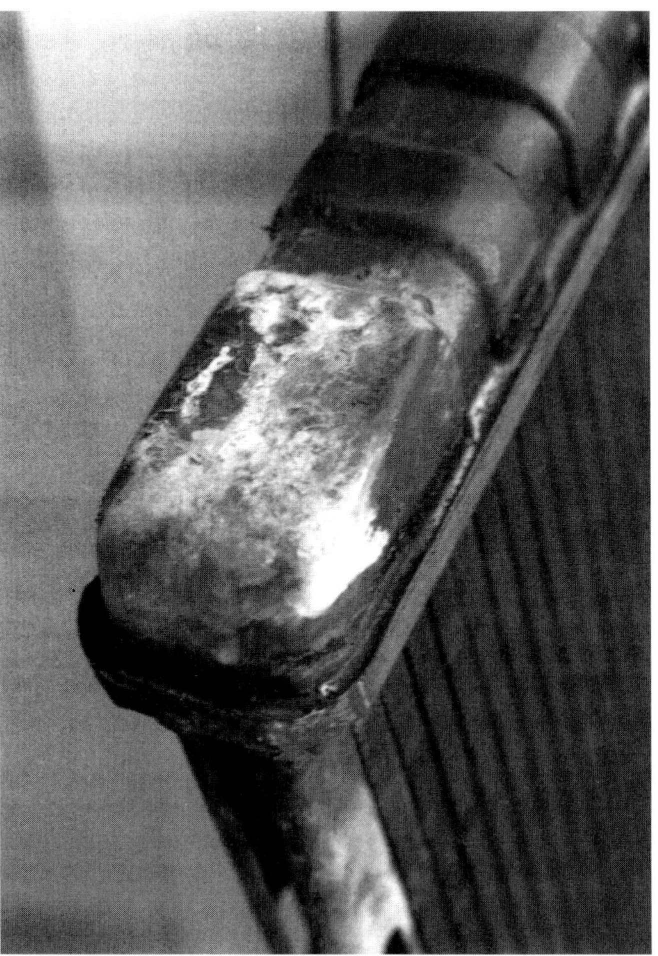

Figure 19-8. *Coolant leaks are characterized by paint removed from the radiator or engine. In most cases, there will be a colored trail (usually green) from the source.*

poor fuel mileage since the ECM cannot enter closed loop operation until the engine is at its normal operating temperature. Overcooling will also cause sludge buildup in the engine internals.

There are two main causes of cooling system problems: leaks and defective components. Leaks can occur at any place in the cooling system. Cooling system components which can become defective include the coolant pump, drive belt, fan, fan clutch, or fan motor, thermostat, radiator, radiator cap, heater core, and hoses.

Checking for Coolant Leaks

A low coolant level is a sign of a coolant leak. Begin checking for coolant leaks by a visual inspection for obvious signs of problems. Leaks usually occur at the water pump seal, radiator seams, and hoses. Leaking coolant will usually stain the area under the leak. In many cases, the leaking coolant will remove the paint from the engine or radiator. See **Figure 19-8.** If the engine has been running long enough to pressurize the system, steam may be seen rising from the leaking area. If the system uses a pressurized coolant recovery system, also check it for leaks.

If a leak cannot be located by visual inspection, fill the cooling system and pressurize it with a radiator pressure tester as in **Figure 19-9A.** If pressure drops, there is a leak somewhere in the system. Check for dripping fluid and trace its source. If the leak cannot be found, it may be inside of the engine, or in the heater core. If the heater core is suspected, check the carpet under the heater core housing. If the leak is inside of the engine, check the engine oil for water. If the engine runs, check for water in the exhaust.

After pressurizing the cooling system, check the radiator cap pressure as shown in **Figure 19-9B.** If the radiator cap will not hold its rated pressure, it should be replaced. Also check the condition of the coolant recovery system. Leaking hoses or a damaged recovery tank should be replaced.

Checking Coolant Pumps and Drive Belts

Coolant pumps usually fail when the shaft bearings wear out. Most bad bearings will cause the pump seal to begin wearing, but some pump bearings become noisy without causing the seal to leak. In rare cases, the impeller will slip or the blades will become filled with deposits. A defective water pump is usually replaced with a new pump.

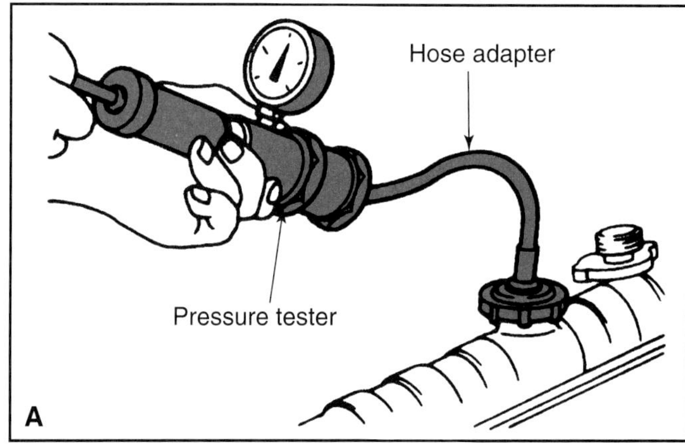

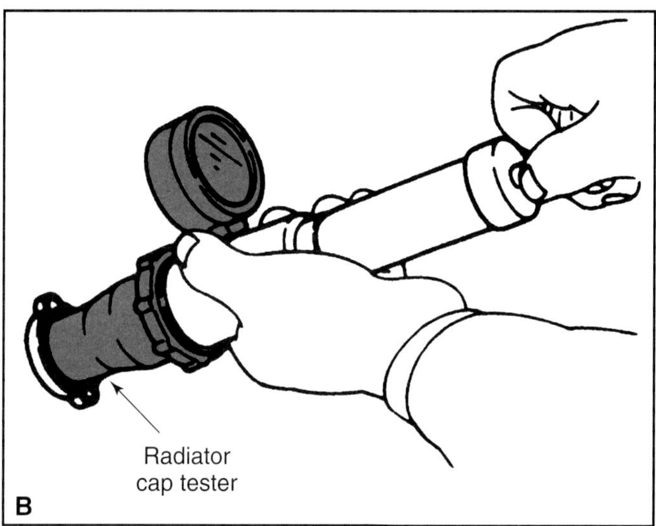

Figure 19-9. *A—A cooling system leak can be found quickly by pressure testing the system. B—When pressure testing the system, do not forget to test the radiator cap. (Subaru)*

The drive belt may wear out and begin slipping, especially when the engine is hot. Slippage will reduce pump output, and may cause overheating, especially at idle. With the engine off, check the belt for tightness and glazing. If the belt is glazed or otherwise worn, it should be replaced. Remember to check both sides of a serpentine belt.

Checking Radiators, Heater Cores, and Hoses

Other than leaks, the most common radiator and heater core problem is clogging. If the cooling system is neglected, or if radiator sealer has been used to seal a cooling system leak, deposits will build up on the radiator tubes. Eventually coolant flow through the tubes is reduced, and some tubes will close entirely. Clogging can also reduce the heating capacity of the radiator core. One way to check the condition of the radiator is to allow the engine to cool off overnight. Then start the engine and allow it to run until the thermostat begins to open (inlet radiator hose starts to get hot). Then allow the engine to run for about two minutes longer. Shut off the engine and feel the rear of the radiator core. The core should be uniformly warm just past the inlet radiator tank, becoming cooler toward the outlet tank. If any part of the core is noticeably cooler than the rest of the core, that section is plugged. Further checking requires that the radiator be removed from the vehicle and the inlet and outlet tanks be removed for core inspection.

Hoses usually do not become clogged, but may become weak and collapse as coolant flows through them. Lower radiator hoses are often collapsed by the suction of the coolant pump. This usually causes overheating at higher engine speeds. If any hose feels soft or is bulged, it should be replaced.

Checking Thermostats

Thermostats can fail open or closed. A thermostat that is stuck closed will cause engine overheating within the first few minutes of engine operation. The usual symptom will be water and steam escaping from the radiator cap and coolant recovery system. If a thermostat sticks open, the engine will take an excessively long time to warm up. Thermostats can be checked for proper operation once they have been removed from the engine. This procedure is covered in the thermostat service section later in this chapter.

Caution: Do not try to correct a cooling system problem by removing the thermostat. Not only does this cause slow warm–up, it will usually not improve the cooling of a hot engine. The coolant will pass through the radiator too quickly to give up its heat. Always locate and solve the real problem instead of removing the thermostat.

Checking Fans, Fan Clutches, and Fan Motors

Proper air flow through the radiator is critical to prevent low speed overheating. The radiator fans must be operating properly. Begin by visually checking the fan for bent, cracked, or missing blades. Replace the fan assembly if it is defective. Do not attempt to repair or straighten any defective fan blades.

Warning: Electric radiator fan motors can start at any time, even when the engine is off. Always remove the fan motor electrical lead or the battery negative cable before working on any electric fan.

Most rear-wheel drive vehicles have a fan clutch installed between the fan pulley and fan. The fan is designed to drive the fan at pulley speed when idling with a hot engine, and to slip at high speeds, and when the engine is cold. If the fan clutch slips at all times, the engine will overheat at idle and low speeds. To check fan clutch operation, allow the engine to warm up thoroughly. Allow the engine to idle, then have an assistant turn off the ignition while you watch the fan assembly. On a hot engine, the fan should spin no more than one half to one revolution after the engine stops. If the fan continues to turn, it should be replaced. Sometimes a fan clutch will jam and fail to slip as the engine speed increases. This can cause a roaring noise at highway speeds. If you cannot turn the fan with the engine off, the fan clutch has jammed.

If you are working on a vehicle with an electric fan that is controlled by the ECM, a scan tool can be used to check the operation of the fan and its control system. A quick visual test is to turn on the air conditioning system. Most electric fans should operate whenever the compressor clutch is engaged. If a scan tool cannot be used, remove the motor lead and check for the presence of power and ground.

An electric fan and its control system can also be checked for proper operation by operating the fan on battery current using a fused jumper wire. If the fan does not operate, it should be replaced. Depending on the type of control system, check the thermostatic switch, fan relay, or ECM fan driver for proper operation.

Fan shrouds must be in place to allow the fan to draw air through the radiator. If these shrouds are missing, the engine will overheat at low speeds, even if the rest of the cooling system is in good condition. In a few cases, missing shrouds can cause overheating at higher speeds, due to air entering the radiator from under the vehicle, preventing air flow through the radiator. If any fan shrouds are missing or damaged, they should be replaced. Certain vehicles also use air dams to force air into the radiator. Air dams can be damaged by road debris or running over obstructions in the road. If a vehicle is equipped with these dams, check them for damage and condition.

Checking the Coolant for Exhaust Gases

Sometimes, combustion gases escape and enter the cooling system, causing overheating. In most cases, this type of leak can be detected by the cylinder leakage test discussed earlier. However, a chemical test, available through most tool manufacturers, can detect the presence of compression gas in the coolant. This tool utilizes a chemical test fluid that changes color when exposed to combustion gases.

Checking Turbocharger Boost Pressure

A general check of turbocharger operation can be made with the dashboard boost gauge. Start the engine and observe the gauge as you gradually accelerate the engine. The gauge should show the turbocharger boost pressure gradually rising to the maximum range and then remaining steady as the engine is accelerated further. If the boost pressure is within the proper range, and no other problems, such as pinging or low power are encountered, the turbocharger is probably okay.

However, most dash gauges are usually not accurate enough to make a precise check of the boost pressure. The dashboard gauge is not accurate to adjust pressure on those units where it is possible. A pressure gauge with a range of 0 to 20 pounds is necessary to accurately check boost pressure. The pressure gauge must be installed on an intake manifold fitting or by using a T-fitting at the wastegate, **Figure 19-10.**

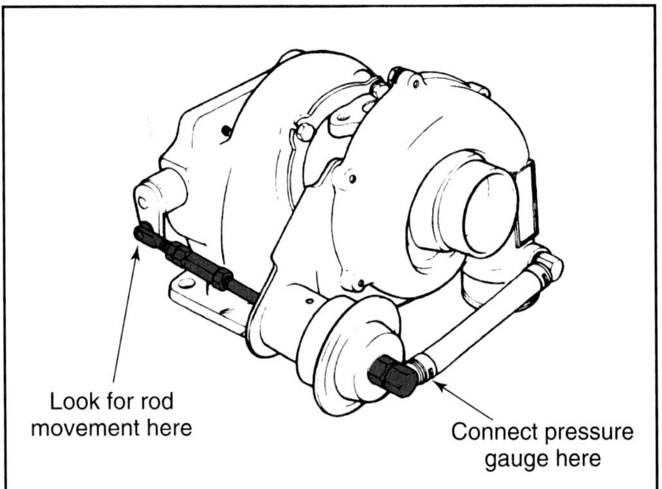

Figure 19-10. *Turbochargers should be checked for external oil leaks and unusual noises. The wastegate rod should have free movement. Be sure to connect the pressure gauge to the right fitting. (Ford)*

Note: Do not attach the pressure gauge to the turbocharger wastegate fitting in such a way that manifold pressure cannot reach the wastegate. Loss of control to the wastegate diaphragm can cause severe damage.

Install a tachometer on the engine and ensure that the engine is fully warmed up. After the pressure gauge and tachometer are installed, start the engine and increase engine speed until the turbocharger begins to operate. Watch the tachometer carefully to ensure that you do not exceed the maximum rpm limit for the engine. Most engines in good condition can speed up to about 4000 rpm without harm. The safe limit for the engine that you are working on should be determined before starting this test.

Increase the engine speed until the limit of boost pressure is reached. The wastegate should open somewhere within the range for maximum boost pressure. If the turbocharger pressures are below normal, either the turbocharger is defective, or the wastegate is opening too soon or stuck partially open. If boost pressures are higher than normal, the wastegate is stuck closed, the linkage is disconnected or jammed, or the actuator diaphragm or manifold hose is leaking. After correcting any problems and rechecking the boost pressure, remove the tachometer and pressure fitting.

Caution: Do not adjust the boost pressure higher than the specifications called for in an attempt to get more power from the engine. This can cause severe engine damage.

If boost pressure is too low, check the exhaust system to ensure that it is not restricted. A restricted exhaust will lower the exhaust gas speed, preventing the turbocharger from turning at sufficient speed. If the exhaust is not restricted, check the turbocharger to ensure that it is in good operating condition. If boost pressure is low because of a damaged turbocharger, it should be replaced.

Checking Superchargers

Begin checking a supercharger by inspecting it for obvious damage. The drive belt should be in good condition and proper tension with no cracks and breaks. Check the supercharger pulley area and seams for evidence of leaks. Since many superchargers contain oil for lubrication, an oil leak can cause exhaust smoking.

Note: Some superchargers will seep some oil from the pulley seal. This is considered to be normal.

Check for intake restrictions such as a clogged air cleaner or kinked intake hose. Check that all hoses and manifold gaskets are in good condition on the pressure side of the supercharger and that all bolts and clamps are tight. If the supercharger has an intercooler, make sure that it is not clogged with debris and is not leaking air. A symptom of a clogged intercooler is detonation at higher engine speeds.

Start the engine, if possible, and listen to the supercharger. Some superchargers will have a low pitched whine, which is normal. If the whine is excessively loud or there are any other unusual noises, locate the source. If the noise is coming from the supercharger, replace it.

The best way to test the supercharger is to determine if it is developing boost pressure. Install a pressure gauge between the supercharger and the cylinder heads or intercooler (if used). Start the engine and check the gauge. Normal boost pressure is lower than turbocharger pressure, ranging from 2-15 psi (13.8-103.4 kPa). Since superchargers are directly driven by the engine, pressure may be present at idle. If the gauge does not show boost pressure as the engine speed is increased, the supercharger or the bypass valve is defective. If pressures are excessive, especially at higher engine speeds, the bypass valve may be stuck closed or misadjusted. Excess pressure will cause severe engine detonation at higher engine speeds.

Note: A supercharged engine should develop vacuum in the intake system between the throttle valves and supercharger intake, just as on a naturally aspirated engine.

If the supercharger is defective, it should be replaced as soon as possible. Since superchargers are expensive, double check your findings to make sure that it is the cause of the problem. Do not operate a defective supercharger for long periods, since it can come apart, allowing metal to enter the engine cylinders, destroying the engine.

Checking Engine Oil Pressure

While low oil pressure will not directly cause a driveability problem, it contributes to engine wear, which can cause many performance problems. Extremely low oil pressure can affect the operation of hydraulic lifters or lash adjusters. A quick way to check for oil pressure is to observe the dashboard gauge or oil light. If the dash indicator or internal engine noise leads you to believe that an oil pressure problem exists, you should perform an oil pressure test by using a gauge connected to the engine's oil system.

Connect the oil pressure gauge to the engine. Usually the connector fitting is the same one used by the oil pressure sending unit, **Figure 19-11.** Next, start the engine and allow it to reach operating temperature. When the engine reaches operating temperature, read the oil pressure on the gauge. It should be at least 20 psi (138 kPa) at idle up to 60 psi (413 kPa) at normal engine cruising speed. If the pressure is low, the oil pump or lubrication system is clogged or the pump is defective. If the pressure is high, a partial restriction or a sticking pressure relief valve may be the cause. On some engines, the oil pump can be easily replaced without major engine disassembly. Most oil pumps are simply replaced after the pan has been removed.

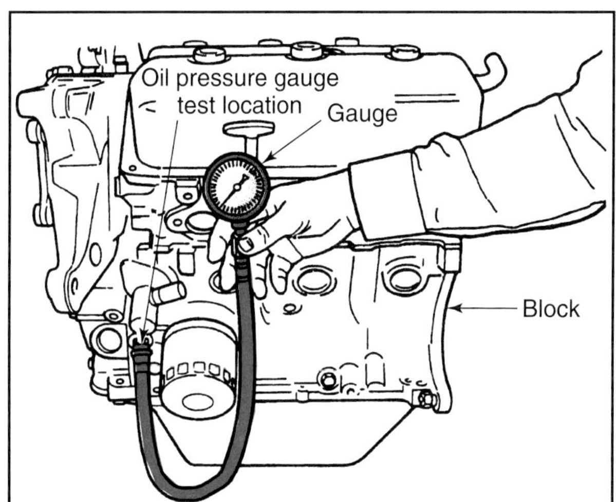

Figure 19-11. *Oil pressure can be easily tested by threading a pressure gauge into the same hole used by the oil pressure sending unit. (Chrysler)*

Preparing to Check Internal Engine Components for Defects

If the external inspection shows no signs of a defect or your earlier tests lead you to believe that the engine has an internal defect, it may be necessary to perform a partial disassembly to test internal parts. If you are preparing to disassemble an engine to find the cause of a driveability problem, you should have eliminated all other external causes. Go back and reinspect or test any part or system if you have *any doubt.*

Note: Be sure to obtain the owner's permission before beginning any engine disassembly.

Checking Valve Train Parts

Valve timing is critical to proper engine performance. If the valves are not precisely timed to the movement of the piston, the engine will not run properly, and the valves may hit the top of the piston. For example, timing gears that are off only one tooth will severely affect driveability. The engine will have low compression, perform sluggishly, hesitate, detonate heavily, or backfire through the intake manifold. If it has jumped several teeth, the engine will not run. One way to quickly check the condition of the timing chain or belt is to check the position of the distributor rotor against the ignition timing marks.

Note: This procedure will not work on engines where the distributor is not driven by the camshaft or on engines with distributorless ignition.

Begin by removing the distributor cap and the number one spark plug. Then crank the engine until the ignition timing marks are on the top dead center position. With the timing marks in this position, the rotor should be under the number one spark plug tower. Next, turn the crankshaft in the opposite direction of its normal rotation. If the crankshaft moves more than six or seven degrees before the rotor begins to move, the chain or belt has excessive slack. For best results, repeat this test several times to ensure that the chain or belt is not loose.

Checking Valve Timing

If the distributor rotor test indicates that the timing is incorrect, the camshaft drive will need to be checked. The camshaft drive gears will always have some type of timing marks. Refer to **Figure 19-12.** To ensure that the valves are properly timed, timing marks are cast or punched into the drive and driven gears. There are three types of camshaft drives: the matching gear type, the gear and chain type, and the gear and belt type. On all three types, valve timing can be checked by removing the timing cover. On some engines, the water pump, vibration damper, and other components must be removed along with the timing cover to expose the camshaft drive mechanism. On some overhead camshaft engines, the timing cover can be removed without disturbing other components.

Once the timing cover is removed, consult the service manual to determine where the mark on the drive (crankshaft) gear should be for making the valve timing check. Then rotate the engine until the mark is in the proper position. Observe the position of the mark on the driven (camshaft) gear. If both gears are in the proper position, the valves are correctly timed. If the marks do not match, the timing is off.

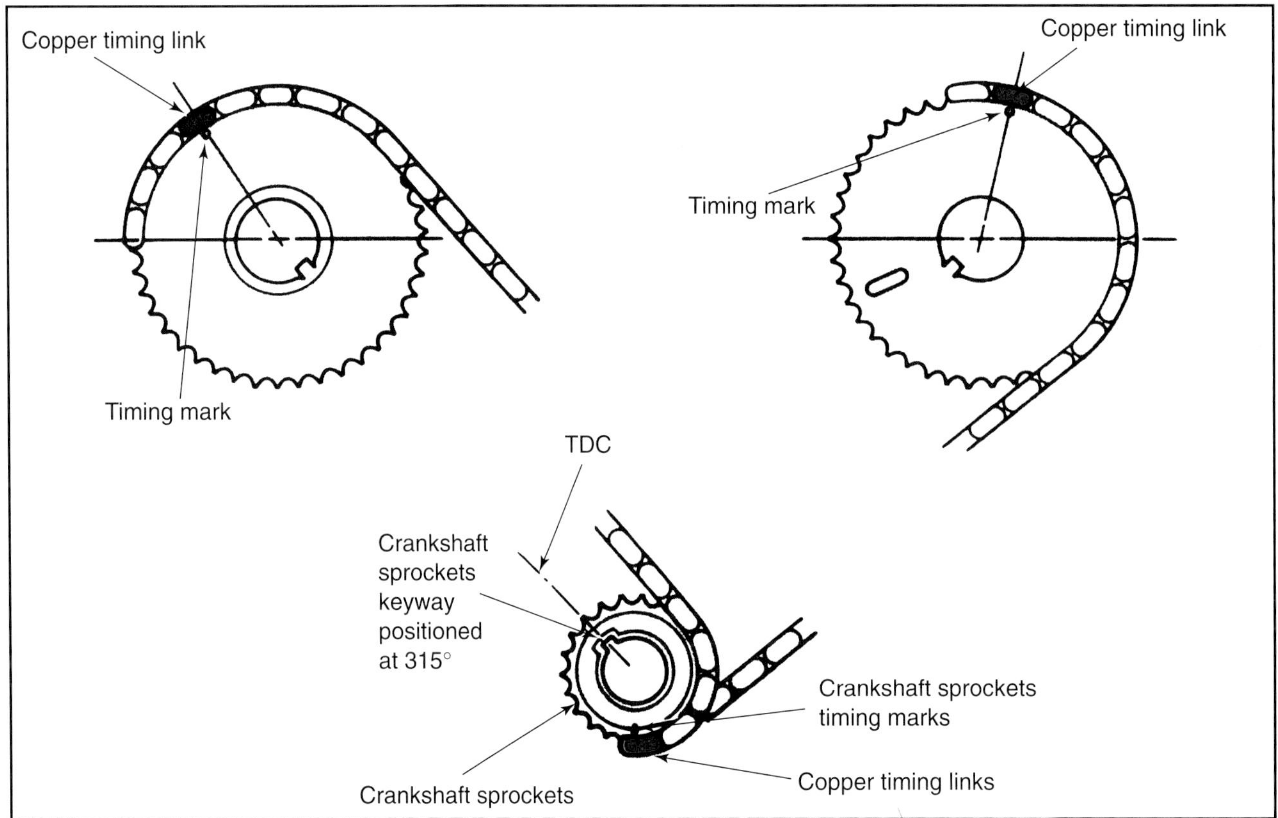

Figure 19-12. *Engine timing can cause many problems if the chain, belt, or gears are off by even one tooth. Make sure all timing marks line up with the engine's reference points. (Ford)*

Checking Overhead Camshaft Timing

On a few overhead camshaft engines, the gears are aligned by matching them to marks cast into the block or head. This type of timing system is usually found on engines with timing belts. Remember that some overhead camshaft engines use a belt-driven distributor. The correct timing of the distributor should also be checked, if equipped. As well as checking the timing of the camshaft, you should also check the condition of the camshaft drive gear.

Checking Chain Drive Systems

A stretched timing chain can cause the ignition timing to be slightly advanced or retarded as well as creating valve train timing problems. Check for a stretched timing chain by turning the camshaft gear to remove all the slack from one side of the chain. Then check the other side for excessive chain deflection, as shown in **Figure 19-13.** More than about .5″ (12.7 mm) of slack will cause retarded valve timing. Timing gear teeth should also be checked for wear or chips. Check all gear attaching bolts for proper torque, and check pressed–on gears for a loose fit on the cam or crankshaft.

The camshaft and crankshaft timing gears should be replaced along with the chain, especially if the camshaft gear is made with molded nylon plastic teeth. Even if it appears to be good, replace these gears since they often fail without warning. Most replacement timing chain sets come with the camshaft and crankshaft gears and should be replaced as a set.

Checking Belt Drive Systems

The gears used on belt drive systems should be checked for wear, scoring, and bearing problems. Check timing belts for missing teeth, cracks in the rubber covering on either side, softness or oil coating, or excessive stretch. The timing belt should be replaced if it shows any signs of damage. See **Figure 19-14.** Bearing checking is especially important on any engine using the timing belt to drive the water pump, distributor, oil pump, or any idler bearings that are not lubricated by engine oil.

Checking Valve Springs, Retainers, and Seals

To service the valve components, the valve cover or covers must be removed. Once the cover(s) are removed, check the springs and retainers for obvious damage. Broken springs or bent retainers should be visible when a close inspection is performed. Also check the rocker arms, rocker mounting studs, pushrods (when used), and other parts for obvious damage or severe wear.

When spark plugs on a particular cylinder repeatedly foul, and engine compression is normal, the most likely cause is a leaking valve stem oil seal, usually called a valve seal. If you suspect that spark plug oil fouling is caused by a defective O-ring type valve seal, it can be checked by the procedure shown in **Figure 19-15.** The suction cup should be placed over the retainer, and a vacuum developed using a vacuum pump. If the seal is good, it will hold vacuum for at least 15 seconds. Vacuum will quickly leak away if the seal is cracked. If the engine uses umbrella type seals, the

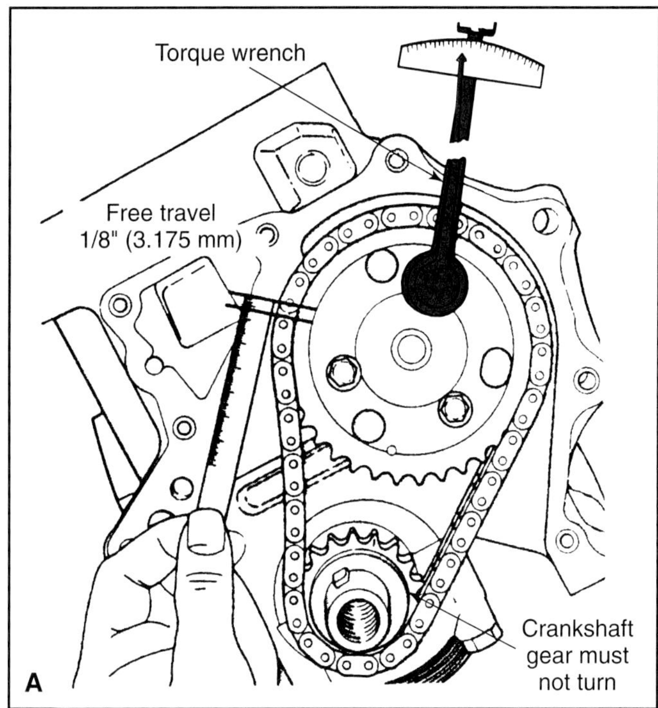

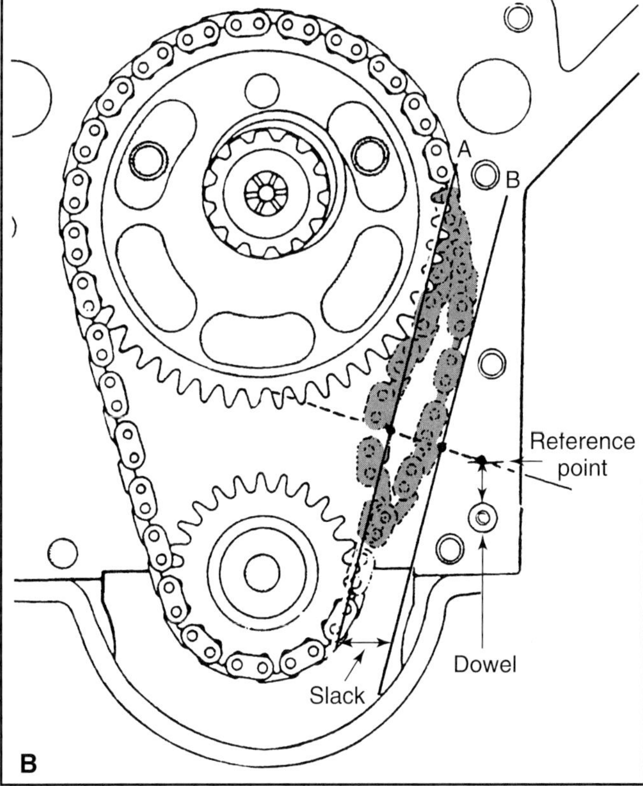

Figure 19-13. *A—Check for timing chain free travel with a ruler. There should be as little deflection as possible. B—The distance between A and B represents the slack in the chain. If a chain has excessive play, it will jump at high speeds. (Chrysler)*

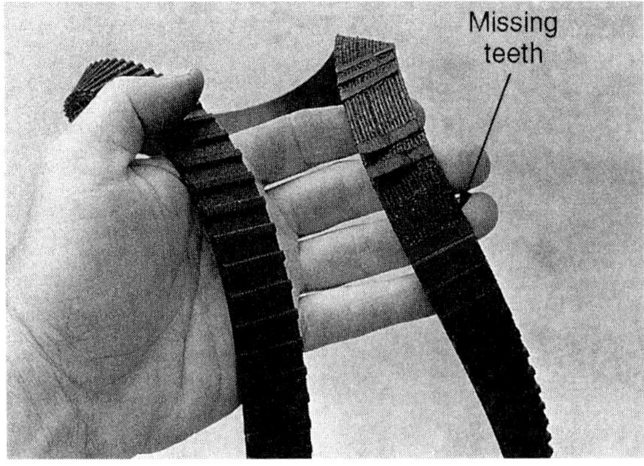

Figure 19-14. *Timing belt condition is critical. This timing belt failed on the engine. If a timing belt fails on most engines, severe damage will result.*

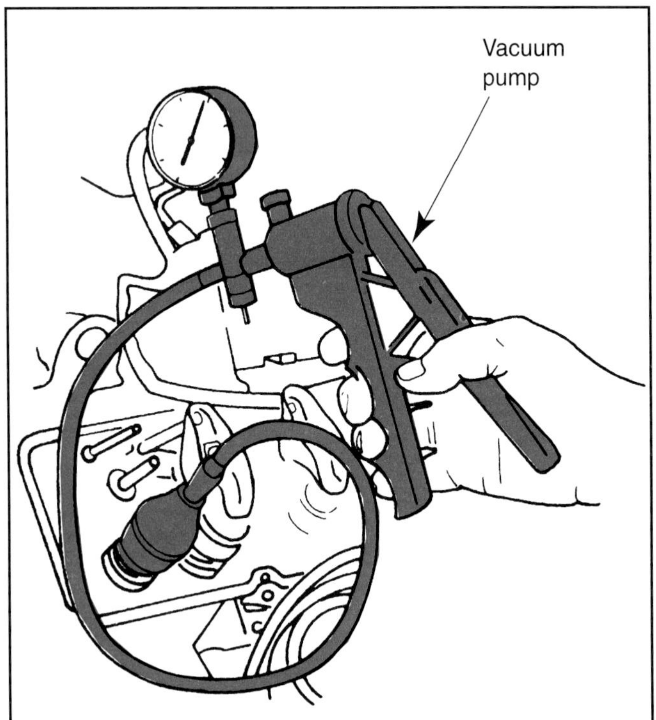

Figure 19-15. *This procedure checks the valve stem seals for leaks. If the engine is equipped with umbrella seals, the valve springs should be removed for a visual inspection. (General Motors)*

retainer and spring must be removed to get a good look at the seal.

Note: Although leaking intake valve seals are the usual cause of oil fouling, do not ignore the exhaust valve seals. Modern camshaft timing often creates a suction at the exhaust valve. If the exhaust valve seal is defective, oil can enter the cylinder through the exhaust valve stem.

The valve spring can be removed without removing the cylinder head. Begin by applying shop air pressure to the cylinder through the spark plug hole. A special adapter must be used to ensure that enough air enters the cylinder to hold both valves in place when the valve retainers are removed. Then a valve spring compressor, **Figure 19-16,** can be used to depress the valve spring and remove the valve keepers. The valve spring can then be removed to expose the umbrella seal. Reassemble the valve springs, retainers, and keepers using new seals.

Caution: Maintain air pressure in the cylinder whenever the valve spring keepers and retainers are disassembled.

After all valve service is completed, reinstall the valve covers. Be sure to use new valve cover gaskets, or gasket forming sealer if specified. Valves often go out of adjustment due to normal wear of the valve train parts. Sometimes the parts become excessively worn, or the excess clearance is caused by a collapsed valve lifter. These problems require parts replacement. In other cases, the excess clearance can be corrected by adjusting the valves. Valve adjustment procedures are covered later in this chapter.

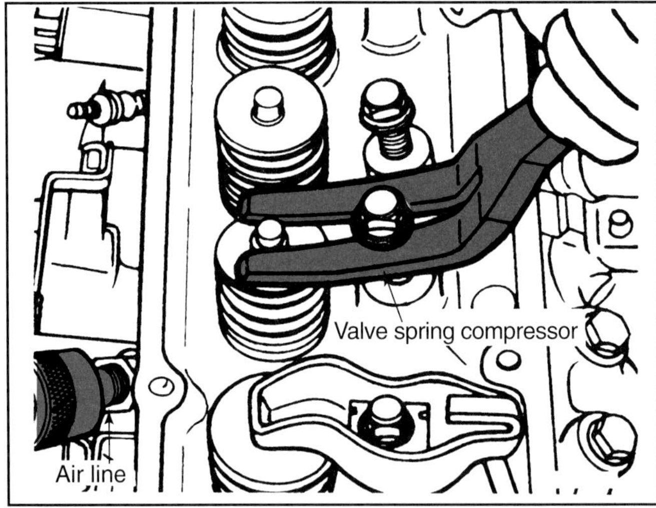

Figure 19-16. *To remove the valve spring, apply air pressure to the cylinder. This prevents the valves from dropping into the cylinder. Then remove the rocker arm and use a valve spring tool to remove the keepers and the spring assembly. Exercise care as the valve spring can come off with considerable force. (Ford)*

Miscellaneous Valve Problems

If the timing belt on an overhead camshaft engine breaks, the valves can be bent when the piston strikes them. Once a valve is bent, it cannot seal against the valve seat, and the engine will not be able to develop compression. Not all overhead camshaft engines will suffer bent valves if the belt breaks, but this is a possibility on many engines, especially if the belt breaks at high engine rpm. The most obvious sign of bent valves is an engine that cranks very easily, due to the lack of compression.

Engines with high mileage, or engines that have been driven for long periods at low speeds, often develop excess carbon deposits on the intake valves. When the engine is cold, these carbon deposits absorb gasoline, making the engine run lean. As the engine heats up, the gasoline is vaporized and causes a rich mixture. Many mysterious cold driveability problems are caused by deposits on the intake valves. Replacing bent valves or removing excess valve deposits requires cylinder head removal.

Checking Valve Adjustment

Engine valve adjustment consists of setting the clearance between the parts that open the valve. These parts are located between the camshaft and the valve stem. Valve adjustment is critical to ensure maximum engine performance with the longest valve life. Valve clearances can be too small, called ***tight clearance,*** or too great, called ***loose clearance.***

If intake valves are too tight, they will not close completely. In this case, the engine will miss and the fuel mixture in the intake manifold will be ignited, usually called "popping back" or backfire through the intake. If the exhaust valves are too tight, the engine will miss, although not as badly as with a tight intake valve. More seriously, tight exhaust valves will eventually melt (usually called burned valves) since they cannot cool off on the valve seat. Tight valves are usually caused by improper adjustment.

In most cases, engine operation will cause wear of the valve train parts, increasing the clearance, and causing loose valves. If the intake or exhaust valves are too loose, they will make a noise usually referred to as clattering. A loose valve train will not open the valve fully, and the engine will not develop full power. The loose setting will also cause excessive valve train shock, leading to increased wear.

Note: Valve clatter should not be confused with the knocking sound produced by detonation or preignition.

Older engines have mechanical lifters that must be adjusted at certain mileage intervals. Most modern engines are equipped with hydraulic valve lifters that can adjust for wear, making periodic manual adjustments unnecessary. Some hydraulic valve systems do not have an adjustment mechanism, and rely on the hydraulic lifter or lash adjuster to maintain the proper clearance. If these systems develop too much clearance, valve train parts must be replaced. Other hydraulic lifter engines are equipped with adjustment mechanisms, allowing the technician to compensate for excessive valve train wear.

Checking Valve Lifters

The valve lifters should also be checked for proper operation. Valve lifters are mechanical, hydraulic or roller, as explained in **Chapter 5.** Mechanical lifters are usually found only on older vehicles. They can be checked for fit in the lifter bore and wear at the point where the lifter contacts the cam lobe. The bottom of the lifter should have a slightly rounded point. This is known as a ***convex face,*** or "crown." If the lifter shows any wear, it should be replaced. Roller lifters do not have a lifter crown. The roller should be checked for wear, pitting, binding, or other defects.

Hydraulic lifters can also be checked for proper crown. Hydraulic lifters are also checked for collapsing. This should be done before disassembling the lifter. Make sure that the lifter is still full of oil, and place it upright on a bench. Push the center section of the lifter with one of the engine pushrods. The center section should move only a slight distance, or not move. If the center section can be moved a great distance down in the lifter body, the lifter is collapsed.

Note: Do not use worn lifters with a new camshaft or new lifters with a worn camshaft. The lifter crown must match the cam lobe taper. Therefore, new lifters or a new cam will not work properly with used mating parts. Always replace the camshaft and lifters as a set.

Checking Camshaft Lobe Lift

Since the camshaft is so vital to proper engine operation, it should be replaced if it fails any of the checks below. If visible, inspect the camshaft for any scoring, pitting, or other obvious wear. If any is found, replace the camshaft. If no obvious defects are found, check the amount of camshaft lift, or ***lobe lift,*** by using a dial indicator. Unless the lobes are badly worn, it is almost impossible to check lobe lift by eye. The steps below are a guideline only. The manufacturer's service manual should also be consulted for lift specifications and exact procedures.

Begin by positioning the point of a dial indicator on the first cam lobe to be checked. On OHV engines, the dial indicator can be connected to the pushrod with the cylinder heads installed on the block, as shown in **Figure 19-17.** Rotate the camshaft slowly in its normal direction of rotation until the indicator registers its lowest reading. This is the lowest point of the cam lobe.

Next, set the dial indicator on zero. Slowly rotate the camshaft until the indicator shows the highest possible reading. This is the high point of the cam lobe. Record the lobe lift reading, and repeat the test on all lobes on the camshaft. Compare the total lift recorded with the specifications in the service manual. If the camshaft readings for all lobes are within specifications, and all other checks are satisfactory, the camshaft can be reused. If any lobe is below specifications, the camshaft should be replaced.

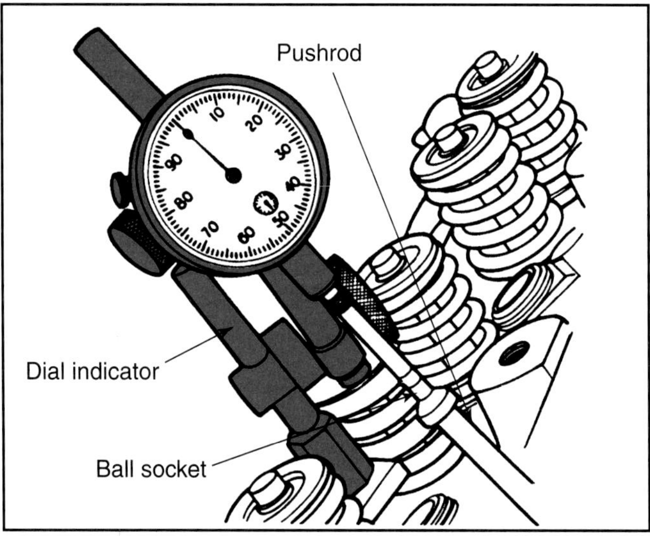

Figure 19-17. *Lobe lift and camshaft condition can be checked by mounting a dial indicator and measuring the distance the pushrod moves. (Ford)*

Engine Service Procedures

The following sections detail adjustment and replacement procedures for some engine components. As explained at the beginning of this chapter, major engine overhaul procedures, such as cylinder reconditioning and piston replacement, are not covered. The parts covered here are components that are typically removed, adjusted, or replaced to correct a driveability problem.

Valve Train Adjustment

On some engines, it is necessary to adjust the valve train clearance, sometimes called ***valve lash,*** for maximum performance. Most manufacturers will indicate in their service manuals if periodic valve train adjustment is necessary. Most valve clearance adjustment procedures are relatively simple, but must be done carefully to avoid damage. Valve clearance is adjusted by one of two methods, depending on the type of valve lifter or lash adjuster. If the lifters are mechanical or if the engine cannot be operated during the adjustment procedure, the lifters are adjusted with feeler gauges with the engine not running. Some valve lifters are adjusted by sound with the engine running. Begin any valve adjustment procedure by determining the type of valve adjustment that must be performed.

Lifter Adjustment, Engine not Running

Remove the necessary engine parts to expose the valve train. If the intake plenum must be removed, place clean shop towels over the intake port openings. Disable the ignition system and bump the starter until the first cylinder is up on its compression stroke. This is easily determined by observing the valves for the number 1 cylinder. When both valves are closed, and the timing marks are aligned, the number 1 cylinder is at the top of its compression stroke. Since the cylinder is on its compression stroke, both valves will be closed. Determine the proper thickness feeler gauge, according to the service manual specifications.

Note: Intake and exhaust valves often have different clearance specifications.

Slide the feeler gauge between the rocker arm and the valve stem, **Figure 19-18.** If the valve slides with only a slight amount of drag, the clearance is correct. If the feeler gauge feels loose, or will not enter the space between the rocker arm and stem, the clearance is wrong. If the clearance is incorrect, locate the valve adjustment mechanism. Loosen the locknut on the adjuster screw, and turn it until the clearance, as measured by the feeler gauge, is correct. Tighten the locknut and recheck the clearance.

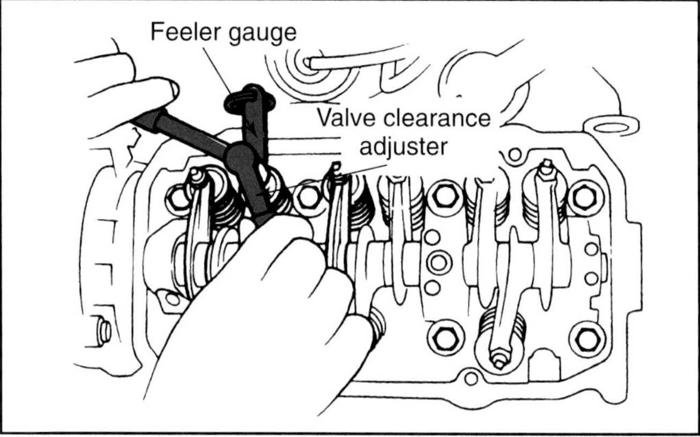

Figure 19-18. *Valve train adjustment is simple if the proper procedure is used. A feeler gauge is almost always used. Some engines require a special tool for valve train adjustment. (Subaru)*

Many manufacturers specify which valves can be adjusted when the number 1 cylinder is at the top of its compression stroke. Turn the engine exactly one turn until the timing marks are again aligned. Then, the other valves can be adjusted. Repeat the procedure on all valves, making sure that each cylinder is on its compression stroke before adjusting the valves for that cylinder. After all valves are adjusted, start the engine and recheck the clearances by inserting the feeler gauge between the rocker and stem as the valves operate. If the feeler gauge cannot be inserted, the valves are too tight. If a rocker arm stops clattering when the gauge is inserted, the setting is too loose.

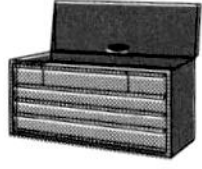

Note: To avoid unnecessary labor, be sure that all adjustments have been made properly before reassembling the engine.

Lifter Adjustment, Engine Running

If the valve lifters are hydraulically operated, start the engine with the valve cover removed. To avoid spilling oil, special clips can be placed over the rocker arms to divert the oil spray from the pushrods. Another method is to place a piece of cardboard next to the rocker arms to deflect the oil back into the engine.

Place a socket and ratchet on the first valve adjuster, **Figure 19-19.** It is easier to remember which valves have been adjusted if you start at the first valve at the front of the head. Back off (loosen) the adjuster until the valve starts to clatter. Begin slowly tightening the adjuster until the clatter stops, then tighten the adjuster an additional amount as called for in the service manual. Repeat this operation for all engine valves.

If any hydraulic valves are still noisy after being adjusted, start the engine and insert a .010″ (.254 mm) feeler gauge between the rocker and stem as the valves operate. If the noise stops when the gauge is inserted, the setting is too loose. This could be caused by incorrect adjustment or worn valve train parts. Readjust or replace parts as necessary, and recheck valve adjustment. Once the valves are adjusted, reinstall the valve cover using a new gasket or sealant as required.

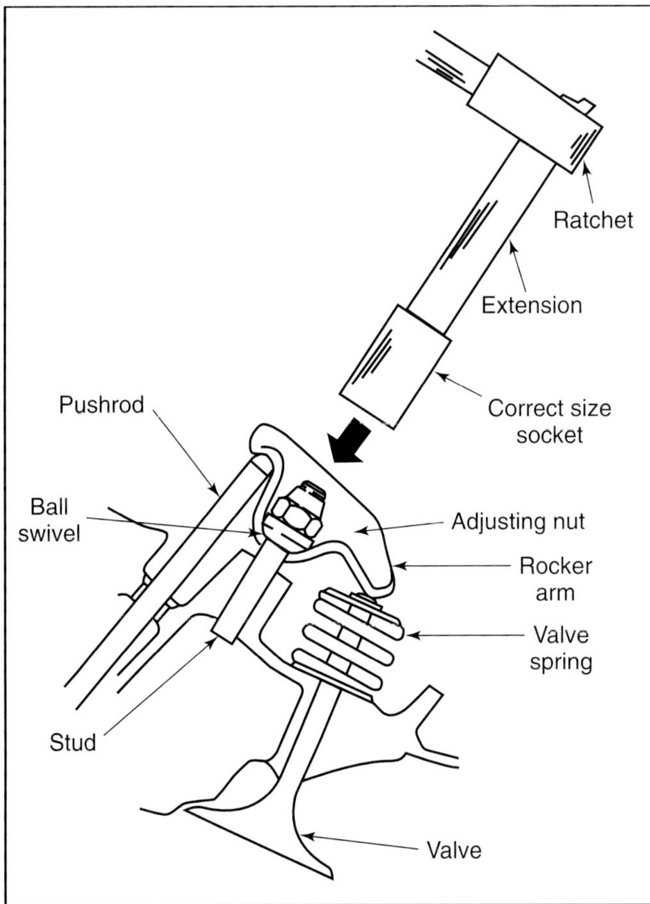

Figure 19-19. *Overhead valve adjustments are usually not necessary unless the rocker arm has been loosened or removed. A socket and ratchet are the only tools needed. This adjustment can also be done with the engine off by turning the pushrod and feeling for tension. Some overhead valve engines adjust the valves automatically when the rocker arms are installed.*

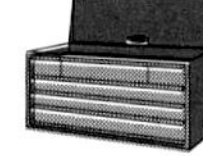

Note: Most engines with hydraulic lifters do not require valve adjustments other than the adjustment performed during initial valve train installation.

Overhead Camshaft Valve Train Adjustment

When the engine has an overhead camshaft, the adjustment procedure may be slightly different. On many overhead camshaft engines, the rocker arm is pushed downward by the camshaft lobe and opens the valve. When an adjustment mechanism is used, it will be on the opposite end of the rocker arm from the valve. Many overhead engines using rocker arms do not have any provision for valve adjustment.

On many overhead camshaft engines the camshaft lobes operate the valves directly, without using rocker arms. Valves on these engines are adjusted by replacing ***shims*** located between the valve and the cam lobe. The illustrations in **Figure 19-20** show how the valve-to-camshaft clearance is checked on this type of engine. If the clearance is excessive, the original shim must be replaced with a thicker shim. The shim thickness is usually referenced in a chart, similar to the one in **Figure 19-20.** The shim is removed and measured with a micrometer. The amount of excess clearance is added to the thickness of the original shim to determine the needed thickness of the replacement shim.

Fan Clutch Replacement

The fan clutch on belt driven fans can be replaced by removing the capscrews holding the fan and clutch assembly to the water pump hub. Fan shrouds may be removed for clearance. It is often easier to leave the fan belt tight until the capscrews have been loosened. After the fan and clutch assembly have been removed and placed on a bench, then remove the screws holding the clutch to the fan blade assembly, and install the new clutch. Then reinstall the clutch and fan assembly on the water pump hub. Be sure to recheck belt tension and shroud clearance before starting the engine.

Caution: Make sure that the fan blade assembly is installed facing in the original direction. If the fan is installed backwards, noise and overheating at cruising speeds may result.

Electric Cooling Fan and Motor Replacement

The radiator fan and motor are usually installed on the engine side of the radiator, as shown in **Figure 19-21.** They can be removed by first disconnecting the motor electrical

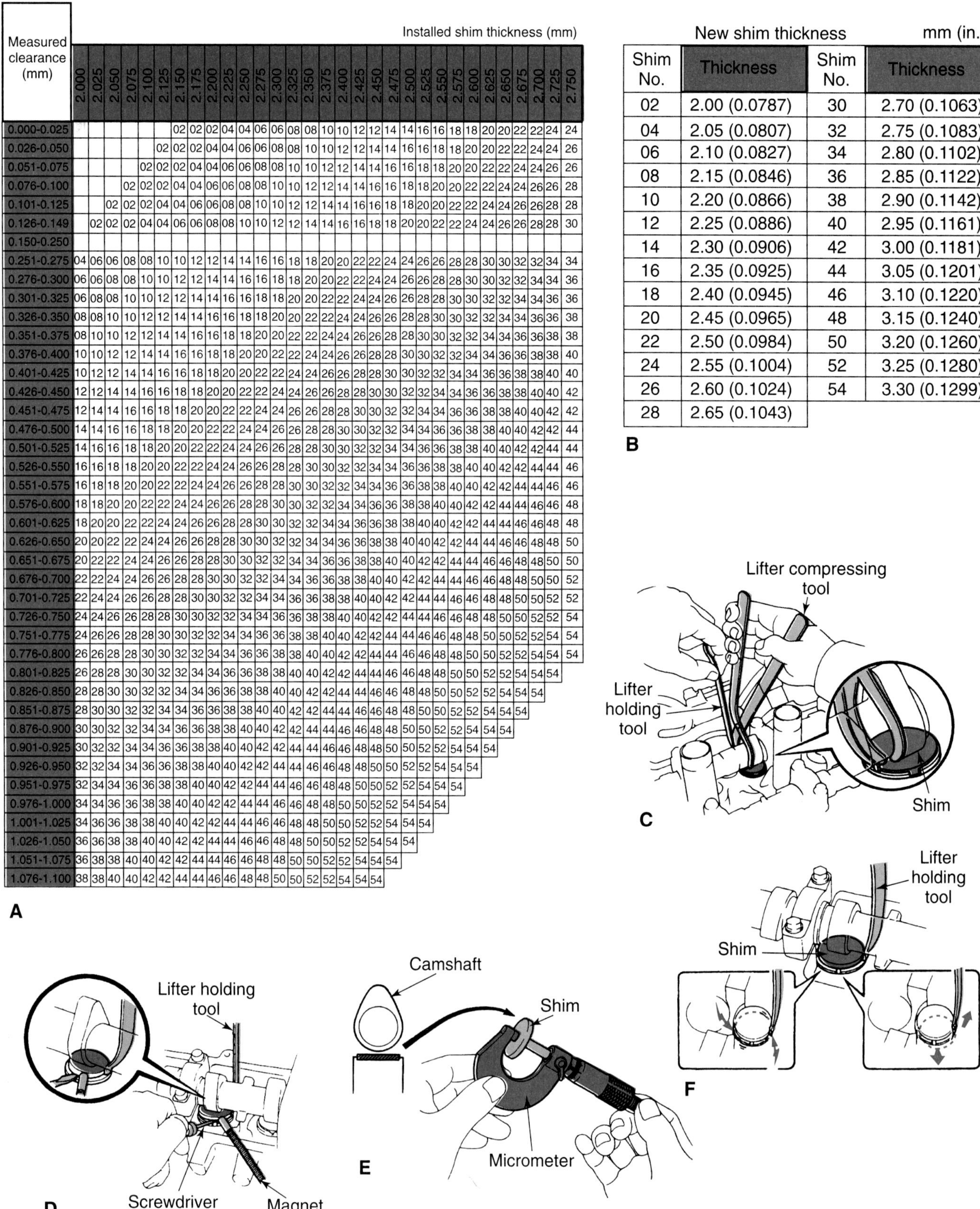

Installed shim thickness (mm)

Measured clearance (mm)	2.000	2.025	2.050	2.075	2.100	2.125	2.150	2.175	2.200	2.225	2.250	2.275	2.300	2.325	2.350	2.375	2.400	2.425	2.450	2.475	2.500	2.525	2.550	2.575	2.600	2.625	2.650	2.675	2.700	2.725	2.750
0.000-0.025							02	02	02	04	04	06	06	08	08	10	10	12	12	14	14	16	16	18	18	20	20	22	22	24	24
0.026-0.050						02	02	02	04	04	06	06	08	08	10	10	12	12	14	14	16	16	18	18	20	20	22	22	24	24	26
0.051-0.075					02	02	02	04	04	06	06	08	08	10	10	12	12	14	14	16	16	18	18	20	20	22	22	24	24	26	26
0.076-0.100				02	02	02	04	04	06	06	08	08	10	10	12	12	14	14	16	16	18	18	20	20	22	22	24	24	26	26	28
0.101-0.125			02	02	02	04	04	06	06	08	08	10	10	12	12	14	14	16	16	18	18	20	20	22	22	24	24	26	26	28	28
0.126-0.149		02	02	02	04	04	06	06	08	08	10	10	12	12	14	14	16	16	18	18	20	20	22	22	24	24	26	26	28	28	30
0.150-0.250																															
0.251-0.275	04	06	06	08	08	10	10	12	12	14	14	16	16	18	18	20	20	22	22	24	24	26	26	28	28	30	30	32	32	34	34
0.276-0.300	06	06	08	08	10	10	12	12	14	14	16	16	18	18	20	20	22	22	24	24	26	26	28	28	30	30	32	32	34	34	36
0.301-0.325	06	08	08	10	10	12	12	14	14	16	16	18	18	20	20	22	22	24	24	26	26	28	28	30	30	32	32	34	34	36	36
0.326-0.350	08	08	10	10	12	12	14	14	16	16	18	18	20	20	22	22	24	24	26	26	28	28	30	30	32	32	34	34	36	36	38
0.351-0.375	08	10	10	12	12	14	14	16	16	18	18	20	20	22	22	24	24	26	26	28	28	30	30	32	32	34	34	36	36	38	38
0.376-0.400	10	10	12	12	14	14	16	16	18	18	20	20	22	22	24	24	26	26	28	28	30	30	32	32	34	34	36	36	38	38	40
0.401-0.425	10	12	12	14	14	16	16	18	18	20	20	22	22	24	24	26	26	28	28	30	30	32	32	34	34	36	36	38	38	40	40
0.426-0.450	12	12	14	14	16	16	18	18	20	20	22	22	24	24	26	26	28	28	30	30	32	32	34	34	36	36	38	38	40	40	42
0.451-0.475	12	14	14	16	16	18	18	20	20	22	22	24	24	26	26	28	28	30	30	32	32	34	34	36	36	38	38	40	40	42	42
0.476-0.500	14	14	16	16	18	18	20	20	22	22	24	24	26	26	28	28	30	30	32	32	34	34	36	36	38	38	40	40	42	42	44
0.501-0.525	14	16	16	18	18	20	20	22	22	24	24	26	26	28	28	30	30	32	32	34	34	36	36	38	38	40	40	42	42	44	44
0.526-0.550	16	16	18	18	20	20	22	22	24	24	26	26	28	28	30	30	32	32	34	34	36	36	38	38	40	40	42	42	44	44	46
0.551-0.575	16	18	18	20	20	22	22	24	24	26	26	28	28	30	30	32	32	34	34	36	36	38	38	40	40	42	42	44	44	46	46
0.576-0.600	18	18	20	20	22	22	24	24	26	26	28	28	30	30	32	32	34	34	36	36	38	38	40	40	42	42	44	44	46	46	48
0.601-0.625	18	20	20	22	22	24	24	26	26	28	28	30	30	32	32	34	34	36	36	38	38	40	40	42	42	44	44	46	46	48	48
0.626-0.650	20	20	22	22	24	24	26	26	28	28	30	30	32	32	34	34	36	36	38	38	40	40	42	42	44	44	46	46	48	48	50
0.651-0.675	20	22	22	24	24	26	26	28	28	30	30	32	32	34	34	36	36	38	38	40	40	42	42	44	44	46	46	48	48	50	50
0.676-0.700	22	22	24	24	26	26	28	28	30	30	32	32	34	34	36	36	38	38	40	40	42	42	44	44	46	46	48	48	50	50	52
0.701-0.725	22	24	24	26	26	28	28	30	30	32	32	34	34	36	36	38	38	40	40	42	42	44	44	46	46	48	48	50	50	52	52
0.726-0.750	24	24	26	26	28	28	30	30	32	32	34	34	36	36	38	38	40	40	42	42	44	44	46	46	48	48	50	50	52	52	54
0.751-0.775	24	26	26	28	28	30	30	32	32	34	34	36	36	38	38	40	40	42	42	44	44	46	46	48	48	50	50	52	52	54	54
0.776-0.800	26	26	28	28	30	30	32	32	34	34	36	36	38	38	40	40	42	42	44	44	46	46	48	48	50	50	52	52	54	54	54
0.801-0.825	26	28	28	30	30	32	32	34	34	36	36	38	38	40	40	42	42	44	44	46	46	48	48	50	50	52	52	54	54	54	
0.826-0.850	28	28	30	30	32	32	34	34	36	36	38	38	40	40	42	42	44	44	46	46	48	48	50	50	52	52	54	54	54		
0.851-0.875	28	30	30	32	32	34	34	36	36	38	38	40	40	42	42	44	44	46	46	48	48	50	50	52	52	54	54	54			
0.876-0.900	30	30	32	32	34	34	36	36	38	38	40	40	42	42	44	44	46	46	48	48	50	50	52	52	54	54	54				
0.901-0.925	30	32	32	34	34	36	36	38	38	40	40	42	42	44	44	46	46	48	48	50	50	52	52	54	54	54					
0.926-0.950	32	32	34	34	36	36	38	38	40	40	42	42	44	44	46	46	48	48	50	50	52	52	54	54	54						
0.951-0.975	32	34	34	36	36	38	38	40	40	42	42	44	44	46	46	48	48	50	50	52	52	54	54	54							
0.976-1.000	34	34	36	36	38	38	40	40	42	42	44	44	46	46	48	48	50	50	52	52	54	54	54								
1.001-1.025	34	36	36	38	38	40	40	42	42	44	44	46	46	48	48	50	50	52	52	54	54	54									
1.026-1.050	36	36	38	38	40	40	42	42	44	44	46	46	48	48	50	50	52	52	54	54	54										
1.051-1.075	36	38	38	40	40	42	42	44	44	46	46	48	48	50	50	52	52	54	54	54											
1.076-1.100	38	38	40	40	42	42	44	44	46	46	48	48	50	50	52	52	54	54	54												

A

New shim thickness mm (in.)

Shim No.	Thickness	Shim No.	Thickness
02	2.00 (0.0787)	30	2.70 (0.1063)
04	2.05 (0.0807)	32	2.75 (0.1083)
06	2.10 (0.0827)	34	2.80 (0.1102)
08	2.15 (0.0846)	36	2.85 (0.1122)
10	2.20 (0.0866)	38	2.90 (0.1142)
12	2.25 (0.0886)	40	2.95 (0.1161)
14	2.30 (0.0906)	42	3.00 (0.1181)
16	2.35 (0.0925)	44	3.05 (0.1201)
18	2.40 (0.0945)	46	3.10 (0.1220)
20	2.45 (0.0965)	48	3.15 (0.1240)
22	2.50 (0.0984)	50	3.20 (0.1260)
24	2.55 (0.1004)	52	3.25 (0.1280)
26	2.60 (0.1024)	54	3.30 (0.1299)
28	2.65 (0.1043)		

B

Figure 19-20. *Overhead camshaft engine valve adjustment procedure. A—Installed shim thickness chart. B—Replacement shim thickness chart. C—Use a lifter compressing tool to carefully depress and hold the lifter. D—Carefully slide the shim out using a magnet and screwdriver. E—Use an outside micrometer to measure the shim thickness. Compare the thickness to the chart in the service manual. F—Slide the replacement shim into place and carefully unload the lifter. (Toyota)* connector to prevent injury from an accidental fan startup. Next, remove any radiator shrouds, engine mounts, or protective covers. Then remove the fasteners holding the fan bracket to the vehicle, and remove the fan and motor as an assembly. To remove the fan motor, first remove the fan locking clip, and slide the fan from the motor shaft, **Figure 19-21A.** Then remove the screws holding the motor to the fan bracket, and pull the motor from the bracket, **Figure 19-21B.**

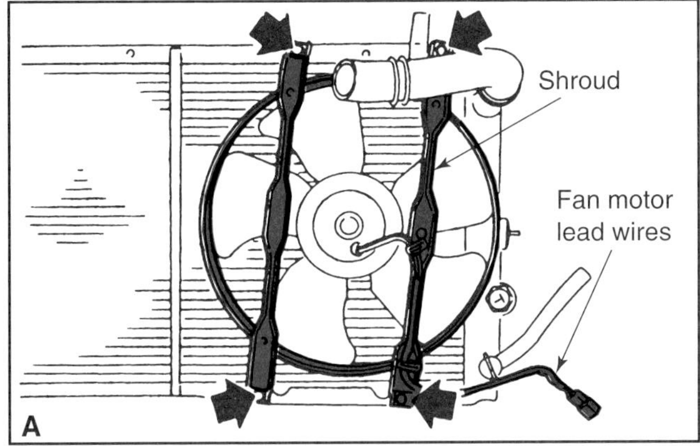

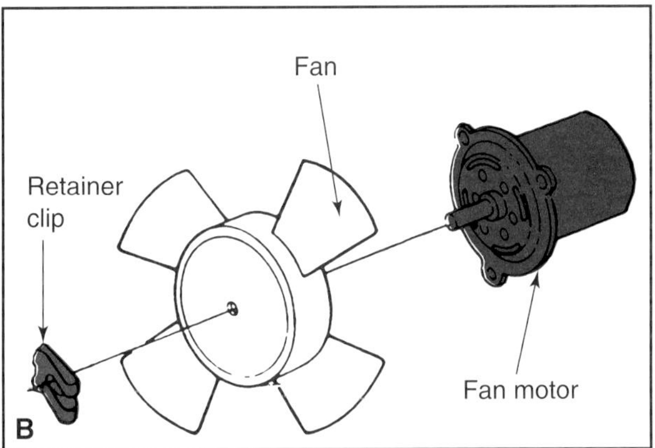

Figure 19-21. *A—Remove the radiator cooling fan by first disconnecting the motor lead and unbolting the frame from the radiator shroud. It may be necessary in some cases to remove other parts to access the fan. B—The fan blade is held to the motor by a clip or bolt. Some bolts used on electric fans have left-hand threads. (Subaru)*

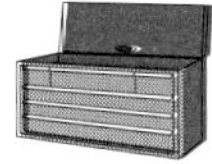

Note: Some cooling fans are attached to the motor by a left-hand thread bolt.

To install the new motor, place it in position on the fan bracket, and install and tighten the mounting screws. Then reinstall the fan on the motor shaft. The entire assembly can then be reinstalled in the vehicle. To prevent injury, reconnect the electrical leads after all other parts are installed. Keep your hands clear of the fan after reinstalling the electrical connector.

Cooling System Thermostat

The engine cooling system thermostat is usually located at the top of the engine at the coolant outlet connected to the upper radiator hose. The thermostat can be replaced easily if a few safety precautions are followed. Start by removing all pressure from the cooling system and draining it to a level below the thermostat. Then remove the upper radiator hose at the thermostat housing. Remove the bypass hose if it is attached to the housing. Remove the thermostat housing attaching bolts, and lightly pry on the housing to loosen the gasket seal. **Figure 19-22** is an exploded view of a typical thermostat and housing assembly.

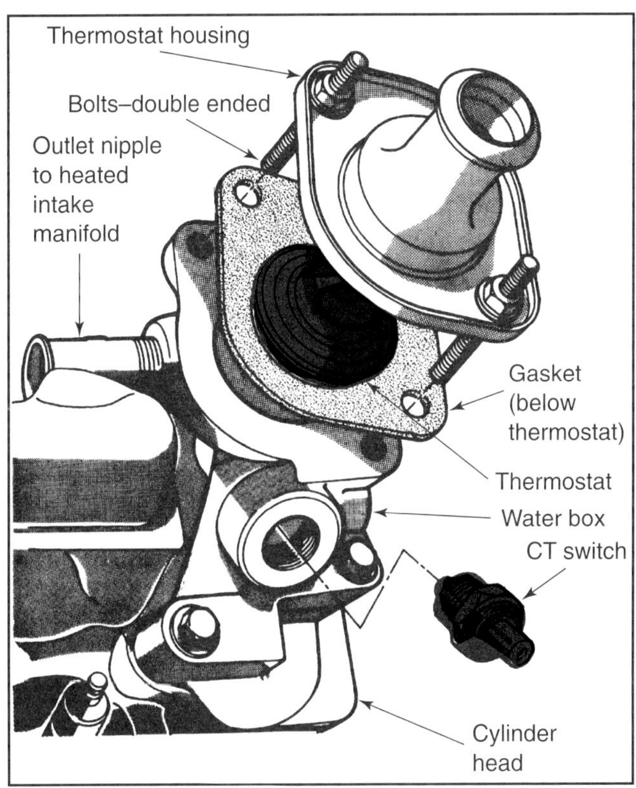

Figure 19-22. *Exploded view of a thermostat housing. The coolant temperature (CT) switch is normally not removed during thermostat service. (Chrysler)*

With the housing removed from the engine, the thermostat will now be visible. Lift the thermostat out of the recess and check it for rust or grease. If rust and grease are found, the cooling system should be thoroughly cleaned and checked for internal and external leaks. Be especially careful to check for cracks in the engine block and heads.

Testing the Thermostat

Once the thermostat has been removed from the engine, it can be checked to see that it opens at the proper temperature. For this test, you will need a water container, thermostat, and heat source. Fill the container with water and suspend the thermostat in the container as shown in **Figure 19-23.** Do not let the thermostat touch the bottom or sides of the pot. Insert a thermometer in the water, and note the opening temperature of the thermostat as the water is heated. The thermostat should start to open within 5° of the rated temperature, and should be completely open after the temperature rises another 15°. Defective thermostats are replaced, not repaired.

Before installing the replacement thermostat, clean all gasket material from the engine and thermostat housing. Then place the thermostat in the recess, ensuring that the temperature sensing element faces into the engine coolant passages. Then reinstall the housing with a new gasket and tighten the attaching bolts. Reinstall the hoses and add coolant per service manual procedures. Recheck the coolant level after the engine has operated with the radiator cap off approximately 10 minutes.

Note: Some vehicles have a bleed valve installed at the thermostat housing. It may be necessary to open this valve to remove all air from the cooling system.

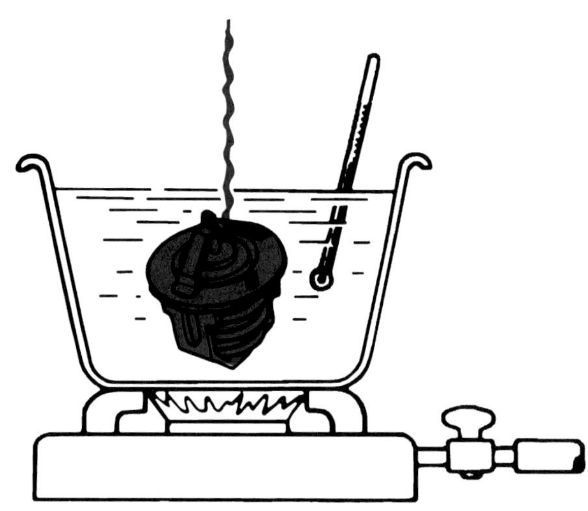

Figure 19-23. *A pot or cup of boiling water can be used to test when a thermostat opens. Suspend the thermostat using mechanic's wire or other suitable material. (Subaru)*

Removing and Installing Intake Manifolds, Plenums, and Exhaust Manifolds

Unless they are damaged by an accident or careless handling, intake and exhaust manifolds and air plenums usually do not require replacement. However, they must often be removed to service other components or to replace gaskets. Leaking gaskets can cause vacuum leaks and exhaust noises. In addition, leaks in intake systems can upset the air fuel ratio, while leaks in the exhaust manifolds can cause incorrect oxygen sensor operation. When removing any manifold, take care not to damage other components.

Intake Plenum Removal and Installation

Often, plenum removal is necessary to access the fuel injectors. Start by disconnecting the negative battery cable and removing any covers that protect the plenum. Drain the coolant to a level below the intake manifold. Remove the throttle body, sensors, output devices, brackets, hoses, and wiring that prevents plenum removal. If necessary, relieve fuel pressure and remove the fuel injector rail. Unbolt the plenum and carefully remove it from the intake manifold, **Figure 19-24.**

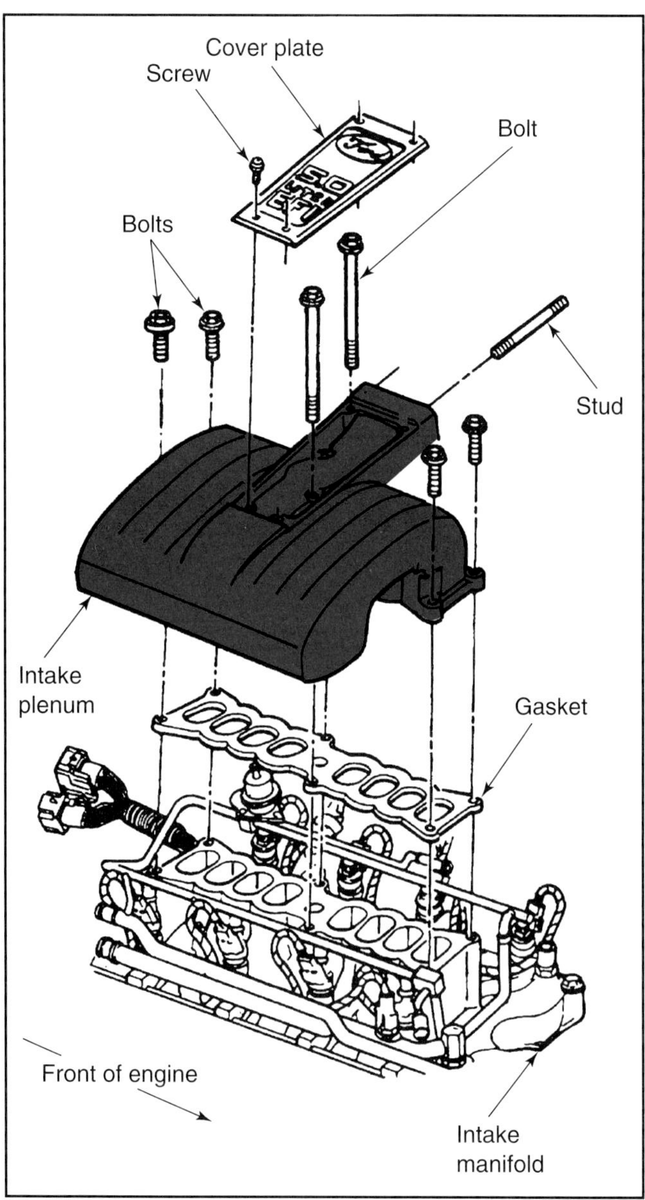

Figure 19-24. *Exploded view of an intake plenum. It is usually necessary to remove the plenum on many engines to access the injectors and fuel rail. (Ford)*

Note: Take care when removing composite plastic intake plenums. They can distort very easily.

To reinstall the plenum, clean all gasket surfaces properly, and remove all gasket material and other debris. Place a new gasket on the manifold and install the plenum and its bolts. Reinstall the throttle body, sensors, output devices, brackets, and the fuel injector rail, if it was removed. Reconnect the battery and refill the coolant. Start the engine and check for leaks. Reset idle speed as needed.

Intake Manifold Removal and Installation

Intake manifolds are normally removed to replace the intake gaskets or as part of major engine work. Disconnect the negative battery cable and remove the air cleaner or any covers. Drain the coolant to a level below the intake manifold and remove the air cleaner. If the engine has an intake plenum, remove this first. Remove the throttle body and fuel injector rail, if the fuel system is a multiport design. Unbolt the intake manifold and lift it from the engine, **Figure 19-25.**

Note: Check the intake manifold mounting surface for warpage with a straightedge. Specifications are located in the service manual.

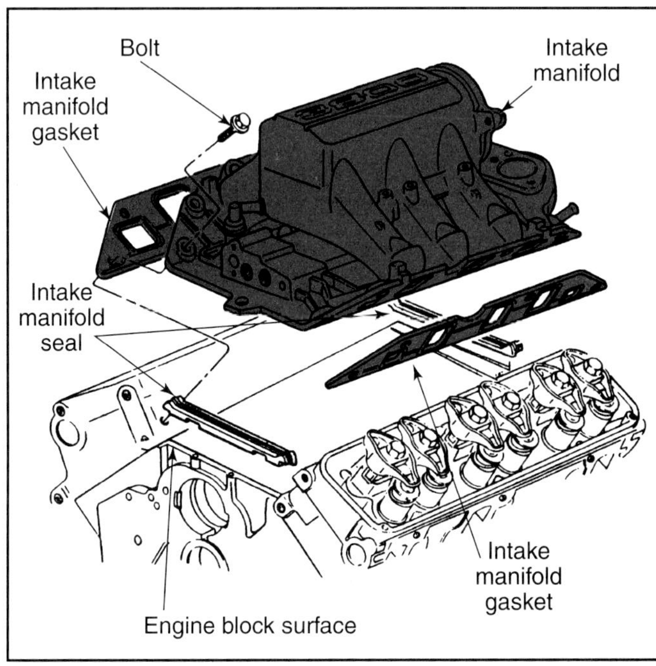

Figure 19-25. *This exploded view shows you the relationship of parts on typical fuel injection intake manifolds. The intake manifold and its gaskets are not normally serviced by driveability technicians. (General Motors)*

Be sure to clean all gasket surfaces properly, and remove all gasket material and other debris before reinstalling the manifold. When reinstalling, be sure that all gaskets are in place and that there are no gaps between parts. In some cases, it may be necessary to coat the manifold bolts with a sealer before installation. Always tighten all manifold and plenum fasteners to the proper torque, and in the proper sequence.

Exhaust Manifold Removal and Installation

Exhaust manifolds are removed due to leakage by either the gasket or the manifold itself. Disconnect the battery's negative cable and remove any oxygen sensors mounted in the exhaust manifold. Remove any other parts that could interfere with exhaust manifold removal. Spray all manifold bolts with a generous amount of penetrating oil. Using a breaker bar and socket, loosen all manifold bolts before using a ratchet for removal. After removing all bolts, carefully lift the manifold from the vehicle, **Figure 19-26.** Inspect the manifold carefully for warpage and leaks. Replace the manifold if excessively warped or leaking. Always replace the exhaust gasket. Install the manifold and gasket together and carefully torque each bolt.

Note: Due to heat stress, it is a good idea to replace exhaust manifold bolts or studs anytime they are removed from the engine.

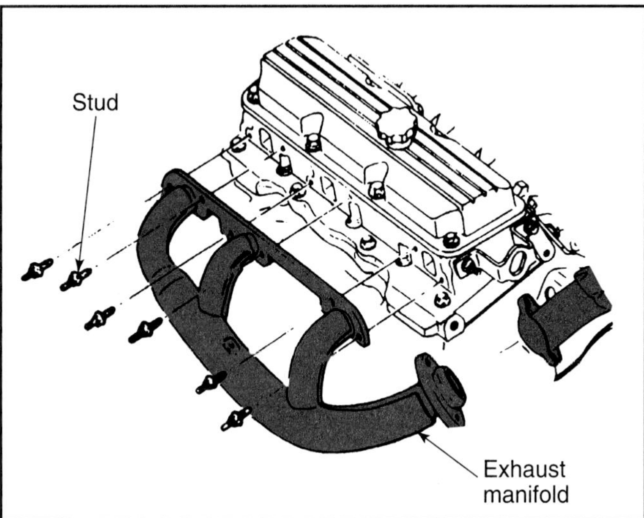

Figure 19-26. *Exhaust manifolds are easier to replace than intakes. However, be sure to spray all bolts with liberal amounts of penetrating oil or you will be repairing broken bolts. (General Motors)*

Replacing Turbochargers

Turbochargers are usually mounted on the side or top of the engine. Since the turbocharger is part of the exhaust system, it should be allowed to cool completely before any of the attaching bolts are removed. Allowing this cool down period will reduce the chance of warping any of the mounting surfaces, as well as reducing the chances of personal injury.

Note: Some turbochargers require extensive engine disassembly for removal. Check the service manual before starting.

The usual turbocharger removal procedure begins with removing any engine parts that obstruct the turbocharger and its mounting hardware. Once the oil lines are exposed, they can be removed, **Figure 19-27A.** In some cases, the wastegate should be removed before removing the turbocharger, while on other engines, the wastegate is separate, and does not have to be removed with the turbocharger. Next, remove the bolts holding the turbocharger to the exhaust and intake runners. The turbocharger can then be removed from the engine as an assembly, **Figure 19-27B.** If the wastegate is installed in the exhaust manifold, always check it for free movement when the turbocharger is removed.

Installing the Turbocharger

To install the replacement turbocharger, place it on the exhaust and intake runners using new gaskets. Then install and tighten the attaching bolts. Next, install the wastegate and linkage if necessary, and reinstall the oil lines. Remove battery power from the ignition coil and crank the engine for about 30 seconds, or until the instrument cluster gauge indicates pressure or the oil light goes out. This will pump engine oil into the turbocharger bearings. Then reconnect the coil wire and start the engine. Monitor engine operation, and check for exhaust or intake manifold leaks. If necessary, adjust the wastegate linkage.

Repairing Turbochargers

Turbochargers are usually replaced instead of rebuilt. However, since a new or remanufactured turbocharger is very expensive, it is sometimes worthwhile to repair minor turbocharger defects, such as leaking oil seals, rather than installing a new unit. Follow the manufacturer's instructions when disassembling and reassembling the turbocharger. When overhauling a turbocharger with an integral wastegate, always make sure that the wastegate and its connecting linkage can move freely.

Note: If the turbocharger bearings were exposed for any reason, lubricate them generously with engine oil before reassembly. After the turbocharger is reassembled, make sure that it turns freely. If it does not, correct the problem before reinstalling the turbocharger on the engine.

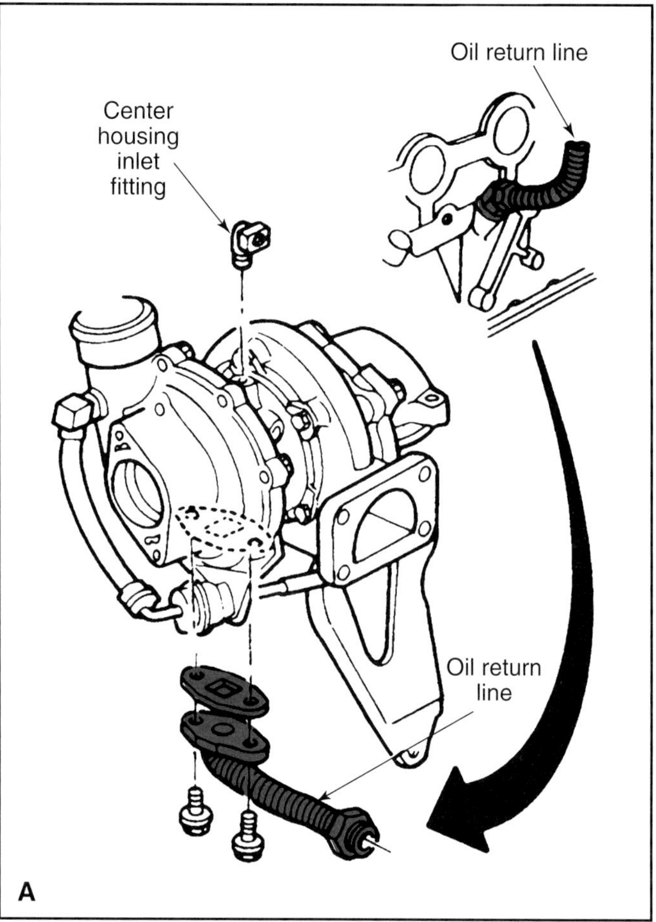

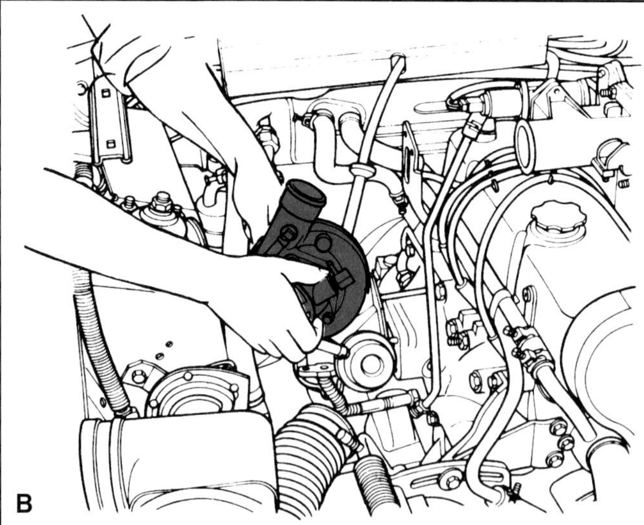

Figure 19-27. *Turbocharger removal is a complex procedure on most vehicles. Make sure the engine is completely cool before beginning removal. A—Remove the oil return line and all connections. B—Carefully remove the turbocharger. The turbocharger is heavier than it looks, so do not underestimate its weight. (Ford)*

Replacing Turbocharger Wastegates

The interior of the wastegate actuator is a diaphragm operated by intake manifold pressure. The diaphragm sometimes ruptures, making the wastegate inoperative. The wastegate linkage cannot be adjusted to compensate for a diaphragm leak. Since it is a sealed unit, the wastegate actuator is usually replaced instead of being overhauled. This is done by disconnecting the linkage and removing the attaching bolts. The new actuator is installed by replacing the bolts and reconnecting the linkage. The linkage can be adjusted to control boost pressure when the engine is operating.

Replacing Superchargers

Defective superchargers are replaced as a unit instead of being serviced, **Figure 19-28**. On a V-type engine,

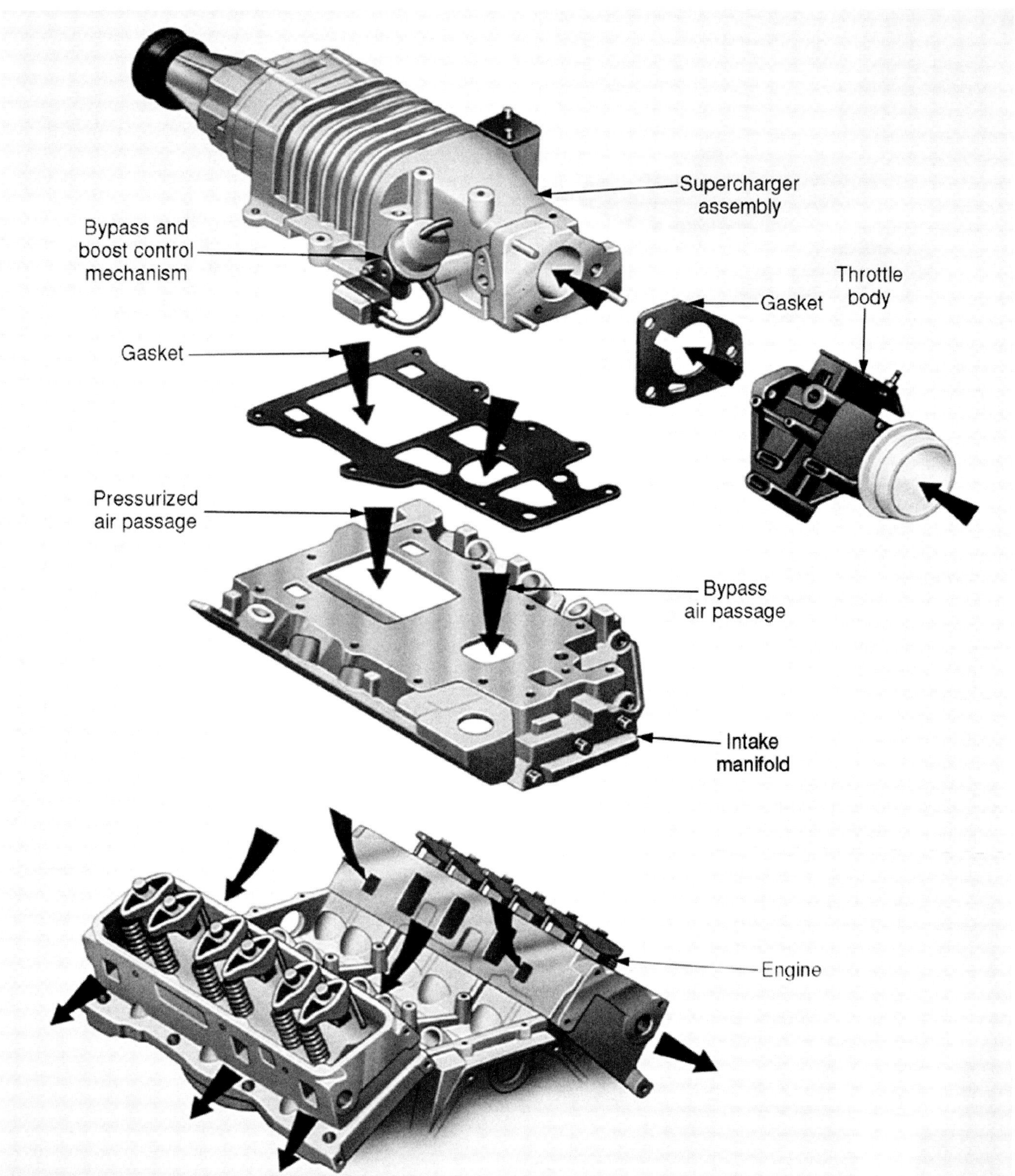

Figure 19-28. *Supercharger are simply replaced as a unit. Due to the cleanliness needed, service in the field is not recommended. (General Motors)*

supercharger removal is similar to intake manifold removal. Remove all intake hoses and linkage as necessary. Disassemble the front of the engine as needed to disconnect the supercharger drive mechanism. Remove all bolts holding the supercharger to the cylinder head or intake manifold and lift the supercharger from the engine. Superchargers are heavy so get someone to help you lift the unit from the engine.

Note: Defective superchargers are returned to the manufacturer for rebuilding. All superchargers have a core deposit and should be turned in when the new unit is obtained.

To install the new supercharger, remove all old gaskets and any sealer from the heads or intake manifold. Place a new gasket on the engine and lower the supercharger onto the gasket. Start the bolts, but do not tighten them. Reattach the drive mechanism, if necessary before torquing. Torque all bolts in the proper sequence. Reattach all hoses, clamps, and oil fittings.

Some manufacturers recommend priming the bearings of a new supercharger before starting. This can be done with a pressure lubricator or by disabling the ignition system and cranking the engine until the oil light goes out. Some superchargers use oil that is self-contained in the unit. Reset idle speed and recheck vehicle operation before returning it to the owner.

Replacing Camshaft Drive Components

The camshaft drive components are mounted in the front of the engine, and can be serviced once the timing cover is removed. There are three main types of camshaft drives: the matching gear type, the gear and chain type, and the gear and belt type. Brief replacement procedures are given for each type. The following procedures are general in nature and the service manual should be consulted when replacing any camshaft drive components. The procedures assume that the engine's original camshaft drive is in need of replacement due to wear and not breakage.

Note: In many cases, if a timing gear, chain, or belt came apart while the engine was running or the engine was cranked after its camshaft drive came apart, severe damage is usually present, requiring extensive engine disassembly and repair or complete engine replacement. Some engines will suffer no damage if its camshaft drive breaks.

Note that most matching gear and gear and chain drive sets can only be replaced after the front timing cover is removed. This requires partial disassembly of the front of the engine. Consult the service manual for the exact removal procedure. Many timing covers used on engines having matching gear or chain and gear driven camshafts also contain the water pump and cooling passages, **Figure 19-29.** The cooling system should be depressurized and drained before the timing cover is removed. The crankshaft pulley hub must also be removed to access the cover. Many crankshaft pulleys can only be removed with a special puller, **Figure 19-30A.** Others are held to the crankshaft with a large bolt, and will slide off once the bolt is removed.

Manufacturers call for replacing the timing cover oil seal whenever the timing cover or crankshaft pulley is removed. **Figure 19-30B** shows a special tool being used to remove the front cover oil seal. If the special tool is not available, the seal can usually be removed by carefully driving it out once the timing cover is removed from the engine. Carefully inspect the crankshaft pulley hub where it contacts the seal. You will often find a groove worn in the pulley where it contacts the seal. If the groove is deeper than about .01 inch (.254 mm), the pulley should be replaced to reduce the possibility of oil leaks.

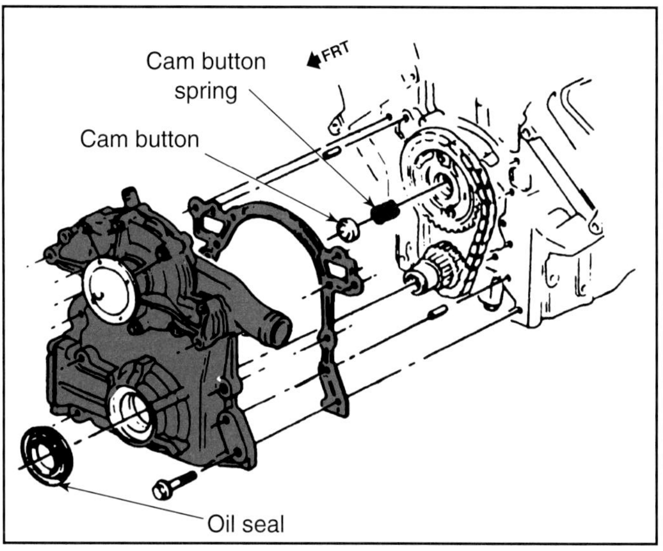

Figure 19-29. Front cover and timing chain assembly. The water pump and other components are often removed with the front cover.

Matching Gear Camshaft Drives

Matching gear camshaft drives, **Figure 19-31**, are removed according to how they are installed on the engine. If the gears are bolted to the cam and crankshafts, remove the attaching bolts and slide the gears from the shafts. Many timing gears are pressed on and may require a special puller tool for removal. On a few overhead cam engines, the camshaft gear must be pressed from the camshaft after the entire assembly is removed from the cylinder head. To install new gears, reverse the removal process. Make sure that the timing marks on new gears are aligned before making the final installation. Before installing the timing cover, crank the engine manually through two revolutions and ensure that the timing marks line up.

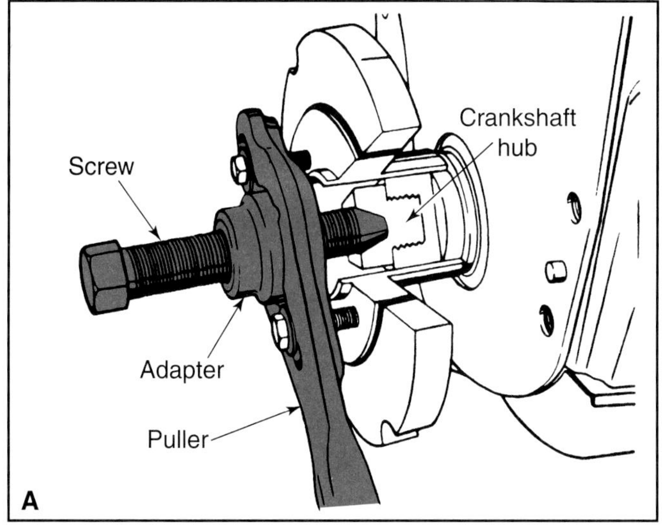

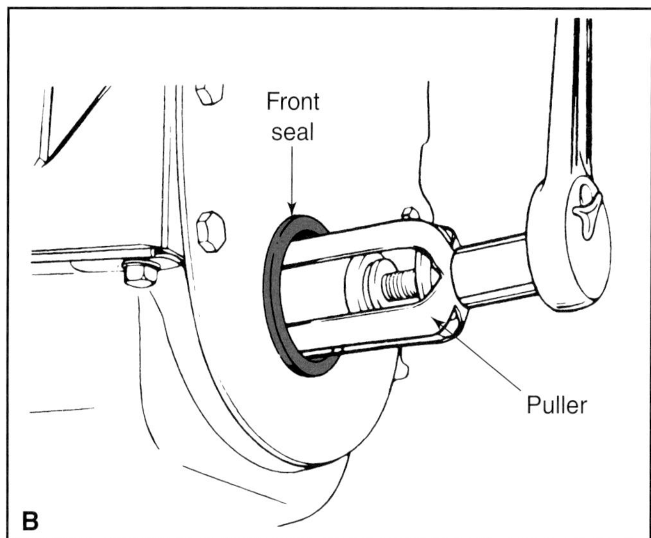

Figure 19-30. *A—A puller is often needed to remove the crankshaft pulley without damage. B—Whenever the crankshaft pulley is replaced, the oil seal should be replaced as well. (General Motors)*

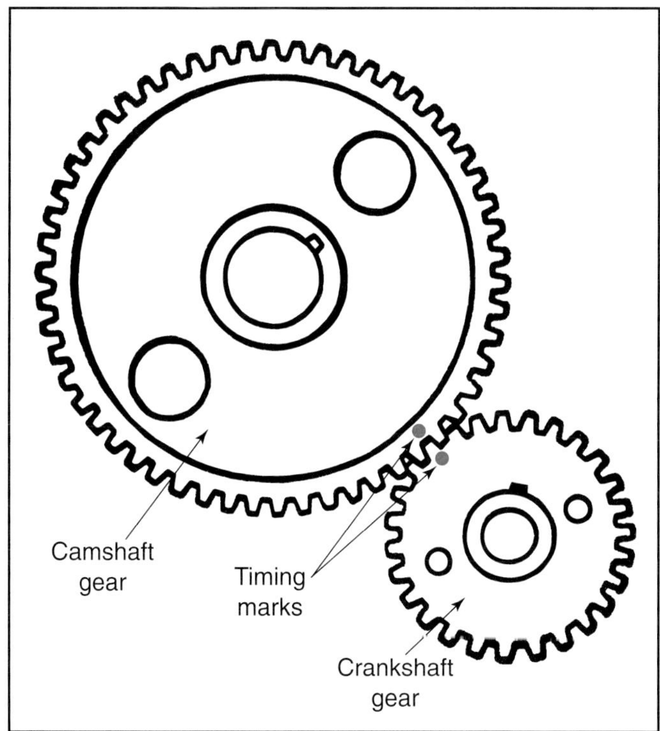

Figure 19-31. *Be sure the timing marks are in line when installing any timing gear, belt, or matching gear set.*

Gear and Chain Camshaft Drives

Gear and chain camshaft drives, **Figure 19-32**, are removed by the same general procedure. Remove the gear attaching screws and pull the timing gears and chain off the cam and crankshaft as a unit. Install the new parts in the same way, making sure that the timing marks are aligned.

Camshaft sprocket
Timing marks
Timing chain
Crankshaft sprocket

Figure 19-32. *Timing setup after installation of a timing gear and chain set. Always replace these as a set. (Chrysler)*

Install the fuel pump eccentric, if used, and any thrust springs or washers. Be sure to follow the service manual for proper removal, installation, and timing procedures.

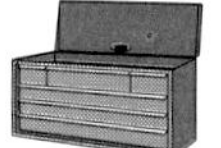

Note: Some overhead camshaft engines require the camshaft(s) to be secured before the timing chain or belt is removed. Check the service manual.

Gear and Belt Camshaft Drives

Gear and belt camshaft drives, **Figure 19-33**, are removed by first removing any belt guards and shielding. Next, remove or back off the belt tensioner. Some belt tensioners are operated by a spring and must be pried away from the belt, while others are released by backing off an adjusting bolt. Once the tension is removed, the belt will slide from the gears. Before installing the new belt, properly align the timing marks on all pulleys. On some engines with a belt-driven distributor, the distributor cap must be removed and the rotor positioned before the belt is installed. Replace any camshaft gears as needed, **Figure 19-34.** Adjust belt tension as needed before starting the engine.

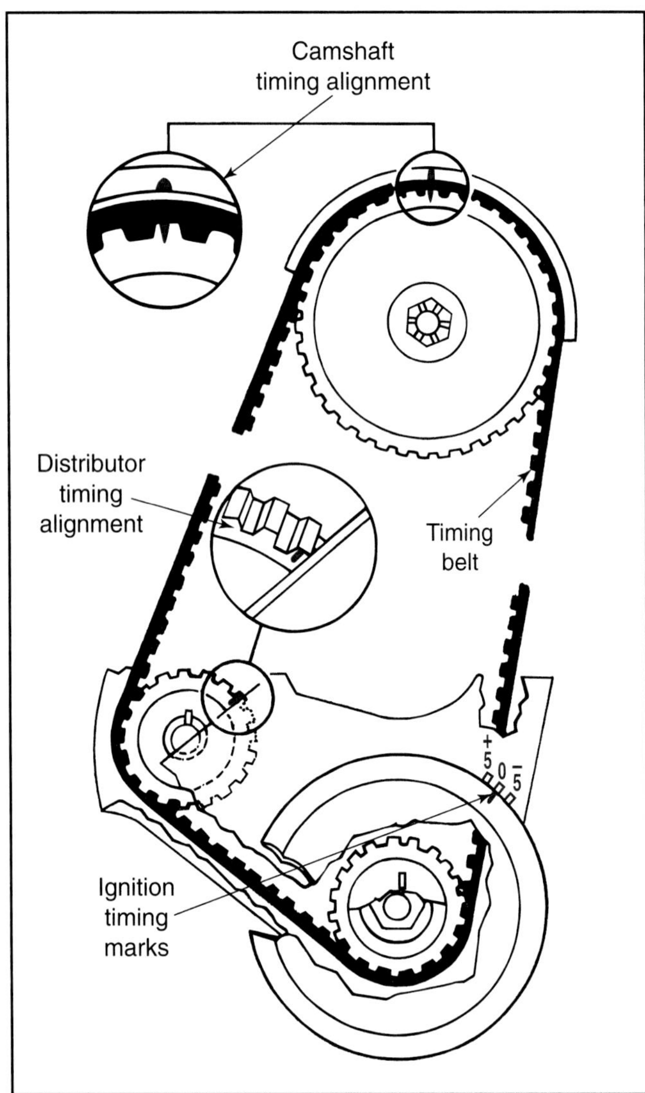

Figure 19-33. *Overhead camshaft engines have timing marks on the engine itself. The gears on this type of engine are rarely replaced. (Ford)*

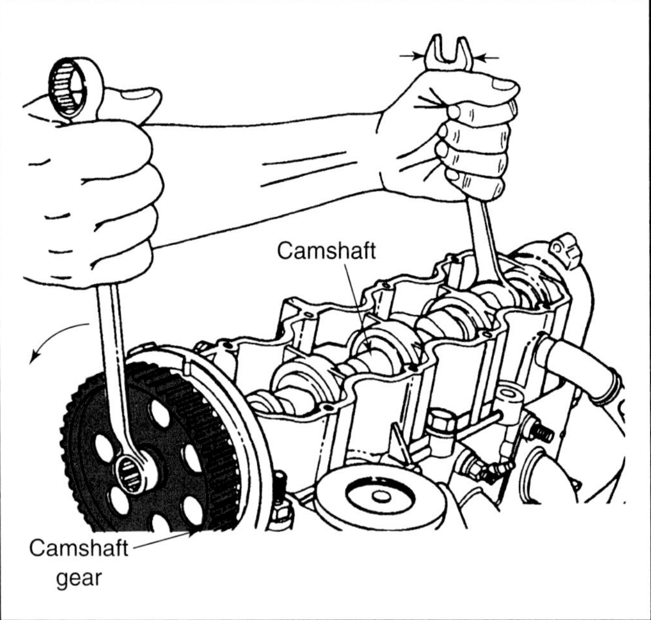

Figure 19-34. *One type of removal procedure for an overhead camshaft gear. Most camshaft gears are pressed on. (Pontiac)*

Follow-Up for Engine Repairs

Check the engine over carefully for any coolant, oil, or vacuum leaks. Road test the vehicle; there should be no performance problems of any type. If there are, investigate each system for something that may be disconnected or improperly routed. Reset anything that is needed. You may want to perform an exhaust gas analysis if the original problem was excess emissions. If needed, clean the engine compartment of any oil or coolant before turning the vehicle over to the owner.

Summary

Before diagnosing an engine problem, check all other engine systems first. After you have determined that the problem is actually in the engine, proceed with engine diagnosis. Begin with a visual inspection for obvious engine problems. This may require some minor disassembly, such as spark plug removal.

Before deciding that a noise is engine related, eliminate all other causes. Then isolate the engine component that is causing the noise. Valve noises can sometimes be fixed without major engine disassembly. Piston, bearing, and crankshaft problems usually require complete disassembly.

Vacuum leaks are a common cause of driveability problems, and should be located. Methods of locating vacuum leaks include spraying oil on suspected leaks or using an ultrasound leak detector. Vacuum gauges can be helpful in locating common engine defects, but must be interpreted carefully. Engine compression readings must also be carefully interpreted to accurately diagnose any compression problems. A wet compression test will isolate the problems to the rings or valves. The cylinder leakage test is a method of determining the exact cause of a compression leak.

The cooling system is a common cause of driveability complaints, and should be carefully inspected for leaks, buildup of deposits, and defective components. A pressure tester can be used to check the system for leaks, and to test the radiator cap. If necessary, the coolant can be checked for the presence of exhaust gases.

Turbocharger boost pressure can be roughly checked using the dashboard gauge, but an accurate test requires the use of a pressure gauge attached to the intake manifold or wastegate. Causes of excessively high or low boost should be investigated. Some wastegates can be adjusted. Supercharger checking procedures are similar to those for turbochargers. Low engine oil pressure can cause driveability problems. Oil pressure should be checked and any problems corrected.

Checking internal engine problems involves at least partial engine disassembly, and the owners' permission should be obtained before beginning. Valve timing can be quickly checked by observing the position of the distributor rotor against the timing marks. The rotor can also be used to check for worn timing chains or belts. More detailed checking of timing mechanism parts requires partial engine disassembly. The valve covers must be removed to check the valve springs and seals. If the valves must be disassembled to access the seals, air pressure can be used to hold the valve in position.

Valve lifters can be checked for excessive wear where they contact the camshaft. Worn lifters must be replaced. The camshaft should also be replaced if lifters are replaced. Hydraulic lifters and lash adjusters should be checked for collapsing and replaced if they are defective. A dial indicator is used to check camshaft lobe condition.

Valves should be adjusted at intervals specified by the manufacturer, or when they become noisy. Some valves do not have a provision for adjustment. Many valves are set by checking clearance with a feeler gauge with the engine not running. Other valves are adjusted by ear with the engine running. Some overhead camshaft engines require replacement of a shim located between the camshaft and valve.

Cooling system fans are easily replaced, but must be installed properly to ensure that the engine does not overheat. Thermostats can be removed and tested. A defective thermostat should be replaced. Intake and exhaust manifolds often require removal to replace other parts. They should be carefully cleaned and resealed before reinstallation. Turbochargers and superchargers are usually replaced as a unit.

Gear, chain, and belt drives often require replacement. Be sure to consult the proper service manual before beginning repairs. Most gear and chain drives require removal of the timing cover to access the timing mechanism. The front seal should be replaced when the cover is removed. When replacing any of the timing gear parts, line up all timing marks. Improper alignment on many overhead camshaft engines can cause severe valve damage.

Know These Terms

Rod knock	Loose clearance
Piston slap	Convex face
Crankshaft end play	Lobe lift
Compression tests	Valve lash
Wet compression tests	Shims
Tight clearance	

Review Questions—Chapter 19

Please do not write in this text. Write your answers on a separate sheet of paper.

1. Noises can be caused by any _____________ part on a vehicle.
2. Which of the following noises does not occur inside of the engine block?
   (A) Rod knocks.
   (B) Valve tapping.
   (C) Piston slap.
   (D) Crankshaft knocks.
3. All of the ______ should be removed to check compression.
4. Oil in the cooling system indicates what conditions are present in the engine?
5. How is a defective water pump usually serviced?
6. To service the valve components, the ______ must be removed.
7. Leaks in an exhaust manifold can cause incorrect ______ operation.
8. If a fan blade assembly is installed backward, what could happen?
   (A) Poor cooling at low speeds.
   (B) Poor cooling at high speeds.
   (C) Noise at high speeds.
   (D) Both B & C.
9. If turbocharger boost pressure is low because of a damaged turbocharger, it should be ______.
10. Make sure that the ______ on new camshaft gears are aligned before final installation.

ASE Certification-Type Questions

1. Removing the drive belts and operating the engine will eliminate the possibility of noise from all of the following, EXCEPT:
   (A) water pump.
   (B) timing chain.
   (C) power steering pump.
   (D) alternator.

2. Technician A says that a slipping fan clutch can cause low speed overheating. Technician B says that missing fan shrouds can cause low speed overheating. Who is right?
   (A) A only.
   (B) B only.
   (C) Both A & B.
   (D) Neither A nor B.

3. A cold engine is started and allowed to idle. After five minutes of operation, water and steam begin violently escaping from the radiator overflow hose. What is the most likely cause?
   (A) Thermostat stuck closed.
   (B) Thermostat stuck open.
   (C) Top radiator hose collapsing.
   (D) Bottom radiator hose collapsing.

4. All of the following statements about cooling system problems are true, EXCEPT:
   (A) ECM output commands are often based on input from the coolant temperature sensor.
   (B) an engine that is pinging excessively may have an overcooling problem.
   (C) the ECM cannot enter closed loop until the engine reaches normal operating temperature.
   (D) overcooling will also cause sludge build-up in the engine internals.

5. The number three cylinder on a four-cylinder engine repeatedly fouls spark plugs. Compression is normal. Technician A says that the intake valve seal could be defective. Technician B says that the exhaust valve could be burned. Who is right?
   (A) A only.
   (B) B only.
   (C) Both A & B.
   (D) Neither A nor B.

6. All of the following statements about overhead valve engines are true, EXCEPT:
   (A) valve adjusters used on overhead camshaft engines are located on the opposite side of the rocker arm from the valve.
   (B) on some overhead camshaft engines the camshaft lobes operate the valves directly, without using rocker arms.
   (C) valves on some overhead camshaft engines are adjusted by replacing shims.
   (D) all overhead valve engines using rocker arms have a provision for valve adjustment.

7. Technician A says that the radiator fan and motor are usually installed on the engine side of the radiator. Technician B says that the fan motor electrical connectors should be removed first when replacing the fan motor. Who is right?
   (A) A only.
   (B) B only.
   (C) Both A & B.
   (D) Neither A nor B.

8. Hydraulic lifters can be checked at these times, EXCEPT:
   (A) when assembled.
   (B) before they are disassembled.
   (C) if they are full of oil.
   (D) when they have bled down.

9. Which of the following must be replaced when the timing cover is removed from the engine?
   (A) Timing cover.
   (B) Water pump.
   (C) Oil seal.
   (D) Crankshaft hub.

10. Technician A says that the timing chain can be checked by removing the timing cover. Technician B says that the timing chain can be checked by removing the distributor cap. Who is right?
    (A) A only.
    (B) B only.
    (C) Both A & B.
    (D) Neither A nor B.

Advanced Certification Questions

The following questions require the use of reference information to find the correct answer. Most of the data and illustrations you will need is in this chapter. Diagnostic trouble codes are listed in **Appendix A.** Failure to refer to the chapter, illustrations, or **Appendix A** for information may result in an incorrect answer.

Cylinder	Normal Compression	Initial Compression	Wet Compression
1	175	100	100
2	175	90	95
3	175	95	95
4	175	105	105

11. A vehicle with a fuel injected inline four-cylinder engine failed a state emissions test with excessively high HC. Results from a compression test are shown in the above figure. All of the following can cause these readings, EXCEPT:
    (A) a blown head gasket.
    (B) excessive piston ring wear.
    (C) worn valves.
    (D) a jumped timing chain.

12. Air escapes from an adjacent spark plug port during a cylinder leakage test. Technician A says that a blown head gasket is the cause. Technician B says that a cracked engine block is the cause. Who is right?
    (A) A only.
    (B) B only.
    (C) Both A & B.
    (D) Neither A nor B.

13. A multiport fuel injected engine has a rough idle. The needle in the vacuum gauge shown above fluctuates at idle and stabilizes as engine speed is increased. Which of the following is the *most likely* cause?
    (A) A worn valve guide.
    (B) A weak valve spring.
    (C) Sticking valves.
    (D) Restricted exhaust.

14. A non-computer control vehicle has a repeated complaint of spark knock. A check of service history and current engine timing shows that it was found to be advanced 3°, 5°, and is currently 4° beyond specifications. Which of the following is the *most likely* cause?
    (A) A ruptured vacuum advance diaphragm.
    (B) A broken timing chain.
    (C) A sticking distributor centrifugal advance mechanism.
    (D) A stretched timing chain.

15. A fuel injected engine overheats in traffic, at stoplights, and has high NO_X. Technician A says that a leaking radiator hose is the cause. Technician B says that a defective electric cooling fan is the cause. Who is right?
    (A) A only.
    (B) B only.
    (C) Both A & B.
    (D) Neither A nor B.

Driveability work takes patience as well as skill and training. You must be able to maintain your composure, even when dealing with the most difficult performance problems. (Fluke)

Chapter 20

Drive Train and Vehicle System Diagnosis

After studying this chapter, you will be able to:

- Check non-engine systems to identify driveability problems.
- Check clutch and manual transmission/transaxle operation.
- Check automatic transmission/transaxle operation.
- Check air conditioner operation.
- Check driveshaft, universal, and constant velocity joint condition.
- Check steering and suspension components.
- Check brake components.
- Replace transmission pressure switches.
- Replace air conditioning pressure switches.

In this chapter, you will learn test procedures to diagnose non-engine systems that can cause driveability problems. With the increased use of computer control and electronics in modern vehicles, non-engine systems are affecting engine performance more than ever. Like **Chapter 19,** this chapter discusses diagnostic routines with very few corresponding repair procedures. This is the last chapter in this text that discusses diagnosis and repair as **Chapter 21** deals with tune-up and maintenance procedures. **Chapters 22** and **23** discuss ASE testing and career opportunities.

Preparing to Check Non-Engine Systems

As mentioned in earlier chapters, a driveability problem may not be engine-related. If you suspect a problem in the drive train, air conditioner, suspension, steering, or brake systems, the tests in this chapter may help to isolate the problem, **Figure 20-1.** Service information for each of these systems could form a textbook, therefore, all procedures are brief. Consult the appropriate service manual for exact procedures.

Visual Inspection

Begin the visual inspection by looking at the vehicle before you drive it. This is a good idea when investigating any problem, since you do not know if the vehicle is safe to drive. Walk completely around the vehicle and look for uneven ride height. Look at the condition of the tires for uneven or unusual wear. Examine the body for any apparent accident damage. Check the vehicle fit and finish as body panels that do not fit well or paint that does not quite match other portions of the vehicle may be a sign of poorly repaired collision damage. If the vehicle is unsafe to drive, do not perform a road test until the vehicle is repaired to a point where it is safe.

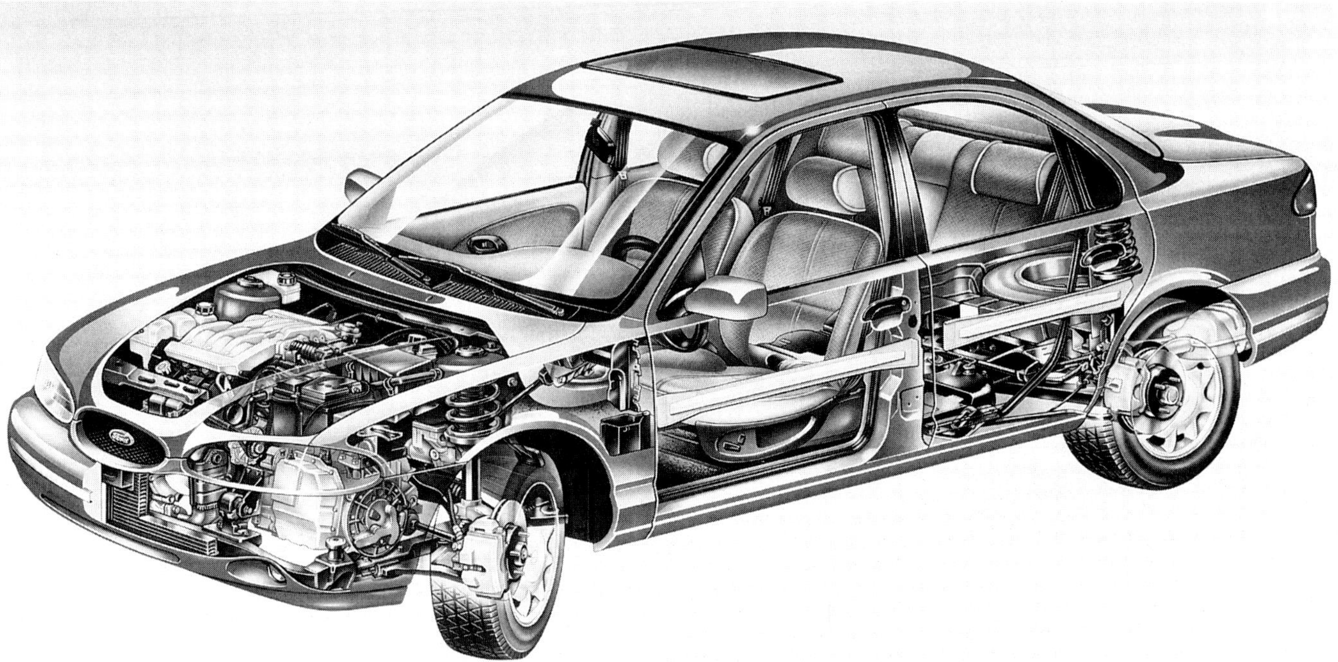

Figure 20-1. *So many systems on a vehicle work with the ECM and the engine that a problem in the transmission, brakes, suspension and steering, and air conditioning can affect driveability. (Ford)*

After the engine is started, check the steering for noise, binding, and excess play. Turn on the air conditioning and listen for excessive clutch cycling. Step on the brake pedal several times; it should be firm and high. If the vehicle is equipped with anti-lock brakes, listen for an ABS system self-test cycle as you begin to drive.

While driving, look and listen for any unusual noises, vibrations, or harshness. Note the operation of the automatic transmission or transaxle. Try to mentally note the approximate speeds that upshifts take place. If the automatic transmissions/transaxle has a lockup torque converter, try to bring the vehicle to highway speed to test the lockup circuitry and converter. Monitor the air conditioning system's performance as there should be little, if any, cycling. When braking, look and listen for noises and vibrations that could indicate a brake problem. When the brakes are released, there should be no drag on the vehicle.

Noise, Vibration, and Harshness

The modern vehicle is heavily insulated to deaden noises and some vibrations from normal vehicle operation. Unfortunately, unitized bodies easily transmit noises that can be heard and felt, even with the great amount of insulation used in today's vehicles.

Noise, vibration, and harshness (NVH) diagnostics is usually associated with drive train and suspension systems. Originally, NVH training was given only to engineers. Today, training in noise, vibration, and harshness diagnostics is becoming increasingly important for all technicians as vehicles become smaller. Most vehicle manufacturers now have dedicated NVH training classes for their technicians. Some automotive programs are beginning to include NVH diagnostics training in their classes. Types of vibrations, noises, and harshness are listed in **Figure 20-2.**

Note: This information is designed to give you a foundation in the principles of noise, vibration, and harshness diagnostics. It is recommended that you receive additional training on this subject, especially if you plan to pursue a career as a drive train or transmission technician.

Principles of Vibration

All vehicles produce some ***vibrations,*** with the majority coming from two sources. Most vibrations come from rotating components, such as the crankshaft, wheels, and drive axles. The other major source of vibrations is the firing impulses produced by the engine during the combustion cycle. It is important to note that some vibrations are part of normal vehicle operation while others are caused by problems, such as defective or imbalanced components. Another factor in NVH diagnostics is that some vibrations, such as normal engine impulses, cannot be completely eliminated. Vibrations also create audible noise.

Creating and Transmitting Noise

Vibrations are the source of almost all ***noise*** complaints since a vibration must usually exist for a noise to occur. Some vibrations begin with a part coming loose, becoming misaligned, or breaking. When a vibrating part touches another part, for example, an exhaust pipe vibrating and touching a heat shield, noise is created. Sometimes noise can be created without any noticeable vibration, however, in most cases a vibration is usually present. Noise can originate from almost every vehicle system, **Figure 20-3.**

Noise, Vibration, and Harshness	
Vibrations That Can Be Felt	**Causes**
Shake—A low frequency vibration that can usually be seen in movement of components. Often referred to as a shake, shimmy, wobble, hop, or shudder.	If vehicle speed related—tires, wheels, brake friction surfaces. If engine speed related—engine.
Buzz—A vibration similar in feel to an electric razor. Can be felt in the hands, feet, or seat of the pants.	Usually related to engine speed—exhaust system, air conditioning compressor, and engine are the most frequent contributors.
Roughness—Vibration similar in feel to an electric saw.	Almost always vehicle speed related—frequently caused by drive train components.
Tingling—High frequency vibration that produces a feeling similar to "pins and needles."	Can be either engine or vehicle speed related.
Audible Vibrations	**Causes**
Boom—Cycling, rhythmic noise similar to thunder or a bass drum. Sometimes driver may complain of "pressure on the ears." Also referred to as a drone, growl, moan, roar, or humming.	Usually vehicle speed related—normally caused by driveline components. May or may not be accompanied by a perceptible vibration.
Moan—A vibration that produces a constant, steady noise similar to a bee or blown air across the top of a soda bottle. Also referred to as a resonance, humming, or buzzing.	Can be either engine or vehicle speed related. Usually caused by the engine or drive train mounts or the exhaust system. May be accompanied by a perceptible vibration.
Howl—Similar to the wind howling.	Can be either engine or vehicle speed related.
Whine—A high–pitched sound, usually prolonged. Similar to vacuum cleaners or mosquitoes.	Almost always vehicle speed related—usually caused by drive train, gear, or tire noise.
Chuckle—A rattling noise similar to a stick or card inside the spokes of a spinning bicycle wheel.	Usually vehicle speed related. Almost always caused by a drive train problem.
Knock—Noise very similar to chuckle, but usually louder. Can also be similar to a knock on a door but with a rhythmic pattern.	Can be engine or vehicle speed related, depending on the affected system. Usually occurs on acceleration or deceleration.
Examples of Harshness	**Causes**
Clunk—A vibration and noise that occurs in response to backlash. Often metallic in sound.	Driveline related, but not speed related. Occurs when something is engaged, such as an automatic transmission.
Chatter—A vibration that can be easily felt, heard, and seen.	Vehicle speed related. Usually indicative of a differential problem. Can be seen frequently on vehicles equipped with limited-slip differentials.
Click—A light repetitive vibration and noise, less distinct than a clunk.	Can be engine or vehicle speed related.

Figure 20-2. *Vibrations, noises, and harshness that can be felt or heard in a vehicle and the systems that produce them.*

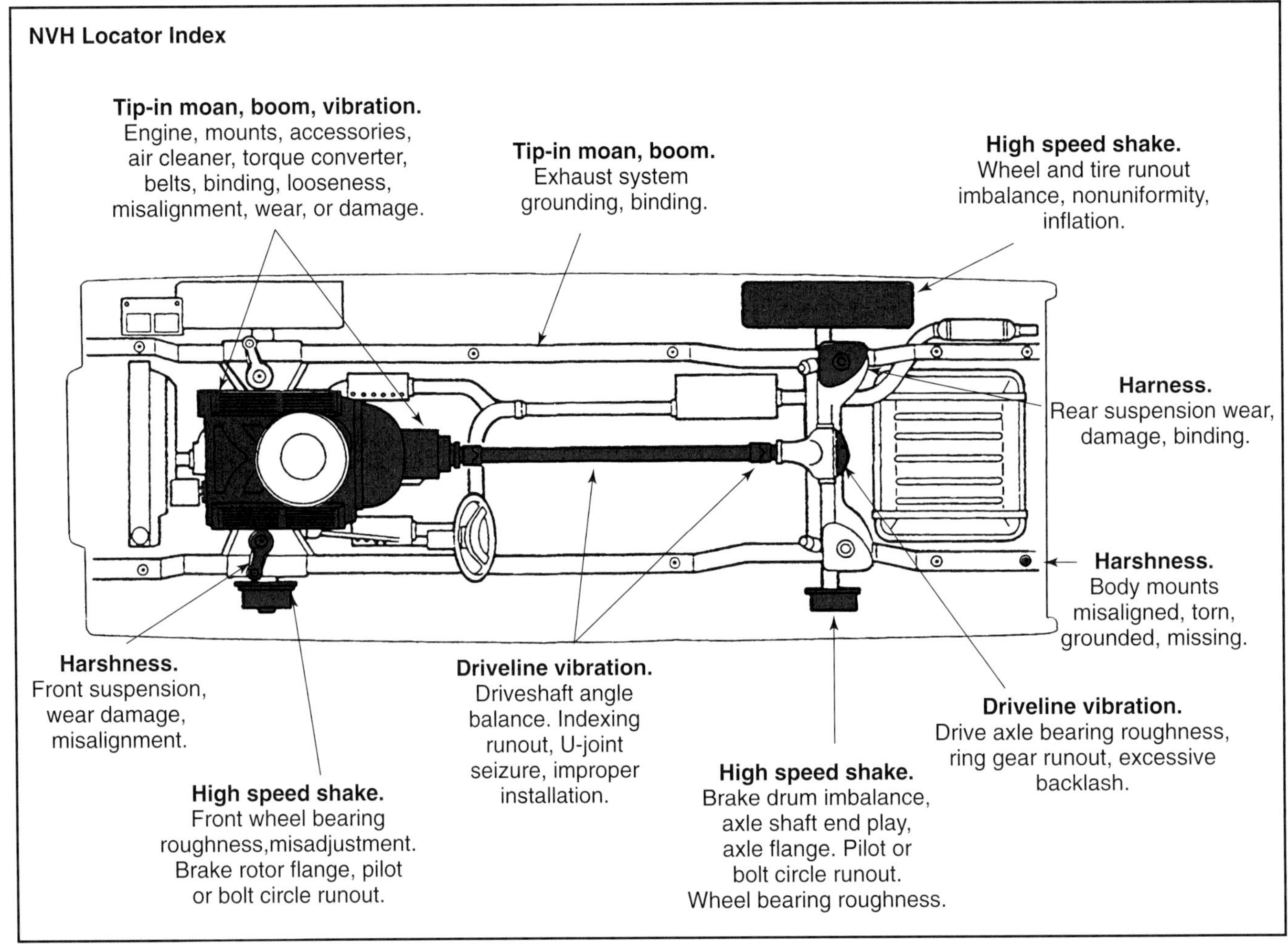

Figure 20-3. *Some typical locations for noise, vibrations, and harshness. These locations are not the only possible transmission sources. (Ford)*

Sources of Harshness

Harshness is when the operation of a component feels hard or more harsh than usual. While harshness is considered a form of vibration, it is more readily felt and heard than a vibration. Harshness is often accompanied by noise. Sources of harshness include the transmission, tires, and rear differential. Harshness is caused by excess play between two components and unusual component wear.

Eliminating Noise, Vibration, and Harshness

To eliminate noise, vibration, or harshness, it is important to understand how they are transmitted. The origin of the vibration is usually called the ***transmitter*** or the *source.* The path the vibration or noise takes is called the ***transfer path*** or *conduit.* The part where the vibration, noise, or harshness is emitted is called the ***receiver,*** *emitter,* or *responder.* An example of this transfer is when a driveshaft becomes imbalanced. The driveshaft transmits vibration through the drive train and unibody to the seats where the driver and passengers can feel it. In this example, the driveshaft is the transmitter, the drive train and the unibody are the transfer paths and the seats are the receiver. Noise, vibrations, and harshness can be measured in frequency or hertz. Specialized equipment is available to measure and convert vibrations to a frequency.

There are many specific methods of eliminating noises and vibrations, depending on the affected system. However, the two main groups are balancing and dampening. ***Balancing*** is performed when weight is added or removed opposite the source to compensate or cancel out the vibration. An example is when weight is added to a wheel to compensate for tire imbalance. ***Dampening*** is usually performed when a vibration cannot be eliminated, such as engine firing impulses. Engine mounts dampen the vibrations created by engine firing impulses. It is very important in NVH correction to find the root cause of the problem. Often, many technicians simply dampen an unusual vibration or noise, leaving the transmitting source uncorrected.

Checking Transmission/Transaxles

In most shops, transmission/transaxle diagnosis and repair is the job of a specialist who usually works on these complex systems exclusively. However, there are some simple tests that you, as a driveability technician, can perform to check these systems for proper operation. The first check is performed during the road test.

Warning: Manual clutches and brake linings contain asbestos, which is a known carcinogen. Cancer and emphysema can result from prolonged exposure to asbestos dust. While this text does not cover replacement of these parts, it does discuss visual inspection. Wear appropriate respiratory protection and minimize creating dust as much as possible while working with these systems.

Checking Manual Clutches

A defective or misadjusted clutch can cause symptoms that could easily be mistaken for engine problems. To check the clutch adjustment, push the clutch pedal with your hand, **Figure 20-4.** You should feel no resistance for approximately the first inch of travel. After 1″ (25.4 mm) of travel, the pedal should be considerably harder to push. The low resistance travel is the clutch linkage free travel, or *free play.* If the free play is not correct, the clutch should be adjusted or replaced if the right amount of free play cannot be obtained.

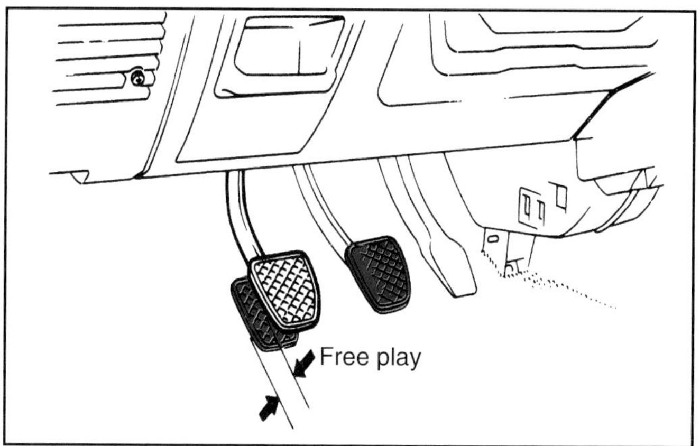

Figure 20-4. *A manual clutch pedal should have as little free play as possible (no more than 1″ (25.4 mm). If play is excessive, clutch wear or slippage is possible.*

If you suspect that the clutch is grabbing or slipping due to oil-soaked or worn out clutch disc linings, the inspection cover should be removed to inspect the clutch facings. Consult the service manual for exact procedures.

Checking Manual Transmissions/Transaxles

Checking the manual transmission/transaxle for problems is fairly simple and is performed by taking the vehicle for a test drive and shifting through all the gears. If the transmission shifts with no noise, vibration, harshness, or binding, the transmission is working properly. If no problems were found with the clutch, the cause of the driveability problem lies elsewhere. It is important to note that most manual clutch and transmission/transaxle problems are usually recognized as drive train problems.

Checking Automatic Transmission/Transaxles

While checking manual transmissions and clutches is fairly simple, testing automatic transmissions/transaxles is more difficult and time-consuming. There are many tests that can be performed on the modern automatic transmission. With the use of computer control to monitor and control transmission operation, the number of necessary tests has increased.

Note: The automatic transmission tests covered here are those normally performed in the course of diagnosing other problems. Consult the service manual for the specific tests for each vehicle

Checking Fluid Level

This is always the first test that should be performed on any automatic transmission. Start by bringing the vehicle up to operating temperature. Place the transmission in park or neutral; be sure to check the service manual for the correct gear. Locate the transmission dipstick and remove it. Wipe it on a clean shop towel, reinsert it, and remove it again. Check the fluid level; it should be full or in a cross-hatched area between the fill and full marks. Examine the fluid for color; it should be red with no bright pink or brown discoloration. Milky looking fluid indicates water or coolant contamination. Take a sample of the fluid and rub it between your fingers. There should be no particles of debris in it. Smell the fluid. A burned smell indicates possible transmission problems, **Figure 20-5.**

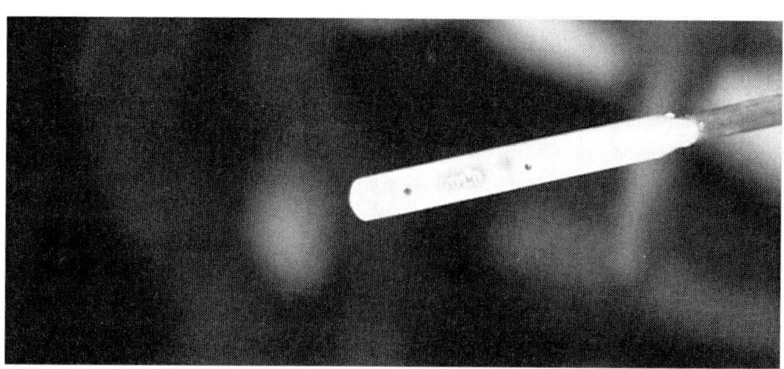

Figure 20-5. *Whenever you investigate any automatic transmission/transaxle problem, check the fluid level first. Many problems are caused by low fluid.*

Checking Automatic Transmission Shift Points

Automatic transmission shift points can be checked to isolate any problems in the transmission control system. The first step in this process is to drive the vehicle and check the shift speeds and feel. The upshift and downshift patterns are usually given in the service manual as a range; for example, the 1-2 shift is supposed to occur between 12-16 mph (5.4-7.2 kph) with a slight throttle opening. If the shift occurs at 14 mph (6.3 kph), it is within the range. However, if it occurs at 22 mph (9.9 kph), it is out of range. Shift feel is more subjective, but the shifts should feel smooth and firm, without excessive harshness or slippage.

Shift charts generally refer to the correct shift points in terms of the miles per hour at which an up or down shift should occur. Note that shift points are slightly different for each transaxle model shown in **Figure 20-6.** The transmission or transaxle model should be known before deciding that the unit is not shifting af the right time. Also note that the shift points are given as a range of speeds. Production tolerances (minor differences between manufactured parts) make it impossible to predict exactly what speed a particular shift will occur. Therefore, the actual shift points can be anywhere within the speed range and still be correct.

The shift point chart in **Figure 20-7** is intended for use with a scan tool. Note that the shift points are given in

Shift Speed Chart

Model	Final drive ratio	1-2 min throttle	2-3 min throttle	3-4 min throttle	4-3 part throttle	3-2 part throttle	4-3 coast down	3-2 coast down	2-1 coast down
AAH, ABH, AFH, ANH, ATH, AWH	3.33	11-14	21-23	38-48	49-55+	36-47	35-52	16-24	8-14
BKH	3.06	13-15	20-22	44-53	55+	49-55+	40-50	16-19	10-13
BJH, BTH	3.33	11-14	21-25	42-50	54-55+	46-52	39-48	17-21	9-12
BRH	3.33	11-14	20-25	41-48	52-55+	44-50	38-46	17-20	9-12
BYH	3.33	12-15	19-21	43-52	55+	45-51	39-49	15-18	10-13
CFH, CTH, CWH	3.33	12-14	21-23	41-50	55+	52-55+	37-48	17-21	10-13
CMH, CRH, CXH	3.33	12-15	21-24	41-51	55+	53-55+	38-49	17-21	10-13
FBH	3.06	13-15	20-22	44-53	55+	49-55+	40-50	16-19	10+
FCH	2.84	12-15	19-21	43-52	55+	45-51	39-49	15-18	10-13
FJH	3.33	11-14	21-25	42-50	53-55+	46-52	39-48	17-21	9-12
FSH	2.84	12-15	19-21	45-55	55+	47-55	41-52	15-18	10-13

Notes:
1. All speeds indicated are in miles per hour. Conversion to kph = mph x 1.609.
2. Shift points will vary slightly due to engine load and vehicle options.
3. Speeds listed with + exceed 55 mph.

Figure 20-6. *Service manuals contain shift charts, such as this, to provide a reference as to the correct shift pattern and speeds for each shift. While test driving, note the speed at each shift and compare the speeds to the shift chart in the appropriate manual. (General Motors)*

		1-2 Shift @ +/– 250 rpm Output Shaft Speed			2-3 Shift @ +/– 200 rpm Output Shaft Speed			3-4 Shift @ +/– 150 rpm Output Shaft Speed			3-1 @ +/– 100 rpm Output Shaft Speed	3-2 @ +/– 100 rpm Output Shaft Speed	3-1 Wide Open Throttle Shift	2-3 Wide Open Throttle Shift
% of TPS		12	25	50	12	25	50	12	25	50				
Trans Cal	Axle													
A	3.08	524	745	1048	927	1209	1854	1290	1572	2378	363	N/A	1290	N/A
B	3.42	514	749	1049	920	1198	1862	1284	1562	2375	364	N/A	1284	N/A
C	3.73	535	744	1046	930	1209	1860	1279	1581	2371	372	763	1279	N/A

Figure 20-7. *This shift chart is designed to be used with information from a scan tool. You would need percentage of throttle position with actual output shaft speed. (General Motors)*

driveshaft revolutions instead of miles per hour. The basic concept of shifts occurring at specific speeds is the same. Using the scan tool, the driveshaft revolutions and shifts (transmission solenoid applications) are displayed next to each other.

If the shift speeds are out of range, or the shift does not feel right, try to adjust the throttle valve linkage and recheck. Always consult the service manual before making any shift linkage adjustments. If the transmission is equipped with a vacuum modulator, check the modulator body neck where it enters the transmission case, as well as the vacuum line for leaks or kinks. The modulator diaphragm should also be checked with a vacuum pump, especially when shift problems are accompanied by an unexplained loss of transmission fluid. See **Figure 20-8.** If the governor is mounted on the side of the case, it can be removed and checked for a stripped gear or stuck valve. Shaft-mounted governors cannot be reached without extensive transmission disassembly.

Note: Many vehicles are equipped with an electric detent (passing gear) switch. If the vehicle will not upshift, check the switch and wiring to ensure that it is not shorted. A short will place the transmission/transaxle in second or third gear at all times.

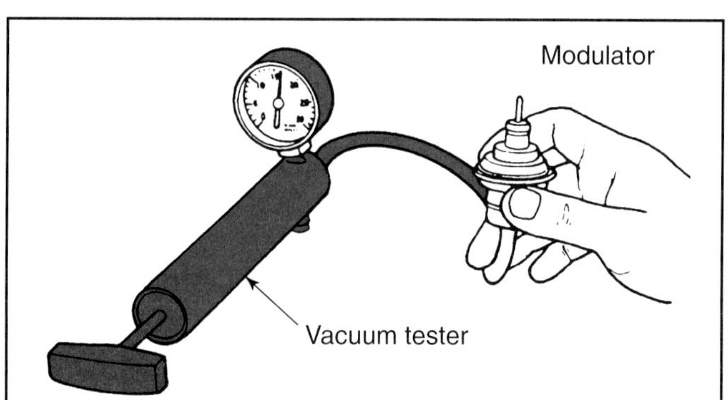

Figure 20-8. *Vacuum modulators can be removed from the transmission case and tested with a vacuum pump. Look for modulator rod movement when vacuum is applied. (Ford)*

Checking Computer-Controlled Transmissions

Before troubleshooting the electronic components of an automatic transmission, make all the basic checks discussed above. Also keep in mind that many electronically controlled automatic transmissions are updated versions of older transmissions, and still contain many hydraulically operated components. Most electronic control systems are designed so that even if an electronic component fails completely, the transmission will continue to move the vehicle without suffering any permanent damage.

Note: A computer-controlled transmission will generally stay in one gear or will start in second gear if the ECM fails.

On some modern automatic transmissions, some shifts may be made by the hydraulic control system, while others are made by solenoids controlled by the ECM. **Figure 20-9** shows one transaxle with a combination of hydraulic and electronic shift controls. Therefore, when confronted with a vehicle that is shifting improperly, always begin diagnosis by determining exactly what shift is being affected. Then determine whether the shift is accomplished hydraulically or electronically.

If the shift problem appears to be caused by the hydraulic system, the transmission may require at least partial teardown to check for stuck valves, plugged governor filters, worn clutches and bands, leaking seals, worn bushings and shafts, or other internal problems. A pressure check, **Figure 20-10,** may help you to determine whether the hydraulic system is operating. Consult the service manual for the installation of the gauge(s) and proper interpretation of pressure readings. Some parts, such as throttle linkage cables, vacuum modulators, and case mounted governors, can be checked and repaired without removing the transmission. However, extensive repairs should not be attempted unless you are thoroughly familiar with the principles and service of automatic transmissions and transaxles.

If the shift problem seems to be caused by the electronic control system, begin by attaching a scan tool and determine whether any transmission related trouble codes have been set. Then proceed to check the indicated systems and components. If no codes have been set, it may be necessary to make detailed tests of various transmission sensors and solenoids until the problem is located. These tests can often be done at the transmission case connector. Also check for disconnected or corroded electrical connectors under the vehicle. Scan tool data may help to isolate a defective component, even if no trouble code has been set. **Figure 20-11** is a list of transmission problems that can be caused by the electronic components.

Checking Vehicle Speed Sensors

The quickest way to check the vehicle speed sensor (VSS) is to drive the vehicle. If the speedometer registers vehicle speed, the sensor is working, **Figure 20-12.** Another way to check the sensor for proper operation is to connect a scan tool to the DLC and monitor vehicle speed during the road test. The scan tool reading of vehicle speed should be within 1 mph of speedometer indicated speed. If the speedometer indicates speed, but the ECM does not, the fault is in the ECM.

Warning: To perform this test, it is necessary to run the vehicle in drive. Be sure to keep clear of the wheels.

To test the VSS, raise the vehicle and connect a jumper wire to the wiring harness from the sensor. Connect a multimeter's positive terminal to the other end of the jumper wire. Set the meter to read AC voltage. Have an assistant

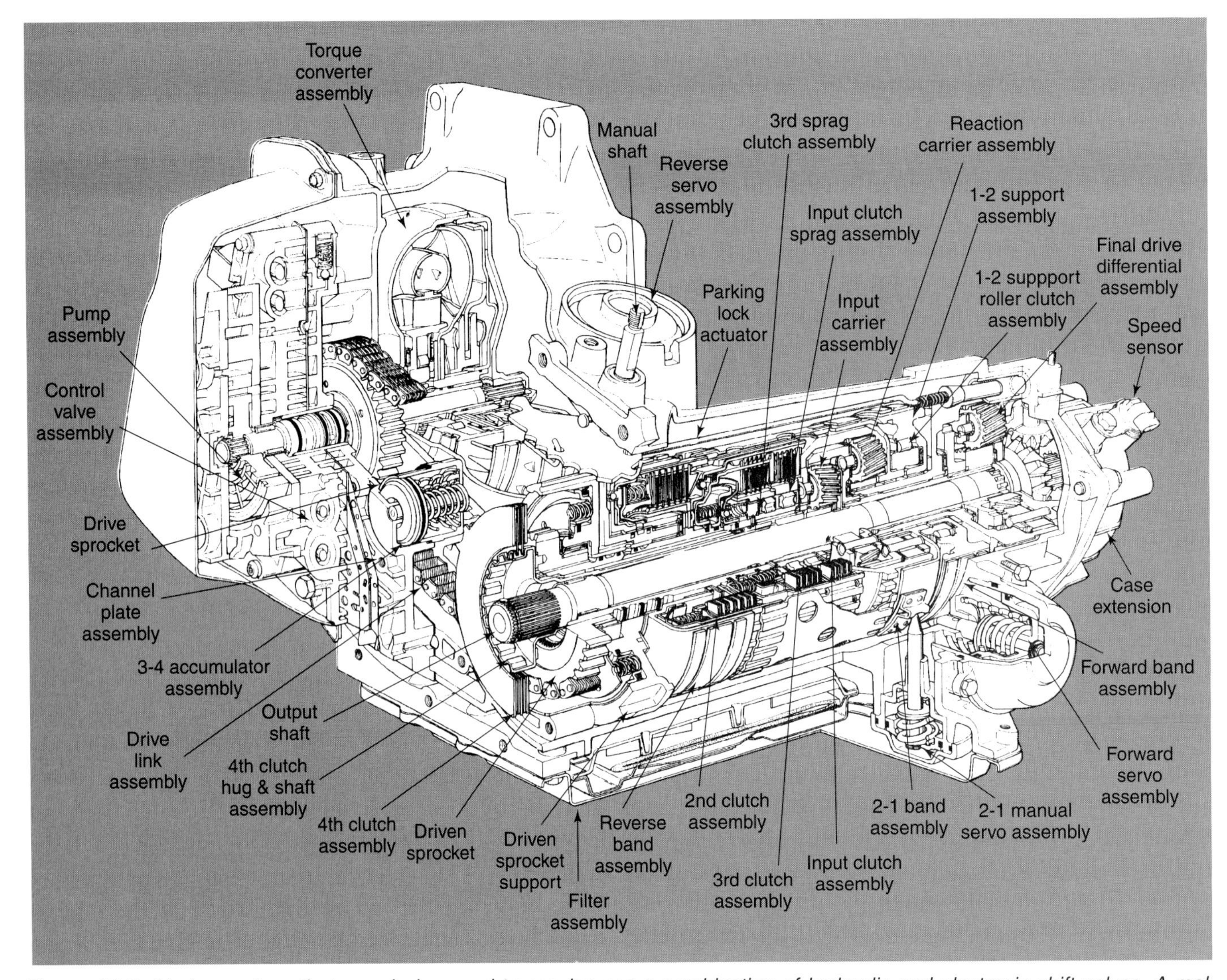

Figure 20-9. *Modern automatic transmissions and transaxles use a combination of hydraulic and electronic shift valves. A malfunction in one of these valves can cause many problems, including slippage, chatter, vibration, and noise. (General Motors)*

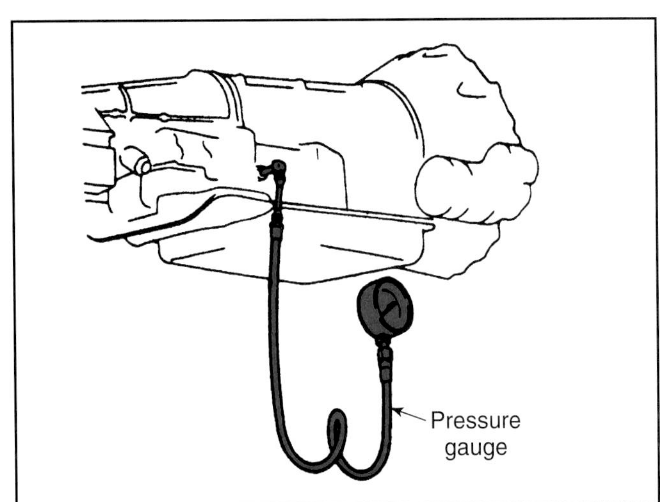

Figure 20-10. *Pressure testing an automatic transmission can reveal the source of a slipping or chatter problem. (Ford)*

shift the vehicle into drive and monitor multimeter output. If the voltage is below approximately .5 volts, check the connection from the ECM. If it is ok, the sensor is defective. If the voltage is above .5 volts, the ECM is defective.

Diagnosing Slippage or Chatter

Slippage is the condition that occurs when engine speed increases without a corresponding increase in vehicle speed. This slippage often takes the form of chatter. ***Chatter*** is a shuddering or jumping effect caused by clutches which are only partially applied. It is usually most noticeable when taking off from stop, but may occur during shifts or when the vehicle is placed under a heavy load, such as climbing a steep hill. It is very easy to mistake chatter for an engine miss.

The severest form of slippage is a complete loss of drive (failure to move when placed in gear). These problems

Component/System	Can Effect
Throttle Position Sensor	• Shift pattern (erratic). • Shift quality (firm or soft). • Engine (rough).
Automatic Transmission Output Shaft Speed Sensor	• Shift pattern (erratic). • TCC Sol Valve apply (at wrong time). • Shift quality (harsh or soft).
Transmission Fluid Pressure (TFP) Manual Valve Position Switch	• TCC Sol Valve apply (no apply if diagnostic code is set). • Shift pattern (no fourth gear in hot mode). • Shift quality (harsh). • Line pressure (high). • Manual downshift (erratic).
Automatic Transmission Fluid Temperature (TFT) Sensor	• TCC Sol Valve control (on or off). • Shift quality (harsh).
Engine Coolant Temperature Sensor	• TCC Sol Valve control (no apply). • Shift quality (harsh).
Shift Solenoid Valves	Gear application (wrong gear, only two gears, no shift).
Brake Switch	• TCC apply (no apply). • No 4th gear if in hot mode.
System Voltage	• Line pressure (high). • Gear application (third gear only). • TCC control (no apply). • No 4th gear if in hot mode.
3-2 Control Shift Solenoid Valve Assembly	• Gear application (third gear only). • 3-2 Downshifts (flare or tie-up).
Pressure Control Solenoid	• Line pressure (high or low). • Shift quality (harsh or soft).
TCC Sol Valve	• TCC Sol Valve apply (no apply). • No 4th gear if in hot mode.
Cruise Control	Delays 3-4 upshift and TCC apply during heavy throttle.
Acceleration Slip Regulation (ASR)	Downshifts.

Figure 20-11. *If any one of a transmission's electronic component malfunctions, it can cause symptoms that look like major engine or transmission problems. Many of these problems would also generate a diagnostic trouble code. (General Motors)*

may be caused by something as simple as low oil level or a clogged filter, or as serious as an inoperative pump, burned clutch or band, or a broken input or output shaft. It is important to figure out the real cause of the problem before attempting any repairs.

If the vehicle suddenly quits moving, the most likely cause is catastrophic failure of an internal transmission part, such as the pump or a shaft. In some cases, splines on a shaft or a hub are stripped off. This is common when a steel shaft is connected to an aluminum hub, such as those used on some planetary gears and torque converter turbines. Correction can only be made by removing and overhauling or replacing the transmission.

If slippage or chatter gradually begins to occur in all gears, and is accompanied by a whining noise that varies with engine speed, the filter may be plugged. If the transmission drops completely out of gear when the vehicle is accelerated, the filter is probably the cause. On other vehicles, the whining noise caused by a clogged filter may disappear when the transmission is placed in park. A pressure gauge can be used to determine whether the transmission pressures are low. Low pressures, however, can be caused by other transmission problems. The simplest way to confirm a filter problem, however, is to remove the oil pan and replace the filter. When the pan is removed, check for excessive deposits of sludge and varnish. If the pan has no more than a light coating of sludge, and only a few metal particles (some metal particle build-up is normal), replace the filter and road test.

Slipping during one or more shifts may be caused by improper linkage adjustment. If the shift is too soft and too low, try adjusting the throttle linkage to give a higher shift. If this adjustment also improves shift quality, misadjusted linkage may be the basic problem. Reset the linkage

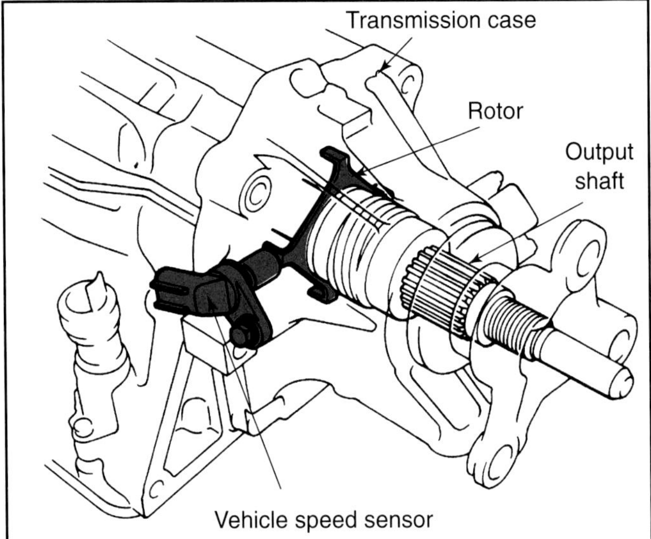

Figure 20-12. *Vehicle speed sensors are usually mounted in the transmission case, near the output shaft. Some vehicles use individual speed sensors mounted at the wheels. (Lexus)*

according to factory procedures and recheck. Keep in mind, however, that prolonged operation of a slipping transmission will severely damage the bands and clutches. Overhaul may be necessary, no matter what the original problem.

Slippage in one gear is usually caused by the failure of the band or clutch that is applied in that gear. If the slippage is accompanied by burned transmission fluid, the band or clutch is probably worn out and glazed. The transmission must be removed and overhauled. If the slippage is being caused by a band that is adjustable, the band adjustment procedure can be performed. However, if the band is already slipping, it is probably too late to cure the problem by adjustment.

Checking Operation of Lockup Torque Converters

Lockup converters are used on almost all modern automatic transmissions. Finding a converter clutch that is stuck in the applied position is easy, since it will stall the engine at idle as soon as the transmission is shifted into a gear other than neutral or park. A clutch that is not applying is more difficult to locate, and may not be felt at all. The simplest way to check the converter lockup clutch is to drive the vehicle with a tachometer or scan tool connected, **Figure 20-13.**

When the converter applies, there should be a noticeable drop in engine rpm with no change on throttle opening. On many automatic transmissions, the clutch application will feel like an extra shift. In some cases, it will be necessary to watch the tachometer closely while accelerating from a stop. There will be an abrupt drop in rpm after each shift, with a smaller drop when the clutch applies. Therefore, a transmission with four speeds should register five changes in rpm on its way from stop to cruising speed.

Diagnosing Lockup Clutch Vibration

Another test is used to verify the cause of a drive train vibration possibly caused by the lockup clutch. Bring the vehicle up to highway speed. When the vibration starts, apply light pressure on the brake pedal, just enough to cause the brake light switch to open, but not apply the brakes. If the vibration stops, pull off the road and shift the transmission into third or disable the overdrive system. Bring the vehicle up to highway speed again. If the vibration does not occur with the clutch disabled, the problem is in the lockup clutch system. However, if the vibration continues, the problem lies elsewhere.

Data Scanned from Vehicle			
Coolant temperature sensor 214°F/101°C	Intake air temperature sensor 88°F/31°C	Manifold absolute pressure sensor 1.1 volts	Throttle position sensor .73 volts
Engine speed sensor 3190 rpm	Oxygen sensor .51 volts	Vehicle speed sensor 57 mph	Battery voltage 13.8 volts
Idle air control valve 21 percent	Evaporative emission canister solenoid Off	Torque converter clutch solenoid Off	EGR valve control solenoid 19 percent
Malfunction indicator lamp Off	Diagnostic trouble codes	Open/closed loop Closed	Fuel pump relay On
Brake light switch On	Cruise control Off	AC compressor clutch On	Knock signal No
Ignition timing (°BTDC)	Base timing: 4		Actual timing: 34

Figure 20-13. *It is possible to monitor converter clutch operation using a scan tool. A tachometer will also work, especially if the vehicle is equipped with one as part of the instrument cluster.*

Note: A misadjusted brake light switch will prevent lockup clutch operation. Also riding the brake pedal will prevent converter clutch engagement.

On some vehicles, one terminal of the diagnostic connector is wired in parallel with the torque converter lockup clutch solenoid. On this type of vehicle, the terminal can be used to separate shifts that are transmission gear changes from the application of the converter clutch which often feels like a shift. This will help the technician separate converter clutch problems from hydraulic problems in the transmission.

To make this test, connect a non-powered test light to the proper terminal, the drive the vehicle. As speed increases, the light will come on when the converter clutch solenoid is energized by the ECM. On some vehicles, the light is on at low speeds and will go off when the converter clutch applies. By matching the shifts to the light operation, the technician can determine which shifts are caused by the converter clutch application, and which shifts are actual transmission gear changes. This procedure can also be accomplished with many scan tools. Set the tool to the powertrain menu and observe the clutch solenoid status as the shifts occur.

Caution: Complex transmission test procedures should be performed by a technician experienced in automatic transmission diagnostics and repair. If transmission tests are performed improperly, transmission damage can result.

Checking Air Conditioner Operation

Since most modern air conditioners control temperatures by cycling the compressor clutch, a malfunction can lead to various driveability symptoms. To check the cycling clutch operation, start the engine and turn on the air conditioner. Set the blower speed to high and open the hood and observe the compressor clutch. A properly operating clutch will cycle on and off anywhere from two to five minutes, depending on the outside heat and sun load. On a very hot day, the clutch may not cycle at all unless the engine speed is increased. A clutch that cycles very rapidly is a sign of a low refrigerant charge. There should be no unusual noise or vibration from the compressor or clutch.

Next, disconnect the A/C blower motor input wire. With the blower not turning, the compressor clutch should begin cycling much more rapidly, as quickly as every 15 seconds. If the clutch does not begin cycling very quickly, there may be a problem with the control system. Check the electric radiator fan(s) for operation. Any fans present should be on when the compressor is on. If the fan(s) are not on, the excessive cycling, as well as possible engine overheating, is caused by a defect in the cooling fan system. If system operation is not as described, consult the correct service manual to isolate the problem. If the air conditioner is the type that does not use a cycling clutch, check the service manual for the proper testing procedures.

ECM Controls for the Air Conditioning System

The air conditioning system in the modern vehicle has many inputs to the ECM. This allows the ECM to adjust engine and drive train operation to compensate for the additional load. In response to a request for air conditioning, the operations performed by the ECM include:

- Delaying compressor clutch engagement for 2-5 seconds when the engine is first started to prevent stalling.
- Increasing idle speed to compensate for compressor load.
- Engaging the radiator fan(s) and on some vehicles, a condenser fan. This allows for better system performance and prevents engine overheating.
- Disengaging the clutch during wide-open throttle or when high power steering pressure is detected, allowing the engine to provide additional power.
- Disengaging the clutch if engine overheating is detected.
- Preventing clutch engagement if low outside ambient air temperatures are detected.
- Preventing clutch engagement if low refrigerant levels are detected.

Note: Some of these inputs are from sensors that are not part of the air conditioning system.

If the ECM did not compensate for compressor load, it could lead to problems, such as stalling at idle or on tight turns, decreased power at wide open throttle, and engine overheating. The ECM prevents clutch engagement at low outside temperatures to increase engine efficiency and prevent system damage. A schematic of an air conditioning control system is shown in **Figure 20-14.**

Air Conditioning Low and High Pressure Switches

Most vehicles use an air conditioning low pressure switch, such as the one in **Figure 20-15.** Some newer vehicles and all vehicles equipped with OBD II monitor this switch for low refrigerant pressure and prevent compressor operation if it senses a low charge. On some vehicles, it will also set a diagnostic trouble code that will prevent compressor clutch engagement and cause the MIL light to illuminate as if a driveability-related problem was present. If a code indicating refrigerant loss is present, a driveability problem is probably not present. Clear the code and begin diagnosing the air conditioning refrigeration system.

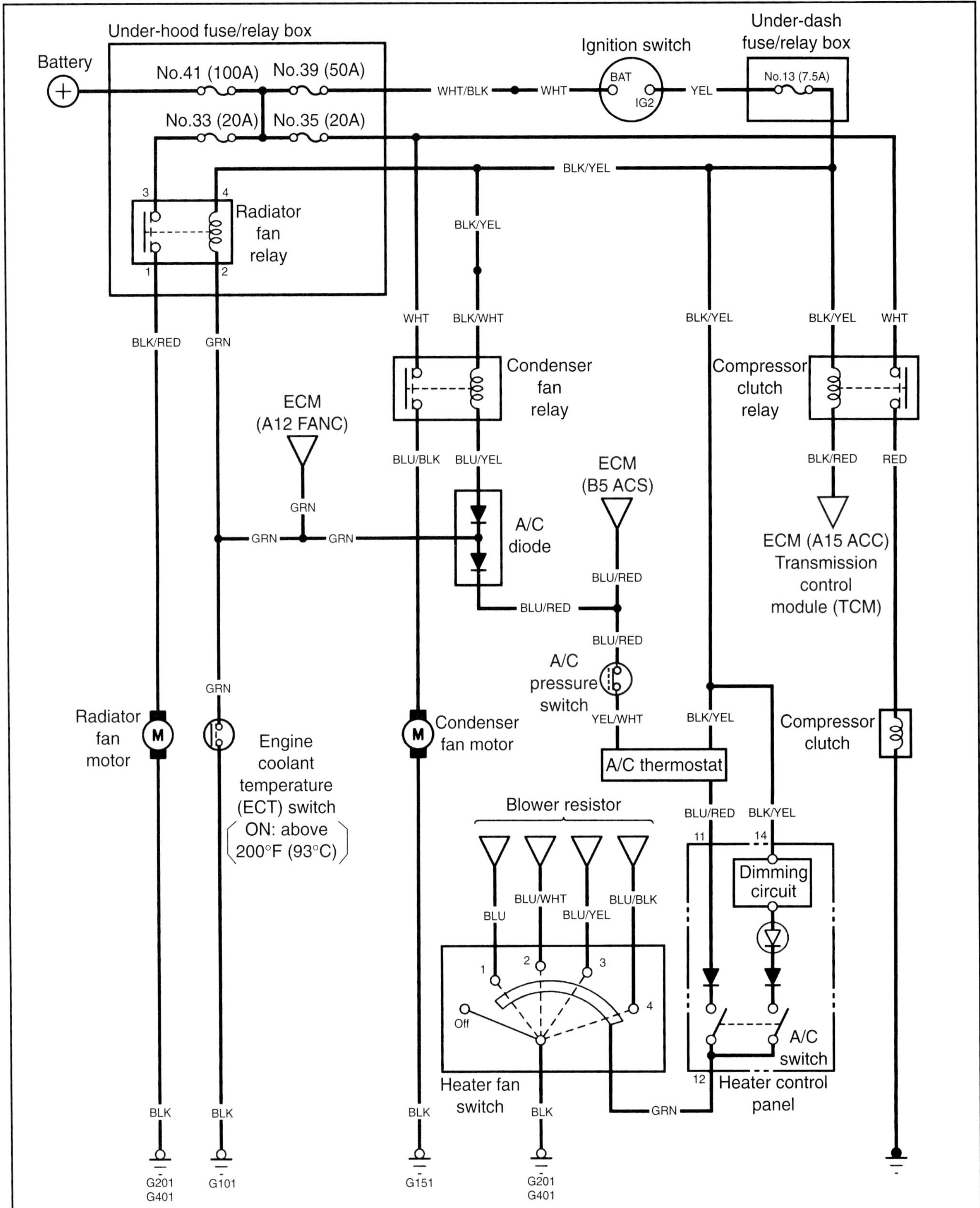

Figure 20-14. *The air conditioning control system has several inputs to and outputs from the ECM. The output from the ECM to the compressor clutch relay allows the ECM to control compressor clutch operation. (Honda)*

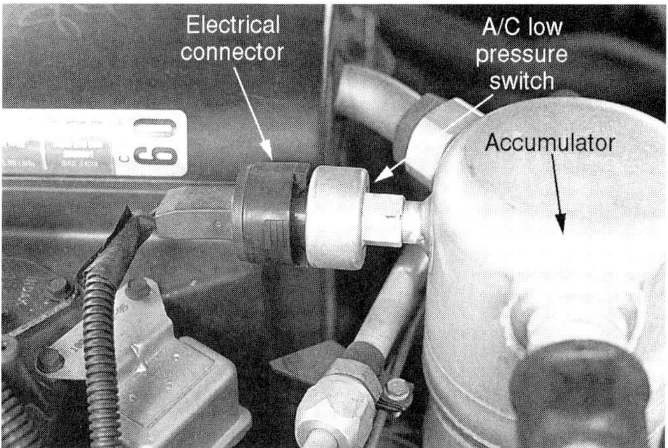

Figure 20-15. The air conditioning system has a low pressure switch mounted in the accumulator or in a line from the evaporator to the compressor. This switch is sometimes monitored by the ECM for low system pressure.

The high pressure switch has been used in air conditioning systems for years. The original function of the high pressure switch was to disengage the compressor when system pressure became too high as a result of a restriction, compressor malfunction, or engine overheating. It functions as part of the control system on earlier vehicles. The ECM in modern vehicles monitors air conditioning high pressure to better control compressor operation. Both of these switches allow the ECM to monitor and control the A/C system for better efficiency as well as protecting the engine, A/C system, and environment from damage.

High and low pressure switches are on-off switches, and can be checked with an ohmmeter or self-powered test light. Both of these switches can be checked with the air conditioner off. When the air conditioner is off, low and high side pressures will equalize throughout the system.

With the air conditioner off, low side pressures will be higher than the low pressure switch setting. Consult the service manual to determine whether the switch is a normally open or normally closed type. Then remove the electrical connector and check across the switch terminals. If the low pressure switch is a normally open type, it should not have continuity when the air conditioner is off. If it does, it is defective. A normally closed low pressure switch should have continuity with the air conditioner off.

High side pressures will be lower than the high side pressure switch setting with the air conditioner off. Therefore, if the high pressure switch is the normally closed type, it should have continuity with the air conditioner off. If the high pressure switch is a normally open type, it will have no continuity with the air conditioner off. If the switch does not have the proper continuity, it should be replaced.

Air Conditioner Cutout Switches

The air conditioner cutout switch is used in some older vehicles to deenergize the air conditioner compressor clutch when the accelerator pedal is fully depressed. Cutting off the compressor clutch eliminates compressor drag when maximum acceleration is needed. If the switch fails to close, the air conditioner clutch will not come on. If the switch fails to open, the clutch will not release when the accelerator is pushed to the floor.

To test this switch, connect a test light to the output terminal of the cutout switch. Turn on the ignition switch, but do not start the engine. Turn on the air conditioner and observe that the light comes on. Then have an assistant push the accelerator pedal to the floor. Check that the light goes out when the throttle is fully open. If the light does not go on and off as specified, adjust the switch by loosening the locknut and turning the switch body until the test light indicates that the switch is operating correctly. If the switch cannot be adjusted, replace it.

Note: The air conditioning cutout switch has been eliminated on most newer vehicles by ECM control of the air conditioning compressor relay.

Checking Driveline Components

There are many types of drivelines in use today, and many kinds of potential problems. Typical driveline problems include noises and vibration. Driveline noises occur when the vehicle is sharply accelerated and decelerated. The usual cause of noises is loose U-joints or CV joints. Driveline vibrations are usually identifiable by the higher frequency (rate of occurrence) than vibrations caused by wheel problems. A driveline vibration will usually seem to occur at about twice the rate of a wheel vibration.

Begin checking the driveline by making sure that there are no loose or dry U-joints (rear-wheel drive vehicles) or loose CV joints (front-wheel drive vehicles). The simplest way to check for looseness is to lift the drive wheels off the ground and try to turn both sides of the suspected joint in opposite directions, **Figure 20-16.** CV joints and U-joints should be firm with no noticeable play. If any looseness is found, the joint should be replaced. Also observe the condition of the CV joint boots. A torn boot means that the CV joint will fail soon from lack of lubrication.

If the joints check out ok, check for bent driveshafts. Most bent driveshafts will be seen to move laterally (up and down) when the vehicle is accelerated with the wheels off the ground. Feeling the entire length of a rear-wheel driveshaft will often reveal a crease caused by the shaft twisting. Rear-wheel driveshaft angles can be checked by the use of an angle gauge, **Figure 20-17.** CV joint design makes front-wheel drive axle angles less critical, and there is usually no procedure for checking them.

Rear-wheel drive vehicles have an axle assembly that can create noises. Whines and roaring noises may be caused by a misadjusted ring and pinion, pinion bearings, or by the outer wheel bearings. Ring and pinion noises are usually high pitched, while pinion bearings are lower in tone, and

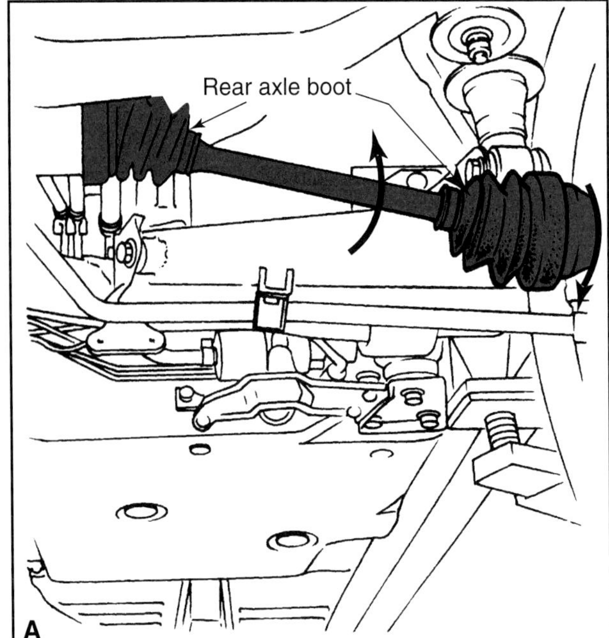

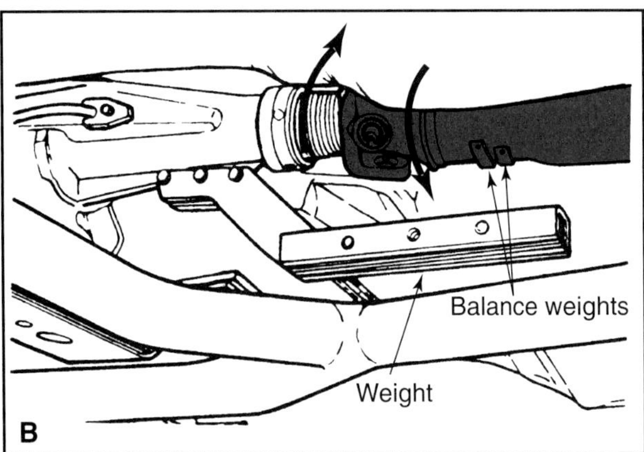

Figure 20-16. *A—Constant velocity joints should have little or no free movement. If the boots are torn or the shaft can be moved excessively, the joint is defective. B—Rear-wheel drive shafts should be able to move slightly in the direction of normal rotation, but no lateral movement with relation to the driveline. (Ford)*

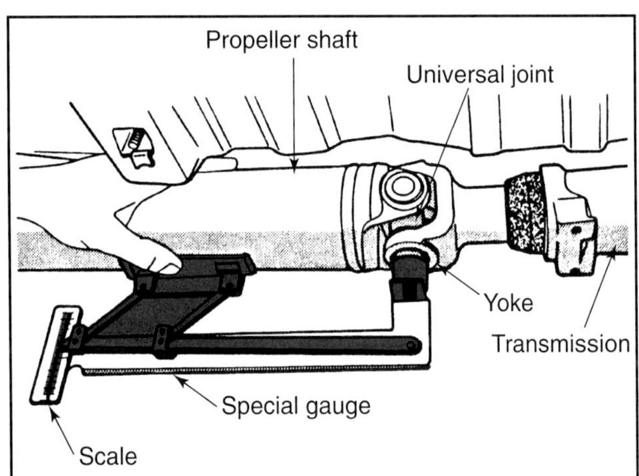

Figure 20-17. *Driveline angle can be measured with an angle gauge. The angle should be within 1° of manufacturers specifications. The vehicle should not be loaded with excess weight when this measurement is taken. (Plymouth)*

vary depending on whether the vehicle is being accelerated or decelerated. The noises made by outer bearings usually vary as the vehicle is turned from side to side. A low oil level can cause some rear axle noises. If the rear axle has a limited slip differential unit, worn clutches or improper lubricant will cause the rear axle to make grumbling and shuddering noises as the vehicle is turned.

Checking Suspension and Steering Components

Often a complaint of engine or drive train vibration is really caused by a steering and suspension problem. The usual causes of vibration in the suspension and steering are loose parts. Front end alignment will not cause vibration, although it may cause pulling and tire wear. Check the front suspension for looseness at each wheel. This is best accomplished by locking the steering wheel and raising the wheels off the ground. Raise the wheels in a way that will unload (take the tension from) the ball joints. On vehicles with springs between the frame and the lower control arm, place the jack under the control arm. If the spring is between the frame or body and the upper control arm, and on MacPherson strut suspensions, place the jack under the frame and allow the lower control arm to hang.

Once the vehicle is properly lifted, vigorously shake the wheel in several places and note wheel movement. This procedure is shown in **Figure 20-18.** If the wheel moves more than roughly .5″ (about 5 mm) in any direction,

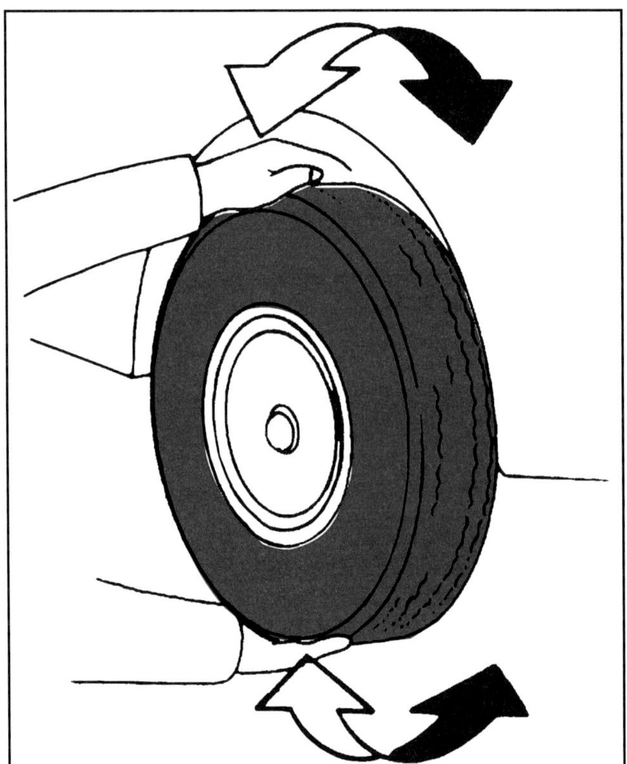

Figure 20-18. *To check the front suspension, grasp the wheel and rock back and forth. Repeat this in several positions. Some play is normal. (Subaru)*

something in the steering or suspension system is excessively loose. As a general rule, excessive movement when the wheel is shaken vertically indicates defective ball joints or control arm bushings. If the wheel moves too much when shaken horizontally, the steering linkage has a loose part. If it moves too much in all directions, the wheel bearings are loose or worn out. Have an assistant shake the wheel as you look under the vehicle. Any suspension or steering part with excessive play should be replaced.

If the vehicle is hard to turn, the problem may be a binding part or misalignment. Check suspected parts for stiffness with the wheels off the ground. Excessive toe in or out can also cause steering problems, as well as pulling. Misaligned rear wheels can cause rear tire cupping, which will make itself known as a roaring noise from the rear axle area. Weak shock absorbers or struts can cause poor rebound or an excessively soft ride, but are usually not misidentified as driveability problems.

Power steering problems consist of leaks or hard steering. If the hard steering complaint is accompanied by a growling or rubbing noise from the power steering pump area, the fluid is probably low. Whenever any power steering problem is encountered, check the fluid level first. If the fluid level is low, add fluid as necessary before proceeding. If the fluid is discolored, the system should be drained and refilled with fresh fluid.

Some power steering systems are equipped with a pressure switch which inputs to the ECM. This switch is normally open. When the steering wheel is in a hard turn, pump pressure to the steering gear rises. This closes the switch, telling the ECM that the power steering system is placing an extra load on the engine. The ECM then raises idle speed, and may turn off the air conditioner compressor clutch. If the switch fails open, the engine may die when the wheels are turned sharply. If the switch fails closed, the idle may be too high, and the air conditioner compressor clutch may not engage.

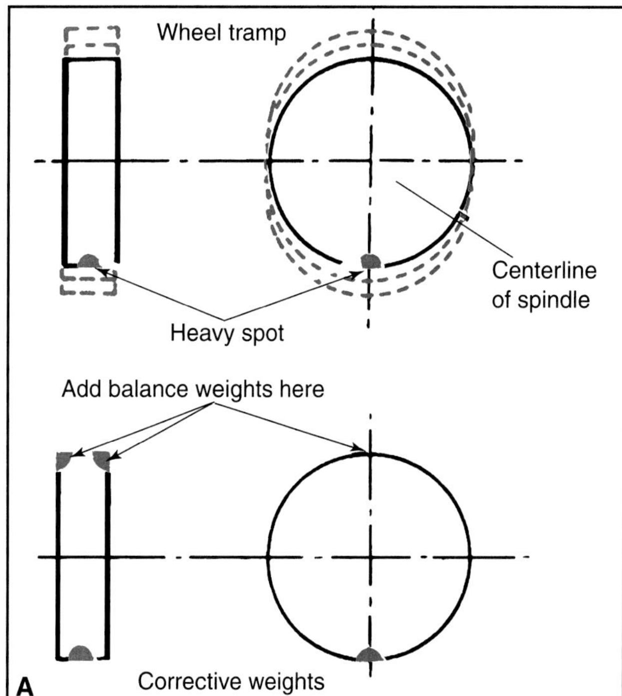

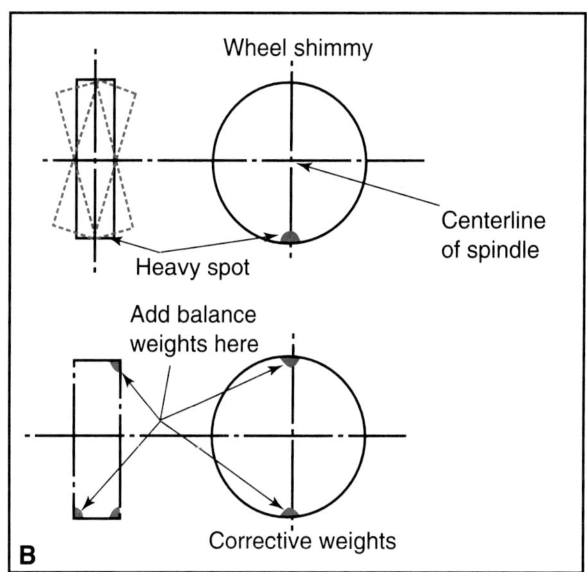

Figure 20-19. *A—Static imbalance will produce an up-and-down motion. This is referred to as wheel tramp. B—Dynamic imbalance will cause a more characteristic side-to-side shake. (Ford)*

Checking Wheel, Hub, and Tire Components

Many vibration problems are caused by imbalanced wheels, bent rims, and defective tires. Of these problems, tire imbalance is the most common. Tire static balance, **Figure 20-19A,** can be checked with a simple bubble balancer, or by spinning the tire with it installed on the axle. If the tire stops in a different position every time it is spun, it is roughly balanced. However, these methods are only approximate, and dynamic balance, **Figure 20-19B,** cannot be checked without a tire balancer. A defective tire usually takes the form of a tread separation. If a tread separation becomes severe, it will usually be visible as a raised spot in the tire tread.

Bent rims can often be spotted by raising the wheel and spinning it. Severe side-to-side motion at the rim is an indication of a bent rim. In many cases, the point of impact can easily be seen. If there is any doubt about the rims' condition, runout can be checked with a dial indicator. In some cases, the wheel hub itself is bent. This must be isolated by removing the rim and checking the hub with a dial indicator. Runout checking procedures are shown in **Figure 20-20.** Bent rims and hubs should be replaced. Do not attempt to straighten a bent rim or hub.

Checking Brake Problems

Brake problems are commonly felt only when the brakes are applied. If the brake system is suspected, check the vehicle rotors for thickness variations. This is usually

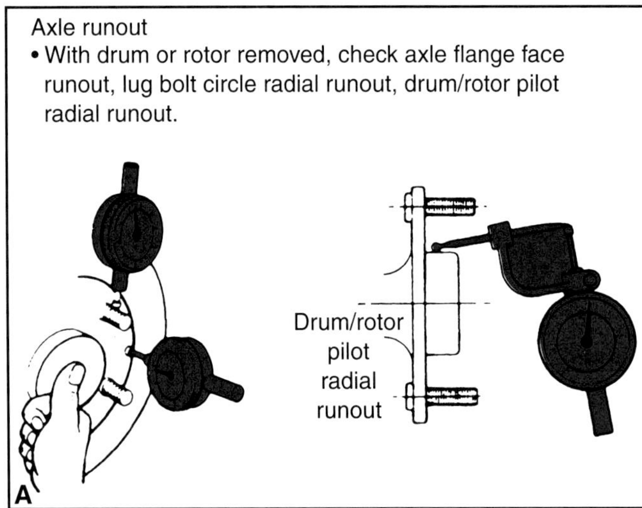

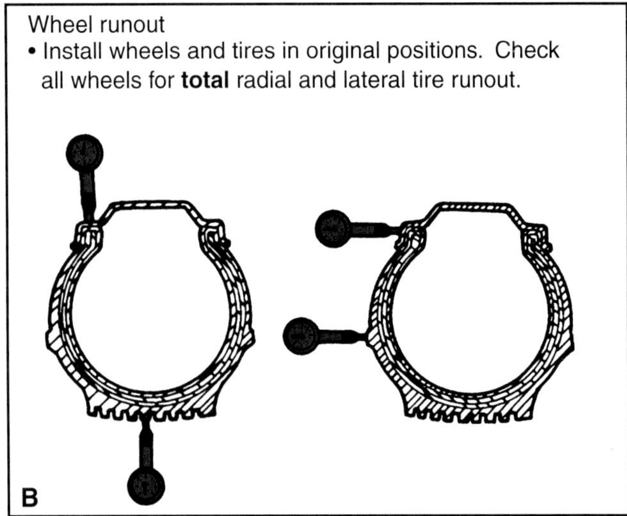

Figure 20-20. *A—Excessive axle runout will produce more noise than vibration unless the bearings are excessively worn or the hub is severely bent. B—Wheel runout can be performed on the vehicle as shown here. A quick check of runout can be made on a tire balancer. (Ford)*

done with a micrometer, as shown in **Figure 20-21.** On drum brakes, check the drums for an out of round condition. This requires a special inside micrometer. If the vehicle pulls in one direction when it is braked, check for severe misadjustment, frozen wheel cylinders or calipers, or brake hoses swollen shut. Anti-lock brake (ABS) problems should occur only when the vehicle is being braked hard. Light brake pedal pulsation when the ABS system is operating is normal. ABS problems will not usually affect engine operation.

Traction Control Systems

If the ABS system also has a traction control system which can modify engine timing or throttle opening, traction control system defects may cause a driveability complaint. Most traction control systems have a servo motor on the throttle body that will control throttle plate position to reduce wheel slippage. This action may be interpreted by the driver as a momentary loss of power.

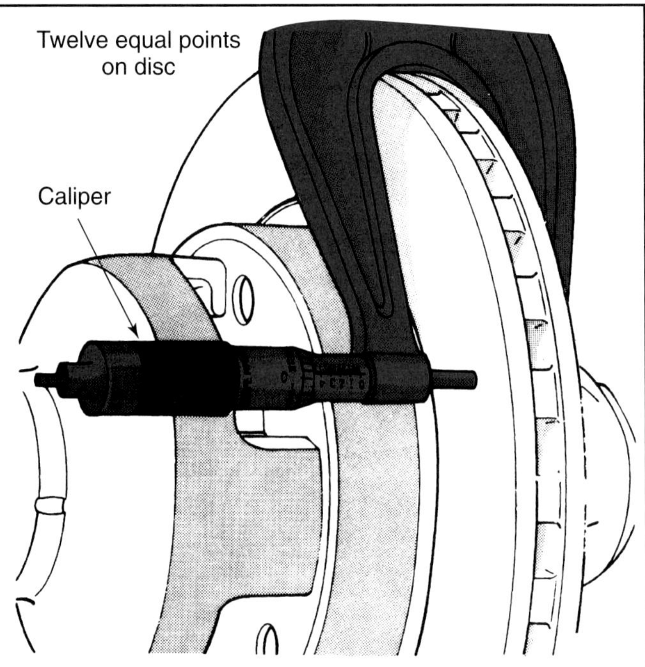

Figure 20-21. *Check the brake rotors for thickness variations with a micrometer. This should be done with the rotors warm as some variations are not evident on a cold part. (Chrysler)*

Correcting Non-Engine Problems that Affect Driveability

Once a problem with a non-engine system is found, it is the responsibility of the technician to repair it. However, in some cases, the driveability technician may not have the training or expertise to properly repair certain systems, such as transmissions. It is also not within the scope of this text to discuss the repair or overhaul of these systems. The procedures discussed here are fairly simple and can be performed by most technicians.

Adjusting Manual Clutch

The modern clutch is a dry plate clutch, which means that it relies on friction between the clutch disc, pressure plate, and flywheel to transmit power. The friction disc is made of the same kind of material as vehicle brake linings, and like brake linings, wears away with use. Eventually, the disc material will wear enough to affect clutch adjustment. Most newer clutches are not adjustable.

Some clutches are equipped with an adjustment device. The clutch linkage can be a rod and lever design, a cable, or a hydraulic system with a master and slave cylinder. Whatever the linkage type, the adjustment usually consists of a threaded adjuster and a lock nut. The adjustment mechanism is usually located on the clutch linkage at the clutch fork.

Before performing any adjustments, check the clutch disc for any foreign material which could cause problems. Oil or grease on the friction disc will cause the clutch to slip or chatter. If the friction disc is contaminated with oil or grease, it should be replaced, and the cause of the original contamination corrected. If the vehicle is driven in deep water, the disc may become wet and slip until it dries out. Do not attempt to adjust a wet clutch assembly.

Follow manufacturer's instructions to adjust the clutch linkage. In general, the clutch pedal should have about 1″ (25.4 mm) of free play before the throwout bearing contacts the pressure plate. Loosen the locknut, turn the adjuster to obtain the proper free play, and then tighten the locknut. If the clutch adjuster has no more threads, the adjustment is used up, and the clutch disc must be replaced.

Adjusting Automatic Transmission Shift Linkage

On vehicles with automatic transmissions, some sort of connection must be provided to allow the transmission to change gears according to the needs of the engine. Automatic transmission shift linkage is usually connected to the fuel system throttle plate assembly that moves an internal transmission valve, called a throttle valve, in response to throttle opening. Movement of the throttle plates causes the throttle valve to move, affecting the other hydraulic components in the transmission. Throttle valve movement will delay upshifts and raise transmission hydraulic pressures under heavy acceleration, and cause early upshifts and lower hydraulic pressures under light acceleration.

In many older transmissions, the throttle valve was operated by intake manifold vacuum through a vacuum device called a modulator, **Figure 20-22.** However, most modern automatic transmissions have direct linkage for firmer upshifts. Using linkage also reduces the chance of late upshifts from a leaking or plugged vacuum line, or from low manifold vacuum due to engine problems. All modern shift linkage is provided with a method of adjustment. If you suspect that shift points are incorrect, too high or too low, check the linkage setting and readjust if necessary.

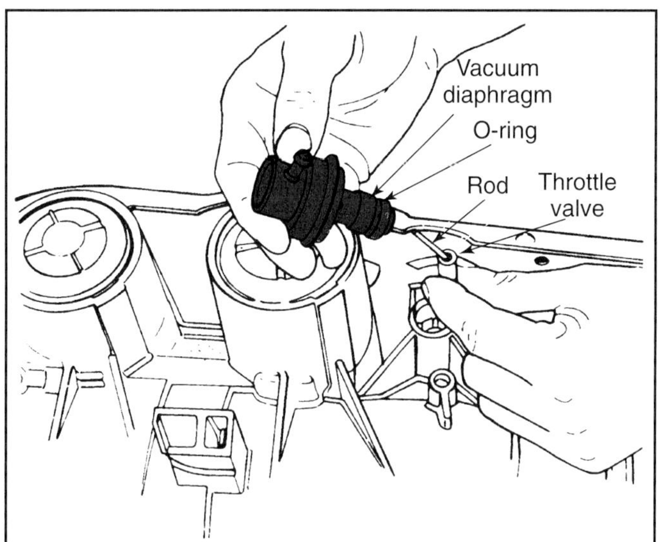

Figure 20-22. *Modulators used on automatic transmissions are controlled by vacuum. However, many modern transmissions do not use modulators and instead, use direct linkage or computer-controlled solenoids. (Ford)*

Note: Any transmission linkage adjustments should be done with the engine shut off.

Several adjustment devices are used on modern transmissions. The simplest type is the threaded adjuster. This type of adjuster is usually used with rod type linkage, although some variations are used with cable linkage. This type is the easiest to adjust, and allows some adjustment flexibility to compensate for engine or transmission wear. Start the adjustment procedure by disconnecting the shift linkage at the throttle valve. Then have an assistant push the accelerator pedal to the floor. Make sure that the throttle plates are fully open. Adjust as necessary. With the throttle plates fully open, pull the shift linkage in the direction of throttle linkage movement, and ensure that the throttle lever and linkage attaching points match closely, as shown in **Figure 20-23.** If they do not match, loosen the linkage locknuts, and turn the threaded adjuster until the attaching points match closely.

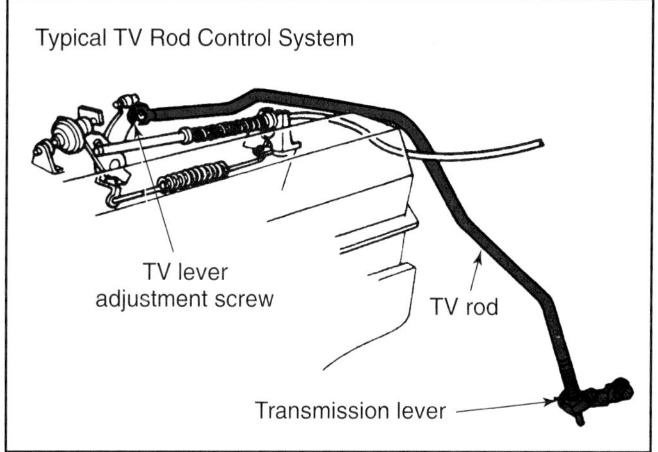

Figure 20-23. *Throttle valve linkages have several points for adjustment. Older vehicles use this type of rod linkage. (Ford)*

Adjusting Cable Linkage

Other transmissions use cable shift linkage. Some cable linkage has a self-adjusting device built into the linkage, **Figure 20-24.** This device consists of a locking mechanism attached to the engine and an adjustable slide built into the shift cable. To set this type of linkage, first check the throttle plate opening, as explained earlier. Then release the locking mechanism, **Figure 20-25A.**

Pull the adjustable slide forward to shorten the linkage as much as possible. While holding the slide, reapply the locking mechanism. Then push the accelerator pedal to the floor. The movement of the pedal pulls the slide to the proper position to adjust the linkage, and no further

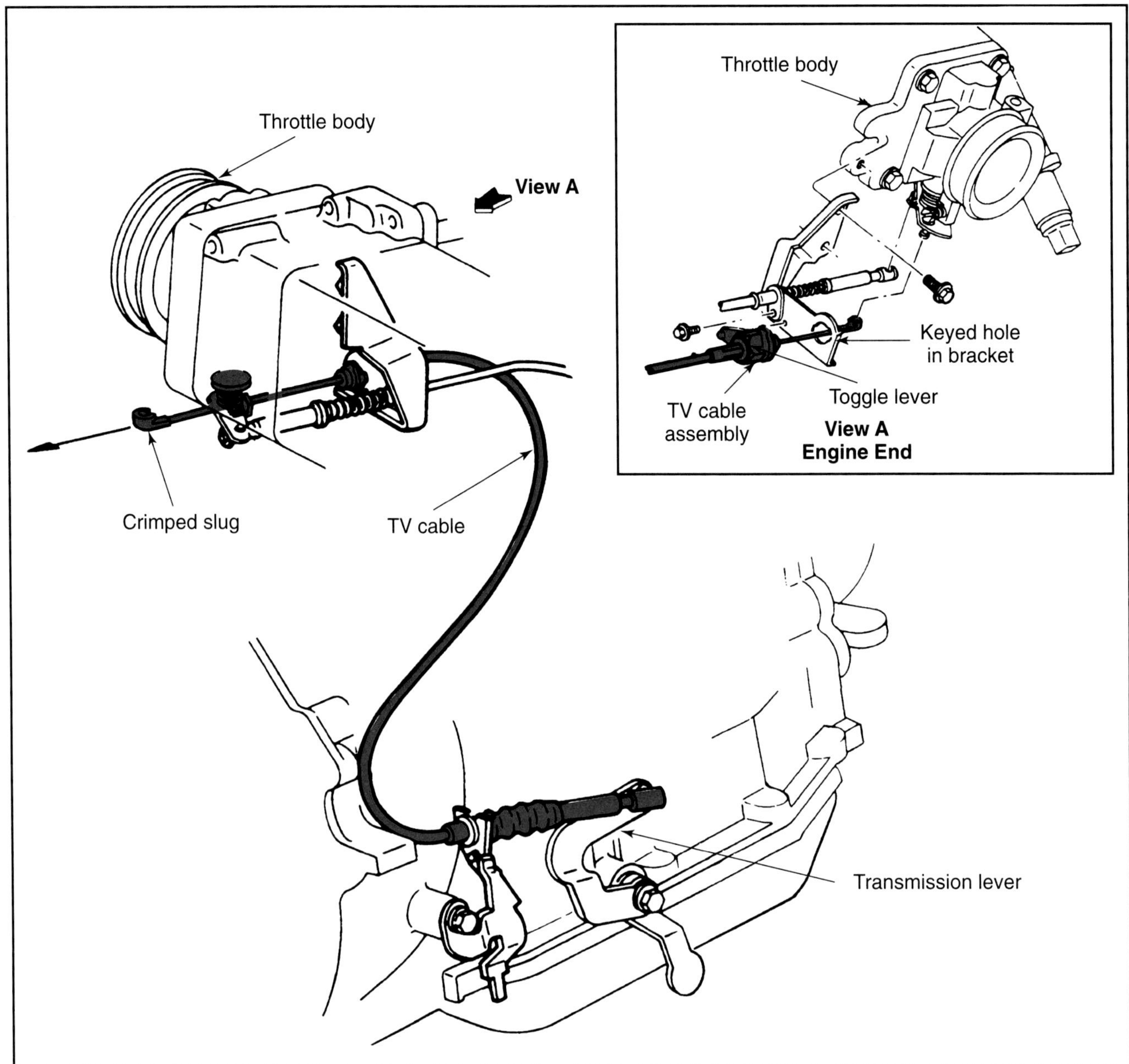

Figure 20-24. *Cable throttle linkage is used on modern vehicles almost exclusively. The adjustment is located near the throttle body. (Ford)*

adjustment is needed. A variation of the self-adjusting cable system uses a slightly different adjustment procedure. Start by releasing the locking mechanism, open the throttle valve to the wide open (WOT) position. With the throttle wide open, relock the mechanism. This sets the linkage and no further adjustment is needed. This is shown in **Figure 20-25B.**

If the transmission still shifts incorrectly, internal transmission work may be needed. This is especially true if the transmission is slipping between shifts, or the fluid appears burned. Automatic transmissions are very complex and further transmission work is beyond the scope of this book. Consult a service manual or text on transmission troubleshooting and repair before proceeding further.

Replacing Torque Converter Lockup Clutch Switch

The lockup clutch is installed in the torque converter of modern automatic transmissions. It is used to eliminate slippage inside the converter by providing a direct mechanical connection between the engine flywheel and the transmission input shaft. If it does not operate, highway gas mileage will be low. If it locks up at low speeds, drive train operation will be rough, and the engine may stall or knock under light acceleration. Therefore, checking the converter electrical controls is critical to proper engine operation. Most modern lockup clutches are controlled through the ECM.

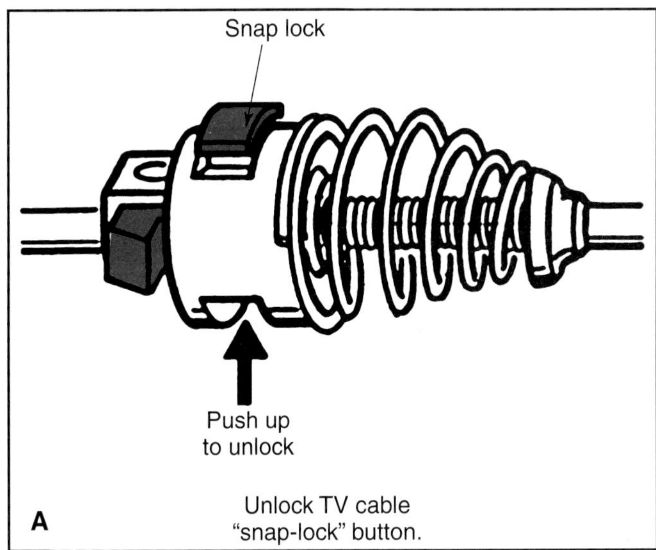

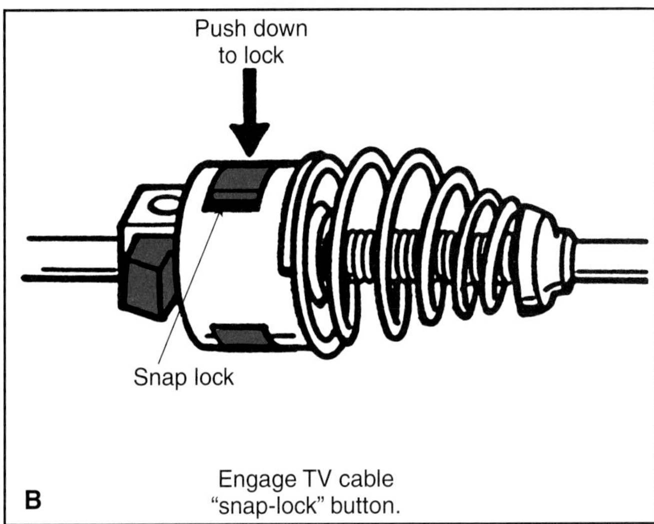

Figure 20-25. *A—To adjust the throttle valve cable, unlock the cable first. This is normally done by unlocking a snap lock or pressing a button on the cable housing. B—Pull the cable to remove the slack and lock the cable. Make sure you do not place too much tension on the cable or the transmission will shift harshly. (General Motors)*

On many older models, two on-off switches control the lockup clutch. One switch is activated by transmission hydraulic pressures in certain gears. This switch is usually mounted on the transmission case. The other switch is controlled by intake manifold vacuum, and is mounted on the engine firewall or inner fender. Follow the manufacturer's instructions to check and adjust these switches.

Replacing Transmission Pressure Switches

Transmission pressure switches are mounted on the transmission case, or installed on the valve body inside the transmission pan. Pressure switches are on-off switches, similar to the familiar engine oil pressure switch. They are installed into the transmission with self-sealing pipe threads. Some transmission pressure switches are shown in **Figure 20-26.**

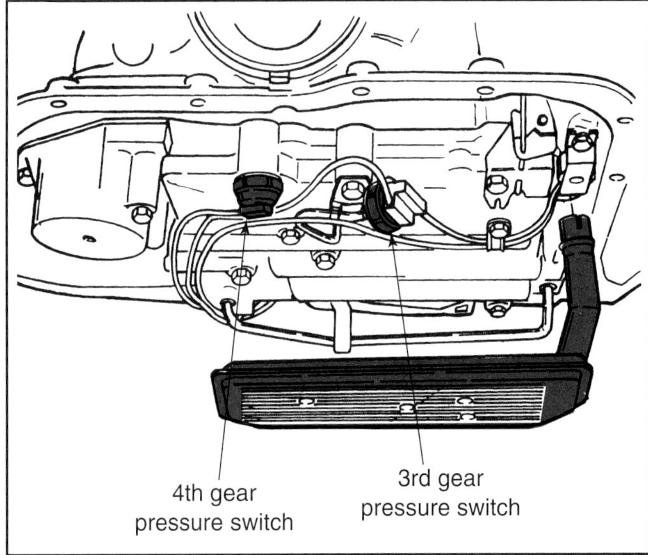

Figure 20-26. *Transmission pressure switches are located in the valve body. Transaxle valve bodies are not this easy to reach, often requiring some disassembly in order to reach the valve body. (General Motors)*

Pressure switches installed on the transmission case can be replaced by removing the electrical connector, and then loosening and unscrewing the switch. Most switches can be removed with a standard socket and/or wrench, while a few can only be removed with a special socket. Before completely removing the switch, position a pan under the transmission to catch any fluid that leaks from the switch port.

If the switch is installed on the valve body, the transmission oil pan must be removed. Loosen the oil pan attaching bolts, and allow the transmission fluid to drain into a pan, as shown in **Figure 20-27.** It is usually easiest to loosen only one side of the pan, so the transmission fluid will drip from only that side. This will reduce the amount of fluid spillage. Then remove the bolts and tilt the transmission pan into the drain pan. Remove the transmission filter, such as the one in **Figure 20-28,** only if it covers the pressure switch. Unscrew the pressure switch after removing the electrical connector.

To install a new pressure switch, install and tighten it into the pressure opening. If the switch is installed on the case, use some non-hardening sealer on the threads. If the switch is inside the pan, no sealer is needed. Then reinstall the electrical connector. If the oil pan was removed, reinstall it using a new gasket. Scrape all old gasket material from the pan and case before reinstallation. You should install a new transmission fluid filter before replacing the pan, especially if the fluid was dirty or overheated.

Refill the transmission with the proper type and amount of transmission fluid, **Figure 20-29,** and start the engine. Run the engine in neutral for about two minutes, then check the

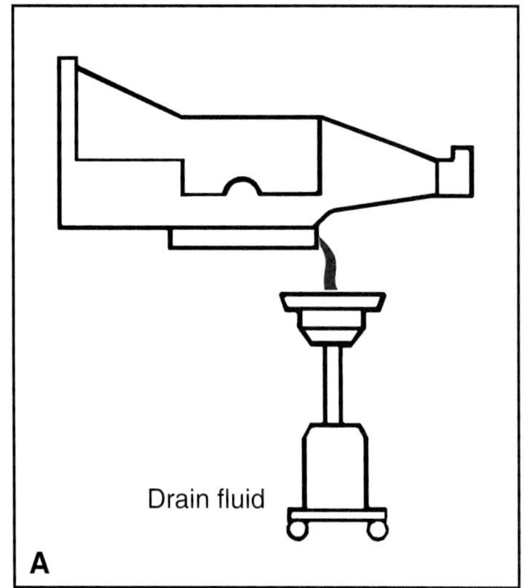

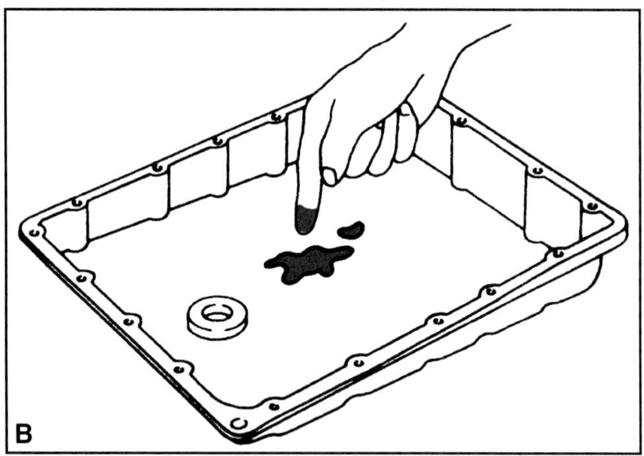

Figure 20-27. *A—Drain the transmission fluid into a pan. Some transmission pans have a drain plug to make this step easier. B—Clean the pan, there will be some metal residue in the bottom. If there is an excessive amount of metal or clutch material in the pan, the transmission must be overhauled or replaced. (Subaru)*

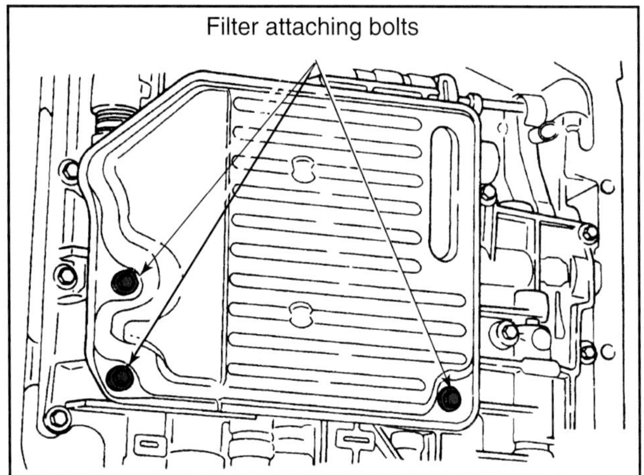

Figure 20-28. *Some filters are bolted to the valve body. A few transmissions use the filter to hold check valves and other devices. Note the position of these as the filter is removed and make sure they are reinstalled in the correct location with the new filter. (Ford)*

Figure 20-29. *Refill the transmission with the correct fluid. Most newer transmissions use Dexron III/Mercon.*

transmission fluid level with the dipstick. If necessary, add more fluid until the reading on the stick is between the add and full marks.

Vehicle Speed Sensors

Vehicle speed sensors are held to the transmission/transaxle by one or more bolts. Position a drain pan below the sensor as some transmission fluid may spill when the sensor is removed. Disconnect the wiring harness from the sensor, unbolt the sensor, and carefully remove the sensor from the transmission.

Make sure the new sensor has a gasket or O-ring. Do not use the old gasket or O-ring. Position the new sensor in the same location as the old sensor and install the retaining bolts. Reconnect the sensor wiring harness and refill the transmission if any fluid leaked.

Air Conditioner Pressure Switch Replacement

Air conditioner high and low pressure switches are serviced by replacing them. Some systems must be discharged to replace the switch. On other systems, the switch mounting ports have Schrader valves, and the switches can be removed and replaced without discharging the system. Always consult the service manual before replacing any air conditioner pressure switch.

Caution: Do not discharge refrigerant into the atmosphere. This is illegal, and also very expensive.

Driveline Service

To replace U-joints or CV joints, the driveshaft must be removed from the vehicle. **Figure 20-30** shows the removal of bolts holding a rear-wheel driveshaft to the rear axle

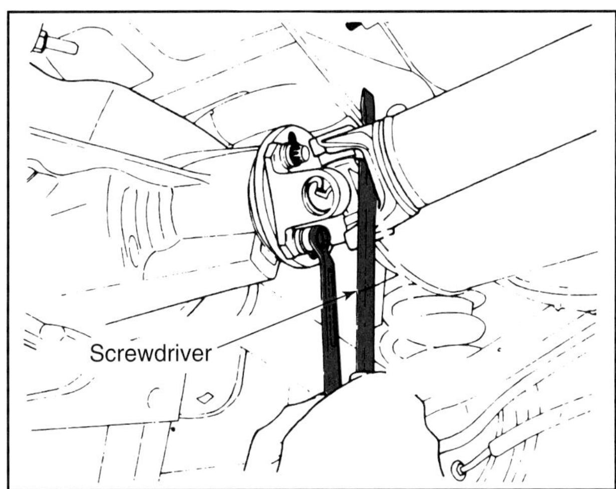

Figure 20-30. *To remove the rear-wheel driveshaft, hold the driveshaft and remove the rear flange or U-bolts with a wrench. It may be necessary to remove other components to remove the driveshaft from the vehicle. (Ford)*

assembly. After these bolts are removed, the front of the driveshaft can be slid out of the transmission and removed from the vehicle. Have a drip pan handy to catch any fluid that leaks from the transmission tailhousing. If the driveshaft has a center support bearing, the center support must be unbolted from the body before the driveshaft can be removed.

Some front-wheel drive CV axle shafts are pressed onto the outer bearing and must be removed with a special puller. Others are slip fits and can be slid out of the bearing once the spindle nut is removed. Spindle nut removal is shown in **Figure 20-31.** To remove most CV axle shafts, the steering knuckle assembly must be removed from the lower control arm. This provides clearance to remove the shaft, **Figure 20-32.** The shaft can then be removed from the transaxle and the vehicle. Some CV axle shafts are bolted to the transaxle output shaft, while others are held in place with a clip on the inner shaft. Most of the models using a clip can be removed by tugging on the shaft, or by carefully prying between the inner CV joint and the transaxle housing. A few designs require the use of a special puller.

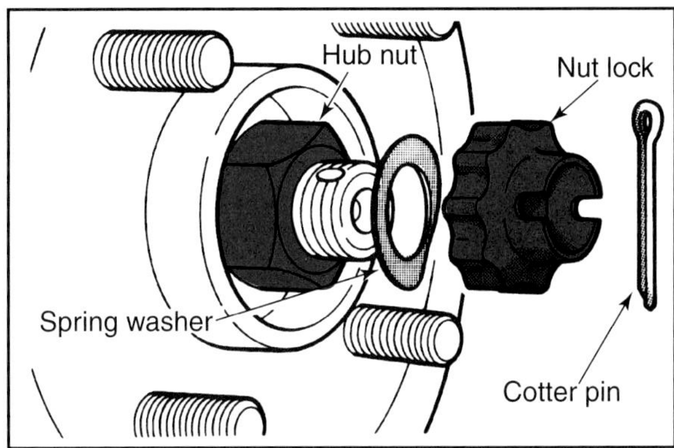

Figure 20-31. *Some CV spindle nuts are held to the bearing with a nut and cotter pin. Others are simply held by a bolt. (Chrysler)*

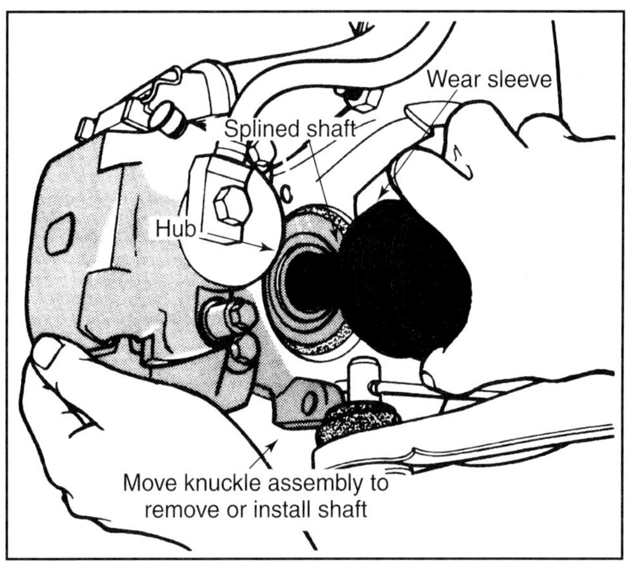

Figure 20-32. *After the nut is removed, the shaft can usually be pulled from the spindle. Note that the spindle has been removed from the ball joint. (Chrysler)*

Once the shaft is out of the vehicle, defective joint or joints can be replaced. Follow the manufacturers joint replacement procedures exactly. Defective U-joints on rear-wheel driveshafts are always replaced. If bent, some rear-wheel driveshafts can be straightened and balanced by shops specializing in driveline service.

If front-wheel drive CV joints are damaged, the entire shaft is often replaced. Some shops specialize in rebuilding CV axles. Any time that a CV axle shaft is bent, it should be replaced. Never try to straighten a CV driveshaft.

Suspension and Steering Service

Suspension parts that can wear out and cause problems include the upper and lower ball joints, strut rod bushings, upper and lower control arm bushings, shock absorbers, and MacPherson strut assemblies when used. Parts that can become bent, usually to collision damage, are the spindles, control arms, strut rods, and the vehicle frame. Replacing suspension parts requires specialized tools and techniques. Do not attempt to repair any suspension part without consulting the proper service manual. Never, under any circumstances, try to straighten any suspension or steering part.

Warning: Many suspension parts are under strong spring tension. Do not attempt to replace any suspension part without first removing the spring tension from the part.

Steering system parts that can wear and cause vibration and handling problems are the inner and outer tie rod ends. Both conventional and rack-and-pinion steering systems have inner and outer tie rod ends. Conventional steering systems also have idler arms, pitman arms, and center links. These parts should be replaced according to the manufacturer's directions. Other steering parts that can cause steering problems are worn gearboxes or rack-and-pinion units, broken steering couplers, loose steering column universal joints (when used), and bent steering linkage.

Note: After any steering and suspension service, the vehicle must be aligned. If alignment settings were changed during parts replacement, severe handling problems and tire wear can result.

Brake Service

Common driveability related brake problems are caused by warped drums and disc brake rotors. Drums and rotors can be machined, or turned, using a brake lathe. **Figure 20-33** shows a rotor being turned in a brake lathe. Only a certain amount of metal can be cut from rotors and drums without dangerously reducing their heat absorbing capabilities. The minimum amount of metal that must be left is stamped on the hub. If a drum or rotor cannot be machined to remove imperfections without removing too much metal, it should be replaced. When the drums or rotors are turned, the pads or shoes are usually replaced. If any brake fluid leaks are found, they should be repaired immediately. Never allow a vehicle to leave the shop with a serious brake defect. Consult the proper manufacturer's service manual for information on brake service.

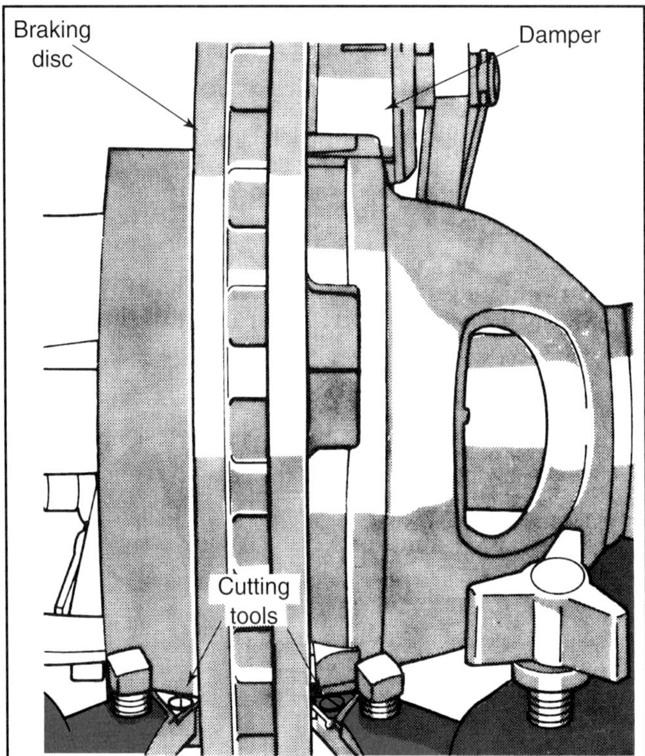

Figure 20-33. *Resurfacing a rotor will usually correct a warped condition. Make sure you do not remove too much metal. (Chrysler)*

Anti-Lock Brake and Traction Control Service

Defects in the anti-lock brake and traction controls can be caused by a hydraulic system problem, or by a defect in the computer control system. The ABS system may be operated by a separate computer, or it may be operated by the ECM. If the ABS or traction control appears to be the source of a problem, consult the proper service manual before attempting any repairs. On many modern vehicles, the ABS control system can be accessed with a scan tool.

Follow-Up for Drive Train and Other Repairs

Check the fluid level of the affected system after repairs are complete and add as needed. Road test the vehicle thoroughly to ensure that no problems are present. Check the ECM for any codes that may have been set during the diagnostic process and clear the ECM's memory. Make sure that the vehicle is clean before releasing it to the owner.

Summary

Always keep in mind that the engine systems are not the only source of trouble. Other problems can be in the manual clutch, automatic transmission shift system, lockup torque converter, air conditioner, brakes, drive train, and suspension systems. Brief checking procedures will enable you to tell whether they are the source of a driveability or performance problem.

Noise, vibration, and harshness diagnosis, usually called NVH diagnosis, is usually concerned with drive train and suspension problems. To perform NVH diagnosis, the technician must be familiar with how noise and vibration are created and transmitted. One method of eliminating NVH complaints is to balance the source of the vibration. If the vibration cannot be eliminated, it can be dampened to keep it from affecting the driver and passengers. If at all possible, the source of the vibration and noise should be eliminated rather than being dampened or suppressed.

Transmissions and transaxles are sources of many driveability complaints. Always check manual clutches for slipping and chatter. Check the manual transmission or transaxle for proper operation. Most automatic transmission problems are more complex than manual transmission problems. Start the automatic transmission diagnosis process by checking fluid level. Then road test and check shift speeds. Refer to shift charts for proper shift speeds. In many cases, linkage can be adjusted to correct shift problems. If the transmission has a vacuum modulator, check it for leaks and proper operation.

Before troubleshooting the electronic components, make all hydraulic checks. Some transmissions have a combination of hydraulic and electronic shift mechanisms. Complaints of slippage usually involve the hydraulic system. Correcting slippage may involve complete teardown, or may be corrected by changing the filter or adjusting the linkage. Check the operation of the lockup torque converter whenever a complaint of vibration or rough shifting is encountered.

Check air conditioner operation for rapid clutch cycling, a sign of low refrigerant levels. Also check the operation of the pressure switches and computer control system, when used. If the vehicle uses a compressor cutout switch, check it at wide open throttle.

The most common driveline problems are loose or dry U-joints or CV joints. Any looseness means the joint should be replaced. If the joints check out OK, check for a bent driveshaft. Rear-wheel drive vehicles can develop noises and vibration in the rear axle assembly. Loose parts in the suspension and steering will cause vibration. Front end alignment is not a cause of vibration. To check for loose parts, raise the front wheel and vigorously shake the wheel in several directions. Any looseness is cause for further investigation.

Bent wheel rims and hubs can be checked with the wheels off the ground. If a bent part is not obvious, a dial indicator can be used to measure runout. Out-of-round brake drums or rotors will cause vibration only when the vehicle is being braked.

Clutch problems can often be corrected by adjustment. There are many kinds of clutch adjusting mechanisms. Some modern clutches are self-adjusting. Automatic transmission shift linkage is adjustable by threaded rods, or a cable system. If the transmission problems cannot be corrected by adjustment, internal transmission work may be needed. The switches that control lockup clutch operation on older vehicles can be adjusted, but most modern lockup clutches are operated by the ECM. Transmission pressure switches and solenoids are easily replaced once the transmission pan is removed.

Air conditioner pressure switches are replaced rather than serviced. Make sure that no refrigerant escapes to the atmosphere when replacing a pressure switch. To replace U-joints or CV joints, the driveshaft must be removed from the vehicle. Some shafts are replaced as a unit. Replacing steering and suspension parts requires special tools and techniques. Do not attempt to repair or replace any steering or suspension part without consulting the proper service manual. After any steering and suspension service, the vehicle must be realigned. Brake drums and rotors can be machined to remove an out-of-round condition.

Know These Terms

Noise, vibration, and harshness (NVH) diagnostics
Vibrations
Noise
Harshness
Transmitter
Transfer path
Receiver
Balancing
Dampening
Shift charts
Slippage
Chatter

Review Questions—Chapter 20

Please do not write in this text. Write your answers on a separate sheet of paper.

1. One inch of free play in the clutch pedal indicates that the clutch is ______.
   (A) correctly adjusted
   (B) slipping
   (C) incorrectly adjusted
   (D) greasy or oily
2. It is normal for automatic transmission shifts to occur ______ under light throttle.
   (A) at very high speeds
   (B) at very low speeds
   (C) within a set range of speeds
   (D) None of the above.
3. Shift points can be anywhere within a certain ________ and still be correct.
4. If an automatic transmission begins slipping all gears, accompanied by a whining noise that varies with engine speed, what could be the cause?
   (A) Clogged filter.
   (B) Stuck shift valve.
   (C) Stuck converter solenoid.
   (D) Low oil level.
5. Air conditioner pressure switches are ______ switches.
6. A driveline vibration will seem to occur at ______ the rate of a wheel vibration.
7. Worn clutches or improper lubricant will cause a limited-slip rear axle to make grumbling and shuddering noises as the vehicle is ______.
8. If a power steering pressure switch fails in the closed position, the ______ may not engage.
9. When will an out-of-round brake rotor cause vibration?
   (A) When the brakes are applied.
   (B) When the brakes are released.
   (C) At all times.
   (D) This condition will not cause vibration.
10. Discharging refrigerant into the atmosphere is _______ and _______.

ASE Certification-Type Questions

1. Technician A says that modern torque converter lockup clutches are usually controlled by the ECM. Technician B says that modern transmissions shift points are controlled by hydraulic pressure. Who is right?
   (A) A only.
   (B) B only.
   (C) Both A & B.
   (D) Neither A nor B.
2. If a transmission band is already slipping, what will solve the problem?
   (A) Adding more fluid.
   (B) Adjusting the band.
   (C) Changing the filter.
   (D) None of the above.
3. When the air conditioner clutch cycles rapidly, the system may have a low ______.
   (A) refrigerant charge
   (B) blower speed
   (C) compressor speed
   (D) None of the above.
4. To check for worn ball joints on a MacPherson strut equipped vehicle, place a jack under the ______.
   (A) ball joint
   (B) frame
   (C) lower control arm
   (D) Both A & B.
5. A torn CV joint boot will cause the CV joint to fail from ______.
   (A) vibration
   (B) lack of lubrication
   (C) acute drive angle
   (D) overspeeding
6. Do not attempt to straighten a bent ______ .
   (A) wheel rim
   (B) driveshaft
   (C) wheel hub
   (D) All of the above.
7. Many suspension parts are under ______ tension.
   (A) spring
   (B) compression
   (C) frame
   (D) None of the above.
8. All of the following statements about diagnosing looseness in a wheel are true, EXCEPT:
   (A) if a wheel moves no more than one inch (about 10 mm) in any direction, everything is tight.
   (B) excessive movement when the wheel is shaken vertically indicates defective ball joints.
   (C) excessive movement when the wheel is shaken horizontally, indicates the steering linkage is loose.
   (D) excessive movement in all directions indicates loose wheel bearings.
9. Technician A says that an ABS system will cause pedal pulsation when the brakes are lightly applied. Technician B says that a warped rotor will pulsate whenever the brakes are applied. Who is right?
   (A) A only.
   (B) B only.
   (C) Both A & B.
   (D) Neither A nor B.
10. If too much metal is removed from a rotor, what can happen?
    (A) It will grab on that side of the vehicle.
    (B) It will not absorb enough heat.
    (C) It will be noisy.
    (D) None of the above.

Advanced Certification Questions

The following questions require the use of reference information to find the correct answer. Most of the data and illustrations you will need is in this chapter. Diagnostic trouble codes are listed in **Appendix A.** Failure to refer to the chapter, illustrations, or **Appendix A** for information may result in an incorrect answer.

11. A fuel injected sedan has a complaint of poor fuel mileage. The scan tool data shown in **Figure 20-13** was captured at highway speed. Technician A says that a defective or misadjusted brake light switch is the problem. Technician B says that an AC compressor clutch switch stuck closed is the problem. Who is right?
    (A) A only.
    (B) B only.
    (C) Both A & B.
    (D) Neither A nor B.

Data Scanned from Vehicle			
Engine temperature 211°F/99°C	Battery voltage 13.1 Volts	Engine speed sensor 1828 RPM	Vehicle speed sensor 9 MPH
Brake light switch Off	Torque converter clutch Off	Shift select position Drive	Throttle position 36 Degrees
Malfunction indicator lamp On	Diagnostic trouble codes P0606,P0700	Engine loop operation Open	Shift solenoid A Off
Shift solenoid B On	Shift solenoid C Off	Shift solenoid D Off	Transmission range Second

12. A multiport fuel injected vehicle's malfunction indicator light is illuminated. Data from a scan tool indicates that diagnostic code P0534 has been set. Which of the following is the *most likely* cause?
    (A) A vehicle speed sensor malfunction.
    (B) An intermittent A/C refrigerant pressure switch malfunction.
    (C) An intermittent malfunction in a transmission shift solenoid.
    (D) An air conditioning refrigerant leak.

13. A vehicle with a computer-controlled transaxle has poor acceleration from a stop. The data above was captured by a scan tool at initial acceleration. Technician A says a defective ECM is the cause. Technician B says that a defective shift solenoid is the cause. Who is right?
    (A) A only.
    (B) B only.
    (C) Both A & B.
    (D) Neither A nor B.

14. A multiport injected engine overheats when the air conditioning system shown in **Figure 20-14** is turned on. The problem could be caused by ______.
    (A) an open A/C diode.
    (B) an open circuit in the BLK/RED wire from ECM terminal A15.
    (C) an open circuit in the BLK/RED wire from the radiator fan relay.
    (D) an open circuit in the GRN wire to the radiator fan relay.

15. An ECM has stored codes P0750 and P0781. Which of the following is the *most likely* cause of these codes?
    (A) A defective 2-3 shift solenoid.
    (B) A defect in the torque converter clutch circuit.
    (C) A defective 1-2 shift solenoid.
    (D) A defective vehicle speed sensor.

Tune-ups are performed on modern engines as part of routine maintenance rather than as a solution to a performance problem. Many engines are not scheduled to receive their first tune-up for 100.000 miles (160,000 km). (Cadillac)

Chapter 21

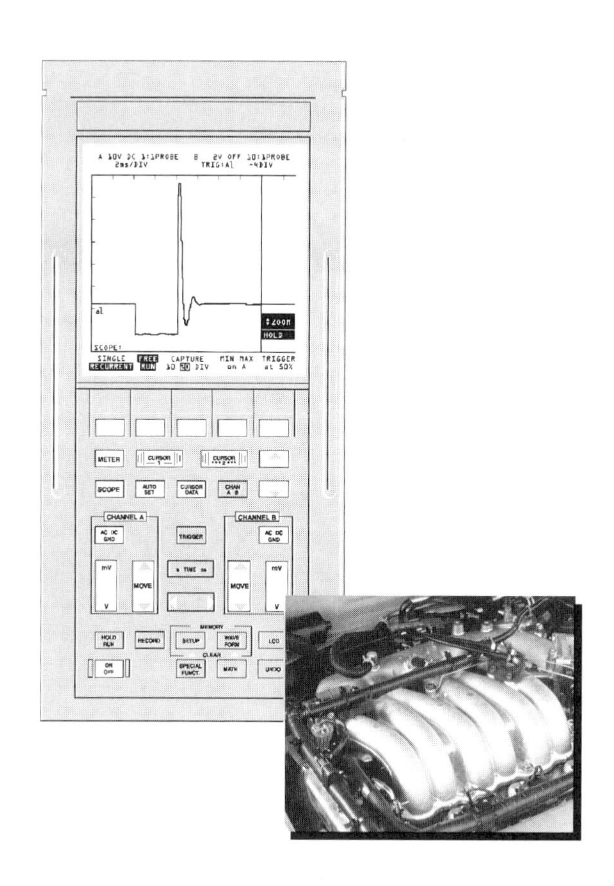

Tune-Up and Maintenance

After studying this chapter, you will be able to:

- ❑ Explain the differences between tune-ups done many years ago and tune-ups done today.
- ❑ Explain how to correctly change spark plugs.
- ❑ Explain how to inspect other ignition system parts.
- ❑ Explain how to set idle speed and mixture.
- ❑ Explain how to replace fuel, air, and emissions filters.
- ❑ Explain how to check turbocharger boost pressure.
- ❑ Explain how to clean injectors.

The concept of the tune-up has changed greatly in the past quarter century. The modern tune-up is less tedious, requiring fewer adjustments, but is more likely to tax the technician's knowledge and diagnostic ability. Instead of replacing ignition or fuel system parts that commonly fail after a set time or mileage, technicians must determine which parts are really defective, and if they are causing the complaint. Unfortunately, many drivers still believe that the tune-up is a sure way to cure performance and driveability problems. Studying this chapter will enable you to understand and contrast the service needs of the older engine with the needs of the modern engine. Knowing the difference will help you to explain to customers that a tune-up may not cure the problem; and what kind of service is really needed. This will reduce misunderstandings and allow you to proceed with the correct repair procedures.

Tune-Ups in the Past

A ***tune-up*** is the process of restoring engine efficiency by troubleshooting problems, and the replacement and/or adjustment of minor parts. In the past, engines required a tune-up about every 12,000 miles (19 200 km), **Figure 21-1.** There were several reasons for this, most involving the shortcomings of the contact point ignition system.

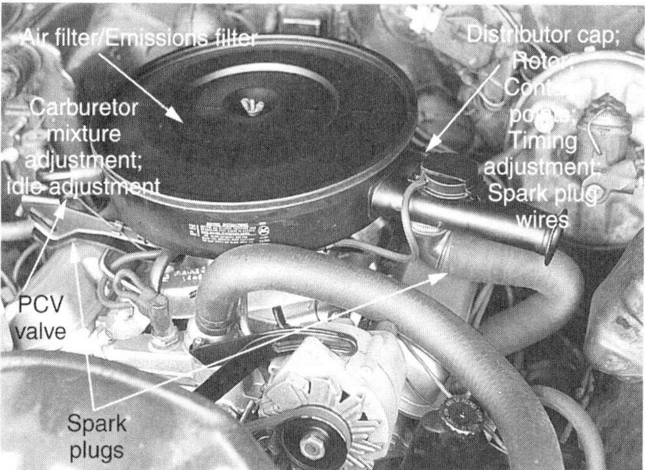

Figure 21-1. *Older vehicles had fewer parts and systems, however, they were more susceptible to performance problems from normal wear, requiring frequent maintenance.*

The contact points in older ignitions usually developed severe pitting and scaling at approximately 12,000–18,000 miles (19 200–28 800 km). At about the same mileage, the point set rubbing block had worn enough to increase dwell and retard ignition timing. The high compression ratios of older engines and use of leaded gasoline would cause the spark plug electrodes to wear rapidly and lead deposits to build up on the insulators. Spark plugs would often begin missing under load around 12,000 miles (19 200 km), sometimes sooner on high compression engines.

Other ignition components could also fail, resulting in a tune-up related problem. It was not uncommon for the insulation around spark plug wires to be burned and crystallized by engine heat. Early resistor wires often corroded internally from a combination of high voltage and moisture. This raised their internal resistance, making plug firing difficult. Distributor caps and rotors were made of an early type of plastic called Bakelite and would develop carbon tracks and cracks from heat and moisture. A big contributing factor to cap and wire problems was the poor fit of many spark plug wire boots, which allowed moisture and dirt to enter.

Condensers would sometimes fail internally, causing complete ignition failure, or go out of calibration, causing point arcing and rapid wear. The distributor shaft and bushings often became worn, causing shaft play and erratic point action. The spark advance mechanisms were a common source of mileage and performance problems. Vacuum spark advance units often leaked vacuum, and centrifugal advance units sometimes rusted and stuck.

Other Tune-Up Related Problems in Older Vehicles

Although most tune-up related problems on older vehicles were caused by the ignition system, many parts of the fuel system could cause trouble. Carburetors often lost their original efficiency because of deposits and wear. The idle speed and mixture required periodic readjustment as wear and deposits slowly changed the carburetor calibration.

In many parts of the country, older engines required adjustments to compensate for differences between winter and summer driving. Carburetor chokes, choke unloader mechanisms, and accelerator pump linkage required readjusting in the spring and fall. Voltage regulators were reset to produce higher charging rates in the winter, and lower in summer. Some manufacturers called for installing a higher temperature cooling system thermostat and removing the air conditioner drive belt in the fall, and installing a lower temperature thermostat and reinstalling the belt in the spring. If a carbureted vehicle was normally driven in high altitudes, it was common for local technicians to replace the jets and/or adjust the air-fuel mixture in order to gain better performance in the thin air.

Tune-Up Intervals

Although tune-ups had to be done often throughout the life of the vehicle, they were relatively easy to do. The parts most likely to cause trouble, such as the spark plugs, contact points, condenser, and rotor, were automatically replaced as part of preventive maintenance, eliminating some diagnosis time. Most adjustments made during a tune-up were simple mechanical adjustments, such as plug gap setting, point setting, or carburetor adjustment. Troubleshooting was relatively easy, usually involving visual checks or very simple performance tests. Since gasoline was inexpensive and there were no emissions checks, most owners took their vehicles in for a tune-up only when they noticed an obvious performance problem, such as a miss.

Tools and Equipment

Very few tools and pieces of test equipment were needed to perform tune-ups on older vehicles. Hand tools needed for a tune-up included a spark plug wrench and a screwdriver for removing and adjusting the points and setting the carburetor. Test equipment consisted of a spark plug gauge, dwell meter, timing light, and sometimes a vacuum gauge. Oscilloscopes were sometimes used, but mostly as time-saving diagnostic tools. Many technicians adjusted point sets by setting the point gap with a feeler gauge or even a matchbook cover. Sometimes the timing was set with a test light connected to the coil negative terminal. With the ignition on and the engine not running, the distributor was turned against rotation until the test light came on, showing that the points had opened. Occasionally, an experienced technician would set an engine's timing by simply listening to the engine idle and adjusting the distributor accordingly.

Modern Tune-Ups

Today, the periodic tune-up is a much different process, performed at much longer intervals, if at all. Tune-ups are no longer a process of changing parts that are known to wear out on a regular basis, but a process of determining whether any parts require service, and adjusting or replacing only those parts. Diagnosis is a much bigger factor in restoring engine efficiency than parts replacement. About the only ignition parts that are still replaced on a regular basis are the spark plugs, **Figure 21-2.** Unleaded gas and higher ignition voltages have extended the useful life of spark plugs from 30,000-40,000 miles (48 000-64 000 km), up to 100,000 miles (160 000 km) in some engines, **Figure 21-3.**

There are no contact points to adjust, wear, or replace. Without the constant pressure of the point rubbing block, distributor shafts and bearings are less likely to wear. The slight wear that could cause point bounce will not affect the operation of electronic ignition systems. While contact points can only open and close in relation to the engine's rotational speed, a solid state electronic ignition module can fire the ignition coil at any speed. This also eliminates the need to check dwell.

Spark plug wire insulation is thicker and more heat resistant than in the past. The wires are equipped with boots

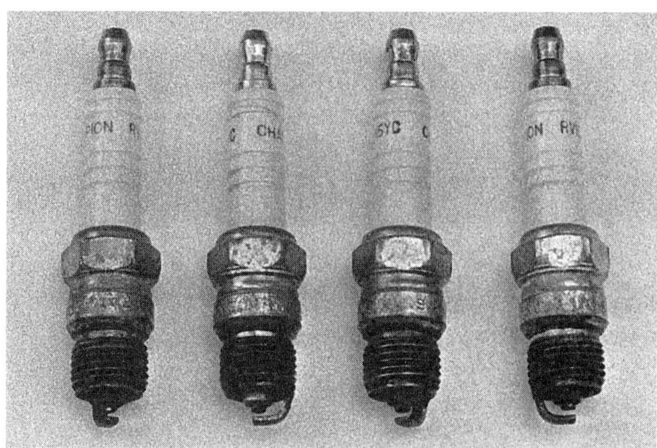

Figure 21-2. *Spark plugs are the most commonly thought of part when a tune-up is mentioned. On some engines, the spark plugs are one of the few items replaced in a tune-up.*

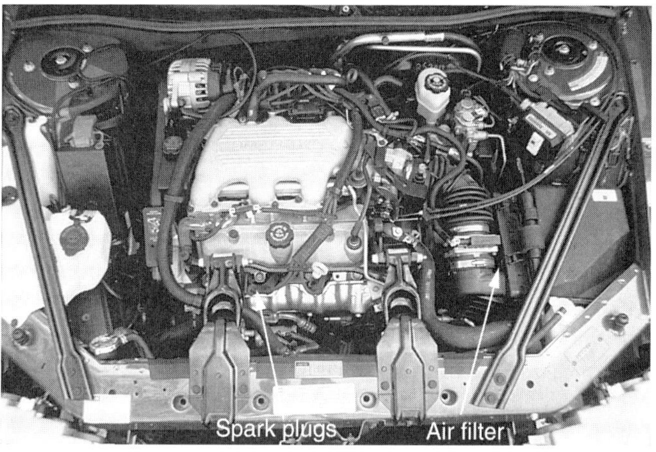

Figure 21-3. *Modern engine designs eliminate the mechanical parts that were the causes of many problems in older vehicles. However, when a problem does occur, it is most likely caused by a defective sensor or output device, rather than a part replaced in a tune-up. (General Motors).*

designed to provide a better seal against dirt and moisture. Distributor caps and rotors are made of better plastics and are much less likely to fail. Many modern engines are distributorless with no cap and rotor and sometimes, no plug wires. Spark advance mechanisms are built into the ECM and will automatically advance the spark as long as the sensors and other computer components operate properly. Since most modern ignition systems are computer-controlled, most engines do not need a timing adjustment during a tune-up. The timing light is used more often as a diagnostic tool than for setting timing during a tune-up.

Carburetors, where they are still used, have sealed idle adjustment screws, with the air-fuel mixture and idle speed adjusted by the ECM. Accelerator pumps, choke mechanisms, and power pistons are also nonadjustable. All of the newest engines are fuel injected and totally controlled by the ECM and related sensors.

Test Procedures and Adjustments

While parts replacement has been simplified, test procedures are more complex. Since most computer control system parts are expensive and do not fail at a set mileage, they must be carefully checked instead of being replaced as part of preventive maintenance. Testing for defective parts, therefore, is much more vital than it was previously. At the same time, testing procedures are much more involved. Instead of simple visual checks and tests, most modern components must be checked for resistance, voltage, or for the presence of waveforms. In many cases, special testers must be used to check individual components, or a similar group of components. In some cases, the parts can only be checked by a process of elimination or by substituting a known good part.

Instead of the simple adjustments on older vehicles, such as point gap or timing, modern adjustments often require a special gauge, tool, or meter. In some cases, the adjustment is similar to older procedures, such as adjusting the pickup coil air gap with a brass feeler gauge. Other electrical devices must be adjusted while monitoring a voltmeter or ohmmeter.

More test equipment is needed to perform the modern tune-up and this equipment is more complex than in the past. In addition to the usual oscilloscopes and timing lights, most modern tune-ups require a scan tool for retrieving trouble codes and interfacing with the ECM and its related sensors and output devices. Other needed equipment includes fuel injection pressure testers, high impedance digital multimeters, testlights, and electronic vacuum leak detectors. Common hand tools are still needed, but more special service tools must be on hand to replace and adjust the sensors, output devices, and fuel injection components. The frequent use of aluminum cylinder heads, intake manifolds, and blocks also makes retorquing a more commonly required tune-up procedure than previously.

The total tool and equipment investment for performing tune-ups is much higher than before. Additionally, much time must be invested in learning about new engine systems. For these reasons, some franchise and independent shops no longer perform tune-up or driveability work. In many shops, the tune-up is called a ***maintenance tune-up*** or an ***emissions tune-up*** and consists of changing the spark plugs and the air, fuel, and emissions filters. This tune-up is often done in response to the emission flags or lights which appear on the dashboards of some cars and light trucks at certain times or mileage intervals.

Tune-Up Procedures

The following section identifies and explains the steps involved in performing a maintenance tune-up on a modern vehicle. It will also briefly cover some of the processes used to perform tune-ups on older engines, and special tune-up procedures that may be necessary on some vehicles.

Note: If the vehicle has a performance problem of any type, you must investigate the complaint as a *driveability problem first*, using the diagnostic tests and methods you learned earlier in this text. In most cases, simply performing the tune-up will not correct the problem.

Replacing Spark Plugs

Even though spark plugs last much longer on newer vehicles, plug replacement is still performed several times during the life of the vehicle. Spark plug replacement is similar on both old and new vehicles. In the past, spark plugs were often taken out, cleaned, and reinstalled. However, it is recommended to simply replace the plugs if they show any signs of wear or deposits.

Caution: If the cylinder head is made of aluminum, allow the engine to cool completely before attempting to remove the spark plugs. Removing the plugs from a hot aluminum head may strip the threads.

Start by removing the plug wire from the first plug to be replaced. It is easy to mix the wires, so remove only one wire at a time. If the plug wire boot is stuck to the plug, try twisting the boot to break it loose from the plug. A special boot removal tool, **Figure 21-4,** will usually remove the boot without damage. While removing the plug wires, note any obvious damage, such as burned insulation or damaged boots.

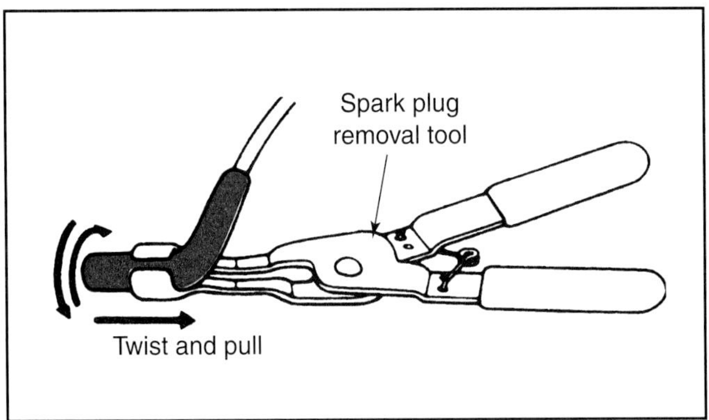

Figure 21-4. *Spark plug removal tools, such as these pliers, prevent spark plug boot damage and burns to your hands from hot metal parts in the close quarters of the modern engine compartment. (Ford)*

Once the wire is removed from the plug, install a proper size spark plug socket or wrench over the plug, and loosen it by turning counterclockwise. If the plug is tight, it may be necessary to give the ratchet or wrench handle a sharp jerk or tug to loosen it from the head. Turn the plug counterclockwise until it is free from the head, then place it to one side. As an aid to diagnosing engine problems, place the plugs in an undisturbed place in the order they came out of the engine. Checking spark plug condition was covered in **Chapter 16.**

Before the new plugs are installed, you may want to test cylinder compression to determine the engine condition. Compression testing was covered in **Chapter 19.** If any cylinders fail the compression test, the engine problem must be corrected before continuing the tune-up. Make sure that the replacement spark plugs are the proper type and heat range. The thread length and heat range should match exactly. Check the plug gap before installing them in the engine.

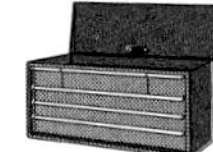

Note: Before installing spark plugs in an aluminum cylinder head, it is a good idea to coat the spark plug threads with anti-seize compound or a drop of clean engine oil.

Exercise care when installing spark plugs. The insulator will not stand much abuse and can crack easily, which can cause misfiring. If the plug wires need to be replaced, do so one at a time as you install the plugs. This decreases the chance of crossing the plug wires. Install the plugs by hand until tight, and turn them an additional 1/8 to 1/4 turn with the ratchet or wrench and reinstall the plug wires. See **Figure 21-5** for tightening information for different types of plugs.

Checking the Distributor Cap and Rotor

Skip this step if the engine has distributorless ignition. Once the new plugs are installed, remove the distributor cap. Check the cap interior and rotor for cracks, oil, rust, or carbon tracks, as shown in **Figure 21-6.** Replace any parts that appear to be worn or damaged.

Checking Vacuum Advance

The vacuum advance, if used, should be checked with a vacuum pump as outlined in **Chapter 16**. If the plate does not move when vacuum is applied, the diaphragm is defective and the vacuum advance unit should be replaced.

Checking the Centrifugal Advance

If the engine has a centrifugal advance mechanism, check it for free movement. Turn the rotor in the same direction as normal distributor rotation, **Figure 21-7.** It should travel about .50″ (12.7 mm) and snap back to its original position when released. If the rotor does not move, the centrifugal advance is stuck, and must be disassembled, cleaned, and any worn parts replaced. If the centrifugal advance is operating properly, lubricate it according to the manufacturer's instructions.

Plug Torque Values			
Plug Thread Type	Torque, Cast Iron Heads	Torque, Aluminum Heads	Turns Past Snug*
10mm (tapered seat)	8-12 ft.-lbs.	8-12 ft.-lbs	1/16-1/4
14 mm (gasketed seat)	25-30 ft.lbs.	18-22 ft. lbs.	1/2-3/4
14 mm (tapered seat)	10-20 ft.-lbs.	7-15 ft.-lbs.	1/16-1/4
18 mm (gasketed seat)	22-28 ft.-lbs.	10-20 ft.-lbs.	1/2-3/4
18 mm (tapered seat)	15-20 ft.-lbs.	10-20 ft.-lbs.	1/16-1/4
*When no torque wrench is available			

Figure 21-5. *When installing spark plugs, make sure that they are torqued to the correct setting. Since many plugs are located in hard-to-reach areas, a set amount of turns past snug is normally used.*

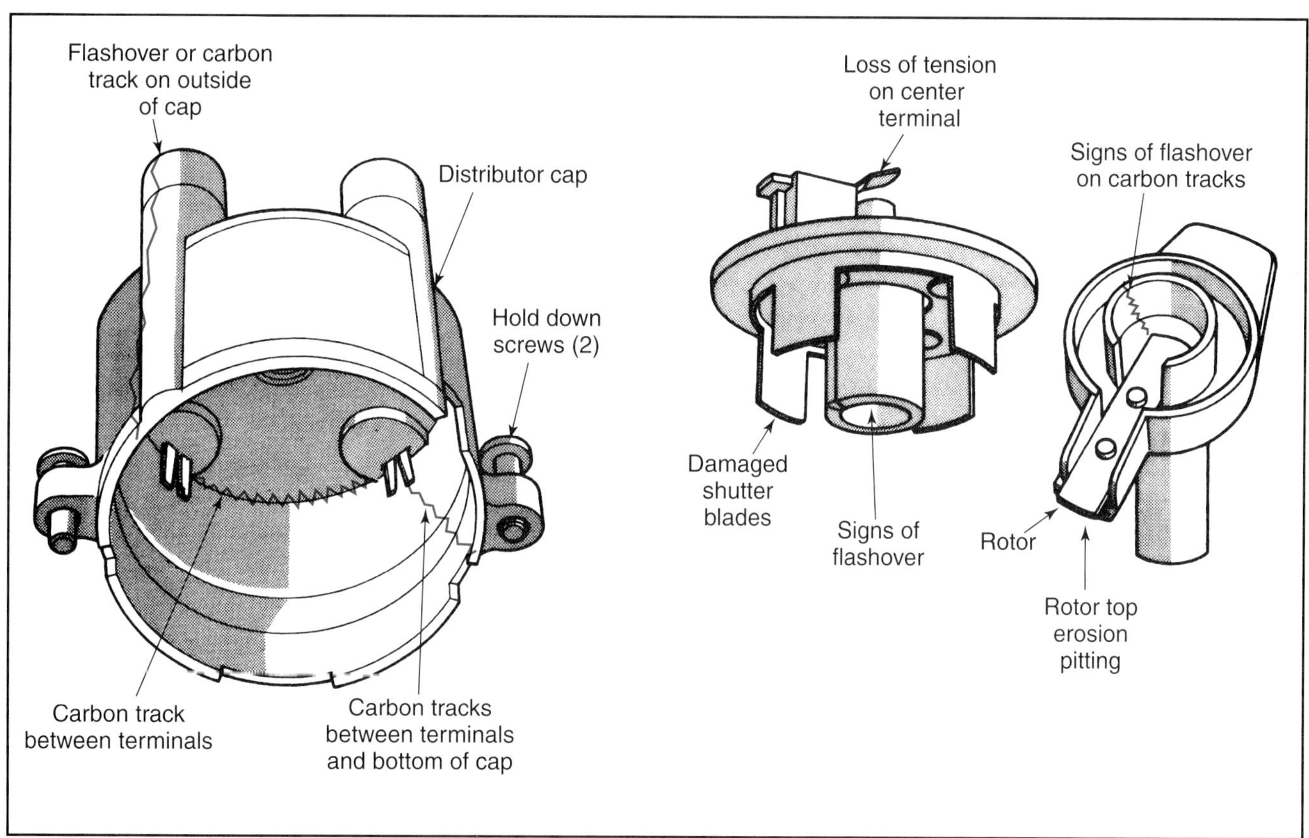

Figure 21-6. *After removing the distributor cap and rotor, inspect both carefully. Shown here are problems typically found on many distributor caps and rotors. Be sure to replace them as a set. (Chrysler)*

Checking and Setting Air Gap

On some electronic ignitions using a pickup coil and trigger wheel, the air gap between the coil and trigger wheel is adjustable. The gap can be adjusted with a set of brass feeler gauges. Steel feeler gauges should not be used as they will be attracted to the pickup coil magnet and drag, giving a false reading. Steel gauges may also magnetize the trigger wheel, causing erratic ignition system operation.

Adjustable pickup coils have an adjustment screw and a hold-down screw, similar to the screw arrangement used on many contact points. The adjustment procedure is similar

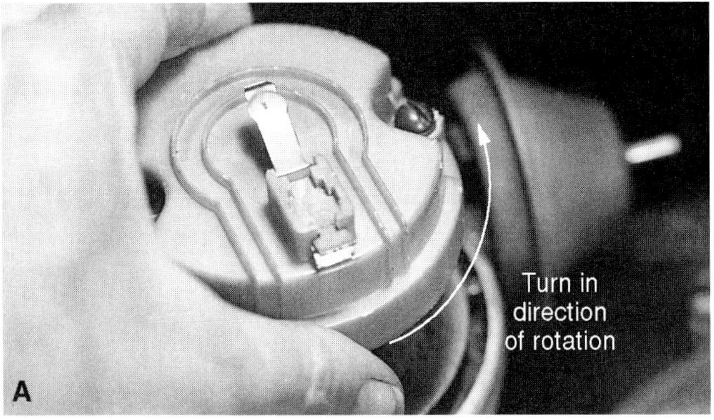

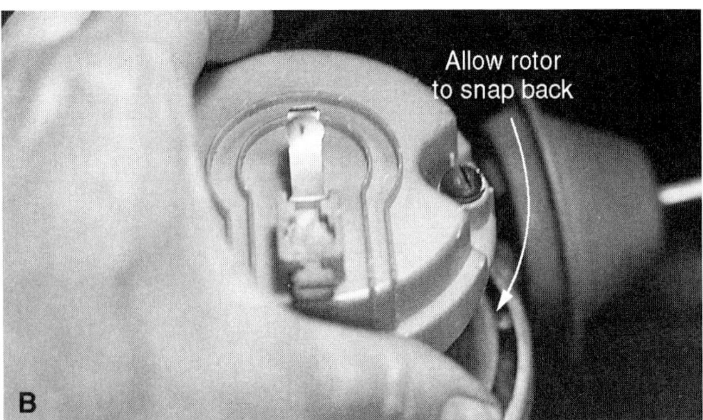

Figure 21-7. *A—To check the centrifugal advance, turn the rotor about .5 in (12.7 mm) in the direction of engine rotation. B—When you release the rotor, the springs in the centrifugal advance should force the rotor to snap back into the original location.*

to that used to set the gap on contact points. Begin by removing the distributor cap, and checking the distributor shaft is not worn.

After the distributor cap is removed, locate the hold-down and adjustment screws. Carefully turn the engine either manually or with the starter until one of the teeth on the trigger wheel is lined up directly across from the pickup coil's stationary tooth. Check the gap by inserting the brass feeler gauge between the two teeth, as shown in **Figure 21-8.** If the air gap is incorrect, slightly loosen the hold-down screw, and turn the adjusting screw to increase or decrease the gap. When the feeler gauge can be pulled between the trigger wheel tooth and stationary tooth with a slight drag, the air gap is adjusted. Tighten the hold-down screw and recheck the air gap. Recheck the gap at several trigger wheel teeth to ensure that the trigger wheel is not out of position on the distributor shaft, and as a double-check that the shaft is not worn.

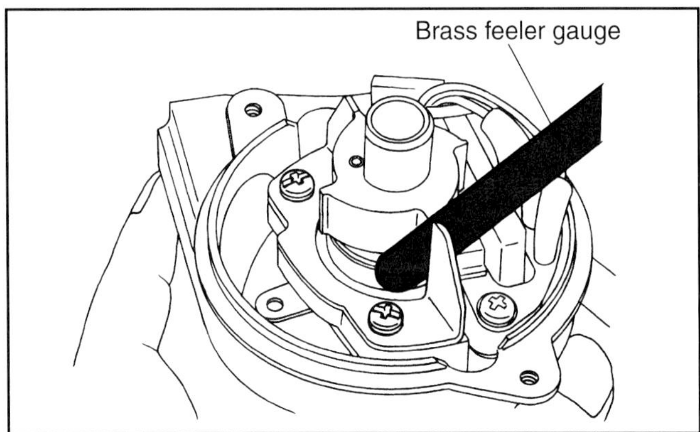

Figure 21-8. *Checking the air gap between the pickup coil and trigger wheel is normally done as part of a repair. However, it should be done during a maintenance tune-up to ensure that no distributor parts have shifted or are worn. (Subaru)*

Changing Filters

Filter replacement is a common part of most tune-ups. The average vehicle is equipped with many filters for removing debris from the engine oil, air, fuel, transmission fluid, and sometimes the power steering fluid and air conditioner refrigerant. The three major types of filters which can affect driveability are the air filter, fuel filter, and emissions filter. These three filters must be changed on a periodic basis to maintain engine performance. This section will discuss the methods of changing these filters.

Air Filters

Changing the ***air filter*** is relatively easy. Filter housings are located at the top of the engine, or off to one side. Begin by removing the wing nut, clips, or screws holding the air cleaner cover in place. Remove the cover and lift the air filter out, **Figure 21-9.** Check the filter and housing for any signs of oil or water, **Figure 21-10.** If the filter and housing is coated with engine oil, do not proceed further with the tune-up as this is evidence of blowby. Advise the driver that the engine is in need of work. A filter that appears to have been saturated with water indicates that the vehicle has been driven through deep water. Advise the driver of the potential engine damage that could be caused by hydrostatic lockup (water drawn and compressed in the engine cylinders).

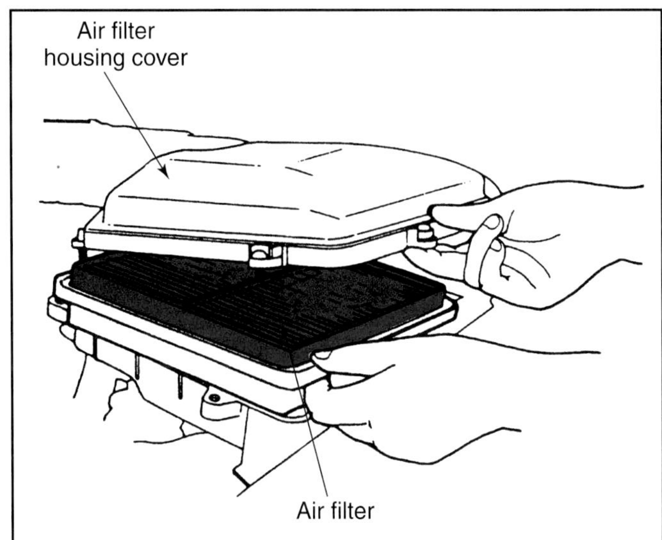

Figure 21-9. *Air filter elements are relatively easy to replace in most cases. Be sure to install the new filter correctly. (Geo)*

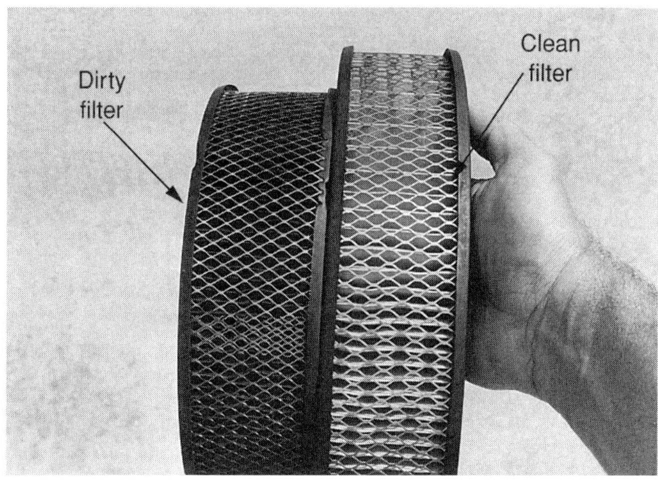

Figure 21-10. *The old filter should be inspected for oil, water, and other contamination that could damage the engine or may be a sign of engine wear. Compare the old filter to the new one to be sure you have the correct replacement.*

Before installing the new element, wipe out the housing to remove any loose dirt, or remove the housing from the vehicle and blow it clean with compressed air. If the emissions filter is inside the air cleaner housing, replace it before reinstalling the cover. This procedure is covered later in this section. Then install the new air cleaner element, and tighten the wing nut or reattach the fasteners.

Caution: Most intake air temperature sensors are located in the air filter housing. Be sure that the sensor terminal is connected after replacing the air filter.

Fuel Filters

Fuel filters are attached to the fuel inlet line with tubing fittings, **Figure 21-11A,** either by hoses and clamps, **Figure 21-11B,** or with special connectors, **Figure 21-11C.** When the fuel filter requires replacement, the first step is to locate the filter. Most modern filters are located on the underside of the vehicle body, and may not be easily spotted. The quickest method when working on a vehicle that you are not familiar with is to trace the fuel line back from the engine.

Once the fuel filter is located, one of several methods is used to remove the filter from the fuel lines, depending on the filter's design. In some cases, the filter is held to the vehicle by a clamp, which should be removed before the lines are removed. Most clamps are held by a single bolt. When replacing filters that are attached by hoses, loosen the clamps and slide the filter and old hoses from the fuel lines. It may be necessary to twist the hoses slightly to loosen them from the lines. When installing the new filter, always use new hoses and clamps.

When replacing filters attached by metal tube fittings, use two wrenches (use a flare-nut wrench on the fuel line) on each fitting to prevent twisting the fuel lines,

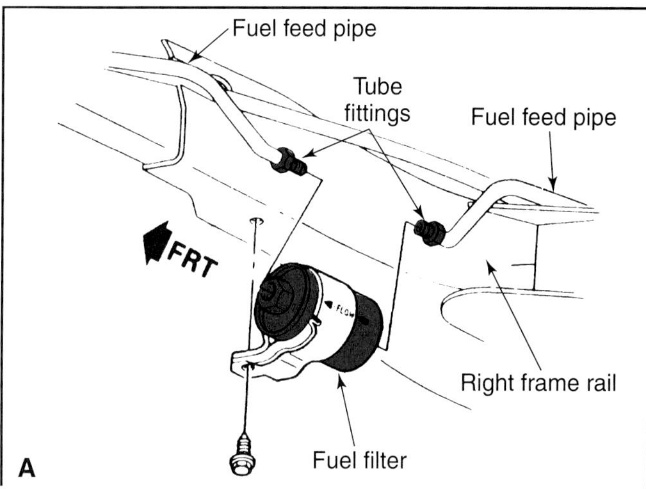

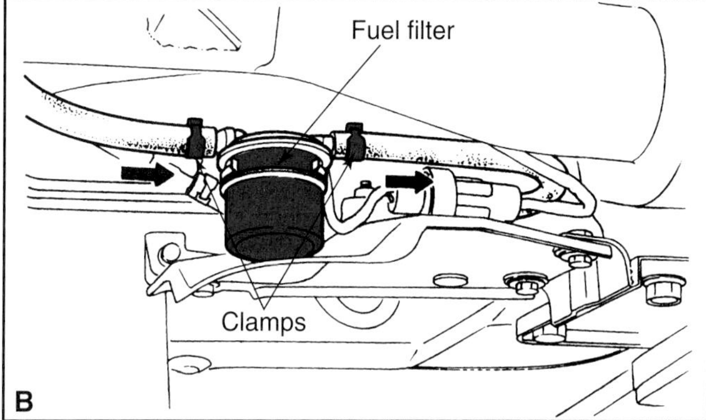

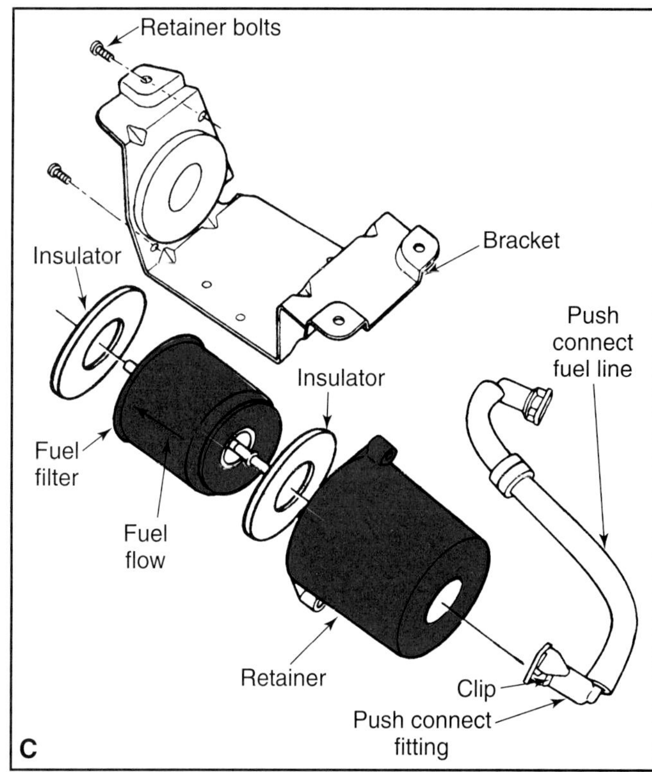

Figure 21-11. *There are several different types of fuel filters used by modern vehicles. A— Tubing connections are used to connect the filter with many fuel systems. B—Most fuel injected vehicles do not use clamps and neoprene fuel lines, but can be found on a few vehicles. C—Push-on fittings are used on almost all vehicles equipped with plastic fuel lines. (General Motors, Subaru, Ford)*

Figure 21-12. This is especially important when changing diesel filters, since the lines are internally coated, and the coating could flake off when the lines are bent. In some cases, it may be necessary to use a liquid Teflon® sealant on the fitting threads when installing the new filter. Be sure that the fittings are not crossthreaded into the filter.

Some filters are attached with special push-on fittings. Carefully press the tab or clip and remove the line from the filter. The removal process for lines using this type of fitting is shown in **Figure 21-13.** Sometimes a special tool must be used to remove them from the filter. Once the new filter is in place, the lines can be reconnected by pushing them onto the filter fittings. Some manufacturers supply new clips with the replacement filter.

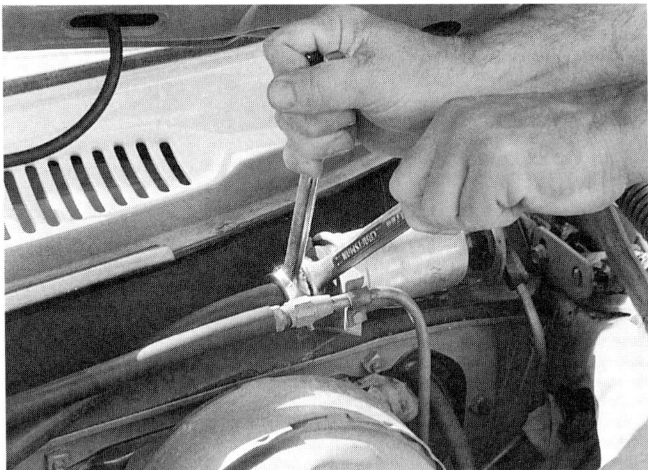

Figure 21-12. *Use a flare-nut wrench when loosening any tubing fitting. This reduces the possibility of rounding off a tube fitting.*

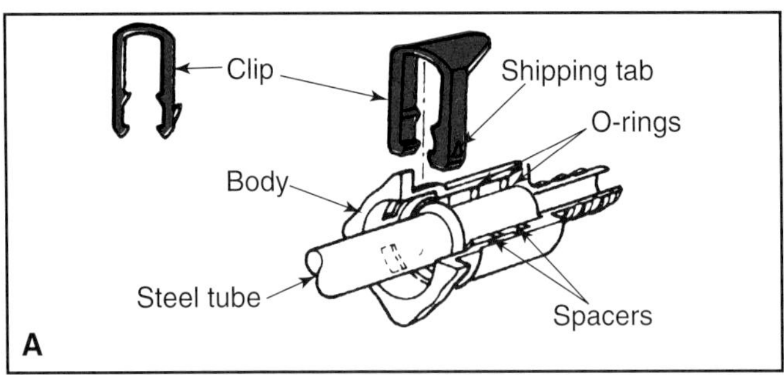

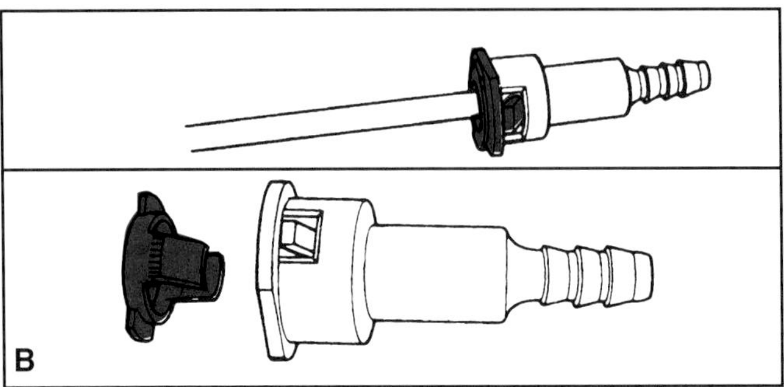

Figure 21-13. *Push-on connectors either have a clip that is removed or a connector that is part of the fitting. Exercise care when handling these connections. (Ford)*

Caution: Do not twist, bend, or kink a plastic fuel line as this can cause a permanent restriction.

Emissions Filters

Emissions filters are used to prevent dust from entering the engine crankcase or fuel system through the PCV system. Most emissions filters are installed as part of the PCV system and should be replaced whenever the PCV valve is replaced, or whenever they appear to be oil or dirt clogged. Most PCV system filters are installed in the air cleaner housing, **Figure 21-14.** They are usually held to the housing by a large sheet metal clip, and can be removed after the clip is pulled from the housing.

Filters are used as part of the evaporative emissions system, and are installed on the vapor canister. They are usually held in place by the canister housing and can be easily pulled out and replaced. In many cases, the evaporative emissions vapor canister is installed in an inaccessible place, and must be removed to replace the filter.

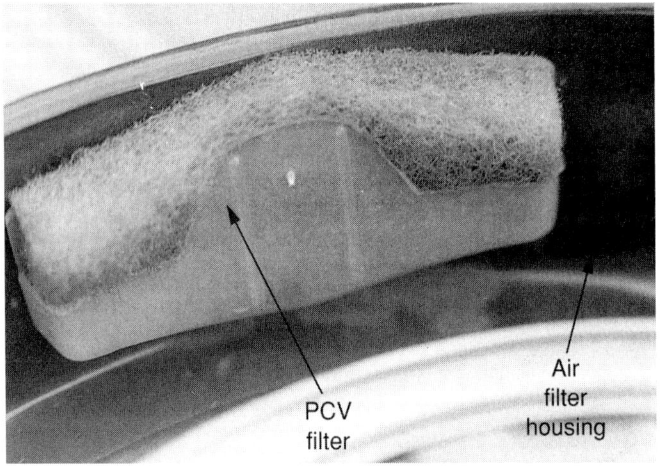

Figure 21-14. *The PCV valve vents gases from the crankcase into the air cleaner. A filter is often located at the entrance to the air cleaner to prevent foreign particles from entering the intake. This filter is the most neglected part in a tune-up.*

Fuel System Maintenance

Like the rest of the engine, the fuel system must receive routine care to ensure maximum performance. In most cases, this involves fuel filter replacement, discussed earlier. Older vehicles equipped with carburetors were often cleaned, adjusted, or rebuilt, depending on the need. Maintenance on fuel injected engines is limited to throttle body cleaning and flushing the fuel injectors.

Throttle Body Cleaning and Maintenance

On both throttle body and multiport fuel injection systems, carbon and gum will build up inside the idle air bypass chamber and just inside the throttle plate opening. The carbon and gum tend to form a ridge around the throttle plate. Eventually the ridge will build up, if not removed, to the point that it will open the throttle plate slightly. This slight opening of the throttle plate can affect overall engine performance by changing the idle speed, forcing the ECM to decrease the idle air control valve opening. This can result in rough idle, stalling, and other performance problems.

To remove this ridge from the throttle plate opening, as well as maintain overall throttle body cleanliness, many technicians use a ***throttle body spray cleaner*** to clean the throttle body. The spray cleaner is also powerful enough to remove the build-up at the throttle opening. A toothbrush can be used to scrub off heavy deposits. Multiport injector throttle bodies can be manually cleaned by removing the intake hose and spraying the throttle body and throttle valve with spray cleaner. Be sure to open the throttle valve to get at deposits behind the valve, **Figure 21-15** .

Caution: Do not spray any type of cleaner through a mass airflow sensor. The chemicals in the spray cleaner will damage the sensor. Remove the sensor before cleaning the throttle body. Do not spray cleaner into the intake manifold or plenum.

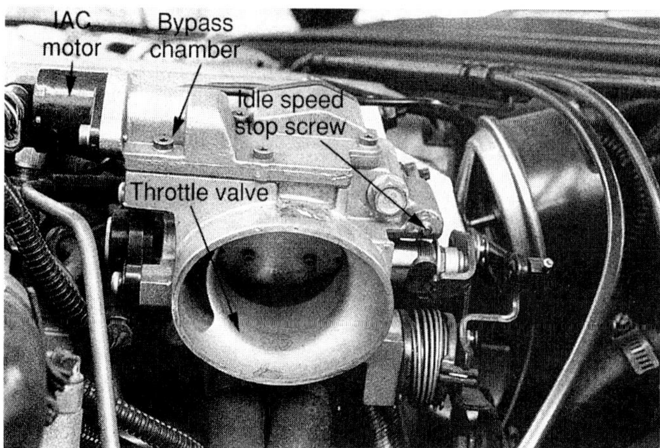

Figure 21-15. *The throttle body should be cleaned whenever a tune-up is performed. This prevents gum and varnish from building on the throttle plate. Do not spray cleaner down into the intake.*

Normally component removal is not necessary, however, it is a good idea to remove the IAC valve and clean the IAC pintle as well as the air bypass area. Some manufacturers do not recommend this procedure, so check the service manual. Try to spray as little cleaner directly into the intake plenum as possible. After all deposits have been removed, reattach any parts that were removed and start the engine. Run the engine at a high idle to purge the intake manifold and engine of cleaner. Further cleaning must be done with the throttle body removed.

Once the throttle body is clean, reinstall all parts and reset the IAC and minimum idle speed according to manufacturer's specifications. If the spray cleaner cannot sufficiently clean the throttle body, it will have to be disassembled and cleaned. This procedure was covered in **Chapter 17.**

Cleaning the Fuel Injectors

The openings inside the fuel injectors are extremely small and may become plugged by dirt or deposits. This can result in a variety of driveability problems. Fuel injectors can be cleaned by removing them from the vehicle, soaking them in cleaner, and blown dry with air pressure. However, this is time-consuming, expensive, and should only be done during a complete throttle body cleaning. In addition, the soaking process may damage internal seals and O-rings not designed to withstand immersion in harsh solvents for long periods. Fuel injector cleaners can be poured into the fuel tank to help prevent the formation of deposits, but most are not strong enough to remove heavy deposits once they are formed. The most positive way to clean the injection system is by using an injector cleaner device, such as that shown in **Figure 21-16.** This procedure was discussed in **Chapter 17.**

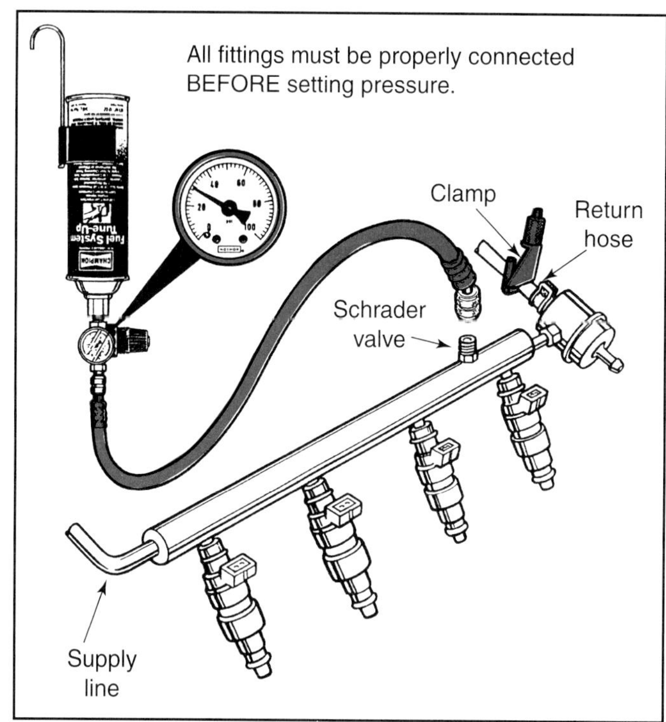

Figure 21-16. *Fuel injectors are often cleaned as part of a tune-up. Remember to check the service manual as some injectors do not need to be cleaned. (Champion Spark Plug)*

Caution: Most manufacturers no longer recommend injector cleaning. Check the service manual before performing any cleaning procedure.

Adjusting Fuel Injection Idle Speed

On newer engines with fuel injection, the ECM controls the opening, or pulse width, of the fuel injectors and is not adjustable. Idle speed is controlled by the ECM through an idle air solenoid or air bleed system. If the IAC valve was removed during throttle body cleaning, the idle speed should be reset. On most engines, this is done by clearing the ECM memory and driving the vehicle. Other engines require a specific idle speed reset procedure, so check the service manual. However, do not adjust the idle speed stop screw unless absolutely necessary to correct a previous adjustment. Idle speed stop screws are usually sealed and used as a fail safe device to maintain a minimum idle if the computer control system fails. In many cases, turning this screw to move the throttle plate will cause the ECM to compensate by adjusting the airflow through the idle air system.

Exhaust Gas Analyzer

An exhaust gas analysis is a good idea whenever any driveability work is completed. As you learned earlier in this text, it is also effective when used to monitor exhaust emissions during mixture adjustments. The basic procedure for exhaust gas analyzer use was covered in **Chapter 18**. The emissions level varies with different states and model years, and the proper specification book should be consulted before adjusting the mixture.

Setting Initial Timing

On older contact point-type systems, the timing was set any time the points were replaced or adjusted. However, the initial ignition timing on an engine equipped with electronic ignition usually does not change. When any other ignition system service is being done, the timing should be checked in case the distributor housing has slipped, or the timing was set incorrectly in the past.

Note: The timing of some modern engines cannot be checked. On these engines, the timing is monitored by pickup devices on the crankshaft or camshaft and controlled by the ECM. Some of these engines do not have timing marks or timing probe receptacles.

Methods of Timing Engines

The timing of most engines can be checked by one of two methods. The most common method is to use a timing light, as explained in **Chapter 16.** A timing meter, also explained earlier, can be used on many engines. Always obtain and review the manufacturer's timing specifications and procedures before setting timing.

Once all preliminary steps have been taken, connect the timing light and clean the timing marks. If the marks are difficult to read, mark them with light colored chalk or paint. Then start the engine and point the timing light at the timing marks, **Figure 21-17.** The timing marks should be at or near the specified position. If the timing appears to be off, recheck all preliminary steps, such as disconnecting hoses or electrical connectors, and setting idle speed, then recheck the timing. If the timing must be adjusted, loosen the distributor hold-down bolt and move the distributor housing until the timing marks are correctly aligned, **Figure 21-18.** Then tighten the hold-down bolt and recheck the timing to ensure the distributor did not slip when the bolt was tightened.

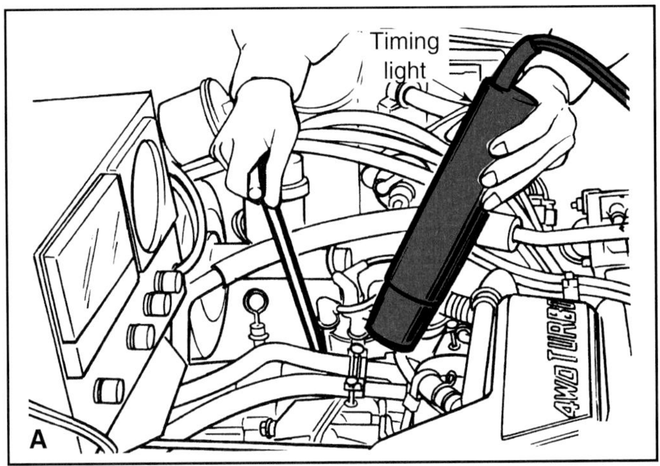

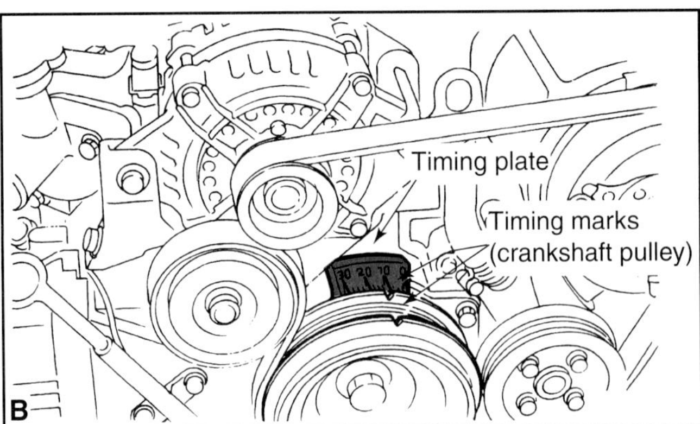

Figure 21-17. *A—A timing meter or timing light can be used to check and set engine timing. B—Before connecting the light or meter, locate and clean the timing marks. Some engines cannot be timed. (Subaru)*

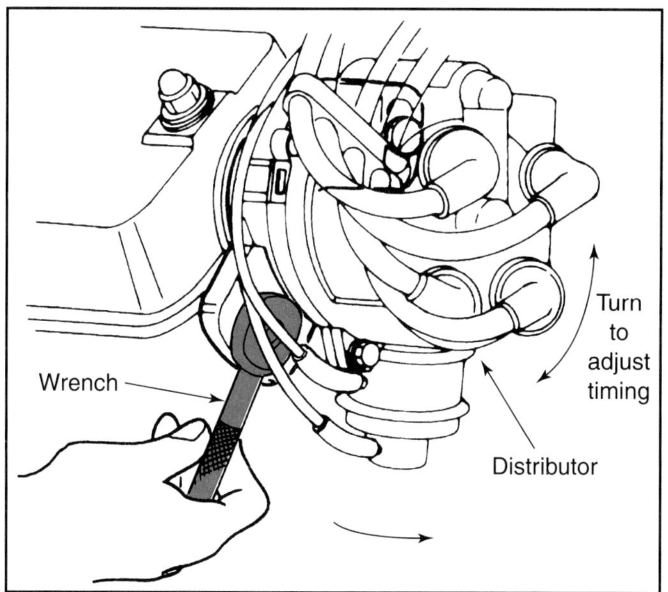

Figure 21-18. *If the timing can be adjusted, loosen the distributor bolt and turn the distributor until the engine is timed properly, and recheck. (Chevrolet)*

Figure 21-19. *Most timing lights are equipped with advance meters, which can be used to check the ignition advance.*

Checking Timing Advance

Correct initial ignition timing and timing advance are critical to engine performance, mileage, and driveability. The timing advance should be checked to ensure that the advance is working, whether it is mechanical, vacuum, or electronic. The timing advance can be checked with a timing light having a built-in advance meter, or with a timing meter on many newer engines. The timing advance of some modern engines cannot be checked because these engines do not have timing marks or timing probe receptacles.

Always refer to the manufacturer's specifications for the amount of advance and engine speed before beginning the timing procedure. The timing on most engines will advance at least 40° at 2500 rpm. Be certain that the initial timing is set correctly, and that any vacuum lines or electrical connectors that were removed to set timing are reconnected. Then connect the timing light or meter. Start the engine and accelerate it to the rpm specified for total advance checking. If you are using a timing light with an advance meter, point it at the timing marks and turn the thumbwheel on the light until the timing marks are aligned on the zero position, then read the advance on the meter, **Figure 21-19.** If you are using a timing meter, read the advance on the meter face. If the timing is not advancing according to specifications, investigate the possible causes using the diagnostic techniques outlined earlier in this text.

PCV Valve Replacement

PCV valve replacement on most vehicles is fairly simple. However, PCV valves on some vehicles are located in an inaccessible area and cannot be replaced without moving one or more parts or wiring. Some PCV valves are located in a recess inside the intake plenum of some multiport fuel injected engines.

To replace the valve, remove any components or wiring necessary to access the valve. Remove the hose from the valve, grasp the valve and pull it out of the engine. If the valve sticks to the grommet, twist the valve while pulling up. After the PCV valve is removed, check the grommet for cracks and leaks. Replace the grommet if any are found. PCV valve replacement is in the reverse order of removal.

Summary

A tune-up is the process of restoring engine efficiency by troubleshooting problems, and the replacement and/or adjustment of minor parts. The tune-up has changed greatly in the past quarter century. In the past, tune-ups had to be done often throughout the life of the vehicle, but they were relatively easy to do. The parts most likely to cause trouble were automatically replaced as part of preventive maintenance. Most adjustments made during a tune-up were simple mechanical adjustments. Troubleshooting was relatively easy, usually involving visual checks or very simple performance tests. Since gasoline was inexpensive and there were no emissions checks, most owners took their vehicles in for a tune-up only when they noticed an obvious performance problem.

The older engine tune-up required very little equipment. Hand tools were limited to a spark plug wrench and a screwdriver for removing and adjusting the points and setting the carburetor. Test equipment consisted of a spark plug gauge, a dwell meter, a timing light, and sometimes a vacuum gauge. Oscilloscopes and exhaust gas analyzers were sometimes used, but mostly for diagnosis.

The modern tune-up is less tedious, requiring much less adjusting and setting than the older version, and is done at much longer intervals, if at all. However, the modern tune-up is more likely to test the technician's knowledge and diagnostic ability. Instead of replacing ignition or fuel system parts that fail on a regular basis, the modern technician must determine which, if any, parts are really defective, and if they are causing the complaint. Diagnosis is a much bigger factor in restoring engine efficiency. About the only tune-up parts that are still replaced on a regular basis are the spark plugs.

While parts replacement has been simplified, test procedures are more complex. Most computer control system parts must be carefully checked instead of being replaced periodically. Testing for defective parts is more vital, and more involved, than previously. Special testers must often be used to check components. In many cases, the parts can only be checked by a process of elimination or by substituting a known good part. Modern adjustments often require a special gauge or meter. In some cases, the adjustment is similar to older procedures, but may involve using a voltmeter, ohmmeter, or other electrical testers.

More test equipment is needed to perform the modern tune-up and this equipment is more complex. In addition to the traditional equipment, modern tune-ups require scan tools, fuel injection pressure testers, high impedance multimeters, and electronic vacuum leak detectors. Common hand tools are still needed, but more special tools must be on hand to test, replace, and adjust the sensors, output devices, and fuel injection components. The total tool and equipment investment for performing tune-ups is much higher than it was previously. Additionally, much time must be invested in learning about new engine systems. For this reason, many shops no longer perform driveability work or tune-ups.

Know These Terms

Tune-up
Maintenance tune-up
Emissions tune-up
Air filter
Emissions filter
Throttle body spray cleaner

Review Questions—Chapter 21

Please do not write in this text. Write your answers on a separate sheet of paper.

1. *True or False?* A tune-up is a sure way to cure performance and driveability problems.
2. Does it take more or less test equipment to perform a tune-up on a newer vehicle? Why?
3. When removing spark plugs from an aluminum cylinder head, be sure that the engine is ______.
4. Always check the _____ of new plugs before installing them in an engine.
5. List at least four items the technician should check after new spark plugs are installed.
6. Before setting the initial timing, you should:
   (A) check the distributor cap and rotor.
   (B) check the centrifugal and vacuum advance mechanisms.
   (C) mark the timing marks with chalk or paint.
   (D) All of the above.
7. When making fuel system adjustments, you should use a(n):
   (A) exhaust gas analyzer.
   (B) multimeter.
   (C) timing light.
   (D) None of the above.
8. Name the three major filters that should be replaced during a tune-up.
9. You should clean fuel injectors by using ______.
   (A) parts cleaning solvent
   (B) a special injector cleaning solution
   (C) carburetor cleaner
   (D) All of the above, depending on the injector design.
10. After cleaning the throttle body, what valve should you reset?

ASE Certification-Type Questions

1. The newest vehicles require a tune-up every ______.
   (A) 12,000 miles (19 200 km)
   (B) 30,000 miles (48 000 km)
   (C) 100,000 miles (160 000 km)
   (D) Both B & C.
2. All of the following parts are replaced during a modern tune-up, EXCEPT:
   (A) spark plugs.
   (B) PCV valve.
   (C) fuel filter.
   (D) fuel injectors.

3. Technician A says that the total tool and equipment investment for performing tune-ups is less than in the past. Technician B says that it is less expensive and more reliable to simply clean the spark plugs and reinstall them. Who is right?
   (A) A only.
   (B) B only.
   (C) Both A & B.
   (D) Neither A nor B.
4. Spark plugs last longer because of all of the following, EXCEPT:
   (A) smaller engines.
   (B) unleaded gas.
   (C) higher ignition voltages.
   (D) higher engine speeds.
5. Technician A says that spark plugs should only be removed from an aluminum head after it has cooled. Technician B says that spark plugs should be removed from an aluminum head while it is hot. Who is right?
   (A) A only.
   (B) B only.
   (C) Both A & B.
   (D) Neither A nor B.
6. The dwell must be checked on an engine with ______.
   (A) contact points
   (B) a Hall-effect switch
   (C) distributorless ignition
   (D) direct ignition
7. All of the following will occur if a throttle body is not cleaned, EXCEPT:
   (A) the idle speed will change.
   (B) the ECM will increase minimum air rate.
   (C) the ECM will decrease minimum air rate.
   (D) a ridge will build up around the throttle plate.
8. Technician A says that it is ok to spray throttle body cleaner through a mass airflow sensor. Technician B says that it is ok to remove an IAC valve to clean the pintle and the bypass port. Who is right?
   (A) A only.
   (B) B only.
   (C) Both A & B.
   (D) Neither A nor B.
9. Technician A says that the idle mixture on a fuel injected engine can be set by the lean drop method. Technician B says that the idle mixture on an older vehicle can be set with an exhaust gas analyzer. Who is right?
   (A) A only.
   (B) B only.
   (C) Both A & B.
   (D) Neither A nor B.
10. The best way to clean fuel injectors is to ______.
   (A) soak them in carburetor cleaner
   (B) pour cleaners in the gas tank
   (C) use a special cleaning solvent
   (D) None of the above.

Advanced Certification Questions

The following questions require the use of reference information to find the correct answer. Most of the data and illustrations you will need is in this chapter. Diagnostic trouble codes are listed in **Appendix A.** Failure to refer to the chapter, illustrations, or **Appendix A** for information may result in an incorrect answer.

11. A fuel injected van with distributorless ignition is in for a maintenance tune-up. During the initial road test, the malfunction indicator light comes on with no noticeable change in vehicle performance. Scan tool information indicates that diagnostic code P0133 was set. All of the following should be done next, EXCEPT:
   (A) clear the ECM memory of stored codes.
   (B) road test the vehicle again.
   (C) visually inspect and adjust the throttle position sensor, if possible.
   (D) visually inspect the oxygen sensor for proper connection.
12. After performing a maintenance tune-up on a multiport fuel injected truck, it has a miss and backfires on acceleration. Which of the following is the *least likely* cause?
   (A) A cracked spark plug insulator.
   (B) Crossed spark plug wires.
   (C) Improper spark plug gap.
   (D) Spark plug heat range too cold.

13. The malfunction indicator light comes on after a maintenance tune-up on a throttle body injected engine. Diagnostic codes P0110, P0111, and P0114 are stored in the ECM's memory. Technician A says that the intake air temperature sensor was disconnected during the tune-up. Technician B says that the manifold absolute pressure (MAP) sensor was disconnected during the tune-up. Who is right?
    (A) A only.
    (B) B only.
    (C) Both A & B.
    (D) Neither A nor B.

14. The air filter on a multiport injected truck engine appears to have been water soaked. Which of the following is the *most likely* cause?
    (A) Driving the vehicle through deep water.
    (B) An engine coolant leak.
    (C) Excessive blowby through the throttle body.
    (D) A leak in the air transfer hose.

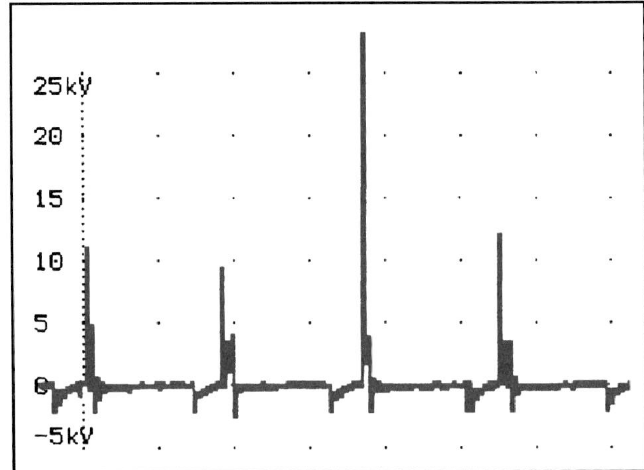

15. A vehicle with a direct ignition system has a rough idle and is in for a maintenance tune-up. The oscilloscope pattern shown above was taken at idle speed. Which of the following is the *most likely* cause?
    (A) A defective ignition coil.
    (B) An open spark plug wire.
    (C) Worn spark plugs.
    (D) Fuel injector wiring harness crossed.

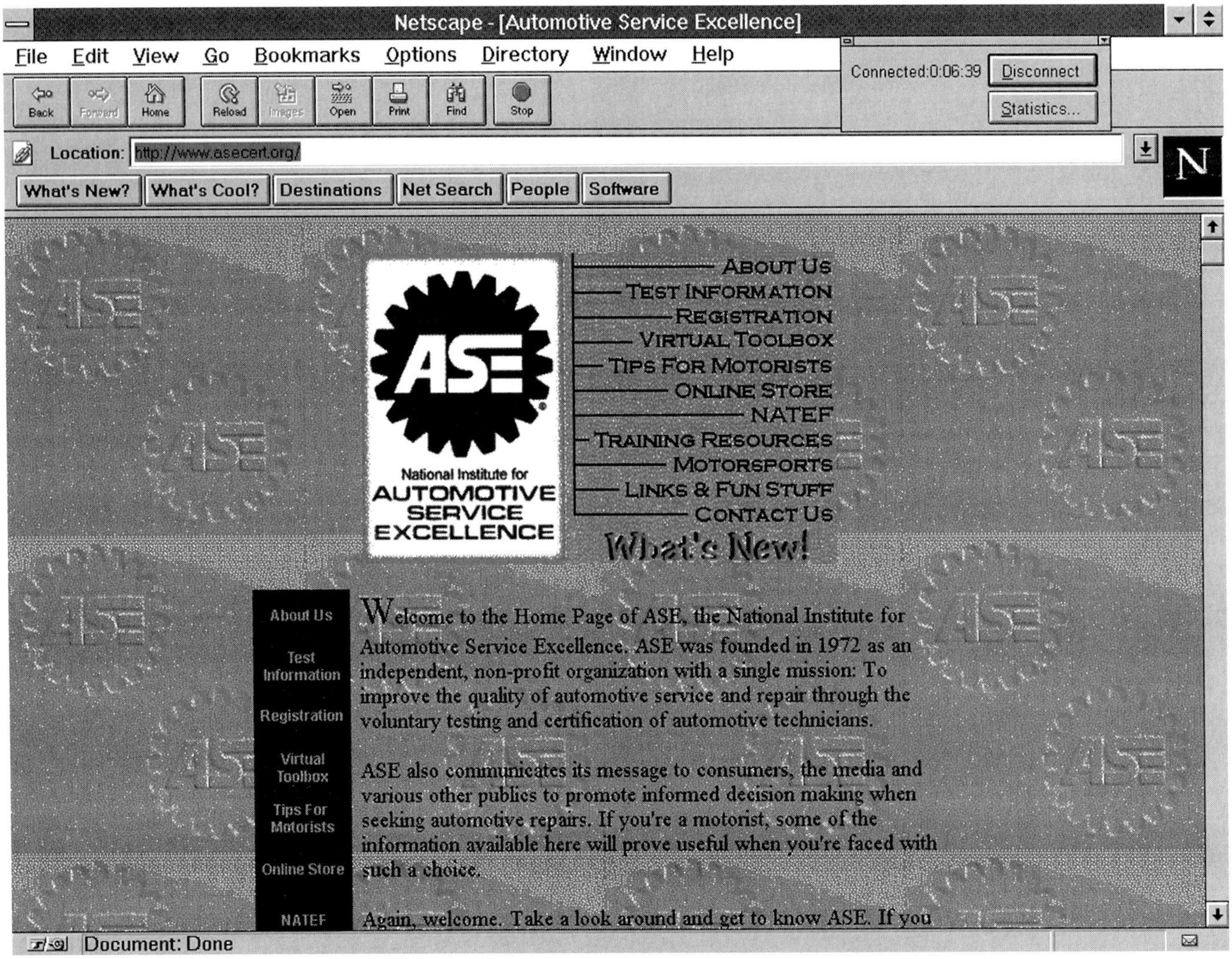

Netscape - [Automotive Service Excellence]

File Edit View Go Bookmarks Options Directory Window Help

Connected:0:06:39 Disconnect Statistics...

Back Forward Home Reload Images Open Print Find Stop

Location: http://www.asecert.org/

What's New? What's Cool? Destinations Net Search People Software

ABOUT US
TEST INFORMATION
REGISTRATION
VIRTUAL TOOLBOX
TIPS FOR MOTORISTS
ONLINE STORE
NATEF
TRAINING RESOURCES
MOTORSPORTS
LINKS & FUN STUFF
CONTACT US

ASE National Institute for AUTOMOTIVE SERVICE EXCELLENCE

What's New!

About Us
Test Information
Registration
Virtual Toolbox
Tips For Motorists
Online Store
NATEF

Welcome to the Home Page of ASE, the National Institute for Automotive Service Excellence. ASE was founded in 1972 as an independent, non-profit organization with a single mission: To improve the quality of automotive service and repair through the voluntary testing and certification of automotive technicians.

ASE also communicates its message to consumers, the media and various other publics to promote informed decision making when seeking automotive repairs. If you're a motorist, some of the information available here will prove useful when you're faced with such a choice.

Again, welcome. Take a look around and get to know ASE. If you

Document: Done

You can find information about ASE certification over the information superhighway at ASE's official World Wide Web site. (ASE)

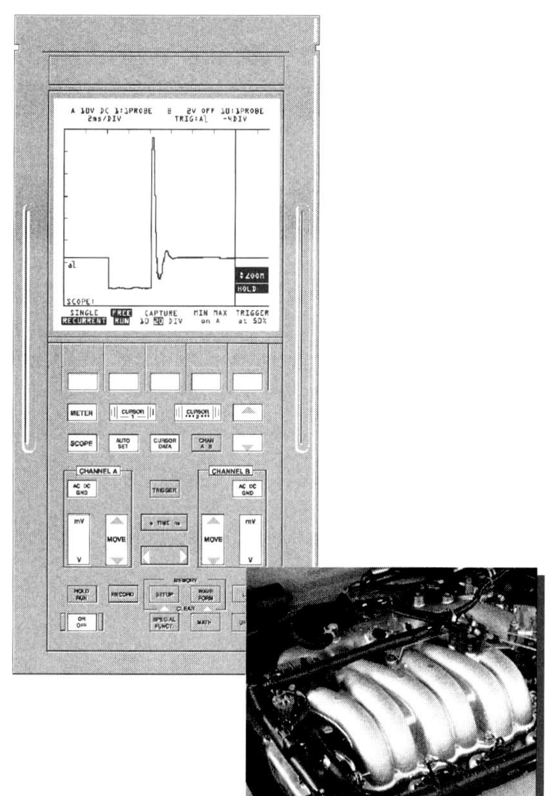

Chapter 22

ASE Certification

After studying this chapter, you will be able to:

- Explain why technician certification is necessary.
- Explain the process of registering for ASE tests.
- Explain how to take the ASE tests.
- Identify typical ASE test questions.
- Explain the content of the ASE Advanced Engine Performance Test.
- Identify the types of reference materials used in the Advanced Engine Performance Test.
- Explain what is done with ASE test results.

This chapter will explain the benefits of National Institute for Automotive Service Excellence (ASE) certification, and the advantages of being ASE certified. This chapter also explains how to apply for and take the ASE tests. When you have finished studying this chapter, you will know the purposes of ASE, the purposes of the ASE tests, the test methods used, and the purpose of the test results.

Reasons for ASE Tests

The concept of setting standards of excellence for skilled jobs is not new. In ancient times, metalworkers, weavers, potters, and other artisans were expected to conform to set standards of product quality. This resulted in the establishment of associations of skilled workers who set standards and enforced rules of conduct. Ancient Egypt, Greece, and Rome had such associations and many medieval industries were regulated by guilds. Many modern labor unions are descended from early associations of skilled workers. Certification processes for aircraft and aerospace workers, and electronics technicians have existed since the beginnings of these industries.

However, this has not been true of the automotive industry. Automobile manufacturing and repair began as a fragmented industry made up of many small vehicle manufacturers and thousands of small repair shops. Although the number of vehicle manufacturers decreased, the number of repair facilities continued to grow in number and in variety. Due to its fragmented, decentralized nature, standards for the automotive repair industry were difficult to establish. For years, there was no unified set of standards for automotive repair knowledge or experience. Anyone could claim to be an automotive technician, no matter how unqualified. This situation resulted in unneeded or improperly done repair work. As a result, a large segment of the public came to regard automotive technicians as unintelligent, dishonest, or both.

This situation changed in 1975, when the ***National Institute for Automotive Service Excellence,*** now called ***ASE,*** was established to provide a certification process for automobile technicians. ASE is a non-profit organization that encourages and promotes high standards of automotive service and repair. ASE does this by providing a series of written tests on various subjects in the automotive repair, medium/heavy truck repair, school bus repair, collision repair/refinishing, engine machinist, and parts specialist areas. Specialized ASE tests in the areas of alternative fuels and advanced engine performance are also available.

These tests are called ***standardized tests,*** which means that the same test in a particular subject is given to everyone throughout the United States. Any person passing one of these tests, and meeting certain work experience requirements, is certified in the subject covered by that test. If a technician can pass all of the tests in the automotive, medium/heavy truck, engine machinist, or collision repair/refinishing areas, he or she is certified as a ***Master Technician*** in that area.

The purposes of the ASE certification test program is to identify and reward skilled and knowledgeable technicians. Periodic recertification provides an incentive for updating skills and also provides guidelines for keeping up with current technology. The test program allows potential employers and the driving public to identify good technicians and helps the technician advance his or her career.

The program is not mandatory on a national level, but many repair shops now hire only ASE certified technicians. Many state emission inspector and manufacturer sponsored technician training programs now require ASE certification for admission and course credit. Some states require ASE certification in order to perform certain system repairs, such as those on the air conditioning system. Close to 400,000 persons are now ASE certified in one or more areas.

ASE Activities

Other activities that ASE is involved in are encouraging the development of effective training programs, conducting research on the best methods of performing instruction, and publicizing the advantages of technician certification. ASE is managed by a board made up of persons from the automotive and truck service industries, motor vehicle manufacturers, state and federal government agencies, schools, other educational groups, and consumer associations.

The advantages that the ASE certification program has brought to the automotive industry include increased respect and trust of automotive technicians, at least of those who are ASE certified. This has resulted in better pay and working conditions for technicians and increased standing in the community. Thanks to ASE, automotive technicians are taking their place next to other skilled artisans.

Applying for the ASE Tests

To apply for an ASE test, begin by obtaining an ***application form*** like the one shown in **Figure 22-1.** To obtain the most current application form, contact ASE at the following address:

ASE

Registration Form
ASE Tests

If extra copies are needed, this form may be reproduced.
Important: Read the Booklet first. Then print neatly when you fill out this form.

1. a. Social Security Number: ___-__-____
1. b. Telephone Number: (During the day) Area Code / Number
1. c. Previous Tests: Have you ever registered for any ASE certification tests before? ❑ Yes ❑ No

2. Last Name: **First** **Middle Initial**

3. Mailing Address:
Number, Street, and Apt. Number
City
State ZIP Code

4. Date of Birth: Month Day Year
5. Sex: ❍ Male ❍ Female

6. Race or Ethnic Group: Blacken one circle. (For research purposes. It will not be reported to anyone and you need not answer if you prefer.)
1 ❍ American Indian 2 ❍ African American 3 ❍ Caucasian/White 4 ❍ Hispanic/Mexican
5 ❍ Oriental/Asian 6 ❍ Puerto Rican 7 ❍ Other (Specify)

7. Education: Blacken only one circle for the highest grade or year you completed.
Grade School and High School (including Vocational): ❍7 ❍8 ❍9 ❍10 ❍11 ❍12
After High School: Trade or Technical (Vocational) School: ❍1 ❍2 ❍3 ❍4
College: ❍1 ❍2 ❍3 ❍4 ❍More

8. Employer: Which of the below best describes your current employer? (Blacken one circle and fill in subcodes from pp. 19-20 as appropriate.)
1 ❍ New Car/Truck Dealer/Distributor, Domestic— / 2 ❍ New Car/Truck Dealer/Distributor, Import— } Enter code for make of car or truck
3 ❍ Service Station—Enter code
4 ❍ Military—Enter Code
5 ❍ Indp't Repair Shop— / 6 ❍ Fleet Repair Shop— } Enter code for vehicle type
7 ❍ Specialty Shop—Enter code
8 ❍ Government/Civil Service
9 ❍ Student
10 ❍ Educator/Instructor
11 ❍ Manufacturer
12 ❍ Independent Collision Shop
13 ❍ Volume Retailer—Enter code
14 ❍ Tire Dealer—Enter code
16 ❍ Utility
17 ❍ Lift Truck Dealer/Repair Shop
18 ❍ Machinist Facility—Enter code
19 ❍ Leasing & Rental Shop—Enter code
15 ❍ Other

9. Test Center: Center Number City State

10. Tests: Blacken the circles (regular) or squares (recertification) for the test(s) you plan to take. (Do not register for a recertification test unless you passed that regular test before.) **Note:** Do not register for more than four regular tests on any single test day.

Regular	Recertification	Regular	Recertification
❍ A1 Auto: Engine Repair	❑ A	❍ A2 Auto: Automatic Trans/Transaxle	❑ E
❍ A8 Auto: Engine Performance	❑ B	❍ A3 Auto: Manual Drive Train and Axles	❑ F
❍ A4 Auto: Suspension & Steering	❑ C	❍ A6 Auto: Electrical/Electronic Systems	❑ G
❍ A5 Auto: Brakes	❑ D	❍ A7 Auto: Heating & Air Conditioning	❑ H
❍ F1 Alt. Fuels: Lt. Vehicle CNG		❍ L1 Adv. Level: Auto Adv. Engine Perf. Spec.	
❍ S4 School Bus: Brakes		❍ T1 Med/Hvy Truck: Gasoline Engines	❑ I
❍ S5 School Bus: Suspension/Steering		❍ T2 Med/Hvy Truck: Diesel Engines	❑ J
		❍ T6 M/Hvy Truck: Elec./Electronic Systems	❑ K
❍ T3 Med/Hvy Truck: Drive Train	❑ L	❍ S6 School Bus: Elec./Electronic Systems	
❍ T4 Med/Hvy Truck: Brakes	❑ M	❍ B2 Body: Painting & Refinish	❑ P
❍ T5 Med/Hvy Truck: Suspension/Steering	❑ N	❍ B3 Body: Nonstructural Analysis	
❍ T8 Med/Hvy Truck: PMI		❍ B4 Body: Structural Analysis	
❍ M1 Machinist: Cylinder Head Specialist	❑ Q	❍ B5 Body: Mechanical & Electrical Components	
❍ M2 Machinist: Cylinder Block Specialist	❑ R		
❍ M3 Machinist: Assembly Specialist	❑ S		

11. Fees: Number of **Regular** tests **(except L1)** marked above______ x $18* = $______
If you are taking the L1 Advanced Level Test, add $40 here + $______
If you marked **any** of the **Recertification** tests, add $20 here + $______
Registration Fee + $ **20**
TOTAL FEE = $______

**Important:* *School Bus Technicians, see page 8 for a money-saving offer! If you qualify, check here ❑ and adjust your fees (Item 11, first line) accordingly.*

❑ MasterCard ❑ VISA Expiration Date
Credit Card # Month Year
Signature of Cardholder

12. Fee Paid By: 1 ❍ Employer 2 ❍ Technician

13. Experience: Blacken one circle. (To substitute training or other appropriate experience, see page 3.)
1 ❍ I certify that I have two years or more of full-time experience (or equivalent) as an automobile, medium/heavy truck, school bus or collision repair/refinish technician or as an engine machinist. (Fill out #14 on opposite side)
2 ❍ I don't yet have the required experience. (Skip 14.)

14. Job History: Required. Provide job history and job details on the opposite side of this form.

15. Authorization: By signing and sending in this form, I accept the terms set forth in the *Registration Booklet* about the tests and the reporting of results.

Signature of Registrant Date

Do not send cash. Use credit card, or enclose a check or money order for the total fee, made payable to ASE/ACT and mail with this form to:
ASE/ACT, P.O. Box 4007, Iowa City, IA 52243

Figure 22-1. *Sample registration form for ASE certification tests. Be sure to fill out all applicable information and include payment of all applicable test fees. (ASE)*

National Institute for Automotive Service Excellence
13505 Dulles Technology Drive, Suite 2
Herndon, VA 22071-3421

ASE tests are given twice each year in the months of May and November. Tests are usually held during a two-week period at night during the work week. The actual test administration is performed by ACT, a non-profit organization experienced in administering standardized tests. The tests are given at designated test centers in over 700 locations in the United States and its territories. If necessary, special test centers can be set up in remote locations. However, there must be enough potential applicants for the establishment of a special test center to be practical.

ASE will send the proper form enclosed in a ***registration booklet*** explaining how to complete the form. When you get the booklet, carefully fill out the form, recording all needed information. You may apply to take as many tests as are being given, fewer tests, or only one test if desired. Work experience, or any substitutes for work experience, should also be included according to the instructions in the registration booklet. If there is any doubt about what should be placed in a particular space, consult the registration booklet.

Anyone may apply for and take an ASE test. However, to become certified, the applicant must have two years experience working as an automobile or truck technician. In some cases, training programs or courses, an apprenticeship program, or time spent performing similar work may be substituted for all or part of the work experience. The one exception to this is the Advanced Engine Performance Specialist test, which requires current certification in Auto Engine Performance and two years documented work experience to be eligible. The reason for this requirement will be explained later in this chapter.

Be sure to determine the closest test center and record its number in the appropriate space. Most test centers are located at local colleges, high schools, or vocational schools. In addition to the application, you must include a check, money order, or credit card number to cover all necessary fees. A fee is charged to register for the test series and a separate fee is charged for each test to be taken. See the latest registration booklet for the current fee structure. In some cases, your employer may pay the registration and test fees or reimburse you for each test that you pass. Check with your employer before sending in your application.

To be accepted for either the Spring or Fall ASE tests, your application and payment must arrive at ASE headquarters at least one month before the test date(s). To ensure that you can take the test at the test center of your choice, send in the application as early as possible.

After sending the application and fees, you will receive an ***admission ticket*** to the test center. This should arrive by mail within two weeks of sending the application. See **Figure 22-2.** If your admission ticket has not arrived, and it is less than two weeks until the first test date, contact ASE using the phone number in the registration booklet. If the desired test center is filled when ASE receives your application, you will be directed to report to the nearest center that has an opening. If it is not possible to go to the alternate test center that was assigned, contact ACT immediately using the phone number in the registration booklet.

Taking the ASE Tests

Be sure to bring your admission ticket with you when reporting to the test center. When you arrive at the test center, you will be asked to produce the admission ticket and a driver's license or other photographic identification. In addition to these items, bring some extra Number 2 pencils. Although pencils will be made available at the test center, extra pencils may save you time if the original pencil breaks.

After you enter the test center and are seated, listen to and follow all instructions given by the test administrators. During the actual test, carefully read all test questions before making a decision as to the proper answer. The ASE tests are designed to measure your knowledge of three things:

- Basic information on how automotive systems and components work.
- Diagnosis and testing of systems and components
- Repairing automotive systems and components.

Each ASE test will contain between 40 and 80 test questions, depending on the subject to be tested. All test questions are multiple-choice, with four possible answers. These types of multiple choice questions are similar to the multiple-choice questions used in this textbook. Questions can be in one or two parts. Samples of these types of test questions are given below.

One-Part Question

1. The throttle body part that controls the entry of air past the throttle valve is the:
   (A) throttle position sensor.
   (B) mass airflow sensor.
   (C) idle air control valve.
   (D) throttle plate adjustment screw.

Notice that the question calls for the best answer out of all of the possibilities. The idle air control valve is the only part that has any control over the flow of air past the throttle valve. Therefore, "C" is correct.

Two-Part Question

This question asks you to read two statements and decide if they are true. Both statements can be true, both can be false, or only one of them can be false.

National Institute for Automotive Service Excellence

ACT, P.O. Box 4007, Iowa City, Iowa 52243, Phone: (319) 337-1433 017910042 T

Admission Ticket

Test Center to which you are assigned:

A

John Smith
123 Main Street
Edens, Il. 60000

REGULAR TESTS (Late arrivals may not be admitted.)		
DATE	REPORTING TIME	TEST(S)
11/14	7:00 PM	A1, A4, A8

RECERTIFICATION TESTS (Late arrivals may not be admitted.)

TEST CODE KEY

A1 Auto: Engine Repair
A2 Auto: Automatic Trans/Transaxle
A3 Auto: Manual Drive Train & Axles
A4 Auto: Suspension & Steering
A5 Auto: Brakes
A6 Auto: Electrical/Electronic Systems
A7 Auto: Heating & Air Conditioning
A8 Auto: Engine Performance
M1 Machinist: Cylinder Head Specialist
M2 Machinist: Cylinder Block Specialist
M3 Machinist: Assembly Specialist
T1 Med/Hvy Truck: Gasoline Engines
T2 Med/Hvy Truck: Diesel Engines
T3 Med/Hvy Truck: Drive Train
T4 Med/Hvy Truck: Brakes
T5 Med/Hvy Truck: Suspension & Steering
T6 Med/Hvy Truck: Elec./Electronic Systems
T8 Med/Hvy Truck: Preventive Main. Inspec.
B2 Coll.: Painting & Refinishing
B3 Coll.: Non-structural Analysis
B4 Coll.: Structural Analysis
B5 Coll.: Mechanical & Elec. Components
B6 Coll.: Damage Analysis & Estimating
P1 Parts: Med/Hvy Truck Parts Specialist
P2 Parts: Automobile Parts Specialist
F1 Alt. Fuels: Lt. Veh. Comprsd. Nat. Gas
L1 Adv. Level: Adv. Engine Perf. Spec.
S1 School Bus: Body Sys. & Spec. Equip.
S4 School Bus: Brakes
S5 School Bus: Suspension & Steering
S6 School Bus: Elec./Electronic Systems

See Notes and Ticketing Rules on reverse side. An asterisk (*) indicates your certification in these areas is expiring.

SPECIAL MESSAGES

-REVIEW ALL INFORMATION ON THIS TICKET. CALL IMMEDIATELY TO REPORT AN ERROR OR IF YOU HAVE QUESTIONS.
-IF YOU MISS ANY EXAMS, FOLLOW THE REFUND INSTRUCTIONS ON THE BACK OF THIS SHEET. THE REFUND DEADLINE IS
-YOU HAVE BEEN ASSIGNED TO AN ALTERNATE TEST CENTER. THE CENTER ORIGINALLY REQUESTED IS FULL.
8010-IL/LOCAL 150 IS LOCATED ON JOLIET AVE, THREE DOORS W. OF LAGRANGE RD ON SOUTH SIDE OF THE ST. ENTER THROUGH BACK DOOR. NO ALCOHOL ON PREMISES.

MATCHING INFORMATION: The information printed in blocks B and C at the right was obtained from your registration form. It will be used to match your registration information and your test information. Therefore, the information at the right must be copied EXACTLY (even if it is in error) onto your answer booklet on the day of the test. If the information is not copied exactly as shown, it may cause a delay in reporting your test results to you.

IF THERE ARE ERRORS: If there are any errors or if any information is missing in block A above or in blocks B and C at the right, you must contact ACT immediately. DO NOT SEND THIS ADMISSION TICKET TO ACT TO MAKE SUCH CORRECTIONS.

Check your tests and test center to be sure they are what you requested. If either is incorrect, call 319/337-1433 immediately. Tests cannot be changed at the test center. **ON THE DAY OF THE TEST,** be sure to bring this admission ticket, positive identification, several sharpened No. 2 pencils, and a watch if you wish to pace yourself.

B FIRST FIVE LETTERS OF LAST NAME

S	M	I	T	H

C SOCIAL SECURITY NUMBER OR ACT IDENTIFICATION NUMBER

1	2	3	4	5	6	7	8	9

SIDE 1

Figure 22-2. *ASE test admission ticket. Bring this ticket along with photographic identification and Number 2 pencils when you go to the test center. (ASE)*

1. Technician A says that the ignition coil changes high voltage into low voltage. Technician B says that the distributor pickup coil produces a low voltage signal. Who is right?

   (A) A only.

   (B) B only.

   (C) Both A and B.

   (D) Neither A nor B.

In this case, the statement of technician A is wrong, since the ignition coil produces high voltage from low voltage. The statement of technician B is correct, since the pickup coil does produce a low voltage signal. Therefore, the correct answer is "B."

Negative Questions

Some questions are called ***negative questions.*** These questions ask you to identify the wrong answer. They will usually have the word "EXCEPT" in the question.

1. The ignition system is comprised of all of the following parts, EXCEPT:

   (A) coil.

   (B) spark plugs.

   (C) fuel injectors.

   (D) ignition module.

Since fuel injectors are the only parts that are not included in the ignition system, the correct answer is "C."

A variation of the negative question will use the words *most* or *least,* such as the following:

1. The computer-controlled engine of a late model car knocks during acceleration. Which of these defects is the *least likely* cause?

   (A) Incorrect timing.

   (B) Defective knock sensor.

   (C) Plugged fuel filter.

   (D) Low octane gasoline.

In this case, the least likely cause of engine knocking is a plugged fuel filter, which is much more likely to cause engine stalling or poor performance instead of knocking. Therefore, the correct answer is "C."

Incomplete Sentence Questions

Some test questions are incomplete sentences, with one of the four possible answers correctly completing the sentence. An example of an incomplete phrase question is given below.

1. The coolant temperature sensor is used to measure ______ temperature.

   (A) exhaust gas

   (B) engine

   (C) incoming air

   (D) ambient (outside) air

Once again the question calls for the best answer. The coolant temperature sensor measures the temperature in the engine by monitoring coolant temperature, so "B" is correct.

ASE Engine Performance Test

The basic ASE Engine Performance test contains approximately 70-80 questions. Some of the questions on this test are similar to those asked on the ASE Engine Repair test. In fact, ASE administered at one time, a combined Engine Repair-Engine Performance test. Recently, ASE has eliminated this test.

The categories in the ASE Engine Performance test include:

- ❑ General Engine Diagnosis
- ❑ Ignition System Diagnosis & Repair
- ❑ Fuel, Air Induction, & Exhaust System Diagnosis & Repair
- ❑ Emissions Control Systems Diagnosis & Repair
- ❑ Computerized Engine Controls Diagnosis & Repair
- ❑ Engine Related Service
- ❑ Engine Electrical Systems Diagnosis & Repair

The fuel section of the test is given at the end and is divided into two sections covering closed loop fuel injection and continuous (import) fuel injection. You are given the choice of which fuel section to take. Only answer the questions pertaining to the fuel section you choose.

ASE Advanced Engine Performance Specialist Test

The ***Advanced Engine Performance Specialist test*** differs from other ASE certification tests. It rigorously tests the technician's experience and ability to use service manuals and other reference materials to properly diagnose emission and driveability complaints, rather than general operation and repair knowledge of a particular system. It is an extension of the standard ASE Engine Performance test. The task list for this test includes 45-50 questions on:

- ❑ General Powertrain Diagnosis
- ❑ Computerized Engine Control Diagnosis
- ❑ Ignition System Diagnosis
- ❑ Fuel System and Air Induction System Diagnosis
- ❑ Emission Control Systems Diagnosis
- ❑ Inspection and Maintenance (I/M) Failure Diagnosis

The Advanced Engine Performance test was created to meet industry demand for a test that complies with Clean Air Act regulations for emissions inspection and maintenance. The questions in this test are more detailed, requiring the use of reference materials supplied in the test booklet. As stated earlier in this chapter, current ASE certification in Engine Performance and two years documented work experience is required for registration. When you register for the advanced test, ASE will send you an admission ticket, along with preparation materials so that you can become familiar with the test format. It is not necessary to take the preparation materials to the test site.

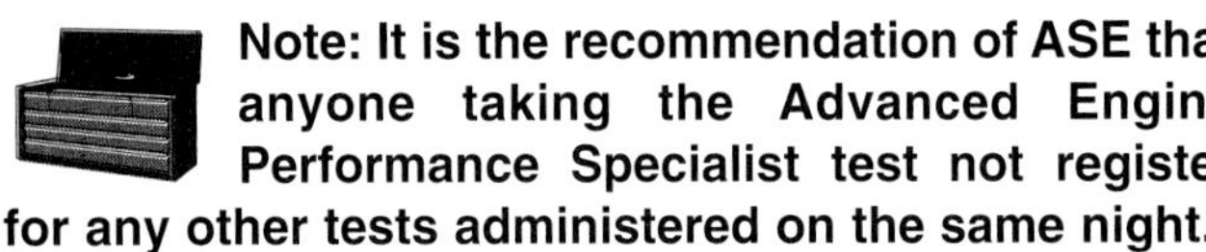

Note: It is the recommendation of ASE that anyone taking the Advanced Engine Performance Specialist test not register for any other tests administered on the same night.

The Composite Vehicle

As mentioned earlier, the Advanced Engine Performance Specialist test contains reference materials to be used to help answer the test questions. The reference materials include charts that show scan tool and exhaust gas analyzer data. A section in the test contains questions that require the technician to refer to a "composite vehicle." This portion tests the technician's ability to use diagrams and schematics to diagnose problems in a specific engine or emissions system.

The ***composite vehicle*** contains a computer, input sensors, output devices, and actuators similar to those that you studied earlier in this text. The composite vehicle resembles the engine component layout illustrations used by some manufacturers in their service manuals. Data for the composite vehicle's sensors, output devices, and operating parameters is included. A corresponding wiring schematic is also provided with the composite vehicle for use in testing the technician's ability to trace electrical problems. **Figure 22-3** and **Figure 22-4** shows a typical composite vehicle and related wiring schematic used in the Advanced Engine Performance Specialist test. The composite vehicle used in the actual test may differ from the one shown here.

Advanced Engine Performance Test Questions

The types of questions on this test are similar in format to those on the standard tests; they are simply more difficult to answer. General system operation, service, and repair knowledge is insufficient to correctly answer most of the test questions. A typical question on the advanced test will reference a chart with scan tool data, test results, or the composite vehicle information. Refer to **Figure 22-5** when dealing with this sample question.

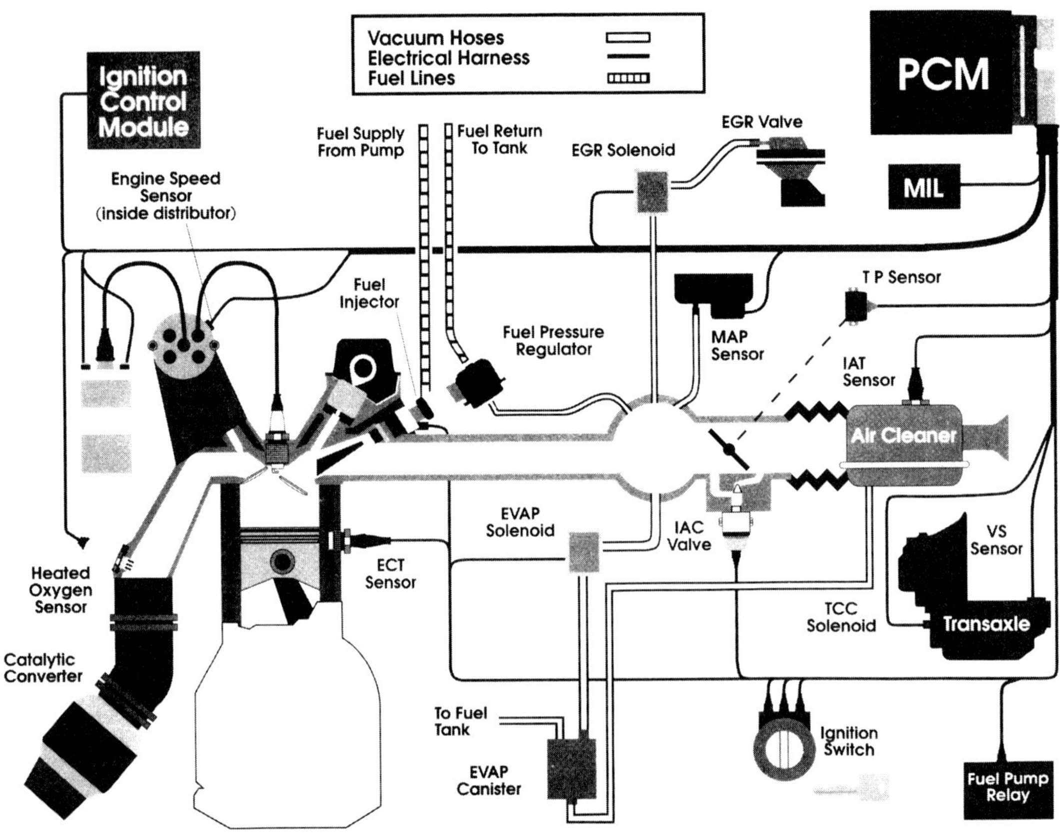

Figure 22-3. *Typical composite vehicle layout used in the Advanced Engine Performance Specialist test. (ASE)*

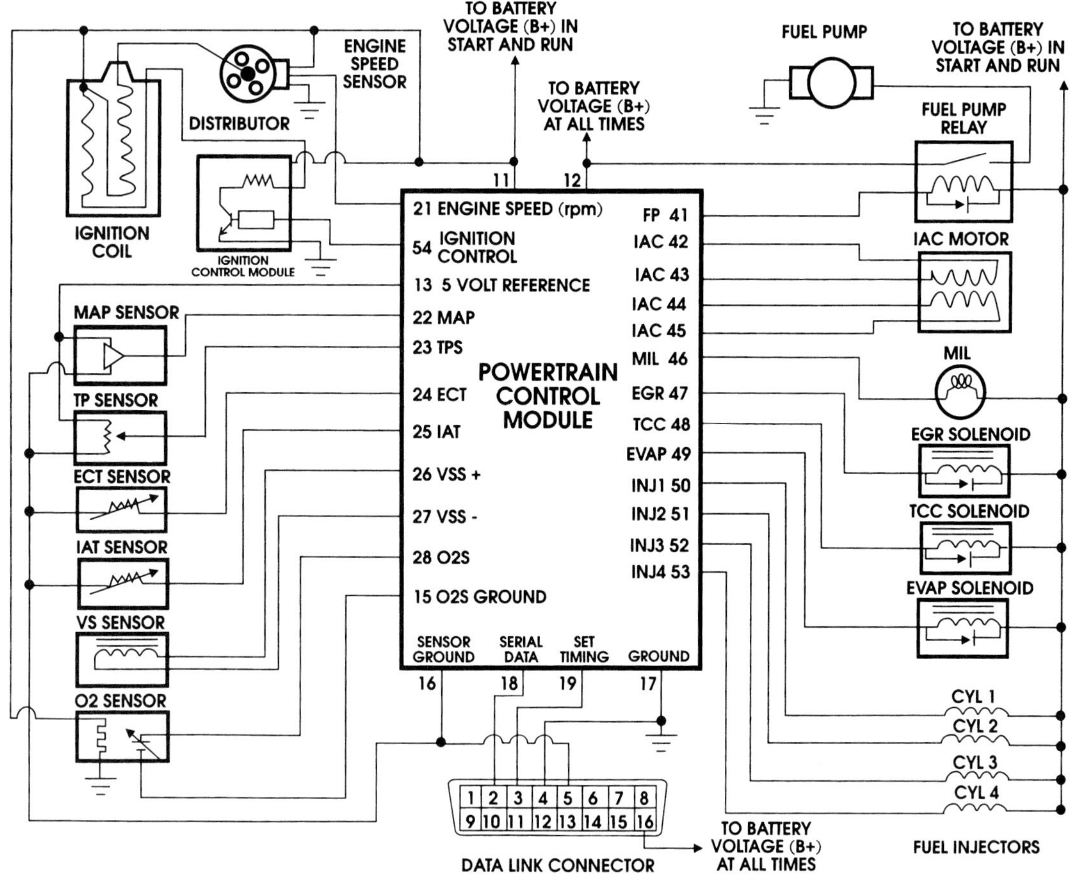

Figure 22-4. *Wiring schematic for the composite vehicle. (ASE)*

Engine Speed	Idle	2000 rpm
HC (ppm)	345	1320
CO (percent)	0.7	1.1
CO_2 (percent)	14.1	15.5
O_2 (percent)	0.2	0.1

Figure 22-5. *Exhaust gas analyzer data chart. Some questions on the Advanced Engine Performance Specialist test use charts similar to this one.*

1. A vehicle with a multiport fuel injected engine has failed a state-mandated emissions test. The exhaust is producing black smoke and foul catalytic converter odor. An exhaust gas analysis has provided the readings shown above. Which of these is the most likely cause?
   (A) Restricted air cleaner.
   (B) Leaking exhaust manifold.
   (C) Low fuel pressure.
   (D) Vacuum leak.

The key to answering this question is correctly interpreting the data and comparing it to normal vehicle operating readings. Answer D would can cause a rich condition. However, this problem would also allow additional oxygen into the exhaust, tricking the oxygen sensor into believing that a lean condition exists. This would show in the exhaust gas analysis as a high O_2 reading. Answers B and C would cause a lean condition and a high O_2 reading. A restricted air cleaner would cause all of the readings shown above, along with the black smoke and foul converter odor, therefore, the correct answer is "A."

A typical question that uses scan tool data and references the composite vehicle information might look like the following question. Refer to **Figure 22-6.**

1. The composite vehicle's engine idles at 1600 rpm. The scan tool data shown above was obtained at idle. Technician A says that a sticking idle air control valve could be the cause. Technician B says that a misadjusted throttle position sensor could be the cause. Who is right?
   (A) A only.
   (B) B only.
   (C) Both A and B.
   (D) Neither A nor B.

Notice that this sample question is like the two-part question you studied earlier, just a bit more complex. This question tests the ability to correctly interpret scan tool

Scan Tool Data			
Engine coolant temperature sensor (CTS) 0.45 volts	Intake air temperature sensor (IAT) 2.4 volts	Manifold absolute pressure sensor (MAP) 1.2 volts	Throttle position sensor (TPS) 0.6 volts
Engine speed sensor (RPM) 1600 rpm	Heated oxygen sensor (HO_2S) 0.3 volts	Vehicle speed sensor (VSS) 0 mph	Battery voltage (B+) 13.4 volts
Idle air control valve (IAC) 87 percent	Evaporative emission canister solenoid (EVAP) Off	Torque converter clutch solenoid (TCC) Off	EGR valve control solenoid (EGR) Off
Malfunction indicator lamp (MIL) On	Diagnostic trouble codes (DTC) P0505, P0507	Open/closed loop Open	Fuel pump relay (FP) On
Ignition timing (°BTDC)	Base timing: 8	Actual timing: 12	

Figure 22-6. *This chart shows typical data displayed by most scan tools. It is very similar to the scan tool charts used in this text and in the Advanced Engine Performance Specialist test along with the composite vehicle.*

information using service manuals as a reference to normal sensor and vehicle operation. The normal sensor values and vehicle operating parameters are included with the composite vehicle reference materials in the test booklet. Using the reference materials, you would find that the stored trouble codes indicate an idle control problem. These codes would be a clue to look at the idle air control value.

An IAC value of 87 percent indicates near maximum idle speed compensation. While a badly misadjusted throttle position sensor could cause an excessively high idle, the high TPS value would be shown in the scan tool data as well as set the two diagnostic trouble codes. You would also find by looking in the composite vehicle reference materials that TPS voltage is close to normal, therefore, "A" is correct.

The last five ASE-type questions in **Chapters 14-21** of this text were designed to closely resemble questions on the Advanced Engine Performance Specialist test. Judging from the difficulty of these sample questions, you can see why only certified, experienced technicians are permitted to take this test.

After completing all the questions in a particular test, recheck all of your answers one time to ensure that you did not miss anything that would change your answer, or that you did not make a careless error or stray mark on the answer sheet. In most cases, rechecking your answers more than once is unnecessary, and often may lead you to change correct answers to incorrect ones. The time allowed for each test is about four hours. However, you may leave after completing you last test and handing in all test material.

Test Results

After taking the test, be prepared to wait from six to eight weeks for the tests to be processed, graded, and mailed. You will receive a confidential ***pass/fail letter*** first, which will report your performance on the tests. The letter will give your preliminary test results, which will state if you have passed, or if more preparation is needed.

A ***test diagnostic report*** will follow in approximately two weeks. The report will show your scores on the test, and whether this score is sufficient for certification. The test questions are also subdivided into general areas to help you to determine any weak areas that require more study. For example, the Auto Engine Performance test questions will be divided into such subsections as ignition system, fuel & exhaust system, emission control system, general engine and electrical diagnosis & service, and computerized engine controls. Also included with the diagnostic report is a verification of certification for all of the tests that you have passed, an ASE shoulder patch, and a pocket card listing all of the areas that you are certified in. See **Figure 22-7.**

Technician Confidentiality

ASE takes the position that all ASE test results are confidential information, and provides them only to the person who took the test. This is done to protect your privacy. The only test information that ASE will release is to confirm to an employer that you are certified in a particular skill area. Test results will be mailed to your home address and will not be provided to anyone else. This is true even if your employer has paid the test fees. If you wish your employer to know exactly how you performed on the tests, you must provide him or her with a copy of your test results.

If you fail a certification test, you can retake it again as many times as you would like. However, you (or your employer) must pay all of the applicable registration and test fees again. You should study all available information in the areas where you did poorly. A copy of the "ASE Preparation Guide" may be helpful to sharpen your skills in these areas. The ASE Preparation Guide is free, and can be obtained by filling out the coupon at the back of the registration booklet.

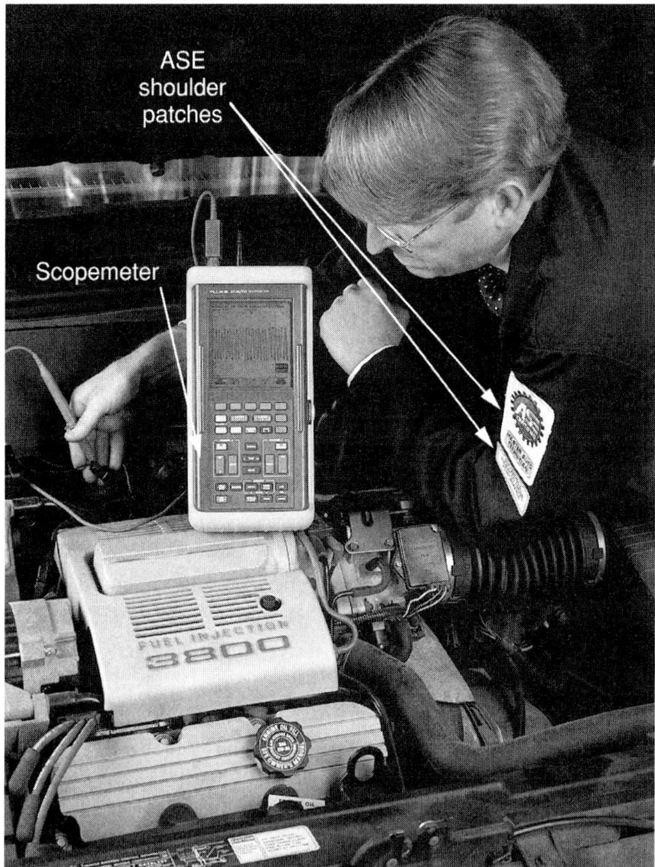

Figure 22-7. *This driveability technician has passed all eight of the ASE automotive certification tests as well as the Advanced Engine Performance Specialist test. Note the advanced engine performance certification strip under the normal ASE certification patch. (Fluke)*

Recertification Tests

Once you have passed the certification test in any area, you must recertify every five years. This assures that your ASE certification remains current and is proof that you have kept up with current technology. Questions on the recertification test are slightly more difficult than those on the regular certification test as they are designed to test the technician's knowledge of current technology.

The process of applying to take the recertification tests is similar to that for the original certification tests. Use the same form and enclose the proper recertification test fees. If you allow a certification to lapse, you must take the regular test to regain your certification.

Summary

The automotive industry was one of the few major industries that did not have testing and certification programs. This resulted in decreased professionalism in the automobile industry, often leading to poor or unneeded repairs, and decreased status and pay for automobile Technicians. The National Institute for Automotive Service Excellence, or ASE, was started in 1975 to help overcome the poor status of the automotive repair industry. ASE tests and certifies automotive technicians in major areas of automotive repair. This has increased the skill level of technicians, resulting in better service, and increased benefits for technicians.

ASE tests are given two times each year, in May and November. Anyone can register to take the tests by filling out the registration form and paying the proper registration and test fees. The registrant must also select the test center that he or she would like to go to. To be considered for certification, the registrant must have two years of hands-on experience as an automotive technician. Proof of this should also be included with the registration form. About two weeks after applying for the test, the technician will receive a test entry ticket which he or she must bring to the test center.

The actual test questions will test your knowledge of general system operation, diagnosing problems, and repair techniques. All of the questions are multiple choice questions with four possible answers.

The Advanced Engine Performance Specialist test was created to fill industry demand for a test that complies with Clean Air Act regulations for emissions inspection and maintenance. This test is designed to evaluate the technician's ability to diagnose engine performance problems using service manuals and other reference materials. The Advanced Engine Performance Specialist test uses the same question format as other ASE tests, however, the questions are more rigorous. Prior certification in Engine Performance and two years documented work experience is required for registration. The questions must be read carefully. The entire test should be gone over one time only to catch careless mistakes.

Test results will arrive within six to eight weeks after the test session. Results are confidential, and will be sent only to the home address of the person who took the test. If a test was passed, and the experience requirement has been met, the technician will be certified for five years. Anyone who fails a test can take it again in the next session. Tests can be taken as many times as necessary. Recertification tests can be taken at the end of the five year certification period.

Know These Terms

National Institute for Automotive Excellence (ASE)
Standardized tests
Master Technician
Application form
Registration booklet
Admission ticket
Negative questions
Advanced Engine Performance Specialist test
Composite vehicle
Pass/fail letter
Test diagnostic report

Review Questions—Chapter 22

Please do not write in this text. Write your answers on a separate sheet of paper.

1. Associations of skilled workers who set standards and enforce rules of conduct have been established since:
   (A) ancient Egypt, Greece, and Rome.
   (B) medieval times.
   (C) 50 years ago.
   (D) 1975.
2. ASE encourages high standards of automotive service and repair by providing a series of ________ tests.
   (A) written
   (B) hands-on
   (C) oral
   (D) computerized
3. When the same test in a particular subject is given to everyone throughout the United States, it is referred to as a ______ test.
   (A) comprehensive
   (B) fair
   (C) standardized
   (D) Both A & B.
4. If a technician can pass all of the tests in the automotive, heavy truck, or collision/refinishing areas, he or she is certified as a:
   (A) trainee.
   (B) master technician.
   (C) general technician.
   (D) knowledgeable technician.
5. ASE tests are held:
   (A) during normal working hours.
   (B) at night and on weekends.
   (C) any time.
   (D) at night during the work week.
6. ASE tests are administered by a non-profit agency called:
   (A) SAT.
   (B) ASE.
   (C) SAE.
   (D) ACT.
7. List three items that you should take with you to the ASE test center.
8. ASE provides test results to:
   (A) the technician who took the test.
   (B) whoever paid for the test.
   (C) the technician's employer.
   (D) Both A & C.
9. The ASE Advanced Engine Performance Specialist test is designed to test what type of skills?
   (A) Ability to use reference materials to diagnose emission and driveability problems.
   (B) Theory of automotive systems.
   (C) General system diagnostic and repair knowledge.
   (D) None of the above.
10. If a technician's certification lapses he or she must:
   (A) take a recertification test.
   (B) take the regular test.
   (C) petition ASE.
   (D) None of the above.

ASE Certification-Type Questions

1. The advantages that ASE certification has brought to automotive technicians include all of the following, EXCEPT:
   (A) increased respect.
   (B) better working conditions.
   (C) lower pay.
   (D) increased standing in the community.
2. ASE tests are given _______ each year.
   (A) once
   (B) twice
   (C) four times
   (D) twelve times

3. ASE tests are designed to measure your knowledge of all of the following, EXCEPT:
   (A) basic information on how automotive systems and components work.
   (B) knowledge of manufacturer specific technology.
   (C) diagnosis and testing of systems and components.
   (D) repairing automotive systems and components.

4. ASE test questions resemble the ones in ______.
   (A) college level courses
   (B) essay-type tests
   (C) verbal examinations
   (D) this book

5. ASE sends all of the following to you after the test, EXCEPT:
   (A) a test diagnostic report.
   (B) a graded copy of your test.
   (C) a confidential pass/fail letter.
   (D) a verification of certification.

6. After you are finished with the last test you should ______.
   (A) turn in your test materials and leave.
   (B) double check your answers once.
   (C) help someone else with their test.
   (D) Both A & B.

7. A technician can retake any certification test ______.
   (A) two times
   (B) four times
   (C) five times
   (D) any number of times

8. Technician A says that anyone taking the ASE Advanced Engine Performance Specialist test must have two years documented work experience and current certification in Engine Repair. Technician B says that it is the recommendation of ASE that anyone taking the ASE Advanced Engine Performance Specialist test register for as many tests as given on that night. Who is right?
   (A) A only.
   (B) B only.
   (C) Both A & B.
   (D) Neither A nor B.

9. The ASE Advanced Engine Performance Specialist Test uses all of the following reference materials, EXCEPT:
   (A) exhaust gas analyzer data.
   (B) schematics.
   (C) service manuals.
   (D) scan tool data charts.

10. Certified technicians must take a recertification test every ______ years.
   (A) 2
   (B) 5
   (C) 10
   (D) There is no set time period.

National Institute for
AUTOMOTIVE SERVICE EXCELLENCE

December 20, XXXX 032527

John Smith
123 Main Street
Edens, Il 60000

Dear ASE Test Taker:

Listed below are the results of your November XXXX ASE Tests. You will soon be receiving a more detailed report.

<u>If your test result is "Pass"</u>, and if you have fulfilled the two-year "hands-on" experience requirement, you will receive a certificate and credential cards for the tests you passed.

<u>If your test result is "More Preparation Needed"</u>, you did not attain a passing score. Check your detailed score report when it arrives. This information may help you prepare for your next attempt.

If you do not receive your detailed report within the next three weeks, please call.

Thank you for participating in the ASE program.

A1	ENGINE REPAIR	PASS
A6	ELECTRICAL/ELECTRONIC SYSTEMS	PASS
A8	ENGINE PERFORMANCE	PASS

123-45-6789

13505 Dulles Technology Drive • Herndon, Virginia 22071-3415 • (703) 713-3800

A sample confidential pass/fail letter. This letter will arrive to you prior to the test diagnostic report, which will give you the actual breakdown of your performance in each area. (ASE)

Chapter 23

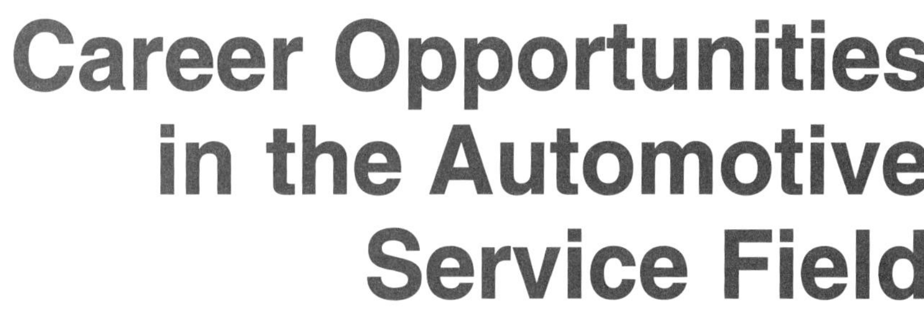

Career Opportunities in the Automotive Service Field

After studying this chapter, you will be able to:

- Identify three main classifications of automotive technicians.
- Identify the major sources of employment in the automotive industry.
- Identify advancement possibilities for automotive technicians.
- Explain how to fill out a job application.
- Explain how to conduct yourself during a job interview.

This chapter will give an overview of career opportunities in the automotive service industry. It includes the types of automotive technicians and what kind of work that they perform. It also includes information on the types of repair outlets as well as the type of work, working conditions, and pay scales you can expect to find in each place. Also included are some of the ways you can locate jobs and become employed in the auto repair industry. It includes information on other types of automotive related jobs that you can enter. Studying this chapter will help you to find and get a job in the automotive service industry.

Automotive Servicing

The business of servicing and repairing cars and trucks has provided employment for millions of people over the last 100 years. It will continue to provide good employment opportunities for many years to come. Like any career, it has its drawbacks, but it also has its rewards.

Most persons in the auto service business work long hours and the diagnosis and repair procedures can be mentally taxing, physically hard, and often hot and dirty. Automotive service has never been a prestige career, however, this is changing as vehicles become more complex and technicians become better trained. The technician often has to deal with difficult, condescending, or sometimes hostile vehicle owners.

The advantages of the auto service business are the opportunity to work with your hands, much less confinement than with many other professions, and the enjoyment of taking something that is not working, and making it work again. Auto repair salaries are usually competitive with those for similar jobs, and it is a secure profession where a good technician can always find work. To ensure that you stay employable, always seek to learn new things and become ASE certified in as many areas as possible.

Levels of Automotive Service Positions

Although the public tends to classify all automotive technicians as "mechanics," there are many types of auto service professionals. The levels of auto service professionals range from the helper or apprentice who changes oil, removes and installs parts, or performs other simple tasks to the certified technician, capable of diagnosing and repairing various automotive systems. Although these levels are unofficial, they tend to hold true throughout the automotive repair industry. It would be possible to further break these levels down into more sublevels, but the general skill classifications will be adequately covered by discussing these three levels.

Helpers

Most ***helpers*** perform many of the common non-repair related tasks around the shop, such as clean up, delivering parts, and driving vehicles. As the helper's skills progress, they may begin to perform the easier types of service and repair jobs, such as installing and balancing tires, changing engine oil and filters, and installing batteries. The skills required of the helper are low, and the pay will be less than that of the other levels. However, the helper position is a good way for many people to start. In fact, many technicians started out doing this kind of automotive service when they were in their teens.

Apprentice

The ***apprentice*** does the service and repair jobs of the helper in many shops. However, they also remove and install parts on vehicles diagnosed by the certified technician. The parts installed are components such as shock absorbers or struts, exhaust system components, and possibly brake master cylinders, alternators, and starters. As the apprentice's skills increase, they may begin to do more complicated repair work and even diagnose vehicle problems under a certified technician's guidance, **Figure 23-1.**

Figure 23-1. *Apprentices work under the guidance of a certified technician, usually performing many of the easier jobs that come into the shop. As an apprentice gains experience, he or she may begin to perform more complex diagnosis and repair jobs. (Fluke)*

Apprentices are paid more than helpers but less than certified technicians. Many apprentices take the opportunity to improve their knowledge and skills and eventually become certified technicians.

Certified Technicians

Certified technicians, **Figure 23-2**, are at the top level, and have the skills to prove it. Most modern technicians are ASE certified in at least one or more automotive areas, and many are certified in more than one test category. The certified technician is usually able to successfully diagnose and repair every area that he or she is certified in and can perform many other service jobs. The certified technician is at the top of the pay scale, also.

Figure 23-2. *Certified technicians have the knowledge and experience to properly diagnose and repair the modern systems in today's vehicles. (Fluke)*

Types of Auto Service Facilities

There are many types of auto service facilities where the aspiring technician can find work. The traditional place to get started in automotive repair, the corner full-service station is all but gone. Most have been replaced by self-service convenience stores that sell fuel and groceries rather than repair vehicles. However, many opportunities to repair vehicles still exist. Even the smallest community has many types of automotive repair facilities. Some of these are discussed in the following sections. In all cases, the technician has to provide his or her own hand and air tools.

New Vehicle Dealers

Most new ***vehicle dealers*** have large, well-equipped service departments to meet the warranty requirements of the vehicle manufacturer, **Figure 23-3.** These service departments are equipped with all of the special tools, test equipment, and service literature needed to service that manufacturer's vehicles. Dealership service departments are also equipped with lifts, parts cleaners, hydraulic presses, brake lathes, and electronic test equipment for efficiently servicing vehicles. Dealers stock all of the most common parts and are usually tied into a factory parts network which allows them to quickly obtain any part. Opportunities for additional training are excellent, with on-site training increasing in popularity.

The rate per hour is competitive between dealerships in the same area and is usually higher than local industry in general. Pay scales at most dealerships are usually based on flat rate time. The number of hours that the technician is paid depends on what work comes in and how fast he or she can complete it. If you can work fast and enough work comes in, the pay can be excellent. Some dealers pay their technicians a combination of a base salary and flat rate hours. Most modern dealers offer benefits packages.

Dealership working conditions are relatively good, and most of the vehicles are new or well cared for older models. Since the dealer must fix any part of the vehicle, the technician can perform a large variety of work. Although many dealer service departments have technicians who work only in specific areas, the trend is toward training all technicians to handle most types of work with specialization in two or three areas. Most repairs will be on the same make of vehicle, although many large dealerships handle more than one make. If you plan on making driveability work one of your specializations, these repairs are now done mostly by the dealer.

The disadvantages are the lack of salary guarantees at most dealers, lower flat rate time for warranty repairs, and fast paced, sometimes hectic working conditions. If you welcome the challenge of being paid by the job and do not mind working under deadlines, a dealership may be the ideal employer. Also check out the local medium duty truck dealerships. Although the work is much heavier, the pay is usually somewhat higher, and working conditions are not as fast paced.

Chain and Department Store Auto Service Centers

Many national chain and department stores have ***auto service centers*** where various types of automotive repairs are performed. These centers often hire technicians for entry-level jobs and may have several classifications above entry-level. Technicians in most of these situations are paid a salary plus commission for work performed. Pay scales for the various classifications of work are competitive and most companies offer generous benefit packages. One advantage of working for large companies such as these is the chance of advancement into other areas, such as sales or management.

One disadvantage of working at the average auto service center is the lack of variety. Most auto centers concentrate on a few types of repairs, such as alignment and brake work, and turn down most other repairs. The work can become monotonous due to the lack of variety. Although the job pressures are usually less than those at dealerships, customers still expect their vehicles to be repaired in a

Figure 23-3. *New vehicle dealerships provide the opportunity to work on the latest models and provide the equipment and service literature to make the job easier. (Land Rover)*

reasonable period of time. However, if you would enjoy the opportunity to work on only one or two areas of automotive repair, this type of job may be ideal for you.

Tire, Muffler, Transmission, and Other Specialty Shops

Tire, muffler, transmission, and other specialty shops can offer good working conditions and good pay, **Figure 23-4.** Most of these shops concentrate on their major specialty, with a few other types of general repair, such as tune-up and brake service. Technicians at these shops are usually paid a competitive salary, plus commission for work performed. A disadvantage of specialty shops is the lack of variety. Since they usually concentrate on a few types of repairs, working at one of these shops can become monotonous.

Many of these shops are franchise operations and the demands of the franchise can create problems. If the prime purpose of the shop is to sell tires, the technician who was hired to do brake repairs may be forced to spend time installing tires. This can be annoying to most technicians who want to be doing the job they were hired for. However, if you are not bothered by this, the specialty shop could be a good work situation.

Figure 23-4. *Auto service chain stores can allow you to work on a few areas of repair. Since they are not supported by a vehicle manufacturer, they do not perform warranty repairs on new vehicles.*

Independent Shops

There are millions of independent auto repair shops. As places to work, they range from excellent to terrible. Many shops are operated by competent and fair managers and have first rate equipment and good working conditions. Some independent shops have almost no equipment, low pay rates, and extremely poor, often dangerous working conditions. The prospective employee should carefully check all aspects of the shop environment before agreeing to work there. Technicians at independent shops are usually paid on a salary plus commission basis.

There are two major classifications of independent repair shops, the general repair shop and the specialty shop. The ***general repair shop*** takes in most types of work, and offers a variety of jobs. General repair shops may avoid some repair jobs which require special equipment, such as automatic transmissions, front end alignment, driveability, or air conditioning service. However, they will usually take a variety of other repair work on different makes and types of vehicles. This can be a good place to work if you like to be involved in many different types of diagnosis and repair.

Specialty repair shops confine their repair work to one area of repair, such as transmissions, driveability and tune-up, or engine repair, **Figure 23-5.** They are fully equipped to handle all aspects of their particular specialty. These shops may occasionally take in other minor repair work while performing a repair or when business in their specialty is slow. These shops can be ideal places to work if you want to concentrate on a specific area of repair. Some specialty shops concentrate on service and repairs for one or a few particular vehicle makes.

Figure 23-5. *Independent shops can sometimes be some of the best places for a technician to work. However, you should be careful when you apply. The appearance of the building, its equipment, and surroundings often indicate the type of work produced by the technicians who work there.*

Note: Many chain, franchise, and independent repair shops no longer perform driveability service. This is due to the fact that most of these shops are not equipped with the necessary speciality tools and equipment needed for driveability work. Many of these shops have had problems with comeback driveability work due to the lack of proper equipment, properly trained technicians, or both. Some shops will not perform maintenance tune-ups, even if the customer requests one.

Fleet Agencies

Most ***fleet agencies,*** such as rental companies, maintain their own repair shops for maintenance and light repairs. The typical fleet repair shop is usually well lighted, clean, and equipped with the same tools and service literature as the local dealership repair shop. Most rental agencies lease their vehicles directly from the manufacturer for a set time and are obligated to return these vehicles in reasonably good condition at the end of the lease.

The work at one of these shops is steady, with little pressure to complete a job by a set deadline. Fleet shops usually pay their technicians a straight salary and provide excellent benefits. Many fleet shops are unionized, which may provide additional pay and benefits. The fleet technician often receives the same type of additional training opportunities as the dealership technician.

However, the opportunities to work in such areas as heavy engine and transmission repair do not exist in most fleet shops. This type of work is normally sent to an outside shop or local dealer. The work can become somewhat monotonous, as most of the repairs are minor and maintenance related. A position at a fleet shop is somewhat difficult to obtain, since there are very few openings due to promotions, retirements, or other causes.

Government Agencies

Many federal, state, and local ***government agencies*** maintain their own vehicles. Government operated repair shops can be good places to work. Pay is straight salary, usually set by law, with no commission. Typical work performed at government shops includes mostly maintenance and light repairs. The work may also include installing radios, lights, cellular telephones, and other special equipment used by a particular agency. Some repairs that are complex or covered under the manufacturer's warranty are sent to a dealership. Although civil service pay scales are somewhat lower than private industry, the benefits are usually excellent. Pay raises, while relatively small, are regular. Most government shops work a 35 or 40 hour week, and have the same holidays as other government agencies.

The working conditions in most of these shops are good, without the stress of deadlines, or having to deal with customers. Hiring procedures are more involved than they are with other auto repair shops. Prospective employees must take examinations, which often have little to do with automotive subjects. US government agencies require a certain level of education, thorough background checks, complete lists of former employers and/or references, and that the employee be a registered voter and/or a US citizen. If you think that you would be interested in working for a government owned repair shop, contact your local state employment agency for the addresses of local, state, and federal employment offices in your area.

Self-Employment

Many persons dream of going into business for themselves. This can be a profitable option for the good technician. However, in addition to mechanical and diagnostic ability, the person with their own business must have a certain type of personality to be successful. This type of person must be able to shoulder responsibilities, handle problems, and look for practical ways to increase business and make a profit. This type of person must maintain a clear idea of what plans, both long and short term, need to be made. A person like this is often called a ***entrepreneur,*** in other words, a person who has the energy and skill to build something out of nothing.

When you have your own business, all of the responsibility for repairs, parts ordering, bookkeeping, debt collection, and other problems are yours. Starting your own shop requires a large investment in tools, equipment, and working space. If the money must be borrowed, you will be responsible for paying it back. However, many people enjoy the feeling of independence, of not having to answer to an employer. If you have the personality to deal with the problems, you may enjoy the feeling of owning your own business.

Another possible method of self-employment is to obtain a franchise from a national service chain. A franchise operation removes some of the headaches of being in business for yourself. Many muffler, tire, transmission, tune-up, and other nationally recognized businesses have local owners. They enjoy the advantages of the franchise affiliation, including national advertising, reliable and reasonable parts supplies, corporate support, and employee benefit programs. Disadvantages include high franchise fees and startup costs, lack of local advertising, and some loss of control of shop operations to the national headquarters.

Other Opportunities in the Auto Service Industry

Many other opportunities are available to the automotive technician. These jobs still involve the servicing of vehicles, without much of the physical work. If you like cars and trucks, but are unsure whether you want to make a career of repairing them, one of these jobs may be for you.

Shop Supervisor

The most likely automotive promotion that you will be offered is ***shop supervisor.*** Many repair facilities are large enough to require one or more supervisors or service managers. If you move into management from the shop floor, your salary will increase, and you will be in a cleaner, physically less demanding position. Many technicians enjoy the management position because it lets them in on the fun part of service, troubleshooting, without actually making the repairs.

The disadvantage of a move to management is that you will no longer be dealing with the concrete principles of machines. Instead, you will deal with the unclear and ever-changing personalities of people. Both the customers and technicians will have problems and attitudes that you will have to deal with. Unlike a vehicle problem, these problems require personality and tact. Sometimes the manager has to give ground, which can be hard for the person who is used to being right, and saying so.

The paperwork load is large for any manager and may not be something that a former technician can get used to. Record keeping requires a good bit of desk time. Automotive record keeping is like balancing a checkbook and writing a term paper every few days. If you do not care to deal with people or keep records, a career in management may not be for you.

Service Advisor

Many people enjoy the challenge of selling. The ***service advisor*** performs a vital service, since repairs will not be performed unless the owner is sold on the necessity of having them done. Service advisors are not necessary in many small independent shops, but are often an important part of large shops, dealership service departments, department store service centers, and specialty repair shops.

The service advisor enjoys a large income and is directly responsible for a large amount of business in the shop. However, selling is a people oriented job, and takes a lot of persuasive ability and diplomacy. Often, a service advisor is the first to endure the wrath of an angry customer if their vehicle still has a problem. If you are not interested in dealing with the public, a service advisor's position may not be for you.

Parts Person

One often overlooked area of the automotive service business is the process of supplying parts. It is as vital as any other area. There are many types of parts outlets, including dealership parts departments, independent parts stores, parts departments in retail stores, and combination parts and service outlets. All of these parts outlets meet the needs of technicians and repair shops, as well as the do-it-yourself needs of the vehicle owner.

Parts persons are trained in the methods of keeping the supply of parts flowing through the system until they reach the ultimate endpoint, the vehicle. This can be a challenging job. Parts must be carefully checked into the parts department and stored so that they can be found again, **Figure 23-6.** When a specific part is needed, it must be located and brought to the person requesting it. If more of the same parts are needed, they must be ordered. If the part is not in stock, it must be special ordered.

The job of the parts person appeals to many people. The job of the parts person does not pay as much as some other areas of automotive service, but rates are comparable with other jobs with the same skill level. If this type of work appeals to you, it may be a good job choice.

Figure 23-6. *A parts department can be a good place to work if you do not like the fast pace of the auto shop.*

Getting a Job

There are many automotive jobs available at all times. The problem is connecting with the job when it is available. There are essentially two hurdles to getting a job: finding a job opening, and successfully applying for the job. Below are some hints for overcoming these hurdles.

Finding Job Openings

Before applying for a job, you must know about a suitable opening. A good place to start is with your instructors. They often have contacts in the local automobile industry and may be able to recommend you to a local company. Another good place to begin your search is the classified section of your local newspaper. Automotive jobs are often advertised in newspapers, especially those for automotive dealerships, specialty and franchise shops, and independent repair shops.

Also visit your local state employment agency or job service, **Figure 23-7.** Most of these agencies keep records of job openings, usually throughout the state, and may be able to connect to a data bank of nationwide job openings. You may be able to connect to some of these job banks at home if you have a computer and on-line access. Some private employment agencies specialize in automotive placement. If there is an agency of this type in your area, arrange for an interview with one of their recruiters.

Visit local repair shops that you are interested in working for. Sometimes these shops have an opening that they have not advertised. If you are interested in working for a chain or department store, most of these stores have a personnel department where you can fill out an application.

Figure 23-7. Your local state employment office has many job listings that are not normally advertised.

Even if no jobs are available, your application will be placed on file in the event a position becomes available in the future.

Applying for the Job

No matter how good your qualifications, you will not get a job if you make a poor impression on the job interviewer. It is important to put your best foot forward when applying for a job.

Start creating a good impression when you fill out the employment application. Type or neatly print when filling out the application. Complete all blanks, and completely explain any blanks in your education or previous employment history. List all of your educational qualifications, including those which many not apply directly to the automotive industry. **Figure 23-8** shows a typical employment application.

When you are called for an interview, try to arrange for a morning interview, since people are most likely to be in a positive mood in the morning. Dress neatly and arrive on time or a little early. When introduced to the interviewer, make an effort to repeat and remember their name. Speak clearly when answering the interviewer's questions. Do not smoke or chew gum during the interview. State your qualifications for the job without bragging or belittling your accomplishments. At the conclusion of the interview, thank the interviewer for his or her time. If you do not hear from the interviewer in a few days, it is permissible to make a brief and polite follow-up call.

Work Ethics

It is extremely tempting and very easy to make money in the automotive business by selling unneeded parts and services, charging for work not done, or performing hurried repairs. Unfortunately, many technicians try to get as much money as possible out of a job, no matter what the actual repair needs. Other technicians work haphazardly to get as many repair jobs as possible completed, and do not really care whether the vehicles are fixed properly. These technicians are one of the primary causes of the driving public's low opinion of automotive technicians. This opinion can only change if every technician acts responsibly and ethically.

After a long and frustrating job, or when business has been slow, even the most honest technician is tempted to make some easy money by cutting a corner or two. You might even get away with it. After all, customers bring their vehicles to you because they don't know how to fix it themselves. This is when you must put yourself in the customers' place. Think about this: if your television, computer, or VCR stops working, you must take it to a shop that specializes in electronic repairs. Unless you are an expert in electronics, you have no way of knowing whether the problem is simple to fix or requires a lot of labor and expensive parts. You must therefore trust a specialist in electronic equipment service to tell you the truth about what is wrong with the unit, fix it as though it was their own, and have it ready when they say they will. The automotive customer is in the same position, trusting you to tell them the truth about what is wrong with their vehicle, and then to fix it to the best of your ability.

There is an unwritten "code of ethics" that all good technicians follow. It can vary from shop to shop, but should be followed to ensure good treatment of the customer. Some of the rules to follow when working with any customer include:

- ❑ Care for each vehicle as if it belonged to you. Do not do anything to a vehicle that you would not be willing to do to your own.
- ❑ Provide a carefully prepared, fairly priced, itemized estimate, which clearly lists all needed repairs and recommended maintenance services.
- ❑ Always obtain customer authorization, in writing or by other satisfactory means, before beginning any repair.
- ❑ Keep the customer informed at all times as to the status of their vehicle.
- ❑ Furnish an itemized invoice for all parts and services. List the complaint, cause, and correction of each problem in all cases. Be sure to identify any parts that are used, rebuilt, or remanufactured. Provide replaced parts for customer inspection upon request.
- ❑ Always strive to fix any problem right the first time.
- ❑ Always be honest and fair.
- ❑ Always maintain high standards for your work, no matter how insignificant the job.
- ❑ Report and correct any and all abuses within the automotive industry.

Dealing with Customers

The driving public's main complaints against technicians are:

- ❑ The original problem is not corrected.
- ❑ The vehicle is not ready when promised.
- ❑ The final bill is much higher than the original estimate.
- ❑ The nature of the problem or the type of repairs is not adequately explained.

Honest Auto Repair **Employment Application**

123 Credibility Street Edens, IL 60000

Date:__________________ Social Security Number:________________________

Name:____

Address:____

Phone:________________ United States citizen?_____________ Can you furnish proof?__________

Employment Desired

Position:__________________ Date you can start:_________________ Expected salary:____________

Are you currently employed?__________ May we inquire of your present employer?_________________

Education

Circle the number for the highest level completed:

High School	Trade/Technical School	Community/Junior College	University
1 2 3 4	1 2	1 2	1 2 3 4

Other:____

Specialized Training or Certifications:____

____

Employment Record

Current/Last employer:__________________________________ From:__________ To:__________

Address:____ Phone:________________

Salary:_____________ Job description:____

Reason for leaving:____

Previous employer:_____________________________________ From:__________ To:__________

Address:____ Phone:________________

Salary:_____________ Job description:____

Reason for leaving:____

Previous employer:_____________________________________ From:__________ To:__________

Address:____ Phone:________________

Salary:_____________ Job description:____

Reason for leaving:____

Figure 23-8. *Getting a job often starts with filling out an employment application.*

Inevitably, some customer-technician friction will occur no matter how much effort is put into avoiding it. Vehicles and people are fallible and things can go wrong. However, complaints can be reduced by ethical conduct and careful service procedures. A lot of friction is caused by a failure to communicate. Never assume that the vehicle owner understands what is involved in a repair operation. Customers have very little knowledge of how much labor is involved to perform a particular repair or the parts, tools, and equipment needed.

To keep your customers happy, always make sure that they know what the cost of a repair will be, and how long the job will take. Explain why your repairs may cost more than the advertised prices at other shops. If you discover the need for additional work after the original estimate, immediately inform the customer about it, and get approval to proceed.

Always provide the customer with an itemized bill for services. Use the repair order to fully and clearly explain just what was done and what parts and labor were involved, **Figure 23-9.** Answer any questions that the customer may have, and explain warranties or guarantees. This helps to satisfy the customer and covers you in the event that the vehicle develops other problems. A clear record of exactly what was done to the vehicle prevents misunderstandings. Keeping accurate records is even more important when the customer declines some needed repairs.

Work Habits

Habits can work for or against you. If you get in the habit of careless diagnosis or sloppy work, these habits will be with you for the rest of your life. However, if you get in the habit of making logical diagnoses and performing careful work, these habits will work to your advantage for your entire career.

The best habit that you can develop is to fix it right the first time. Unfortunately, it is much easier to fall into bad habits, because at the time they seem like the path of least resistance, the easiest thing to do. It is important to avoid the path of least resistance by always trying to do things right. After doing the right thing enough times, it will become a habit.

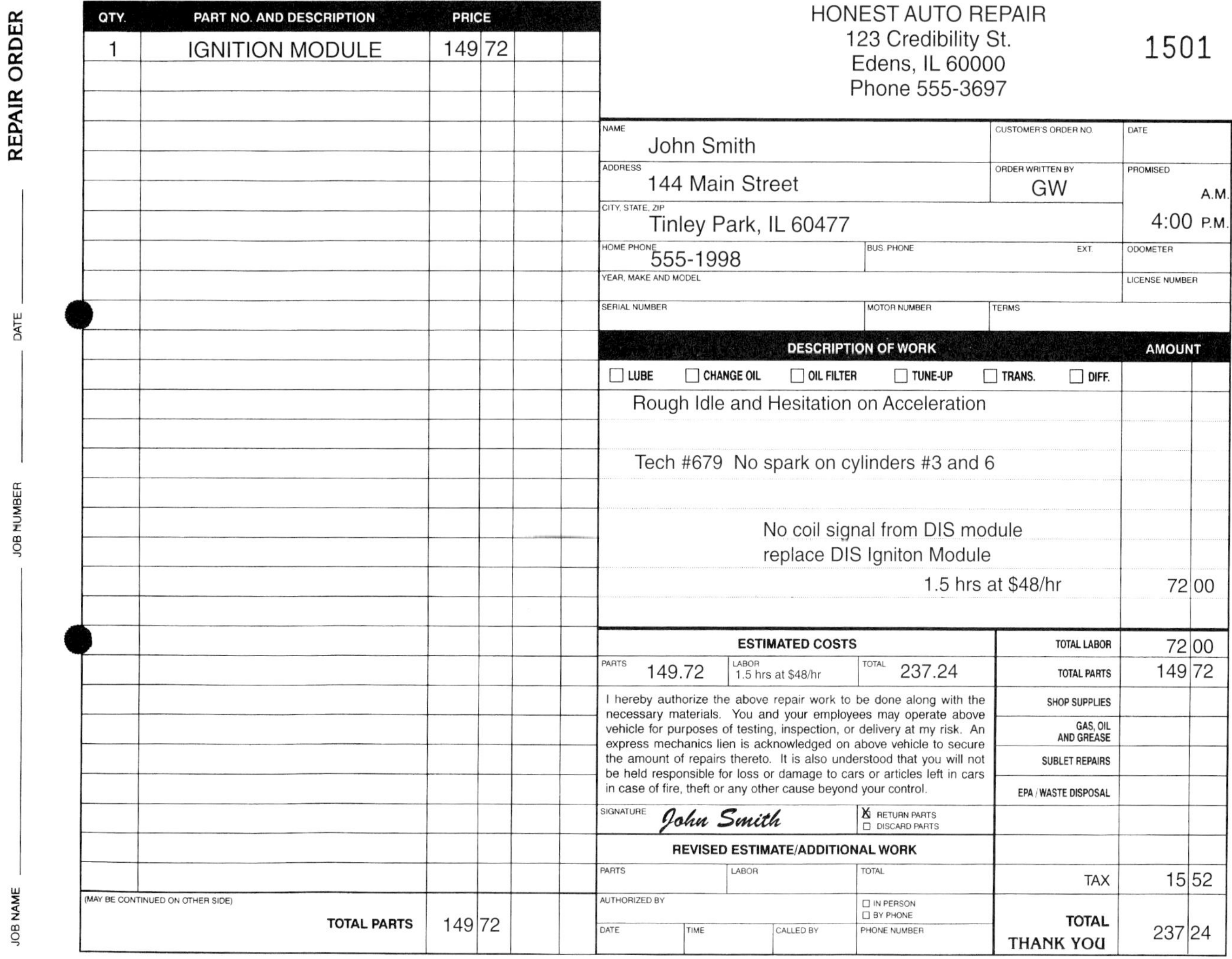

REPAIR ORDER

DATE

JOB NUMBER

JOB NAME

QTY.	PART NO. AND DESCRIPTION	PRICE
1	IGNITION MODULE	149 72
(MAY BE CONTINUED ON OTHER SIDE)	TOTAL PARTS	149 72

HONEST AUTO REPAIR
123 Credibility St.
Edens, IL 60000
Phone 555-3697

1501

NAME: John Smith
CUSTOMER'S ORDER NO.
DATE
ADDRESS: 144 Main Street
ORDER WRITTEN BY: GW
PROMISED: 4:00 P.M.
CITY, STATE, ZIP: Tinley Park, IL 60477
HOME PHONE: 555-1998
BUS. PHONE
EXT.
ODOMETER
YEAR, MAKE AND MODEL
LICENSE NUMBER
SERIAL NUMBER
MOTOR NUMBER
TERMS

DESCRIPTION OF WORK	AMOUNT
☐ LUBE ☐ CHANGE OIL ☐ OIL FILTER ☐ TUNE-UP ☐ TRANS. ☐ DIFF.	
Rough Idle and Hesitation on Acceleration	
Tech #679 No spark on cylinders #3 and 6	
No coil signal from DIS module	
replace DIS Igniton Module	
1.5 hrs at $48/hr	72 00

ESTIMATED COSTS

PARTS: 149.72 LABOR: 1.5 hrs at $48/hr TOTAL: 237.24

I hereby authorize the above repair work to be done along with the necessary materials. You and your employees may operate above vehicle for purposes of testing, inspection, or delivery at my risk. An express mechanics lien is acknowledged on above vehicle to secure the amount of repairs thereto. It is also understood that you will not be held responsible for loss or damage to cars or articles left in cars in case of fire, theft or any other cause beyond your control.

SIGNATURE: John Smith
☒ RETURN PARTS
☐ DISCARD PARTS

REVISED ESTIMATE/ADDITIONAL WORK

PARTS LABOR TOTAL
AUTHORIZED BY
☐ IN PERSON
☐ BY PHONE
DATE TIME CALLED BY PHONE NUMBER

TOTAL LABOR	72 00
TOTAL PARTS	149 72
SHOP SUPPLIES	
GAS, OIL AND GREASE	
SUBLET REPAIRS	
EPA / WASTE DISPOSAL	
TAX	15 52
TOTAL THANK YOU	237 24

Figure 23-9. *When filling out a repair order, remember to list the complaint, what caused it, and what you did to correct it. Also list all parts that were replaced and any adjustments made.*

Summary

The automotive service industry provides employment for many people, and will continue to do so. Automotive service has some disadvantages, such as long hours; hard work, both mentally and physically; lack of status; and difficulties in dealing with the public. Advantages include interesting work, the security of a guaranteed career, and the enjoyment of diagnosing and correcting problems. Always stay employable by learning new things and obtaining ASE certification.

The three general classes of technicians are the helper, the apprentice, and the certified technician. The helper does the simplest tasks, such as tire changing and lubrication. Many helpers move up into the other classes after a short time. The apprentice installs new parts, such as shock absorbers and strut assemblies, and eventually moves into other repairs and some diagnostics. The certified technician performs the most complex diagnosis and repair jobs on vehicles, and makes the most money.

There are many places to work as an automotive technician. Among the most popular are new car and truck dealers, auto centers affiliated with department or chain stores, specialty shops, independent repair shops, and government agencies. Some people prefer to have their own businesses, either as independent owners, or as part of a franchise system.

Other opportunities in the automotive service field include moving into management as a foreman or service manager, or into sales as a service advisor or specialty sales person. Another often overlooked employment possibility is in the automotive parts business.

To obtain a job in the automotive business, first locate possible job openings. Try the local newspaper, state job service, and repair shops in your area. Most department or other large stores with auto service centers have personnel departments where you can fill out a job application. To get a job, you must make a good impression, no matter how qualified you are. Fill out all job applications carefully and neatly, listing your qualifications honestly. When invited to a job interview, dress neatly, arrive on time, and be courteous. Answer all questions without over or understating your abilities and experience. Follow up the interview with a brief phone call within a few days.

Know These Terms

Helper
Apprentice
Certified technician
Vehicle dealers
Auto service centers
General repair shop
Specialty repair shop
Fleet agencies
Government agencies
Entrepreneur
Shop supervisor
Service advisor
Parts person

Review Questions—Chapter 23

Please do not write in this text. Write your answers on a separate sheet of paper.

1. List some of the disadvantages of working in the automotive service field.
2. List some of the advantages of the automotive service field.
3. List the three general classes of automotive technicians.
4. Technicians in chain and department store automotive centers are paid by what type of system?
   (A) Salary only.
   (B) Salary plus commission.
   (C) Commission for work performed only.
   (D) Both B & C.
5. ______ confine their repair work to one area of repair, such as transmissions, driveability and tune-up, or engine repair.
6. A major disadvantage of moving up to management is that you will be dealing with ______ instead of ______.
7. In what types of automotive repair shops are service advisors needed?
8. Parts outlets meet the needs of which of the following groups?
   (A) Technicians.
   (B) Repair shops.
   (C) Vehicle owners.
   (D) All of the above.
9. There are two major hurdles to getting a job. Name them.
10. The time to start making a good impression on a possible employer is:
    (A) when you fill out the employment application.
    (B) during the job interview.
    (C) in a follow up letter or call.
    (D) as soon as you get the job.

ASE Certification-Type Questions

1. Places where the prospective automotive technician can get started in the automobile industry include all of the following, EXCEPT:
   (A) specialty stores.
   (B) self-serve convenience stores.
   (C) chain stores.
   (D) new vehicle dealers.

2. Automotive service jobs are advertised in all of the following, EXCEPT:
   (A) local newspapers.
   (B) flyers and handbills.
   (C) state employment agency data banks.
   (D) certain private employment agencies.
3. A technician being paid by the flat rate method is being ______.
   (A) paid by the amount of repair work that he/she does
   (B) paid by the number of hours spent in the shop
   (C) paid a salary based on 35 or 40 hours per week
   (D) paid a percentage of the profit that the shop makes
4. Large new vehicle dealerships may handle several ______.
   (A) makes of vehicle
   (B) types of repair
   (C) methods of technician payment
   (D) Both A & B.
5. A general repair shop can be a good place to work if you like ______.
   (A) working on only one type of vehicle
   (B) one kind of repair work
   (C) a variety of work
   (D) monotonous work
6. Advantages of working for a government agency includes all of the following, EXCEPT:
   (A) steady pay.
   (B) excellent benefits.
   (C) demanding deadlines.
   (D) regular raises.
7. Technician A says that, when seeking a job as a driveability technician, independent shops and chain stores are good places to look. Technician B says that vehicle dealerships and shops that specialize in driveability work are good places to look. Who is right?
   (A) A only.
   (B) B only.
   (C) Both A & B.
   (D) Neither A nor B.
8. When you own your own automotive repair business, you will have all of the following, EXCEPT:
   (A) responsibility for debt collection.
   (B) profits.
   (C) loss of control of shop operations.
   (D) independence from answering to an employer.
9. Needed parts can be special ordered if they are ______.
   (A) no longer manufactured
   (B) not in stock
   (C) not vital to vehicle operation
   (D) are cheaper
10. The best time for a job interview is ______.
   (A) the morning
   (B) late afternoon
   (C) outside normal working hours
   (D) whenever it is convenient for you

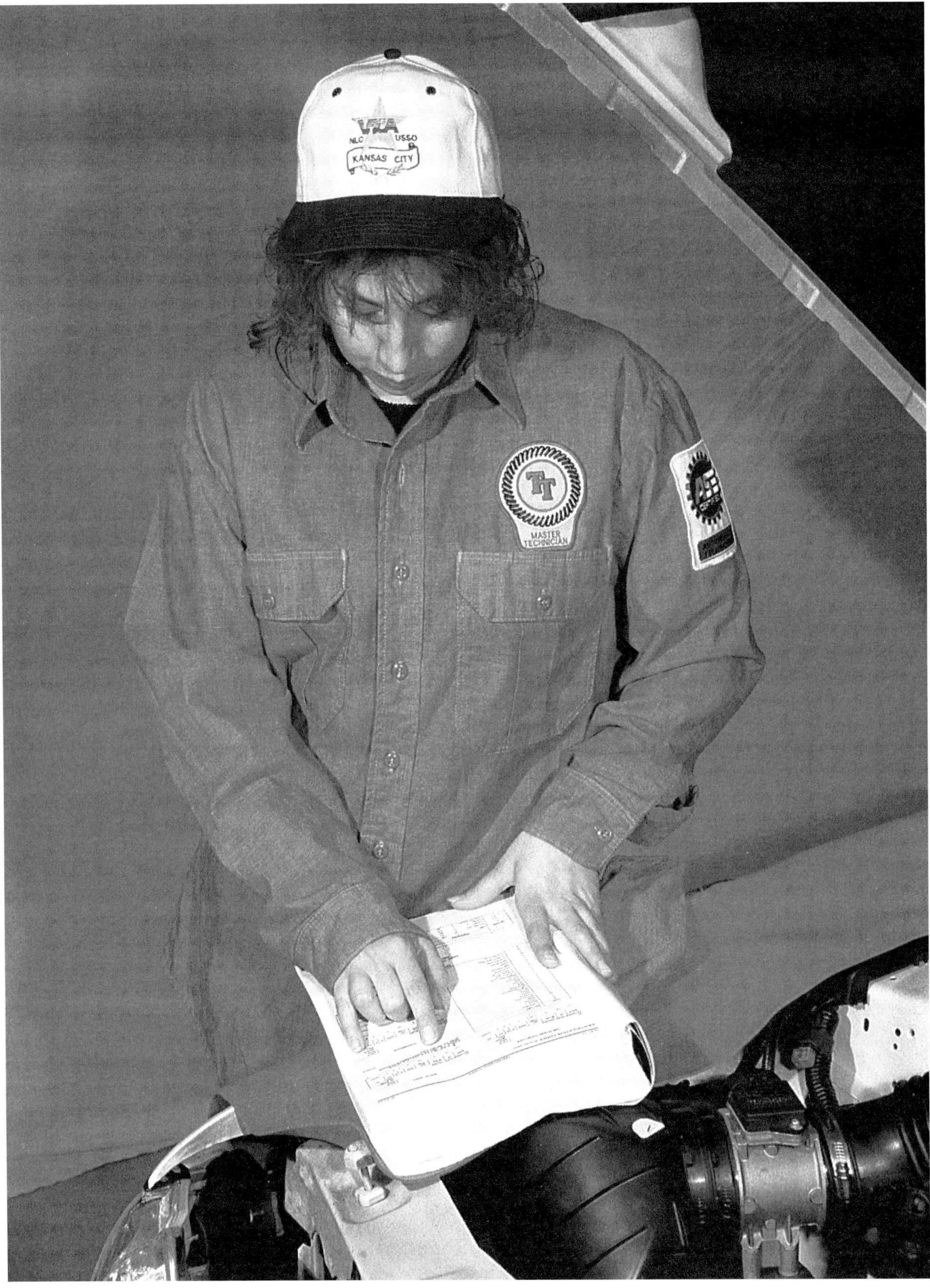

You should always look in a service manual after retrieving trouble codes from a vehicle's computer control system. (Jack Klasey)

Appendix A

Diagnostic Codes

This appendix is provided to you for use in answering some of the review questions in this text and as a convenient reference for codes you may encounter while diagnosing vehicles in the shop. The list of codes shown here was compiled from several sources and was current at the time of publication. Goodheart-Willcox Publisher cannot be held responsible for any errors or omissions. Any codes generated by a vehicle's computer control system should be checked in the appropriate service manual.

OBD II Trouble Codes

These are the generic SAE codes generated by OBD II diagnostic systems. Some late model OBD I ECMs will provide these codes to some scan tools along with the normal two digit code. Manufacturer specific codes (codes that begin with a P1 alpha-numeric designator) should be looked up in the service manual.

P01XX Fuel and Air Metering

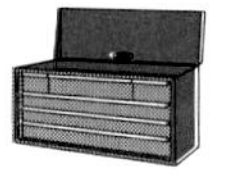

Note: For systems with single O_2 sensors, use codes for Bank 1 sensor. Bank 1 contains cylinder #1. Sensor 1 is closest to the engine.

P0100 Mass or Volume Airflow Circuit Malfunction
P0101 Mass or Volume Airflow Circuit Range/Performance Problem
P0102 Mass or Volume Airflow Circuit Low Input
P0103 Mass or Volume Airflow Circuit High Input
P0104 Mass or Volume Airflow Circuit Intermittent
P0105 Manifold Absolute Pressure/Barometric Pressure Circuit Malfunction
P0106 Manifold Absolute Pressure/Barometric Pressure Circuit Range/Performance Problem
P0107 Manifold Absolute Pressure/Barometric Pressure Circuit Low Input
P0108 Manifold Absolute Pressure/Barometric Pressure Circuit High Input
P0109 Manifold Absolute Pressure/Barometric Pressure Circuit Intermittent
P0110 Intake Air Temperature Circuit Malfunction
P0111 Intake Air Temperature Circuit Range/Performance Problem
P0112 Intake Air Temperature Circuit Low Input
P0113 Intake Air Temperature Circuit High Input
P0114 Intake Air Temperature Circuit Intermittent
P0115 Engine Coolant Temperature Circuit Malfunction
P0116 Engine Coolant Temperature Circuit Range/Performance Problem
P0117 Engine Coolant Temperature Circuit Low Input
P0118 Engine Coolant Temperature Circuit High Input
P0119 Engine Coolant Temperature Circuit Intermittent
P0120 Throttle/Pedal Position Sensor/Switch A Circuit Malfunction
P0121 Throttle/Pedal Position Sensor/Switch A Circuit Range/Performance Problem

P0122 Throttle/Pedal Position Sensor/Switch A Circuit Low Input
P0123 Throttle/Pedal Position Sensor/Switch A Circuit High Input
P0124 Throttle/Pedal Position Sensor/Switch A Circuit Intermittent
P0125 Insufficient Coolant Temperature for Closed Loop Fuel Control
P0126 Insufficient Coolant Temperature for Stable Operation
P0130 Oxygen Sensor Circuit Malfunction (Bank 1 Sensor 1)
P0131 Oxygen Sensor Circuit Low Voltage (Bank 1 Sensor 1)
P0132 Oxygen Sensor Circuit High Voltage (Bank 1 Sensor 1)
P0133 Oxygen Sensor Circuit Slow Response (Bank 1 Sensor 1)
P0134 Oxygen Sensor Circuit No Activity Detected (Bank 1 Sensor 1)
P0135 Oxygen Sensor Heater Circuit Malfunction (Bank 1 Sensor 2)
P0136 Oxygen Sensor Circuit Malfunction (Bank 1 Sensor 2)
P0137 Oxygen Sensor Circuit Low Voltage (Bank 1 Sensor 2)
P0138 Oxygen Sensor Circuit High Voltage (Bank 1 Sensor 2)
P0139 Oxygen Sensor Circuit Slow Response (Bank 1 Sensor 2)
P0140 Oxygen Sensor Circuit No Activity Detected (Bank 1 Sensor 2)
P0141 Oxygen Sensor Heater Circuit Malfunction (Bank 1 Sensor 2)
P0142 Oxygen Sensor Circuit Malfunction (Bank 1 Sensor 3)
P0143 Oxygen Sensor Circuit Low Voltage (Bank 1 Sensor 3)
P0144 Oxygen Sensor Circuit High Voltage (Bank 1 Sensor 3)
P0145 Oxygen Sensor Circuit Slow Response (Bank 1 Sensor 3)
P0146 Oxygen Sensor Circuit No Activity Detected (Bank 1 Sensor 3)
P0147 Oxygen Sensor Heater Circuit Malfunction (Bank 1 Sensor 3)
P0150 Oxygen Sensor Circuit Malfunction (Bank 2 Sensor 1)
P0151 Oxygen Sensor Circuit Low Voltage (Bank 2 Sensor 1)
P0152 Oxygen Sensor Circuit High Voltage (Bank 2 Sensor 1)
P0153 Oxygen Sensor Circuit Slow Response (Bank 2 Sensor 1)
P0154 Oxygen Sensor Circuit No Activity Detected (Bank 2 Sensor 1)
P0155 Oxygen Sensor Heater Circuit Malfunction (Bank 2 Sensor 1)
P0156 Oxygen Sensor Circuit Malfunction (Bank 2 Sensor 1)
P0157 Oxygen Sensor Circuit Low Voltage (Bank 2 Sensor 2)
P0158 Oxygen Sensor Circuit High Voltage (Bank 2 Sensor 2)
P0159 Oxygen Sensor Circuit Slow Response (Bank 2 Sensor 2)
P0160 Oxygen Sensor Circuit No Activity Detected (Bank 2 Sensor 1)
P0161 Oxygen Sensor Heater Circuit Malfunction (Bank 2 Sensor 2)
P0162 Oxygen Sensor Circuit Malfunction (Bank 2 Sensor 2)
P0163 Oxygen Sensor Circuit Low Voltage (Bank 2 Sensor 3)
P0164 Oxygen Sensor Circuit High Voltage (Bank 2 Sensor 3)
P0165 Oxygen Sensor Circuit Slow Response (Bank 2 Sensor 3)
P0166 Oxygen Sensor Circuit No Activity Detected (Bank 2 Sensor 3)
P0167 Oxygen Sensor Heater Circuit Malfunction (Bank 2 Sensor 3)
P0170 Fuel Trim Malfunction (Bank 1)
P0171 System too Lean (Bank 1)
P0172 System too Rich (Bank 1)
P0173 Fuel Trim Malfunction (Bank 1)
P0174 System too Lean (Bank 2)
P0175 System too Rich (Bank 2)
P0176 Fuel Composition Sensor Circuit Malfunction
P0177 Fuel Composition Sensor Circuit Range/Performance
P0178 Fuel Composition Sensor Circuit Low Input
P0179 Fuel Composition Sensor Circuit High Input
P0180 Fuel Temperature Sensor A Circuit Malfunction
P0181 Fuel Temperature Sensor A Circuit Range/Performance
P0182 Fuel Temperature Sensor A Circuit Low Input
P0183 Fuel Temperature Sensor A Circuit High Input
P0184 Fuel Temperature Sensor A Circuit Intermittent
P0185 Fuel Temperature Sensor B Circuit Malfunction
P0186 Fuel Temperature Sensor B Circuit Range/Performance
P0187 Fuel Temperature Sensor B Circuit Low Input
P0188 Fuel Temperature Sensor B Circuit High Input
P0189 Fuel Temperature Sensor B Circuit Intermittent
P0190 Fuel Rail Pressure Sensor Circuit Malfunction
P0191 Fuel Rail Pressure Sensor Circuit Range/Performance
P0192 Fuel Rail Pressure Sensor Circuit Low Input
P0193 Fuel Rail Pressure Sensor Circuit High Input
P0194 Fuel Rail Pressure Sensor Circuit Intermittent
P0195 Engine Oil Temperature Sensor Malfunction
P0196 Engine Oil Temperature Sensor Range/Performance
P0197 Engine Oil Temperature Sensor Low
P0198 Engine Oil Temperature Sensor High
P0199 Engine Oil Temperature Sensor Intermittent

P02XX Fuel and Air Metering

P0200 Injector Circuit Malfunction
P0201 Injector Circuit Malfunction–Cylinder 1
P0202 Injector Circuit Malfunction–Cylinder 2
P0203 Injector Circuit Malfunction–Cylinder 3
P0204 Injector Circuit Malfunction–Cylinder 4
P0205 Injector Circuit Malfunction–Cylinder 5
P0206 Injector Circuit Malfunction–Cylinder 6
P0207 Injector Circuit Malfunction–Cylinder 7
P0208 Injector Circuit Malfunction–Cylinder 8
P0209 Injector Circuit Malfunction–Cylinder 9
P0210 Injector Circuit Malfunction–Cylinder 10
P0211 Injector Circuit Malfunction–Cylinder 11
P0212 Injector Circuit Malfunction–Cylinder 12
P0213 Cold Start Injector 1 Malfunction
P0214 Cold Start Injector 2 Malfunction
P0215 Engine Shutoff Solenoid Malfunction
P0216 Injection Timing Control Circuit Malfunction
P0217 Engine Over Temperature Condition
P0218 Transmission Over Temperature Condition
P0219 Engine Overspeed Condition
P0220 Throttle/Pedal Position Sensor/Switch B Circuit Malfunction
P0221 Throttle/Pedal Position Sensor/Switch B Circuit Range/Performance Problem
P0222 Throttle/Pedal Position Sensor/Switch B Circuit Low Input
P0223 Throttle/Pedal Position Sensor/Switch B Circuit High Input
P0224 Throttle/Pedal Position Sensor/Switch B Circuit Intermittent
P0225 Throttle/Pedal Position Sensor/Switch C Circuit Malfunction
P0226 Throttle/Pedal Position Sensor/Switch C Circuit Range/Performance Problem
P0227 Throttle/Pedal Position Sensor/Switch C Circuit Low Input
P0228 Throttle/Pedal Position Sensor/Switch C Circuit High Input
P0229 Throttle/Pedal Position Sensor/Switch C Circuit Intermittent
P0230 Fuel Pump Primary Circuit Malfunction
P0231 Fuel Pump Secondary Circuit Low
P0232 Fuel Pump Secondary Circuit High
P0233 Fuel Pump Secondary Circuit Intermittent
P0235 Turbocharger Boost Sensor A Circuit Malfunction
P0236 Turbocharger Boost Sensor A Circuit Range/Performance
P0237 Turbocharger Boost Sensor A Circuit Low
P0238 Turbocharger Boost Sensor A Circuit High
P0239 Turbocharger Boost Sensor B Circuit Malfunction
P0240 Turbocharger Boost Sensor B Circuit Range/Performance
P0241 Turbocharger Boost Sensor B Circuit Low
P0242 Turbocharger Boost Sensor B Circuit High
P0243 Turbocharger Wastegate Solenoid A Malfunction
P0244 Turbocharger Wastegate Solenoid A Range/Performance
P0245 Turbocharger Wastegate Solenoid A Low
P0246 Turbocharger Wastegate Solenoid A High
P0247 Turbocharger Wastegate Solenoid B Malfunction
P0248 Turbocharger Wastegate Solenoid B Range/Performance
P0249 Turbocharger Wastegate Solenoid B Low
P0250 Turbocharger Wastegate Solenoid B High
P0251 Injection Pump A Rotor/Cam Malfunction
P0252 Injection Pump A Rotor/Cam Range/Performance
P0253 Injection Pump A Rotor/Cam Low
P0254 Injection Pump A Rotor/Cam High
P0255 Injection Pump A Rotor/Cam Intermittent
P0256 Injection Pump B Rotor/Cam Malfunction
P0257 Injection Pump B Rotor/Cam Range/Performance
P0258 Injection Pump B Rotor/Cam Low
P0259 Injection Pump B Rotor/Cam High
P0260 Injection Pump B Rotor/Cam Intermittent
P0261 Cylinder 1 Injector Circuit Low
P0262 Cylinder 1 Injector Circuit High
P0263 Cylinder 1 Contribution/Balance Fault
P0264 Cylinder 2 Injector Circuit Low
P0265 Cylinder 2 Injector Circuit High
P0266 Cylinder 2 Contribution/Balance Fault
P0267 Cylinder 3 Injector Circuit Low
P0268 Cylinder 3 Injector Circuit High
P0269 Cylinder 3 Contribution/Balance Fault
P0270 Cylinder 4 Injector Circuit Low
P0271 Cylinder 4 Injector Circuit High
P0272 Cylinder 4 Contribution/Balance Fault
P0273 Cylinder 5 Injector Circuit Low
P0274 Cylinder 5 Injector Circuit High
P0275 Cylinder 5 Contribution/Balance Fault
P0276 Cylinder 6 Injector Circuit Low
P0277 Cylinder 6 Injector Circuit High
P0278 Cylinder 6 Contribution/Balance Fault
P0279 Cylinder 7 Injector Circuit Low
P0280 Cylinder 7 Injector Circuit High
P0281 Cylinder 7 Contribution/Balance Fault
P0282 Cylinder 8 Injector Circuit Low
P0283 Cylinder 8 Injector Circuit High
P0284 Cylinder 8 Contribution/Balance Fault
P0285 Cylinder 9 Injector Circuit Low
P0286 Cylinder 9 Injector Circuit High
P0287 Cylinder 9 Contribution/Balance Fault
P0288 Cylinder 10 Injector Circuit Low
P0289 Cylinder 10 Injector Circuit High
P0290 Cylinder 10 Contribution/Balance Fault
P0291 Cylinder 11 Injector Circuit Low
P0292 Cylinder 11 Injector Circuit High
P0293 Cylinder 11 Contribution/Balance Fault
P0294 Cylinder 12 Injector Circuit Low
P0295 Cylinder 12 Injector Circuit High
P0296 Cylinder 12 Contribution/Balance Fault

P03XX Ignition System or Misfire

Note: Bank 1 contains cylinder #1.

P0300 Random/Multiple Cylinder Misfire Detected
P0301 Cylinder 1 Misfire Detected
P0302 Cylinder 2 Misfire Detected
P0303 Cylinder 3 Misfire Detected
P0304 Cylinder 4 Misfire Detected
P0305 Cylinder 5 Misfire Detected
P0306 Cylinder 6 Misfire Detected
P0307 Cylinder 7 Misfire Detected
P0308 Cylinder 8 Misfire Detected
P0309 Cylinder 9 Misfire Detected
P0310 Cylinder 10 Misfire Detected
P0311 Cylinder 11 Misfire Detected
P0312 Cylinder 12 Misfire Detected
P0320 Ignition/Distributor Engine Speed Input Circuit Malfunction
P0321 Ignition/Distributor Engine Speed Input Circuit Range/Performance
P0322 Ignition/Distributor Engine Speed Input Circuit No Signal
P0323 Ignition/Distributor Engine Speed Input Circuit Intermittent
P0325 Knock Sensor 1 Circuit Malfunction (Bank 1 or Single Sensor)
P0326 Knock Sensor 1 Circuit Range/Performance (Bank 1 or Single Sensor)
P0327 Knock Sensor 1 Circuit Low Input (Bank 1 or Single Sensor)
P0328 Knock Sensor 1 Circuit High Input (Bank 1 or Single Sensor)
P0329 Knock Sensor 1 Circuit Input Intermittent (Bank 1 or Single Sensor)
P0330 Knock Sensor 2 Circuit Malfunction (Bank 2)
P0331 Knock Sensor 2 Circuit Range/Performance (Bank 2)
P0332 Knock Sensor 2 Circuit Low Input (Bank 2)
P0333 Knock Sensor 2 Circuit High Input (Bank 2)
P0334 Knock Sensor 2 Circuit Input Intermittent (Bank 2)
P0335 Crankshaft Position Sensor A Circuit Malfunction
P0336 Crankshaft Position Sensor A Circuit Range/Performance
P0337 Crankshaft Position Sensor A Circuit Low Input
P0338 Crankshaft Position Sensor A Circuit High Input
P0339 Crankshaft Position Sensor A Circuit Intermittent
P0340 Camshaft Position Sensor Circuit Malfunction
P0341 Camshaft Position Sensor Circuit Range/Performance
P0342 Camshaft Position Sensor Circuit Low Input
P0343 Camshaft Position Sensor Circuit High Input
P0344 Camshaft Position Sensor Circuit Intermittent
P0350 Ignition Coil Primary/Secondary Circuit Malfunction
P0351 Ignition Coil A Primary/Secondary Circuit Malfunction
P0352 Ignition Coil B Primary/Secondary Circuit Malfunction
P0353 Ignition Coil C Primary/Secondary Circuit Malfunction
P0354 Ignition Coil D Primary/Secondary Circuit Malfunction
P0355 Ignition Coil E Primary/Secondary Circuit Malfunction
P0356 Ignition Coil F Primary/Secondary Circuit Malfunction
P0357 Ignition Coil G Primary/Secondary Circuit Malfunction
P0358 Ignition Coil H Primary/Secondary Circuit Malfunction
P0359 Ignition Coil I Primary/Secondary Circuit Malfunction
P0360 Ignition Coil J Primary/Secondary Circuit Malfunction
P0361 Ignition Coil K Primary/Secondary Circuit Malfunction
P0362 Ignition Coil L Primary/Secondary Circuit Malfunction
P0370 Timing Reference High Resolution Signal A Malfunction
P0371 Timing Reference High Resolution Signal A Too Many Pulses
P0372 Timing Reference High Resolution Signal A Malfunction
P0373 Timing Reference High Resolution Signal Intermittent/Erratic Pulses
P0374 Timing Reference High Resolution Signal A No Pulse
P0375 Timing Reference High Resolution Signal B Malfunction
P0376 Timing Reference High Resolution Signal B Too Many Pulses
P0377 Timing Reference High Resolution Signal B Too Few Pulses
P0378 Timing Reference High Resolution Signal B Intermittent/Erratic Pulses
P0379 Timing Reference High Resolution Signal B No Pulse
P0380 Glow Plug/Heater Circuit Malfunction
P0381 Glow Plug/Heater Indicator Circuit Malfunction
P0385 Crankshaft Position Sensor B Circuit Malfunction
P0386 Crankshaft Position Sensor B Circuit Range/Performance
P0387 Crankshaft Position Sensor B Circuit Low Input
P0388 Crankshaft Position Sensor B Circuit High Input
P0389 Crankshaft Position Sensor B Circuit Intermittent

P04XX Auxiliary Emission Controls

Note: Bank 1 contains cylinder #1.

P0400 Exhaust Gas Recirculation Flow Malfunction
P0401 Exhaust Gas Recirculation Flow Insufficient Detected
P0402 Exhaust Gas Recirculation Flow Excessive Detected
P0403 Exhaust Gas Recirculation Circuit Malfunction
P0404 Exhaust Gas Recirculation Circuit Range/Performance
P0405 Exhaust Gas Recirculation Sensor A Circuit Low
P0406 Exhaust Gas Recirculation Sensor A Circuit High
P0407 Exhaust Gas Recirculation Sensor B Circuit Low
P0408 Exhaust Gas Recirculation Sensor B Circuit High
P0410 Secondary Air Injection System Malfunction
P0411 Secondary Air Injection System Incorrect Flow Detected
P0412 Secondary Air Injection System Switching Valve A Circuit Malfunction
P0413 Secondary Air Injection System Switching Valve A Circuit Open
P0414 Secondary Air Injection System Switching Valve A Circuit Shorted
P0415 Secondary Air Injection System Switching Valve B Circuit Malfunction
P0416 Secondary Air Injection System Switching Valve B Circuit Open
P0417 Secondary Air Injection System Switching Valve B Circuit Shorted
P0420 Catalyst System Efficiency Below Threshold (Bank 1)
P0421 Warm Up Catalyst Efficiency Below Threshold (Bank 1)
P0422 Main Catalyst Efficiency Below Threshold (Bank 1)
P0423 Heated Catalyst Efficiency Below Threshold (Bank 1)
P0424 Heated Catalyst Temperature Below Threshold (Bank 1)
P0430 Catalyst System Efficiency Below Threshold (Bank 2)
P0431 Warm Up Catalyst Efficiency Below Threshold (Bank 2)
P0432 Main Catalyst Efficiency Below Threshold (Bank 2)
P0433 Heated Catalyst Efficiency Below Threshold (Bank 2)
P0434 Heated Catalyst Temperature Below Threshold (Bank 2)
P0440 Evaporative Emission Control System Malfunction
P0441 Evaporative Emission Control System Incorrect Purge Flow
P0442 Evaporative Emission Control System Leak Detected (Small Leak)
P0443 Evaporative Emission Control System Purge Control Valve Circuit Malfunction
P0444 Evaporative Emission Control System Purge Control Valve Circuit Open
P0445 Evaporative Emission Control System Purge Control Valve Circuit Shorted
P0450 Evaporative Emission Control System Pressure Sensor Malfunction
P0451 Evaporative Emission Control System Pressure Sensor Range/Performance
P0452 Evaporative Emission Control System Pressure Sensor Low Input
P0453 Evaporative Emission Control System Pressure Sensor High Input
P0454 Evaporative Emission Control System Pressure Sensor Intermittent
P0455 Evaporative Emission Control System Leak Detected (Gross Leak)
P0460 Fuel Level Sensor Circuit Malfunction
P0461 Fuel Level Sensor Circuit Range/Performance
P0462 Fuel Level Sensor Circuit Low Input
P0463 Fuel Level Sensor Circuit High Input
P0464 Fuel Level Sensor Circuit Intermittent
P0465 Purge Flow Sensor Circuit Malfunction
P0466 Purge Flow Sensor Circuit Range/Performance
P0467 Purge Flow Sensor Circuit Low Input
P0468 Purge Flow Sensor Circuit High Input
P0469 Purge Flow Sensor Circuit Intermittent
P0470 Exhaust Pressure Sensor Malfunction
P0471 Exhaust Pressure Sensor Range/Performance
P0472 Exhaust Pressure Sensor Low
P0473 Exhaust Pressure Sensor High
P0474 Exhaust Pressure Sensor Intermittent
P0475 Exhaust Pressure Control Valve Malfunction
P0476 Exhaust Pressure Control Valve Range/Performance
P0477 Exhaust Pressure Control Valve Low
P0478 Exhaust Pressure Control Valve High
P0479 Exhaust Pressure Control Valve Intermittent

P05XX Vehicle Speed, Idle Control, and Auxiliary Inputs

P0500 Vehicle Speed Sensor Malfunction
P0501 Vehicle Speed Sensor Range/Performance
P0502 Vehicle Speed Sensor Circuit Low Input
P0503 Vehicle Speed Sensor Intermittent/Erratic/High
P0505 Idle Control System Malfunction
P0506 Idle Control System RPM Lower Than Expected
P0507 Idle Control System RPM Higher Than Expected
P0510 Closed Throttle Position Switch Malfunction
P0530 A/C Refrigerant Pressure Sensor Circuit Malfunction
P0531 A/C Refrigerant Pressure Sensor Circuit Range/Performance
P0532 A/C Refrigerant Pressure Sensor Circuit Low Input
P0533 A/C Refrigerant Pressure Sensor Circuit High Input
P0534 Air Conditioner Refrigerant Charge Loss
P0550 Power Steering Pressure Sensor Circuit Malfunction
P0551 Power Steering Pressure Sensor Circuit Range/Performance

P0552 Power Steering Pressure Sensor Circuit Low Input
P0553 Power Steering Pressure Sensor Circuit High Input
P0554 Power Steering Pressure Sensor Circuit Intermittent
P0560 System Voltage Malfunction
P0561 System Voltage Unstable
P0562 System Voltage Low
P0563 System Voltage High
P0565 Cruise Control On Signal Malfunction
P0566 Cruise Control Off Signal Malfunction
P0567 Cruise Control Resume Signal Malfunction
P0568 Cruise Control Set Signal Malfunction
P0569 Cruise Control Coast Signal Malfunction
P0570 Cruise Control Acceleration Signal Malfunction
P0571 Cruise Control/Brake Switch A Circuit Malfunction
P0572 Cruise Control/Brake Switch A Circuit Low
P0573 Cruise Control/Brake Switch A Circuit High
P0574 through **P0580** Reserved for Cruise Control System Codes

P06XX Computer and Auxiliary Outputs

P0600 Serial Communication Link Modification
P0601 Internal Control Module Memory Check Sum Error
P0602 Control Module Programming Error
P0603 Internal Control Module Keep Alive Memory (KAM) Error
P0604 Internal Control Module Random Access Memory (RAM) Error
P0605 Internal Control Module Read Only Memory (ROM) Error
P0606 PCM Processor Fault

P07XX Transmission

P0700 Transmission Control System Malfunction
P0701 Transmission Control System Range/Performance
P0702 Transmission Control System Electrical
P0703 Torque Converter/Brake Switch B Circuit Malfunction
P0704 Clutch Switch Input Circuit Malfunction
P0705 Transmission Range Sensor Circuit Malfunction (PRNDL Input)
P0706 Transmission Range Sensor Circuit Range/Performance
P0707 Transmission Range Sensor Circuit Low Input
P0708 Transmission Range Sensor Circuit High Input
P0709 Transmission Range Sensor Circuit Intermittent
P0710 Transmission Fluid Temperature Sensor Circuit Malfunction
P0711 Transmission Fluid Temperature Sensor Circuit Range/Performance
P0712 Transmission Fluid Temperature Sensor Low Input
P0713 Transmission Fluid Temperature Sensor Circuit High Input
P0714 Transmission Fluid Temperature Sensor Circuit Intermittent
P0715 Input/Turbine Speed Sensor Circuit Malfunction
P0716 Input/Turbine Speed Sensor Circuit Range/Performance
P0717 Input/Turbine Speed Sensor Circuit No Signal
P0718 Input/Turbine Speed Sensor Circuit Intermittent
P0719 Torque Converter/Brake Switch B Circuit Low
P0720 Output Speed Sensor Circuit Malfunction
P0721 Output Speed Sensor Circuit Range/Performance
P0722 Output Speed Sensor Circuit No Signal
P0723 Output Speed Sensor Circuit Intermittent
P0724 Torque Converter/Brake Switch B Circuit High
P0725 Engine Speed Input Circuit Malfunction
P0726 Engine Speed Input Circuit Range/Performance
P0727 Engine Speed Input Circuit No Signal
P0728 Engine Speed Input Circuit Intermittent
P0730 Incorrect Gear Ratio
P0731 Gear 1 Incorrect Ratio
P0732 Gear 2 Incorrect Ratio
P0733 Gear 3 Incorrect Ratio
P0734 Gear 4 Incorrect Ratio
P0735 Gear 5 Incorrect Ratio
P0736 Reverse Incorrect Ratio
P0740 Torque Converter Clutch Circuit Malfunction
P0741 Torque Converter Clutch Circuit Performance or Stuck Off
P0742 Torque Converter Clutch Circuit Stuck On
P0743 Torque Converter Clutch Circuit Electrical
P0744 Torque Converter Clutch Circuit Intermittent
P0745 Pressure Control Solenoid Malfunction
P0746 Pressure Control Solenoid Performance or Stuck Off
P0747 Pressure Control Solenoid Stuck On
P0748 Pressure Control Solenoid Electrical
P0749 Pressure Control Solenoid Intermittent
P0750 Shift Solenoid A Malfunction
P0751 Shift Solenoid A Performance or Stuck Off
P0752 Shift Solenoid A Stuck On
P0753 Shift Solenoid A Electrical
P0754 Shift Solenoid A Intermittent
P0755 Shift Solenoid B Malfunction
P0756 Shift Solenoid B Performance or Stuck Off
P0757 Shift Solenoid B Stuck On
P0758 Shift Solenoid B Electrical
P0759 Shift Solenoid B Intermittent
P0760 Shift Solenoid C Malfunction
P0761 Shift Solenoid C Performance or Stuck Off
P0762 Shift Solenoid C Stuck On
P0763 Shift Solenoid C Electrical
P0764 Shift Solenoid C Intermittent
P0765 Shift Solenoid D Malfunction
P0766 Shift Solenoid D Performance or Stuck Off
P0767 Shift Solenoid D Stuck On
P0768 Shift Solenoid D Electrical
P0769 Shift Solenoid D Intermittent
P0770 Shift Solenoid E Malfunction
P0771 Shift Solenoid E Performance or Stuck Off
P0772 Shift Solenoid E Stuck On
P0773 Shift Solenoid E Electrical
P0774 Shift Solenoid E Intermittent
P0780 Shift Malfunction
P0781 1-2 Shift Malfunction

P0782 2-3 Shift Malfunction
P0783 3-4 Shift Malfunction
P0784 4-5 Shift Malfunction
P0785 Shift/Timing Solenoid Malfunction
P0786 Shift/Timing Solenoid Range/Performance
P0787 Shift/Timing Solenoid Low
P0788 Shift/Timing Solenoid High
P0789 Shift/Timing Solenoid Intermittent
P0790 Normal/Performance Switch Circuit Malfunction

OBD I Codes

This list of codes was current at the time of publication. In some cases, codes were combined or eliminated to avoid redundancies. You should check any codes against the appropriate service manual.

BMW

1989 and later 3–Series

1 Airflow meter
2 Oxygen sensor
3 Coolant temperature sensor
4 Throttle position sensor (TPS)

1989 and later 5 and 7–Series

1000 End of diagnosis
1211 Electronic control unit (ECU)
1215 Airflow sensor
1221 Oxygen sensor
1222 Oxygen sensor regulation
1223 Coolant temperature sensor
1224 Air temperature sensor
1231 Battery voltage out of range
1232 Idle switch
1233 Full throttle switch
1251 Fuel injectors (final stage 1)
1252 Fuel injectors (final stage 2)
1261 Fuel pump relay
1262 Tank vent
1264 Oxygen sensor heating relay
1444 No faults in memory

Chrysler, Dodge, and Plymouth – domestic cars and light trucks

*Check Engine Lamp on
**Check Engine Lamp on (California only)
88 Start of test
11 Engine not cranked since battery was disconnected (Dakota pick-up 2.5L models)
11 Engine not cranked since battery was disconnected/No reference signal
12 Memory standby power lost
13* MAP (manifold absolute pressure) sensor vacuum circuit – slow or no change in MAP sensor input and/or output
14* MAP (manifold absolute pressure) sensor electrical circuit – high or low voltage
15** Vehicle speed/distance sensor circuit
16* Loss of battery voltage
16 Knock sensor
17 Engine running too cold
21** Oxygen sensor circuit
22* Coolant temperature sensor unit–high or low voltage
23 Throttle body air temperature sensor circuit–high or low voltage
24* Throttle position sensor circuit – high or low voltage
25** ISC (idle speed control) motor driver circuit
26* Peak injector current has not been reached or injector circuits have high resistance
27* Fuel injector control circuit or injector output circuit not responding.
31* Canister purge solenoid circuit failure
32** EGR (exhaust gas recirculation) system open; short in transducer solenoid or failure; power loss to PCM during diagnostic test
33 Air conditioning clutch cutout relay circuit
34 Speed control vacuum or vent control solenoid circuits–an open or shorted circuit at the EGR solenoid
35 Cooling fan relay, high speed fan or low speed fan control relays
35 Idle switch circuit; cooling fan relay circuit
36* Air switching solenoid circuit (non-turbo) or wastegate solenoid circuit on turbocharged models
37 Part throttle unlock solenoid driver circuit (automatic transmission only) or shift indicator light circuit (lockup converter)
41 Charging system excess or lack of field current
42 Automatic shutdown relay driver circuit (ASD)
43 Ignition coil control circuit or spark interface circuit
44 Loss of FJ2 to logic board/battery temperature out of range or failure in the SMEC/SBEC
45 Overboost shut-off circuit on MAP sensor reading above overboost limit detected/overdrive solenoid (A-500 or A-518 automatic transmission)
46* Charging system voltage too high
47 Charging system voltage too low
51** Oxygen sensor indicates lean
52** Oxygen sensor indicates rich
53 Module internal problem; SMEC/SBEC failure; internal engine controller fault condition detected
54 Problem with the distributor synchronization circuit
55 End of code output
61* BARO solenoid failure
62 Emissions reminder light mileage is not being updated
63 EEPROM write denied–controller failure
64 Flexible fuel (methanol) sensor indicates concentration sensor input more than the acceptable voltage
64 Flexible fuel (methanol) sensor indicates concentration sensor input less than the acceptable voltage
65 Manifold tune valve solenoid circuit open or shorted
66 Communication problem between TCM and PCM

Ford Motor Company

1983–89 V8 w/MCU

11 System OK
12 RPM out of spec
25 Knock sensor
41 Fuel mixture lean
42 Fuel mixture rich
44 Thermactor
45 Thermactor not bypassing
46 Thermactor not bypassing
51 Hi-low vacuum switch
53 Dual temperature switch
54 Mid-temperature switch
55 Mid-vacuum switch
61 & 65 Mid-vacuum switch closed
62 Barometric switch

Two- and three-digit EEC IV Codes

Test condition: I–ignition on/engine off, R–engine running, C–continuous

Code Displayed	Test Condition	Condition
11	I, R, C	System OK, test sequence complete
12	R	Idle speed control out of range
13	I, R, C	Normal idle out of range
14	I, C	Ignition profile pickup erratic
15	I	ROM test failure
15	C	Power interrupt to keep alive memory
16	R	Erratic idle oxygen sensor out of range throttle not closing
17	R	Curb idle out of range
18	C	No ignition signal to ECM
18	R	SPOUT circuit open
19	I	No power to processor
19	C	CID sensor failure
21	I, R	Coolant temperature out of range
21	I, R, C	Coolant temperature sensor out of range
22	I, R, C	MAP, BARO out of range
23	I, R, C	Throttle position signal out of range
24	I, R	Air charge temperature low, improper sensor installation
25	R	Knock not sensed in test
26	I, R	Vane mass airflow sensor contamination or circuit
27	C	Vehicle speed sensor or circuit
28	I, R	Vane air temperature sensor adjustment or circuit
29	C	No continuity in vehicle speed sensor circuit
31	I, R, C	Canister or EGR valve control system
32	I, R, C	Canister or EGR valve control system
33	R, C	Canister or EGR valve not operating properly
34	I, R, C	Canister or EGR control circuit
35	I, R, C	EGR pressure feedback, regulator circuit
38	C	Idle control circuit
39	C	Automatic overdrive circuit
41	C	Oxygen sensor signal (ex. 5.0L SEFI); fuel pressure out of range (5.0L SEFI)
41	R	Lean mixture (ex. 5.0L SEFI); injectors out of balance (5.0L SEFI)
42	R	Fuel pressure out of range (5.0L SEFI)
42	R, C	Fuel mixture rich (ex. 5.0L SEFI)
43	C	Lean fuel mixture at wide open throttle
43	R	Engine too warm for test
44	R	Air management system inoperative
45	R	Thermactor air diverter circuit
45	C	DIS coil pack, 1 circuit failure
46	R	Thermactor air bypass circuit
46	C	DIS coil pack, 2 circuit failure
47	R	Low flow of unmetered air at idle
48	R	High flow of unmetered air at idle
48	C	DIS coil pack, 3 circuit failure
49	C	SPOUT signal defaulted to 10 degrees
51	I, C	Coolant temperature sensor out of range
52	I, R	Power steering pressure switch out of range
53	I, C	Throttle sensor input out of range
54	I, C	Vane airflow sensor, air change temperature sensor
55	R	Charging system under voltage (1983-88 ex.3.8L TBI)
55	R	Open ignition key power circuit (1983-88 3.8L TBI only, 1989-90 all)
56	I, R, C	Vane mass airflow sensor or circuit
57	C	Transmission neutral pressure switch circuit
58	I	CFI; idle control circuit, EFI; vane airflow circuit
58	R	Idle control/motor or circuit
58	C	Vane air temperature sensor or circuit
59	I, C	Transmission throttle pressure switch circuit (ex. 3.0L SHO, 3.8L supercharger)
61	I, C	Coolant temperature switch out of range
62	I	Transmission circuit fault
63	I, C	Throttle position sensor circuit
64	I, C	Vane air temperature, air charge temperature sensor
65	C	Fuel control system not switching to closed loop
66	I, C	No vane mass airflow sensor signal
67	I,R,C	Neutral drive switch or circuit
67	C	A/C clutch switch circuit
68	I, R, C	Idle tracking switch (TBI); Vane air temperature circuit (EFI)
69	I, C	Vehicle speed sensor circuit
71	C	Idle tracking switch (TBI); Electrical interference (EFI)
72	R	No MAP or mass airflow sensor change
72	C	System power circuit, electrical interference
73	I, R	Throttle position sensor or circuit
74	R	Brake on/off ground circuit fault
75	R	Brake on/off power circuit fault
76	R	No vane airflow change
77	R	Throttle "goose" test not performed
78	C	Power circuit
79	I	A/C clutch circuit
81	I	Thermactor air circuit, turbo boost circuit

82	I	Thermactor air circuit, integrated controller circuit
82	I	Supercharger bypass circuit
83	I	EGR control circuit (3.8L CFI, 2.3L EFI ex. turbo only)
83	I	Cooling fan circuit (2.3L turbo, 2.5L, 3.0L, 3.8L EFI only) (1983-88, 1989-90 SHO)
83	I, C	EGR solenoid circuit (1989-90 ex. SHO)
84	I, R	EGR control circuit
85	I, R	Canister purge circuit (ex. 2.3L turbo); turbo transmission shift control circuit (2.3L)
85	C	Excessive fuel pressure or flow
85	I	Canister purge circuit
86	C	Low fuel pressure of flow
87	I, R, C	Fuel pump circuit
88	I	Clutch converter circuit (2.3L turbo) others; integrated controller
89	I	Lock-up solenoid
91	R, C	Oxygen sensor, fuel pressure, injector balance
92	R	Fuel mixture rich, fuel pressure high
93	I	Throttle plate position or TPS sensor
94	R	Secondary air inoperative
95	I, C	Fuel pump circuit
95	R	Thermactor air diverter circuit
96	R	Thermactor air bypass circuit
98	R	Repeat test sequence (1983-88)
99	R	System has not learned to control idle
99	R	system has not learned to control idle
111	ORC	System OK, testing complete
112	OC	Air Charge Temperature sensor indicates 254˚F/circuit grounded
113	OC	ACT indicates–40˚F/circuit grounded
114	OR	ACT out of self-test range
116	OR	Engine Coolant Temperature out of self-test range
117	OC	ECT indicates 254˚F/circuit grounded
118	OC	ECT indicates –40˚F/circuit grounded
121	ORC	Throttle Position (TP) circuit voltage higher or lower than expected
121	OC	Throttle Position (TP) circuit voltage below minimum
123	OC	Throttle Position (TP) circuit voltage above maximum
124	C	Throttle Position (TP) sensor voltage higher than expected
125	C	Throttle Position (TP) sensor voltage lower than expected
129	R	Insufficient Mass Air Flow (MAF) change during Dynamic Response test
157	C	Mass Air Flow (MAF) sensor voltage below minimum
158	OC	Mass Air Flow (MAF) sensor voltage above maximum
159	OR	Mass Air Flow (MAF) sensor higher or lower than expected during KOEO, KOER
167	R	Insufficient Throttle Position (TP) change during Dynamic Response test
171	C	Fuel system at adaptive limits, oxygen sensor (HEGO) unable to switch
172	R,C	Lack of HEGO switches, indicates lean

173	R,C	Lack of HEGO switches, indicates rich
179	C	Fuel system at lean adaptive limit at part throttle, system rich
181	C	Fuel system at rich adaptive limit at part throttle, system lean
182	C	Fuel system at lean adaptive limit at idle, system rich
183	C	Fuel system at rich adaptive limit at idle, system lean
184	C	Mass Air Flow (MAP) higher than expected
185	C	Mass Air Flow (MAP) lower than expected
186	C	Injector pulse width higher than expected
187	C	Injector pulse width lower than expected
211	C	Profile Ignition Pickup (PIP) circuit fault
212	C	Loss of Ignition Diagnostic Monitor (IDM) input to computer/SPOUT circuit grounded
213	R	SPOUT circuit open
214	C	Cylinder ID circuit failure
215	C	Coil 1 primary circuit failure
216	C	Coil 2 primary circuit failure
226	O	Identification Diagnostic Monitor (IDM) signal not received
326	R,C	Pressure Feedback EGR (PFE) circuit voltage low
327	ORC	Pressure Feedback EGR (PFE) circuit below minimum voltage
332	R,C	EGR valve opening not detected
335	O	Pressure Feedback EGR (PFE) sensor voltage out of range
336	R,C	Exhaust pressure high/PFE circuit voltage high
337	ORC	PFE circuit above maximum voltage
338	C	Engine coolant temperature lower than normal
339	C	Engine coolant temperature higher than normal
341	O	Octane Adjust (OCT ADJ) circuit open
411	R	RPM out of specified range during self-test low rpm check
412	R	RPM out of specified range during self-test high rpm check
452	R	Insufficient input from Vehicle Speed Sensor (VSS)
511	O	Read Only Memory (ROM) (computer) test failure
512	C	Keep Alive Memory (KAM) (computer) test failure
513	O	Computer internal voltage failure
522	O	Vehicle not in Park or Neutral during KOEO/Neutral Drive Switch (NDS) circuit open
528	C	Clutch Engage Switch (CES) circuit failure
536	R,C	Brake On/Off (600) circuit failure or not actuated during self-test
538	R	Brief wide open throttle not sensed during self-test or operator error
539	O	Air conditioning on or defrost on during self-test
542	OC	Fuel pump secondary circuit failure
543	OC	Fuel pump secondary circuit failure
556	OC	Fuel pump relay primary circuit failure
558	O	EGR Vacuum Regulator (EVR) circuit failure
563	O	High Electro Drive Fan (HEDF) circuit failure

Code Displayed	Test Condition	Condition
564	O	Electro Drive Fan (EDF) circuit failure
565	O	Canister Purge (CANP) circuit failure
621	OC	Shift solenoid 1 circuit failure
622	OC	Shift solenoid 2 circuit failure
634	C	Error in Transmission Select Switch (TSS) circuits
636	OR	Transmission Oil Temperature (TOT) sensor out of self-test circuits
637	C,C	TOT sensor voltage above maximum
638	OC	TOT sensor voltage below minimum
639	R,C	Insufficient input from Transmission Speed Sensor (TSS)
641	OC	Shift solenoid 3 circuit failure
643	OC	Converter clutch control circuit failure
998	R	Hard fault is present—Failure Mode Effects Management (FMEM)
No Codes		Unable to initiate self-test or unable to output codes

General Motors

Note: Any VIN letter reference is the eighth letter in the vehicle identification number. This letter refers to the type of engine.

12 No tach reference to ECM, no codes stored
13 Oxygen sensor circuit
13 Left oxygen sensor open circuit (VIN J)
13 Oxygen sensor not ready (left sensor, Allante)
14 Coolant sensor circuit; high temperature indicated, voltage low
14 Shorted coolant sensor circuit (Cadillac)
15 Coolant sensor circuit; low temperature indicated, voltage high
15 Open coolant sensor circuit (Cadillac)
16 System overvoltage
16 DIS fault line (VIN J)
16 Missing 2X reference circuit (1992 2.3 L)
17 RPM signal problem
17 Spark reference circuit
17 Right oxygen sensor not ready (Allante)
17 Shorted crank signal circuit (Cadillac TBI)
18 Open crank signal circuit (Cadillac)
19 Crankshaft position sensor
19 Shorted fuel pump circuit (Cadillac)
20 Open fuel pump circuit (Cadillac)
21 Shorted throttle position sensor circuit (Cadillac)
21 Throttle position sensor voltage high
22 Open throttle position sensor circuit (Cadillac)
22 Throttle position sensor voltage low
23 EST/bypass circuit shorted or open (Cadillac)
23 Intake air temperature (IAT) circuit voltage high, low temperature indicated
23 Open or ground MC solenoid (carbureted)
24 Vehicle speed sensor circuit
25 Intake air temperature (IAT) circuit voltage low, high temperature indicated
25 Modulated displacement failure (Cadillac)
26 Quad driver failure
26 Shorted throttle switch circuit (Cadillac)
27 Open throttle switch circuit (Cadillac)
27 Second gear switch
28 Third or fourth gear circuit (Cadillac)
28 Third gear switch
29 Fourth gear switch
30 ISC circuit problem (Cadillac TBI, 4.1L 4.5L 4.9L 6.0L)
30 RPM error (Cadillac MFI, 4.1L 4.5L 4.9L 6.0L)
31 Park/Neutral switch
31 Canister purge solenoid (carbureted, Cadillac, 4-cyl., 5.0L, 5.7L)
31 Shorted MAP sensor circuit (Cadillac, 4.1L 4.5L 4.9L 6.0L)
31 Wastegate solenoid (Pontiac, Buick Turbo 1.8L)
31 Camshaft sensor circuit (VIN J, VIN T)
32 Barometric or altitude sensor
32 EGR system fault
32 Open MAP sensor circuit (Cadillac, 4.1L 4.5L 4.9L 6.0L)
33 MAP sensor voltage high, low vacuum (4-cyl. 2.0L) (Cadillac, 4-cyl., 5.0L, 5.7L)
33 MAF sensor frequency high (V6 2.8L MFI) (Cadillac, 4-cyl., 5.0L, 5.7L)
34 MAP sensor voltage low, high vacuum
34 MAF sensor frequency low
35 Idle speed or idle air control circuit error
35 Shorted barometric sensor circuit (Cadillac, 4.1L 4.5L 4.9L 6.0L)
36 MAF circuit burnoff
36 Closed throttle airflow too high (VIN A)
36 DIS missing or extra EST signal (VIN J)
36 Open barometric sensor circuit (Cadillac, 4.1L 4.5L 4.9L 6.0L)
36 Transaxle shift control problem (1992)
37 Shorted MAT sensor circuit (Cadillac, 4.1L 4.5L 4.9L 6.0L)
38 Brake switch
38 Open MAT sensor circuit (Cadillac, 4.1L 4.5L 4.9L 6.0L)
39 Torque converter clutch circuit
40 Power steering pressure switch circuit (Cadillac, 4.1L 4.5L 4.9L 6.0L)
41 Cylinder select error
41 Incorrect or faulty Mem-cal
41 Cam sensor circuit (3.8 L Code 3, C) (Cadillac, 4.1L 4.5L 4.9L 6.0L)
41 Missing 1X reference circuit (1992 2.3L)
41 No distributor reference pulses to ECM
42 EST circuit grounded or open
42 Left oxygen sensor lean (Cadillac MFI, 4.1L 4.5L 4.9L 6.0L)
43 ESC signal low or high voltage
43 Left oxygen sensor rich (Cadillac MFI, 4.1L 4.5L 4.9L 6.0L)
44 Air-fuel mixture too lean
44 Lean exhaust signal (VIN J, left sensor; right sensor, Cadillac MFI, 4.1L 4.5L 4.9L 6.0L)
45 Air-fuel mixture rich

45 Air-fuel mixture rich (Cadillac, 4-cyl., 5.0L, 5.7L)
45 Rich exhaust signal (VIN J, left sensor; right sensor, Cadillac MFI, 4.1L 4.5L 4.9L 6.0L)
46 Power steering pressure switch
46 Right to left bank fueling imbalance (Cadillac, 4.1L 4.5L 4.9L 6.0L)
46 Vehicle anti-theft system (VATS)
47 ECM/BCM data problem (VIN C)
48 EGR system fault (Cadillac, 4.1L 4.5L 4.9L 6.0L)
48 Misfire (VIN C)
50 Second gear pressure switch (Cadillac, 4.1L 4.5L 4.9L 6.0L)
51 Faulty PROM, MEM-CAL, ECM or installation
52 ECM memory reset indicator (Cadillac, 4.1L 4.5L 4.9L 6.0L)
52 CALPAK error
52 Engine oil temperature sensor low (VIN J, VIN 8)
53 Distributor signal interrupt (Cadillac, 4.1L 4.5L 4.9L 6.0L)
53 EGR vacuum valve sensor (carbureted) (Cadillac, 4-cyl., 5.0L, 5.7L)
53 System overvoltage
54 Fuel pump circuit voltage low
54 Shorted MC solenoid (carbureted) (Cadillac, 4-cyl., 5.0L, 5.7L)
55 ECM error
55 Fuel lean monitor (VIN J)
55 TPS misadjusted (Cadillac, 4.1L 4.5L 4.9L 6.0L)
56 Vacuum sensor circuit signal voltage low (VIN J)
58 Vehicle anti-theft system (Cadillac, 4.1L 4.5L 4.9L 6.0L)
58 PASS fuel enable circuit
60 Transmission not in drive (Cadillac, 4.1L 4.5L 4.9L 6.0L)
61 Cruise control circuit (Allante) set & resume engaged simultaneously (Cadillac, 4.1L 4.5L 4.9L 6.0L)
61 Cruise vent solenoid circuit (1992 3.8L)
61 Degraded oxygen sensor
61 Secondary port throttle valve system (VIN J)
62 Cruise control circuit (Allante);vehicle speed exceeds maximum limit (Cadillac, 4.1L 4.5L 4.9L 6.0L)
62 Cruise vacuum circuit (3.8L)
62 Engine oil temperature circuit high (VIN J, VIN 8)
62 Transmission gear switch signal
63 Cruise control circuit, vehicle speed & set speed tolerance exceeded (Cadillac, 4.1L 4.5L 4.9L 6.0L)
63 MAP sensor voltage high (3.8L)
63 EGR flow problem (3.8L, VIN C)
64 EGR flow problem (3.8L, VIN C)
64 Cruise control circuit, vehicle acceleration exceeds maximum limit (Cadillac, 4.1L 4.5L 4.9L 6.0L)
64 Right oxygen sensor circuit lean (VIN J)
64 MAP sensor voltage low (3.8L)
65 Cruise control circuit (Allante); coolant temperature exceeds maximum limit (Cadillac, 4.1L 4.5L 4.9L 6.0L)
65 Servo position sensor
65 Cruise servo position sensor (1992 3.8L)
65 EGR flow problem (3.8L, VIN C)
65 Fuel injector circuit low current (Quad 4 VIN A, D)
66 A/C low pressure switch
66 Cruise control circuit; engine RPM exceeds maximum limit (Cadillac, 4.1L 4.5L 4.9L 6.0L)
66 Power mode switch (VIN J)
67 Cruise control circuit; switch circuit shorted (Cadillac, 4.1L 4.5L 4.9L 6.0L)
67 Cruise switch circuit
68 Cruise control circuit; switch circuit shorted (Cadillac, 4.1L 4.5L 4.9L 6.0L)
68 Cruise system problem
69 A/C head pressure switch circuit
70 Intermittent TPS (Cadillac, 4.1L 4.5L 4.9L 6.0L)
71 Intermittent MAP (Cadillac, 4.1L 4.5L 4.9L 6.0L)
73 Intermittent coolant sensor (Cadillac, 4.1L 4.5L 4.9L 6.0L)
74 Intermittent MAT (Cadillac, 4.1L 4.5L 4.9L 6.0L)
75 Intermittent speed sensor (Cadillac, 4.1L 4.5L 4.9L 6.0L)
80 Fuel system rich (Cadillac, 4.1L 4.5L 4.9L 6.0L)
85 Throttle body service required (Cadillac, 4.1L 4.5L 4.9L 6.0L)
96 Torque converter overstress (Cadillac, 4.1L 4.5L 4.9L 6.0L)
97 P/N D/R engaged problem (Cadillac, 4.1L 4.5L 4.9L 6.0L)
97 P/N D/R engagement problem (Cadillac, 4.1L 4.5L 4.9L 6.0L)

Honda

1988 and later models

0 Faulty ECU
1 Oxygen sensor or circuit
2 Faulty ECU
3/5 Manifold absolute pressure (MAP) sensor or circuit
4 Crank angle sensor or circuit
6 Coolant temperature sensor or circuit
7 Throttle angle sensor or circuit
8 TDC position/crank angle sensor circuit
9 Crank angle sensor or circuit
10 Intake air temperature sensor or circuit
11 No particular symptom shown or system does not operate–faulty ECU
12 Exhaust gas recirculation (EGR) failure
13 Atmosphere pressure sensor circuit
14 Electronic air control valve (EACV)
15 No ignition output signal–possible faulty igniter
16 Fuel injector circuit
17 Vehicle speed sensor or circuit
19 Lock-up control solenoid valve (automatic transmission)
20 Electric load detector–possible open or grounded circuit in ECU wiring
21 VTEC spool solenoid valve circuit (Civic & Civic Del-Sol)
22 VTEC oil pressure switch circuit (Civic & Civic Del-Sol)
23 Knock sensor (Prelude)
30 A/T control unit ECM fuel injector signal "A" (Accord and Prelude)
31 A/T control unit and ECM circuit (Accord and Prelude)
41 Heated oxygen sensor–heater circuit
43 Fuel supply system circuit (except D15Z1 engine)
48 Heated oxygen sensor circuit (D15Z1 engine)

Mazda

2 Crank position sensor–NE sensor
4 Crank position sensor–G sensor
5/7 Left/right side knock sensor
6 Speedometer sensor
9 Coolant temperature sensor
10 Intake air temperature sensor
11 Intake air temperature sensor
12 Throttle sensor–full range
13 Pressure sensor
14 Atmospheric pressure sensor (replace ECU)
15/23 Left/right side oxygen sensor
16 EGR switch side oxygen
17 Oxygen sensor–inaccurate
17/24 Left/right side feedback system
18 Throttle sensor–narrow range
20 Metering oil pump position sensor
23 Fuel temperature sensor
25 Solenoid valve–pressure regulator control
26 Metering oil pump stepper motor
27 Metering oil pump
28 Solenoid valve–EGR
29 Solenoid valve–EGR vent
30 Solenoid valve–split air bypass
31 Solenoid valve–relief 1
32 Solenoid valve–switching
33 Solenoid valve–port air bypass
34 Solenoid valve–idle speed control
36/37 Right/left side oxygen sensor heater
37 Metering oil pump
39 Solenoid valve–relief 2
40 Solenoid valve–purge control
41 Variable inertia charging system (DOHC only)
42 Solenoid valve–turbo precontrol
43 Solenoid valve–wastegate control
44 Solenoid valve–turbo control
45 Solenoid valve–charge control
46 Solenoid valve–charge relief control
50 Solenoid valve–double throttle control
51 Fuel pump relay
54 Air pump relay
65 Air conditioning signal
71 Injector–front secondary
73 Injector–rear secondary
76 Slip lock-off signal
77 Torque reduced signal

Mercedes-Benz

CIS-E

1 No system malfunction
2 Throttle valve switch
3 Coolant temperature sensor
4 Airflow sensor position indicator
5 Oxygen sensor
6 Not used
7 TD signal
8 Altitude correction capsule
9 Electro-hydraulic actuator (EHA)
10 Throttle valve switch, idle speed contact
12 EGR temperature sensor (California 420 & 560 series only)

Bosch CIE–E III

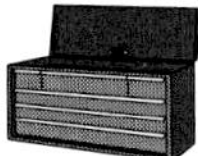

Note: If ICU is replaced, check ignition timing. If trouble code warning light blinks while driving, knock regulation is at maximum control limit. [1] – California only.

Code	Location of Fault	Problem
1111	ICU or FICU	Defective memory circuits in control unit
2121	Idle Switch	Switch stuck closed or problem in wiring to switch
2122	Engine Speed Signal	No engine speed signal from ICU Term. No.17 or hall sender to FICU Term No. 30
2123	Full Throttle Switch	Switch stuck closed or problem
2132	No Data Transmitted	Disconn. or open wire between ICU Term No. 5 from FICU to ICU & FICU Term. No. 1 or ICU Term No. 3 & FICU Term. No. 13 or Defective Control Units
2141	Knock Regulation	Engine or ignition knock is causing timing to be retarded (max. limit)
2142	Knock Sensor	Defective sensor or sensor wiring
2223	Altitude Sensor	No Signal from sensor
2232[1]	Air Sensor	No Signal from Sensor to FICU or break in wire between FICU Term. No. 21 & ICU Term. No. 8
2233	Reference Voltage	No reference voltage from ICU Term. No. 212 air sensor & altitude sensor & FICU Term. No. 26
2312	Cool. Temp. Sensor	No signal from sensor
2341	Oxygen Sensor	Oxygen sensor control operating at rich or lean limit
2342	Oxygen Sensor	No signal from sensor
4431	Idle Stab. Valve	Problem in wiring to idle stab. valve
4444	No Faults Stored	
0000	End of Diagnosis	

Nissan

1990 300ZX, Stanza; 1991–92 All ex. Trucks & Maxima SOCH: After obtaining codes, turn mode switch clockwise after two seconds, counterclockwise.

1990 Others, 1991–92 Trucks & Maxima SOHC: After obtaining codes, turn mode switch fully clockwise. After LEDs flash four times, turn mode switch counterclockwise and turn ignition switch off.

1597cc ex. Turbo, 1983 1957cc all, 1984–86 California:

LED off: No problem in system

LED on: Flashes slowly

Fuel mixture solenoid

Vacuum switching solenoids (1.5L)

Anti-diesel solenoid

Flashes quickly

Oxygen sensor

Engine temperature sensor

1983–86 1488cc Turbo, 1974cc FI, 2960cc
1984–87 1809cc Turbo
1986 2389cc FI:

11 Crank angle sensor
12 Airflow meter
13 Engine temperature sensor
14 Vehicle speed sensor
21 Ignition signal
22 Fuel pump circuit
23 Throttle valve or idle switch
24 Neutral/park switch
31 AC speed-up valve (with AC)
31 System OK (without AC)
32 Starter signal
33 Oxygen sensor
34 Detonation sensor
41 Fuel temperature sensor
42 Altitude sensor
42 Throttle sensor (1986 Pickup)
43 Battery voltage high/low
43 Injector (1986 Pickup)
44 System OK

1597cc Federal & 1987 1597cc Calif.

Red LED off, Green LED on: No problem in system

Red LED on, Green LED flashing quickly:

Vacuum sensor
Altitude sensor
Engine temperature sensor
Air temperature sensor

Red LED on, Green LED flashing slowly:

Mixture heater relay
Fuel mixture solenoid
Coasting richer solenoid
Anti-diesel solenoid
Idle speed control valve

1987–92 ex. 1987 1809cc Turbo:

11 Crank angle sensor
12 Airflow meter
13 Water temperature sensor
14 Vehicle speed sensor
15 Mixture ratio control
21 Ignition signal
22 Fuel pump circuit
23 Idle switch
24 Neutral/clutch switch
25 Idle speed control
31 ECU
32 EGR
33 Oxygen sensor 1987–89 (1987–91 300 ZX left side)
33 EGR sensor, 1990–91 others
34 Detonation sensor
35 Exhaust temperature sensor
41 Air temperature sensor
42 TPS (4-cyl) or fuel temperature sensor (V6)
43 Throttle sensor
44, 55 System OK
45 Injector leak
51 Injector circuit
53 Oxygen sensor (300ZX right side)
54 AT control unit to ECU problem
55 No malfunction

Saturn

11 Transaxle codes present
12 Diagnostic check only
13 Oxygen sensor circuit
14 Coolant sensor/high temperature
15 Coolant sensor/low temperature
17 PCM fault pull-up resistor
19 6X signal fault (1992 and 1993 models only)
21 Throttle position sensor voltage high.
22 Throttle position sensor voltage low
23 IAT circuit low
24 VSS circuit–no signal
25 IAT circuit-temperature out of high range
26 Quad driver output fault
32 EGR system fault
33 MAP circuit–voltage out of range high
34 MAP circuit–voltage out of range low
35 Idle air control (IAC)–rpm out of range
41 Ignition control circuit open or shorted
42 Bypass circuit–open or shorted
41 and **42** IC control circuit grounded/bypass open
43 Knock sensor circuit–open or shorted
44 Oxygen sensor indicates lean exhaust
45 Oxygen sensor indicates rich exhaust
46 Power steering pressure circuit open or shorted (1991 models only)
49 High idle indicates vacuum leak
51 PCM memory error
55 A/D error
81 ABS message fault (1993 vehicles only)
82 PCM internal communication fault

Subaru

14 Duty solenoid valve control system
15 CFC system
21 Water temperature
22 VLC solenoid control system
23 Pressure sensor system
24 Idle-up solenoid valve control system
25 FCV solenoid valve control system
32 Oxygen sensor system
33 Car speed sensor system
35 Purge control solenoid valve control system
52 Clutch switch system (FWD) model only
62 Idle-up system
63 Idle-up system

Toyota

1989–91 All ex. Van

1987–88 Pickup w/FI

1986–88 Celica, Corolla w/FI, Supra 3.0L w/o Super Monitor

1985–87 Camry, MR2, Cressida w/o Super Monitor

With engine at operating temperature, turn ignition switch on and place a jumper between cavities T and E1 (1985–88) or TE1 and E1 (1989–92) on the Check Engine connector located by the airflow meter or the strut tower. "Check Engine" light with flash codes.

1987–88 FX16

1985–89 Van

1985–86 Pickup w/FI, Supra 2.8L

1985 Celica, Corolla GTS

1983–84 Camry

With engine at operating temperature, turn ignition on and jumper the cavities on the two wire yellow connector in the engine compartment. "Check Engine" light will flash codes.

1985–88 Cressida, Supra w/Super Monitor

Turn ignition on and push "Select" & "Input" buttons simultaneously and hold for 3 seconds. Pause, then hold "Set" button for 3 seconds. Codes will appear on screen if any are stored.

1983–84, Cressida, Supra

With the engine at operating temperature, turn the ignition switch on and place a jumper between the cavities of the Check Engine connector located by the distributor. Connect an analog VOM to the EFI service connector located by the air intake pickup hose at right side of radiator. Connect plus (+) to upper cavity and minus (–) to lower-left cavity. Needle on voltmeter will pulse between 2.5 & 5 volts if no codes are stored. Codes that are stored will be displayed as pulses on the meter.

All models

To clear memory, remove STOP or EFI fuse from fuse block for 30 seconds.

1983–95 Camry, Van

1986 Camry Canada, Van

1 System OK
2,3 Airflow meter signal
4 Engine temperature sensor circuit
5 Oxygen sensor
6 Ignition signal circuit
7 Throttle position circuit

1988 Pickup Turbo

1987 Van

1985–87 MR2, Pickup w/FI, Corolla GTS

1986 Camry U.S., Celica 2S Eng.

1 System OK
2 Airflow meter signal
3 Ignition signal
4 Engine temperature sensor circuit
5 Oxygen sensor
6 RPM circuit
7 Throttle position circuit
8 Intake air temperature circuit
10 Starter signal
11 AC switch
12 Knock sensor
13 Turbocharger pressure

1989–92 All

1988 Corolla, MR2, FX, Van, Pickup ex. Turbo

1987–88 Camry

1986–88 Celica 3S eng.

1983–88 Supra, Cressida

No code System OK
11 ECU circuit
12, 13 RPM signal circuit
14 Ignition signal
16 Transmission electronic control
21 Oxygen sensor circuit (left side, 2958cc)
22 Engine temperature sensor circuit
23 & 24 Intake air temperature circuit
25 Air-fuel ratio lean

26 Air-fuel ratio rich
28 Oxygen sensor circuit (right side, 2958cc)
31 Vacuum switches (1988–89 1.6L 2V Calif.)
31 Vacuum sensor (1989–92 1.5L FI, 1.6L FI, 2.2L)
31, 32 Airflow meter circuit (others)
34 Turbocharger pressure signal
35 Turbocharger pressure sensor signal
35 AAC sensor signal (ex. Turbo)
41 Throttle position sensor circuit
43 Starter signal
51 Neutral start or AC switch
52 Knock sensor circuit
53 Knock sensor circuit in ECU
71 EGR
72 Fuel cut solenoid

Carburetors were used effectively for many years before fuel injection became the preferred method of fuel delivery. Before the advent of electronic fuel injection, the Corvette was equipped over its many years of production with a carburetor or mechanical fuel injection. (Chevrolet)

Appendix B

Carburetion

After studying this appendix, you will be able to:

- Discuss the operation of carburetors.
- Describe the different types of carburetors.
- List the various carburetor circuits and systems.
- Remove, rebuild, and install a carburetor.
- Adjust the various carburetor systems.

This appendix discusses the principles, operation, and service of carburetors. Long before electronics made fuel injection possible on the modern vehicle, carburetors were the fuel delivery system of choice on gasoline powered engines. Although carburetors have been completely replaced by fuel injection systems on all vehicles built since 1993, there are still millions of vehicles with carburetors in use. However, this number will decrease as time and mileage claim these older vehicles. It is not necessary for you to know this material to learn the systems and concepts in this text. This material was placed here primarily as a reference for you when servicing carbureted engines.

Carburetors

The ***carburetor*** is a device for vaporizing gasoline, mixing it with air, and delivering it to the engine. There are many sizes and types of carburetors, and most carburetors on vehicles built after 1980 are controlled by the ECM. However, all carburetors have the same basic components.

Carburetors are assembled from several pewter or aluminum castings, which have been precision machined to attach to each other without large gaps or leaks. Internal passages are drilled in the castings for fuel and air flow. Gaskets and seals are used to prevent gas or air leaks. Specialized valves, springs, vacuum diaphragms, and linkages are attached to the main castings.

All modern carburetors contain several basic systems or ***circuits.*** There are also systems that are used on some carburetors and not others. Carburetor systems are operated by pressure differences (vacuum), gravity, calibrated springs, mechanical linkages, or in newer carburetors, motors or solenoids controlled by the ECM. Newer carburetors also contain sensors for providing various inputs to the ECM.

Main Metering System

Every carburetor has at least one large internal passage for air and gas, called a ***throat*** or ***barrel.*** Incoming air and gasoline are mixed in the throat before entering the intake manifold and traveling to the engine cylinders. Carburetors can have one, two, or four throats. The throats that are used for normal driving are called the primary throats or ***primary system.*** If the carburetor has a second set of throats for extra power, these are called the secondary throats or ***secondary system.***

One area in the carburetor throat is smaller than the rest of the throat. This restricted area is called the ***venturi.*** The venturi area of a two-barrel carburetor is shown in **Figure B-1.** Incoming air speeds up as it flows through the restricted area. Speeding up the air causes the air pressure

Figure B-1. *The incoming gasoline is mixed with air in the carburetor throat. This is the throat and venturi set-up for a two-barrel carburetor.*

to drop, forming a vacuum in the venturi. This is called ***venturi vacuum.*** Venturi vacuum pulls gasoline into the venturi through internal passages, mixing it with air below the venturi. The venturi and the related fuel passages are called the ***main metering system.*** **Figure B-2** is a diagram of a main metering system. This system supplies the proper amount of fuel to the engine at cruising speeds. The main metering system is unable to operate at low speeds, since the air flow through the venturi is too slow to create a vacuum.

The internal fuel passages of the carburetor contain a calibrated restriction called the ***main jet.*** The main jet determines how much gas can be pulled into the carburetor throat through the main metering system, and therefore, the air-fuel ratio at cruising speeds. Jet size affects power, mileage, and driveability at higher speeds. At low speeds, gas flow is lower than the restricting capacity of the jets. Some jets can be replaced with larger or smaller units while others are fixed and cannot be removed.

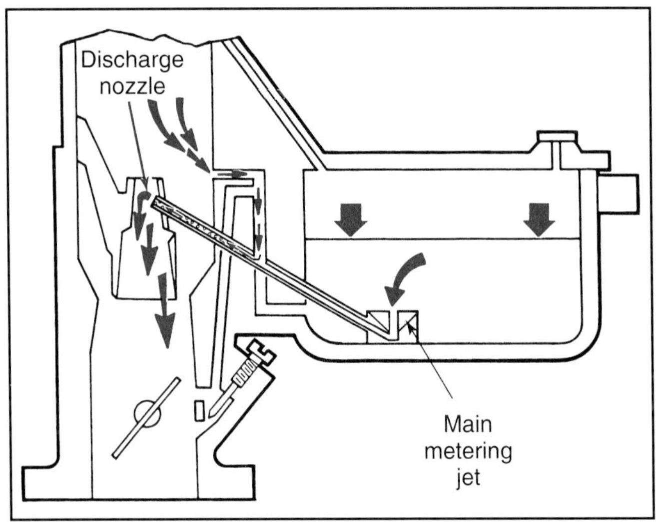

Figure B-2. *The main metering system supplies fuel for basic carburetor operation. Other systems are needed for optimum performance. (Ford)*

Power System

Under heavy loads, the engine requires a richer mixture to develop full power. The ***power system,*** or ***power valve,*** is a fuel enrichment valve, operated by engine manifold vacuum and a spring. It richens the fuel mixture by bypassing the main jet, allowing more gas to enter the main metering system when the engine is under heavy loads. The power valve is controlled by manifold vacuum opposing spring pressure. High manifold vacuum can hold the valve closed against spring pressure.

When the engine is under a heavy load, manifold vacuum drops off, and the spring pulls the valve open. When the valve opens, extra gasoline can enter the engine. A variation of this is the ***step-up rod*** system. In this system, a tapered needle is placed in the main jet. Manifold vacuum pushes the needle into the main jet, restricting fuel flow. Under heavy loads, the vacuum drops off, and spring tension can pull the needle from the jet. The needle is tapered to match fuel flow to the increase in air flow. The further the needle is pulled from the jet, the more fuel can flow. On a few systems, the needle is operated by linkage from the throttle plates. **Figure B-3** illustrates the most common types of power valves.

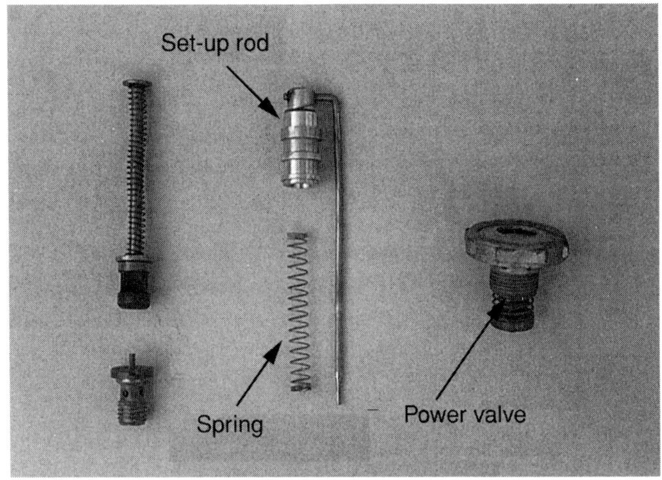

Figure B-3. *Power valves and the power circuit give the carburetor the ability to supply additional fuel for full power performance.*

Idle and Off-Idle Systems

The main metering system operates poorly at idle and low speeds. To maintain the proper air-fuel ratio at low speeds, two other systems are built into the carburetor. Both rely on intake manifold vacuum, which is created by the cylinders attempting to draw air past the closed throttle plate. Intake manifold vacuum is present whenever the engine is idling or under light load.

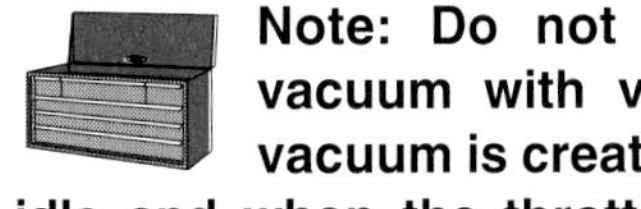

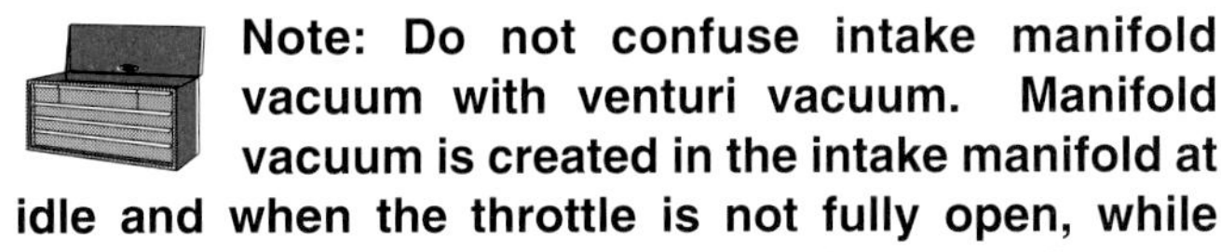

Note: Do not confuse intake manifold vacuum with venturi vacuum. Manifold vacuum is created in the intake manifold at idle and when the throttle is not fully open, while venturi vacuum is developed only in the carburetor venturi.

The ***idle system*** consists of internal carburetor passages that connect the fuel bowl to the carburetor throat below the throttle plate. Refer to **Figure B-4.** Intake manifold vacuum pulls gasoline from the bowl through the internal passages. The passages travel upward from the bowl to calibrated air bleeds. Air is drawn through the bleed passages which helps to vaporize the gasoline before it enters the intake manifold below the throttle plates. The idle system is usually equipped with one adjustment device per throat, called an ***idle needle,*** **Figure B-5.** The needle is threaded and can be turned to adjust idle mixture, although the screws are usually sealed on late model vehicles.

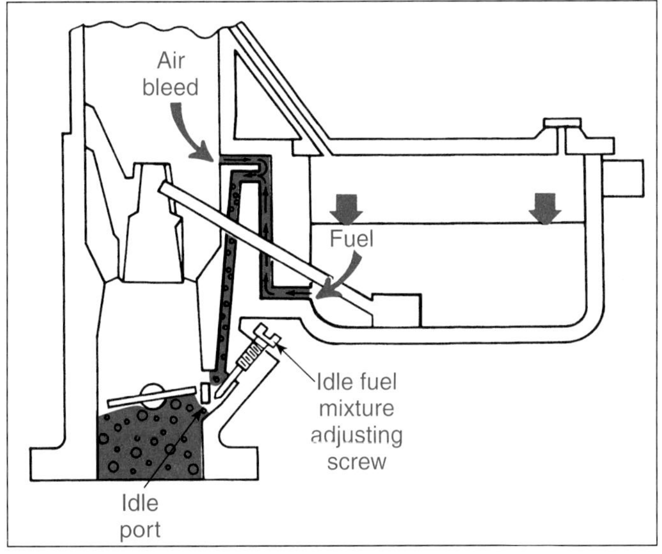

Figure B-4. *The idle circuit provides the right air-fuel mixture for idle operation. The main metering circuit does not perform well at idle. (Ford)*

The ***off-idle system,*** **Figure B-6,** consists of a slot or transfer port in the side of the carburetor, connected to the idle system fuel passages. As the carburetor throttle plate opens, more of the slot is exposed to manifold vacuum, pulling more fuel into the intake manifold. The off-idle system maintains the proper air-fuel ratio when the throttle opening is between idle and cruising speeds. The idle and off-idle systems send fuel into the engine as long as there is manifold vacuum, even at cruising speeds. The main metering system is calibrated to accept this extra fuel and still maintain the correct air-fuel ratio.

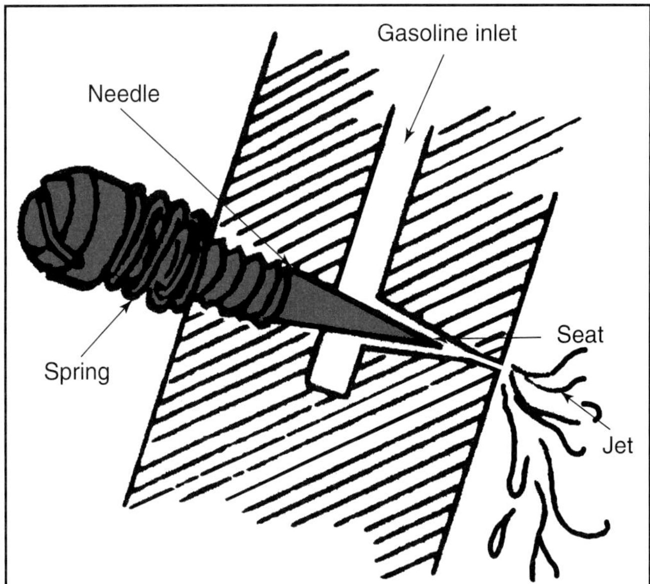

Figure B-5. *The idle mixture can be adjusted by turning an idle needle. However, the idle mixture on most carburetors cannot be adjusted.*

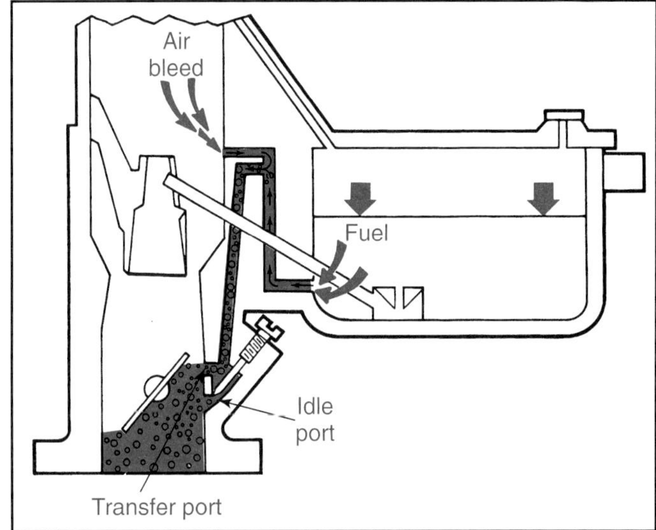

Figure B-6. *Off-idle circuits consist of an extra slot in the throat which provides fuel when the throttle plate is opened. (Ford)*

Throttle Plate

The ***throttle plate*** contains the throttle valves, which are opened and closed to control the amount of air and fuel that enters the engine, and therefore, how fast the engine runs. An adjustment screw attached to the throttle linkage holds the throttle valve slightly open to allow the engine to idle. Carburetor idle speed on many modern vehicles is controlled by an idle speed motor or solenoid, operated by the ECM. **Figure B-7** shows a typical throttle valve used on a two-barrel carburetor. Every carburetor throat contains a throttle valve.

Figure B-7. *The throttle plate serves as the control for the amount of air and fuel that enters the engine. It also forms the base of most carburetors.*

Float System

Air-fuel ratios are partially determined by the level of gasoline in the carburetor bowl. If the level is too low, the venturi vacuum will be able to pull less gas into the venturi, and the fuel mixture will be lean. If the level is too high, the mixture will be rich.

Fuel bowl level is controlled by the ***float system.*** The float system consists of a lightweight float, hinged on one end. The float rides on the surface of the gasoline in the carburetor bowl. Some floats are hollow, while others are made of a solid plastic material that is lighter than gasoline. The float is attached to the carburetor casting through a hinge. The float controls a fuel inlet valve called a ***needle and seat.*** The needle is attached to the float and moves with it, while the seat is threaded or pressed into the carburetor casting.

The float system operates whenever the engine is running, as shown in **Figure B-8.** As fuel flows through the other

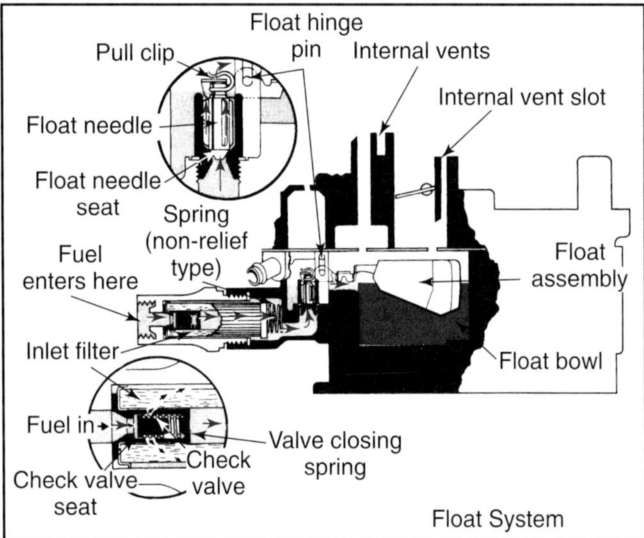

Figure B-8. *The float regulates how much fuel enters the carburetor. In some respects, it serves the same purpose as the fuel pressure regulator in the fuel injection system. (General Motors)*

carburetor circuits to the engine, the fuel level in the bowl drops. The float drops with the fuel level, and allows the needle to move away from the seat, opening the passageway from the fuel pump. Fuel flows into the bowl until the fuel level rises high enough to cause the float to close the needle and seat. The process repeats continuously as the engine uses fuel.

Bowl Vents

The float bowl is vented to allow air to replace the gasoline as it is used. An internal bowl vent connects the bowl with the air horn, while an external bowl vent is connected to the evaporative emissions system. External ***bowl vents*** reduce the effects of gasoline vaporizing in the bowl during engine idle, or when a warm engine is shut off during hot weather. The evaporative emissions system is discussed in **Chapter 10.**

Accelerator Pump

The ***accelerator pump*** is a device that overcomes the tendency of the engine to hesitate when it is accelerated quickly. When quick acceleration is necessary, the driver presses the accelerator pedal, which opens the throttle plates, and allows large amounts of air into the engine. Gasoline is heavier than air, and the main metering system and off-idle system must catch up with the increased air flow. Since the carburetor cannot immediately deliver enough fuel to match the increased air flow, the engine will receive air without fuel, and hesitate, or possibly stall.

The accelerator pump is used to inject extra gasoline into the engine when the carburetor throttle is opened quickly. The accelerator pump is a small pump that injects gasoline into the carburetor throat. The accelerator pump is operated by linkage connected to the throttle valve. When the throttle is opened, the pump squirts fuel into the carburetor throat. Check valves ensure that the fuel flows in the proper direction and that air is not drawn into the system when the throttle is released. Accelerator pumps are either the plunger type, **Figure B-9,** or the diaphragm type.

Choke

The ***choke*** is a flat metal plate that pivots on a shaft and is used to richen the carburetor mixture when the engine is cold. It resembles the throttle plate, but is always mounted on the top of the carburetor throat at the air horn. An engine always operates best when the air-fuel ratio is consistent, whether it is hot or cold. However, when an engine is cold, the gasoline is harder to vaporize, and often recondenses on the cold manifold and combustion chamber surfaces. Since liquid gasoline will not burn, the engine will stumble and hesitate, even though it is receiving the same amount of gasoline as when it is hot.

To compensate for this condensation, the choke is closed to increase the amount of gas flowing through the engine. When the choke plate is closed, air flow through the

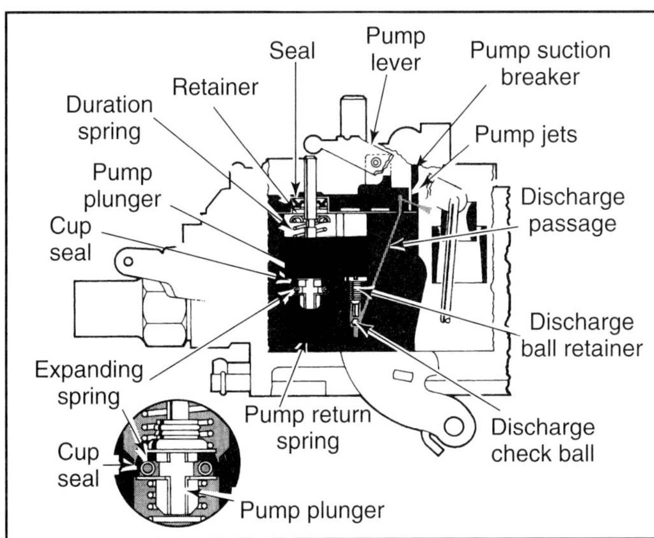

Figure B-9. *Accelerator pumps inject extra fuel into the carburetor throat when the driver depresses the accelerator pedal. (General Motors)*

carburetor air horn is restricted, creating a vacuum in the upper air passage. The vacuum causes extra fuel to be drawn through the other carburetor systems, richening the mixture. Some chokes are operated by a thermostatic coil spring. The coil spring is made of two different metals, laminated together. The different expansion rates of the two metals causes it to coil up when cold and uncoil as it heats up. When the engine is cold, the choke "coils" (tightens) and closes the choke through linkage. As the engine warms up, the choke "uncoils" (loosens) and allows the choke to open. See **Figure B-10.**

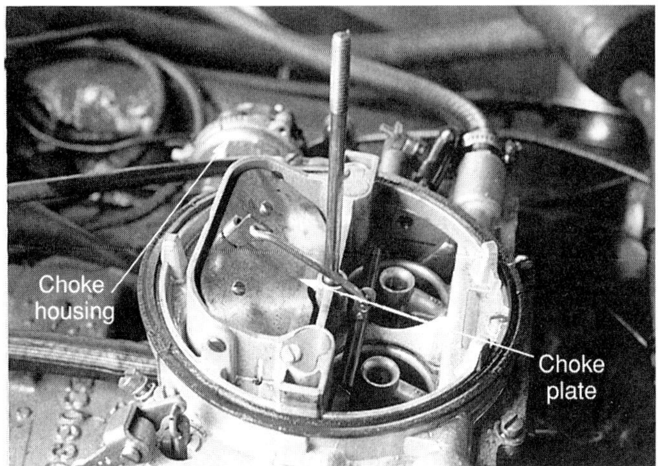

Figure B-10. *The choke limits the amount of air that can enter the engine when cold, richening the air-fuel mixture.*

The choke coil is often installed on the carburetor casting. Some choke coils are installed on the intake manifold. Since they are separate from the carburetor, they are called *divorced chokes.* Most chokes have a manifold vacuum operated device which opens the choke as soon as the engine starts. It is called a vacuum break or ***choke pull-off.*** The choke pull-off may be located in the choke housing, or may be a separate diaphragm. **Figure B-11** shows a divorced choke and choke pull-off.

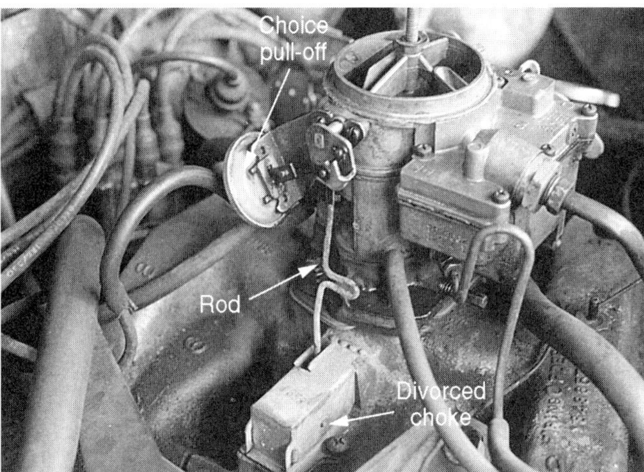

Figure B-11. *Some choke devices are separate from the carburetor. However, this system is not used on newer carburetors.*

Since the richer mixture provided by the choke contributes to exhaust emissions, modern engines are designed so that the choke is opened within a few moments of starting. Devices such as thermostatic air cleaners, coolant heated manifolds, and manifold heat control valves are used for heating the fuel intake system to improve cold driveability and reduce exhaust emissions.

Secondary System

The secondary system is an additional carburetor throat or throats, equipped with throttle valves, as shown in **Figure B-12.** On carburetors having a secondary system, the throats used for normal driving are called the primary

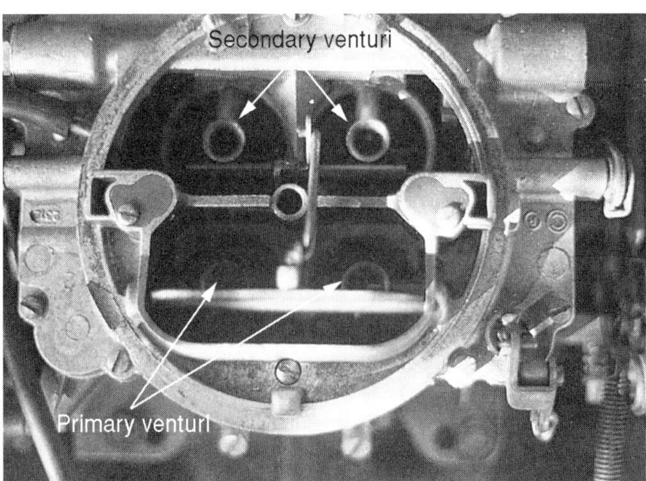

Figure B-12. *The secondary throats are larger than the primaries. Secondary systems open just after the primary valves to provide maximum power.*

system. The secondary system allows extra air and gas into the engine when more power is needed. When the accelerator is pressed to the floor, the secondary throttle valves are opened by mechanical linkage. When the secondary throttle valves open, extra air enters the engine through the secondary throats. Venturi systems in the secondary throat deliver extra gasoline. A carburetor with a secondary system allows the engine to run smoothly and get good mileage at low speeds and still perform well when extra power is needed. Although once found only on high performance cars, carburetors with secondary systems are commonly used on smaller engines.

Hot Idle Compensator

The ***hot idle compensator*** is a carburetor device which opens to allow more air into the engine when carburetor temperatures become excessive. It does this to prevent percolation by using a bi-metal strip attached to a metering valve, **Figure B-13.** Percolation is excess fuel vaporization caused by liquid gasoline boiling in the carburetor bowl and being drawn into the carburetor air horn. To reduce percolation, the hot idle compensator's metering valve opens before carburetor temperatures reach the point where percolation could occur. The extra air flowing through the compensator port leans out the fuel mixture and increases idle speed. Increasing idle speed also increases the speed of the water pump and radiator fan to reduce engine temperatures. Although hot idle compensators were once common, they are used less widely on newer carburetors since they complicate the job of precisely controlling the carburetor air-fuel ratio.

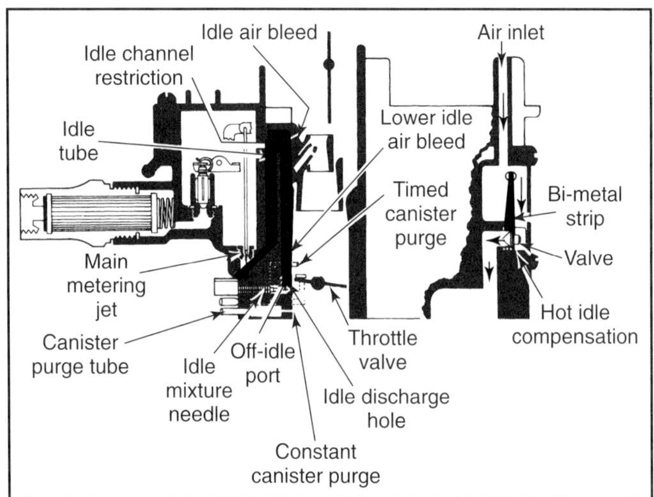

Figure B-13. *The hot idle compensator prevents fuel percolation and vapor lock when the engine is hot by allowing more cool air into the carburetor. (AC-Delco)*

Other Carburetor Devices

To improve engine driveability and performance, other solenoids, motors and devices were added to the carburetor. Some of these devices were used before computer controls were added. Many of these devices are explained in the following sections.

Electric Choke

As stated earlier, on most carburetors, a choke is used to lean the air-fuel mixture on cold engines. Older vehicles used a driver-operated choke or a mechanical choke operated by engine temperature. Newer carburetors use an ***electric choke.*** The electric choke housing resembles other choke housings, but contains a heating element made of high resistance electrical wire, as shown in **Figure B-14.**

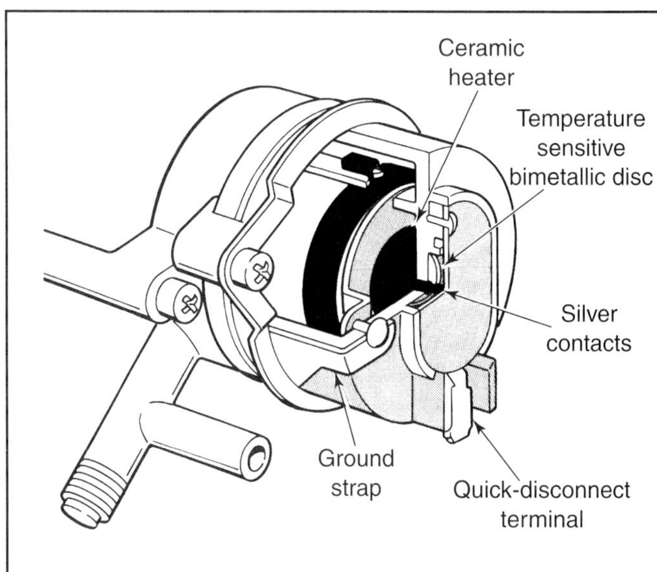

Figure B-14. *Most modern carburetors use electric chokes. An electric choke provides better control of choke opening and closing. (Ford)*

When current passes through the element, the resistance in the wire creates heat, which quickly uncoils the choke spring. The choke also contains a thermostatic switch. When the engine is cold, the switch closes, heating the choke coil. When the temperature becomes high enough, the switch opens, cutting off the current to the coil. Some electric chokes operate from a terminal attached to the alternator's stator terminal. Therefore, the only time the choke is being heated is when the engine is actually running. Other manufacturers accomplish the same thing by running the choke current through a pressure switch. With this system, the coil receives current only after the engine has started and developed oil pressure.

Idle Control Solenoids

Idle control solenoids are electric solenoids used to control the engine's idle speed. The solenoid plunger contacts part of the throttle linkage, holding the throttle plate open when the solenoid is energized. Two kinds of idle control solenoids are used on modern engines. The ***anti-dieseling solenoid*** is energized when the engine is running, and is used to set the idle speed. When the ignition is turned off, the solenoid is de-energized and the throttle return spring completely closes the throttle plate, cutting the engine off from any incoming air or fuel. This is done to prevent run-on (dieseling), which is a common problem on some modern carbureted engines. A typical anti-dieseling solenoid is shown in **Figure B-15.**

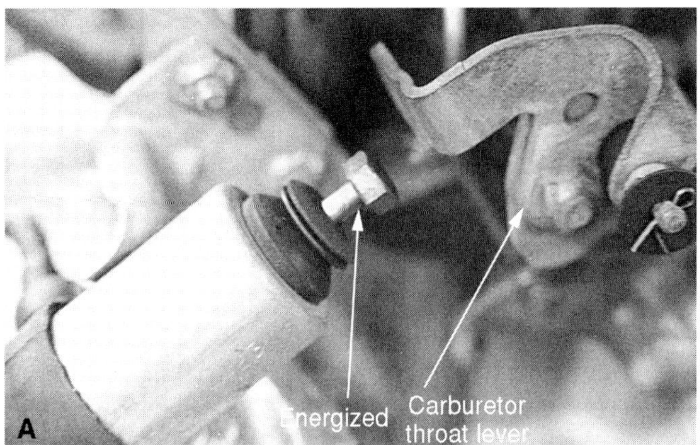

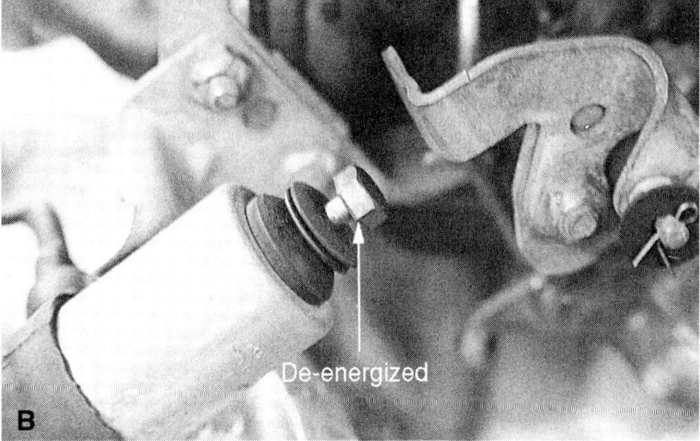

Figure B-15. *The anti-dieseling solenoid forces the throttle valve closed when the engine is shut off. A—Anti-dieseling solenoid energized. B—De-energized solenoid.*

The ***air conditioning speed-up solenoid***, **Figure B-16**, is used to compensate for the drag of the air conditioner compressor. It is inactive when the air conditioner compressor is off. When the compressor clutch is energized, the same electrical circuit energizes the solenoid. The solenoid opens the throttle, increasing engine idle.

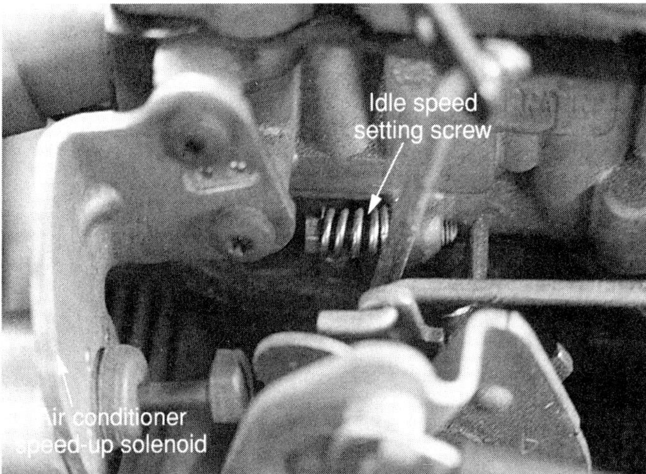

Figure B-16. *The air conditioning speed-up solenoid will open the throttle slightly to increase idle speed to compensate for the air conditioning compressor load.*

Idle Control Motors

A few engines are equipped with ***idle control motors,*** which control the engine idle by use of a small motor attached to a worm drive plunger. The motor extends or retracts the plunger in response to commands from the computer. The idle speed motor may also contain a throttle position switch. Idle control motors are used on some throttle body fuel injection systems.

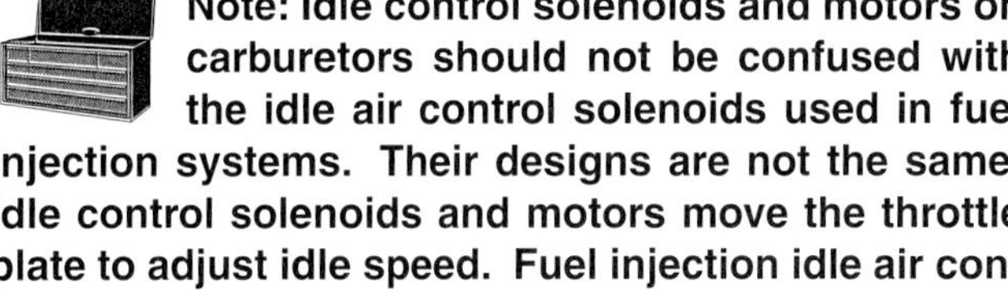

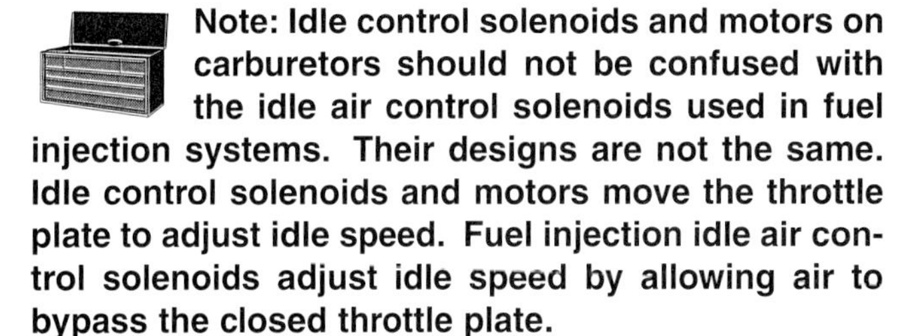

Note: Idle control solenoids and motors on carburetors should not be confused with the idle air control solenoids used in fuel injection systems. Their designs are not the same. Idle control solenoids and motors move the throttle plate to adjust idle speed. Fuel injection idle air control solenoids adjust idle speed by allowing air to bypass the closed throttle plate.

Variable Venturi Carburetors

Some older engines made use of a type of carburetor called the ***variable venturi carburetor.*** In these carburetors, the size of the venturi was varied to match air flow through the carburetor. Air pressure between the throttle plate and venturi plunger operated a diaphragm to open or close the venturi as needed. Fuel flow through the main jet was controlled by a tapered needle, similar to that in the power valve of many conventional carburetors. The variable venturi carburetor was equipped with many of the same devices as a fixed venturi carburetor. It had a fuel bowl with a float system, throttle valves, and often an accelerator pump, idle passages, and a cold enrichment system operated by a choke coil spring.

Computer-Controlled Feedback Carburetors

The carburetor mixture control solenoid has no effect on the choke, accelerator pump, power valve, or secondary system, although it can partially compensate for over-rich mixtures in these systems by leaning out the systems it does control. The solenoid can be installed in the carburetor passages so that it affects the air-fuel ratio by restricting the flow of fuel or by restricting the amount of air that mixes with the fuel.

The ECM in a carburetor control system relies on inputs from the throttle position sensor, MAP sensor, oxygen sensor, and other sensors to determine the proper solenoid pulse length. It controls the solenoid by transmitting an output command in the form of a pulsed signal. The percentage of the pulse that is "on" (current flowing) or "off" (current not flowing) determines the mixture control solenoid operation and, therefore, the air-fuel ratio.

Throttle Position Sensor

On feedback carburetors, a ***throttle position sensor*** is installed on the throttle shaft. This sensor sends a throttle opening reading to the ECM. The ECM uses this reading to make adjustments to the ignition and fuel system. A few throttle position sensors are installed in the accelerator pump linkage. See **Figure B-17.** Testing a carburetor's throttle position sensor is the same as on a fuel injection system.

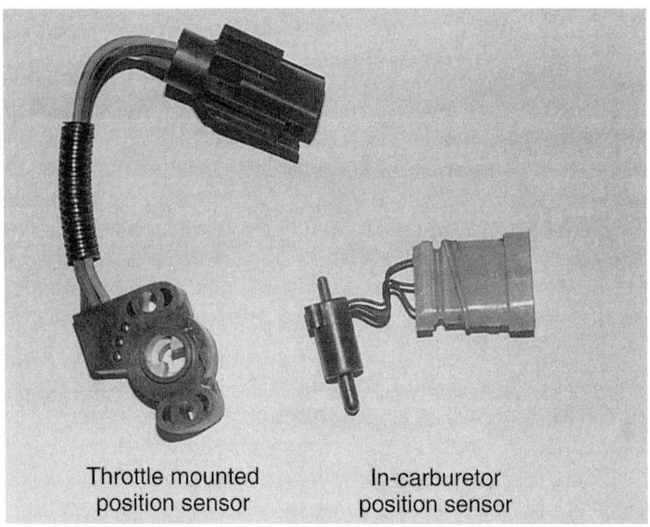

Figure B-17. *Throttle position sensors can be mounted on the throttle shaft or inside the carburetor bowl.*

Mixture Control Solenoid

The air-fuel ratios on modern carburetors are adjusted automatically by a ***mixture control solenoid.*** The mixture control solenoid is an electric solenoid controlled by the ECM. Carburetors with these solenoids have the same circuits as older carburetors. The solenoid is a more precise way of controlling the fuel and air flow through the main metering and idle systems. The mixture control solenoid controls carburetor mixtures by either controlling fuel flow through the internal passages of the carburetor, or by controlling air flow through a vacuum port in the carburetor. Refer to **Figure B-18.**

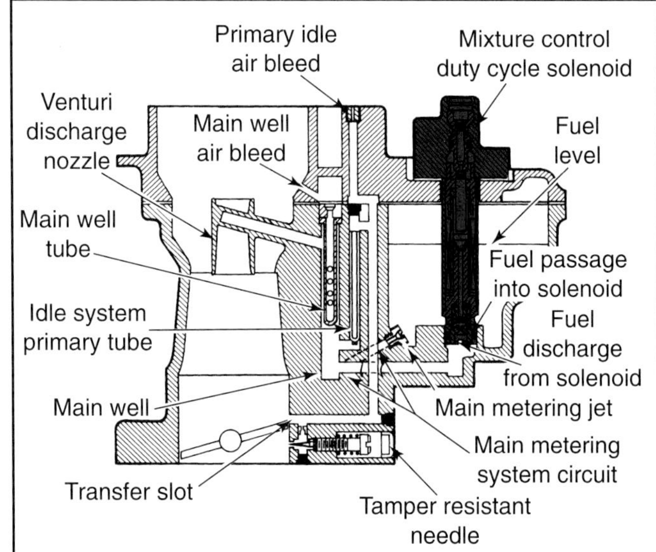

Figure B-18. *Mixture control solenoids control the fuel flow and in some respect, the air-fuel mixture in the engine. This solenoid allows for some computer control of the carburetor. (Chrysler)*

The solenoid turns on and off, opening and closing an internal passage in the carburetor. Some solenoids operate as much as ten times per second. The ECM controls the length of the solenoid pulse, based on inputs from various engine sensors. Typical inputs are engine temperature, manifold vacuum, throttle opening, and exhaust gas oxygen. **Figure B-19** shows the installation of a mixture control solenoid on a commonly used carburetor.

Carburetor air-fuel ratios are modified by the ECM and the mixture control solenoid. The only systems that the solenoid controls are the main metering, and sometimes the idle and off-idle systems. On older engines, the idle mixture adjustment screws can be adjusted to check the mixture setting. If turning the screws improves the air-fuel ratio, they are misadjusted, and may be the cause of emissions problems, as well as an idle or cruising speed problem. If the idle cannot be adjusted, look for vacuum leaks at the carburetor or manifold caused by loose manifold bolts, leaking gaskets, and misplaced or leaking vacuum hoses. On newer carburetors, the idle mixture cannot be easily adjusted. The carburetor may require an overhaul if it will not idle properly.

Mechanical Fuel Pumps

Mechanical fuel pumps are usually used on engines with carburetors. To test the pump pressure, remove the fuel line at the carburetor and install the fuel pressure gauge in

Figure B-19. *All mixture control solenoids are installed inside the carburetor. Some can be removed without disassembling the carburetor. However, this solenoid design requires carburetor air horn removal.*

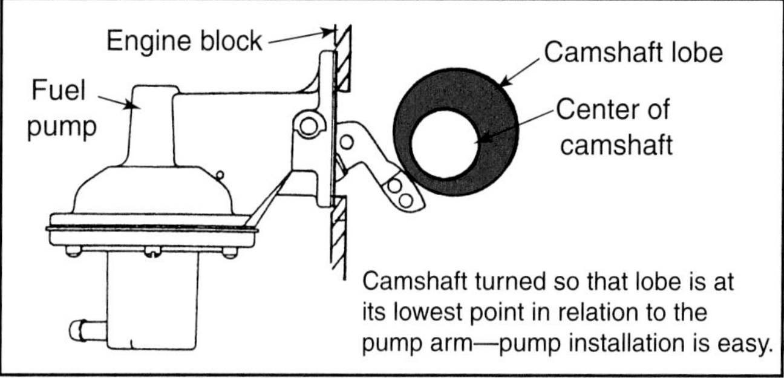

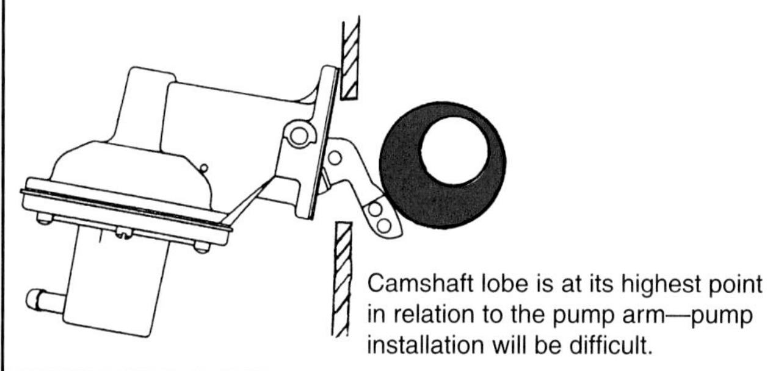

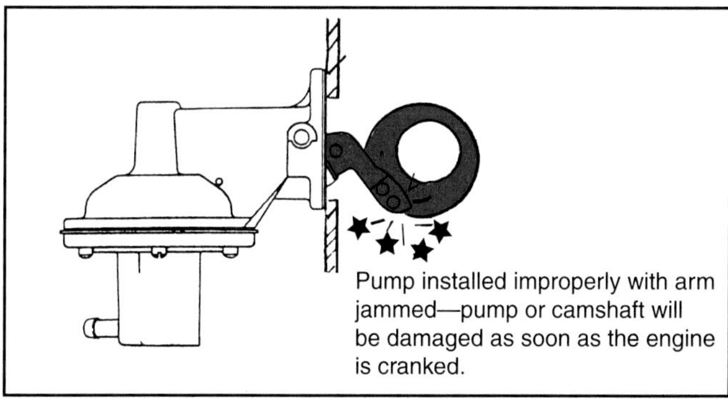

Figure B-20. *Mechanical fuel pumps are often driven by a lobe on or bolted to the front of the camshaft. These illustrations show the fuel pump position during installation.*

the line. A T-fitting should be used so that the engine can still receive gasoline. Start the engine and allow it to idle, then read the pressure gauge. Check the gauge reading against the vehicle manufacturer's specifications. On most engines, the pressure should be between 4-7 psi (27.6-48.26 kPa).

If the pressure is less than specified, check to see if any part of the fuel line, such as from the pump to the carburetor, or from the tank to the pump, is restricted. Look for kinks in metal lines, or evidence of swelling on neoprene rubber hoses. If any of the lines are restricted, clean or replace them. Also check all filters if they were not checked previously.

If defects in the fuel lines or filters are not restricting fuel flow, the pump is probably defective. On some engines with pumps driven by a camshaft lobe or pushrod, it is possible for the cam lobe or the fuel pump pushrod to be worn enough to reduce pump output.

Mechanical Fuel Pump Replacement

Mechanical fuel pumps are installed on the engine and operated by a camshaft lobe directly, **Figure B-20,** or through a pushrod, **Figure B-21.** Begin the replacement process by removing the fuel lines, then the bolts holding the pump to the engine. Clean the mounting surface of all old gasket and/or sealant material. Using a flashlight, inspect the pump pushrod and/or cam. If they are excessively worn, they should be replaced. Before installing the new pump, compare the old and new pump arms. They should match closely, or the new pump is not the correct replacement.

Note: If the fuel pump pushrod or cam interferes with pump installation, manually turn the crankshaft one revolution. This will allow additional clearance between the pushrod or cam and the fuel pump arm. If the pushrod will not remain in its recess, lightly coat the pushrod with all-purpose grease and reinstall.

Place the pump on the engine mounting surface, using a new gasket and sealer if required. In most cases, the pump arm will contact the camshaft or pushrod, preventing the pump body from completely seating on the engine mounting surface. To ensure that the pump is installed correctly, manually turn the engine to remove as much tension as possible from the pump arm. If the pump is operated by a pushrod, make sure that the pushrod is in the proper position. Then install the pump mounting bolts and tighten them

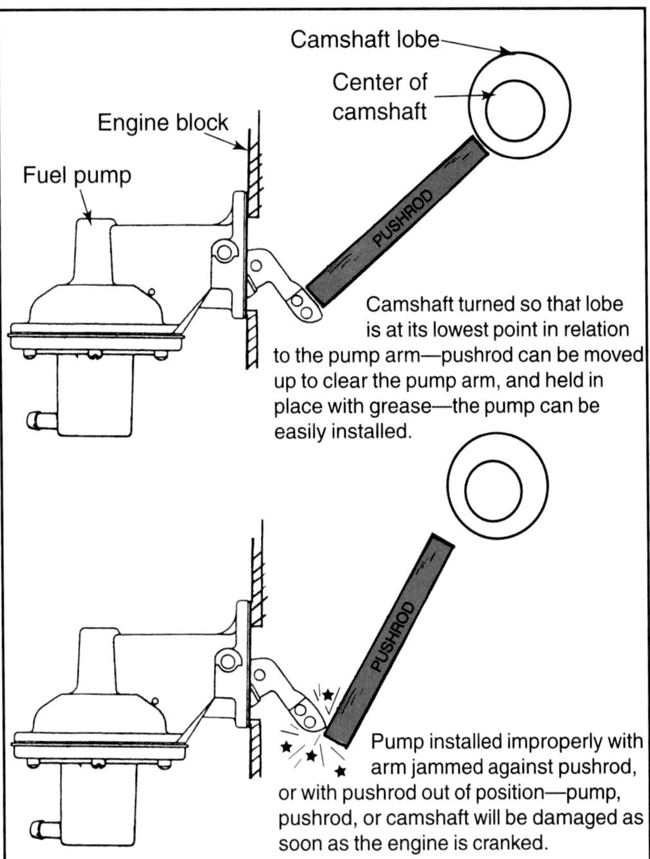

Figure B-21. *When installing mechanical fuel pumps driven by push rods, make sure the pushrod is as far inside the engine as possible. Fuel pump damage will result or installation will be difficult if the push rod is not properly positioned.*

evenly until the pump and engine surfaces are flush. Reinstall the lines and crank the engine until the fuel lines refill with gasoline. This may take as much as 60 seconds of cranking in 15 second intervals.

Carburetor Diagnosis and Service

The carburetor can be checked for obvious damage or defects. Before removing the air cleaner, check that the thermostatic air flap is free and operating properly. On a running engine, the flap should be open when the air cleaner housing is hot, and closed when the air cleaner housing is cold.

One of the most common problem areas on carburetors, both old and new, is the choke. Check choke adjustment according to the service manual. Some chokes on newer engines are not adjustable, especially when the carburetor is computer-controlled.

Move the choke plate and ensure that it moves freely and does not stick at any point. Also check the choke vacuum break to ensure that it opens the choke slightly when the engine starts, and is not set to open the choke too much. Make sure that the choke pull-off diaphragm does not leak, and that the vacuum hose to the diaphragm is not disconnected or leaking. While inspecting the choke system, check the fast (cold) idle speed setting, and adjust as needed.

Check the accelerator pump by moving the throttle while observing the carburetor throat. A stream of gasoline should squirt from the accelerator pump outlet nozzles. If you cannot see a stream, either the carburetor is out of gas or the accelerator pump is defective. Check the fuel pump and filters as explained later in this section. If gas is reaching the carburetor, but is not being pumped into the throat, the accelerator pump is defective. Either the pump plunger is worn out, or one of the check balls is leaking or out of position. If the carburetor is a feedback type, make sure all connections are in place and clean.

Checking Mixture Control Solenoids

To check mixture control (MC) solenoids, you will need a multimeter capable of recording duty cycles or pulse widths. As an alternative, a dwell meter can be used on many vehicles. The engine must be at operating temperature and in closed loop mode for this test. Connect the meter to the mixture control solenoid as shown in **Figure B-22.** On some vehicles, the meter is connected to a green lead from the vehicle wiring harness.

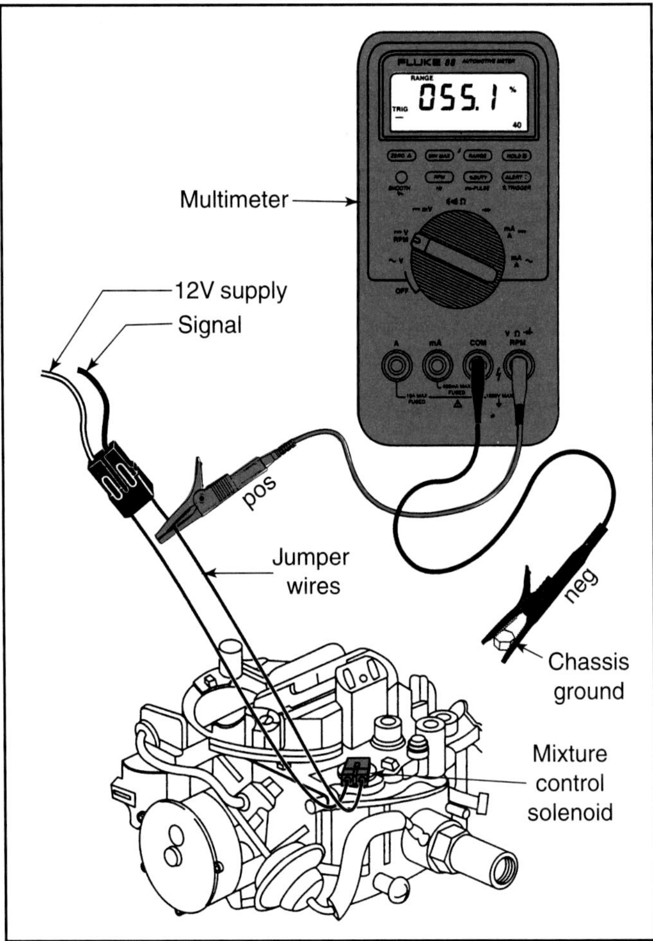

Figure B-22. *Digital meters capable of recording duty cycles or pulse widths can be used to check mixture control solenoids. Connect the meter as shown here. Scan tools can also be used to check MC solenoid operation. (Fluke)*

Note: Some scan tools can be used to check mixture control solenoid operation.

Set the multimeter to the *duty cycle* position, or set the dwell meter to the *six-cylinder* position. Start the engine and observe the duty cycle reading. It should fluctuate within a range of about 20% with the ECM in closed loop mode. A reading of 50% on the multimeter, or 30° of dwell on a dwell meter, is normal on most systems.

To test MC solenoid operation, create an artificial rich mixture using propane enrichment, but be careful not to stall the engine by adding too much. Propane enrichment is discussed later. Do not use carburetor cleaner or gasoline to richen the mixture. The duty cycle should rise as the ECM commands full lean to compensate for the rich mixture.

Turn off the propane enrichment valve and create an artificial lean mixture by removing a vacuum hose and covering it with your thumb. As you remove your thumb from the hose, the duty cycle will reduce as the ECM tries to compensate for the vacuum leak. Do not let in too much air or the engine may stall. If all of these tests show that the computer control system and MC solenoid are operating properly, remove the meter and propane enrichment device and reconnect the vacuum hose.

Figure B-23. *Some electric chokes are adjustable by loosening screws and turning the cover. Others are secured by rivets, which must be drilled out to loosen the cover. (Ford)*

Choke Adjustment

As long as carburetors are in existence, the choke will be a source of driveability problems. The choke system is the single greatest cause of cold driveability problems on carbureted engines. It is difficult to precisely control the accuracy of the choke closing in relation to engine temperature, outside air temperature, and air-fuel ratio needs. It is difficult to perfectly control the choke, even with computer systems.

Adjusting the choke for the best average operating conditions is critical to proper cold engine performance. Many chokes are sealed, either with a riveted cover, or aligning lugs that can only be installed one way. Other chokes can be adjusted. Factory service procedures call for aligning the marks on the choke plate with the proper marks on the stationary housing, or for bending a linkage rod to a specified dimension. Always follow the factory procedures to make a preliminary choke adjustment. Refer to **Figure B-23.** In many cases, this adjustment will solve the cold engine problem.

Often, however, the original factory adjustment cannot take into account the loss of tension on the choke spring, gradual wear, or dirt buildup in the carburetor or engine. Therefore, it may become necessary to adjust the choke to non-factory specifications, or to replace the choke coil assembly.

Before setting the choke, always make sure that the engine ignition and compression systems are in good condition. Also check the operation of the EFE system and the thermostatic air cleaner, and make sure that they do not have any defects. In many cases, a satisfactory adjustment can be made by setting a cold choke to close completely, with no additional tension. In other cases, the adjustment procedure will take considerable trial-and-error setting, possibly over several days and temperature changes.

Vacuum Break Adjustment

The vacuum break is a simple adjustment that is often overlooked. A misadjusted vacuum break can cause stalling when a cold engine is first started. To adjust the vacuum break on some carburetors, a section of the linkage can be bent, as shown in **Figure B-24.** On other carburetors, the vacuum break linkage contains an adjustment screw, **Figure B-25.** To adjust the vacuum break, start by cooling the choke coil, or allowing the engine to stand for at least six hours. If the vehicle has an electric choke, do not turn the ignition switch to the on position during the cool down period.

Once the choke coil is cold, open the throttle, and allow the choke plate to close completely. Apply vacuum to the vacuum break without starting the engine, or turning on the ignition switch. The easiest way to do this is with a hand-held vacuum pump. Observe whether the vacuum break opens the choke the required amount. If not, make adjustments and reapply vacuum to recheck until adjusted properly.

Float Adjustment

The carburetor float assembly can be checked by removing the carburetor air horn. Carburetor disassembly is covered in more detail later. Once the float is exposed,

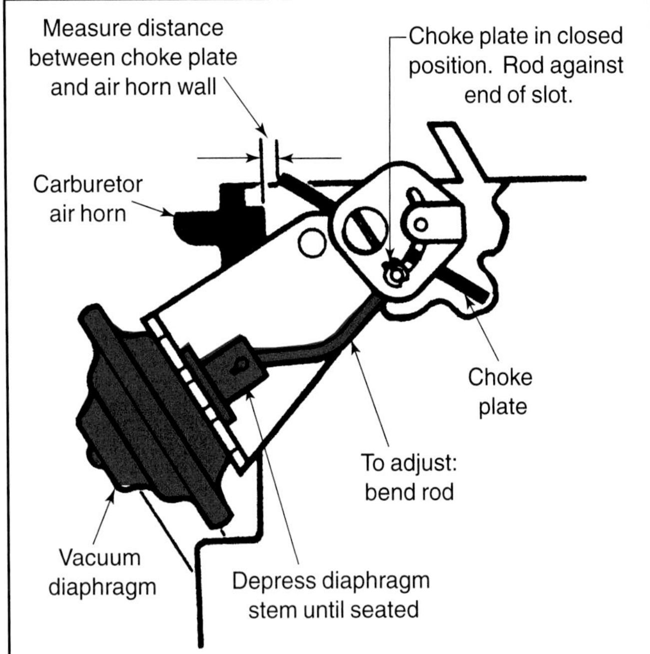

Figure B-24. *To adjust the vacuum break, as well as other adjustments, it may be necessary to bend part of the linkage. (Ford)*

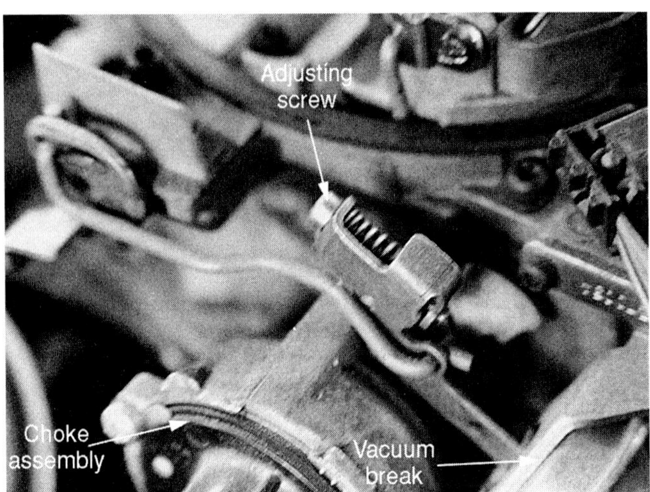

Figure B-25. *Some vacuum breaks have an adjustment screw. Exercise care when adjusting any carburetor screw.*

check for a defective needle and seat or a bent or leaking carburetor float. A worn out needle and seat can be spotted by a wear ring around the needle point where it contacts the seat. A leaking hollow float will make a sloshing sound when shaken. Solid floats can be checked for obvious damage and compared against a known good float if they appear to have absorbed gasoline.

If the needle, seat, and float are in good condition, check and adjust the float setting. While the carburetor is partially disassembled, check the float bowl for signs of dirt or water. If possible, check the power valve. The power valve could have a leaking diaphragm, be stuck, or have weak return springs, depending on the type used in the carburetor.

It is possible to check the float level on some carburetors without removing the carburetor air horn. Insert a ***float gauge*** in a vent slot on the air horn, **Figure B-26.** In some cases, a drill bit can be used to substitute for the gauge. Do not press down on the gauge as float damage or flooding will result. The marks on the gauge will line up with the top of the casting. Check the measurement against float level specifications. If the float setting is off, remove the air horn and adjust the float. To set either type of float, the carburetor does not have to be removed from the engine.

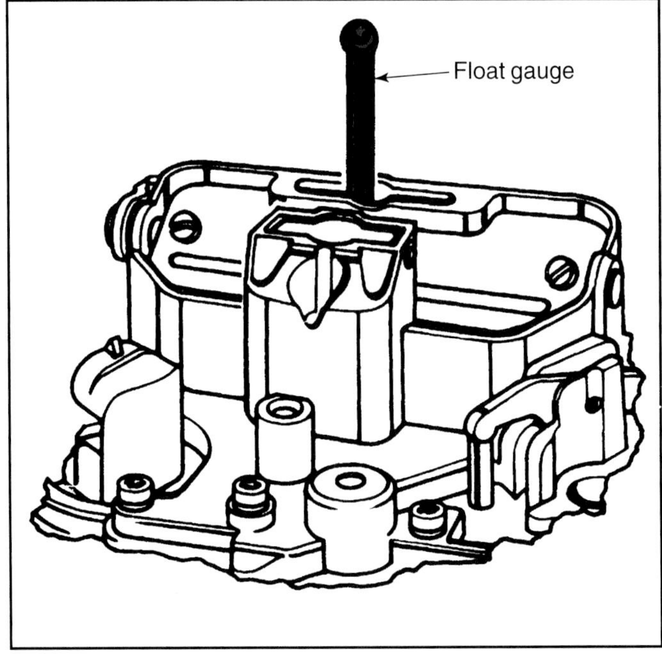

Figure B-26. *The float level on some carburetors can be checked by using a float gauge. (General Motors)*

To adjust a bowl cover mounted float, remove the hinge pin and the float. If the needle and seat were worn, replace them before adjusting the float. If the needle and seat are okay, reassemble the float and measure the distance between the top of the float and the cover gasket. This is shown in **Figure B-27.** Compare the reading with the

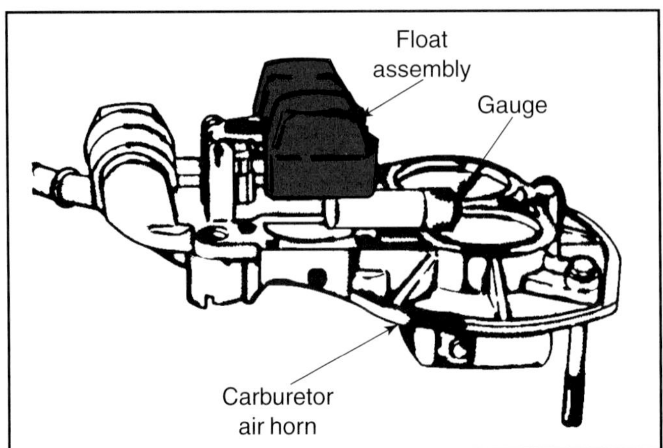

Figure B-27. *When checking the float with the air horn removed, check the distance between the horn and the float with the casting turned upside down. This checks the maximum height of the float when the carburetor bowl is full of fuel. (Chrysler)*

manufacturer's specifications. If the reading is not correct, bend the float arm until the proper reading is obtained.

Next, turn the bowl cover over and measure the ***float drop.*** This is usually measured as the distance from the cover gasket to the top of the float, as shown in **Figure B-28.** If the float drop is not correct, bend the metal projection at the rear of the float assembly until the correct float drop is obtained. Recheck the float setting and readjust as necessary. Then reinstall the bowl cover on the carburetor. Start the engine and check carburetor operation.

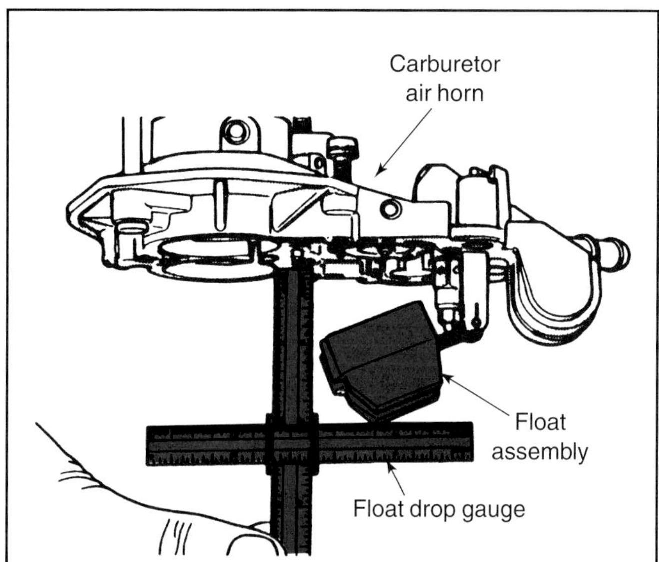

Figure B-28. *Use a float drop gauge to check the distance the float drops. This ensures the float will not drop too low, which can result in a lean mixture. (Chrysler)*

To adjust a bowl mounted float, remove the bowl cover. Note that there is no float drop setting, since the float drop limit is reached when the float contacts the bottom of the bowl. Reach into the bowl and remove the float hinge pin and float. Inspect the needle and seat for wear. If the needle and seat are worn, replace them before adjusting the float. If the needle and seat are in good condition, reassemble the float.

While holding the hinge pin and needle seat down firmly, press on the float hinge on the opposite side of the hinge pin. Measure from the top of the float to the top of the bowl casting. See **Figure B-29.** If the measurement is incorrect, bend the float arm until the proper setting is reached. Reinstall the bowl cover on the carburetor. On some models, the hinge pin is bowed and must be opened slightly to fit tightly and allow the float to work properly. Start the engine and check carburetor operation.

Adjusting Throttle Position Sensors

Figure B-30 shows a typical throttle position sensor installed in a four-barrel carburetor and operated by the accelerator pump linkage. Obtain the correct voltage specifications for the sensor, and set the voltmeter to the proper range. Then use a probe to penetrate the correct wires, and measure the input and output voltages. If the input voltage is incorrect, check the input wiring and make repairs as necessary. If the output voltage is incorrect, first ensure that the throttle plates are in the fully closed position. Then remove the plug covering the sensor adjustment screw and turn the screw until the voltmeter shows the proper voltage. Then reinstall a new protective plug over the adjustment screw.

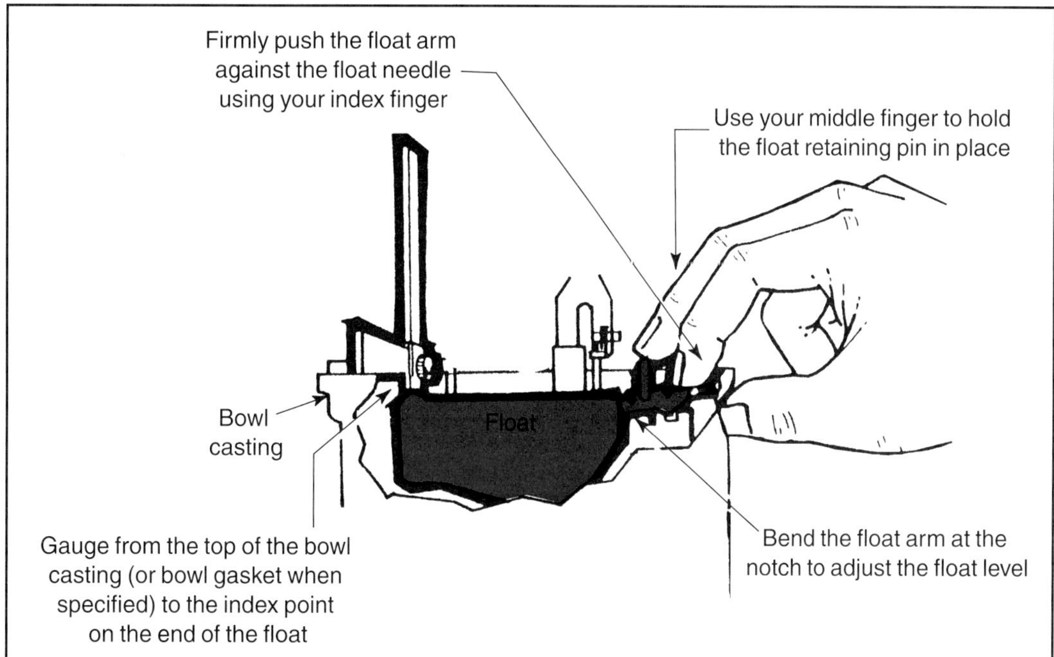

Figure B-29. *If the float adjustment is not correct, bend the float arm and recheck. (Chrysler)*

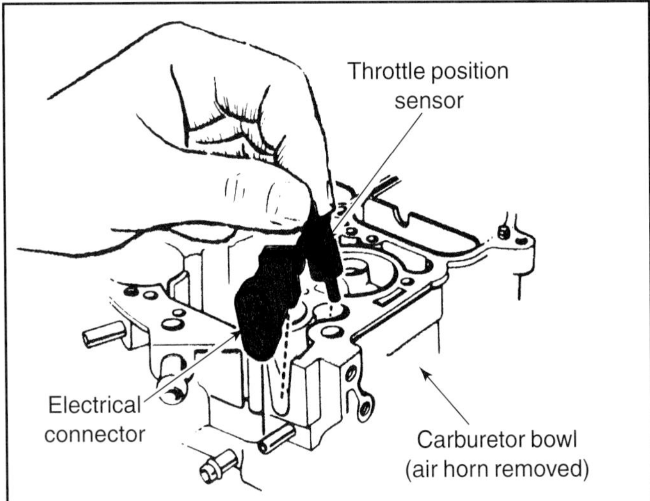

Figure B-30. *Typical installation of an internal throttle position sensor. (General Motors)*

Carburetor Overhaul

This section contains a general procedure for carburetor overhaul. Always consult the proper service manual to obtain specific carburetor rebuilding information. Carburetor overhaul kits also contain exploded views and the settings needed for the particular carburetor serviced by that kit.

Removing the Carburetor

The following procedure can be used to remove most carburetors. Before working on a carburetor having any type of electrical connections, disconnect the battery's negative cable first. Remove the air cleaner or air duct to the carburetor air horn. Check the air filter and determine if it needs replacement. Then disconnect any vacuum lines that attach to the carburetor. Also disconnect any electrical connectors at the mixture solenoid, bowl vent, throttle position sensor, or idle speed control. **Figure B-31** shows some typical connections.

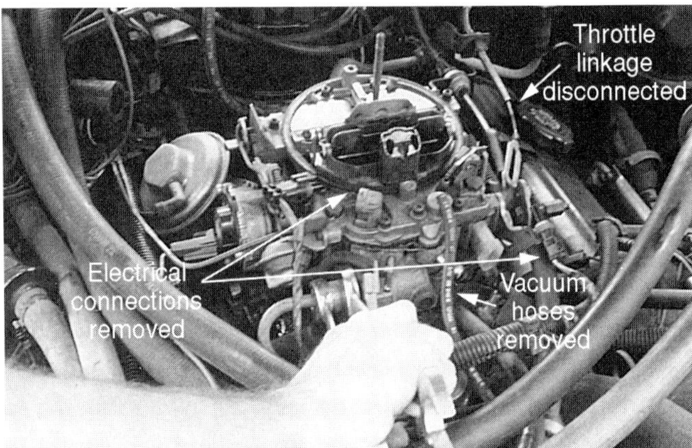

Figure B-31. *Make sure you remove all lines, connectors, and linkage before unbolting the carburetor from the engine.*

Next, remove the throttle linkage, any automatic transmission linkage, and the return springs at the carburetor. If the carburetor has a divorced choke, remove the choke link rod from the choke shaft. Then remove the fuel line from the carburetor inlet.

After all other lines and fittings are disconnected, remove the fasteners that hold the carburetor to the intake manifold. Some carburetors are installed by capscrews, while others are held by nuts that thread onto studs in the manifold. Lift the carburetor off the intake manifold and place clean rags or other covering over the manifold opening. Keep the carburetor in an upright position so that the fuel will not spill out. The fuel in the bowl can be checked for water or dirt when the carburetor is taken apart.

Disassembling the Carburetor

The following procedures can be used to completely disassemble a typical carburetor. Note that many of the most common service adjustments and parts replacements can be performed without completely disassembling the carburetor.

Closely inspect the carburetor for dirt, water, sediment, or other foreign matter as you disassemble it. The gasoline in the bowl can be poured into a clean container and checked for contaminants before disposal. While disassembling the carburetor, also check for damage or improper assembly from previous overhauls.

If the carburetor has an attached choke, remove the choke cover screws and retainers and remove the cover and thermostatic coil. Some choke covers are held to the housing with rivets. Do not drill out the rivets unless you are sure that the choke requires readjustment or replacement. Also remove the choke cover gasket and separator plate from the carburetor recess.

Remove the fast idle cam and levers from the choke shaft. Levers can be removed from the rods by turning them until the slot in the lever passes over the tang on the link rod. Unless it is absolutely necessary, do not remove the choke plate from its shaft, as it is hard to reinstall in the proper position. If the choke uses an external vacuum break, remove it now.

If the accelerator pump is located in the bowl cover, disconnect the pump rod from the throttle lever. If the mixture control solenoid is installed through the cover, remove it before removing the cover. If the bowl cover contains other electrical parts, they should be removed before further carburetor disassembly. If the gas filter is installed in the bowl cover, remove the inlet nut and discard the filter. Next, remove the bowl cover screws and lift the cover from the carburetor body. See **Figure B-32.** If the gasket is stuck, lightly pry on the cover to loosen it. Then place the cover upside down on a clean surface.

If the float is hinged from the bowl cover, slide the hinge pin out and lift the float assembly from the cover. Check the float for dents or leaks and check the hinge and pin for wear or binding. Remove the float needle from the float or seat as applicable. Then remove the needle seat and gasket with a screwdriver or socket wrench. Check the needle and seat for wear.

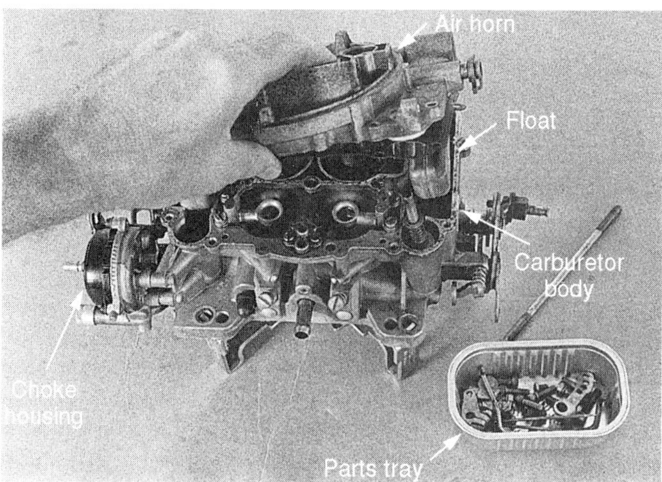

Figure B-32. *Keep all parts organized as you disassemble the carburetor. This will simplify reassembly.*

Remove the power piston, if it is installed in the bowl cover. On carburetors having the accelerator pump in the bowl cover, remove the pump retainer clip and remove the pump. After all other parts are removed from the bowl cover, remove the gasket and lay it aside so that the replacement gasket can be matched with it. Disassemble the bowl and throttle body. If the accelerator pump was installed in the cover, remove the pump plunger return spring from the pump well. On models with a bowl mounted accelerator pump, remove the retaining screws and remove the pump cover, diaphragm, and spring. Then remove the main jets and power valve.

If the float assembly is located in the bowl, remove the float, needle, and seat. If the fuel filter is installed in the bowl, remove it. Then remove the screws on top of the venturi cluster, and remove the cluster, gasket, and main well inserts. In some cases, it may be necessary to lightly pry on the venturi cluster to free the parts from the gasket. If the bowl assembly contains the accelerator pump discharge ball and spring, remove them. Also remove the mixture control solenoid and throttle position sensor if they are installed in the bowl.

Some carburetor chokes have an internal vacuum break piston. Detach the linkage, free the piston from its bore, then pull the piston and shaft from the carburetor. Remove the choke housing attaching screws, and remove the choke housing and gasket. If the choke or vacuum break assembly has heavy carbon deposits, there is an exhaust gas leak into the choke heater passages. The intake manifold should be checked for leaks and replaced if necessary.

Turn the carburetor upside down and remove the screws holding the bowl to the throttle plate. Remove the throttle plate and gasket. Lay the gasket to one side for comparison with the replacement gasket. The bowl and throttle plate are a single casting on some carburetors. Remove the idle adjusting needles and springs from the throttle plate. It may be necessary to remove caps installed at the factory before the idle needles can be removed. Cut two small V-shaped notches in the throttle plate above the caps and use a small chisel to remove the notches and the caps. Remove the idle speed motor or solenoid, if they are mounted on the bowl or throttle plate.

Carburetor Cleaning and Inspection

Most carburetor problems are caused by gum and varnish build up in the internal passages, restricting air and fuel flow. Other problems are caused by dirt, water, or carbon buildup on the inside or outside passages or moving parts of the carburetor. Dirt on the carburetor parts will prevent their inspection for defects. Therefore, successful carburetor rebuilding depends upon thorough cleaning.

Soak the metal parts of the carburetor in fresh cleaning solvent. Do not leave the castings in the cleaning solvent for more than 20 minutes, or the solvent will dissolve the metal coating that is placed on the castings to prevent internal leaks.

Caution: Do not place non-metal parts or metal parts containing non-metal components in the solvent. They will be ruined immediately.

After the carburetor parts have been in the solvent, remove them, wash the parts with water to neutralize the solvent and blow clean with compressed air. Do not attempt to clean jets or internal passages with drill bits or wires. They will enlarge the openings and change carburetor calibration. Once the passages are blown out, blow off all external areas until they are dry. Do not use rags to dry parts since they will leave lint.

Warning: Always wear eye protection and observe all safety precautions when using compressed air. Immediately wash off any solvent that splashes on your skin or clothing.

After the parts are thoroughly cleaned, check moving parts for wear. Moving parts include the throttle and choke shaft bores in the throttle body, cover castings, idle adjusting needle(s), internal and external linkages, fast idle cam steps, and the accelerator pump bore. Also check the bottom of the bowl for corrosion caused by water. Replace all defective parts. Before beginning reassembly, note the carburetor model number, and order all needed parts. The model number is cast on the carburetor bowl or on a tag affixed to the carburetor.

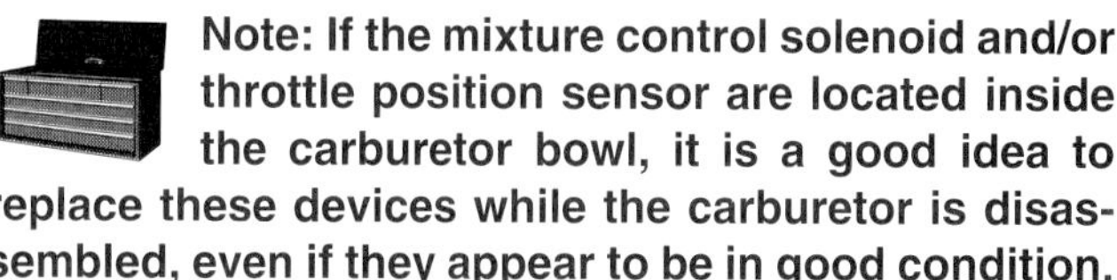

Note: If the mixture control solenoid and/or throttle position sensor are located inside the carburetor bowl, it is a good idea to replace these devices while the carburetor is disassembled, even if they appear to be in good condition.

Carburetor Assembly

Start by ensuring that you have the proper overhaul kit for the carburetor. Use the part numbers on the carburetor casting or tag to ensure that you have the correct kit. The kit should contain new gaskets and O-rings, needle and seat, accelerator pump, and other parts that are normally replaced, **Figure B-33.** Other parts, such as floats and vacuum break assemblies, must be purchased separately.

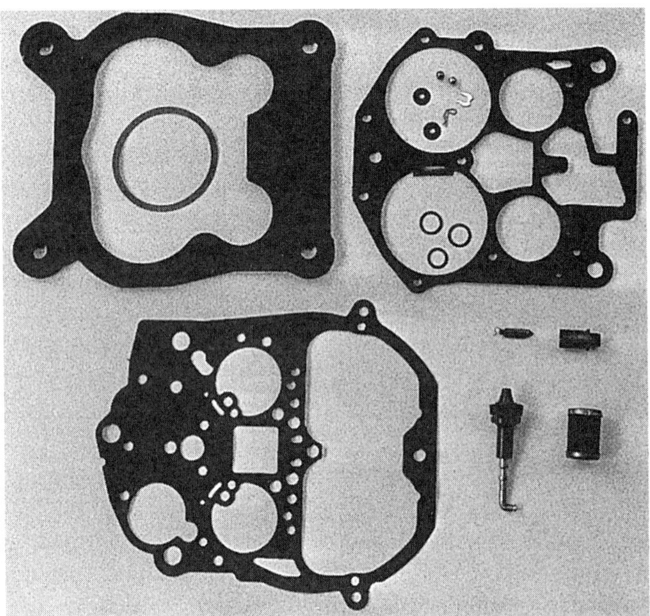

Figure B-33. *Carburetor repair kits have the gaskets, O-rings, and other parts needed to restore proper carburetor operation. Other parts such as throttle position sensors and mixture control solenoids must be purchased separately.*

Begin by reassembling the throttle plate. If the idle adjusting needles and springs were removed, install them into the throttle plate and seat them by hand. Do not use a screwdriver to force the needle against the seat, or it will be damaged. Back the screws out two turns after seating them. This will serve as a starting idle mixture adjustment. Install the idle speed solenoid if it is attached to the throttle plate.

Turn the carburetor bowl upside down and place a new throttle plate gasket in position. Place the throttle plate over the bowl and install the screws securely. If the carburetor has an attached choke, install the choke housing using a new choke-to-bowl gasket. Place the vacuum break piston in the choke housing and slide it into the bore. Attach any linkage screws inside the choke housing. Turn the bowl over and reinstall the internal parts. Locate the correct size accelerator pump inlet check ball and drop it into the inlet hole. Then replace the accelerator pump return spring and retainer. If the accelerator pump is in the bowl, install it now. On plunger-type accelerator pumps, the plunger seal is the only part that will be changed, and the pump must be disassembled to install the new seal.

If the accelerator pump check ball is installed under the venturi cluster, install it along with its spring and retainer. Then install the venturi cluster and gasket. Reinstall the main metering jets, power valve, and any electrical components that go inside the bowl. If the float is installed in the bowl, install the seat, needle, and float and adjust to specifications.

After the bowl components are reinstalled, reassemble the bowl cover. If the accelerator pump and linkage are installed in the cover, reinstall them. Then reinstall the needle, seat, and gasket, and the power piston if applicable. Install the cover gasket, then place the float and needle in position. Insert the float hinge pin and adjust the float level and float drop according to the instructions given in the service manual or sheet provided with the overhaul kit. Float settings were covered earlier in this appendix.

After you are certain that the float is accurately adjusted, lower the bowl cover onto the bowl. Make sure that the cover gasket is correctly positioned and that the accelerator pump plunger is correctly installed in its bore and moves freely. Also make sure that all external linkages that cannot be replaced after the cover is on, are installed. Then install and tighten the cover screws evenly. Install any other electrical devices as necessary.

Install any linkages that were removed during disassembly. Install the choke plate if it was removed, ensuring that the choke valve does not bind in the air horn before tightening the screws. If the carburetor has an attached choke, place the choke coil, gasket, and cover into position and install the choke coil and cover. Rotate the cover until the index marks on the cover and housing are lined up according to specifications. After making the adjustment, observe the choke. If the choke is closed, install the retainers and screws on the choke housing.

Note: Some manufacturers recommend filling the bowl with clean gasoline when assembling a carburetor.

If not done earlier, replace the fast idle cam and choke linkage. If the carburetor has an external vacuum break diaphragm, install and adjust it now. Install the air cleaner gasket and air cleaner mounting stud. If you filled the bowl with gasoline, operate the throttle lever several times and check the discharge from the accelerator pump jets. If the stream of gasoline is steady, the carburetor is ready to install.

Carburetor Installation

Clean the intake manifold gasket surface, and install a new gasket. If the manifold has heat crossover passages, check them for carbon buildup, and clean if needed. Then lower the carburetor onto the manifold. If the original carburetor was installed with washers, reinstall them. Install and tighten the carburetor hold-down bolts or nuts.

Reconnect the carburetor throttle and transmission linkage, if necessary. Reconnect all vacuum lines, fuel lines, and electrical connections. Attach a vacuum gauge to the thermostatic air cleaner port on the intake manifold. If the battery was disconnected, reconnect it and start the engine.

The engine should start easily without backfiring if the carburetor was filled with gasoline before installation. Allow the engine to warm up, and readjust the choke and vacuum break if they appear to be misadjusted.

After the engine is warm, adjust the idle speed and mixture as necessary. If no problems are encountered, install the air cleaner assembly and hoses, and then road test the vehicle. If any problems are encountered, review the rebuilding procedures to determine the cause.

Adjusting Carburetor Idle Speed and Mixture

The carburetor idle mixture must meet the requirements of local emission control laws. The idle mixture must be adjusted to produce no more than the legal amount of CO, NO_X, and HC. The mixture must be adjusted with the help of an exhaust gas analyzer, or by the propane enrichment method. Some older vehicles are set by the ***lean drop method.***

On newer vehicles with carburetors, the idle mixture cannot be set without removing caps from the carburetor. Most modern carburetors have sealed idle mixture screws, which can only be adjusted by removing the carburetor and drilling out a seal plug. On these carburetors, idle mixture and speed adjustments are not necessary, or desirable. The following sections explain how to make idle speed and mixture adjustments on newer carburetors. If the engine will not idle properly on an engine with a computer-controlled carburetor, check for vacuum leaks and defective parts before adjusting the mixture.

Lean Drop Method

The following idle speed and mixture setting procedure applies to older carbureted engines only. Check the manufacturer's service manual for adjustment procedures and the location of the idle speed and mixture screws. **Figure B-34** shows some typical idle mixture screw locations. If possible, always leave the air cleaner and all hoses installed during the idle adjustment procedures. If the engine idle can be adjusted, an adjustment procedure will be given on the emissions decal installed in the engine compartment. Before adjusting the idle, read the information on the decal to determine if any special procedures must be performed first. Examples are setting timing, disconnecting electrical devices, placing the transmission in drive or neutral, and turning the air conditioner on or off. Before starting the adjustment procedure, ensure that the engine is warm, and the choke is completely open.

Figure B-34. *Not all idle screws are as easy to reach as these. Most are located in the throttle plate and are somewhat difficult to reach when the carburetor is installed.*

Begin by connecting a tachometer and a vacuum gauge to the engine. Set the parking brake and place the transmission in neutral. Check that the carburetor is on the lowest step of the fast idle cam. Then start the engine and allow it to idle for at least one minute to stabilize the mixture. After one minute, adjust the idle speed screw to obtain the correct engine speed. If idle speed is controlled by a solenoid or motor, use the speed setting procedure listed on the emissions decal or in the service manual. If the idle speed is too high, and you cannot lower it with the adjustment screw, recheck the fast idle cam and choke linkage, and ensure that the accelerator linkage or cable is not binding. Also check that the throttle return spring has enough tension to completely close the throttle, and that the mounting bolts are tight.

Next, remove the idle screw restrictor caps. Turn the idle mixture screw clockwise until it is seated lightly. On most carburetors, turning the screw clockwise will move the mixture needle in to lean out the mixture. Turning the screw counterclockwise will move the needle out to richen the mixture. Turning the screw completely in should cause the engine to die. If it does not, the carburetor is running too rich, and should be overhauled. If the carburetor has two screws, perform this operation on one screw at a time. As either screw is turned in, the engine will begin to run roughly and may stall.

Caution: Do not seat the screws tightly, since this will distort the calibrated surfaces of the needle and the carburetor.

After lightly seating the mixture screws, turn them out about two turns to make a preliminary adjustment. Then restart the engine and adjust the mixture to produce the highest vacuum reading and idle speed. If the carburetor has two mixture screws, adjust each screw separately, ensuring that the total number of turns for each screw is within one turn of each other. Reset the idle speed to specifications, and readjust the mixture if necessary. Then turn the mixture screw in (lean) until the idle speed drops off 50 rpm, or as directed by the service manual or the underhood emissions decal. If the carburetor has two screws, turn them in equally. The manifold vacuum will drop off slightly. Then raise the idle speed to the specified setting using the speed adjusting screw.

Check the adjustment by putting the transmission in drive, if it has an automatic transmission, and turning on the air conditioner. If the engine does not stall or idle roughly, the mixture adjustment is satisfactory. Install new idle screw restrictor caps, remove all test equipment, and reinstall any engine components that were removed.

Propane Enrichment Method

The following procedure is used to set the idle mixture by the ***propane enrichment method.*** To perform this adjustment, you will need a propane enrichment device, **Figure B-35.** Inspect the propane enrichment device to make sure that the cylinder is at least half full and that both propane control valves are fully closed. Then place the propane bottle upright in the engine compartment, away from engine or exhaust system heat. Before starting, ensure that the engine is warm and the choke is fully open.

Begin by reading the procedure on the engine compartment emissions decal or in the service manual. Apply the parking brake and place the transmission in neutral. Turn off the air conditioner and connect a tachometer to the engine. A vacuum gauge is not necessary for this procedure. Start the engine and allow it to idle. After about one minute, check the tachometer to ensure that the idle speed is correct. Reset the idle speed if necessary. Then remove the PCV valve from the valve cover and allow the valve to draw in outside air. Disconnect a vacuum hose at the intake manifold and install the supply hose from the propane bottle in its place. If the air cleaner was removed, reinstall it before proceeding.

With the engine running at idle, open the main control valve of the propane enrichment device, and then slowly open the propane metering valve until the engine reaches its maximum idle speed. Do not move the throttle. When the amount of propane added becomes excessive, engine idle speed will begin to drop. Continue to adjust the metering valve for the highest engine rpm. Once the highest rpm is obtained, leave the propane metering valve at that setting. Then adjust the idle speed screw downward to the rpm specified on the engine compartment emissions decal or service manual.

Increase engine speed to about 2500 rpm for 15 seconds, then let the engine return to idle. Adjust the propane metering valve to get the highest engine idle speed. If this adjustment changed the idle speed, readjust it to specifications. Turn off the main propane valve and allow engine speed to drop as the rich mixture caused by the propane leans out. After about one minute, slowly adjust the mixture screw(s) until the specified idle speed is reached.

Raise the engine speed again to about 2500 rpm for 15 seconds, then release the throttle. Allow the engine to idle for about one minute and recheck engine speed. Reopen the main propane valve and adjust the propane metering valve to get the highest engine rpm. If the maximum speed is more than 25 rpm higher or lower than the specified propane rpm, repeat the adjustment procedure. Once the engine speed does not vary by more than 25 rpm, turn off both propane valves. Then stop the engine and remove the propane supply hose. Reinstall the manifold vacuum hose and PCV hose.

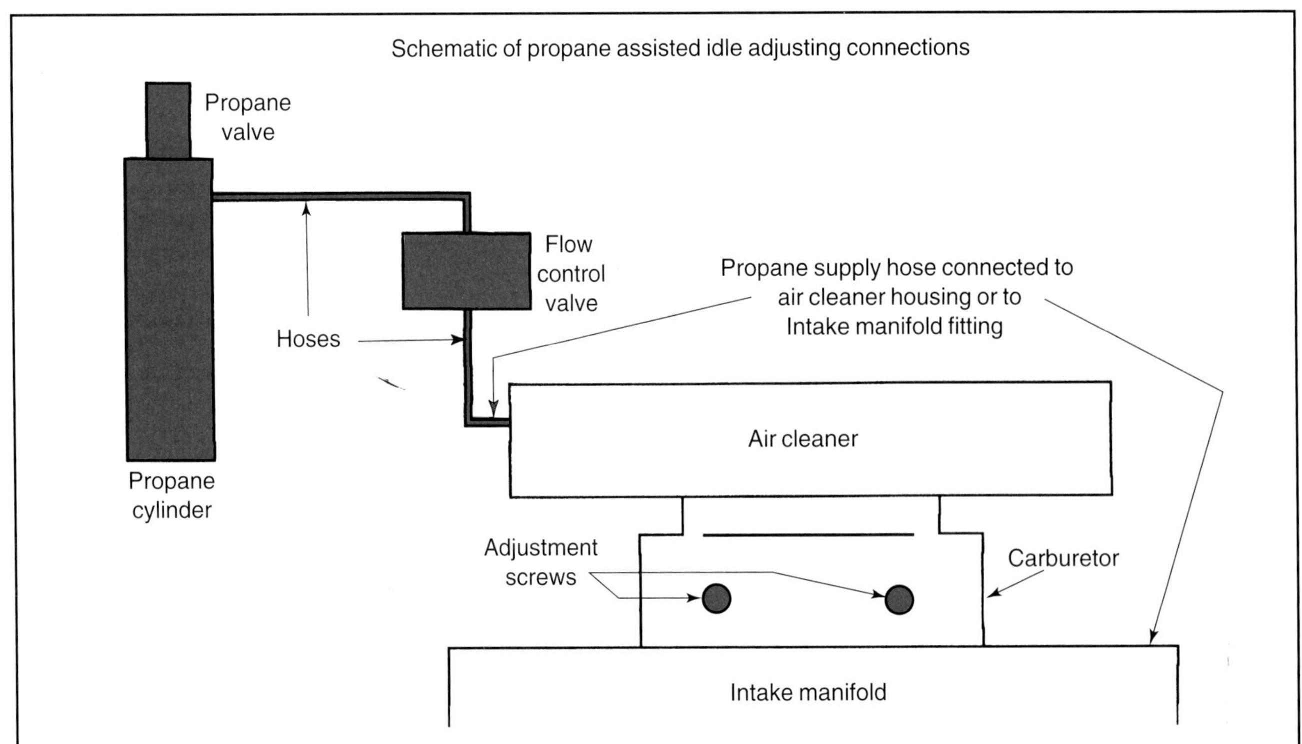

Figure B-35. *The propane enrichment method uses this setup to supply propane to the engine. This method is not normally performed on feedback carburetors.*

Adjustment With an Exhaust Gas Analyzer

To adjust the idle with an exhaust gas analyzer, start and calibrate the analyzer. Then place the exhaust probe of the analyzer in the vehicle tailpipe. The analyzer should be in its manual mode when used for calibration. Adjust the mixture to obtain the optimum level of emissions, which is much lower than that allowed by most local laws. The basic procedure for exhaust gas analyzer use is covered in **Chapter 18.** The emissions level varies with different model years, and the proper specification book should be consulted before adjusting the mixture.

Note: Although most states do not require emissions testing on vehicles made before 1975, it is a good idea to use an exhaust gas analyzer when adjusting the air-fuel mixture on these vehicles.

Once the vehicle air-fuel ratio at idle is correct, allow the exhaust gas analyzer's readings to stabilize. Briefly open the throttle, then allow it to return to idle. Observe the exhaust gas analyzer. The mixture should briefly become much richer, then return to the original settings. This will confirm that the accelerator pump is working properly

If the vehicle is equipped with a computer control system, the following checks can be made. Start by creating a rich mixture. This can be done by partially closing the choke plate. The throttle plate should remain closed. Do not restrict the air flow enough to stall the engine, just enough to create a rich mixture. Do not spray gasoline or carburetor cleaner into the intake to richen the mixture, as this will overload the oxygen sensor. When the air flow is restricted, the exhaust gas analyzer should show a momentary rich mixture, and then begin to lean out as the ECM begins to compensate for the rich mixture signal from the oxygen sensor.

When this check is complete, remove the air flow restriction and operate the engine at high idle for one minute to clear any extra gasoline out of the intake passages. Allow the exhaust gas analyzer readings to stabilize. Create a lean mixture by removing the PCV hose or another vacuum hose. The exhaust gas analyzer should show a momentary lean mixture, then begin to move to the rich position as the ECM begins to compensate for the lean signal from the oxygen sensor. When the test is complete, reconnect the hose.

Follow-Up for Carburetor Repairs

Road test the vehicle thoroughly. There should be no noticeable odors or fumes. Recheck the ECM for trouble codes that may have been set during the repair process. If possible, keep the vehicle overnight. This will allow any gaskets that were replaced to swell and seal more effectively. The idle settings, speed, and mixture can then be readjusted, if necessary. By keeping the vehicle overnight, any cold start circuits, devices, or carburetor choke settings can be rechecked with the engine cold. Perform an exhaust gas analysis to ensure the vehicle complies with local emissions standards.

Summary

All carburetors contain the same basic systems, whether they are one, two, or four throat models. Carburetors are composed of many different systems which operate at different speeds, loads, and temperatures. All carburetors contain main metering, throttle valve, float, idle and off-idle, accelerator pump, power, and choke systems. Some carburetors contain other systems to meet different requirements, such as secondary throats for more power. Most modern carburetors are controlled by the ECM, and contain mixture control solenoids. Most modern carburetors are equipped with a throttle position sensor which feeds information into the ECM.

Some carburetor components, such as the choke and accelerator pump, can be examined to determine their condition without disassembling the carburetor. Air, fuel, and emission filters can be checked to determine the need for replacement. Some carburetor components, such as the choke and accelerator pump, can be examined to determine their condition without disassembling the carburetor.

The choke system is the single greatest cause of cold engine problems on carbureted engines. It is difficult to precisely control the accuracy of the choke closing in relation to the engine temperature, outside air temperature, and air/fuel ratio needs. It is difficult to perfectly control the choke, even with electronic control systems.

Adjusting the choke for the best average operating conditions is critical to proper cold engine performance. Factory adjustment cannot take into account the loss of tension on the choke spring, and gradual wear or dirt buildup in the carburetor or engine. Therefore, it may become necessary to adjust the choke to non-factory specifications, or to replace the choke coil assembly.

The vacuum break is a simple adjustment that is often overlooked. A misadjusted vacuum break can cause stalling when a cold engine is first started. On some carburetors, the vacuum break linkage contains an adjustment screw.

Newer vehicle carburetors have smaller main jets, reduced accelerator pump output, and power valve springs which do not allow the power valve to open until the engine is under very heavy loads. The idle mixture screws are sealed to prevent readjustment. Most carburetors are controlled by the ECU through a mixture control solenoid installed on the carburetor.

There are two types of carburetor floats. One is hinged in the bowl cover. Gas flows down through the cover casting, the needle and seat, and into the bowl. The second type is hinged in the bowl. Gas flows through the bowl casting, the needle and seat, and into the bowl.

It is important to have the proper overhaul kit for the carburetor. The part numbers on the carburetor casting or tag should be checked to ensure that you have the correct kit. The carburetor overhaul kit contains exploded views and settings needed for the particular carburetor serviced by that kit. Most common carburetor service adjustments and parts replacements can be performed without completely disassembling the carburetor.

Most carburetor problems are caused by gum and varnish building up inside the internal passages, restricting air and fuel flow. Other carburetor problems are caused by dirt, water, or carbon buildup on the inside or outside passages, or moving parts of the carburetor. Dirt on the carburetor parts will prevent their inspection for defects. Successful carburetor rebuilding depends upon thorough cleaning.

After the parts are thoroughly cleaned, moving parts should be checked for wear. Also, the bottom of the bowl should be checked for corrosion caused by water. If a problem is found, the part should be replaced. If possible, the vehicle should be kept overnight after the overhaul. This allows the gaskets to swell and seal more effectively. The idle speed and mixture can then be readjusted. The choke setting can also be rechecked after the engine cools off overnight.

Know These Terms

Carburetor
Circuits
Throat
Barrel
Primary system
Secondary system
Venturi
Venturi vacuum
Main metering system
Main jet
Power system
Power valve
Step-up rod
Idle system
Idle needle
Off-idle system
Throttle plate
Float system
Needle and seat
Bowl vents
Accelerator pump
Choke
Choke pull-off
Hot idle compensator
Electric choke
Idle control solenoids
Anti-dieseling solenoid
Air conditioning speed-up solenoid
Idle control motors
Variable venturi carburetor
Throttle position sensor
Mixture control solenoid
Float gauge
Float drop
Lean drop method
Propane enrichment method

Review Questions—Appendix B

Please do not write in this text. Write your answers on a separate sheet of paper.

1. The most common cause of cold engine problems on carbureted engines is the ______.
2. A misadjusted vacuum break can cause stalling when a _______ engine is first started.
3. If the carburetor ______ and ______ are worn, replace them before adjusting the float.
4. What are the two main reasons that the carburetor float bowl is vented?
5. The off-idle system maintains the proper air-fuel ratio when the throttle opening is between ____ and ____ speeds.
6. The bowl cover must be turned over to measure the float ______.
7. The fuel in a carburetor bowl can be checked for _____ or ____ when the carburetor is taken apart.
8. You should adjust the carburetor idle speed and mixture when the engine is ______.
9. The ______ determines how much gas can be pulled into the carburetor throat through the main metering system.
10. The main metering system is unable to operate at ______ speeds, since the air flow through the venturi is too slow to create a vacuum.

ASE Certification-Type Questions

1. In the carburetor, incoming air and gasoline are mixed in the ______ .
   (A) throat
   (B) bowl
   (C) choke
   (D) jets
2. The area in the carburetor throat that is smaller than the rest of the throat is the ______ .
   (A) air horn
   (B) throttle valve
   (C) choke plate
   (D) venturi

3. Technician A says that most power valves are controlled by manifold vacuum and spring pressure. Technician B says that most power valves are controlled by venturi vacuum and spring pressure. Who is right?
   (A) A only.
   (B) B only.
   (C) Both A & B.
   (D) Neither A nor B.
4. If the level of gasoline in the carburetor bowl is low, the air-fuel mixture will be ______ .
   (A) lean
   (B) rich
   (C) stoichiometric
   (D) lean or rich, depending on the relevant air temperature
5. The mixture control solenoid controls carburetor mixtures by controlling ______ .
   (A) fuel flow
   (B) idle speed
   (C) air flow
   (D) Both A & C.
6. Technician A says that some chokes are operated by engine heat. Technician B says that some chokes are operated by electricity. Who is right?
   (A) A only.
   (B) B only.
   (C) Both A & B.
   (D) Neither A nor B.
7. On many modern carburetors, the ______ is not adjustable.
   (A) idle speed
   (B) accelerator pump
   (C) choke
   (D) All of the above.
8. When adjusting either type of float, the ________ is removed.
   (A) bowl cover
   (B) needle and seat
   (C) carburetor bowl
   (D) Both A & B.
9. To ensure that you order the proper carburetor overhaul kit, use the part numbers on ______.
   (A) the carburetor kit
   (B) the carburetor casting or tag
   (C) the engine block
   (D) the VIN number
10. Technician A says that carburetor parts should be soaked in cleaning solvent for no more than 60 minutes. Technician B says that carburetor parts should be soaked in cleaning solvent for no more than 30 minutes. Who is right?
   (A) A only.
   (B) B only.
   (C) Both A & B.
   (D) Neither A nor B.

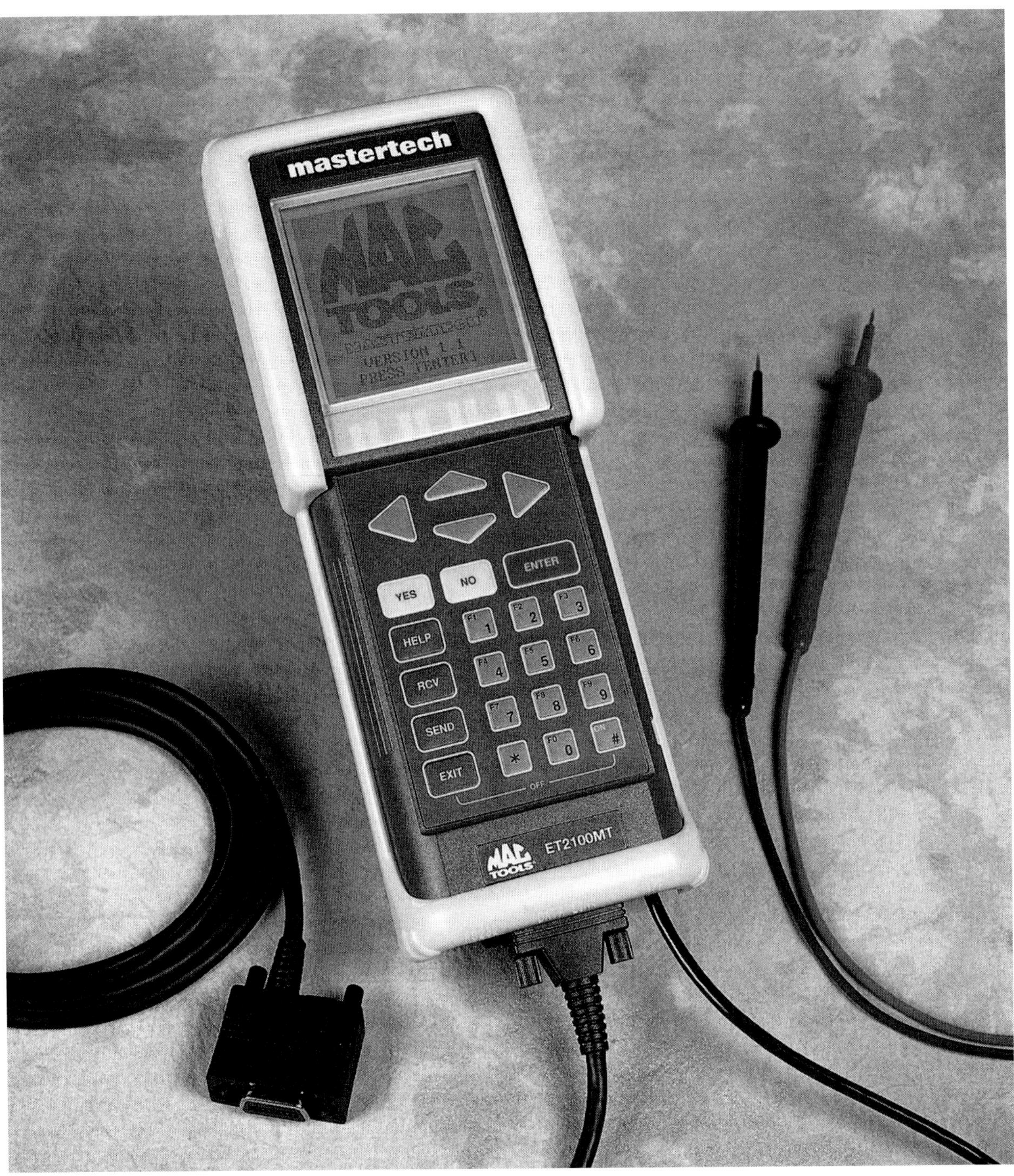

The newest scan tools are equipped with multimeters and other test tools. (Mac Tools)

Appendix C

Using Scan Tools

The scan tool is one of the modern technician's most powerful tools. It provides the ability to monitor and control engine, transmission, brake, steering, air bag, and body system functions from a package no larger than a book. The scan tool allows you to monitor system functions while the vehicle is being driven. Traditional engine analyzers cannot do this due to their size.

This appendix takes you through a step-by-step description of the most common scan tool functions and capabilities. This appendix is not specific to any one type of scan tool, but describes the features common to most scan tools you will encounter. It is recommended that you read the scan tool's manual before you use it or if you ever have any questions as to the purpose of a specific scan tool function.

Introduction to Scan Tools

A common misconception about ***scan tools*** is that they are foolproof, **Figure C-1.** However, this statement is far from the truth. Many technicians, especially those who have had very limited exposure to scan tools, either make a mistake interpreting the scan tool data or fail to use the tool properly. Modern scan tools are capable of generating so much information, and can control so many systems, that even experienced technicians may feel overwhelmed. Knowing how scan tools work, how to use them, and the information they generate properly can make the diagnostic process that much easier.

Scan Tools Types

Scan tools can be broken down into three groups: dedicated, multi-system, and generic. ***Dedicated scan tools*** work with the computer systems of only one manufacturer. They allow the technician to check and clear computer codes and monitor some sensor operations. However, they have extremely limited or no test capabilities. When coupled with selector switches or special harness connectors, dedicated scan tools are able to access the computerized systems of one manufacturer's vehicles.

Multi-system scan tools can retrieve trouble codes from the various systems used on different lines of vehicles. They are sometimes referred to as multi-line scan tools. These scan tools are able to present visually the sensor information coming into the vehicle's ECM, the operational status of output devices, and has some limited test capabilities. They consist of a scan tool, diagnostic connectors, and software packages, or cartridges. The cartridges are designed to interface with the various computer systems in each manufacturer's line of vehicles. Different connectors and cartridges are available for each system and line of vehicle.

Generic scan tools are able to access all the computer systems of more than one manufacturer. These tools are very powerful, with diagnostic capabilities equal to or

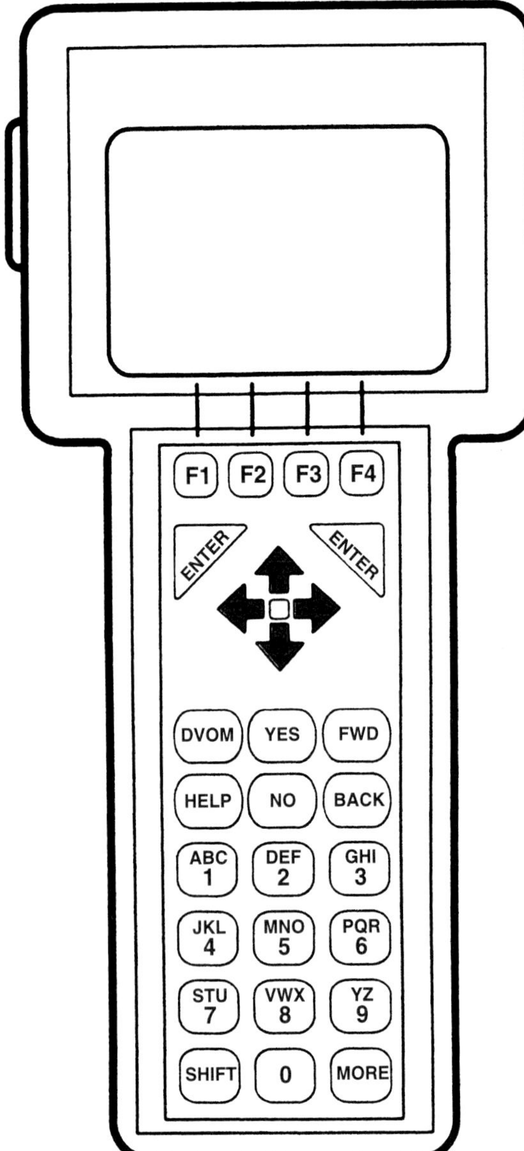

Figure C-1. *Scan tools come in a variety of sizes and capabilities. This one is typical of a generic scan tool design. (Chrysler)*

greater than some engine analyzers. Some have multimeters and other test equipment built into them. Generic scan tool programming can be updated by downloading new vehicle and system specifications. They have the capacity for future expansion for even greater diagnostic power.

Bi-Directional Scan Tools

To accurately monitor and control vehicle systems, scan tools must be able to perform more than one function. Modern scan tools can transmit and receive information and are referred to as ***bi-directional scan tools.*** A bi-directional scan tool can receive sensor inputs and control actuator operation, often at the same time. Since early scan tools did not do much more than read trouble codes, there was no need for bi-directional capability.

Scan Tool Interface

Most scan tools have a ***liquid crystal display (LCD)*** that can vary in definition(clarity and quality of images), size, and shape. All the information generated by the scan tool, as well as menus and command functions are displayed on this screen. Some scan tools have a light emitting diode (LED) or low-definition liquid crystal display. Newer scan tools have high-definition LCD displays that can show graphs and waveforms, just like an engine analyzer's oscilloscope. To control the tool's functions, a keypad and other controls are located beside or just under the display.

Scan Tool Keypad and Controls

The ***keypad*** controls the scan tool's functions and allows the user to input information and make selections from a menu provided by the scan tool's software, **Figure C-2.** The keypad on most scan tools use touch sensitive keys, laminated for protection against dirt, oil, and grease. A few scan tools have raised keys. The scan tool may produce an audible "click" when a key is pressed.

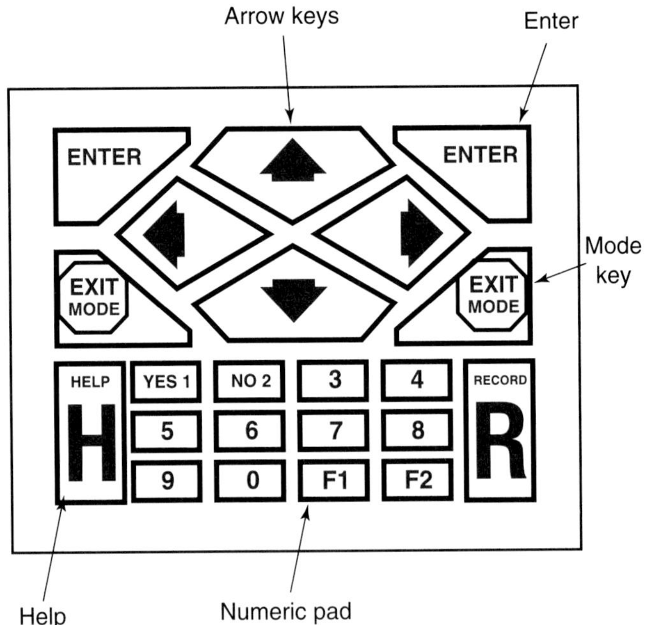

Figure C-2. *The keypad is the command center for the scan tool. Using the keys, you can step through the various menu screens provided by the scan tool software. (OTC)*

Except for some early dedicated scan tools, almost all scan tools will have two or more keys labeled *Enter* and *Exit,* or *Yes* and *No.* These are referred to as action or *command keys.* These keys are used to enter information, exit menus, and give a response to questions prompted by the scan tool. Some scan tools are equipped with both sets of command keys.

The keypad will probably also have *up* and *down arrow* keys. This allows the user to scroll through menu selections. Some keypads also have *left* and *right arrow* keys that permit switching between menu functions or screens. These keys are referred to as *selection keys.* Some scan tools have a thumbwheel that allows the user to scroll through the information shown on the display.

The keypad may also have a key labeled *More* along with or in place of one or both sets of arrow keys. The More key is used when there are more selections available than can be shown on the display. If you need to go back because you made a mistake or have to change scan tool settings, you can use the Exit or arrow keys.

The scan tool keypad often has *numeric keys,* similar to a calculator. Scan tool functions can also be activated using these numeric keys. The numeric keys are also used to enter vehicle information, such as part or all of the VIN number. Some keypads have letters above the numbers like a touch-tone telephone.

Some scan tools have *function keys,* labeled F1, F2, etc., in place of or in addition to the numeric keys. They are similar in style to the function keys on the top or side of a personal computer keyboard. In most cases, these keys serve the same purpose as numeric keys or provide additional tool capabilities. Scan tools may have additional keys, such as *Help* and so-called "soft keys," whose functions can vary, depending on the scan tool's software, capabilities, and manufacturer.

The scan tool may have one or more indicator LEDs built into the keypad. These LEDs indicate that the scan tool is receiving power or if a vehicle system has passed or failed a particular test. There may be more LEDs on a particular scan tool with different functions than described here. This is why it is important that you read the user's manual before connecting the scan tool to the data link connector.

Cables and Adapters

The scan tool needs compatible cables and adapters that will allow it to access any data link connector, **Figure C-3.** Dedicated scan tools will only have the cables and adapters needed for the vehicle computer systems it is designed to access. Cables are needed to connect the scan tool to the data link connector and power. Additional cables may be needed to extend the scan tool into the vehicle compartment as well as downloading new information into the scan tool.

Multi-system and generic scan tools will have adapters that allow them to access any system for which they have programming. In the case of multi-system scan tools, the necessary adapters will come with the program cartridge. Scan tool cables and adapters often come in a kit along with the tool and user manual. As newer vehicles with standardized 16-pin data link connector replace older vehicles, fewer adapters will be needed.

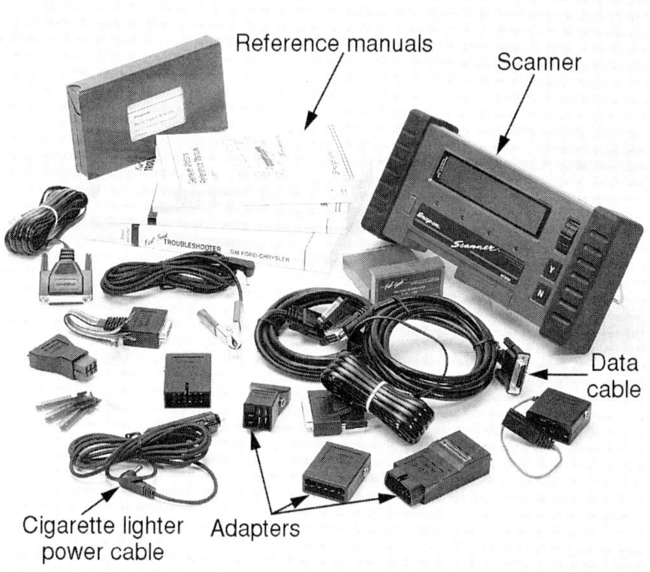

Figure C-3. *The wide variety of vehicles on the road requires many cables and adapters. (Snap-on Tools)*

Scan Tool Built-in Terminals

Depending on the type of scan tool, it can have a variety of terminals to connect to other external devices, such as computerized analyzers. All scan tools have one terminal that is used by the cable that connects the scan tool to the data link connector. This is usually either a 24- or 36-pin terminal. Newer scan tools have more than one built-in terminal. These can include, but are not necessarily limited to:

- Telephone jack—allows the scan tool to communicate with a computer via modem connection.
- RS-232 serial port—used for connecting to a printer, computer, or other serial communications device.
- RS-485 port—allows the scan tool to communicate with a computer or engine analyzer.
- Multimeter terminals—allows the scan tool to be used as a multimeter, volt/amp tester, or oscilloscope.
- DIN port—used in place of a 24- or 36-pin terminal or to connect additional cables.

Cartridges and Software

The scan tool, like a computer, must have software in order to function. The first scan tools did no more than read trouble codes. As manufacturers included greater self-diagnostic capability on more vehicle systems, technicians needed additional tools that could monitor and test all these systems. Tool manufacturers began offering scan tools that used small software cartridges, often no larger than a deck of cards.

While these program cartridges allowed scan tools to monitor and control more system functions, there were many drawbacks. As new vehicle models came out or if a mid-year change was made to a vehicle system, the old program cartridges could not be updated and had to be replaced. The program cartridge used to diagnose engine performance problems could not be used to diagnose anti-lock brake and other on-board computer systems.

Scan tool manufacturers finally offered a solution to this problem with generic cartridges, sometimes referred to as ***mass storage cartridges***, that can store up to 1 megabyte of data. Mass storage cartridges allowed the user to update the cartridge. This is done by connecting the scan tool to a computerized analyzer or personal computer (PC) and downloading an updated program from a CD-ROM disk to the cartridge through the scan tool.

The latest scan tools have self-contained programming on fixed ***personal computer memory card industry association (PCMCIA)*** cards, that are about the size of a charge card. They are capable of storing up to 10 megabytes of programming and vehicle data per card. No external program cartridges are needed for the scan tool to operate. Most of these scan tools also have slots for one or more PCMCIA cards. This type of scan tool can be updated by downloading new information as needed from a personal computer or computerized analyzer. The software will have a program or version number and date when it was created or downloaded.

Using Scan Tools

Using scan tools safely and properly go hand-in-hand with being able to correctly interpret the information generated. If using the scan tool results in damage to the tool or vehicle, fixing the driveability problem will become a secondary issue.

The first step is to locate the scan tool. It should be located in a plastic box or carrying case, along with all the necessary cables, adapters, and in some cases, software cartridges. If the tool is not located there, it is probably being used by another technician in your shop. If you find yourself using the scan tool frequently, you might want to investigate purchasing your own scan tool.

Rules for Scan Tools Usage

The following are several hints for using scan tools properly and safely. A scan tool may look fairly harmless, but can create a very dangerous situation if it is used or placed in the vehicle improperly.

- ❑ Make sure the cables are located away from the accelerator, brake pedal, and all moving parts.
- ❑ Do not shut off the ignition key unless prompted to do so by the scan tool. Any scanned information may be lost.
- ❑ Do not disconnect the scan tool cable from the data link connector during any programming or learn procedure. Damage to the ECM may result.
- ❑ Do not remove the scan tool cartridge or program card from the scan tool while power is on.
- ❑ Do not expose the scan tool to excessive or direct sunlight for extended periods as it can damage the display screen.
- ❑ Do not hang the scan tool from the rear view mirror. This will place it in a location where the cables or the scan tool itself may become caught in the steering wheel. The tool's weight may pull the rear view mirror off of its mounting, possibly damaging the mirror or windshield.
- ❑ If the data link connector is located under the hood, use an extension cable that allows the scan tool to be located inside the passenger compartment or use a chassis dynamometer.
- ❑ Be aware of any special operating conditions that exist while the scan tool is connected to the data link connector. For example, some ABS systems are disabled anytime a scan tool is connected to the data link connector. The scan tool will usually warn you if certain system or vehicle functions will be disabled.
- ❑ Do not drive and monitor scan tool functions. You will not be able to devote full attention to either the scan tool display or the road. If it is necessary to monitor scan tool functions during a road test, have another technician or shop foreman drive the vehicle for you.
- ❑ Do not perform any actuator or scan tool intrusive tests while driving the vehicle. Doing so may disable a vital safety system, such as the brakes. Only use the scan tool's system monitor and snapshot features during a road test.
- ❑ Do not perform any system learn functions during a road test.
- ❑ If you must leave the vehicle for more than a few minutes, disconnect the scan tool from battery power. Scan tools draw a great amount of power and can drain the battery if the engine is not running.

Connecting the Scan Tool

Locate the vehicle's data link connector. It will most likely be located in the passenger compartment, or in the engine compartment. Check the service manual for the exact location as some vehicles use several data link connectors for various on-board computer systems. Look at the connector and determine the cable and/or adapter needed. Unpack the scan tool and connect the proper cable and adapter, if needed. If the scan tool requires one, load the correct program cartridge for the vehicle or system, **Figure C-4.**

Carefully plug in the scan tool data cable to the data link connector. The cable should connect to the data link connector with little effort. If resistance is encountered, remove the cable and check for bent pins. Double check to be sure you are using the correct adapter. Connect the scan tool to battery power, **Figure C-5.** Most scan tools draw power from the cigar lighter or directly from the battery.

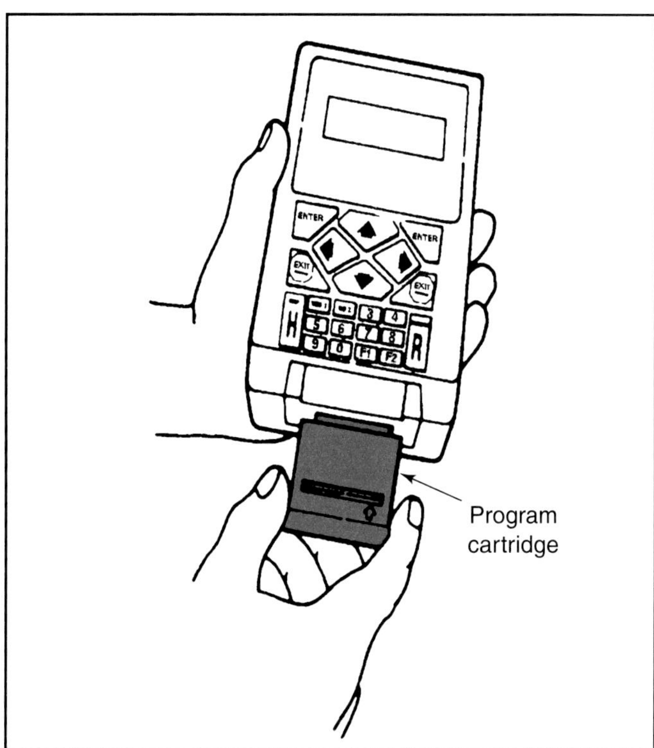

Figure C-4. *Make sure you plug in the program cartridge before connecting the scan tool to power. The cartridge should fit securely, but not tight in its slot. (OTC)*

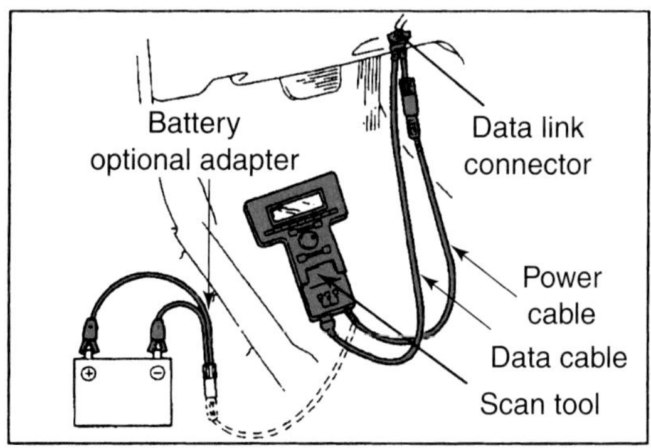

Figure C-5. *A properly connected scan tool will have cable connections to power and the vehicle's data link connector.*

Once connected to power, the scan tool display should light, along with any indicator LEDs. In some cases, the scan tool will perform a ***power-on self-test (POST),*** which verifies that the scan tool is operating properly. This test should only take a few moments. After the self-test, **Figure C-6,** the display should indicate:

- The type of scan tool and copyright information.
- The data cartridge or program number in use.
- In some cases, the ID number for the scan tool's internal processor.

GW Tool Corporation
Copyright 1998

Version 2.0, Program 1017CJSZ

Starting Internal Self-Diagnostics

Figure C-6. *When the scan tool receives power, the first screen should indicate the scan tool make, copyright information, and what program or cartridge is installed.*

Inputting Vehicle Data

The next step in most cases will be to input data about the vehicle into the scan tool. The scan tool will request that you enter the vehicle make, model, engine type, transmission, and if certain options are installed. In some cases, the scan tool will be able to read this information automatically from the vehicle's on-board computers. Remember, if you make a mistake at any point, use the arrow or exit keys to back up.

In most cases, the scan tool will receive this information by asking you to enter certain letters and numbers from the VIN number, **Figure C-7.** If this number is not on the repair order, it should be located on the dash, on a sticker inside one of the door jams, or in the engine compartment.

Once the scan tool knows the make, model, and options on the vehicle, it will present the first of many menu screens. From here, you will be able to select the system, monitor, or test you wish to perform.

Scan Tool Menu Functions

The first display that you should see is a main menu screen that will list several choices, based on the scan tool's programming. In most cases, the menu is a list of general scan tool functions. Use the arrow keys to select the function that you need, in this case, powertrain diagnostics. Then press the *Enter* or *Yes* key to proceed to the next screen, **Figure C-8.**

The next menu screen will ask the specific computerized system you want to diagnose. It will list all the computerized systems available on that particular vehicle make and model. Once again, you will scroll to and select the system where the problem is most likely located.

Once the selection of a system is made, a third menu will ask you which particular test, or feature you would like to use. This menu is normally referred to as the *Diagnostic menu.* It is from this menu where you will perform the majority of your selections, **Figure C-9.** In most cases, the first step should be to see if any trouble codes have been stored in the ECM's volatile memory.

Diagnostic Trouble Codes

As you learned earlier, reading trouble codes was the primary function of the first scan tools. The first step you

VIN Digit Identifier		
1	Country of Origin	1=United States 2=Canada 3=Mexico 4=United States Built J=Japan K=Korean
2	Make	Letter identifying the vehicle's manufacturer.
3	Vehicle type	Letter indentifying vehicle type (passenger car, truck, etc.)
4	Passenger safety*	Denotes on-board safety systems installed.
5	Vehicle line	Indicates the vehicle model.
6	Series	Indicates trim level (sedan, sport, luxury, etc.).
7	Body style	Indicates body style (2dr, 4dr, etc.).
8	Engine	Number or letter indicating the type of engine.
9	Check digit	Used to separate model year, assembly plant, and assembly sequence numbers.
10	Model year	Letter corresponding with the vehicle's year of make.
11	Assembly plant	Letter or number used by manufacturer to denote point of assembly.
12-17	Assembly sequence	Numbers denoting sequence in which vehicle was assembled.

*System may be denoted by the seventh VIN digit on some manufacturers' vehicles.

Figure C-7. *Each number in the vehicle identification number (VIN) describes the vehicle. Numbers or letters are used to describe the vehicle's make, model, manufacturer, year and order of production.*

System Option Menu

1. Powertrain
2. Body Control
3. ABS/TCS
4. Air Bags
5. Steering/Suspension
6. Cruise Control
7. Anti-Theft System
8. Climate Control

Figure C-8. *By using the arrow keys, you can select menu options. A shaded box often appears over each selection as you scroll through the menu box.*

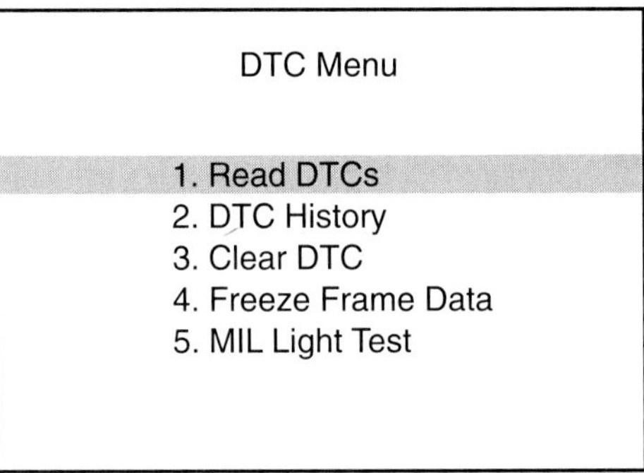

Figure C-10. *Depending on the scan tool, there are many submenu, such as this one.*

Diagnostic Menu

1. Read DTCs
2. System Tests
3. Input/Output Monitor
4. Actuator Tests
5. System Monitors
6. Anti-Theft System
7. Custom Display
8. Miscellaneous

Figure C-9. *The diagnostic menu will give you a host of choices in which to begin your scan tool diagnosis. In most cases, the first choice should be to read trouble codes or DTCs.*

should take once you reach the diagnostic menu is to check the ECM for any stored codes, **Figure C-10.** Scroll to the *Read Trouble Codes* choice and enter.

Reading Diagnostic Trouble Codes

Trouble codes are usually either two-digit numbers or five-digit alpha-numeric codes. Some scan tools may present trouble codes in both formats, if applicable. However, do not attempt to correlate these trouble codes on your own. When the scan tool reads codes, they will give the code number, as well as an explanation of the problem. If there is more than one code stored, the scan tool will give them in numeric order, from lowest to highest. Write each code and the problem as indicated by the scan tool, **Figure C-11.**

After all the codes are presented, the scan tool will ask if you want to see the trouble code history. The history indicates the number of times each code has been set within a certain number of keystarts or engine warm-ups. Even if only one code has been stored, select this function, as it will give you an indication if the code is hard or intermittent. Select *DTC History* and enter.

Code P0132
Oxygen Sensor High Voltage (B1 S1)

Code P0171
System Too Rich, Bank 1

Want To See DTC History?
1=Yes 2=No

Figure C-11. *Trouble codes will be listed in ascending numerical order, no matter how many times the code has been set. The number of times the code was set in memory is shown in the history.*

Interpreting Diagnostic Trouble Code History

The trouble code history will indicate the number of times each code has appeared, usually in the last 40-80 keystarts or engine warm-ups. Each DTC history is presented in order, starting with the most frequently occurring code, **Figure C-12.** The history will also give you a good indication as to whether a particular code is hard or intermittent. In cases where several codes have been stored, it also indicates which sensor, actuator, or system you should look at first. When this information is presented, write this information with the applicable code numbers.

Code History

Code P0132 36 of 40

Code P0171 4 of 40

Figure C-12. *Code history shows trouble codes in descending order, by the number of times the code was set in memory during a specified number of keystarts or engine warmups.*

For example, if a code has appeared 32 times in the last 34 keystarts, it is probably a hard code. However, if it has only appeared 5 times in the last 34 keystarts, it is most likely a soft or intermittent code. Even with the DTC history function, you should still clear the ECM memory and road test the vehicle. The code could have been stored the last time the vehicle was in for service. This also verifies that the conditions that set a code are still present, which can prevent lost diagnostic time looking for a problem that may have been corrected at an earlier date. However, if the vehicle is OBD II compliant, do not erase the codes until the ECM freeze frame data and failure record have been checked.

Using the Freeze Frame Data

Vehicles equipped with OBD II computer systems have an additional feature that saves or freezes sensor and output device data when certain trouble codes are stored. This information is called ***freeze frame data.***

Freeze frame data can reduce the time needed to diagnose a driveability problem by saving a picture of ECM sensor/actuator data when the code was set. In the past, it was necessary to reset the ECM and take the vehicle on a road test with the scan tool set to snapshot the ECM when the problem took place. The drawback to this is if the problem did not occur, no data would be captured. Freeze frame shows the data present when the original problem took place, **Figure C-13.**

Freeze Frame

Short-Term Fuel Trim: 155
Long-Term Fuel Trim: 146
Open/Closed Loop: Open
Engine Load: 0%
Coolant Temperature: 106°C
Engine RPM: 842 RPM
Vehicle Speed: 0 MPH
FF caused by DTC: Yes

Figure C-13. *The ECM in OBD II equipped vehicles has a freeze frame feature, which will store sensor input and actuator outputs for certain systems whenever a type A or B code is set in memory.*

Type A and B codes will store data in the freeze frame, however, type C and D codes will not. Most ECMs have enough memory to store one or two freeze frames. In all cases, a freeze frame for a type A code will overwrite the information for a type B code, no matter how many times the type B code has occurred.

The monitored systems that will cause freeze frame data to be stored are:

- ❑ Fuel trim monitor.
- ❑ Misfire monitor.
- ❑ Catalyst monitor.
- ❑ Oxygen sensor monitors and heaters.
- ❑ Secondary air monitor.
- ❑ EVAP system leak detection system.
- ❑ Purge flow monitor.

In some cases, the freeze frame data can be downloaded into the scan tool and kept for comparison to other test readings, or for subsequent downloading to a computer mainframe database.

Failure Record

In addition to the freeze frame, some vehicles have a ***failure record,*** which records the vehicle's sensor and actuator information when a trouble code is set. Unlike the freeze frame, the failure record will save information when type C and D codes are set. ECMs can keep several failure records in memory. If a code resets more than once, the information from the most recent code will be saved.

If the code is not a type A code, the vehicle operating information will most likely be located in the failure record. As with freeze frame data, write down the code, sensor inputs, and output device status. After you have obtained the codes, their history, freeze frame and/or failure record data, you can proceed to erase the codes.

Using Scan Tools to Clear Trouble Codes

After viewing the trouble codes and/or history information, the scan tool will ask if you wish to clear the stored codes. In some cases, there may be a separate menu selection titled *Clear DTC* or *Clear Stored Codes.* Make sure you have written the code information, trouble code history, and freeze frame information, where applicable. Select *Yes* or *Enter* when prompted to clear codes, **Figure C-14.** After a second, the scan tool will tell you if the codes have been cleared. In some cases, the MIL light may flash while the scan tool is clearing the ECM memory.

Clear Trouble Codes

Note: All Trouble Code Information
Including History And Freeze Frame
Will Be Erased.

Do You Wish To Clear Codes?

Press Yes To Clear
Press No To Continue

Figure C-14. *After writing down the trouble code information, clear the ECM memory and road test the vehicle or operate the system to see if the code resets.*

Clearing the stored trouble codes also clears the code history, failure record, and freeze frame data. After the scan tool indicates the codes have been cleared, recheck the ECM memory to ensure that the codes have been cleared or a potential hard code has not reset. If no codes are set, drive the vehicle or activate the system that caused the original code(s). Recheck the ECM for codes if the MIL lamp illuminates or after a few minutes of driving or system operation.

Using the Snapshot Feature

If the stored trouble code is intermittent or you cannot duplicate the conditions that set the code in the shop, especially on OBD I equipped vehicles, some scan tools will allow you to automatically take a picture of ECM functions as the code sets in memory. This is often referred to as a ***snapshot feature.*** This is similar to the freeze frame data stored in the ECMs of OBD II vehicles.

To set the scan tool to make a snapshot of ECM data, first clear the ECM of any stored trouble codes. Then, select the *snapshot* feature. In most cases, you will have the choice of selecting one of several options for snapshot data capture, **Figure C-15.** You should select either the *Automatic Capture* feature or *Capture When Code Sets.* There is usually a *Manual Capture,* which can be used, however, this requires someone to constantly monitor the scan tool and vehicle operation. In most cases, the problem may occur too quickly for a person to manually activate the scan tool snapshot.

Once the scan tool is set-up, road test the vehicle. Try to simulate the actual conditions that set the code as closely as possible. When the problem occurs, the scan tool will capture the data from the sensors. Drive the vehicle back to the shop, but do not shut the vehicle off until you have written down the captured information.

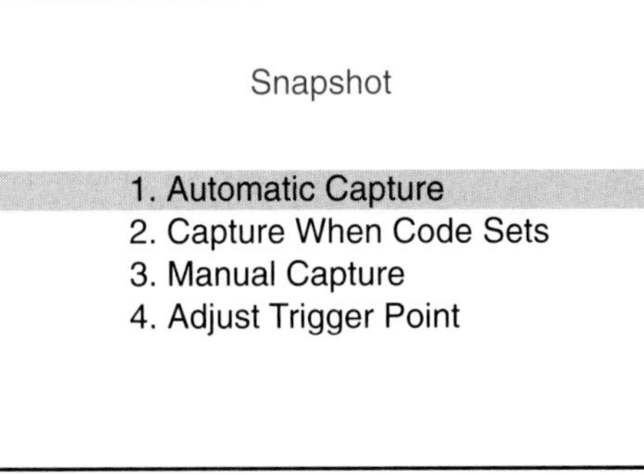

Figure C-15. *The snapshot feature is available on some scan tools. You should use automatic capture, since it will allow you to concentrate on the vehicle's operation rather than the scan tool.*

Sensor and Actuator Display

The sensor display feature will show you the major systems the ECM monitors. In most cases, you will have three or four selections, **Figure C-16.** The *Input Sensor* function will show only the sensor inputs to the ECM for that particular system. Choosing *Actuator Tests* will show the ECM outputs to the actuators and will allow you to control their operation. Some scan tools combine the input and output selections into one function. The other selection is the *Custom Display,* which allows you to choose which individual functions you want to see.

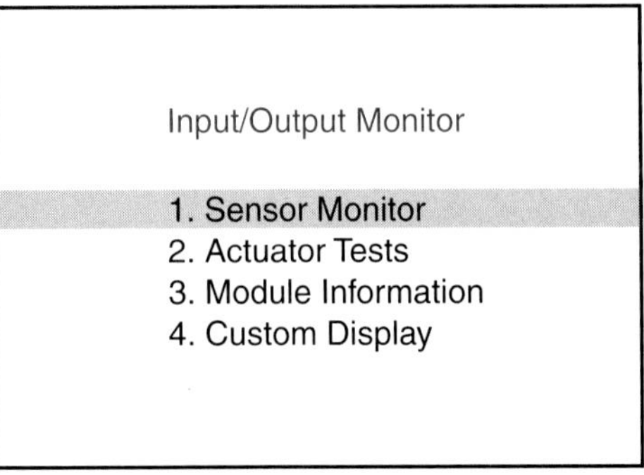

Figure C-16. *There are menu options that will allow you to see the sensor inputs and ECM outputs, as they occur.*

Sensor/Actuator Display

When you select the Sensor/Actuator display function, the scan tool will show you the inputs of all vehicle sensors and its outputs to the actuators, in real time, **Figure C-17.** From these menu screens is where much of your diagnostic information will be found. Use the More key or the arrow keys to scroll through the different sensor and actuator readings.

Sensor Monitor
Coolant Temperature Sensor: 101°C
Intake Air Temperature Sensor: 28°C
MAP Sensor Voltage: 2.66 Volts
Barometric Pressure: 30.4 in Hg
Throttle Angle: 0%
TPS Voltage: .43 Volts
Engine RPM: 788 RPM
Vehicle Speed: 0 MPH
1 of 4

Sensor Monitor
Short-Term Fuel Trim: 132
Long-Term Fuel Trim: 129
Injector Pulse Width: 2 mSec
Oxygen Sensor B1, S1: .583 Volts
Oxygen Sensor B1, S2: .594 Volts
Oxygen Sensor B2, S1: .601 Volts
Oxygen Sensor B1, S3: .451 Volts
Open/Closed Loop: Closed
2 of 4

Sensor Monitor
Engine Run Time: 6:24
Closed Loop Time: 2.03
Battery Voltage: 13.6 Volts
Battery Temperature: 52°C
IAC Motor: 34%
Knock Sensor: No
Cooling Fan 1: Yes
Cooling Fan 2: No
3 of 4

Sensor Monitor
A/C Request: Yes
A/C Compressor: On
Power Steering: Normal
Cruise Control: Off
Brake Switch: Off
MIL Light: Off
Spark Timing: 8°BTDC
Spark Advance: 12°
4 of 4

Figure C-17. *When looking at sensor information, there is usually more information than can be shown on one screen. Note the number of screens at the lower-left hand corner. In most cases, you would use the arrow keys or More key to scroll through this information.*

Depending on the size of the display screen, up to eight sensor/actuator readings will be shown. It is possible to select individual readings for closer scrutiny. This is explained in the next section.

Selecting Custom Displays

One of the most important features of any scan tool is the ability to ***customize*** the display to show information as you need it. In other words, to only show those sensors, actuators, and other readings needed to effectively diagnose the vehicle's problem. For example, if you are trying to find out why an engine is running rich, you probably do not want to look at information about the torque converter clutch or the transmission gear position.

Selecting a custom display is the easy part. Knowing what sensors, actuators, and readings to display is the difficult part. It takes time, experience, and knowledge of the particular vehicle's systems to know which readings to display, **Figure C-18.**

Custom Display
Open/Closed Loop: Closed
Short-Term Fuel Trim: 131
Oxygen Sensor B1, S1: .508 Volts
Oxygen Sensor B2, S1: .544 Volts
VOID
VOID
VOID
VOID

Figure C-18. *Customizing the scan tool display eliminates sensors and actuators that are probably not affecting the driveability problem.*

To create a customized display, select the *Custom Display* feature from the menu and choose the sensor(s) and actuator(s) that you want to monitor. Most scan tools will allow you to selectively display at least two sensor readings at a time. **Figure C-19** shows some combinations used in driveability work. It is to your advantage to show as many devices as the scan tool display screen will allow. These device readings will stay as long as you have custom display selected.

Using the Scan Tool to Control Actuator Devices

As well as monitoring sensor inputs, the scan tool can test devices controlled by the ECM. This gives you the ability to test the circuits of certain ECM-controlled output devices, reducing or in some cases, eliminating the need to check the circuit using a meter or test light.

Custom Display Selection	Custom Display Selection
Any suspect sensor **and** Open/Closed loop	Checks if defective sensor is affecting closed loop operation.
Any suspect sensor **and** MIL lamp	Checks if defective sensor is setting a DTC.
Short-term fuel trim **and** Oxygen sensor(s)	Checks for lazy oxygen affecting fuel trim.
Short- or Long-term fuel trim **and** Open/Closed loop	Used to check if fuel trim problem is affecting closed loop operation.
Injector pulse width **and** Short- or Long-term fuel trim	Checks if injector pulse width is changing in relation to fuel trim.
Oxygen sensor heater **and** Oxygen sensor	Checks heater and oxygen sensor circuits.
Oxygen sensor(s) **and** Catalyst monitor	Checks if catalytic converter is working properly.
Mass airflow sensor or airflow **and** injector pulse width	Checks if airflow is being detected/calculated and ECM is adjusting pulse width to match.
MAP or BARO sensor **and** Injector pulse width	Checks if ECM is adjusting pulse width to match changes in engine vacuum.
Throttle position sensor **and** Throttle angle	Ensures correct TPS adjustment in relation to throttle angle.
Spark advance or spark or Knock retard **and** Knock sensor	Checks if ECM is adjusting ignition timing in response to engine knock (tap on block with hammer).
Oxygen sensor(s), Coolant temperature, Intake air temperature, Engine run time **and** Open/Closed loop	Checks how quickly each sensor meets parameters for closed loop operation.
Open/Closed loop, MIL, **and** Engine run time	Used to check ECM operations when a code for excessive open loop operation is stored.
EGR solenoid **and** EGR position	Checks if EGR solenoids are responding to ECM commands.
Engine speed **and** Transmission gear solenoids	Checks transmission shift points in relation to engine speed.
Cruise control **and** Brake switch	Ensures cruise control disengages when brakes are applied.
Vehicle speed sensor **and** Transmission gear solenoids	Used to check when each gear solenoid applies in relation to vehicle speed.
Transmission position switch **and** Transmission gear solenoids	Used to check when each gear solenoid applies in relation to transmission gear (manually shifted).
Engine speed, Vehicle speed sensor, Brake switch, **and** Torque converter clutch	Used to verify torque converter clutch operation.
Coolant temperature sensor **and** Cooling fan	Verifies if ECM is commanding fans on in relation to engine temperature.
Cooling fan **and** A/C clutch	Checks if ECM is commanding fan on with the A/C clutch.
Power steering switch **and** A/C clutch	Ensures A/C clutch is disengaged when steering turned to lock.

Figure C-19. *One of the hard parts of learning how to use a scan tool is knowing which readout combinations are useful. This is a list of a few scan tool readout combinations used by driveability technicians.*

Vehicle Systems

There are several ECM controlled systems that can be tested or controlled using a scan tool. These include, but are not limited to:

- ❑ A/C compressor.
- ❑ Cooling fans.
- ❑ Engine idle control.
- ❑ Fuel injectors.
- ❑ Ignition coils.
- ❑ EGR solenoids.
- ❑ Evaporative emissions purge valve.
- ❑ MIL lamp.
- ❑ Fuel pump relay.
- ❑ Torque converter clutch.
- ❑ Shift indicator lamp (manual transmissions only).
- ❑ Tachometer output.
- ❑ All computer-controlled relays and solenoids.

By selecting the *Actuator Test* function, you can select which actuator to manually switch on and off, **Figure C-20.** The three most frequently used are engine idle control, fuel injectors, and ignition coils. Fuel injectors and ignition coil control is often combined into one selection called cylinder power balance, which will be discussed later.

Actuator Test Menu

1. Cylinder Power Balance
2. EVAP Purge System
3. EGR Solenoids
4. Idle Control
5. Clear Flood
6. OBD II Drive Cycle
7. Idle Learn
8. ECM Programming

1 of 2

Figure C-20. *Scan tools can control actuator operation, overriding the vehicle's ECM. To protect the engine, vehicle, and technician, some actuator tests can only be performed for a finite period of time by the scan tool's software.*

Engine Idle

You can use the scan tool to control the idle speed on fuel injected engines. The scan tool can automatically test idle air control valve operation by varying idle speed on a random basis. Make sure that the vehicle is in park and that the parking brake is set before you begin this test.

Select the function for engine idle control. The scan tool should give you the choice of either automatic or manual control. It is recommended that manual control be used, **Figure C-21.** If the engine's idle air control system has a defective motor, automatic control may result in the engine operating at an undesirable speed.

Increase idle speed in increments of 100 rpm. Most scan tools will not allow idle speeds above 2000-3000 rpms. Slowly increase engine speed to 1500-2000 rpms. Then, carefully reduce engine speed in 100 rpm increments to normal idle speed. Make sure the engine is at or near idle speed before exiting this test function. Do not simply back out of this test function, as the idle speed may remain where it was manually assigned.

Manual Idle Control Menu

Press 1 Or Up Arrow To Increase Engine Idle
Press 2 Or Down Arrow To Decrease Engine Idle.

Figure C-21. *One of the actuator functions the scan tool can control is the idle air control motor, which varies idle speed. This can be done manually or automatically in some cases.*

Power Balance Tests

Among the actuator tests that can be performed to the fuel system include cylinder ***power balance tests.*** Power balance tests can be performed to isolate a misfiring cylinder and to determine if it is caused by the fuel or ignition system or some other problem. Most scan tools have the capability of performing power balance tests automatically. Some scan tools give you the ability to disable any injector or cylinder manually.

A power balance test is performed by shutting off one injector at a time. Select the menu function for cylinder ignition or injector testing and start the test, **Figure C-22.** Depending on your selection, the scan tool will shut off the ignition spark or fuel injector to one cylinder for a few seconds. If the idle speed does not change when the cylinder is disabled, the cause of the misfire is located in that cylinder.

Cylinder Power Balance Test

Select System To Disable

1. Ignition
2. Fuel Injectors

Figure C-22. *A cylinder power balance test can disable either the fuel injectors or the ignition spark at each cylinder. This allows the user to isolate the cause of a weak or misfiring cylinder.*

Using Scan Tools to Learn ECM and Engine System Functions

Most newer ECMs use chips that are erased and reprogrammed rather than replaced. They can be programmed with the vehicle's operating parameters as well as reprogrammed to correct certain driveability problems. The scan tool can be used to program these chips as well as to reset certain ECM and vehicle functions.

ECM Programming

ECM programming is done by downloading the new operating information through the scan tool. The scan tool is first programmed with the vehicle information from a computer or computerized analyzer. Then, the information is downloaded from the scan tool into the ECM. Follow the instruction in the service manual whenever you are programming or reprogramming an ECM.

Using a scan tool to program an ECM is called ***indirect programming.*** Information on ECM programming is located in **Chapter 15.** Actual programming details vary between manufacturers, but the basic procedure begins with placing the ECM in the programming mode. The scan tool menu is accessed using the keypad. Most scan tools will use a high capacity cartridge or PCMCIA card to store the information.

Connections for ECM Programming

To begin ECM programming, make sure that the battery is fully charged. Recharge the battery if necessary. Connect the scan tool to the ECM data link connector. Make any other vehicle connections as needed before proceeding with ECM programming. Do not disconnect the scan tool from the DLC during programming.

Once the vehicle information is entered, go to the programming software and follow the directions as prompted. It may be necessary to turn the ignition switch on or off during the connection and programming procedure. The next step is to determine the type of programming that is needed.

It is possible on some vehicles to program more than one ECM. When programming a new ECM, make sure you only program that ECM, **Figure C-23.** While most ECMs have internal circuitry that protects them from accidental programming, be careful not to program the wrong one. Do not attempt to program an ECM with information from the old ECM or an ECM from another vehicle. An erasure may need to be performed on a new ECM before initial programming can take place.

ECM Programming

Select ECM To Program
1. Powertrain
2. Body Control
3. ABS/TCS
4. Air Bags
5. Steering/Suspension

Figure C-23. *It is possible to reprogram the ECMs on certain vehicles. Make sure you are programming the correct ECM.*

Allow sufficient time for programming to take place. Do not touch any connections until you are sure that the programming sequence is complete. After programming is finished, turn the ignition switch to the position called for and disconnect the computer or scan tool. After programming is complete, check ECM and control system operation.

Reprogramming ECMs

If you are reprogramming an ECM, determine the date that the ECM's programming was downloaded or the current calibration number, **Figure C-24.** If the information installed is the latest version, no further action is required.

Before reprogramming the ECM, you must first erase the existing information. To erase the old programming, select the erase function and follow the scan tool's directions. After this step is complete, download the new information into the ECM. On some systems, the erasure step is not necessary as the new program will automatically overwrite the old information as it loads into the ECM.

ECM Information

1997 3.6L SMFI V-6
Unleaded Fuel
Federal Emissions
Auto O/D Trans, FWD
Dual fan, Cycling A/C
ECM part #12345678
Program #58824B
Program Date 6/4/97

Figure C-24. *The ECM program numbers and date of programming can be found under ECM Information, Vehicle Information, or Module Information.*

Idle Speed Learn

Some vehicles require a scan tool to set the vehicle's idle speed. This is also known as setting the minimum air rate. If this is not done after certain service procedures, the engine will have an unstable idle, or surge. The engine may also stall at times for no apparent reason.

To perform the idle learn, clear the ECM memory by disconnecting battery power. After battery power has been restored, connect the scan tool and turn the ignition key to run, but do not start the engine. Enter the vehicle's information and select the *Idle Learn* function. Follow the scan tool's instructions. It may be necessary to allow the engine to idle to operating temperature before the scan tool will set idle speed. Once the idle speed is set, shut off the vehicle and disconnect the scan tool.

Scan Tool Maintenance

Like all other tools, a scan tool must be treated with respect if it is to provide years of trouble-free service. You will not realize how valuable a scan tool is until you do not have it, because it is out for repair due to abuse or neglect.

Many scan tools come with a leather or plastic sleeve or glove. Keep the scan tool in this glove at all times. This will protect it from damage as well as to help reduce the amount of contact with dirt, oil, grease, and fluids. Most scan tools come with a plastic carrying case. Return the scan tool to this case when you are finished with it, along with all cables and scan tool cartridges (if equipped). Make sure you check and replace the battery, if equipped.

Try to keep the scan tool as clean as possible. Wipe it clean of dirt, oil, and grease using a dry shop towel after each use. Do not expose the scan tool to water or other fluids. Do not use an abrasive cleanser to clean the scan tool. Clean the display screen and keypad with a nonabrasive glass cleaner applied to a soft, static-free cloth.

Clean the cables with a mild soap solution, never immerse the scan tool in water. While most scan tools have laminated surfaces, they are not waterproof.

Some scan tools are equipped with one or more cooling fans. These fans have a small filter that should be removed and cleaned using warm soapy water. If the filter cannot be cleaned, replace the filter. If the scan tool has a protective glove or sleeve, it should also be cleaned at this time.

When you are finished using the scan tool, put it back in its case. Remove any external program cartridges, cables, and any adapters and place them in their respective compartments. Close the case and return it to the tool room or location where it is normally stored. This way, you or any other technician in the shop can easily find the scan tool the next time it is needed.

Common Scan Tool Usage Problems

As with any electronic tool, you will no doubt encounter problems in using scan tools. If you run into any problems, refer to the scan tool's user manual. It contains many pages with hints on solving problems encountered while using the scan tool. Some common problems and their causes are listed in the following paragraphs.

Scan Tool Appears to Have No Power

This is the most common problem you will encounter. It is usually caused by a poor external power connection at the cigar lighter or battery. Make sure your source of scan tool power is clean and tight. If the scan tool draws power from the data link connector, make sure the connection at the data link connector is tight and that the pins are straight and clean.

If the connection is good and the tool still does not activate, check the cigar lighter socket for power. If there is no power at the socket, the most likely cause is a blown fuse. If the cigar lighter socket is full of ash, clean it as it may be the cause of the incomplete circuit or blown fuse. Make sure the battery voltage is not below 9.5 volts. If necessary, recharge the battery before conducting any diagnostic tests. The least probable cause is a problem inside the scan tool.

No Data Message Appears in the Scan Tool Display

A frequent cause of *No Data* messages is a poor connection between the scan tool and the ECM, **Figure C-25.** The location of the poor connection is usually at the data link connector. Check the connection at the data link connector. Make sure all pins in the data link connector, adapter, and cable are clean, straight, and tight. Check the data link connector to ensure that you are using the correct adapter. Also make sure you are connected to the right vehicle connector.

No Data

Check Scan Tool Connection To DLC

Figure C-25. *If a No Data message appears on the scan tool screen, check the cable connection at the data link connector. In most cases, the problem is a poor connection.*

If the connection appears to be good, make sure the ignition key is on. In most cases, the ignition key must be turned to run before the ECM will send information to the scan tool. A third cause involves poor scan tool program cartridge installation. Make sure the program cartridge is installed correctly and you are using the right one for the particular vehicle.

Scan Tool Displays an Error Message

A scan tool is capable of displaying many error messages, related and unrelated to vehicle operation. Often, the scan tool is not the cause of the error message. The problem is most likely caused by a problem in the vehicle or an error by the user. Some of the more common messages are:

- No Response—the scan tool is unable to communicate with the computer system selected or is caused by connecting the scan tool to the wrong vehicle connector, **Figure C-26.**

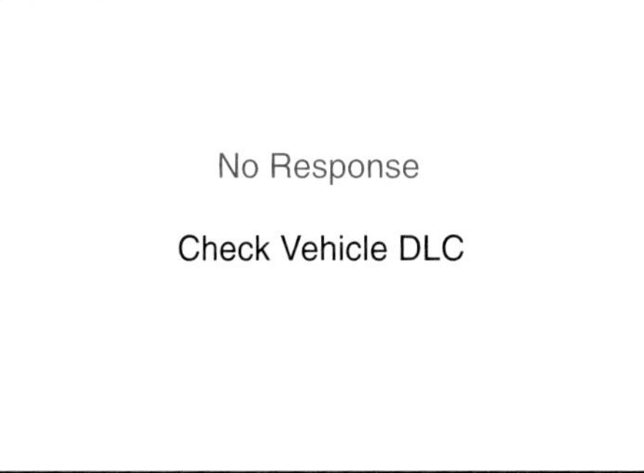

Figure C-26. *A No Response message indicates an inability to communicate with the ECM.*

- No Bus Bias—caused by the scan tool's inability to communicate with a controller. Usually caused by a fault in the vehicle's data bus wiring.
- Not Available—will appear if you select an item or system that is not on the vehicle or is not supported by the scan tool.
- Unable to Access System—the scan tool cannot communicate with the vehicle's controller. Check the connection at the data link connector. Also check the ECM fuse to ensure the ECM is receiving power. If there is power to the ECM, scan tool, and the connection to the data link connector is good, the plug-in connectors at the ECM are loose or the ECM is defective, **Figure C-27.**

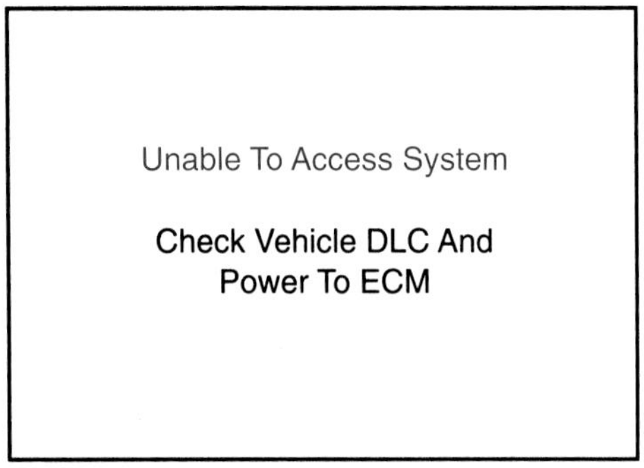

Unable To Access System

Check Vehicle DLC And
Power To ECM

Figure C-27. *A defective ECM, or no power to the ECM will cause an Unable To Access System message to appear in the scan tool display.*

- Scan Tool Displays Only Symbols and ASCII Code—if the data cartridge is correctly installed, the problem is caused by an internal defect in the scan tool or program cartridge, **Figure C-28**. Try another program cartridge for another vehicle system or try the scan tool on another vehicle. Also, try reprogramming the scan tool data cartridge or card. If it still does not work, send the scan tool to a certified repair facility.

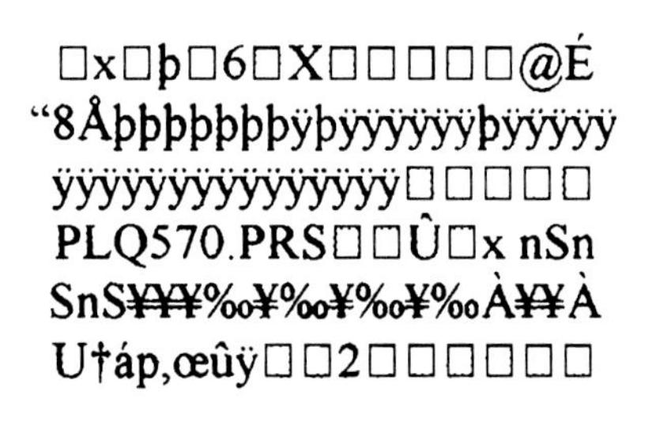

Figure C-28. *If you see a display similar to this when you plug in the scan tool, check the program cartridge or card. Most likely, the problem is in the scan tool itself.*

Other Scan Tool Functions

The latest scan tools are becoming more like miniaturized engine analyzers than traditional stand-alone scan tools. Some have multimeters incorporated into them. With adapters, they can function as high-amperage volt-amp testers and can be used to check starting and charging systems. A few scan tools have oscilloscope functions and will generate waveforms similar to those on an engine analyzer's scope.

With special infared probes, some scan tools can function as thermometers capable of reading the high temperatures generated by engine and exhaust system parts. Future scan tools will probably have exhaust gas analyzer and other diagnostic functions currently performed by much larger equipment. Scan tools have been in existence less than 20 years, but have undergone as much evolution as the vehicles they are designed to daignose. As technology improves, scan tools will no doubt become more powerful.

Useful Tables

CONVERSION CHART

METRIC/U.S. CUSTOMARY UNIT EQUIVALENTS

Multiply:	by:	to get:	Multiply:	by:	to get:
ACCELERATION					
feet/sec^2	x 0.3048	= meters/sec^2 (m/s^2)	x 3.281	= feet/sec^2	
inches/sec^2	x 0.0254	= meters/sec^2 (m/s^2)	x 39.37	= inches/sec^2	
ENERGY OR WORK (watt–second = joule = newton–meter)					
foot–pounds	x 1.3558	= joules (J)	x 0.7376	= foot–pounds	
calories	x 4.187	= joules (J)	x 0.2388	= calories	
Btu	x 1055	= joules (J)	x 0.000948	= Btu	
watt–hours	x 3600	= joules (J)	x 0.0002778	= watt–hours	
kilowatt–hrs.	x 3.600	= megajoules (MJ)	x 0.2778	= kilowatt-hrs	
FUEL ECONOMY AND FUEL CONSUMPTION					
miles/gal	x 0.42514	= kilometers/liter (km/L)	x 2.3522	= miles/gal	
Note:					
235.2/(mi/gal) = liters/100km					
235.2/(liters/100 km) = mi/gal					
LIGHT					
footcandles	x 10.76	= lumens/$meter^2$ (lm/m^2)	x 0.0929	= footcandles	
PRESSURE OR STRESS (newton/sq meter = pascal)					
inches Hg(60°F)	x 3.377	= kilopascals (kPa)	x 0.2961	= inches Hg	
pounds/sq in	x 6.895	= kilopascals (kPa)	x 0.145	= pounds/sq in	
inches H_2O(60°F)	x 0.2488	= kilopascals (kPa)	x 4.0193	= inches H_2O	
bars	x 100	= kilopascals (kPa)	x 0.01	= bars	
pounds/sq ft	x 47.88	= pascals (Pa)	x 0.02088	= pounds/sq ft	
POWER					
horsepower	x 0.746	= kilowatts (kW)	x 1.34	= horsepower	
ft–lbf/min	x 0.0226	= watts (W)	x 44.25	= ft–lbf/min	
TORQUE					
pounds–inches	x 0.11298	= newton–meters (N-m)	x 8.851	= pound–inches	
pound–feet	x 1.3558	= newton–meters (N-m)	x 0.7376	= pound–feet	
VELOCITY					
miles/hour	x 1.6093	= kilometers/hour (km/h)	x 0.6214	= miles/hour	
feet/sec	x 0.3048	= meters/sec (m/s)	x 3.281	= feet/sec	
kilometers/hr	x 0.27778	= meters/sec (m/s)	x 3.600	= kilometers/hr	
miles/hour	x 0.4470	= meters/sec (m/s)	x 2.237	= miles/hour	

COMMON METRIC PREFIXES

mega	(M)	= 1 000 000	or 10^6	centi	(c)	= 0.01	or 10^{-2}
kilo	(k)	= 1 000	or 10^3	milli	(m)	= 0.001	or 10^{-3}
hecto	(h)	= 100	or 10^2	micro	(μ)	= 0.000 001	or 10^{-6}

INCHES 0 1 2 3 4 5

METRIC/U.S. CUSTOMARY UNIT EQUIVALENTS

Multiply:	by:	to get:	Multiply:	by:	to get:
LINEAR					
inches	x 25.4	= millimeters (mm)	x 0.03937	= inches	
feet	x 0.3048	= meters (m)	x 3.281	= feet	
yards	x 0.9144	= meters (m)	x 1.0936	= yards	
miles	x 1.6093	= kilometers (km)	x 0.6214	= miles	
inches	x 2.54	= centimeters (cm)	x 0.3937	= inches	
microinches	x 0.0254	= micrometers (μm)	x 39.37	= microinches	
AREA					
$inches^2$	x 645.16	= $millimeters^2$(mm^2)	x 0.00155	= $inches^2$	
$inches^2$	x 6.452	= $centimeters^2$(cm^2)	x 0.155	= $inches^2$	
$feet^2$	x 0.0929	= $meters^2$(m^2)	x 10.764	= $feet^2$	
$yards^2$	x 0.8361	= $meters^2$(m^2)	x 1.196	= $yards^2$	
$acres^2$	x 0.4047	= hectares (10^4m^2)			
		ha	x 2.471	= acres	
$miles^2$	x 2.590	= $kilometers^2$ (km^2)	x 0.3861	= $miles^2$	
VOLUME					
$inches^3$	x 16387	= $millimeters^3$ (mm^3)	x 0.000061	= $inches^3$	
$inches^3$	x 16.387	= $centimeters^3$ (cm^3)	x 0.06102	= $inches^3$	
$inches^3$	x 0.01639	= liters (L)	x 61.024	= $inches^3$	
quarts	x 0.94635	= liters (L)	x 1.0567	= quarts	
gallons	x 3.7854	= liters (L)	x 0.2642	= gallons	
$feet^3$	x 28.317	= liters (L)	x 0.03531	= $feet^3$	
$feet^3$	x 0.02832	= $meters^3$ (m^3)	x 35.315	= $feet^3$	
fluid oz	x 29.57	= milliliters (mL)	x 0.03381	= fluid oz	
$yards^3$	x 0.7646	= $meters^3$ (m^3)	x 1.3080	= $yards^3$	
teaspoons	x 4.929	= milliliters (mL)	x 0.2029	= teaspoons	
cups	x 0.2366	= liters (L)	x 4.227	= cups	
MASS					
ounces (av)	x 28.35	= grams (g)	x 0.03527	= ounces (av)	
pounds (av)	x 0.4536	= kilograms (kg)	x 2.2046	= pounds (av)	
tons (2000 lb)	x 907.18	= kilograms (kg)	x 0.001102	= tons (2000 lb)	
tons (2000 lb)	x 0.90718	= metric tons (t)	x 1.1023	= tons (2000 lb)	
FORCE					
ounces—f (av)	x 0.278	= newtons (N)	x 3.597	= ounces—f (av)	
pounds—f (av)	x 4.448	= newtons (N)	x 0.2248	= pounds—f (av)	
kilograms—f	x 9.807	= newtons (N)	x 0.10197	= kilograms—f	

TEMPERATURE

32 98.6 212

°F -40 0 40 80 120 160 200 240 280 320 °F

°C -40 -20 0 20 40 60 80 100 120 140 160 °C

°Celsius = 0.556 (°F – 32) °F = (1.8° C) + 32

millimeters 0 10 20 30 40 50 60 70 80 90 100 110 120 130

TAP/DRILL CHART

COARSE STANDARD THREAD (N.C.) Formerly U.S. Standard Thread					FINE STANDARD THREAD (N.F.) Formerly S.A.E. Thread				
Sizes	Threads Per Inch	Outside Diameter at Screw	Tap Drill Sizes	Decimal Equivalent of Drill	Sizes	Threads Per Inch	Outside Diameter at Screw	Tap Drill Sizes	Decimal Equivalent of Drill
1	64	.073	53	0.0595	0	80	.060	3/64	0.0469
2	56	.086	50	0.0700	1	72	.073	53	0.0595
3	48	.099	47	0.0785	2	64	.086	50	0.0700
4	40	.112	43	0.0890	3	56	.099	45	0.0820
5	40	.125	38	0.1015	4	48	.112	42	0.0935
6	32	.138	36	0.1065	5	44	.125	37	0.1040
8	32	.164	29	0.1360	6	40	.138	33	0.1130
10	24	.190	25	0.1495	8	36	.164	29	0.1360
12	24	.216	16	0.1770	10	32	.190	21	0.1590
1/4	20	.250	7	0.2010	12	28	.216	14	0.1820
5/16	18	.3125	F	0.2570	1/4	28	.250	3	0.2130
3/8	16	.375	5/16	0.3125	5/16	24	.3125	I	0.2720
7/16	14	.4375	U	0.3680	3/8	24	.375	Q	0.3320
1/2	13	.500	27/64	0.4219	7/16	20	.4375	25/64	0.3906
9/16	12	.5625	31/64	0.4843	1/2	20	.500	29/64	0.4531
5/8	11	.625	17/32	0.5312	9/16	18	.5625	0.5062	0.5062
3/4	10	.750	21/32	0.6562	5/8	18	.625	0.5687	0.5687
7/8	9	.875	49/64	0.7656	3/4	16	.750	11/16	0.6875
1	8	1.000	7/8	0.875	7/8	14	.875	0.8020	0.8020
1 1/8	7	1.125	63/64	0.9843	1	14	1.000	0.9274	0.9274
1 1/4	7	1.250	1 7/64	1.1093	1 1/8	12	1.125	1 3/64	1.0468
					1 1/4	12	1.250	1 11/64	1.1718

BOLT TORQUING CHART

METRIC STANDARD

Grade of Bolt		5D	.8G	10K	12K	
Min. Tensile Strength		71,160 P.S.I.	113,800 P.S.I.	142,200 P.S.I.	170,679 P.S.I.	
Grade Markings on Head		5D	8G	10K	12K	Size of Socket or Wrench Opening
Metric		Foot Pounds				Metric
Bolt Dia.	U.S. Dec Equiv.					Bolt Head
6mm	.2362	5	6	8	10	10mm
8mm	.3150	10	16	22	27	14mm
10mm	.3937	19	31	40	49	17mm
12mm	.4720	34	54	70	86	19mm
14mm	.5512	55	89	117	137	22mm
16mm	.6299	83	132	175	208	24mm
18mm	.709	111	182	236	283	27mm
22mm	.8661	182	284	394	464	32mm

SAE STANDARD/FOOT POUNDS

Grade of Bolt	SAE 1 & 2	SAE 5	SAE 6	SAE 8		
Min. Tensile Strength	64,000 P.S.I.	105,000 P.S.I.	133,000 P.S.I.	150,000 P.S.I.		
Markings on Head					Size of Socket or Wrench Opening	
U.S. Standard	Foot Pounds				U.S. Regular	
Bolt Dia.					Bolt Head	Nut
1/4	5	7	10	10.5	3/8	7/16
5/16	9	14	19	22	1/2	9/16
3/8	15	25	34	37	9/16	5/8
7/16	24	40	55	60	5/8	3/4
1/2	37	60	85	92	3/4	13/16
9/16	53	88	120	132	7/8	7/8
5/8	74	120	167	180	15/16	1.
3/4	120	200	280	296	1-1/8	1-1/8

DECIMAL CONVERSION CHART

FRACTION	INCHES	M/M	FRACTION	INCHES	M/M
1/64	.01563	.397	33/64	.51563	13.097
1/32	.03125	.794	17/32	.53125	13.494
3/64	.04688	1.191	35/64	.54688	13.891
1/16	.6250	1.588	9/16	.56250	14.288
5/64	.07813	1.984	37/64	.57813	14.684
3/32	.09375	2.381	19/32	.59375	15.081
7/64	.10938	2.778	39/64	.60938	15.478
1/8	.12500	3.175	5/8	.62500	15.875
9/64	.14063	3.572	41/64	.64063	16.272
5/32	.15625	3.969	21/32	.65625	16.669
11/64	.17188	4.366	43/64	.67188	17.066
3/16	.18750	4.763	11/16	.68750	17.463
13/64	.20313	5.159	45/64	.70313	17.859
7/32	.21875	5.556	23/32	.71875	18.256
15/64	.23438	5.953	47/64	.73438	18.653
1/4	.25000	6.350	3/4	.75000	19.050
17/64	.26563	6.747	49/64	.76563	19.447
9/32	.28125	7.144	25/32	.78125	19.844
19/64	.29688	7.541	51/64	.79688	20.241
5/16	.31250	7.938	13/16	.81250	20.638
21/64	.32813	8.334	53/64	.82813	21.034
11/32	.34375	8.731	27/32	.84375	21.431
23/64	.35938	9.128	55/64	.85938	21.828
3/8	.37500	9.525	7/8	.87500	22.225
25/64	.39063	9.922	57/64	.89063	22.622
13/32	.40625	10.319	29/32	.90625	23.019
27/64	.42188	10.716	59/64	.92188	23.416
7/16	.43750	11.113	15/16	.93750	23.813
29/64	.45313	11.509	61/64	.95313	24.209
15/32	.46875	11.906	31/32	.96875	24.606
31/64	.48438	12.303	63/64	.98438	25.003
1/2	.50000	12.700	1	1.00000	25.400

SOME COMMON ABBREVIATIONS

U.S CUSTOMARY		METRIC	
UNIT	ABBREVIATION	UNIT	ABBREVIATION
inch	in.	kilometer	km
feet	ft.	hectometer	hm
yard	yd.	dekameter	dam
mile	mi.	meter	m
grain	gr.	decimeter	dm
ounce	oz.	centimeter	cm
pound	lb.	millimeter	mm
teaspoon	tsp.	cubic centimeter	cm^3
tablespoon	tbsp.	kilogram	kg
fluid ounce	fl. oz.	hectogram	hg
cup	c.	dekagram	dag
pint	pt.	gram	g
quart	qt.	decigram	dg
gallon	gal.	centigram	cg
cubic inch	in^3	milligram	mg
cubic foot	ft^3	kiloliter	kl
cubic yard	yd^3	hectoliter	hl
square inch	in^2	dekaliter	dl
square foot	ft^2	liter	L
square yard	yd^2	centiliter	cl
square mile	mi^2	milliliter	ml
Fahrenheit	F	square kilometer	km^2
barrel	bbl.	hectare	ha
fluid dram	fl. dr.	are	a
board foot	bd. ft.	centare	ca
rod	rd.	tonne	t
dram	dr.	Celsius	C
bushel	bu.		

Acknowledgments

The author would like to thank the following firms and individuals for their assistance in the preparation of **Auto Engine Performance & Driveability**.

Acura Automobile Division; Audi of America; Automotive Diagnostics; Balco Inc.; BMW of North America; Robert Bosch Corporation; Buick Motor Division; Cadillac Motor Division; Champion Spark Plug Div. of Cooper Industries; Chevrolet Motor Division; Chrysler Corporation; Clayton Industries; EMI-Tech; Ferret Instruments; John Fluke Manufacturing Company; Ford Motor Company; GMC Trucks; General Motors Corporation, Service Technology Group; American Honda Motor Company; Infiniti North America; Land Rover of North America; Lexus Automobile Division; Mac Tools; Mazda Motor of America; Mercedes-Benz of North America; Mitsubishi Motors of America; National Institute for Automotive Service Excellence; Nissan North America; Oldsmobile Motor Division; OTC Division of SPX Corp.; Pontiac Motor Division; Saab Cars USA; Saturn Corporation; Snap-On Tools; Subaru of America; Toyota Motor Sales.

The author would also like to thank the following persons and organizations who provided vehicles, parts and test equipment for the photographs as well as other items used throughout this text.

Paul Anderson; Brent Blake, International Automotive Technicians Network; David Bryson, Sears Auto Center, Greenville, SC; James Cledras; James Daley; Corey Glassman, John Fluke Corporation; Marsh Hayward; Danny McCown, Century Lincoln Mercury/BMW, Greenville, SC; Herbert and Dorothy MacMillian; Palemetto Chapter, Pontiac–Oakland Club International; Oliver Scheurmann; Martin W. Stockel; Martin T. Stockel; C.W. Weeks; Natchitoches Junior College, Natchitoches, LA; Michael Wood.

Index

E

K

L

M

N

O

T